**Cell Signaling and
Growth Factors in Development**

Edited by
K. Unsicker and
K. Krieglstein

Further Titles of Interest

R.A. Meyers
Encyclopedia of Molecular Cell Biology and Molecular Medicine
2005, second edition
3-527-30542-4

D. Wedlich
Cell Migration in Development and Disease
2005
ISBN 3-527-30587-4

A. Ridley
Cell Motility
2004
ISBN 0-470-84872-3

G.S. Stein, A.B. Pardee
Cell Cycle and Growth Control
2004, second edition
ISBN 0-471-25071-6

W.W. Minuth, R. Strehl, K. Schumacher
Tissue Engineering
2005
ISBN 3-527-31186-6

E.A. Nigg
Centrosomes in Development and Disease
2004
ISBN 3-527-30980-2

M. Schliwa
Molecular Motors
2003
ISBN 3-527-30594-7

Cell Signaling and Growth Factors in Development

From Molecules to Organogenesis

Edited by
Klaus Unsicker and
Kerstin Krieglstein

WILEY-
VCH

WILEY-VCH Verlag GmbH & Co. KGaA

Editor

Prof. Dr. Klaus Unsicker
University of Heidelberg
Interdisciplinary Center for Neurosciences (IZN)
Department of Neuroanatomy
Im Neuenheimer Feld 307
69120 Heidelberg
Germany

Prof. Dr. Kerstin Krieglstein
University of Göttingen
Medical School
Department of Neuroanatomy
Kreuzbergring 36
37075 Göttingen
Germany

Coverillustration
The simplified signaling network (see Fig. 15.5 from Susan Mackem) and the mouse embryo section represent the starting point and goal of the fascinating way from molecules to organo-genesis (mouse embryo section reprinted from »The Atlas of Mouse Development« edited by M.H. Kaufmann, cover page, 1992 with permission from Elsevier).

■ This book was carefully produced. Nevertheless, authors, editor and publisher do not warrant the information contained therein to be free of errors. Readers are advised to keep in mind that statements, data, illustrations, procedural details or other items may inadvertently be inaccurate.

Library of Congress Card No.: applied for
A catalogue record for this book is available from the British Library

**Bibliographic information published by
Die Deutsche Bibliothek**

Die Deutsche Bibliothek lists this publication in the Deutsche Nationalbibliografie; detailed bibliographic data is available in the Internet at http://dnb.dbb.de.

© 2006 WILEY-VCH Verlag GmbH & Co. KGaA, Weinheim
All rights reserved (including those of translation in other languages). No part of this book may be reproduced in any form – by photoprinting, microfilm, or any other means – nor transmitted or translated into machine language without written permission from the publishers. Registered names, trademarks, etc. used in this book, even when not specifically marked as such, are not to be considered unprotected by law.

Printed in the Federal Republic of Germany.
Printed on acid-free paper.
Typesetting   pagina GmbH, Tübingen
Printing   betz-druck GmbH, Darmstadt
Bookbinding   J. Schäffer GmbH i. G., Grünstadt
ISBN-13:   978-3-527-31034-0
ISBN-10:   3-527-31034-7

Contents

Volume 1

Preface   *XXVII*

List of Contributors   *XXIX*

Color Plates   *XXXVII*

I   **Cell Signaling and Growth Factors in Development**

1   **Stem Cells**
    Christian Paratore and Lukas Sommer
1.1   Introduction   *3*
1.2   Maintenance of Stemness in Balance with Stem Cell Differentiation   *5*
1.2.1   Wnt Signaling   *7*
1.2.2   Wnt Signaling Regulates ‚Stemness' in ESCs   *8*
1.2.3   Wnt Signaling in Hematopoietic Stem Cells   *8*
1.2.4   Wnt Signal Activation in the Skin   *10*
1.2.5   Multiple Roles of Canonical Wnt Signaling in Neural Stem Cells   *11*
1.2.6   Aberrant Wnt signal activation in carcinogenesis   *13*
1.3   The Notch Signaling Pathway   *13*
1.3.1   Notch Signaling During Hematopoiesis   *14*
1.3.2   Notch1 Functions as a Tumor Suppressor in Mouse Skin   *15*
1.3.3   Notch Signaling in the Nervous System and its Role in Neural Differentiation and Stem Cell Maintenance   *15*
1.3.4   Aberrant Notch Signaling   *18*
1.4   Signaling Pathway of the TGFβ Family Members   *18*
1.4.1   BMP Signaling in ESCs   *20*

Cell Signaling and Growth Factors in Development. Edited by K. Unsicker and K. Krieglstein
Copyright © 2006 WILEY-VCH Verlag GmbH & Co. KGaA, Weinheim
ISBN 3-527-31034-7

1.4.2	The Influence of TGFβ Family Members on MSC Differentiation	20
1.4.3	Tgfβ Factors Act Instructively on NCSC Differentiation	21
1.4.4	Aberrant Growth Regulation by Mutations in the Tgfβ Signaling Pathway	22
1.5	Shh Signaling	23
1.5.1	Hematopoiesis and T-cell Maturation	24
1.5.2	The Role of Shh in the Nervous System	24
1.5.3	Shh Signaling in Tumorigenesis	25
1.6	Conclusions	26

2 Germ Cells
Pellegrino Rossi, Susanna Dolci, Donatella Farini, and Massimo De Felici

2.1	Introduction	39
2.2	Primordial Germ Cells	41
2.2.1	Growth Factors in PGC Commitment	41
2.2.2	Growth Factors in PGC Proliferation	42
2.2.2.1	The KL/KIT System	42
2.2.2.2	LIF/OSM/IL6 Superfamily	43
2.2.2.3	TGFβ Superfamily and Neuropeptides	44
2.3	Oocytes	45
2.3.1	Growth Factors for Fetal Oocytes	45
2.3.2	Growth Factors for Growing and Mature Oocytes	46
2.3.3	Growth Factors in the Initiation of Follicle and Oocyte Growth	48
2.3.4	From Pre-antral to Ovulatory Follicles: Oocytes take Control	50
2.4	Male Germ Cells	52
2.4.1	GDNF	53
2.4.2	BMPs	55
2.4.3	KL/Kit System	56
2.4.4	Control of Meiotic Progression of Spermatocytes	58

3 Implantation and Placentation
Susana M. Chuva de Sousa Lopes, Christine L. Mummery, and Sólveig Thorsteinsdóttir

3.1	Introduction	73
3.1.1	Formation of the Trophectoderm, the Precursor of the Invading Trophoblast	74
3.1.2	Initial Blastocyst-Uterine Interaction	74
3.1.3	Differentiation of Extra-embryonic Lineages During Implantation and Gastrulation	76

3.1.4	Steps towards a Functional Placenta	78
3.1.5	Comparison of Mouse and Human Implantation and Placentation	79
3.2	Molecular Mechanisms and Biological Effects	80
3.2.1	Preparation of the Blastocyst for Implantation (E3.5–E4.5)	80
3.2.2	Uterine-Embryonic Signaling and Adhesion (E4.5)	81
3.2.3	The Invasion of the Uterus (E4.5–E7.5)	85
3.2.3.1	Decidualization	85
3.2.3.2	Anchoring of Trophoblast Cells and Differentiation during Peri-implantation	86
3.2.4	The Formation and Development of the Chorio-allantoic Placenta (E8.5–E16.5)	88
3.3	Clinical Relevance in Humans	90
3.4	Summary	92
	Acknowledgments	93

4   **Cell Movements during Early Vertebrate Morphogenesis**
Andrea Münsterberg and Grant Wheeler

4.1	Introduction	107
4.2	History: Classic Experiments in the Study of Gastrulation	108
4.3	Gastrulation and Neurulation in Different Vertebrate Species	109
4.3.1	Amphibians: *Xenopus laevis*	109
4.3.2	The Teleost, Zebrafish	111
4.3.3	Amniotes: Chick and Mouse	112
4.4	Mechanistic Aspects, Molecules and Molecular Networks	114
4.4.1	Wnt Signaling	115
4.4.2	Convergent Extension and the Regulation of Cytoskeletal Dynamics	118
4.4.3	Fibroblast Growth Factors (FGF)	119
4.4.4	Regulation of Epithelial Mesenchymal Transition (EMT)	122
4.4.5	Cadherins and Protocadherins	123
4.4.6	Extracellular Matrix and Integrin Receptors	124
4.4.7	Platelet Derived Growth Factor (PDGF)	125
4.5	Conclusion and Outlook	126

5   **Head Induction**
Clemens Kiecker

5.1	Introduction	141
5.2	Classical Concepts for Head Induction	141
5.3	Tissues Involved in Anterior Specification	145
5.3.1	The Gastrula Organizer	145

5.3.2	Primitive and Definitive Anterior Endoderm	146
5.3.3	Ectodermal Signaling Centers	149
5.3.4	Summary	149
5.4	Signaling Pathways Involved in Anterior Specification	149
5.4.1	BMP Signaling	150
5.4.2	FGF Signaling	153
5.4.3	IGF Signaling	154
5.4.4	Nodal Signaling	155
5.4.5	Retinoid Signaling	157
5.4.6	Wnt/β-Catenin Signaling	158
5.4.7	Noncanonical Wnt Signaling	160
5.4.8	Summary	161
5.5	Mechanistic Models for Head Induction	162
5.5.1	The Two- and Three-Inhibitor Models	162
5.5.2	The Organizer-Gradient Model Dualism	162
5.6	Transcriptional Regulators in Anterior Specification	163
5.6.1	ANF/Hesx1	164
5.6.2	Blimp1	164
5.6.3	FoxA2/HNF3β	164
5.6.4	Goosecoid	165
5.6.5	Hex	165
5.6.6	Lim1	165
5.6.7	Otx2	166
5.6.8	Six3	167
5.6.9	Transcriptional Repression in Anterior Specification	167
5.6.10	Summary	167
5.7	Conclusions and Outlook	168
	Abbreviations	169
	Acknowledgments	170
	Notes added in proof	170

6 Anterior-Posterior Patterning of the Hindbrain: Integrating Boundaries and Cell Segregation with Segment Formation and Identity
Angelo Iulianella and Paul A. Trainor

6.1	Introduction	189
6.2	Hindbrain Development	190
6.2.1	Segmentation into Rhombomeres and a Blueprint for Craniofacial Development	190
6.2.2	Segment- and Boundary-restricted *Hox* Gene Expression	193
6.2.3	Altering the *Hox* code: *Hox* Gene Loss- and Gain-of-Function	194
6.2.4	Initiating *Hox* Gene Expression in the Hindbrain	196

6.2.5	Interactions between FGF and Retinoid Signaling *199*	
6.2.6	*Krox20* and *Kreisler* Regulate Paralogous Groups 2 and 3 in the Vertebrate Hindbrain *201*	
6.2.7	Establishing the *Hox* Code in the Vertebrate Hindbrain: the Role of Auto- and Cross-regulation *204*	
6.2.8	A Mechanism for Establishing and Maintaining Hindbrain Segmentation *206*	
6.2.9	Coupling Cell Lineage Restriction with Discrete Domains of *Hox* Gene Expression *209*	
6.3	Conclusions *211*	
	Acknowledgments *211*	

7	**Neurogenesis in the Central Nervous System**	
	Véronique Dubreuil, Lilla Farkas, Federico Calegari, Yoichi Kosodo, and Wieland B. Huttner	
7.1	Introduction *229*	
7.2	Intrinsic Regulation of Neurogenesis *231*	
7.2.1	Neuronal Determination *231*	
7.2.1.1	Proneural Genes *231*	
7.2.1.2	Regulation of Proneural Activity *232*	
7.2.1.3	Proliferative Determinants *236*	
7.2.2	Neuronal Differentiation *236*	
7.3	Extrinsic Regulation of Neurogenesis *237*	
7.3.1	Wnt Factors *237*	
7.3.1.1	Wnt Pathway *238*	
7.3.1.2	Propagation of the Wnt Signal *238*	
7.3.1.3	Expression and Function during Neurogenesis *239*	
7.3.1.4	Effectors of the Wnt Pathway *240*	
7.3.1.5	Cross-talk between Signaling Pathways *241*	
7.3.2	Hedgehog Factors *241*	
7.3.2.1	Shh Pathway *241*	
7.3.2.2	Expression and Function during Neurogenesis *242*	
7.3.2.3	Eye Differentiation as a Model for Shh Activity *243*	
7.3.2.4	Cross-talk between Signaling Pathways *244*	
7.3.3	Fibroblast Growth Factors *244*	
7.3.3.1	Ligands, Receptors and Expression during Neurogenesis *244*	
7.3.3.2	Function during Neurogenesis *245*	
7.3.3.3	Cross-talk between Signaling Pathways *246*	
7.3.4	Transforming Growth Factors-α, Neuregulins and Epidermal Growth Factors *246*	
7.3.4.1	Ligands, Receptors and Expression during Neurogenesis *247*	
7.3.4.2	Function during Neurogenesis *247*	
7.3.5	Transforming Growth Factors-β *248*	

7.3.5.1   Ligands, Receptors and Expression during Neurogenesis   248
7.3.5.2   Function in Neurogenesis   249
7.3.5.3   Cross-talk between Signaling Pathways   249
7.3.5.4   BMP Family   250
7.3.6   Other Factors   251
7.3.6.1   Neurotrophins   252
7.3.6.2   Neurokines: LIF, CNTF   253
7.3.6.3   PDGF   253
7.3.6.4   GH   254
7.4   Cell Cycle Regulation and Neuronal Fate Determination   254
7.4.1   Cell Cycle of Neuroepithelial Cells   255
7.4.1.1   Interkinetic Nuclear Migration and Cell Division   255
7.4.1.2   Length of Cell Cycle of Neural Stem Cells   255
7.4.2   Cell-fate Determinants Influencing the Cell Cycle   256
7.4.2.1   Extrinsic Cell-fate Determinants Regulate the Cell Cycle   256
7.4.2.2   Intrinsic Cell-fate Determinants Regulate the Cell Cycle   257
7.4.3   Cell Cycle Regulators Influencing Cell Fate   258
7.4.3.1   Cell Cycle Regulators   258
7.4.3.2   Cell Cycle Regulators Regulate Cell-fate Determination   258
7.4.4   Model of Cell Cycle Lengthening   259
7.5   Neuron-generating Asymmetric Cell Division   260
7.5.1   Background   260
7.5.2   Asymmetric Cell Divisions of Neuroblasts in the *Drosophila* CNS   260
7.5.2.1   Cell Lineage   260
7.5.2.2   Apical-Basal NB Polarity   261
7.5.2.3   Control of Spindle Orientation   262
7.5.2.4   Cell-fate Determinants   262
7.5.3   The Neuronal Cell Lineage in the Mammalian CNS   263
7.5.3.1   Neuron-generating Divisions and Cell-fate Asymmetry   263
7.5.3.2   Cell Polarity in the Mammalian Neuroepithelium   264
7.5.3.3   Neuron-generating Division and Asymmetric Cell Division in Mammals   264
7.5.3.4   Asymmetric Distribution of Cell-fate Determinants in Mammals   266
7.6   Conclusions   266

8   Generating Cell Diversity
Ajay Chitnis
8.1   Introduction   287
8.2   Establishing Early Compartments of the Embryo   288
8.2.1   Establishing the Three Germ Layers   288
8.2.1.1   Nodals Function as Morphogens to Define Germ Layers   288

8.2.1.2	Maternal Factors Regulate Nodal Signaling	290
8.2.2	Establishing the Dorsal-Ventral Axis	290
8.2.2.1	Formation of the Vertebrate Organizer	290
8.2.2.2	β-Catenin-dependent Dorsalizing Factors Position the Organizer	291
8.2.2.3	The Vertebrate Dorsal Organizer is a Self-organizing Tissue	291
8.2.2.4	The BMP Gradient Determines Compartments of the Ectoderm	292
8.2.2.5	Neural Genes Prevent Differentiation and Maintain a Neural State	293
8.2.3	Establishing the Rostral-Caudal (Anterior-Posterior) Axis	293
8.2.3.1	The Blastoderm Margin is an Early Source of Posteriorizing Factors	293
8.2.3.2	Rostral-Caudal Patterning of the Anterior Neuroectoderm	294
8.2.3.3	Secondary Organizers and Compartment Boundaries	295
8.2.3.4	Rostral-Caudal Patterning of the Hindbrain	296
8.2.3.5	Establishing a Rostral-Caudal FGF Gradient	296
8.2.3.6	The FGF Gradient Regulates Mesoderm and Neuroectoderm Fate	297
8.3	Early Neurogenesis in the Neural Plate	298
8.3.1	Proneural Genes Define Neurogenic Domains	298
8.3.2	Notch Signaling Regulates Differentiation within Neurogenic Domains	300
8.3.3	Cell Fate Decisions Regulated by Notch	300
8.3.4	A Brief Review of Notch Signaling	301
8.3.5	Factors that Influence the Outcome of Notch Signaling	302
8.3.5.1	Fringe Glycosyl Transferases	302
8.3.5.2	E3 Ubiquitin Ligases	303
8.3.5.3	Numb and Biasing Notch Signaling in the SOP	303
8.3.6	Determination of Neurogenic and Non-neurogenic Domains in the Neuroectoderm	304
8.4	Determination of Neuronal Identities in the Spinal Cord	305
8.4.1	Cellular Organization of the Spinal Cord	305
8.4.2	Specification of Dorsal Neurons	306
8.4.3	The Notochord and Floor Plate Establish a Hedgehog Gradient	308
8.4.4	Class I and Class II Transcription Factors Respond to a Hedgehog Gradient to Define Discrete Compartments of Neuron Progenitors in the Ventral Cord	309
8.4.5	Specification of Motor Neuron Fate	310
8.4.6	Organization of the Motor Neurons and the LIM Code	311
8.5	Concluding Remarks	313

8.5.1	Gradients, Compartments, Boundaries and Local Interactions	*313*
8.5.2	Three Principles of Transcription Factor Regulation	*314*
8.5.3	Temporal Regulation of Cell Diversity	*314*

9 The Molecular Basis of Directional Cell Migration
Hans Georg Mannherz

9.1	Introduction	*321*
9.2	How and What Moves Cells	*322*
9.3	The Cytoskeleton	*323*
9.3.1	Microtubules	*325*
9.3.2	Microfilaments	*325*
9.3.3	Intermediate Filaments	*327*
9.4	Motile Membrane Extensions containing Microfilaments	*328*
9.4.1	The Cytoskeletal Components of Transient Actin-Powered Protrusive Organelles	*328*
9.4.2	Lamellipodial Protrusion	*330*
9.5	Modulation of the Structure of Transient Protusive Membrane Extensions	*333*
9.6	Formation of Substratum Contacts	*334*
9.7	Cell Body Traction	*336*
9.8	Rear Detachment	*337*
9.9	The Role of Microtubules in Cell Migration	*338*
9.10	The Role of Intermediate Filaments in Cell Migration	*338*
9.11	Modulation of the Cytoskeletal Organization by External Signals	*338*
9.12	The Cytoskeletal Organization of Neuronal Growth Cones	*341*
9.13	Conclusions	*343*
	Acknowledgement	*344*

10 Cell Death in Organ Development
Kerstin Krieglstein

10.1	Introduction	*347*
10.2	Programmed Cell Death	*347*
10.3	Genetics of Programmed Cell Death	*348*
10.4	Cell Death Receptors	*349*
10.5	Growth Factor Deprivation	*349*
10.6	Programmed Cell Death in Developing Organs	*350*
10.6.1	Cavitation of the Early Embryo during Implantation	*350*
10.6.2	Cardiovascular Development	*350*
10.6.3	Renal Development	*351*

10.6.4	Nervous System Development 351
10.6.5	Inner Ear Development 352
10.6.6	Removal of Interdigital Tissue and Separation of Digits 352
10.6.7	Removal of Mal-instructed Cells of the Immune System 353
10.7	Concluding Remarks 353
	Acknowledgments 353

Volume 2

II Cell Signaling and Growth Factors in Organogenesis

11 Dorso-Ventral Patterning of the Vertebrate Central Nervous System
Elisa Martí, Lidia García-Campmany, and Paola Bovolenta

11.1	Introduction 361
11.2	Generating Cell Diversity in the Dorsal Neural Tube 363
11.2.1	The Neural Crest 363
11.2.1.1	Cellular and Molecular Inducers of Neural Crest 364
11.2.1.2	Molecular Identity of Newly-induced Neural Crest Cells 366
11.2.2	The Roof Plate 367
11.2.3	Signals from the Roof Plate Pattern the Dorsal Spinal Cord 369
11.2.3.1	Transcriptional Code Defining Dorsal Spinal Cord Interneurons 370
11.3	Generating Cell Diversity in the Ventral Neural Tube 371
11.3.1	The Floor Plate 371
11.3.2	Sonic Hedgehog Secreted from the Notochord and the Floor Plate Patterns the Ventral Spinal Cord 373
11.4	Generation of D-V Patterning in the Anterior Neural Tube Follows Rules Similar to those Used in the Spinal Cord 374
11.4.1	The Eye 375
11.4.1.1	D-V Patterning of the Optic Vesicle 376
11.4.1.2	D-V Patterning of the Optic Cup 379
11.4.2	D-V Patterning of the Telencephalon 381
11.5	Conclusions and Perspectives 384

12		**Novel Perspectives in Research on the Neural Crest and its Derivatives**
		Chaya Kalcheim, Matthias Stanke, Hermann Rohrer, Kristjan Jessen, and Rhona Mirsky
12.1		General Introduction 395
12.2		Novel Techniques to Investigate NC Development and Fate 396
12.2.1		Cellular Tracers 396
12.2.2		Molecules Expressed by NC Cells 397
12.2.3		Investigating the Function of Molecules Expressed by NC Cells 397
12.3		Mechanisms of Specification and Emigration of NC Cells 399
12.3.1		Specification of the NC 399
12.3.1.1		Transcription Factors in NC Formation 400
12.3.2		The Delamination of NC Progenitors from the Neural Tube 401
12.3.2.1		A Balance between BMP and its Inhibitor Noggin Regulates NC Delamination in the Trunk 402
12.3.2.2		BMP-dependent Genes and NC Delamination 402
12.3.2.3		The Role of the Cell Cycle in NC Delamination 403
12.4		Peripheral Neuronal Lineages: Pluripotentiality and Early Restrictions of Migratory NC 404
12.4.1		Evidence from Cell Lineage Analysis *In Vivo* and *In Vitro* 404
12.4.2		NC Stem Cells and the PNS Lineage 406
12.5		Molecular Control of Neuron Development in Peripheral Ganglia 408
12.5.1		Extrinsic Signals in Sensory, Sympathetic, Parasympathetic and Enteric Neuron Development 408
12.5.1.1		Sensory Ganglia 408
12.5.1.2		Sympathetic Ganglia 408
12.5.1.3		Parasympathetic Ganglia 410
12.5.1.4		Enteric Ganglia 410
12.5.2		Transcriptional Control of Peripheral Neuron Development 411
12.5.2.1		Sensory Neurons 411
12.5.2.2		Sympathetic and Parasympathetic Neurons 413
12.5.2.3		Enteric Neurons 414
12.6		Growth Cone Navigation and Innervation of Peripheral Targets 414
12.6.1		Semaphorins 415
12.6.2		GDNF-family Ligands 415
12.6.3		Hepatocyte Growth Factor (HG) 416
12.6.4		Macrophage Stimulating Protein (MSP) 416
12.6.5		TGF-β/BMP Family 416

12.7	Factors Controlling Neuronal Survival in the PNS	417
12.8	Glial Cell Development from the NC-Peripheral Glial Lineages: Fate Choice and Early Developmental Events	417
12.8.1	The Main Cell Types	417
12.8.2	Molecular Markers of Early Glial Development	419
12.8.3	Transcription Factors that Control the Emergence of Glia	421
12.8.4	Inductive Signals Involved in Glial Specification	422
12.8.5	Developmental Plasticity of Early Glia	423
12.9	Signals that Control Schwann Cell-precursor Survival and Schwann Cell Generation	423
12.9.1	Schwann Cell Precursors	423
12.9.2	Schwann Cell Generation	425
	Acknowledgments	426

13 Eye Development
Filippo Del Bene and Joachim Wittbrodt

13.1	Vertebrate Eye Development: An Overview	449
13.2	Patterning of the Anterior Neural Plate	449
13.2.1	Posteriorizing Factors	450
13.2.2	Repression of Posteriorizing Factors is required for Head Formation	451
13.2.3	Subdivision of the Anterior Neural Plate	451
13.3	Transcription Factors Function in the Establishment of Retinal Identity	453
13.4	Patterning of the Eye Field and its Medio-lateral Separation	456
13.5	Differentiation of Neuroretina and Pigmented Retinal Epithlium	458
13.5.1	Trigger from the Ventral Midline	458
13.5.2	Retinal Polarity	458
13.5.3	Intrinsic Signals in the Neuroretina	460
13.5.4	Six3 and Geminin	461
13.5.5	Chx10	462
13.6	Differentiation of Retinal Cell Types	463
13.6.1	Retinal Fate Determination by Basic Helix-Loop-Helix (bHLH) and Homeobox Transcription Factors	465
13.6.2	Retinal Ganglion Cells	466
13.6.3	Amacrine Cells	467
13.6.4	Bipolar Cells	467
13.6.5	Horizontal Cells	468
13.6.6	Photoreceptor Cells	468
13.6.7	Maintenance of Retinal Stem Cells and Generation of Müller Glia	468

14   Mammalian Inner Ear Development: Of Mice and Man
Bernd Fritzsch and Kirk Beisel
14.1   Introduction   487
14.1.1   An Outline of Mammalian Ear Evolution   487
14.1.2   Human Deafness Related Mutations   488
14.2   Morphological and Cellular Events in Mammalian Ear Development   491
14.2.1   Molecular Basis of Ear Placode Induction   493
14.2.2   Molecular Basis of Ear Morphogenesis   494
14.2.2.1 FGFs   495
14.2.2.2 The EYA/SIX/DACH Complex   497
14.2.2.3 GATA3   499
14.2.3   Molecular Biology of Otoconia, Cupula and Tectorial Membrane Formation and Maintenance   500
14.2.4   Molecular Basis of Mammalian Ear Histogenesis   503
14.2.4.1 Sensory Neuron Development   504
14.2.4.2 Hair Cell Development   507
14.3   Postnatal Maturation   510
14.4   Aging, Hair Cell Loss and the Molecular Biology of Hair Cell Regeneration   510

15   Limb Development
Susan Mackem
15.1   Introduction   523
15.1.1   Morphological Landmarks of the Limb   524
15.1.2   Embryological Elucidation of Inductive Interactions in Limb Development   524
15.2   The Major Signaling Pathways in Limb Development   528
15.2.1   Retinoids   528
15.2.2   Fgfs   530
15.2.3   Wnts   535
15.2.4   Bmps   541
15.2.5   Shh   547
15.2.6   Other Players   551
15.3   Limb Initiation and Formation of the Limb "Organizers"   553
15.3.1   Axial Cues for Limb Initiation and AER Induction   553
15.3.2   Establishment of Early DV Polarity and AER Formation in Limb Bud Ectoderm   556
15.3.3   AER Maturation and Maintenance   558
15.3.4   ZPA Formation   560
15.4   Function of Organizers in Regulating Pattern and Growth along the Limb Axes   562
15.4.1   The AER-Fgfs, RA, and the PD Axis   562

15.4.2	Dorsal Ectodermal Wnt7a and the DV Axis	566
15.4.3	ZPA-Shh and the AP Axis	568
15.5	Coordination of Patterning, Outgrowth and Differentiation: Positive Co-regulation and Feedback Loops Synchronize Output from Limb Organizers, and Antagonism between Signaling Pathways Contributes to Polarity	571
15.6	Ongoing Late Regulation and Realization of Pattern	573
15.6.1	Condensation and Segmentation of Skeletal Elements	573
15.6.2	Apoptosis and Sculpting the Final Limb Form	576
15.7	Potential Biomedical Applications and Future Directions	578

16 Skeletal Development
William A. Horton

16.1	Introduction	619
16.2	Skeletal Morphogenesis	620
16.2.1	Pre-condensation	620
16.2.2	Membranous Skeletal Development	620
16.2.3	Endochondral Skeletal Development	621
16.3	Skeletal Growth	624
16.3.1	Growth Plate Structure and Organization	624
16.3.2	Regulation of Linear Growth	626
16.3.2.1	Ihh-PTHrP Circuit	626
16.3.2.2	FGF-FGFR3 Circuit	627
16.3.2.3	CNP-GC-B Circuit	628
16.3.2.4	VEGF-MMP9 Circuit	628
16.4	Therapeutic Considerations	629
16.5	Conclusions	631

17 Musculature and Growth Factors
Petra Neuhaus, Herbert Neuhaus, and Thomas Braun

17.1	Introduction	641
17.2	Induction of Primary Myogenesis by Cell Signaling and Growth Factors	642
17.2.1	Wnt Molecules	643
17.2.2	Shh, Ihh	646
17.2.3	BMPs	647
17.2.4	FGFs	649
17.3	Tour Guides: How Growth Factors Guide Migration of Muscle Precursor Cells During Embryonic Development	651
17.3.1	HGF	651
17.3.2	FGFs	652
17.4	Control of Muscle Size and Muscle Fiber Diversity by Local and Circulating Growth Factors	653

17.4.1	Control of Muscle Size	653
17.4.1.1	Myostatin	653
17.4.1.2	IGFs	654
17.4.2	Muscle Fiber Diversity	656
17.4.2.1	Shh	656
17.4.2.2	Wnt	657
17.5	Growth Factors in Muscle Wasting and Skeletal Muscle Regeneration	659
17.5.1	Mediation of Muscle Wasting by Cell Signaling Molecules	659
17.5.2	Activation of Muscle Satellite Cells by Growth Factors	660
17.5.2.1	HGF	662
17.5.2.2	FGF	662
17.6	Does Growth Factor Mediated Recruitment of Uncommitted Stem Cells Contribute to Skeletal Muscle Cell Regeneration?	664

18	**Skin Development**	
	Lydia Sorokin and Leena Bruckner-Tuderman	
	Abbreviations	679
18.1	Introduction	679
18.2	Morphology of the Skin and Development of Hair Follicles and Other Adnexal Structures	681
18.2.1	Epidermis	681
18.2.2	Non-Keratinocyte Cell Types in the Epidermis	683
18.2.3	Hair Follicle	686
18.2.4	Sebaceous Glands	687
18.2.5	Nails	688
18.2.6	Eccrine and Apocrine Sweat Glands	688
18.3	Cell Adhesion and the Role of Adhesion Molecules in Development	689
18.3.1	Cell-Cell Adhesion	689
18.3.2	Cell-Extracellular Matrix Adhesion in the Epidermis	690
18.3.2.1	Hemidesmosomes	690
18.3.2.2	Inter-Hemidesmosomal Spaces	693
18.3.3	Role of Laminins in Epidermal Basement Membrane Development	695
18.4	Development of the Dermal Matrix	698
18.5	Epithelial-Mesenchymal Interactions and Signaling	700
18.6	Epidermal Stem Cells	703
18.7	Pathological Skin Conditions Caused by Developmental Abnormalities	705
18.8	Future Considerations	706
	Acknowledgments	707

19 Tooth Development
Xiu-Ping Wang and Irma Thesleff

19.1 Introduction 719
19.2 Signaling Pathways and their Networks Regulating Tooth Development 722
19.2.1 Fibroblast Growth Factors 722
19.2.2 TGFβ Family 725
19.2.3 Sonic Hedgehog 729
19.2.4 Wnt Family 730
19.2.5 Tumor Necrosis Factors 732
19.2.6 Other Growth Factors 733
19.2.7 Integrations between Growth Factor Signal Pathways 734
19.3 Biological Effects 735
19.3.1 Formation of Dental Placodes 736
19.3.2 Epithelial Cell Proliferation 736
19.3.3 Patterning of the Tooth Crown 738
19.3.4 Cell Differentiation 738
19.4 Biomedical Application 739
19.5 Summary 742

20 Gastrointestinal Tract
Daniel Ménard, Jean-François Beaulieu, François Boudreau, Nathalie Perreault, Nathalie Rivard, and Pierre H. Vachon

20.1 Introduction 755
20.2 Development of the Gastrointestinal Mucosa 756
20.2.1 Specialization of Epithelium and Functional Units 756
20.2.1.1 Esophagus 756
20.2.1.2 Stomach 756
20.2.1.3 Small Intestine 756
20.2.1.4 Colon 757
20.2.2 Mesenchymal-Epithelial Interactions 757
20.2.2.1 Growth Factors and Morphogens 758
20.2.2.2 Hox Genes 759
20.2.2.3 Epimorphin/Syntaxin 2 760
20.2.2.4 Basement Membrane Proteins 760
20.2.2.5 Mesodermal Transcription Factors 761
20.3 Gastric Cell Proliferation and Differentiation 761
20.3.1 Functional Compartmentalization of Gastric Glands 761
20.3.2 Hormones and Growth Factors 762
20.3.3 ECM and Integrins 763
20.4 Intestinal Cell Proliferation and Differentiation 765
20.4.1 Hormones and Growth Factors 765
20.4.2 ECM and Integrins 766

20.4.2.1 Development 768
20.4.2.2 Anteroposterior (AP) Axis 769
20.4.2.3 Crypt-Villus Axis 769
20.4.3 Cell Signaling Pathways 770
20.3.4.1 ERK-MAP Kinase Cascade 770
20.4.3.2 p38–MAP Kinase Cascade 771
20.4.3.3 Wnt Pathway 772
20.4.3.4 Phosphatidylinositol 3–Kinase Signaling Pathway 772
20.4.4 Transcription Factors 774
20.4.4.1 Transcription Factors Involved in the Determination of Intestinal Epithelial Cell Lineage 775
20.4.4.2 Transcriptional Regulators of Intestinal-specific Genes 776
20.4.5 Cell Survival, Apoptosis and Anoikis 778
20.4.5.1 Crypt-Villus Axis Distinctions 779
20.4.5.2 Proximal-Distal Axis Distinctions 780
20.4.5.3 Intestinal Cell Survival and Death: Differences and Differentiation 781
20.5 Biomedical Applications 782

21 Cell Signaling and Growth Factors in Lung Development
David Warburton, Saverio Bellusci, Pierre-Marie Del Moral, Stijn DeLanghe, Vesa Kaartinen, Matt Lee, Denise Tefft, and Wei Shi
21.1 Introduction 791
21.1.1 The Stereotypic Branch Pattern of Respiratory Organs 791
21.1.2 Transduction of Candidate Growth Factor Peptide Ligand Signals 792
21.1.3 Examples of Peptide Growth Factor Signaling Pathways 792
21.2 Growth Factors and Lung Development 792
21.2.1 Candidate Growth Factors in Lung Development 792
21.2.2 Growth Factor-mediated Epithelial-Mesenchymal Interactions and Lung Development 793
21.2.3 FGF10 793
21.2.4 The Role of BMP4 794
21.2.5 The Role of the Vasculature and VEGF Signaling 794
21.2.6 Postnatal Lung Development 795
21.2.7 The Influence of Peptide Growth Factor Signaling on the Correct Organization of the Matrix 795
21.3 Growth Factor Signaling Pathways in Lung Development 795
21.3.1 Critical Signaling Pathways in Lung Development 795
21.3.1.1 BMPs 795
21.3.1.2 Activin Receptor-like Kinases 796
21.3.1.3 ALKs in Pulmonary Development 796
21.3.1.4 ALKs and the Pulmonary Vasculature 797

21.3.1.5	ALKs, Pulmonary Fibrosis and Inflammation	797
21.3.2	FGF Signaling Promotes Outgrowth of Lung Epithelium	798
21.3.3	FGF2b	798
21.3.4	FGF10	799
21.3.5.	FGF10 Activity	799
21.3.5.1.	Regulation of FGF Signaling	800
21.3.5.2.	Mice and Humans Possess Several *Spry* Genes (*Mspry1–4* and *Hspry1–4*)	800
21.3.5.3.	Sprouty Binds Downstream Effector Complexes	800
21.3.5.4.	FGFs as Tyrosine Kinase Receptors	801
21.3.5.5.	Relationship between FGF Signaling and Spry during Development	801
21.3.5.6.	Spry2 and Spry4 Share a Common Inhibitory Mechanism	801
21.3.4	Sonic Hedgehog, Patched and Hip	802
21.3.4.1	Expression and Activation of TGFβ Family of Peptides	802
21.3.4.2.	Developmental Specificity of the TGFβ1 Overexpression Phenotype	803
21.3.4.3.	TGFβ Signaling	804
21.3.4.4.	The Bleomycin-induced Model of Lung Fibrosis	804
21.3.5.	VEGF Isoform and Cognate Receptor Signaling in Lung Development	805
21.3.5.1.	Both Humans and Mice have Three Different VEGF Isoforms	805
21.3.5.2.	*Vegf* Knockout Mice have a Lethal Phenotype within the Early Stages of Embryonic Development (E8.5–E9)	805
21.3.5.3.	The Role of VEGF in Maintaining Alveolar Structure	806
21.3.5.4.	VEGF-C and VEGF-D	806
21.3.5.5.	VEGF Isoforms Induce Vasculogenesis, Angiogenesis and Lymphoangiogenesis	806
21.3.6.	Wnt Signaling	806
21.3.7.	Inactivation of the β-Catenin Gene	808
21.3.8.	Dickkopf Regulates Matrix Function	808
21.4	Regulation of Signaling Networks	809
21.4.1.	Growth Factor Tyrosine Kinase and TGF-β Pathways	809
21.4.2.	Mutual Regulation of Intracellular Signaling Networks	809
21.4.3.	Regulation of TGF-β signaling by EGF Signaling	810
21.4.4.	Calcium Signaling and the Mitochondrial Apoptosis Pathway	810
21.5	Developmental Modulation of Growth Factor Signaling by Adapter Proteins	810
21.6	Morphogens and Morphogenetic Gradients	811
21.6.1.	Coordination of Growth Factor Morphogenetic Signals to Determine Lung Development	811

21.6.2.	Action of Morphogens	*811*
21.6.3.	Other Morphogen Gradient Systems	*812*
21.6.4.	The APR Model	*812*
21.6.5.	The Modified Turing Model	*813*
21.6.6.	Expression of *Fgf10*	*813*
21.6.7.	Tip-splitting Event	*813*
21.6.8.	The Value of Hypothetical Models	*814*
21.6.9.	Retinoic Acid Receptors	*815*
21.7.	Conclusion	*815*

22 Molecular Genetics of Liver and Pancreas Development
Tomas Pieler, Fong Cheng Pan, Solomon Afelik, and Yonglong Chen

22.1	Introduction	*823*
22.2	From the Fertilized Egg to Primitive Endodermal Precursor Cells	*825*
22.3	Commitment to Pancreas and Liver Fates in Xenopus	*826*
22.4	Liver and Pancreas Specification in Mouse, Chicken and Zebrafish	*826*
22.5	Proliferation and Differentiation of Functionally Distinct Pancreatic and Hepatic Cell Populations	*829*
22.6	Transdifferentiation of Pancreas and Liver	*830*
22.7	Pancreas and Liver Regeneration	*830*
22.8	Generation of Pancreatic and Hepatic Cells from Pluripotent Embryonic Precursor Cells	*832*

23 Molecular Networks in Cardiac Development
Thomas Brand

23.1	Introduction	*841*
23.2	Implications of Studying Heart Development for Adult Cardiology	*841*
23.3	Model Organisms	*842*
23.3.1	The Mouse Embryo (*Mus musculus*)	*842*
23.3.2	The Chick Embryo (*Gallus gallus*)	*843*
23.3.3	The Frog Embryo (*Xenopus laevis*)	*843*
23.3.4	The Zebrafish Embryo (*Danio rerio*)	*843*
23.3.5	Lower Chordates	*844*
23.3.6	*Drosophila melanogaster*	*844*
23.3.7	*Homo sapiens*	*844*
23.4	Anatomical Description of Heart Development	*845*
23.5	Cardiac Induction	*846*
23.5.1	The Role of BMP2 in Heart Induction	*847*
23.5.2	Canonical Wnt Signaling Interferes with Heart Formation in Vertebrates	*849*

23.5.3	FGF Cooperates with BMP2 *850*	
23.5.4	Cripto *851*	
23.5.5	Shh *852*	
23.5.6	Notch Signaling Interferes with Myocardial Specification *852*	
23.6	Transcription Factor Families Involved in Early Heart Induction *853*	
23.6.1	The NK Family of Homeobox Genes *853*	
23.6.2	The GATA Family of Zinc-finger Transcription Factors *854*	
23.6.3	Serum Response Factor *855*	
23.6.4	Synergistic Interaction of Cardiac Transcription Factors *855*	
23.7	Tubular Heart Formation *856*	
23.8	Left-Right Axis Development *857*	
23.8.1	Looping Morphogenesis *857*	
23.8.1.1	Mechanisms of L-R Axis Determination *857*	
23.8.1.2	Generation of Initial Asymmetry *858*	
23.8.1.3	Transfer of L-R Asymmetry to the Organizer Tissue *858*	
23.8.1.4	The Nodal Flow Model *859*	
23.8.1.5	Transfer of L-R Asymmetry to the Lateral Plate *861*	
23.8.1.6	Asymmetric Organ Morphogenesis *862*	
23.8.2	A-P Axis Formation in the Heart *864*	
23.8.3	Dorso-ventral Polarity of the Heart Tube *864*	
23.9	Chamber Formation *865*	
23.9.1	Analysis of Growth Patterns in the Heart *865*	
23.9.2	Reprogramming of Gene Expression at the Onset of Chamber Development *865*	
23.9.3	The Secondary or Anterior Heart Field (AHF) *866*	
23.9.4	The Right Ventricle is a Derivative of the Anterior Heart Field *868*	
23.9.5	T-box Genes Pattern the Cardiac Chambers and are Involved in Septum Formation *869*	
23.9.6	Cell-Cell Interaction in Chamber Formation *869*	
23.9.6.1	Formation of Compact and Trabecular Layer *869*	
23.9.6.2	The Epicardium Controls Ventricular Compact Layer Formation *870*	
23.9.6.3	Epicardial Cells Form the Coronary Vasculature after Epithelial Mesenchymal Transition *871*	
23.10	Outflow Tract Patterning and the Role of the Neural Crest *872*	
23.10.1	Tbx1 is Mutated in the DiGeorge Syndrome *873*	
23.10.2	The Neural Crest Cells have an Early Role in the Heart Tube *873*	
23.11	Signals Governing Valve Formation *874*	
23.11.1	The Tgfβ Superfamily and Valve Formation *874*	
23.11.2	AV Cushion Formation Requires Hyaluronic Acid *875*	
23.11.3	NFAT2 probably Mediates VEGF Signaling during Cushion Formation *876*	

- 23.11.4 Wnt and Notch/Delta Signaling Pathways and Cardiac Valve Formation   876
- 23.12 Epigenetic Factors   877
- 23.13 Outlook   877
  - Acknowledgments   879

24 Vasculogenesis
Georg Breier

- 24.1 Introduction   909
- 24.2 Modes of Blood Vessel Morphogenesis: Vasculogenesis, Angiogenesis and Arteriogenesis   909
- 24.3 Endothelial Cell Differentiation and Hematopoiesis   910
- 24.4 Adult Arteriogenesis and Vasculogenesis   911
- 24.5 Endothelial Cell Growth and Differentiation Factors   912
- 24.6 VEGF and VEGF Receptors   913
- 24.7 Other VEGF Family Members: Involvement in Angiogenesis and Lymphangiogenesis   914
- 24.7.1 Angiopoietins and Ties in Angiogenesis and Lymphangiogenesis   915
- 24.7.2 Ephrins, Notch, and Arteriovenous Differentiation   916
- 24.8 Hypoxia-inducible Factors and Other Endothelial Transcriptional Regulators   917
- 24.9 Outlook   917

25 Inductive Signaling in Kidney Morphogenesis
Hannu Sariola and Kirsi Sainio

- 25.1 Early Differentiation of the Kidney   925
- 25.2 Regulation of Ureteric Bud Branching   928
- 25.3 Genes Affecting Early Nephrogenesis   928
- 25.4. Signaling Molecules, their Receptors, and the Integrins   930
- 25.4.1 Glial Cell Line-derived Neurotrophic Factor   930
- 25.4.2 Fibroblast Growth Factors   932
- 25.4.3 Leukemia Inhibitory Factor   932
- 25.4.4 Wnt Proteins   933
- 25.4.5. Bone Morphogenetic Proteins   933
- 25.4.6. Sonic Hedgehog   934
- 25.4.7 Formins   934
- 25.4.8 Hepatocyte Growth Factor and Met Receptor Tyrosine Kinase   934
- 25.4.9. Integrins   935
- 25.4.10. Other Molecules   935
- 25.5 Transcription Factors   935

25.5.1	Wilms' Tumor Gene 1	935
25.5.2	Pax–2	936
25.5.3	Six-Eya Complex	936
25.5.4	Emx–2	937
25.5.5	Lim–1	937
25.5.6	Fox Genes	937
25.6	Future Perspectives	938

26 Molecular and Cellular Pathways for the Morphogenesis of Mouse Sex Organs

Humphrey Hung-Chang Yao

26.1	Introduction	947
26.2	Building the Foundation: Establishment of the Urogenital Ridge	948
26.2.1	Gonadogenesis in Mice	949
26.2.2	Molecular Mechanisms of Gonadogenesis	949
26.3	Parting the Way: Sexually Dimorphic Development of the Gonad	950
26.3.1	Embryonic Testis	951
26.3.1.1	Sry: The Master Switch from the Y Chromosome	951
26.3.1.2	Sox9: The Master Switch Downstream of Sry	953
26.3.1.3	Specification of Sertoli Cell Lineage	954
26.3.1.4	Cellular Events Triggered by *Sry*	955
26.3.2	Embryonic Ovary	958
26.3.2.1	The Quest for the Ovary-determining Gene	958
26.3.2.2	Female Germ Cells: The Key to Femaleness	960
26.4	Dimorphic Development of the Reproductive Tracts	961
26.4.1	Initial Formation of the Wolffian and Müllerian Ducts	961
26.4.2	Sexually Divergent Development of Reproductive Tracts	963
26.4.3	Patterning of the Reproductive Tracts	964
26.5	Morphogenesis of the External Genitalia	966
26.6	Summary	967

Index 979

Preface

Developmental biology investigates and describes, in the broadest sense, changes in organisms from their origins to birth, and on into adulthood, or even to death. Developmental changes are characterized by increasing complexity and specialization of cells and organs. Research in developmental biology has always been a combination of experiment and description. Although mostly morphologically oriented in its beginnings, developmental biology has experienced exponential growth in the 20th century following its fruitful merger with molecular biology and genetics during the 1980s and 1990s.

What we are currently witnessing is the integration of molecular cell biology into developmental biology, with the goal of eventually understanding all developmental events at the level of and as the result of cell biological events. Following this trend, we may ultimately expect to understand development at the resolutional level of molecular machines, individual molecules, and even atoms. The elucidation of signaling between developing cells and the analysis of intracellular signaling networks represent an intermediate but important and indispensable step in the attainment of this goal.

This book addresses representative examples of inter- and intracellular signaling in development involving growth and transcription factors. The various chapters cover general aspects of development such as e.g. stem cells, implantation and placentation, generation of cell diversity, cell migration, cell death, anterior-posterior and dorso-ventral patterning. Chapters concerning the development of specific structures and organs address, *inter alia*, the nervous system, gastrointestinal, respiratory, and reproductive tracts, cardiovascular and excretory systems, sense organs, and limb, skeletal and muscle development.

We thank all the authors for having taken on the burden of writing these extensive reviews and sharing with the community their profound knowledge of their specialist subjects. We are particularly grateful to those authors who submitted their chapters early and who have shown great patience in awaiting the completed volumes.

We also thank all the reviewers, who through their critically assessments have helped the authors to maximize the quality of their chapters.

Cell Signaling and Growth Factors in Development. Edited by K. Unsicker and K. Krieglstein
Copyright © 2006 WILEY-VCH Verlag GmbH & Co. KGaA, Weinheim
ISBN 3-527-31034-7

Special thanks go to Ursel Lindenberger for secretarial assistance in Heidelberg (Neuroanatomy Secretary Heidelberg) and to Dr Andreas Sendtko and the Wiley team for having initiated and promoted the project.

Heidelberg and Göttingen
September 2005

Klaus Unsicker and Kerstin Krieglstein

List of Contributors

Solomon Afelik
Georg-August-University of
Goettingen
Goettingen Center for Molecular
Biosciences
Department of Developmental
Biochemistry
Justus-von-Liebig-Weg 11
37077 Göttingen
Germany

Jean-François Beaulieu
CIHR Group on Functional
Development and Physiopathology of
the Digestive Tract
Department of Anatomy and Cell
Biology
Faculté de médecine et des sciences de
la santé
Université de Sherbrooke
Sherbrooke (Québec) J1H 5N4
Canada

Kirk W. Beisel
Creighton University
Department of Biomedical Sciences
723 N 18th Street
Omaha, NE 68178
USA

Saverio Bellusci
University of Southern California Keck
School of Medicine
Developmental Biology Program
Saban Research Institute
Childrens Hospital Los Angeles
4650 Sunset Boulevard
Los Angeles CA 90027
USA

François Boudreau
CIHR Group on Functional
Development and Physiopathology of
the Digestive Tract
Department of Anatomy and Cell
Biology
Faculté de médecine et des sciences de
la santé
Université de Sherbrooke
Sherbrooke (Québec) J1H 5N4
Canada

Paola Bovolenta
Instituto Cajal de Neurobiología
(CSIC)
Av. Doctor Arce 37
28002 Madrid
Spain

Thomas Brand
University of Wuerzburg
Molecular Developmental Biology
Cell and Developmental Biology
(Zoology I)
Am Hubland
97074 Würzburg
Germany

Thomas Braun
Department of Physiological
Chemistry
Martin-Luther-University
Halle-Wittenberg
Hollystr. 1
06114 Halle
Germany

Georg Breier
Medical Faculty Carl Gustav Carus
University of Technology Dresden
Institute of Pathology
Fetscherstr. 74
01307 Dresden
Germany

Leena Bruckner-Tuderman
Department of Dermatology
University of Freiburg
Hauptstr. 7
79104 Freiburg
Germany

Federico Calegari
Max Planck Institute of Molecular Cell
Biology and Genetics
Pfotenhauerstrasse 108
01307 Dresden
Germany

Yonglong Chen
Georg-August-University Goettingen
Goettingen Center for Molecular
Biosciences
Department of Developmental
Biochemistry
Justus-von-Liebig-Weg 11
37077 Göttingen
Germany

Ajay Chitnis
Unit on Vertebrate Neural
Development
Laboratory of Molecular Genetics
National Institute of Child Health and
Human Development
National Institutes of Health
6 Center Drive
Bethesda, MD20892
USA

Susana M. Chuva de Sousa Lopes
Hubrecht Laboratory
Netherlands Institute for
Developmental Biology
Uppsalalaan 8
3584 CT Utrecht
The Netherlands

Massimo De Felici
Section of Histology and Embryology
Department of Public Health and Cell
Biology
University of Rome "Tor Vergata"
Via Montpellier 1
00133 Rome
Italy

Stijn DeLanghe
Developmental Biology Program
Saban Research Institute
Childrens Hospital Los Angeles
4650 Sunset Boulevard
Los Angeles CA 90027
USA

List of Contributors | XXXI

Filippo Del Bene
European Molecular Biology
Laboratory
EMBL
Developmental Biology Programme
Meyerhofstrasse 1
69117 Heidelberg
Germany

Pierre-Marie Del Moral
University of Southern California
School of Dentistry
Developmental Biology Program
Saban Research Institute
Childrens Hospital Los Angeles
4650 Sunset Boulevard
Los Angeles CA 90027
USA

Susanna Dolci
Section of Human Anatomy
Department of Public Health and Cell
Biology
University of Rome "Tor Vergata"
Via Montpellier 1
00133 Rome
Italy

Véronique Dubreuil
Max Planck Institute of Molecular Cell
Biology and Genetics
Pfotenhauerstrasse 108
01307 Dresden
Germany

Donatella Farini
Section of Histology and Embryology
Department of Public Health and Cell
Biology
University of Rome "Tor Vergata"
Via Montpellier 1
00133 Rome
Italy

Lilla Farkas
Max Planck Institute of Molecular Cell
Biology and Genetics
Pfotenhauerstrasse 108
01307 Dresden
Germany

Bernd Fritzsch
Creighton University
Department of Biomedical Sciences
723 N 18th Street
Omaha, NE 68178
USA

Lidia García-Campmany
Instituto de Biología Molecular de
Barcelona (CSIC)
Parc Cientific de Barcelona
Josef Samitier 1–5
08028 Barcelona
Spain

William A. Horton
Director, Research Center
Shriners Hospital for Children
3101 S. W. Sam Jackson Park Road
Portland, OR 97239
USA

Wieland B. Huttner
Max Planck Institute of Molecular Cell
Biology and Genetics
Pfotenhauerstrasse 108
01307, Dresden
Germany

Angelo Iulianella
Stowers Institute for Medical Research
1000 E. 50th Street,
Kansas City, MO 64110
USA

Kristjan Jessen
Department of Anatomy and
Developmental Biology
University College London
Gower Street
WC1E 6BT London
UK

Vesa Kaartinen
University of Southern California Keck
School of Medicine
Developmental Biology Program
Saban Research Institute
Childrens Hospital Los Angeles
4650 Sunset Boulevard
Los Angeles CA 90027
USA

Chaya Kalcheim
Department of Anatomy and Cell
Biology
Hebrew University-Hadassah Medical
School
P.O. Box 12272
Jerusalem 91120
Israel

Clemens Kiecker
MRC Centre for Developmental
Neurobiology
Guy's Hospital Campus
King's College
University of London
London SE1 1UL
UK

Yoichi Kosodo
Max Planck Institute of Molecular Cell
Biology and Genetics
Pfotenhauerstrasse 108
01307 Dresden
Germany

Kerstin Krieglstein
Department of Neuroanatomy
Medical Faculty
University of Göttingen
Kreuzbergring 36
37075 Göttingen
Germany

Matt Lee
University of Southern California
School of Dentistry
Center for Craniofacial Molecular
Biology
Health Sciences Campus
Los Angeles, CA 90033
USA

Susan Mackem
National Cancer Institute
Laboratory of Pathology
Building 10, Room 2A33
9000 Rockville Pike
Bethesda, MD 20892–1500
USA

Hans Georg Mannherz
Department of Anatomy and Cell
Biology
Ruhr-University
Universitätsstr. 150
44780 Bochum
Germany

Elisa Martí
Instituto de Biología Molecular de
Barcelona (CSIC)
Parc Cientific de Barcelona
Josep Samitier 1–5
08028 Barcelona
Spain

Daniel Ménard
Département d'Anatomie et de
Biologie Cellulaire
Faculté de Médecine
Université de Sherbrooke
Sherbrooke
Québec, J1H 5N4
Canada

Rhona Mirsky
Department of Anatomy and
Developmental Biology
University College London
Gower Street
WC1E 6BT London
UK

Christine L. Mummery
Hubrecht Laboratory
Netherlands Institute for
Developmental Biology
Uppsalalaan 8
3584 CT Utrecht
The Netherlands

Andrea Münsterberg
School of Biological Sciences
Developmental Biology Research
Groups
University of East Anglia
Norwich NR4 7TJ
UK

Herbert Neuhaus
Department of Physiological
Chemistry
Martin-Luther-University
Halle-Wittenberg
Hollystr. 1
06114 Halle
Germany

Petra Neuhaus
Department of Physiological
Chemistry
Martin-Luther-University
Halle-Wittenberg
Hollystr. 1
06114 Halle
Germany

Fong Cheng Pan
Georg-August-University of
Goettingen
Goettingen Center for Molecular
Biosciences
Department of Developmental
Biochemistry
Justus-von-Liebig-Weg 11
37077 Göttingen
Germany

Christian Paratore
Institute of Cell Biology
Swiss Federal Institute of Technology
ETH-Hoenggerberg HPM E47
8093 Zürich
Switzerland

Nathalie Perreault
CIHR Group on Functional
Development and Physiopathology of
the Digestive Tract
Department of Anatomy and Cell
Biology
Faculté de médecine et des sciences de
la santé
Université de Sherbrooke
Sherbrooke (Québec) J1H 5N4
Canada

Tomas Pieler
Georg-August-University of Goettingen
Goettingen Center for Molecular Biosciences
Department of Developmental Biochemistry
Justus-von-Liebig-Weg 11
37077 Göttingen
Germany

Nathalie Rivard
CIHR Group on Functional Development and Physiopathology of the Digestive Tract
Department of Anatomy and Cell Biology
Faculté de médecine et des sciences de la santé
Université de Sherbrooke
Sherbrooke (Québec) J1H 5N4
Canada

Hermann Rohrer
Max-Planck-Institut für Hirnforschung
Abteilung Neurochemie
Deutschordenstrasse 46
60528 Frankfurt
Germany

Pellegrino Rossi
Section of Human Anatomy
Department of Public Health and Cell Biology
University of Rome "Tor Vergata"
Via Montpellier 1
00133 Rome
Italy

Kirsi Sainio
Institute of Biomedicine
University of Helsinki
P.O. Box 63
00014 Helsinki
Finland

Hannu Sariola
Institute of Biomedicine
University of Helsinki
P.O. Box 63
00014 Helsinki
Finland

Wei Shi
University of Southern California Keck School of Medicine
Developmental Biology Program
Saban Research Institute
Childrens Hospital Los Angeles
4650 Sunset Boulevard
Los Angeles CA 90027
USA

Lukas Sommer
Institute of Cell Biology
Swiss Federal Institute of Technology
ETH-Hoenggerberg HPM E38
8093 Zürich
Switzerland

Lydia Sorokin
Department of Experimental Pathology
Lund University Hospital
Sölvegatan 8B
22362 Lund
Sweden

Matthias Stanke
Max-Planck-Institut für Hirnforschung
Abteilung Neurochemie
Deutschordenstrasse 46
60528 Frankfurt
Germany

Denise Tefft
University of Southern California Keck
School of Medicine
Developmental Biology Program
Saban Research Institute
Childrens Hospital Los Angeles
4650 Sunset Boulevard
Los Angeles CA 90027
USA

Irma Thesleff
Institute of Biotechnology
University of Helsinki
P.O. Box 63
00014 Helsinki
Finland

Sólveig Thorsteinsdóttir
Department of Animal Biology and
Centre for Environmental Biology
Faculty of Sciences
University of Lisbon
Lisbon
Portugal
and
Gulbenkian Institute of Science
Oeiras
Portugal

Paul A. Trainor
Stowers Institute for Medical Research
1000 E. 50th Street,
Kansas City, MO 64110
USA

Xiu-Ping Wang
Institute of Biotechnology
University of Helsinki
P.O. Box 63
00014 Helsinki
Finland

Grant Wheeler
School of Biological Sciences
University of East Anglia
Norwich NR4 7TJ
UK

Pierre H. Vachon
CIHR Group on Functional
Development and Physiopathology of
the Digestive Tract
Department of Anatomy and Cell
Biology
Faculté de médecine et des sciences de
la santé
Université de Sherbrooke
Sherbrooke (Québec) J1H 5N4
Canada

David Warburton
University of Southern California Keck
School of Dentistry
Vice Chair, Department of Surgery,
Leader, Developmental Biology
Program,
Saban Research Institute
Childrens Hospital Los Angeles
4650 Sunset Boulevard
Los Angeles CA 90027
USA

Joachim Wittbrodt
European Molecular Biology
Laboratory
EMBL
Developmental Biology Programme
Meyerhofstrasse 1
69117 Heidelberg
Germany

Humphrey Hung-Chang Yao
Department of Veterinary Biosciences
University of Illinois
3806 VMBSB
2001 South Lincoln Avenue
Urbana, IL 61802
USA

Color Plates

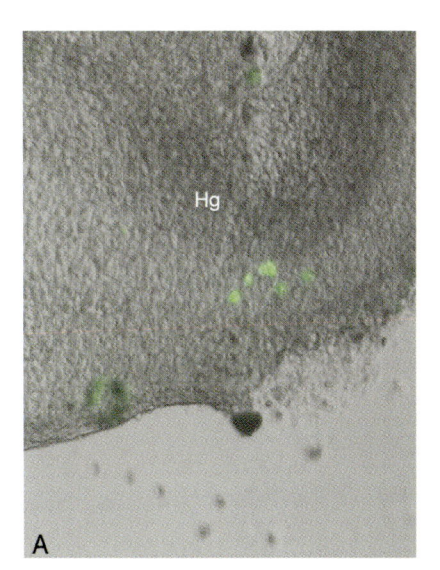

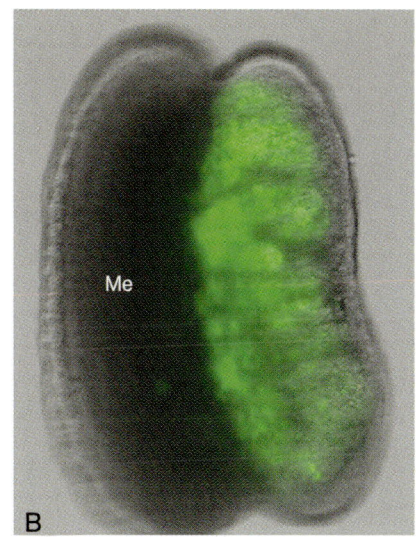

Fig. 2.2
Kit/GFP PGCs in migration towards the hindgut (Hg) (A) and within the testis (B) of mouse embryo of 9.5 and 12.5 dpc. Me, mesonephros. This figure also appears on page 43.

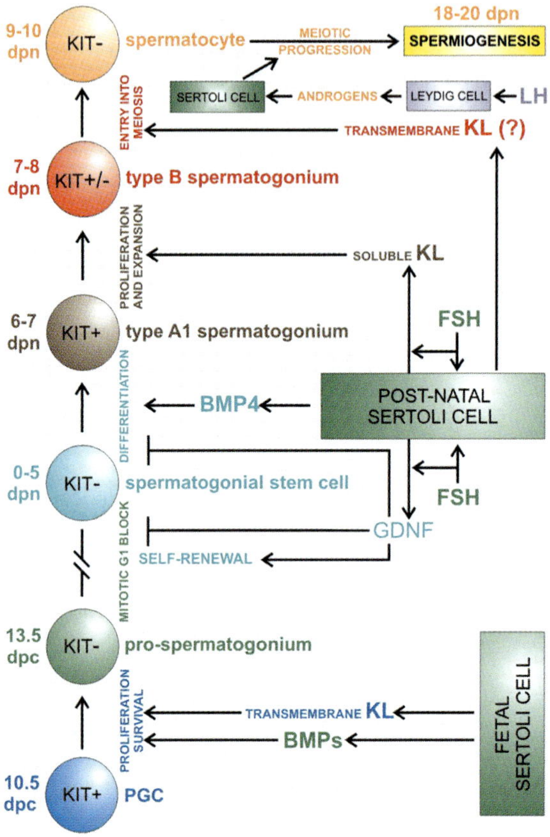

Fig. 2.5
Schematic representation of the best characterized paracrine GFs that act directly on mouse male germ cells, thus controlling progression of spermatogenesis and hormonal regulation of their expression in the somatic cell compartment. Different colours mark male germ cells at various developmental stages and the specific action of the indicated GFs on differentiative and proliferative events. Fluctuations during mitotic stages of expression of the *c-kit* gene, encoding the receptor for KL, are also indicated. This figure also appears on page 54.

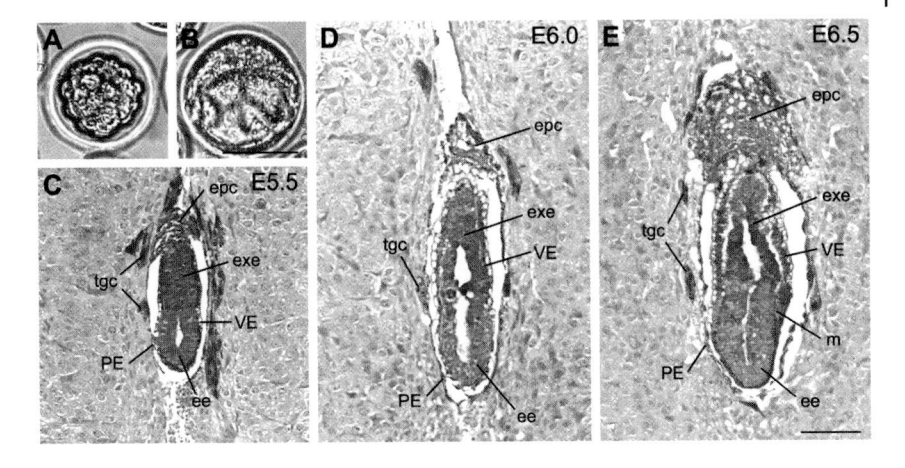

Fig. 3.1
Mouse embryonic development from morula to gastrula. (A) After compaction, the morula continues cleaving without cellular growth. (B) The blastocyst forms with the expansion of the blastocoelic cavity resulting from vectorial fluid transport by trophectoderm cells (dark blue, polar trophectoderm; light blue, mural trophectoderm). The ICM cells (yellow) and the primitive endoderm (orange) form distinct populations inside the blastocyst. (C-E) After implantation the ICM develops into the epiblast or embryonic ectoderm (yellow), the primitive endoderm gives rise to visceral endoderm (orange) and parietal endoderm (pink), the mural trophectoderm differentiates into the primary trophoblast giant cells (light blue) and the polar trophectoderm forms the ectoplacental cone (dark blue) and the extra-embryonic ectoderm (green). Although the tissue organization remains similar after implantation, the embryo starts growing in size. Gastrulation begins with the invagination of nascent mesoderm (gray) at the posterior side of the embryo. ee, epiblast or embryonic ectoderm; epc, ectoplacental cone; exe, extra-embryonic ectoderm; PE, parietal endoderm; m, nascent mesoderm; tgc, trophoblast giant cells; VE, visceral endoderm. Scale bar: 50 μm (A and B) and 100 μm (C-E). This figure also appears on page 75.

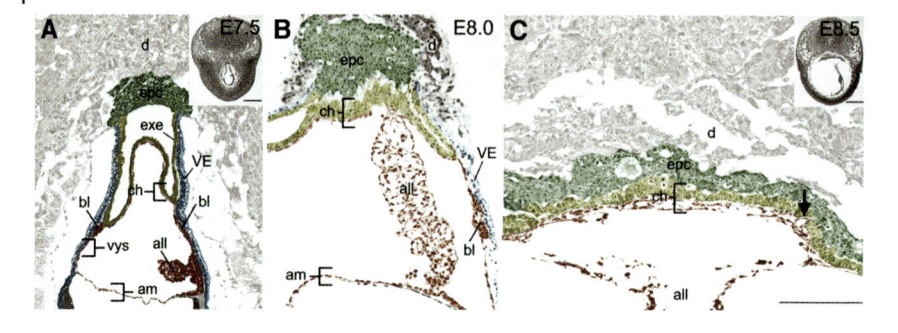

Fig. 3.2
Chorio-allantoic fusion. (A) By E7.5, the extra-embryonic mesoderm (brown) contributes to the formation of four extra-embryonic structures: the amnion (together with ectoderm), the visceral yolk sac (together with visceral endoderm (blue)), the chorion (together with extra-embryonic ectoderm (yellow)) and the allantois. (B) The allantois is a finger-like structure, which grows across the exocoelomic cavity towards the chorion. It is formed by a sheath of mesothelium and an inner core of vascularizing mesoderm. At this stage the ectoplacental cavity becomes progressively reduced as the chorion squeezes against the ectoplacental cone (green). (C) At E8.5 the allantois spreads across the chorion surface and blood vessels (arrow) immediately sprout to vascularize the budding chorio-allantoic placenta. Note that although both the embryo and extra-embryonic structures increase tremendously in size, the size of the deciduum remains constant (see inserts in A and C). all, allantois; am, amnion; bl, blood island; ch, chorion; d, deciduum; epc, ectoplacental cone; exe, extra-embryonic ectoderm; VE, visceral endoderm; vys, visceral yolk sac. Scale bar: 250 μm (A-C) and 500 μm (inserts in A and C). This figure also appears on page 77.

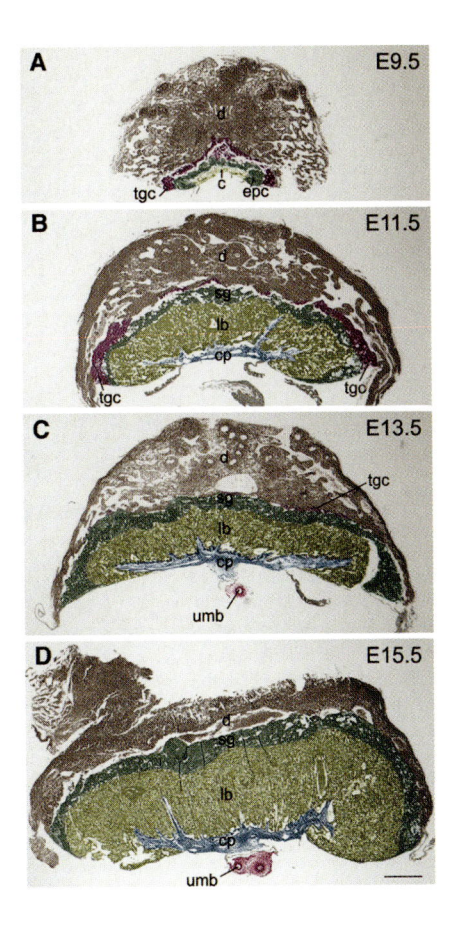

Fig. 3.3
Placentation. (A) An increasing number of trophoblast giant cells (magenta) and ectoplacental cone cells (green) form the placenta of the E9.5 embryo. The allantois vasculature sprouts into folds in the chorion (yellow). (B) Two days later, the labyrinth (yellow) has developed greatly between the spongiotrophoblast layer (green) and the chorionic plate (blue). The trophoblast giant cells (magenta) become localized at the border of the placenta. (C and D) At E13.5 and E15.5, the structure of the placenta remains essentially the same, but the labyrinth (yellow) continues to expand. Note that the umbilical cord is also visible (pink). c, chorion; cp, chorionic plate; d, deciduum; epc, ectoplacental cone; lb, labyrinth; sg, spongiotrophoblast; tgc, trophoblast giant cells; umb, umbilical cord. Scale bar: 500 μm. This figure also appears on page 78.

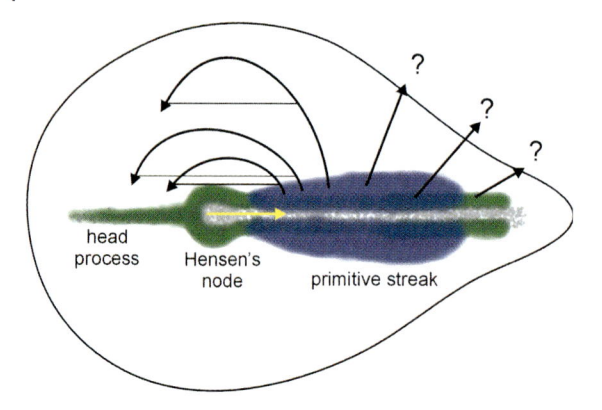

Fig. 4.2
Model suggesting that cell movement is directed by repulsion from the streak mediated by FGF8 and attraction towards the midline in response to FGF4 (reprinted with permission from [83]). The FGF4 expression pattern is indicated in green, and FGF8 expression is indicated in blue. The black arrows show the cell movement trajectories observed. The yellow arrow indicates the direction of node regression. It was suggested that cell movement away from the streak is the result of negative chemotaxis in response to FGF8 produced by cells in the middle primitive streak, while the inward movement of the cells emerging from the anterior streak is the result of attraction by FGF4 secreted by cells in the head process and notochord. The cells in the caudal part of the embryo are initially repelled by FGF8 and subsequently attracted by an unknown signal present in the region between the area opaca and the area pellucida. This figure also appears on page 121.

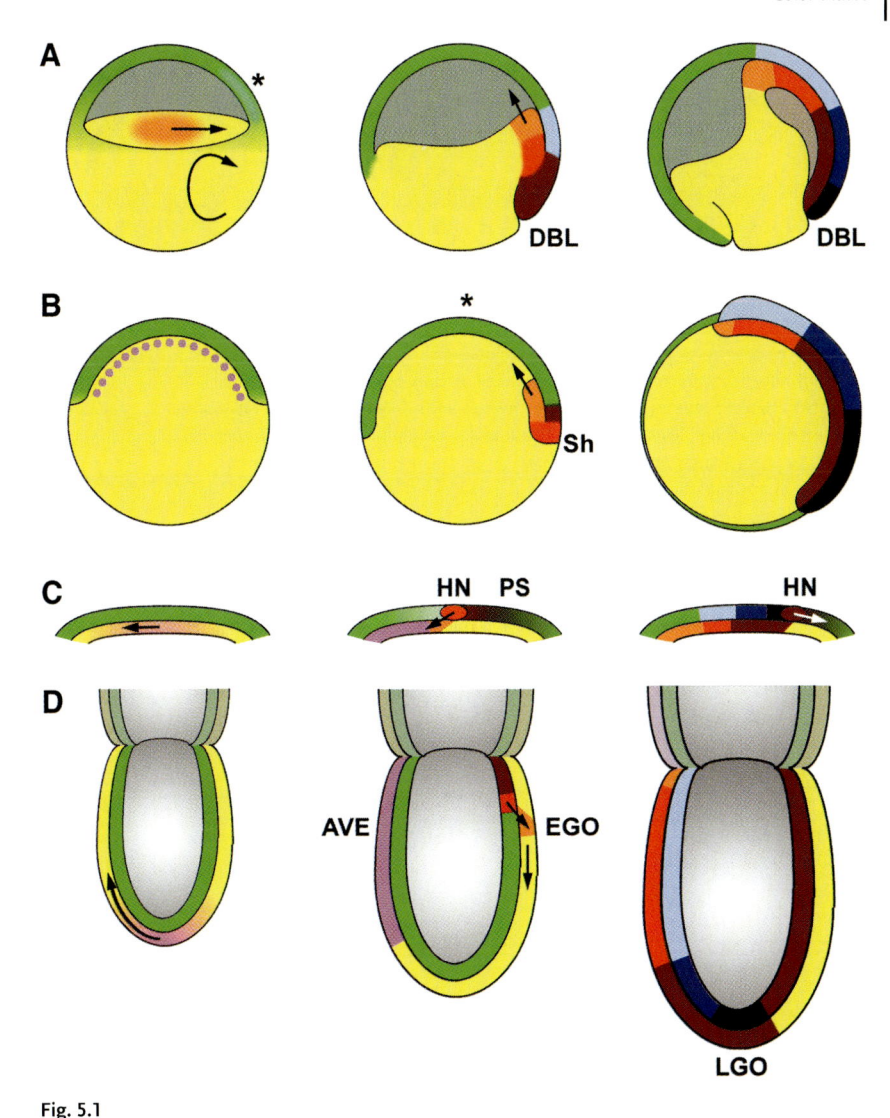

Fig. 5.1
Schematic depiction of early vertebrate embryos at comparable stages of development. Definitive AE is shown in orange, PME in red, CM in brown. Anterior, medial and posterior neuroectoderm are shown in light blue, dark blue and black, respectively. Left-hand panels: pre-gastrula; middle panels: early gastrula; right-hand panels: late gastrula. (A) *Xenopus* embryo, dorsal side is to the right, animal/anterior pole points upwards. Note "vegetal rotation" and BCNE (asterisk) at pre-gastrula stage. (B) Zebrafish embryo, dorsal side is to the right, animal/anterior points upwards. Note YSL (pink dots) in left panel. Row–1 is indicated by asterisk in early gastrula. Sh, shield. (C) Chick embryo, anterior is to the left, dorsal side points upwards. Note anterior movement of hypoblast in left and posterior regression of Hensen's node (HN) in right-hand panel. PS, primitive streak. (D) Mouse embryo, anterior is to the left, distal tip of egg cylinder points down. Note movement of AVE from distal (pre-gastrula) to anterior pole (early gastrula) and its subsequent displacement by PME (late gastrula). Adapted from [226]. This figure also appears on page 142.

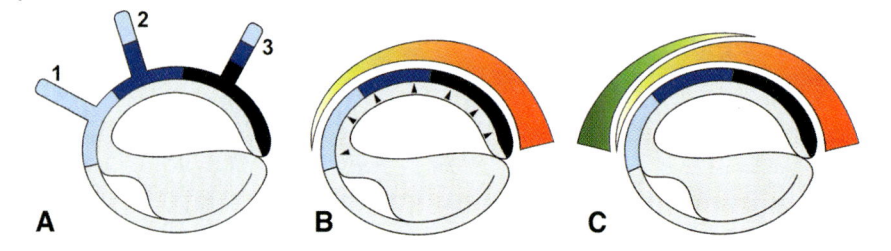

Fig. 5.4
Gradient models in AP patterning of amphibian embryos. Anterior, medial and posterior neuroectoderm are shown in light blue, dark blue and black, respectively. (A) Nieuwkoop's experiment: ectodermal folds differentiate into anterior neuroectoderm when implanted into the anterior neural plate (1), into anterior neuroectoderm distally and medial neural structures basally when implanted medially (2) and into an AP range of neural tissues from distal to proximal when implanted into the posterior neural plate (3). (B) Nieuwkoop's activation-transformation model. Black arrowheads indicate the activating signal and the orange wedge the transforming signal. (C) The Finnish two-inducer model. Archencephalic (green) and spinocaudal inducer (orange) form opposing AP gradients. Adapted from [236]. This figure also appears on page 145.

Fig. 6.1
The genetic regulation of hindbrain segment and identity. (A) The subdivision of the hindbrain into seven rhombomere segments exerts a profound influence over patterning of the branchiomotor nerves, the cranial ganglia and the migration of neural crest cells into the adjacent branchial arches. This registration is crucial for providing the blueprint for craniofacial development. (B) A large number of transcription factors, membrane proteins and secreted factors are expressed dynamically in restricted compartments or segments during hindbrain development. These genes are crucial not only for compartment formation and the restriction of cell movements but also for the identity and differentiation of individual segments. This figure also appears on page 192.

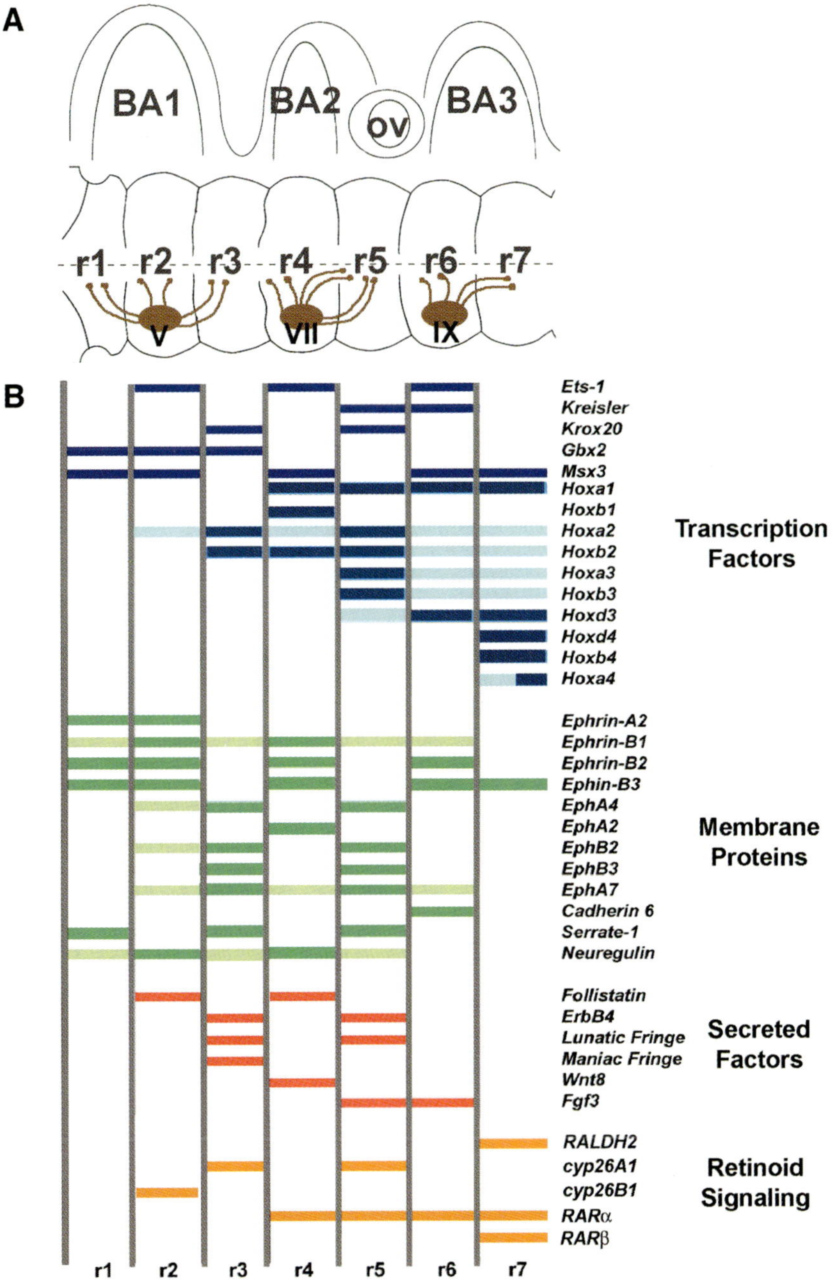
A
BA1
BA2
OV
BA3
r1 r2 r3 r4 r5 r6 r7
V
VII
IX
B
Ets-1
Kreisler
Krox20
Gbx2
Msx3
Hoxa1
Hoxb1
Hoxa2
Hoxb2
Hoxa3
Hoxb3
Hoxd3
Hoxd4
Hoxb4
Hoxa4
Transcription Factors
Ephrin-A2
Ephrin-B1
Ephrin-B2
Ephin-B3
EphA4
EphA2
EphB2
EphB3
EphA7
Cadherin 6
Serrate-1
Neuregulin
Membrane Proteins
Follistatin
ErbB4
Lunatic Fringe
Maniac Fringe
Wnt8
Fgf3
Secreted Factors
RALDH2
cyp26A1
cyp26B1
RARα
RARβ
Retinoid Signaling
r1 r2 r3 r4 r5 r6 r7

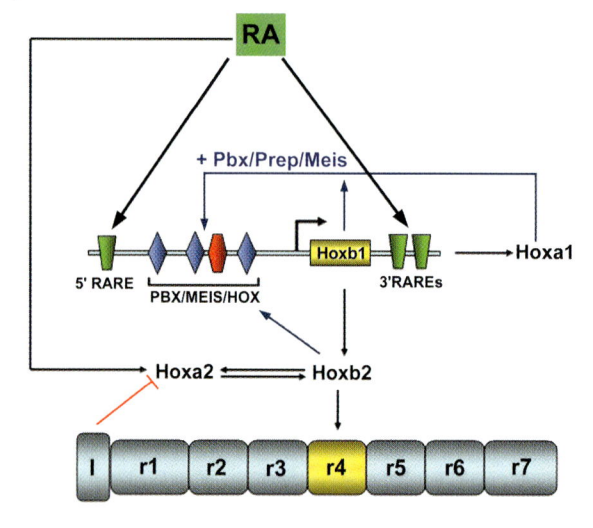

Fig. 6.2
Establishing individual rhombomere (r) 4 identity. The complex genetic regulation of r4 identity commences with the induction of Hoxb1 expression in the neuroepithelium in response to RA (retinoic acid) mediated by 3′ RAREs (retinoic acid response elements) and through para-regulation by Hoxa1 and PBX/PREP/MEIS complex binding. 5′ RAREs then serve to repress Hoxb1 expression in r3 and r5 thus restricting the domain of Hoxb1 expression to r4. In addition, FGF signaling from centers such as the isthmus may repress anterior Hox gene expression in the hindbrain. The maintenance of Hoxb1 expression in r4 is achieved through auto-regulation and possibly by the cross-regulation of Hoxb2 which is mediated through PBX/MEIS/HOX tripartite complex binding. Hoxb1 cross-regulates Hoxb2 which in turn is involved in a para-regulatory mechanism with Hoxa2. These interactions are essential for the maintenance of r4 identity and subsequently for appropriate differentiation. This figure also appears on page 199.

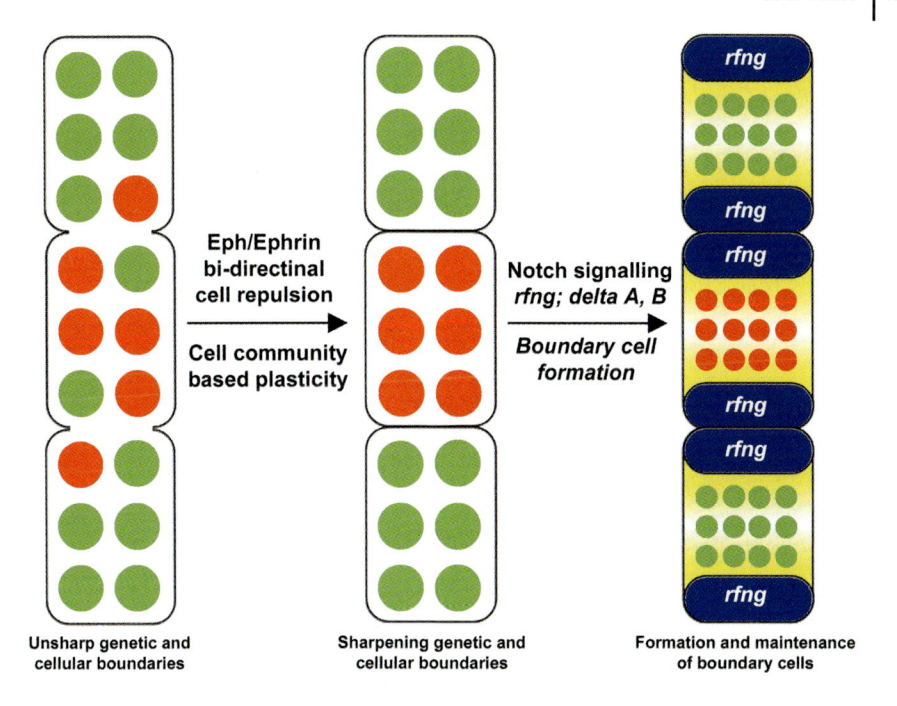

Fig. 6.3

The morphological generation of hindbrain segmentation. Distinct rhombomere cell populations begin to sort out through bidirectional cell repulsion and differential cell adhesion which is mediated by the *Eph/ephrin* family. Superimposed on this segregation is cellular-based plasticity which helps to ensure that cells which potentially might become positioned inappropriately adopt the genetic profile of their neighbors. As a consequence, the initially diffuse gene expression boundaries sharpen in response to this cell sorting and plasticity. Concomitant with this process, differential Notch signaling mediated by *rfng* and *delta* leads to the formation of boundary cells which exhibit low rates of proliferation and establishes the rhombomeres as units of cell lineage restriction. This figure also appears on page 209.

Fig. 9.1

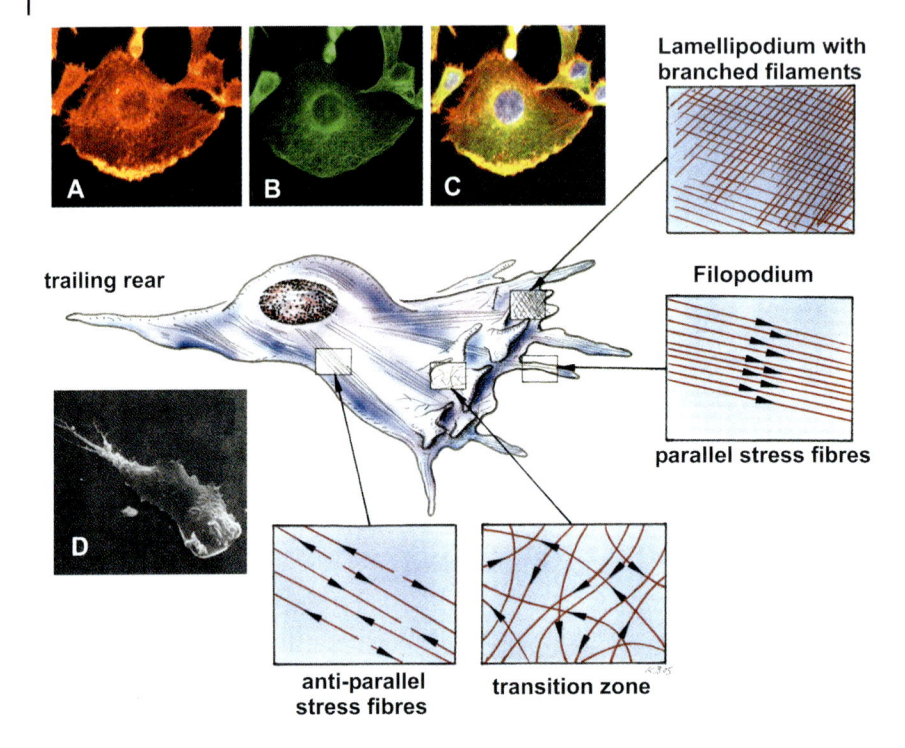

shape and F-actin organization in migrating cells. (A) Fibroblastic cell fluorescently stained for F-actin, (B) for tubulin and (C) merger of A and B with additional chromatin stain (blue). Note in (A) the different forms of F-actin organization: a homogenously stained rim in the region of the lamellipodium (bottom of the cell) and straight bundled stress fibers in the cell body. (Center) Schematic diagram of a migrating fibroblastic cell and its different forms of F-actin organization. (D) Scanning electron micrograph of a migrating fibroblast. Note the triangular shape with a small lamellipodium on the right lower side. The different organizational forms of actin and their location are depicted in the schematic diagram of a migrating cell. Special regions are magnified in the four insets representing the lamellipodial region containing a branched F-actin network, a filopodium with parallel stress fibers, the transition zone with an open network, and bundles of anti-parallel orientated actin filaments forming the stress fibers crossing the cell body from front to rear. This figure also appears on page 323.

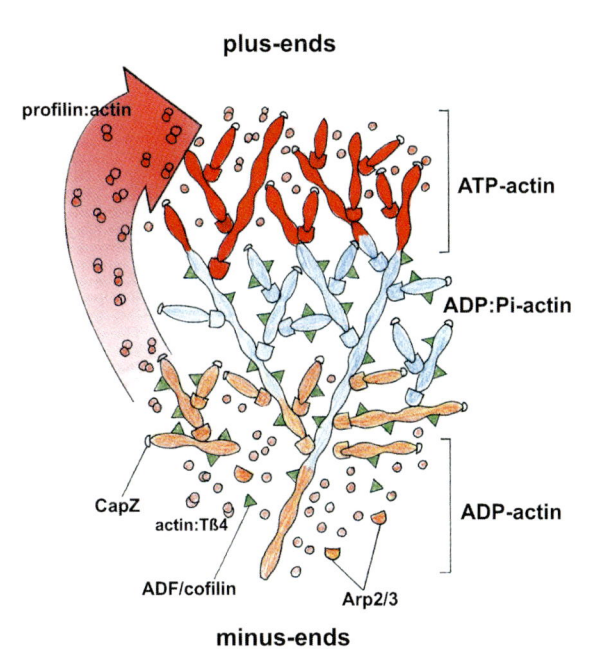

Fig. 9.4
Schematic representation of the branched F-actin network within the lamellipodium. Plus-ends on top are located at the lamellipodial periphery. Here new actin subunits are added to existing filaments or new filaments are nucleated. Within the newly added filaments actin contains bound ATP which is subsequently hydrolyzed to ADP and inorganic phosphate (Pi) both of which remain bound to actin. Towards the minus-ends the Pi is released and the actin contains bound ADP. Minus-end dissociation is stimulated by ADF/cofilin (green triangles), thereafter ADF/cofilin is replaced by either profilin (circles) or thymosin β4 (half moons). The profilin : actin complex is reshuffled to the plus-ends for further outgrowth; the actin : thymosin β4 complex represents the storage form of G-actin. This figure also appears on page 329.

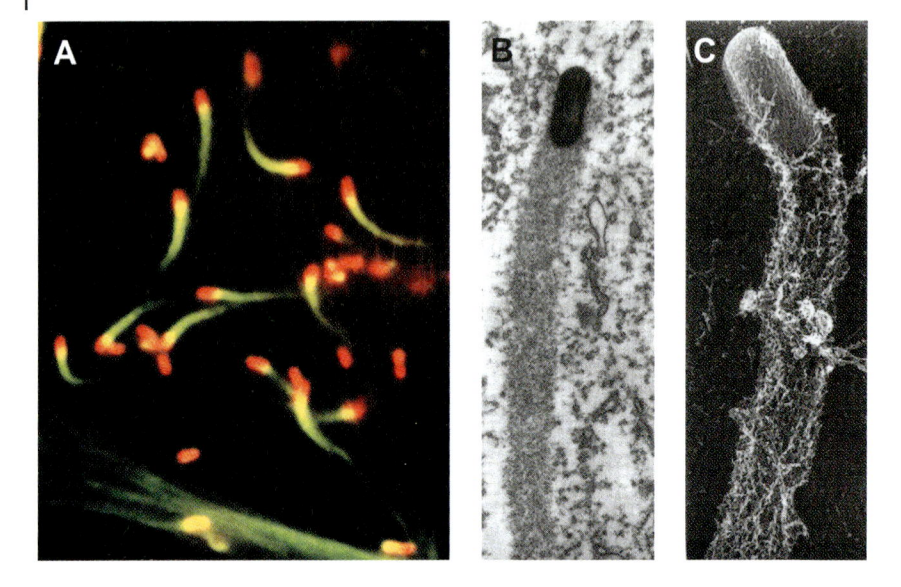

Fig. 9.5
Comet tails induced by the intracellular bacteria *Listeria monocytogenes*. (A) Fluorescent staining of a listeria-infected cell. The bacteria are stained red and the F-actin comet tail in green. Points of F-actin attachment to the bacteria are yellow due to overlap of both stains (from [8]). Different electron micrographs of listeria comet tails: (B) transmission EM at lower magnification [9] and (C) visualized by the deep-etch technique [10]. This figure also appears on page 331.

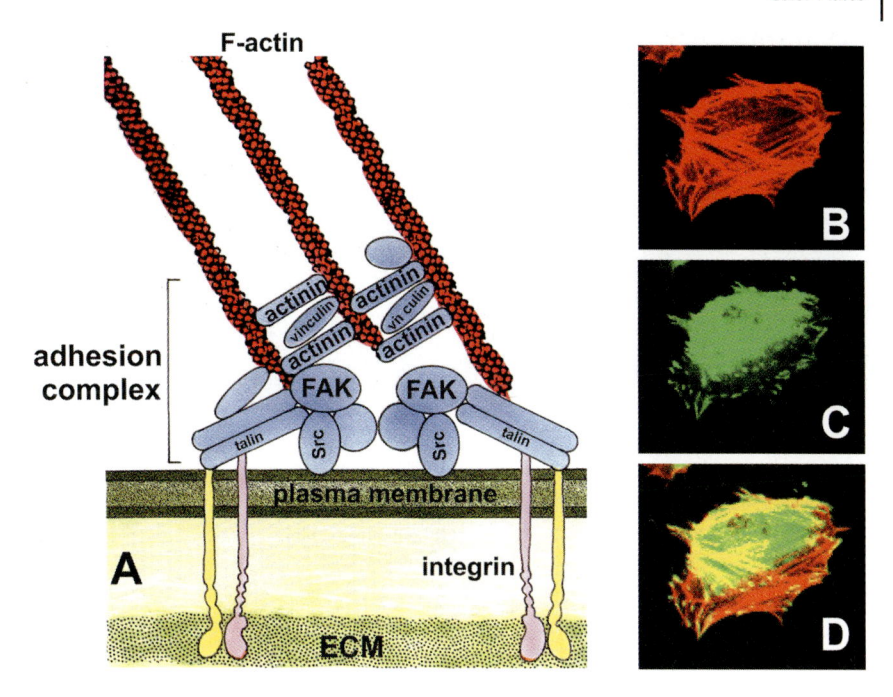

Fig. 9.6
Organization and composition of focal contacts underneath the plasma membrane. (A) Diagram of the organization of a focal contact to the extracellular matrix (ECM). Heterodimeric integrin as transmembrane ECM-receptors and the organization of a parallel bundle of F-actin with a number of associated proteins essential for the attachment of the actin filaments to the plasma membrane (see text for details). Open circles represent various components depending on the type of contact. (B) Fibroblastic cell stained for F-actin (red), (C) for vinculin (green); note the punctate or streaky pattern marking focal contacts from which F-actin bundles radiate (yellow in (D) by the superposition of both stains). This figure also appears on page 335.

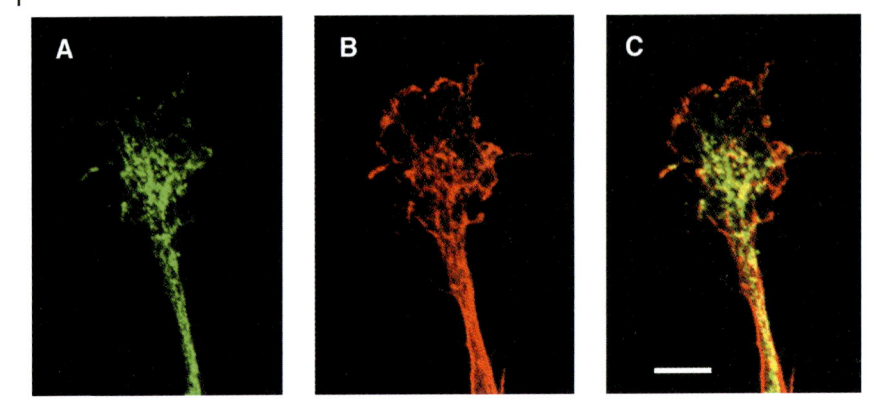

Fig. 9.8
Growth cone of explanted embryonic chicken (E 10) dorsal root ganglia cells fluorescently stained with anti-tubulin (A), anti-actin (B), and overlay of both stains (C). Bar corresponds to 5 μm (kindly provided by Dr. Carsten Theiss, Department of Cytology, Bochum). This figure also appears on page 342.

Part I
Cell Signaling and Growth Factors in Development

1
Stem Cells

Christian Paratore and Lukas Sommer

1.1
Introduction

Stem cells are the founder cells for every organ, tissue and cell in the body. They are undifferentiated cells that can give rise to several lineages of differentiated cell types. In addition, stem cells are able to self-renew and thus to produce undifferentiated descendents, some of which are stem cells again (Fig. 1.1). These features allow stem cells to fulfill their multiple functions, namely to provide enough cells during organogenesis, to control tissue homeostasis and, in addition, to ensure regeneration and repair, at least of certain tissues. It is because of these characteristics that stem cells are a prime target of applied research that seeks to treat degenerative diseases by cell replacement therapies. So far, researchers have used embryonic, fetal, and adult stem cells as a source from which to generate various specialized cell types. Any disease caused by tissue degeneration can be a potential candidate for stem cell therapies, including Parkinson's and Alzheimer's disease, stroke, spinal cord injury, heart diseases, burns, and many more. However, to realize the clinical potential of stem cells, it is crucial to have a deeper insight into the mechanisms regulating stem cell self-renewal and their ability to produce the correct cell type at the appropriate time and location in correct numbers. In this chapter, we review how extracellular signals influence stem cell behavior. This overview can by no means provide an exhaustive list of all signal transduction pathways reported to act on stem cells. Rather, we try to illustrate aspects of stem cell development by discussing some specific signals affecting stem cell proliferation, fate decision, and differentiation.

The stem cells with the broadest range of potential are cells isolated from the inner cell mass (ICM) of the blastocyst. These embryonic stem cells (ESCs) are pluripotent and able to respond to morphogenic signals and to differentiate into any desired cell type of the three germ layers. In culture, ESCs can be propagated almost indefinitely, demonstrating their unlimited potential with respect to growth and differentiation. Additionally, the developmental and therapeutic potential of adult stem cells isolated from various tissues is also being investigated. The bone marrow (BM) is composed of the non-adherent hematopoietic and adherent stromal cell compartment. The adherent BM stromal cell fraction contains pluripotent mesenchymal

Cell Signaling and Growth Factors in Development. Edited by K. Unsicker and K. Krieglstein
Copyright © 2006 WILEY-VCH Verlag GmbH & Co. KGaA, Weinheim
ISBN 3-527-31034-7

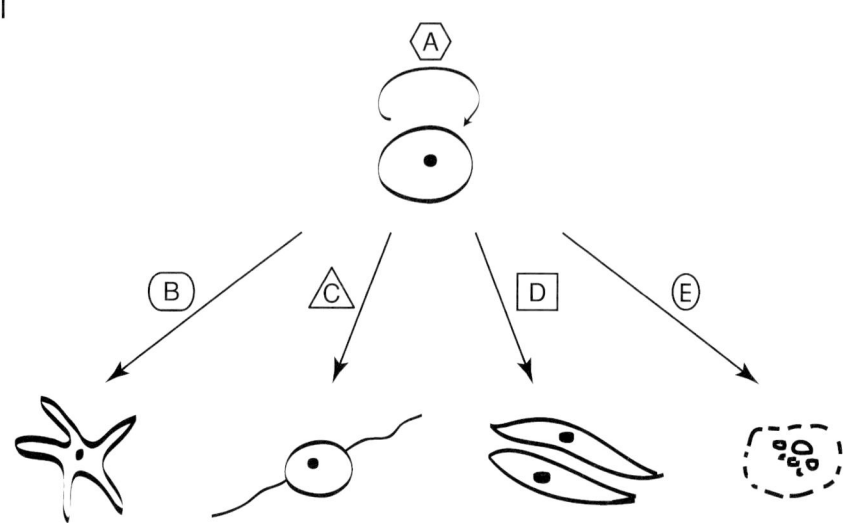

Fig. 1.1
Stem cell fates are regulated in a signal-dependent manner. Stem cells are multipotent, that is, they are able to generate many different derivatives. In addition, stem cells have the capacity to self-renew. At any time-point, neighboring cells, growth factors, and extracellular matrix components that adjust the balance between self-renewal, differentiation, or apoptosis influence the fate of stem cells. This decision is regulated by numerous cell-intrinsic and cell-extrinsic factors (identified here by A-E), some of which maintain the cells as stem cells whereas others induce cell death, or differentiation into various lineages. Usually, a combination of factors involving distinct signaling cascades is linked to a cell-specific output.

stem cells (MSCs) that can be induced to differentiate into various mesenchymal lineages as well as into most somatic lineages including derivatives of the brain [1–3]. Apart from bone marrow-derived stem cells, multipotent adult stem cells from the adult dermis [4], muscle, and brain [5] have been described to generate cells representing derivatives of multiple germ layers. These results have been explained by the capability of the cells to trans-differentiate. The term trans-differentiation describes the conversion of a cell type of a specific tissue lineage into a cell type of another lineage, involving reprogramming of gene expression due to altered microenvironmental cues. It has been hypothesized that tissue injury increases the rate at which bone marrow-derived stem cells trans-differentiate [3, 6]. These results have been debated, however, and it has been suggested that trans-differentiation events – if they occur at all – are rare and that the appearance of donor cell markers in host tissues might arise by other mechanisms. First, transplanted cells might undergo fusion with endogenous differentiated cells [7, 8]. In fact, the ability to fuse is characteristic of many cell types, such as myoblasts, hepatocytes, and others ([9] and references therein). Alternatively, cells from a given lineage might de-differentiate into a more naive state that allows the cell to re-differentiate along new lineages. Finally, a very rare pluripotent stem cell might persist until adulthood, and upon transplantation would be able to generate a broad variety of cells representing derivatives of all three germ layers, depending on its environment. Thus, when elucidating the potential of stem cells in culture or *in vivo*, it is not sufficient to analyze the expression of appropriate lineage markers; rather, possible fusion events have to be excluded, and the purity of the stem cells has to be considered in order to rule out their contamination by additional cells with other potentials. This can be achieved by clonal analysis of prospectively identified cells that, if possible, have been minimally manipulated (for example without culturing) before use. The ultimate proof that a given stem cell can adopt a certain fate lies in the demonstration of its functional integration into the tissue.

1.2
Maintenance of Stemness in Balance with Stem Cell Differentiation

Many stem cells reside in a spatially restricted compartment called a niche. This niche provides an environment that supports the survival of the multipotent stem cell without induction of differentiation. Neighboring differentiated cell types secrete factors and provide a milieu of extracellular matrix that allows stem cells to self-renew and to maintain the capacity to respond to differentiation programs (Fig. 1.2). Physical contact between stem cells and their non-stem cell neighbors in the niche is critical in keeping the stem cells within this compartment and in maintaining stem cell character. Often, stem cells within the niche are quiescent or slow-cycling, but proliferation might be induced by injury. Niches have been described, for example, for germ cells, in the bulge of the hair follicle, the bone marrow, the crypt of an intestinal villus, and the subventricular zone of the brain (reviewed in [10]). It is still a matter of investigation which factors control stemness, that is, the maintenance of stem cell properties. It is likely that various signaling pathways are involved, including Notch, bone morphogenetic proteins (BMPs), transforming growth factor β (TGFβ), and Wnt signaling (see below). Several groups have applied microarray technology with the goal of identifying genes that control stemness. The transcriptional profiles of ESCs, hematopoietic stem cells (HSCs), neural stem cells (NSCs), and neural crest stem cells (NCSCs) have been compared and analyzed [11–13]. However, only very few genes were found to be commonly expressed in all stem cells, and it appears to be difficult to define a valid genetic fingerprint that determines stemness of all stem cells or even of a specific stem cell subtype. This could be explained by the usage of different microarray chips, technical difficulties, or the purity of the analyzed cells. Furthermore, the data might reflect substantial intrinsic differences between different types of stem cells.

Cell-intrinsic properties determine how a stem cell interprets the signals present in its environment. At each cell division, stem cells have to choose between self-renewal and differentiation. The mechanisms determining how quiescent or slow-cycling stem cells are induced to start proliferation or differentiation are still largely unknown. One possibility might be that the stem cells, which are slowly cycling "fill"

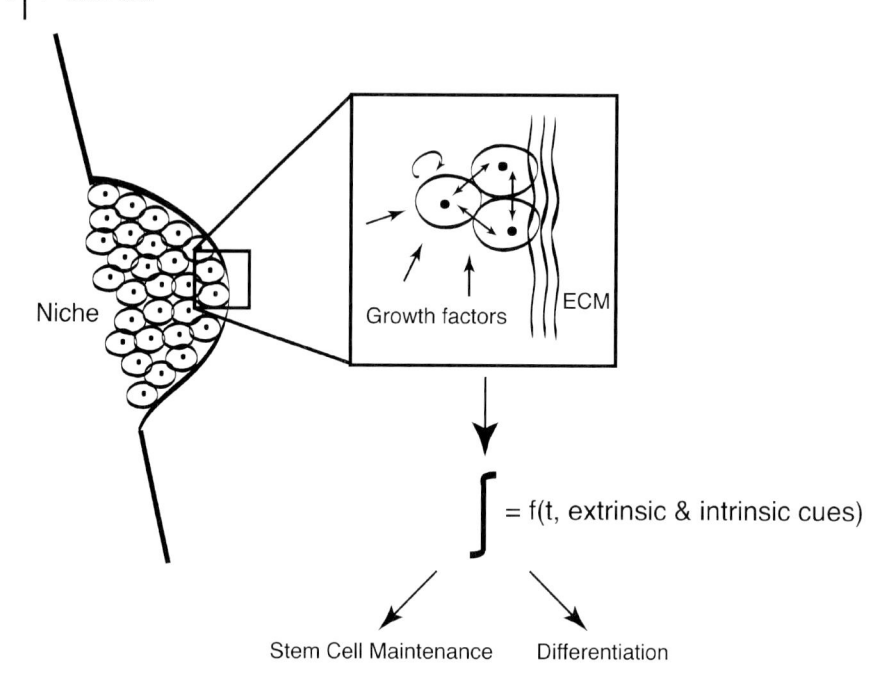

Fig. 1.2
Signaling in the niche. The niche provides an environment that attracts stem cells and keeps them in an undifferentiated state by supporting self-renewing cell divisions. Accordingly, differentiation may be initiated when the stem cell leaves the niche. The balance of quiescence, self-renewal, and cell commitment is influenced by secreted growth factors that initiate intracellular signaling cascades and activate distinct sets of transcription factors. Further, the extracellular matrix (ECM) plays an important role in retaining the stem cells in the niche. Thus, self-renewal versus lineage specification and differentiation are the result of the capacity of a stem cell to integrate multiple signals that vary with location and time.

the niche and subsequently leave it. Outside the niche the cells are exposed to an environment that is permissive or even inductive for differentiation. Alternatively, stress or injury might change the extrinsic signaling in a way that induces differentiation. Stem cells may undergo symmetrical divisions to generate identical twins to self-renew or to differentiate, or they may undergo asymmetric cell divisions, yielding one differentiated progeny and one stem cell daughter [14, 15]. Therefore, the total number of stem cells represents a dynamic balance between symmetric and asymmetric cell divisions in the niche. In addition, the stem cell number is controlled via programmed cell death. Due to the exponential expansion of a single progenitor cell, elimination of stem cells or precursors by programmed cell death at early stages will have a marked effect on the final number of terminally-differentiated cells. Again, the balance between maintenance and depletion of the progenitor pool size has to be tightly controlled by the extracellular environment. Extrinsic factors could actively promote cell death, or the withdrawal of trophic support by growth factors that act as survival factors might induce cell death [16].

In principle, all stem cells and precursors respond to multiple growth factors, and their effects can be modulated by extracellular matrix components (reviewed in [17]). Several different integrins that bind to the extracellular matrix seem to be differentially involved in the regulation of proliferation, cell migration and differentiation. Binding to extracellular matrix proteins such as laminin activates an intracellular signaling pathway via phosphatidylinositol3 (PI3) kinase and Akt kinase [18]. In sum, stem cell development is controlled by the combinatorial activity of multiple factors, acting in signaling networks (Fig. 1.2, [10, 19]). The composition of such networks is dynamic, changing with time and location. To unravel the players involved, researchers have to elucidate the contribution of individual signal transduction pathways, knowing that this contribution is likely to be modulated by the cross-talk with other pathways.

1.2.1
Wnt Signaling

Wnt proteins are important regulators of cell proliferation and differentiation [20]. The Wnt signaling pathway involves proteins that directly participate in both gene transcription and cell adhesion. Nineteen Wnt genes with diverse functions exist in mammalian genomes. Wnt molecules are secreted lipid-modified signaling proteins [21] that bind to Frizzled receptors on the cell surface. Several cytoplasmic components transduce the signal to β-catenin (Armadillo in Drosophila), which enters the nucleus and forms a complex with a high mobility group (HMG) box-containing DNA binding protein such as TCF (T cell factor) and LEF (lymphoid enhancer factor). This complex activates many different target genes and is modulated by cross-talk of Wnt/ß-catenin signaling with various other signal transduction pathways including signaling by Notch, TGFβ factors, FGFs, and Shh [22–24]. In addition, many proteins have been identified that interact with TCF and mediate repression. One such repressor is the Groucho protein in Drosophila (known as TLE in vertebrates). Groucho binds to TCF, repressing the expression of downstream target genes [25].

The central player of the canonical Wnt signaling pathway is β-catenin, which in the absence of Wnt is degraded in the cytoplasm. Excess β-catenin is phosphorylated by glycogen synthase kinase 3β (GSK3β) and then targeted for proteosome-mediated degradation. In the presence of Wnt signaling, Dishevelled (Dsh) becomes activated, which leads to the uncoupling of β-catenin from the degradation pathway by inhibition of GSK3β activity. This results in the accumulation of β-catenin, which enters the nucleus and interacts with partners such as TCF/LEF. Therefore, stabilization of β-catenin and its accumulation in the cytoplasm is a crucial step in canonical Wnt-dependent target gene expression. Apart from GSK3β, several proteins are instrumental in tightly regulating β-catenin levels in the cell, including adenomatous polyposis coli (APC) and Axin/Conductin. In addition to its function in the above-described Wnt signaling pathway, β-catenin plays a role in the structural organization and function of cadherins. β-Catenin binds to the cytoplasmic domain of type I cadherins, linking cadherins through α-catenin to the actin cytoskeleton [24, 26, 27].

1.2.2
Wnt Signaling Regulates ‚Stemness' in ESCs

The mechanisms controlling multipotency and differentiation of ESCs are of fundamental interest. So far, several factors have been described that affect cell-fate decisions and self-renewal of ESCs, including Oct4, Fgf4, Nanog and Sox2 [28–32]. The self-renewal capacity of mouse ESCs can be maintained by growth factors provided by feeder cells or exogenously [33]. In such cultures, ESCs from the inner cell mass of blastocysts remain multipotent and can be propagated nearly indefinitely. Various signaling pathways have been implicated in regulating the self-renewal capacity and multipotency of ESCs. One signal described in this process is the leukemia inhibitory factor (LIF) that is produced by feeder layers of inactivated mouse fibroblasts on which ESCs have been maintained in culture. LIF activates the Janus kinase (JAK) as well as signal transducer and activator of transcription–3 (Stat–3). However, while activation of Stat–3 is sufficient to maintain self-renewal of mouse ESC, LIF has no effect on human ESC self-renewal [34]. Large-scale gene expression profiling of undifferentiated human ESCs revealed that the main components of the canonical Wnt signaling pathway are expressed [34, 35]. Intriguingly, overexpression of Wnt1 or of stabilized β-catenin and lack of APC in ESCs results in the inhibition of neural differentiation and in activation of downstream targets of Wnt signaling such as cyclins, c-myc and BMP [35, 36]. Moreover, treatment of ESCs with a specific synthetic pharmacological inhibitor of GSK3β activates the canonical Wnt pathway and allows both mouse and human ESCs to remain undifferentiated [37]. Such drug-treated cells display sustained expression of transcription factors including Nanog and Oct–3/4, which are important in controlling the pluripotent state of ESCs. Finally, mutations in APC associated with increased tumor incidence result in increased doses of ß-catenin and interfere with the differentiation of ESCs into the three germ layers [38]. Taken together, canonical Wnt signaling has emerged as a crucial factor in regulating ESC maintenance.

1.2.3
Wnt Signaling in Hematopoietic Stem Cells

The hematopoietic stem cell is a multipotent cell in the bone marrow that has the capacity to provide for the life-long production of all blood lineages. The mechanisms regulating HSC lineage decisions and self-renewal *in vivo* have been difficult to define. However, it was possible to establish the importance of the hematopoietic microenvironment through the use of long-term bone marrow culture systems in which maintenance of HSCs at low frequencies is supported by culturing hematopoietic cells on stroma. Subsequently, candidate stem cell factors have been identified by direct addition of purified factors to *in vitro* cultures of HSC populations followed by transplantation of the cultured cells. Many early acting cytokines such as interleukin–3 (IL–3), IL–6, and Kit ligand stimulate proliferation of committed progenitor cells while allowing only limited expansion of HSCs capable of long-term multi-lineage repopulation [39]. In contrast, both conditioned media from cells expressing Wnt proteins and, more recently, addition of purified Wnt3a have been demonstrated to induce self-renewal of HSCs [21, 39]. Overexpression of β-catenin in long-term *in vitro* cultures leads to expansion of HSCs in an immature state, indicating an involvement of the canonical Wnt pathway in this process [40]. On the other hand, applying soluble inhibitors that prevent Wnt proteins from binding and activating the Frizzled receptors reduces HSC growth *in vitro*. Similarly, ectopic expression of Axin, which increases β-catenin degradation, has an inhibitory effect on growth of HSCs and on cell survival. Wnt signaling in HSCs might act through Notch (see below) and the transcription factor HoxB4, both of which have been shown to be involved in self-renewal of HSCs and are upregulated in response to Wnt in HSCs. Thus, Wnt signal activation and the nuclear functions of β-catenin

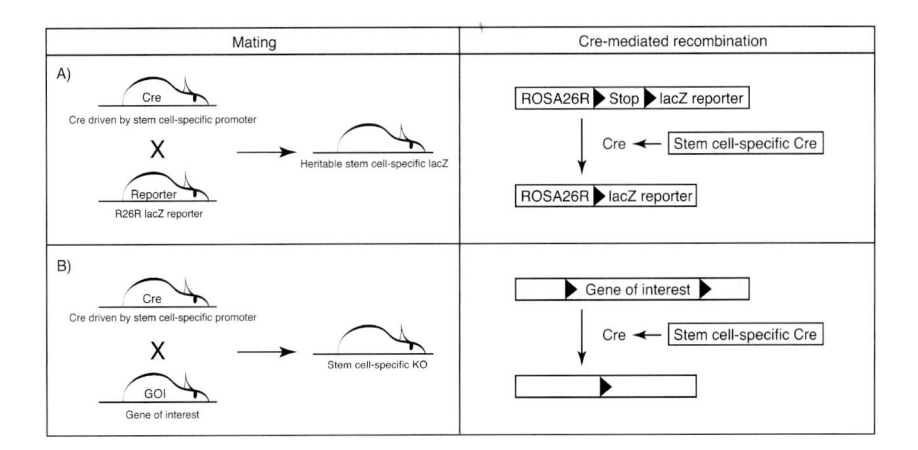

Fig. 1.3

In vivo fate mapping and conditional gene ablation in mice. Mice that express Cre recombinase from a stem cell-specific promoter mated with a reporter mouse line such as Rosa26R [203] produce mice that have heritable lacZ expression (A). Rosa26R reporter mice have the lacZ gene preceded by a transcriptional stop cassette that is flanked by loxP sites. All cells in which the Cre recombinase has been active, as well as their descendants, are lacZ-positive, allowing fate mapping. Conditional gene ablation is performed by using mice that carry alleles in which the gene of interest is flanked by loxP sites (B). Stem cell-specific expression of Cre eliminates the gene of interest. Additionally, inducible forms of Cre recombinase can be activated by injection or feeding of tamoxifen [204], allowing not only cell type-specific but also stage-specific gene manipulation.

enable HSCs to proliferate and to limit their differentiation potential, thereby sustaining self-renewal in long-term culture and functional reconstitution of hematopoietic lineages *in vivo*. However, conditional ablation of β-catenin using the cre/loxP technology (Fig. 1.3) in hematopoietic stem cells did not impair hematopoiesis and lymphopoiesis, suggesting that β-catenin is not required for self-renewal and development of hematopoietic stem cells under physiological conditions [41].

1.2.4
Wnt Signal Activation in the Skin

In the skin, cells in the basal layer proliferate, leave this layer, stop dividing and undergo terminal differentiation. Cells in the outermost layer of the skin are cornified and continually shed from the surface of the epidermis. Therefore, throughout the entire lifespan of the individual new differentiated cells must be produced. However, not all dividing cells within the basal layer are stem cells. As a stem-cell daughter is fated to undergo final differentiation, it first divides a small number of times as a transit-amplifying cell and thereby amplifies the number of terminally-differentiating cells generated by each stem cell. Skin stem cells reside in specific niches (bulge) of hair follicles and are bipotent, as they give rise to both keratinocytes of the hair follicle and the interfollicular epidermis [42, 43]. The niche is characterized by a variety of extracellular matrix proteins such as β1 integrins that are expressed at higher levels in human interfollicular epidermal stem cells than in transit-amplifying cells. An association between Wnt signaling and skin stem cell development was suggested as a result of the finding that cultured human epidermal stem cells with high levels of β1 integrins also displayed a higher level of β-catenin than transit-amplifying cells. Indeed, modulation of β-catenin activity affects the proportion of epidermal stem cells in culture [44], and mice expressing stable β-catenin under the control of an epidermal keratin promoter display excess skin epithelium and develop excess fur caused by postnatal hair follicle morphogenesis [45]. These data suggest that canoncial Wnt signal activation might maintain the stem cell character of adult epidermal cells.

However, Wnt/β-catenin might have further roles in the skin, promoting hair lineage proliferation and differentiation: Activation of c-myc, a possible downstream target of β-catenin, stimulates the exit from the stem cell compartment and cells turn into transit-amplifying cells [46, 47]. Moreover, *in vivo* manipulation of genes encoding Wnt signaling components indicates an essential role of Wnt signaling in fate decision processes of skin stem cells [48–50]. In particular, β-catenin-deficient stem cells fail to differentiate into follicular keratinocytes and instead adopt an epidermal fate [48]. Thus in the skin, canonical Wnt signaling can apparently elicit different cellular responses. How this is regulated remains to be determined, but it is conceivable that alterations in TCF/LEF transcription factors interacting with β-catenin are involved in controlling the fate of skin stem cells [49].

1.2.5
Multiple Roles of Canonical Wnt Signaling in Neural Stem Cells

Another tissue in which Wnt signaling has pleiotropic effects, presumably depending on location and developmental stage, is the nervous system (Fig. 1.4). In the central nervous system (CNS), gene deletion studies have demonstrated that Wnt signaling requires neural progenitor proliferation and hippocampal development [51] and the expansion of dorsal neural cells including the neural crest [52]. Conversely, Wnt signal activation by overexpression of Wnt or of constitutively activated β-catenin impairs neuronal differentiation and increases the progenitor pool, resulting in a massive enlargement of neural tissue in certain areas of the brain [53–55]. To address the question of whether the observed phenotypes are due to effects on multipotent, self-renewing neural stem cells or on transient amplifying progenitors, researchers rely on the availability of neural stem cell cultures. Using such systems, an increase in secondary neurosphere formation (indicating self-renewal activity of sphere-forming cells) has been reported from β-catenin-overexpressing cells derived from the ganglionic eminence [56]. In contrast, Wnt proteins were found to promote maturation and proliferation of neural progenitors from the cortex, apparently without affecting secondary or tertiary sphere formation [57]. These differences might be due to region-specific or context-dependent responses to Wnt signaling. Indeed, it has recently been reported that Wnts promote neuronal differentiation of neural stem cells at later stages of cortical development, while at early stages they control the expansion of neural stem cells [58].

The first example of Wnt promoting stem cell-fate decisions rather than proliferation in the nervous system was provided by studies carried out with neural crest stem cells (NCSCs). During vertebrate development, these cells delaminate from the closing dorsal neural tube and emigrate to various locations within the embryo to give rise to neuronal and glial cell types of most of the peripheral nervous system (PNS), and to several non-neural structures including pigmented cells, smooth muscle cells in the outflow tract of the heart, and craniofacial bones, cartilage, and connective tissues [59]. A variety of signals has been described that influence cell-fate decisions of NCSCs in culture [60, 61]. BMP signaling causes NCSCs to form autonomic neurons; TGFβ promotes smooth muscle-like cells and, under certain conditions, autonomic neurogenesis; and neuregulin induces a glial phenotype. Several studies also reported that Wnt signaling plays a role at multiple stages of neural crest development. *In vivo*, Wnt signaling is involved in early neural crest induction and expansion [52, 62, 63]. Furthermore, both in avian cell cultures and in zebrafish *in vivo*, activation of the Wnt signaling pathway in neural crest cells promotes the formation of pigment cells [64, 65], while neural crest cells deficient in β-catenin fail to produce melanoblasts during development [66]. In addition, NCSCs lacking β-catenin fail to generate sensory neuronal precursors, and mutant neural crest cells are unable to aggregate in dorsal root ganglia (DRG) or to generate sensory neurons and satellite glia. Cell culture analysis revealed that NCSCs without β-catenin emigrate and proliferate normally but are unable to acquire a sensory neuronal fate. In a complementary set of gain-of-function experiments, Wnt/β-

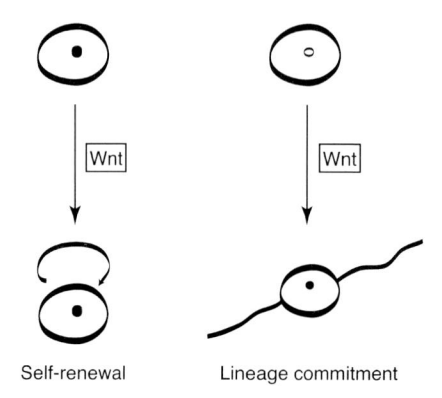

Fig. 1.4
Differences in Wnt signal interpretation. Different types of stem cells (depicted by different colored nuclei) respond differentially to Wnt signaling. In certain stem cell types Wnt signaling promotes stem cell self-renewal, while in other stem cells it regulates lineage commitment. Therefore, there are cell-intrinsic differences between stem cell types. Such differences are presumably due to distinct determinants that change spatially and over time. Furthermore, the microenvironment modulates Wnt activity by additional factors (indicated by X or Y). Thus, the biological activity of Wnt in a particular environment is influenced by the convergence of Wnt signaling with other signal transduction pathways.

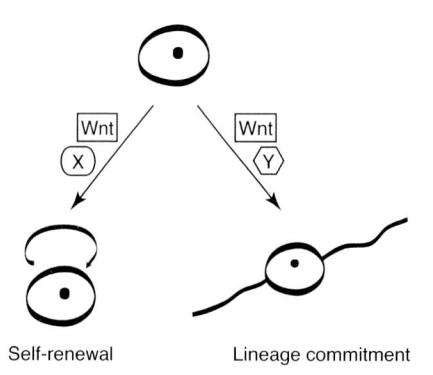

catenin signal activation was shown to regulate sensory fate decisions in emigrating NCSCs while having little effect on the stem cell population size [67]. In particular, as shown by *in vivo* fate mapping of mutant cells (Fig. 1.3), NCSCs expressing a constitutively active form of β-catenin produce sensory neurons at the expense of other crest derivatives, some at ectopic cranial locations of the embryo that are usually devoid of neural derivatives of the neural crest. At locations of normal sympathetic ganglia formation, sensory rather than sympathetic neurons are generated in these mutant embryos. Clonal analysis of cultured cells further demonstrated that canonical Wnt signaling induces sensory neurogenesis by acting instructively on early NCSCs. Thus, in contrast to other types of stem cells, Wnt signaling does not control proliferation but rather promotes sensory fate decision in multipotent NCSCs (Fig. 1.4) [67].

1.2.6
Aberrant Wnt signal activation in carcinogenesis

Given that Wnt signaling is a crucial growth factor for many types of stem cells, its activity needs to be highly controlled to ensure proper organogenesis and tissue homeostasis. Indeed, deregulation of the Wnt signaling pathway affects cell-fate decision, adhesion, and migration, and results in induction and progression of several forms of cancer, indicating that cancers may be a consequence of dysregulation of stem cell programs. Accordingly, Wnt/β-catenin signaling is not only essential for the homeostasis of the intestinal epithelium [68] but sustained β-catenin activity has also been directly implicated in the formation of colon carcinoma [69, 70]. Thus, mutations resulting in increased β-catenin levels have been found in genes encoding β-catenin itself, APC, or Axin, [71, 72], and the presence of constitutively active TCF/β-catenin complexes in the nucleus is characteristics of some cancers [73]. The accompanying inappropriate activation of Wnt target genes is considered to be a critical, early event in the course of carcinogenesis. Thus, understanding how canonical Wnt signaling regulates cellular processes during normal development will likely yield important insights into the regulatory mechanisms involved in cancer progression in the adult.

1.3
The Notch Signaling Pathway

The Notch/Delta signaling pathway is highly conserved across species and is involved in cell-fate specification both in vertebrate and invertebrate development [74–76]. The Notch proteins are cell surface receptors that consist of a single transmembrane polypeptide with a ligand-binding extracellular domain containing several tandem epidermal growth factor (EGF)-like repeats. Mammals have four Notch receptors encoded by four different genes. Notch receptors are activated by Delta-like ligands (Dll–1, –3, and –4) and Serrate-like ligands (Jagged–1 and –2) presented by neighboring cells. The signaling pathway is initiated by ligand binding, which induces a proteolytic cleavage of the Notch intracellular domain (NICD). Once released from the plasma membrane, NICD translocates to the nucleus where it binds the transcriptional regulator CSL (CBF1/Suppressor of Hairless/Lag1), DNA-binding proteins [77], and the Mastermind (Mam)/Lag3 co-activator [78]. In an inactivate state, CSL associates with transcriptional co-repressors that inhibit target gene expression [79]. However, when the cleaved intracellular domain of Notch enters the nucleus, co-repressors are replaced, co-activators recruited, and expression of members of the Hairy enhancer of Split (HES) and HES-related (HERP) genes is initiated. HES proteins that act as transcriptional repressors belong to the basic helix-loop-helix (bHLH) family of transcription factors. These proteins are involved in several lineage-specification processes and mediate many of the primary effects of Notch activation. The HES proteins inhibit the expression of lineage-specifying bHLH genes, such as Mash1 and Neurogenins (which regulate neurogenesis), MyoD (involved in myogenesis), and E2A (involved in B lymphopoiesis) [80]. Additionally, there exists CSL-independent signaling activities, although the mechanisms of these signaling pathways remain to be further elucidated (for review see [81]).

Notch signaling pathways are used in a variety of developmental contexts. The different output of Notch signaling strongly depends on the cellular context. Therefore, different target genes are expressed in different cells upon ligand stimulation [82]. Further, Notch signaling is modulated at several levels by extracellular, cytoplasmic, and nuclear proteins. At the extracellular level, Notch receptors and perhaps Notch ligands undergo posttranslational modification, as for example glycosylation by Fringe proteins [83, 84]. Fringes selectively alter the sensitivity of the Notch receptor to activation by different ligands [85], while several proteins, such as Numb inhibit Notch signaling by targeting cytoplasmic or nuclear NICD for ubiquitination and proteosomal degradation [86, 87].

Although most of our initial understanding of the Notch signaling pathway came from studies in worms and flies, Notch signaling has by now been shown to play several roles in vertebrates, ranging from controlling cell lineage decisions to pattern formation [76]. Notch receptors and ligands are widely expressed in the developing vertebrate embryo, and the generation of mutants of Notch ligands or receptors demonstrated important functions in cell-fate decisions in tissue derived from all three primary germ layers.

1.3.1
Notch Signaling During Hematopoiesis

Notch signaling is involved in many aspects of hematopoiesis. Both Notch receptors and ligands are widely expressed in the hematopoietic system, corroborating the important role of Notch signaling in hematopoiesis. For example, forced expression of Notch1 in HSCs can promote their ability to self-renew and suppress their differentiation into myeloid, erythroid, or lymphoid lineages [88]. In addition to Notch signaling, the Sonic hedgehog and Wnt signaling pathways have been implicated in adult HSC expansion and self-renewal [40, 89]. In the future it will be important to understand how these pathways interact and regulate the size of the HSC pool *in vivo*. However, although many gain-of-function experiments supported the idea that Notch signaling is involved in HSC maintenance, conditional loss-of-function approaches for Notch1 [90] and Notch2 [91] failed to demonstrate the role of Notch signaling in adult HSCs.

Notch1 function is best characterized in T/B lymphoid cell-fate specification [90, 92]. Loss-of-function and gain-of-function experiments revealed that Notch1 signaling is required for the determination of T cells from a progenitor that is capable of forming both T and B cells. Ablation of Notch1 function results in a hypotrophic thymus and bone marrow progenitors are instructed to develop into B cells at the expense of T cells after entering the thymus [93]. In addition, gain-of-function experiments in which NICD or Delta are expressed in bone marrow precursors leads to ectopic T cell development in the bone marrow and blocks B cell development [94, 95]. Thus, Notch signaling must be absent or negatively regulated during B-cell development in the bone marrow. In sum, Notch signaling is critical for T cell- versus B cell-fate specification.

1.3.2
Notch1 Functions as a Tumor Suppressor in Mouse Skin

In the skin, the role of Notch signaling has been assessed by tissue-specific conditional gene ablation. Disruption of *RBP-J*, encoding a Notch signaling component, resulted in increased epidermal cell formation from hair follicle stem cells at the expense of hair cells [96]. A keratinocyte-specific conditional ablation of Notch1 results in deregulation of the normal balance between growth and differentiation [97]. Withdrawal from the cell cycle is a prerequisite of terminal keratinocyte differentiation [98]. It has been shown that activated Notch1 causes the arrest of keratinocyte growth via increased expression of the cyclin-dependent kinase inhibitor p21. Therefore, inactivation of Notch1 in young mice induces hyperproliferation of the basal epidermal layer and deregulates expression of multiple differentiation markers. This suggests a role for Notch1 as a critical integrator of signals which controls the induction of keratinocyte growth arrest and early versus late stages of differentiation in the epidermis [97]. Furthermore, the role of Notch1 in adult mice has been investigated by applying the same loss-of-function approach using tissue-specific inducible gene ablation [99]. Surprisingly, these studies indicated that long-term Notch1 deficiency leads to epidermal and corneal hyperplasia followed by the development of skin tumors in various parts of the body. These results are unexpected as Notch signaling has previously been associated with maintaining proliferative cell populations and with cancer progression [100, 101]. It appears that loss of Notch1 signaling in the epidermis of mouse skin de-represses Wnt signaling and leads to increased levels of free, signaling competent β-catenin [99]. In contrast, forced Notch1 signaling in the epidermis and primary keratinocytes represses β-catenin signaling. This supports the hypothesis that Notch1 signaling inhibits β-catenin-mediated signaling in keratinocytes, and acts as a tumor suppressor in the skin.

1.3.3
Notch Signaling in the Nervous System and its Role in Neural Differentiation and Stem Cell Maintenance

There is substantial evidence to show that in the CNS Notch signaling regulates neural differentiation and stem cell maintenance. Activation of Notch signaling has been associated with the inhibition of neuronal differentiation, whereas repression of Notch activity promotes neurogenesis (reviewed in [74, 102]). Several studies have suggested that interference with Notch signaling leads to premature neurogenesis and a depletion of the neural stem cell pool [103–108]. For instance, mice deficient in Hes1, one of the downstream signaling effectors, display a decrease in the number of embryonic neural progenitor cells and commitment to the neuronal lineage is accelerated [104]. Moreover, conditional deletion of Notch1 in the cerebellar primordium *in vivo* results in upregulation of neuronal markers concomitant with reduced expression of the progenitor marker nestin [108].

Similarly, Notch activation inhibits oligodendrocyte differentiation in culture [109] while conditional ablation of Notch1 in oligodendrocyte precursors leads to their precocious maturation [110]. In turn, constitutive activation of Notch either renders multipotent progenitor cells permissive for cues inducing gliogenesis or instructs such cells to adopt a glial fate. In the retina, Notch1 and Hes1 are expressed in retinal progenitor cells and downregulated in differentiating and mature neurons [111, 112]. Forced expression of a constitutively activated Notch1 gene in rat retinal progenitor cells blocks the normal differentiation of the neuronal cell types and promotes formation of an unidentified cell type [112]. More recently these results have been complemented by forced expression of Hes1 or activated Notch1 in progenitor cells, which promotes formation of cells expressing Müller glia markers [113]. A possible mechanism by which Notch1–HES signaling exerts its function might be the repression of Mash1, a bHLH transcription factor required for neurogenesis [114]. Another study has shown that Notch activation in telencephalic progenitors promotes radial glia development [115]. In general, it remains to be elucidated whether activation of the Notch signaling pathway simply inhibits one fate (e.g. neurogenesis) leading to the promotion of a default pathway, or whether it directly promotes a specific fate.

Morrison and colleagues found that expression of activated Notch *in vivo* inhibits neuronal differentiation in the PNS. In addition, NCSCs in which the Notch signaling pathway is activated by soluble Delta are driven into the glial lineage *in vitro* [116]. Thereby, Notch instructs NCSCs to adopt a glial fate, even if exposure to Delta is only transient. It has been hypothesized that Notch signal activity is highly context-dependent and that the influence of Notch signaling is modified by additional signals. Addition of soluble Delta together with BMP2 revealed that neural crest cells become progressively more gliogenic and less neurogenic during development [117]. The decrease in sensitivity to the instructive neurogenic signal BMP2 as well as the increase in sensitivity to the anti-neurogenic and gliogenic signal Delta correlate with an increase in the ratio of expression of Notch1 to that of the Notch antagonist Numb [117]. Therefore, cells from distinct origins or isolated at different time-points display individual intrinsic properties that facilitate cell type-specific interpretation of Notch signaling, ranging from inhibition of differentiation and maintenance of the cells as progenitors to active instruction of progenitor cells to generate a particular cell lineage.

To further elucidate the role of Notch signaling in CNS stem cells, various research groups have made use of neurosphere cell culture systems. The self-renewing capacity of neural stem cells can be assessed in such neurosphere cultures [118]. In this assay, neural cells are cultured clonally and examined for their ability to form cell clusters (spheres). The differentiation potential of progenitors within spheres can be demonstrated by dissociation and subsequent differentiation in appropriate culture media. Self-renewal capacity can be addressed by serial subcloning experiments in which the generation of secondary spheres from dissociated primary spheres is monitored (Fig. 1.5). Upon ablation of Notch1, cells derived from the forebrain are unable to generate neurospheres, indicating a depletion of the neural stem cell pool [119]. In particular, homozygous disruption of Notch1 or CSL in mice disturbs the self-renewal capacity of the stem cells while promoting neuronal and glial differentiation. Likewise, the sphere-forming capacity is reduced in Hes1– and Hes5–deficient cells from the embryonal telencephalon, confirming the reduced self-renewal capacity of mutant NSCs [107]. It has been proposed that Notch signaling is primarily involved in symmetric divisions of neural stem cells within the CNS. Therefore, as a consequence of attenuated Notch signaling, fewer symmetrical and self-renewing divisions of mutant neural stem cells take place, concomitant with an increase in neuronal and astroglial differentiation of the neural progenitor cells.

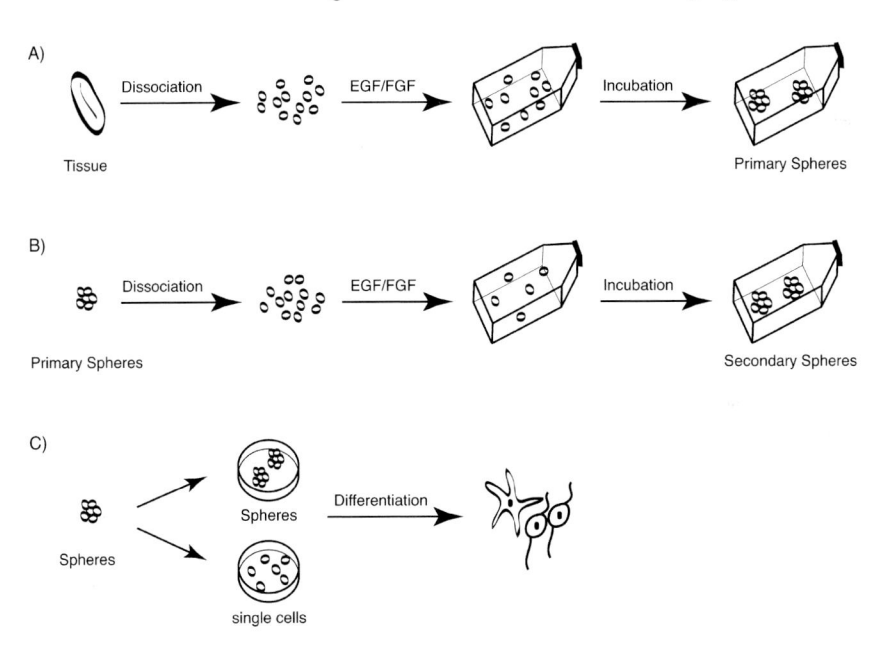

Fig. 1.5
The neurosphere assay demonstrates the self-renewal capacity and differentiation potential of stem cells. Stem cells from several different locations have been isolated and cultured as single floating cells in the presence of the growth factors EGF and FGF (A). Under these conditions the stem cells start to form floating cell aggregates termed neurospheres (first described in [118]). These primary spheres can be dissociated into single cells and again cultured clonally. The formation of secondary spheres proves the existence of stem cells that display the capacity for self-renewal (B). The differentiation potential of sphere-derived cells can be shown in adhesive cultures (C). The spheres are plated as entire spheres or as single cells, and after culturing in differentiation medium the cellular composition can be assessed.

This finding can be seen as an alternative to the idea that Notch signaling is directly and instructively involved in cell-fate decisions of neuronal and glial differentiation in the mammalian CNS [115]. Notch-dependent self-renewal is presumably mediated by endothelial cells that in co-culture promote the expansion of neuroepithelial cells and *in vivo* are thought to provide a vascular stem cell niche [120].

1.3.4
Aberrant Notch Signaling

Given its complex functions in normal tissue development and homeostasis, it is not surprising that aberrant Notch signaling gives rise to some human diseases. These include cerebral autosomal dominant arteriopathy with subcortical infarcts and leukoencephalopathy (CADASIL; [121]), together with several human cancers. CADASIL is an autosomal dominant disorder mainly affecting the arteries of the brain. The cause of the disease is a systemic arteriopathy that is associated with mutations of Notch3 resulting in destruction of arteriolar vascular smooth muscle cells. Another known disease is the Alagille syndrome (AGS), which is caused by Jagged–1 mutations. It is an autosomal dominant disease characterized by defects in liver, heart, skeleton and eye [122–124]. Most of the patients suffering from AGS carry a mutation in the Jagged–1 gene or the entire gene is deleted. The molecular mechanism underlying the disease is largely unknown. It is hypothesized that in addition to Notch other environmental or genetic factors are involved, such as upstream and downstream modulators of Notch signaling.

Aberrant activation of Notch signaling promotes neoplastic transformation of many cell types, which might be explained by Notch inhibiting other signaling pathways [125]. So far, many human and murine cancers including certain neuroblastomas, and mammary, skin, cervical and prostate cancers, are correlated with alterations in expression of Notch proteins and/or ligands. Often, the causal relationships still remain to be proven. Nevertheless, the elucidation of the Notch signaling pathway might allow the manipulation of Notch signaling via delivery of soluble Notch ligands or other strategies, in order to establish possible therapeutic anti-cancer treatments in the future.

1.4
Signaling Pathway of the TGFβ Family Members

Members of the TGFβ superfamily play a role in many aspects of embryonic development and adult homeostasis by affecting cell proliferation, differentiation and migration. The family includes TGFβ isoforms, BMPs, activins, and growth and differentiation factors (GDFs). Originally, they were identified as proteins capable of inducing ectopic cartilage and bone in mammals [126]. TGFβ family members are secreted dimeric cytokines that bind to type II single transmembrane receptors with intrinsic serine/threonine kinase activity, binding is followed by ligand-induced heterodimerization of type I and type II receptors [127]. Subsequently, the type I receptor is phosphorylated by the type II receptor and intracellular signaling is propagated by phosphorylation of specific Smad proteins that translocate to the nucleus where they control the transcription of target genes. There are eight vertebrate Smads that can be separated into three functional groups: common partner Smads (Co-Smads), receptor-regulated Smads (R-Smads) and inhibitory Smads (I-Smads). Smad2 and Smad3 are R-Smads that become phosphorylated and activated by TGFβ and activin receptors, whereas Smad1, Smad5 and Smad8 are activated in response to BMP or other ligands [128, 129]. Once activated, R-Smads are released from the receptor complex and form a heterotrimeric complex with the Co-Smad Smad4. Finally, activated R-Smad/Co-Smad complexes efficiently translocate into the nucleus, and in conjunction with other nuclear factors, regulate transcription of target genes. I-Smads (i.e. Smad6 and Smad7) can negatively regulate TGFβ signaling on several levels by binding to type I receptors and thereby preventing R-Smads from being activated by type I receptors. Additionally, I-Smads inhibit signaling by competing with Co-Smad interaction and by targeting the receptors for degradation. Smad proteins mediate transcriptional activation or repression depending on their associated partners. R-Smads and Smad4 are expressed in most cell types whereas the expression of the inhibitory Smad6 and Smad7 is highly regulated by extracellular signals. The level of the Smad pool is mainly regulated in a signaling-independent manner. Smad ubiquitination-related factor 1 (Smurf1) is an E3 ubiquitin ligase that catalyzes the transfer of the ubiquitin moiety to its target substrates. Smurfs appear to regulate BMP signaling by targeting non-activated Smad1 and –5 for protein degradation, thereby preventing spurious activation of the pathway. Additionally, the ubiquitin-proteasome pathway not only regulates the steady-state levels of R-Smads, but is also involved in the degradation of activated R-Smads. Smads also function as adaptors that recruit Smurfs to target proteins and thereby control the level of Smad-associating proteins. Two highly conserved negative regulators of Smad transcriptional function are c-Ski and SnoN, which are members of the Ski family of proto-oncogenes. Both antagonize TGFβ signaling through direct interaction with Smad4 and R-Smads [130, 131]. Smad signaling is terminated by either dephosphorylation or by ubiquitination and proteosome-mediated degradation of activated R-Smads.

Although there are only few receptors and Smads, a great versatility of signaling is possible by combinatorial interactions of type I and II receptors, oligomeric interaction complexes formed with Smads, and specific transcription factors whose levels change temporally and spatially depending on the cellular context. Differences in stability of signaling components and their subcellular localization may also affect the cellular response. In addition to Smad-mediated transcription, TGFβ also activates other signaling cascades such as the mitogen-activated protein kinase (MAPK) pathway. Some of these pathways regulate Smad activation, but others might induce Smad-independent responses [132, 133]. Additionally, other signaling pathways help to define the responses to TGFβ factors in a context-dependent manner.

1.4.1
BMP Signaling in ESCs

It is still matter of investigation how many growth factors and signaling pathways are involved in ESC self-renewal. Recently, gene expression profiling suggested that in addition to Wnt signaling as previously mentioned, BMP4 might support ESC self-renewal [134]. Qi and coworkers showed that BMP4 inhibits MAPK pathways in ESCs. MAPK pathways are crucial for signal transduction of many mitogens including LIF, BMPs, and FGFs [135, 136]. Changes in the balance of MAPK activity might determine whether the cells remain undifferentiated or whether differentiation is induced. Furthermore, BMP4 acts synergistically with LIF to promote self-renewal of ESCs [134, 137]. Introduction of the inhibitory Smad family members Smad6 and Smad7 into ESCs in order to antagonize BMP signaling reduced the self-renewal capacity of ESCs and induced differentiation [137]. This is accomplished by the induction of Id proteins through BMP/Smad signaling. Id proteins are negative regulatory helix-loop-helix factors that prevent the transcriptional activity of bHLH factors such as MyoD and Mash1 [138, 139]. Therefore, the suppression of ESC differentiation by BMP4 is likely achieved via induction of Id genes. In summary, in ESCs the two signaling pathways initiated by LIF and BMP act in combination and are highly controlled in order to sustain self-renewal.

1.4.2
The Influence of TGFβ Family Members on MSC Differentiation

MSCs isolated from bone marrow have the capacity to differentiate into a variety of cell types such as bone, cartilage, muscle and fat tissue [2, 3, 140]. MSCs can be isolated from the adult, and therefore it is likely that these stem cells participate in regeneration and repair. Members of the TGFβ superfamily have important roles in regulating the differentiation of mesenchymal cells. BMPs can induce differentiation of mesenchymal cells into cells with chondroblastic and osteoblastic phenotypes. Furthermore, mesenchymal cell lines respond to multiple isoforms of BMP, including BMP2 and BMP7 [141–144]. BMPs induce specific transcription factors, such as Sox9, Dlx5, and c-fos that are known to determine the commitment of mesenchymal cells into chondrogenic or osteogenic lineages. This process, in which chondrogenic differentiation precedes osteogenesis, occurs in several steps that are dose- and time-dependent [145]. Furthermore, TGFβ and activin provide competence for the early stages of chondroblastic differentiation, but at late stages in the osteoblastic differentiation pathway TGFβ acts as an inhibitor. BMPs and TGFß also block differentiation into the myogenic lineage. TGFβ inhibits muscle formation via direct interaction of Smad3 with MyoD [146], a bHLH transcription factor that plays an important role in myogenesis. Similarly, TGFβ is also an inhibitor of adipogenesis mainly via Smad3. In sum, the TGFβs and the BMPs exert several functions demonstrating positive and negative effects on bone development. The cross-talk between TGFβ and BMP signaling has not yet been fully elucidated but temporal expression and the dosage of the individual factors are important.

1.4.3
Tgfβ Factors Act Instructively on NCSC Differentiation

Members of the TGFβ superfamily have multiple functions during neural development, including lineage commitment, proliferation, survival, apoptosis, differentiation, and morphogenesis [147, 148]. In the CNS, BMP signaling is involved in the patterning of the neural tube, regulation of apoptosis, survival and maturation. In the PNS, factors of the BMP subclass together with other factors play a role in neural crest induction [149–152]. At later stages of PNS development, BMP2/4 promote autonomic neurogenesis *in vitro* and *in vivo* [153–155]. TGFβ family members have been shown to act instructively on NCSCs. BMP2 promotes a neuronal and, to a lesser extent, a smooth muscle-like fate in clonal cultures of multipotent progenitors derived from neural crest, sciatic nerve, dissociated DRG, and enteric nervous system [156]. Likewise, single progenitor cells are instructed by TGFβ to exclusively adopt a non-neural fate. Cardiac neural crest gives rise to smooth muscle cells in the outflow tract of the heart [157, 158] where TGFβ isoforms are expressed [159, 160]. TGFβ2 null mice exhibit developmental cardiac defects but it is not clear whether the deficiency is in lineage determination, migration, or maturation of crest cells [161]. BMP2 and BMP4 are expressed in the dorsal aorta close to areas of autonomic neurogenesis [153, 155]. BMP2 induces and maintains the basic helix-loop-helix transcription factor Mash1 that is crucial for autonomic neuronal differentiation [162–164]. Thus, only if BMP2 expression persists in the environment is a neural crest-derived cell able to adopt an autonomic fate. The *in vivo* expression pattern of BMP and TGFβ are consistent with the role of these factors in regulating cell-fate decisions in the developing PNS. In chicken embryos, a requirement for BMP signaling in autonomic neurogenesis has been demonstrated using the BMP agonist Noggin [154].

In vivo, however, progenitor cells of the PNS are exposed to multiple signals during migration and at sites of differentiation. Thus, it is conceivable that distinct signaling pathways act on a multipotent progenitor by modulating each other, thereby producing biological effects that are different from those elicited by the individual signals alone. In neural crest cultures, BMP2 and TGFβ act co-dominantly, while these TGFβ family members are dominant over other signals such as NRG1 [165]. In contrast, the gliogenic activity of Notch signaling suppresses induction of neurogenesis by BMP2 [116]. Additionally, cell-cell interactions termed community effects, influence lineage decisions [166]. Cell clusters of neural crest-derived progenitors, in contrast to single cells, display a reduced non-neural potential when exposed to TGFβ factors (Fig. 1.6). Although individual progenitor cells have the potential to give rise to non-neural smooth muscle-like cells in response to TGFβ factors, neurogenesis or, at slightly higher doses of TGFβ, apoptosis is promoted at the expense of the non-neural fate in progenitor communities in the presence of these instructive signals [167, 168]. Thus, the community effect reveals a synergy between TGFβ signaling and signal transduction pathways provided by short-range cell-cell interactions. Conceivably, this allows the fine tuning of the cell-fate decision and programmed cell death, which is an important process in development to control cell numbers and patterning [169–171]. The molecular basis of these effects has yet to be identified, but cellular interactions via cell-cell contact, local accumulation of secreted signals, or gap junctions are presumably involved [166]. In general, members of the TGFβ superfamily can undertake a variety of different, context-dependent functions in developmental systems. TGFβ signaling in early NCSC development represents an example of how TGFβ signal transduction pathways are able to operate as part of a signaling network which integrates multiple environmental cues that a cell is exposed to.

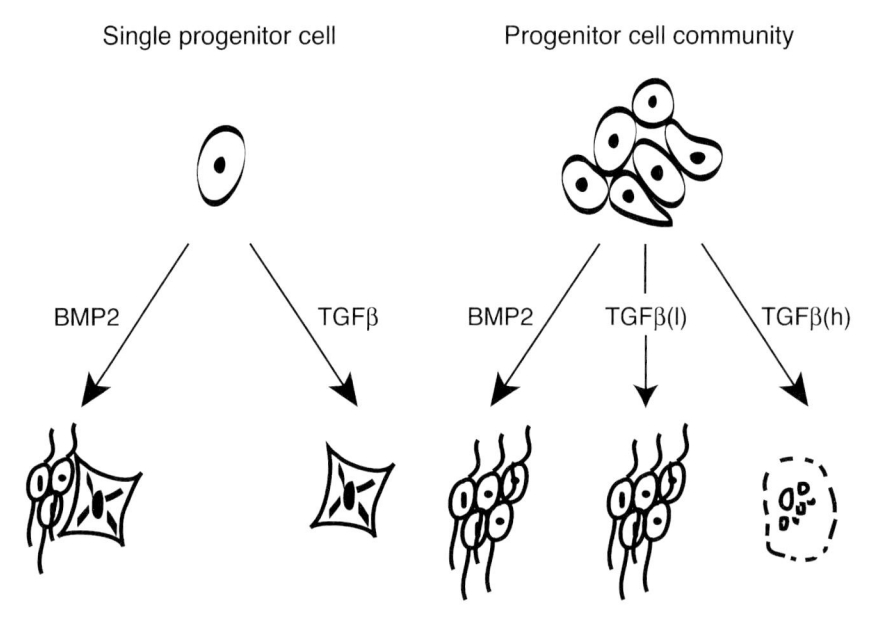

Fig. 1.6
Context-dependent TGFβ signaling in neural crest-derived progenitor cells. In response to BMP2, single progenitor cells can produce neuronal as well as a non-neural, smooth muscle-like progeny. TGFβ promotes only a non-neural fate in neural crest-derived single progenitors. In contrast, the non-neural cells are completely suppressed by short-range cell-cell interactions provided by progenitor cell communities. Instead, members of the TGFβ factor family induce neurogenesis in such communities. Additionally, higher doses of TGFβ promote cell death as an alternative fate. Thus, the fate of stem and progenitor cells is influenced by multiple signals that act in combination and at changing concentrations.

1.4.4
Aberrant Growth Regulation by Mutations in the Tgfβ Signaling Pathway

The disruption of the TGFβ signaling pathway has been implicated in the progression of several human diseases. For instance, TGFβ signaling has been shown to be involved in various forms of cancer such as breast, pancreatic, colon and lung cancer. TGFβ signaling displays tumor suppressor activities, as it acts as an inhibitor of cell growth and an inducer of apoptosis that regulates the homeostasis of rapidly proliferating tissues, such as renewing epithelia and blood cells. On the other hand, TGFβ also has pro-oncogenic activities that can lead to enhanced epithelial to mesenchymal transition (EMT), growth stimulation, increased motility, and invasiveness. The TGFβ-mediated growth arrest in many cells can be attributed to the downregulation of c-myc. This repression is achieved by the binding of a Smad complex to a TGFβ-inhibitory element in the c-myc promoter [172]. A second important event that leads to TGFβ-induced growth arrest is the induction of two major cell cycle inhibitors, the cyclin-dependent kinase (CDK) inhibitors p15 and p21, directly via Smad-dependent transcriptional activation [173, 174]. The components of the TGFβ signaling pathway that are most commonly mutated in human cancers are Smad4 (originally termed "deleted in pancreatic carcinoma, locus4" or DPC4) and Smad2 [132]. Furthermore, the TGFβ system and the Ras/MAPK pathways interact in tumorigenesis. TGFβ is able to activate the MAPK pathways directly, and interacts with these pathways when they are activated by other cues. Many pro-oncogenic responses to TGFβ seem to be either Smad-independent, or require cooperation of Smad with alternative pathways. Smad7 is upregulated in human pancreatic cancer, and its overexpression leads to a loss of TGFβ-induced growth inhibition [175]. Thus, proteins that interact with Smads and modulate their activity might be direct targets of oncogenic change.

1.5
Shh Signaling

Sonic hedgehog (Shh) is a member of the Hedgehog (Hh) family of secreted signaling proteins carrying out diverse functions during vertebrate development. Originally, Shh was identified as a regulator of cell-fate determination and body segment polarity. In some contexts, Hh signals act as morphogens in a dose-dependent manner, in others as mitogens regulating cell proliferation. In many contexts, the Shh network functions as a "cell-fate switch" where the cell state is changed at a critical threshold level. For example, Shh is secreted from the notochord and organizes the developing neural tube by forming a concentration gradient. The distinct levels of Shh establish distinct regions of homeodomain transcription factor domains along the dorso-ventral axis, thereby specify neuronal identity [176–178].

A key component of the Shh signaling pathway is the 12–transmembrane domain receptor Ptc (patched in Drosophila), which acts as a key inhibitory regulator of the constitutively active G-protein coupled receptor component Smoothened (Smo). Binding of Shh inactivates Ptc and allows Smo to become active, which leads to transcription of downstream target genes of the Gli family and Ptc itself [179, 180]. There are three Gli proteins that interpret the Shh signal in a combinatorial fashion by having both activator and repressor activities [181–183]. Further, Ptc also regulates the movement of Hh through tissue, as binding of Hh limits the spread of Hh from its source. The ability of Shh to exert its function is regulated by a series of posttranslational processes. The approximately 45–kDa Hh precursor molecule undergoes an autoproteolytic cleavage that removes the C-terminal end. During this cleavage a cholesterol moiety is covalently attached to the remaining active N-terminal fragment [184]. Additionally, the protein is palmitolyated at the N-terminal end [185]. These lipid modifications of Hh may play a role in targeting it to rafts, and may affect the ability of Shh to activate reporter constructs in cultured cells and target genes *in vivo* [186, 187].

1.5.1
Hematopoiesis and T-cell Maturation

As already mentioned in previous sections, factors regulating the pool of HSCs are still a matter of active research. Bhardwaj et al. showed in 2001 that Hh and its putative receptors, Ptc and Smo, along with the downstream transcription factors Gli1, Gli2, and Gli3, are expressed in primitive human blood cells and stromal cells of the hematopoietic microenvironment. Blocking of endogenously produced Hh or addition of exogenous soluble Hh can control the proliferation of uncommitted human hematopoietic cells [89]. Furthermore, Shh signaling influences T-cell differentiation, which depends on interactions between the thymic epithelium and developing thymocytes in the thymus [188]. It has been shown that Hh signaling is already active during early thymocyte development. Shh is produced by the thymic epithelium, and its receptors Ptc and Smo are expressed by thymocytes. Inhibition of Shh increases the differentiation of thymocytes and treatment with Shh inhibits their differentiation [189].

1.5.2
The Role of Shh in the Nervous System

During embryonic development Gli genes are expressed in proliferative zones of the brain. BrdU incorporation experiments demonstrated a mitogenic effect of Shh on nestin-positive progenitors [190]. Furthermore, neurosphere assays using embryonic neocortical progenitors showed that Shh signaling is required for normal proliferation and self-renewal [191]. In particular, cells isolated from the cortex of Shh-deficient animals produced neurospheres at a much lower frequency as compared to control cells. Therefore, Shh signaling provides a mechanism regulating the number of stem cells in the developing mouse neocortex.

It has been reported that until adulthood localized zones of active neurogenesis persist in the brain. Neurogenesis in the adult mammalian brain takes place in the SVZ of the lateral ventricular walls of the forebrain and in the subgranular layer of the dentate gyrus of the hippocampus [192]. Stem cells in these zones are periventricular astrocytes [193, 194] that are induced by inductive signals to produce new neurons. Two recent reports by Lai et al. [195] and Machold et al. [196] showed that Shh signaling is involved in cell proliferation in adult neurogenic niches. Lai and colleagues report that Shh signaling regulates the proliferation of progenitor cells in the adult rat hippocampus, which can be blocked by applying an inhibitor of Shh signaling in the subgranular zone [195]. Moreover, the removal of Shh signaling results in a reduced number of neural progenitors in both the postnatal subventricular zone and the hippocampus [196]. Shh may directly regulate the cell cycle, as it upregulates the expression of type D and E cyclins [197]. Therefore, Shh appears to act on adult multipotent hippocampal progenitor cells by inducing proliferation. Consistent with this idea, an Hh agonist increases the proliferation and Gli1 expression in the SVZ and dentate gyrus [196]. Finally, the requirement for Shh in the maintenance of telencephalic stem cells has been assessed by the neurosphere assay, revealing that progenitors from the SVZ with impaired Shh signaling have a reduced potential to generate neurospheres. The combined data suggest that Shh is required for the maintenance of telencephalic stem cell niches in the adult brain. Possibly, Shh signaling acts at a certain concentration range together with other growth factors to establish an environment in which the stem cells are able to persist and to proliferate.

1.5.3
Shh Signaling in Tumorigenesis

Aberrant Shh signaling is thought to contribute to the neoplastic transformation of cells arising from two different cell types of ectodermal origin in the embryo: the epithelial cell of the skin (Gorlin syndrome; basal cell carcinomas, BCCs) and the neural precursors in the brain (gliomas, medullablastoma; [197]). Consistent with this idea, overexpression of Gli1 in the CNS of tadpoles as well as in the tadpole skin leads to tumor formation [190, 198]. Further, cyclopamine, a plant-derived drug that selectively inhibits the Hh-Gli pathway by suppressing the activity of Smo, is able to inhibit brain tumor growth [190]. Additionally, mice that carry a mutation in the patched gene are susceptible to medullablastoma formation [199, 200]. In humans, analysis of many sporadic brain tumors showed expression of three Gli genes [190]. In particular, inappropriate activation of the Shh-Gli pathway has been associated with familial brain tumors such as primitive neuroectodermal tumors (PNETs) of the cerebellum or medullablastoma. Medullablastomas represent the most common malignant brain tumors of childhood [201]. They form a heterogeneous group of tumors believed to arise from immature precursor cells of the cerebellar granule cells. Normally, Shh, which is produced by the Purkinje neurons, controls the growth of the cerebellum and promotes proliferation of granule neuron precursors in the external germinal layer (EGL) of the cerebellum. It is assumed that medullablastomas arise when granule neuron precursors inappropriately maintain Shh-Gli signal activation [190, 202]. More generally, it has been proposed that stem cells displaying sustained Shh signal activity might be responsible for the development of some tumors. Not surprisingly, the role of stem cells in tumorigenesis and of the signaling pathways involved has become a major focus of cancer research.

1.6
Conclusions

In recent years, stem cell research has made considerable progress and several of the signaling pathways that influence stem cell development have been brought to light. We have to be aware, however, that a complex orchestra of signaling cascades rather than individual signaling pathways controls stem cell specification, expansion, and differentiation. Distinct signaling pathways might activate, inhibit or modulate each other, thereby eliciting different biological responses. Moreover, the combination of signals involved likely changes in a spatiotemporal manner. Therefore, it will be a challenge for the future to identify the crucial key points in the signaling network that determines the fate of a particular stem cell type at a specific time-point and location. The use of functional genomics and proteomics should provide several candidate molecules. Cell culture experiments are helpful in the elucidation of the function of such candidate factors (and factor combinations), because they allow one to study the influence of multiple factors on cell-fate decisions in defined but changeable contexts. Furthermore, generating animal models carrying multiple mutations, possibly stem cell-specific and inducible, will be necessary to better understand signal integration by stem cells *in vivo*.

References

1 Mezey, E., et al. Turning blood into brain: cells bearing neuronal antigens generated in vivo from bone marrow. *Science*, **2000**; *290(5497)*: 1779–1782.
2 Jiang, Y., et al. Neuroectodermal differentiation from mouse multipotent adult progenitor cells. *Proc. Natl Acad. Sci. USA*, **2003**; *100 (Suppl 1)*: 11854–11860.
3 Jiang, Y., et al. Pluripotency of mesenchymal stem cells derived from adult marrow. *Nature*, **2002**; *418(6893)*: 41–49.
4 Toma, J.G., et al. Isolation of multipotent adult stem cells from the dermis of mammalian skin. *Nature Cell Biol.*, **2001**; *3(9)*: 778–784.
5 Jiang, Y., et al. Multipotent progenitor cells can be isolated from postnatal murine bone marrow, muscle, and brain. *Exp. Hematol.*, **2002**; *30(8)*: 896–904.
6 LaBarge, M.A. and H.M. Blau, Biological progression from adult bone marrow to mononucleate muscle stem cell to multinucleate muscle fiber in response to injury. *Cell*, **2002**; *111(4)*: 589–601.
7 Wagers, A.J., et al. Little evidence for developmental plasticity of adult hematopoietic stem cells. *Science*, **2002**; *297(5590)*: 2256–2259.
8 Frisen, J., Stem cell plasticity? *Neuron*; **2002**; *35(3)*: 415–418.
9 Wagers, A.J. and I.L. Weissman, Plasticity of adult stem cells. *Cell*, **2004**; *116(5)*: 639–648.

10 Fuchs, E., T. Tumbar, and G. Guasch, Socializing with the neighbors: stem cells and their niche. *Cell*, **2004**; *116(6)*: 769–778.
11 Ivanova, N.B., et al. A stem cell molecular signature. *Science*, **2002**; *298(5593)*: 601–604.
12 Ramalho-Santos, M., et al. »Stemness«: transcriptional profiling of embryonic and adult stem cells. *Science*, **2002**; *298(5593)*: 597–600.
13 Buchstaller, J., et al. Efficient isolation and gene expression profiling of small numbers of neural crest stem cells and developing Schwann cells. *J. Neurosci.*, **2004**; *24(10)*: 2357–2365.
14 Lu, B., L. Jan, and Y.N. Jan, Control of cell divisions in the nervous system: symmetry and asymmetry. *Annu. Rev. Neurosci.*, **2000**; *23*: 531–556.
15 Cai, L., N.L. Hayes, and R.S. Nowakowski, Synchrony of clonal cell proliferation and contiguity of clonally related cells: production of mosaicism in the ventricular zone of developing mouse neocortex. *J Neurosci.*, **1997**; *17(6)*: 2088–2100.
16 Raoul, C., B. Pettmann, and C.E. Henderson, Active killing of neurons during development and following stress: a role for p75(NTR) and Fas? *Curr. Opin. Neurobiol.*, **2000**; *10(1)*: 111–117.
17 Dutton, R. and P.F. Bartlett, Precursor cells in the subventricular zone of the adult mouse are actively inhibited from differentiating into neurons. *Dev. Neurosci.*, **2000**; *22(1–2)*: 96–105.
18 Armulik, A., T. Velling, and S. Johansson, The integrin beta1 subunit transmembrane domain regulates phosphatidylinositol 3–kinase-dependent tyrosine phosphorylation of Crk-associated substrate. *Mol. Biol. Cell*, **2004**; *15(6)*: 2558–2567.
19 Sommer, L. and M. Rao, Neural stem cells and regulation of cell number. *Prog. Neurobiol.*, **2002**; *66(1)*: 1–18.
20 Cadigan, K.M. and R. Nusse, Wnt signaling: a common theme in animal development. *Genes Dev.*, **1997**; *11(24)*: 3286–3305.
21 Willert, K., et al. Wnt proteins are lipid-modified and can act as stem cell growth factors. *Nature*, **2003**; *423*: 448–452.
22 De Strooper, B. and W. Annaert, Where Notch and Wnt signaling meet. The presenilin hub. *J. Cell Biol.*, **2001**; *152(4)*: F17–F20.
23 Hecht, A. and R. Kemler, Curbing the nuclear activities of beta-catenin. Control over Wnt target gene expression. *EMBO Rep.*, **2000**; *1(1)*: 24–28.
24 Nelson, W.J. and R. Nusse, Convergence of Wnt, beta-catenin, and cadherin pathways. *Science*, **2004**; *303(5663)*: 1483–1487.
25 Brantjes, H., et al. All Tcf HMG box transcription factors interact with Groucho-related co-repressors. *Nucleic Acids Res.*, **2001**; *29(7)*: 1410–1419.
26 Jamora, C. and E. Fuchs, Intercellular adhesion, signalling and the cytoskeleton. *Nature Cell Biol.*, 2002. 4(4): p. E101–E108.
27 Gumbiner, B.M., Regulation of cadherin adhesive activity. *J. Cell Biol.*, **2000**; *148(3)*: 399–404.
28 Nichols, J., et al. Formation of pluripotent stem cells in the mammalian embryo depends on the POU transcription factor Oct4. *Cell*, **1998**; *95(3)*: 379–391.

29 Feldman, B., et al. Requirement of FGF–4 for postimplantation mouse development. *Science*, **1995**; *267(5195)*: 246–24
30 Niwa, H., J. Miyazaki, and A.G. Smith, Quantitative expression of Oct–3/4 defines differentiation, dedifferentiation or self-renewal of ES cells. *Nature Genet.*, **2000**; *24(4)*: 372–376.
31 Mitsui, K., et al. The homeoprotein Nanog is required for maintenance of pluripotency in mouse epiblast and ES cells. *Cell*, **2003**; *113(5)*: 631–642.
32 Avilion, A.A., et al. Multipotent cell lineages in early mouse development depend on SOX2 function. *Genes Dev.*, **2003**; *17(1)*: 126–140.
33 Smith, A.G., Embryo-derived stem cells: of mice and men. *Annu. Rev. Cell Dev. Biol.*, **2001**; *17*: 435–462.
34 Sato, N., et al. Molecular signature of human embryonic stem cells and its comparison with the mouse. *Dev. Biol.*, **2003**; *260(2)*: 404–413.
35 Aubert, J., et al. Functional gene screening in embryonic stem cells implicates Wnt antagonism in neural differentiation. *Nature Biotechnol.*, **2002**; *20(12)*: 1240–1245.
36 Haegele, L., et al. Wnt signalling inhibits neural differentiation of embryonic stem cells by controlling bone morphogenetic protein expression. *Mol. Cell. Neurosci.*, **2003**; *24(3)*: 696–708.
37 Sato, N., et al. Maintenance of pluripotency in human and mouse embryonic stem cells through activation of Wnt signaling by a pharmacological GSK–3–specific inhibitor. *Nature Med.*, **2004**; *10(1)*: 55–63.
38 Kielman, M.F., et al. Apc modulates embryonic stem-cell differentiation by controlling the dosage of beta-catenin signaling. *Nature Genet.*, **2002**; *32(4)*: 594–605.
39 Austin, T.W., et al. A role for the Wnt gene family in hematopoiesis: expansion of multilineage progenitor cells. *Blood*, **1997**; *89(10)*: 3624–3635.
40 Reya, T., et al. A role for Wnt signalling in self-renewal of haematopoietic stem cells. *Nature*, **2003**; *423*: 409–414.
41 Cobas, M., et al. β-Catenin Is Dispensable for Hematopoiesis and Lymphopoiesis. *J. Exp. Med.*, **2004**; *199(2)*: 221–229.
42 Oshima, H., et al. Morphogenesis and renewal of hair follicles from adult multipotent stem cells. *Cell*, **2001**; *104(2)*: 233–245.
43 Alonso, L. and E. Fuchs, Stem cells in the skin: waste not, Wnt not. *Genes Dev.*, **2003**; *17(10)*: 1189–1200.
44 Zhu, A.J. and F.M. Watt, beta-catenin signalling modulates proliferative potential of human epidermal keratinocytes independently of intercellular adhesion. *Development*, **1999**; *126(10)*: 2285–2298.
45 Gat, U., et al. *De novo* hair follicle morphogenesis and hair tumors in mice expressing a truncated beta-catenin in skin. *Cell*, **1998**; *95(5)*: 605–614.
46 Gandarillas, A. and F.M. Watt, c-Myc promotes differentiation of human epidermal stem cells. *Genes Dev.*, **1997**; *11(21)*: 2869–2882.
47 Arnold, I. and F.M. Watt, c-Myc activation in transgenic mouse epidermis results in mobilization of stem cells and differentiation of their progeny. *Curr. Biol.*, **2001**; *11(8)*: 558–568.

48 Huelsken, J., et al. beta-Catenin controls hair follicle morphogenesis and stem cell differentiation in the skin. *Cell*, **2001**; *105*(4): 533–545.
49 Merrill, B.J., et al. Tcf3 and Lef1 regulate lineage differentiation of multipotent stem cells in skin. *Genes Dev.*, **2001**; *15*(13): 1688–1705.
50 Niemann, C., et al. Expression of DeltaNLef1 in mouse epidermis results in differentiation of hair follicles into squamous epidermal cysts and formation of skin tumours. *Development*, **2002**; *129*(1): 95–109.
51 Lee, S.M., et al. A local Wnt–3a signal is required for development of the mammalian hippocampus. *Development*, **2000**; *127*(3): 457–467.
52 Ikeya, M., et al. Wnt signalling required for expansion of neural crest and CNS progenitors. *Nature*, **1997**; *389*(6654): 966–970.
53 Zechner, D., et al. beta-Catenin signals regulate cell growth and the balance between progenitor cell expansion and differentiation in the nervous system. *Dev. Biol.*, **2003**; *258*(2): 406–418.
54 Megason, S.G. and A.P. McMahon, A mitogen gradient of dorsal midline Wnts organizes growth in the CNS. *Development*, **2002**; *129*(9): 2087–2098.
55 Chenn, A. and C.A. Walsh, Regulation of cerebral cortical size by control of cell cycle exit in neural precursors. *Science*, **2002**; *297*(5580): 365–369.
56 Israsena, N., et al. The presence of FGF2 signaling determines whether beta-catenin exerts effects on proliferation or neuronal differentiation of neural stem cells. *Dev. Biol.*, **2004**; *268*(1): 220–231.
57 Viti, J., A. Gulacsi, and L. Lillien, Wnt regulation of progenitor maturation in the cortex depends on Shh or fibroblast growth factor 2. *J. Neurosci.*, **2003**; *23*(13): 5919–5927.
58 Hirabayashi, Y., et al. The Wnt/beta-catenin pathway directs neuronal differentiation of cortical neural precursor cells. *Development*, **2004**; *131*(12): 2791–2801.
59 Le Douarin, N.M. and C. Kalcheim, *The Neural Crest*. 2nd ed. Cambridge, UK: Cambridge University Press, **1999**.
60 Le Douarin, N.M. and E. Dupin, Multipotentiality of the neural crest. *Curr. Opin. Genet. Dev.*, **2003**; *13*(5): 529–536.
61 Paratore, C., et al. Cell-intrinsic and cell-extrinsic cues regulating lineage decisions in multipotent neural crest-derived progenitor cells. *Int. J. Dev. Biol.*, **2002**; *46*(1): 193–200.
62 Saint-Jeannet, J.P., et al. Regulation of dorsal fate in the neuraxis by Wnt–1 and Wnt–3a. *Proc. Natl. Acad. Sci. USA*, **1997**; *94*(25): 13713–13718.
63 Garcia-Castro, M.I., C. Marcelle, and M. Bronner-Fraser, Ectodermal Wnt function as a neural crest inducer. *Science*, **2002**; *297*(5582): 848–851.
64 Dorsky, R.I., R.T. Moon, and D.W. Raible, Control of neural crest cell fate by the Wnt signalling pathway. *Nature*, **1998**; *396*(6709): 370–373.
65 Jin, E.J., et al. Wnt and BMP signaling govern lineage segregation of melanocytes in the avian embryo. *Dev. Biol.*, **2001**; *233*(1): 22–37.

66 Hari, L., et al. Lineage-specific requirements of ß-catenin in neural crest development. *J. Cell Biol.*, **2002**; *159*: 867–880.

67 Lee, H.Y., et al. Instructive role of Wnt/beta-catenin in sensory fate specification in neural crest stem cells. *Science*, **2004**; *303*(*5660*): 1020–1023; published online 8 January 2004 (10.1126/science.1091611).

68 Pinto, D., et al. Canonical Wnt signals are essential for homeostasis of the intestinal epithelium. *Genes Dev.*, **2003**; *17*(*14*): 1709–1713.

69 Hata, K., et al. Tumor formation is correlated with expression of beta-catenin-accumulated crypts in azoxymethane-induced colon carcinogenesis in mice. *Cancer Sci.*, **2004**; *95*(*4*): 316–320.

70 van de Wetering, M., et al. The beta-catenin/TCF–4 complex imposes a crypt progenitor phenotype on colorectal cancer cells. *Cell*, **2002**; *111*(*2*): 241–250.

71 Kikuchi, A., Tumor formation by genetic mutations in the components of the Wnt signaling pathway. *Cancer Sci.*, **2003**; *94*(*3*): 225–229.

72 Lustig, B. and J. Behrens, The Wnt signaling pathway and its role in tumor development. *J. Cancer Res. Clin. Oncol.*, **2003**; *129*(*4*): 199–221.

73 Polakis, P., Wnt signaling and cancer. *Genes Dev.*, **2000**; *14*(*15*): 1837–1851.

74 Artavanis-Tsakonas, S., M.D. Rand, and R.J. Lake, Notch signaling: cell fate control and signal integration in development. *Science*, **1999**; *284*(*5415*): 770–776.

75 Harper, J.A., et al. Notch signaling in development and disease. *Clin. Genet.*, **2003**; *64*(*6*): 461–472.

76 Lai, E.C., Notch signaling: control of cell communication and cell fate. *Development*, **2004**; *131*(*5*): 965–973.

77 Bray, S. and M. Furriols, Notch pathway: making sense of suppressor of hairless. *Curr. Biol.*, **2001**; *11*(*6*): R217–R221.

78 Petcherski, A.G. and J. Kimble, LAG–3 is a putative transcriptional activator in the C. elegans Notch pathway. *Nature*, **2000**; *405*(*6784*): 364–368.

79 Kao, H.Y., et al. A histone deacetylase corepressor complex regulates the Notch signal transduction pathway. *Genes Dev.*, **1998**; *12*(*15*): 2269–2277.

80 Iso, T., L. Kedes, and Y. Hamamori, HES and HERP families: multiple effectors of the Notch signaling pathway. *J. Cell Physiol.*, **2003**; *194*(*3*): 237–255.

81 Martinez Arias, A., V. Zecchini, and K. Brennan, CSL-independent Notch signalling: a checkpoint in cell fate decisions during development? *Curr. Opin. Genet. Dev.*, **2002**; *12*(*5*): 524–533.

82 Cooper, M.T., et al. Spatially restricted factors cooperate with notch in the regulation of Enhancer of split genes. *Dev. Biol.*, **2000**; *221*(*2*): 390–403.

83 Haines, N. and K.D. Irvine, Glycosylation regulates Notch signalling. *Nature Rev. Mol. Cell Biol.*, **2003**; *4*(*10*): 786–797.

84 Haltiwanger, R.S. and P. Stanley, Modulation of receptor signaling by glycosylation: fringe is an O-fucose-beta1,3–N-acetylglucosaminyltransferase. *Biochim. Biophys. Acta*, **2002**; *1573*(*3*): 328–335.

85 Hicks, C., et al. Fringe differentially modulates Jagged1 and Delta1 signalling through Notch1 and Notch2. *Nature Cell Biol.*, **2000**; *2(8)*: 515–520.
86 Nie, J., et al. LNX functions as a RING type E3 ubiquitin ligase that targets the cell fate determinant Numb for ubiquitin-dependent degradation. *EMBO J.*, **2002**; *21(1–2)*: 93–102.
87 McGill, M.A. and C.J. McGlade, Mammalian numb proteins promote Notch1 receptor ubiquitination and degradation of the Notch1 intracellular domain. *J. Biol. Chem.*, **2003**; *278(25)*: 23196–23203.
88 Stier, S., et al. Notch1 activation increases hematopoietic stem cell self-renewal *in vivo* and favors lymphoid over myeloid lineage outcome. *Blood*, **2002**; *99(7)*: 2369–2378.
89 Bhardwaj, G., et al. Sonic hedgehog induces the proliferation of primitive human hematopoietic cells via BMP regulation. *Nature Immunol.*, **2001**; *2(2)*: 172–180.
90 Radtke, F., et al. Deficient T cell fate specification in mice with an induced inactivation of Notch1. *Immunity*, **1999**; *10(5)*: 547–558.
91 Saito, T., et al. Notch2 is preferentially expressed in mature B cells and indispensable for marginal zone B lineage development. *Immunity*, **2003**; *18(5)*: 675–685.
92 Wilson, A., et al. Cutting edge: an essential role for Notch–1 in the development of both thymus-independent and -dependent T cells in the gut. *J. Immunol.*, **2000**; *165(10)*: 5397–5400.
93 Wilson, A., H.R. MacDonald, and F. Radtke, Notch 1–deficient common lymphoid precursors adopt a B cell fate in the thymus. *J. Exp. Med.*, **2001**; *194(7)*: 1003–1012.
94 Pui, J.C., et al. Notch1 expression in early lymphopoiesis influences B versus T lineage determination. *Immunity*, **1999**; *11(3)*: 299–308.
95 Dorsch, M., et al. Ectopic expression of Delta4 impairs hematopoietic development and leads to lymphoproliferative disease. *Blood*, **2002**; *100(6)*: 2046–2055.
96 Yamamoto, N., et al. Notch/RBP-J signaling regulates epidermis/hair fate determination of hair follicular stem cells. *Curr. Biol.*, **2003**; *13(4)*: 333–338.
97 Rangarajan, A., et al. Notch signaling is a direct determinant of keratinocyte growth arrest and entry into differentiation. *EMBO J.*, **2001**; *20(13)*: 3427–3436.
98 Missero, C., et al. The absence of p21Cip1/WAF1 alters keratinocyte growth and differentiation and promotes ras-tumor progression. *Genes Dev.*, **1996**; *10(23)*: 3065–3075.
99 Nicolas, M., et al. Notch1 functions as a tumor suppressor in mouse skin. *Nature Genet.*, **2003**; *33(3)*: 416–421.
100 Milner, L.A. and A. Bigas, Notch as a mediator of cell fate determination in hematopoiesis: evidence and speculation. *Blood*, **1999**; *93(8)*: 2431–2448.
101 Pear, W.S., et al. Exclusive development of T cell neoplasms in mice transplanted with bone marrow expressing activated Notch alleles. *J. Exp. Med.*, **1996**; *183(5)*: 2283–2291.
102 Beatus, P. and U. Lendahl, Notch and neurogenesis. *J. Neurosci. Res.*, **1998**; *54(2)*: 125–136.

103 Ishibashi, M., et al. Targeted disruption of mammalian hairy and Enhancer of split homolog–1 (HES–1) leads to up-regulation of neural helix-loop-helix factors, premature neurogenesis, and severe neural tube defects. *Genes Dev.*, **1995**; *9*: 3136–3148.

104 Nakamura, Y., et al. The bHLH gene hes1 as a repressor of the neuronal commitment of CNS stem cells. *J. Neurosci.*, **2000**; *20(1)*: 283–293.

105 de la Pompa, J.L., et al. Conservation of the Notch signalling pathway in mammalian neurogenesis. *Development*, **1997**; *124(6)*: 1139–1148.

106 Ohtsuka, T., et al. Hes1 and Hes5 as notch effectors in mammalian neuronal differentiation. *EMBO J.*, **1999**; *18(8)*: 2196–2207.

107 Ohtsuka, T., et al. Roles of the basic helix-loop-helix genes Hes1 and Hes5 in expansion of neural stem cells of the developing brain. *J. Biol. Chem.*, **2001**; *276(32)*: 30467–30474.

108 Lutolf, S., et al. Notch1 is required for neuronal and glial differentiation in the cerebellum. *Development*, **2002**; *129(2)*: 373–385.

109 Wang, S., et al. Notch receptor activation inhibits oligodendrocyte differentiation. *Neuron*, **1998**; *21*: 63–75.

110 Genoud, S., et al. Notch1 control of oligodendrocyte differentiation in the spinal cord. *J. Cell Biol.*, **2002**; *158(4)*: 709–718.

111 Tomita, K., et al. Mammalian hairy and Enhancer of split homolog 1 regulates differentiation of retinal neurons and is essential for eye morphogenesis. *Neuron*, **1996**; *16(4)*: 723–734.

112 Bao, Z.Z. and C.L. Cepko, The expression and function of Notch pathway genes in the developing rat eye. *J. Neurosci.*, **1997**; *17(4)*: 1425–1434.

113 Furukawa, T., et al. rax, Hes1, and notch1 promote the formation of Muller glia by postnatal retinal progenitor cells. *Neuron*, **2000**; *26(2)*: 383–394.

114 Nieto, M., et al. Neural bHLH genes control the neuronal versus glial fate decision in cortical progenitors. *Neuron*, **2001**; *29(2)*: 401–413.

115 Gaiano, N., J.S. Nye, and G. Fishell, Radial glial identity is promoted by Notch1 signaling in the murine forebrain. *Neuron*, **2000**; *26(2)*: 395–404.

116 Morrison, S.J., et al. Transient Notch activation initiates an irreversible switch from neurogenesis to gliogenesis by neural crest stem cells. *Cell*, **2000**; *101(5)*: 499–510.

117 Kubu, C.J., et al. Developmental changes in Notch1 and numb expression mediated by local cell-cell interactions underlie progressively increasing delta sensitivity in neural crest stem cells. *Dev. Biol.*, **2002**; *244(1)*: 199–214.

118 Reynolds, B.A. and S. Weiss, Generation of neurons and astrocytes from isolated cells of the adult mammalian central nervous system. *Science*, **1992**; *255(5052)*: 1707–1710.

119 Hitoshi, S., et al. Notch pathway molecules are essential for the maintenance, but not the generation, of mammalian neural stem cells. *Genes Dev.*, **2002**; *16(7)*: 846–858.

120 Shen, Q., et al. Endothelial cells stimulate self-renewal and expand neurogenesis of neural stem cells. *Science*, **2004**; *304(5675)*: 1338–1340.

121 Joutel, A., et al. Notch3 mutations in CADASIL, a hereditary adult-onset condition causing stroke and dementia. *Nature*, **1996**; *383(6602)*: 707–710.
122 Emerick, K.M., et al. Features of Alagille syndrome in 92 patients: frequency and relation to prognosis. *Hepatology*, **1999**; *29(3)*: 822–829.
123 Li, L., et al. Alagille syndrome is caused by mutations in human Jagged1, which encodes a ligand for Notch1. *Nature Genet.*, **1997**; *16(3)*: 243–251.
124 Oda, T., et al. Mutations in the human Jagged1 gene are responsible for Alagille syndrome. *Nature Genet.*, **1997**; *16(3)*: 235–242.
125 Weng, A.P. and J.C. Aster, Multiple niches for Notch in cancer: context is everything. *Curr. Opin. Genet. Dev.*, **2004**; *14(1)*: 48–54.
126 Wozney, J.M., et al. Novel regulators of bone formation: molecular clones and activities. *Science*, **1988**; *242(4885)*: 1528–1534.
127 Shi, Y. and J. Massague, Mechanisms of TGF-beta signaling from cell membrane to the nucleus. *Cell*, **2003**; *113(6)*: 685–700.
128 Miyazawa, K., et al. Two major Smad pathways in TGF-beta superfamily signalling. *Genes Cells*, **2002**; *7(12)*: 1191–1204.
129 Zwijsen, A., K. Verschueren, and D. Huylebroeck, New intracellular components of bone morphogenetic protein/Smad signaling cascades. *FEBS Lett.*, **2003**; *546(1)*: 133–139.
130 Liu, X., et al. Ski/Sno and TGF-beta signaling. *Cytokine Growth Factor Rev.*, **2001**; *12(1)*: 1–8.
131 Wang, W., et al. Ski represses bone morphogenic protein signaling in Xenopus and mammalian cells. *Proc. Natl Acad. Sci. USA*, **2000**; *97(26)*: 14394–14399.
132 de Caestecker, M.P., et al. The Smad4 activation domain (SAD) is a proline-rich, p300–dependent transcriptional activation domain. *J. Biol. Chem.*, **2000**; *275(3)*: 2115–2122.
133 Mulder, K.M., Role of Ras and Mapks in TGFbeta signaling. *Cytokine Growth Factor Rev.*, **2000**; *11(1–2)*: 23–35.
134 Qi, X., et al. BMP4 supports self-renewal of embryonic stem cells by inhibiting mitogen-activated protein kinase pathways. *Proc. Natl Acad. Sci. USA*, **2004**; *101(16)*: 6027–6032.
135 Cobb, M.H., MAP kinase pathways. *Prog. Biophys. Mol. Biol.*, **1999**; *71(3–4)*: 479–500.
136 Kyriakis, J.M. and J. Avruch, Mammalian mitogen-activated protein kinase signal transduction pathways activated by stress and inflammation. *Physiol. Rev.*, **2001**; *81(2)*: 807–869.
137 Ying, Q.L., et al. BMP induction of Id proteins suppresses differentiation and sustains embryonic stem cell self-renewal in collaboration with STAT3. *Cell*, **2003**; *115(3)*: 281–292.
138 Jen, Y., H. Weintraub, and R. Benezra, Overexpression of Id protein inhibits the muscle differentiation program: *in vivo* association of Id with E2A proteins. *Genes Dev.*, **1992**; *6(8)*: 1466–1479.
139 Lyden, D., et al. Id1 and Id3 are required for neurogenesis, angiogenesis and vascularization of tumour xenografts. *Nature*, **1999**; *401(6754)*: 670–677.

140 Grigoriadis, A.E., J.N. Heersche, and J.E. Aubin, Differentiation of muscle, fat, cartilage, and bone from progenitor cells present in a bone-derived clonal cell population: effect of dexamethasone. *J. Cell Biol.*, **1988**; *106*(6): 2139–2151.

141 Katagiri, T., et al. The non-osteogenic mouse pluripotent cell line, C3H10T1/2, is induced to differentiate into osteoblastic cells by recombinant human bone morphogenetic protein–2. *Biochem. Biophys. Res. Commun.*, **1990**; *172*(1): 295–299.

142 Wang, E.A., et al. Bone morphogenetic protein–2 causes commitment and differentiation in C3H10T1/2 and 3T3 cells. *Growth Factors*, **1993**; *9*(1): 57–71.

143 Puleo, D.A., Dependence of mesenchymal cell responses on duration of exposure to bone morphogenetic protein–2 *in vitro*. *J. Cell Physiol.*, **1997**; *173*(1): 93–101.

144 Ducy, P., et al. Osf2/Cbfa1: a transcriptional activator of osteoblast differentiation. **Cell**, **1997**; *89*(5): 747–754.

145 Shea, C.M., et al. BMP treatment of C3H10T1/2 mesenchymal stem cells induces both chondrogenesis and osteogenesis. J. Cell Biochem., **2003**; *90*(6): 1112–1127.

146 Liu, D., B.L. Black, and R. Derynck, TGF-beta inhibits muscle differentiation through functional repression of myogenic transcription factors by Smad3. *Genes Dev.*, **2001**; *15*(22): 2950–2966.

147 Hogan, B.L., Bone morphogenetic proteins: multifunctional regulators of vertebrate development. *Genes Dev.*, **1996**; *10*(13): 1580–1594.

148 Mehler, M.F., et al. Bone morphogenetic proteins in the nervous system. *Trends Neurosci.*, **1997**; *20*(7): 309–317.

149 Liem, K.F., et al. Dorsal differentiation of neural plate cells induced by BMP-mediated signals from epidermal ectoderm. *Cell*, **1995**; *82*: 969–979.

150 Mujtaba, T., M. Mayer-Proschel, and M.S. Rao, A common neural progenitor for the CNS and PNS. *Dev. Biol.*, **1998**; *200*(1): 1–15.

151 Selleck, M.A., et al. Effects of Shh and Noggin on neural crest formation demonstrate that BMP is required in the neural tube but not ectoderm. *Development*, **1998**; *125*(24): 4919–4930.

152 Wilson, P.A. and A. Hemmati-Brivanlou, Induction of epidermis and inhibition of neural fate by Bmp–4. *Nature*, **1995**; *376*: 331–333.

153 Reissmann, E., et al. Involvement of bone morphogenetic protein–4 and bone morphogenetic protein–7 in the differentiation of the adrenergic phenotype in developing sympathetic neurons. *Development*, **1996**; *122*(7): 2079–2088.

154 Schneider, C., et al. Bone morphogenetic proteins are required in vivo for the generation of sympathetic neurons. *Neuron*, **1999**; *24*(4): 861–870.

155 Shah, N., A. Groves, and D.J. Anderson, Alternative neural crest cell fates are instructively promoted by TGFβs superfamily members. *Cell*, **1996**; *85*: 331–343.

156 Sommer, L., Context-dependent regulation of fate decisions in multipotent progenitor cells of the peripheral nervous system. *Cell Tissue Res.*, **2001**; *305*: 211–216.

157 Kirby, M.L. and K.L. Waldo, Neural crest and cardiovascular patterning. *Circ. Res.*, **1995**; *77(2)*: 211–215.
158 Olson, E.N. and D. Srivastava, Molecular pathways controlling heart development. *Science*, **1996**; *272(5262)*: 671–676.
159 Millan, F.A., et al. Embryonic gene expression patterns of TGF beta 1, beta 2 and beta 3 suggest different developmental functions *in vivo*. *Development*, **1991**; *111(1)*: 131–143.
160 Pelton, R.W., et al. *In situ* hybridization analysis of TGF beta 3 RNA expression during mouse development: comparative studies with TGF beta 1 and beta 2. *Development*, **1990**; *110(2)*: 609–620.
161 Sanford, L.P., et al. TGFbeta2 knockout mice have multiple developmental defects that are non-overlapping with other TGFbeta knockout phenotypes. *Development*, **1997**; *124(13)*: 2659–2670.
162 Guillemot, F., et al. Mammalian achaete-scute homolog–1 is required for the early development of olfactory and autonomic neurons. *Cell*, **1993**; *75*: 463–476.
163 Sommer, L., et al. The cellular function of MASH1 in autonomic neurogenesis. *Neuron*, **1995**; *15*: 1245–1258.
164 Lo, L., L. Sommer, and D.J. Anderson, MASH1 maintains competence for BMP2–induced neuronal differentiation in post-migratory neural crest cells. *Curr. Biol.*, **1997**; *7*: 440–450.
165 Shah, N.M. and D.J. Anderson, Integration of multiple instructive cues by neural crest stem cells reveals cell-intrinsic biases in relative growth factor responsiveness. *Proc. Natl Acad. Sci. USA*, **1997**; *94(21)*: 11369–11374.
166 Gurdon, J.B., P. Lemaire, and K. Kato, Community effects and related phenomena in development. *Cell*, **1993**; *75(5)*: 831–834.
167 Hagedorn, L., U. Suter, and L. Sommer, P0 and PMP22 mark a multipotent neural crest-derived cell type that displays community effects in response to TGF-β family factors. *Development*, **1999**; *126(17)*: 3781–3794.
168 Hagedorn, L., et al. Autonomic neurogenesis and apoptosis are alternative fates of progenitor cell communities induced by TGFbeta. *Dev. Biol.*, **2000**; *228(1)*: 57–72.
169 Deshmukh, M. and E.M. Johnson, Jr., Programmed cell death in neurons: focus on the pathway of nerve growth factor deprivation-induced death of sympathetic neurons. *Mol. Pharmacol.*, **1997**; *51(6)*: 897–906.
170 Raff, M., Cell suicide for beginners. *Nature*, **1998**; *396(6707)*: 119–122.
171 Silos-Santiago, I., et al. Molecular genetics of neuronal survival. *Curr. Opin. Neurobiol.*, **1995**; *5(1)*: 42–49.
172 Chen, C.R., Y. Kang, and J. Massague, Defective repression of c-myc in breast cancer cells: A loss at the core of the transforming growth factor beta growth arrest program. *Proc. Natl Acad. Sci. USA*, **2001**; *98(3)*: 992–999.
173 Feng, X.H., X. Lin, and R. Derynck, Smad2, Smad3 and Smad4 cooperate with Sp1 to induce p15(Ink4B) transcription in response to TGF-beta. *EMBO J.*, **2000**; *19(19)*: 5178–5193.
174 Pardali, K., et al. Role of Smad proteins and transcription factor Sp1 in p21(Waf1/Cip1) regulation by transforming growth factor-beta. *J. Biol. Chem.*, **2000**; *275(38)*: 29244–29256.

175 Kleeff, J., et al. The TGF-beta signaling inhibitor Smad7 enhances tumorigenicity in pancreatic cancer. *Oncogene*, **1999**; *18*(39): 5363–5372.

176 Ericson, J., et al. Sonic hedgehog induces the differentiation of ventral forebrain neurons: a common signal for ventral patterning within the neural tube. *Cell*, **1995**; *81*(5): 747–756.

177 Ericson, J., et al. Two critical periods of Sonic Hedgehog signaling required for the specification of motor neuron identity. *Cell*, **1996**; *87*(4): 661–673.

178 Jacob, J. and J. Briscoe, Gli proteins and the control of spinal-cord patterning. *EMBO Rep.*, **2003**; *4*(8): 761–765.

179 Ingham, P.W. and A.P. McMahon, Hedgehog signaling in animal development: paradigms and principles. *Genes Dev.*, **2001**; *15*(23): 3059–3087.

180 Nybakken, K. and N. Perrimon, Hedgehog signal transduction: recent findings. *Curr. Opin. Genet. Dev.*, **2002**; *12*(5): 503–511.

181 Lee, J., et al. Gli1 is a target of Sonic hedgehog that induces ventral neural tube development. *Development*, **1997**; *124*(13): 2537–2552.

182 Ruiz i Altaba, A., Gli proteins encode context-dependent positive and negative functions: implications for development and disease. *Development*, **1999**; *126*(14): 3205–3216.

183 Ruiz i Altaba, A., Combinatorial Gli gene function in floor plate and neuronal inductions by Sonic hedgehog. *Development*, **1998**; *125*(12): 2203–2212.

184 Porter, J.A., K.E. Young, and P.A. Beachy, Cholesterol modification of hedgehog signaling proteins in animal development. *Science*, **1996**; *274*(5285): 255–259.

185 Pepinsky, R.B., et al. Identification of a palmitic acid-modified form of human Sonic hedgehog. *J. Biol. Chem.*, **1998**; *273*(22): 14037–14045.

186 Peters, C., et al. The cholesterol membrane anchor of the Hedgehog protein confers stable membrane association to lipid-modified proteins. *Proc. Natl Acad. Sci. USA*, **2004**; *101*(23): 8531–8536.

187 Chen, M.H., et al. Palmitoylation is required for the production of a soluble multimeric Hedgehog protein complex and long-range signaling in vertebrates. *Genes Dev.*, **2004**; *18*(6): 641–659.

188 Boyd, R.L., et al. The thymic microenvironment. *Immunol. Today*, **1993**; *14*(9): 445–459.

189 Outram, S.V., et al. Hedgehog signaling regulates differentiation from double-negative to double-positive thymocyte. *Immunity*, **2000**; *13*(2): 187–197.

190 Dahmane, N., et al. The Sonic Hedgehog-Gli pathway regulates dorsal brain growth and tumorigenesis. *Development*, **2001**; *128*(24): 5201–5212.

191 Palma, V. and A. Ruiz i Altaba, Hedgehog-GLI signaling regulates the behavior of cells with stem cell properties in the developing neocortex. *Development*, **2004**; *131*(2): 337–345.

192 Gage, F.H., Stem cells of the central nervous system. *Curr. Opin. Neurobiol.*, **1998**; *8*(5): 671–676.

193 Doetsch, F., et al. Subventricular zone astrocytes are neural stem cells in the adult mammalian brain. *Cell*, **1999**; *97(6)*: 703–716.
194 Capela, A. and S. Temple, LeX/ssea–1 is expressed by adult mouse CNS stem cells, identifying them as nonependymal. *Neuron*, **2002**; *35(5)*: 865–875.
195 Lai, K., et al. Sonic hedgehog regulates adult neural progenitor proliferation in vitro and *in vivo*. *Nature Neurosci.*, **2003**; *6(1)*: 21–27.
196 Machold, R., et al. Sonic hedgehog is required for progenitor cell maintenance in telencephalic stem cell niches. *Neuron*, **2003**; *39(6)*: 937–950.
197 Ruiz, I.A.A., V. Palma, and N. Dahmane, Hedgehog-Gli signalling and the growth of the brain. *Nature Rev. Neurosci.*, **2002**; *3(1)*: 24–33.
198 Dahmane, N., et al. Activation of the transcription factor Gli1 and the Sonic hedgehog signalling pathway in skin tumours. *Nature*, **1997**; *389(6653)*: 876–881.
199 Goodrich, L.V., et al. Altered neural cell fates and medulloblastoma in mouse patched mutants. *Science*, **1997**; *277(5329)*: 1109–1113.
200 Hahn, H., et al. Patched target Igf2 is indispensable for the formation of medulloblastoma and rhabdomyosarcoma. *J. Biol. Chem.*, **2000**; *275(37)*: 28341–28344.
201 Chintagumpala, M., S. Berg, and S.M. Blaney, Treatment controversies in medulloblastoma. *Curr. Opin. Oncol.*, **2001**; *13(3)*: 154–159.
202 Wechsler-Reya, R.J. and M.P. Scott, Control of neuronal precursor proliferation in the cerebellum by Sonic Hedgehog. *Neuron*, **1999**; *22(1)*: 103–114.
203 Soriano, P., Generalized lacZ expression with the ROSA26 Cre reporter strain. *Nature Genet.*, **1999**; *21(1)*: 70–71.
204 Feil, R., et al. Ligand-activated site-specific recombination in mice. *Proc. Natl Acad. Sci. USA*, **1996**; *93*: 10887–10890.

2
Germ Cells

Pellegrino Rossi, Susanna Dolci, Donatella Farini, and Massimo De Felici

2.1
Introduction

Germ cell development in mammalian gametogenesis involves a complex sequence of events regulated by a large variety of molecules. Although hormones have always been considered to be the primary regulators of gametogenesis, it is now well established that growth factors (GFs) in concert and against or in some case autonomously of hormones, may control or modulate directly or indirectly many, if not all, of the processes of germ cell development. A number of GFs produced by gonadal somatic cells and germ cells have been identified and appear to act locally on the autocrine and paracrine hormones to regulate gametogenesis. Bi-directional communication between gonadal somatic cells and germ cells is critical for gonad function and fertility. An increasing body of evidence has led to the proposition that several GFs are important mediators of such communication and crucial in controlling and determining female and male fertility.

In this chapter, we will focus mainly on GFs acting directly on female and male germ cells and on those produced by the germ cells themselves. Germ cells produce and are themselves a target for GFs which are also active in other cell types; however, some GFs have been identified which are almost exclusively produced by germ cells. As in other cell types, GFs often have pleiotropic effects on germ cells. Depending on the developmental stage, they may influence germ cell proliferation, differentiation and/or survival. Such effects are often seen in *in vitro* studies and in some cases have been confirmed *in vivo*, for example by genetic ablation. Indeed, several aspects of the action(s) and molecular pathways of GFs in germ cells are still poorly understood due to difficulties intrinsic to the study of germ cells, either *in vivo* or *in vitro*.

The main stages of mammalian gametogenesis are shown in Fig. 2.1 and can be summarized as follows. The precursors of gametes, known as primordial germ cells (PGCs), are first detectable in extra-embryonic regions during the early embryonic period (for a review see [1]). PGCs migrate into the undifferentiated gonads (gonadal ridges, GRs) where they are known generically as gonocytes or as oogonia and prespermatogonia when the GRs begin to differentiate into ovaries or testes respectively. PGCs and gonocytes are characterized by intense mitotic proliferation that

---

Cell Signaling and Growth Factors in Development. Edited by K. Unsicker and K. Krieglstein
Copyright © 2006 WILEY-VCH Verlag GmbH & Co. KGaA, Weinheim
ISBN 3-527-31034-7

ceases when oogonia enter into meiosis in the fetal ovary and prespermatogonia into G1/G0 arrest in the fetal testis. In the female, during the fetal period or shortly after birth, meiotic germ cells, named primary oocytes, are arrested at the end of prophase I in a stage called diplotene or dyctiate. At this stage an oocyte is surrounded by a single layer of pre-granulosa cells forming a primordial follicle. During the female's reproductive life ovulated eggs are derived from this pool of primordial follicles. When follicles leave the resting pool they undergo a primordial to primary follicle transition and will grow and develop until either ovulation occurs or follicles undergo atresia. In the male, prespermatogonia remain quiescent until after birth, when they are reactivated and start the process of spermatogenesis. Spermatogenesis is initiated when prespermatogonia differentiate into spermatogonia. While some spermatogonia become self-renewing spermatogonial stem cells, most after several rounds of mitotic divisions differentiate into meiotic spermatocytes. These give rise to haploid spermatids which eventually become spermatozoa. Spermatogenesis occurs within testicular structures called seminiferous tubules in which Sertoli cells represent the main somatic cell constituents.

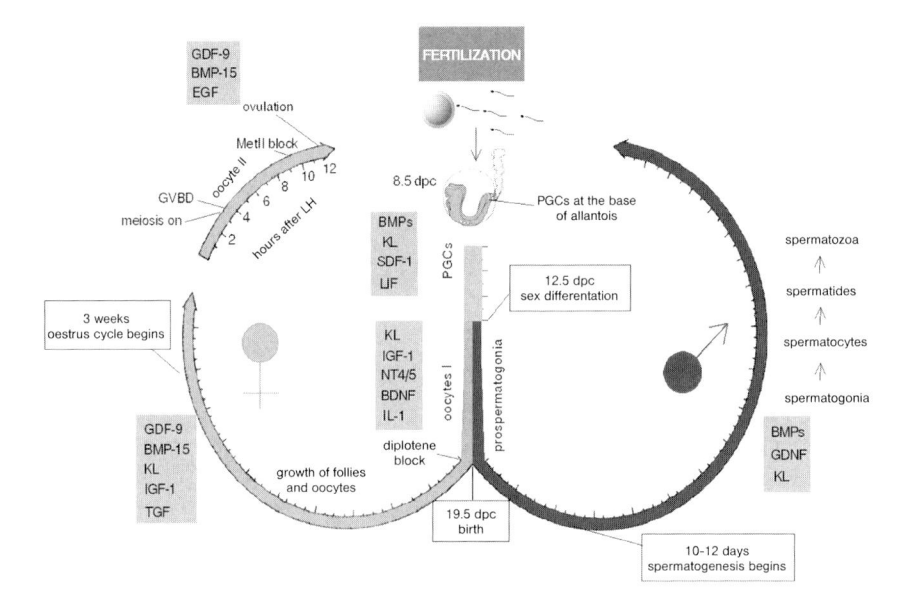

Fig. 2.1
Schematic representation of the main events of mouse gametogenesis. Some of the growth factors playing a role in the germ cell development during the indicated period are shown.

2.2
Primordial Germ Cells

Primordial germ cells are first identified at around 7.25 days post coitum (dpc) at the head-fold stage as a cluster of about 45 alkaline phosphatase (AP)-expressing cells at the base of the allantois [2]. As early as they can be identified as AP positive cells, PGCs express genes such as *Oct4* and *Nanog*, which are known to be essential for inner cell mass and ES pluripotency. PGCs then migrate from the base of allantois through the hindgut [3] and the dorsal mesentery to colonize GRs by 11.5 dpc [4]. At this stage, the number of PGCs increases about 60–fold up to about 3000 cells [4] due to active proliferation. After gonadal colonization, PGCs continue to proliferate within the developing gonads and initiate the process of sexual differentiation. By 13.5 dpc, there are approximately 20,000 PGCs present in both male and female embryos. In the male, germ cells undergo mitotic arrest (prespermatogonia) within the seminiferous cords, while in the female PGCs enter meiosis (oocytes) and progress up to the diplotene stage of meiosis I, when they become arrested until puberty [5].

2.2.1
Growth Factors in PGC Commitment

Lineage analysis has shown that the PGCs are derived before gastrulation from a subpopulation of proximal epiblast cells adjacent to the trophoblast-derived extra-embryonic ectoderm. Specification of germ cell lineage within the epiblast depends on bone morphogenetic protein 4 (BMP4), a member of the transforming growth factor-β (TGFβ) superfamily. On several genetic backgrounds, all of the *Bmp4* null (homozygous) mutants fail to generate PGCs, and *Bmp4* heterozygous embryos have a reduced number of PGCs (50 % of wild-type) at various developmental stages. Furthermore, in *in vitro* experiments it has been shown that BMP4 treatment enables the recruitment of pluripotent cells to a PGC phenotype by a multi-step process involving an initial pre-commitment of epiblast cells and a following stage of PGC phenotypic determination [6]. Among the *Bmp* superfamily members, *Bmp2* and *Bmp8b* are also required for PGC generation, and null animals for either gene display significantly reduced number of PGCs [7, 8]. Among the signal transduction molecules involved in BMP signaling it has been shown that Smad1 and Smad5 are both essential for PGC formation and knock-out embryos for either of the two genes display severe reduction of primordial germ cells [9, 10].

2.2.2
Growth Factors in PGC Proliferation

2.2.2.1 The KL/KIT System

Several genes are known to regulate PGC proliferation/survival during this period (for a review, see [1]), however the mechanism of action has been clearly demonstrated for only a few of them. One of these genes encodes the tyrosine kinase-coupled receptor Kit, whose mutations or deletions affect PGC development. The *c-kit* gene maps to chromosome 5 in the mouse, and several deletions which map within the *c-kit* locus result in a severe phenotype in homozygous animals and at least white spotting of the mouse fur in the heterozygous animals (*White spotting*, W). In many W mutations, homozygous mice display anemia, defects of pigmentation and sterility, due to the early loss of hemopoietic precursors, melanoblasts and PGCs. In the early embryo *c-kit* expression is shared by PGCs, neural crest cells and hemopoietic stem cells. In all the three cell lineages *c-kit* transcription is regulated by at least six DNAse-hypersensitive sites (HS) located either within the proximal 5′ region of the gene or within the first intron [11]. Transgenic embryos which carry a GFP construct driven by the six HS sites show specific GFP expression in PGCs (Fig. 2.2), hemopoietic precursors and neural crest derivatives. An overlapping phenotype to the W mutations is produced by deletions within the *Steel locus* (*Sl*, chromosome 10) which encodes for the kit ligand (*Kl*) gene. KL (also called Stem Cell Factor, SCF; Steel Factor, SF, SLF; Mast Cell Growth Factor, MGF), is produced either as a potentially soluble or membrane-bound form by the somatic cell types which surround PGCs during their migratory period up to gonadal colonization and its expression temporally overlaps *c-kit* expression. Several data demonstrate that KL is essential for PGC proliferation/survival either *in vivo* or *in vitro* (for reviews see [1, 12]). Our knowledge about the growth factor requirements for PGCs relies mostly on *in vitro* culture studies. PGCs can be grown for up to 7–10 days when cultured onto permissive feeder layers such as STO or TM4 cells, but not when cultured on COS or CV–1 cells, which do not express KL. Furthermore, when PGCs are cultured in the absence of the membrane-bound form of KL, they rapidly disappear from the culture, mirroring the *in vivo* phenotype in which mice lacking the membrane-bound form of KL (*Steel-Dickie*) have few PGCs and become sterile in adult life [13].

KL signaling in responsive cells is mediated by KIT receptor dimerization and autophosphorylation. The activated receptor is then able to phosphorylate several substrates, thus initiating different signaling pathways, which include the phosphatidylinositol 3–kinase (PI3–K)/AKT/mTOR/p70^{S6K}, Ras/mitogen-activated protein kinase kinase (MEK)/mitogen-activated protein kinase (MAPK), the Janus kinase (JAK)/signal transducer and activator of transcription (STAT), and the Src signaling pathways (for reviews see [14–16]). It has recently been shown that one of the key effectors of KIT signaling in PGCs is AKT [17, 18]. In contrast to spermatogonia (see below), activation of AKT in PGCs appears to be independent of PI3–K, whereas it is sensitive to Src and MAPK inhibitors [17]. However, in short term assay the PI3–K inhibitor LY294002 abolishes KL-dependent AKT phosphorylation [18], suggesting multiple alternative pathways for AKT activation in PGCs. AKT-activated

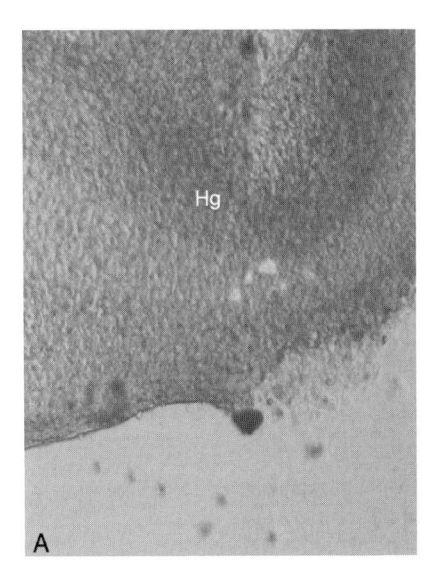

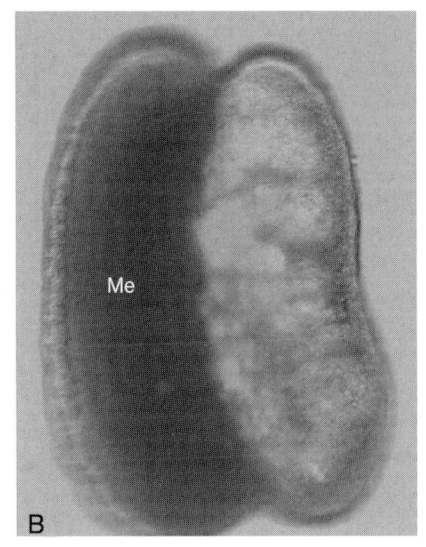

Fig. 2.2
Kit/GFP PGCs in migration towards the hindgut (Hg) (A) and within the testis (B) of mouse embryo of 9.5 and 12.5 dpc. Me, mesonephros. (This figure also appears with the color plates.)

pathways in PGCs involve mTOR/FRAP and $p70^{S6K}$ signaling and pharmacological inhibition of MEK/MAPK, Src, and mTOR/FRAP inhibits PGC proliferation. The tumor suppressor gene *Pten*, which encodes a lipid phosphatase that inhibits PI3–K activity, also plays a crucial role in the regulation of PGC proliferation and differentiation. Deletion or inactivation of the *Pten* gene in PGCs results in constitutive AKT activation which is responsible for testicular teratomas in male newborn mice and increased EGC (Embryonic Germ Cell) production in both sexes [18, 19].

2.2.2.2 LIF/OSM/IL6 Superfamily

PGC growth in culture is not solely dependent on KL production by the feeder layer. Indeed, STO cell culture medium contains diffusible factors, which in combination with KL, can stimulate PGC proliferation. We and others have found that one of these factors is Leukaemia Inhibitory Factor (LIF) [20, 21]. LIF is a cytokine first discovered by its ability to inhibit proliferation of mouse monocytic leukemia cells and to induce them to differentiate into macrophages [22]. LIF also has anti-differentiation activity in ES cells and can be used to maintain them in the undifferentiated state in culture. Based on its secondary structure and genomic organization, LIF has been included in a family of cytokines which include Interleukin–6 (IL–6), Ciliary Neurotrophic Factor (CNTF) and Oncostatin M (OSM) [23]. LIF is able to bind to two types of receptor, a low affinity receptor (LIFR) and gp130, which is the shared receptor for members of the IL–6 family of cytokines. Low-affinity LIF receptors are expressed by PGCs in the developing gonad and it has been shown that at least in culture gp130 plays a pivotal role in mediating signal transduction activated by LIF in PGCs [24, 25]. Addition of LIF to the culture medium promotes survival and reduces apoptosis of PGCs [20, 26], it also supports PGC proliferation when added in combination with KL [21]. In the presence of Basic Fibroblast Growth Factor (bFGF) and KL, LIF is able to prolong PGC proliferation in culture, leading to the derivation of embryonic stem cell-like cells, termed EG cells which have been shown to behave as pluripotent stem cells when transplanted into host blastocysts [27, 28]. Surprisingly, inactivation of gp130 specifically in PGCs by the Cre-lox system results in a slight decrease in male PGCs only, suggesting that it is dispensable for PGC survival/proliferation [29].

Stimulation of PGC proliferation by LIF in the presence of KL and bFGF may be due to the direct effect of bFGF on the PGCs [30]. Indeed, receptors for bFGF have been described in PGCs and also in other cell types present in the cultures [30]. It should also be emphasized that a combination of KL and LIF with retinoic acid or forskolin, both of which have been shown to be potent mitogens for PGC, also promotes proliferation of PGC and formation of EG cells in culture [25].

2.2.2.3 TGFβ Superfamily and Neuropeptides

Among the TGFβ superfamily, BMP4 has been shown to promote proliferation of PGCs in culture [6], and we have found that type IA BMP receptors and Smad5 are expressed by PGCs during their migratory period through gonadal colonization [31] (Fig. 2.2). In contrast, TGFβ1 and activin have been demonstrated to reduce *in vitro* proliferation of PGCs in culture. TGFβ1 inhibits proliferation of PGCs taken from 8.5–day-old embryos and cultured on embryonic fibroblast feeder layers and it has also been suggested to be one of the chemoattractant molecules for migratory PGCs [32].

Neuropeptides of the secretin-glucagon vasoactive intestinal polypeptide-GH-releasing hormone family such as PACAP–27 and PACAP–38 have been shown to stimulate *in vitro* proliferation of mouse primordial germ cells. These neuropeptides are able to bind to PGCs and gonadal somatic cells through the type I PACAP receptor and activate adenylate cyclase to increase intracellular levels of cAMP [33]. Interestingly, dibutyryl cAMP and forskolin, two agents known to enhance the level of intracellular cAMP markedly increase the number of migratory and post-migratory PGCs in culture by stimulating their proliferation [34]. A scheme of the pathways potentially involved in the control of PGC proliferation/survival under the control of GFs is shown in Fig. 2.3.

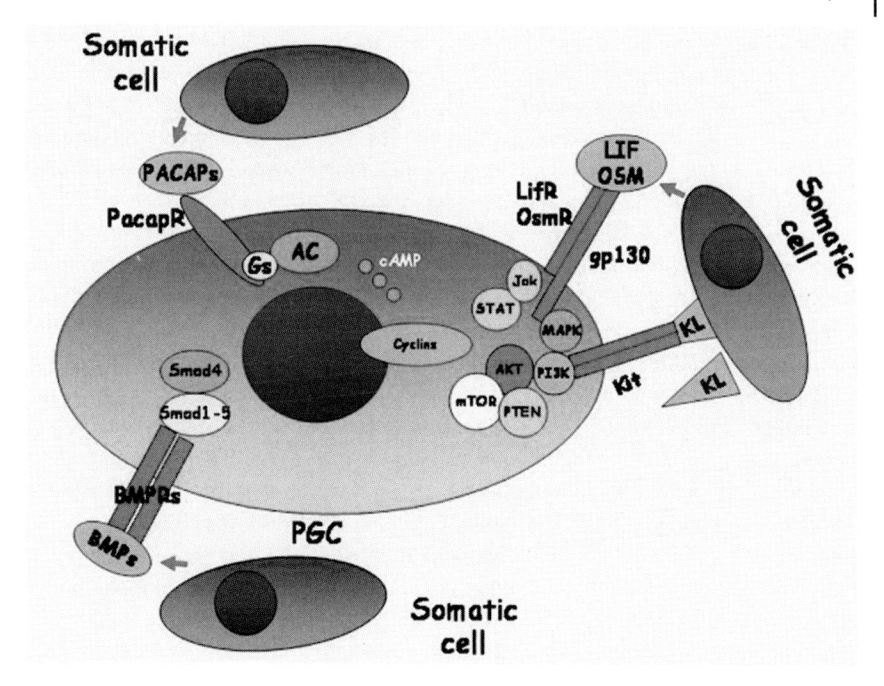

Fig. 2.3
Schematic representation of possible molecular pathways activated by certain GFs (KL, BMPs and PACAPs) which are reported to influence mouse PGC development.

2.3
Oocytes

2.3.1
Growth Factors for Fetal Oocytes

During fetal life germ cells enter meiosis as primary oocytes in the female. Oocytes pass through leptotene, zygotene and pachytene stages before arresting in diplotene at about the time of birth. However, most oocytes (around 70–90% depending on the species) fail to complete their pre-follicular development and a minority are enclosed in the primordial follicles that comprise the ovarian reserve at birth. In fact, most oocytes are lost through programmed cell death [35, 36]. Hormones do not seem to be involved in such processes. Very little information is available with regard to whether GFs regulate the complex processes of meiotic prophase I. It seems certain that germ cells do not require GFs to undergo meiosis in the fetal ovary since this process, at least in the mouse, is likely to be autonomous to the germ cell and to be the default differentiation pattern for PGCs. Using *in vitro* organ culture it has been reported that a combination of GFs (KL, LIF, and Insulin Growth Factor–1,

IGF–1) was able to promote the progression of meiosis in mouse fetal ovaries [37]. Using a similar experimental approach, Morita et al. [38] have previously shown that the same GF cocktail was able to reduce mouse oocyte apoptosis through PI3–K-mediated signaling, which is a classical pathway for the activation of the receptor tyrosine kinase. A significant anti-apoptotic effect has also been reported to be exerted on isolated mouse fetal oocytes in culture by KL and IGF–1, alone or in combination but not by LIF ([39, 40] our unpublished observations). In accordance with these results are the findings that mouse oocytes undergoing diplotene arrest, upregulate the KL receptor Kit both *in vivo* and *in vitro*, while expressing constant levels of the IGF–1 receptor throughout meiotic prophase I [40–43]. The observation that a subpopulation of somatic cells of the fetal ovary express KL and that oocytes at the end of prophase I co-express both KL and Kit at mRNA and protein levels, suggests that the KL/Kit system plays either a paracrine or autocrine role in oocytes during this period [42]. According to Morita et al. [44], fetal oocytes also express type 1 interleukin receptor–1 (ILR–1) and the addition of IL–1α or IL–1β to the culture medium significantly increases oocyte survival in fetal ovaries cultured *in vitro*. It has also been proposed that neurotrophins, namely neurotrophin 4/5 (NT4/5) and brain-derived neurotrophic factor (BDNF), play a role in maintaining the survival of mouse and human fetal oocytes [45–47]. NT4/5 and BDNF are present in the fetal ovary where the oogonia and oocytes preferentially express the truncated isoform of the favored receptor TrkB. Ablation of these receptors leads to severe depletion of primordial follicles at birth [46, 47].

Taken together these results indicate that apoptosis of fetal oocytes may partly depend on the availability of trophic growth factors produced by the surrounding gonadal cells. Whereas some growth factor (e.g. IGF–1, neurotrophins) might be necessary throughout the entire prophase I, others (e.g. KL) might be required at specific stages. An intriguing possibility to address in future studies is whether GFs can influence some of the events of meiotic recombination and whether this is associated with their anti-apoptotic action on fetal oocytes. To this end the molecular pathways activated by GFs in fetal oocytes should be investigated. However, with the exception of the involvement of PI3–K activation in the anti-apoptotic action of certain GFs on fetal oocytes in culture (see above), no information is so far available on this topic.

2.3.2
Growth Factors for Growing and Mature Oocytes

Peripheral endocrine hormones and local paracrine and autocrine factors contribute in a coordinated fashion to the process of recruitment, development or atresia, selection and ovulation of follicles after birth. It is well known that folliculogenesis is dependent at least in its later stages, on FSH and LH and that the early stages of the follicular growth can occur without gonadotropins. This has led to a follicle classification system based on functional rather than morphological criteria [48, 49]. Follicles can be quiescent (primordial), committed to growth (pre-antral and antral), ovulatory or atretic. After antrum formation, follicles become dependent on FSH and LH stimulation for continuing growth and development. The targets of FSH and LH within the follicle are granulosa and theca cells, respectively.

Follicles are believed to leave the primordial pool in an ordered sequence. Once initiated, follicle growth ends with degeneration (atresia) or formation of dominant ovulatory follicles.

However, the mechanisms responsible for the employment and initiation of growth of primordial follicles remain little understood, but GFs are certainly involved in such processes. GFs acting on granulosa cells and/or oocytes at different stages of maturation have been described and it is now well known that the oocyte expresses a number of GF receptors and is able to produce several GFs (Tables 2.1, 2.2 and Fig. 2.4).

Tab. 2.1
Growth factors reported to be expressed at mRNA and/or protein levels in mammalian oocytes.

Growth factor	Species	Oocytes Fetal	Growing	Fully grown and/or ovulated
GDF–9	Mouse, rat, human, sheep, bovine, pig, ovine, marsupial	–	+	+
BMP–15	Mouse, rat, human, sheep, bovine, pig, ovine, marsupial	–	+	+
BMP–6	Mouse, bovine	–	+	+
bFGF	Rat, human, bovine	–	+	+
FGF–8	Mouse, rat	–	+	+
TGFβ2	Mouse, human	ND	ND	+
TGFα	Bovine, human	+	+	+
IGF–1	Rat, bovine	–	+	+
EGF	Human, pig, bovine	+	+	+
KL	Mouse, sheep	+	+	–
LIF	Rat	–	+	+

ND, not determined.

Tab. 2.2
Growth factor receptors reported to be expressed at mRNA and/or protein levels in mammalian oocytes

Growth factor receptor	Species	Oocytes Fetal	Growing	Fully grown and/or ovulated
c-Kit	Mouse, rat, human, sheep	+	+	+
ILR–1	Mouse	+	ND	ND
BMP-R1–R2	Rat, bovine, sheep	ND	+	+
IFNγ-R1–2	Mouse	ND	ND	+
IGF–1–R	Mouse, rat, human, bovine	+	+	+
GH-R	Bovine	?	ND	+
Act-RII-IIB	Mouse, rat, human	±	+	+
TrkB,TrkB-T1	Mouse, human	+	+	ND
EGF-R	Mouse, sheep	ND	+	+

ND, not determined.

2.3.3
Growth Factors in the Initiation of Follicle and Oocyte Growth

Primordial follicles consist of an oocyte surrounded by a single layer of granulosa cells. The initiation of follicle growth is characterized by the proliferation of granulosa cells, changes in their shape from flattened to cuboidal, enlargement of the oocyte, and formation of the zona pellucida. There is evidence to suggest that oocyte growth follows, rather than precedes, the changes in the granulosa cells. The onset of follicle and oocyte growth follows a geographically-determined pattern, in which the first follicles to grow are situated at the inner part of the ovary cortex. It is reasonable to speculate that early follicle growth is regulated by locally produced stimulatory and/or inhibitory signals. These signals might come from the follicles themselves and/or from the surrounding cells. Early studies [50] suggested that the cohort of growing follicles exerts an inhibitory effect on the primordial follicle pool. It was reported that the number of growing follicles was inversely correlated with the number of small, non-growing follicle pools. A decrease in such pools caused a larger fraction of follicles to begin to grow. In the human ovary, initiation of the follicle growth phase seems to accelerate towards the menopause as the pool of small, non-growing follicles decreases [51–53]. A similar relationship was also seen in rats whose ovaries were depleted of small oocytes by treatment with busulphan. An alternative explanation is that the number of follicles entering the growing phase is constant over time and may not be dependent on the pool of small oocytes [54]. Thus, when the stockpile of primordial follicles is large, the percentage undergoing

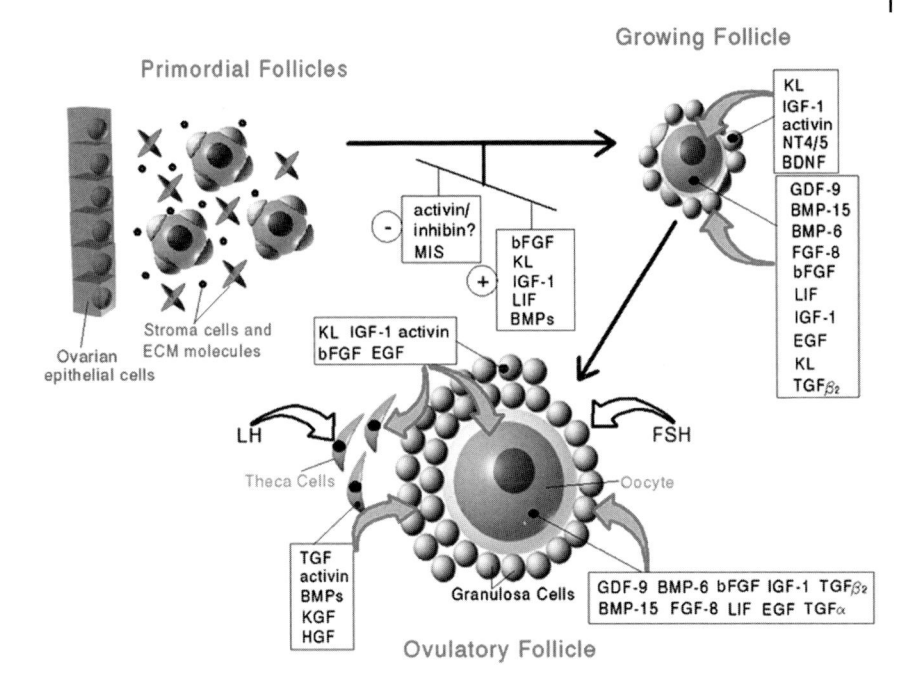

Fig. 2.4
Schematic representation of the involvement of GFs in three stages of follicle and oocyte development in mammals. The recruitment of primordial follicles to the growing phase is likely to result from the action of inhibitory and stimulatory growth factors. During the growing phase and until the formation of the ovulatory follicle, bidirectional communication between the granulosa cells and oocytes is established and is mediated by GFs; with the formation of theca cells these cells also participate in such communication. FSH and LH hormones play a crucial role after antrum formation (not shown in the figure).

growth is low; but when the stockpile is small, the percentage of primordial follicles entering the growth phase is large. In support of this supposition, the results of *in vitro* cultures have shown that the number of oocytes that begin to grow within fragments of fetal ovaries is unrelated to the number of small oocytes, provided that the fragments contain more than a certain minimum number of small oocytes. However, the identity and origin of the signal(s) that positively or negatively influence the release of primordial follicles from the resting pool still remains to be elucidated. Studies of bovine and baboon follicle development *in vitro* support the view that the ovarian stroma exerts inhibitory effects on the development of primordial follicles [55, 56]. Moreover, there is evidence to indicate that ovarian epithelium and extracellular matrix are the source of GFs which regulate this process (e.g. activin [57]; bFGF [58–60]; Müllerian inhibiting factor/anti-Müllerian hormone, MIS/AMH [61–63]; KL [64–66]; LIF [67]). From this evidence it is possible to speculate that the fate of primordial follicles i.e. growth or remaining quiescent, is regulated by the balance between the local stimulatory and inhibitory actions of GFs on granulosa cells (Fig. 2.2).

The crucial role of one of these GFs, namely KL, in the initiation of follicle growth is supported by several reports. Mice with naturally occurring mutations in *Kl* (*Steel¹*, *Steel^panda* and *Steel^con*) can initially form apparently normal primordial follicles which are unable to progress beyond the primary stages [64, 68, 69], and a similar, albeit less penetrant, phenotype is seen with mutations in the KL receptor c-Kit [70]. In addition, Yoshida et al. [71] reported that administering anti-c-Kit antibody for the first 5 days after birth completely blocked the onset of primordial follicle growth and the survival of developing primary follicles. There is also evidence that KL is able to directly stimulate the first stage of oocyte growth [65, 40]. In several mammalian species including humans, KL is produced by the granulosa cells whereas its receptor c-Kit is present in oocytes and theca cells [72–77]. It is possible that when a sufficient number of cuboidal granulosa cells are formed, the local levels of KL become sufficiently high to stimulate the c-Kit receptors present on the oocyte. Interestingly, KL is present in large amount in cords and clusters of somatic cells in the central region of the mouse ovary where the oocytes start to grow [73, 42]. We can speculate that after initiation of growth, the oocyte begins to produce the GFs necessary to promote further proliferation and differentiation of granulosa cells (see below). However, other factors which differ from KL and are as yet are unidentified, are likely to be involved in promoting oocyte growth [78].

2.3.4
From Pre-antral to Ovulatory Follicles: Oocytes take Control

Among the local ovarian factors regulating folliculogenesis, there is growing evidence that members of the TGFβ superfamily play an important role. These members include activins, TGFβ, growth differentiation factor–9 (GDF–9), BMPs and perhaps MIS.

Activin is produced in the ovary and the expression pattern of mRNA for the active subunits (activin A and B) changes during folliculogenesis. Numerous functional studies support a role for activin in inhibiting or stimulating folliculogenesis at various stages mainly through a balance with inhibin and actions on granulosa cells (for a review, see [79, 80]). However, the expression of activin receptors II and IIB has been detected in growing and mature oocytes of various mammalian species (mouse [81–83]; rat [84]; bovine [85, 86]; human [87]), suggesting also a direct action on the oocyte. In support of such role, *in vitro* studies in rodents and primates have revealed that activin can accelerate *in vitro* meiotic maturation of oocytes [88–90]. Moreover, when tested in a bovine *in vitro* maturation system, activin A enhanced the post-fertilization developmental competence of the *in vitro*-matured denuded oocytes [91].

TGFβ is expressed in follicular cells mainly by theca cells and has proliferative and cytodifferentiation actions on granulosa cells (for reviews see [79, 80]). Interestingly, mouse oocytes express TGFβ2 mRNA [92] and bovine oocyte TGFα protein [93], supporting the notion that the oocyte exerts paracrine action on the surrounding follicle cells. Indeed one of the most exciting discoveries in the past few years is that the process of folliculogenesis is controlled by GFs secreted by the oocyte itself

(Table 2.1). In 1993, McPherron and his co-workers discovered GDF–9 and showed that it is selectively synthesized in mouse oocytes [94, 95]. Definitive evidence that GDF–9 is obligatory for folliculogenesis and fertility came from the loss-of-function studies of Dong et al. [96]. In addition to the above-mentioned GFs, another seven GFs have been identified in mammalian oocytes (Tab. 2.1): bone morphogenetic protein–15 (BMP–15/GDF–9B) [97, 98], bone morphogenetic protein–6 (BMP–6) [99], fibroblast growth factor–8 (FGF–8) [100], IGF–1 [101, 102], bFGF [59, 103, 60] LIF [67] and epidermal growth factors (EGF) [104]. Robust evidence that these oocyte GFs play critical roles in follicle development and female fertility in all mammalian species studied, has been presented only for the two members of the TGFβ superfamily GDF–9 and BMP–15. GDF–9 and BMP–15 mRNA are expressed specifically in the mouse and human oocyte of small pre-antral follicles, but not in the primordial follicles [99, 95, 97, 105]. However, in cows and pigs, low levels of GDF–9 expression have been detected in the oocytes of primordial follicles [106] indicating species-dependent variability in GDF–9 expression at earlier stages of follicular development. Expression of both GDF–9 and BMP–15 persists in the oocytes at all subsequent stages of folliculogenesis. Recent results indicate that the expression of these GFs in the mouse is regulated by the germ cell nuclear factor (GCNF) [107], a member of the orphan nuclear receptor superfamily of ligand-activated transcription factors expressed specifically by germ cells [108]. Experiments with knockout mice have demonstrated that in the absence of GDF–9, folliculogenesis is blocked at the primary pre-antral stage [96]. A similar phenotype has been reported in homozygous sheep carrying a naturally-occurring X-linked mutation in BMP–15 [109]. GDF–9 and BMP–15 appear to be classic granulosa mitogens (for reviews see [110, 79, 80]). Moreover, in some experiments, recombinant GDF–9 was able to downregulate the expression of KL mRNA in granulosa cells [111]. Fully grown, but not growing oocytes, exert a similar effect on granulosa cells and negatively regulate the ability of pre-antral granulosa cells to stimulate oocyte growth in culture [112]. Furthermore, in $GDF-9^{null}$ mice, ovarian KL mRNA levels are markedly increased [113]. Since KL promotes oocyte growth and oocytes are likely to produce and secrete GDF–9 from primary follicles onward, Joyce et al. [111], have postulated the following regulatory feedback loop. During early follicular development, the growing oocyte promotes or at least allows KL production by granulosa cells. KL produced by granulosa cells in turn stimulates continued oocyte growth. In antral follicles, GDF–9 secretion from the oocyte becomes effective at suppressing KL production by granulosa cells. As a consequence the stimulus for oocyte growth is reduced and eventually halted. To further complicate such a bidirectional loop, Otsuka and Shimaski [114] reported that in the rat BMP–15 produced by the oocyte stimulates KL expression in granulosa cells; KL however, is an inhibitor of BMP–15 expression.

GFs might also play a role in the maintenance of the meiotic block in the oocyte at the diplotene stage of prophase I and in the resumption of meiosis. Although enzymes typically involved in the downstream signaling of GF receptor activation, such as PI3–K and MAPKs, seem to play a role in regulating some of the pathways involved in meiotic resumption/maturation of fully grown mammalian oocytes [115, 116], there is no robust evidence to show that GFs act directly on the mature oocyte. In the rat, human chorionic gonadotropin-induced meiotic resumption in oocytes is accompanied by a loss of expression of KL in cumulus cells [117]. Furthermore, KL inhibits the progression of meiosis in denuded rat oocytes in culture [118]. With the possible exception of activin (see above), no other data regarding the direct action of GFs on the oocyte in such processes have been reported. In fact, EGF is a potent stimulant of rodent and porcine oocyte maturation, but its action is on the granulosa and cumulus cells rather than on the oocyte [119–121].

Oocyte-derived factors are essential not only to regulate follicle development but also for efficient ovulation. For example, oocytes secrete factors that promote cumulus expansion, induce hyaluronic acid synthesis and pentraxin3, inhibit plasminogen activator production and suppress FSH-induced LH receptor mRNA synthesis (for reviews see [122–125]). *In vitro* experiments in rodents have shown that recombinant GDF–9, but not BMP–15 or BMP–6, can mimic many, but not all, of these oocyte-specific actions [113]. In particular, GDF–9 stimulates the expression of hyluronan synthase 2 (HAS2), cyclo-oxygenase 2 (COX–2), and steroid acute regulatory protein (StAR). The physiological relevance of GDF–9–stimulated HAS2 and StAR appears to be associated with cumulus expansion and progesterone production, respectively [113].

BMP4 and BMP7 are among other BMPs expressed in the ovary particularly in thecal cells of rat and bovine Graafian follicles [126, 127] and BMP3 is expressed in human granulosa cells [128]. Furthermore, the functional BMP receptors BMPR-IA, IB and -II are expressed in the granulosa cells and oocytes of rats [126], sheep [129] and cattle [127]. Such BMPs appear to influence granulosa cell sensitivity to FSH and IGF and BMP4 and BMP7 at least, seem to inhibit luteinization by modulating steroidogenesis [126].

MIS/AMH is a product of granulosa cells and in contrast to activin and GDF–9, it has been shown to negatively regulate the growth of small follicles [130]. However, McGee et al. [131] showed that MIS can enhance the FSH-induced increase in follicular diameter and cell number when added to rat pre-antral follicles *in vitro* suggesting that this factor acts on folliculogenesis in a stage-dependent manner.

2.4
Male Germ Cells

In the fetal testis PGCs differentiate into prespermatogonia which undergo mitotic arrest in the G1/G0 stage. In mice active proliferation of quiescent spermatogonial precursor cells re-starts after birth. Proliferation and differentiation of mitotic spermatogonia in the pre-pubertal testis depend on the production of secreted growth factors by the nursing Sertoli cells within the seminiferous tubules. After the first wave of spermatogenesis at puberty, the cycle and the wave of the seminiferous epithelium are established. In the adult mouse testis, undifferentiated spermatogonial stem cells (As spermatogonia) have been described as single cells that are able both to renew themselves and to produce more differentiated Apr (paired) spermatogonia. The Apr cells then divide into Aal (aligned) spermatogonia that further differentiate into A1 spermatogonia (for a review see [132]). The appearance of A1 spermatogonia coincides with the re-expression of the *c-kit* gene, which encodes the receptor for the Kit ligand (KL, [41, 133–135]), and which had been previously downregulated in the male fetal testis at the time of mitotic quiescence. Kit is a tyrosine kinase receptor that mediates proliferation/survival signals in type A spermatogonia [136–138]. Upon Kit expression, spermatogonia become sensitive to KL produced by Sertoli cells [136, 139] and undergo a finite number of proliferative cycles (forming the A2–A4, intermediate, and B spermatogonia), before entering meiosis as preleptotene spermatocytes. Thus the temporal appearance of Kit expression and of KL sensitivity in spermatogonia at around 6–7 days postnatum (dpn) [41, 133, 138] marks the switch from the Aal spermatogonia to the A1–B differentiating cell types.

Postnatal spermatogenesis is under the control of FSH and LH, whose targets within the seminiferous tubules are somatic cells (Sertoli and Leydig, respectively). In response to LH, Leydig cells produce androgens, whose receptors are also present in Sertoli cells, but not in germ cells, as is the case with FSH receptors. As a consequence, all the hormonal effects on germ cells are mediated by somatic cells, which, in response, produce signals which stimulate germ cell proliferation and differentiation. These signals are important for resumption of mitotic divisions in the germ cell compartment of the seminiferous epithelium, as well as for the subsequent progression of the differentiation program: entry into meiosis, control of the meiotic divisions; spermiogenesis after haploidization of the DNA content of the future male gamete, with the final production of mature spermatozoa. FSH action in Sertoli cells is required for quantitatively normal spermatogenesis, and appears to mainly regulate the proliferation of germ cells at mitotic stages of differentiation [140]. Sertoli cell proliferation is directly regulated by FSH at puberty [141], thus establishing the somatic nursing compartment for spermatogenesis. Moreover, at least two paracrine growth factors acting on the spermatogonial compartment are known to be produced by Sertoli cells under the regulation of FSH: glial cell line-derived neurotrophic factor (GDNF), acting on the spermatogonial stem cells [142], and KL, acting on differentiating type A spermatogonia [136, 138]. Another important paracrine factor secreted by Sertoli cells, at least during the first wave of spermatogenesis, is BMP4, which also acts directly on the spermatogonial compartment [31]. The Sertoli cell-mediated action of androgens appears to regulate further stages of germ cell differentiation, both at meiotic and post-meiotic stages [140], but little is currently known about the cellular mediators produced by Sertoli cells. A schematic representation of the best characterized paracrine GFs controlling mouse male germ cell development is shown in Fig. 2.5.

2.4.1
GDNF

GDNF-family ligands interact with membrane receptors designated GFRαs which, in turn, mediate stimulation of the Ret receptor tyrosine kinase. Ret and GFRα–1

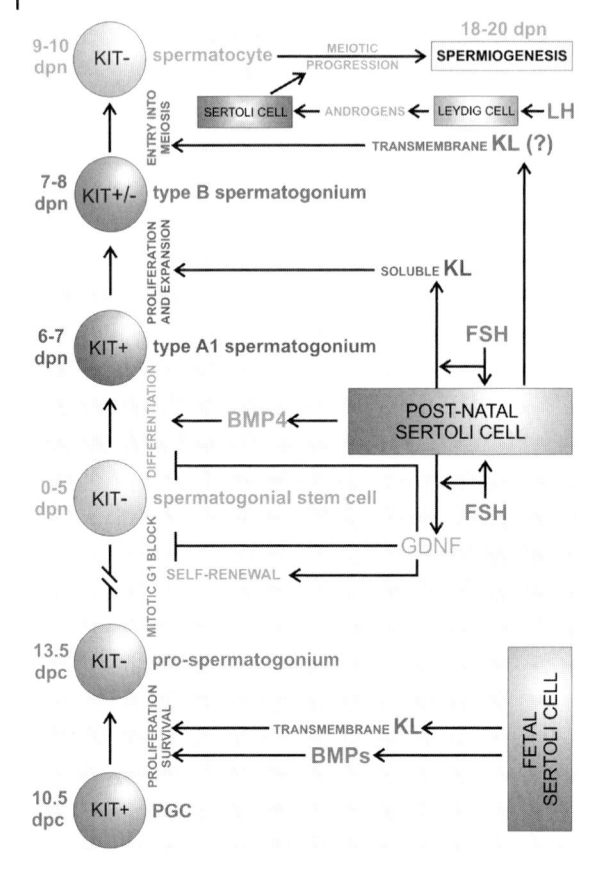

Fig. 2.5
Schematic representation of the best characterized paracrine GFs that act directly on mouse male germ cells, thus controlling progression of spermatogenesis and hormonal regulation of their expression in the somatic cell compartment. Different colours mark male germ cells at various developmental stages and the specific action of the indicated GFs on differentiative and proliferative events. Fluctuations during mitotic stages of expression of the *c-kit* gene, encoding the receptor for KL, are also indicated. (This figure also appears with the color plates.)

(the GDNF receptor) are expressed by mitotic germ cells, while GDNF is expressed by Sertoli cells, and *in vitro* addition of GDNF stimulates DNA synthesis in enriched mitotic populations of germ cells [143]. *In vivo* proliferation of undifferentiated spermatogonia, some of which are germinal stem cells, is accelerated by high concentrations of glial cell line-derived neurotrophic factor (GDNF), while FSH stimulation has been reported to increase GDNF mRNA levels in cultured Sertoli cells [144]. Transgenic loss-of-function and overexpression models show that the concentration of GDNF produced by Sertoli cells regulates cell-fate decisions of undifferentiated spermatogonial cells that include the stem cells for spermatogenesis [142]. Gene-targeted mice with one *Gdnf-null* allele show depletion of stem cell reserves, whereas mice overexpressing GDNF show accumulation of undifferentiated spermatogonia. They are unable to respond properly to differentiation signals and undergo apoptosis upon treatment with retinoic acid. Thus, GDNF contributes to the paracrine regulation of spermatogonial self-renewal and negatively regulates their differentiation [142].

2.4.2
BMPs

One of the factors controlling the differentiation of spermatogonia into the Kit-expressing stage in mice is BMP4, a growth factor belonging to the TGFβ-BMP superfamily. BMP4 is expressed at high levels in pre-puberal Sertoli cells, and its receptors Alk3 and BMPIIR are specifically expressed in mitotic spermatogonia during the first week postnatum (dpn) [31]. Furthermore, BMP4 is able to induce DNA synthesis in undifferentiated spermatogonia from 4 dpn. BMP4 action is mediated by a rapid nuclear translocation of Smad4, which associates with Smad5. Upon nuclear translocation, the Smad4/Smad5 complexes are able to recruit the transactivating factor CBP and to bind Smad-responsive DNA sequences [31]. Alk3 and Smad5 are exclusively expressed in the germline compartment of the postnatal testis. Finally, BMP4 is able to induce *c-kit* expression in Kit-negative spermatogonia, thus conferring KL sensitivity to these cells. Thus, the BMP4/Alk3/Smad5 pathway is expressed and operates in the postnatal testis and is involved in the regulation of spermatogonial differentiation [31]. Since spermatogonial stem cells are able to renew themselves and at the same time to progress through differentiation (i.e. to the Kit-dependent stages of proliferation), BMP4 may be one of the factors that regulates this process. Alternatively, BMP4 may act on a subset of spermatogonia which have lost their stem cell features but will still become Kit-positive cells. The exact physiological role played by BMP4 in postnatal spermatogenesis cannot be assessed with currently available animal models since *Bmp4−/−* mice die before birth. However, the dramatic reduction in BMP4 expression in adult Sertoli cells suggest that its action might be important mainly during the first wave of spermatogenesis, and that its action on spermatogonial differentiation in adult mice might be replaced by the action of other paracrine factors produced by the testicular environment once the cycle and the wave of the seminiferous epithelium have been established. Interestingly, another member of the TGFβ-BMP superfamily, BMP8b, also appears to play a role in mouse spermatogenesis. BMP8b has been shown to be produced by meiotic and post-meiotic germ cells, rather than by Sertoli cells, and exerts proliferative and anti-apoptotic effects on both spermatogonia and spermatocytes, as suggested by analysis of the phenotype of knockout animals [145]. However, the infertile phenotype of *Bmp8b−/−* animals is not highly penetrant, suggesting that other factors, such as BMP8a, might compensate for its loss within the seminiferous tubules in adult mice.

2.4.3
KL/Kit System

Re-expression of Kit in differentiating spermatogonia has led to the hypothesis that the KL/Kit interaction is required for the proliferation and/or survival of these cells. Several lines of evidence support this hypothesis. Early studies performed in *W* mutant mice suggested that Kit plays a critical role in the differentiation from type A to type B spermatogonia [146]. Moreover, a peculiar *Steel* mutation, Sl^{17H}, resulting in a splicing defect in the KL cytoplasmic tail in the homozygous condition, induces sterility in males but not in females due to the loss of spermatogonia during postnatal development [147].

The first evidence to suggest that Kit plays an important role in mitotically dividing male germ cells came from the observation that addition of the soluble form of KL to *in vitro*-cultured male germ cells from 7–8–day-old mice at mitotic stages of differentiation, stimulates DNA synthesis selectively in type A, but not in type B spermatogonia [136]. Selective induction of DNA synthesis in type A spermatogonia by soluble KL has been confirmed using an *in vitro* tissue culture system for stage-defined seminiferous tubules from the adult rat testis [148]. The important role of the KL/Kit system in the maintenance of the germ cell line in the mouse postnatal testis is also suggested by *in vivo* and *in vitro* experiments in which the interaction between KL and Kit was blocked by an antibody directed against the extracellular region of the receptor. Under these conditions, Kit-expressing type A spermatogonia, but not Kit-negative spermatogonial stem cells, are depleted [133] and unable to proliferate [149], and differentiating spermatogonia show increased levels of apoptosis [150]. The role of soluble KL in promoting survival and/or proliferation of rat and porcine spermatogonia has also been reported [151, 152]. Compelling evidence for an essential role of KL in postnatal spermatogenesis came from the observation that a mutation in the Kit docking site for the p85 subunit of PI3–K, artificially introduced by a knock-in strategy, causes a dramatic reduction of the spermatogonial population in the pre-pubertal testis by blocking proliferation of differentiating spermatogonia at day 8 postnatum [153, 154] whereas self-renewal of undifferentiated spermatogonial stem cells remains intact [155]. Addition of KL to *in vitro*-cultured differentiating spermatogonia results in a transient activation of extracellular signal-regulated kinases (Erk)1/2 as well as of PI3–K-dependent AKT kinase [138]. These events are followed by a rapid redistribution of cyclin D3, which becomes predominantly nuclear, whereas its total cellular concentration does not change. Nuclear accumulation of cyclin D3 is coupled to the transient activation of the associated kinase activity, assayed using retinoblastoma protein (Rb) as a substrate. These events are followed by a transient accumulation of cyclin E, stimulation of the associated histone H1–kinase activity, a delayed accumulation of cyclin A2, and Rb hyper-phosphorylation. All the events associated with KL-induced cell cycle progression are inhibited by the addition of either a PI3–K inhibitor or a mitogen-activated protein-kinase kinase (MEK) inhibitor, indicating that both MEK and PI3–K are essential for the Kit-mediated proliferative response. In contrast, the anti-apoptotic effect of KL is not influenced by the separate addition of either MEK or PI3–K inhibitors. Thus, KL effects on mitogenesis and survival in Kit-expressing spermatogonia rely on different signal transduction pathways [138].

In the descendants of telomerase null mutant mice the discovery of sterility which is brought about by the lack of proliferation of the spermatogonia, has highlighted the role of telomerase activity in mouse spermatogenesis [156]. Since the proliferation of spermatogonia is under the control of KL, the possible role of the latter in the regulation of telomerase activity has been explored [157]. KL was found to induce telomerase activity in both Kit-positive proliferating male primordial germ cells (PGCs) and in postnatal differentiating type A spermatogonia, and this increase was found to be dependent on PI3–K activity [157]. These data suggest that the induction of telomerase by KL may contribute to the self-renewing potential of mitotic male germ cells.

Expression of mRNA for KL is induced by the pituitary hormone FSH in pre-pubertal mouse Sertoli cells cultured *in vitro*, as a result of the increase in cAMP levels [136, 139]. Stage-dependent induction of KL mRNA expression by FSH has also been observed in the adult rat testis [158]. Maximal levels of induction of KL mRNA are observed in the stages of development of the seminiferous epithelium which show the maximal sensitivity to FSH stimulation, and in which type A spermatogonia are actively dividing. Interestingly, the soluble and transmembrane forms of KL are differentially expressed during testis development. Sertoli cells from pre-pubertal mice express mRNA encoding for both the soluble and transmembrane form, but mRNA encoding for the soluble form is expressed at higher levels in coincidence with the beginning of the spermatogenic process, and the two transcripts are expressed at equivalent levels in the adult testis [136]. Moreover, FSH and/or cAMP analogs, in addition to increasing KL mRNA levels, also modify the splicing pattern of the two isoforms in cultured mouse Sertoli cells in favor of the mRNA encoding for the soluble form [136]. The splicing pattern of KL mRNA in Sertoli cells might also be influenced by local changes of pH in the seminiferous epithelium [159]. FSH and/or cAMP regulation of KL expression in rat Sertoli cells is due, at least in part, to transcriptional activation [158], and cAMP-responsive regions of the KL promoter have been identified in the 5′ flanking regions of the human [160], rat [161] and mouse [162] KL genes.

The possibility that KL also plays a role in male meiosis was originally raised by the observation that intraperitoneal injection of antibodies directed against the extracellular portion of Kit, together with inducing apoptosis of spermatogonial cells, also induced a striking increase in apoptosis of the dividing spermatocytes at stage XII of the mouse seminiferous epithelium [150]. Since the injected antibody cannot pass through the blood-testis barrier, clearly this observation cannot be ascribed to a direct effect of the antibody on the spermatocytes, but it probably reflects an alteration of complex local paracrine networks which regulate meiotic progression of germ cells as a consequence of the functional alteration of other Kit-expressing testicular cells, namely Leydig cells, which in turn might produce other paracrine factors, such as androgens, thus regulating meiotic divisions of spermatocytes (see below). It has been reported that the transmembrane form of KL may play a role in trans-meiotic progression in cultures of spermatogenic cells co-cultured with the

KL-expressing 15–P1 cell line [163]. This has been argued on the basis of the observation that trans-meiotic progression in this *in vitro* system was inhibited by the soluble form of KL or by antibodies directed against the extracellular portion of the Kit receptor. Since *c-kit* is not expressed after the entry of male germ cells into meiosis [134, 164], it could be hypothesized that whereas soluble KL promotes the mitotic cell cycle in dividing type A spermatogonia, transmembrane KL promotes entry of type B spermatogonia into the meiotic cell cycle. In agreement with this hypothesis is the finding that the highest levels of the transmembrane form of KL are detected immunohistochemically in stages VII-VIII of the mouse seminiferous epithelium, and that, at these stages, a peculiar pattern of expression is observed over the whole Sertoli cell membrane rather than just in the proximity of the basal layer of germ cells [163]. Since the transition from type B spermatogonia to preleptotene spermatocytes actually occurs at stages VII-VIII, it is conceivable that the transmembrane form of KL might regulate the entry of mitotic germ cells into the first meiotic prophase.

The *in vitro* generation of spermatocytes and spermatids from telomerase-immortalized mouse type A spermatogonial cells cultured in the presence of KL has been reported [165]. This differentiation can apparently occur in the absence of supportive cells and/or other growth factors. If these data can be confirmed in other laboratories, this or other immortalized differentiating spermatogonial cell lines may serve as a powerful tool in the elucidation of the molecular mechanisms of spermatogenesis beyond the stage of spermatogonial divisions.

2.4.4
Control of Meiotic Progression of Spermatocytes

Very little is known about the factors which regulate the entry of male germ cells into meiosis in the postnatal testis. In yeast, the switch from the mitotic to the meiotic cell cycle is triggered by metabolic stimuli such as nutrient starvation [166] and it is possible that local environmental changes in the seminiferous tubules which modify nutrient availability (i.e. the formation of the blood-testis barrier by Sertoli cells at puberty) might influence the homologous switch in mammalian male germ cells by changing their metabolism. Interestingly, the main difference between the fetal gonads in the two sexes is the intensive development of blood vessels in the testis, in which germ cells arrest at mitotic stages of differentiation, versus the ovary, in which oogonia undergo meiosis in the cortical region, in which blood vessels are rare.

There is also little information regarding the factors which regulate meiotic progression of spermatocytes and the subsequent morphogenetic events of spermiogenesis. It is widely accepted that whereas the action of FSH on Sertoli cells mainly regulates spermatogonial proliferation, the action of androgens on the same cells is important to regulate later stages of differentiation, namely meiotic divisions and maturation of spermatozoa [140]. Compelling evidence for the Sertoli cell-mediated role of androgens in regulating meiotic progression of spermatocytes has been obtained through conditional knock-out mouse models. Male mice in which the androgen receptor gene was selectively disrupted in Sertoli cells are sterile because spermatogenesis is blocked at the middle-late pachytene and diplotene stages of meiosis, which strongly suggests that androgen-regulated Sertoli cell-derived signals are required by spermatocytes so that they can undergo meiotic divisions [167, 168].

Even though the nature of these Sertoli cell factors, which are likely to be GFs *in vivo*, are currently unknown, much progress has been made in dissecting the molecular pathways which regulate meiotic progression of *in vitro*-cultured spermatocytes artificially triggered by treatment with okadaic acid (OA), a selective inhibitor of protein Ser/Thr phosphatases PP1 and PP2A. This treatment triggers the first meiotic G2/M transition of middle-late pachytene spermatocytes [169]. Chromosome condensation induced in these cells by OA treatment requires the activation of the MAPK pathway, independently from MPF activity [170]. The kinase activity of p90Rsk2 is modulated by the MAPK pathway in spermatocytes and activated ERK1/2 and p90Rsk2 associate with condensing meiotic chromatin. Moreover, p90Rsk2 phosphorylates and activates the Ser/Thr kinase Nek2 through phosphorylation of its N-terminal catalytic domain [171]. Nek2 is the murine homolog of the Ser/Thr kinase NIMA, which, in *Aspergillus nidulans* is associated with chromatin and is involved in chromosome condensation during the mitotic cell cycle [172]. Thus Nek2 might be involved in the mechanisms leading to DNA condensation during progression through the first spermatogenic meiotic division [171]. Nek2 physically interacts both *in vitro* and *in vivo* with chromatin structural proteins of the class A High Mobility Group (HMG) family (HMGA1 e HMGA2), and this interaction requires the C-terminal regulatory domain of Nek2, but is not dependent on the activation state of the enzyme [173]. Furthermore, both recombinant and endogenous Nek2 are able to phosphorylate HMGA1/2 *in vitro*. Endogenous HMGA2 is strongly phosphorylated in pachytene spermatocytes during the first G2/M transition, and this phosphorylation is completely abolished by pretreatment with U0126, a specific inhibitor of the MEK-ERK pathway. Phosphorylation by Nek2 is able to reduce the affinity of HMGA2 for chromatin both *in vitro* and *in vivo* [173]. Since *hmga2–/–* mice are sterile and show a dramatic impairment of spermatogenesis [174], it is possible that the functional interaction between HMGA2 and Nek2 plays a crucial role in the correct process of chromatin condensation in meiosis. These results suggest a model in which, during progression through the first spermatogenic meiotic division, the activation of the MEK/ERK/p90Rsk2 pathway by an unknown environmental signal leads to phosphorylation and activation of Nek2, which is constitutively associated with the chromatin through the interaction with the non-histone protein HMGA2. Activated Nek2 then phosphorylates HMGA2 provoking its release from DNA, and thus allowing the association of other proteins which are important for chromosome condensation [173]. Moreover, this event might help to regulate changes in gene expression patterns that characterize specific developmental events during the transition to spermiogenesis [164].

References

1 De Felici, M., Scaldaferri, M.L., Lobascio, M., Iona, S., Nazzicone, V., Klinger, F.G., Farini, D. *In vitro* approaches to the study of primordial germ cell lineage and proliferation. *Hum. Reprod. Update* **2004**; *10*: 197–206.

2 Anderson, R., Copeland, T.K., Scholer, H., Heasman, J., Wylie, C. The onset of germ cell migration in the mouse embryo. *Mech. Dev.* **2000**; *91*: 61–68.

3 Tsang, T.E., Khoo, P.L., Jamieson, R.V., Zhou, S.X., Ang, S.L., Behringer, R., Tam, P.P.L. The allocation and differentiation of mouse primordial germ cells *Int. J. Dev. Biol.* **2001**; *45*: 549–556.

4 Tam, P.P., Snow, M.H. Proliferation and migration of primordial germ cells during compensatory growth in mouse embryos. *J. Embryol. Exp. Morphol.* **1981**; *64*: 133–147.

5 McLaren, A. Mammalian germ cells: birth, sex, and immortality. *Cell. Struct. Funct.* **2001**; *26*: 119–122.

6 Pesce, M., Gioia Klinger, F., De Felici, M. Derivation in culture of primordial germ cells from cells of the mouse epiblast: phenotypic induction and growth control by Bmp4 signalling *Mech. Dev.* **2002**; *112*: 15–24.

7 Ying, Y., Zhao, G.Q. Mammalian germ cells: birth, sex, and immortality. *Dev. Biol.* **2001**; *232*: 484–492.

8 Ying, Y., Liu, X.M., Marble, A., Lawson, K.A., Zhao, G.Q. Requirement of Bmp8b for the generation of primordial germ cells in the mouse. *Mol. Endocrinol.* **2000**; *14*: 1053–1063.

9 Tremblay, K.D., Dunn, N.R., Robertson, E.J. Mouse embryos lacking Smad1 signals display defects in extra-embryonic tissues and germ cell formation. *Development* **2001**; *18*: 3609–3621.

10 Chang, H., Matzuk, M.M. Smad5 is required for mouse primordial germ cell development. *Mech. Dev.* **2001**; *104*: 61–67.

11 Cairns, L.A., Moroni, E., Levantini, E., Giorgetti, A., Klinger, F.G., Ronzoni, S., Tatangelo, L., Tiveron, C., De Felici, M., Dolci, S., Magli, M.C., Giglioni, B., Ottolenghi, S. Kit regulatory elements required for expression in developing hematopoietic and germ cell lineages. *Blood* **2003**; *102*: 3954–3962.

12 De Felici, M. Regulation of primordial germ cell development in the mouse. *Int. J. Dev. Biol.* **2000**; *44*: 575–580.

13 Dolci, S., Williams, D.E., Ernst, M.K., Resnick, J.L., Brannan, C.I., Lock, L.F., Lyman, S.D., Boswell, H.S., Donovan, P.J. Requirement for mast cell growth factor for primordial germ cell survival in culture. *Nature* **1991**; *352*: 809–811.

14 Blume-Jensen, P., Janknecht, R., Hunter, T. The kit receptor promotes cell survival via activation of PI 3–kinase and subsequent Akt-mediated phosphorylation of Bad on Ser136. *Curr. Biol.* **1998**; *8*: 779–782.

15 Rameh, L.E., Cantley, L.C. The role of phosphoinositide 3–kinase lipid products in cell function. *J. Biol. Chem.* **1999**; *274*: 88347–8350.

16 Ueda, S., Mizuki, M., Ikeda, H., Tsujimura, T., Matsumura, I., Nakano, K., Daino, H., Honda, I.Z., Sonoyama, J., Shibayama,

H., Sugahara, H., Maschii, T., Kanakura, Y. Critical roles of c-Kit tyrosine residues 567 and 719 in stem cell factor-induced chemotaxis: contribution of src family kinase and PI3–kinase on calcium mobilization and cell migration. *Blood* **2002**; *99*: 3342–3349.

17 De Miguel, M.P., Cheng, L., Holland, E.C., Federspiel, M.J., Donovan, P.J. Dissection of the c-Kit signaling pathway in mouse primordial germ cells by retroviral-mediated gene transfer. *Proc. Natl Acad. Sci. USA* **2002**; *99*: 10458–10463.

18 Moe-Behrens, G.H., Klinger, F.G., Eskild, W., Grotmol, T., Haugen, T.B., De Felici, M. Akt/PTEN signaling mediates estrogen-dependent proliferation of primordial germ cells *in vitro*. *Mol. Endocrinol.* **2003**; *17*: 2630–2638.

19 Kimura, T., Suzuki, A., Fujita, Y., Yomogida, K., Lomeli, H., Asada, N., Ikeuchi, M., Nagy, A., Mak, T.W., Nakano, T. Conditional loss of PTEN leads to testicular teratoma and enhances embryonic germ cell production. *Development* **2003**; *130*: 1691–1700.

20 De Felici, M., Dolci, S. Leukemia inhibitory factor sustains the survival of mouse primordial germ cells cultured on TM_4 feeders layers. *Dev. Biol.* **1991**; *147*: 281–284.

21 Matsui, Y., Toksoz, D., Nishikawa, S., Williams, D., Zsebo, K., Hogan, B.L. Effect of Steel factor and leukaemia inhibitory factor on murine primordial germ cells in culture. *Nature* **1991**; *353*: 750–752.

22 Metcalf, D. Haemopoietic growth factors. *Med. J.* **1988**; *148*: 516–519.

23 Bazan, J.F. Neuropoietic cytokines in the hematopoietic fold. *Neuron* **1991**; *7*: 197–208.

24 Kishimoto, T., Taga, T., Akira, S. Cytokine signal transduction. *Cell* **1994**; *76*: 253–262.

25 Koshimizu, U., Taga, T., Watanabe, M., Saito, M., Shirayoshi, Y., Kishimoto, T., Nakatsuji, N. Functional requirement of gp130–mediated signaling for growth and survival of mouse primordial germ cells in vitro and derivation of embryonic germ (EG) cells. *Development* **1997**; *122*: 1235–1242.

26 Pesce, M., Farrace, M.G., Piacentini, M, Dolci, S., De Felici, M. Stem cell factor and leukemia inhibitory factor promote primordial germ cell survival by suppressing programmed cell death (apoptosis). *Development* **1993**; *118*: 1089–1094.

27 Matsui, Y., Zsebo, K., Hogan, B.M.L. Derivation of pluripotential embryonic stem cells from murine primordial germ cells in culture. *Cell* **1992**; *70*: 841–847.

28 Resnick, J.L., Bixler, L.S., Cheng, L., Donovan, P.J. Long-term proliferation of mouse primordial germ cells in culture. *Nature* **1992**; *359*: 550–551.

29 Molyneaux, K.A., Schaible K., Wylie, C. The chemokine SDF1/CXCL12 and its receptor CXCR4 regulate mouse germ cell migration and survival. *Development* **2003**; *130*: 4287–4294.

30 Resnick, J.L., Ortiz, M., Keller, J.R., Donovan, P.J. Role of fibroblast growth factors and their receptors in mouse primordial germ cell growth. *Biol. Reprod.* **1998**; *59*: 1224–1229.

31 Pellegrini, M., Grimaldi, P., Rossi, P., Geremia, R., Dolci, S. Developmental expression of BMP4/ALK3/SMAD5 signaling

32 Godin, I., Wylie, C.C. TGFβ1 inhibits proliferation and has a chemotropic effect on mouse primordial germ cells in culture *Development* **1991**; 113: 1451–1457.

33 Pesce, M., Canipari, R., Ferri,G.L., Siracusa, G., De Felici, M. Pituitary adenylate cyclase-activating polypeptide (PACAP) stimulates adenylate cyclase and promotes proliferation of mouse primordial germ cells. *Development* **1996**; *122*: 215–221.

34 De Felici, M., Dolci, S., Pesce, M. Proliferation of mouse primordial germ cells *in vitro*: a key role for cAMP. *Dev. Biol.* **1993**; *157*: 277–280.

35 Pesce, M., Farrace, M.G., Amendola, A., Piacentini, M., De Felici, M. Steel factor regulation of apoptosis in mouse primordial germ cells. In: *Cell Death in Reproductive Physiology*, J. Tilly (Ed.), Serono Symposium USA, **1997**.

36 Morita, Y., Tilly, J.L. Sphingolipid regulation of female gonadal cell apoptosis. *Ann. NY Acad. Sci.* **2000**; *905*: 209–220.

37 Lyrakou, S.L, Hultén, M.A., Hartshorne, G.M. Growth factors promote meiosis in mouse fetal ovaries *in vitro*. *Molec. Hum. Reprod.* **2002**; *8*: 906–911.

38 Morita, Y., Manganaro, T.F., Tao, X., Martimbeau, S., Donahoe, P.K., Tilly, Y.L. Requirement for phosphatidylinositol–3'-kinase in cytokine-mediated germ cell survival during fetal oogenesis in the mouse. *Endocrinology* **1999**; *140*: 941–949.

39 De Felici, M., Di Carlo, A., Pesce, M., Iona, S., Farrace, M.G., Piacentini, M., Bcl–2 and Bax regulation of apoptosis in germ cells during prenatal oogenesis in the mouse embryo. *Cell Death Diff.* **1999**; *6*: 908–915.

40 Klinger, F.G., De Felici, M. In vitro development of growing oocytes from fetal mouse oocytes: stage-specific regulation by stem cell factor and granulosa cells. *Dev. Biol.* **2002a**; *44*: 85–95.

41 Manova, K., Nocka, K., Besmer, P., Bachvarova, R.F. Gonadal expression of c-kit encoded at the W locus of the mouse. *Development* **1990**; *110*: 1057–1069.

42 Doneda, L., Klinger, F.G., La Rizza, L., De Felici, M. KL/KIT co-expression in mouse fetal oocytes. *Int. J. Dev. Biol.* **2002**; *46*: 1015–1021.

43 Klinger, F.G., De Felici, M. Antiapoptotic effects of SCF and IGF–1 on fetal mouse oocytes. *Dev. Biol.* **2002b**; 247: Abstr. 292.

44 Morita, Y., Maravedi, D.V., Bergeron, L., Wang, S., Perez, G.I., Tsutsumi, O., Taketani, Y., Asano, M., Horai, R., Korsmeyer, S.J., Iwakura, Y., Yuan, J., Tilly, J.L. Caspase–2 deficiency prevents programmed germ cell death resulting from cytokine insufficiency but not meiotic defects caused by loss of ataxia telangiectasia-mutated (Atm) gene function. *Cell Death Differ.* **2001**; *8*: 614–620.

45 Anderson, R.A., Robinson, L.L., Brooks, J., Spears, N. Neurotropins and their receptors are expressed in the human fetal ovary. *J. Clin. Endocrinol. Metab.* **2002**; *87*: 890–897.

46 Spears, N., Molinek, M.D., Robinson, L.L., Fulton, N., Cameron, H., Shimoda, K., Telfer, E.E., Anderson, R.A., Price, D.J.

The role of neurotrophin receptors in female germ-cell survival in mouse and human. *Development* **2003**; *130*: 5481–5491.

47 Paredes, A., Romero, C., Dissen, G.A., DeChiara, T.M., Reichardt, L., Cornea, A., Ojeda, S.R., Xu, B. TrkB receptors are required for follicular growth and oocyte survival in the mammalian ovary. *Dev. Biol.* **2004**; *267*: 430–449.

48 Hirshfield, A.N. Development of follicles in the mammalian ovary. *Int. Rev. Cytol.* **1991**; 124: 43–101.

49 Scaramuzzi, R.J., Adams, N.R., Baird, D.T., Campbell, B.K., Downing, J.A., Findlay, J.K., Henderson, K.M., Martin, G.B., McNatty, K.P., McNeilly, A.S. A model for follicle selection and the determination of ovulation rate in the ewe. *Reprod. Fertil. Dev.* **1993**; *5*: 459–478.

50 Krarup, T., Pedersen, T., Faber, M. Regulation of oocyte growth in the mouse ovary. *Nature* **1969**; *224*: 187–188.

51 Gougeon, A., Chainy, G.B. Morphometric studies of small follicles in ovaries of women at different ages. *J. Reprod. Fertil.* **1987**; *81*: 433–442.

52 Faddy, M.J., Gosden, R.G., Gougeon, A., Richardson, S.J., Nelson, J.F. Accelerated disappearance of ovarian follicles in mid-life: implications for forecasting menopause. *Hum. Reprod.* **1992**; *7*: 1342–1346.

53 Faddy, M.J., Gosden, R.G. A mathematical model of follicle dynamics in the human ovary. *Hum. Reprod.* **1995**; *10*: 770–775.

54 Hirshfield, A.N. Relationship between the supply of primordial follicles and the onset of follicular growth in rats. *Biol. Reprod.* **1994**; *50*: 421–428.

55 Wandji, S.A., Srsen, V., Voss, A.K., Eppig, J.J., Fortune, JE. Initiation *in vitro* of growth of bovine primordial follicles. *Biol. Reprod.* **1996**; *55*: 942–948.

56 Wandji, S.A., Srsen, V., Nathanielsz, P.W., Eppig, J.J., Fortune, J.E. Initiation of growth of baboon primordial follicles *in vitro*. *Hum. Reprod.* **1997**; *12*: 1993–2001.

57 Oktay, K., Karlikaya, G., Akman, O., Ojakian, G.K., Oktay, M. Interaction of extracellular matrix and activin-A in the initiation of follicle growth in the mouse ovary. *Biol. Reprod.* **2000**; *63*: 457–461.

58 Gospodarowicz, D., Plouet, J., Fujii, D.K. Ovarian germinal epithelial cells respond to basic fibroblast growth factor and express its gene: implications for early folliculogenesis. *Endocrinology* **1989**; *125*: 1266–1276.

59 van Wezel, I.L., Umapathysivam, K., Tilley, W.D., Rodgers, R.J. Immunohistochemical localization of basic fibroblast growth factor in bovine ovarian follicles. *Mol. Cell. Endocrinol.* **1995**; *115*: 133–140.

60 Nilsson, E.E., Parrott, J.A., Skinner, M.K. Basic fibroblast growth factor induces primordial follicle development and initiates folliculogenesis. *Mol. Cell. Endocrinol.* **2001**; *175*: 123–130.

61 Baarends, W.M., Hoogerbrugge, J.W., Post, M., Visser, J.A., De Rooij, D.G., Parvinen, M., Themmen, A.P., Grootegoed, J.A. Anti-mullerian hormone and anti-mullerian hormone

type II receptor messenger ribonucleic acid expression during postnatal testis development and in the adult testis of the rat. *Endocrinology* **1995**; *136*: 5614–5622.

62 Durlinger, A.L., Kramer, P., Karels, B., de Jong, F.H., Uilenbroek, J.T., Grootegoed, J.A., Themmen, A.P. Control of primordial follicle recruitment by anti-Mullerian hormone in the mouse ovary. *Endocrinology* **1999**; *140*: 5789–5796.

63 Durlinger, A.L., Gruijters, M.J., Kramer, P., Karels, B., Ingraham, H.A., Nachtigal, M.W., Uilenbroek, J.T., Grootegoed, J.A., Themmen, A.P. Anti-Mullerian hormone inhibits initiation of primordial follicle growth in the mouse ovary. *Endocrinology* **2002**; *143*: 1076–1084.

64 Huang, E.J., Manova, K., Packer, A.I., Sanchez, S., Bachvarova, R.F., Besmer, P. The murine steel panda mutation affects kit ligand expression and growth of early ovarian follicles. *Dev. Biol.* **1993**; *157*: 100–109.

65 Packer, A.I., Hsu, Y.C., Besmer, P., Bachvarova, R.F. The ligand of the c-kit receptor promotes oocyte growth. *Dev. Biol.* **1994**; *161*: 194–205.

66 Parrott, J.A., Skinner, M.K. Kit ligand actions on ovarian stromal cells: effects on theca cell recruitment and steroid production. *Mol. Reprod. Dev.* **2000**; *55*: 55–64.

67 Nilsson, E.E., Kezele, P., Skinner, M.K. Leukemia inhibitory factor (LIF) promotes the primordial to primary follicle transition in rat ovaries. *Mol. Cell. Endocrinol.* **2002**; *188*: 65–73.

68 Kuroda, H., Terada, N., Nakayama, H., Matsumoto, K., Kitamura, Y. Infertility due to growth arrest of ovarian follicles in Sl/Slt mice. *Dev. Biol.* **1988**; *126*: 71–79.

69 Bedell, M.A., Brannan, C.I., Evans, E.P., Copeland, N.G., Jenkins, N.A., Donovan, P.J. DNA rearrangements located over 100 kb 5' of the Steel (Sl)-coding region in Steel-panda and Steel-contrasted mice deregulate Sl expression and cause female sterility by disrupting ovarian follicle development. *Genes Dev.* **1995**; *9*: 455–470.

70 Kissel, H., Timokhina, I., Hardy, M.P., Rothschild, G., Tajima, Y., Soares, V., Angeles, M., Whitlow, S.R., Manova, K., Besmer, P. Point mutation in kit receptor tyrosine kinase reveals essential roles for kit signaling in spermatogenesis and oogenesis without affecting other kit responses. *EMBO J.* **2000**; *19*: 1312–1326.

71 Yoshida, H., Takakura, N., Kataoka, H., Kunisada, T., Okamura, H., Nishikawa, S.I. Stepwise requirement of c-kit tyrosine kinase in mouse ovarian follicle development. *Dev. Biol.* **1997**; *184*: 122–137.

72 Manova, K., Nocka, K., Besmer, P., Bachvarova, R.F. Gonadal expression of c-kit encoded at the W locus of the mouse. *Development* **1990**; *110*: 1057–1069.

73 Manova, K., Huang, E.J., Angels, M., De Leon, V., Sanchez, S., Pronovost, S.M., Besmer, P., Bachvarova, R.F. The expression pattern of the c-kit ligand in gonads of mice supports a role for the c-kit receptor in oocyte growth and in proliferation of spermatogonia. *Dev. Biol.* **1993**; *157*: 85–99.

74 Horie, K., Takakura, K., Taii, S., Narimoto, K, Noda, Y., Nishikawa, S., Nakayama, H., Fujita, J., Mori, T. The expression of

c-kit protein during oogenesis and early embryonic development. *Biol. Reprod.* **1991**; *45*: 547–552.

75 Motro, B., Bernstein, A. Dynamic changes in ovarian c-kit and Steel expression during the estrous reproductive cycle. *Dev. Dyn.* **1993**; *197*: 69–79.

76 Clark, D.E., Tisdall, D.J., Fidler, A.E., McNatty, K.P. Localization of mRNA encoding c-kit during the initiation of folliculogenesis in ovine fetal ovaries. *J. Reprod. Fertil.* **1996**; *106*: 329–335.

77 Horie, K., Fujita, J., Takakura, K., Kanzaki, H., Suginami, H., Iwai, M., Nakayama, H., Mori, T. The expression of c-kit protein in human adult and fetal tissues. *Hum. Reprod.* **1993**; *8*: 1955–1962.

78 Cecconi, S., Rossi, G., De Felici, M., Colonna, R. Mammalian oocyte growth *in vitro* is stimulated by soluble factor(s) produced by preantral granulosa cells and by Sertoli cells. *Mol. Reprod. Dev.* **1996**; *44*: 540–546.

79 Findlay, J.K., Drummond, A.E., Dyson, M.L., Baillie, A.J., Robertson, D.M., Eithier, J.F. Recruitment and development of the follicle; the roles of the transforming growth factor-beta superfamily *Mol. Cell. Endocrinol.* **2002**; *191*: 35–43.

80 Knight, P.G., Glister, C. Local roles of TGF-beta superfamily members in the control of ovarian follicle development. *Animal Reprod. Sci.* **2003**; *78*: 165–183.

81 Wu, T.C., Jih, M.H., Wang, L., Wan, Y.I. Expression of activin receptor II and IIB mRNA isoforms in mouse reproductive organs and oocytes *Mol. Reprod. Dev.* **1994**; *38*: 9–15.

82 Manova, K., De Leon, V., Angeles, M., Kalantry, S., Giarre, M., Attisano, L., Wrana, J., Bachvarova, R.F. mRNAs for activin receptors II and IIB are expressed in mouse oocytes and in the epiblast of pregastrula and gastrula stage mouse embryos *Mech. Dev.* **1995**; *49*: 3–11.

83 Sidis, Y., Fujiwara, T., Leykin, L., Isaacson, K., Toth, T., Schneyer, A.L. Characterization of inhibin/activin subunit, activin receptor, and follistatin messenger ribonucleic acid in human and mouse oocytes: evidence for activin's paracrine signaling from granulosa cells to oocytes. *Biol. Reprod.* **1998**; *59*: 807–812.

84 Cameron, V.A., Nishimura, E., Mathews, L.S., Lewis, K.A., Sawchenko, P.E., Vale, W.W. Hybridization histochemical localization of activin receptor subtypes in rat brain, pituitary, ovary, and testis. *Endocrinology* **1994**; *134*: 799–808.

85 Hulshof, S.C.J., Figueiredo, J.R., Beckers, J.F., Bevers, M.M., Vanderstichele, H., van der Hurk, R. Isolation and characterization of preantral follicles from foetal bovine ovaries *Theriogenology* **1997**; *48*: 133–142.

86 Izadyar, F., Dijkstra, G., Van Tol, H.T.A. Van den Eijnden-van Raaij, A.J.M., Van den Hurk, R., Colenbrander, B., Bevers, M.M. Immunohistochemical localization and mRNA expression of activin, inhibin, follistatin, and activin receptor in bovine cumulus-oocyte complexes during *in vitro* maturation. *Mol. Reprod. Dev.* **1998**; *49*:186–195.

87 Martins da Silva, S.J., Bayne, R.A.L., Cambray, N., Hartley, P.S., McNeilly, A.S., Anderson, R.A. Expression of activin sub-

units and receivers in the developing human ovary: activin A promotes germ cell survival and proliferation before primordial follicle formation. *Dev. Biol.* **2004**; *266*: 334–345.

88 Alak, B.M., Smith, G.D., Woodruff, T.K., Stouffer, R.L., Wolf, D.P. Enhancement of primate oocyte maturation and fertilization *in vitro* by inhibin A and activin A. *Fertil. Steril.* **1996**; *55*: 646–653.

89 Sadatsuki, M., Tsutsumui, O., Yamada, R., Maramatsu, M., Taketani, Y. Local regulatory effects of activin A and follistatin on meiotic maturation of rat oocytes. *Biochem. Biophys. Res. Commun.* **1993**; *196*: 388–395.

90 Alak, B.M., Coskun, S., Friedman, C.I., Kennard, E.A., Kim, M.H., Seifer, D.B. Activin A stimulates meiotic maturation of human oocytes and modulates granulosa cell steroidogenesis *in vitro*. *Fertil. Steril.* **1998**; *70*: 1126–1130.

91 Silva, C.C., Knight, P.G. Modulatory actions of activin-A and follistatin on the developmental competence of in vitro-matured bovine oocytes. *Biol. Reprod.* **1998**; *58*: 558–565.

92 Schmid, P., Cox, D., van der Putten, H., McMaster, G.K., Bilbe, G. Expression of TGF-beta s and TGF-beta type II receptor mRNAs in mouse folliculogenesis: stored maternal TGF-beta 2 message in oocytes. *Biochem. Biophys. Res. Commun.* **1994**; *201*: 649–656.

93 Glister, C., Groome, N.P., Knight, P.G. Oocyte-mediated suppression of follicle-stimulating hormone- and insulin-like growth factor-induced secretion of steroids and inhibin-related proteins by bovine granulosa cells *in vitro*: possible role of transforming growth factor alpha. *Biol. Reprod.* **2003**; *68*: 758–765.

94 McPherron, A.C., Lee, S.J. GDF-3 and GDF-9: two new members of the transforming growth factor-beta superfamily containing a novel pattern of cyst *J. Biol. Chem.* **1993**; *268*: 3444–3449.

95 McGrath, S.A., Esquela, A.F., Lee, S.J. Oocyte-specific expression of growth/differentiation factor-9. *Mol. Endocrinol.* **1995**; *9*: 131–136.

96 Dong, J., Albertini, D.F., Nishimori, K., Kumar, T.R., Lu, N., Matzuk, M.M. Growth differentiation factor-9 is required during early ovarian folliculogenesis. *Nature* **1996**; *383*: 531–535.

97 Dube, J.L. The bone morphogenetic protein 15 gene is X-linked and expressed in oocytes. *Mol. Endocrinol.* **1998**; *12*: 1809–1817.

98 Laitinen, M., Vuojolainen, K., Jaatinen, R., Ketola, I., Aaltonen, J., Lehtonen, E., Heikinheimo, M., Ritvos, O. A novel growth differentiation factor-9 (GDF-9) related factor is co-expressed with GDF-9 in mouse oocytes during folliculogenesis *Mech. Dev.* **1998**; *78*: 135–140.

99 Lyons, K.M., Pelton, R.W., Hogan, B.M.L. Patterns of expression of murine Vgr-1 and BMP-2a RNA suggest that transforming growth factor-beta-like genes coordinately regulate aspects of embryonic development. *Genes Dev.* **1989**; *3*: 1657–1668.

100 Valve, E. FGF-8 is expressed during specific phases of rodent oocyte and spermatogonium development. *Biochem. Biophys. Res. Commun.* **1997**; *232*: 173–177.

101 Zhou, J., Wang J., Penny D., Monget P., Arraztoa J.A., Fogelson L.J., Bondy C.A. Insulin-like growth factor binding protein 4 expression parallels luteinizing hormone receptor expression and follicular luteinization in the primate ovary. *Biol. Reprod.* **2003**; *69*: 22–29.
102 Vendola, K., Zhou, J., Wang, J., Famuyiwa, O.A., Bievre, M., Bondy, C.A. Androgens promote oocyte insulin-like growth factor I expression and initiation of follicle development in the primate ovary. *Biol. Reprod.* **1999**; *61*: 353–357.
103 Yamamoto, S., Konishi, I., Nanbu, K. Komatsu, T., Mandai, M., Kuroda, H., Matsushita, K., Mori, T. Expression of vascular endothelial growth factor (VEGF) during folliculogenesis and corpus luteum formation in the human ovary *Gynecol. Endocrinol.* **1997**; *11*: 223–230.
104 Singh, B., Kennedy, T.G., Tekpetey, F.R., Armstrong, D.T. Gene expression and peptide localization for epidermal growth factor receptor and its ligands in porcine luteal cells. *Mol. Cell. Endocrinol.* **1995**; *113*: 137–143.
105 Aaltonen, J., Laitinen, M.P., Vuojolainen, K., Jaatinen, R., Horelli-Kuitunen, N., Seppa, L., Louhio, H., Tuutri, T., Sjoberg, J., Butzow, R., Hovata, O., Dale, L., Ritvos, O. Human growth differentiation factor 9 (GDF–9) and its novel homolog GDF–9B are expressed in oocytes during early folliculogenesis. *J. Clin. Endocrinol. Metab.* **1999**; *84*: 2744–2750.
106 Bodensteiner, K.J., Clay, C.M., Moeller, C.L., Sawyer, H.R. Molecular cloning of the ovine Growth/Differentiation factor–9 gene and expression of growth/differentiation factor–9 in ovine and bovine ovaries. *Biol. Reprod.* **1999**; *50*: 421–428.
107 Lan, Z.J., Gu, P., Xu, X., Jackson, K.J., Demayo, F.J., O'Malley, B.W., Cooney, A.J. GCNF-dependent repression of BMP–15 and GDF–9 mediates gamete regulation of female fertility. *EMBO J.* **2003**; *15*: 4070–4081.
108 Chen, F., Cooney, A.J., Wang, Y., Law, S.W., O'Malley, B.W. Cloning of a novel orphan receptor (GCNF) expressed during germ cell development. *Mol. Endocrinol.* **1994**; *8*: 1434–1444.
109 Galloway, S.M., McNatty, K.P., Cambridge, L.M., Laitinen, M.P., Juengel, J.L., Jokiranta, T.S., McLaren, R.J., Luiro, K., Dodds, K.G., Montgomery, G.W., Beattie, A.E., Davis, G.H., Ritvos, O. Mutations in an oocyte-derived growth factor gene (BMP15) cause increased ovulation rate and infertility in a dosage-sensitive manner *Nature Gen.* **2000**; *25*: 279–283.
110 Erickson, G.F., Shimasaki, S. The role of the oocyte in folliculogenesis. *TEM* **2000**; *11*: 193–198.
111 Joyce, I.M., Clark, A.T., Pendola, F.L., Eppig, J.J. Comparison of recombinant growth differentiation factor–9 and oocyte regulation of KIT ligand messenger ribonucleic acid expression in mouse ovarian follicles. *Biol. Reprod.* **2000**; *63*: 1669–1675.
112 Cecconi, S., Rossi, G. Mouse antral oocytes regulate preantral granulosa cell ability to stimulate oocyte growth *in vitro*. *Dev. Biol.* **2001**; *233*: 186–191.
113 Elvin, J.A., Yan, C.N., Wang, P., Nishimori, K., Matzuk, M.M. Molecular characterization of the follicle defects in the growth

114 Otsuka, F., Shimasaki, S. A negative feedback system between oocyte bone morphogenetic protein 15 and granulosa cell kit ligand: its role in regulating granulosa cell mitosis. *Proc. Natl Acad. Sci. USA* **2002**; *99*: 8060–8065.

differentiation factor 9–deficient ovary. *Mol. Endocrinol.* **1999**; *13*: 1018–1034.

115 Anas, M.K., Shimada, M., Terada, T. Possible role for phosphatidylinositol 3–kinase in regulating meiotic maturation of bovine oocytes *in vitro*. *Theriogenology* **1998**; *50*: 347–356.

116 Fan, H.Y., Sun, Q.Y. Protein kinase C and mitogen-activated protein kinase cascade in mouse cumulus cells: cross talk and effect on meiotic resumption of oocyte. *Biol. Reprod.* **2004**; *70*: 535–547.

117 Ismail, R.S., Dubè, M., Vanderhyden, B.C. Hormonally regulated expression and alternative splicing of kit ligand may regulate kit-induced inhibition of meiosis in rat oocytes. *Dev. Biol.* **1997**; *184*: 333–342.

118 Ismail, R.S., Okawara, Y., Fryer, J.N., Vanderhyden, B.C. Ismail, R.S Hormonal regulation of the ligand for c-kit in the rat ovary and its effects on spontaneous oocyte meiotic maturation. *Mol. Reprod. Dev.* **1996**; *43*: 458–469.

119 Singh, B., Zhang, X., Armstrong, D.T. Porcine oocytes release cumulus expansion-enabling activity even though porcine cumulus expansion *in vitro* is independent of the oocyte. *Endocrinology* **1993**; *132*: 1860–1862.

120 Dekel, N, Sherizly, I. Epidermal growth factor induces maturation of rat follicle-enclosed oocytes. *Endocrinology* **1985**; *116*: 406–409.

121 Downs, S.M., Daniel, S.A.J., Eppig, J.J. Induction of maturation in cumulus cell-enclosed mouse oocytes by follicle-stimulating hormone and epidermal growth factor: evidence for a positive stimulus of somatic cell origin. *J. Exp. Zool.* **1988**; *245*: 86–96.

122 Salustri, A., Hascall, V.C., Campioni, A., Yanagishita, M. Oocytes-Granulosa cell interactions. In: *The Ovary*, Adashi E.Y, Leung P.C.K. (Eds), Raven Press, New York, USA, **1993**.

123 Salustri, A. Paracrine actions of oocytes in the mouse preovulatory follicle. *I. J. Dev. Biol.* **2000**; *44*: 591–597.

124 Matzuk, M.M., Burns, K.H., Viveiros, M.M., Eppig, J.J. Intercellular communication in the mammalian ovary: oocytes carry the conversation. *Science* **2002**; *296*: 2178–2180.

125 Elvin, J.A., Yan, C., Matzuk, M.M. Oocyte-expressed TGF-beta superfamily members in female fertility. *Mol. Cell Endocrinol.* **2000**; *159*: 1–5.

126 Shimasaki, S., Zachow, R.J., Li, D., Kim, H., Iemura, S., Ueno, N., Sampath, K., Chang, R.J., Erickson, G.F. A functional bone morphogenetic protein system in the ovary. *Proc. Natl. Acad. Sci USA* **1999**; *96*: 7282–7287.

127 Glister, C., Knight, P.G. Immunocytochemical evidence for a functional bone morphogenetic protein (BMP) signalling system in bovine antral follicles. *Reproduction Abstract Series 29* **2002**; *5*: Abstr 4.

128 Jaatinen, R., Laitinen, M.P., Vuojolainen, K., Aaltonen, J., Louhio, H., Heikinheimo K., Lehtonen, E., Ritvos, O. Localization

of growth differentiation factor–9 (GDF–9) mRNA and protein in rat ovaries and cDNA cloning of rat GDF–9 and its novel homolog GDF–9B. *Mol. Cell Endocrinol* **1999**; *156*: 189–193.

129 Souza, C.J., Campbell, B.K., McNeilly, A.S., Baird, D.T. Effect of bone morphogenetic protein 2 (BMP2) on oestradiol and inhibin A production by sheep granulosa cells, and localization of BMP receptors in the ovary by immunohistochemistry. *Reproduction* **2002**; *123*: 363–369.

130 Durlinger, A.L., Kramer, P.., Karels, B., de Jong, F.H., Uilenbroek, J.T., Grootegoed, J.A., Themmen, A.P. Control of primordial follicle recruitment by anti-Mullerian hormone in the mouse ovary *Endocrinology* **1999**; *140*: 5789–5796.

131 McGee, E.A., Smith, R., Spears, N., Nachtigal, M.W., Ingraham, H., Hsueh, A.J. Mullerian inhibitory substance induces growth of rat preantral ovarian follicles. *Biol. Reprod.* **2001**; *64*: 293–298.

132 de Rooij, D.G. Proliferation and differentiation of spermatogonial stem cells. *Reproduction* **2001**; *121*: 347–354.

133 Yoshinaga, K., Nishikawa, S. Ogawa, M., Hayashi, S., Kunisada, T., Fujimoto, T., Nishikawa, S.I. Role of c-kit in mouse spermatogenesis: identification of spermatogonia as a specific site of c-kit expression and function. *Development* **1991**; *113*: 689–699.

134 Sorrentino, V., Giorgi, M., Geremia, R., Besmer, P., Rossi, P. Expression of the c-kit proto-oncogene in the murine male germ cells. *Oncogene* **1991**; *6*: 149–151.

135 Schrans-Stassen, B. H., van de Kant, H. J., de Rooij D. G., van Pelt, A.M. Differential expression of c-kit in mouse undifferentiated and differentiating type A spermatogonia. *Endocrinology* **1999**; *140*: 5894–5900.

136 Rossi, P., Dolci, S., Albanesi, C., Grimaldi, P., Ricca, R., Geremia, R. Follicle-stimulating hormone induction of steel factor (SLF) mRNA in mouse Sertoli cells and stimulation of DNA synthesis in spermatogonia by soluble SLF. *Dev. Biol.* **1993**; *155*: 68–74.

137 Feng, L.X., Ravindranath, N., Dym, M. Stem cell factor/c-kit up-regulates cyclin D3 and promotes cell cycle progression via the phosphoinositide 3–kinase/p70 S6 kinase pathway in spermatogonia. *J. Biol. Chem.* **2000**; *275*: 25572–25576.

138 Dolci, S., Pellegrini, M., Di Agostino, S., Geremia, R., Rossi, P. Signaling through extracellular signal-regulated kinase is required for spermatogonial proliferative response to stem cell factor *J. Biol. Chem.* **2001**; *276*: 40225–40233.

139 Rossi, P., Albanesi, C., Grimaldi, P., Geremia, R. Expression of the mRNA for the ligand of c-kit in mouse Sertoli cells. *Biochem. Biophys. Res. Commun.* **1991**; *176*: 910–914.

140 McLachlan, R. I., O'Donnell, L., Meachem, S. J., Stanton, P. G., de Kretser, D. M., Pratis, K., Robertson, D. M. Identification of specific sites of hormonal regulation in spermatogenesis in rats, monkeys, and man. *Recent Prog. Horm. Res.* **2002**; *57*: 149–179.

141 Allan, C. M., Garcia, A., Spaliviero, J., Zhang, F. P., Jimenez, M., Huhtaniemi, I., Handelsman, D. J. Complete Sertoli cell

proliferation induced by follicle-stimulating hormone (FSH) independently of luteinizing hormone activity: evidence from genetic models of isolated FSH action. *Endocrinology* **2004**; *145*: 1587–1593.

142 Meng, X., Lindahl, M., Hyvonen, M. E., Parvinen, M., de Rooij, D. G., Hess, M. W., Raatikainen-Ahokas, A., Sainio, K., Rauvala, H., Lakso, M., Pichel, J. G. Westphal, H., Saarma, M., Sariola, H. Regulation of cell fate decision of undifferentiated spermatogonia by GDNF. *Science* **2000**; *287*: 1489–1493.

143 Viglietto, G., Dolci, S., Bruni, P., Baldassarre, G., Chiariotti, L., Melillo, R. M., Salvatore, G., Chiappetta, G., Sferratore, F., Fusco, A., Santoro, M. Glial cell line-derived neutrotrophic factor and neurturin can act as paracrine growth factors stimulating DNA synthesis of Ret-expressing spermatogonia. *Int. J. Oncol.* **2000**; *16*: 689–694.

144 Tadokoro, Y., Yomogida, K., Ohta, H., Tohda, A., Nishimune, Y. Homeostatic regulation of germinal stem cell proliferation by the GDNF/FSH pathway. *Mech. Dev.* **2002**; *113*: 29–39.

145 Zhao, G. Q., Deng, K., Labosky, P.A., Liaw, L., Hogan, B.L. The gene encoding bone morphogenetic protein 8B is required for the initiation and maintenance of spermatogenesis in the mouse. *Genes Dev.* **1996**; *10*: 1657–1669.

146 Koshimizu, U., Sawada, K., Tajima, Y., Watanabe, D. and Nishimune, Y. White-spotting mutations affect the regenerative differentiation of testicular germ cells: demonstration by experimental cryptorchidism and its surgical reversal. Biol. Reprod. **1991**; *45*: 642–648.

147 Brannan, C. I., Bedell, M. A., Resnick, J. L., Eppig, J. J., Handel, M. A., Williams, D. E., Lyman, S. D., Donovan, P. J., Jenkins, N. A., Copeland, N.G. Developmental abnormalities in Steel17H mice result from a splicing defect in the steel factor cytoplasmic tail. Genes Dev. **1992**; *6*: 1832–1842.

148 Hakovirta, H., Yan, W., Kaleva, M., Zhang, F., Vanttinen, K., Morris, P. L., Soder, M., Parvinen, M., Toppari, J. Function of stem cell factor as a survival factor of spermatogonia and localization of messenger ribonucleic acid in the rat seminiferous epithelium. Endocrinology **1999**; *140*: 1492–1498.

149 Tajima, Y., Sawada, K., Morimoto, T. and Nishimune, Y. Switching of mouse spermatogonial proliferation from the c-kit receptor-independent type to the receptor-dependent type during differentiation. J. Reprod. Fertil. **1994**; *102*: 117–122.

150 Packer, A. I., Besmer, P., Bachvarova, R. F. Kit ligand mediates survival of type A spermatogonia and dividing spermatocytes in postnatal mouse testes. Mol. Reprod. Dev. **1995**; *42*: 303–310.

151 Allard, E.K., Blanchard, K.T., Boekelheide, K. Exogenous stem cell factor (SCF) compensates for altered endogenous SCF expression in 2,5–hexanedione-induced testicular atrophy in rats. Biol. Reprod. **1996**; *55*: 185–193.

152 Dirami, G., Ravindranath, N., Pursel, V., Dym M. Effects of stem cell factor and granulocyte macrophage-colony stimulating factor on survival of porcine type A spermatogonia cultured in KSOM. Biol. Reprod. **1999**; *61*: 225–230.

153 Kissel, H., Timokhina, I., Hardy, M.P., Rothschild, G., Tajima, Y., Soares, V., Angeles, M., Whitlow, S.R., Manova, K., Besmer, P. Point mutation in kit receptor tyrosine kinase reveals essential roles for kit signaling in spermatogenesis and oogenesis without affecting other kit responses. *EMBO J.* **2000**; *19*: 1312–1326.

154 Blume-Jensen, P., Jiang, G., Hyman, R., Lee, K.F., O'Gorman, S., Hunter, T. Kit/stem cell factor receptor-induced activation of phosphatidylinositol 3′-kinase is essential for male fertility. *Nat. Genet.* **2000**; *24*: 157–162.

155 Ohta, H., Yomogida, K., Dohmae, K., Nishimune, Y. Regulation of proliferation and differentiation in spermatogonial stem cells: the role of c-kit and its ligand SCF. *Development* **2000**; *127*: 2125–2131.

156 Lee, H.W., Blasco, M.A., Gottlieb, G.J., Horner, J.W. 2nd, Greider, C.W., DePinho, R.A. Essential role of mouse telomerase in highly proliferative organs. *Nature* **1998**; *392*: 569–574.

157 Dolci S, Levati L, Pellegrini M, Faraoni I, Graziani G, Di Carlo A, Geremia R. Stem cell factor activates telomerase in mouse mitotic spermatogonia and in primordial germ cells . *J. Cell Sci.* **2002**; *115*: 1643–1649.

158 Yan, W., Linderborg, J., Suominen, J., Toppari, J. Stage-specific regulation of stem cell factor gene expression in the rat seminiferous epithelium. Endocrinology **1999**; *140*: 1499–1504.

159 Mauduit, C., Chatelain, G., Magre, S., Brun, G., Benahmed, M., Michel, D. J. Regulation by pH of the alternative splicing of the stem cell factor pre-mRNA in the testis. Biol. Chem. **1999**; *274*: 770–775.

160 Taylor, W.E., Najmabadi, H., Strathearn, M., Jou, N-T., Liebling, M., Rajavashisth, T., Chanani, N., Phung, L., Bhasin, S. Human stem cell factor promoter deoxyribonucleic acid sequence and regulation by cyclic 3′,5′-adenosine monophosphate in a Sertoli cell line. *Endocrinology* **1996**; *137*: 5407–5414.

161 Jiang, C., Hall, S.J., Boekelheide, K. Cloning and characterization of the 5′ flanking region of the stem cell factor gene in rat Sertoli cells. *Gene* **1997**; *185*: 285–290.

162 Grimaldi, P., Capolunghi, F., Geremia, R., Rossi, P. Cyclic adenosine monophosphate (cAMP) stimulation of the kit ligand promoter in sertoli cells requires an Sp1–binding region, a canonical TATA box, and a cAMP-induced factor binding to an immediately downstream GC-rich element. *Biol. Reprod.* **2003**; *69*: 1979–1988.

163 Vincent, S., Segretain, D., Nishikawa, S., Nishikawa, S.I., Sage, J., Cuzin, F., Rassoulzadegan, M. Stage-specific expression of the Kit receptor and its ligand (KL) during male gametogenesis in the mouse: a Kit-KL interaction critical for meiosis. *Development* **1998**; *125*: 4585–4593.

164 Rossi, P., Dolci, S., Sette, C., Capoluoghi, F., Pellegrini, M., Loiarro, M., Di Agostino, S., Baronetto, M.P., Grimaldi, P., Merico, D., Martegani, E., Geremia, R. Analysis of the gene expression profile of mouse male meiotic germ cells. *Gene Expr. Patterns* **2004**; *4*: 267–281.

165 Feng, L.X., Chen, Y., Dettin, L., Pera, R.A., Herr, J.C., Goldberg, E., Dym, M. Generation and *in vitro* differentiation of a spermatogonial cell line. *Science* **2002**; *297*: 392–395.

166 Honigberg, S.M., Purnapatre, K. Signal pathway integration in the switch from the mitotic cell cycle to meiosis in yeast. *J. Cell Sci.* **2003**; *116*: 2137–2147.

167 De Gendt, K., Swinnen, J.V., Saunders, P.T., Schoonjans, L., Dewerchin, M., Devos, A., Tan, K., Atanassova, N., Claessens, F., Lecureuil, C., Heyns, W., Carmeliet, P., Guillou, F., Sharpe, R.M., Verhoeven, G. A Sertoli cell-selective knockout of the androgen receptor causes spermatogenic arrest in meiosis. *Proc. Natl. Acad. Sci. U S A* **2004**; *101*: 1327–1332.

168 Chang, C., Chen, Y.T,, Yeh, S.D., Xu, Q., Wang, R.S., Guillou, F., Lardy, H., Yeh, S. Infertility with defective spermatogenesis and hypotestosteronemia in male mice lacking the androgen receptor in Sertoli cells *Proc. Natl. Acad. Sci USA* **2004**; *101*: 6876–6881.

169 Wiltshire, T., Park, C., Caldwell, K.A., Handel, M.A. Induced premature G2/M-phase transition in pachytene spermatocytes includes events unique to meiosis *Dev. Biol.* **1995**; *169*: 557–567.

170 Sette, C., Barchi, M., Bianchini, A., Conti, M., Rossi, P., Geremia, R. Activation of the mitogen-activated protein kinase ERK1 during meiotic progression of mouse pachytene spermatocytes. *J. Biol. Chem.* **1999**; *274*: 33571–33579.

171 Di Agostino, S., Rossi, P., Geremia, R., Sette, C. The MAPK pathway triggers activation of Nek2 during chromosome condensation in mouse spermatocytes. *Development* **2002**; *129*: 1715–1727.

172 Fry, A.M., Nigg, E.A. Cell cycle. The NIMA kinase joins forces with Cdc2. *Curr. Biol.* **1995**; *5*: 1122–1125.

173 Di Agostino, S., Fedele, M., Chieffi, P., Fusco, A., Rossi, P., Geremia, R., Sette, C. Phosphorylation of high-mobility group protein A2 by Nek2 kinase during the first meiotic division in mouse spermatocytes. *Mol. Biol. Cell.* **2004**; *15*: 1224–1132.

174 Chieffi, P., Battista, S., Barchi, M., Di Agostino, S., Pierantoni, G.M., Fedele, M., Chiariotti, L., Tramontano, D., Fusco, A. HMGA1 and HMGA2 protein expression in mouse spermatogenesis. *Oncogene* **2002**; *21*: 3644–3650.

3
Implantation and Placentation
Susana M. Chuva de Sousa Lopes, Christine L. Mummery, and Sólveig Thorsteinsdóttir

3.1
Introduction

The eutherian embryo is unique among mammals (eutherians, monotremata and marsupials) in that it not only implants like all mammals, but it develops a placenta, a specialized organ that allows the embryo to develop within and in close physical association with its mother. This provides safety from potential external dangers, and ensures a constant supply of vital components from the maternal blood. To achieve maternal link, the embryo spends a considerable amount of its initial developmental effort preparing and establishing the connections with the maternal environment. This involves the production of various cell types that play key roles in implantation and placentation, whilst supporting normal development of the embryo itself. Implantation thus involves complex cell-cell communication at two levels: (1) *within the embryo*, where early cell differentiation and morphogenetic arrangements progressively set aside the extra-embryonic lineages while also supporting the cells of the embryo proper and (2) *between the embryo and the maternal environment*, ensuring a tight coordination of events leading to successful implantation and placentation. The preparation of the embryo and the two-way interaction between the embryo and the uterus are prerequisites for successful development in mammals.

Gene ablation studies in mice targeting growth factors, their receptors, adhesion molecules and transcription factors have highlighted the complexity of events involved in implantation and placentation. In this chapter, we focus on the mouse as a model system for understanding these processes and show that progress in the field of mouse implantation has also provided insight into human implantation and in particular placental disorders.

3.1.1
Formation of the Trophectoderm, the Precursor of the Invading Trophoblast

Development of the mammalian embryo from fertilization to implantation is termed pre-implantation development and is characterized by a relative independence from the maternal environment. The fertilized egg completes meiosis and is pushed towards the uterus by cilia in the oviduct. While in the oviduct, the embryo remains within a glycoprotein coat, the zona pellucida, which was synthesized during oogenesis. The zona pellucida not only plays a role in fertilization (reviewed in [1]) but also keeps the blastomeres of the early embryo together and prevents implantation in the oviduct.

Preparation of the embryo for implantation starts at the mid–8–cell stage with "compaction" (reviewed in [2, 3]): increased E-cadherin-mediated adhesion between all blastomeres, results in them flattening against each other, forming a smooth, spherical ball of cells, the compacted *morula*. This polarizes the blastomeres into basolateral and apical domains. During the next cleavage, some blastomeres divide along the apical/basal axis, forming two polar outer cells, while others divide into one outer and one inner blastomere, the latter being non-polarized [4]. Thus two different lineages become established in the morula (Fig. 3.1A): (1) the outer cells form the first epithelium of the embryo, the trophectoderm, the precursor of the *trophoblast lineage*; (2) the inner cells remain non-polarized and pluripotent, giving rise to the *inner cell mass (ICM)*. At the 32–cell stage, "cavitation" starts as the trophectoderm cells, now forming a tight junction seal separating the inner cells from the maternal environment, secrete fluid directionally into the morula. This leads to the formation of the fluid-filled blastocoel cavity and the embryo is designated a *blastocyst* (Fig. 3.1B). As the blastocyst expands, the ICM cells become positioned at one end of this cavity. Trophectoderm cells in contact with the ICM are termed *polar trophectoderm*, while those lining the blastocoel cavity are called *mural trophectoderm*.

3.1.2
Initial Blastocyst-Uterine Interaction

The blastocyst reaches the uterus on embryonic day (E) 3.5 in the mouse. Shortly after, a trypsin-like protease, strypsin/ISP1 [5, 6], expressed by the trophectoderm creates a hole in the zona pellucida and the expanding blastocyst escapes from its glycoprotein coat, an event termed "hatching". The hatched blastocyst is initially not adhesive, but becomes competent for attachment when the uterus reaches a receptive stage, which occurs at E4.5 in the mouse. If the uterus does not become receptive, the blastocyst remains non-adhesive and enters a metabolically inactive state. This state of delayed implantation obeys seasonal cues in some mammals, is induced in certain species if the female is lactating and can be induced in mice by treatments that alter the balance of progesterone and estrogen (reviewed in [7]). However, if the uterus enters a receptive stage, the blastocyst becomes competent for attachment, as a result of cross-talk between them.

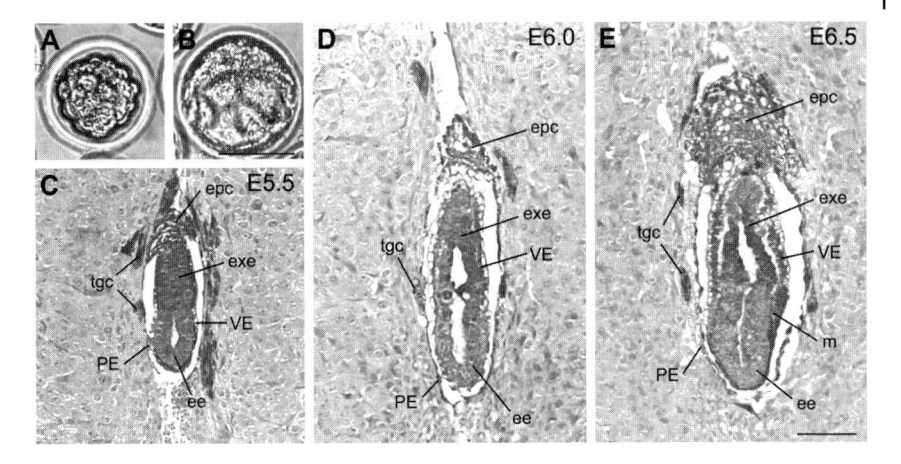

Fig. 3.1
Mouse embryonic development from morula to gastrula. (A) After compaction, the morula continues cleaving without cellular growth. (B) The blastocyst forms with the expansion of the blastocoelic cavity resulting from vectorial fluid transport by trophectoderm cells (dark blue, polar trophectoderm; light blue, mural trophectoderm). The ICM cells (yellow) and the primitive endoderm (orange) form distinct populations inside the blastocyst. (C-E) After implantation the ICM develops into the epiblast or embryonic ectoderm (yellow), the primitive endoderm gives rise to visceral endoderm (orange) and parietal endoderm (pink), the mural trophectoderm differentiates into the primary trophoblast giant cells (light blue) and the polar trophectoderm forms the ectoplacental cone (dark blue) and the extra-embryonic ectoderm (green). Although the tissue organization remains similar after implantation, the embryo starts growing in size. Gastrulation begins with the invagination of nascent mesoderm (gray) at the posterior side of the embryo. ee, epiblast or embryonic ectoderm; epc, ectoplacental cone; exe, extra-embryonic ectoderm; PE, parietal endoderm; m, nascent mesoderm; tgc, trophoblast giant cells; VE, visceral endoderm. Scale bar: 50 μm (A and B) and 100 μm (C-E). (This figure also appears with the color plates.)

In the mouse, the uterine lumen forms a fold that surrounds the blastocyst, bringing it in close contact with the uterine epithelium. The presence of an apposed blastocyst activates a decidual response in the uterine wall [8]. This starts with an increase in vascular permeability within the uterine stroma, immediately followed by blastocyst attachment to the uterine epithelium. Soon after, apoptosis is induced in this epithelium and it is shed, facilitating trophoblast penetration and invasion by displacement and phagocytosis of the uterine epithelial and stromal cells [9, 10]. Between E5.5 and E6.0, the stromal cells encapsulating the embryo stop proliferating and differentiate becoming polyploid [11]. These cells constitute the primary decidual zone (PDZ). By E8.0, the PDZ undergoes apoptosis enlarging the implantation chamber and is replaced by the differentiating stromal cells that surround the PDZ and form the secondary decidual zone (SDZ). Similarly to the PDZ, the SDZ also degenerates later in development.

3.1.3
Differentiation of Extra-embryonic Lineages During Implantation and Gastrulation

The invading trophoblasts are formed by cells quite different from the epithelial trophectoderm. The differentiation from one into the other involves changes that are reminiscent of an epithelial to mesenchymal transition (reviewed in [3]). The first step occurs as the mural trophectoderm differentiates into primary *trophoblast giant cells* (Fig. 3.1C-E). These cells undergo endoreduplication of their DNA becoming polyploid and also become motile with different adhesion characteristics from their trophectoderm precursor. These primary trophoblast giant cells form the so-called *yolk sac placenta* that surrounds the early post-implantation embryo. This non-vascular placenta is also present in bats and armadillos and is believed to exchange materials between mother and conceptus during the first one-third of gestation, until the formation of the chorio-allantoic placenta later in development [12].

Cells of the polar trophectoderm remain proliferative and undifferentiated due to signals emanating from the ICM (reviewed in [3]) and give rise to the *extra-embryonic ectoderm* and the *ectoplacental cone* (Fig. 3.1C-E), which displace the ICM into the blastocoel cavity as they grow. A second wave of trophoblast giant cells differentiates from the ectoplacental cone and migrates to surround the conceptus, thus contributing to the expansion of the yolk sac placenta. These *secondary trophoblast giant* cells also endoreduplicate their DNA and differentiate into a highly invasive cell type. Meanwhile, a stem cell population continues to proliferate in the ectoplacental cone, giving rise first to more secondary trophoblast giant cells and later contributing to the chorio-allantoic placenta.

At the late blastocyst stage, two cell populations are set aside in the ICM: the cells facing the blastocoel cavity are termed *primitive endoderm* (or *hypoblast*), an extra-embryonic lineage, while the remaining cells of the ICM form the *primitive ectoderm* (or *epiblast*; Fig. 3.1B). Two cell populations develop from the primitive endoderm (reviewed in [13]): cells leaving the ICM as individual mesenchymal cells and migrating on the basal side of the trophectoderm eventually lining the blastocoel cavity are termed *parietal endoderm* (PE; Fig. 3.1C-E). PE cells are responsible for synthesizing a thick basement membrane, *Reichert's membrane*, deposited between the PE and the trophectoderm. Reichert's membrane is a transient structure unique to rodents and bats that not only contributes to the regulation of nutrient and waste exchange between embryo and mother, but also protects the embryo against the immune system of the mother. Together, the PE, trophectoderm and Reichert's membrane are designated the *parietal yolk sac*, the structure that directly contacts the uterine tissue. Primitive endoderm cells that remain in contact with the ICM, differentiate into epithelial *visceral endoderm* (VE), which initially remains in contact with the ICM/epiblast and extra-embryonic ectoderm (Fig. 3.1C-E). The anterior visceral endoderm (AVE) plays an important role in determining the anterior-posterior axis of the developing embryo (reviewed in [14]). Gastrulation, epiblast cells give rise to the *definitive endoderm* of the embryo proper displacing the VE towards the extra-embryonic ectoderm. Meanwhile, extra-embryonic mesodermal cells derived from the posterior primitive streak (Fig. 3.1E) colonize the basal side of the visceral endoderm, differentiating into extra-embryonic mesothelium and blood islands (Fig. 3.2A and B). These cells together with the VE and its basement membrane form the *visceral yolk sac* (Fig. 3.2A). The visceral yolk sac is the first site of hematopoiesis in the conceptus providing blood cells (Fig. 3.2A and B) to the early embryo as the heart begins rhythmic contraction around E8.5, but this "extra-embryonic" blood (with primitive nucleated red blood cells) is then gradually replaced by blood produced by the fetal liver and aorta-gonad-mesonephros region [15, 16]. The extra-embryonic mesoderm also contributes to the *chorion* (with the extra-embryonic ectoderm), the *amnion* (with the embryonic ectoderm) and the *allantois* located at the posterior side of the embryo that grows into the exocoelomic cavity and eventually develops into the umbilical cord (see Fig. 3.2A and B).

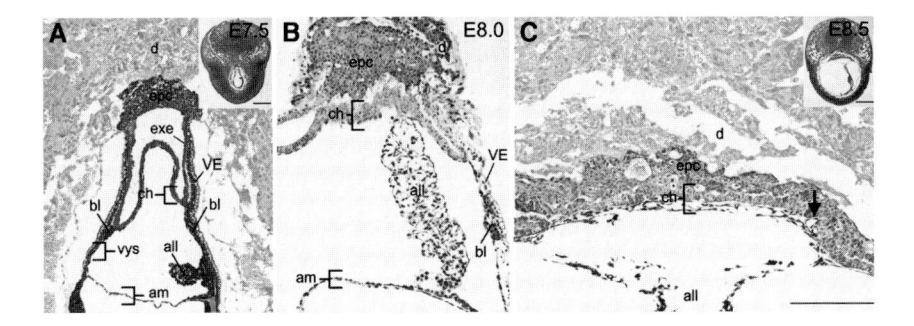

Fig. 3.2
Chorio-allantoic fusion. (A) By E7.5, the extra-embryonic mesoderm (brown) contributes to the formation of four extra-embryonic structures: the amnion (together with ectoderm), the visceral yolk sac (together with visceral endoderm (blue)), the chorion (together with extra-embryonic ectoderm (yellow)) and the allantois. (B) The allantois is a finger-like structure, which grows across the exocoelomic cavity towards the chorion. It is formed by a sheath of mesothelium and an inner core of vascularizing mesoderm. At this stage the ectoplacental cavity becomes progressively reduced as the chorion squeezes against the ectoplacental cone (green). (C) At E8.5 the allantois spreads across the chorion surface and blood vessels (arrow) immediately sprout to vascularize the budding chorio-allantoic placenta. Note that although both the embryo and extra-embryonic structures increase tremendously in size, the size of the deciduum remains constant (see inserts in A and C). all, allantois; am, amnion; bl, blood island; ch, chorion; d, deciduum; epc, ectoplacental cone; exe, extra-embryonic ectoderm; VE, visceral endoderm; vys, visceral yolk sac. Scale bar: 250 μm (A-C) and 500 μm (inserts in A and C). (This figure also appears with the color plates.)

3.1.4
Steps towards a Functional Placenta

The first step in the formation of the *chorio-allantoic placenta* occurs around E8.5 when the allantois, elongated across the exocoelomic cavity, makes close contact with the *chorionic plate*, the chorion flattened against the ectoplacental cone (Fig. 3.2B and C). This event is called chorio-allantoic fusion. A few hours after first contact, folds appear in the chorionic epithelium and blood vessels sprouting from the allantois invade the spaces formed by the folds (Fig. 3.2C). Over the next days, chorionic trophoblast cells and fetal blood vessels undergo extensive branching, referred to as "vascular invasion" of the chorion, forming the placental *labyrinth layer* (Fig. 3.3). However, the chorionic plate remains as a defined structure, adjacent to the highly vascularized placental labyrinth (Fig. 3.3; reviewed in [17]).

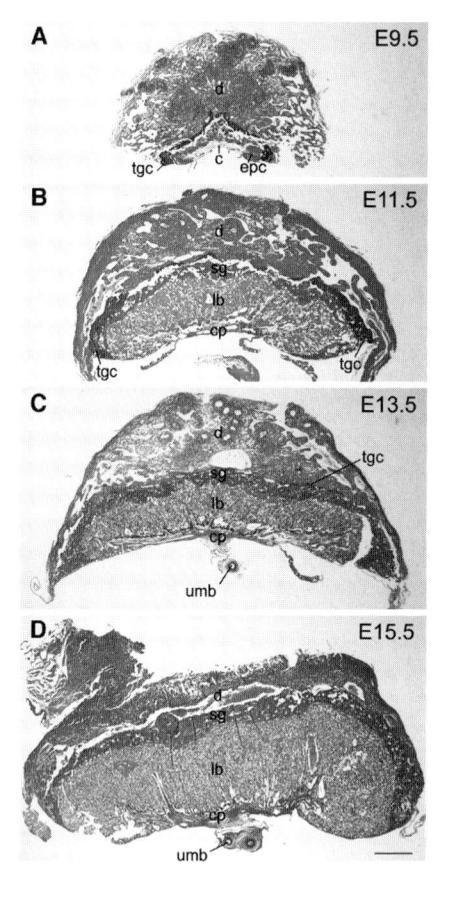

Fig. 3.3
Placentation. (A) An increasing number of trophoblast giant cells (magenta) and ectoplacental cone cells (green) form the placenta of the E9.5 embryo. The allantois vasculature sprouts into folds in the chorion (yellow). (B) Two days later, the labyrinth (yellow) has developed greatly between the spongiotrophoblast layer (green) and the chorionic plate (blue). The trophoblast giant cells (magenta) become localized at the border of the placenta. (C and D) At E13.5 and E15.5, the structure of the placenta remains essentially the same, but the labyrinth (yellow) continues to expand. Note that the umbilical cord is also visible (pink). c, chorion; cp, chorionic plate; d, deciduum; epc, ectoplacental cone; lb, labyrinth; sg, spongiotrophoblast; tgc, trophoblast giant cells; umb, umbilical cord. Scale bar: 500 μm. (This figure also appears with the color plates.)

For branching morphogenesis and labyrinth development to proceed normally, the labyrinthine trophoblast (derived from the chorion) responsible for the maintenance of the maternal space has to differentiate (reviewed in [17]). In the mouse, two separate *syncytiotrophoblast* cell layers develop to surround the lumen of the maternal blood sinus in the labyrinth. These distinct concentric cell layers are formed from the fusion of two or more precursor cells into a multinucleated cell type, the syncytiotrophoblast. Furthermore, a mononuclear trophoblast cell type is present inside the lumen of these branches containing the maternal blood. Between one and four large maternal arterial blood vessels, known as the central arterial canals cross the *spongiotrophoblast layer*, a compact non-syncytial cell layer located between the trophoblast giant cells (forming the outer layer of the placenta at the interface with the maternal deciduum) and the placental labyrinth (Fig. 3.3B-D) [18, 19]. Interestingly, the maternal endothelium is progressively removed and as the maternal blood enters the placental labyrinth the small branches formed by the syncytiotrophoblast cell layers become the maternal "blood vessels" themselves [18]. Transport of gases and nutrients thus occurs in the placental labyrinth from the maternal blood through the syncytiotrophoblast cells and fetal blood vessels into the fetal blood, while embryonic waste products are transported in the opposite direction. The chorio-allantoic placenta starts functioning around E10.0 and reaches full maturation at E17.5, meeting the ever-growing demands of the fetus.

3.1.5
Comparison of Mouse and Human Implantation and Placentation

The mechanisms used by mammals for implantation are species dependent (reviewed in [7]). Firstly, murine blastocysts adhere to the uterine wall through the mural trophectoderm, whilst human blastocysts attach via the polar trophectoderm. Secondly, uterine invasion is also initially different in mice and humans. In mice, the uterine wall directly surrounding the blastocyst undergoes apoptosis and is phagocytosed by the invasive trophoblast cells facilitating penetration, while human trophoblasts squeeze between epithelial cells forcing penetration through the uterine wall. Although mice rely during the first one-third of gestation on a yolk sac placenta, they also develop a chorio-allantoic placenta as the demand for glucose increases during organogenesis. Surprisingly, the chorio-allantoic placenta of mice and humans is rather similar compared with that of other mammals. Both mouse and human placentae form a single unit of attachment with the uterus and are therefore classified as *discoid*, although the human placenta consists of separate branching units clustered together. The degree of trophoblastic invasion of the uterus and erosion of the maternal tissue is also similar in mice and humans so that both placentae are classified as *hemochorial*. In neither species is there a maternal uterine epithelial tissue interfacing with the outer layer of the placenta. In addition, the endothelial cells of maternal capillaries are lost and therefore the maternal blood contacts fetal syncytiotrophoblast cells directly (reviewed in [17]).

Mouse and human placentae are structured in layers. The interface layer is formed in both species by polyploid cells (trophoblast giant cells and *extravillous cytotrophoblast* cells, respectively) although murine giant cells are less invasive and have a higher DNA content. The middle layer of the placenta in mice is formed by spongiotrophoblast cells and in humans by the *column cytotrophoblast* cells. Cells of this middle layer can differentiate into cells of the interface layer. The labyrinth layer in the mouse is the functional equivalent of the *chorionic villi* in humans (reviewed in [17]). In addition, placentae in both species have a chorionic plate from which the umbilical cord sprouts to make physical contact with the fetus.

3.2
Molecular Mechanisms and Biological Effects

3.2.1
Preparation of the Blastocyst for Implantation (E3.5–E4.5)

During the first days of development, the "free-floating" embryo depends for survival on uterine fluid from the mother for nutrition since it lacks a blood supply, but the embryo is nevertheless practically independent of maternal growth factors and can undergo cell division, apoptosis and differentiation autonomously. In mice there are no targeted deletions of growth factors or their respective receptors that compromise blastocyst development, with the exception of transforming growth factor-β1 (TGFβ1) where ablation in a predominantly CF1 genetic background causes developmental arrest at the 2- to 4-cell stage [20]. Deletion of other cytokines however, does cause mild abnormal phenotypes, such as reduced cell number (colony stimulating factor–1 (CSF–1) and granulocyte-macrophage colony stimulating factor (GM-CSF)) or increased apoptosis (TGFα) in the blastocyst [21–23]. In spite of this apparent growth factor independence, the rate of development *in vitro* is enhanced by growth factor addition to the culture media (reviewed in [24]). At this stage (human and mouse) the reproductive tract and the embryo itself are rich sources of growth factors (and their respective receptors) from various families, including insulin-like growth factor (IGF), epidermal growth factor (EGF), fibroblast growth factor (FGF), platelet derived growth factor (PDGF), TGFα, TGFβ and tumour necrosis factor (TNF). Thus, although gene targeting studies indicate that growth factors (produced either by the mother or the blastocyst) are not essential during pre-implantation development, it remains possible that maternal mRNA and proteins present in the embryo during its first cell divisions (and not targeted using conventional knockout techniques) could play important roles during early pre-implantation development. Support for this comes from studies showing a profound effect of ectopic dominant negative FGF or TGFβ receptors on early mouse embryos [25, 26].

In pre-implantation embryos, cell-cell interactions mediated by various adhesion and junction molecules are extremely important for maintenance of structure and function. Deletion of the cell adhesion molecule E-cadherin [27, 28] or its intracellular binding partner, α-E-catenin [29] is lethal at the blastocyst stage. In those mutant embryos, compaction takes place due to the presence of maternal protein

[27, 28], but cell-cell adhesion is then progressively lost and blastocyst expansion does not occur. Furthermore, the small GTPase RhoA is important in the recruitment of the actin cytoskeleton to sites of adhesion during compaction [30]. A number of different connexins, components of gap junctions, are expressed during pre-implantation development [31]. However, deletion of single connexins in the mouse has no consequences for blastocyst formation, possibly due to redundancy between connexins [32]. The formation of tight junctions and desmosomes between trophectoderm cells results in the permeability seal necessary for vectorial fluid transport and cavitation. This involves a large number of molecules, including ZO–1, occludin, cingulin (tight junctions) and plakoglobin, desmoplakin and desmoglein (desmosomes) [33]. The importance of each of these proteins for pre-implantation development remains unclear. Although before implantation the embryo assembles a continuous basement membrane under the trophectoderm as well as between the cells of the ICM [34, 35], the presence of this membrane is apparently not essential for successful development to the blastocyst stage [36]. Likewise, the embryo can develop in the absence of the zona pellucida.

The concerted expression of the transcription factors Oct4 (*Pou5f1*) and Nanog determines cellular identity in the expanded blastocyst; both factors are therefore crucial during pre-implantation development. Oct4 is present in the morula and becomes restricted to the ICM as the blastocyst forms, its primary function being to prevent differentiation to trophectoderm cells [37]. Independently, Nanog becomes expressed only in the ICM and is necessary for maintenance of pluripotency; in its absence, the ICM becomes endoderm [38, 39]. Interestingly, apart from serving as a transcriptional repressor of genes that stimulate trophectoderm differentiation, Oct4 can (together with the transcription factor, Sox2) activate FGF4 expression [37]. FGF signaling in turn stimulates cell proliferation both in pre-implantation [25] and early post-implantation development [40].

To date, a number of trophectoderm-specific proteins have been identified including CSF–1 receptor (*c-fms*), TGFβ2, trophoblast interferon τ and Na^+/K^+-ATPase which are highly expressed in mural trophectoderm cells. However, none of these molecules appears essential for trophectoderm differentiation (reviewed in [41, 42]).

3.2.2
Uterine-Embryonic Signaling and Adhesion (E4.5)

After hatching from the zona pellucida the blastocyst can attach to the uterine wall. Implantation is complex and highly regulated (the major events are summarized in Fig. 3.4). It is still not fully understood at the molecular level, mainly because of the difficulty in studying it *in vivo* and the lack of appropriate *in vitro* models. Pre-implantation embryonic development and preparation of the uterine environment are regulated independently and blastocysts generated *in vitro* can successfully implant in the uterus of non-pregnant females treated with the correct hormone cocktails, both in mice and humans. In the mouse, both estrogen and progesterone, produced by the ovary (follicles and corpora lutea, respectively), are necessary to trigger a receptive state in the uterus ("window of implantation"), for a limited period of time (24–36 h) while catecholestrogen, a metabolite of estrogen activates the blastocyst by inducing the expression of perlecan and EGF-receptor (ErbBs) (reviewed in [43, 44]). These two activation events lead to the dialogue necessary for successful implantation. Disturbing the balance between estrogen and progesterone renders the uterus non-receptive for blastocyst attachment. This occurs, for example, during lactation as suckling keeps the secretion of prolactin high and gonadotropin low, leading to delayed implantation (diapause). As soon as the levels of estrogen/progesterone reach the right balance, implantation can take place (reviewed in [7]). Moreover, deficiency in the prolactin receptor (*Prlr*–/–) leads to abnormal levels of progesterone and consequently failure of implantation [45, 46].

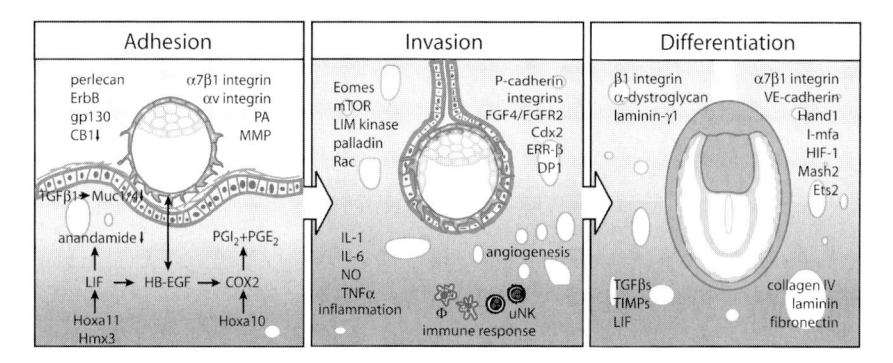

Fig. 3.4
Major factors and processes involved in mouse implantation. Implantation in the mouse can be divided into three major phases: adhesion between the blastocyst and the uterine lumen, invasion of the uterus by the trophoblast and decidualization and, finally, differentiation of the yolk sac placenta. The deciduum/uterus, with their respective blood vessels, is represented in blue, the uterine epithelium in pink, the ECM on the surface of the uterine epithelium (adhesion) and Reichert's membrane (differentiation) in light brown, the trophectoderm and its derivatives in green, the ICM and derived tissues in yellow, the primitive endoderm and its derivatives in orange. All abbreviations used are explained in the text.

The activated blastocyst triggers expression of heparin-binding (HB)-EGF in the surrounding uterine epithelium 6–7 h before the initial attachment reaction [47]. HB-EGF then acts on the blastocyst promoting ErbB1 phosphorylation [47]. Leukemia inhibitor factor (LIF) is also expressed by uterine cells surrounding the embryo at the time and site of implantation suggesting a role for maternal LIF. Indeed, *Lif*-deficient blastocysts are able to implant when placed in the uterus of wild-type pseudo-pregnant females, but all blastocysts fail to implant in *Lif*-deficient mothers [48]. Accordingly, inactivation of the LIF co-receptor *gp130* (genetic background dependent), but not the LIF receptor *Lif-r*, results in implantation failure [49–51], suggesting that although both LIF receptors are expressed by the blastocyst, gp130 mediates the major signal in response to maternal LIF during implantation. Interestingly, maternal LIF probably also plays a role in inducing HB-EGF, as HB-EGF expression is completely lost in the uterus of *Lif*-deficient mice [52].

In the uterine wall, HB-EGF induces the production of cyclooxygenase–2 (COX–2), an enzyme restricted to the implantation site and essential for both decidualization and implantation [53, 54]. COX–2, together with COX–1, are the rate-limiting enzymes in the biosynthesis of prostaglandins, converting arachidonic acid into PGH_2, an intermediate compound that can subsequently be turned into different types of prostaglandins [55]. In particular, prostacyclin (PGI_2) and its non-canonical binding partner peroxisome proliferator-activated receptor-δ (PPARδ), together with prostaglandin E_2 (PGE_2) play important roles in both decidualization and implantation [56]. Furthermore, an increasing number of factors are being implicated in cellular and molecular responses of the uterus during implantation, including EGF, interleukin, FGF, IGF, bone morphogenetic protein (BMP), Wnt (Wingless/Integrated) and Hedgehog (Hh) family members, although their exact roles are still unclear [41, 57, 58]. BMP2, 4 and 5 and nodal, all members of the TGFβ superfamily, are involved in embryo spacing in the uterus [58, 59]. The homeobox genes *Hoxa10* and *Hoxa11* as well as *Hmx3* (a non-classical homeobox gene) have also been implicated in the decidual response leading to a successful implantation. Hoxa10 probably acts independently of LIF, as *Hoxa10*–deficient mice show normal LIF levels in the uterus and vice versa [52, 60], however the levels of COX2 are reduced [61] resulting in implantation defects. In contrast, females deficient in either *Hoxa11* or *Hmx3* show downregulation of uterine LIF and blastocysts fail to implant, mimicking the situation observed in *Lif*-deficient uteri [62–64]. The implantation phenotype is completely rescued when embryos from homozygous females are placed into wild-type receptive uteri [64]. Cannabinoid signaling has also been implicated in implantation. During the window of implantation, both the high levels of anandamide (endocannabinoid) in the uterus and the cannabionoid receptor CB1 in the embryo are coordinately downregulated [65] suggesting that high signaling levels are not compatible with implantation. In agreement with this, low concentrations of anandamide stimulate blastocyst outgrowth *in vitro* while high concentrations are inhibitory [66]. Interestingly, the levels of anandamide and CB1 remain high in *Lif*-deficient uteri and in dormant blastocysts during delayed implantation [65].

During the window of implantation, both the blastocyst and the uterus become transiently adhesive. The molecules involved during this crucial step are still largely unknown. Mucin1 (Muc1) and Muc4, high molecular-weight glycoproteins are present at the surface of the uterine epithelium providing a physical barrier to infections and outside agents by reducing cellular adhesion [67, 68]. Muc1 and Muc4 are downregulated during the window of implantation, facilitating attachment of the blastocyst to the uterine wall and although expressed in most tissues of the female reproductive tract, hormonal regulation is restricted to the uterine epithelium [69, 70]. Interestingly, downregulation of Muc4 by increased progesterone levels appears to involve a paracrine action of TGFβ1, since progesterone upregulates TGFβ1 expression and exogenous TGFβ1 downregulates Muc4 in cultured rat uterine epithelial cells [71]. A protein implicated in the adhesion between uterus and blastocyst is perlecan (*Hspg2*), a heparin-sulfate proteoglycan that is strongly upregulated at the surface of attachment-competent blastocysts. However, *Hspg2*–deficient mice show no implantation defects [72, 73]. As mentioned above, HB-EGF is synthesized in the uterine epithelium just prior to implantation and, interestingly, its transmembrane form (as opposed to the secreted form) clearly stays associated with the uterine epithelium and may bind to ErbBs and perlecan on the blastocyst surface (reviewed in [7, 43, 44]).

Prior to attachment, the mural trophectoderm cells differentiate from epithelial cells to the invasive and highly polarized primary trophoblast giant cells (reviewed in [3]). The change in polarity involves the expression and redistribution of specific integrin complexes, including $\alpha v \beta 3$, $\alpha 5 \beta 1$ and $\alpha 7 \beta 1$ from the basolateral to the apical membrane [74, 75]. Finally, protrusive activity is initiated, an event thought to be regulated by the T-box transcription factor, Eomesodermin (*Eomes*) (reviewed in [3]). $\alpha 7 \beta 1$ integrin, a receptor for the basement membrane glycoprotein laminin, appears to play a major role in implantation and early yolk sac placenta formation. Its presence in the invading trophoblast in mice and *in vitro* assays using trophoblast cell lines suggest that $\alpha 7 \beta 1$–laminin interactions not only play an active role in invasion but that they may also regulate primary trophoblast differentiation [76]. Although a trophoblast phenotype has not been described in $\alpha 7$ integrin (*Itga7*) null mouse embryos, interestingly, about half of the homozygous embryos die around E12.5 [77]. αv Integrins interact with the extracellular matrix molecule osteopontin, which is induced in the uterine stroma by progesterone during the implantation window [78]. Inactivation of αv integrin (*Itgav*) results in a placental phenotype in most homozygous embryos, but obvious defects are only detected at later stages of placental development (see Section 3.2.4). It is not clear whether the fibronectin receptor $\alpha 5 \beta 1$ integrin is essential for implantation *per se*. However, extra-embryonic and embryonic vascular development is impaired in $\alpha 5$ integrin (*Itga5*) null embryos which die around E10.0 [79].

When in contact with the uterine wall, the trophoblast begins secreting proteases, including plasminogen activators (PAs) and the matrix metalloproteinases (MMPs), collagenase and stromelysin [80, 81], which digest extracellular matrix (ECM) proteins and thereby facilitate attachment. Furthermore, ECM digestion products stimulate cell motility, probably contributing to the shift in behavior observed in the trophoblast giant cells. ADAMs, a class of molecules with metalloprotease and disintegrin domains, might also be important during implantation. ADAMs have been reported to bind integrins, including $\alpha v \beta 3$ and $\alpha 5 \beta 1$, interact with heparin-sulfate proteoglycans and ECM proteins, and cleave HB-EGF and amphiregulin (reviewed in [82]). However, to date no defects in implantation have been reported in mice deficient in any protease or their inhibitors, although there may be functional redundancy.

Prior to implantation, the mural trophectoderm cells start endoreduplicating their DNA and rapidly become unable to divide. However, polar trophectoderm cells remain proliferative (and diploid) contributing to expansion of the trophoblast giant cell population (reviewed in [3]). It is thought that the polar trophectoderm cells remain multipotent due to signals secreted by the ICM [83]. Accordingly, the introduction of a second ICM inside a blastocyst results in a second zone of proliferative trophectoderm, while the removal of the ICM results in the differentiation of the trophectoderm in giant cells [84, 85]. FGF signaling seems to play a crucial role here by regulating the onset of differentiation to trophoblast giant cells (see Section 3.2.3.2).

3.2.3
The Invasion of the Uterus (E4.5–E7.5)

3.2.3.1 Decidualization

The decidual response that starts only hours after the initial attachment of the blastocyst to the uterine epithelium has many similarities with an acute inflammatory response (see Fig. 3.4). This includes the recruitment of non-antigen specific immune response cells such as the specialized uterine natural killer cells (uNK) and macrophages (Φ), while T- and B-cells are virtually absent (antigen specific). The production of inflammatory factors such as interleukin (IL)–1, IL–6, nitric oxide (NO) and TNFα, which in turn increase vascular permeability and stimulate the formation of new blood vessels (angiogenesis) in the deciduum, also takes place (reviewed in [86]). Nevertheless and quite remarkably, the maternal immune system does not reject the embryo, which is in fact a semi-allogenic "graft". The embryo may avoid the immune attack by producing indoleamine 2,3–dioxygenase (IDO), an enzyme that degrades the amino acid tryptophan, but IDO is not detected in the embryo until E7.5 [87]. Depletion of tryptophan prevents activated T-cells from replicating, thereby suppressing their action in the proximity of the embryo (reviewed in [88]). Alternatively, during pregnancy, the maternal T-cells acquire a transient and reversible state of tolerance specific for paternal histocompatability antigens. Accordingly, during this period, the mother can be inoculated with paternal tumor cells and not reject them, as would normally occur [89].

Cellular changes also take place as the uterine epithelium enters apoptosis and disappears (phagocytosed by the trophoblast cells thereby facilitating invasion), while stromal mesenchymal cells rapidly proliferate and undergo postmitotic differentiation to form polyploid decidual cells. Important for stromal cell polyploidy during decidualization is the upregulation of cyclin D3 by HB-EGF [90] and the subsequent interaction of cyclin D3 with p21 and cdk6 [91]. Female mice lacking the IL–11 receptor (*IL–11Rα*) form only very small deciduae and as a consequence, the chorio-allantoic placenta does not develop and the embryos degenerate [92, 93]. As mentioned above, females deficient in *Hoxa10*, *Hoxa11* and *Cox2* also exhibit specific defects in decidualization. Interestingly, the decidual reaction is completely independent of the presence and/or attachment of the embryo to the uterine wall, but can be induced in pseudo-pregnant females using artificial stimuli, such as intraluminal oil infusion. Paria and colleagues [94] demonstrated that it is progesterone alone, in the absence of estrogen that triggers decidualization. On the other hand, tight junctions formed between decidual cells (as the uterine epithelium is lost) in the primary decidual zone are specifically triggered by the blastocyst, but not by artificial stimuli or isolated ICM. This tight junction barrier appears impermeable for immunoglobulins, therefore playing an important role in avoiding the maternal immune system [95].

Another important function of the deciduum is to control excessive invasion by the trophectoderm. It does this by producing among other factors TGFβs, tissue inhibitors of metalloproteinases (TIMPs) and LIF, which then limit trophectoderm invasion by reducing the activity of proteases and favoring the synthesis and deposition of ECM components, including type IV collagen, laminin and fibronectin (reviewed in [41, 86]). It is of note that mouse embryos transplanted to non-uterine tissues are tumorigenic and will form teratocarcinomas [96].

3.2.3.2 Anchoring of Trophoblast Cells and Differentiation during Peri-implantation

In mice, the mural trophoblast cells acquire a pseudomalignant invasive character (see Fig. 3.4 for a summary of the major events involved). However, mouse trophoblasts are far less invasive than human trophoblasts *in vivo*, invading primarily by directed phagocytosis of apoptotic epithelial and decidual cells, instead of active migration. The mouse primary trophoblast giant cells form protrusions that facilitate the phagocytosis of large particles and whole cells [10] and thus proteins involved in actin-filament assembly, such as LIM kinase, palladin and probably GTPase Rac are upregulated (reviewed in [3]). Exposure of the embryo to amino acids leads to activation of mTOR, a serine-threonine kinase involved in initiating changes in protrusive activity [97].

During peri-implantation, the migrating trophoblast giant cells downregulate E-cadherin and transiently upregulate P-cadherin altering their cell-cell adhesion properties [98–100]. This switch may be regulated by the transcription factor Snail [101, 102]. Later, VE-cadherin replaces P-cadherin [3], adherens junctions and desmosomes are formed both between trophoblast giant cells and between trophoblast giant cells and maternal decidual cells, while tight junctions are completely absent between trophoblast giant cells but evident between decidual cells [9, 12, 95, 103]. The repertoire of integrin α-subunits expressed by the trophoblast cells enlarges after implantation now providing more receptors for laminins (α1β1, α2β1, α3β1, α6β1, α7β1) and collagens (α1β1, α2β1) in addition to the αv integrins present previously [74]. Interestingly, although the α7β1 integrin seems to play a pivotal role in the uterine invasion by trophoblast giant cells [76], absence of β1 integrin first affects the ICM which degenerates around E5.5 [104, 105]. It is thus possible that αv integrins [74] and non-integrin laminin receptors such as dystroglycan [106] support the initial phases of trophoblast invasion when β1 integrins are absent.

The formation of Reichert's membrane from laminins, type IV collagen and perlecan secreted by the parietal endoderm is crucial for the integrity of the maternal-embryo interface. Therefore inactivation of genes involved in differentiation and/or migration of parietal endoderm cells, including *Lamc1* (laminin-γ1), *Itgb1* (β1 integrin) and/or leading to a defective Reichert's membrane, including *Dag1* (dystroglycan) are lethal to the embryo during peri-implantation [36, 104, 107].

A number of genes including *Fgf4* and FGF receptor 2 (*Fgfr2*) are specifically involved in allocation of the trophoblast lineage. During peri-implantation, *Fgf4* is expressed specifically by the ICM/epiblast cells [108]. Moreover, trophoblast cells in the vicinity of the FGF4–producing ICM remain in a proliferative state, while cells away from the FGF4 source differentiate into trophoblast giant cells. Interestingly, the trophectoderm cells adjacent to the ICM are exposed to a laminin isoform (laminin 10/11) that is not present on the mural trophectoderm (only in contact with laminin 1) and laminin 10/11 stimulates the proliferation of trophoblast cells *in vitro* whereas laminin 1 does not [76]. Furthermore, it has recently been shown that human ECV304 cells (an endothelial cell line) proliferate better in the presence of FGFs if grown on laminin 10/11 [109]. This raises the interesting possibility that signals generated by FGFs and laminin 10/11 cooperate in promoting cell proliferation of the polar trophectoderm *in vivo*. Trophoblast cells in culture rapidly acquire a giant cell-like phenotype when FGF4 is removed from the culture media [83]. Tanaka and colleagues [83] demonstrated that the proliferative (polar) trophoblast cells, which form the ectoplacental cone and extra-embryonic ectoderm, constitute a population of trophoblast stem cells and succeeded in isolating a trophoblast stem cell line. This line is diploid, immortal, expresses markers of extra-embryonic ectoderm (and not of other lineages) and similarly to (polar) trophoblast cells rapidly differentiates into giant trophoblast cells when FGF4 is removed from the culture medium. Embryos deficient in *Fgf4* undergo implantation and trigger decidualization, but die shortly thereafter [40] and embryonic stem cells (as opposed to trophoblast stem cells) can be obtained from the ICM of *Fgf4*–deficient blastocysts, suggesting that FGF4 indeed acts primarily in trophoblast differentiation at this stage. In support of this proposal, it has been found that embryos deficient in *Fgfr2* die a few hours post-implantation [110] and overexpression of dominant negative FGF receptor 2 (FGFR2) blocks trophoblast proliferation during pre-implantation [25].

Two genes downstream of the FGF signaling pathway also involved in allocation of the trophoblast lineage are the transcription factors *Eomes* (see Section 3.2.2.) and *Cdx2*. Both genes are involved in patterning the embryo during gastrulation, but targeted deletion of either gene causes peri-implantation death, primarily due to failure in trophoblast differentiation [111, 112]. The estrogen-receptor-related receptor β (ERR-β) is expressed in a subset of extra-embryonic ectoderm cells and is involved in maintenance of trophoblast stem cells, as embryos lacking this receptor are deficient in diploid trophoblast cells, but trophoblast giant cells are overabundant and death occurs around E10.5 [113]. DP1, a member of the E2F/DP transcription factor family also seems to be required for maintenance of trophoblast-derived tissues [114]. Furthermore, genes that specifically regulate ectoplacental cone function (*Ets2*) and trophoblast giant cell differentiation (*Hand1* and *I-mfa*) have been identified by targeted deletion in mice. *Ets2*–deficient embryos implant and form normal extra-embryonic tissues, except that ectoplacental cone proliferation and differentiation are reduced leading to embryonic lethality at E8.5 [115]. *Hand1*, also known as *Hxt*, codes for a basis helix-loop-helix (bHLH) transcription factor highly expressed in trophoblast giant cells during implantation [116] and its targeted deletion results in the absence of trophoblast giant cells [117]. Similarly, on a C57BL/6 genetic background, deletion of the bHLH inhibitor *I-mfa* leads to a severe reduction in trophoblast giant cells [118]. Interestingly, both Hand1 and I-mfa seem to drive differentiation to trophoblast giant cells, while negatively regulating spongiotrophoblast cell formation. *Mash2*, another bHLH family member is necessary for the allocation of ectoplacental cone cells to the spongiotrophoblast cell lineage and in its absence this lineage is not formed, while there is an expansion of trophoblast giant cells [119]. Both Hand1 and I-mfa interfere with Mash2 activity, suggesting that the balance between these proteins may determine the formation of spongiotrophoblast or trophoblast giant cells [118, 120]. Finally, another interesting regulator of trophoblast cell fate is the oxygen level. Low levels of oxygen activate hypoxia-inducible factor–1 (HIF–1), a dimeric transcription factor composed of the bHLH-PAS protein HIF–1α and the arylhydrocarbon receptor nuclear translocator (ARNT). HIF–1 promotes differentiation into the spongiotrophoblast lineage since inactivation of *Arnt* leads to a reduction of the spongiotrophoblast layer and an expansion of the trophoblast giant cell component, with no alteration in ERR-β or Mash2 expression [121]. Thus oxygen is another important regulator of trophoblast development.

3.2.4
The Formation and Development of the Chorio-allantoic Placenta (E8.5–E16.5)

Implantation can only be considered successful if it leads to the formation of the definitive placenta that supports the conceptus to birth. The formation of the chorio-allantoic placenta starts at E8.5 with chorio-allantoic fusion (Fig. 3.2). Mutations that interfere with the proper formation of the allantois, often lead to defects in chorio-allantoic fusion and subsequent failure in the formation of the chorio-allantoic placenta (see Fig. 3.5 for a summary of the major factors involved). Therefore, mutations in genes that affect mesoderm formation in general, for example *Brachyury T*, *Bmp2*, *Bmp4*, *Bmp5/Bmp7* and *Smad1*, also exhibit defects in allantois formation and fusion [122–126]. Mutations in either *Vcam1* (vascular cell-adhesion molecule 1) or *Itga4* (α4 integrin), result in specific defects in the adhesion of the allantois (normally expresses Vcam1) to the chorion (normally expresses α4β1) in about half of mutant embryos [127–129]. Interestingly, inactivation of *Wnt7b*, normally expressed in the chorion, leads to a loss of α4 immunoreactivity in the chorion and a subsequent failure in chorio-allantoic fusion [130], suggesting a role for autocrine Wnt signaling in regulating the adhesion properties of the chorionic trophoblast. Although the interaction between Vcam1 and α4β1 integrin is important during chorio-allantoic fusion, in half of null embryos this process proceeds normally and Vcam1 and α4β1 integrin expression is normal in other mutants with similar specific chorio-allantoic fusion defects, including deficiency in MJR, a DnaJ-related co-chaperone [131].

After close apposition of the allantois and chorion has taken place, the chorionic trophoblast and the blood vessels of the allantois form folds that subsequently undergo branching morphogenesis, much like that which occurs in the lung and

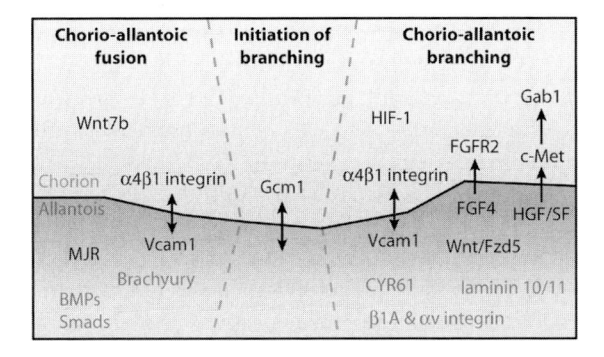

Chorio-allantoic fusion	**Initiation of branching**	**Chorio-allantoic branching**

```
                                                              Gab1
                                                               ↑
                    Wnt7b                   HIF-1
                                                      FGFR2   c-Met
Chorion    α4β1 integrin    Gcm1    α4β1 integrin      ↑       ↑
Allantois       ↕             ↓          ↑            FGF4   HGF/SF
           MJR  Vcam1                  Vcam1   Wnt/Fzd5
                Brachyury              CYR61        laminin 10/11
           BMPs
           Smads                       β1A & αv integrin
```

Fig. 3.5
Major molecular interactions involved in early chorio-allantoic placental development in the mouse. The chorio-allantoic placenta forms through the coordinated morphogenesis of the chorionic trophoblast cells and the allantois-derived fetal blood vessels. Adhesion of the allantois to the chorion, an event called chorio-allantoic fusion, occurs through the binding of Vcam1 to the α4β1 integrin and Wnt7b expression in the chorion is necessary for α4 expression to occur. However, MJR is also necessary for chorio-allantoic fusion. Furthermore, all genes involved in the development of the mesoderm during gastrulation (including Brachyury, BMPs and Smads) play a role in allantois development and are thus essential for this event to occur. After contact with the allantois, the transcription factor Gcm1, expressed in the chorionic trophoblast, plays a key role in the initiation of branching morphogenesis. The branching of the chorionic trophoblast and the invasion of the allantois-derived fetal blood vessels into the spaces generated is a complex process involving many regulatory factors (see [14] for a thorough review). Signaling events mediated by FGFs and HGF/SF secreted by the allantois and transduced by their respective receptors in the chorionic trophoblast (FGFR2 and c-Met) are essential for normal branching morphogenesis, whereas Wnt signaling through Fzd5 appears to be important for the development of the fetal blood vasculature. Labyrinthine development is also dependent on HIF−1 expression in the trophoblast as well as continuing Vcam1–α4β1 interactions. The presence of αv-containing integrins, the β1A integrin variant, CYR61 and the laminin α5 chain (present in laminin 10/11) is also important for labyrinthine development, although in these cases it is not yet clear whether these molecules play a role in trophoblast branching or in morphogenesis of the allantois-derived vasculature. All abbreviations used are explained in the text.

kidney (reviewed in [17]). Initiation of this branching and the subsequent differentiation of the trophoblast cells into syncytiotrophoblast, are dependent on the presence of the transcription factor Gcm1 (glial-cells missing−1) in the chorionic trophoblast cells [132, 133]. Interestingly, Gcm1 is expressed in the chorion before the allantois makes contact, but branching is only initiated after contact. This suggests that a signaling event initiated through chorio-allantoic contact is necessary for maintaining (or complementing) Gcm1 expression, which then regulates the initiation of branching and trophoblast differentiation (Fig. 3.5).

Although Gcm1 is essential for the initiation of chorio-allantoic morphogenesis, a number of other molecules and signaling pathways are necessary for successful branching morphogenesis and the development of the labyrinth, those of major importance being summarized in Fig. 3.5. The best characterized are the FGF, HGF/SF (hepatocyte growth factor/scatter factor), and Wnt pathways (reviewed in [17]). As described above, FGF signaling plays a role in normal trophoblast cell proliferation. *Fgfr2*–null mouse embryos show an inhibition in the expansion of this cell population and consequently have a severely underdeveloped labyrinth [134]. FGFR2 signaling could also be important for branching since lung branching is impaired in *Fgfr2* mutants [135]. Inactivation of *Hgf/Sf*, its receptor *c-Met*, or *Gab1* a downstream effector of c-Met, also has dramatic effects on trophoblast branching in the labyrinth [136–138]. Hypoxia is also a regulator of labyrinth development, since *Arnt* deficient placentae show a reduction in this layer, the primary defect being in the trophoblast component [121]. Thus, in all these cases labyrinth development seems to be impaired due to a failure in trophoblast development. In contrast, embryos with disruptions in Wnt signaling that lead to an abnormal phenotype in labyrinth development display an apparently normal trophoblast component but impairment in the development of the allantois-derived embryonic vasculature instead [139, 140].

Cell adhesion is also important during labyrinth development. The *Vcam1*–null embryos that undergo normal chorio-allantoic fusion show a defect in the vascular development of the labyrinth [127] suggesting a role for Vcam1–α4β1 integrin interactions at these stages as well. Inactivation of *Itgav* (αv integrin) leads to abnormal labyrinth development and mid-gestation lethality in 80 % of null embryos [141]. Inactivation of *Gja7* (connexin45), *Gjb2* (connexin26), *Lama5* (laminin α5 chain, present in laminin 10/11) or *Cyr61*, an ECM/integrin-associated protein also reduces labyrinth branching [142–144] or nutrient transport [145]. Finally, replacement of the β1A integrin subunit by the striated muscle-specific β1D variant *in vivo* leads to impaired labyrinth development in 66 % of placentae [146]. Further studies should address whether it is the trophoblast or vascular compartment that is affected in these integrin/ECM mutants. Only then will it be possible to hypothesize how transcription factors and growth factor signaling collaborate with cell adhesion events to produce the materno-fetal interface.

3.3
Clinical Relevance in Humans

Furthering our understanding of mouse implantation and placental development, as described in the above sections, should provide novel insights into human implantation and placental function and provide new strategies for improving the rates successful pregnancy. For example, LIF receptors are expressed on early human embryos [147] and the LIF mRNA concentration peaks in human endometrium at the time of implantation [148] as in the mouse, so defects in either could affect implantation. Moreover, it has been shown that cultured endometrial explants from infertile women produced less LIF than those from fertile women [149, 150] and therefore administration of LIF to infertile women may increase the chances of successful implantation. HB-EGF is also expressed in both humans and mice during the implantation window [151, 152] and soluble HB-EGF has been shown to improve the development of *in vitro* human embryos [153]. However, the IGF signaling pathway for example which regulates trophoblast invasion in humans (and primates) is not necessary for implantation in mice (reviewed in [154]). Embryos null for IGFs and the IGF type I receptor do, however, show embryonic growth restriction [155, 156]. It is therefore important to realize that although many similarities exist between implantation in mice and humans, there are also significant differences.

The regulation of components of the extracellular matrix also appears to be crucial for successful implantation both in mice and humans. Aberrant expression of αvβ3 integrin was noted in women with retarded endometrial development or endometriosis, whilst the absence of α4β1 integrin, which recognizes fibronectin present on the fetal trophoblast, may result in incomplete maternal recognition of the embryo [157, 158]. Furthermore, although the expression of Muc1 is increased in the human uterus during the receptive stage [159] in contrast to mice [69], careful analysis may reveal a local reduction of Muc1 expression at the site of blastocyst attachment. If that is the case, directing the levels of Muc1 in the uterus may help control the rate of implantation in humans.

Several genes necessary for mouse placental development are expressed in an analogous manner in humans; these include homologs of *Mash2, Hand1, Gcm1, Hgf/Sf* and *c-Met* [160–164]. Although to date there is little functional data concerning these genes in human placental development, what is known is consistent with mouse studies (reviewed in [17]). Evidence from genetic studies in mice indicates that different aspects of placental development are controlled by distinct genetic pathways. For example mutants affected in villi/labyrinth morphogenesis are not usually affected in invasive trophoblast (giant cell) differentiation. Likewise missed abortion (pregnancy losses in the first 2 months of gestation), some types of intrauterine growth retardation and pre-eclampsia (acute toxemia at the time of birth) are often considered as part of a spectrum but in fact they may be distinct pathologies affecting different placental components. Pre-eclampsia is often associated with defective differentiation and invasion of extravillous cytotrophoblast cells [165], while in missed abortion [166–168] and severe cases of early onset intrauterine growth retardation [169] the placenta is characterized by reduced branching of the chorionic villi and underlying vasculature. Compared to the mouse, placental tissues in humans are highly invasive and have a much greater potential to become malignant so that implantation in humans is a strictly controlled process in space and time. Any disequilibrium between the factors favoring the invasive properties of the trophoblast and factors limiting its invasion can lead to pathological states. Placental tissue remaining after birth can result in the formation of highly malignant tumors. Uncontrolled invasion is characteristic of ectopic pregnancies, placenta accreta, hydatiform moles and choriocarcinomas. Insufficient invasion may be implicated in severe hypertensive diseases such as pre-eclampsia, as described above.

It is clear that there are many common, yet unresolved, complications of human pregnancy that would benefit from a better understanding of cellular assignment and the underlying cause of implantation and placental failure in mice. Once the diseases are classified by cellular criteria, it will be much easier to define the molecular pathways causing the defects and perhaps devise strategies based on immunological intervention to control pregnancy loss and pre- and postnatal pathological conditions.

3.4
Summary

Implantation is a complex process involving developmental changes in both the implanting embryo and the receiving uterus. These changes must be synchronized in space and time for successful pregnancy to occur. Thus the preparation of the embryo and the development of uterine receptivity should occur within a window of time that permits successful communication and contact between the two. After initial contact occurs, both sides collaborate to form the materno-fetal interface that permits the survival of the embryo and development to term. Implantation and placentation vary considerably among the different groups of mammals and, therefore, it is not possible to put forth a summary of events that applies to all mammals. However, gene targeting studies in mice have (often unexpectedly) revealed implantation and placentation phenotypes in a long list of gene ablations. A careful analysis of these phenotypes, coupled with physiological and cell biological studies have therefore put the mouse at center stage in implantation research. These studies have highlighted the involvement of numerous transcription factors, members of several growth factor families (and their respective receptors) as well as cell-cell and cell-ECM adhesion molecules although an understanding of the causal relationship between all these factors is still far from clear. In this chapter we review the different phases of implantation in the mouse, including the preparation of the pre-implantation embryo and the establishment of uterine receptivity, initial contact and invasion of the embryo into the uterine environment, and the progressive establishment of the functional materno-fetal interface, the chorio-allantoic placenta. We highlight the major factors and signaling events involved in these different steps, showing the variety of molecular interactions. We also show that the mouse is a useful model for studying human implantation and in particular placentation. In some cases, human homologs of key mouse genes exist and are expressed in an analogous manner. In other cases, the careful analysis of certain mouse implantation and placental phenotypes have revealed that disorders previously grouped into the same category are in fact caused by the malfunctioning of very different genes, processes or cell populations, and have thus proven useful in the understanding of human disorders as well. Further understanding of mouse implantation and placentation is thus not only interesting from the point of view of cell and developmental biology, but also has clear implications for human reproductive biology.

Acknowledgments

We are grateful to J. Korving, M. van Rooijen and A. Brouwer for technical assistance and B. A. J. Roelen for useful comments on the manuscript. S. M. C. S. L. was supported by Fundação para a Ciência e Tecnologia, Portugal and FEDER (SFRH/BD/827/2000).

References

1 Jungnickel, M. K., Sutton, K. A. and Florman, H. M. In the beginning: lessons from fertilization in mice and worms. *Cell* 2003, *114*, 401–404.
2 Fleming, T. P., Sheth, B. and Fesenko, I. Cell adhesion in the preimplantation mammalian embryo and its role in trophectoderm differentiation and blastocyst morphogenesis. *Front Biosci* 2001, *6*, D1000–D1007.
3 Sutherland, A. Mechanisms of implantation in the mouse: differentiation and functional importance of trophoblast giant cell behavior. *Dev Biol* 2003, *258*, 241–251.
4 Fleming, T. P. A quantitative analysis of cell allocation to trophectoderm and inner cell mass in the mouse blastocyst. *Dev Biol* 1987, *119*, 520–531.
5 Perona, R. M. and Wassarman, P. M. Mouse blastocysts hatch *in vitro* by using a trypsin-like proteinase associated with cells of mural trophectoderm. *Dev Biol* 1986, *114*, 42–52.
6 O'Sullivan, C. M., Liu, S. Y., Karpinka, J. B. and Rancourt, D. E. Embryonic hatching enzyme strypsin/ISP1 is expressed with ISP2 in endometrial glands during implantation. *Mol Reprod Dev* 2002, *62*, 328–334.
7 Carson, D. D., Bagchi, I., Dey, S. K., Enders, A. C., Fazleabas, A. T., Lessey, B. A. and Yoshinaga, K. Embryo implantation. *Dev Biol* 2000, *223*, 217–237.
8 Psychoyos, A., Greep, R. O., Astwood, E. G., Geiger, S. R. Endocrine control of egg implantation. In *Handbook of Physiology*, Greep R. O., Astwood, E. G., Geiger, S. R. (eds). American Physiological Society: Washington, 1973, 187–215.
9 Bevilacqua, E. M. and Abrahamsohn, P. A. Trophoblast invasion during implantation of the mouse embryo. *Arch Biol Med Exp (Santiago)* 1989, *22*, 107–118.
10 Rassoulzadegan, M., Rosen, B. S., Gillot, I. and Cuzin, F. Phagocytosis reveals a reversible differentiated state early in the development of the mouse embryo. *EMBO J* 2000, *19*, 3295–3303.
11 Ansell, J. D., Barlow, P. W. and McLaren, A. Binucleate and polyploid cells in the decidua of the mouse. *J Embryol Exp Morphol* 1974, *31*, 223–227.
12 Bevilacqua, E. M. and Abrahamsohn, P. A. Ultrastructure of trophoblast giant cell transformation during the invasive stage of implantation of the mouse embryo. *J Morphol* 1988, *198*, 341–351.

13 Gardner, R. L. Origin and differentiation of extra-embryonic tissues in the mouse. *Int Rev Exp Pathol* **1983**, *24*, 63–133.

14 Beddington, R. S. and Robertson, E. J. Axis development and early asymmetry in mammals. *Cell* **1999**, *96*, 195–209.

15 Medvinsky, A. L., Samoylina, N. L., Muller, A. M. and Dzierzak, E. A. An early pre-liver intraembryonic source of CFU-S in the developing mouse. *Nature* **1993**, *364*, 64–67.

16 Baron, M. H. Embryonic origins of mammalian hematopoiesis. *Exp Hematol* **2003**, *31*, 1160–1169.

17 Rossant, J. and Cross, J. C. Placental development: lessons from mouse mutants. *Nature Rev Genet* **2001**, *2*, 538–548.

18 Adamson, S. L., Lu, Y., Whiteley, K. J., Holmyard, D., Hemberger, M., Pfarrer, C. and Cross, J. C. Interactions between trophoblast cells and the maternal and fetal circulation in the mouse placenta. *Dev Biol* **2002**, *250*, 358–373.

19 Redline, R. W. and Lu, C. Y. Localization of fetal major histocompatibility complex antigens and maternal leukocytes in murine placenta. Implications for maternal-fetal immunological relationship. *Lab Invest* **1989**, *61*, 27–36.

20 Kallapur, S., Ormsby, I. and Doetschman, T. Strain dependency of TGFβ1 function during embryogenesis. *Mol Reprod Dev* **1999**, *52*, 341–349.

21 Pollard, J. W. Role of colony-stimulating factor–1 in reproduction and development. *Mol Reprod Dev* **1997**, *46*, 54–60; discussion 60–61.

22 Robertson, S. A., Sjoblom, C., Jasper, M. J., Norman, R. J. and Seamark, R. F. Granulocyte-macrophage colony-stimulating factor promotes glucose transport and blastomere viability in murine preimplantation embryos. *Biol Reprod* **2001**, *64*, 1206–1215.

23 Brison, D. R. and Schultz, R. M. Apoptosis during mouse blastocyst formation: evidence for a role for survival factors including transforming growth factor α. *Biol Reprod* **1997**, *56*, 1088–1096.

24 Hardy, K. and Spanos, S. Growth factor expression and function in the human and mouse preimplantation embryo. *J Endocrinol* **2002**, *172*, 221–236.

25 Chai, N., Patel, Y., Jacobson, K., McMahon, J., McMahon, A. and Rappolee, D. A. FGF is an essential regulator of the fifth cell division in preimplantation mouse embryos. *Dev Biol* **1998**, *198*, 105–115.

26 Roelen, B. A., Goumans, M. J., Zwijsen, A. and Mummery, C. L. Identification of two distinct functions for TGFβ in early mouse development. *Differentiation* **1998**, *64*, 19–31.

27 Larue, L., Ohsugi, M., Hirchenhain, J. and Kemler, R. E-cadherin null mutant embryos fail to form a trophectoderm epithelium. *Proc Natl Acad Sci USA* **1994**, *91*, 8263–8267.

28 Riethmacher, D., Brinkmann, V. and Birchmeier, C. A targeted mutation in the mouse E-cadherin gene results in defective preimplantation development. *Proc Natl Acad Sci USA* **1995**, *92*, 855–859.

29 Torres, M., Stoykova, A., Huber, O., Chowdhury, K., Bonaldo, P., Mansouri, A., Butz, S., Kemler, R. and Gruss, P. An α-E-

catenin gene trap mutation defines its function in preimplantation development. *Proc Natl Acad Sci USA* **1997**, *94*, 901–906.

30 Clayton, L., Hall, A. and Johnson, M. H. A role for Rho-like GTPases in the polarisation of mouse eight-cell blastomeres. *Dev Biol* **1999**, *205*, 322–331.

31 Houghton, F. D., Barr, K. J., Walter, G., Gabriel, H. D., Grummer, R., Traub, O., Leese, H. J., Winterhager, E. and Kidder, G. M. Functional significance of gap junctional coupling in preimplantation development. *Biol Reprod* **2002**, *66*, 1403–1412.

32 Kidder, G. M. and Winterhager, E. Intercellular communication in preimplantation development: the role of gap junctions. *Front Biosci* **2001**, *6*, D731–D736.

33 Collins, J. E. and Fleming, T. P. Epithelial differentiation in the mouse preimplantation embryo: making adhesive cell contacts for the first time. *Trends Biochem Sci* **1995**, *20*, 307–312.

34 Leivo, I., Vaheri, A., Timpl, R. and Wartiovaara, J. Appearance and distribution of collagens and laminin in the early mouse embryo. *Dev Biol* **1980**, *76*, 100–114.

35 Thorsteinsdóttir, S. Basement membrane and fibronectin matrix are distinct entities in the developing mouse blastocyst. *Anat Rec* **1992**, *232*, 141–149.

36 Smyth, N., Vatansever, H. S., Murray, P., Meyer, M., Frie, C., Paulsson, M. and Edgar, D. Absence of basement membranes after targeting the *LAMC1* gene results in embryonic lethality due to failure of endoderm differentiation. *J Cell Biol* **1999**, *144*, 151–160.

37 Nichols, J., Zevnik, B., Anastassiadis, K., Niwa, H., Klewe-Nebenius, D., Chambers, I., Scholer, H. and Smith, A. Formation of pluripotent stem cells in the mammalian embryo depends on the POU transcription factor Oct4. *Cell* **1998**, *95*, 379–391.

38 Mitsui, K., Tokuzawa, Y., Itoh, H., Segawa, K., Murakami, M., Takahashi, K., Maruyama, M., Maeda, M. and Yamanaka, S. The homeoprotein Nanog is required for maintenance of pluripotency in mouse epiblast and ES cells. *Cell* **2003**, *113*, 631–642.

39 Chambers, I., Colby, D., Robertson, M., Nichols, J., Lee, S., Tweedie, S. and Smith, A. Functional expression cloning of Nanog, a pluripotency sustaining factor in embryonic stem cells. *Cell* **2003**, *113*, 643–655.

40 Feldman, B., Poueymirou, W., Papaioannou, V. E., DeChiara, T. M. and Goldfarb, M. Requirement of FGF-4 for postimplantation mouse development. *Science* **1995**, *267*, 246–249.

41 Cross, J. C., Werb, Z. and Fisher, S. J. Implantation and the placenta: key pieces of the development puzzle. *Science* **1994**, *266*, 1508–1518.

42 Slager, H. G., Lawson, K. A., van den Eijnden-van Raaij, A. J., de Laat, S. W. and Mummery, C. L. Differential localization of TGFβ2 in mouse preimplantation and early postimplantation development. *Dev Biol* **1991**, *145*, 205–218.

43 Paria, B. C., Song, H. and Dey, S. K. Implantation: molecular basis of embryo-uterine dialogue. *Int J Dev Biol* **2001**, *45*, 597–605.

44 Paria, B. C., Reese, J., Das, S. K. and Dey, S. K. Deciphering the cross-talk of implantation: advances and challenges. *Science* **2002**, *296*, 2185–2188.

45 Ormandy, C. J., Camus, A., Barra, J., Damotte, D., Lucas, B., Buteau, H., Edery, M., Brousse, N., Babinet, C., Binart, N. and Kelly, P. A. Null mutation of the prolactin receptor gene produces multiple reproductive defects in the mouse. *Genes Dev* **1997**, *11*, 167–178.

46 Grosdemouge, I., Bachelot, A., Lucas, A., Baran, N., Kelly, P. A. and Binart, N. Effects of deletion of the prolactin receptor on ovarian gene expression. *Reprod Biol Endocrinol* **2003**, *1*, 12.

47 Das, S. K., Wang, X. N., Paria, B. C., Damm, D., Abraham, J. A., Klagsbrun, M., Andrews, G. K. and Dey, S. K. Heparin-binding EGF-like growth factor gene is induced in the mouse uterus temporally by the blastocyst solely at the site of its apposition: a possible ligand for interaction with blastocyst EGF-receptor in implantation. *Development* **1994**, *120*, 1071–1083.

48 Stewart, C. L., Kaspar, P., Brunet, L. J., Bhatt, H., Gadi, I., Kontgen, F. and Abbondanzo, S. J. Blastocyst implantation depends on maternal expression of leukaemia inhibitory factor. *Nature* **1992**, *359*, 76–79.

49 Ware, C. B., Horowitz, M. C., Renshaw, B. R., Hunt, J. S., Liggitt, D., Koblar, S. A., Gliniak, B. C., McKenna, H. J., Papayannopoulou, T., Thoma, B. et al. Targeted disruption of the low-affinity leukemia inhibitory factor receptor gene causes placental, skeletal, neural and metabolic defects and results in perinatal death. *Development* **1995**, *121*, 1283–1299.

50 Yoshida, K., Taga, T., Saito, M., Suematsu, S., Kumanogoh, A., Tanaka, T., Fujiwara, H., Hirata, M., Yamagami, T., Nakahata, T., et al. Targeted disruption of gp130, a common signal transducer for the interleukin 6 family of cytokines, leads to myocardial and hematological disorders. *Proc Natl Acad Sci USA* **1996**, *93*, 407–411.

51 Ernst, M., Inglese, M., Waring, P., Campbell, I. K., Bao, S., Clay, F. J., Alexander, W. S., Wicks, I. P., Tarlinton, D. M., Novak, U., et al. Defective gp130–mediated signal transducer and activator of transcription (STAT) signaling results in degenerative joint disease, gastrointestinal ulceration, and failure of uterine implantation. *J Exp Med* **2001**, *194*, 189–203.

52 Song, H., Lim, H., Das, S. K., Paria, B. C. and Dey, S. K. Dysregulation of EGF family of growth factors and COX–2 in the uterus during the preattachment and attachment reactions of the blastocyst with the luminal epithelium correlates with implantation failure in LIF-deficient mice. *Mol Endocrinol* **2000**, *14*, 1147–1161.

53 Chakraborty, I., Das, S. K., Wang, J. and Dey, S. K. Developmental expression of the cyclo-oxygenase–1 and cyclo-oxygenase–2 genes in the peri-implantation mouse uterus and their differential regulation by the blastocyst and ovarian steroids. *J Mol Endocrinol* **1996**, *16*, 107–122.

54 Lim, H., Paria, B. C., Das, S. K., Dinchuk, J. E., Langenbach, R., Trzaskos, J. M. and Dey, S. K. Multiple female reproductive failures in cyclooxygenase 2–deficient mice. *Cell* **1997**, *91*, 197–208.

55 Smith, W. L. and Dewitt, D. L. Prostaglandin endoperoxide H synthases–1 and –2. *Adv Immunol* **1996**, *62*, 167–215.
56 Lim, H., Gupta, R. A., Ma, W. G., Paria, B. C., Moller, D. E., Morrow, J. D., DuBois, R. N., Trzaskos, J. M. and Dey, S. K. Cyclo-oxygenase-2–derived prostacyclin mediates embryo implantation in the mouse via PPARδ. *Genes Dev* **1999**, *13*, 1561–1574.
57 Kapur, S., Tamada, H., Dey, S. K. and Andrews, G. K. Expression of insulin-like growth factor-I (IGF-I) and its receptor in the peri-implantation mouse uterus, and cell-specific regulation of IGF-I gene expression by estradiol and progesterone. *Biol Reprod* **1992**, *46*, 208–219.
58 Paria, B. C., Ma, W., Tan, J., Raja, S., Das, S. K., Dey, S. K. and Hogan, B. L. Cellular and molecular responses of the uterus to embryo implantation can be elicited by locally applied growth factors. *Proc Natl Acad Sci USA* **2001**, *98*, 1047–1052.
59 Pfendler, K. C., Yoon, J., Taborn, G. U., Kuehn, M. R. and Iannaccone, P. M. *Nodal and Bone morphogenetic protein 5* interact in murine mesoderm formation and implantation. *Genesis* **2000**, *28*, 1–14.
60 Benson, G. V., Lim, H., Paria, B. C., Satokata, I., Dey, S. K. and Maas, R. L. Mechanisms of reduced fertility in *Hoxa–10* mutant mice: uterine homeosis and loss of maternal *Hoxa–10* expression. *Development* **1996**, *122*, 2687–2696.
61 Lim, H., Ma, L., Ma, W. G., Maas, R. L. and Dey, S. K. *Hoxa–10* regulates uterine stromal cell responsiveness to progesterone during implantation and decidualization in the mouse. *Mol Endocrinol* **1999**, *13*, 1005–1017.
62 Hsieh-Li, H. M., Witte, D. P., Weinstein, M., Branford, W., Li, H., Small, K. and Potter, S. S. *Hoxa 11* structure, extensive antisense transcription, and function in male and female fertility. *Development* **1995**, *121*, 1373–1385.
63 Gendron, R. L., Paradis, H., Hsieh-Li, H. M., Lee, D. W., Potter, S. S. and Markoff, E. Abnormal uterine stromal and glandular function associated with maternal reproductive defects in *Hoxa–11* null mice. *Biol Reprod* **1997**, *56*, 1097–1105.
64 Wang, W., Van De Water, T. and Lufkin, T. Inner ear and maternal reproductive defects in mice lacking the *Hmx3* homeobox gene. *Development* **1998**, *125*, 621–634.
65 Paria, B. C., Song, H., Wang, X., Schmid, P. C., Krebsbach, R. J., Schmid, H. H., Bonner, T. I., Zimmer, A. and Dey, S. K. Dysregulated cannabinoid signaling disrupts uterine receptivity for embryo implantation. *J Biol Chem* **2001**, *276*, 20523–20528.
66 Wang, J., Paria, B. C., Dey, S. K. and Armant, D. R. Stage-specific excitation of cannabinoid receptor exhibits differential effects on mouse embryonic development. *Biol Reprod* **1999**, *60*, 839–844.
67 DeSouza, M. M., Surveyor, G. A., Price, R. E., Julian, J., Kardon, R., Zhou, X., Gendler, S., Hilkens, J. and Carson, D. D. MUC1/episialin: a critical barrier in the female reproductive tract. *J Reprod Immunol* **1999**, *45*, 127–158.
68 Hilkens, J., Ligtenberg, M. J., Vos, H. L. and Litvinov, S. V. Cell membrane-associated mucins and their adhesion-modulating property. *Trends Biochem Sci* **1992**, *17*, 359–363.

69 Surveyor, G. A., Gendler, S. J., Pemberton, L., Das, S. K., Chakraborty, I., Julian, J., Pimental, R. A., Wegner, C. C., Dey, S. K. and Carson, D. D. Expression and steroid hormonal control of Muc–1 in the mouse uterus. *Endocrinology* **1995**, *136*, 3639–3647.

70 Idris, N. and Carraway, K. L. Sialomucin complex (Muc4) expression in the rat female reproductive tract. *Biol Reprod* **1999**, *61*, 1431–1438.

71 Carraway, K. L. and Idris, N. Regulation of sialomucin complex/Muc4 in the female rat reproductive tract. *Biochem Soc Trans* **2001**, *29*, 162–166.

72 Costell, M., Gustafsson, E., Aszodi, A., Morgelin, M., Bloch, W., Hunziker, E., Addicks, K., Timpl, R. and Fässler, R. Perlecan maintains the integrity of cartilage and some basement membranes. *J Cell Biol* **1999**, *147*, 1109–1122.

73 Arikawa-Hirasawa, E., Watanabe, H., Takami, H., Hassell, J. R. and Yamada, Y. Perlecan is essential for cartilage and cephalic development. *Nature Genet* **1999**, *23*, 354–358.

74 Sutherland, A. E., Calarco, P. G. and Damsky, C. H. Developmental regulation of integrin expression at the time of implantation in the mouse embryo. *Development* **1993**, *119*, 1175–1186.

75 Schultz, J. F. and Armant, D. R. β1– and β3–class integrins mediate fibronectin binding activity at the surface of developing mouse peri-implantation blastocysts. Regulation by ligand-induced mobilization of stored receptor. *J Biol Chem* **1995**, *270*, 11522–11531.

76 Klaffky, E., Williams, R., Yao, C. C., Ziober, B., Kramer, R. and Sutherland, A. Trophoblast-specific expression and function of the integrin α7 subunit in the peri-implantation mouse embryo. *Dev Biol* **2001**, *239*, 161–175.

77 Mayer, U., Saher, G., Fässler, R., Bornemann, A., Echtermeyer, F., von der Mark, H., Miosge, N., Poschl, E. and von der Mark, K. Absence of integrin α7 causes a novel form of muscular dystrophy. *Nature Genet* **1997**, *17*, 318–323.

78 Johnson, G. A., Burghardt, R. C., Bazer, F. W. and Spencer, T. E. Osteopontin: roles in implantation and placentation. *Biol Reprod* **2003**, *69*, 1458–1471.

79 Yang, J. T., Rayburn, H. and Hynes, R. O. Embryonic mesodermal defects in α5 integrin-deficient mice. *Development* **1993**, *119*, 1093–1105.

80 Strickland, S., Reich, E. and Sherman, M. I. Plasminogen activator in early embryogenesis: enzyme production by trophoblast and parietal endoderm. *Cell* **1976**, *9*, 231–240.

81 Brenner, C. A., Adler, R. R., Rappolee, D. A., Pedersen, R. A. and Werb, Z. Genes for extracellular-matrix-degrading metalloproteinases and their inhibitor, TIMP, are expressed during early mammalian development. *Genes Dev* **1989**, *3*, 848–859.

82 White, J. M. ADAMs: modulators of cell-cell and cell-matrix interactions. *Curr Opin Cell Biol* **2003**, *15*, 598–606.

83 Tanaka, S., Kunath, T., Hadjantonakis, A. K., Nagy, A. and Rossant, J. Promotion of trophoblast stem cell proliferation by FGF4. *Science* **1998**, *282*, 2072–2075.

84 Gardner, R. L. and Johnson, M. H. An investigation of inner cell mass and trophoblast tissues following their isolation from the mouse blastocyst. *J Embryol Exp Morphol* **1972**, *28*, 279–312.
85 Ilgren, E. B. Review article: control of trophoblastic growth. *Placenta* **1983**, *4*, 307–328.
86 Duc-Goiran, P., Mignot, T. M., Bourgeois, C. and Ferre, F. Embryo-maternal interactions at the implantation site: a delicate equilibrium. *Eur J Obstet Gynecol Reprod Biol* **1999**, *83*, 85–100.
87 Munn, D. H., Zhou, M., Attwood, J. T., Bondarev, I., Conway, S. J., Marshall, B., Brown, C. and Mellor, A. L. Prevention of allogeneic fetal rejection by tryptophan catabolism. *Science* **1998**, *281*, 1191–1193.
88 Moffett, J. R. and Namboodiri, M. A. Tryptophan and the immune response. *Immunol Cell Biol* **2003**, *81*, 247–265.
89 Tafuri, A., Alferink, J., Möller, P., Hämmerling, G. J. and Arnold, B. T cell awareness of paternal alloantigens during pregnancy. *Science* **1995**, *270*, 630–633.
90 Tan, Y., Li, M., Cox, S., Davis, M. K., Tawfik, O., Paria, B. C. and Das, S. K. HB-EGF directs stromal cell polyploidy and decidualization via cyclin D3 during implantation. *Dev Biol* **2004**, *265*, 181–195.
91 Tan, J., Rajá, S., Davis, M. K., Tawfik, O., Dey, S. K. and Das, S. K. Evidence for coordinated interaction of cyclin D3 with p21 and cdk6 in directing the development of uterine stromal cell decidualization and polyploidy during implantation. *Mech Dev* **2002**, *111*, 99–113.
92 Bilinski, P., Roopenian, D. and Gossler, A. Maternal IL–11Rα function is required for normal decidua and fetoplacental development in mice. *Genes Dev* **1998**, *12*, 2234–2243.
93 Robb, L., Li, R., Hartley, L., Nandurkar, H. H., Koentgen, F. and Begley, C. G. Infertility in female mice lacking the receptor for interleukin 11 is due to a defective uterine response to implantation. *Nature Med* **1998**, *4*, 303–308.
94 Paria, B. C., Tan, J., Lubahn, D. B., Dey, S. K. and Das, S. K. Uterine decidual response occurs in estrogen receptor-α-deficient mice. *Endocrinology* **1999**, *140*, 2704–2710.
95 Wang, X., Matsumoto, H., Zhao, X., Das, S. K. and Paria, B. C. Embryonic signals direct the formation of tight junctional permeability barrier in the decidualizing stroma during embryo implantation. *J Cell Sci* **2004**, *117*, 53–62.
96 Stevens, L. C. The development of transplantable teratocarcinomas from intratesticular grafts of pre- and postimplantation mouse embryos. *Dev Biol* **1970**, *21*, 364–382.
97 Martin, P. M., Sutherland, A. E. and Van Winkle, L. J. Amino acid transport regulates blastocyst implantation. *Biol Reprod* **2003**, *69*, 1101–1108.
98 Damjanov, I., Damjanov, A. and Damsky, C. H. Developmentally regulated expression of the cell-cell adhesion glycoprotein cell-CAM 120/80 in peri-implantation mouse embryos and extraembryonic membranes. *Dev Biol* **1986**, *116*, 194–202.
99 Nose, A. and Takeichi, M. A novel cadherin cell adhesion molecule: its expression patterns associated with implantation and

organogenesis of mouse embryos. *J Cell Biol* **1986**, *103*, 2649–2658.

100 Parast, M. M., Aeder, S. and Sutherland, A. E. Trophoblast giant-cell differentiation involves changes in cytoskeleton and cell motility. *Dev Biol* **2001**, *230*, 43–60.

101 Batlle, E., Sancho, E., Franci, C., Dominguez, D., Monfar, M., Baulida, J. and Garcia De Herreros, A. The transcription factor Snail is a repressor of *E-cadherin* gene expression in epithelial tumour cells. *Nature Cell Biol* **2000**, *2*, 84–89.

102 Cano, A., Perez-Moreno, M. A., Rodrigo, I., Locascio, A., Blanco, M. J., del Barrio, M. G., Portillo, F. and Nieto, M. A. The transcription factor Snail controls epithelial-mesenchymal transitions by repressing E-cadherin expression. *Nature Cell Biol* **2000**, *2*, 76–83.

103 Bevilacqua, E. M., Katz, S. and Abrahamsohn, P. A. Contact between trophoblast and antimesometrial decidual cells in the mouse. *Microsc Electron Biol Celular* **1985**, *9*, 45–49.

104 Stephens, L. E., Sutherland, A. E., Klimanskaya, I. V., Andrieux, A., Meneses, J., Pedersen, R. A. and Damsky, C. H. Deletion of β1 integrins in mice results in inner cell mass failure and peri-implantation lethality. *Genes Dev* **1995**, *9*, 1883–1895.

105 Fässler, R. and Meyer, M. Consequences of lack of β1 integrin gene expression in mice. *Genes Dev* **1995**, *9*, 1896–1908.

106 Yotsumoto, S., Fujiwara, H., Horton, J. H., Mosby, T. A., Wang, X., Cui, Y. and Ko, M. S. Cloning and expression analyses of mouse dystroglycan gene: specific expression in maternal decidua at the peri-implantation stage. *Hum Mol Genet* **1996**, *5*, 1259–1267.

107 Williamson, R. A., Henry, M. D., Daniels, K. J., Hrstka, R. F., Lee, J. C., Sunada, Y., Ibraghimov-Beskrovnaya, O. and Campbell, K. P. Dystroglycan is essential for early embryonic development: disruption of Reichert's membrane in *Dag1*–null mice. *Hum Mol Genet* **1997**, *6*, 831–841.

108 Rappolee, D. A., Basilico, C., Patel, Y. and Werb, Z. Expression and function of FGF–4 in peri-implantation development in mouse embryos. *Development* **1994**, *120*, 2259–2269.

109 Genersch, E., Ferletta, M., Virtanen, I., Haller, H. and Ekblom, P. Integrin αvβ3 binding to human α5–laminins facilitates FGF–2– and VEGF-induced proliferation of human ECV304 carcinoma cells. *Eur J Cell Biol* **2003**, *82*, 105–117.

110 Arman, E., Haffner-Krausz, R., Chen, Y., Heath, J. K. and Lonai, P. Targeted disruption of fibroblast growth factor (FGF) receptor 2 suggests a role for FGF βsignaling in pregastrulation mammalian development. *Proc Natl Acad Sci USA* **1998**, *95*, 5082–5087.

111 Russ, A. P., Wattler, S., Colledge, W. H., Aparicio, S. A., Carlton, M. B., Pearce, J. J., Barton, S. C., Surani, M. A., Ryan, K., Nehls, M. C., et al. Eomesodermin is required for mouse trophoblast development and mesoderm formation. *Nature* **2000**, *404*, 95–99.

112 Chawengsaksophak, K., James, R., Hammond, V. E., Kontgen, F. and Beck, F. Homeosis and intestinal tumours in *cdx2* mutant mice. *Nature* **1997**, *386*, 84–87.

113 Luo, J., Sladek, R., Bader, J. A., Matthyssen, A., Rossant, J. and Giguere, V. Placental abnormalities in mouse embryos lacking the orphan nuclear receptor ERRβ. *Nature* **1997**, *388*, 778–782.
114 Kohn, M. J., Bronson, R. T., Harlow, E., Dyson, N. J. and Yamasaki, L. *Dp1* is required for extra-embryonic development. *Development* **2003**, *130*, 1295–1305.
115 Yamamoto, H., Flannery, M. L., Kupriyanov, S., Pearce, J., McKercher, S. R., Henkel, G. W., Maki, R. A., Werb, Z. and Oshima, R. G. Defective trophoblast function in mice with a targeted mutation of Ets2. *Genes Dev* **1998**, *12*, 1315–1326.
116 Cross, J. C., Flannery, M. L., Blanar, M. A., Steingrimsson, E., Jenkins, N. A., Copeland, N. G., Rutter, W. J. and Werb, Z. *Hxt* encodes a basic helix-loop-helix transcription factor that regulates trophoblast cell development. *Development* **1995**, *121*, 2513–2523.
117 Riley, P., Anson-Cartwright, L. and Cross, J. C. The Hand1 bHLH transcription factor is essential for placentation and cardiac morphogenesis. *Nature Genet* **1998**, *18*, 271–275.
118 Kraut, N., Snider, L., Chen, C. M., Tapscott, S. J. and Groudine, M. Requirement of the mouse *I-mfa* gene for placental development and skeletal patterning. *EMBO J* **1998**, *17*, 6276–6288.
119 Guillemot, F., Nagy, A., Auerbach, A., Rossant, J. and Joyner, A. L. Essential role of Mash–2 in extraembryonic development. *Nature* **1994**, *371*, 333–336.
120 Scott, I. C., Anson-Cartwright, L., Riley, P., Reda, D. and Cross, J. C. The HAND1 basic helix-loop-helix transcription factor regulates trophoblast differentiation via multiple mechanisms. *Mol Cell Biol* **2000**, *20*, 530–541.
121 Adelman, D. M., Gertsenstein, M., Nagy, A., Simon, M. C. and Maltepe, E. Placental cell fates are regulated *in vivo* by HIF-mediated hypoxia responses. *Genes Dev* **2000**, *14*, 3191–3203.
122 Rashbass, P., Cooke, L. A., Herrmann, B. G. and Beddington, R. S. A cell autonomous function of Brachyury in T/T embryonic stem cell chimaeras. *Nature* **1991**, *353*, 348–351.
123 Ying, Y. and Zhao, G. Q. Cooperation of endoderm-derived BMP2 and extraembryonic ectoderm-derived BMP4 in primordial germ cell generation in the mouse. *Dev Biol* **2001**, *232*, 484–492.
124 Winnier, G., Blessing, M., Labosky, P. A. and Hogan, B. L. Bone morphogenetic protein–4 is required for mesoderm formation and patterning in the mouse. *Genes Dev* **1995**, *9*, 2105–2116.
125 Solloway, M. J. and Robertson, E. J. Early embryonic lethality in *Bmp5;Bmp7* double mutant mice suggests functional redundancy within the 60A subgroup. *Development* **1999**, *126*, 1753–1768.
126 Tremblay, K. D., Dunn, N. R. and Robertson, E. J. Mouse embryos lacking Smad1 signals display defects in extra-embryonic tissues and germ cell formation. *Development* **2001**, *128*, 3609–3621.
127 Gurtner, G. C., Davis, V., Li, H., McCoy, M. J., Sharpe, A. and Cybulsky, M. I. Targeted disruption of the murine *VCAM1* gene: essential role of VCAM–1 in chorioallantoic fusion and placentation. *Genes Dev* **1995**, *9*, 1–14.

128 Kwee, L., Baldwin, H. S., Shen, H. M., Stewart, C. L., Buck, C., Buck, C. A. and Labow, M. A. Defective development of the embryonic and extraembryonic circulatory systems in vascular cell adhesion molecule (VCAM–1) deficient mice. *Development* **1995**, *121*, 489–503.

129 Yang, J. T., Rayburn, H. and Hynes, R. O. Cell adhesion events mediated by α4 integrins are essential in placental and cardiac development. *Development* **1995**, *121*, 549–560.

130 Parr, B. A., Cornish, V. A., Cybulsky, M. I. and McMahon, A. P. Wnt7b regulates placental development in mice. *Dev Biol* **2001**, *237*, 324–332.

131 Hunter, P. J., Swanson, B. J., Haendel, M. A., Lyons, G. E. and Cross, J. C. Mrj encodes a DnaJ-related co-chaperone that is essential for murine placental development. *Development* **1999**, *126*, 1247–1258.

132 Anson-Cartwright, L., Dawson, K., Holmyard, D., Fisher, S. J., Lazzarini, R. A. and Cross, J. C. The glial cells missing–1 protein is essential for branching morphogenesis in the chorioallantoic placenta. *Nat Genet* **2000**, *25*, 311–314.

133 Schreiber, J., Riethmacher-Sonnenberg, E., Riethmacher, D., Tuerk, E. E., Enderich, J., Bosl, M. R. and Wegner, M. Placental failure in mice lacking the mammalian homolog of glial cells missing, GCMa. *Mol Cell Biol* **2000**, *20*, 2466–2474.

134 Xu, X., Weinstein, M., Li, C., Naski, M., Cohen, R. I., Ornitz, D. M., Leder, P. and Deng, C. Fibroblast growth factor receptor 2 (FGFR2)-mediated reciprocal regulation loop between FGF8 and FGF10 is essential for limb induction. *Development* **1998**, *125*, 753–765.

135 Arman, E., Haffner-Krausz, R., Gorivodsky, M. and Lonai, P. *Fgfr2* is required for limb outgrowth and lung-branching morphogenesis. *Proc Natl Acad Sci USA* **1999**, *96*, 11895–11899.

136 Uehara, Y., Minowa, O., Mori, C., Shiota, K., Kuno, J., Noda, T. and Kitamura, N. Placental defect and embryonic lethality in mice lacking hepatocyte growth factor/scatter factor. *Nature* **1995**, *373*, 702–705.

137 Bladt, F., Riethmacher, D., Isenmann, S., Aguzzi, A. and Birchmeier, C. Essential role for the *c-met* receptor in the migration of myogenic precursor cells into the limb bud. *Nature* **1995**, *376*, 768–771.

138 Sachs, M., Brohmann, H., Zechner, D., Muller, T., Hulsken, J., Walther, I., Schaeper, U., Birchmeier, C. and Birchmeier, W. Essential role of Gab1 for signaling by the c-Met receptor *in vivo*. *J Cell Biol* **2000**, *150*, 1375–1384.

139 Monkley, S. J., Delaney, S. J., Pennisi, D. J., Christiansen, J. H. and Wainwright, B. J. Targeted disruption of the *Wnt2* gene results in placentation defects. *Development* **1996**, *122*, 3343–3353.

140 Ishikawa, T., Tamai, Y., Zorn, A. M., Yoshida, H., Seldin, M. F., Nishikawa, S. and Taketo, M. M. Mouse Wnt receptor gene *Fzd5* is essential for yolk sac and placental angiogenesis. *Development* **2001**, *128*, 25–33.

141 Bader, B. L., Rayburn, H., Crowley, D. and Hynes, R. O. Extensive vasculogenesis, angiogenesis, and organogenesis precede lethality in mice lacking all αv integrins. *Cell* **1998**, *95*, 507–519.

142 Kruger, O., Plum, A., Kim, J. S., Winterhager, E., Maxeiner, S., Hallas, G., Kirchhoff, S., Traub, O., Lamers, W. H. and Willecke, K. Defective vascular development in connexin 45–deficient mice. *Development* **2000**, *127*, 4179–4193.

143 Miner, J. H., Cunningham, J. and Sanes, J. R. Roles for laminin in embryogenesis: exencephaly, syndactyly, and placentopathy in mice lacking the laminin α5 chain. *J Cell Biol* **1998**, *143*, 1713—1723.

144 Mo, F. E., Muntean, A. G., Chen, C. C., Stolz, D. B., Watkins, S. C. and Lau, L. F. CYR61 (CCN1) is essential for placental development and vascular integrity. *Mol Cell Biol* **2002**, *22*, 8709–8720.

145 Gabriel, H. D., Jung, D., Butzler, C., Temme, A., Traub, O., Winterhager, E. and Willecke, K. Transplacental uptake of glucose is decreased in embryonic lethal connexin26–deficient mice. *J Cell Biol* **1998**, *140*, 1453–1461.

146 Cachaço, A. S., Chuva de Sousa Lopes, S. M., Kuikman, I., Bajanca, F., Abe, K., Baudoin, C., Sonnenberg, A., Mummery, C. L. and Thorsteinsdóttir, S. Knock-in of integrin β1D affects primary but not secondary myogenesis in mice. *Development* **2003**, *130*, 1659–1671.

147 van Eijk, M. J., Mandelbaum, J., Salat-Baroux, J., Belaisch-Allart, J., Plachot, M., Junca, A. M. and Mummery, C. L. Expression of leukaemia inhibitory factor receptor subunits LIFRβ and gp130 in human oocytes and preimplantation embryos. *Mol Hum Reprod* **1996**, *2*, 355–360.

148 Charnock-Jones, D. S., Sharkey, A. M., Fenwick, P. and Smith, S. K. Leukaemia inhibitory factor mRNA concentration peaks in human endometrium at the time of implantation and the blastocyst contains mRNA for the receptor at this time. *J Reprod Fertil* **1994**, *101*, 421–426.

149 Delage, G., Moreau, J. F., Taupin, J. L., Freitas, S., Hambartsoumian, E., Olivennes, F., Fanchin, R., Letur-Konirsch, H., Frydman, R. and Chaouat, G. In vitro endometrial secretion of human interleukin for DA cells/leukaemia inhibitory factor by explant cultures from fertile and infertile women. *Hum Reprod* **1995**, *10*, 2483–2488.

150 Hambartsoumian, E. Endometrial leukemia inhibitory factor (LIF) as a possible cause of unexplained infertility and multiple failures of implantation. *Am J Reprod Immunol* **1998**, *39*, 137–143.

151 Leach, R. E., Khalifa, R., Ramirez, N. D., Das, S. K., Wang, J., Dey, S. K., Romero, R. and Armant, D. R. Multiple roles for heparin-binding epidermal growth factor-like growth factor are suggested by its cell-specific expression during the human endometrial cycle and early placentation. *J Clin Endocrinol Metab* **1999**, *84*, 3355–3363.

152 Yoo, H. J., Barlow, D. H. and Mardon, H. J. Temporal and spatial regulation of expression of heparin-binding epidermal growth factor-like growth factor in the human endometrium: a possible role in blastocyst implantation. *Dev Genet* **1997**, *21*, 102–108.

153 Martin, K. L., Barlow, D. H. and Sargent, I. L. Heparin-binding epidermal growth factor significantly improves human blas-

tocyst development and hatching in serum-free medium. *Hum Reprod* **1998**, *13*, 1645–1652.

154 Nayak, N. R. and Giudice, L. C. Comparative biology of the IGF system in endometrium, decidua, and placenta, and clinical implications for fetal growth and implantation disorders. *Placenta* **2003**, *24*, 281–296.

155 Liu, J. P., Baker, J., Perkins, A. S., Robertson, E. J. and Efstratiadis, A. Mice carrying null mutations of the genes encoding insulin-like growth factor I (*Igf–1*) and type 1 IGF receptor (*Igf1r*). *Cell* **1993**, *75*, 59–72.

156 Baker, J., Liu, J. P., Robertson, E. J. and Efstratiadis, A. Role of insulin-like growth factors in embryonic and postnatal growth. *Cell* **1993**, *75*, 73–82.

157 Nardo, L. G., Bartoloni, G., Di Mercúrio, S. and Nardo, F. Expression of $\alpha v \beta 3$ and $\alpha 4 \beta 1$ integrins throughout the putative window of implantation in a cohort of healthy fertile women. *Acta Obstet Gynecol Scand* **2002**, *81*, 753–758.

158 Skrzypczak, J., Mikolajczyk, M. and Szymanowski, K. Endometrial receptivity: expression of $\alpha 3 \beta 1$, $\alpha 4 \beta 1$ and $\alpha v \beta 1$ endometrial integrins in women with impaired fertility. *Reprod Biol* **2001**, *1*, 85–94.

159 Hey, N. A., Graham, R. A., Seif, M. W. and Aplin, J. D. The polymorphic epithelial mucin MUC1 in human endometrium is regulated with maximal expression in the implantation phase. *J Clin Endocrinol Metab* **1994**, *78*, 337–342.

160 Alders, M., Hodges, M., Hadjantonakis, A. K., Postmus, J., van Wijk, I., Bliek, J., de Meulemeester, M., Westerveld, A., Guillemot, F., Oudejans, C., Little, P. and Mannens, M. The human Achaete-Scute homologue 2 (ASCL2,HASH2) maps to chromosome 11p15.5, close to IGF2 and is expressed in extravillus trophoblasts. *Hum Mol Genet* **1997**, *6*, 859–867.

161 Janatpour, M. J., Utset, M. F., Cross, J. C., Rossant, J., Dong, J., Israel, M. A. and Fisher, S. J. A repertoire of differentially expressed transcription factors that offers insight into mechanisms of human cytotrophoblast differentiation. *Dev Genet* **1999**, *25*, 146–157.

162 Knofler, M., Meinhardt, G., Vasicek, R., Husslein, P. and Egarter, C. Molecular cloning of the human Hand1 gene/cDNA and its tissue-restricted expression in cytotrophoblastic cells and heart. *Gene* **1998**, *224*, 77–86.

163 Nait-Oumesmar, B., Copperman, A. B. and Lazzarini, R. A. Placental expression and chromosomal localization of the human *Gcm1* gene. *J Histochem Cytochem* **2000**, *48*, 915–922.

164 Somerset, D. A., Li, X. F., Afford, S., Strain, A. J., Ahmed, A., Sangha, R. K., Whittle, M. J. and Kilby, M. D. Ontogeny of hepatocyte growth factor (HGF) and its receptor (c-met) in human placenta: reduced HGF expression in intrauterine growth restriction. *Am J Pathol* **1998**, *153*, 1139–1147.

165 de Groot, C. J., O'Brien, T. J. and Taylor, R. N. Biochemical evidence of impaired trophoblastic invasion of decidual stroma in women destined to have preeclampsia. *Am J Obstet Gynecol* **1996**, *175*, 24–29.

166 Meegdes, B. H., Ingenhoes, R., Peeters, L. L. and Exalto, N. Early pregnancy wastage: relationship between chorionic vas-

cularization and embryonic development. *Fertil Steril* **1988**, *49*, 216–220.
167 Ornoy, A., Salamon-Arnon, J., Ben-Zur, Z. and Kohn, G. Placental findings in spontaneous abortions and stillbirths. *Teratology* **1981**, *24*, 243–252.
168 van Lijnschoten, G., Arends, J. W. and Geraedts, J. P. Comparison of histological features in early spontaneous and induced trisomic abortions. *Placenta* **1994**, *15*, 765–673.
169 Krebs, C., Macara, L. M., Leiser, R., Bowman, A. W., Greer, I. A. and Kingdom, J. C. Intrauterine growth restriction with absent end-diastolic flow velocity in the umbilical artery is associated with maldevelopment of the placental terminal villous tree. *Am J Obstet Gynecol* **1996**, *175*, 1534–1542.

4
Cell Movements during Early Vertebrate Morphogenesis

Andrea Münsterberg and Grant Wheeler

4.1
Introduction

As an embryo develops, the various cells and tissues have to move in relation to each other to create a three-dimensional organism. Two of the most extreme examples of these morphogenetic movements are gastrulation and neurulation. During gastrulation, the blastula-stage embryo profoundly rearranges its form. Cells from the surface of the blastula move into their correct position on the inside which sets up the future body plan. Among different vertebrates there are similarities in the underlying mechanisms of morphogenetic movements, both at the cellular and the molecular level.

Classical fate mapping experiments have shown that the blastula is made up of presumptive endoderm, mesoderm and ectoderm. After gastrulation these tissues become completely rearranged leading to an embryo with the endoderm and mesoderm on the inside and ectoderm on the outside. For these morphogenetic movements to occur the cells undergo changes in adhesion and motility, however, often there is little accompanying increase in cell number or total cell mass.

During neurulation, which closely follows and overlaps with gastrulation, the neural plate ectoderm begins to fold at the edges, rise up and fold towards the midline where the edges fuse to form the neural tube. The neural tube then descends below the epidermis. Concomitant with gastrulation and neurulation movements cells become destined to be for example dorsal or ventral mesoderm, or neural versus epidermal. Because these processes occur at the same time, morphogenesis and patterning/cell-fate choices of mesoderm and neuroectoderm cells are closely linked and the signals that control morphogenesis become integrated with those that control cell-fate specification. This chapter will concentrate on the morphogenetic processes and the embryological experiments investigating the tissues that drive these movements. We will then look at some of the molecules and signaling pathways that have been implicated in the control of these important events in vertebrate embryogenesis.

Cell Signaling and Growth Factors in Development. Edited by K. Unsicker and K. Krieglstein
Copyright © 2006 WILEY-VCH Verlag GmbH & Co. KGaA, Weinheim
ISBN 3-527-31034-7

4.2
History: Classic Experiments in the Study of Gastrulation

Even without a detailed understanding of cellular behavior during gastrulation embryologists have been fascinated by this complex process for a long time and have speculated as to what central force might be coordinating the whole event. Here we will briefly cover the emergence of embryology as a field of scientific study and highlight the history of research into gastrulation in particular. For further reading we refer to some excellent reviews that focus on the historical perspective [1, 2] and to text books on developmental biology [3, 4].

With the advent of microscopy scientists were able to begin to look at the earliest events in development. However, it was not until the 19th century that careful analyses of cell movements during embryogenesis were begun and the field of experimental embryology was born. Scientists began to describe the processes of gastrulation as early as the 1820s when it was noted that the lower white hemisphere of an amphibian egg was progressively covered by the pigmented upper half. In 1874 Haeckel [5] proposed the name "Gastrula" to describe this early stage in development. In 1888 Roux [6] began experimenting with embryos by cleaving them with hot needles to see how this would perturb their development. The first fate maps using vital dyes were published in the 1920s [1] and have been continually refined to this day using dyes, tritium and protein markers such as HRP, β-gal and GFP [7–19].

Some of the most important experiments shaping developmental biology during the 20th century were those carried out by Hans Spemann and Hilde Mangold. They demonstrated that transplantation of the dorsal lip of the blastopore to the ventral side of an amphibian blastula embryo resulted in the formation of a secondary body axis consisting of tissue derived from the host and donor [20]. This small piece of tissue, subsequently called the "organizer", is thus able to coordinate a cascade of events, including secondary inductive events that lead to the formation of a complete embryo. After the discovery of the amphibian organizer by Spemann and Mangold, experiments by Waddington in the 1950s in the chick led to the identification of Hensen's node as having similar organizer properties [21]. Transplantation experiments showed that Hensen's node could induce an ectopic axis that included neural tube and somites, which were largely host derived. In the mouse a similar region called the node has also been identified. Grafts from transgenically marked mouse node or from a node labeled with DiI, to a posterolateral location in a host embryo at the same developmental stage results in the induction of a second neural axis and the formation of ectopic somites [22]. These experiments demonstrated that the mammalian node is equivalent to Spemann's organizer and Henson's node, sharing the same general properties. The organizer/node is clearly important in orchestrating the morphogenetic movements of gastrulation and neurulation. In addition, the organizer is the source of inductive signals, which continually influence the cells undergoing movement and migration to specify different mesodermal, endodermal and neural fates.

In the second half of the 20th century, other important milestones concerning gastrulation and neurulation have included the discovery that morphogenesis is driven by differential cell affinity or adhesion [23]. The Nieuwkoop center, which induces the Spemann organizer, was identified [24]. The use of chick quail chimeras for fate mapping studies in avian embryos was pioneered by LeDouarin [25] and it was observed that shaping of the neural plate and closure of the neural groove require convergent extension movements [26]. Over the last 20 years in addition to producing detailed fate maps Keller and colleagues have also described many of the other morphogenetic movements associated with gastrulation, specifically in amphibians [27, 28]. Schoenwolf and colleagues have provided insights into the cellular basis of neurulation in the chick embryo [29] and fate maps of the mouse primitive streak embryos were provided by a series of papers [30–36]. Recent, molecular analysis has revealed that many of the same genes are expressed in the organizer of different vertebrates.

4.3
Gastrulation and Neurulation in Different Vertebrate Species

4.3.1
Amphibians: *Xenopus laevis*

The primary model organism for studying vertebrate gastrulation has been the amphibian embryo, in particular *Xenopus laevis*. The cellular mechanisms underlying *Xenopus* gastrulation have been elucidated over a number of years using very precise dye marking techniques, combined with electron microscopy, time lapse video microscopy and analysis of cell behavior in tissue explants [27, 37, 38]. The main processes include involution, the rolling in of the endoderm and mesoderm at the blastopore lip, which is driven by bottle cells, followed by tissue separation of the mesoderm from the ectoderm [39], convergent extension of the mesoderm and epiboly which is the spreading of the ectoderm as the endoderm and mesoderm move inside. At this point we would like to refer to some excellent descriptions and detailed illustrations of this process in a recent review by Ray Keller [27] and also to an assortment of developmental biology text books for an overview [3, 4, 40].

During involution the involuting marginal zone (IMZ) which is localized approximately opposite the site of sperm entry on the future dorsal side of the embryo, turns inward and back on itself. Involution begins with the formation of a local depression, the blastoporal groove. This is driven by the apical constriction at this position of the so-called bottle cells, which as a result become wedge shaped and cause an inward bending of the epithelial sheet. After this initiation event, most of the morphogenetic movements and changes of form are caused by vegetal rotation movements [27, 41]. It has been shown that in *Xenopus*, an active distortion of the vegetal cell mass leads to a dramatic expansion of the blastocoel floor and a concomitant turning around of the marginal zone. Analysis of explant slices of the gastrula has shown that cell rearrangement drives the vigorous inward surging of the vegetal region into the blastocoel. Because these movements occur in explants they are thought to be active and intrinsic to the vegetal endoderm. Thus the prospective endoderm, which was previously thought to move passively, seems to provide the main driving force for the internalization of the mesendoderm during the first half of gastrulation. Subsequently the cells move directionally towards the animal pole using the fibronectin-containing extracellular matrix, produced by the overlying blastocoel roof as a substrate [42–46]. Even though the internalizing mesendodermal cell mass is brought into contact with the blastocoel roof the two tissues remain separated by the cleft of Brachet. This is important for the movement of the two tissues past each other. The property of internalized cells to remain separated on the surface of the blastocoel roof substratum has been investigated in some detail [39, 47, 48] and some of the factors involved will be discussed below.

The eventual closure of the blastopore and the anterior-posterior extension of the body axis is brought about largely by a specific type of cell movement behavior: convergence and extension. In this process, the cells in the involuting marginal zone intercalate both radially and mediolaterally. Together, this results in a narrowing and elongation of the tissue along the future anterior-posterior body axis. These movements are active in dorsal mesodermal tissue and posterior neural tissue and drive the second half of gastrulation and neurulation. Explants of the mesodermal and neural tissues converge and extend in culture, when they are unattached to a substrate or to other parts of the embryo. Recent studies mainly in *Xenopus* and Zebrafish embryos have implicated a role for non-canonical Wnt signaling in some of these processes ([49] and see below).

As the mesendoderm becomes internalized at the blastopore the ectoderm spreads over the whole gastrula during epiboly, which involves the thinning and spreading of the blastocoel roof in all directions. Epiboly is mediated by spreading and division of cells in the superficial layer and the radial intercalation of mesenchymal cells from the deeper layers resulting in the formation of fewer layers, which occupy a larger area [37]. Epiboly starts at the animal cap at the blastula stage, before the fibronectin-containing matrix on the inner surface of the blastocoel roof is laid down [37, 50] and it is unclear what adhesive forces are involved in this early phase of radial intercalation. In the absence of an involuting IMZ the animal cap bears numerous folds, suggesting that spreading during epiboly is active and force generating [27, 42]. Continued radial intercalation results in the final stages of animal cap epiboly and extension of the IMZ, and this has been shown to be dependent on cell-matrix interactions involving integrins [51]. This process can be simulated by a recently developed computer model [52].

Gastrulation is closely linked with the process of neurulation, the emergence of the neural plate and the induction of neural tissue from the ectoderm, since both depend on the activity of the Spemann organizer. As the cells from the dorsal blastopore lip involute they signal the overlying ectoderm to form neural plate tissue. Cells overlying the dorsal mesoderm undergo shape changes, they become elongated and columnar; cell migration, intercalation and convergent extension gives rise to a pear-shaped neural plate. Cell shape changes drive the rolling up of the plate, with the edges of the plate fusing dorsally to give rise to a tube. Neurulation also involves changes in cell adhesion, protrusive activity and the mechanical influences of the tissues outside the plate itself [29, 53–56]. By the end of gastrulation regional specification of the neural plate has already begun and chapter 6 will discuss patterning of the neural plate along the anterior-posterior axis.

4.3.2
The Teleost, Zebrafish

Over the past few decades the Zebrafish embryo, *Danio rerio*, has become another extremely attractive vertebrate model organism in which to study the cellular and molecular basis of early embryogenesis. This is for a number of reasons, including its transparency, speed of development, ease of manipulation and the wealth of available mutants [57–61]. The teleost embryo develops as a blastoderm, which sits on top of a large uncleaved yolk cell and consists of an outer epithelial sheet and a multilayered deep layer. The blastoderm, which is also referred to as the epiblast, undergoes epiboly and eventually encloses the entire yolk. In fundulus embryos, it was shown that during epiboly the blastoderm is pulled vegetally due to a tight attachment of cells at the margin to the yolk syncytial layer by a junctional complex, which is contiguous with the actin cytoskeleton of the yolk syncytial and cytoplasmic layers [62]. In addition, microtubules have been implicated in epiboly [63, 64]. As in *Xenopus*, the deep cells undergo radial intercalation to form a larger but thinner area during epiboly. Gastrulation begins at 50–70 % epiboly with cells from the edge of the blastoderm involuting or ingressing to form the so-called germ ring. The cells of the lateral germ ring migrate directionally towards the future dorsal side where they form a thickened region, the embryonic shield, which is equivalent to the dorsal blastopore lip in *Xenopus*. The directed migration of lateral cells towards the dorsal midline involves polarized protrusive activity across an as yet uncharacterized substratum and has not been observed in amphibian embryos. As the cells approach the paraxial region they align and adopt a medio-lateral elongated shape characteristic of intercalating cells. In the embryonic shield, cells of the epiblast involute to form the dorsal hypoblast. Both layers undergo convergence and extension movements and similar to amphibians, an elongated, narrow axis is formed, underlying the neural plate [65]. At the present time the nature of the interface between the epiblast and hypoblast in Zebrafish has not been characterized. In *Xenopus* a fibronectin-containing matrix is laid down and serves as a substrate for mesendoderm migration during gastrulation. This boundary also mediates the separation behavior of the different layers.

The large-scale mutant screens undertaken in Zebrafish gave rise to a collection of mutants that affect early morphogenesis. Some of the mutants have in the meantime been analyzed in more detail (see below); however, others remain to be characterized still further. Mutants described include those with impaired or incomplete epiboly, those in which notochord formation is affected or which cause distinct effects on cell fates in the gastrula. The mutations may affect the formation, maintenance or function of the dorsal organizer, or affect cellular rearrangements during gastrulation including the major morphogenetic processes, epiboly, convergence and extension, and tail morphogenesis [66–70]. The mutants provide an exciting opportunity to dissect the control of gastrulation in vertebrates and to unravel the cellular and molecular machinery involved.

4.3.3
Amniotes: Chick and Mouse

In addition to the embryo proper, the amniote egg will also give rise to extra-embryonic tissues and membranes to provide the embryo with an environment that supports its development. Because of this there is added variation compared to the early morphogenesis of amphibians and fish.

The newly laid avian egg consists of a large uncleaved yolk cell with the overlying blastoderm or blastodisc of the developing embryo. The blastodisc sits above the subgerminal cavity and is attached to the yolk along its edge. The center of the disc, the area pellucida, has a relatively transparent appearance since it is only a few cell layers thick, its cells have less yolk and they do not lie directly on the yolk mass. The area opaca, a band of cells along the edge of the blastodisc, consists of yolkier cells. At the periphery of the area opaca cells merge into a syncytial area with the yolk. The orientation of the anterior-posterior axis and the location of the future posterior end of the embryo are determined by the effects of gravity during the egg's passage through the oviduct [71]. How gravity breaks the radial symmetry of the blastodisc is currently not known. At the posterior end of the blastodisc a thickened, crescent-shaped structure, Koller's sickle, forms below the epiblast. Cells advancing anteriorly from Koller's sickle together with cells delaminating from the epiblast form a loose tissue layer beneath the area pellucida called the hypoblast.

It was thought initially that the hypoblast influences the organization of the epiblast; however reinvestigation of these experiments showed that when the hypoblast was rotated relative to the epiblast, the axis of the embryo remained unaffected. This suggests that the hypoblast does not act by inducing the epiblast to form a primitive streak [72, 73] and indicates that the epiblast has an axial bias independent of the hypoblast.

In the chick embryo, experiments to identify a region that is equivalent to the Nieuwkoop center of the amphibian embryo involve the precise marking and grafting of cells and have been challenging. The hypoblast, Koller's sickle and the posterior marginal zone have been investigated for their ability to induce the organizer. Cell marking experiments showed that the posterior side of the marginal zone and the posterior region of the epiblast layer have the ability to initiate the embryonic axis. A graft of a particular posterior blastoderm region can initiate an ectopic streak, and is able to recruit other neighboring cells to the developing ectopic streak. The posteriolateral part of the marginal zone region also has such abilities, which are inhibited during normal development. Thus, the cells in the posterior marginal zone including Koller's sickle, have organizer properties which induce the mesoderm and determine the initiation site of gastrulation in the chick embryo [74, 75].

In the chick embyro, two secreted factors, cWnt8c and the TGF-β superfamily member cVg1, have been shown to cooperate in inducing the primitive streak. cWnt8C is expressed in the marginal zone, around the circumference of the embryo, and cVg1 is expressed in the posterior part of the marginal zone [76]. Misexpression of Vg1 in the anterior marginal zone induces an ectopic primitive streak and recapitulates the morphological and molecular changes associated with normal primitive streak formation [77].

Experiments using bromodeoxyuridine (BrdU) incorporation to assess the rate of cell proliferation in the chick stage XIII blastoderm, detected a relatively high level of labeled cells around the posterior region of the area opaca, the marginal zone, Koller's sickle and the epiblast. It was suggested therefore that directional axis formation and the ability to initiate an embryonic axis could be attributed to a region of proliferation in the posterior side of a stage XIII blastoderm [78].

Gastrulation in the chick is dominated by two major movements, first the extension of the primitive streak followed by regression of Hensen's node. The details of how the primitive streak forms from epithelial epiblast cells are still not very well understood, but it involves the lateral to medial movement of cells towards the midline which results in a thickening at the posterior midline. The primitive streak extends anteriorly from the epiblast above Koller's sickle [55, 79] and forms the midline of the early embryo, defining the anterior-posterior axis. The anterior end of the extending primitive streak forms a raised, thickened area called Hensen's node, the avian organizer.

Two mechanisms have been suggested for primitive streak extension. This could be driven by convergence and extension of a region of the epiblast overlying the anterior part of Koller's sickle and/or by oriented cell division [55, 79, 80]. There is evidence to suggest that indeed both mechanisms are used. During primitive streak thickening the epiblast cells columnarize, undergo an epithelial to mesenchymal transition (EMT) and ingress to form deeper layers of the primitive streak. Throughout gastrulation, presumptive endodermal and mesodermal cells in the epiblast move medially towards the streak where they undergo EMT and subsequently migrate as mesenchymal cells into the deeper layers and outwards, giving rise to embryonic endoderm and mesoderm. In contrast, in *Xenopus* only the cells at the leading edge of the involuted marginal zone become migratory as the mesoderm extends. Presumptive neural and non-neural ectoderm remains in the epiblast. Classic fate mapping experiments indicated that the position of cells within the streak correlates with their eventual medio-lateral position in the embryo, with anterior streak cells giving rise to medial tissues and more posterior streak cells contributing to more lateral tissues [12, 18, 81, 82]. This was confirmed recently by observing GFP-labeled cells directly as they move out from the primitive streak using long-term video microscopy [83]. This study also uncovered opposing roles for different members of the fibroblast growth factor family (FGF) in controlling movement behavior of primitive streak cells. Once the primitive streak has fully extended, cells leave Hensen's node anteriorly, they converge and give rise to the extending notochord. In addition, notochord extension is driven by division of cells both within the node and in the notochord [84]. The node then begins to regress through the primitive streak, the cellular and molecular mechanisms that drive this regression movement is not understood at present, although it is possible that the convergence and extension of the notochord actively pushes the node in a posterior direction. The primitive streak itself, even though not required for regression movements, may still contribute to node regression and may pull the node backwards [55, 79]. There is some evidence to suggest that FGF receptor signalling may be required for this process [83].

Early embryogenesis of mammalian embryos is difficult to study due to their intrauterine development. On the other hand, the power of mouse reverse genetics has provided a wealth of information regarding the molecular networks involved in controlling morphogenesis. The early mouse blastocyst (equivalent to the *Xenopus* blastula and the chick blastodisc) is highly asymmetric and contains a large blastocoel with the inner cell mass (ICM) on one side. The ICM will give rise to the epiblast of the embryo, and is surrounded by trophectoderm cells and visceral endoderm cells. The mouse epiblast is not a flat disc as in the case of the chick but has an unusual cup shape with the dorsal side on the inner surface of the cup. Extensive cell marking experiments indicate that gastrulation movements in the mouse are similar to those in the chick [17, 32, 36]; however, the details remain to be investigated. Posterior epiblast cells move towards the midline and form a primitive streak, which extends in the future midline from posterior to anterior and has a node region at its anterior end. Cells move through the streak to establish definitive mesoderm and endoderm layers. The node is an organizing center and has the ability to induce a secondary axis [22]. Additional signals are needed for the formation of most anterior structures and in mammals the visceral endoderm, an extra-embryonic cell layer that covers the egg cylinder stage embryo prior to gastrulation, is required for head formation [30]. The visceral endoderm plays an active role in guiding early development and cells in the anterior visceral endoderm function as an early organizer, prior to the formation of the primitive streak, by specifying the fate of underlying embryonic tissues [85].

4.4
Mechanistic Aspects, Molecules and Molecular Networks

The previous paragraphs have summarized the cell rearrangements and movements involved in vertebrate gastrulation and even though biologists have had an active interest in this complex process for a century, not much was known about the molecular control of these events. However, the past decade has seen a rapid advance in our knowledge and in the following sections we will consider what is currently known about the molecular basis of these events. Not unexpectedly very different kinds of factors are involved. These range from secreted peptide signaling molecules to their cognate receptors, transduction cascades and transcription factors to cytoskeletal proteins, cell adhesion molecules, extracellular matrix components and receptors.

4.4.1
Wnt Signaling

The Wnt family of secreted glycoproteins is expressed throughout vertebrate development and these glycoproteins are involved in many cellular functions such as cell fate, polarity, adhesion, proliferation and differentiation. Recent work has shown that Wnt signaling and activation of the downstream pathway component, disheveled, regulate convergent extension in vertebrates.

Since the discovery of the frizzled, seven-pass transmembrane proteins as Wnt receptors, rapid progress has been made in unravelling the downstream signal transduction pathways activated by this family of ligands and it has been found that Wnts can signal through a number of different mechanisms (Figure 4.1). The best understood is the "canonical" pathway in which Wnt signaling stabilizes cytoplasmic β-catenin leading to direct changes in the transcription of target genes [86]. This pathway is usually implicated in changes of cell fate and is for example, involved in the induction of organizer cell fate by inducing siamois [87–90]. A number of β-catenin-independent, or "non-canonical" pathways have been characterized more recently. These include the planar cell polarity pathway (PCP), which is similar to a pathway known to regulate cell and tissue polarity in Drosophila [91, 92]. In vertebrates, the PCP pathway has been found to play a role in convergent extension movements in both gastrulation and neurulation, it has also been implicated in tissue separation during morphogenesis (for reviews see [49, 93, 94]). The PCP pathway overlaps with the canonical Wnt signaling pathway in that it requires frizzled receptors and the cytoplasmic phosho-protein disheveled, which locates to the membrane when the receptor becomes activated by ligand binding [95]. However, the PCP pathway diverges downstream of disheveled [96, 97] and responses to non-canonical Wnt signaling include an increase in intracellular Ca^{2+}, activation of rho, rac and/or cdc42, heterotrimeric G protein activation upstream of disheveled, and activation of the JNK pathway [94]. Thus, the non-canonical pathway seems to have many different branches whose activation is context dependent.

The mammalian homolog of Drosophila Naked Cuticle (Nkd) has been shown to interact directly with disheveled [98]. mNkd acts in a cell-autonomous manner to inhibit the canonical Wnt pathway and to stimulate c-Jun-N-terminal kinase activity. Expression of mNkd disrupts convergent extension in Xenopus and it may act as a switch to direct disheveled activity towards the PCP pathway, and away from the canonical Wnt pathway. In contrast, the casein kinase lepsilon stimulates Wnt signaling and inhibits planar cell polarity in Drosophila by reducing the activity of disheveled in the JNK pathway [99].

Wnt ligands thought to signal via non-canonical pathways include Wnt11, Wnt5a and Wnt4, and overexpression of these by RNA injection in early Xenopus embryos causes defective convergent extension [100, 101]. A dominant negative form of disheveled was shown to affect axis elongation in Xenopus and these experiments implicated Wnt/dsh signaling in morphogenesis [102]. Further analysis at the cellular level showed that the dominant negative disheveled resulted in a lack of polarized cell protrusions, which are required for cell intercalation during axis extension [95].

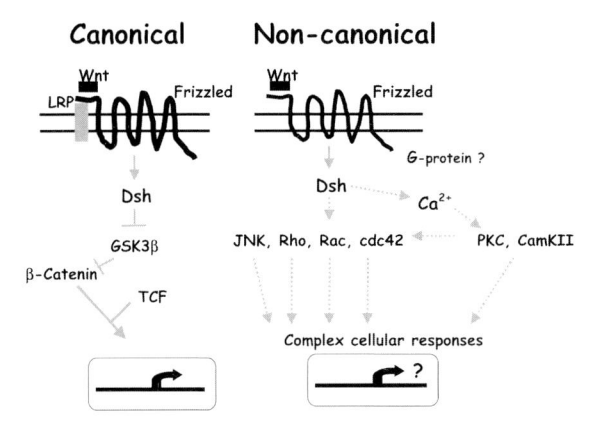

Fig. 4.1
Canonical and non-canonical Wnt signaling pathways. (A) A simplified scheme of the canonical Wnt signaling pathway. In this pathway a Wnt ligand binds to a Frizzled (Fz) receptor and a co-receptor of the low density lipoprotein recptor-related protein (LRP) family. This leads to the inhibition of Glycogen synthase kinase beta (GSK3β) by disheveled. This leads to an accumulation of β-catenin in the cytoplasm. In the absence of Wnt, GSK3β phosphoylates β-catenin thus targeting it for degradation. Stabilized β-catenin is able to associate with transcription factors of the ternary complex factor (TCF)/lymphoid enhancer factor 1 (LEF1) family. (B) The vertebrate non-canonical pathway also requires Frizzled receptors and like the canonical pathway involves the cytoplasmic signal transduction protein disheveled (Dsh). The pathway however differs in requiring Dsh to be localized to the cell membrane via its DEP domain. From Dsh the main signaling involves the small GTPases of the Rho family. The precise roles of Rho versus Rac and Cdc42 remain unclear, as does the potential role of the JNK pathway. Dsh can also stimulate calcium flux and the activation of the calcium-sensitive kinases PKC and CamKII, suggesting that the Wnt/calcium pathway, which had been previously thought of as distinct from PCP-like signaling, may be part of a common non-canonical pathway. Wnt/calcium signaling possibly works via hetero-trimeric G-proteins to mobilize intracellular Ca^{2+}. How this interacts with Dsh is unclear at this time. Whereas the canonical pathway mainly leads to changes in gene expression, non-canonical signaling can cause many different responses in the cell including changes in cell polarity, cell adhesion, cytoskeleton, as well as possibly changes in gene expression. The two pathways are also known to interact with each other.

It was subsequently shown that disheveled is activated specifically by Wnt11 to control convergent extension movements [103, 104]. In Zebrafish silberblick mutants, which lack a functional Wnt11 gene, and in *Xenopus* embryos expressing a dominant negative Wnt11, morphogenesis is defective and convergent extension movements are strongly inhibited. It was shown that β-catenin-dependent signaling is not involved, since activation of this pathway could not rescue the phenotype in these mutants. In addition, axis elongation in animal caps treated with dominant negative Wnt11 could be restored by providing a form of disheveled that can no longer interact with the β-catenin-dependent pathway. Interestingly, these morphogenetic changes are not accompanied by changes in cell fate [95, 103, 104]. In this context Wnt11 seems to act through the frizzled7 (Xfz7) receptor whose zygotic expression strongly increases at the beginning of gastrulation and is predominantly localized to the presumptive neuroectoderm and deep cells of the involuting mesoderm [105]. Misexpression of Xfz7 in the dorsal equatorial region affects convergent extension movements and delays mesodermal involution without affecting expression of mesodermal marker genes.

Another Zebrafish mutant, pipetail, which is an allele of Wnt5, has defects in tail extension indicating that Wnt5 is required for elongation in posterior regions of the gastrula and modulation of calcium flux may be involved [106]. Furthermore, the silberblick/pipetail double mutant shows a more severe phenotype than each single mutant suggesting some redundancy between the two Wnt ligands, Wnt11 and Wnt5 [107, 108]. In Zebrafish, transmission of the Wnt11 signal during convergent extension is potentiated by a heparin sulfate proteoglycan encoded by the Knypek gene (Kny) [109]. In Knypek mutant embryos, cells fail to elongate and to align medio-laterally. Molecular and genetic analysis indicates that Knypek and Wnt11 act in the same pathway and suggests that Knypek may bind and stabilize Wnt11 on the cell surface. Further studies in Zebrafish also demonstrated that no-tail (brachyury or T) interacts with Knypek and pipetail to regulate posterior body development. The authors propose that tail-forming movements employ mechanisms that regulate gastrulation together with mechanisms unique to the posterior body [110].

Other genes known to be involved in *Drosophila* PCP signaling [111] also seem to function in vertebrate gastrulation. These include strabismus/van gogh, a membrane protein, prickle, a LIM domain protein and the seven-pass transmembrane cadherin, flamingo. A combination of mutant analysis and knock-down experiments in Zebrafish and *Xenopus* using morpholinos, which resulted in similar convergent extension phenotypes, indicated that all these components are involved [94, 112–115]. A recent confocal analysis of live Zebrafish embryos looked at the patterns of cell division orientation in the dorsal epiblast at gastrulation. This study reported that non-canonical Wnt signaling, including Wnt11, disheveled and strabismus, is involved in establishing a non-randon animal-vegetal polarity of cell division. The authors suggested that this process has an important function in elongation of tissues along the anterior-posterior axis [116].

Another potential player on the scene is the vertebrate homolog of *Drosophila* nemo, nemo-like kinase (Nlk), which inhibits the DNA-binding ability of β-catenin/Tcf complexes by phosphorylation, thereby blocking activation of Wnt targets. Nlk is involved in cell-fate specification and overexpression and morpholino experiments in Zebrafish showed that nlk affects ventro-lateral mesoderm formation and patterning [117]. In addition, Nlk strongly enhances convergent/extension phenotypes associated with wnt11/silberblick, suggesting a role in modulating cell movements and providing an example of cases where modulation of Wnt signaling affects both cell-fate decisions and morphogenesis.

A recent report identified an important function for signaling through a p75 neurotrophin receptor-related transmembrane protein, NRH1, in the regulation of convergent extension movements [118]. NRH1 is expressed in marginal zone tissues of the *Xenopus* gastrula and in the posterior ectoderm of the neurula. NRH1 activated downstream effectors of the Wnt/planar cell polarity pathway: small GTPases and the cascade of MKK7–JNK independent of disheveled. Decreasing NRH1 function inhibits convergent extension movements. The phenotype could be rescued by co-injection of downstream effectors, suggesting that NRH1 functions as a positive modulator of planar cell polarity signaling downstream of Xdsh.

The function of the non-canonical Wnt pathway has been conserved in amniote gastrulation movements as shown by genetic analysis in mice. For example, loop-tail mice carry a semi-dominant mutation in a strabismus homolog (Vangl2) and display failures in neural tube closure, which may be due to defective convergent extension of the neural plate [119]. Alternatively, it has been suggested that loop-tail normally restricts the lateral extent of floor plate differentiation by negatively regulating the expression of Shh [120]. Mice mutant in a homolog of flamingo (Celsr1) also display a neural tube phenotype [121]). Together these studies provide evidence for the involvement of a planar cell polarity pathway in vertebrate neurulation. Interestingly, the planar polarity of inner ear hair cells is also affected in Celsr1 and loop-tail mice, reminiscent of planar polarity of bristles in the fly [121, 122]. Stereocilia are also affected in mice mutant in a gene homologous to *Drosophila* scribble, identifying this gene as a member of the PCP pathway [123].

The mouse protein tyrosine kinase 7 (PTK7), which encodes an evolutionarily conserved transmembrane protein with tyrosine kinase homology, has also been identified as a novel regulator of the PCP pathway in vertebrates. Mutations in PTK7 disrupt neural tube closure and stereociliary bundle orientation. Furthermore, PTK7 interacts genetically with Vangl2, the mouse homolog of Van Gogh. In *Xenopus*, PTK7 is required for neural convergent extension and neural tube closure [124].

Non-canonical Wnt signaling is also important for maintaining proper separation of the germ layers during gastruation. Morpholino mediated knock-down of *Xenopus* frizzled7 leads to gastrulation defects [48] due to the loss of separation between the involuting mesoderm from the overlying ectoderm. Recent work shows that this is mediated in part by paraxial protocadherin (PAPC), a homophilic adhesion protein that mediates selective cell-cell adhesion and cell sorting [47]. PAPC can interact with the *Xenopus* frizzled7 receptor to modulate the activity of the Rho GTPase and c-jun N-terminal kinase, two effectors of PCP signaling. Knock-down experiments show that this function of PAPC is essential for the regulation of tissue separation during gastrulation. This is particularly interesting because of the known links between Wnt signaling and cell adhesion [125, 126].

4.4.2
Convergent Extension and the Regulation of Cytoskeletal Dynamics

Convergent extension movements depend on specific and coordinated changes of the cytoskeleton, a pre-requisite for cell motility. How Wnt/disheveled signaling is leading to these morphogenetic changes is at present the subject of intense research activity. An increasing number of molecular players are being uncovered, including some that are known regulators of the cytoskeleton, however their distinct roles and the different contexts in which they are required remain to be investigated.

One of the first reports suggesting that Wnts signal through key regulators of the actin cytoskeleton to mediate gastrulation movements demonstrated that the effects of Xfz7 and Xwnt11 over-expression on convergent extension in activin-treated animal caps can be reversed by co-expression of a dominant negative mutant of the small GTPase Cdc42 [105]. The Rho family of GTPases, which include rho, rac and cdc42, regulate many essential cellular processes, including actin dynamics and cell adhesion and it has been shown that *Xenopus* Cdc42 is expressed in tissues undergoing extensive morphogenetic changes, such as the deep layers of involuting mesoderm and posterior neuroectoderm during gastrulation [127]. Over-expression of either wild-type or dominant-negative Cdc42 results in changes in cell adhesiveness and interferes with convergent extension movements without affecting mesodermal cell specification. Dominant negative Cdc42 rescues PKC-alpha- or XWnt5a-mediated inhibition of convergent extension, suggesting that XCdc42 acts downstream of the Wnt/Ca^{2+} signaling pathway [127]. Further studies investigating the regulation of Cdc42 found that its activity increases as a response to Wnt11 expression in animal cap ectoderm. This activation is dependent on PKC, consistent with the possibility that Cdc42 is a molecular target of Wnt11/Xfz7 signals [128]. Investigating the control of convergent extension movements by Rac and Rho during *Xenopus* gastrulation revealed that they have both distinct and overlapping roles in regulating the motility of axial mesoderm cells [129]. Time-lapse recording of mesoderm explants showed that cell protrusions and lamellipodia formation were controlled by the two GTPases. The authors suggested that Rho and Rac operate in distinct pathways that are integrated to control cell motility during convergent extension. This is consistent with other work which shows that Wnt-mediated activation of Rac is independent of Rho. Furthermore, inhibition of Rac function results in gastrulation defects without affecting Wnt/Frizzled activation of the Rho or β-catenin pathways [130]. It has also been shown that myristoylated alanine-rich C kinase substrate (MARCKS) plays a role in regulating the cortical actin formation required for dynamic morphogenetic movements. MARCKS is an actin-binding, membrane-associated protein and is involved in the control of cell morphology, motility, adhesion and protrusive activity in embryonic cells. Blocking MARKS during *Xenopus* gastrulation results in impaired morphogenetic movements, including convergent extension mediated by the non-canonical Wnt pathway [131].

4.4.3
Fibroblast Growth Factors (FGF)

Fibroblast growth factor (FGF)-mediated signals have long been implicated in controlling cell behavior and gastrulation movements in a number of different animal model systems. These include the invertebrate *Drosophila*, where FGF plays a role in the spreading of mesoderm cells over the ectoderm [132]. In vertebrates, FGF was first identified as a factor involved in mesoderm induction. However, experiments using a dominant-negative form of the FGF receptor FGFR1 showed that FGF is additionally involved in controlling convergent extension in Xenopus embryos [133, 134]. Activity of the FGF signaling pathway is regulated by a number of negative feed back regulators including sprouty, sef, spred and MKP3. It was shown, again in Xenopus, that sprouty–2 inhibits FGF-dependent gastrulation movements without affecting mesoderm induction and patterning, providing evidence that these two processes can be separated [135].

Recent work in chick embryos, which used live imaging of GFP-labeled primitive streak cells to establish their movement trajectories, established that cell movement patterns during gastrulation are controlled by positive and negative chemotaxis in response to FGF4 and FGF8. Based on the expression patterns of FGF4 and FGF8 in early chick embryos it was hypothesized that FGF8 acts as a chemorepellent to drive cells away from the primitive streak, while FGF4 from the forming head process and node attracts the cells back towards the midline once the node starts to regress (Figure 4.2) [83]. Misexpression of a dominant negative receptor, dnFGFR1c, resulted in severe malformations, inhibition of cell movements and loss of responsiveness towards FGFs. Some of the phenotypes observed were reminiscent of those observed in Xenopus embryos [133], with an expansion of the head at the expense of somite formation and tail structures. Furthermore, dnFGFR1 expression within the node inhibited node regression while regression took place normally when FGF signaling was blocked in primitive streak cells [83]. Primitive streak extension also requires FGF signaling within cells of the node. It was found in chick embyros, that when cells in the node are electroporated with a dnFGFR1 construct, the streak does not elongate properly. FGF8 may also be involved in the continued posterior elongation of the embryo, driven initially by regression of the node and subsequently by the caudal movement of the tail bud. This hypothesis is based on the finding that a gradient of FGF8 protein is set up by mRNA decay, resulting in the highest levels of FGF8 being maintained at the posterior most end of the embryo [136]. In addition, this work uncovered a novel mechanism by which a morphogen gradient might be established and could have wide implication for morphogenesis and patterning.

Investigation of the phenotypes of knockout mice also demonstrated the importance of FGF signaling in mammalian gastrulation. Both FGF8 and FGFR1 mutants show severe defects in gastrulation movements at the primitive streak stage [137–139]. The finding that FGF8 knockout mice show an accumulation of mesoderm cells in the streak is entirely consistent with the observations in chick embyos [139]. Similarly, mice that lack the FGFR1 receptor die post-implantation and chimeric analysis has revealed that signaling through this receptor is required for cells to undergo the epithelial to mesenchymal transition (EMT) [138, 140]. In order to allow cells to leave the streak E-cadherin must be downregulated, resulting in EMT [141], FRFR1 null mice fail to do so and therefore cells are trapped in the streak [137]. FGF signals mediated by the FGFR1 receptor regulate the expression of snail (sna), a zinc finger containing a transcriptional repressor, thereby controlling E-cadherin expression in the streak. These authors also proposed that ectopic E-cadherin expression may indirectly affect the activity of canonical Wnt signaling in the streak, by sequestering cytoplasmic β-catenin, which would then no longer be available to bind

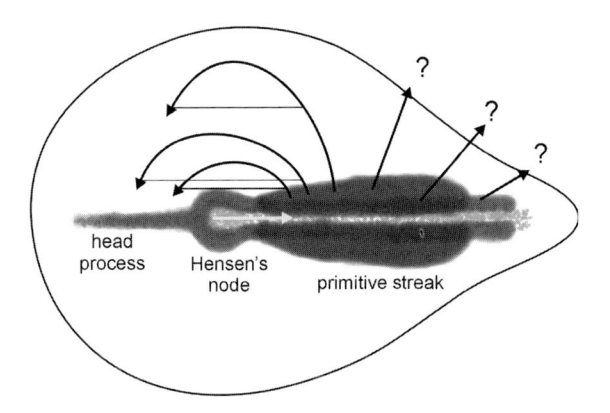

Fig. 4.2
Model suggesting that cell movement is directed by repulsion from the streak mediated by FGF8 and attraction towards the midline in response to FGF4 (reprinted with permission from [83]). The FGF4 expression pattern is indicated in green, and FGF8 expression is indicated in blue. The black arrows show the cell movement trajectories observed. The yellow arrow indicates the direction of node regression. It was suggested that cell movement away from the streak is the result of negative chemotaxis in response to FGF8 produced by cells in the middle primitive streak, while the inward movement of the cells emerging from the anterior streak is the result of attraction by FGF4 secreted by cells in the head process and notochord. The cells in the caudal part of the embryo are initially repelled by FGF8 and subsequently attracted by an unknown signal present in the region between the area opaca and the area pellucida. (This figure also appears with the color plates.)

to Tcf/Lef family transcription factors and activate Wnt target genes. This suggests a possible link between Wnt and FGF pathways and would provide a molecular mechanism to integrate cell adhesion, morphogenetic cell movements and cell-fate specification during mouse gastrulation. FGFR1–mediated signaling also regulates mesodermal cell-fate specification and patterning by controlling Brachyury and Tbx6 gene expression.

It has been shown in *Xenopus*, chick and mouse that the expression of many Ets genes is associated with mesenchymal-epithelial interactions and changes in extracellular matrix proteins [142–144]. Many Ets genes are expressed in migratory cells, for example neural crest and endothelial cells. They are also transcribed in embryonic areas affected by epithelio-mesenchymal transitions (see 4.4.4) and may coordinate changes in the expression of adhesion molecules and modulate extracellular matrix composition [145].

In mice, Fgf8 appears to be the principal ligand required for mesodermal development, as mouse Fgf8 mutants do not form mesoderm [139]. In Zebrafish however, fgf8 mutants have only mild defects in posterior mesodermal development. This suggested that it is not the only Fgf ligand involved in the development of this tissue. Using morpholino-based gene inactivation it was shown that fgf24 is required together with fgf8 to promote posterior mesoderm development [146]. In addition, genetic evidence suggests that as in other vertebrates these Fgf signaling components interact with the T-box transcription factors, no tail (brachyury) and spadetail (Tbx16) to regulate mesoderm development.

4.4.4
Regulation of Epithelial Mesenchymal Transition (EMT)

As described above, FGFR1 receptors regulate the expression of the snail transcriptional repressor, thereby controlling E-cadherin expression in the streak. In mouse embryos that are homozygous for a null mutation in the snail gene a mesoderm layer forms, however, it is morphologically abnormal [147]. Some mesodermal cells retain epithelial characteristics with apical basal polarity and adherens junctions between cells. The mesoderm continues to express E-cadherin in the absence of snail function, confirming that E-cadherin is a target for snail. Previous studies in cultured cells and in metastatic carcinomas had shown that snail is indeed a direct transcriptional repressor of E-cadherin expression, and is able to trigger an epithelial mesenchymal transition when overexpressed in a number of epithelial cell lines [148]. EMT is concomitant with the aquisition of migratory and invasive behavior and epithelial cells that ectopically express Snail adopt a fibroblast-like phenotype and acquire tumorigenic and invasive properties [149]. Consistent with these findings, incubation of chick embryos with antisense oligonucleotides to the snail family gene, slug, results in defects in delamination of both the mesoderm and neural crest [150].

In Zebrafish, the signal transducer and activator of transcription 3 (Stat3) has been found to be required for anterior migration of dorsal mesendodermal cells and for the convergence of neighboring paraxial cells [151]. Loss-of-function experiments result in a shortened anterior-posterior axis; however embryos lacking Stat3 activity show no defects in early cell-fate specification. Interestingly, Stat3 is activated in the dorsal organizer by the maternal Wnt/beta-catenin pathway, thus linking signals involved in the specification of cell fates with the coordination of gastrulation movements. Subsequent experiments identified LIV1, a breast-cancer-associated zinc transporter protein, as a downstream target of Stat3 [152]. LIV1 is essential for the nuclear localization of Snail and thus mediates the role of Stat3 in the EMT of Zebrafish gastrula organizer cells and their anterior migration. This work further elucidates the molecular network of signals and transcription factors controlling EMT.

Studies in *Drosophila* have shown that pebble (pbl), which encodes a Rho-family GTP exchange factor (GEF), is required in mesodermal cells during the transition from an epithelial to mesenchymal morphology during gastrulation [153]. Cells mutant for pebble fail to undergo the normal epithelial-mesenchymal transition (EMT) and dorsal migration that follows ventral furrow formation. Pebble is also required for cytokinesis; however, rescue experiments demonstrate that this can be separated from its function in EMT. Both, EMT and cytokinesis depend on actin organization and consistent with this, the GTP exchange function of pebble is required for both processes. As yet there is no information regarding the function of a pebble ortholog in vertebrate gastrulation movements, but considering the importance of the Rho family of GTPases this is only a matter of time.

4.4.5
Cadherins and Protocadherins

Cell adhesion molecules were first predicted to be involved in the phenomenon of "cell sorting" by different embryonic cell types into separate tissues by Steinberg [154]. Experiments by Takeichi and colleagues supported this idea by showing that different cadherins expressed on a similar cell can mediate cell sorting [155]. Even the amount of cadherin being expressed can lead to sorting [156]. Cadherins are a large family of cell adhesion molecules which have many recognized roles in development including regulating cell and tissue polarity, cell sorting, cell migration and cell rearrangements [157]. They are defined by having a varying number of tandem repeats of a CD domain (~ 110 amino acids long) ranging from 1–2 to 34. Ca^{2+} binds to the linker region between these repeats and is necessary for their function. Within the cadherin family are a number of subfamilies some of which have been implicated in morphogenesis, including the classical cadherins, proto-cadherins, Fat-like cadherins and seven-pass transmembrane cadherins.

During convergent extension movements the cells are in constant contact with each other and are in effect pulling on each other to help drive their movement. Thus, cell movements during gastrulation and neurulation require dynamic changes in cell adhesive properties. Experiments in *Xenopus* showed that activin decreases the adhesive function of C-cadherin (a classical cadherin) molecules on the surface of blastomeres. This suggested that decreased cadherin-mediated cell-cell adhesion is associated with increased morphogenetic movement [158]. Injection of RNA encoding a truncated C-cadherin into the prospective dorsal involuting marginal zone, caused defects in gastrulation without affecting cell specification [59] and it was shown that animal cap elongation requires the reduction of C-cadherin-mediated adhesion [160].

In *Xenopus* three protocadherins have been described which are expressed in early development. NF-protocadherin (NFPC) is expressed in the embryonic ectoderm and mediates cell adhesion [161]. It was shown recently that the cellular protein TAF1, previously identified as a histone-associated protein, binds the NFPC cytoplasmic domain and TAF1 can rescue the ectodermal disruptions caused by a dominant-negative NFPC construct [162]. Axial protocadherin (AXPC) is expressed in the axial mesoderm and together with another family member, paraxial protocadherin (PAPC) mediates the sorting of different mesodermal cells [163, 164]. Paraxial protocadherin (PAPC) plays an important role in gastrulation movements. Overexpression of PAPC can promote convergent extension perhaps by providing the traction needed for cell motility. A dominant-negative form of PAPC inhibits the elongation of animal cap explants without altering mesodermal patterning [163]. Over-expression studies suggested that tissue separation may require modulation of cadherin function and that this is controlled by the paired-class homeodomain transcription factors Mix.1 and goosecoid (gsc) [39]. Studies in mouse and *Xenopus* also implicated the transcription factor Lim1 in the regulation of PAPC and gastrulation movements. Expression of PAPC is lost in the nascent mesoderm of Lim1(−/−) mouse embryos and in the organizer of Lim1–depleted *Xenopus* embryos [165]. It was shown recently that PAPC functionally interacts with the Wnt/PCP pathway in the control of convergence and extension movements [166]. In this context XPAPC functions as a signaling molecule that signals through the small GTPases Rho A and Rac 1 and c-jun N-terminal kinase (JNK) to coordinate cell polarity of the involuting mesoderm. In addition, PAPC also has a function in tissue separation during gastrulation (see Section 4.4.1) where it was also shown that PAPC and Xfz7 can directly interact with each other [47]. Interestingly, despite common downstream components PAPC cannot compensate for the loss of Xfz7 receptor function suggesting they have non-redundant functions in convergent extension and tissue separation [47, 166].

In Zebrafish paraxial protocadherin (PAPC) is expressed in trunk mesoderm undergoing morphogenesis. Microinjection studies using a dominant-negative construct suggest that it is required for proper dorsal convergence movements [167]. Genetic studies show that PAPC is a close downstream target of spadetail, a T-box transcription factor previously implicated in mesodermal morphogenetic movements [66, 167]. In axial mesoderm the floating head (flh) homeobox gene is required to repress the expression of both, spadetail and PAPC, thereby promoting notochord differentiation and suppressing paraxial mesoderm. In the mouse Not, an ortholog of flh, is also involved in posterior notochord development [168].

N-cadherin (Ncad) is a classical cadherin that is implicated in several aspects of vertebrate embryonic development, including neural tube formation. Characterization of Zebrafish parachute (pac) mutations, which correspond to an N-cadherin homolog, revealed that convergent cell movements during neurulation are severely compromised [169]. In parachute mutant embryos, neuroectodermal cell adhesion is altered and β-catenin stabilization/localization is affected, as a result many neurons become progressively displaced along the dorsoventral and the anteroposterior axes.

4.4.6
Extracellular Matrix and Integrin Receptors

Integrins are a family of heterodimeric cell surface proteins responsible for cell attachments to the extracellular matrix (ECM) [170]. From the early stages in development interactions between the ECM and integrin receptors are crucial for normal morphogenesis. Perturbation of these interactions results in disruption of gastrulation [42, 171–175] and neurulation [176, 177]. In addition to their role in adhesion, integrins have also been demonstrated to play a role in cell signaling, growth, differentiation and regulation of gene expression [178]. The 17 known α and eight known β subunits pair to form a variety of cell surface receptors which can bind to many different proteins in the ECM including fibronectin (Fn), the laminins and the collagens. Mouse knockouts of integrin subunits expressed during early development, such as α5 and β1, have shown them to be important in morphogenesis [175, 179, 180]. Molecules that are involved in integrin-mediated cell signaling, which are found at the sites of ECM-integrin contact, have also been shown to cause morphogenetic defects. For example, the mouse knockouts of FAK and paxillin are similar to those of Fn [181, 182].

In vertebrates, fibronectin (Fn) is a major constituent of the ECM that is assembled into extracellular matrices [183, 184]. Fibronectin-null mice are embryonic lethal and have defects in mesodermal derivatives [185]. Over recent years the role of Fn during gastrulation has been further refined in *Xenopus*. It has been shown that Fn is required for the cellular rearrangements that drive epiboly in the marginal zone at gastrulation [51, 186]. The fibronectin fibril matrix on the blastocoel roof of the *Xenopus* gastrula also determines the direction of mesoderm cell migration. It has been shown that activin and FGF affect fibronectin matrix assembly [187]. In addition, injection of monoclonal antibodies directed against the central cell-binding domain of fibronectin combined with confocal and time-lapse microscopy showed that integrin-based interactions with Fn are necessary for the establishment of cell polarity in the deep layers of the dorsal marginal zone and blastocoel roof [186]. Interestingly, integrin-based interactions with Fn were sufficient to cause membrane localization of disheveled-GFP suggesting that integrin and Wnt signaling pathways interact to regulate radial intercalation in *Xenopus* embryos. Furthermore, integrin-ECM interactions regulate cadherin-mediated cell adhesion, and by modulating cell intercalation affect convergent extension [51]. Analysis of the role of the integrin alpha cytoplasmic tail demonstrated that activin induction-dependent cell spreading, mesoderm cell and explant motility, and the ability to assemble Fn matrix on the blastocoel roof varied with specific alpha subunit tail sequences [188].

4.4.7
Platelet Derived Growth Factor (PDGF)

PDGF signaling is involved in a variety of processes including embryonic development of the nervous system, heart and lung development and angiogenesis. At the cellular level PDGF signaling targets the regulation of proliferation and the control of cell motility and migration. In the mouse, knockouts have suggested that PDGF has a function in mesoderm migration but how it does this is unclear [189].

As in the mouse, PDGF and the PDGF receptor are expressed in a complementary pattern in the *Xenopus* gastrula embryo. PDGFA is expressed in the blastocoel roof (BCR) and the PDGFRα is expressed in *Xenopus* mesoderm at gastrulation. Disruption of PDGF function with a dominant-negative PDGFRα leads to aberrant movement of the mesoderm [190] and treatment of mesoderm aggregates with PDGFAA stimulates spreading on Fn *in vitro* [191]. Recent observations show that interfering with PDGF function randomizes mesoderm migration on conditioned substrate *in vitro* and disrupts the orientation of migratory cells in the embryo. PDGF signaling is therefore playing a role in the guidance of migrating mesoderm during gastrulation and disruption results in reduced head formation and axis elongation [192]. Platelet Derived Growth Factor (PDGF) functions upstream of phosphoinositide 3–kinases (PI3K) and in Zebrafish it was shown that PI3K are required in mesendodermal cells for process formation and cell polarization at the onset of gastrulation [193]. In addition, protein kinase B (PKB), a downstream effector of PI3K activity, localizes to the leading edge of migrating mesendodermal cells indicating its function in directed cell migration [193].

4.5
Conclusion and Outlook

Despite a much improved insight into the cellular and molecular mechanisms that control vertebrate morphogenesis there are still many issues that remain poorly understood. For example how cell cycle progression and morphogenetic movements, which are incompatible within invaginating cells, are coordinated is only beginning to be investigated in vertebrates [194]. Evidence from *Drosophila* suggests that there seems to be an interaction between tribbles, a protein affecting the activity of the cell cycle regulator, string (cdc25) and the snail pathway. How the two pathways interact to promote gastrulation is not understood at present [195–198]. The potential role of tribbles in vertebrate gastrulation has only been investigated in *Xenopus*, where a homolog, Xtrb2, functions in the coordination of synchronous cell divisions during the stages of blastula formation [199]. It was also shown that inhibition of cell cycle progression is necessary for convergent extension movements in *Xenopus* [200, 201]. Another burning question is how morphogenesis is coordinated with axial patterning and a recent paper has provided some insight showing that the establishment of anterior-posterior polarity involves graded activin-like signaling and is directly linked to convergent extension movements [202, 203]. This chapter has summarized some of the molecular mechanisms involved in morphogenesis and the next goal will be to link these pathways together into an integrated network.

References

1 Beetschen, J. C. Amphibian gastrulation: history and evolution of a 125–year-old concept. *Int J Dev Biol* **2001**; 45: 771–795.
2 Schoenwolf, G. C. Cutting, pasting and painting: experimental embryology and neural development. *Nat Rev Neurosci* **2001**; 2: 763–771.
3 Gilbert, S. F. *Developmental Biology*. Sunderland: Sinauer Associates, Inc., **2003**.

4 Wolpert, L. *Principles of Development*. Oxford: Oxford University Press, **2002**.
5 Haeckel, E. Die GastreaTheorie, die phylogenetische Klassification des Tierreichs und die Homologie der Keimblätter. *Jen Zts Naturwiss* **1874**; *8*: 1–55.
6 Sander, K. "Mosaic Work" and "assimilating effects" in embryogenesis: Wilhelm Roux's conclusions after disabling frog blastomeres. *Roux's Arch Dev Biol* **1991**; *200*: 265–267.
7 Dale, L. and Slack, J. M. Fate map for the 32-cell stage of Xenopus laevis. *Development* **1987**; *99*: 527–551.
8 Garcia-Martinez, V., Alvarez, I. S. and Schoenwolf, G. C. Locations of the ectodermal and nonectodermal subdivisions of the epiblast at stages 3 and 4 of avian gastrulation and neurulation. *J Exp Zool* **1993**; *267*: 431–446.
9 Keller, R. E. Vital dye mapping of the gastrula and neurula of Xenopus laevis. I. Prospective areas and morphogenetic movements of the superficial layer. *Dev Biol* **1975**; *42*: 222–241.
10 Keller, R. E. Vital dye mapping of the gastrula and neurula of Xenopus laevis. II. Prospective areas and morphogenetic movements of the deep layer. *Dev Biol* **1976**; *51*: 118–137.
11 Kumano, G. and Smith, W. C. Revisions to the Xenopus gastrula fate map: implications for mesoderm induction and patterning. *Dev Dyn* **2002**; *225*: 409–421.
12 Hatada, Y. and Stern, C. D. A fate map of the epiblast of the early chick embryo. *Development* **1994**; *120*: 2879–2889.
13 Lane, M. C. and Smith, W. C. The origins of primitive blood in Xenopus: implications for axial patterning. *Development* **1999**; *126*: 423–434.
14 Lane, M. C. and Sheets, M. D. Rethinking axial patterning in amphibians. *Dev Dyn* **2002**; *225*: 434–447.
15 Lawson, K. A. Fate mapping the mouse embryo. *Int J Dev Biol* **1999**; *43*: 773–775.
16 Lawson, K. A. and Pedersen, R. A. Clonal analysis of cell fate during gastrulation and early neurulation in the mouse. *Ciba Found Symp* **1992**; *165*: 3–21; discussion 21–26.
17 Psychoyos, D. and Stern, C. D. Fates and migratory routes of primitive streak cells in the chick-embryo. *Development* **1996**; *122*: 1523–1534.
18 Schoenwolf, G. C., Garcia-Martinez, V. and Dias, M. S. Mesoderm movement and fate during avian gastrulation and neurulation. *Dev Dyn* **1992**; *193*: 235–248.
19 Smith, J. L., Gesteland, K. M. and Schoenwolf, G. C. Prospective fate map of the mouse primitive streak at 7.5 days of gestation. *Dev Dyn* **1994**; *201*: 279–289.
20 Spemann, H. and Mangold, H. Ueber induktion von Embryonalanlagen durch implantation artfremder Organisatoren. *Wilhelm Roux Arch. Entwicklungsmech* **1924**; *100*: 599–638.
21 Boettger, T., Knoetgen, H., Wittler, L. and Kessel, M. The avian organizer. *Int J Dev Biol* **2001**; *45*: 281–287.
22 Beddington, R. S. Induction of a second neural axis by the mouse node. *Development* **1994**; *120*: 613–620.
23 Townes, P. L. and Holtfreter, J. Directed movements and selective adhesion of embryonic amphibian cells. *J Exp Zool* **1955**; *128*: 53–120.

24 Nieuwkoop, P. D. The formation of the mesoderm in urodele amphibians. I. Induction by the endoderm. *Wilhelm Roux Arch Entwicklungsmech Org* **1969**; *162*: 341–373.

25 Le Douarin, N. A biological cell labeling technique and its use in expermental embryology. *Dev Biol* **1973**; *30*: 217–222.

26 Jacobson, A. G. and Gordon, R. Changes in the shape of the developing vertebrate nervous system analyzed experimentally, mathematically and by computer simulation. *J Exp Zool* **1976**; *197*: 191–246.

27 Keller, R., Davidson, L. A. and Shook, D. R. How we are shaped: the biomechanics of gastrulation. *Differentiation* **2003**; *71*: 171–205.

28 Shook, D. R., Majer, C. and Keller, R. Pattern and morphogenesis of presumptive superficial mesoderm in two closely related species, *Xenopus laevis* and *Xenopus tropicalis*. *Dev Biol* **2004**; *270*: 163–185.

29 Colas, J. F. and Schoenwolf, G. C. Towards a cellular and molecular understanding of neurulation. *Dev Dyn* **2001**; *221*: 117–145.

30 Beddington, R. S. and Robertson, E. J. Anterior patterning in mouse. *Trends Genet* **1998**; *14*: 277–284.

31 Quinlan, G. A., Davidson, B. P. and Tam, P. P. Lineage allocation during early embryogenesis. Mapping of the neural primordia and application to the analysis of mouse mutants. *Methods Mol Biol* **2001**; *158*: 227–250.

32 Tam, P. P. and Beddington, R. S. The formation of mesodermal tissues in the mouse embryo during gastrulation and early organogenesis. *Development* **1987**; 99: 109–126.

33 Tam, P. P. Regionalisation of the mouse embryonic ectoderm: allocation of prospective ectodermal tissues during gastrulation. *Development* **1989**; *107*: 55–67.

34 Tam, P. P. and Behringer, R. R. Mouse gastrulation: the formation of a mammalian body plan. *Mech Dev* **1997**; *68*: 3–25.

35 Tam, P. P., Kanai-Azuma, M. and Kanai, Y. Early endoderm development in vertebrates: lineage differentiation and morphogenetic function. *Curr Opin Genet Dev* **2003**; *13*: 393–400.

36 Wilson, V. and Beddington, R. S. Cell fate and morphogenetic movement in the late mouse primitive streak. *Mech Dev* **1996**; *55*: 79–89.

37 Keller, R. E. The cellular basis of epiboly: an SEM study of deep-cell rearrangement during gastrulation in *Xenopus laevis*. *J Embryol Exp Morphol* **1980**; *60*: 201–234.

38 Keller, R. E., Danilchik, M., Gimlich, R. and Shih, J. The function and mechanism of convergent extension during gastrulation of *Xenopus laevis*. *J Embryol Exp Morphol* **1985**; *89*(Suppl.): 185–209.

39 Wacker, S., Grimm, K., Joos, T. and Winklbauer, R. Development and control of tissue separation at gastrulation in Xenopus. *Dev Biol* **2000**; *224*: 428–439.

40 Wilt, F. H. and Hake, S. C. *Principles of Developmental Biology*. New York, London: W.W. Norton & Company, **2004**.

41 Winklbauer, R. and Schurfeld, M. Vegetal rotation, a new gastrulation movement involved in the internalization of the mes-

oderm and endoderm in Xenopus. *Development* **1999**; *126*: 3703–3713.

42 Boucaut, J. C., Darribere, T., Boulekbache, H. and Thiery, J. P. Prevention of gastrulation but not neurulation by antibodies to fibronectin in amphibian embryos. *Nature* **1984a**; *307*: 364–367.

43 Boucaut, J. C., Darribere, T., Poole, T. J., Aoyama, H., Yamada, K. M. and Thiery, J. P. Biologically active synthetic peptides as probes of embryonic development: a competitive peptide inhibitor of fibronectin function inhibits gastrulation in amphibian embryos and neural crest cell migration in avian embryos. *J Cell Biol* **1984b**; *99*: 1822–1830.

44 Davidson, L. A., Hoffstrom, B. G., Keller, R. and DeSimone, D. W. Mesendoderm extension and mantle closure in *Xenopus laevis* gastrulation: combined roles for integrin alpha(5)beta(1), fibronectin, and tissue geometry. *Dev Biol* **2002**; *242*: 109–129.

45 Winklbauer, R. Mesodermal cell migration during Xenopus gastrulation. *Dev Biol* **1990**; *142*: 155–168.

46 Winklbauer, R. and Keller, R. E. Fibronectin, mesoderm migration, and gastrulation in Xenopus. *Dev Biol* **1996**; *177*: 413–426.

47 Medina, A., Swain, R. K., Kuerner, K. M. and Steinbeisser, H. Xenopus paraxial protocadherin has signaling functions and is involved in tissue separation. *EMBO J* **2004**; *23*: 3249–3258.

48 Winklbauer, R., Medina, A., Swain, R. K. and Steinbeisser, H. Frizzled-7 signalling controls tissue separation during Xenopus gastrulation. *Nature* **2001**; *413*: 856–860.

49 Keller, R. Shaping the vertebrate body plan by polarized embryonic cell movements. *Science* **2002**; *298*: 1950–1954.

50 Winklbauer, R. and Stoltz, C. Fibronectin fibril growth in the extracellular matrix of the Xenopus embryo. *J Cell Sci* **1995**; *108*(Pt 4): 1575–1586.

51 Marsden, M. and DeSimone, D. W. Integrin-ECM interactions regulate cadherin-dependent cell adhesion and are required for convergent extension in Xenopus. *Curr Biol* **2003**; *13*: 1182–1191.

52 Longo, D., Peirce, S. M., Skalak, T. C., Davidson, L., Marsden, M., Dzamba, B. and DeSimone, D. W. Multicellular computer simulation of morphogenesis: blastocoel roof thinning and matrix assembly in *Xenopus laevis*. *Dev Biol* **2004**; *271*: 210–222.

53 Copp, A. J., Greene, N. D. and Murdoch, J. N. The genetic basis of mammalian neurulation. *Nat Rev Genet* **2003**; *4*: 784–793.

54 Lawson, A., Anderson, H. and Schoenwolf, G. C. Cellular mechanisms of neural fold formation and morphogenesis in the chick embryo. *Anat Rec* **2001a**; *262*: 153–168.

55 Lawson, A. and Schoenwolf, G. C. New insights into critical events of avian gastrulation. *Anat Rec* **2001**; *262*: 238–252.

56 Lowery, L. A. and Sive, H. Strategies of vertebrate neurulation and a re-evaluation of teleost neural tube formation. *Mech Dev* **2004**; *121*: 1189–1197.

57 Kimmel, C. B. Genetics and early development of zebrafish. *Trends Genet* **1989**; *5*: 283–288.

58 Kimmel, C. B., Ballard, W. W., Kimmel, S. R., Ullmann, B. and Schilling, T. F. Stages of embryonic development of the zebrafish. *Dev Dyn* **1995**; *203*: 253–310.

59 Nusslein-Volhard, C. Of flies and fishes. *Science* **1994**; *266*: 572–574.

60 Stemple, D. L. and Driever, W. Zebrafish: tools for investigating cellular differentiation. *Curr Opin Cell Biol* **1996**; *8*: 858–864.

61 Solnica-Krezel, L., Stemple, D. L. and Driever, W. Transparent things: cell fates and cell movements during early embryogenesis of zebrafish. *Bioessays* **1995**; *17*: 931–939.

62 Betchaku, T. and Trinkaus, J. P. Contact relations, surface activity, and cortical microfilaments of marginal cells of the enveloping layer and of the yolk syncytial and yolk cytoplasmic layers of fundulus before and during epiboly. *J Exp Zool* **1978**; *206*: 381–426.

63 Solnica-Krezel, L. and Driever, W. Microtubule arrays of the zebrafish yolk cell: organization and function during epiboly. *Development* **1994**; *120*: 2443–2455.

64 Strahle, U. and Jesuthasan, S. Ultraviolet irradiation impairs epiboly in zebrafish embryos: evidence for a microtubule-dependent mechanism of epiboly. *Development* **1993**; *119*: 909–919.

65 Glickman, N. S., Kimmel, C. B., Jones, M. A. and Adams, R. J. Shaping the zebrafish notochord. *Development* **2003**; *130*: 873–887.

66 Hammerschmidt, M., Pelegri, F., Mullins, M. C., Kane, D. A., Brand, M., van Eeden, F. J., Furutani-Seiki, M., Granato, M., Haffter, P., Heisenberg, C. P. et al. Mutations affecting morphogenesis during gastrulation and tail formation in the zebrafish, Danio rerio. *Development* **1996**; *123*: 143–151.

67 Kane, D. A., Hammerschmidt, M., Mullins, M. C., Maischein, H. M., Brand, M., van Eeden, F. J., Furutani-Seiki, M., Granato, M., Haffter, P., Heisenberg, C. P. et al. The zebrafish epiboly mutants. *Development* **1996**; *123*: 47–55.

68 Odenthal, J., Haffter, P., Vogelsang, E., Brand, M., van Eeden, F. J., Furutani-Seiki, M., Granato, M., Hammerschmidt, M., Heisenberg, C. P., Jiang, Y. J. et al. Mutations affecting the formation of the notochord in the zebrafish, Danio rerio. *Development* **1996**; *123*: 103–115.

69 Solnica-Krezel, L., Stemple, D. L., Mountcastle-Shah, E., Rangini, Z., Neuhauss, S. C., Malicki, J., Schier, A. F., Stainier, D. Y., Zwartkruis, F., Abdelilah, S. et al. Mutations affecting cell fates and cellular rearrangements during gastrulation in zebrafish. *Development* **1996**; *123*: 67–80.

70 Stemple, D. L., Solnica-Krezel, L., Zwartkruis, F., Neuhauss, S. C., Schier, A. F., Malicki, J., Stainier, D. Y., Abdelilah, S., Rangini, Z., Mountcastle-Shah, E. et al. Mutations affecting development of the notochord in zebrafish. *Development* **1996**; *123*: 117–128.

71 Eyal-Giladi, H. Establishment of the axis in chordates: facts and speculations. *Development* **1997**; *124*: 2285–2296.

72 Khaner, O. Axis determination in the avian embryo. *Curr Top Dev Biol* **1993**; *28*: 155–180.

73 Khaner, O. The rotated hypoblast of the chicken embryo does not initiate an ectopic axis in the epiblast. *Proc Natl Acad Sci USA* **1995**; *92*: 10733–10737.

74 Bachvarova, R. F., Skromne, I. and Stern, C. D. Induction of primitive streak and Hensen's node by the posterior marginal zone in the early chick embryo. *Development* **1998**; *125*: 3521–3534.
75 Khaner, O. The ability to initiate an axis in the avian blastula is concentrated mainly at a posterior site. *Dev Biol* **1998**; *194*: 257–266.
76 Skromne, I. and Stern, C. D. Interactions between Wnt and Vg1 signalling pathways initiate primitive streak formation in the chick embryo. *Development* **2001**; *128*: 2915–2927.
77 Skromne, I. and Stern, C. D. A hierarchy of gene expression accompanying induction of the primitive streak by Vg1 in the chick embryo. *Mech Dev* **2002**; *114*: 115–118.
78 Zahavi, N., Reich, V. and Khaner, O. High proliferation rate characterizes the site of axis formation in the avian blastula-stage embryo. *Int J Dev Biol* **1998**; *42*: 95–98.
79 Lawson, A., Colas, J. F. and Schoenwolf, G. C. Classification scheme for genes expressed during formation and progression of the avian primitive streak. *Anat Rec* **2001b**; *262*: 221–226.
80 Wei, Y. and Mikawa, T. Formation of the avian primitive streak from spatially restricted blastoderm: evidence for polarized cell division in the elongating streak. *Development* **2000**; *127*: 87–96.
81 Garcia-Martinez, V. and Schoenwolf, G. C. Positional control of mesoderm movement and fate during avian gastrulation and neurulation. *Dev Dyn* **1992**; *193*: 249–256.
82 Selleck, M. A. and Stern, C. D. Fate mapping and cell lineage analysis of Hensen's node in the chick embryo. *Development* **1991**; *112*: 615–626.
83 Yang, X., Dormann, D., Münsterberg, A. E. and Weijer, C. J. Cell movement patterns during gastrulation in the chick are controlled by positive and negative chemotaxis mediated by FGF4 and FGF8. *Dev Cell* **2002**; *3*: 425–437.
84 Charrier, J. B., Teillet, M. A., Lapointe, F. and Le Douarin, N. M. Defining subregions of Hensen's node essential for caudalward movement, midline development and cell survival. *Development* **1999**; *126*: 4771–4783.
85 Bielinska, M., Narita, N. and Wilson, D. B. Distinct roles for visceral endoderm during embryonic mouse development. *Int J Dev Biol* **1999**; *43*: 183–205.
86 Huelsken, J. and Behrens, J. The Wnt signalling pathway. *J Cell Sci* **2002**; *115*: 3977–3978.
87 Brannon, M. and Kimelman, D. Activation of Siamois by the Wnt pathway. *Dev Biol* **1996**; *180*: 344–347.
88 Fagotto, F., Guger, K. and Gumbiner, B. M. Induction of the primary dorsalizing center in Xenopus by the Wnt/GSK/beta-catenin signaling pathway, but not by Vg1, Activin or Noggin. *Development* **1997**; *124*: 453–460.
89 Fan, M. J. and Sokol, S. Y. A role for Siamois in Spemann organizer formation. *Development* **1997**; *124*: 2581–2589.
90 Laurent, M. N., Blitz, I. L., Hashimoto, C., Rothbacher, U. and Cho, K. W. The Xenopus homeobox gene twin mediates Wnt induction of goosecoid in establishment of Spemann's organizer. *Development* **1997**; *124*: 4905–4916.

91 Adler, P. N. Planar signaling and morphogenesis in Drosophila. *Dev Cell* **2002**; *2*: 525–535.

92 Axelrod, J. D. and McNeill, H. Coupling planar cell polarity signaling to morphogenesis. *Sci World J* **2002**; *2*: 434–454.

93 Tada, M., Concha, M. L. and Heisenberg, C. P. Non-canonical Wnt signalling and regulation of gastrulation movements. *Semin Cell Dev Biol* **2002**; *13*: 251–260.

94 Veeman, M. T., Axelrod, J. D. and Moon, R. T. A second canon. Functions and mechanisms of beta-catenin-independent Wnt signaling. *Dev Cell* **2003**; *5*: 367–377.

95 Wallingford, J. B., Rowning, B. A., Vogeli, K. M., Rothbacher, U., Fraser, S. E. and Harland, R. M. Dishevelled controls cell polarity during Xenopus gastrulation. *Nature* **2000**; *405*: 81–85.

96 Axelrod, J. D., Miller, J. R., Shulman, J. M., Moon, R. T. and Perrimon, N. Differential recruitment of Dishevelled provides signaling specificity in the planar cell polarity and Wingless signaling pathways. *Genes Dev* **1998**; *12*: 2610–2622.

97 Capelluto, D. G., Kutateladze, T. G., Habas, R., Finkelstein, C. V., He, X. and Overduin, M. The DIX domain targets dishevelled to actin stress fibres and vesicular membranes. *Nature* **2002**; *419*: 726–729.

98 Yan, D., Wallingford, J. B., Sun, T. Q., Nelson, A. M., Sakanaka, C., Reinhard, C., Harland, R. M., Fantl, W. J. and Williams, L. T. Cell autonomous regulation of multiple Dishevelled-dependent pathways by mammalian Nkd. *Proc Natl Acad Sci USA* **2001**; *98*: 3802–3807.

99 Cong, F., Schweizer, L. and Varmus, H. Casein kinase Iepsilon modulates the signaling specificities of dishevelled. *Mol Cell Biol* **2004**; *24*: 2000–2011.

100 Du, S. J., Purcell, S. M., Christian, J. L., McGrew, L. L. and Moon, R. T. Identification of distinct classes and functional domains of Wnts through expression of wild-type and chimeric proteins in Xenopus embryos. *Mol Cell Biol* **1995**; *15*: 2625–2634.

101 Moon, R. T., Campbell, R. M., Christian, J. L., McGrew, L. L., Shih, J. and Fraser, S. Xwnt–5A: a maternal Wnt that affects morphogenetic movements after overexpression in embryos of *Xenopus laevis*. *Development* **1993**; *119*: 97–111.

102 Sokol, S. Y. Analysis of Dishevelled signalling pathways during Xenopus development. *Curr Biol* **1996**; *6*: 1456–1467.

103 Heisenberg, C. P., Tada, M., Rauch, G. J., Saude, L., Concha, M. L., Geisler, R., Stemple, D. L., Smith, J. C. and Wilson, S. W. Silberblick/Wnt11 mediates convergent extension movements during zebrafish gastrulation. *Nature* **2000**; *405*: 76–81.

104 Tada, M. and Smith, J. C. Xwnt11 is a target of Xenopus Brachyury: regulation of gastrulation movements via Dishevelled, but not through the canonical Wnt pathway. *Development* **2000**; *127*: 2227–2238.

105 Djiane, A., Riou, J., Umbhauer, M., Boucaut, J. and Shi, D. Role of frizzled 7 in the regulation of convergent extension movements during gastrulation in *Xenopus laevis*. *Development* **2000**; *127*: 3091–3100.

106 Westfall, T. A., Brimeyer, R., Twedt, J., Gladon, J., Olberding, A., Furutani-Seiki, M. and Slusarski, D. C. Wnt–5/pipetail

functions in vertebrate axis formation as a negative regulator of Wnt/beta-catenin activity. *J Cell Biol* **2003**; *162*: 889–898.
107 Carreira-Barbosa, F., Concha, M. L., Takeuchi, M., Ueno, N., Wilson, S. W. and Tada, M. Prickle 1 regulates cell movements during gastrulation and neuronal migration in zebrafish. *Development* **2003**; *130*: 4037–4046.
108 Kilian, B., Mansukoski, H., Barbosa, F. C., Ulrich, F., Tada, M. and Heisenberg, C. P. The role of Ppt/Wnt5 in regulating cell shape and movement during zebrafish gastrulation. *Mech Dev* **2003**; *120*: 467–476.
109 Topczewski, J., Sepich, D. S., Myers, D. C., Walker, C., Amores, A., Lele, Z., Hammerschmidt, M., Postlethwait, J. and Solnica-Krezel, L. The zebrafish glypican knypek controls cell polarity during gastrulation movements of convergent extension. *Dev Cell* **2001**; *1*: 251–264.
110 Marlow, F., Gonzalez, E. M., Yin, C., Rojo, C. and Solnica-Krezel, L. No tail co-operates with non-canonical Wnt signaling to regulate posterior body morphogenesis in zebrafish. *Development* **2004**; *131*: 203–216.
111 Rawls, A. S. and Wolff, T. Strabismus requires Flamingo and Prickle function to regulate tissue polarity in the Drosophila eye. *Development* **2003**; *130*: 1877–1887.
112 Heisenberg, C. P. Wnt signalling: Refocusing on Strabismus. *Curr Biol* **2002**; *12*: R657–R659.
113 Heisenberg, C. P. and Tada, M. Wnt signalling: a moving picture emerges from van Gogh. *Curr Biol* **2002a**; *12*: R126–R128.
114 Heisenberg, C. P. and Tada, M. Zebrafish gastrulation movements: bridging cell and developmental biology. *Semin Cell Dev Biol* **2002b**; *13*: 471–479.
115 Park, M. and Moon, R. T. The planar cell-polarity gene stbm regulates cell behaviour and cell fate in vertebrate embryos. *Nat Cell Biol* **2002**; *4*: 20–25.
116 Gong, Y., Mo, C. and Fraser, S. E. Planar cell polarity signalling controls cell division orientation during zebrafish gastrulation. *Nature* **2004**; *430*: 689–693.
117 Thorpe, C. J. and Moon, R. T. nemo-like kinase is an essential co-activator of Wnt signaling during early zebrafish development. *Development* **2004**; *131*: 2899–2909.
118 Sasai, N., Nakazawa, Y., Haraguchi, T. and Sasai, Y. The neurotrophin-receptor-related protein NRH1 is essential for convergent extension movements. *Nat Cell Biol* **2004**; *6*: 741–748.
119 Kibar, Z., Vogan, K. J., Groulx, N., Justice, M. J., Underhill, D. A. and Gros, P. Ltap, a mammalian homolog of Drosophila Strabismus/Van Gogh, is altered in the mouse neural tube mutant Loop-tail. *Nat Genet* **2001**; *28*: 251–255.
120 Murdoch, J. N., Doudney, K., Paternotte, C., Copp, A. J. and Stanier, P. Severe neural tube defects in the loop-tail mouse result from mutation of Lpp1, a novel gene involved in floor plate specification. *Hum Mol Genet* **2001**; *10*: 2593–2601.
121 Curtin, J. A., Quint, E., Tsipouri, V., Arkell, R. M., Cattanach, B., Copp, A. J., Henderson, D. J., Spurr, N., Stanier, P., Fisher, E. M. et al. Mutation of Celsr1 disrupts planar polarity of inner ear hair cells and causes severe neural tube defects in the mouse. *Curr Biol* **2003**; *13*: 1129–1133.

122 Dabdoub, A., Donohue, M. J., Brennan, A., Wolf, V., Montcouquiol, M., Sassoon, D. A., Hseih, J. C., Rubin, J. S., Salinas, P. C. and Kelley, M. W. Wnt signaling mediates reorientation of outer hair cell stereociliary bundles in the mammalian cochlea. *Development* **2003**; *130*: 2375–2384.

123 Montcouquiol, M., Rachel, R. A., Lanford, P. J., Copeland, N. G., Jenkins, N. A. and Kelley, M. W. Identification of Vangl2 and Scrb1 as planar polarity genes in mammals. *Nature* **2003**; 423, 173–177.

124 Lu, X., Borchers, A. G., Jolicoeur, C., Rayburn, H., Baker, J. C. and Tessier-Lavigne, M. PTK7/CCK–4 is a novel regulator of planar cell polarity in vertebrates. *Nature* **2004**; *430*: 93–98.

125 Fanto, M. and McNeill, H. Planar polarity from flies to vertebrates. *J Cell Sci* **2004**; *117*: 527–533.

126 Nelson, W. J. and Nusse, R. Convergence of Wnt, beta-catenin, and cadherin pathways. *Science* **2004**; *303*: 1483–1487.

127 Choi, S. C. and Han, J. K. Xenopus Cdc42 regulates convergent extension movements during gastrulation through Wnt/Ca2+ signaling pathway. *Dev Biol* **2002**; *244*: 342–357.

128 Penzo-Mendez, A., Umbhauer, M., Djiane, A., Boucaut, J. C. and Riou, J. F. Activation of Gbetagamma signaling downstream of Wnt–11/Xfz7 regulates Cdc42 activity during Xenopus gastrulation. *Dev Biol* **2003**; *257*: 302–314.

129 Tahinci, E. and Symes, K. Distinct functions of Rho and Rac are required for convergent extension during Xenopus gastrulation. *Dev Biol* **2003**; *259*: 318–335.

130 Habas, R., Dawid, I. B. and He, X. Coactivation of Rac and Rho by Wnt/Frizzled signaling is required for vertebrate gastrulation. *Genes Dev* **2003**; *17*: 295–309.

131 Iioka, H., Ueno, N. and Kinoshita, N. Essential role of MARCKS in cortical actin dynamics during gastrulation movements. *J Cell Biol* **2004**; *164*: 169–14.

132 Wilson, R. and Leptin, M. Fibroblast growth factor receptor-dependent morphogenesis of the Drosophila mesoderm. *Philos Trans R Soc Lond B Biol Sci* **2000**; *355*: 891–895.

133 Amaya, E., Musci, T. J. and Kirschner, M. W. Expression of a dominant negative mutant of the FGF receptor disrupts mesoderm formation in Xenopus embryos. *Cell* **1991**; *66*: 257–270.

134 Kroll, K. L. and Amaya, E. Transgenic Xenopus embryos from sperm nuclear transplantations reveal FGF signaling requirements during gastrulation. *Development* **1996**; *122*: 3173–3183.

135 Nutt, S. L., Dingwell, K. S., Holt, C. E. and Amaya, E. Xenopus Sprouty2 inhibits FGF-mediated gastrulation movements but does not affect mesoderm induction and patterning. *Genes Dev* **2001**; *15*: 1152–1166.

136 Dubrulle, J. and Pourquie, O. fgf8 mRNA decay establishes a gradient that couples axial elongation to patterning in the vertebrate embryo. *Nature* **2004**; *427*: 419–422.

137 Ciruna, B. and Rossant, J. FGF signaling regulates mesoderm cell fate specification and morphogenetic movement at the primitive streak. *Dev Cell* **2001**; *1*: 37–49.

138 Ciruna, B. G., Schwartz, L., Harpal, K., Yamaguchi, T. P. and Rossant, J. Chimeric analysis of fibroblast growth factor recep-

tor–1 (Fgfr1) function: a role for FGFR1 in morphogenetic movement through the primitive streak. *Development* **1997**; *124*: 2829–2841.
139. Sun, X., Meyers, E. N., Lewandoski, M. and Martin, G. R. Targeted disruption of Fgf8 causes failure of cell migration in the gastrulating mouse embryo. *Genes Dev* **1999**; *13*: 1834–1846.
140. Yamaguchi, T. P., Harpal, K., Henkemeyer, M. and Rossant, J. fgfr–1 is required for embryonic growth and mesodermal patterning during mouse gastrulation. *Genes Dev* **1994**; *8*: 3032–3044.
141. Burdsal, C. A., Damsky, C. H. and Pedersen, R. A. The role of E-cadherin and integrins in mesoderm differentiation and migration at the mammalian primitive streak. *Development* **1993**; *118*: 829–844.
142. Fafeur, V., Tulasne, D., Queva, C., Vercamer, C., Dimster, V., Mattot, V., Stehelin, D., Desbiens, X. and Vandenbunder, B. The ETS1 transcription factor is expressed during epithelial-mesenchymal transitions in the chick embryo and is activated in scatter factor-stimulated MDCK epithelial cells. *Cell Growth Differ* **1997**; *8*: 655–665.
143. Maroulakou, I. G. and Bowe, D. B. Expression and function of Ets transcription factors in mammalian development: a regulatory network. *Oncogene* **2000**; *19*: 6432–6442.
144. Meyer, D., Durliat, M., Senan, F., Wolff, M., Andre, M., Hourdry, J. and Remy, P. Ets–1 and Ets–2 proto-oncogenes exhibit differential and restricted expression patterns during *Xenopus laevis* oogenesis and embryogenesis. *Int J Dev Biol* **1997**; *41*: 607–620.
145. Remy, P. and Baltzinger, M. The Ets-transcription factor family in embryonic development: lessons from the amphibian and bird. *Oncogene* **2000**; *19*: 6417–6431.
146. Draper, B. W., Stock, D. W. and Kimmel, C. B. Zebrafish fgf24 functions with fgf8 to promote posterior mesodermal development. *Development* **2003**; *130*: 4639–4654.
147. Carver, E. A., Jiang, R., Lan, Y., Oram, K. F. and Gridley, T. The mouse snail gene encodes a key regulator of the epithelial-mesenchymal transition. *Mol Cell Biol* **2001**; *21*: 8184–8188.
148. Batlle, E., Sancho, E., Franci, C., Dominguez, D., Monfar, M., Baulida, J. and Garcia De Herreros, A. The transcription factor snail is a repressor of E-cadherin gene expression in epithelial tumour cells. *Nat Cell Biol* **2000**; *2*: 84–89.
149. Cano, A., Perez-Moreno, M. A., Rodrigo, I., Locascio, A., Blanco, M. J., del Barrio, M. G., Portillo, F. and Nieto, M. A. The transcription factor snail controls epithelial-mesenchymal transitions by repressing E-cadherin expression. *Nat Cell Biol* **2000**; *2*: 76–83.
150. Nieto, M. A., Sargent, M. G., Wilkinson, D. G. and Cooke, J. Control of cell behavior during vertebrate development by Slug, a zinc finger gene. *Science* **1994**; *264*: 835–839.
151. Yamashita, S., Miyagi, C., Carmany-Rampey, A., Shimizu, T., Fujii, R., Schier, A. F. and Hirano, T. Stat3 Controls Cell Movements during Zebrafish Gastrulation. *Dev Cell* **2002**; *2*: 363–375.

152 Yamashita, S., Miyagi, C., Fukada, T., Kagara, N., Che, Y. S. and Hirano, T. Zinc transporter LIVI controls epithelial-mesenchymal transition in zebrafish gastrula organizer. *Nature* **2004**; *429*: 298–302.

153 Smallhorn, M., Murray, M. J. and Saint, R. The epithelial-mesenchymal transition of the Drosophila mesoderm requires the Rho GTP exchange factor Pebble. *Development* **2004**; *131*: 2641–2651.

154 Steinberg, M. S. Reconstruction of tissues by dissociated cells. Some morphogenetic tissue movements and the sorting out of embryonic cells may have a common explanation. *Science* **1963**; *141*: 401–408.

155 Takeichi, M. Cadherin cell adhesion receptors as a morphogenetic regulator. *Science* **1991**; *251*: 1451–1455.

156 Godt, D. and Tepass, U. Drosophila oocyte localization is mediated by differential cadherin-based adhesion. *Nature* **1998**; *395*: 387–391.

157 Tepass, U., Truong, K., Godt, D., Ikura, M. and Peifer, M. Cadherins in embryonic and neural morphogenesis. *Nat Rev Mol Cell Biol* **2000**; *1*: 91–100.

158 Brieher, W. M. and Gumbiner, B. M. Regulation of C-cadherin function during activin induced morphogenesis of Xenopus animal caps. *J Cell Biol* **1994**; *126*: 519–527.

159 Lee, C. H. and Gumbiner, B. M. Disruption of gastrulation movements in Xenopus by a dominant-negative mutant for C-cadherin. *Dev Biol* **1995**; *171*: 363–373.

160 Zhong, Y., Brieher, W. M. and Gumbiner, B. M. Analysis of C-cadherin regulation during tissue morphogenesis with an activating antibody. *J Cell Biol* **1999**; *144*: 351–359.

161 Bradley, R. S., Espeseth, A. and Kintner, C. NF-protocadherin, a novel member of the cadherin superfamily, is required for Xenopus ectodermal differentiation. *Curr Biol* **1998**; *8*: 325–334.

162 Heggem, M. A. and Bradley, R. S. The cytoplasmic domain of Xenopus NF-protocadherin interacts with TAF1/set. *Dev Cell* **2003**; *4*: 419–429.

163 Kim, S. H., Yamamoto, A., Bouwmeester, T., Agius, E. and Robertis, E. M. The role of paraxial protocadherin in selective adhesion and cell movements of the mesoderm during Xenopus gastrulation. *Development* **1998**; *125*: 4681–4690.

164 Kuroda, H., Inui, M., Sugimoto, K., Hayata, T. and Asashima, M. Axial protocadherin is a mediator of prenotochord cell sorting in Xenopus. *Dev Biol* **2002**; *244*: 267–277.

165 Hukriede, N. A., Tsang, T. E., Habas, R., Khoo, P. L., Steiner, K., Weeks, D. L., Tam, P. P. and Dawid, I. B. Conserved requirement of Lim1 function for cell movements during gastrulation. *Dev Cell* **2003**; *4*: 83–94.

166 Unterseher, F., Hefele, J. A., Giehl, K., De Robertis, E. M., Wedlich, D. and Schambony, A. Paraxial protocadherin coordinates cell polarity during convergent extension via Rho A and JNK. *EMBO J* **2004**; *23*: 3259–3269.

167 Yamamoto, A., Amacher, S. L., Kim, S. H., Geissert, D., Kimmel, C. B. and De Robertis, E. M. Zebrafish paraxial protocad-

herin is a downstream target of spadetail involved in morphogenesis of gastrula mesoderm. *Development* **1998**; *125*: 3389–3397.

168 Abdelkhalek, H. B., Beckers, A., Schuster-Gossler, K., Pavlova, M. N., Burkhardt, H., Lickert, H., Rossant, J., Reinhardt, R., Schalkwyk, L. C., Muller, I. et al. The mouse homeobox gene Not is required for caudal notochord development and affected by the truncate mutation. *Genes Dev* **2004**; *18*: 1725–1736.

169 Lele, Z., Folchert, A., Concha, M., Rauch, G. J., Geisler, R., Rosa, F., Wilson, S. W., Hammerschmidt, M. and Bally-Cuif, L. parachute/n-cadherin is required for morphogenesis and maintained integrity of the zebrafish neural tube. *Development* **2002**; *129*: 3281–3294.

170 Hynes, R. O. Integrins: versatility, modulation, and signaling in cell adhesion. *Cell* **1992**; *69*: 11–25.

171 Darribere, T., Guida, K., Larjava, H., Johnson, K. E., Yamada, K. M., Thiery, J. P. and Boucaut, J. C. *In vivo* analyses of integrin beta 1 subunit function in fibronectin matrix assembly. *J Cell Biol* **1990**; *110*: 1813–1823.

172 Darribere, T., Yamada, K. M., Johnson, K. E. and Boucaut, J. C. The 140-kDa fibronectin receptor complex is required for mesodermal cell adhesion during gastrulation in the amphibian *Pleurodeles waltlii*. *Dev Biol* **1988**; *126*: 182–194.

173 Ramos, J. W. and DeSimone, D. W. Xenopus embryonic cell adhesion to fibronectin: position-specific activation of RGD/synergy site-dependent migratory behavior at gastrulation. *J Cell Biol* **1996**; *134*: 227–240.

174 Ramos, J. W., Whittaker, C. A. and DeSimone, D. W. Integrin-dependent adhesive activity is spatially controlled by inductive signals at gastrulation. *Development* **1996**; *122*: 2873–2883.

175 Yang, J. T., Rayburn, H. and Hynes, R. O. Embryonic mesodermal defects in alpha 5 integrin-deficient mice. *Development* **1993**; *119*: 1093–1105.

176 Lallier, T. E. and DeSimone, D. W. Separation of neural induction and neurulation in Xenopus. *Dev Biol* **2000**; *225*: 135–150.

177 Lallier, T. E., Whittaker, C. A. and DeSimone, D. W. Integrin alpha 6 expression is required for early nervous system development in *Xenopus laevis*. *Development* **1996**; *122*: 2539–2554.

178 Giancotti, F. G. and Ruoslahti, E. Integrin signaling. *Science* **1999**; *285*: 1028–1032.

179 Yang, J. T., Bader, B. L., Kreidberg, J. A., Ullman-Cullere, M., Trevithick, J. E. and Hynes, R. O. Overlapping and independent functions of fibronectin receptor integrins in early mesodermal development. *Dev Biol* **1999**; *215*: 264–277.

180 Fassler, R. and Meyer, M. Consequences of lack of beta 1 integrin gene expression in mice. *Genes Dev* **1995**; *9*: 1896–1908.

181 Hagel, M., George, E. L., Kim, A., Tamimi, R., Opitz, S. L., Turner, C. E., Imamoto, A. and Thomas, S. M. The adaptor protein paxillin is essential for normal development in the mouse and is a critical transducer of fibronectin signaling. *Mol Cell Biol* **2002**; *22*: 901–915.

182 Ilic, D., Furuta, Y., Kanazawa, S., Takeda, N., Sobue, K., Nakatsuji, N., Nomura, S., Fujimoto, J., Okada, M. and Yama-

moto, T. Reduced cell motility and enhanced focal adhesion contact formation in cells from FAK-deficient mice. *Nature* **1995**; *377*: 539–544.

183 DeSimone, D. W., Norton, P. A. and Hynes, R. O. Identification and characterization of alternatively spliced fibronectin mRNAs expressed in early Xenopus embryos. *Dev Biol* **1992**; *149*: 357–369.

184 Hynes, R. O. *Fibronectins*. New York: Springer, **1990**.

185 George, E. L., Georges-Labouesse, E. N., Patel-King, R. S., Rayburn, H. and Hynes, R. O. Defects in mesoderm, neural tube and vascular development in mouse embryos lacking fibronectin. *Development* **1993**; *119*: 1079–1091.

186 Marsden, M. and DeSimone, D. W. Regulation of cell polarity, radial intercalation and epiboly in Xenopus: novel roles for integrin and fibronectin. *Development* **2001**; *128*: 3635–3647.

187 Nagel, M. and Winklbauer, R. Establishment of substratum polarity in the blastocoel roof of the Xenopus embryo. *Development* **1999**; *126*: 1975–1984.

188 Na, J., Marsden, M. and DeSimone, D. W. Differential regulation of cell adhesive functions by integrin alpha subunit cytoplasmic tails *in vivo*. *J Cell Sci* **2003**; *116*: 2333–2343.

189 Soriano, P. The PDGF alpha receptor is required for neural crest cell development and for normal patterning of the somites. *Development* **1997**; *124*: 2691–2700.

190 Ataliotis, P., Symes, K., Chou, M. M., Ho, L. and Mercola, M. PDGF signalling is required for gastrulation of *Xenopus laevis*. *Development* **1995**; *121*: 3099–3110.

191 Symes, K. and Mercola, M. Embryonic mesoderm cells spread in response to platelet-derived growth factor and signaling by phosphatidylinositol 3–kinase. *Proc Natl Acad Sci USA* **1996**; *93*: 9641–9644.

192 Nagel, M., Tahinci, E., Symes, K. and Winklbauer, R. Guidance of mesoderm cell migration in the Xenopus gastrula requires PDGF signaling. *Development* **2004**; *131*: 2727–2736.

193 Montero, J. A., Kilian, B., Chan, J., Bayliss, P. E. and Heisenberg, C. P. Phosphoinositide 3–kinase is required for process outgrowth and cell polarization of gastrulating mesendodermal cells. *Curr Biol* **2003**; *13*: 1279–1289.

194 Duncan, T. and Su, T. T. Embryogenesis: coordinating cell division with gastrulation. *Curr Biol* **2004**; *14*: R305–R307.

195 Grosshans, J. and Wieschaus, E. A genetic link between morphogenesis and cell division during formation of the ventral furrow in Drosophila. *Cell* **2000**; *101*: 523–531.

196 Johnston, L. A. The trouble with tribbles. *Curr Biol* **2000**; *10*: R502–R504.

197 Mata, J., Curado, S., Ephrussi, A. and Rorth, P. Tribbles coordinates mitosis and morphogenesis in Drosophila by regulating string/CDC25 proteolysis. *Cell* **2000**; *101*: 511–522.

198 Seher, T. C. and Leptin, M. Tribbles, a cell-cycle brake that coordinates proliferation and morphogenesis during Drosophila gastrulation. *Curr Biol* **2000**; *10*: 623–629.

199 Saka, Y. and Smith, J. C. A Xenopus tribbles orthologue is required for the progression of mitosis and for development of the nervous system. *Dev Biol* **2004**; *273*: 210–225.

200 Leise, W. F., 3rd and Mueller, P. R. Inhibition of the cell cycle is required for convergent extension of the paraxial mesoderm during Xenopus neurulation. *Development* **2004**; *131*: 1703–1715.

201 Murakami, M. S., Moody, S. A., Daar, I. O. and Morrison, D. K. Morphogenesis during Xenopus gastrulation requires Wee1–mediated inhibition of cell proliferation. *Development* **2004**; *131*: 571–580.

202 Keller, R. Developmental biology: heading away from the rump. *Nature* **2004**; *430*: 305–306.

203 Ninomiya, H., Elinson, R. P. and Winklbauer, R. Antero-posterior tissue polarity links mesoderm convergent extension to axial patterning. *Nature* **2004**; *430*: 364–367.

5
Head Induction

Clemens Kiecker

5.1
Introduction

The vertebrate head is a complex structure that consists of derivatives of all three germ layers and is characterized by an accumulation of sensory organs, an oral apparatus and the brain, arguably the most intriguing vertebrate organ. It forms at the anterior end of the organism hence its induction is closely linked to the establishment of the anteroposterior (AP, head-to-tail; see Abbreviations) body axis and occurs early in embryogenesis. Thus, head formation depends on both the establishment of anterior identity as well as neural induction and, in fact, "anterior neural induction" and "head induction" are often used interchangeably. This view may seem somewhat simplistic, but it has proved useful in many experimental systems since the induction of anterior neural markers is the earliest reliable indication of head formation. Head-like structures are also found in non-vertebrates and it appears that basic AP patterning genes are conserved between all bilaterians [1, 2]. It has been proposed that the AP axis is the primordial axis of multicellular animals that corresponds to the single axis of polarity in radially symmetrical organisms such as *Hydra* [3]. Here, unifying concepts of how anterior identity is established in vertebrate embryos will be reviewed.

5.2
Classical Concepts for Head Induction

Fundamental progress in understanding head induction was made in the 1920s when Hilde Mangold and Hans Spemann performed transplantation experiments using amphibian embryos and demonstrated that tissue from the dorsal blastopore lip (DBL, a structure that becomes visible early in embryogenesis at the onset of gastrulation; Fig. 5.1A) is able to induce a secondary embryonic axis including a head when grafted to the ventral side of a host embryo (Fig. 5.2A). Transplantations between pigmented and unpigmented newt species revealed that only few cells of

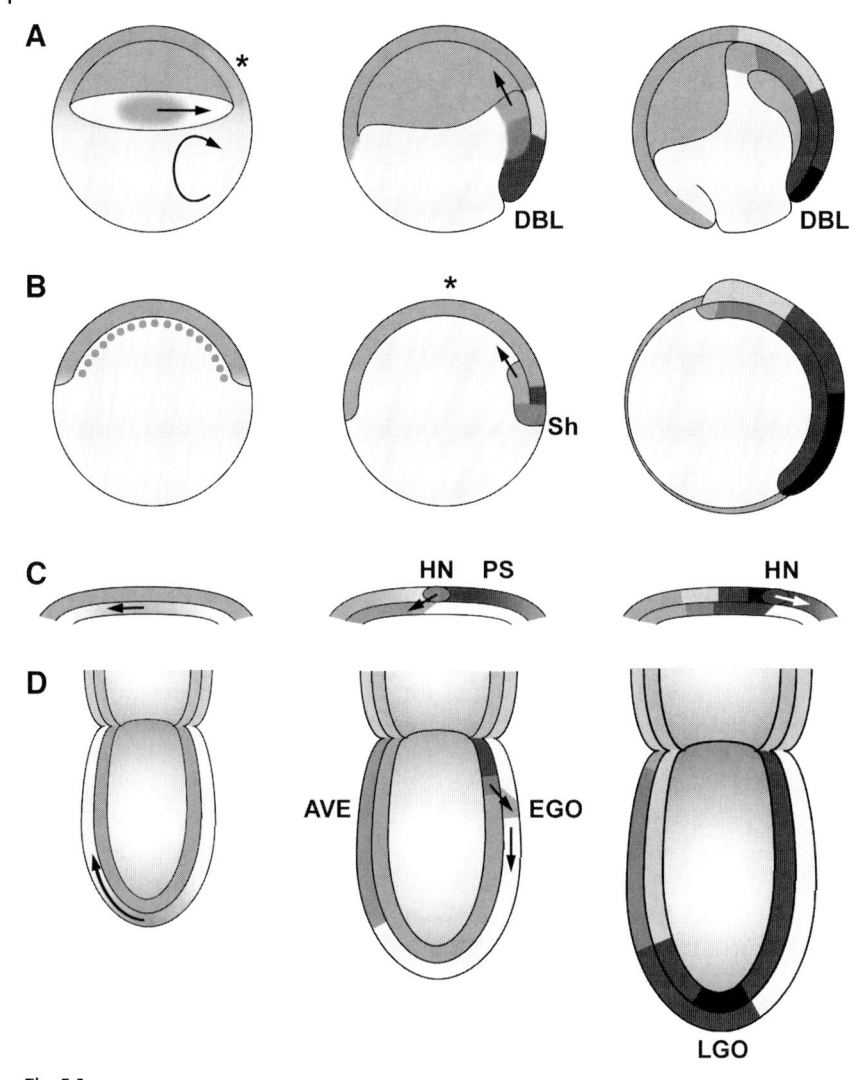

Fig. 5.1
Schematic depiction of early vertebrate embryos at comparable stages of development. Definitive AE is shown in orange, PME in red, CM in brown. Anterior, medial and posterior neuroectoderm are shown in light blue, dark blue and black, respectively. Left-hand panels: pre-gastrula; middle panels: early gastrula; right-hand panels: late gastrula. (A) *Xenopus* embryo, dorsal side is to the right, animal/anterior pole points upwards. Note "vegetal rotation" and BCNE (asterisk) at pre-gastrula stage. (B) Zebrafish embryo, dorsal side is to the right, animal/anterior points upwards. Note YSL (pink dots) in left panel. Row–1 is indicated by asterisk in early gastrula. Sh, shield. (C) Chick embryo, anterior is to the left, dorsal side points upwards. Note anterior movement of hypoblast in left and posterior regression of Hensen's node (HN) in right-hand panel. PS, primitive streak. (D) Mouse embryo, anterior is to the left, distal tip of egg cylinder points down. Note movement of AVE from distal (pre-gastrula) to anterior pole (early gastrula) and its subsequent displacement by PME (late gastrula). Adapted from [226]. (This figure also appears with the color plates.)

the experimentally-induced axis were derived from the graft while most of it consisted of host-derived tissue, indicating that the grafted cells had recruited and organized surrounding tissue to form an axis [4]. These observations coined the term "gastrula organizer" for the amphibian DBL and earned Spemann the Nobel Prize for Medicine in 1935. In the 1930s, Spemann extended his experiments by using grafts from embryos at different stages of gastrulation. He found that DBLs from young gastrulas induced complete ectopic axes including heads while DBLs from advanced gastrulas induced only posterior structures such as trunk and tail, suggesting that the organizer progressively loses its head-inducing potential during gastrulation (Fig. 5.2B) [5]. Cell labeling studies have shown that the DBL region is not a fixed population of cells, but a dynamic structure, through which mesodermal and endodermal cells invaginate during gastrulation [6]. Thus, Spemann's observations are best explained by a population of cells with head-inducing potential that is present in the early, but not the late, DBL.

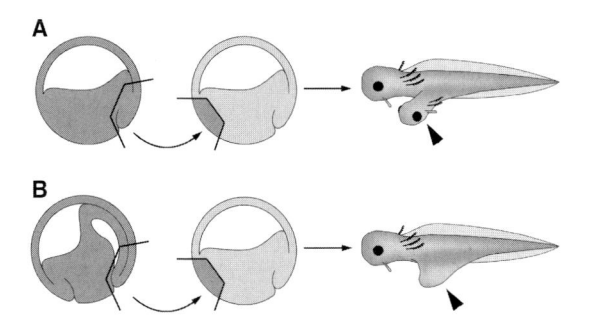

Fig. 5.2
The Spemann/Mangold organizer experiment. (A) Grafting DBL tissue from an early amphibian gastrula to the ventral side of a host embryo results in induction of a secondary body axis with head. (B) Grafting DBL tissue from a late gastrula only induces secondary trunk structures. Arrowheads mark ectopic inductions. Adapted from [276].

The early gastrula DBL is fated to form the axial mesendoderm (AME) that, from anterior to posterior, consists of anterior endoderm (AE), prechordal mesendoderm (PME) and chordamesoderm (CM). Spemann's colleague Otto Mangold dissected the archenteron roof of early neurula embryos (containing the AME) into four transverse segments and tested their inductive potential in gastrula hosts (Fig. 5.3). The anterior-most segments (containing AE) only induced ectopic balancers and parts of the oral apparatus while segments containing PME and anterior CM frequently induced ectopic head structures such as balancers, eyes, otic vesicles, fore- and midbrain. Segments containing central or posterior CM only induced ectopic trunk and tail structures, respectively. Mangold concluded that induction by the organizer and its derivatives is regionally specific, with tissue between PME and anterior CM emitting the strongest head-inducing signals [7]. Taken together, Spemann's and Mangold's observations suggest that the gastrula organizer progressively loses its head-inducing potential because PME and anterior CM precursors are present in the early gastrula DBL but leave this region soon after the onset of gastrulation. They proposed that the gastrula organizer consists of distinct head- and trunk-inducing regions.

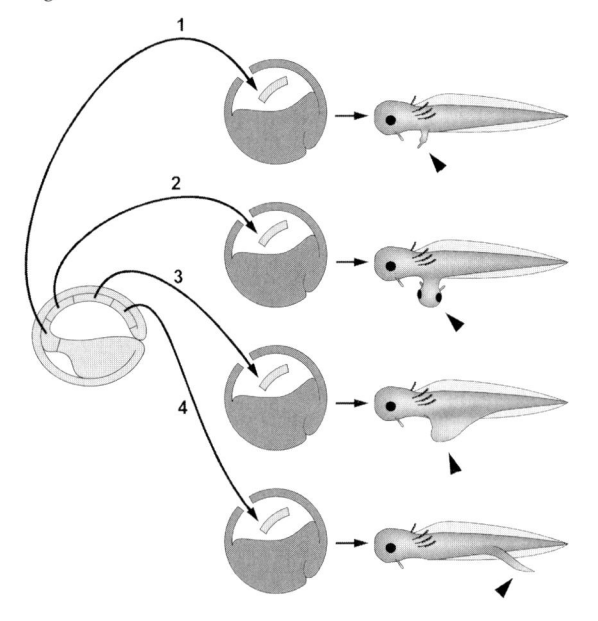

Fig. 5.3
Regionally specific induction by organizer derivatives. The archenteron roof of an amphibian neurula (light gray, anterior to the left, dorsal to the top) was cut into four consecutive AP stripes that were grafted into host gastrulas (dark gray). The anterior-most explant induced ectopic balancers (1), the second induced additional heads (2), the third induced extra trunk structures (3) and the posterior-most explant induced ectopic tails (4). Arrowheads mark ectopic inductions. Adapted from [276].

While Spemann's organizer model is able to explain the binary decision of head versus trunk induction, it is not sufficient to describe patterning along the entire AP axis. For example, how does the central nervous system become subdivided anteroposteriorly into forebrain, midbrain, hindbrain and spinal cord? In the 1950s, alternative models were developed that explain axial patterning by dose-dependent activities of diffusible signaling molecules (morphogens). Pieter Nieuwkoop and coworkers implanted ectodermal folds at different AP levels into the prospective neural plate of amphibian embryos. When Nieuwkoop analyzed these embryos at later stages, he found that folds implanted at the forebrain level had differentiated exclusively into anterior neural structures while folds implanted at more posterior levels had differentiated into posterior neural tissue at their base and into forebrain distally (Fig. 5.4A). Nieuwkoop proposed that AP neural patterning occurs in two steps (Fig. 5.4B): first, a vertical signal from the AME induces the overlying ectoderm to become anterior neuroectoderm (activation); second, a graded signal within the plane of the neuroectoderm that increases from anterior to posterior confers progressively posterior identity to the neural plate (transformation) [8]. According to this model, a head forms where the activating signal is present but the transforming signal is absent. Similarly, the "Finnish school" of embryologists suggested that AP patterning is regulated by two opposing gradients of an archencephalic (forebrain-type) and a spinocaudal (posteriorizing and mesoderm-inducing) inducer and that different AP levels are defined by their ratio of concentrations (Fig. 5.4C) [9]. We will see later that both qualitative organizer and quantitative gradient models are not necessarily contradictory and that they can be reconciled in light of our current knowledge of the molecular mechanisms mediating AP specification.

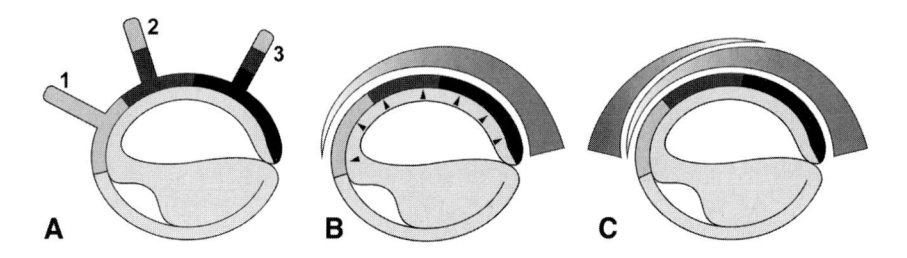

Fig. 5.4
Gradient models in AP patterning of amphibian embryos. Anterior, medial and posterior neuroectoderm are shown in light blue, dark blue and black, respectively. (A) Nieuwkoop's experiment: ectodermal folds differentiate into anterior neuroectoderm when implanted into the anterior neural plate (1), into anterior neuroectoderm distally and medial neural structures basally when implanted medially (2) and into an AP range of neural tissues from distal to proximal when implanted into the posterior neural plate (3). (B) Nieuwkoop's activation-transformation model. Black arrowheads indicate the activating signal and the orange wedge the transforming signal. (C) The Finnish two-inducer model. Archencephalic (green) and spinocaudal inducer (orange) form opposing AP gradients. Adapted from [236]. (This figure also appears with the color plates.)

5.3
Tissues Involved in Anterior Specification

5.3.1
The Gastrula Organizer

Structures equivalent to Spemann's organizer have been identified in chick (Hensen's node), fish (shield), rabbit and mouse (node). All these organizer regions induce axial duplications when transplanted to ectopic locations in host embryos and they all give rise to AE, PME and CM (Fig. 5.1) [10–14]. Inductive effects are even observed when organizer tissue is grafted between different species suggesting that the underlying mechanisms are evolutionarily conserved [13, 15–18].

While the gastrula organizer induces embryonic axes including heads in amphibians [4], chick [17] and fish [19], the situation has been more controversial in the mouse. Here, grafts of either the late gastrula organizer (LGO) [14] or the early gastrula organizer (EGO) [20] (Fig. 5.1D) were only able to induce secondary axes lacking anterior structures, suggesting that the mouse organizer on its own is not sufficient to induce head structures. Meanwhile this view has been challenged by the demonstration that both mouse and rabbit nodes are able to induce forebrain markers in chick hosts [18] and a recent study has demonstrated that the mouse node is able to induce secondary axes including anterior structures when transplanted at mid-gastrula stage (MGO) in mice [21]. The seemingly different head-inducing abilities of different vertebrate organizers are thus likely to represent a temporal phenomenon rather than a qualitative functional difference: the EGO may not yet have acquired full head-inducing potential while the LGO may have lost it already (in a similar manner to the late gastrula DBL in amphibians). Similarly, neither the early anterior primitive streak [22] nor the node after emigration of the head process [23] (Fig. 5.1C) are able to induce anterior structures in chick.

The vertebrate gastrula organizer and its derivatives are sufficient to induce head structures, but are they also required for head formation *in vivo*? If organizer formation in the frog *Xenopus* is inhibited by UV irradiation [24] or by blocking the function of the homeobox gene *siamois* (*sia*) [25, 26], the embryos develop without axial characteristics. Microsurgical ablation of PME from chick or *Xenopus* at early gastrula stages as well as removal of anterior AME from mouse embryos all result in head truncations [17, 27, 28]. In contrast, removal of PME from older chick embryos does not reduce the anterior extent of the central nervous system although dorsoventral (DV) defects including cyclopia become apparent [29]. Ablation of the prospective shield region from late blastula zebrafish embryos prevents anterior neural induction [30] and zebrafish mutants of the homeobox gene *bozozok* (*boz*, the functional equivalent of *sia*) show severe anterior defects [31]. However, the anterior extent of the neuraxis is not reduced if the shield is removed at gastrula stages, although these embryos develop DV defects of the neural tube and are often cyclopic [19]. This suggests that the fish organizer is only required for anterior neural induction before gastrulation. In conclusion, the organizer is required for head formation in all model vertebrates although this requirement may differ temporally. Anterior specification seems to occur particularly early in zebrafish. In addition, other signaling centers may compensate for a lack of gastrula organizer activity in fish (see Section 5.3.3).

5.3.2
Primitive and Definitive Anterior Endoderm

Although vertebrate organizers are more comparable in terms of their head-inducing potential than previously thought, the incomplete axial inductions described above sparked a hunt for other anterior-inducing centers. The anterior visceral endoderm (AVE) of the mouse embryo, an extra-embryonic tissue induced at the distal tip of the early egg cylinder that expresses the homeobox gene *Hex*, shifts anteriorly until it lies next to the prospective anterior neural plate at the onset of gastrulation (Fig. 5.1D) [32]. AVE ablation results in anterior defects [33], yet, the AVE alone is not sufficient to induce anterior character when transplanted to ectopic locations [20] (however, anterior neural markers are induced in chick hosts by rabbit AVE [17]). Co-culture of AVE and ectodermal explants *in vitro* does not result in induction of anterior, but in suppression of posterior marker genes in the ectoderm, suggesting that AVE functions permissively rather than instructively [34].

The requirement for the AVE in anterior specification prompted a search for equivalent structures in other vertebrates. The pre-gastrula chick embryo consists of an upper layer, that gives rise to the embryo proper (epiblast), and an extra-embryonic lower layer (hypoblast; Fig. 5.1C). It has been suggested that the anterior hypoblast is equivalent to the mammalian AVE: it is an extra-embryonic structure that moves anteriorly until it underlies the prospective anterior neural plate before the onset of gastrulation and it expresses chick *Hex* [35, 36]. In transplantation assays, anterior hypoblast is not able to stably induce forebrain markers in chick epiblast (unlike rabbit AVE) [17]. However, it transiently induces early neural markers and is able to direct the movement of epiblast cells [36]. Furthermore, it is required for the establishment of anterior identity as demonstrated by ablation and explant assays [17, 37]. Thus, both mammalian AVE and avian anterior hypoblast seem to perform comparable roles in counteracting posteriorizing signals, thereby promoting anterior fates.

Various attempts have been made to identify equivalents of the AVE in anamniotes. A unifying characteristic of AVE and hypoblast are their anteriorly directed pre-gastrulation movements (Fig. 5.1C and D). In *Xenopus*, endodermal cells perform extensive rotational movements before the onset of gastrulation (Fig. 5.1A). This "vegetal rotation" transports yolky cells from the center of the embryo to the dorsal side into the organizer region where they form the AE, an embryonic structure fated to give rise to the liver and parts of the foregut [38]. During gastrulation, AE cells form the leading edge of the tissue that moves anteriorly towards the animal pole of the frog embryo. Strikingly, these cells express the *Xenopus* ortholog of *Hex* suggesting that they may constitute a structure equivalent to the AVE [39]. This hypothesis is supported by the common expression of several organizer effectors in AVE, anterior hypoblast and frog AE (Table 5.1). In particular, Cerberus (Cer), that is able to induce ectopic heads in *Xenopus*, is expressed in all three tissues [36, 40–43]. Yet, amphibian AE shows only low head-inducing potential in transplantation assays [7, 40] although it is able to induce telencephalic and cement gland marker genes in explant assays [39, 44, 45]. In contrast to AVE and hypoblast, there is no evidence for a requirement for the amphibian AE in head formation. Selective removal of either the AE or the PME from dorsal segments of early *Xenopus* gastrulas has revealed that the PME, but not the AE, is essential for head induction in this assay [27]. This finding does not rule out an involvement of the AE in anterior specification at earlier stages. Alternatively, AE may regenerate quickly in operated embryos. In fact, ablation experiments in chick point towards a requirement for definitive AE in anterior development [37, 46].

Tab. 5.1 Regional expression of selected organizer effectors.

Function	Factor	Expressed in							
		EGO	AE	PME	CM	A/H/Y	ANB	ANE	BCNE
BMP inhibitors	Chordin	+	–	+	+	–	–	Only in zebrafish	+
	Noggin	+	–	+	+	–	–	Only in zebrafish	+
	Follistatin	+	–	–	+	–	–	–	–
	Tsukushi	+	?	+	+	Weak	–	–	?
	Xnr3	Only SBE	–	–	–	–	–	–	+
BMPs	ADMP	+	–	+/–	+	–	–	–	?
Nodal inhibitors	Antivin/Lefty1	+	+/–	+	+	+	–	–	–
RA metabolism	CYP26	–	–	+	–	+	+	+	?
Wnt inhibitors	Dkk1	+	+	+	–	+	–	–	–
	Frzb1	+	+	+	–	–	–	–	–
	Frzb2/Crescent	+	+	+	–	?	–	–	–
	Xsfrp2	+	–	–	–	–	–	+	?
	Tlc	–	–	–	–	–	+	–	–
Wnts	Wnt3a	+	–	–	+	–	–	–	–
BMP, Nodal, Wnt inhibitor	Cer	+	+	–	–	+	–	–	–
Transcription factors	Anf1/Hesx1	+	+	+	–	+	+	–	–
	Blimp1	+	+	+	–	+	–	–	–
	FoxA2/Hnf3β	+	+	+	+	+	–	–	?
	Goosecoid	+	–	+	–	–	–	–	–
	Hex	–	+	–	–	+	–	–	–
	Lim1	+	+	+	+/–	+	–	–	–
	Otx2	+	+	+	+/–	+	–	+	?
	Six3	–	–	–	–	–	–	+	–
	Sia, Boz	+	–	–	–	–	–	–	+

A/H/Y = AVE, anterior hypoblast, dorsal YSL; ANE = anterior neuroectoderm. For references see text. Note absence of BMP inhibitors from AE, A/H/Y, ANB and absence of Wnt inhibitors from CM.

In addition to the yolky cells of the leading edge there is another *Hex*-expressing population of cells in *Xenopus*: the suprablastoporal endoderm (SBE), an epithelium that covers the early organizer and gives rise to the archenteron roof and later to the gut lining. Removal of this tissue results in anterior truncations and SBE is able to induce axial marker genes in an explant recombination assay [47]. Taken together, there is some evidence for a requirement for definitive endodermal tissues in anterior specification in both chick and frog embryos, but their head-inducing potential is rather low.

In zebrafish, hundreds of extra-embryonic yolk cell nuclei underlie the dome-shaped blastula embryo and form the so-called yolk syncytial layer (YSL; Fig. 5.1B). Since these nuclei undergo concerted pre-gastrulation movements and zebrafish

Hex is expressed in the dorsal part of the YSL, it has been proposed that this dorsal YSL is equivalent to the chick anterior hypoblast and the mouse AVE [48, 49]. However, no functional data support or weaken this hypothesis to date.

5.3.3
Ectodermal Signaling Centers

A single stripe of cells at the anterior border of the neural plate of gastrulating zebrafish embryos (row–1) is also involved in anterior specification (Fig. 5.1B). Surgical removal of the row–1 results in anterior truncations and, conversely, row–1 cells are able to induce anterior fates in posterior neuroectoderm [50]. In mice, co-transplantation of anterior neural boundary (ANB) with AVE and EGO tissue results in enhanced induction of anterior neural markers suggesting the presence of a comparable signaling center in mammals [20].

A population of dorsal ectodermal cells is required for brain formation in *Xenopus* before the onset of gastrulation, although they have no inducing activity in grafting experiments. This region is characterized by the expression of two neural inducers and has been named blastula Chordin- and Noggin-expressing center (BCNE; Fig. 5.1A) [51].

5.3.4
Summary

The gastrula organizer is necessary and sufficient for head induction in all vertebrate model organisms and head-inducing potential resides in its anterior derivatives. In zebrafish, anterior neural specification may occur comparably early and is supported by signaling from the ANB, making AME dispensable for this process. In amniotes, anterior extra-embryonic endoderm is also required for head induction but it does not act as an inducer on its own, suggesting a rather permissive function. Embryonic tissues involved in head/anterior neural induction are summarized in Table 5.2.

5.4
Signaling Pathways Involved in Anterior Specification

The identification of factors that underlie inductions by the organizer was regarded as the Holy Grail of embryology since its discovery by Spemann. Over the last two decades, molecular biology has led to the characterization of genes and signaling pathways controlling embryonic development, resulting in two surprising conclusions: (1) embryogenesis is governed by a relatively limited set of signaling pathways that is utilized in a reiterative fashion and (2) inductions by the organizer are mostly

5 Head Induction

Tab. 5.2 Head-organizer activity of vertebrate gastrula tissues, deduced from microsurgical studies.

	Chick		Mouse		Xenopus		Zebrafish	
	Required	Sufficient	Required	Sufficient	Required	Sufficient	Required	Sufficient
EGO	Regenerates after ablation	No [22]	?	Only with AVE and ANB [20]	Regenerates after ablation; inhibition of organizer formation: no axis (see text)	Yes [4]	Yes (pre-gastrula) [30]	?
MGO		Yes [10]	?	Yes [21]			No [19]	Yes [11]
LGO	Unlikely, head induction has occurred	No [23]	No [277]	No [14]	Unlikely, head induction has occurred	No [5]		Unlikely
AE	Yes [37,46]	?	Yes [28]	?	Yolky AE: no [27]; SBE: yes [47]	Very weak [7]; SBE in explant assay [47]	No [19]	Yes [19]
PME	Yes [17]	Yes [29]		?	Yes [27]	Yes [7]		
AVE, ant. hypoblast, dorsal YSL	Yes [17,37]	transient induction [36]	Yes [33]	Only with ANB and EGO [20]	–	–	?	?
ANB	?	?	?	Only with AVE and EGO [20]	?	?	Yes [50]	Yes [50]
BCNE	–	–	–	–	Yes (pre-gastrula) [51]	No [51]	?	?

mediated by inhibitory, rather than instructive, signals. As detailed below, both notions prove true for the mechanisms underlying head induction.

5.4.1
BMP Signaling

One of the best characterized roles of the gastrula organizer in axis induction is the inhibition of signaling by bone morphogenetic proteins (BMPs), members of the transforming growth factor β (TGFβ) superfamily [52]. Six of the factors secreted by the organizer are BMP antagonists (Table 5.1): Noggin, Chordin, Follistatin, Cer and Tsukushi are all able to bind BMPs and sequester them in the extracellular space while Nodal-related 3 may interfere with BMP receptor complex formation [53–58]. A large number of studies mostly performed in amphibian embryos has established that DV patterning of the gastrula is governed by a BMP signaling gradient that

5.4 Signaling Pathways Involved in Anterior Specification

results from the interplay of widely expressed BMPs and organizer-derived inhibitors (Fig. 5.5A). In the mesoderm, high levels of BMP signaling specify ventral-most fates (blood, lateral plate mesoderm); lower levels, intermediate fates (somites); and absence of signaling, dorsal-most fates (PME, CM) [59, 60]. In the ectoderm, high BMP levels promote epidermis formation, lower levels specify the neural plate margin and the neural plate forms where BMPs are absent. The idea that neuralization results from BMP inhibition is known as the "default model" for neural induction since it implies that ectodermal cells will form neuroectoderm by default, as long as they are not exposed to anti-neuralizing BMPs (Fig. 5.5B) [59, 61]. The key role of BMP signaling in early DV patterning is supported by copious genetic data in zebrafish [62, 63].

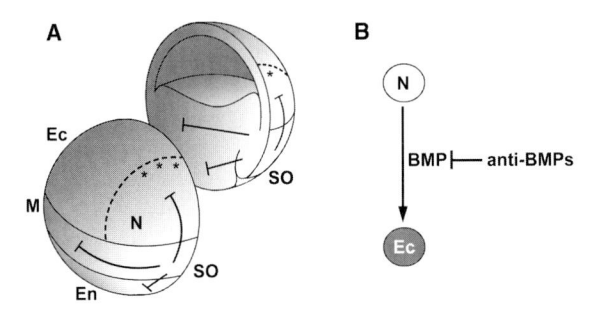

Fig. 5.5
DV patterning of the amphibian gastrula. (A) Schematic depiction of hemisected amphibian gastrula (animal pole points upwards). Spemann's organizer (SO) antagonizes BMPs in ectoderm (Ec), mesoderm (M) and endoderm (En), resulting in a gradient of BMP activity throughout the whole embryo (gray). Low BMP levels specify the neural plate (N) within the dorsal Ec; intermediate levels specify the neural plate margin (dashed line) including the anterior margin (asterisks). (B) Default model for neural induction: BMPs inhibit neural (N) and promote epidermal fates (Ec) in ectodermal cells. Neural induction occurs if BMP activity is blocked by BMP antagonists or in the absence of all signals (default state).

Localized expression of various BMP inhibitors on the ventral side of *Xenopus* embryos gives rise to the induction of secondary body axes. However, these ectopic axes typically lack head structures, comparable to late gastrula DBL grafts (see Section 5.2) suggesting that BMP inhibition is a function of the trunk organizer [56, 58, 64–67]. Yet, it has been demonstrated recently that axes with heads can be obtained by ventral co-expression of several BMP antagonists [68]. These findings indicate that BMP inhibition is necessary and, under certain conditions, sufficient for head induction but that ectopic expression of single BMP antagonists may not be enough to block the full spectrum of head-inhibiting BMP signals active on the ventral side of *Xenopus* embryos.

Global overexpression of BMPs in fish or frog results in ventralized embryos lacking axial characteristics. However, head truncations develop at only mildly elevated BMP levels, supporting the contention that BMPs specifically antagonize anterior specification [69–72]. In line with these findings, a genetic compound knockout of *Chordin* and *Noggin* in mouse leads to severe head truncations [73]. At present, it remains unclear whether increased BMP signaling affects neuroectoderm [74], AME [71] or both. Furthermore, it is still unknown at which developmental stage the headless phenotype becomes manifest; a preliminary analysis of marker gene expression in *Chordin$^{-/-}$;Noggin$^{-/-}$* embryos has indicated that anterior neural fates are induced at early stages but fail to be maintained [73]. Both *Chordin* and *Noggin* are expressed in the AME, but not in the AVE (Table 5.1), confirming that the AVE alone does not act as a head organizer but that AME, AVE and maybe other tissues need to synergize in anterior specification. Milder anterior defects become apparent in Chordin-deficient mice carrying one functional allele of Noggin (*Chordin$^{-/-}$; Noggin$^{+/-}$*), possibly due to impaired PME and ANB formation [75]. Similarly, forebrain defects arise in mice lacking the BMP cofactor Twisted Gastrulation [76, 77].

Despite the evidence that forebrain development requires inhibition of BMP signaling, several studies suggest that intermediate BMP levels may be necessary to induce structures derived from the anterior margin of the neural plate such as the telencephalon and the *Xenopus* cement gland [74, 78]. BMPs flank the neural plate not only laterally but also anteriorly and it is therefore likely that graded BMP activities regulate marginal versus medial rather than just DV fates within the neural plate (Fig. 5.5A) [74]. It is tempting to speculate that the formation of the ANB itself may depend on appropriate levels of BMP signaling [79, 80].

Besides their broad ventrolateral expression domains, BMPs are also expressed in a small patch of dorsoanterior cells of the organizer [81–84]. To date, there is no explanation for the expression of these ventralizing and posteriorizing factors in tissues with dorsalizing and anterior-inducing activity. At first sight, the expression of the BMP-related ADMP (anti-dorsalizing morphogenetic protein) in the organizer of frog, chick and fish embryos is similarly counterintuitive [85–88]. While it has been proposed that ADMP serves to position the organizer in chick [86], functional experiments in frog and fish suggest a role in organizer patterning where ADMP promotes posterior at the expense of anterior organizer identity [88–90].

In *Xenopus*, a positive role in head formation was recently uncovered for BMP3 and BMP3b, two divergent BMP family members. BMP3 enhances anterior structures while BMP3b results in ectopic head formation following overexpression. These differential effects are likely to result from differing biochemical functions: BMP3b inhibits both Nodals (see Section 5.4.4) and ventralizing BMPs whereas BMP3 only antagonizes the latter [91].

In conclusion, BMPs play a primarily antagonistic role during vertebrate head induction, although some anterior structures may require intermediate levels of BMP signaling. BMPs counteract the establishment of anterior identity at multiple levels: (1) they inhibit organizer formation by ventralizing the mesoderm, (2) they antagonize neural fates, (3) ADMP posteriorizes the AME and defines the trunk versus the head organizer and (4) they may exert a direct posteriorizing influence on neuroectoderm.

5.4.2
FGF Signaling

Fibroblast growth factors (FGFs) can signal via different intracellular pathways the most important of which is arguably the mitogen-activated protein kinase (MAPK) pathway [92]. Overexpression of an intracellularly truncated, dominant-negative form of a *Xenopus* FGF receptor (XFD) results in severe gastrulation defects and embryos with posterior truncations, strongly suggesting a role for FGF signaling in AP axis formation [93, 94]. Indeed, overexpression of FGFs results in headless *Xenopus* and zebrafish embryos [93, 95, 96], posterior neural markers are progressively induced by increasing doses of FGFs in frog neuroectoderm [97] and the promoter of the *Xenopus* homeobox gene *caudal homolog 3*, a regulator of posterior development, contains FGF response elements [98]. Similarly, posterior neural markers are induced in chick following implantation of FGF-soaked heparin beads [99–101]. Activation of the FGF pathway can be detected in *Xenopus* gastrula embryos in regions giving rise to the posterior nervous system using an antibody directed against phosphorylated MAPK [102, 103]. Therefore, it has been proposed that FGFs are prime candidates for Nieuwkoop's transforming signal [97]. Yet, FGFs fail to induce various posterior neural markers in chick neural plate explants [104] and *Xenopus* embryos expressing XFD display a relatively normal AP pattern in the anterior neural plate [94, 102, 105], suggesting that FGF signaling alone is not sufficient for neural posteriorization. Furthermore, there is no genetic evidence in mouse or zebrafish to support a role for FGFs as transforming signals.

Electroporation of XFD into the chick node results in anteriorized embryos and time-lapse analysis has revealed that electroporated cells leave the node prematurely during its posterior regression [106]. This observation has led to the proposition that FGFs are not transforming signals themselves but that they define a stem cell niche within the node where cells are exposed to the true transformer. Cells that remain in the node for longer are exposed to these signals for a longer time and acquire more posterior fates.

To further complicate the role of FGFs in anterior specification, they may not only antagonize but also promote anterior fates, depending on their site of expression: FGF8 is expressed in the ANB of all vertebrates [107] and FGF3 is expressed even earlier in zebrafish row–1 cells [108]. However, FGFs from the ANB are more likely to be involved in forebrain patterning [45] rather than its induction, since all forebrain primordia are present in embryos with compromised FGF signaling. The various roles of FGF signaling during neural patterning have been extensively reviewed [79, 80, 109–111].

Despite the large body of evidence supporting the default model for neural induction (see Section 5.4.1), there is an ongoing debate as to whether BMP inhibition really is the initial step in this process or whether it serves to stabilize neural character in cells that have already received other neural-inducing signals. It has been suggested that FGFs are one such signal, priming ectodermal cells to acquire neural fate [112–116]. A recent study in *Xenopus* has revealed that activated MAPK phosphorylates and thereby inactivates the intracellular BMP transducer Smad1 [117].

Thus, neural induction by BMP inhibition and by FGF-MAPK signaling are integrated at the level of Smad1, reconciling the default model and the FGF-centric models for neuralization and providing an explanation for why FGFs are able to induce neural fates even in the presence of exogenous BMPs [114]: they feed into the BMP pathway intracellularly, downstream of BMP ligands (Fig. 5.6). The ability of FGF8 to induce secondary embryonic axes in zebrafish may be a consequence of this intracellular anti-BMP activity [118].

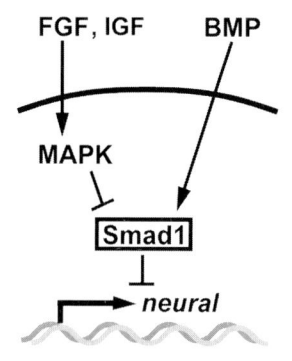

Fig. 5.6
Neural inducing and inhibiting signals are integrated at the level of Smad1. After [119].

Taken together, FGFs play positive and negative roles during anterior specification. First, they are involved in mesoderm induction and neuralization while, at later stages, they act in a posteriorizing fashion. Subsequently, FGFs act from local signaling centers such as the ANB and the midbrain-hindbrain boundary (MHB) to pattern the developing nervous system.

5.4.3
IGF Signaling

Insulin-like growth factors (IGFs) have been implicated in the regulation of growth, life span and homeostasis as well as in neoplastic transformation. Surprisingly, IGF signaling has also been associated with head and neural induction in *Xenopus*: overexpression of different IGFs or of an IGF-binding protein promotes anterior development at the expense of that of the trunk, localized activation of the IGF pathway in ventral regions of the frog embryo leads to induction of ectopic head-like structures and IGFs induce anterior neural marker genes in ectodermal explants. Conversely, inhibition of IGF receptor signaling results in embryos displaying anterior truncations and blocks neural induction by BMP inhibitors [119, 120]. These observations indicate that IGFs act as head inducers in a dual manner: (1) they promote anterior versus posterior fates and (2) they are neural inducers. The neural-inducing activity of IGFs is best explained by their ability to activate the MAPK signaling axis which results in inactivation of the BMP transducer Smad1, as demonstrated for FGFs (Fig. 5.6) [117].

5.4.4
Nodal Signaling

Nodals, like BMPs, are members of the TGFβ superfamily of signaling factors [52]. Over the last few years, it has been well established that activation of the Nodal pathway is crucial for mesoderm and endoderm formation in vertebrates [121]. It has been suggested that Nodals act in a dose-dependent manner in *Xenopus* with high levels of signaling specifying dorsal and low levels specifying ventral mesodermal fates [122]. This gradient can be visualized at blastula stages using antibodies against phosphorylated Smad2, an intracellular transducer of Nodal signals [103, 123]. Similarly, graded Nodal signaling gives rise to the patterning of the AP axis of the zebrafish organizer: high levels specify anterior (AE, PME) and lower levels specify trunk fates (CM) [124, 125]. Since the anterior AME (and/or its precursors) is required for anterior specification, one would expect that embryos in which Nodal signaling is impaired would develop anterior defects. Indeed, a large number of different mouse knockouts with reduced levels of Nodal signaling have been generated and all of them display anterior defects to varying degrees [121]. Head truncations are also observed following expression of a dominant-negative version of Nodal-related 2 in frog [126].

Which tissues are affected in Nodal pathway knockouts? Conditional inactivation of Smad2 or Smad4 in the epiblast (that gives rise to the embryo proper) has revealed an essential role for Nodal signaling in the formation of the anterior primitive streak (AE and PME), consistent with the role of Nodal signaling in AP organizer patterning proposed in zebrafish. Unsurprisingly, these mice develop anterior truncations [127–129]. In addition, anterior development is severely impaired in mouse chimeras in which extra-embryonic tissues are Nodal-deficient while the embryo itself is rescued (by injection of wild-type embryonic stem cells into mutant blastocysts), indicating that a Nodal signal originating from the AVE is required for head formation [130]. Furthermore, formation of the AVE itself requires Smad2 and Smad4 function [131, 132]. Taken together, Nodal signaling is required for anterior specification in the mouse at three distinct levels (Fig. 5.7A): (1) in extra-embryonic tissues for AVE development, (2) from the AVE to the epiblast for head induction and (3) within the epiblast for the establishment of the organizer.

Zebrafish double mutants for the *Nodal*-related genes *cyclops* and *squint* (*cyc;sqt*) as well as mutants for the essential Nodal cofactor *one-eyed pinhead* (*oep*), in which both maternal and zygotic contributions of this gene are deleted (*MZoep*), do not form mesoderm or an organizer but anterior neural fates are induced and AP neural patterning is surprisingly unaffected [133, 134]. This observation conflicts with the anterior truncations observed in Nodal-deficient mice and has been interpreted as an argument against an essential role for the organizer in AP axis establishment.

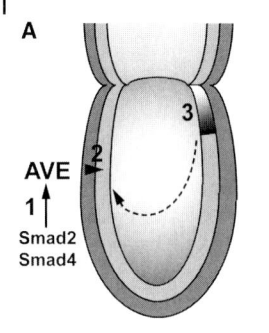

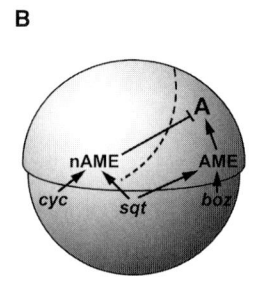

Fig. 5.7
Nodal signaling during vertebrate anterior specification. (A) In the mouse embryo, Smad2 and Smad4 are essential for AVE formation (1), a Nodal signal from the AVE induces anterior character in the adjacent epiblast (2) and graded Nodal activity forms and patterns the organizer at the opposite side (3). (B) In zebrafish, *sqt* and *boz* synergize in AME induction, thereby inducing anterior identity, while *cyc* and *sqt* induce nonaxial mesendoderm (nAME), a source of posteriorizing signals.

However, early β-catenin signaling (see Section 5.4.6) transiently induces the expression of some organizer factors (Chordin, Noggin) that may be sufficient for anterior neural induction in *cyc;sqt* mutants [135], a view supported by recent studies in *Xenopus* [51, 136]. Thus, Nodals are required to maintain, but not to induce, the expression of dorsalizing factors. Alternatively, signaling from the ANB may compensate for the lack of an organizer in *cyc;sqt* mutant fish [50]. Most importantly though, not only the organizer but also large parts of the lateral, ventral and posterior mesoderm do not develop in *cyc;sqt* or *MZoep* mutants and these tissues are known sources of posteriorizing signals [137]. Hence, both anterior-inducing centers and their antagonizing signals are absent from these embryos. An elegant study using different crosses of *cyc*, *sqt* and *boz* mutants has provided genetic evidence for this view: while *boz;sqt* double mutants show severe anterior truncations that are more pronounced than in *boz* single mutants [135], anterior structures are rescued in *boz;sqt;cyc* triple mutants [138]. This result indicates that *boz* and *sqt* cooperate in organizer formation while *cyc* is upstream of an antagonistic (posteriorizing) signal (Fig. 5.7B).

In *Xenopus* and zebrafish, overexpression of Nodal-related factors during gastrulation results in anterior truncations. Thus, Nodal signaling needs to be inhibited in order to maintain anterior identity at later stages [57, 72, 125]. In fact, several Nodal inhibitors are expressed in anterior signaling centers (Table 5.1). *Xenopus* AE expresses the multifunctional inhibitor Cer that sequesters BMP, Nodal and Wnt ligands extracellularly [40, 57]. Depletion of endogenous Cer activity shows no effect on its own, but it aggravates the anterior defects of embryos with elevated BMP, Nodal or Wnt levels [91, 139]. A Cer-related factor with a comparable inhibitory spectrum, Coco, has recently been identified that is expressed in the ectoderm and in the marginal zone during gastrulation in *Xenopus* [140]. The chick ortholog of *Cer* is expressed first in the hypoblast and later in the AE and possibly the PME [36]. A genetic loss-of-function of mouse *Cerberus-like* (*Cer-l*, expressed in AVE and AE) does not seem to affect development [141–144], but simultaneous depletion of *Cer-l* and the Nodal antagonist *Lefty1* elicits the formation of excessive mesoderm and multiple primitive streaks. This phenotype is rescued by removal of one allele of *Nodal* (*Cer-l$^{-/-}$;Lefty1$^{-/-}$;Nodal$^{+/-}$*), providing genetic evidence for dose-dependent Nodal signaling [145]. Complementary effects have been observed in a gain-of-function study in chick [146], culminating in a model in which Nodal inhibitors from the AVE/hypoblast restrict the position of the primitive streak to the posterior end of the embryo and thereby determine early AP polarity of the embryo [147]. A study in chick embryos, suggesting that the AE promotes PME fates by secreting different TGFβs, somewhat conflicts with the idea that BMPs and Nodals have to be inhibited for anterior development [148]. At this point, the answer to how these factors escape the various TGFβ inhibitors expressed in the anterior of the embryo remains elusive.

In summary, the Nodal pathway plays multiple and sometimes opposing roles at different stages during head development. This complicates the interpretation of the phenotypes in various Nodal gain- and loss-of-function situations. However, over the last few years it has become clear that, early in development, Nodal signaling is essential to generate the tissues required for anterior specification (AVE, organizer) while later, during gastrulation, it needs to be inhibited to maintain anterior identity. The implication of the Nodal pathway in head development has even attracted the attention of clinical researchers: mutations in the human genes *TGIF*, encoding a transcription factor that mediates Nodal signals, and *TDGF1*, an ortholog of zebrafish *oep*, result in holoprosencephaly, a common developmental forebrain defect with a variable phenotype ranging from mild facial abnormalities to cyclopia and anophthalmia [149, 150].

5.4.5
Retinoid Signaling

Retinoic acid (RA) was the first substance considered as a potential transforming signal because RA treatment of vertebrate embryos resulted in a loss of head structures accompanied by an upregulation of posterior, at the expense of anterior, neural marker genes [151]. This effect can be mimicked by overexpression of a cellular RA binding protein, a constitutively active form of RA receptor (RAR) α1 or the RA-producing enzyme retinaldehyde dehydrogenase 2 (RALDH2) in *Xenopus* embryos [152–154].

RALDH2 is expressed in the posterior-lateral mesoderm in all vertebrates examined [154–158]. Mouse and zebrafish mutants for *RALDH2* as well as RAR knockout mice display defects in the posterior hindbrain and the spinal cord while the anterior part of the CNS is properly patterned [155, 157–160]. Comparable defects are observed in *Xenopus* after overexpression of the RA-degrading enzyme CYP26 [161] which is expressed in the anterior neuroectoderm of vertebrate embryos [156, 161, 162]. Surprisingly, head development is largely normal in mice carrying loss-of-function mutations in *CYP26* while they display a broad range of other defects [163, 164]. A recent study in *Xenopus* has provided evidence that, in the absence of RA, RARs act as transcriptional repressors (presumably by recruiting co-repressors) and that this constitutive repression of RA target genes is necessary for head formation [165]. There is evidence that a gradient of RA patterns the hindbrain [151, 166], although rescue experiments in which hindbrain patterning in RA-deficient quail embryos is largely restored by exogenous, global application of RA, are somewhat inconsistent with a requirement for a finely tuned endogenous gradient [167].

In conclusion, RA antagonizes head formation and is required for the establishment of posterior identity although some aspects of early anterior patterning may require a transient RA activity [168].

5.4.6
Wnt/β-Catenin Signaling

The canonical Wnt/β-catenin pathway is activated by secreted factors of the Wnt1 class in vertebrates. These Wnts bind to a receptor complex consisting of members of the Frizzled (Fz) family of seven-pass transmembrane proteins and the low density lipoprotein receptor-related proteins (LRPs) 5 or 6. Intracellularly, Wnt1–type signals are transduced via the adaptor protein Dishevelled (Dsh) which inhibits glycogen synthase kinase 3β (GSK3β) by a poorly-understood mechanism involving the adaptor protein Axin. In the absence of Wnt activity, GSK3β phosphorylates the multifunctional protein β-catenin and thereby targets it for degradation by the proteasome. Thus, cytoplasmic levels of β-catenin are low in cells that do not receive a Wnt signal. Upon pathway activation, GSK3β is inhibited and β-catenin accumulates in the cytoplasm, then translocates into the nucleus and associates with transcriptional cofactors such as members of the LEF/TCF (lymphoid enhancer-binding factor/T cell-specific transcription factor) family to initiate the transcription of target genes [169, 170].

In addition to Nodal activation, vertebrate organizer formation depends on an instructive Wnt/β-catenin-type signal [171, 172]. The competence of *Xenopus* and zebrafish embryos to respond to Wnt signals changes dramatically at the midblastula transition (MBT) when zygotic transcription starts. After the MBT, Wnt/β-catenin signaling exerts a posteriorizing activity and antagonizes the head organizer. Overexpression of class 1 Wnts, the Wnt transducers β-catenin or TCF3, or treatment with the GSK3 inhibitor Li^+ at these stages results in repression of anterior development [72, 173–187]. A posteriorizing role for Wnt/β-catenin signaling has also been suggested by a study using transgenic mice ectopically expressing chick Wnt8c and by treatment of gastrulating chick embryos with Li^+ or Wnt3a protein [188–190]. Wnt/β-catenin signaling is not only sufficient to repress anterior development but is also required for posterior development: overexpression of Wnt antagonists in *Xenopus* or zebrafish as well as treatment of chick embryos with a Wnt inhibitor result in the formation of embryos with enlarged head and shortened trunk structures [67, 190–201]. Zebrafish embryos that lack Wnt8 function develop pronounced trunk defects [202]. Furthermore, mice carrying mutations in *Wnt3a*, *Wnt5a*, *LRP6*, *TCF1* and *LEF1* or *TCF1* and *TCF4* show posterior truncations, providing genetic evidence of a requirement for Wnts in posteriorization [203–207]. The change of embryonic competence towards Wnt/β-catenin signaling from organizer induction to posteriorization is based on an exchange of nuclear cofactors rather than the employment of a different signal transduction pathway [182, 208].

Wnt antagonists are expressed in all tissues involved in anterior induction (Table 5.1). Inhibition of the extracellular Wnt antagonist Dickkopf1 (Dkk1) or knockdown of the Dkk1 co-receptor Kremen in *Xenopus* yields embryos with microcephaly and cyclopia, indicating that this Wnt inhibitor is essential for head formation [67, 181, 209]. A requirement for Wnt inhibition by Cer has been suggested by the increased sensitivity to Wnt signals in Cer-depleted *Xenopus* embryos [139]. Both *Cer-l* and *Dkk1* are expressed in the mouse AVE, but only the latter is also expressed in the anterior AME [41–43, 196, 210]. While genetic loss-of-function of *Cer-l* does not seem to affect normal development (see Section 5.4.4) [141–144, 211], $Dkk1^{-/-}$ mice are headless. The analysis of chimeras has revealed that *Dkk1* function is required in the AME, but not in the AVE, for head formation [212] and experiments in frog have shown that Dkk1 is required for PME formation [181]. Thus, if Wnt inhibition from the AVE is also essential during anterior specification, it may be mediated redundantly and it would be interesting to test whether anterior development is impaired in mice lacking both Cer-l and Dkk1 in the AVE.

Head truncations are also observed in the zebrafish mutants *headless* and *masterblind* in which the intracellular Wnt antagonists TCF3 and Axin1, respectively, are inactivated and in mice deficient for ICAT (inhibitor of β-catenin and TCF), providing genetic support for the contention that intracellular Wnt inhibition is essential for anterior specification [213–216]. XsalF, the *Xenopus* ortholog of the *Drosophila* zinc-finger transcription factor Spalt, is required for the expression of two Wnt antagonists, TCF3 and GSK3β in forebrain and midbrain anlagen and its knockdown results in anterior defects [217].

Experiments in chick, *Xenopus* and zebrafish embryos have shown that Wnt/β-catenin signaling is necessary and sufficient to posteriorize neuroectoderm directly and in a dose-dependent fashion, suggesting that Wnts may act as a transforming signal in Nieuwkoop's sense [190, 218, 219]. Importantly, an increasing AP gradient of nuclear localization of β-catenin can be detected transiently in the emerging neural plate of the gastrulating *Xenopus* embryo, providing evidence for a Wnt signaling gradient *in vivo* [219]. It is likely that this gradient is shaped by the differential expression of Wnt receptors and downstream targets [187, 220]. A Wnt-unrelated extracellular protein, WISE (Wnt modulator in surface ectoderm), has recently been identified that is able to activate or inhibit the Wnt pathway in a context-dependent manner [221]. Genetic support for a requirement for graded Wnt signaling is provided by different allelic combinations of $Wnt3a^-$ and *vestigial tail*, a hypomorphic mutation of *Wnt3a*, that display varying posterior truncations depending on the dose of Wnt3a [222].

In zebrafish, the ANB secretes the Wnt antagonist Tlc [186]. Local expression of Tlc rescues telencephalic marker gene expression after ANB ablation, indicating that

Wnt inhibition is the major function of the ANB in anterior neural induction. *Wnt8b* is expressed between the posterior forebrain and the midbrain primordia later in gastrulation and it has been shown that the antagonism between Wnt8b and ANB-derived Tlc is crucial for the patterning of this area [186, 187]. These observations suggest that Wnt/β-catenin signaling posteriorizes the neural plate in two sequential steps: early in gastrulation it regulates patterning along the entire AP axis, while later it modulates AP patterning in a more localized fashion within the anterior neural plate [80]. This notion is supported by selective rescue experiments in the mildly posteriorized zebrafish mutant *colgate,* encoding an as yet unidentified Wnt antagonist, where depletion of Wnt8 rescues early, global posteriorization while depletion of Wnt8b rescues regional posteriorization in the anterior neuroectoderm [223]. Two phases of posteriorization by Wnts have also been postulated in chick [224].

In conclusion, Wnt/β-catenin signaling regulates anterior specification during at least three phases: (1) it is necessary to induce the gastrula organizer which is required for anterior specification, (2) it counteracts the head organizer and regulates AP patterning by graded signaling and (3) it regulates AP regionalization between fore- and midbrain. It is of note that the "head-inducing" IGF pathway counteracts Wnt/β-catenin signaling at the level of GSK3β in *Xenopus* embryos (see Section 5.4.3) [119, 120]. Despite the large body of evidence supporting a role for Wnt/β-catenin signaling in posteriorization it is also clear that this pathway alone is not able to account for all aspects of AP patterning. The competence of different parts of the neural plate to respond to Wnts is regionally restricted [219], presumably by other diffusible signals [162] and by a pre-pattern of transcription factors (see Section 5.6.8) [220, 224].

5.4.7
Noncanonical Wnt Signaling

The *silberblick* (*slb*) mutant in zebrafish is characterized by mild cyclopia and other head defects. A mutation in the *Wnt11* locus has been identified as the cause of this phenotype. AP marker gene expression is unaltered but convergence and extension movements, the driving force behind axial elongation during gastrulation, are impaired in *slb* mutants [225]. The *slb* phenotype is therefore best explained by the failure of the AME to elongate, resulting in a posterior misallocation of this tissue that is a known source of midline patterning signals [226]. A large number of follow-up studies in fish and frog embryos have revealed a noncanonical Wnt pathway that acts via Fzs and Dsh, but independently from Axin, GSK3β and β-catenin, and is essential for convergence and extension [227]. Overexpression of the Wnt inhibitor Crescent/Frzb2 (Crs) in *Xenopus* embryos elicits a phenotype comparable to *slb* [228, 229] even though the neural plate is anteriorized in these embryos, suggesting that Crs antagonizes both Wnt1– and Wnt11–type signals. Expression of a dominant-negative form of the small GTPase RhoA, a link between noncanonical Wnt signaling and cytoskeletal organization [230], has also been shown to affect head development [231]. Although noncanonical Wnt signaling is not directly involved in anterior specification, the head defects resulting from its inhibition emphasize the importance of understanding the embryo as a dynamic structure in which the relative movement of tissues is as important as instructive signaling because sources of signals and receiving tissues are constantly relocated.

5.4.8
Summary

Several conserved signaling pathways are involved in vertebrate anterior specification. Most of them are employed in a reiterative fashion and elicit different, sometimes even opposing, phenotypical effects in successive phases of development. This complicates the interpretation of gain- and loss-of-function experiments and highlights the requirement to modulate gene function in specific tissues and defined time-windows. Often, both the AP and the DV embryonic axes appear to be controlled by a given diffusible signal. While some signals may in fact exert separable effects on AP and DV, it is important to keep in mind that the embryonic axes unfold during gastrulation and that their assignment to a pre-gastrula embryo is artificial nomenclature, as reflected by an ongoing debate about axial allocation in *Xenopus* [232].

Early in embryogenesis, instructive Nodal and β-catenin signaling are required to induce the gastrula organizer. Nodal signaling is also essential for AVE formation in mice. Strikingly, virtually all molecular pathways examined counteract head formation at only slightly later stages, during gastrulation: BMP, FGF, Nodal, RA and Wnt/β-catenin signaling all antagonize head organizer activity and posteriorize or ventralize embryos. It is the inhibition of these pathways, rather than instructive signaling, that is required for anterior specification. Even IGF, seemingly the only instructive head-inducing signal, antagonizes BMP and Wnt/β-catenin signaling intracellularly. Likewise, neural induction appears to be largely based on inhibition rather than on instructive signaling, as suggested by the default model for neural induction, since the neural-inducing activities of FGFs and IGFs may be mediated, at least in part, by inhibition of the BMP transducer Smad1.

Several of the pathways counteracting head organizer activity have been proposed to act like Nieuwkoop's transforming signal by imparting AP identity in a dose-dependent fashion. However, none of these pathways is sufficient on its own to establish the refined AP pattern observed later. It is therefore likely that several transforming signals act in concert to set up posterior domains [162, 176, 178, 184]. Signaling pathways often interact on multiple levels and it remains to be established which signals posteriorize which tissues directly and which of the observed effects are mediated indirectly through the induction of secondary signals.

5.5
Mechanistic Models for Head Induction

5.5.1
The Two- and Three-Inhibitor Models

The gastrula organizer expresses BMP, Nodal and Wnt inhibitors (Table 5.1) [60, 226]. While inhibitors of BMP and Nodal signaling are expressed in AE, PME and CM, Wnt inhibitors are exclusively expressed anteriorly, suggesting that Wnt inhibition is a distinguishing feature of the head organizer. Ectopic expression of various BMP antagonists on the ventral side of *Xenopus* embryos induces secondary trunks lacking anterior structures (see Section 5.4.1) while co-expression of BMP and Wnt antagonists elicits the formation of ectopic axes with head structures. Based on these observations, the following two-inhibitor model for head induction was proposed: the head organizer simultaneously represses both BMP and Wnt signaling while the trunk organizer antagonizes only BMPs [67]. This model has been strengthened by genetic data in zebrafish where the lack of BMP antagonists in the *boz;chordino* double mutant leads to a loss of both head and trunk structures [233] and in mice where the simultaneous reduction of BMP and Wnt inhibitors in $Dkk1^{+/-};Noggin^{+/-}$ doubles results in head truncations that are not observed in single heterozygotes [234]. Similarly, the simultaneous reduction of Chordin, Dkk1 and Noggin levels in *Xenopus* embryos leads to stronger anterior truncations than depletion of either factor on its own [234]. Simultaneous inhibition of BMP and Wnt signaling is also required for the formation of head muscles after gastrulation [235]. This raises the interesting possibility that the combined inhibition of BMPs and Wnts acts as a conserved module that is used reiteratively during vertebrate head development.

Cer acts as a head inducer when expressed ectopically in *Xenopus* embryos [40] and has been shown to antagonize factors of the BMP, Nodal and Wnt families [57]. A three-inhibitor model for head induction was proposed where the inhibition of all three pathways is required for head induction. This hypothesis is complemented by a recent study demonstrating that the simultaneous activation of the three pathways induces tails in zebrafish [72]. There is experimental evidence for both the two- and the three-inhibitor model and they are fully compatible in light of the extensive cross-talk between signaling pathways during embryogenesis (Fig. 5.8A).

5.5.2
The Organizer-Gradient Model Dualism

While Spemann postulated distinct head- and trunk-inducing regions within the gastrula organizer, Nieuwkoop proposed a posteriorizing gradient of a transforming agent. Both models are now matched by their molecular counterparts: the two-inhibitor model corresponds to Spemann's idea of head and trunk organizers while posteriorizing FGF, RA and Wnt gradients represent Nieuwkoop's transforming signal. Both concepts have often been regarded as mutually exclusive and there has

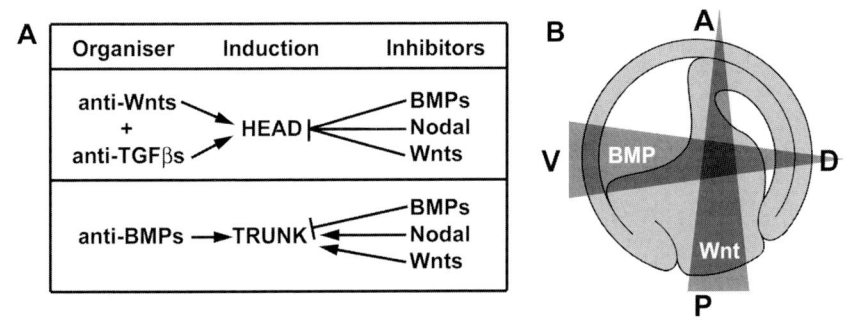

Fig. 5.8
Molecular interpretation of classical models for embryonic patterning. (A) The two/three inhibitor model for head induction corresponds to Spemann's head and trunk organizer. (B) Revised double gradient model: the DV axis of the amphibian gastrula is patterned by a BMP gradient, the AP axis by a Wnt gradient. Adapted from [219].

been a tendency to dismiss the organizer-centric view in favor of one that is gradient based, as the latter appears to be more suitable for explaining patterning into multiple AP regions. However, the molecular mechanisms described above suggest that Spemann's and Nieuwkoop's models do not contradict each other but merely represent two different views of AP patterning. The interplay between posteriorizing signals and their antagonists, secreted by anterior organizing centers, establishes a gradient of posteriorizing activity. A head will form at the pole of the embryo where this activity is lowest. Thus, like an electron can act as a wave or a particle, AP patterning appears to be regulated by organizers or gradients, depending on the experimental approach [236]. Simultaneous inhibition of BMP and Wnt signaling, as suggested by the qualitative two-inhibitor model, results in the formation of orthogonal BMP and Wnt gradients regulating DV and AP patterning, respectively. Our current molecular view of embryonic axis formation can thus be interpreted as a reincarnation of the classical quantitative two-gradient model of Saxén and Toivonen (Fig. 5.8B) [9].

5.6
Transcriptional Regulators in Anterior Specification

The integration of the multiple extracellular signals that regulate early development often occurs at a transcriptional level. Several transcriptional regulators involved in establishing anterior identity have been identified and will be reviewed in this section. Transcriptional regulators specific for signaling pathways (Smads, TCFs) have been discussed in Section 5.4.

5.6.1
ANF/Hesx1

In mouse, the homeobox gene *Hesx1* is first expressed in the AVE and later on also in the anterior AME and the ANB [237]. $Hesx1^{-/-}$ mice display forebrain truncations suggesting that this gene is involved in establishing anterior identity [238]. However, forebrain markers are initially induced but fail to be maintained at later stages in these mice arguing against an early role for *Hesx1*. An analysis of chimeras (see Section 5.4.4) has shown that *Hesx1* is required for anterior specification in the epiblast but not in the AVE. The anterior defects observed in the knockout are most likely due to an indirect effect since expression of FGF8 in the ANB is impaired in these mice [237].

The *Xenopus* ortholog of Hesx1, Xanf, is expressed in the organizer, acts as a repressor and is able to induce truncated secondary axes following misexpression [239]. Yet, an essential role in organizer function is not supported by loss-of-function experiments using dominant-negative versions of Xanf [240].

5.6.2
Blimp1

The transcriptional repressor Blimp1 is expressed in PME and AE in *Xenopus* as well as in AE and AVE in mouse. A genetic knockout of this gene has not been described, but gain- and loss-of-function experiments in *Xenopus* have revealed a potential role for Blimp1 in anterior specification, possibly by promoting AE-specific gene expression. Co-expression of Blimp1 and Chordin induces ectopic heads while overexpression of an activator form of Blimp1 results in head truncations [241].

5.6.3
FoxA2/HNF3β

The winged helix transcription factor FoxA2 (hepatocyte nuclear factor 3β, HNF3β) is expressed in the mouse organizer as well as in the AVE and in the corresponding tissues in chick [242, 243]. Mice lacking *FoxA2* function fail to form an organizer, are headless and show pronounced DV neural patterning defects [242]. Organizer formation and anterior defects are largely rescued if FoxA2 is expressed in extra-embryonic tissues [244]. However, this gene is also required in embryonic tissues for forebrain and AE maintenance at later stages [245].

5.6.4
Goosecoid

The homeobox gene *Goosecoid* (*Gsc*) was the first organizer-specific gene described [246]. During gastrulation, it is specifically expressed in the PME and it is a target of both the early β-catenin signal that leads to organizer induction [247, 248] and the dorsalizing Nodal pathway [122, 134, 249]. Ectopic expression of *Gsc* in ventral *Xenopus* blastomeres results in induction of ectopic axes including heads, and the expression of antimorphic Gsc forms interferes with head formation [250, 251]. However, genetic ablation of *Gsc* in mouse does not affect gastrulation or head formation, possibly due to genetic redundancy with related genes [252, 253]. Yet, $Gsc^{-/-}$ organizer tissue shows a reduced neural-inducing capacity in a transplantation assay, indicating that in the organizer Gsc is upstream of neural inducers that are not essential in the context of the whole embryo [254]. In $Gsc^{-/-};FoxA2^{+/-}$ compound mice, signaling from the ANB and the embryonic midline are reduced, resulting in embryos with reduced forebrain structures [255].

5.6.5
Hex

Hex is the earliest marker of the AVE in mouse and it is also expressed in the mouse AE, the chick hypoblast, the yolky AE and the SBE in *Xenopus* as well as in the dorsal YSL in zebrafish (see Section 5.3.2) [32, 35, 36, 39, 47, 48]. Functional experiments in frog and fish have demonstrated that *Hex* is a target of both the early dorsalizing β-catenin and Nodal signals [256] and that its overexpression results in induction of several dorsal and anterior markers (such as *Cer*) while others are repressed (*Gsc*, *Chordin*, but see [48]). Removing *Hex* function leads to downregulation of *Cer* expression and to a truncation of head structures in *Xenopus* [47, 257]. The changes in gene expression following modulation of Hex levels suggest that it may promote AE versus PME fates. $Hex^{-/-}$ mice are characterized by forebrain truncations similar to those observed in *Hesx1* mutants and the analysis of chimeras has shown that *Hex* function is required in the AE [237]. An additional requirement in the AVE has not yet been ruled out. Similar to *Hesx1*, forebrain markers are intially induced in *Hex* mutant mice, suggesting that it is required in the maintenance phase rather than in the early induction phase of anterior development.

5.6.6
Lim1

Lim1 is expressed in both AVE and AME in the mouse (and in the corresponding chick tissues [243]) and its genetic ablation results in mice lacking forebrain and midbrain [258]. Interestingly, chimeric mice lacking Lim1 function only in AVE or AME are headless, supporting the view that neither structure is sufficient on its own for head formation [259]. *Lim1* genetically interacts with *FoxA2*, as double mutants show a more severe phenotype than each single mutant, and this interaction is required in the AVE [260].

In *Xenopus* and zebrafish, *Lim1* is expressed in the organizer and acts downstream of Nodal and *sia* [249, 261]. Misexpression experiments using mutant forms of Lim1 have suggested that Lim1 is able to mediate some organizer activities and its depletion results in embryos with axial (including head) defects. A molecular analysis of this phenotype has led to the suggestion that Lim1 functions by coordinating cell movements during gastrulation [262]. Thus, the Lim1 loss-of-function phenotype may be another example for AP patterning defects that are caused by disturbed tissue movements during gastrulation, as discussed for noncanonical Wnt signaling (see Section 5.4.7).

5.6.7
Otx2

The homeobox gene *Otx2* is expressed in the AVE, the AME and the anterior neuroectoderm in the gastrulating mouse embryo and a comparable expression pattern is observed in chick [263, 264]. The AVE fails to move anteriorly in *Otx2*–deficient mice and extensive genetic analyses have indicated that extra-embryonic expression of *Otx* genes is sufficient to rescue the early induction of anterior neural tissue, even if the epiblast is *Otx2*–negative. However, these transiently rescued embryos still develop anterior truncations, presumably because anterior fates are initially induced but not maintained at later stages [264–266].

Explant assays have revealed a requirement for Otx2 in the anterior neuroectoderm where it may act as a competence factor that allows the anterior neural plate to respond to signals from the ANB [266, 267]. Obviously, this does not rule out that Otx2 is also required in the AME, but a conditional knockout of this gene to address this question has not yet been described.

In *Xenopus*, Otx2 is expressed in the DBL and ectopic expression induces ectopic cement glands and rudimentary axes [268]. Furthermore, Otx2 is able to anteriorize AME and it has been shown to inhibit convergence and extension movements [269, 270].

Otx2 expression is downregulated by the posteriorizing factors discussed in Section 5.4 [263, 269]. During gastrulation and neurulation, the neural *Otx2* expression domain is delimited posteriorly by the expression domain of another homeobox gene, *Gbx2*, and the position of the MHB is anticipated by the border between these two expression domains. The MHB itself acts as an important local signaling center (see Section 5.4.2) that regulates the development of both midbrain and hindbrain and different mouse mutants have indicated that its position critically depends on Otx2 [110, 111, 264, 271]. Thus, *Otx2* is a precedent for a gene that links early AP patterning to the establishment of secondary organizing centers, resulting in progressive refinement of AP subregionalization.

5.6.8
Six3

The homeobox gene *Six3* is expressed in anterior neuroectoderm and its inactivation results in forebrain truncations [272]. Overexpression of Six3 rescues the anterior defects characteristic for the zebrafish *headless* mutant (see Section 5.4.6), suggesting that Six3 acts by inhibiting Wnt/β-catenin signaling. Electroporation experiments in chick have revealed a mutual repression between Six3 and Wnt1 in the midbrain [272].

Furthermore, it has been suggested that Six3, similar to Otx2, acts as a competence factor regulating the response of the anterior neural plate to inductive signals from secondary signaling centers such as the ANB. As for *Otx2* and *Gbx2*, *Six3* expression is bound posteriorly by expression of the homeobox gene *Irx3*. Hence, the neural plate is subdivided anteroposteriorly by a code of mutually complementary expression domains of homeobox transcription factors that may act by establishing different regions of competence [273]. This "competence code" is likely to be a direct readout of the gradients regulating early AP patterning [224].

5.6.9
Transcriptional Repression in Anterior Specification

At least two posteriorizing pathways act by derepressing the transcription of posterior genes instead of inducing them instructively. In the absence of Wnt signals, zebrafish TCF3 interacts with co-repressors of the Groucho family to repress target genes [213, 220, 274]. Upon Wnt activation, β-catenin displaces the co-repressors and converts the transcriptional complex into an activator form [274]. Similarly, posterior genes are repressed by unliganded RARs that recruit co-repressors such as SMRT (silencing mediator of RAR and thyroid hormone receptor) and N-CoR (nuclear receptor co-repressor) and this effect is relieved once RA binds to its receptor [165]. Thus, a double repression mechanism, extracellular inhibition of posteriorizing growth factors and intracellular inhibition of the transcription of their target genes, may constitute a reiterated motif in anterior development.

5.6.10
Summary

Several transcription factors are essential for anterior specification and they are expressed in nested anterior domains. For many of them, the analysis of chimeric mice has allowed the requirements in embryonic and extra-embryonic tissues to be distinguished. Transcription factors regulate anterior specification (1) by defining domains of cellular competence to respond to extracellular signals or (2) by acting as downstream mediators of these signals. Our understanding of which signal results in induction or repression of which target gene is only in its beginnings, but the increasing use of large-scale gene expression profiling will help to unravel the complex interactions between extracellular signals and transcriptional regulation in the near future.

5.7
Conclusions and Outlook

The vertebrate gastrula organizer plays a central role in the establishment of anterior identity. At early stages, organizer formation depends on Nodal and β-catenin signaling but only slightly later, during gastrulation, the embryo undergoes a dramatic change in competence to respond to these signals and both Nodal and Wnt/β-catenin signaling as well as BMP, FGF and RA signaling counteract anterior development. The organizer secretes inhibitors of BMPs, Nodals and Wnts and the combinatorial inhibition of these three classes of signals is required and sufficient for head formation.

The extra-embryonic endoderm in amniotes (AVE, hypoblast) is a signaling center that influences the anterior region of the embryo before the organizer has reached it and it is essential, albeit not sufficient, for head induction. Chimeric approaches in mouse have pointed towards a sequential requirement for AVE and organizer tissues (see Sections 5.4.4, 5.6.1, 5.6.3, 5.6.5, 5.6.7). This indicates that anterior specification in amniotes occurs in two steps: first, AVE induces a labile anterior character and, second, the MGO and its derivatives reinforce this fate decision and are required for its maintenance. It is noteworthy that, despite their different origins, both the AVE and the organizer seem to function by similar molecular mechanisms as they antagonize the same classes of posteriorizing signals via largely overlapping sets of inhibitors (Table 5.1).

Nieuwkoop proposed that a graded signal is responsible for the pattern of the AP axis in the amphibian embryo. Despite our recent evidence that posteriorizing factors can act dose-dependently to specify more than a binary cell-fate decision, it remains to be established how different extracellular concentrations of a factor are transduced via the cell membrane to elicit distinct intracellular responses. Most signaling molecules are able to bind to a spectrum of receptors and the differential distribution of each of these may influence the extracellular distribution of the ligand as well as regulate the strength of intracellular signal transduction. Furthermore, it is as yet unclear how different classes of posteriorizing signals interact. Some may act directly by specifying posterior cell fates while others may just modulate the cellular competence to respond to the real transformers. Finally, it remains to be established how the extensive movements single cells and whole tissues undergo during the unfolding of the vertebrate body plan influence the distribution of signals. Since the gastrulating embryo is a highly dynamic structure, patterning gradients may not only form by diffusion of signaling factors and their inhibitors but also by the relative movement of signaling sources and receiving tissues.

5.7 Conclusions and Outlook

Are the establishment of anterior identity and neural induction separable processes or are they linked, as suggested by Nieuwkoop? Recent experiments using chick explants seem to favour the first hypothesis [37] and this is in agreement with the observation that the key players in neuralization and AP specification differ (for example, BMP/Wnt gradient). Organizer grafts have revealed the presence of an inherent AP pattern in the uncommitted ectoderm of the zebrafish embryo [275]. Thus, it is possible that neural induction simply reveals an AP character that has already been established before gastrulation. This idea is also supported by the emergence of an AP neural pattern in zebrafish embryos lacking both anteriorizing and posteriorizing signals (see Section 5.4.4) [138].

An attempt to combine tissue interactions, signals and effectors resulting in anterior specification in a comprehensive model for head induction is shown in Fig. 5.9.

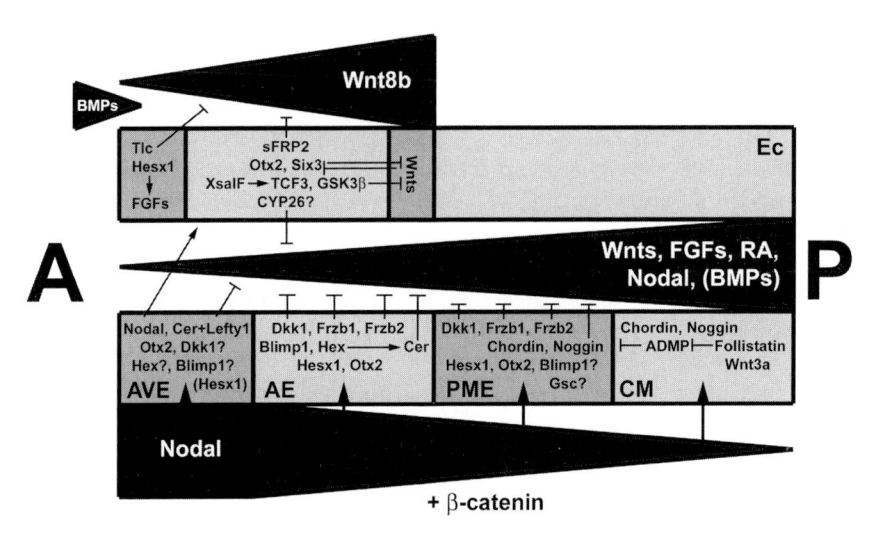

Fig. 5.9
Model for the establishment of the AP axis in vertebrates. Graded Nodal activity synergizes with β-catenin to induce and regionalize the AME (and the AVE in amniotes). The differential expression of molecular effectors in AVE, AME derivatives and overlying ectoderm (Ec), results in the formation of a gradient of transforming activity that patterns the AP axis. A secondary AP Wnt gradient regulates patterning between forebrain and midbrain while intermediate BMP levels may be required for the induction of some anterior ectodermal structures. No specific function has been characterized in this tissue.

Abbreviations

ADMP, anti-dorsalizing morphogenetic protein; AE, anterior endoderm (definitive, embryonic); AME, axial mesendoderm; AP, anteroposterior; AVE, anterior visceral endoderm (extra-embryonic); BCNE, blastula Chordin- and Noggin-expressing center; BMP, bone morphogenetic protein; *boz*, *bozozok*; Cer, Cerberus; Cer-l, Cerberus-like; CM, chordamesoderm; *cyc, cyclops*; Crs, Crescent; DBL, dorsal blastopore lip; Dkk, Dickkopf; Dsh, Dishevelled; DV, dorsoventral; FGF, fibroblast growth factor; Fz, Frizzled; Gsc, Goosecoid; GSK3β, glycogen synthase kinase 3β; HNF3β, hepatocyte nuclear factor 3β; ICAT, inhibitor of β-catenin and TCF; IGF, insulin-like growth factor; LEF, lymphoid enhancer-binding factor; LRP, lipoprotein receptor-related protein; MAPK, mitogen-activated protein kinase; MBT, mid-blastula transition; MHB, midbrain-hindbrain boundary; *MZoep*, maternal and zygotic *one-eyed pinhead*; *oep, one-eyed pinhead*; PME, prechordal mesendoderm; RA, retinoic acid; RALDH2, retinaldehyde dehydrogenase 2; RAR, retinoic acid receptor; SBE, suprablastoporal endoderm; *sia, siamois*; SMRT, silencing mediator of RAR and thyroid hormone receptor; *sqt, squint*; TCF, T cell-specific transcription factor; TGFβ, transforming growth factor β; WISE, Wnt modulator in surface ectoderm; XFD, *Xenopus* dominant-negative FGF receptor; YSL, yolk syncytial layer

Acknowledgments

I thank Esther Bell and Andrew Lumsden for helpful comments on the manuscript and stimulating discussions and apologize to all those researchers whose work I could not cite due to space constraints.

Notes added in proof

While this chapter was under review, further evidence for the essential role of BMP signaling in DV patterning of the vertebrate gastrula embryo was provided. Simultaneous loss of BMP7 and Tsg in mice results in underdeveloped ventroposterior structures (Zakin et al. *Development* **2005**; *132*: 2489–2499). Similarly, ventral structures are reduced and the neural plate becomes expanded progressively in *Xenopus* embryos with reduced levels of BMP2, BMP4, BMP7 and Tsg (Reversade et al. *Development* **2005**; *132*: 3381–3392). Conversely, no dorsal axis develops in frog embryos in which Chordin, Noggin and Follistatin are absent (Khokha et al. *Dev Cell* **2005**; *8*: 401–411). These studies indicate that the rather weak effects on DV patterning that have been observed following the impairment of single BMPs or BMP inhibitors are likely to result from extensive functional redundancy. At the same time, several studies now support an early role for FGF signaling in neural induction that reaches beyond intracellular BMP antagonism (Linker & Stern. *Development* **2004**; *131*: 5671–5681; Delaune et al. *Development* **2005**; *132*: 299–310). Unexpectedly, neuralisation of dissociated ectoderm cells – previously regarded as evidence for the default model of neural induction – appears to result from an activation of the FGF-MAPK pathway, rather than the dilution of BMPs (Kuroda et al. *Genes Dev* **2005**; *19*: 1022–1027). Finally, *blimp1* has been found to be disrupted in the zebrafish *u-boot* and *narrowminded* mutants (Baxendale et al. *Nat Genet* **2004**; *36*: 88–93; Hernandez-Lagunas et al. *Dev Biol* **2005**; *278*: 347–357). However, no major anterior defects are observed in these two mutants. Surprisingly, embryos treated with a morpholino against *blimp1* show enhanced dorsoanterior features suggesting that *blimp1* counteracts, rather than promotes, dorsalising signals (Wilm & Solnica-Krezel. *Development* **2005**; *132*: 393–404).

References

1 Reichert H. Conserved genetic mechanisms for embryonic brain patterning. *Int J Dev Biol* **2002**; *46*: 81–87.
2 Lowe CJ, Wu M, Salic A, Evans L, Lander E, Stange-Thomann N, Gruber CE, Gerhart J, Kirschner M. Anteroposterior patterning in hemichordates and the origins of the chordate nervous system. *Cell* **2003**; *113*: 853–865.
3 Meinhardt H. Models for the generation of the embryonic body axes: ontogenetic and evolutionary aspects. *Curr Opin Genet Dev* **2004**; *14*: 446–454.
4 Spemann H, Mangold H. Über Induktion von Embryonalanlagen durch Implantation artfremder Organisatoren. *Arch Mikrosk Anat Entwicklungsmechan* **1924**; *100*: 599–638.
5 Spemann H. Über den Anteil von Implantat und Wirtskeim an der Orientierung und Beschaffenheit der induzierten Embryonalanlage. *Roux' Arch Entwicklungsmech* **1931**; *123*: 389–517.
6 Vogt W. Gestaltungsanalyse am Amphibienkeim mit örtlicher Vitalfärburg. II. Gastrulation und Mesodermbildung bei Urodelen und Anuren. *Wilhelm Roux' Arch Entwicklungsmech Org* **1929**; *120*: 384–706.
7 Mangold O. Über die Induktionsfähigkeit der verschiedenen Bezirke der Neurula von Urodelen. *Naturwissenschaften* **1933**; *21*: 761–766.
8 Nieuwkoop PD, Botterenbrood EC, Kremer A et al. Activation and organization of the central nervous system in amphibians. *J Exp Zool* **1952**; *120*: 1–108.
9 Saxén L, Toivonen S. The two-gradient hypothesis in primary induction: The combined effect of two types of inducers mixed in different ratios. *J Embryol Exp Morph* **1961**; *9*: 514–533.
10 Waddington CH. Induction by the primitive streak and its derivatives in the chick. *J Exp Zool* **1933**; *10*: 38–46.
11 Oppenheimer JM. Transplantation experiments on developing teleosts (Fundulus and Perca). *J Exp Zool* **1936**; *72*: 409–437.
12 Waddington CH. Organizers in mammalian development. *Nature* **1936**; *138*: 125.
13 Blum M, Gaunt SJ, Cho KW, Steinbeisser H, Blumberg B, Bittner D, De Robertis EM. Gastrulation in the mouse: the role of the homeobox gene *goosecoid*. *Cell* **1992**; *69*: 1097–1106.
14 Beddington RS. Induction of a second neural axis by the mouse node. *Development* **1994**; *120*: 613–620.
15 Kintner CR, Dodd J. Hensen's node induces neural tissue in *Xenopus* ectoderm. Implications for the action of the organizer in neural induction. *Development* **1991**; *113*: 1495–1505.

16 Hatta K, Takahashi Y. Secondary axis induction by heterospecific organizers in zebrafish. *Dev Dyn* **1996**; *205*: 183–195.

17 Knoetgen H, Viebahn C, Kessel M. Head induction in the chick by primitive endoderm of mammalian, but not avian origin. *Development* **1999**; *126*: 815–825.

18 Knoetgen H, Teichmann U, Wittler L, Viebahn C, Kessel M. Anterior neural induction by nodes from rabbits and mice. *Dev Biol* **2000**; *225*: 370–380.

19 Saudé L, Woolley K, Martin P, Driever W, Stemple DL. Axis-inducing activities and cell fates of the zebrafish organizer. *Development* **2000**; *127*: 3407–3417.

20 Tam PP, Steiner KA. Anterior patterning by synergistic activity of the early gastrula organizer and the anterior germ layer tissues of the mouse embryo. *Development* **1999**; *126*: 5171–5179.

21 Kinder SJ, Tsang TE, Wakamiya M, Sasaki H, Behringer RR, Nagy A, Tam PP. The organizer of the mouse gastrula is composed of a dynamic population of progenitor cells for the axial mesoderm. *Development* **2001**; *128*: 3623–3634.

22 Lemaire L, Roeser T, Izpisua-Belmonte JC, Kessel M. Segregating expression domains of two *goosecoid* genes during the transition from gastrulation to neurulation in chick embryos. *Development* **1997**; *124*: 1443–1452.

23 Storey KG, Crossley JM, De Robertis EM, Norris WE, Stern CD. Neural induction and regionalisation in the chick embryo. *Development* **1992**; *114*: 729–741.

24 Gerhart J, Danilchik M, Doniach T, Roberts S, Rowning B, Stewart R. Cortical rotation of the *Xenopus* egg: consequences for the anteroposterior pattern of embryonic dorsal development. *Development* **1989**; *107*: 37–51.

25 Fan MJ, Sokol SY. A role for Siamois in Spemann organizer formation. *Development* **1997**; *124*: 2581–2589.

26 Kessler DS. Siamois is required for formation of Spemann's organizer. *Proc Natl Acad Sci USA* **1997**; *94*: 13017–13022.

27 Schneider VA, Mercola M. Spatially distinct head and heart inducers within the *Xenopus* organizer region. *Curr Biol* **1999**; *9*: 800–809.

28 Camus A, Davidson BP, Billiards S, Khoo P, Rivera-Perez JA, Wakamiya M, Behringer RR, Tam PP. The morphogenetic role of midline mesendoderm and ectoderm in the development of the forebrain and the midbrain of the mouse embryo. *Development* **2000**; *127*: 1799–1813.

29 Pera EM, Kessel M. Patterning of the chick forebrain anlage by the prechordal plate. *Development* **1997**; *124*: 4153–4162.

30 Grinblat Y, Gamse J, Patel M, Sive H. Determination of the zebrafish forebrain: induction and patterning. *Development* **1998**; *125*: 4403–4416.

31 Fekany K, Yamanaka Y, Leung T, Sirotkin HI, Topczewski J, Gates MA, Hibi M, Renucci A, Stemple D, Radbill A, Schier AF, Driever W, Hirano T, Talbot WS, Solnica-Krezel L. The zebrafish *bozozok* locus encodes Dharma, a homeodomain protein essential for induction of gastrula organizer and dorsoanterior embryonic structures. *Development* **1999**; *126*: 1427–1438.

32. Thomas PQ, Brown A, Beddington RS. *Hex*: a homeobox gene revealing peri-implantation asymmetry in the mouse embryo and an early transient marker of endothelial cell precursors. *Development* **1998**; *125*: 85–94.
33. Thomas P, Beddington R. Anterior primitive endoderm may be responsible for patterning the anterior neural plate in the mouse embryo. *Curr Biol* **1996**; *6*: 1487–1496.
34. Kimura C, Yoshinaga K, Tian E, Suzuki M, Aizawa S, Matsuo I. Visceral endoderm mediates forebrain development by suppressing posteriorizing signals. *Dev Biol* **2000**; *225*: 304–321.
35. Yatskievych TA, Pascoe S, Antin PB. Expression of the homebox gene *Hex* during early stages of chick embryo development. *Mech Dev* **1999**; *80*: 107–109.
36. Foley AC, Skromne I, Stern CD. Reconciling different models of forebrain induction and patterning: a dual role for the hypoblast. *Development* **2000**; *127*: 3839–3854.
37. Chapman SC, Schubert FR, Schoenwolf GC, Lumsden A. Anterior identity is established in chick epiblast by hypoblast and anterior definitive endoderm. *Development* **2003**; *130*: 5091–5101.
38. Winklbauer R, Schürfeld M. Vegetal rotation, a new gastrulation movement involved in the internalization of the mesoderm and endoderm in *Xenopus*. *Development* **1999**; *126*: 3703–3713.
39. Jones CM, Broadbent J, Thomas PQ, Smith JC, Beddington RS. An anterior signaling centre in *Xenopus* revealed by the homeobox gene *XHex*. *Curr Biol* **1999**; *9*: 946–954.
40. Bouwmeester T, Kim S, Sasai Y, Lu B, De Robertis EM. Cerberus is a head-inducing secreted factor expressed in the anterior endoderm of Spemann's organizer. *Nature* **1996**; *382*: 595–601.
41. Belo JA, Bouwmeester T, Leyns L, Kertesz N, Gallo M, Follettie M, De Robertis EM. Cerberus-like is a secreted factor with neuralizing activity expressed in the anterior primitive endoderm of the mouse gastrula. *Mech Dev* **1997**; *68*: 45–57.
42. Biben C, Stanley E, Fabri L, Kotecha S, Rhinn M, Drinkwater C, Lah M, Wang CC, Nash A, Hilton D, Ang SL, Mohun T, Harvey RP. Murine cerberus homologue mCer–1: a candidate anterior patterning molecule. *Dev Biol* **1998**; *194*: 135–151.
43. Shawlot W, Deng JM, Behringer RR. Expression of the mouse *cerberus-related* gene, *Cerr1*, suggests a role in anterior neural induction and somitogenesis. *Proc Natl Acad Sci USA* **1998**; *95*: 6198–6203.
44. Bradley L, Wainstock D, Sive H. Positive and negative signals modulate formation of the *Xenopus* cement gland. *Development* **1996**; *122*: 2739–2750.
45. Lupo G, Harris WA, Barsacchi G, Vignali R. Induction and patterning of the telencephalon in *Xenopus laevis*. *Development* **2002**; *129*: 5421–5436.
46. Withington S, Beddington R, Cooke J. Foregut endoderm is required at head process stages for anteriormost neural patterning in chick. *Development* **2001**; *128*: 309–320.
47. Smithers LE, Jones CM. *Xhex*-expressing endodermal tissues are essential for anterior patterning in *Xenopus*. *Mech Dev* **2002**; *119*: 191–200.

48 Ho CY, Houart C, Wilson SW, Stainier DY. A role for the extraembryonic yolk syncytial layer in patterning the zebrafish embryo suggested by properties of the *hex* gene. *Curr Biol* **1999**; *9*: 1131–1134.

49 D'Amico LA, Cooper MS. Morphogenetic domains in the yolk syncytial layer of axiating zebrafish embryos. *Dev Dyn* **2001**; *222*: 611–624.

50 Houart C, Westerfield M, Wilson SW. A small population of anterior cells patterns the forebrain during zebrafish gastrulation. *Nature* **1998**; *391*: 788–792.

51 Kuroda H, Wessely O, De Robertis EM. Neural Induction in *Xenopus*: Requirement for Ectodermal and Endomesodermal Signals via Chordin, Noggin, β-Catenin, and Cerberus. *PLoS Biol* **2004**; *2*: E92.

52 Shi Y, Massagué J. Mechanisms of TGF-β signaling from cell membrane to the nucleus. *Cell* **2003**; *113*: 685–700.

53 Zimmerman LB, De Jesus-Escobar JM, Harland RM. The Spemann organizer signal noggin binds and inactivates bone morphogenetic protein 4. *Cell* **1996**; *86*: 599–606.

54 Piccolo S, Sasai Y, Lu B, De Robertis EM. Dorsoventral patterning in *Xenopus*: inhibition of ventral signals by direct binding of chordin to BMP–4. *Cell* **1996**; *86*: 589–598.

55 Fainsod A, Deissler K, Yelin R, Marom K, Epstein M, Pillemer G, Steinbeisser H, Blum M. The dorsalizing and neural inducing gene *follistatin* is an antagonist of BMP–4. *Mech Dev* **1997**; *63*: 39–50.

56 Hansen CS, Marion CD, Steele K, George S, Smith WC. Direct neural induction and selective inhibition of mesoderm and epidermis inducers by Xnr3. *Development* **1997**; *124*: 483–492.

57 Piccolo S, Agius E, Leyns L, Bhattacharyya S, Grunz H, Bouwmeester T, De Robertis EM. The head inducer Cerberus is a multifunctional antagonist of Nodal, BMP and Wnt signals. *Nature* **1999**; *397*: 707–710.

58 Ohta K, Lupo G, Kuriyama S, Keynes R, Holt CE, Harris WA, Tanaka H, Ohnuma S. Tsukushi functions as an organizer inducer by inhibition of BMP activity in cooperation with Chordin. *Dev Cell* **2004**; *7*: 347–358.

59 De Robertis EM, Kuroda H. Dorsal-Ventral Patterning and Neural Induction in *Xenopus* Embryos. *Annu Rev Cell Dev Biol* **2004**; *20*: 285–308.

60 Niehrs C. Regionally specific induction by the Spemann-Mangold organizer. *Nat Rev Genet* **2004**; *5*: 425–434.

61 Munoz-Sanjuan I, Hemmati-Brivanlou A. Neural induction, the default model and embryonic stem cells. *Nat Rev Neurosci* **2002**; *3*: 271–280.

62 Schier AF. Axis formation and patterning in zebrafish. *Curr Opin Genet Dev* **2001**; *11*: 393–404.

63 Hammerschmidt M, Mullins MC. Dorsoventral patterning in the zebrafish: bone morphogenetic proteins and beyond. *Results Probl Cell Differ* **2002**; *40*: 72–95.

64 Smith WC, Knecht AK, Wu M, Harland RM. Secreted noggin protein mimics the Spemann organizer in dorsalizing *Xenopus* mesoderm. *Nature* **1993**; *361*: 547–549.

65 Hemmati-Brivanlou A, Kelly OG, Melton DA. Follistatin, an antagonist of activin, is expressed in the Spemann organizer and displays direct neuralizing activity. *Cell* **1994**; *77*: 283–295.

66 Sasai Y, Lu B, Steinbeisser H, Geissert D, Gont LK, De Robertis EM. *Xenopus* chordin: a novel dorsalizing factor activated by organizer-specific homeobox genes. *Cell* **1994**; *79*: 779–790.

67 Glinka A, Wu W, Onichtchouk D, Blumenstock C, Niehrs C. Head induction by simultaneous repression of Bmp and Wnt signalling in *Xenopus*. *Nature* **1997**; *389*: 517–519.

68 Yamamoto TS, Takagi C, Hyodo AC, Ueno N. Suppression of head formation by Xmsx–1 through the inhibition of intracellular nodal signaling. *Development* **2001**; *128*: 2769–2779.

69 Dale L, Howes G, Price BM, Smith JC. Bone morphogenetic protein 4: a ventralizing factor in early *Xenopus* development. *Development* **1992**; *115*: 573–585.

70 Jones CM, Lyons KM, Lapan PM, Wright CV, Hogan BL. DVR–4 (bone morphogenetic protein–4) as a posterior-ventralizing factor in *Xenopus* mesoderm induction. *Development* **1992**; *115*: 639–647.

71 Sedohara A, Fukui A, Michiue T, Asashima M. Role of BMP–4 in the inducing ability of the head organizer in *Xenopus laevis*. *Zool Sci* **2002**; *19*: 67–80.

72 Agathon A, Thisse C, Thisse B. The molecular nature of the zebrafish tail organizer. *Nature* **2003**; *424*: 375–376.

73 Bachiller D, Klingensmith J, Kemp C, Belo JA, Anderson RM, May SR, McMahon JA, McMahon AP, Harland RM, Rossant J, De Robertis EM. The organizer factors Chordin and Noggin are required for mouse forebrain development. *Nature* **2000**; *403*: 658–661.

74 Barth KA, Kishimoto Y, Rohr KB, Seydler C, Schulte-Merker S, Wilson SW. Bmp activity establishes a gradient of positional information throughout the entire neural plate. *Development* **1999**; *126*: 4977–4987.

75 Anderson RM, Lawrence AR, Stottmann RW, Bachiller D, Klingensmith J. Chordin and noggin promote organizing centers of forebrain development in the mouse. *Development* **2002**; *129*: 4975–4987.

76 Petryk A, Anderson RM, Jarcho MP, Leaf I, Carlson CS, Klingensmith J, Shawlot W, O'Connor MB. The mammalian *twisted gastrulation* gene functions in foregut and craniofacial development. *Dev Biol* **2004**; *267*: 374–386.

77 Zakin L, De Robertis EM. Inactivation of mouse *Twisted gastrulation* reveals its role in promoting Bmp4 activity during forebrain development. *Development* **2004**; *131*: 413–424.

78 Gammill LS, Sive H. Coincidence of otx2 and BMP4 signaling correlates with *Xenopus* cement gland formation. *Mech Dev* **2000**; *92*: 217–226.

79 Wilson SW, Rubenstein JL. Induction and dorsoventral patterning of the telencephalon. *Neuron* **2000**; *28*: 641–651.

80 Wilson SW, Houart C. Early steps in the development of the forebrain. *Dev Cell* **2004**; *6*: 167–181.

81 Kishimoto Y, Lee KH, Zon L, Hammerschmidt M, Schulte-Merker S. The molecular nature of zebrafish *swirl*: BMP2 func-

tion is essential during early dorsoventral patterning. *Development* **1997**; *124*: 4457–4466.

82 Martinez-Barbera JP, Toresson H, Da Rocha S, Krauss S. Cloning and expression of three members of the zebrafish *Bmp* family: *Bmp2a*, *Bmp2b* and *Bmp4*. *Gene* **1997**; *198*: 53–59.

83 Nikaido M, Tada M, Saji T, Ueno N. Conservation of BMP signaling in zebrafish mesoderm patterning. *Mech Dev* **1997**; *61*: 75–88.

84 Hartley KO, Hardcastle Z, Friday RV, Amaya E, Papalopulu N. Transgenic *Xenopus* embryos reveal that anterior neural development requires continued suppression of BMP signaling after gastrulation. *Dev Biol* **2001**; *238*: 168–184.

85 Moos MJ, Wang S, Krinks M. Anti-dorsalizing morphogenetic protein is a novel TGF-β homolog expressed in the Spemann organizer. *Development* **1995**; *121*: 4293–4301.

86 Joubin K, Stern CD. Molecular interactions continuously define the organizer during the cell movements of gastrulation. *Cell* **1999**; *98*: 559–571.

87 Dickmeis T, Rastegar S, Aanstad P, Clark M, Fischer N, Korzh V, Strähle U. Expression of the *anti-dorsalizing morphogenetic protein* gene in the zebrafish embryo. *Dev Genes Evol* **2001**; *211*: 568–572.

88 Lele Z, Nowak M, Hammerschmidt M. Zebrafish *admp* is required to restrict the size of the organizer and to promote posterior and ventral development. *Dev Dyn* **2001**; *222*: 681–687.

89 Dosch R, Niehrs C. Requirement for anti-dorsalizing morphogenetic protein in organizer patterning. *Mech Dev* **2000**; *90*: 195–203.

90 Willot V, Mathieu J, Lu Y, Schmid B, Sidi S, Yan YL, Postlethwait JH, Mullins M, Rosa F, Peyrieras N. Cooperative action of ADMP- and BMP-mediated pathways in regulating cell fates in the zebrafish gastrula. *Dev Biol* **2002**; *241*:59–78.

91 Hino J, Nishimatsu S, Nagai T, Matsuo H, Kangawa K, Nohno T. Coordination of BMP–3b and cerberus is required for head formation of *Xenopus* embryos. *Dev Biol* **2003**; *260*:138–157.

92 Böttcher RT, Niehrs C. Fibroblast growth factor signalling during early vertebrate development. *Endocrine Rev* **2005**; *26*: 63–77.

93 Griffin K, Patient R, Holder N. Analysis of FGF function in normal and *no tail* zebrafish embryos reveals separate mechanisms for formation of the trunk and the tail. *Development* **1995**; *121*: 2983–2994.

94 Kroll KL, Amaya E. Transgenic *Xenopus* embryos from sperm nuclear transplantations reveal FGF signaling requirements during gastrulation. *Development* **1996**; *122*: 3173–3183.

95 Pownall ME, Tucker AS, Slack JM, Isaacs HV. eFGF, *Xcad3* and *Hox* genes form a molecular pathway that establishes the anteroposterior axis in *Xenopus*. *Development* **1996**; *122*: 3881–3892.

96 Christen B, Slack JM. FGF-8 is associated with anteroposterior patterning and limb regeneration in *Xenopus*. *Dev Biol* **1997**; *192*: 455–466.

97 Doniach T. Basic FGF as an inducer of anteroposterior neural pattern. *Cell* **1995**; *83*: 1067–1070.

98 Haremaki T, Tanaka Y, Hongo I, Yuge M, Okamoto H. Integration of multiple signal transducing pathways on Fgf response elements of the *Xenopus caudal* homologue Xcad3. *Development* 2003; *130*: 4907–4917.

99 Henrique D, Tyler D, Kintner C, Heath JK, Lewis JH, Ish-Horowicz D, Storey KG. cash4, a novel *achaete-scute* homolog induced by Hensen's node during generation of the posterior nervous system. *Genes Dev* 1997; *11*: 603–615.

100 Alvarez IS, Araujo M, Nieto MA. Neural induction in whole chick embryo cultures by FGF. *Dev Biol* 1998; *199*: 42–54.

101 Storey KG, Goriely A, Sargent CM, Brown JM, Burns HD, Abud HM, Heath JK. Early posterior neural tissue is induced by FGF in the chick embryo. *Development* 1998; *125*: 473–484.

102 Curran KL, Grainger RM. Expression of activated MAP kinase in *Xenopus laevis* embryos: evaluating the roles of FGF and other signaling pathways in early induction and patterning. *Dev Biol* 2000; *228*: 41–56.

103 Schohl A, Fagotto F. β-catenin, MAPK and Smad signaling during early *Xenopus* development. *Development* 2002; *129*: 37–52.

104 Muhr J, Graziano E, Wilson S, Jessell TM, Edlund T. Convergent inductive signals specify midbrain, hindbrain, and spinal cord identity in gastrula stage chick embryos. *Neuron* 1999; *23*: 689–702.

105 Godsave SF, Durston AJ. Neural induction and patterning in embryos deficient in FGF signaling. *Int J Dev Biol* 1997; *41*: 57–65.

106 Mathis L, Kulesa PM, Fraser SE. FGF receptor signalling is required to maintain neural progenitors during Hensen's node progression. *Nat Cell Biol* 2001; *3*: 559–566.

107 Shimamura K, Rubenstein JL. Inductive interactions direct early regionalization of the mouse forebrain. *Development* 1997; *124*: 2709–2718.

108 Walshe J, Mason I. Unique and combinatorial functions of Fgf3 and Fgf8 during zebrafish forebrain development. *Development* 2003; *130*: 4337–4349.

109 Liu A, Joyner AL. Early anterior/posterior patterning of the midbrain and cerebellum. *Annu Rev Neurosci* 2001; *24*: 869–896.

110 Rhinn M, Brand M. The midbrain-hindbrain boundary organizer. *Curr Opin Neurobiol* 2001; *11*: 34–42.

111 Wurst W, Bally-Cuif L. Neural plate patterning: upstream and downstream of the isthmic organizer. *Nat Rev Neurosci* 2001; *2*: 99–108.

112 Hongo I, Kengaku M, Okamoto H. FGF signaling and the anterior neural induction in *Xenopus*. *Dev Biol* 1999; *216*: 561–581.

113 Harland R. Neural induction. *Curr Opin Genet Dev* 2000; *10*: 357–362.

114 Streit A, Berliner AJ, Papanayotou C, Sirulnik A, Stern CD. Initiation of neural induction by FGF signalling before gastrulation. *Nature* 2000; *406*: 74–78.

115 Wilson SI, Edlund T. Neural induction: toward a unifying mechanism. *Nat Neurosci* 2001; *4*: 1161–1168.

116 Stern CD: Induction and initial patterning of the nervous system – the chick embryo enters the scene. *Curr Opin Genet Dev* **2002**; *12*: 447–451.

117 Pera EM, Ikeda A, Eivers E, De Robertis EM. Integration of IGF, FGF, and anti-BMP signals via Smad1 phosphorylation in neural induction. *Genes Dev* **2003**; *17*: 3023–3028.

118 Fürthauer M, Thisse C, Thisse B. A role for FGF–8 in the dorsoventral patterning of the zebrafish gastrula. *Development* **1997**; *124*: 4253–4264.

119 Pera EM, Wessely O, Li SY, De Robertis EM. Neural and head induction by insulin-like growth factor signals. *Dev Cell* **2001**; *1*: 655–665.

120 Richard-Parpaillon L, Heligon C, Chesnel F, Boujard D, Philpott A: The IGF pathway regulates head formation by inhibiting Wnt signaling in *Xenopus*. *Dev Biol* **2002**; *244*: 407–417.

121 Schier AF. Nodal signaling in vertebrate development. *Annu Rev Cell Dev Biol* **2003**; *19*: 589–621.

122 Agius E, Oelgeschläger M, Wessely O, Kemp C, De Robertis EM. Endodermal Nodal-related signals and mesoderm induction in *Xenopus*. *Development* **2000**; *127*: 1173–1183.

123 Lee MA, Heasman J, Whitman M. Timing of endogenous activin-like signals and regional specification of the *Xenopus* embryo. *Development* **2001**; *128*: 2939–2952.

124 Gritsman K, Talbot WS, Schier AF. Nodal signaling patterns the organizer. *Development* **2000**; *127*: 921–932.

125 Thisse B, Wright CV, Thisse C. Activin- and Nodal-related factors control antero-posterior patterning of the zebrafish embryo. *Nature* **2000**; *403*: 425–428.

126 Osada SI, Wright CV. *Xenopus* nodal-related signaling is essential for mesendodermal patterning during early embryogenesis. *Development* **1999**; *126*: 3229–3240.

127 Vincent SD, Dunn NR, Hayashi S, Norris DP, Robertson EJ. Cell fate decisions within the mouse organizer are governed by graded Nodal signals. *Genes Dev* **2003**; *17*: 1646–1662.

128 Chu GC, Dunn NR, Anderson DC, Oxburgh L, Robertson EJ. Differential requirements for Smad4 in TGFβ-dependent patterning of the early mouse embryo. *Development* **2004**; *131*: 3501–3512.

129 Dunn NR, Vincent SD, Oxburgh L, Robertson EJ, Bikoff EK. Combinatorial activities of Smad2 and Smad3 regulate mesoderm formation and patterning in the mouse embryo. *Development* **2004**; *131*: 1717–1728.

130 Varlet I, Collignon J, Robertson EJ. *nodal* expression in the primitive endoderm is required for specification of the anterior axis during mouse gastrulation. *Development* **1997**; *124*: 1033–1044.

131 Sirard C, de la Pompa JL, Elia A, Itie A, Mirtsos C, Cheung A, Hahn S, Wakeham A, Schwartz L, Kern SE, Rossant J, Mak TW. The tumor suppressor gene *Smad4/Dpc4* is required for gastrulation and later for anterior development of the mouse embryo. *Genes Dev* **1998**; *12*: 107–119.

132 Waldrip WR, Bikoff EK, Hoodless PA, Wrana JL, Robertson EJ. Smad2 signaling in extraembryonic tissues determines ante-

rior-posterior polarity of the early mouse embryo. *Cell* **1998**; *92*: 797–808.

133 Feldman B, Gates MA, Egan ES, Dougan ST, Rennebeck G, Sirotkin HI, Schier AF, Talbot WS. Zebrafish organizer development and germ-layer formation require nodal-related signals. *Nature* **1998**; *395*: 181–185.

134 Gritsman K, Zhang J, Cheng S, Heckscher E, Talbot WS, Schier AF. The EGF-CFC protein one-eyed pinhead is essential for nodal signaling. *Cell* **1999**; *97*: 121–132.

135 Shimizu T, Yamanaka Y, Ryu SL, Hashimoto H, Yabe T, Hirata T, Bae YK, Hibi M, Hirano T. Cooperative roles of Bozozok/Dharma and Nodal-related proteins in the formation of the dorsal organizer in zebrafish. *Mech Dev* **2000**; *91*: 293–303.

136 Wessely O, Agius E, Oelgeschläger M, Pera EM, De Robertis EM. Neural induction in the absence of mesoderm: β-catenin-dependent expression of secreted BMP antagonists at the blastula stage in *Xenopus*. *Dev Biol* **2001**; *234*: 161–173.

137 Woo K, Fraser SE. Specification of the zebrafish nervous system by nonaxial signals. *Science* **1997**; *277*: 254–257.

138 Sirotkin HI, Dougan ST, Schier AF, Talbot WS. *bozozok* and *squint* act in parallel to specify dorsal mesoderm and anterior neuroectoderm in zebrafish. *Development* **2000**; *127*: 2583–2592.

139 Silva AC, Filipe M, Kuerner KM, Steinbeisser H, Belo JA. Endogenous Cerberus activity is required for anterior head specification in *Xenopus*. *Development* **2003**; *130*: 4943–4953.

140 Bell E, Munoz-Sanjuan I, Altmann CR, Vonica A, Brivanlou AH. Cell fate specification and competence by Coco, a maternal BMP, TGFβ and Wnt inhibitor. *Development* **2003**; *130*: 1381–1389.

141 Simpson EH, Johnson DK, Hunsicker P, Suffolk R, Jordan SA, Jackson IJ. The mouse *Cer1* (*Cerberus related* or *homologue*) gene is not required for anterior pattern formation. *Dev Biol* **1999**; *213*: 202–206.

142 Belo JA, Bachiller D, Agius E, Kemp C, Borges AC, Marques S, Piccolo S, De Robertis EM. Cerberus-like is a secreted BMP and nodal antagonist not essential for mouse development. *Genesis* **2000**; *26*: 265–270.

143 Shawlot W, Min Deng J, Wakamiya M, Behringer RR. The *cerberus-related* gene, *Cerr1*, is not essential for mouse head formation. *Genesis* **2000**; *26*: 253–258.

144 Borges AC, Marques S, Belo JA. The BMP antagonists cerberus-like and noggin do not interact during mouse forebrain development. *Int J Dev Biol* **2001**; *45*: 441–444.

145 Perea-Gomez A, Vella FD, Shawlot W, Oulad-Abdelghani M, Chazaud C, Meno C, Pfister V, Chen L, Robertson E, Hamada H, Behringer RR, Ang SL. Nodal antagonists in the anterior visceral endoderm prevent the formation of multiple primitive streaks. *Dev Cell* **2002**; *3*: 745–756.

146 Bertocchini F, Stern CD. The hypoblast of the chick embryo positions the primitive streak by antagonizing nodal signaling. *Dev Cell* **2002**; *3*: 735–744.

147 Yamamoto M, Saijoh Y, Perea-Gomez A, Shawlot W, Behringer RR, Ang SL, Hamada H, Meno C. Nodal antagonists regu-

late formation of the anteroposterior axis of the mouse embryo. *Nature* **2004**; *428*: 387–392.

148 Vesque C, Ellis S, Lee A, Szabo M, Thomas P, Beddington R, Placzek M. Development of chick axial mesoderm: specification of prechordal mesoderm by anterior endoderm-derived TGFβ family signalling. *Development* **2000**; *127*: 2795–2809.

149 Gripp KW, Wotton D, Edwards MC, Roessler E, Ades L, Meinecke P, Richieri-Costa A, Zackai EH, Massagué J, Muenke M, Elledge SJ. Mutations in *TGIF* cause holoprosencephaly and link NODAL signalling to human neural axis determination. *Nat Genet* **2000**; *25*: 205–208.

150 de la Cruz JM, Bamford RN, Burdine RD, Roessler E, Barkovich A, J., Donnai D, Schier AF, Muenke M. A loss-of-function mutation in the CFC domain of *TDGF1* is associated with human forebrain defects. *Hum Genet* **2002**; *110*: 422–428.

151 Maden M. Retinoid signalling in the development of the central nervous system. *Nat Rev Neurosci* **2002**; *3*: 843–853.

152 Dekker EJ, Vaessen MJ, van den Berg C, Timmermans A, Godsave S, Holling T, Nieuwkoop P, Geurts van Kessel A, Durston A. Overexpression of a cellular retinoic acid binding protein (xCRABP) causes anteroposterior defects in developing *Xenopus* embryos. *Development* **1994**; *120*: 973–985.

153 Blumberg B, Bolado JJ, Moreno TA, Kintner C, Evans RM, Papalopulu N. An essential role for retinoid signaling in anteroposterior neural patterning. *Development* **1997**; *124*: 373–379.

154 Chen Y, Pollet N, Niehrs C, Pieler T. Increased XRALDH2 activity has a posteriorizing effect on the central nervous system of *Xenopus* embryos. *Mech Dev* **2001**; *101*: 91–103.

155 Niederreither K, Subbarayan V, Dollé P, Chambon P. Embryonic retinoic acid synthesis is essential for early mouse post-implantation development. *Nat Genet* **1999**; *21*: 444–448.

156 Swindell EC, Thaller C, Sockanathan S, Petkovich M, Jessell TM, Eichele G. Complementary domains of retinoic acid production and degradation in the early chick embryo. *Dev Biol* **1999**; *216*: 282–296.

157 Begemann G, Schilling TF, Rauch GJ, Geisler R, Ingham PW. The zebrafish *neckless* mutation reveals a requirement for *raldh2* in mesodermal signals that pattern the hindbrain. *Development* **2001**; *128*: 3081–3094.

158 Grandel H, Lun K, Rauch GJ, Rhinn M, Piotrowski T, Houart C, Sordino P, Kuchler AM, Schulte-Merker S, Geisler R, Holder N, Wilson SW, Brand M. Retinoic acid signalling in the zebrafish embryo is necessary during pre-segmentation stages to pattern the anterior-posterior axis of the CNS and to induce a pectoral fin bud. *Development* **2002**; *129*: 2851–2865.

159 Niederreither K, Vermot J, Schuhbaur B, Chambon P, Dollé P. Retinoic acid synthesis and hindbrain patterning in the mouse embryo. *Development* **2000**; *127*: 75–85.

160 Wendling O, Ghyselinck NB, Chambon P, Mark M. Roles of retinoic acid receptors in early embryonic morphogenesis and hindbrain patterning. *Development* **2001**; *128*: 2031–2038.

161 Hollemann T, Chen Y, Grunz H, Pieler T. Regionalized metabolic activity establishes boundaries of retinoic acid signalling. *EMBO J* **1998**; *17*: 7361–7372.

162 Kudoh T, Wilson SW, Dawid IB. Distinct roles for Fgf, Wnt and retinoic acid in posteriorizing the neural ectoderm. *Development* **2002**; *129*: 4335–4346.

163 Abu-Abed S, Dollé P, Metzger D, Beckett B, Chambon P, Petkovich M. The retinoic acid-metabolizing enzyme, CYP26A1, is essential for normal hindbrain patterning, vertebral identity, and development of posterior structures. *Genes Dev* **2001**; *15*: 226–240.

164 Sakai Y, Meno C, Fujii H, Nishino J, Shiratori H, Saijoh Y, Rossant J, Hamada H. The retinoic acid-inactivating enzyme CYP26 is essential for establishing an uneven distribution of retinoic acid along the anterio-posterior axis within the mouse embryo. *Genes Dev* **2001**; *15*: 213–225.

165 Koide T, Downes M, Chandraratna RA, Blumberg B, Umesono K. Active repression of RAR signaling is required for head formation. *Genes Dev* **2001**; *15*: 2111–2121.

166 Dupé V, Lumsden A. Hindbrain patterning involves graded responses to retinoic acid signalling. *Development* **2001**; *128*: 2199–2208.

167 Gale E, Zile M, Maden M. Hindbrain respecification in the retinoid-deficient quail. *Mech Dev* **1999**; *89*: 43–54.

168 Halilagic A, Zile MH, Studer M. A novel role for retinoids in patterning the avian forebrain during presomite stages. *Development* **2003**; *130*: 2039–2050.

169 Wodarz A, Nusse R. Mechanisms of Wnt signaling in development. *Annu Rev Cell Dev Biol* **1998**; *14*: 59–88.

170 Moon RT, Bowerman B, Boutros M, Perrimon N. The promise and perils of Wnt signaling through β-catenin. *Science* **2002**; *296*: 1644–1646.

171 Sokol SY. Wnt signaling and dorso-ventral axis specification in vertebrates. *Curr Opin Genet Dev* **1999**; *9*: 405–410.

172 Weaver C, Kimelman D. Move it or lose it: axis specification in *Xenopus*. *Development* **2004**; *131*: 3491–3499.

173 Christian JL, Moon RT. Interactions between Xwnt-8 and Spemann organizer signaling pathways generate dorsoventral pattern in the embryonic mesoderm of *Xenopus*. *Genes Dev* **1993**; *7*: 13–28.

174 Kelly GM, Erezyilmaz DF, Moon RT. Induction of a secondary embryonic axis in zebrafish occurs following the overexpression of β-catenin. *Mech Dev* **1995**; *53*: 261–273.

175 Kelly GM, Greenstein P, Erezyilmaz DF, Moon RT. Zebrafish *wnt8* and *wnt8b* share a common activity but are involved in distinct developmental pathways. *Development* **1995**; *121*: 1787–1799.

176 McGrew LL, Lai CJ, Moon RT. Specification of the anteroposterior neural axis through synergistic interaction of the Wnt signaling cascade with noggin and follistatin. *Dev Biol* **1995**; *172*: 337–342.

177 Fredieu JR, Cui Y, Maier D, Danilchik MV, Christian JL. Xwnt-8 and lithium can act upon either dorsal mesodermal or neurectodermal cells to cause a loss of forebrain in *Xenopus* embryos. *Dev Biol* **1997**; *186*: 100–114.

178 McGrew LL, Hoppler S, Moon RT. Wnt and FGF pathways cooperatively pattern anteroposterior neural ectoderm in *Xenopus*. *Mech Dev* **1997**; *69*: 105–114.

179 Saint-Jeannet JP, He X, Varmus HE, Dawid IB. Regulation of dorsal fate in the neuraxis by Wnt–1 and Wnt–3a. *Proc Natl Acad Sci USA* **1997**; *94*: 13713–13718.

180 McGrew LL, Takemaru K, Bates R, Moon RT. Direct regulation of the *Xenopus engrailed–2* promoter by the Wnt signaling pathway, and a molecular screen for Wnt-responsive genes, confirm a role for Wnt signaling during neural patterning in *Xenopus*. *Mech Dev* **1999**; *87*: 21–32.

181 Kazanskaya O, Glinka A, Niehrs C. The role of *Xenopus* dickkopf1 in prechordal plate specification and neural patterning. *Development* **2000**; *127*: 4981–4992.

182 Darken RS, Wilson PA. Axis induction by wnt signaling: Target promoter responsiveness regulates competence. *Dev Biol* **2001**; *234*: 42–54.

183 Domingos PM, Itasaki N, Jones CM, Mercurio S, Sargent MG, Smith JC, Krumlauf R. The Wnt/β-catenin pathway posteriorizes neural tissue in *Xenopus* by an indirect mechanism requiring FGF signalling. *Dev Biol* **2001**; *239*: 148–160.

184 Gamse JT, Sive H. Early anteroposterior division of the presumptive neurectoderm in *Xenopus*. *Mech Dev* **2001**; *104*: 21–36.

185 Hamilton FS, Wheeler GN, Hoppler S. Difference in XTcf–3 dependency accounts for change in response to β-catenin-mediated Wnt signalling in *Xenopus* blastula. *Development* **2001**; *128*: 2063–2073.

186 Houart C, Caneparo L, Heisenberg C, Barth K, Take-Uchi M, Wilson S. Establishment of the telencephalon during gastrulation by local antagonism of Wnt signaling. *Neuron* **2002**; *35*: 255–265.

187 Kim SH, Shin J, Park HC, Yeo SY, Hong SK, Han S, Rhee M, Kim CH, Chitnis AB, Huh TL. Specification of an anterior neuroectoderm patterning by Frizzled8a-mediated Wnt8b signalling during late gastrulation in zebrafish. *Development* **2002**; *129*: 4443–4455.

188 Pöpperl H, Schmidt C, Wilson V, Hume CR, Dodd J, Krumlauf R, Beddington RS. Misexpression of *Cwnt8C* in the mouse induces an ectopic embryonic axis and causes a truncation of the anterior neuroectoderm. *Development* **1997**; *124*: 2997–3005.

189 Roeser T, Stein S, Kessel M. Nuclear β-catenin and the development of bilateral symmetry in normal and LiCl-exposed chick embryos. *Development* **1999**; *126*: 2955–2965.

190 Nordström U, Jessell TM, Edlund T. Progressive induction of caudal neural character by graded Wnt signaling. *Nat Neurosci* **2002**; *5*: 525–532.

191 Hoppler S, Brown JD, Moon RT. Expression of a dominant-negative Wnt blocks induction of *MyoD* in *Xenopus* embryos. *Genes Dev* **1996**; *10*: 2805–2817.

192 Pierce SB, Kimelman D. Overexpression of *Xgsk–3* disrupts anterior ectodermal patterning in *Xenopus*. *Dev Biol* **1996**; *175*: 256–264.

193 Leyns L, Bouwmeester T, Kim SH, Piccolo S, De Robertis EM. Frzb–1 is a secreted antagonist of Wnt signaling expressed in the Spemann organizer. *Cell* **1997**; *88*: 747–756.

194 Wang S, Krinks M, Lin K, Luyten FP, Moos M, Jr. Frzb, a secreted protein expressed in the Spemann organizer, binds and inhibits Wnt–8. *Cell* **1997**; *88*: 757–766.

195 Deardorff MA, Tan C, Conrad LJ, Klein PS. Frizzled–8 is expressed in the Spemann organizer and plays a role in early morphogenesis. *Development* **1998**; *125*: 2687–2700.

196 Glinka A, Wu W, Delius H, Monaghan AP, Blumenstock C, Niehrs C. Dickkopf–1 is a member of a new family of secreted proteins and functions in head induction. *Nature* **1998**; *391*: 357–362.

197 Hsieh JC, Kodjabachian L, Rebbert ML, Rattner A, Smallwood PM, Samos CH, Nusse R, Dawid IB, Nathans J. A new secreted protein that binds to Wnt proteins and inhibits their activities. *Nature* **1999**; *398*: 431–436.

198 Fekany-Lee K, Gonzalez E, Miller-Bertoglio V, Solnica-Krezel L. The homeobox gene *bozozok* promotes anterior neuroectoderm formation in zebrafish through negative regulation of BMP2/4 and Wnt pathways. *Development* **2000**; *127*: 2333–2345.

199 Hashimoto H, Itoh M, Yamanaka Y, Yamashita S, Shimizu T, Solnica-Krezel L, Hibi M, Hirano T. Zebrafish Dkk1 functions in forebrain specification and axial mesendoderm formation. *Dev Biol* **2000**; *217*: 138–152.

200 Shinya M, Eschbach C, Clark M, Lehrach H, Furutani-Seiki M. Zebrafish Dkk1, induced by the pre-MBT Wnt signaling, is secreted from the prechordal plate and patterns the anterior neural plate. *Mech Dev* **2000**; *98*: 3–17.

201 Mercurio S, Latinkic B, Itasaki N, Krumlauf R, Smith JC. Connective-tissue growth factor modulates WNT signalling and interacts with the WNT receptor complex. *Development* **2004**; *131*: 2137–2147.

202 Lekven AC, Thorpe CJ, Waxman JS, Moon RT. Zebrafish *wnt8* encodes two wnt8 proteins on a bicistronic transcript and is required for mesoderm and neurectoderm patterning. *Dev Cell* **2001**; *1*: 103–114.

203 Takada S, Stark KL, Shea MJ, Vassileva G, McMahon JA, McMahon AP. Wnt–3a regulates somite and tailbud formation in the mouse embryo. *Genes Dev* **1994**; *8*: 174–189.

204 Yamaguchi TP, Takada S, Yoshikawa Y, Wu N, McMahon AP. T (*Brachyury*) is a direct target of Wnt3a during paraxial mesoderm specification. *Genes Dev* **1999**; *13*: 3185–3190.

205 Galceran J, Farinas I, Depew MJ, Clevers H, Grosschedl R. Wnt3a$^{-/-}$-like phenotype and limb deficiency in Lef1(–/–)Tcf1(–/–) mice. *Genes Dev* **1999**; *13*: 709–717.

206 Pinson KI, Brennan J, Monkley S, Avery BJ, Skarnes WC. An LDL-receptor-related protein mediates Wnt signalling in mice. *Nature* **2000**; *407*: 535–538.

207 Gregorieff A, Grosschedl R, Clevers H. Hindgut defects and transformation of the gastro-intestinal tract in Tcf4(–/–)/Tcf1(–/–) embryos. *EMBO J* **2004**; *23*: 1825–1833.

208 Wheeler GN, Hamilton FS, Hoppler S. Inducible gene expression in transgenic *Xenopus* embryos. *Curr Biol* **2000**; *10*: 849–852.

209 Davidson G, Mao B, del Barco Barrantes I, Niehrs C. Kremen proteins interact with Dickkopf1 to regulate anteroposterior CNS patterning. *Development* **2003**; *129*: 5587–5596.

210 Pearce JJ, Penny G, Rossant J. A mouse *cerberus/Dan*-related gene family. *Dev Biol* **1999**; *209*: 98–110.

211 Stanley EG, Biben C, Allison J, Hartley L, Wicks IP, Campbell IK, McKinley M, Barnett L, Koentgen F, Robb L, Harvey RP. Targeted insertion of a *lacZ* reporter gene into the mouse *Cer1* locus reveals complex and dynamic expression during embryogenesis. *Genesis* **2000**; *26*: 259–264.

212 Mukhopadhyay M, Shtrom S, Rodriguez-Esteban C, Chen L, Tsukui T, Gomer L, Dorward DW, Glinka A, Grinberg A, Huang SP, Niehrs C, Belmonte JC, Westphal H. *Dickkopf1* is required for embryonic head induction and limb morphogenesis in the mouse. *Dev Cell* **2001**; *1*: 423–434.

213 Kim CH, Oda T, Itoh M, Jiang D, Artinger KB, Chandrasekharappa SC, Driever W, Chitnis AB. Repressor activity of *Headless/Tcf3* is essential for vertebrate head formation. *Nature* **2000**; *407*: 913–916.

214 Heisenberg CP, Houart C, Take-Uchi M, Rauch GJ, Young N, Coutinho P, Masai I, Caneparo L, Concha ML, Geisler R, Dale TC, Wilson SW, Stemple DL. A mutation in the Gsk3–binding domain of zebrafish Masterblind/Axin1 leads to a fate transformation of telencephalon and eyes to diencephalon. *Genes Dev* **2001**; *15*: 1427–1434.

215 van de Water S, van de Wetering M, Joore J, Esseling J, Bink R, Clevers H, Zivkovic D. Ectopic Wnt signal determines the eyeless phenotype of zebrafish *masterblind* mutant. *Development* **2001**; *128*: 3877–3888.

216 Satoh K, Kasai M, Ishidao T, Tago K, Ohwada S, Hasegawa Y, Senda T, Takada S, Nada S, Nakamura T, Akiyama T. Anteriorization of neural fate by inhibitor of β-catenin and T cell factor (ICAT), a negative regulator of Wnt signaling. *Proc Natl Acad Sci USA* **2004**; *101*: 8017–8021.

217 Onai T, Sasai N, Matsui M, Sasai Y. *Xenopus* XsalF: anterior neuroectodermal specification by attenuating cellular responsiveness to Wnt signaling. *Dev Cell* **2004**; *7*: 95–106.

218 Erter CE, Wilm TP, Basler N, Wright CV, Solnica-Krezel L. *Wnt8* is required in lateral mesendodermal precursors for neural posteriorization in vivo. *Development* **2001**; *128*: 3571–3583.

219 Kiecker C, Niehrs C. A morphogen gradient of Wnt/β-catenin signalling regulates anteroposterior neural patterning in *Xenopus*. *Development* **2001**; *128*: 4189–4201.

220 Dorsky RI, Itoh M, Moon RT, Chitnis A. Two *tcf3* genes cooperate to pattern the zebrafish brain. *Development* **2003**; *130*: 1937–1947.

221 Itasaki N, Jones CM, Mercurio S, Rowe A, Domingos PM, Smith JC, Krumlauf R. Wise, a context-dependent activator and inhibitor of Wnt signalling. *Development* **2003**; *130*: 4295–4305.

222 Greco TL, Takada S, Newhouse MM, McMahon JA, McMahon AP, Camper SA. Analysis of the *vestigial tail* mutation demonstrates that *Wnt–3a* gene dosage regulates mouse axial development. *Genes Dev* **1996**; *10*: 313–324.

223 Nambiar RM, Henion PD. Sequential antagonism of early and late Wnt-signaling by zebrafish *colgate* promotes dorsal and anterior fates. *Dev Biol* **2004**; *267*: 165–180.

224 Braun MM, Etheridge A, Bernard A, Robertson CP, Roelink H. Wnt signaling is required at distinct stages of development for the induction of the posterior forebrain. *Development* **2003**; *130*: 5579–5587.

225 Heisenberg CP, Tada M, Rauch GJ, Saudé L, Concha ML, Geisler R, Stemple DL, Smith JC, Wilson SW. Silberblick/Wnt11 mediates convergent extension movements during zebrafish gastrulation. *Nature* **2000**; *405*: 76–81.

226 Kiecker C, Niehrs C. The role of prechordal mesendoderm in neural patterning. *Curr Opin Neurobiol* **2001**; *11*: 27–33.

227 Veeman MT, Axelrod JD, Moon RT. A second canon. Functions and mechanisms of β-catenin-independent Wnt signaling. *Dev Cell* **2003**; *5*: 367–377.

228 Bradley L, Sun B, Collins-Racie L, LaVallie E, McCoy J, Sive H. Different activities of the frizzled-related proteins frzb2 and sizzled2 during *Xenopus* anteroposterior patterning. *Dev Biol* **2000**; *227*: 118–132.

229 Pera EM, De Robertis EM. A direct screen for secreted proteins in *Xenopus* embryos identifies distinct activities for the Wnt antagonists Crescent and Frzb–1. *Mech Dev* **2000**; *96*: 183–195.

230 Habas R, Kato Y, He X. Wnt/Frizzled activation of Rho regulates vertebrate gastrulation and requires a novel Formin homology protein Daam1. *Cell* **2001**; *107*: 843–854.

231 Wünnenberg-Stapleton K, Blitz IL, Hashimoto C, Cho KW. Involvement of the small GTPases XRhoA and XRnd1 in cell adhesion and head formation in early *Xenopus* development. *Development* **1999**; *126*: 5339–5351.

232 Lane MC, Sheets MD. Rethinking axial patterning in amphibians. *Dev Dyn* **2002**; *225*: 434–447.

233 Gonzalez EM, Fekany-Lee K, Carmany-Rampey A, Erter C, Topczewski J, Wright CV, Solnica-Krezel L. Head and trunk in zebrafish arise via coinhibition of BMP signaling by *bozozok* and *chordino*. *Genes Dev* **2000**; *14*: 3087–3092.

234 del Barco Barrantes I, Davidson G, Gröne HJ, Westphal H, Niehrs C. *Dkk1* and *noggin* cooperate in mammalian head induction. *Genes Dev* **2003**; *17*: 2239–2244.

235 Tzahor E, Kempf H, Mootoosamy RC, Poon AC, Abzhanov A, Tabin CJ, Dietrich S, Lassar AB. Antagonists of Wnt and BMP signaling promote the formation of vertebrate head muscle. *Genes Dev* **2003**; *17*: 3087–3099.

236 Kiecker C, Niehrs C. The role of Wnt signaling in vertebrate head induction and the organizer-gradient model dualism. In: *Wnt Signaling in Development* (Kühl M. Ed.). Georgetown, TX: Landes Bioscience/Eurekah.com, **2003**; 71–89.

237 Martinez-Barbera JP, Beddington RS. Getting your head around *Hex* and *Hesx1*: forebrain formation in mouse. *Int J Dev Biol* **2001**; *45*: 327–336.

238 Dattani MT, Martinez-Barbera JP, Thomas PQ, Brickman JM, Gupta R, Martensson IL, Toresson H, Fox M, Wales JK, Hind-

marsh PC, Krauss S, Beddington RS, Robinson IC. Mutations in the homeobox gene *HESX1/Hesx1* associated with septo-optic dysplasia in human and mouse. *Nat Genet* **1998**; *19*: 125–133.

239 Zaraisky AG, Ecochard V, Kazanskaya OV, Lukyanov SA, Fesenko IV, Duprat AM. The homeobox-containing gene *XANF–1* may control development of the Spemann organizer. *Development* **1995**; *121*: 3839–3847.

240 Ermakova GV, Alexandrova EM, Kazanskaya OV, Vasiliev OL, Smith MW, Zaraisky AG. The homeobox gene, *Xanf–1*, can control both neural differentiation and patterning in the presumptive anterior neurectoderm of the *Xenopus laevis* embryo. *Development* **1999**; *126*: 4513–4523.

241 de Souza FS, Gawantka V, Gomez AP, Delius H, Ang SL, Niehrs C. The zinc finger gene *Xblimp1* controls anterior endomesodermal cell fate in Spemann's organizer. *EMBO J* **1999**; *18*: 6062–6072.

242 Ang SL, Rossant J. $HNF–3\beta$ is essential for node and notochord formation in mouse development. *Cell* **1994**; *78*: 561–574.

243 Chapman SC, Schubert FR, Schoenwolf GC, Lumsden A. Analysis of spatial and temporal gene expression patterns in blastula and gastrula stage chick embryos. *Dev Biol* **2002**; *245*: 187–199.

244 Dufort D, Schwartz L, Harpal K, Rossant J. The transcription factor $HNF3\beta$ is required in visceral endoderm for normal primitive streak morphogenesis. *Development* **1998**; *125*: 3015–3025.

245 Hallonet M, Kaestner KH, Martin-Parras L, Sasaki H, Betz UA, Ang SL. Maintenance of the specification of the anterior definitive endoderm and forebrain depends on the axial mesendoderm: a study using $HNF3\beta/Foxa2$ conditional mutants. *Dev Biol* **2002**; *243*: 20–33.

246 Cho KW, Blumberg B, Steinbeisser H, De Robertis EM. Molecular nature of Spemann's organizer: the role of the *Xenopus* homeobox gene *goosecoid*. *Cell* **1991**; *67*: 1111–1120.

247 Brannon M, Kimelman D. Activation of *Siamois* by the Wnt pathway. *Dev Biol* **1996**; *180*: 344–347.

248 Carnac G, Kodjabachian L, Gurdon JB, Lemaire P. The homeobox gene *Siamois* is a target of the Wnt dorsalisation pathway and triggers organiser activity in the absence of mesoderm. *Development* **1996**; *122*: 3055–3065.

249 Toyama R, O'Connell ML, Wright CV, Kuehn MR, Dawid IB. Nodal induces ectopic *goosecoid* and *lim1* expression and axis duplication in zebrafish. *Development* **1995**; *121*: 383–391.

250 Ferreiro B, Artinger M, Cho K, Niehrs C. Antimorphic goosecoids. *Development* **1998**; *125*: 1347–1359.

251 Latinkic BV, Smith JC. *Goosecoid* and *mix.1* repress *Brachyury* expression and are required for head formation in *Xenopus*. *Development* **1999**; *126*: 1769–1779.

252 Rivera-Perez JA, Mallo M, Gendron-Maguire M, Gridley T, Behringer RR. *Goosecoid* is not an essential component of the mouse gastrula organizer but is required for craniofacial and rib development. *Development* **1995**; *121*: 3005–3012.

253 Yamada G, Mansouri A, Torres M, Stuart ET, Blum M, Schultz M, De Robertis EM, Gruss P. Targeted mutation of the murine *goosecoid* gene results in craniofacial defects and neonatal death. *Development* **1995**; *121*: 2917–2922.

254 Zhu L, Belo JA, De Robertis EM, Stern CD. *Goosecoid* regulates the neural inducing strength of the mouse node. *Dev Biol* **1999**; *216*: 276–281.

255 Filosa S, Rivera-Perez JA, Gomez AP, Gansmuller A, Sasaki H, Behringer RR, Ang SL. *Goosecoid* and *HNF–3β* genetically interact to regulate neural tube patterning during mouse embryogenesis. *Development* **1997**; *124*: 2843–2854.

256 Zorn AM, Butler K, Gurdon JB. Anterior endomesoderm specification in *Xenopus* by Wnt/β-catenin and TGF-β signalling pathways. *Dev Biol* **1999**; *209*: 282–297.

257 Brickman JM, Jones CM, Clements M, Smith JC, Beddington RS. Hex is a transcriptional repressor that contributes to anterior identity and suppresses Spemann organiser function. *Development* **2000**; *127*: 2303–2315.

258 Shawlot W, Behringer RR. Requirement for *Lim1* in head-organizer function. *Nature* **1995**; *374*: 407–408.

259 Shawlot W, Wakamiya M, Kwan KM, Kania A, Jessell TM, Behringer RR. *Lim1* is required in both primitive streak-derived tissues and visceral endoderm for head formation in the mouse. *Development* **1999**; *126*: 4925–4932.

260 Perea-Gomez A, Shawlot W, Sasaki H, Behringer RR, Ang S. *HNF3β* and *Lim1* interact in the visceral endoderm to regulate primitive streak formation and anterior-posterior polarity in the mouse embryo. *Development* **1999**; *126*: 4499–4511.

261 Kodjabachian L, Karavanov AA, Hikasa H, Hukriede NA, Aoki T, Taira M, Dawid IB. A study of *Xlim1* function in the Spemann-Mangold organizer. *Int J Dev Biol* **2001**; *45*: 209–218.

262 Hukriede NA, Tsang TE, Habas R, Khoo PL, Steiner K, Weeks DL, Tam PP, Dawid IB. Conserved requirement of *Lim1* function for cell movements during gastrulation. *Dev Cell* **2003**; *4*: 83–94.

263 Bally-Cuif L, Gulisano M, Broccoli V, Boncinelli E. *c-otx2* is expressed in two different phases of gastrulation and is sensitive to retinoic acid treatment in chick embryo. *Mech Dev* **1995**; *49*: 49–63.

264 Simeone A, Puelles E, Acampora D. The *Otx* family. *Curr Opin Genet Dev* **2002**; *12*: 409–415.

265 Acampora D, Avantaggiato V, Tuorto F, Briata P, Corte G, Simeone A. Visceral endoderm-restricted translation of *Otx1* mediates recovery of *Otx2* requirements for specification of anterior neural plate and normal gastrulation. *Development* **1998**; *125*: 5091–5104.

266 Rhinn M, Dierich A, Shawlot W, Behringer RR, Le Meur M, Ang SL. Sequential roles for *Otx2* in visceral endoderm and neuroectoderm for forebrain and midbrain induction and specification. *Development* **1998**; *125*: 845–856.

267 Tian E, Kimura C, Takeda N, Aizawa S, Matsuo I. *Otx2* is required to respond to signals from anterior neural ridge for forebrain specification. *Dev Biol* **2002**; *242*: 204–223.

268 Pannese M, Polo C, Andreazzoli M, Vignali R, Kablar B, Barsacchi G, Boncinelli E. The *Xenopus* homologue of *Otx2* is a maternal homeobox gene that demarcates and specifies anterior body regions. *Development* **1995**; *121*: 707–720.

269 Andreazzoli M, Pannese M, Boncinelli E. Activating and repressing signals in head development: the role of *Xotx1* and *Xotx2*. *Development* **1997**; *124*: 1733–1743.

270 Morgan R, Hooiveld MH, Pannese M, Dati G, Broders F, Delarue M, Thiery JP, Boncinelli E, Durston AJ. Calponin modulates the exclusion of *Otx*-expressing cells from convergence extension movements. *Nat Cell Biol* **1999**; *1*: 404–408.

271 Liu A, Joyner AL. Early anterior/posterior patterning of the midbrain and cerebellum. *Annu Rev Neurosci* **2001**; *24*: 869–896.

272 Lagutin OV, Zhu CC, Kobayashi D, Topczewski J, Shimamura K, Puelles L, Russell HR, McKinnon PJ, Solnica-Krezel L, Oliver G. Six3 repression of Wnt signaling in the anterior neuroectoderm is essential for vertebrate forebrain development. *Genes Dev* **2003**; *17*: 368–379.

273 Kobayashi D, Kobayashi M, Matsumoto K, Ogura T, Nakafuku M, Shimamura K. Early subdivisions in the neural plate define distinct competence for inductive signals. *Development* **2002**; *129*: 83–93.

274 Roose J, Molenaar M, Peterson J, Hurenkamp J, Brantjes H, Moerer P, van de Wetering M, Destrée O, Clevers H. The *Xenopus* Wnt effector XTcf–3 interacts with Groucho-related transcriptional repressors. *Nature* **1998**; *395*: 608–612.

275 Koshida S, Shinya M, Mizuno T, Kuroiwa A, Takeda H. Initial anteroposterior pattern of the zebrafish central nervous system is determined by differential competence of the epiblast. *Development* **1998**; *125*: 1957–1966.

276 Gilbert SF. *Developmental Biology* (6th edn). Sunderland, MA: Sinauer Associates, Inc. Publishers, **2000**.

277 Davidson BP, Kinder SJ, Steiner K, Schoenwolf GC, Tam PP. Impact of node ablation on the morphogenesis of the body axis and the lateral asymmetry of the mouse embryo during early organogenesis. *Dev Biol* **1999**; *211*: 11–26.

6
Anterior-Posterior Patterning of the Hindbrain: Integrating Boundaries and Cell Segregation with Segment Formation and Identity

Angelo Iulianella and Paul A. Trainor

6.1
Introduction

Morphologically, the formation of the primitive streak at the commencement of gastrulation signals the initiation of anterior-posterior patterning during vertebrate development. One of the important consequences of the morphogenetic movement of cells during gastrulation is to organize populations of cells in the correct position that are destined to contribute to a particular embryonic structure. Segmentation is an underlying feature of vertebrate embryonic development and describes the process in which repeated units of progenitor tissue are generated. Shortly after the completion of gastrulation, segmentation becomes clearly evident in the vertebrate head, particularly in the nervous system in the form of serially repeated brain vesicles, cranial ganglia and branchial motor nerves. Despite an initial homology, individual segments and their derivatives become distinguished as they differentiate to confer regional character along a body axis. In this chapter we discuss the potential mechanisms that lead to the segmentation and anterior-posterior patterning of the vertebrate nervous system with particular emphasis on hindbrain.

Gastrulation is a morphogenetic process that leads to the formation of the mesoderm and the generation of a triblastic embryo composed of endoderm (the innermost layer), mesoderm (middle layer) and ectoderm (outer layer) and this occurs between 6.5 and 7.5 days post coitum (dpc) of mouse embryonic development. It is from the ectoderm that the neuroepithelium and subsequently the major tissues of the central and peripheral nervous systems are derived. Shortly after the completion of gastrulation and the induction of the neural plate, the process of neurulation takes place in which the neural plate develops into a tubular structure called the neural tube. The anterior portion of the neural tube will become the brain whereas the posterior portion of the neural tube will develop into the spinal cord. Differential growth and morphogenesis within the neural tube results in individual regions of the neuroepithelium expanding at different rates such that the anterior neural tube becomes subdivided into three major brain vesicles: the forebrain (prosencephalon), the midbrain (mesencephalon) and the hindbrain (rhombencephalon). The forebrain later becomes further subdivided into the anterior telencephalon and the more

---

Cell Signaling and Growth Factors in Development. Edited by K. Unsicker and K. Krieglstein
Copyright © 2006 WILEY-VCH Verlag GmbH & Co. KGaA, Weinheim
ISBN 3-527-31034-7

caudal diencephalon. The telencephalon gives rise to the cerebral cortex, basal ganglia, hippocampus, amygdala and olfactory bulb. The diencephalon will generate the thalamus, hypothalamus, subthalamus and epithalamus. Similar to the forebrain, the hindbrain is further regionalized into the anterior metencephalon and the posterior myelencephalon. The metencephalon ultimately gives rise to the pons and cerebellum, the part of the brain responsible for co-coordinating movements, posture and balance. The myelencephalon eventually forms the medulla oblongata, the nerves of which regulate respiratory, gastrointestinal and cardiovascular movements. In contrast to the forebrain and hindbrain, the midbrain is thought not to be subdivided further morphologically and its lumen becomes the cerebral aqueduct.

6.2
Hindbrain Development

6.2.1
Segmentation into Rhombomeres and a Blueprint for Craniofacial Development

Of all the regions of the developing nervous system, the hindbrain has been the focus of intense attention because even after its subdivision into the metencephalon and myelencephalon it undergoes a third round of division into seven contiguous compartments which are termed rhombomeres [1]. These transverse periodic neuroepithelial bulges are clearly distinguishable by 9–9.5 days in mouse and 1.5–2 days in chick during embryonic development (Fig. 6.1). Each rhombomere constitutes a compartment of cell lineage restriction and each territory exhibits sharp restricted domains of gene expression [1–6]. Although transient in nature, rhombomeric segmentation persists particularly in the ventricular zone until at least 9 days of chick embryonic development [7]. Presumably this ensures a continuity of the segmental cues that specify neuroepithelial cells in the hindbrain and as such underlies the development of individual rhombomeres into well-defined regions of the mature adult brain [8].

Segmentation and the formation of compartments is an integral component of embryonic development and the segmental organization of the hindbrain presages the periodic organization of neurons and cranial motor nerves (Fig. 6.1A) [5, 6, 9–11]. The first subset of neurons to form in the hindbrain and extend axons are the reticular neurons and they do so in an alternate manner within the hindbrain arising firstly in rhombomere 4 (r4) and then shortly after in rhombomeres 2 and 6 [5, 9]. Subsequently, reticular neurons develop in the odd numbered rhombomeres, such that reticular neuron formation conforms to a periodicity pattern.

The formation and disposition of motor neurons also conforms to a segment periodicity pattern (Fig. 6.1A). The cell bodies of individual cranial nerves exhibit a precise relationship to specific rhombomeres such that the motor nerves of the first three branchial arches (V, trigeminal, VII, facio-acoustic and IX, glosso-pharyngeal) are respectively derived from nuclei that are confined within r2, r4 and r6 [5, 12, 13].

Each individual nerve is then subsequently augmented by neurons developing in the caudally adjacent rhombomere. Despite the dual rhombomere origin for each of the branchiomotor nerves, the axons only project from the neural tube through exit points contained within the even numbered rhombomeres. This clearly illustrates how the metameric pattern of cranial nerves and reticular neurons are underpinned by the segmental organization of the hindbrain [11, 14].

In addition to patterning the cranial nerves, the segmental organization of the hindbrain also influences the patterns of cranial neural crest cell migration (Fig. 6.1A) [15–18]. Neural crest cells are a transient, migratory, stem cell-like population that arises at the lateral edges of the neural plate along almost the entire neuraxis. Forming at the junction between the neuroectoderm and ectoderm, neural crest cells are essential for proper craniofacial morphogenesis as they give rise to the majority of the bone, cartilage, connective and peripheral nerve tissue in the head [19–24]. In the mouse the first population of neural crest cells to emigrate from the neural tube do so from the caudal midbrain and rostral hindbrain at the 5–6–somite stage (8.25–8.5 dpc), long before closure of the neural tube. In contrast, the commencement of neural crest migration in the chick at the 6–7–somite stage coincides with closure of the neural tube [19, 25, 26]. The patterns of neural crest cell induction and migration in mouse and chick embryos are very similar and the duration of emigration from all axial levels typically lasts between 9 and 12 h.

The pattern of hindbrain neural crest migration consists of three broad yet segregated streams of cells lateral to rhombomeres r2, r4 and r6, each of which populates the adjacent first, second and third branchial arches respectively in keeping with their craniocaudal order [15, 17]. Although each rhombomere has an inherent capacity to generate neural crest cells [27], significantly fewer neural crest cells delaminate from rhombomeres 3 and 5 compared to the even-numbered rhombomeres [15, 17]. Rather than migrating laterally however, the relatively small number of neural crest cells derived from r3 and r5 migrate rostrally or caudally, joining the even-numbered rhombomere neural crest streams as they fill the branchial arches [17, 28]. Disrupting rhombomere boundary formation and segmentation affects the migration of cranial neural crest cells [29] which highlights the profound impact that the segmental organization of the hindbrain has on cranial neural crest cell development.

The segmental organization of the hindbrain therefore is a conserved strategy that is used by vertebrates to provide the essential ground plan or foundations for establishing many of the critical features of craniofacial development. The precise register that exists between rhombomeres, the patterns of innervation of the branchial arches and the organization of sensory and motor neurons in the hindbrain raises the issue of how the identity of individual rhombomeres is established. In this chapter we describe the coupling between segmentation, anterior-posterior positional specification and the segmental restriction of cell movement that is essential for the establishment and maintenance of compartments with distinct identities during hindbrain development and which in turn is critical for normal craniofacial development.

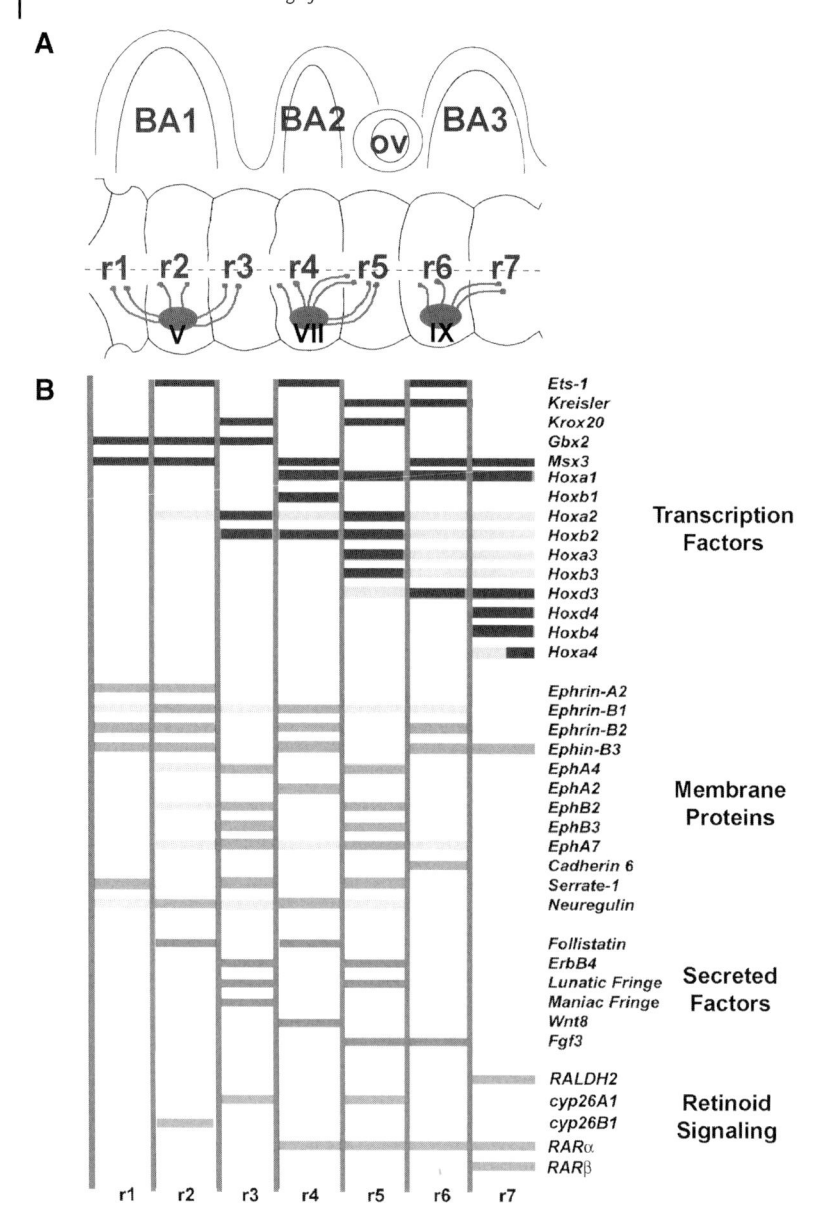

Fig. 6.1

The genetic regulation of hindbrain segment and identity. (A) The subdivision of the hindbrain into seven rhombomere segments exerts a profound influence over patterning of the branchiomotor nerves, the cranial ganglia and the migration of neural crest cells into the adjacent branchial arches. This registration is crucial for providing the blueprint for craniofacial development. (B) A large number of transcription factors, membrane proteins and secreted factors are expressed dynamically in restricted compartments or segments during hindbrain development. These genes are crucial not only for compartment formation and the restriction of cell movements but also for the identity and differentiation of individual segments. (This figure also appears with the color plates.)

6.2.2
Segment- and Boundary-restricted *Hox* Gene Expression

A remarkable feature of the vertebrate hindbrain is that the morphological appearance of segments occurs concomitantly with the acquisition of unique gene expression profiles in the rhombomeres (Fig. 6.1B). Among the most critical regulatory proteins involved in the formation and patterning of the hindbrain are members of the *Hox* transcription factor family. *Hox* genes are involved in hierarchical and reciprocal gene interactions that lead to the establishment of unique expression signatures delineating the different rhombomeres.

There are 39 *Hox* genes in mice and humans, and 48 in teleosts, organized in distinct clusters on separate chromosomes. In mammals, *Hox* genes are found in four separate clusters (*HoxA-HoxD*) which are believed to have evolved via duplication from a single ancestral vertebrate *Hox* complex. *Hox* genes are arranged in 13 highly related paralogous groups within each cluster, but no single cluster contains all 13 paralogs, presumably due to evolutionary gene loss. The zebrafish genome has undergone an additional duplication event giving rise to three more *Hox* clusters [30]. However, extensive gene loss in teleosts including the abolition of an entire D cluster, have resulted in a total of 48 *Hox* genes. Irrespective of the number of *Hox* genes in a species, their conserved clustered organization underpins a spatio-temporal colinearity which is critical for anterior-posterior patterning during embryonic development. The essence of spatiotemporal colinearity is that the more 3' an individual *Hox* gene is positioned in a cluster, the earlier it is activated and the more anteriorly it is expressed.

Hox gene expression in the mouse is initiated during gastrulation in the mesoderm tissue that has ingressed through the primitive streak and is then found shortly thereafter in the overlying nascent neuroepithelium [31]. At E7.5, during the earliest stages of hindbrain development, *Hoxa1* and *Hoxb1* expression in the neuroepithelium is initially broad and extends anteriorly to the presumptive r3/r4 boundary [32]. While *Hoxa1* expression rapidly diminishes, the expression domain of *Hoxb1* (in zebrafish *Hoxb1a* and *Hoxb1b*) [33] is refined, becoming restricted to r4 (Fig. 6.1B). *Hoxa2* is expressed in the neuroepithelium with an anterior limit that corresponds to the r1/r2 boundary [34]. Within this broad domain, *Hoxa2* exhibits intensified expression in r3 and r5 as the segments form [35]. Similarly, *Hoxb2* which displays a broad domain of expression extending anteriorly to the r2/r3 boundary, also exhibits upregulation in r3 through r5 [36]. The anterior border of *Hoxa3* and *Hoxb3* expression is set at the r4/r5 boundary and elevated levels are observed in r5 and r6 for *Hoxa3* and r5 for *Hoxb3* [32]. The group 4 *Hox* genes are the caudalmost expressing *Hox* members in the hindbrain. The anterior boundary of *Hoxa4*, *Hoxb4*, and *Hoxd4* is set in the caudal hindbrain at r6/r7, while *Hoxc4* is expressed even more posteriorly at the poorly-defined boundary between r7 and the spinal cord [37]. In teleosts, *Hoxa4a* (formerly *Hoxx4*) displays a similar pattern to *Hoxc4* in the mouse, being restricted anteriorly to r7 [38, 39]. To date no hindbrain expression has been observed for members of the *Hox* paralogous groups 5–13. Therefore during embryonic development, only members of the *Hox* paralogous groups 1–4 are expressed in the hindbrain such that each rhombomere displays a unique combination of *Hox* genes. This molecular signature or "combinatorial *Hox* code" specifies anterior-posterior segment identity and subsequently patterns the posterior CNS and craniofacial region.

6.2.3
Altering the *Hox* code: *Hox* Gene Loss- and Gain-of-Function

Given the highly conserved nature of hindbrain segmentation and the correlation with *Hox* gene expression domains, it was hypothesized that alterations to the *Hox* code would exhibit a profound effect on hindbrain development. Mutational studies in both mice and zebrafish have subsequently confirmed the crucial regulatory roles played by *Hox* genes during hindbrain development. For instance, the loss of mouse *Hoxb1* function coincides with the transformation of r4 to an r2–like character [40]. This results in the aberrant patterning of the VIIth cranial nerve, such that motor neurons derived from the r4 territory of *Hoxb1* mutants now display characteristics of trigeminal neurons (Vth cranial nerve), which normally arise from r2 [41, 42]. This clearly demonstrates that segment-specific expression of *Hox* genes in the hindbrain imparts regional specialization in the CNS.

Similarly, removal of *Hoxa1* function in the mouse leads to the loss of facial and abducens motor nerves as well as inner ear defects which are consistent with a perturbation of both the neurons and neural crest cells that originate in r4 and r5, where *Hoxa1* is expressed. Furthermore *Hoxa1* null mutant mice also exhibit other patterning defects ranging from a reduction in the r4 and r5 territories to a complete absence of r5 and an associated reduction of the otic vesicle. $Hoxa1^{-/-}$ mutant embryos also display a partial transformation of r3 to an r2–like identity, which may reflect the importance of *Hoxa1* in establishing r2–specific identity through cross-regulation of *Hoxb2* [43]. In zebrafish, the combined knockdown of the r4–restricted paralogs *Hoxb1a* and *Hoxb1b* resulted in the loss of the VIIth cranial nerve, with *Hoxb1a* playing a more critical role in the patterning of r4 derivatives [44]. Extensive synergy has also been noted for the function of the group 1 paralogs in the mouse. *Hoxa1/Hoxb1* double null mutants exhibit a complete loss of the r4 territory, leading to the spectacular disarray of craniofacial motor neuron ganglia VII to XI as well as the complete absence of neural crest-derived second branchial arch structures [45, 46].

While the loss of function of group 1 *Hox* genes leads to perturbations including anteriorization of the r4/r5 region, in contrast, the overexpression of individual *Hox* genes can lead to posteriorization of the hindbrain. *Hoxa1* overexpression results in the posterior transformation of r2 to an r4 identity [47]. Similar gain-of-function phenotypes have also been observed by the regional overexpression of *Hoxb1* and *Hoxa2* in the chick hindbrain [48]. When *Hoxb1* was mis-expressed in the hindbrain neuroepithelium anterior to the r3/r4 boundary (where it is endogenously expressed), the resulting r2 territory expressed molecular markers as well as motor neuron migration tracts characteristic of r4 [48]. Interestingly, the ectopic expression of *Hoxb1* or *Hoxa2* in r1, which is normally devoid of *Hox* gene expression, leads to the ectopic production of branchiomotor neurons of either facial or trigeminal character, respectively [49]. Thus, *Hox* genes behave in a cell-autonomous manner to specify the distinct segmental identity of neural cell types and character in the hindbrain.

The group 2 *Hox* gene, *Hoxa2* does not appear to play a crucial role in the hindbrain patterning as mouse null mutants and hypomorphs (where transcriptional activity is as low as 20 %) do not display any overt hindbrain segmentation defects [50]. *Hoxa2* mutants do however exhibit mild alterations in the projection of motor neurons from r2 and r3, demonstrating a role for *Hoxa2* in the dorso-ventral patterning of the rhombencephalic neural tube [51]. In contrast, the second branchial arch neural crest-derived structures are severely affected in the null and hypomorphic mutants. $Hoxa2^{-/-}$ mice exhibit mirror-image duplications of Meckel's cartilage and associated skeletal elements, suggesting that *Hoxa2* functions as a homeotic selector gene for the fate of second branchial arch neural crest cells. An homologous phenotype in the second arch cartilages was noted for the morpholino knockdown of the zebrafish group 2 *Hox* genes [52].

Hoxb2 is not essential for the initial patterning of the hindbrain, but is required for the subsequent maintenance of r4 identity and the proper anterior-posterior and dorso-ventral patterning of r4 neurons [53]. In the *Hoxb2* null mutants, migration of the r4 branchiomotor nerves is reduced which compromises facial motor ganglion development. This phenotype is reminiscent of the *Hoxb1* null mutation, although less severe, suggesting *cis*-regulation in the *HoxB* cluster is crucial for the proper patterning of r4.

In contrast to the individual null mutants, *Hoxa2*/*Hoxb2* double display hindbrain segmentation abnormalities in the form of aberrantly specified boundaries between r1–r4 [54]. Although the cranial sensory nerves derived from this region were largely normal the phenotype suggests a synergistic interaction between group 2 *Hox* members during hindbrain segmentation. These genes also function synergistically to dorso-ventrally pattern the anterior hindbrain. Similar functional redundancy between paralogous *Hox* genes was also observed in zebrafish, where the combined knockdown of both *Hoxa2* and *Hoxb2* was required to reveal a role for these genes in second branchial arch patterning [52].

The generation of compound mutants involving the group 1 and group 2 *Hox* genes demonstrates the importance of *Hox* gene cross-regulation in the patterning of the hindbrain. For example, *Hoxa1*/*Hoxa2* double mutants display abnormal specification of r2–5 and their derivatives which are similar but significantly more severe than the *Hoxa1* single mutants [55]. The defects are compounded by the abolition of *Hoxb1* expression and the inability to maintain a proper r3/r4 boundary and territories.

Loss of function of the group 3 *Hox* genes highlights crucial roles for these genes in patterning the caudal hindbrain and r5/r6 in particular as expected from their expression patterns. *Hoxa3* mutants display deficiencies and fusions of the IXth and Xth cranial ganglia as well as abnormal neural crest derivatives of the posterior hindbrain, including the throat cartilages, thyroid and parathyroid glands [56].

Hoxb3 mutants exhibit similar defects in the cranial ganglia, although at a lower penetrance [57, 58], while in contrast *Hoxd3* mutants do not display any overt defects in the neurogenic derivatives of the caudal hindbrain [59, 60]. The generation of double group 3 null mutants increased the severity of the defects observed in the single nulls, demonstrating functional redundancy between the different paralogs, particularly in the patterning of the IXth cranial nerve [57, 60]. The loss of *Hoxa3* and *Hoxb3* also results in the ectopic expression of *Hoxb1* in r6 which is indicative of an anterior transformation of r6 to an r4 identity. This transformation consequently leads to the formation of r4–like facial branchiomotor neurons in the mutant r6 territory, [61]. Furthermore, the overexpression of *Hoxa3* in the anterior hindbrain region of chick embryos led to the ectopic formation of somatic motor neurons in r1–r4, which normally develop only within r5–r7 [62]. These studies suggests that the *Hox* group 3 genes function as homeotic selectors in establishing and regulating posterior (r5–r7) hindbrain fate.

In summary, *Hox* genes play key roles in the patterning of the hindbrain. Not only do *Hox* genes play key roles in the segmentation process itself but more importantly they cell autonomously impart specific molecular identities and morphological fates to naive neuroepithelium.

6.2.4
Initiating *Hox* Gene Expression in the Hindbrain

The initiation of *Hox* gene expression is dependent upon signaling from the vitamin A derivative retinoic acid (RA). RA is enriched in the caudal hindbrain in various vertebrates [63–68] and *Raldh2* (Retinal dehydrogenase 2), a major component of the vitamin A biosynthetic pathway, is expressed throughout the paraxial mesoderm including the occipital and cervical somites which lie adjacent to the caudal hindbrain [69–71]. In contrast, enzymes that degrade RA are expressed in the hindbrain neuroepithelium (Fig. 6.1B). For instance, *cyp26A1* transcripts localize to r2 and *cyp26B1* is expressed in r2–r6 in the presumptive hindbrain [70, 72–74]. This suggests the presence of a sharp retinoid gradient in the posterior hindbrain originating from the paraxial mesoderm that patterns the neurectoderm through the regulation of *Hox* genes. Consistent with this idea, grafting paraxial tissue or RA-soaked beads underneath the anterior hindbrain resulted in the ectopic induction of caudally expressed *Hox* genes, such as *Hoxa3*, *Hoxb4* and *Hoxd4*. Consequently this led to the formation of motor neurons characteristic of r5–r7 in the rostral hindbrain [62, 75]. The exogenous application of RA during vertebrate gastrulation potently induces and anteriorizes *Hox* gene expression and leads to posterior homeotic transformations along the anterior-posterior axis [76–81]. Furthermore, the response of *Hox* genes to RA is colinear such that 3′ *Hox* genes are activated earlier and at lower concentrations than genes at the 5′ end of a cluster (reviewed in [82]). This again reflects the importance of the complex organization of the *Hox* gene transcription factor family and its direct relationship to anterior-posterior patterning.

Retinoic acid exerts its effects by binding to and activating a heterodimer comprised of a retinoic acid receptor (RAR) and a retinoid X receptor (RXR), which are members of the steroid hormone superfamily of ligand-inducible transcriptional activators (reviewed in [83]). At least two members of the *RAR*s are expressed abundantly in the hindbrain, with an anterior limit at the r3/r4 border for *RARα* and r6/r7 for *RARβ* (Fig. 6.1B) [84–91], while *RXRα* and *RXRβ* are expressed throughout the hindbrain neurectoderm [91]. The RAR/RXR heterodimer is a sequence-specific transcriptional activator that recognizes DNA sequences consisting of direct repeats called retinoic acid response elements (RAREs) in the regulatory regions of target genes [83, 92, 93]. The consensus sequence for RAREs is highly variable but consists of a core hexapeptide motif (A/G)G(G/T)TCA separated by one to five non-conserved nucleotides, with the five-nucleotide spacing being the most common [83, 94].

RAREs have been identified in the regulatory regions of several *Hox* genes, particularly those belonging to paralog groups 1–4, confirming that the regulation of *Hox* gene transcription by retinoids is direct [95–103]. Importantly, the structure and position of the RAREs is remarkably well conserved in all the vertebrates studied, highlighting the central role played by retinoid signaling in *Hox* gene regulation [94]. In the mouse, an RARE located in the 3′ regulatory region of the *Hoxa1* gene is required to set its anterior expression limit at the presumptive r3/r4 boundary and mediates its response to RA (Fig. 6.2) [95, 104, 105]. For *Hoxb1*, two RAREs located in the 3′ end of the gene mediate its expression during gastrulation and serve to restrict its anterior expression limit to the r3/r4 boundary [98, 101, 103]. Interestingly, another RARE located in the 5′ regulatory region of *Hoxb1* is essential for restricting *Hoxb1* transcripts to r4 at later stages by repressing its expression in r3 and r5, demonstrating that RAREs can mediate both inductive and repressive effects on target genes (Fig. 6.2) [99, 101]. Mice deficient for the 3′ RAREs from *Hoxa1* or *Hoxb2* have confirmed their critical roles in regulating endogenous *Hoxa1* and *Hoxb1* expression in the hindbrain [46, 100, 105].

RAREs have not yet been characterized for the group 2 and group 3 genes, although genome analysis has revealed putative RARE-type direct repeats in the regulatory region of the group 3 paralogous group [94, 106]. However, Mazanares et al. [106] identified an RA-responsive RARE that lies between *Hoxb3* and *Hoxb4* which may mediate the effects of retinoids on both genes. RA signaling is directly implicated in the regulation of the group 4 *Hox* genes as they all possess critical RAREs in their regulatory regions. In the case of *Hoxb4*, an RARE is required for its expression in the neuroepithelium up to the r6/r7 boundary, and this enhancer mediates the retinoid signal coming from the paraxial mesoderm underlying the caudal hindbrain [107]. Similarly, RAREs present in the *Hoxa4* and *Hoxd4* promoter regions establish the proper anterior limit of expression in the caudal hindbrain and mediate the RA-inducibility of these genes [97, 108–112]. The response of group 4 paralogs to exogenous RA is slightly different to group 1 *Hox* genes. *Hoxa1* and *Hoxb1* are rapidly induced by RA treatment between 7.5–8.0 dpc, however there is no response of the group 4 *Hox* genes at this stage. Conversely, if RA is administered between 8.5 and 9.5 dpc, *Hoxa4*, *Hoxb4* and *Hoxd4* are activated but the group 1 *Hox* genes exhibit no response. The mechanism underlying the differential response of individual *Hox* genes to RA remains to be understood, however this is at least partially regulated by the sequence specificity of the RARE. Converting the sequence of a *Hoxb4* RARE to that of a *Hoxb1* RARE resulted in a corresponding switch from a *Hoxb4–* to a *Hoxb1*–like anterior expression boundary. This suggests that these individual motifs do indeed interpret positional information and furthermore supports the concept that RAREs are critical components important for defining the boundaries of *Hox* gene expression in the hindbrain [107].

The hindbrain is highly sensitive to retinoid signal and the manipulation of the levels of retinoids leads to dramatic perturbations in rhombencephalic development. The mutation of *Raldh2*, an enzyme that catalyzes the formation of RA, leads to the truncation of the posterior hindbrain and is associated with the loss or downregulation of *Hoxa1*, *Hoxb1*, *Hoxa3*, *Hoxb3*, and *Hoxd4* expression [113, 114]. Similarly, a mutation in the zebrafish *raldh2* gene (known as the *neckless* mutant), leads to a severe downregulation of *Hoxb4* which results in branchiomotor nerve defects in the caudal hindbrain [115, 116]. Exogenously applied RA can rescue some of the defects associated with mutations in *raldh2*, demonstrating that retinoids are required for the proper development of the caudal hindbrain [114–116]. A similar disruption of normal hindbrain patterning and *Hox* group 1, 2 and 4 expression has been observed in quails and rats raised on a vitamin A-deficient diet (VAD) [117–119]. Further support for the importance of RA signaling in the caudal hindbrain comes from studies using the pan-RAR antagonist BMS493, which causes a dose-dependent reduction of posterior rhombomeres along with the reduction or loss of caudally-expressed *Hox* genes in the hindbrain [120, 121]. Interestingly, the most severe form of retinoid deprivation observed in these studies involved the truncation of the hindbrain posterior to r4, while the remaining rhombomere territories were enlarged resembling those of the anterior-most hindbrain [120].

Null mutations in the retinoid receptors result in severe hindbrain abnormalities, demonstrating that these transcription factors are key mediators of RA signaling in the hindbrain. *RARα/RARγ* double null mutant mouse embryos display hindbrain patterning defects as severe as those observed in VAD studies and *Raldh2* mutants, while *RARα/RARβ* double null mutants lack rhombomeres caudal to r6 and display an enlarged r5 territory [121, 122]. Therefore, the loss of retinoid signaling leads to a truncation of the caudal hindbrain and an expansion of anterior hindbrain fate.

As discussed earlier, a corollary to these observations is that excess retinoid signaling in the form of exogenous RA posteriorizes hindbrain fate. Similarly, the inactivation of *cyp26A1*, a RA-catabolizing enzyme expressed in r2, results in the posterior transformation of the anterior hindbrain, as assayed by the ectopic induction of *Hoxb1* rostral to its normal expression domain in r4 [123, 124]. These mutants also displayed aberrant migration of the trigeminal nerves reflecting the altered specification of the anterior hindbrain. Collectively, these studies demonstrate the key roles played by retinoid signaling in regulating anterior-posterior patterning of the hindbrain. RA diffusing from the cranial paraxial mesoderm underlying the posterior hindbrain acts to specify the posterior hindbrain by setting the anterior expression limits of the group 1–4 *Hox* genes.

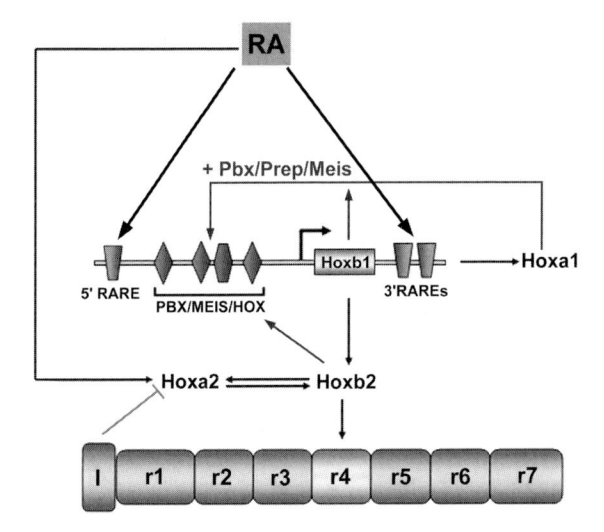

Fig. 6.2
Establishing individual rhombomere (r) 4 identity. The complex genetic regulation of r4 identity commences with the induction of *Hoxb1* expression in the neuroepithelium in response to RA (retinoic acid) mediated by 3′ RAREs (retinoic acid response elements) and through para-regulation by *Hoxa1* and PBX/PREP/MEIS complex binding. 5′ RAREs then serve to repress *Hoxb1* expression in r3 and r5 thus restricting the domain of *Hoxb1* expression to r4. In addition, FGF signaling from centers such as the isthmus may repress anterior *Hox* gene expression in the hindbrain. The maintenance of *Hoxb1* expression in r4 is achieved through auto-regulation and possibly by the cross-regulation of *Hoxb2* which is mediated through PBX/MEIS/HOX tripartite complex binding. *Hoxb1* cross-regulates *Hoxb2* which in turn is involved in a para-regulatory mechanism with *Hoxa2*. These interactions are essential for the maintenance of r4 identity and subsequently for appropriate differentiation. (This figure also appears with the color plates.)

6.2.5
Interactions between FGF and Retinoid Signaling

In addition to retinoids, FGFs also play central roles in establishing the anterior-posterior patterning of the hindbrain. Blocking FGF signaling disrupts the establishment of a hindbrain expression program in zebrafish resulting in extensive patterning defects in r3–5 [125, 126]. Conversely, exogenous FGF signaling is able to induce ectopic expression of 5′ *HoxB* genes (*Hoxb6–Hoxb9*) in the caudal hindbrain, (which is anterior to their normal expression domains), via a mechanism that requires the activation of members of the *Cdx* family of transcriptional factors [127–129]. These same 5′ *Hox* members are in turn refractory to retinoid treatment [127]. In contrast, *Hox* genes at the 3′ end of the complex which can be activated by RA, fail to respond to exogenous FGF signals. This suggests that the *HoxB* complex can be subdivided into two separate regulatory domains, a retinoid-responsive group and an FGF-dependent group, that together produce the striking colinearity of *Hox* gene expression along the anterior-posterior axis of the spinal cord and hindbrain.

An important signaling center for hindbrain patterning is the isthmus, which is the junction between the midbrain and hindbrain. The isthmus influences the fate of cells in the anterior hindbrain by restricting the anterior expression limits of *Hox* gene expression and the activity of the isthmus appears to be mediated by FGF8, a secreted growth factor that activates tyrosine kinase signaling (Fig. 6.2). Blocking FGF8 signaling *in vivo* in the chick isthmus using antibodies leads to an anterior expansion of *Hoxa2* into r1 and the loss of r1–specific identity and fate [130]. In *acerebellar* (*ace*) zebrafish embryos, which lack Fgf8 function, the isthmus is missing and cell types characteristic of r1, the most anterior hindbrain region, are also lacking [131–134]. In the mouse however, the conditional inactivation of the *Fgf8* gene affects more anterior CNS structures, such as the midbrain and cerebellum [135]. Currently it is thought that perhaps *Fgf15* or other members of the FGF family can compensate for the lack of *Fgf8* function in the mouse. The repressive influence of the isthmus was also demonstrated by the posterior transposition of the isthmus into the hindbrain which downregulated hindbrain *Hox* gene expression [136]. Beads soaked with FGF8–protein can mimic the effects of the isthmus and inhibit *Hox* gene expression when implanted into *Hox*-positive regions of the hindbrain [136]. Collectively, these results imply that in addition to activating 5' *Hox* genes, FGF signaling can also repress 3' *Hox* gene expression. FGF signaling may therefore antagonize the inductive effects of RA on *Hox* gene expression and collectively this serves to restrict the expression domains of 3' *Hox* genes to an anterior limit that corresponds with the r1/2 boundary in the hindbrain.

In addition to the isthmus playing a role in patterning the anterior hindbrain, r4 has also been hypothesized to function as a signaling center within the zebrafish hindbrain based on the localized expression of *Fgf3* and *Fgf8*. Time-lapse analysis revealed that r4 develops prior to other rhombomeres in zebrafish and that neuronal differentiation also initiates in r4 before occurring in adjacent segments [137]. Transplanting r4 cells or ectopically expressing *Fgf3* or *Fgf8* induces the expression of r5/r6 markers. Genetic knockdown of either *Fgf3* or *Fgf8* however has only mild effects on the patterning of r5 and r6 which is suggestive of functional redundancy between *Fgf3* and *Fgf8*. In contrast the combined inhibition of both genes using morpholinos, or *Fgf3* morpholinos in the *ace*/*Fgf8* mutant background, results in a dramatic loss of r5 and r6 development as well as a reduction of more anterior rhombomeres and the abolition of *Hoxa2* expression [137–139]. As a consequence *Fgf3*/*Fgf8* knockdown embryos exhibited severe segmentation defects which was evident in highly disorganized axonal projections as well as the abolition of reticulospinal neurons characteristic of r1–r3 and r5–r7 [137, 138]. Furthermore, globally blocking FGF signaling disrupts the establishment of a hindbrain expression program in zebrafish resulting in extensive patterning defects in r3–5. This defect was associated primarily with the loss of expression of *Krox20* and *Kreisler/mafB*, two crucial regulators of r3–r5 territories which are discussed below [126]. These results demonstrate that at least in zebrafish, r4 constitutes another localized signaling center, this time within the hindbrain. In this context *Fgf3* and *Fgf8* function cooperatively to activate *Hox* gene expression and coordinate the proper segmentation of adjacent rhombomeres together with the subsequent differentiation of their derivatives during anterior-posterior patterning of the hindbrain.

Recent evidence now indicates that the RA and FGF signaling pathways interact to regulate *Hox* gene expression and that this complex cross-regulatory relationship is important for the proper patterning of the anterior-posterior axis in vertebrates. RA is able to antagonize FGF signaling in the posterior of the embryo by downregulating its expression in nascent neuroepthelium and paraxial mesoderm arguing that it acts in a repressive manner [140]. In contrast, in *Raldh2* mutants, the expression of *Fgf3* in r4 is lost suggesting that RA acts in this context as an inducer [113]. In *Xenopus* embryos FGF signaling is required for the proper expression of several retinoid signaling components such as *RARα, Raldh2,* and *cyp26A*, while at the same time FGF receptors (*FGFR1* and *FGFR4)* have been hypothesized to function as targets of *RARα* [141]. Collectively this argues for functional interdependence between the FGF and RA signaling pathways during anterior-posterior patterning. Furthermore, these results highlight the presence of an elaborate mechanism driven by FGFs and RA that is critical for establishing precise *Hox* gene expression boundaries within rhombomeres and specifying regional fate within the hindbrain.

6.2.6
Krox20 and *Kreisler* Regulate Paralogous Groups 2 and 3 in the Vertebrate Hindbrain

The conserved expression of *Hox* genes suggest that hindbrain patterning is under exquisite regulatory control. We have summarized key observations implicating a role for RA and FGF signaling in helping initiate *Hox* gene expression in the hindbrain, however in the group 2 *Hox* genes, additional regulatory inputs are required for their elevated levels of expression in r3 and r5. Studies in a number of vertebrates have revealed an essential role for *Krox20*, a zinc-finger transcription factor, in regulating *Hox* genes in the odd-numbered rhombomeres. *Krox20* is dynamically expressed in r3 and r5 in all vertebrates studied (Fig. 6.1B) [142–147]. The mutation of *Krox20* in the mouse leads to a progressive loss of r3 and r5 territories, and results in the complete absence of the derivatives of these rhombomeres [29, 146, 148–151]. In addition, the segmental organization of the cranial nerves is highly perturbed as *Krox20*$^{-/-}$ embryos show fusions and misrouting of the trigeminal, facial, and vestibular ganglia [148, 152]. Chimeric analysis of *Krox20* mutants demonstrated that in the absence of *Krox20*, cells of the presumptive r3/r5 territories adopt a fate characteristic of the even-numbered rhombomeres [151]. Conversely, the overexpression of *Krox20* in the chick hindbrain is sufficient to convert the fate of the even-numbered rhombomeres to that of r3/r5 [153]. The loss of a single *Krox20* allele is sufficient to exacerbate the *Hoxb1* null phenotype, leading to a loss of r3 fate, which was not previously observed for the *Hoxb1* mutation [154]. These experiments indicate that *Krox20* functions cell-autonomously to specify r3 and r5 fate in the hindbrain.

The upregulation of *Hoxa2* and *Hoxb2* in odd-numbered rhomobomeres suggested direct regulation by *Krox20*. Indeed, *Krox20* binding sites have been identified in the 5′ regulatory region of both *Hoxa2* [35, 155] and *Hoxb2* [36, 156] and analyses in transgenic mice have demonstrated the critical role for these sites in the upregulation of *Hoxa2* and *Hoxb2* in odd-numbered rhombomeres. Consistent with this phenomenon, the overexpression of *Krox20* in the chick hindbrain was sufficient to induce *Hoxa2* expression in even-numbered rhombomeres [153], and the ectopic expression of *Krox20* in r4 of transgenic mice induced the expression of a *Hoxa2* reporter gene [35]. Conversely in *Krox20* null mutant mice, the upregulation of *Hoxa2* and *Hoxb2* transcripts in r3 and r5 was compromised [35, 36, 156, 157]. Although *Krox20* sites have been highly conserved during the course of vertebrate evolution, on their own they are incapable of directing group 2 *Hox* expression in the r3 and r5 regions, suggesting that they must cooperate with additional *cis*-regulatory elements that have yet to be characterized [155, 157, 158]. For the *Hoxa2* regulatory region, at least five additional motifs are required along with the *Krox20* binding sites for proper enhancer activity [157, 158]. Importantly, the organization and number of these sites vary greatly between the different group 2 paralogs and among different vertebrates, suggesting the dynamic evolution of *Krox20* co-factor recruitment for group 2 *Hox* expression [158].

Underpinning the upregulation of *Hoxa3* and *Hoxb3* in the caudal hindbrain is a Maf basic domain-leucine zipper transcription factor family member called *Kreisler (Kr)/mafB* [159, 160]. *Kreisler* is one of the earliest markers of the presumptive r5 territory, after which it is also expressed in r6, but *Kreisler* expression rapidly declines in these regions as the segment boundaries are sharpened (Fig. 6.1B) [159, 161]. In zebrafish, the *Kreisler/MafB* gene is known as *valentino* (*val*) and its transcripts are initially expressed in a broad domain that is ultimately subdivided into r5 and r6 as the definitive rhombomeres are formed [162]. The mutation of this gene in the zebrafish and mouse confirm its conserved role in caudal hindbrain development [160–164]. *Valentino/MafB* zebrafish mutants do not form boundaries between r4–r7 as they lack r5 and r6 territories altogether, and mosaic analysis confirms the cell-autonomous role for this gene as the mutants cells are excluded from r5 and r6 in wild-type hindbrains [162, 164]. In mammals, *Kreisler/MafB* mutants lack segmentation posterior to the r3/r4 boundary, however the phenotype is not as severe as that in zebrafish. Despite having lost r5 identity entirely, a number of r6 markers are still present [159, 160, 163, 165]. Consequently, *Kreisler* mutants display aberrant inner ear development and exhibit deficiencies in both the neuronal and the neural crest derivatives of the r5/r6 region [160, 161, 163, 165].

In both the mouse and zebrafish, *Kreisler/val/MafB* functions upstream of the group 3 *Hox* genes in r5 and r6 [166, 167]. The regulatory regions of the murine *Hoxa3* and *Hoxb3* genes contain essential *Kreisler/MafB* binding sites that direct their expression in r5 and r6 [166–168]. *Kreisler* mutants lack the expression of *Krox20*, *Hoxa2*, *Hoxb2*, *Hoxb3* and *Hoxb4* in r5, consistent with a loss of this territory [166, 167]. *Hoxa3* fails to be upregulated in r6 of *Kreisler* mice, suggesting the compromised maintenance of this segment. Similarly, zebrafish *val* mutants also fail to upregulate *Hoxa3* and *Hoxb3* in r5 and r6 [39]. The differences between the zebrafish and mouse *Kreisler* mutation are at least in part due to the fact that *Kreisler* is not a null mutation but rather affects the regulatory elements that direct its expression to r5 and r6, while *valentino* is a null allele [159, 160, 164]. In addition, the *MafB* target gene *Hoxb3* is expressed in r5 and r6 in teleosts, whereas it is expressed only in r5 in the mouse, and therefore can account for the greater severity of *MafB* loss in zebrafish versus the mouse. Thus, *Kreisler/val/MafB* acts as a true segmental gene that is crucial for the segmental expression of *Hoxa3* in r5 and *Hoxb3* in r5 and r6.

Krox20 also provides critical inputs in the regulation of the group 3 *Hox* members in r5 interacting with *Kreisler* [150, 169]. *Hoxb3* expression is downregulated in r5 of *Krox20* mutant hindbrains [150], and the *Hoxb3* regulatory region contains both *Kreisler* and *Krox20* binding sites that are essential for enhancer activity in transgenic mice [169]. Thus, *Krox20* and *Kreisler* cooperate in establishing the unique identities of segments in the vertebrate hindbrain by activating the group 3 *Hox* genes in r5.

Another regulator of r5/r6 patterning, is *variant hepatocyte nuclear factor 1 (vhnf1)*, which was recently identified in zebrafish and found to encode a divergent homeobox family member [170]. *Vhnf1* mutants, like *val*, display a loss of r5 and r6 territories and an absence of *Krox20* and *val/MafB* expression in these rhombomeres. *Vhnf1* transcripts localize to the caudal hindbrain with a sharp anterior limit at the r4/r5 boundary, coinciding with the location of the FGF hindbrain signaling center described above. Interestingly, *vhnf1* is able to repress *Hoxb1a* expression and restrict it to a narrow domain identifying the r4 boundary, and it can suppress the anterior transformation of the caudal hindbrain by overexpression of *Hox* paralog group 1 [171, 172]. Moreover, *vhnf1* requires intact FGF signaling to activate *val/MafB* expression and therefore specify r5/r6 identity [171]. This synergy between FGF signals and *vhnf1* is required to restrict r4 fate and allow the caudal hindbrain to be further regionalized into distinctive r5 and r6 fates.

Similar to the interaction between *vhnf1* and FGF signaling, retinoid signaling influences both *Krox20* and *Kreisler* expression in the hindbrain. RA excess in both amphibians and mammals can lead to a dramatic reduction in *Krox20* expression [173–175]. In *RARα/RARγ* double mutants, *Raldh2* mutants, retinoid antagonist-treated mouse embryos, and VAD quail embryos, *Krox20* is no longer expressed in distinct stripes in r3 and r5 but is highly downregulated and often appears as a diffuse signal throughout the caudal hindbrain [114, 117, 120, 121]. *Kreisler* expression is also lost in the hindbrains of antagonist-treated embryos and double *RAR* mutants [120, 121].

Collectively these results demonstrate that an extraordinarily intricate coordination of *Krox20*, *Kreisler/valentino*, *vhnf1*, retinoic acid and FGF signaling is required to precisely govern the establishment of *Hox* gene expression domains and anterior-posterior patterning during hindbrain segmentation.

6.2.7
Establishing the *Hox* Code in the Vertebrate Hindbrain: the Role of Auto- and Cross-regulation

Among the few well-characterized *Hox* gene targets are the *Hox* genes themselves (Fig. 6.2). A unique feature of the *Hox* complexes in vertebrates is the clustered organization of individual *Hox* members which may have facilitated the *cis*-and *trans*-regulatory interactions that act to reinforce the segment-specific gene expression program in the hindbrain. *Hox* gene expression occurs in two phases, an initiation phase followed by a maintenance phase. As was discussed above, retinoid and FGF signals are crucial in initiating *Hox* gene expression in the hindbrain, after which complex cross-regulation within the *Hox* complexes helps refine and maintain the rhombomeric *Hox* code, along with key inputs from other factors such as *Krox20* and *Kreisler/val/MafB* (Fig. 6.2).

In the case of *Hoxb1*, transcription is initiated with a potent input from RA signaling, after which strong r4 expression is maintained through the use of a highly conserved auto-regulatory region in the 5′ regulatory region of *Hoxb1* [176]. This region contains three related bipartite binding motifs comprised of overlapping *Hox* and *Pbx/Meis* sites, the latter belonging to the TALE class of *Extradenticle/Homothorax*-like homeobox gene family [177]. Thus, upon the initiation of expression by retinoids, *Hoxb1* regulates its own expression in the hindbrain through these auto-regulatory enhancers (Fig. 6.2). A similar cross-regulatory motif has been identified in the critical 5′ regulatory region that mediates *Hoxb2* expression in r4 [43]. *Hoxb1* can to bind to the *Pbx/Hox* motif of *Hoxa2 in vitro* and the mutation of this site leads to a loss of reporter activity in r4 in transgenic mice [43]. An interesting corollary to this observation is that *Hoxa2* lacks any *Pbx/Hox* binding motifs in its enhancer, thus demonstrating the importance to cross-regulatory interactions in establishing the rhombomeric *Hox* code. Consistent with this notion, the loss of either *Hoxb2* or *Hoxb1* leads to a partial transformation of r4 to an r2–like identity, highlighting the dynamic *cis*-regulatory interactions governing *Hox* gene expression [41, 53, 54].

Additional enhancers that mediate the auto- and cross-regulatory interactions have also been described for the group 4 *Hox* members [107, 178, 179]. A highly conserved neural enhancer capable of driving *Hox* paralog group 4 expression in the hindbrain up to the r6/r7 boundary is located in the 3′ flanking regions of *Hoxb4* and *Hoxd4* [107, 109, 179–182]. *Cis*-interactions are essential for the proper expression patterns of *Hoxb4*, as the mutation of a shared *Pbx/Hox* enhancer located in between the *Hoxb4* and *Hoxb3* genes abolished their expression in the hindbrain [179]. Thus, like *Hoxb1*, auto-regulation through *Pbx/Hox* bipartite motifs reinforces *Hoxb4* expression in the caudal hindbrain following the initiation of its expression by RA.

Auto-regulatory elements consisting of two *Pbx/Hox* binding sites are also essential for directing *Hoxa3* and *Hoxb3* expression in the hindbrain [183, 184]. The regulation of *Hox* genes through these enhancers can be complex as both *Hoxb3* and *Hoxb4* can bind to and regulate the *Hoxb3* enhancer. The auto- and cross-regulations of the *Hox* group 1–4 members illustrate the highly dynamic interactions involved in establishing robust gene regulatory networks that impart specific segment identity in the vertebrate hindbrain.

There are at least four members of the *Pbx* gene family that collectively are widely expressed, through the developing embryo, including the presumptive hindbrain [185–189]. The importance of *Pbx* as a binding partner in regulating *Hox* gene expression has most clearly been demonstrated in the hindbrains of zebrafish. The loss of function of *Pbx4* in the zebrafish *lazarus* (*lzr*) mutant causes a severe deficiency in hindbrain patterning resembling the loss of multiple *Hox* genes in the mouse [188, 190]. The facial cranial nerve in *Pbx4/lzr* mutants fails to migrate posteriorly to r4 and the axons of trigeminal ganglia project abnormally into the second instead of the first branchial arch, resembling *Hoxb1* and *Hoxa2* loss of functions, respectively. Moreover, *Pbx4/lzr* is needed to potentiate the effects of overexpression of *Hox* paralog groups 1 and 2 in the zebrafish hindbrain [188, 190]. The combined loss of both *lzr/Pbx4* and *Pbx2* leads to a dramatic anteriorization of the entire zebrafish hindbrain to an r1–like identity. These results indicate that *Hox* genes impart positional information and are required to establish progressively posterior cell fates in the hindbrain. In mice, the loss of *Pbx* genes does not lead to obvious hindbrain phenotypes, presumably due to redundancies among the four members [191–196]. The generation of double knockouts however should be very informative.

MEIS is a member of the TALE-homeobox protein family which constitutes another family of *Hox* co-factors related to the *Pbx* genes that are expressed during gastrulation and in the presumptive hindbrain neuroepithelium of all vertebrates studied [197–203]. Similar to *Pbx*, *Meis3* overexpression in teleosts increases the severity of *Hox* gain-of-function phenotypes and blocking *Meis* function recapitulates the *lzr/Pbx4* mutant phenotype [203, 204]. MEIS proteins appear to regulate *Hox* expression by forming trimeric complexes with *Pbx/Hox* [203–207]. Moreover, MEIS proteins can directly bind to the *Hoxb2* r4 enhancer and participate as obligate members with the *Pbx/Hox* complex to regulate expression [205], but they can also presumably form trimeric complexes without binding to DNA, since the *Hoxb1* auto-regulatory enhancer does not require *Meis* binding [208]. A divergent *Meis* family member *Prep1.1* also functions in hindbrain development at least in part through the stabilization of *Pbx/Hox* complexes in the nuclei of target cells [209, 210]. Removing the *Pbx*-binding region of *Prep1.1* or sequestering it to the cytoplasm is sufficient to perturb *Hox* gene expression in r3 and r4 as well as dramatically affect the development of neural derivatives of these territories. Interestingly, retinoids appear to provide a further input into *Hox* gene regulation as a number of studies have shown their ability to upregulate *Meis* and *Pbx* genes *in vitro* and *in vivo* [211–216].

6.2.8
A Mechanism for Establishing and Maintaining Hindbrain Segmentation

The sequence and timing of rhombomere boundary formation has been best described in avian embryos [217]. In contrast to the sequential formation of somites in the trunk, the formation of rhombomeres during hindbrain development does not occur in a strict anterior-posterior order. Rhombomere boundaries arise in a specific sequence starting with the boundary between rhombomeres 5 and 6 at stage 9 –. This is followed shortly thereafter by the establishment of the r3/4 boundary at stage 9, the r2/3 boundary at stage 9 +, the r4/5 boundary at stage 10, the r1/2 boundary at stage 11 + and the formation of the r6/7 boundary at stage 12 – [217]. Thus by the time the chick embryo has 10 somites, boundaries delineating rhombomeres 2, 3, 4, 5 and 6 are clearly identifiable.

The sequence of boundary formation results in r3 being the first hindbrain segment to be fully delineated morphologically in the chick and it is tempting to speculate that this is also true genetically as the early expression of *Krox20* in r3 coincides with its formation. However, the maturation of even-numbered rhombomeres precedes that of odd-numbered rhombomeres such that the pattern of reticular neuron and trigeminal motor neuron formation occurs in r2 prior to r3. Despite the identification of a number of families of genes with rhombomere-restricted domains of expression in the hindbrain, many more remain to be uncovered and it is expected that their patterns of expression will reflect the differential maturation of specific hindbrain regions.

Current hypotheses speculate that rhombomere boundaries represent specialized domains. At the outset, boundaries are regions characterized by low cell density where cells exhibit reduced rates of division and interkinetic nuclear migration [5, 218]. Lineage analyses in the avian hindbrain demonstrated that there is restricted cell movement across rhombomere boundaries indicating that rhombomeres function as units of cell lineage restriction [4]. When single cells were fluorescently labeled prior to boundary formation, clonal derivatives frequently spanned more than one segment. However when an individual cell was labeled at the time of boundary formation or shortly thereafter, the resulting cellular clones were restricted to a single segment. This restriction in the anterior-posterior movement of segment precursors persists until late stages of neurogenesis such that individual rhombomeres give rise to unique regions of the mature adult brain [7, 8]. Thus from the moment that boundaries are recognizable morphologically, rhombomeres represent compartments similar to insect segments [219].

The function of the rhombomere boundary was initially explored in detail through boundary ablation experiments performed in avians. Mechanical removal of rhombomere boundaries invariably results in their reconstruction and subsequently normal segment morphology and axonal colonization [217]. It is important to note however that this is not true in many genetic ablation studies which often lead to the absence of rhombomere boundaries and disrupted nerve patterning [29, 146, 148–151]. Hence this argues that rhombomere boundaries do not merely represent a physical barrier. If this was the case cells from adjacent rhombomeres should associate seamlessly. Instead the implication is that cells from adjacent rhombomeres must recognize surface differences among the cells which allows for the recreation of a missing boundary. The juxtaposition of adjacent rhombomeres such as r3 and r4, for example, always leads to the establishment of new boundaries [217]. This is also true when rhombomeres lying three or more segments apart are brought into contact with each other. Combining tissues of identical origins such as r3 and r3 or from similar origins such as r3 and r5, never results in the establishment of a boundary. Similarly, even-numbered rhombomere combinations (r2 and r4 or r4 and r6) also do not establish a boundary when juxtaposed. These results suggested alternating cell surface properties within the odd- versus even-numbered rhombomeres that constitute the hindbrain and the possibility of a hierarchy of selective cell adhesion.

This idea was subsequently tested *in vitro* in experiments designed to assess the extent of mixing or segregation exhibited by cells from different rhombomeres in the belief that they would display selective adhesive characteristics. Cells of r3 origin were observed to intermingle more appreciably with r3 cells than r5 cells and the miscibility of these cell populations was reflected in the absence of boundary formation between them. In contrast, r4 cells do not mix with r3 cells demonstrating a tight correlation between boundary formation and a lack of cell mixing at the interface [220]. Hence self/self (r3/3 or r4/4) cell combinations mix more freely than those from alternate segments (r3/5 or r2/4) which in turn are more miscible than combinations of cells from adjacent segments (r3/4 or r4/5). This implies the existence of a hierarchy within the hindbrain whereby individual rhombomeres are characterized by distinct repertoires of cell surface molecules [220]. This region-specific segregation however appears to be calcium dependent as the inhibition of Ca^{2+}-dependent adhesion molecules abolishes this selectivity allowing dissimilar rhombomeric cell populations to mix freely with each other [221].

The best candidates for being molecular mediators of this phenomenon are the Eph single pass tyrosine kinase transmembrane receptors and membrane bound Ephrin proteins [222, 223]. Eph receptors and Ephrin proteins have emerged in parallel with cell adhesion molecules as key regulators of the establishment and maintenance of compartmental organization (Fig. 6.3). In particular, recent work has indicated that Eph receptors can enhance cell adhesion. The expression patterns of *Eph* receptors and *ephrins* are consistent with the possibility that they restrict cell movements between hindbrain segments. *EphA4*, *EphB2*, and *EphB3* are expressed at high levels in r3 and r5, co-localizing with *Krox20*, whereas *ephrinB1*, *ephrinB2* and *ephrinB3* are expressed at high levels in r2/r4 and r6 (Fig. 6.1B). As a consequence of this complementary expression, interactions between EphA4 and EphB receptors with ephrinB proteins are likely to occur at and define rhombomere interfaces within the hindbrain.

Initial clues to the function of Eph receptors was obtained in zebrafish from the ectopic expression of the truncated EphA4 receptor which lacked the kinase domain [224]. Cells with an r3 or r5 identity were now present in r2, r4 and r6. Truncated EphA4 acted in a dominant-negative fashion blocking endogenous Eph receptors and in addition it also functioned as a ligand to ectopically activate ephrinB proteins in the odd-numbered rhombomeres. Similar results were also obtained when *ephrinB2* was ectopically overexpressed. This phenomenon is indicative of the bi-directional signaling activities of Eph receptors and ephrin proteins. To resolve whether rhombomeric cells were undergoing a change in identity or simply exhibiting more substantial intermingling, eight-cell stage zebrafish embryos were injected with *ephrinB2* RNA [222]. Cells expressing *ephrinB2* in r3/r5 were found to be locally restricted to the rhombomere boundaries whereas in r2/4/6 they were scattered throughout the segment.

Therefore the mosaic expression of Eph receptors or ephrin proteins is sufficient for cell sorting and segment formation occurs as a consequence of bidirectional signaling-mediated cellular repulsion. This implies that the interactions between endogenous Eph receptors and ephrins at rhombomere boundaries create a zone with lower cell-cell affinities compared with non-boundary regions [225]. In an *in vitro* fishball assay, where cells expressing *ephrinB2* were juxtaposed with cells expressing *EphB2* and/or *EphA4*, a restriction in the intermingling of these two cell populations was observed [223]. However, if the Eph receptor or the ephrin is omitted from one of the cell populations, then cell intermingling occurs. This implies that activation of any endogenous EphB receptor or ephrinB protein is insufficient to restrict intermingling and promote cellular segregation. It is the bidirectional signaling between two cell populations that restricts their miscibility. Furthermore, the larger intercellular spaces observed at rhombomere boundaries [5] may be due to the cell repulsion mediated by Eph-ephrin interactions which as a consequence prevents the stable cell contacts required for gap junction assembly [225]. Through Lucifer yellow labeling and cellular transfer analyses it was demonstrated that gap junctions are not formed in the presence of Eph-ephrin bidirectional signaling between two cell populations. Therefore it is currently believed that the activation of an Eph receptor or ephrin protein each triggers a repulsion or de-adhesion response. There is accumulating evidence that Eph receptors and ephrin proteins could regulate the function of cell adhesion molecules [226–228] and it can be envision therefore that the coordination of cellular communication and a restriction in cell miscibility is crucial for establishing compartments during hindbrain segmentation.

Mechanisms governing the formation and maintenance of boundary cells have until recently, remained elusive. Notch signaling however, has now been uncovered as a critical regulator of rhombomere boundary cells. In the zebrafish hindbrain, *radical fringe* (*rfng*) is expressed in boundary cells and delta genes are expressed adjacent to boundaries consistent with sustained activation of Notch in boundary cells [229]. Mosaic expression experiments reveal that the activation of the Notch/Su(H) pathway regulates cell affinity properties that segregate cells to boundaries in the zebrafish hindbrain. Activation of the Notch pathway correlates with the transient suppression of neuronal differentiation at rhombomere boundaries and is required to maintain boundary cells [229]. These findings reveal that at hindbrain boundaries, there is a coupling of two roles of Notch activation found in other tissue contexts, the regulation of cell affinity and inhibition of neurogenesis, and such a coupling may be important for these cells to act as a stable signaling center.

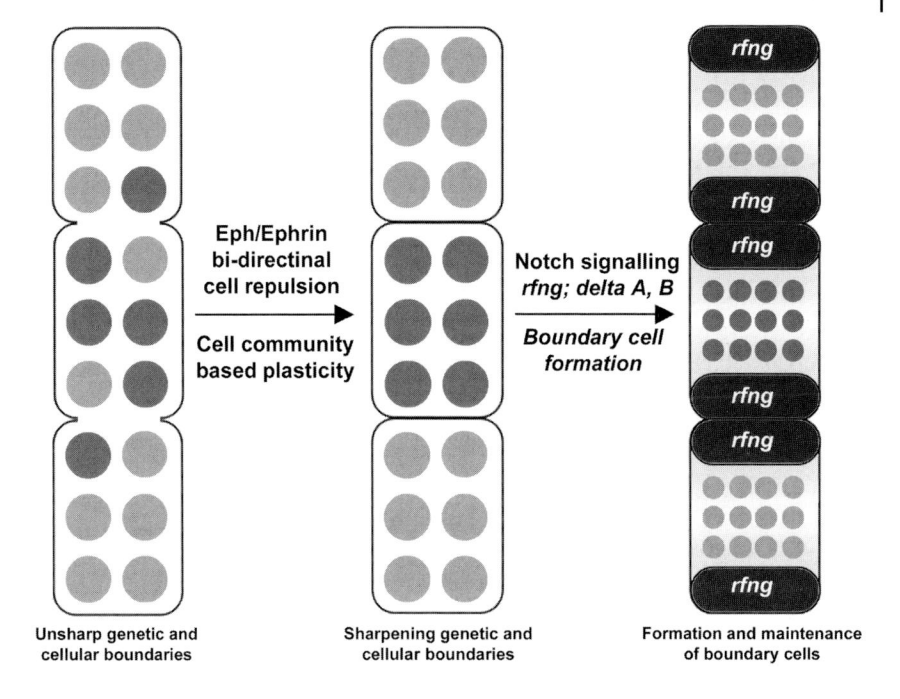

Fig. 6.3
The morphological generation of hindbrain segmentation. Distinct rhombomere cell populations begin to sort out through bidirectional cell repulsion and differential cell adhesion which is mediated by the *Eph/ephrin* family. Superimposed on this segregation is cellular-based plasticity which helps to ensure that cells which potentially might become positioned inappropriately adopt the genetic profile of their neighbors. As a consequence, the initially diffuse gene expression boundaries sharpen in response to this cell sorting and plasticity. Concomitant with this process, differential Notch signaling mediated by *rfng* and *delta* leads to the formation of boundary cells which exhibit low rates of proliferation and establishes the rhombomeres as units of cell lineage restriction. (This figure also appears with the color plates.)

6.2.9
Coupling Cell Lineage Restriction with Discrete Domains of *Hox* Gene Expression

During the early phase of hindbrain development, gene expression boundaries are initially generally diffuse. The gene expression boundaries however become refined and sharpened concomitantly with the generation of rhombomere boundaries. The restriction of intermingling between rhombomeres in the developing hindbrain is therefore crucial for the establishment and maintenance of segment identity, both morphologically and genetically. This is underscored by numerous gain- and loss-of-function analyses in several vertebrates which have demonstrated the functional importance of *Hox* genes during hindbrain development. For example, in transplantations of neural plate stage tissue within the hindbrain, the ectopic tissue acquired the *Hox* gene identity and neuroanatomy appropriate to its new location [230]. In contrast, when rhombomeres are transplanted to ectopic locations within the hindbrain they generally display molecular and cellular autonomy [231]. Therefore during normal rhombomere development, an early period of plasticity is followed by a definitive commitment to specific segmental fates and this is accompanied by the establishment of a unique genetic address of *Hox* gene expression that is maintained through subsequent cell divisions. Consequently it was hypothesized that a direct correlation existed between the commitment to rhombomere specific fates and the sharp restricted domains of *Hox* gene expression in the hindbrain and that this process was coordinated with the restrictions in cell mixing that promoted the generation of rhombomere compartments.

Models which argue for autonomous *Hox* gene expression in rhombomeres are largely based on experiments performed in avian embryos which generally involved the manipulation of large blocks of tissue such as pairs of rhombomeres or entire hindbrain neural tubes. More recent analyses in mouse [232] and zebrafish [233] embryos have now revealed an added level of complexity in the mechanisms linking cell movement restrictions with the sharpening of gene expression domains and the generation of segment identity. These studies centered upon the development of techniques for transplanting small groups of genetically-marked cells within the mouse and zebrafish hindbrains which surprisingly uncovered a considerable degree of neuroepithelial plasticity with respect to the patterns of *Hox* gene expression in individual rhombomere cells. When small groups of cells were isolated from rhombomeres 3, 4 or 5 and heterotopically transplanted into r2, the majority of the transplanted cells remained as a cohort in their new location and autonomously maintained their original anterior-posterior *Hox* gene identity [232]. However, within these cohorts, a small number of cells often became separated from the primary graft and mixed with their new neighbors. These transplanted cells exhibited clear cellular plasticity by failing to maintain their appropriate *Hox* gene (*Hoxb1*, *Hoxb2*, *Hoxa2*) expression patterns. In zebrafish embryos, not only did transplanted cells switch off inappropriate gene expression but they were also observed to activate *Hox* genes (*Hoxa2*, *Hoxa3*) that were characteristic of their new location [233]. Furthermore it was also revealed that in grafts of small numbers of cells, outlying cells within the graft but more than two cell diameters from the center, were able to switch their fate. Therefore individual rhombomere cells maintain an inherent plasticity with respect to *Hox* gene expression and cell fate [234]. The observation that cell fate and identity is autonomously maintained in rhombomere cells that remain as a cohort but not in cells that intermingle with the surrounding populations indicates that cell-community effects are important for reinforcing regional identity.

The mechanism for generating hindbrain segmentation and establishing *Hox* gene expression domains is therefore an intricately coordinated multi-step process (Fig. 6.3). Initially repulsive cell signaling between the Eph receptors and ephrin proteins located on adjacent cells, serves to segregate cells with differential cell adhesion affinities. This appears to be a calcium-dependent cell adhesion response [221] and is the first step in establishing distinct rhombomere territories as units of cell lineage restriction. Subsequently, interactions between adjacent segments induce the formation of boundary cells in which neurogenesis is inhibited. The regulation of cell affinity differences between boundary and non-boundary cells is governed by differential Notch activation [229]. These two processes are tightly linked since the inhibition of Eph-ephrin signaling can disrupt boundary cell identity and fate [224]. The integration of these signaling pathways probably stabilizes interactions at the interfaces of rhombomere boundaries helping to ensure their function as units of cell lineage and fate restriction. Superimposed onto this already complex mechanism is the plasticity exhibited by individual hindbrain cells such that as rhombomere boundaries are forming, dispersed cells can change their fates [232]. Taken together these different processes including cell repulsion, boundary cell induction and community-induced plasticity collectively provide a mechanism for coordinating the progressive generation of precise rhombomeric domains of gene expression in the hindbrain.

6.3 Conclusions

The morphological and molecular segmentation of the hindbrain functions to maintain an appropriate anterior-posterior register between the neural tube, neural crest and branchial arches, which underscores the functional importance of the blueprint provided by the hindbrain during craniofacial morphogenesis. The establishment of unique rhombomeric identities requires the expression of members of the *Hox* paralog groups 1–4 along the anterior-posterior axis of the vertebrate hindbrain. The *Hox* genes are transcription factors whose functions are essential for not only setting the distinct rhombomeric territories of the hindbrain, but for also influencing the fate of the descendants of the hindbrain segments. Initiating this "*Hox* code" requires both localized RA and FGF signaling, along with *Krox20* and *Kreisler*, and the subsequent establishment and maintenance of *Hox* gene expression depends on elaborate cross- and auto-regulatory loops within the *Hox* complexes. In this way, an initially uniform neuroepithelium acquires the regionalized anterior-posterior character that contributes so importantly to the structure and innervation of the vertebrate head.

Acknowledgments

Angelo Iulianella is the recipient of a Canadian Institutes of Health Research postdoctoral fellowship. Paul Trainor is supported by a March of Dimes Basil O'Connor Starter Scholar award and grant R01 DE016082–01 from the National Institutes of Dental and Craniofacial Research. The authors also acknowledge the support of the Stowers Institute for Medical Research.

References

1 S. Vaage, The segmentation of the primitive neural tube in chick embryos (*Gallus domesticus*). *Adv Anat Embryol Cell Biol* **1969**; *41*: 1–88.
2 E. Birgbauer, J. Sechrist, M. Bronner-Fraser, S. Fraser, Rhombomeric origin and rostrocaudal reassortment of neural crest cells revealed by intravital microscopy. *Development* **1995**; *121*: 935–945.
3 E. Birgbauer, S. E. Fraser, Violation of cell lineage restriction compartments in the chick hindbrain. *Development* **1994**; *120*: 1347–1356.
4 S. Fraser, R. Keynes, A. Lumsden, Segmentation in the chick embryo hindbrain is defined by cell lineage restrictions. *Nature* **1990**; *344*: 431–435.
5 A. Lumsden, R. Keynes, Segmental patterns of neuronal development in the chick hindbrain. *Nature* **1989**; *337*: 424–428.
6 A. Lumsden, R. Krumlauf, Patterning the vertebrate neuraxis. *Science* **1996**; *274*: 1109–1115.
7 R. Wingate, A. Lumsden, Persistence of rhombomeric organisation in the postsegmental avian hindbrain. *Development* **1996**; *122*: 2143–2152.
8 F. Marín, L. Puelles, Morphological fate of rhombomeres in quail/chick chimeras:a segmental analysis of hindbrain nuclei. *European J. Neuroscience* **1995**; *7*: 1714–1738.
9 J. D. Clarke, A. Lumsden, Segmental repetition of neuronal phenotype sets in the chick embryo hindbrain. *Development* **1993**; *118*: 151–162.
10 J. D. W. Clarke, L. Erskine, A. Lumsden, Differential progenitor dispersal and the spatial origin of early neurons can explain the predominance of single-phenotype clones in the chick hindbrain. *Developmental Dynamics* **1998**; *212*: 14–26.
11 R. Keynes, R. Krumlauf, Hox genes and regionalization of the nervous system. *Annual Review of Neuroscience* **1994**; *17*: 109–132.
12 H. Marshall, S. Nonchev, M. H. Sham, I. Muchamore, A. Lumsden, R. Krumlauf, Retinoic acid alters hindbrain Hox code and induces transformation of rhombomeres 2/3 into a 4/5 identity. *Nature* **1992**; *360*: 737–741.
13 E. M. Carpenter, J. M. Goddard, O. Chisaka, N. R. Manley, M. R. Capecchi, Loss of *Hoxa–1 (Hox–1.6)* function results in the reorganization of the murine hindbrain. *Development* **1993**; *118*: 1063–1075.
14 R. Keynes, A. Lumsden, Segmentation and the origins of regional diversity in the vertebrate central nervous system. *Neuron* **1990**; *4*: 1–9.
15 A. Lumsden, N. Sprawson, A. Graham, Segmental origin and migration of neural crest cells in the hindbrain region of the chick embryo. *Development* **1991**; *113*: 1281–1291.
16 G. Serbedzija, S. Fraser, M. Bronner-Fraser, Vital dye analysis of cranial neural crest cell migration in the mouse embryo. *Development* **1992**; *116*: 297–307.

17 J. Sechrist, G. N. Serbedzija, T. Scherson, S. E. Fraser, M. Bronner-Fraser, Segmental migration of the hindbrain neural crest does not arise from its segmental generation. *Development* **1993**; *118*(3): 691–703.
18 P. A. Trainor, P. P. L. Tam, Cranial paraxial mesoderm and neural crest of the mouse embryo-codistribution in the craniofacial mesenchyme but distinct segregation in the branchial arches. *Development* **1995**; *121*: 2569–2582.
19 N. Le Douarin, *The Neural Crest*. Cambridge: Cambridge University Press, **1983**.
20 D. Noden, The role of the neural crest in patterning of avian cranial skeletal, connective, and muscle tissues. *Dev Biol* **1983**; *96*: 144–165.
21 G. F. Couly, A. Grapin-Bottom, P. Coltey, N. M. Le Douarin, The regeneration of the cephalic neural crest, a problem revisited: the regenerating cells originate from the contralateral or from the anterior and posterior neural folds. *Development* **1996**; *122*: 3393–3407.
22 G. Couly, A. Grapin-Botton, P. Coltey, B. Ruhin, N. M. Le Douarin, Determination of the identity of the derivatives of the cephalic neural crest: incompatibilty between *Hox* gene expression and lower jaw development. *Development* **1998**; *125*: 3445–3459.
23 P. Hunt, P. Ferretti, R. Krumlauf, P. Thorogood, Restoration of normal Hox code and branchial arch morphogenesis after extensive deletion of hindbrain neural crest. *Dev Biol* **1995**; *168*: 584–597.
24 J. R. Saldivar, J. W. Sechrist, C. E. Krull, S. Ruffin, M. Bronner-Fraser, Dorsal hindbrain ablation results in the rerouting of neural crest migration and the changes in gene expression, but normal hyoid development. *Development* **1997**; *124*: 2729–2739.
25 K. Tosney, The segregation and early migration of cranial neural crest cells in the avian embryo. *Dev Biol* **1982**; *89*: 13–24.
26 S. Horstadius, *The Neural Crest*. London: Oxford University Press, **1950**.
27 P. A. Trainor, D. Sobieszczuk, D. Wilkinson, R. Krumlauf, Signalling between the hindbrain and paraxial tissues dictates neural crest migration pathways. *Development* **2002**; *129*: 433–442.
28 P. M. Kulesa, S. E. Fraser, Neural crest cell dynamics revealed by time-lapse video microscopy of whole embryo chick explant cultures. *Dev Biol* **1998**; *204*: 327–344.
29 P. J. Swiatek, T. Gridley, Perinatal lethality and defects in hindbrain development in mice homozygous for a targeted mutation of the zinc finger gene *Krox–20*. *Genes Dev* **1993**; *7*: 2071–2084.
30 A. Amores, A. Force, Y.-L. Yan, L. Joly, C. Amemiya, A. Fritz, R. Ho, J. Langeland, V. Prince, Y.-L. Wang, M. Westerfield, M. Ekker, J. Postlehwait, Zebrafish *hox* clusters and vertebrate genome evolution. *Science* **1998**; *282*: 1711–1714.
31 J. Deschamps, E. van den Akker, S. Forlani, W. De Graaff, T. Oosterveen, B. Roelen, J. Roelfsema, Initiation, establishment and maintenance of Hox gene expression patterns in the mouse. *Int J Dev Biol* **1999**; *43*: 635–650.

32 D. G. Wilkinson, S. Bhatt, M. Cook, E. Boncinelli, R. Krumlauf, Segmental expression of Hox–2 homeobox-containing genes in the developing mouse hindbrain. *Nature* **1989**; *341*: 405–409.

33 D. Alexandre, J. Clarke, E. Oxtoby, Y.-L. Yan, T. Jowett, N. Holder, Ectopic expression of *Hoxa–1* in the zebrafish alters the fate of the mandibular arch neural crest and phenocopies a retinoic acid -induced phenotype. *Development* **1996**; *122*: 735–746.

34 P. Hunt, M. Gulisano, M. Cook, M. Sham, A. Faiella, D. Wilkinson, E. Boncinelli, R. Krumlauf, A distinct *Hox* code for the branchial region of the head. *Nature* **1991**; *353*: 861–864.

35 S. Nonchev, C. Vesque, M. Maconochie, T. Seitanidou, L. Ariza-McNaughton, M. Frain, H. Marshall, M. H. Sham, R. Krumlauf, P. Charnay, Segmental expression of Hoxa–2 in the hindbrain is directly regulated by Krox–20. *Development* **1996**; *122*: 543–554.

36 M. H. Sham, C. Vesque, S. Nonchev, H. Marshall, M. Frain, R. Das Gupta, J. Whiting, D. Wilkinson, P. Charnay, R. Krumlauf, The zinc finger gene *Krox–20* regulates *Hoxb–2 (Hox2.8)* during hindbrain segmentation. *Cell* **1993**; *72*: 183–196.

37 S. J. Gaunt, R. Krumlauf, D. Duboule, Mouse homeo-genes within a subfamily, Hox–1.4, –2.6 and –5.1, display similar anteroposterior domains of expression in the embryo, but show stage- and tissue-dependent differences in their regulation. *Development* **1989**; *107*: 131–141.

38 V. E. Prince, L. Joly, M. Ekker, R. K. Ho, Zebrafish hox genes: genomic organization and modified colinear expression patterns in the trunk. *Development* **1998**; *125*: 407–420.

39 V. E. Prince, C. B. Moens, C. B. Kimmel, R. K. Ho, Zebrafish *hox* genes: expression in the hindbrain region of wild-type and mutants of the segmentation gene, *valentino*. *Development* **1998**; *125*: 393–406.

40 J. Goddard, M. Rossel, N. Manley, M. Capecchi, Mice with targeted disruption of *Hoxb1* fail to form the motor nucleus of the VIIth nerve. *Development* **1996**; *122*: 3217–3228.

41 A. Gavalas, C. Ruhrberg, J. Livet, C. E. Henderson, R. Krumlauf, Neuronal defects in the hindbrain of Hoxa1, Hoxb1 and Hoxb2 mutants reflect regulatory interactions among these Hox genes. *Development* **2003**; *130*: 5663–5679.

42 B. R. Arenkiel, P. Tvrdik, G. O. Gaufo, M. R. Capecchi, Hoxb1 functions in both motor neurons and in tissues of the periphery to establish and maintain the proper neuronal circuitry. *Genes & Development* **2004**; *18*: 1539–1552.

43 M. Maconochie, S. Nonchev, M. Studer, S.-K. Chan, H. Pöpperl, M.-H. Sham, R. Mann, R. Krumlauf, Cross-regulation in the mouse *HoxB* complex: the expression of *Hoxb2* in rhombomere 4 is regulated by *Hoxb1*. *Genes Dev* **1997**; *11*: 1885–1896.

44 J. M. McClintock, M. A. Kheirbek, V. E. Prince, Knockdown of duplicated zebrafish hoxb1 genes reveals distinct roles in hindbrain patterning and a novel mechanism of duplicate gene retention. *Development* **2002**; *129*: 2339–2354.

45 A. Gavalas, P. Trainor, L. Ariza-McNaughton, R. Krumlauf, Synergy between Hoxa1 and Hoxb1: the relationship between arch patterning and the generation of cranial neural crest. *Development* **2001**; *128*: 3017–3027.
46 A. Gavalas, M. Studer, A. Lumsden, F. Rijli, R. Krumlauf, P. Chambon, *Hoxa1* and *Hoxb1* synergize in patterning the hindbrain, cranial nerves and second pharyngeal arch. *Development* **1998**; *125*: 1123–1136.
47 M. Zhang, H. J. Kim, H. Marshall, M. Gendron-Maguire, D. A. Lucas, A. Baron, L. J. Gudas, T. Gridley, R. Krumlauf, J. F. Grippo, Ectopic Hoxa–1 induces rhombomere transformation in mouse hindbrain. *Development* **1994**; *120*: 2431–2442.
48 E. Bell, R. Wingate, A. Lumsden, Homeotic transformation of rhombomere identity after localized *Hoxb1* misexpression. *Science* **1999**; *284*: 21682171.
49 S. Jungbluth, E. Bell, A. Lumsden, Specification of distinct motor neuron identities by the singular activities of individual Hox genes. *Development* **1999**; *126*: 2751–2758.
50 M. Gendron-Maguire, M. Mallo, M. Zhang, T. Gridley, Hoxa–2 mutant mice exhibit homeotic transformation of skeletal elements derived from cranial neural crest. *Cell* **1993**; *75*: 1317–1331.
51 A. Gavalas, M. Davenne, A. Lumsden, P. Chambon, F. Rijli, Role of *Hoxa–2* in axon pathfinding and rostral hindbrain patterning. *Development* **1997**; *124*: 3693–3702.
52 M. P. Hunter, V. E. Prince, Zebrafish hox paralogue group 2 genes function redundantly as selector genes to pattern the second pharyngeal arch. *Dev Biol* **2002**; *247*: 367–389.
53 J. Barrow, M. Capecchi, Targeted disruption of the *Hoxb2* locus in mice interferes with expression of *Hoxb1* and *Hoxb4*. *Development* **1996**; *122*: 3817–3828.
54 M. Davenne, M. Maconochie, R. Neun, J.-F. Brunet, P. Chambon, R. Krumlauf, F. Rijli, *Hoxa2* and *Hoxb2* control dorsoventral patterns of neuronal development in the rostral hindbrain. *Neuron* **1999**; *22*: 677–691.
55 J. R. Barrow, H. S. Stadler, M. R. Capecchi, Roles of Hoxa1 and Hoxa2 in patterning the early hindbrain of the mouse. *Development* **2000**; *127*: 933–944.
56 O. Chisaka, M. Capecchi, Regionally restricted developmental defects resulting from targeted disruption of the mouse homeobox gene *hox1.5*. *Nature* **1991**; *350*: 473–479.
57 N. Manley, M. Capecchi, *Hox* group 3 paralogous genes act synergistically in the formation of somitic and neural crest-derived structures. *Dev Biol* **1997**; *192*: 274–288.
58 N. Manley, M. Capecchi, Hox group 3 paralogs regulate the development and migration of the thymus, thyroid and parathyroid glands. *Dev Biol* **1998**; *195*: 1–15.
59 B. Condie, M. Capecchi, Mice homozygous for a targeted disruption of *Hoxd–3(Hox–4.1)* exhibit anterior transformations of the first and second cervical vertebrae, the atlas and axis. *Development* **1993**; *119*: 579–595.
60 B. G. Condie, M. R. Capecchi, Mice with targeted disruptions in the paralogous genes *Hoxa–3* and *Hoxd–3* reveal synergistic interactions. *Nature* **1994**; *370*: 304–307.

61 G. O. Gaufo, K. R. Thomas, M. R. Capecchi, Hox3 genes coordinate mechanisms of genetic suppression and activation in the generation of branchial and somatic motoneurons. *Development* **2003**; *130*: 5191–5201.

62 S. Guidato, F. Prin, S. Guthrie, Somatic motorneurone specification in the hindbrain: the influence of somite-derived signals, retinoic acid and Hoxa3. *Development* **2003**; *130* 2981–2896.

63 J. Rossant, R. Zirngibl, D. Cado, M. Shago, V. Goguere, Expression of retinoic acid response element-hsp *lacZ* transgene defines specific domains of transcriptional activity during mouse embryogenesis. *Genes Dev* **1991**; *5*: 1333–1344.

64 B. L. M. Hogan, C. Thaller, G. Eichle, Evidence that Hensen's node is a site of retinoic acid synthesis. *Nature* **1992**; *359*: 237–241.

65 M. Colbert, E. Linney, A. LaMantia, Local sources of retinoic acid coincide with retinoid-mediated transgene activity during embryonic development. *Proc Natl Acad Sci USA* **1993**; *90*: 6572–6576.

66 P. McCaffery, U. C. Drager, Hot spots of retinoic acid synthesis in the developing spinal cord. *Proc Natl Acad Sci USA* **1994**; *91*: 7194–7197.

67 C. Horton, M. Maden, Endogenous distribution of retinoids during normal development and teratogenesis in the mouse embryo. *Dev Dyn* **1995**; *202*: 312–323.

68 M. Maden, E. Sonneveld, P. T. van der Saag, E. Gale, The distribution of endogenous retinoic acid in the chick embryo: implications for developmental mechanisms. *Development* **1998**; *125*: 4133–4144.

69 K. Niederreither, P. McCaffery, U. C. Drager, P. Chambon, P. Dollé, Restricted expression and retinoic acid-induced down-regulation of the retinaldehyde dehydrogenase type 2 (RALDH–2) gene during mouse development. *Mech Dev* **1997**; *62*: 67–78.

70 A. Blentic, E. Gale, M. Maden, Retinoic acid signalling centres in the avian embryo identified by sites of expression of synthesising and catabolising enzymes. *Dev Dyn* **2003**; *227*: 114–127.

71 T. Oosterveen, F. Meijlink, J. Deschamps, Expression of retinaldehyde dehydrogenase II and sequential activation of 5' Hoxb genes in the mouse caudal hindbrain. *Gene Expr Patt* **2004**; *4*: 243–247.

72 H. Fujii, T. Sato, S. Kaneko, O. Gotoh, Y. Fujii-Kuriyama, K. Osawa, S. Kato, H. Hamada, Metabolic inactivation of retinoic acid by a novel P450 differentially expressed in developing mouse embryos. *EMBO J* **1997**; *16*: 4163–4173.

73 G. MacLean, S. Abu-Abed, P. Dollé, A. Tahayato, P. Chambon, M. Petkovich, Cloning of a novel retinoic-acid metabolizing cytochrome P450, Cyp26B1, and comparative expression analysis with Cyp26A1 during early murine development. *Mech Dev* **2001**; *107*: 195–201.

74 S. Reijntjes, E. Gale, M. Maden, Expression of the retinoic acid catabolising enzyme CYP26B1 in the chick embryo and its regulation by retinoic acid. *Gene Expr Patt* **2003**; *3*: 621–627.

75 N. Itasaki, J. Sharpe, A. Morrison, R. Krumlauf, Reprogramming *Hox* expression in the vertebrate hindbrain: Influence of paraxial mesoderm and rhombomere transposition. *Neuron* **1996**; *16*: 487–500.

76 A. Durston, J. Timmermans, W. Hage, H. Hendriks, N. de Vries, M. Heideveld, P. Nieuwkoop, Retinoic acid causes an anteroposterior transformation in the developing central nervous system. *Nature* **1989**; *340*: 140–144.

77 A. Simeone, D. Acampora, L. Arcioni, P. W. Andrews, E. Boncinelli, F. Mavilio, Sequential activation of HOX2 homeobox genes by retinoic acid in human embryonal carcinoma cells. *Nature* **1990**; *346*: 763–766.

78 M. Kessel, P. Gruss, Homeotic transformations of murine prevertebrae and concomitant alteration of Hox codes induced by retinoic acid. *Cell* **1991**; *67*: 89–104.

79 R. A. Conlon, J. Rossant, Exogenous retinoic acid rapidly induces anterior ectopic expression of murine *Hox–2* genes *in vivo*. *Development* **1992**; *116*: 357–368.

80 P. Kolm, H. Sive, Regulation of the *Xenopus* labial homeodomain genes, *HoxA1* and *HoxD1*: activation by retinoids and peptide growth factors. *Dev Biol* **1995**; *167*: 34–49.

81 P. Kolm, V. Apekin, H. Sive, *Xenopus* hindbrain patterning requires retinoid signaling. *Dev Biol* **1997**; *192*: 1–16.

82 H. Marshall, A. Morrison, M. Studer, H. Pöpperl, R. Krumlauf, Retinoids and Hox genes. *FASEB J* **1996**; *10*: 969–978.

83 P. Chambon, A decade of molecular biology of retinoic acid receptors. *FASEB J* **1996**; *10*: 940–954.

84 E. Ruberte, P. Dolle, A. Krust, A. Zelent, G. Morriss-Kay, P. Chambon, Specific spatial and temporal distribution of retinoic acid receptor gamma transcripts during mouse embryogenesis. *Development* **1990**; *108*: 213–222.

85 E. Ruberte, P. Dolle, P. Chambon, G. Morriss-Kay, Retinoic acid receptors and cellular retinoid binding proteins II. Their differential pattern of transcription during early morphogenesis in mouse embryos. *Development* **1991**; *111*: 45–60.

86 E. Ruberte, P. Kastner, P. Dolle, A. Krust, P. Leroy, C. Mendelsohn, A. Zelent, P. Chambon, Retinoic acid receptors in the embryo. *Sem Dev Biol* **1991**; *2*: 153–159.

87 E. Ruberte, V. Friederich, G. Morriss-Kay, P. Chambon, Differential distribution patterns of CRABP-I and CRABP-II transcripts during mouse embryogenesis. *Development* **1992**; *115*: 973–989.

88 E. Ruberte, V. Friederich, P. Chambon, G. Morriss-Kay, Retinoic acid receptors and cellular retinoid binding proteins. III. Their differential transcript distribution during mouse nervous system development. *Development* **1993**, *118*. 267–282.

89 C. Mendelsohn, E. Ruberte, M. Le Meur, G. Morriss-Kay, P. Chambon, Developmental analysis of the retinoic acid-inducible RAR-beta 2 promoter in transgenic animals. *Development* **1991**; *113*: 723–734.

90 C. Mendelsohn, S. Larkin, M. Mark, M. LeMeur, J. Clifford, A. Zelent, P. Chambon, RAR beta isoforms: distinct transcriptional control by retinoic acid and specific spatial patterns of

promoter activity during mouse embryonic development. *Mech Dev* **1994**; *45*: 227–241.

91 P. Dollé, V. Fraulob, P. Kastner, P. Chambon, Developmental expression of murine *retinoid X receptor (RXR)* genes. *Mech Dev* **1994**; *45*: 91–104.

92 D. J. Mangelsdorf, R. M. Evans, The RXR heterodimers and orphan receptors. *Cell* **1995**; *83*: 841–850.

93 D. J. Mangelsdorf, C. Thummel, M. Beato, P. Herrlich, G. Schütz, K. Umesono, B. Blumberg, P. Kastner, M. Mark, P. Chambon, R. M. Evans, The Nuclear Receptor Superfamily: The Second Decade. *Cell* **1995**; *83*: 835–839.

94 G. Mainguy, P. M. In der Rieden, E. Berezikov, J. M. Woltering, R. H. Plasterk, A. J. Durston, A position-dependent organisation of retinoid response elements is conserved in the vertebrate Hox clusters. *Trends Genet* **2003**; *19*: 476–479.

95 A. W. Langston, L. J. Gudas, Identification of a retinoic acid responsive enhancer 3' of the murine homeobox gene *Hox–1.6*. *Mech Dev* **1992**; *38*: 217–228.

96 M. Moroni, M. Vigano, F. Mavilio, Regulation of the human *HOXD4* gene by retinoids. *Mech Dev* **1993**; *44*: 139–154.

97 H. Pöpperl, M. Featherstone, Identification of a retinoic acid response element upstream of the murine *Hox–4.2* gene. *Mol Cell Biol* **1993**; *13*: 257–265.

98 H. Marshall, M. Studer, H. Pöpperl, S. Aparicio, A. Kuroiwa, S. Brenner, R. Krumlauf, A conserved retinoic acid response element required for early expression of the homeobox gene *Hoxb–1*. *Nature* **1994**; *370*: 567–571.

99 M. Studer, H. Pöpperl, H. Marshall, A. Kuroiwa, R. Krumlauf, Role of a conserved retinoic acid response element in rhombomere restriction of *Hoxb–1*. *Science* **1994**; *265*: 1728–1732.

100 M. Studer, A. Gavalas, H. Marshall, L. Ariza-McNaughton, F. Rijli, P. Chambon, R. Krumlauf, Genetic interaction between *Hoxa1* and *Hoxb1* reveal new roles in regulation of early hindbrain patterning. *Development* **1998**; *125*: 1025–1036.

101 T. Ogura, R. Evans, A retinoic acid-triggered cascade of *Hoxb–1* gene activation. *Proc Natl Acad Sci USA* **1995**; *92*: 387–391.

102 T. Ogura, R. Evans, Evidence for two distinct retinoic acid response pathways for *Hoxb–1* gene regulation. *Proc Natl Acad Sci USA* **1995**; *92*: 392–396.

103 A. Langston, J. Thompson, L. Gudas, Retinoic acid-responsive enhancers located 3' of the HoxA and the HoxB gene clusters. *J Biol Chem* **1997**; *272*: 2167–2175.

104 M. Frasch, X. Chen, T. Lufkin, Evolutionary-conserved enhancers direct region-specific expression of the murine *Hoxa–1* and *Hoxa–2* loci in both mice and *Drosophila*. *Development* **1995**; *121*: 957–974.

105 V. Dupé, M. Davenne, J. Brocard, P. Dollé, M. Mark, A. Dierich, P. Chambon, F. Rijli, *In vivo* functional analysis of the *Hoxa1* 3' retinoid response element (3' RARE). *Development* **1997**; *124*: 399–410.

106 M. Manzanares, H. Wada, N. Itasaki, P. A. Trainor, R. Krumlauf, P. W. Holland, Conservation and elaboration of Hox gene regulation during evolution of the vertebrate head. *Nature* **2000**; *408*: 854–857.

107 A. Gould, N. Itasaki, R. Krumlauf, Initiation of rhombomeric *Hoxb4* expression requires induction by somites and a retinoid pathway. *Neuron* **1998**; *21*: 39–51.

108 A. Morrison, M. Moroni, L. Ariza-McNaughton, R. Krumlauf, F. Mavilio, *In vitro* and transgenic analysis of a human *HOXD4* retinoid-responsive enhancer. *Development* **1996**; *122* 1895–1907.

109 A. Morrison, L. Ariza-McNaughton, A. Gould, M. Featherstone, R. Krumlauf, *HOXD4* and regulation of the group 4 paralog genes. *Development* **1997**; *124*: 3135–3146.

110 A. Packer, D. Crotty, V. Elwell, D. Wolgemuth, Expression of the murine Hoxa4 gene requires both autoregulation and a conserved retinoic acid response element. *Development* **1998**; *125*: 1991–1998.

111 F. Zhang, H. Pöpperl, A. Morrison, E. Kovàcs, V. Prideaux, L. Schwarz, R. Krumlauf, J. Rossant, M. Featherstone, Elements both 5' and 3' to the murine *Hoxd4* gene establish anterior borders of expression in mesoderm and neuroectoderm. *Mechanisms of Development* **1997**; *67*: 49–58.

112 F. Zhang, E. Nagy Kovacs, M. S. Featherstone, Murine hoxd4 expression in the CNS requires multiple elements including a retinoic acid response element. *Mech Dev* **2000**; *96*: 79–89.

113 K. Niederreither, V. Subbarayan, P. Dollé, P. Chambon, Embryonic retinoic acid synthesis is essential for early mouse post-implantation development. *Nature Genet* **1999**; :1, 444–448.

114 K. Niederreither, J. Vermot, B. Schuhbaur, P. Chambon, P. Dollé, Retinoic acid synthesis and hindbrain patterning in the mouse embryo. *Development* **2000**; *127*: 75–85.

115 G. Begemann, T. F. Schilling, G. J. Rauch, R. Geisler, P. W. Ingham, The zebrafish neckless mutation reveals a requirement for raldh2 in mesodermal signals that pattern the hindbrain. *Development* **2001**; *128*: 3081–3094.

116 H. Grandel, K. Lun, G. J. Rauch, M. Rhinn, T. Piotrowski, C. Houart, P. Sordino, A. M. Kuchler, S. Schulte-Merker, R. Geisler, N. Holder, S. W. Wilson, M. Brand, Retinoic acid signalling in the zebrafish embryo is necessary during pre-segmentation stages to pattern the anterior-posterior axis of the CNS and to induce a pectoral fin bud. *Development* **2002**; *129*: 2851–2865.

117 M. Maden, E. Gale, I. Kostetskii, M. Zile, Vitamin A deficient quail embryos have half a hindbrain and other neural defects. *Curr Biol* **1996**; *6*: 417–426.

118 E. Gale, M. Zile, M. Maden, Hindbrain respecification in the retinoid-deficient quail. *Mech Dev* **1999**; *89*: 43–54.

119 J. C. White, M. Highland, M. Kaiser, M. Clagett-Dame, Vitamin A-deficiency in the rat embryo results in anteriorization of the posterior hindbrain which is prevented by maternal consumption of retinoic acid or retinol. *Dev Biol* **2000**; *220*: 263–284.

120 V. Dupé, A. Lumsden, Hindbrain patterning involves graded responses to retinoic acid signalling. *Development* **2001**; *128*: 2199–2208.

121 O. Wendling, N. B. Ghyselinck, P. Chambon, M. Mark, Roles of retinoic acid receptors in early embryonic morphogenesis and hindbrain patterning. *Development* **2001**; *128*: 2031–2038.

122 V. Dupé, N. Ghyselinck, O. Wendling, P. Chambon, M. mark, Key roles of retinoic acid receptors alpha and beta in the patterning of the caudal hindbrain, pharyngeal arches and otocyst in the mouse. *Development* **1999**; *126*: 5051–5059.

123 S. Abu-Abed, P. Dolle, D. Metzger, B. Beckett, P. Chambon, M. Petkovich, The retinoic acid-metabolizing enzyme, CYP26A1, is essential for normal hindbrain patterning, vertebral identity, and development of posterior structures. *Genes Dev* **2001**; *15*: 226–240.

124 Y. Sakai, C. Meno, H. Fujii, J. Nishino, H. Shiratori, Y. Saijoh, J. Rossant, H. Hamada, The retinoic acid-inactivating enzyme CYP26 is essential for establishing an uneven distribution of retinoic acid along the anterio-posterior axis within the mouse embryo. *Genes Dev* **2001**; *15*: 213–225.

125 F. Marin, P. Charnay, Hindbrain patterning: FGFs regulate Krox20 and mafB/kr expression in the otic/preotic region. *Development* **2000**; *127*: 4925–4935.

126 N. M. Roy, C. G. Sagerstrom, An early Fgf signal required for gene expression in the zebrafish hindbrain primordium. *Brain Res Dev Brain Res* **2004**; *148*: 27–42.

127 S. Bel-Vialar, N. Itasaki, R. Krumlauf, Initiating Hox gene expression: In the early chick neural tube differential sensitivity to FGF and RA signalling subdivides the Hoxb genes into two distinct groups. *Development* **2002** *129*: 5103–5115.

128 L. W. Gamer, C. V. Wright, Murine Cdx–4 bears striking similarities to the Drosophila caudal gene in its homeodomain sequence and early expression pattern. *Mech Dev* **1993**; *43*: 71–81.

129 F. Beck, T. Erler, A. Russell, R. James, Expression of Cdx–2 in the mouse embryo and placenta: possible role in patterning of the extra-embryonic membranes. *Dev Dyn* **1995**; *204*: 219–227.

130 C. Irving, I. Mason, Signalling by fgf8 from the isthmus patterns the anterior hindbrain and establishes the anterior limit of *Hox* gene expression. *Development* **2000**; *127*: 177–186.

131 M. Brand, C. P. Heisenberg, Y. J. Jiang, D. Beuchle, K. Lun, M. Furutani-Seiki, M. Granato, P. Haffter, M. Hammerschmidt, D. A. Kane, R. N. Kelsh, M. C. Mullins, J. Odenthal, F. J. van Eeden, C. Nusslein-Volhard, Mutations in zebrafish genes affecting the formation of the boundary between midbrain and hindbrain. *Development* **1996**; *123*: 179–190.

132 S. Guo, J. Brush, H. Teraoka, A. Goddard, S. W. Wilson, M. C. Mullins, A. Rosenthal, Development of noradrenergic neurons in the zebrafish hindbrain requires BMP, FGF8, and the homeodomain protein soulless/Phox2a. *Neuron* **1999**; *24*: 555–566.

133 F. Reifers, H. Bohli, E. C. Walsh, P. H. Crossley, D. Y. Stainier, M. Brand, Fgf8 is mutated in zebrafish acerebellar (ace) mutants and is required for maintenance of midbrain-hindbrain boundary development and somitogenesis. *Development* **1998**; *125*: 2381–2395.

134 J. Jaszai, F. Reifers, A. Picker, T. Langenberg, M. Brand, Isthmus-to-midbrain transformation in the absence of midbrain-hindbrain organizer activity. *Development* **2003**; *130*: 6611–6623.

135 C. L. Chi, S. Martinez, W. Wurst, G. R. Martin, The isthmic organizer signal FGF8 is required for cell survival in the prospective midbrain and cerebellum. *Development* **2003**; *130*: 2633–2644.
136 C. Irving, I. Mason, Regeneration of isthmic tissue is the result of a specific and direct interaction between rhombomere 1 and midbrain. *Development* **1999**; *126*: 3981–3989.
137 L. Maves, W. Jackman, C. B. Kimmel, FGF3 and FGF8 mediate a rhombomere 4 signaling activity in the zebrafish hindbrain. *Development* **2002**; *129*: 3825–3837.
138 J. Walshe, H. Maroon, I. M. McGonnell, C. Dickson, I. Mason, Establishment of hindbrain segmental identity requires signaling by FGF3 and FGF8. *Curr Biol* **2002**; *12*: 1117–1123.
139 E. L. Wiellette, H. Sive, Early requirement for fgf8 function during hindbrain pattern formation in zebrafish. *Dev Dyn* **2004**; *229*: 393–399.
140 R. Diez del Corral, I. Olivera-Martinez, A. Goriely, E. Gale, M. Maden, K. Storey, Opposing FGF and retinoid pathways control ventral neural pattern, neuronal differentiation, and segmentation during body axis extension. *Neuron* **2003**; *40*: 65–79.
141 J. Shiotsugu, Y. Katsuyama, K. Arima, A. Baxter, T. Koide, J. Song, R. A. Chandraratna, B. Blumberg, Multiple points of interaction between retinoic acid and FGF signaling during embryonic axis formation. *Development* **2004**; *131*: 2653–2667.
142 D. G. Wilkinson, S. Bhatt, P. Chavrier, R. Bravo, P. Charnay, Segment-specific expression of a zinc-finger gene in the developing nervous system of the mouse. *Nature* **1989**; *337*: 461–465.
143 M. A. Nieto, L. C. Bradley, D. G. Wilkinson, Conserved segmental expression of *Krox–20* in the vertebrate hindbrain and its relationship to lineage restriction. *Development Supplement* **1991**; *2*: 59–62.
144 L. C. Bradley, A. Snape, S. Bhatt, D. G. Wilkinson, The structure and expression of the *Xenopus Krox–20* gene: conserved and divergent patterns of expression in the rhombomeres and neural crest. *Mech Dev* **1992**; *40*: 73–84.
145 E. Oxtoby, T. Jowett, Cloning of the zebra fish Krox–20 gene (Krx–20) and its expression during hindbrain development. *Nucleic Acids Res* **1993**; *21*: 1087–1095.
146 S. Schneider-Maunoury, P. Topilko, T. Seitanidou, G. Levi, M. Cohen-Tannoudji, S. Pournin, C. Babinet, P. Charnay, Disruption of *Krox–20* results in alteration of rhombomeres 3 and 5 in the developing hindbrain. *Cell* **1993**; *75*: 1199–1214.
147 O. Voiculescu, P. Charnay, S. Schneider-Maunoury, Expression pattern of a Krox–20/Cre knock-in allele in the developing hindbrain, bones, and peripheral nervous system. *Genesis* **2000**; *26*: 123–126.
148 S. Schneider-Maunoury, T. Seitanidou, P. Charnay, A. Lumsden, Segmental and neuronal architecture of the hindbrain of *Krox–20* mouse mutants. *Development* **1997**; *124*: 1215–1226.
149 T. D. Jacquin, V. Borday, S. Schneider-Maunoury, P. Topilko, G. Ghilini, F. Kato, P. Charnay, J. Champagnat, Reorganization of pontine rhythmogenic neuronal networks in Krox–20 knockout mice. *Neuron* **1996**; *17*: 747–758.

150 T. Seitanidou, S. Schneider-Maunoury, C. Desmarquet, D. Wilkinson, P. Charnay, Krox20 is a key regulator of rhombomere-specific gene expression in the developing hindbrain. *Mech Dev* 1997; 65: 31–42.

151 O. Voiculescu, E. Taillebourg, C. Pujades, C. Kress, S. Buart, P. Charnay, S. Schneider-Maunoury, Hindbrain patterning: Krox20 couples segmentation and specification of regional identity. *Development* 2001; 128: 4967–4978.

152 S. Garel, M. Garcia-Dominguez, P. Charnay, Control of the migratory pathway of facial branchiomotor neurones. *Development* 2000; 127: 5297–5307.

153 F. Giudicelli, E. Taillebourg, P. Charnay, P. Gilardi-Hebenstreit, Krox–20 patterns the hindbrain through both cell-autonomous and non cell-autonomous mechanisms. *Genes Dev* 2001; 15: 567–580.

154 F. Helmbacher, C. Pujades, C. Desmarquet, M. Frain, F. Rijli, P. Chambon, P. Charnay, Hoxa1 and Krox20 synergize to control the development of rhombomere 3. *Development* 1998; 125: 4739–4748.

155 S. Nonchev, M. Maconochie, C. Vesque, S. Aparicio, L. Ariza-McNaughton, M. Manzanares, K. Maruthainar, A. Kuroiwa, S. Brenner, P. Charnay, R. Krumlauf, The conserved role of Krox–20 in directing Hox gene expression during vertebrate hindbrain segmentation. *Proc Natl Acad Sci USA* 1996; 93: 9339–9345.

156 C. Vesque, M. Maconochie, S. Nonchev, L. Ariza-McNaughton, A. Kuroiwa, P. Charnay, R. Krumlauf, Hoxb–2 transcriptional activation by Krox–20 in vertebrate hindbrain requires an evolutionary conserved cis-acting element in addition to the Krox–20 site. *EMBO J* 1996; 15: 5383–5896.

157 M. K. Maconochie, S. Nonchev, M. Manzanares, H. Marshall, R. Krumlauf, Differences in Krox20–dependent regulation of Hoxa2 and Hoxb2 during hindbrain development. *Dev Biol* 2001; 233: 468–481.

158 S. Tumpel, M. Maconochie, L. M. Wiedemann, R. Krumlauf, Conservation and diversity in the cis-regulatory networks that integrate information controlling expression of Hoxa2 in hindbrain and cranial neural crest cells in vertebrates. *Dev Biol* 2002; 246: 45–56.

159 S. P. Cordes, G. S. Barsh, The mouse segmentation gene kr encodes a novel basic domain-leucine zipper transcription factor. *Cell* 1994; 79: 1025–1034.

160 I. J. McKay, I. Muchamore, R. Krumlauf, M. Maden, A. Lumsden, J. Lewis, The kreisler mouse: a hindbrain segmentation mutant that lacks two rhombomeres. *Development* 1994; 120: 2199–2211.

161 M. Manzanares, P. Trainor, S. Nonchev, L. Ariza-McNaughton, J. Brodie, A. Gould, H. Marshall, A. Morrison, C.-T. Kwan, M.-H. Sham, D. Wilkinson, R. Krumlauf, The role of kreisler in segmentation during hindbrain development. *Dev Biol* 1999; 211: 220–237.

162 C. B. Moens, S. P. Cordes, M. W. Giorgianni, G. S. Barsh, C. B. Kimmel, Equivalence in the genetic control of hindbrain segmentation in fish and mouse. *Development* 1998; 125: 381–391.

163 M. A. Frohman, G. R. Martin, S. Cordes, L. P. Halamek, G. S. Barsh, Altered rhombomere-specific gene expression and hyoid bone differentiation in the mouse segmentation mutant *kreisler (kr)*. *Development* **1993**; *117*: 925–936.

164 C. B. Moens, Y.-L. Yan, B. Appel, A. G. Force, C. B. Kimmel, *valentino*: a zebrafish gene required for normal hindbrain segmentation. *Development* **1996**; *122*: 3981–3990.

165 I. McKay, J. Lewis, A. Lumsden, Organization and development of facial motor neurons in the *kreisler* mutant mouse. *Eur J Neurosci* **1997**; *9*: 1499–1506.

166 M. Manzanares, S. Cordes, L. Ariza-McNaughton, V. Sadl, K. Maruthainar, G. Barsh, R. Krumlauf, Conserved and distinct roles of *kreisler* in regulation of the paralogous *Hoxa3* and *Hoxb3* genes. *Development* **1999**; *126*: 759–769.

167 M. Manzanares, S. Cordes, C.-T. Kwan, M.-H. Sham, G. Barsh, R. Krumlauf, Segmental regulation of *Hoxb3* by *kreisler*. *Nature* **1997**; *387*: 191–195.

168 M. Manzanares, S. Bel-Vialer, L. Ariza-McNaughton, E. Ferretti, H. Marshall, M. K. Maconochie, F. Blasi, R. Krumlauf, Independent regulation of initiation and maintenance phases of *Hoxa3* expression in the vertebrate hindbrain involves auto and cross-regulatory mechanisms. *Development* **2001**; *128*: 3595–3607.

169 M. Manzanares, J. Nardelli, P. Gilardi-Hebenstreit, H. Marshall, F. Giudicelli, M. T. Martinez-Pastor, R. Krumlauf, P. Charnay, Krox20 and kreisler co-operate in the transcriptional control of segmental expression of Hoxb3 in the developing hindbrain. *EMBO J* **2002**; *21*: 365–376.

170 Z. Sun, N. Hopkins, vhnf1, the MODY5 and familial GCKD-associated gene, regulates regional specification of the zebrafish gut, pronephros, and hindbrain. *Genes Dev* **2001**; *15*: 3217–3229.

171 E. L. Wiellette, H. Sive, vhnf1 and Fgf signals synergize to specify rhombomere identity in the zebrafish hindbrain. *Development* **2003**; *130*: 3821–3829.

172 S. K. Choe, C. G. Sagerstrom, Paralog group 1 hox genes regulate rhombomere 5/6 expression of vhnf1, a repressor of rostral hindbrain fates, in a meis-dependent manner. *Dev Biol* **2004**; *271*: 350–361.

173 G. M. Morriss-Kay, P. Murphy, R. E. Hill, D. R. Davidson, Effects of retinoic acid excess on expression of Hox–2.9 and Krox–20 and on morphological segmentation in the hindbrain of mouse embryos. *EMBO J*. **1991**; *10*: 2985–2995.

174 N. Papalopulu, J. Clarke, L. Bradley, D. Wilkinson, R. Krumlauf, N. Holder, Retinoic acid causes abnormal development and segmental patterning of the anterior hindbrain in Xenopus embryos. *Development* **1991**; *113*: 1145–1159.

175 H. Wood, G. Pall, G. Morriss-Kay, Exposure to retinoic acid before or after the onset of somitogenesis reveals separate effects on rhombomeric segmentation and 3' HoxB gene expression domains. *Development* **1994**; *120*: 2279–2285.

176 H. Pöpperl, M. Bienz, M. Studer, S. Chan, S. Aparicio, S. Brenner, R. Mann, R. Krumlauf, Segmental expression of *Hoxb1* is

controlled by a highly conserved autoregulatory loop dependent upon *exd/Pbx*. *Cell* **1995**; *81*: 1031–1042.

177 R. S. Mann, M. Affolter, Hox proteins meet more partners. *Curr Opin Genet Dev* **1998**; *8*: 423–429.

178 A. Gould, Functions of mammalian *Polycomb*-group and *trithorax*-group related genes. *Curr Opin Genet Dev* **1997**; *7*: 488–494.

179 A. Gould, A. Morrison, G. Sproat, R. White, R. Krumlauf, Positive cross-regulation and enhancer sharing: two mechanisms for specifying overlapping *Hox* expression patterns. *Genes Dev* **1997**; *11*: 900–913.

180 J. Whiting, H. Marshall, M. Cook, R. Krumlauf, P. W. J. Rigby, D. Stott, R. K. Allemann, Multiple spatially specific enhancers are required to reconstruct the pattern of *Hox–2.6* gene expression. *Genes Dev* **1991**; *5*: 2048–2059.

181 S. Aparicio, A. Morrison, A. Gould, J. Gilthorpe, C. Chaudhuri, P. W. J. Rigby, R. Krumlauf, S. Brenner, Detecting conserved regulatory elements with the model genome of the Japanese puffer fish *Fugu rubripes*. *Proc Natl Acad Sci USA* **1995**; *92*: 1684–1688.

182 A. Morrison, C. Chaudhuri, L. Ariza-McNaughton, I. Muchamore, A. Kuroiwa, R. Krumlauf, Comparative analysis of chicken *Hoxb–4* regulation in transgenic mice. *Mech Dev* **1995**; *53*: 47–59.

183 C. T. Kwan, S. L. Tsang, R. Krumlauf, M. H. Sham, Regulatory analysis of the mouse Hoxb3 gene: multiple elements work in concert to direct temporal and spatial patterns of expression. *Dev Biol* **2001**; *232*: 176–190.

184 M. Manzanares, S. Bel-Vialar, L. Ariza-McNaughton, E. Ferretti, H. Marshall, M. M. Maconochie, F. Blasi, R. Krumlauf, Independent regulation of initiation and maintenance phases of Hoxa3 expression in the vertebrate hindbrain involve auto- and cross-regulatory mechanisms. *Development* **2001**; *128*: 3595–3607.

185 M. P. Kamps, C. Murre, X.-H. Sun, D. Baltimore, A new homeobox gene contributes the DNA binding domain of the t(1;19) translocation protein in pre-B ALL. *Cell* **1990**; *60*: 547–555.

186 J. Nourse, J. D. Mellentin, N. Galili, J. Wilkinson, E. Stanbridge, S. D. Smith, M. L. Cleary, Chromosomal translocation t(1;19) results in synthesis of a homeobox fusion mRNA that codes for a potential chimeric transcription factor. *Cell* **1990**; *60*: 535–545.

187 K. Monica, N. Galili, J. Nourse, D. Saltman, M. Cleary, *PBX2* and *PBX3*, new homeobox genes with extensive homology to the human proto-oncogene *PBX1*. *Mol Cell Biol* **1991**; *11*: 6149–6157.

188 H. Popperl, H. Rikhof, H. Chang, P. Haffter, C. B. Kimmel, C. B. Moens, lazarus is a novel pbx gene that globally mediates hox gene function in zebrafish. *Mol Cell* **2000**; *6*: 255–267.

189 K. Wagner, A. Mincheva, B. Korn, P. Lichter, H. Popperl, Pbx4, a new Pbx family member on mouse chromosome 8, is expressed during spermatogenesis. *Mech Dev* **2001**; *103*: 127–131.

190 K. L. Cooper, W. M. Leisenring, C. B. Moens, Autonomous and nonautonomous functions for Hox/Pbx in branchiomotor neuron development. *Dev Biol* **2003**; *253*: 200–213.

191 J. F. DiMartino, L. Selleri, D. Traver, M. T. Firpo, J. Rhee, R. Warnke, S. O'Gorman, I. L. Weissman, M. L. Cleary, The Hox cofactor and proto-oncogene Pbx1 is required for maintenance of definitive hematopoiesis in the fetal liver. *Blood* **2001**; *98*: 618–626.

192 L. Selleri, M. J. Depew, Y. Jacobs, S. K. Chanda, K. Y. Tsang, K. S. Cheah, J. L. Rubenstein, S. O'Gorman, M. L. Cleary, Requirement for Pbx1 in skeletal patterning and programming chondrocyte proliferation and differentiation. *Development* **2001**; *128*: 3543–3557.

193 L. Selleri, J. DiMartino, J. van Deursen, A. Brendolan, M. Sanyal, E. Boon, T. Capellini, K. S. Smith, J. Rhee, H. Popperl, G. Grosveld, M. L. Cleary, The TALE homeodomain protein Pbx2 is not essential for development and long-term survival. *Mol Cell Biol* **2004**; *24*: 5324–5331.

194 S. K. Kim, L. Selleri, J. S. Lee, A. Y. Zhang, X. Gu, Y. Jacobs, M. L. Cleary, Pbx1 inactivation disrupts pancreas development and in Ipf1–deficient mice promotes diabetes mellitus. *Nature Genet* **2002**; *30*: 430–435.

195 C. A. Schnabel, R. E. Godin, M. L. Cleary, Pbx1 regulates nephrogenesis and ureteric branching in the developing kidney. *Dev Biol* **2003**; *254*: 262–276.

196 C. A. Schnabel, L. Selleri, M. L. Cleary, Pbx1 is essential for adrenal development and urogenital differentiation. *Genesis* **2003**; *37*: 123–130.

197 F. Biemar, N. Devos, J. A. Martial, W. Driever, B. Peers, Cloning and expression of the TALE superclass homeobox Meis2 gene during zebrafish embryonic development. *Mech Dev* **2001**; *109*: 427–431.

198 T. Zerucha, V. E. Prince, Cloning and developmental expression of a zebrafish meis2 homeobox gene. *Mech Dev* **2001**; *102*: 247–250.

199 T. Nakamura, D. A. Largaespada, J. D. Shaughnessy, N. A. Jenkins, N. G. Copeland, Cooperative activation of *Hoxa* and *Pbx1*–related genes in murine myeloid leukaemias. *Nature Genet* **1996**; *12*: 149–153.

200 F. Cecconi, G. Proetzel, G. Alvarez-Bolado, D. Jay, P. Gruss, Expression of Meis2, a Knotted-related murine homeobox gene, indicates a role in the differentiation of the forebrain and the somitic mesoderm. *Dev Dyn* **1997**; *210*: 184–190.

201 S. Steelman, J. J. Moskow, K. Muzynski, C. North, T. Druck, J. C. Montgomery, K. Huebner, I. O. Daar, A. M. Buchberg, Identification of a conserved family of Meis1–related homeobox genes. *Genome Res* **1997**; *7*: 142–156.

202 A. Salzberg, S. Elias, N. Nachaliel, L. Bonstein, C. Henig, D. Frank, A Meis family protein caudalizes neural cell fates in Xenopus. *Mech Dev* **1999**; *80*: 3–13.

203 N. Vlachakis, S. K. Choe, C. G. Sagerstrom, Meis3 synergizes with Pbx4 and Hoxb1b in promoting hindbrain fates in the zebrafish. *Development* **2001**; *128*: 1299–1312.

204 A. J. Waskiewicz, H. A. Rikhof, R. E. Hernandez, C. B. Moens, Zebrafish Meis functions to stabilize Pbx proteins and regulate hindbrain patterning. *Development* **2001**; *128*: 4139–4151.

205 Y. Jacobs, C. A. Schnabel, M. L. Cleary, Trimeric association of Hox and TALE homeodomain proteins mediates Hoxb2 hindbrain enhancer activity. *Mol Cell Biol* **1999**; *19*: 5134–5142.

206 C. P. Chang, Y. Jacobs, T. Nakamura, N. A. Jenkins, N. G. Copeland, M. L. Cleary, Meis proteins are major in vivo DNA binding partners for wild-type but not chimeric Pbx proteins. *Mol Cell Biol* **1997**; *17*: 5679–5687.

207 P. S. Knoepfler, M. P. Kamps, The highest affinity DNA element bound by Pbx complexes in t(1;19) leukemic cells fails to mediate cooperative DNA-binding or cooperative transactivation by E2a-Pbx1 and class I Hox proteins – evidence for selective targetting of E2a-Pbx1 to a subset of Pbx-recognition elements. *Oncogene* **1997**; *14*: 2521–2531.

208 E. Ferretti, H. Marshall, H. Pöpperl, M. Maconochie, R. Krumlauf, F. Blasi, Segmental expression of *Hoxb2* in r4 requires two distinct sites that facilitate co-operative interactions and ternary complex formation between Perp, Pbx and Hox proteins. *Development* **2000**; *127*: 155–166.

209 G. Deflorian, N. Tiso, E. Ferretti, D. Meyer, F. Blasi, M. Bortolussi, F. Argenton, Prep1.1 has essential genetic functions in hindbrain development and cranial neural crest cell differentiation. *Development* **2004**; *131*: 613–627.

210 S. K. Choe, N. Vlachakis, C. G. Sagerstrom, Meis family proteins are required for hindbrain development in the zebrafish. *Development* **2002**; *129*: 585–595.

211 P. Qin, R. Cimildoro, D. M. Kochhar, K. J. Soprano, D. R. Soprano, PBX, MEIS, and IGF-I are potential mediators of retinoic acid-induced proximodistal limb reduction defects. *Teratology* **2002**; *66*: 224–234.

212 P. Qin, J. M. Haberbusch, K. J. Soprano, D. R. Soprano, Retinoic acid regulates the expression of PBX1, PBX2, and PBX3 in P19 cells both transcriptionally and post-translationally. *J Cell Biochem* **2004**; *92*: 147–163.

213 M. Oulad-Abdelghani, C. Chazaud, P. Bouillet, V. Sapin, P. Chambon, P. Dolle, Meis2, a novel mouse Pbx-related homeobox gene induced by retinoic acid during differentiation of P19 embryonal carcinoma cells. *Dev Dyn* **1997**; *210*: 173–183.

214 P. S. Knoepfler, M. P. Kamps, The Pbx family of proteins is strongly upregulated by a post- transcriptional mechanism during retinoic acid-induced differentiation of P19 embryonal carcinoma cells. *Mech Dev* **1997**; *63*: 5–14.

215 N. Mercader, E. Leonardo, M. E. Piedra, A. C. Martinez, M. A. Ros, M. Torres, Opposing RA and FGF signals control proximodistal vertebrate limb development through regulation of Meis genes. *Development* **2000**; *127*: 3961–3970.

216 C. Dibner, S. Elias, R. Ofir, J. Souopgui, P. J. Kolm, H. Sive, T. Pieler, D. Frank, The Meis3 protein and retinoid signaling interact to pattern the Xenopus hindbrain. *Dev Biol* **2004**; *271*: 75–86.

217 S. Guthrie, A. Lumsden, Formation and regeneration of rhombomere boundaries in the developing chick hindbrain. *Development* **1991**; *112*: 221–229.

218 S. C. Guthrie, M. Butcher, A. Lumsden, Patterns of cell division and interkinetic nuclear migration in the chick embryo hindbrain. *J Neurobiol* **1991**; *22*: 742–754.

219 A. Garcia-Bellido, P. Ripoll, G. Morata, Developmental compartmentalisation of the wing disk of Drosophila. *Nature* **1973**; *245*: 251–253.

220 S. Guthrie, V. Prince, A. Lumsden, Selective dispersal of avian rhombomere cells in orthotopic and heterotopic grafts. *Development* **1993**; *118*: 527–538.

221 A. Wizenmann, A. Lumsden, Segregation of rhombomeres by differential chemoaffinity. *Mol Cell Neurosci* **1997**; *9*: 448–459.

222 Q. Xu, G. Mellitzer, V. Robinson, D. Wilkinson, In vivo cell sorting in complementary segmental domains mediated by *Eph* receptors and *ephrins*. *Nature* **1999**; *399*: 267–271.

223 G. Mellitzer, Q. Xu, D. Wilkinson, Eph receptors and ephrins restrict cell intermingling and communication. *Nature* **1999**; *400*: 77–81.

224 Q. Xu, G. Alldus, N. Holder, D. G. Wilkinson, Expression of truncated *Sek–1* receptor tyrosine kinase disrupts the segmental restriction of gene expression in the *Xenopus* and zebrafish hindbrain. *Development* **1995**; *121*: 4005–4016.

225 Q. Xu, G. Mellitzer, D. G. Wilkinson, Roles of Eph receptors and ephrins in segmental patterning. *Phil Trans R Soc Lond B Biol Sci* **2000**; *355*: 993–1002.

226 R. S. Winning, J. B. Scales, T. D. Sargent, Disruption of cell adhesion in Xenopus embryos by Pagliaccio, an Eph-class receptor tyrosine kinase. *Dev Biol* **1996**; *179*: 309–319.

227 A. Zisch, E. Pasquale, The Eph family: a multude of receptors that mediate cell recognition signals. *Cell Tissue Res* **1997**; *290*: 217–226.

228 T. L. Jones, L. D. Chong, J. Kim, R. H. Xu, H. F. Kung, I. O. Daar, Loss of cell adhesion in *Xenopus laevis* embryos mediated by the cytoplasmic domain of XLerk, an erythropoietin-producing hepatocellular ligand. *Proc Natl Acad Sci USA* **1998**; *95*: 576–581.

229 Y. C. Cheng, M. Amoyel, X. Qiu, Y. J. Jiang, Q. Xu, D. G. Wilkinson, Notch activation regulates the segregation and differentiation of rhombomere boundary cells in the zebrafish hindbrain. *Dev Cell* **2004**; *6*: 539–550.

230 A. Grapin-Botton, M.-A. Bonnin, L. Ariza-McNaughton, R. Krumlauf, N. M. LeDouarin, Plasticity of transposed rhombomeres: Hox gene induction is correlated with phenotypic modifications. *Development* **1995**; *121*: 2707–2721.

231 S. Guthrie, I. Muchamore, A. Kuroiwa, H. Marshall, R. Krumlauf, A. Lumsden, Neuroectodermal autonomy of *Hox–2.9* expression revealed by rhombomere transpositions. *Nature* **1992**; *356*: 157–159.

232 P. Trainor, R. Krumlauf, Plasticity in mouse neural crest cells reveals a new patterning role for cranial mesoderm. *Nature Cell Biol* **2000**; *2*: 96–102.

233 T. Schilling, Plasticity of zebrafish Hox expression in the hindbrain and cranial neural crest hindbrain. *Dev Biol* **2001**; *231*: 201–216.

234 P. Trainor, R. Krumlauf, Patterning the cranial neural crest: Hindbrain segmentation and *Hox* gene plasticity. *Nature Rev Neurosci* **2000**; *1*: 116–124.

7
Neurogenesis in the Central Nervous System

Véronique Dubreuil, Lilla Farkas, Federico Calegari, Yoichi Kosodo, and Wieland B. Huttner

Abbreviations

CNS, central nervous system; NSC, neural stem cells; bHLH, basic helix-loop-helix; PNS, peripheral nervous system; EGF, epidermal growth factor; NIC, Notch intracellular domain; DSL, Delta/Serrate/lag–2; CSL, CBF–1/Suppressor of hairless/Lag–1; pRb, retinoblastoma protein; Wg, Wingless; Hh, Hedgehog; Shh, Sonic Hedgehog; BMP, bone morphogenetic protein; LEF/TCF, lymphocyte enhancer factor/T-cell factor; VZ, ventricular zone; SVZ, subventricular zone; TGF-α/β, transforming growth factor α/β; EGFR, epidermal growth factor receptor; FGF, fibroblast growth factors; FGFR, fibroblast growth factor receptor; HS, heparan sulfate proteoglycans; MHB, mid-hindbrain boundary; RA, retinoic acid; ORN, olfactory receptor neurons; PDGF, platelet-derived growth factor; BDNF, brain-derived neurotrophic factor; NGF, nerve growth factors; CNTF, ciliary neurotrophic factor; LIF, leukocyte inhibitory factor; GH, growth hormone; BrdU, bromodeoxyuridine; CDK, cyclin dependent kinase; CKI, CDK inhibitors; NB, neuroblasts; GMC, ganglion mother cell; aPKC, atypical protein kinase C; Pins, partner of Inscuteable; PON, partner of Numb

7.1
Introduction

Neurons, astrocytes and oligodendrocytes of the central nervous system (CNS) arise from a common pool of proliferating multipotent progenitor cells also called neural stem cells (NSC). In the neural tube, these cells are arranged in a polarized epithelium, the neuroepithelium, also referred to as the ventricular zone (VZ, the layer facing the lumen of the neural tube; Fig. 7.1). Neurons are the first cell type to be born from these neural progenitors, with gliogenesis following neurogenesis. In addition to this temporal order there is also spatial order, with neurogenesis proceeding in a rostral-to-caudal direction from the hindbrain to the spinal cord and in a caudal-to-rostral direction from the midbrain to the forebrain. The production of

Cell Signaling and Growth Factors in Development. Edited by K. Unsicker and K. Krieglstein
Copyright © 2006 WILEY-VCH Verlag GmbH & Co. KGaA, Weinheim
ISBN 3-527-31034-7

neurons requires the complex integration of environmental and intrinsic cues at the cellular level to control the balance between proliferation, differentiation, and survival.

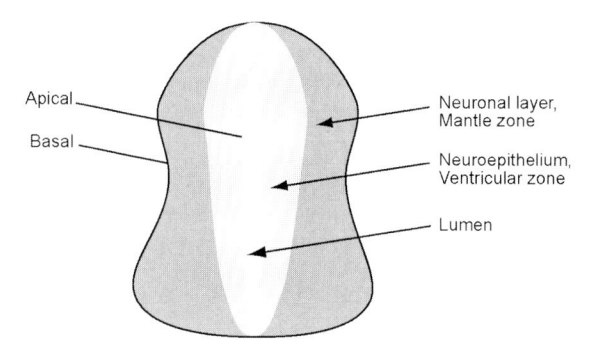

Fig. 7.1
The neural tube. The progenitor cells (dividing NSC) are arranged in a single cell layer facing the lumen of the neural tube. These cells are polarized: the apical side faces the lumen and the basal side faces the cortical layer. During neuronal differentiation, the young neurons escape from the VZ and accumulate in the cortical layer.

In response to differentiation signals, which may originate extracellularly as well as intracellularly, NSC generate neurons. They exit the cell cycle (the post-mitotic state being a common characteristic of neurons), migrate to the appropriate neuronal layer, and change their gene expression pattern. This early step of neuronal differentiation is then followed by further steps of neuronal maturation, including neurite outgrowth, axon pathfinding, synapse formation, etc.

The neuronal determination of a progenitor cell (its selection from a pool of NSC), constitutes a generic aspect of neurogenesis that can be applied to all types of neurons. However, it should be remembered that the NSC, and their progeny, are under strict spatio-temporal cues that influence the identity of the resulting neurons. Generic differentiation and neuronal subtype specification are therefore tightly linked and cannot readily be dissociated. This chapter will address only the generic aspects of neurogenesis. The principal focus will concern mammalian neurogenesis but whenever appropriate, data from non-mammalian model species will be included. Indeed, the fundamental studies in the fly have established most of the present concepts about neurogenesis, and the developmental principles are strikingly conserved between species.

The first part of this chapter will discuss the major intracellular players and the complex intrinsic regulation involved in the neuronal determination of progenitors. Then, the most important extracellular factors that influence neurogenesis, and their respective signaling pathways, will be reviewed. The last two parts of this chapter will specifically address the cellular mechanisms underlying the switch of progenitors to neurogenesis. One concerns the role of the length of the progenitor cell cycle on the determination process. The other concerns asymmetric cell division of neural progenitors, which has been pioneered in the fly and for which there is increasing evidence also with regard to mammalian neurogenesis.

7.2
Intrinsic Regulation of Neurogenesis

In *Drosophila*, as in mammals, NSC are arranged in an epithelium. When neurogenesis starts, only a fraction of these progenitors is instructed to become neurons. This primary step in neurogenesis is crucial, first, to avoid the complete loss of the progenitors, and second, to allow time-dependent factors to influence cell fate. Consequently, the decision to become a neuron is strongly regulated at the cellular level. As neurogenesis begins, positive regulators instructing neuronal differentiation are switched on, and are counteracted by factors involved in the maintenance of the undifferentiated, proliferative state. The balance between inducing and repressing factors determines the onset and extent of neurogenesis.

Neurogenesis is a process that is highly conserved between invertebrates and vertebrates. Orthologs of most of the molecular players involved, both factors promoting and inhibiting neuronal differentiation, have been found in the various species analyzed [1–3].

7.2.1
Neuronal Determination

7.2.1.1 Proneural Genes

The major intrinsic regulators of neurogenesis, the proneural genes, were first cloned in *Drosophila* in the late 1970s [4]. Homologs to these genes, the *Drosophila* *achaete-scute* complex and *atonal*, were then found in vertebrates [1, 5, 6]. Their initial characterization, through the analysis of different *Drosophila* mutants, demonstrated their requirement during neurogenesis in the peripheral nervous system (PNS). Loss-of-function mutants were found to exhibit a loss of external sense organs, while gain-of-function mutations induce supernumerary sensory bristles [7].

Biochemical characterization and expression

The proneural genes code for transcription factors with a characteristic basic helix-loop-helix (bHLH) domain; the basic domain is involved in DNA binding and the HLH domain in protein-protein interactions (homo- or heterodimerization with other HLH proteins) [8]. Proneural genes are neural-specific transcriptional activators. Upon heterodimerization with ubiquitously expressed bHLH factors (daughterless in *Drosophila* and E2A or E2-2 in vertebrates), their binding to the E-box consensus sequence leads to gene transactivation [9]. Proneural genes in vertebrates are classified according to their *Drosophila* counterparts: achaete-scute complex homolog-like (Ascl), atonal homolog (atoh) and, in addition, the neurogenins.

As observed in *Drosophila*, the expression domains of the proneural genes in vertebrates delimit a competent region where neurogenesis can take place. This characteristic expression in the neuroepithelium is in accordance with an early function in the selection of the future neuron.

Gain- and loss-of-function phenotypes in vertebrates
As is the case in *Drosophila*, gain-of-function experiments (ectopic or overexpression) with proneural genes in vertebrates induce supernumerary neurons, in accordance with their expected neurogenic activity [10–12]. These results strongly support the inherent ability of these transcription factors to promote neuronal differentiation of proliferating cells.

In addition, proneural genes are also required during the neuronal differentiation process. Indeed, individual gene inactivation for different proneural genes in mouse embryos triggers a loss of some specific neuronal progenitors [13–16].

Effects induced by proneural activity
So far, the direct target genes of the proneural bHLH transcription factors are poorly characterized. Despite the description of the consensus DNA-binding motif, little is known about the DNA specificity of each proneural gene.

The proneural genes are able to promote neuronal differentiation by integrating different aspects of neuronal development. First, they are involved in the cell cycle withdrawal of progenitor cells. This is achieved indirectly by the release of lateral inhibition (via the activation of *delta*), which maintains the proliferative state (see below). More directly, the expression of different proneural genes *in vitro* has been shown to upregulate cell cycle inhibitors [17]. Second, they trigger the sequence of events leading to neuronal differentiation. The proneural genes are able to induce a cascade of transcriptional regulators (mostly from the bHLH class) whose presence is critical for proper neuronal differentiation (see below). Finally, proneural genes influence neural fate determination of the progenitor cells by repressing non-neuronal fate. Notably, they can bias the neuronal versus glial choice; various studies have shown that proneural genes can repress glial differentiation [18–20]. Concerning this last point, it is remarkable that the expression of proneural genes is nevertheless observed during glial differentiation [21–23]. This may suggest that the neuronal potential of the proneural factor changes with time or that neural progenitors become unresponsive with time to neuronal identity cues.

It is worth mentioning that proneural genes also play a pivotal role in the identity of the neurons produced from their expression domains, probably by interacting with positional information cues that are present throughout the neural tube [6].

7.2.1.2 Regulation of Proneural Activity
As described above, neurogenesis is highly dependent on proneural activity. A tight control of the proneural genes is therefore a prerequisite for the proper generation of neurons. The regulation of proneural activity occurs both at the transcriptional and protein level.

Lateral inhibition

Proneural activity is principally regulated by a cell-cell signaling pathway between progenitor cells in the neuroectoderm. This pathway involves the two transmembrane proteins Delta (ligand) and Notch (receptor), which are important cell fate determinants in numerous developmental systems [2, 24]. In *Drosophila*, the Delta-Notch pathway is the key for the lateral inhibition process involved in the selection of a neuronal precursor cell. The term "lateral inhibition" refers to the negative cross-regulation between neighboring cells that is required to establish the precise pattern of neurons in the fly, or to maintain the pool of progenitors in vertebrates (Fig. 7.2).

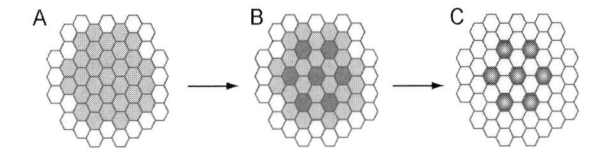

Fig. 7.2
Neuronal precursor selection and lateral inhibition. Inside the proneural group, all the cells are equivalent, and have the capacity to become neuronal precursors. A disequilibrium in the equivalence group perturbs the Delta-Notch signaling and some scattered cells start to emerge by repressing their neighbors via the lateral inhibition pathway. Selected cells adopt the neuronal fate while the surrounding cells adopt ectodermal identity (in *Drosophila* PNS) or remain in an undifferentiated state (in the CNS).

Extensively studied during the development of the fly sense organs, lateral inhibition refers to the process in which one cell is selected from a group of equivalent cells. The expression of a proneural gene (at a low level) defines this equivalence group (or proneural cluster). In these cells, Notch and Delta are also present, each cell having the capacity to send a signal via Delta, but also to receive a signal via Notch. Downstream of Notch signaling are transcriptional repressors that can silence the proneural genes such that the activation of Notch impairs neuronal differentiation. Importantly, the proneural genes induce the upregulation of *Delta*, leading to the activation of Notch in the adjacent cells which triggers proneural repression and *Delta* down-regulation, thereby generating a negative feedback loop in the signaling pathway. At equilibrium, signals sent and received by each cell would lead to mutual inhibition. By inducing an imbalance in the signaling, the auto-amplification of the signal due to the negative feedback in the signaling cascade would induce an all or none response (progenitor fate versus neuronal precursor fate). For example, an increase in proneural activity in one cell (cell A) would induce Delta and activate Notch in the surrounding cells. In response, the proneural genes would decrease and thus *Delta* expression would be reduced in the neighbors. As a consequence, cell A would amplify its own proneural and Delta expression while repressing their expression in the surrounding cells (Fig. 7.2).

Delta-notch pathway

The cell biology of Delta-Notch signaling is very complex: both ligand and receptor are tightly regulated [25] by glycosylation [26], ubiquitination [27], endocytosis [28] and proteolysis [29]. Notch and Delta are one-pass transmembrane proteins rich in epidermal growth factor (EGF) repeats. During biosynthetic transport to the cell surface, Notch undergoes a first constitutive cleavage in its ectodomain by the furin protease in the trans-Golgi network. The two proteolytic fragments remain associated and form an active heterodimeric Notch receptor presented at the plasma membrane. Upon Delta-Notch heterodimerization (via the EGF repeats), Notch is subject to additional proteolytic cleavages leading to its activation. Binding of Notch ligands (Delta or Jagged, DSL) activates extracellular proteases that cleave the extracellular part of Notch and render the C-terminal domain (still attached to the membrane) susceptible to gamma-secretase (presenilin and nicastrin are part of the γ-secretase complex). This last cleavage leads to the release of the Notch intracellular domain (NIC), which enters the nucleus. NIC then forms an activator complex by recruiting the CBF–1/Suppressor of hairless/Lag–1 (CSL) transcription factors and the protein mastermind (Fig. 7.3A).

In the absence of NIC, CSL represses its transcriptional targets. Notch activation leads to the formation of a transcriptional activator complex (NIC/CSL), which directly transactivates *Hes* gene promoters. *Hes* expression is therefore a powerful read-out of Notch activity.

Hes genes encode transcription factors of the bHLH class with repressor activity also called negative bHLH. They are able to heterodimerize via their C-terminal WRPW domain with the co-repressor Groucho/TLE (*Drosophila*/vertebrates) and to repress target genes [30]. The principal targets of Hes factors are the proneural genes. In addition to transcriptional repression of the proneural genes, Hes factors can also block the proneural proteins by forming non-active heterodimers with the proneural bHLH proteins, thereby inhibiting their activity [31] (Fig. 7.3B).

Notch activity associated with CSL factors is known as the canonical Notch pathway. Some studies, however, suggest the existence of other Notch pathways, which are independent of CSL activity [32, 33].

Notch signaling during mammalian neurogenesis

In mammals, all four *Notch* genes are expressed in the proliferative region at specific stages of nervous system development. Manipulations of *Notch* expression or of genes in the Delta-Notch pathway show neuronal phenotypes.

Notch1 inactivation results in early lethality. Nevertheless, it is possible to determine that in the absence of *Notch1*, there is a premature expression of proneural genes and a premature generation of neurons [34]. Consistent with the previous results, the conditional inactivation of *Notch1* in the midbrain also results in neurogenesis defects [35]. In complement with the inactivation experiments, overexpression of NIC in mice decreases neuronal production [36].

In accordance with the function of the Notch pathway during neurogenesis, manipulations of the different proteins in the signaling pathway also induce neuronal phenotypes. Impairing Notch processing using a presenilin mutant [37] or inacti-

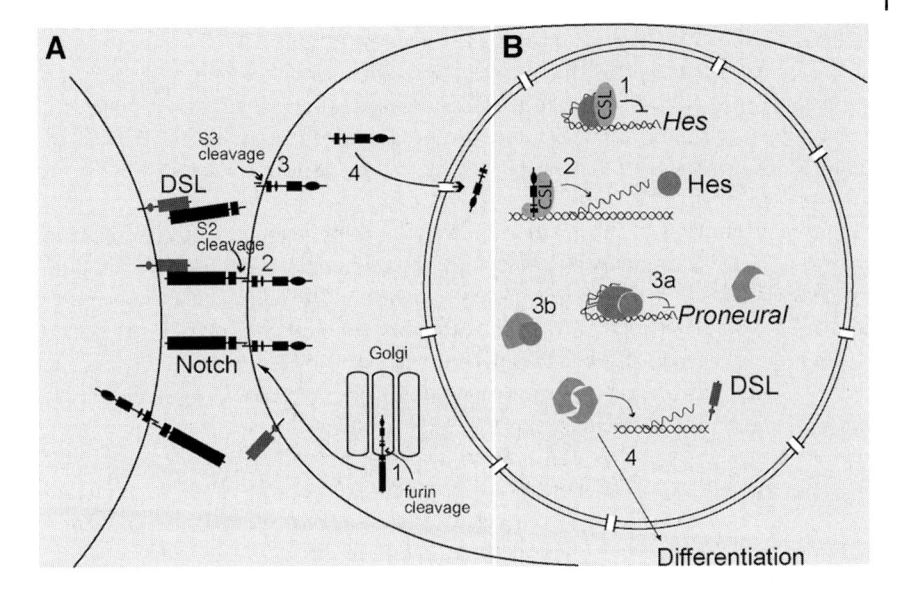

Fig. 7.3
Notch signaling. (A) Notch processing. 1, Notch is cleaved during its processing in the Golgi apparatus and appears in the plasma membrane as a heterodimer. 2, Upon DSL (Delta/Serrate/Lag-2) ligand activation Notch is processed by the TACE/ADAM proteases, which triggers the release of the extracellular domain. 3, Notch intracellular fragment (NIC) is accessible to γ-secretase activity, is cleaved, released in the cytoplasm and 4, translocates into the nucleus. (B) Transcriptional interactions. 1, CSL (CBF1/RBP-Jx/Su(H)/Lag-1) factors recruit co-repressor complexes and repress *Hes* genes. 2, In presence of NIC, the activator complex NIC/CSL/mastermind transactivates Hes target genes. Hes acts at two levels to repress the proneural activity: 3a, Hes represses proneural genes by forming a complex with the co-repressor groucho/TLE, and 3b, Hes forms an inactive complex with the bHLH factors. 4, In the absence of proneural repression, the neuronal determination program is activated and leads, most notably, to the upregulation of *Delta* (DSL).

vating *Hes1* and *Hes5* (the principal target genes of the Notch pathway during neurogenesis) produces defects in neurogenesis similar to those observed in the *Notch* inactivation mutants. In addition, *Hes* genes, when overexpressed, can repress neuronal differentiation [38] while their inactivation causes premature neuronal differentiation [39]. The general effects on neurogenesis described above have also been associated with defects in the expression of the crucial regulators of neurogenesis, the proneural genes.

To conclude, lateral inhibition, by exerting tight control of the proneural activity via its principal effectors, the Hes repressors, is a critical requirement in the regulation of neuronal determination.

7.2.1.3 Proliferative Determinants

Proliferation of neuroepithelial cells is not only regulated by the Delta-Notch signaling pathway but also by various factors expressed in the progenitors themselves. Among these proliferative factors, the genes of the Id family are one class of interesting transcription factor that probably act again by fine-tuning the proneural activity.

Id proteins encoded by four *Id* genes in vertebrates (homolog to *Drosophila extramacrochaete*) are transcription factors with an HLH domain but no DNA binding domain. They are thus able to form heterodimers with bHLH proteins via the HLH domain. Because they lack the basic DNA-binding domain, the heterodimer generated is unable to bind DNA and therefore, act as a dominant-negative complex [40].

In the neural tube, *Id* gene expression is specifically restricted to the proliferative region, and inversely correlated with neurogenesis, suggesting a role in proliferation of NSC. Id proteins are involved in the control of cell differentiation and cell cycle progression in different cell lineages. During neurogenesis, they repress proneural activity and regulate the cell cycle. Inactivation studies in mice show a size reduction of the neural tube associated with premature differentiation, in accordance with a role in the inhibition of differentiation [41]. Moreover, overexpression of Id2 induces an overgrowth of the cortex in addition to a block of cortical differentiation [42]. Apart from the block of proneural activity that would be associated with the natural dominant effect of Id on bHLH factors, it has also been shown that the effect on cortical differentiation could be reverted by co-expressing pRb, the retinoblastoma protein, suggesting a direct link with cell cycle regulation.

Neurogenesis, and more precisely neuronal determination as the first chronological event leading to neuronal generation, is strongly dependent on proneural gene expression and function. As shown in the above section, various levels of regulation are required to shape proneural activity.

7.2.2
Neuronal Differentiation

Once the neural progenitor has given rise to a neuron, a cascade of differentiation genes is switched on. The proneural genes, as transcriptional activators, are able to start this neuronal differentiation program that is characterized by the sequential activation of different genes (mostly transcription factors) essential to implement the neuronal phenotype.

One of the first genes transiently turned on by the proneural factors is *Delta* (as described above), which is followed by the expression of other members of the *bHLH* family of transcription factors whose expression is specific to the neural tissue. Among these genes, NSCL1, NeuroD and Math3 are transiently expressed in the young post-mitotic neuron that is en route to the neuronal layer. Depending on the region of the nervous system observed, different sequences of gene expression are activated in the young neurons. For example NeuroM is turned on before NeuroD in the spinal cord, but after it in the dorsal root ganglia [43].

The ectopic expression of bHLH genes, like that of these proneural genes, is characterized by a neurogenic activity. However, their expression pattern is not consistent with an early role in neurogenesis and gain- and loss-of-function analyses suggest rather that they have roles in the maintenance of the differentiated state, in cell cycle exit, and even in survival of specific neuronal populations [44–46].

Other transcription factors are also critical to the differentiation process. Zinc-finger transcription factors (EBF family, Zic family) seem to be able to coordinate neuronal differentiation and could function to reinforce the identity of committed neurons or to promote neuron survival [47, 48].

Very few of the effector targets of these transcription factors have been identified but it can be assumed that many important determinants are mobilized to transform the proliferative progenitor into a differentiated cell. Various biological modifications (cell cycle exit, cellular mobility, cell shape remodeling) are involved in this transformation and are still to be characterized.

Various regulators not cited above are also involved in the process of neurogenesis and participate in the overall spatial and temporal patterning of neurogenesis. The gradual acquisition of a generic neuronal identity is dependent upon complex transcriptional regulation as suggested in this section. Each step being highly controlled by different factors necessary for both maintaining the pool of progenitor cells and stimulating neuronal generation.

7.3
Extrinsic Regulation of Neurogenesis

This chapter will focus only on the effect of different extracellular factors on neurogenesis in the CNS. A review of different signaling factors acting on the balance of proliferation-differentiation is presented here, with the principal aim being to demonstrate their function in neurogenesis, either as inhibitors or as activators of differentiation.

Most of the known signaling pathways seem to be involved in the establishment of the nervous system, providing the extracellular cues for the spatio-temporal control of neurogenesis. Some of these factors are known to act in the maintenance of NSC proliferation, which is critical for neurogenesis, either directly stimulating proliferation, promoting survival, or both. Other factors are also actively required to promote neuronal differentiation. In both cases, the deregulation of the pathway associated with these factors often leads to neuronal generation defects.

7.3.1
Wnt Factors

The Wnt (Wingless/Int–1) proteins play diverse roles during development, acting on cell proliferation, cell polarity and cell fate determination [49]. Wnt family members stimulate stem cell proliferation in various tissues including the CNS [50–53].

7.3.1.1 Wnt Pathway

Members of the Wnt family are secreted signaling glycoproteins related to *Drosophila* Wingless, Wg (reviewed in [54]). Intracellular signaling of the Wnt pathway involves at least three diverse branches. First, the beta-catenin pathway also called the canonical Wnt pathway, which activates target genes in the nucleus. Second the planar cell polarity pathway, which involves jun N-terminal kinase (JNK) and cytoskeletal rearrangements. Third, the Wnt/Ca^{2+} pathway (for a recent update see [55]). The core components of these pathways are the Frizzled seven-pass transmembrane receptors and the single transmembrane co-receptors LRP (low-density lipoprotein receptor related proteins). Intracellularly, the canonical Wnt pathway leads to the stabilization of cytosolic β-catenin, which is otherwise degraded by a huge protein complex. This complex consists of the casein kinase, the glycogen synthase kinase 3beta (GSK3beta), the scaffold proteins axin and diversin, and the tumor suppressor gene product APC. This complex triggers phosphorylation of β-catenin and its degradation by the ubiquitin-proteasome pathway. In the presence of Wnt, Dishevelled blocks β-catenin degradation, allowing the stabilized β-catenin to enter the nucleus where it associates with the transcription factors LEF/TCF (lymphocyte enhancer factor/T-cell factor). The complex β-catenin/LEF/TCF activates the Wnt target genes.

Extracellular antagonists of the Wnt signaling pathway have been characterized in different developmental systems as Wnt signal modulators. The antagonists can be divided into two broad classes depending on their mode of action (for review see [56]). Both classes of molecules prevent ligand-receptor interactions, but by different mechanisms. Members of the first class, which include the sFRP (secreted Frizzled-related protein) family, WIF (Wnt inhibitory factor)–1 and Cerberus, bind directly to Wnt proteins. The second class of factors comprises certain members of the Dickkopf (Dkk) family and act by binding to one component of the Wnt receptor complex. The two classes of antagonist differentially inhibit the canonical and non-canonical Wnt pathway, and they can also antagonize each other [56].

7.3.1.2 Propagation of the Wnt Signal

Wnt signaling molecules are secreted proteins able to function at a distance from their source of production [57]. During *Drosophila* development Wg can act as a long-range morphogen directly inducing transcriptional targets in a concentration-dependent manner [58, 59]. However, whether these molecules are classical morphogens can be questioned particularly in regard to the concentration-dependent responses they induce (for review see [60]). In *Drosophila*, Wnt are proposed to act as long-range signals (20–30 cell diameters away from their site of synthesis [59]). In vertebrates, however, Wnt factors have been shown to bind with high affinity to proteoglycans of the extracellular matrix thereby restricting the likelihood of their diffusion [61].

The mechanism used by Wg to reach its target is still unclear. So far wingless has been proposed to spread by restricted diffusion, cell delivery and planar transcytosis (reviewed [62]). First, Wg-carrying vesicles could be inherited during proliferation and transported with the moving cell [63]. Second, Wg-carrying vesicles could be actively transported through cells by planar transcytosis [64]. To support the previous hypotheses, vesicles containing the Wg protein have been found in Wg-responsive cells in *Drosophila* embryonic imaginal disks. The restricted diffusion model explains the formation of the unstable Wg gradient on the basolateral surface of the wing imaginal disk epithelium [65]. The restricted diffusion model suggests that spreading of Wg is constrained by interactions with cell surface or extracellular matrix. Dynamin-dependent endocytosis has also been involved in the secretion of Wg, but not in the spreading of Wg into the extracellular space. Endocytosis seems to contribute to the shaping of the gradient by removing extracellular Wg [65]. The Wg morphogen gradient forms by rapid movement of ligand through the extracellular space, and depends on continuous secretion and rapid turnover.

7.3.1.3 Expression and Function during Neurogenesis

mRNA expression of six Wnt (Wnt1–6) homologs has been analyzed in detail in embryonic and adult mouse tissues [66]. The most prominent member of the family, Wnt1, is expressed along the entire antero-posterior axis of the neural tube in the dorsal midline, and in a belt-like fashion at the midbrain-hindbrain boundary. Wnt7a is expressed in the germinal layer of the embryonic mouse cortex [67], as are several frizzled receptors [68]. Wnt7b is expressed in the cortical plate, particularly by deep layer neurons [69, 70].

Analyses of inactivated mutants for one or more genes of the Wnt signaling pathway tend to corroborate the idea that the Wnt pathway acts in the maintenance of cellular proliferation through its mitogenic action. For example, in Wnt1 mutant mice the midbrain is deleted [71, 72], as is the hippocampus in Wnt3a mutants [52] and LEF1 mutants [73]. The Wnt1/Wnt3a double mutants show a reduction of the caudal diencephalon, the rostral hindbrain and the cranial and spinal ganglia [74]. In the case of mice lacking Wnt3a, caudo-medial cortical progenitors appear to be specified normally, but then they under-proliferate. By mid-gestation, the hippocampus also presents a strong phenotype, being either absent or constituted by a few residual cells [52]. These findings suggest that Wnt1 and Wnt3a play broad, semi-redundant roles in growth control of the neural progenitors rather than simply specifying regional cell fates as it is broadly documented [75]. Consistent with this view, ectopic expression of Wnt1 in transgenic mice causes an overgrowth of the neural tube without altering the primary patterning of cell identities along the dorso-ventral axis [76].

Wnt factors not only act during brain development, but also during spinal cord neurogenesis. In the spinal cord, neurogenesis proceeds in a ventro-dorsal wave, opposed to the dorso-ventral mitogenic gradient. Different Wnt ligands are expressed in the spinal cord. They can be classified depending on their expression and mitogenic activity: Wnt1 and Wnt3a, which are restricted to the dorsal midline of the spinal cord, have a mitogenic activity, while more broadly expressed Wnt factors (Wnt3, Wnt4, Wnt7a, Wnt7b) do not. The mitogenic Wnt factors form a dorsal to ventral concentration gradient that correlates with the growth gradient established in the neural tube as it grows; the proliferation rate is highest dorsally while the differentiation rate is highest ventrally [53]. In order to describe the morphogenesis of the spinal cord, a "mitogen gradient model" has been proposed. The predictions from this model are in good correlation with results from mutant analyses. Loss of function of mitogenic Wnt factors like the double mutant Wnt1/Wnt3a leads to a marked reduction in the number of spinal cord neural precursors, with the phenotype being more pronounced in the dorsal part of the spinal cord [74]. Mouse mutants of non-mitogenic Wnts (inactivation of Wnt4 or Wnt7a/Wnt7b) do not show such phenotypes as their neural tubes grow normally [53]. Wnt3 mutant mice die at gastrulation, therefore, the possible effects on spinal cord neurogenesis of these embryos cannot be analyzed [77].

7.3.1.4 Effectors of the Wnt Pathway

As described above, Wnt pathways converge on the β-catenin/LEF-TCF transcriptional complex. Some of the target genes have been characterized and their deregulation confirms that they act in the Wnt pathway. Consistent with the Wnt mitogenic effect, the constitutive activation of β-catenin in mouse neural precursors shows that they re-enter the cell cycle rather than differentiate [78–80]. The expression of stabilized β-catenin in embryonic or adult mice leads to an expansion of the progenitor pool, a horizontal enlargement of the VZ and a massive folding of the cortex. This dramatic disorganization of the neocortical layering occurs despite the presence of increased apoptosis [78, 79].

Emx2, a homeobox gene, is regulated in the forebrain through LEF/TCF enhancer sites, strongly suggesting that it is a target gene of the Wnt canonical pathway [81]. Interestingly, *Emx2* has been shown to control proliferation and promote symmetric cell divisions and multipotency of cortical progenitors [82, 83]. The overgrowth induced by constitutive β-catenin expression also suggests that progenitor cells are biased toward a symmetric proliferative division, indicating that the activity of β-catenin could be regulated through the regulation of *Emx2*. The stimulation of proliferation induced by β-catenin could also be more direct. In fact, the canonical Wnt signaling pathway has also been shown to regulate some cell cycle regulators. More precisely, in spinal neural progenitors, the β-catenin/TCF complex can promote cell cycle progression and repress cell cycle exit through the transcriptional activation of *cyclin D1* and *cyclin D2* [53].

In addition to its function as a key mediator of the intracellular Wnt pathway, β-catenin interacts with cadherins and cytoskeletal components and is therefore involved in cell adhesion, cell polarity [84, 85] and in mechanisms determining the symmetry of cell division [86]. Mutations or immunogenic inactivation of zebrafish N-cadherins, which bind β-catenin, disturb the integrity of the neuroepithelial VZ [87, 88]. Along these lines, the conditional inactivation of β-catenin in mouse from E10.5 onwards, triggers a dramatic disorganization of the telencephalic neuroepithelium. This defect is also associated with disruption of interkinetic nuclear migration, loss of adherens junctions, block of neuronal radial migration and a decrease in cell proliferation after E15.5 [89]. In newborns, a premature disassembly of the radial glial scaffold and an increased number of astrocytes are also detected in the cortex [89].

7.3.1.5 Cross-talk between Signaling Pathways

Neurogenesis in the cortical region of mammals requires a first step of progenitor maturation, which is characterized by the transition of progenitors from the VZ to the subventricular zone (SVZ), where they continue to divide before differentiating. Wnt ligands, in addition to their mitogenic activity, are involved in this maturation step in a common pathway involving FGF2, Shh and BMP4 [90].

The canonical Wnt pathway may cross-talk with the transforming growth factor β (TGF-β)/bone morphogenetic protein (BMP) signaling or other cytokine signaling through the mitogen-activated protein kinases (MAPK) pathway [91]. In fact, it seems that some of the MAPK can counteract Wnt signaling through phosphorylation of the β-catenin/LEF-TCF complex [91]. A similar negative regulation between the Wnt and TGF-β/BMP pathway has also been described in *Drosophila*. In this case, decapentaplegic (BMP homolog) triggers the repression of *Wg* by inactivation of Hh [92].

7.3.2
Hedgehog Factors

The Sonic Hedgehog (Shh) signaling pathway is highly conserved and involved in various developmental processes. The most studied function of this molecule concerns the establishment of the dorso-ventral patterning in the CNS. In patterning, Shh acts as a classical morphogen; different cellular fates can be induced by Shh depending on the concentration of Shh received which is thus dependent on the distance of the Shh-sensitive cell from the source of Shh. Besides its role in patterning, Shh also plays a role in proliferation of neural precursors and axonal growth.

7.3.2.1 Shh Pathway

Shh is a member of the Hh family of signaling molecules (reviewed in [93]). Shh is a secreted molecule that is also found at the cell surface. It acts as a morphogen, and triggers different cell fates in a concentration-dependent manner. The protein follows different maturation steps during its synthesis. First, an autocatalytic cleavage releases the N-terminal active peptide that is then modified by addition of a cholesterol molecule in the C-terminus and a palmitoyl group in the N-terminus [94]. Different hypotheses have already been proposed to explain the mode of diffusion of Shh, however, it is still unclear how it works as a long-range signal [95]. It has recently been shown that the megalin endocytic-receptor can also bind Shh suggesting that endocytosis could be part of the mechanism of Shh diffusion in tissue [96, 97].

Shh binds its receptor Patched (12–pass transmembrane protein), which releases the Smoothened seven-pass transmembrane protein and triggers the activation of the Gli zinc finger transcription factors [94].

7.3.2.2 Expression and Function during Neurogenesis

Shh is produced by two ventral midline signaling centers: the notochord, the axial mesoderm underlying the neural tube, and the floor plate, a specialized population of cells at the ventral midline of the neural tube [98]. As development proceeds, Shh appears in more rostral locations: in the zona limitans interthalamica, and in the medial ganglionic eminence in the basal telencephalon. During late development and in adulthood Shh is found in the cerebellar cortex and the optic tectum [99, 100]. Shh is also expressed by differentiated cells of the mouse cortex in a layer-specific manner [101]. The Gli transcription factors are also expressed in the CNS; they are notably found in progenitor populations of the cortex and of the spinal cord [101, 102].

The cerebellar cortex originates from two distinct germinal layers, a typical VZ and an external germinal layer. The latter exists only transiently and contains granule cell progenitors. After clonal expansion, the granule cells exit the cell cycle and migrate through the Purkinje cell layer to their final location. Shh is expressed in migrating and settled Purkinje neurons and acts as a potent mitogen to expand the granule cell progenitor pool [103–105]. Granule cell maturation and migration therefore occur in a Shh-rich environment, which is in contradiction to Shh being a mitogenic factor for the granule cells. In this context, modulation of the Shh response seems to be dependent on the extracellular matrix. Laminin glycoproteins present in the proliferative region stimulate Shh-induced proliferation, while vitronectins, encountered when the cells migrate, down-regulate the mitogenic response to Shh [106, 107]. The vitronectin effect could be mediated by its ability to phosphorylate the cAMP responsive element binding protein (CREB) that is alone able to trigger the differentiation of the granule cells despite the presence of Shh [106].

Manipulations of Shh expression have been used to define the function of the pathway during neurogenesis and have shown that Shh activity can regulate patterning, proliferation and fate determination. Ectopic Shh expression in the mouse dorsal spinal cord increases the proliferation rate of precursors at early stages [108]. In the same study, Shh-responsive cells were found to be post-mitotic later in development, suggesting that the mitogenic competence of Shh varies with time [108]. Additionally, and still in line with the mitogenic effect of Shh, misexpression of Shh in the embryonic telencephalon leads not only to a ventralization of the tissue [109] and the appearance of supernumerary oligodendrocytes [110] but also to an abnormal proliferation of the neural precursors and pronounced hypertrophy of the telencephalic region [101, 111]. In the postnatal telencephalon, Shh signaling both promotes proliferation and maintains the neural progenitor state [112]. In adult brain regions, it has been suggested that it is involved in maintaining stem cell niches [112].

In general agreement with the gain-of-function phenotypes, the Shh null mutants have defects in dorso-ventral patterning, ventral fate specification and cell proliferation. In these mutants the telencephalon is dysmorphic, strongly reduced in size (up to 90 %) and appears as a single fused vesicle [113, 114]. Oligodendrocyte differentiation markers are also missing in Shh mutant mice, in accordance with its function in the specification of ventral identity [115].

The diverse results presented so far show that Shh action is crucial for the building of the vertebrate brain through the regulation of stem cell numbers, the control of precursor proliferation and the organization of cellular identity [101, 103–105, 116]. Overall or local changes in Shh levels or its reception, may have contributed to the evolution of sizes and shapes of the brain such as expansion of the primate neocortex, or the tectum in birds, and the cerebellum of electrosensitive fish. The price for such plasticity may be tumorigenesis. Many cancers arise from constitutive Shh signaling in various tissues, including the brain (reviewed in [117, 118]).

7.3.2.3 Eye Differentiation as a Model for Shh Activity

After formation of the optic cup, multipotent retinal precursors give rise to all major cell types in the retina [119]. Neuronal differentiation is usually initiated in the central retina and subsequently spreads into the periphery [120]. Shh, the related tiggywinkle and desert hh, the Shh receptors, Patched 1 and 2 are expressed in differentiating eye cells [121, 122]. The production of Shh in the ventral diencephalic midline accounts for the early patterning in vertebrates whereas endogenously expressed members of the Shh signaling pathway control both proliferation and differentiation of retinal precursor cells [121–123].

In zebrafish, neuronal differentiation is initiated in the ventro-nasal rather than the central retina [124–127]. Shh secreted by the differentiated retinal ganglion cells drives the wave of neurogenesis from the centre to the periphery of the retina. Propagation of the neurogenic wave involves the induction of its own expression in the uncommitted cells and the activation of the Ras-MAP kinase pathway [123].

A neurogenic wave sweeps across the *Drosophila* eye imaginal disk and generates ommatidia behind the progressing morphogenetic furrow in a manner similar to that which occurs in vertebrates [128, 129]. Hh initiates neurogenesis in adjacent undifferentiated cells through induction of the expression of *atonal*, a proneural bHLH transcription factor. Newly differentiated neurons in turn secrete Hh which initiates further neuronal differentiation and progression of the morphogenetic furrow [130].

7.3.2.4 Cross-talk between Signaling Pathways

The exposure of cells to various signaling factors can be either sequential or concomitant. An example of the first concerns the generation of different cell types through the sequential action of Shh and BMP. During development of the cerebral cortex, gamma-aminobutyric acid (GABA)ergic neurons and oligodendrocytes are generated from a common neural stem cell population located in the VZ of the ventral forebrain. Most of these reach their final position after tangential migration. During their maturation, these progenitors therefore experience different signaling cues dependent upon their position. First, ventral Shh restricts the progenitors to GABAergic neuronal and oligodendrocytic programs by the induction of bHLH transcription factors like Olig2 and Mash1 [131]. Subsequent exposure to different levels of BMP determines the neuronal versus glial choice. Neuronal differentiation is implemented by activation of the BMP pathway, while the presence of a BMP antagonist (noggin) allows oligodendrocyte differentiation [132].

In the mouse neocortex, Shh and EGF signaling may synergize through the regulation of the EGF receptor (EGFR) to control precursor proliferation [133]. A negative interaction between the Shh pathway and BMP pathways also seems to occur at the level of Gli activity [134]. It has been shown that the effector of the BMP pathway, the smad protein, can bind the Gli3 thereby providing the means to counteract Shh activity. The Gli proteins also seem to be modulated by some components of the Wnt pathway [135].

7.3.3
Fibroblast Growth Factors

Fibroblast growth factors (FGF) are widely involved in developmental processes and are present in almost all mammalian tissues. In the nervous system, FGF are not only involved in early neurogenesis, but also in axon growth, neuroprotection, and synaptic plasticity.

7.3.3.1 Ligands, Receptors and Expression during Neurogenesis

FGF ligands are monomeric molecules that form multimers under association with heparan sulfate proteoglycans (HS). The cluster of FGF ligands triggers the activation of FGF receptor tyrosine kinases (FGFR) [136, 137]. Activation of the receptor by autophosphorylation in turn allows the transient assembly of multiple intracellular signaling complexes.

Ten out of 23 FGF family members are expressed in the brain, and four receptors (FGFR) have been identified so far. The analysis of the expression pattern of different FGF shows that some of them are expressed in relevant brain regions during neurogenesis suggestive of an active role during this process [138]. Among all the FGF, FGF1 (acidic FGF), FGF2 (basic FGF), FGF6 and FGF7 are expressed in the proliferating mouse neuroepithelium [138, 139]. Moreover, FGFR isoforms are also detected in the VZ at early developmental stages showing that, indeed, FGF signaling can be integrated at the level of neural progenitors [140]. In addition, heparin and HS, which are thought to play a role in the ligand presentation, are secreted by neuroepithelial cells. Looking at various developmental times, different glycosylation states of HS have been detected and have been associated to a modulation of HS affinity for various FGF. FGF1 and FGF2, which are expressed sequentially in the brain when neurogenesis starts, could therefore be modulated by the particular form of HS present in the tissue [138].

7.3.3.2 Function during Neurogenesis

In vitro, FGF2, like EGF is known and widely used to stimulate the mitotic activity of various stem cells dissociated from embryonic and adult mouse brain regions. Because both neurons and oligodendrocytes can be obtained from NSC stimulated by FGF2, it is thought that this action is not neuronal specific [138, 141]. However, it has also been suggested that FGF2 can preferentially stimulate the proliferation of neuronal progenitors from spinal cord progenitor cultures [142]. A proper analysis of the dose-effect of FGF2 has also demonstrated that low doses of FGF2 can contribute to enrich the culture in neuronal-committed precursors [140]. Indeed, cortical progenitor cells in culture are able to differentiate preferentially into neurons in the presence of FGF2. This effect seems to be mediated by the MEK-C/EBP-ERK pathway, as FGF2 is able to stimulate the phosphorylation of MEK kinases in these NSC cultures. Additionally, blocking or enhancing ERK phosphorylation (a MEK substrate) affects neurogenesis promoted by FGF2 on behalf of gliogenesis [143].

A mitogenic effect promoted by FGF has clearly been shown *in vitro*, raising the question of the relevance of this effect *in vivo* and of its ability to increase the pool of neuronal progenitors. The presence of components of the signaling pathway at the appropriate times during development is already providing some evidence for such an activity *in vivo*. Mice lacking *FGF2* show a reduction in neuronal density in the neocortex suggesting a role for FGF2 during corticogenesis [144]. Along the same lines of evidence, injection of anti-FGF2 antibodies decreases the number of neural cells produced while intra-luminal injection of FGF2 generated additional neurons and glial cells [145]. Together, these results show the *in vivo* mitogenic activity of FGF2 and its importance in stimulating the proliferative capacity of neural progenitors and, as a consequence, in regulating the number of neurons produced in the cortical region. A similar effect also seems to be triggered by FGF8 in another specific domain of the brain, again arguing for the involvement of FGF in the regional modulation of cellular growth. FGF8 is expressed at the midbrain-hindbrain junction (MHB) where it acts as a mediator of the organizer activity which regulates the proper patterning of the adjacent midbrain region. Besides its patterning function, FGF8 also seems to regulate the overall growth of the midbrain region [146]. At the MHB, additional FGF are also expressed, FGF18 and FGF17, and inactivation of *Fgf17* leads to midbrain tissue loss due to a decrease in precursor cell proliferation [147].

7.3.3.3 Cross-talk between Signaling Pathways

The FGF pathway interacts with other signaling pathways to integrate intrinsic information or other important cues. Interactions between the Notch pathway and the growth factors FGF1 and FGF2 have been described *in vitro* while examining the mechanisms of action of FGF on neuronal differentiation [148]. E10 neural progenitor cells from the forebrain, sensitive to the mitogen activity of FGF (FGF1, FGF2 or FGF8 for instance) increase *Notch1* expression and decrease *Delta* expression. Moreover, by interfering with Notch activity (blocking or activating) the inhibition of neuronal production normally induced by FGF is perturbed. As both pathways are required to block neurogenesis, this study suggests that FGF activity is mediated via the Notch pathway, at least in part.

Two recent studies from the laboratory of K. Storey provide a nice example of exogenous influences on neurogenesis, and at the same time reveal the opposing interactions of FGF and retinoic acid (RA) pathways on neurogenesis [149, 150]. The spinal cord is surrounded by paraxial mesoderm. As development proceeds, both tissues appear to differentiate concomitantly in a rostro-caudal sequence: the spinal cord generates neurons and the mesoderm forms somites. The authors have shown that the paraxial mesoderm is able to influence neuronal differentiation in the adjacent neuroectoderm. The undifferentiated mesoderm or presomitic mesoderm, expresses FGF while the somites (differentiated mesoderm) express RA. On the other hand, RA receptors are expressed in the spinal cord at the level of the somites in the region where neurogenesis begins. The results suggest that FGF signaling from the presomitic mesoderm (but also from the caudal neural plate) is involved in the maintenance of the undifferentiated state in the neuroectoderm and that a RA signal coming from the somites can promote neuronal differentiation. The effects of RA are partly due to its ability to down-regulate *FGF* expression (*Fgf8*), but a more active process also seems to be involved, as repressing the FGF pathway alone is not sufficient to promote neurogenesis. FGF8 can also regulate RA production thereby providing cross-regulation feedback that may be essential to the coordination of both neurogenesis and somitogenesis.

Altogether, the *in vivo* and *in vitro* results concerning FGF function suggest that it has a role in the regulation of the proliferative pool of progenitors, providing a means by which to regulate the timing of neurogenesis and the number of neural cells produced.

7.3.4
Transforming Growth Factors-α, Neuregulins and Epidermal Growth Factors

Although they are not as widely expressed as the FGF family of ligands and receptors, TGF-α and EGF are also expressed in patterns that are suggestive of a role in regulating the proliferation of precursor populations in the developing and adult nervous system.

7.3.4.1 Ligands, Receptors and Expression during Neurogenesis

TGF-α, EGF and neuregulins signal via a common receptor, EGFR a tyrosine kinase receptor encoded by the *ErbB* gene. The proteins alternately referred to as neuregulins, neu differentiation factors (NDF), glial growth factor (GGF), and acetylcholine receptor-inducing activity (ARIA) are encoded by a single differentially-spliced gene. Neuregulin receptors are encoded by the *ErbB–2–4* genes (reviewed in [151]). EGF, like FGF, is a monomeric ligand that binds to EGFR and thereby triggers receptor dimerization and activation.

TGF-α is expressed in the proliferating cells of the developing rat basal ganglia by E13, in the germinal zone of the midbrain by E15, and in the VZ of the medial ganglionic eminence by E17. EGFR mRNA is found in the germinal zone of the midbrain and the external granule layer of the cerebellum by E15 [152]. Postnatally, EGFR continues to be expressed in regions undergoing active neurogenesis including the cerebellar granule layer and the SVZ [153]. EGFR is also expressed in the granule layer and in proliferating cells of the dentate gyrus as observed by immunohistochemistry [154].

7.3.4.2 Function during Neurogenesis

Much of the evidence from *in vitro* studies demonstrates that EGFR ligands regulate proliferation of distinct precursor populations. For example, precursors from fetal rodent striatum that are expanded using EGF as a mitogen stay multipotent and can give rise to the three major cell types of the CNS: neurons, astrocytes, and oligodendrocytes [155, 156]. Similarly, stem cells isolated from the adult SVZ proliferate in response to EGF [157–159]. Transitory amplification of the cells of the adult SVZ shows that these cells retain stem cell competence under the influence of EGF signaling [160]. The same mitogenic effect is also observed using TGF-α. For example, retinal progenitor cells from embryonic and postnatal rats maintained as explants or in monolayer culture proliferate in response to TGF-α dependent on the maturation stage of the cells [159, 161, 162]. Finally, EGF is a mitogen for dissociated precursors from postnatal rat olfactory epithelium [163, 164].

In contrast to the proliferative effects of FGF2 on early VZ progenitors, EGF is a potent mitogen for the late multipotent progenitors of the embryonic and adult SVZ [156, 165]. The transition from VZ to SVZ in the mouse cerebral cortex is correlated with the upregulation of EGFR [152, 153], which confers mitotic responsiveness to EGF family ligands [166]. The EGF-responsive population represents a subset of SVZ cells and is itself heterogeneous, including stem cells and more restricted precursors. Maturation to an EGF-responsive state requires positive and negative regulators such as the BMP or FGF2 [167]. In addition, TGF-α also seems to have a mitogenic function. Adult mice with a targeted deletion in the *TGF-α* gene show diminished proliferation of precursors within the SVZ, consistent with a proliferative function during neurogenesis [168]. Moreover, mice lacking functional EGFR have defects in cortical neurogenesis, which may be associated with the role of EGFR ligands in proliferation, differentiation, migration or survival of neural precursor populations [169].

Besides their role in proliferation, EGFR ligands appear to play a role in regulating differentiation of precursor populations. Rat retinal progenitor cells expressing exogenous EGFR *in vivo* following infection with a retrovirus encoding the human EGFR, differentiate preferentially into Mueller glial cells [162]. In explant cultures, this glial differentiation occurs at the expense of rod photoreceptor cell differentiation, suggesting that activation of EGFR in retinal precursors regulates such lineage decisions. Long-term administration of EGF to the lateral ventricles of adult rats does not appear to induce the generation of dentate gyrus or olfactory bulb granule neurons, and instead induces generation of astrocytes within the SVZ [159]. Evidence also suggests roles for other members of the EGF family in regulating lineage decisions during neural development. Targeted deletion of neuregulin [170–172] or of the genes encoding ErbB2 [173], ErbB3 [174], or ErbB4 [175] resulted in mice having profound defects in CNS or PNS populations of neurons or glia. Among the most prominent defects described in these studies, the dramatic reduction in Schwann cells in mice lacking ErbB3 indicates a role for neuregulin signaling in the generation of this lineage from the neural crest.

The different ligands signaling through the EGFR seem to have different roles in neurogenesis. On one hand they were shown to stimulate proliferation and used as mitogen factors, on the other they seem to be involved in the gliogenic lineage decision. These two aspects of EGF signaling which are in apparent contradiction may be reconciled by considering the chronology of neuronal and glial differentiation.

7.3.5
Transforming Growth Factors-β

Members of this family play critical roles in regulating developmental processes, so it is not surprising that they function during the development of the nervous system. However, identifying their precise roles in regulating the generation of neuronal populations has been a difficult problem.

7.3.5.1 Ligands, Receptors and Expression during Neurogenesis

The TGF-β superfamily of ligands is extremely large, with over 40 members identified in different organisms from *Drosophila* to mammals. The superfamily includes TGF-β members, activins and BMP. TGF-β members are secreted peptide growth factors forming dimers. They exert their effects through a class of heterodimeric receptors with serine-threonine protein kinase activity [176]. Once activated, the receptors phosphorylate the Smad proteins (Smad2 and –3 are phosphorylated by TGF-β and activin family members while Smad1, –5 and –8 are phosphorylated by BMP). This triggers the formation of a Smad complex, which enters the nucleus, recruits cofactors and regulates target genes.

TGF-β1, -β2, and –β3 are expressed in the developing rodent brain and spinal cord [177–180], but, in contrast to the expression patterns of TGF-α and FGF family members, they are highly expressed in regions of neuronal differentiation and not in proliferative zones. In the postnatal rat brain, TGF-β2 is also expressed in the hippocampus, dentate gyrus, and cerebellum [181].

7.3.5.2 Function in Neurogenesis

The expression pattern of TGF-β argues against a general role in early progenitor proliferation. However, some TGF-β members have been shown to regulate proliferation in certain neural populations. TGF-β3 is mitogenic for rat retinal precursors *in vitro* and enhances the mitogenic effects of EGF and acidic FGF [182]. In the same culture, the presence of TGF-β3 increases the number of retinal neurons, suggesting that it acts on the precursors of retinal neurons [182]. This latest result and others are suggestive of a role for TGF-βs in the commitment of the progenitor or in the promotion of differentiation. For example, despite the presence of the mitogenic EGF, TGF-β2 can induce the generation of olfactory neurons *in vitro* from olfactory epithelium cultures [163]. In addition, TGF-β2 inhibits the proliferation of cerebellar precursors [181, 183], although this effect can be modulated with the culture conditions [183]. TGF-β is also an anti-proliferative signal for pluripotent neural crest cells and for committed melanogenic cells; the TGF-β-mediated anti-proliferative activity dominates over the FGF–2/neurotrophin-mediated mitogenic signal and enhances sensory and adrenergic neurogenesis [184].

7.3.5.3 Cross-talk between Signaling Pathways

Neurogenesis continues throughout adult life in the mammalian olfactory epithelium. This is a very dynamic process involving proliferation, differentiation and cell death. It is highly likely that not only different autocrine, but also paracrine signals, are responsible for its regulation.

Numerous *in vitro* studies concerning the olfactory receptor neurons (ORN) suggest that TGF-β promotes the maturation and/or differentiation of olfactory progenitors. However, the physiological relevance of these effects as well as the exact mechanisms of action can be questioned. Nonetheless, there is emerging evidence that FGF2, TGFβ–2 and platelet-derived growth factor (PDGF) act sequentially on precursor cells and immature neurons during adult olfactory epithelium neurogenesis [185].

Both mature and immature ORN express the TGF-β type II receptor (TGFβ-RII), suggesting that these cells could effectively respond to a TGF-β signal [186]. In the olfactory epithelium of TGF-α-overexpressing transgenic mice, a reduction in the terminal differentiation of ORN is observed [186]. This differentiation defect is associated with a reduction of TGFβ-RII protein levels. These results indicate that interactions between TGF-α and TGF-β signaling pathways are responsible for the correct differentiation of ORN *in vivo* [186].

7.3.5.4 BMP Family

In the nervous system, BMP activity is well characterized in the establishment of dorso-ventral patterning [187, 188]. Much evidence now indicates that in addition to specifying regional patterning within the developing neural tube, members of the BMP family members can also regulate neuronal and glial differentiation.

Expression and function during neurogenesis

Close to the spinal cord, BMP are produced by the epithelial ectoderm overlying the neural tube and also in the dorsal-most part of the spinal cord itself where they play a role in the generation of dorsal cell types [189]. Consistent with this hypothesis, exposure of neural tube explants to BMP4, –5, –7 or related TGF-β members (Dsl1, Activin A or Activin B) induced the generation of spinal cord interneurons [189]. The BMP factors also seem to be important cues which bias progenitor identity toward other specific neuronal fates. For example, BMP9 is a major differentiating factor for cholinergic CNS neurons [190].

BMP can also trigger neuronal differentiation of neocortical precursors from the VZ in different culture systems [132, 191]. This effect on progenitor cells is highly stage and dose dependent. At early stages in the mouse (E13) BMP inhibit proliferation and promote cell death. Later on (E16), they induce neuronal or astroglial differentiation at low doses, but promote cell death at high concentrations [132, 192]. During perinatal development BMP signaling enhances astrogliogenesis and blocks oligodendrocyte differentiation [132]. Repression of BMP activity using its potent natural inhibitor Noggin, which acts by high affinity binding to BMP and prevents BMP binding to cell surface receptors, confirms the activity of BMP in the differentiation of neocortical neurons. Accordingly, Noggin inhibits neuronal differentiation triggered by BMP [193]. Interestingly, Noggin protein expression is detected from E15 in the developing cortex, suggesting a balanced regulation of neocortical neurogenesis mediated by the interaction of Noggin and BMP *in vivo* [193]. In the adult SVZ the balance between Noggin and BMP also seems to regulate neuronal-glial production but in the opposite manner, suggesting that various progenitor populations differ in their sensitivity to the BMP-Noggin balance [194]. In this case, Noggin is expressed by ependymal cells adjacent to the SVZ, whereas BMP and their receptors are expressed by SVZ cells. BMP signaling enhances glial differentiation and the counteracting Noggin promotes neuronal differentiation by inhibiting BMP activity [194].

At late developmental stages, and in rostral regions, BMP activity is involved in the generation of astrocytes [188]. Isolated mouse cerebellar granule cell precursors from the outermost proliferative zone of the external germinal layer can differentiate into astroglial cells when exposed to BMP [195]. The astroglial differentiation induced by BMP appears at the expense of oligodendrogenesis and neurogenesis and BMP are in fact potent inhibitors of oligodendrocyte specification, as has been shown in the chick spinal cord [196]. Exposure of proliferating precursors isolated from fetal striatal SVZ to BMP also induces their differentiation into astrocytes and decreases the proportion of cells differentiating into oligodendrocytes or neurons [197]. Along the same lines of evidence, BMP2 exposure of telencephalic neural progenitors in culture also promotes astrocytic differentiation at the expense of neuronal differentiation [198].

In vivo, overexpression of BMP4 in transgenic mice directs progenitor cells to commit to the astrocytic rather than to the oligodendrocyte lineage [199]. In this study, differentiation of radial glial cells into astrocytes was accelerated, suggesting that radial glial cells are a source of at least some of the supernumerary astrocytes.

The early requirement of BMP signaling during embryogenesis has so far impaired the establishment of direct genetic evidence for BMP function *in vivo*. BMP2 and BMP4 mutant mice die before the majority of neural development has occurred [200]. Mice with mutations affecting either BMPRIA or BMPRII also arrest early in development which also prevents any conclusions being drawn about the function of the BMP pathway during neurogenesis [201, 202].

Effectors of the BMP pathway

It has been shown that two genes induced by BMP could mediate the apoptosis observed in response to BMP. These two putative downstream mediators are the transcription factor msx2 and the cell cycle regulator p21 (CIP1/WAF1). The inhibition of their induction blocks BMP-induced apoptosis, however, they are not sufficient to induce apoptosis on their own [203].

The role of BMP signaling in the choice of astrocytic cell fate has been associated with the upregulation of some negative HLH factors: Id1, Id3, and Hes5. The results show that HLH proteins could mediate the BMP2 dependence of the neurogenic fate of these cells [198].

Cross-talk between signaling pathways

In addition to the direct binding of noggin, chordin, and follistatin (BMP antagonists) that modulates BMP activity, other extracellular factors can also influence the BMP pathway by converging and interfering with its downstream effectors. EGF and FGF for instance, modulate the BMP pathway by altering Smad protein activity, a major effector of the BMP pathway, thereby achieving differential activities [204–206]. As already mentioned BMP also exerts opposing action to Wnt activity and to Shh in the spinal cord.

TGF-β superfamily members play various roles during neural differentiation. The effects that are generated in their presence seem to be highly dependent on the environment and, therefore, highly modulated spatially and temporally. Contrary to the factors described so far, they do not have mitogenic activity. They are mostly involved in cell-fate decisions and may act as instructive or permissive signals to bias the cellular identity adopted by progenitors.

7.3.6
Other Factors

Other signaling molecules and growth factors present interesting expression patterns and activity, which could be associated with their putative role in establishing the neurogenesis gradient during development. Although still fragmentary, the available information regarding the effects of such factors will be mentioned in this section.

7.3.6.1 Neurotrophins

Neurotrophins have been involved in neuronal maturation, mostly as survival factors. This activity is described in the neurotrophic hypothesis, which suggests that neuronal survival is tightly linked to the limited amount of neurotrophic factors that the neuron encounters when the axon reaches its target [207]. This late developmental role of neurotrophins in neuronal maturation supplements their earlier functions during neurogenesis, which are of relevance to this chapter.

Ligands, receptors and expression during neurogenesis

Brain-derived neurotrophic factor (BDNF), the nerve growth factors (NGF) and neurotrophins 3 and 4/5 (NT3/4), commonly known as neurotrophins are secreted ligands that bind the receptor tyrosine kinases of high affinity: TrkA (NGF receptor), TrkB (BDNF and NT4/5 receptor) and TrkC (NT3 receptor). Members of the neurotrophin family also bind to the low affinity receptor $p75^{NTR}$.

Despite having a late developmental role in neuronal survival and axonal pathfinding, the neurotrophins could also be involved in earlier processes of neuronal differentiation. This is notably suggested by the expression pattern of these factors, as well as their receptors, at early developmental stages. For example, NT3 and its receptor are detected in the developing cortex when neurogenesis is initiated [208]. In addition, after plating out mouse cortical stem cells, the presence of FGF2, BDNF, NT3 and the receptors (TrkB and TrkC) is detected in the culture [209].

Function in neurogenesis

While in the PNS NT3 is a powerful mitogen, no such evidence has been found for a similar role in the CNS. Instead, *in vitro* experiments suggest that NT3 is able to promote neuronal differentiation [141]. Indeed, while FGF2 treatment of cortical stem cell cultures stimulates progenitor proliferation, the addition of anti-NT3 antibody reduces the number of neurons generated without directly affecting proliferation. Other pieces of evidence suggest that *in vitro*, NT3 and BDNF can induce neuronal differentiation of hippocampal cells in culture [210]. In another set of experiments, NGF in combination with FGF2 has been shown to be involved in the activation of proliferation of precursor cells derived from the rat striatum [211].

In vivo analysis of the expression of BDNF and its receptor TrkB suggests that they play a role in optic differentiation. Both messengers are found to be highly expressed in the retinal neuroepithelium, and disruption of BDNF signaling by dominant-negative receptors significantly blocks the normal differentiation of neurons. Two hypotheses have been developed: BDNF may have a specific survival function for the differentiated neurons, or it may be involved more directly in the progress of differentiation [212].

Effectors of the neurotrophin pathway

Neurotrophins are able to stimulate a number of well-characterized intracellular pathways likely to transduce the signal to numerous cellular functions [213]. For example, upon *in vitro* stimulation with BDNF, NT3 or NT4, phosphorylation of MAP kinases is detected, indicating an effective activation of the neurotrophin receptor [209].

At least two studies suggest that the neurotrophin pathway is connected to regulation of proneural activity. The first observation concerns the regulation of Hes1 by NGF in PC12 cells. When treated with NGF, PC12 cells express neuronal markers and start to extend neurites, and a similar effect has been observed when Hes1 activity was blocked using a dominant-negative form [214]. In order to find the link between the two phenomena, the authors observed that exposing the cells to NGF leads to phosphorylation of the DNA-binding site of Hes1, thereby rendering Hes1 inactive. Moreover, overexpression of Hes1 blocks NGF activity, suggesting that neuronal differentiation induced by NGF requires the inhibition of Hes1 activity [214]. The second observation concerns the induction of the proneural genes *Mash1* and *Math1* in NSC culture stimulated with NGF, BDNF or NT3 [215]. The combination of a mitotic signal (FGF2) with neurotrophins can induce *Math1* or *Mash1* expression but the cells could not differentiate until the mitogenic factor was removed [215], indicating that the sequence of action of different factors could be important for the proper regulation of neurogenesis.

7.3.6.2 Neurokines: LIF, CNTF

Ciliary neurotrophic factor (CNTF) and Leukocyte Inhibitory Factor (LIF) are two cytokines that act via the glycoprotein 130–linked receptor, LIFRβ and CNTFRα, the latter being specific for CNTF. They have been shown to have pleiotropic actions on different cell types. In the developing CNS, they promote differentiation or survival of astrocytes, oligodendrocytes or neurons but they also act earlier during development to stimulate the renewal of the NSC. One of the first indications of such a role is suggested by the early expression of CNTFRα in the proliferative region of the telencephalon at E14 [216].

In vitro, CNTF/LIF signaling through LIFR activation has been involved in the maintenance of NSC proliferation emanating from embryonic and adult regions undergoing neurogenesis [217].

The mechanism mediating the cytokine action on NSC may require Notch signaling. Indeed, addition of CNTF to forebrain NSC cultures can increase Notch1 expression and Notch1 processing [216].

7.3.6.3 PDGF

Concerning PDGF, ligands and receptors are expressed in cortical NSC suggesting that it may be a good candidate for the control of neurogenesis. Moreover, activated receptors are also detected in NSC indicating that signaling is effectively taking place [218]. In addition, exposure of NSC cultures to PDGF for a short time, stimulates neuronal differentiation.

7.3.6.4 GH

GH may also be required as a neurogenic factor. In a study of the intracellular regulator of cytokine signaling SOCS2 (an inhibitor of JAK/STAT signaling), which is also a mediator of GH signaling, effects on GH activity and neuronal development have been correlated [219]. SOCS2 is expressed in neuroepithelial cells, and $Socs2^{-/-}$ mutant brains have fewer neurons than wild-type brains suggesting a requirement for this protein during normal neuronal differentiation. In addition to the brain phenotype, there is a reduction in the number of Ngn1–expressing cells. These effects can also be correlated to the deregulation of GH activity: SOCS2 being normally involved in the inhibition of STAT5 activation triggered by GH.

To conclude, many extrinsic factors are required to fine-tune the development of the nervous system. Most of them act locally providing mitogenic or differentiating cues to neural progenitors, and at the same time they can also convey positional information to bias neural fate. Hence, there is no simple consensus model of the factors responsible for the neurogenic switch.

7.4
Cell Cycle Regulation and Neuronal Fate Determination

Neurogenesis is defined by the acquisition of neuronal features, one of which is the post-mitotic state. The transition from a proliferative state to a quiescent mitotic state is therefore a highly regulated process occurring concomitantly with neurogenesis.

The maintenance of proliferation in the NSC pool is crucial for the establishment of neural diversity. Understanding the fluctuations of the NSC cell cycle therefore provides a basis on which to understand the balance between proliferation and differentiation.

During neurogenesis, the relationship between cell cycle regulation and cell fate determination is controlled by a cohort of extra- and intracellular signals. The balance between proliferation and differentiation appears to be regulated from both sides. First, various factors known to regulate neurogenesis seem to control the cell cycle. Second, the cell cycle parameters themselves influence neural differentiation. Regulation of cell cycle length appears to be part of the mechanism leading to neuronal production.

7.4.1
Cell Cycle of Neuroepithelial Cells

7.4.1.1 Interkinetic Nuclear Migration and Cell Division

Neural progenitor cells are characterized by their arrangement in a single cell layer and their epithelial character. In vertebrates, with regard to the cell cycle, the nucleus of these cells oscillates between the ventricular and the cortical surface (in the apical-basal plan of cell polarity), in a movement called interkinetic nuclear migration. The nucleus divides on the apical side of the neuroepithelium then moves basally during the G1 phase. In the basal region, the S phase occurs and the nucleus returns to its apical position during the G2 phase in order to divide again [220].

In the neuroepithelium three types of cell division have been described with respect to the fate of the daughter cells. Symmetric, proliferative divisions generate two progenitors. Differentiative divisions can be subdivided into symmetric and asymmetric neuron-generating divisions [221–226]. Before neurogenesis, NSC proliferate and increase the pool of NSC by symmetric, proliferative division. At the beginning of neurogenesis, some cells switch to differentiative divisions that generate either a neuron and a NSC, in the case of symmetric neuron-generating division or two neurons, in the case of symmetric neuron-generating division. During the neurogenic process, the number of differentiative divisions increases.

7.4.1.2 Length of Cell Cycle of Neural Stem Cells

Cell cycle length during neurogenesis

The first correlation between a change in cell cycle parameters and mammalian neurogenesis dates from the 1990s when Caviness' group analyzed the cell cycle of neuroepithelial cells using the method of cumulative bromodeoxyuridine (BrdU) labeling. Neuroepithelial cells have been found to increase their average cell cycle length in temporal and spatial correlation with neurogenesis [227, 228]. At the onset of neocortical neurogenesis (E11), when the vast majority of neuroepithelial cells undergo proliferative divisions, the average length of a cell cycle is about 8 h. During the development of the CNS, the mean cell cycle length of neuroepithelial cells increases to reach 18 h at the end of the neurogenic interval (E16), when most of the cells have switched from proliferative to neuron-generating divisions [227]. This increase in cell cycle length is selectively due to a lengthening of the G1 phase of the cell cycle, whereas the length of the other phases stays constant i.e. about 4 and 2 h for the S and G2/M phase, respectively. The length of the G1 phase of neuroepithelial cells quadruplicates: from about 3 h, prior to the onset of neurogenesis, to more than 12 h at the end of the neurogenic interval [227].

The cell cycle of subpopulations of progenitors
By taking a closer look at what happens in the neuroepithelium at a fixed time point, contradictory studies report homogeneous or heterogeneous cell cycle length. The issue of cell cycle homogeneity of NSC in the VZ is therefore quite controversial.

Despite the supposed presence of different populations of progenitors, measures of cell cycle length in the mouse neocortex have suggested that the cell cycle is homogeneous [229]. However, other studies suggest that a difference in cell cycle length could, in fact, exist. For example, differences in BrdU incorporation have been observed between distinct co-existing subpopulations of radial glial cells, suggesting that if not length, at least some parameters of the cell cycle vary between radial glial cell subpopulations [230]. This discrepancy could, in part, be attributed to the lack, in the first study, of direct markers allowing the discrimination of various subpopulations. Analysis of the cell cycle length of proliferative dividing cells and neuron-generating cells in the neuroepithelium has shown that these two co-existing cell populations have different cell cycle lengths [226, 231]. The neuron-generating cells have a significantly longer cell cycle compared to the proliferating neuroepithelial cells.

7.4.2
Cell-fate Determinants Influencing the Cell Cycle

Extrinsic and intrinsic factors involved in the determination and progression of neurogenesis act in part through the control of some important cell cycle regulators (for reviews see [232–234]). Most if not all of these regulators of neurogenesis seem to act on the cell cycle either directly or indirectly (see Sections 7.1 and 7.2).

7.4.2.1 Extrinsic Cell-fate Determinants Regulate the Cell Cycle
Shh, for example, is a well-characterized morphogen, which is involved in neuronal patterning and differentiation. It has also been shown to act as a potent mitogen affecting cell cycle progression. Using the GAL4/UAS system, inducible expression of Shh in mice inhibits the generation of post-mitotic neurons while it increases the proliferation rate of neuroepithelial cells [108]. Similar results have also been obtained in the developing cerebellum. In this case, synthesis and secretion of Shh by Purkinje cells inhibits differentiation of granule cells. Again, inhibition of differentiation has been associated with an increase in cell proliferation [105]. The exact molecular mechanisms by which this cell-fate determinant is able to influence cell cycle progression are not yet clear. Some data suggest, however, that Shh could directly regulate certain cell cycle regulators. Genomic screens have recently shown that activation of the Shh signaling pathway leads to an increase in cell cycle regulators such as cyclinD and N-myc [235].

Another example concerns the Delta-Notch signaling pathway involved in cell-fate determination (see Section 7.1.1.2). In addition to its activity in the switch between proliferation and differentiation of NSC, the Delta-Notch pathway influences cell cycle progression. Activation of the Notch receptor by Delta is alone sufficient to inhibit neuronal differentiation and, conversely, its repression promotes it. Interestingly, upregulation and downregulation of the Notch pathway have been shown to increase and inhibit NSC proliferation respectively [236, 237]. This regulation of cell proliferation seems to take place during the G1–to-S transition by the control of p21 and p27 expression, two G1 cell cycle inhibitors [238].

In vitro analyses of the effect of bFGF and Neurotrophin3 (NT3) shows that the cell cycle length correlates, and may determine, differentiation [239]. Dissociated mouse cortical progenitors submitted to bFGF or NT3 exposure, proliferate and differentiate respectively. This effect was correlated to shortening and increasing the length of G1 phase respectively.

7.4.2.2 Intrinsic Cell-fate Determinants Regulate the Cell Cycle

Some transcription factors have also been involved in the cell-fate decision and cell cycle regulation. The first example concerns proneural genes, which can directly regulate the transcription of some cell cycle inhibitors. Their overexpression in cell culture is effectively associated with an increase in *p27* expression and followed by cell cycle exit [17].

In another example, overexpression of the Phox2b transcription factor in the vertebrate neural tube is both necessary and sufficient to induce neuronal differentiation (probably via the regulation of proneural genes) and, to inhibit cell proliferation [240, 241]. The mechanisms by which Phox2b influences cell cycle progression have not yet been analyzed and, in particular, it is not known if its action occurs during a particular phase of the cell cycle.

Many other examples could be added (see Section 7.1 and [232–234]) either as extracellular factors or intracellular factors influencing the cell cycle. The few studies reported here provide some typical examples of cell cycle regulation. The preceding examples suggest that the coordination of the cell differentiation program and cell cycle is dependent upon cell-fate determinants, which can influence directly or indirectly cell cycle regulators [232, 242]. This view considers the inhibition of cell cycle progression as a consequence of a cell-fate change and many studies seem to support this hypothesis.

In addition to the dominant idea that cell-fate determinants are upstream of cell cycle regulation, emerging data argues that cell cycle parameters by themselves may influence cell fate.

7.4.3
Cell Cycle Regulators Influencing Cell Fate

7.4.3.1 Cell Cycle Regulators

The main group of cell cycle regulators is the cyclin dependent kinases (CDK) family. The activity of various members of the CDK family constitutes the "thermostat" of cell cycle progression. The general principle is that activation of CDK leads to, and is essential for, progression through the cell cycle. An increase in CDK activity leads to an increase in cell proliferation and, conversely, the inactivation of CDK lengthens or even blocks, cell cycle progression.

The activity of CDK is essentially regulated by three mechanisms. First, CDK are activated after binding with their respective cyclin partners whose synthesis and degradation are tightly controlled. They are then activated through phosphorylation of threonine residues by the CDK activating kinases. Finally, the cyclin/CDK complexes are inhibited via interaction with the CDK inhibitors (CKI) such as p16, p21 and p27 (for comprehensive review see [243, 244]).

7.4.3.2 Cell Cycle Regulators Regulate Cell-fate Determination

Manipulating the activity of cell cycle regulators involved in the control of G1 progression, either *in vitro* or *in vivo*, triggers changes not only in cell cycle parameters, but also in the cell fate of neuronal precursors [232–234, 242].

One significant report in this area concerns the proliferation arrest of pheochromocitoma cells (PC12) and the acquisition of a post-mitotic neuronal phenotype under NGF exposure. This change in cell fate has been correlated with inhibition of CDK2 activity. Interestingly, when CDK2 is directly inhibited by antisense probes or by treatment with the specific CDK2 inhibitor olomoucine, PC12 cells acquire a neuronal phenotype independent of NGF stimulation [245]. Another *in vitro* example of the influence of cell cycle regulators on cell fate is the overexpression of p73, a known inhibitor of CDK activity. Similarly to PC12 cells, which can be activated to acquire a neuronal phenotype by NGF, retinoblastoma cells can be induced to differentiate by RA. In retinoblastoma cells treated with RA, neuronal differentiation is associated with upregulation of p73 expression. Overexpression of p73 and the consequent inhibition of cell cycle progression is alone sufficient to induce neuronal differentiation in the absence of RA [246].

The activity of p27 (another CKI) has also been implicated in neural fate determination. In both neuronal and glial progenitor cells, the cytoplasmic concentration of p27 increases over time. The "intrinsic timer" model proposes that gradual accumulation of p27 after each cell cycle constitutes the core of the clock, thereby controlling the timing of differentiation. Reaching the appropriate concentration of p27 would eventually lead to a block in the cell cycle and differentiation of progenitor cells [247–249]. However, in oligodendrocyte differentiation, the effect of p27 on cell-fate determination has been shown to be independent of its ability to control the cell cycle [250]. Moreover, inducible overexpression of p27 in the neuroepithelium of transgenic mouse embryos seems to increase neuronal production without any visible effects on the length of the cell cycle of neuroepithelial cells [251].

In line with these observations, *in vivo* analyses strengthen the role of cell cycle regulators in the differentiation process. Indirect evidence suggests that slowing down the cell cycle induces neuronal differentiation during mouse neurogenesis. Expression of the anti-proliferative gene *TIS21* during the G1 phase [252, 253] correlates with neurogenesis and is confined to a subpopulation of neuron-generating neuroepithelial cells [226, 254]. The presence of TIS21 before the differentiative division suggests indeed that cell cycle inhibition occurs before the progenitor exits the cell cycle. As more direct evidence, the overexpression of PC3 (the rat homolog of TIS21) is alone sufficient to enhance neurogenesis, at the same time inhibiting the extent of neuroepithelial cell proliferation [255]. In agreement with the previous observations, the inhibition of cell cycle progression by olomoucine is alone sufficient to trigger premature neurogenesis in mouse embryos developing in whole embryo culture [256].

As described in this section, the inhibition of cell proliferation may lead to a change in cell fate. Until now it has often been hypothesized that blocking the cell cycle is the cause of differentiation [248]. However, *in vivo* data suggest that lengthening rather than blocking the cell cycle occurs upstream of cell differentiation [231].

7.4.4
Model of Cell Cycle Lengthening

Considering the proliferation to differentiation switch as the cell-fate change of interest, the regulation of cell cycle kinetics has been shown to be essential. The crucial phase of the cell cycle involved is the G1 phase. In all cases reported, shortening of G1 phase has been associated with inhibition of differentiation, while inhibition of G1 progression correlates with stimulation of neurogenesis. Lengthening of the G1 phase, which occurs in neuronal progenitors, is an upstream event preceding neuronal differentiation.

The lengthening model proposes that an increase in length of the critical phase of the cell cycle leads to differentiation. The underlying mechanism postulated by this model is that a critical activity dependent upon time triggers a cell-fate change. To illustrate this model, the regulation of activity of a critical cell-fate determinant may require its accumulation or its degradation. For example, a factor specifically synthesized during the G1 phase and whose activity is dependent upon a certain threshold would therefore be dependent on the length of the G1 phase to be effective. In other words, a short cell cycle would not allow the cell-fate determinant-mediated effects, thus preventing cell-fate change.

With regard to the asymmetric distribution of some cell-fate determinants during mitosis (see Section 7.4), this model would explain how an asymmetric distribution of cell-fate determinants may or may not lead to an asymmetric cell fate of the daughter cells. This would be consistent with the observation that lengthening of the cell cycle occurring at the onset of neurogenesis correlates with asymmetric neuron-generating cell divisions whereas a further lengthening, occurring at later stages of neurogenesis, correlates with symmetric neuron-generating cell divisions [223, 227, 257].

7.5 Neuron-generating Asymmetric Cell Division

7.5.1 Background

Observations of the NSC in mammals have indicated the existence of two kinds of divisions leading to self-renewal: the symmetric, proliferating division generating two NSC and the asymmetric, neuron-generating division giving rise to one neuron and one self-renewing progenitor. This discrimination has been based on the cell fate of the progeny.

The past 20 years have seen the emergence and the characterization of the mechanisms underlying the symmetry versus asymmetry of cell fates. The mechanisms involved seem mostly to be intrinsic to the cell lineage and concern the symmetry of inheritance of fate determinants by the daughter cells. This inheritance is crucial to bias the identity of the receiving cell.

Asymmetric cell division can hence be defined on a cell biological basis or on a cell-fate basis. In the first case, the asymmetry of division can already be identified in the mitotic mother cell while in the second case analysis of the daughter cells is required.

Cell-fate asymmetry has been studied in the nervous system of diverse organisms as an essential mechanism to regulate the generation of neurons from NSC. In *Drosophila*, the mechanisms underlying this fate asymmetry have been extensively analyzed providing the evidence for the determinant-distribution asymmetry model. In mammals, some data has emerged pointing to the existence of similar mechanisms involved in the establishment of cell-fate asymmetry. In parallel, the increasing progress in live-imaging has produced many descriptive studies on cell lineages, and in particular the mammalian neural progenitor lineage, providing a more detailed characterization of cell-fate asymmetry.

7.5.2 Asymmetric Cell Divisions of Neuroblasts in the *Drosophila* CNS

7.5.2.1 Cell Lineage
Effective genetic tools used in *Drosophila* have allowed researchers to analyze the molecular mechanisms of the asymmetric cell division involved in the generation of neurons in the CNS. The molecular machinery was first identified in neurogenesis in *Drosophila* and were later found in other developmental models.

In the *Drosophila* CNS, neurons arise from neuronal precursors called neuroblasts (NB) arranged in a polarized neuroectodermal epithelium. As neurogenesis starts, selected NB leave the epithelium and migrate to a basal position in a process called delamination. Shortly after delamination, NB divide asymmetrically along the apical-basal axis of cell polarity [258]. The daughter cell which is closer to the apical surface remains as a NB, while the other daughter becomes a small ganglion mother cell (GMC). NB division can be considered as a stem cell-like division, as the NB produced continue to divide, while GMC divide only once more to produce a pair of neurons or glial cells [259, 260]. In *Drosophila*, all neuronal and glial lineages arising from NB have been traced and identified [258, 261–263].

7.5.2.2 Apical-Basal NB Polarity

The molecular machinery involved in NB asymmetric division has been analyzed in detail. The maintenance of apical-basal cell polarity and the tight control of spindle orientation during division constitute prerequisites for the asymmetric cell division of NB (for reviews see [264, 265]).

Prior to delamination, NB are parts of the neuroectodermal epithelium, and similar to the neuroectodermal cells, they are characterized by epithelial polarity features. They are notably connected to the adjacent cells by the junctional complex of the zonula adherens. When NB delaminate (in a basal movement), they lose their cell contacts and move inside the embryo. NB translocation is accompanied by a loss of the apical stalk in the neuroectodermal epithelium and by a rounding up of NB after delamination (for reviews see [264, 266]). Despite the striking cell shape remodeling, the apical-basal polarity is maintained during the delamination process.

The maintenance of polarity which constitutes an essential step in asymmetric division, has been attributed to a well-conserved complex of proteins: the PAR–6/PAR–3/aPKC (atypical protein kinase C) complex. This complex was originally characterized in *C. elegans*, in studies investigating the establishment of zygote cell division and polarity. In the zygote, the distribution of the Par/aPKC protein complex defines the anterior domain of the cell cortex, and is indispensable for asymmetric division (for review see [267, 268]). Homologs of these proteins have also been found in *Drosophila*. They are located in the subapical region of the neuroectoderm, in the apical stalk of NB during delamination, and in the apical cell cortex after NB have fully delaminated. As in *C. elegans*, mutant phenotypes of the genes encoding *bazooka* (*Drosophila PAR–3* homolog), *Drosophila aPKC* and *DmPAR–6* lead to a loss of apical-basal polarity in epithelial cells and in NB. The PAR/aPKC complex seems, therefore, to be a key component of cell polarity which has been conserved between species and used to establish the polarity of various cell types (see review [264, 268, 269]).

7.5.2.3 Control of Spindle Orientation

Given that both *Drosophila* neuroectodermal cells which divide symmetrically, and NB which divide asymmetrically, use a related molecular mechanism to maintain their cellular polarity, the distinction between symmetric and asymmetric division should require additional mechanisms. One prominent difference between neuroectodermal and NB cell division is the orientation of their plane of division. Neuroectodermal cell divisions occur in the plane of the neuroectoderm while NB divisions occur along the apical-basal axis. The orientation of the mitotic spindle therefore, constitutes a crucial step in the generation of the asymmetric distribution of factors segregated along the apical-basal axis.

The protein Inscuteable is the major player in this process. Inscuteable is not expressed in neuroectodermal cells, and the protein is first detected in the apical stalk of NB during delamination. In delaminated NB, the localization of Inscuteable to the apical cell cortex is re-established at each cell cycle between late interphase and anaphase [266]. In the absence of the protein, the mitotic spindle in NB fails to rotate [270], and NB divide in random orientations [271]. The connection between apical-basal polarity and mitotic spindle orientation is established via the protein Bazooka which is required for apical localization of Inscuteable [272, 273]. Inscuteable recruits Pins (Partner of Inscuteable) and the associated G protein αi subunit. The apical co-localization of Inscuteable, Pins, and $G\alpha i$ is interdependent and essential for proper asymmetric cell division. So far, however, the mechanism underlying the regulation of spindle orientation by the complex Inscuteable/Pins/$G\alpha i$ is not clear, and no direct interactions have been found with microtubules or the centrosome that would account for the activity of the Inscuteable/Pins/$G\alpha i$ complex [264].

7.5.2.4 Cell-fate Determinants

Cell-fate determinants are mostly proteins, Prospero, Miranda, Staufen, Numb and Partner of Numb (PON) but also mRNA (*Prospero*). These determinants are localized in a basal crescent in NB, and inherited by the GMC. Miranda, Staufen and PON are adaptor proteins required to recruit Prospero (protein and mRNA) and Numb at the basal cell cortex. Miranda binds to Prospero and Staufen [274, 275]. Staufen, an RNA binding protein, segregates Prospero mRNA by binding to its 3' untranslated region [276, 277]. Prospero, a homeobox transcription factor, is required for the transcription of GMC-specific genes but also for the downregulation of NB-specific genes in the GMC [278, 279]. Numb is transported to the GMC during metaphase by the action of PON [280]. Numb acts by repressing Notch activity in the GMC, however, the mechanisms of Numb activity are not yet clear [265, 268, 281].

The tight localization of the protein complexes described so far is crucial for the asymmetric segregation of cell-fate determinants. For example, the apical protein Inscuteable is required for the proper localization of Numb, Prospero, and Miranda. All three proteins still localize asymmetrically without Inscuteable, but their crescents are formed at random positions around the cell cortex and are no longer correlated with the spindle poles [268, 270]. In addition to the apical complexes, the protein products of the tumor suppressor genes *lethal giant larvae* (*lgl*), *discs large* (*dlg*) and *scribble* (*scrib*) are essential for the basal localization of cell-fate determinants [282, 283]. While the distribution of these proteins is not asymmetric, the activity of Lgl is nonetheless restricted to the basal cortex by phosphorylation triggered by the apical DaPKC protein [284].

7.5.3
The Neuronal Cell Lineage in the Mammalian CNS

7.5.3.1 Neuron-generating Divisions and Cell-fate Asymmetry

The pool of neural progenitors that constitutes the wall of the neural tube is exponentially amplified by symmetric, proliferative division before neurogenesis. When neurogenesis starts, the neural progenitors switch from a symmetric to an asymmetric mode of division. After neurogenesis, an identity switch towards glial cell type differentiation takes place. It is assumed that the completion of neural differentiation is characterized by another switch from an asymmetric to a symmetric mode of division giving rise to two differentiated cells. However, this simplistic model has been re-evaluated by the analysis of clonal progeny of mammalian NSC (*in vitro* and *in vivo*). It has been shown that neurons come from two co-existing types of division: asymmetric neuron-generating divisions and symmetric neuron-generating divisions, which generate two neurons [221, 224–226].

In the mammalian nervous system, the fate asymmetry resulting from the asymmetric division of neuronal progenitors has been traditionally deduced from lineage studies. Analyses using single-labeled neuroepithelial cells *in vivo* coupled to characterization of the progeny (morphology and localization) have established the first basis for the progenitor versus neuron fate asymmetry of the daughter cells [224, 285, 286].

Recent progress in imaging and particularly in time-lapse recording systems in intact living tissue, have produced convincing evidence and large amounts of data concerning distinct daughter fates. According to this strategy, time-lapse studies of dividing neuroepithelial cells or radial glia cells (which can be considered as a subpopulation of neuroepithelial cells, see [264, 287]) have been followed in slice cultures of mammalian embryonic brain. Asymmetric morphology, marker localization (as for static analysis), and asymmetric behavior of the daughter cells have been taken as evidence for an asymmetric, neuronal versus progenitor, fate. Kriegstein and colleagues followed GFP-labeled precursor cells and their progeny *in vivo* for up to 3 days. They showed that the radial glial cell is indeed a neuronal precursor. In addition, radial glial cells maintain their pial processes (also called the "basal process") throughout cell division, although it becomes extremely thin at certain time points during mitosis [288]. Similar results have been reported after examination of the morphology and cell cycle of differently labeled radial glial cells (dye labeling from the pial surface in a cortical slice culture) [289]. Up to 140 h of recording of the embryonic radial glial cells have led to the detailed description of patterns of cell division and migration [225].

The neuron-generating division has also been specifically recorded and characterized using GFP knock-in mice. In this study, *GFP* has been inserted into the *TIS21* locus, encoding an anti-proliferative gene, whose mRNA is selectively expressed in neuron-generating cells, but not proliferative neuroepithelial cells nor in neurons [226, 254]. Two kinds of neuron-generating division have been observed [225, 226]: apical divisions resulting in asymmetric neuron-progenitor daughter fate, and basal divisions resulting in symmetric neuron-neuron daughter fate.

The mechanisms that could explain neuronal fate are still not as well understood as in *Drosophila*. The co-existence of two types of neuron-generating progenitors raises the question of the existence of two different mechanisms involved in neuronal production in the mammalian CNS. The basally dividing progenitor generating two neurons is reminiscent of the GMC in *Drosophila* and suggests that similar factors and asymmetric cell division could be involved in its generation. This would also suggest the existence of NB-like neuroepithelial cells, which are, if they exist, indistinguishable from the proliferative progenitors.

The mechanisms of fate asymmetry resulting from the apical neuron-generating division seem to depend, as in *Drosophila*, upon the segregation of progenitor versus neuronal determinants (see the Section 7.5.3.2).

7.5.3.2 Cell Polarity in the Mammalian Neuroepithelium

Asymmetric distribution of some cellular determinants has also been described in mammals. The data collected so far in mammals suggest that, on one hand, some features may be shared with the fly mode of neurogenesis, and on the other hand, distinct mechanisms may be involved.

Like other epithelia, mammalian neuroepithelial cells have an apical-basal polarity. The epithelial characteristics of the neuroepithelial cells change with development. During the neural plate stage, neuroepithelial cells show the typical features of epithelial cells. They are polarized, with distinct apical and basolateral surfaces [290], the polarity is maintained by functional tight junctions [291], and their basal plasma membrane is in contact with a basal lamina [292]. With neural tube closure and the onset of neurogenesis, neuroepithelial cells start to lose some of these epithelial features. They switch from E-cadherin to N-cadherin expression [293], functional tight junctions are lost [291], and the polarized delivery of certain membrane proteins to the apical and basolateral plasma membrane is downregulated [290].

7.5.3.3 Neuron-generating Division and Asymmetric Cell Division in Mammals

It has been proposed that the type of division neuroepithelial cells undergo critically depends on the orientation of the cleavage plane relative to the apical-basal cell axis, as for *Drosophila* NB divisions. Vertical cleavage planes (i.e. parallel to the apical-basal axis) would give rise to symmetric, proliferative divisions while horizontal cleavage planes (i.e. perpendicular to the apical-basal axis) would give rise to asymmetric, neuron-generating divisions [294–296]. However, the proportion of horizontal cleavages observed is not sufficient to account for the neurons produced (particularly during the peak of neurogenesis): there are too few horizontal cleavage planes [83, 297, 298]. It has therefore been suggested that neuron-generating as well as proliferative divisions of the neuroepithelial cells may have vertical cleavage planes.

Given the particularly elongated shape of the neuroepithelial cells, the small apical domain of the cell could easily be asymmetrically distributed with a vertical cleavage plane. Such an unequal partition of the apical plasma membrane to one daughter could lead to an asymmetric distribution of determinants in the small apical region (as previously described in *Drosophila*) [299]. To test this hypothesis, the position of the apical plasma membrane (Fig. 7.4, white box) relative to the cleavage plane (Fig. 7.4, broken line) has been examined in proliferative or neuron-generating divisions using the *TIS21*–GFP knock-in mice to discriminate between the two progenitors [226]. The switch from proliferative to neuron-generating divisions correlates with the symmetric versus asymmetric inheritance of the apical membrane rather than with the vertical versus horizontal rotation of the cleavage plane [300]. Hence, this result suggests that some determinants, being in this case the apical plasma membrane, are asymmetrically distributed and correlated to a specific cell fate in mammalian neuron-generating divisions.

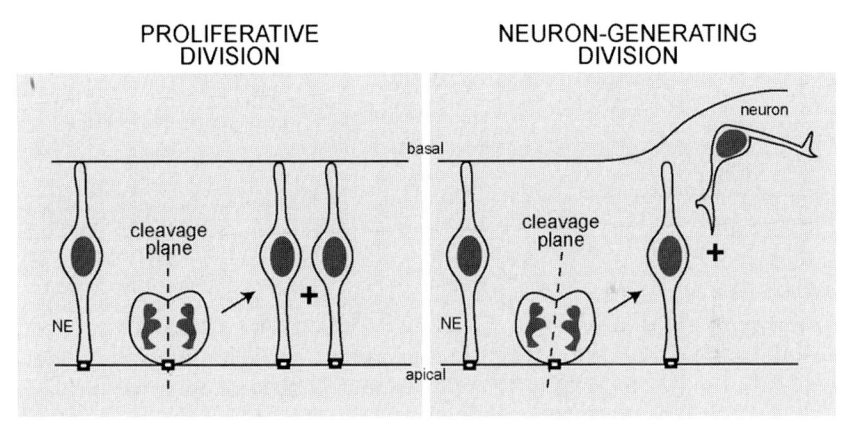

Fig. 7.4
Model of the distribution of apical plasma membranes in proliferative or neuron-generating dividing neuroepithelial (NE) cells in the developing mouse embryo. Orientation of the cleavage plane (broken line) is almost parallel to the apical-basal axis in both cases, but the position of the apical plasma membrane (white box) in relation to the cleavage plane is equal (*left*) or unequal (*right*). In the case of neuron-generating asymmetric division, the daughter cell which inherits the apical plasma membrane will remain a neuroepithelial cell, while the other (lacking the apical plasma membrane) will become a neuron.

7.5.3.4 Asymmetric Distribution of Cell-fate Determinants in Mammals

So far, very little is known about the mechanisms triggering asymmetric division in mammals. Homologs of the Par3/Par6/DaPKC complex do exist and their distribution in the apical region (junctional distribution) has been described [301]. Moreover, the mammalian homolog of *Drosophila* Bazooka (mPAR–3) is distributed equally or unequally in symmetric, proliferative or asymmetric, neuron-generating dividing neuroepithelial cells, respectively [300]. mPAR–3, and likely its binding partners mPar6 and aPKC, could therefore be involved in the control of apical-basal polarity and the symmetry of cell division as in *Drosophila*. One possible cell-fate determinant, mammalian Numb, has been shown to localize to an apical crescent of neuroepithelial cells [302, 303], however, the role of Numb in mammalian neurogenesis still needs to be determined [264, 304].

The comparison between *Drosophila* and mammals is highly suggestive of a strong conservation of the mechanisms involved in neuron generation. Even the apparent contradiction concerning the cleavage plane orientation may be explained by the atypical morphology of the mammalian neuroepithelial cells [300]. Comprehension of the mechanism underlying the production of neurons in either *Drosophila* or mammals represents an important step in the understanding of neurogenesis.

7.6 Conclusions

Spatial and temporal control of neurogenesis needs to be tightly coordinated to permit the proper generation of neurons in the proper place, at the proper time.

Many regulators of neurogenesis are either involved in the promotion of differentiation or in the regulation of proliferation. The neurogenic switch, revealed at the level of the cell cycle division, requires the mobilization of various factors and mechanisms in the neural stem cells. As highlighted in this chapter, some common core mechanisms are involved in neuronal differentiation. Transcriptional regulation, cell cycle control and asymmetric distribution of cell-fate determinants are among these common critical steps involved in the control of neuronal generation. However, a consensus concerning the events upstream of the neurogenic switch has not yet been established.

Extensive cross-regulation between the distinct pathways and mechanisms regulating neurogenesis make it complicated to fully understand the chronology of events triggering neuronal differentiation. Furthermore, the balance between the diverse signals may be as important as their chronology. Finally, local variations in the integration of these signals may provide the flexibility required to build the central nervous system.

References

1 Lee, J.E. Basic helix-loop-helix genes in neural development. *Curr Opin Neurobiol* **1997**; *7*: 13–20.
2 Lewis, J. Notch signalling and the control of cell fate choices in vertebrates. *Semin Cell Dev Biol* **1998**; *9*: 583–589.
3 Artavanis-Tsakonas, S., Rand, M.D., and Lake, R.J. Notch signaling: cell fate control and signal integration in development. *Science* **1999**; *284*: 770–776.
4 Ghysen, A., and Richelle, J. Determination of sensory bristles and pattern formation in Drosophila. II. The achaete-scute locus. *Dev Biol* **1979**; *70*: 438–452.
5 Kageyama, R., and Nakanishi, S. Helix-loop-helix factors in growth and differentiation of the vertebrate nervous system. *Curr Opin Genet Dev* **1997**; *7*: 659–665.
6 Bertrand, N., Castro, D.S., and Guillemot, F. Proneural genes and the specification of neural cell types. *Nat Rev Neurosci* **2002**; *3*: 517–530.
7 Ghysen, A., Dambly-Chaudiere, C., Jan, L.Y., and Jan, Y.N. Cell interactions and gene interactions in peripheral neurogenesis. *Genes Dev* **1993**; *7*: 723–733.
8 Murre, C., McCaw, P.S., Vaessin, H., Caudy, M., Jan, L.Y., Jan, Y.N., Cabrera, C.V., Buskin, J.N., Hauschka, S.D., Lassar, A.B., et al. Interactions between heterologous helix-loop-helix proteins generate complexes that bind specifically to a common DNA sequence. *Cell* **1989**; *58*: 537–544.
9 Cabrera, C.V., and Alonso, M.C. Transcriptional activation by heterodimers of the achaete-scute and daughterless gene products of Drosophila. *EMBO J* **1991**; *10*: 2965–2973.
10 Blader, P., Fischer, N., Gradwohl, G., Guillemont, F., and Strahle, U. The activity of neurogenin1 is controlled by local cues in the zebrafish embryo. *Development* **1997**; *124*: 4557–4569.
11 Turner, D.L., and Weintraub, H. Expression of achaete-scute homolog 3 in Xenopus embryos converts ectodermal cells to a neural fate. *Genes Dev* **1994**; *8*: 1434–1447.
12 Mizuguchi, R., Sugimori, M., Takebayashi, H., Kosako, H., Nagao, M., Yoshida, S., Nabeshima, Y., Shimamura, K., and Nakafuku, M. Combinatorial roles of olig2 and neurogenin2 in the coordinated induction of pan-neuronal and subtype-specific properties of motoneurons. *Neuron* **2001**; *31*: 757–771.
13 Cau, E., Gradwohl, G., Fode, C., and Guillemot, F. Mash1 activates a cascade of bHLH regulators in olfactory neuron progenitors. *Development* **1997**; *124*: 1611–1621.
14 Ben-Arie, N., Bellen, H.J., Armstrong, D.L., McCall, A.E., Gordadze, P.R., Guo, Q., Matzuk, M.M., and Zoghbi, H.Y. Math1 is essential for genesis of cerebellar granule neurons. *Nature* **1997**; *390*: 169–172.
15 Scardigli, R., Schuurmans, C., Gradwohl, G., and Guillemot, F. Crossregulation between Neurogenin2 and pathways specifying neuronal identity in the spinal cord. *Neuron* **2001**; *31*: 203–217.

16 Kay, J.N., Finger-Baier, K.C., Roeser, T., Staub, W., and Baier, H. Retinal ganglion cell genesis requires lakritz, a Zebrafish atonal Homolog. *Neuron* **2001**; *30*: 725–736.

17 Farah, M.H., Olson, J.M., Sucic, H.B., Hume, R.I., Tapscott, S.J., and Turner, D.L. Generation of neurons by transient expression of neural bHLH proteins in mammalian cells. *Development* **2000**; *127*: 693–702.

18 Koyano-Nakagawa, N., Wettstein, D., and Kintner, C. Activation of Xenopus genes required for lateral inhibition and neuronal differentiation during primary neurogenesis. *Mol Cell Neurosci* **1999**; *14*: 327–339.

19 Sun, Y., Nadal-Vicens, M., Misono, S., Lin, M.Z., Zubiaga, A., Hua, X., Fan, G., and Greenberg, M.E. Neurogenin promotes neurogenesis and inhibits glial differentiation by independent mechanisms. *Cell* **2001**; *104*: 365–376.

20 Nieto, M., Schuurmans, C., Britz, O., and Guillemot, F. Neural bHLH genes control the neuronal versus glial fate decision in cortical progenitors. *Neuron* **2001**; *29*: 401–413.

21 Kondo, T., and Raff, M. Basic helix-loop-helix proteins and the timing of oligodendrocyte differentiation. *Development* **2000**; *127*: 2989–2998.

22 Cai, L., Morrow, E.M., and Cepko, C.L. Misexpression of basic helix-loop-helix genes in the murine cerebral cortex affects cell fate choices and neuronal survival. *Development* **2000**; *127*: 3021–3030.

23 Wang, S., Sdrulla, A., Johnson, J.E., Yokota, Y., and Barres, B.A. A role for the helix-loop-helix protein Id2 in the control of oligodendrocyte development. *Neuron* **2001**; *29*: 603–614.

24 Frisen, J., and Lendahl, U. Oh no, Notch again! *Bioessays* **2001**; *23*: 3–7.

25 Schweisguth, F. Regulation of Notch signaling activity. *Curr Biol* **2004**; *14*: R129–R138.

26 Haines, N., and Irvine, K.D. Glycosylation regulates Notch signaling. *Nat Rev Mol Cell Biol* **2003**; *4*: 786–797.

27 Lai, E.C. Protein degradation: four E3s for the notch pathway. *Curr Biol* **2002**; *12*, R74–78.

28 Kramer, H. RIPping Notch apart: a new role for endocytosis in signal transduction? *Sci STK* **2000**; *2000*: PE1.

29 Weinmaster, G. Notch signal transduction: a real rip and more. *Curr Opin Genet Dev* **2000**; *10*: 363–369.

30 Fisher, A., and Caudy, M. The function of hairy-related bHLH repressor proteins in cell fate decisions. *Bioessays* **1998**; *20*: 298–306.

31 Davis, R.L., and Turner, D.L. Vertebrate hairy and Enhancer of split related proteins: transcriptional repressors regulating cellular differentiation and embryonic patterning. *Oncogene* **2001**; *20*: 8342–8357.

32 Brennan, K., and Gardner, P. Notching up another pathway. *Bioessays* **2002**; *24*: 405–410.

33 Martinez Arias, A., Zecchini, V., and Brennan, K. CSL-independent Notch signaling: a check point in cell fate decisions during development? *Curr Opin Genet Dev* **2002**; *12*: 524–433.

34 De la Pompa, J.L., Wakeham, A., Correia, K.M., Samper, E., Brown, S., Aguilera, R.J., Nakano, T., Honjo, T., Mak, T.W.,

Rosant, J., and Conlon, R. Conservation of the notch signalling pathway in mammalian neurogenesis. *Development* **1997**; *124*: 1139–1148.

35 Lutolf, S., Radtke, F., Aguet, M., Suter, U., and Taylor, V. Notch1 is required for neuronal and glial differentiation in the cerebellum. *Development* **2002**; *129*: 373–385.

36 Lardelli, M., Williams, R., Mitsiadis, T., and Lendahl, U. Expression of the Notch 3 intracellular domain in mouse central nervous system progenitor cells is lethal and leads to disturbed neural tube development. *Mech Dev* **1996**; *59*: 177–190.

37 Shen, J., Bronson, R.T., Chen, D.F., Xia, W., Selkoe, D.J., and Tonegawa, S. Skeletal and CNS defects in Presenilin–1–deficient mice. *Cell* **1997**; *89*: 629–639.

38 Ohtsuka, T., Sakamoto, M., Guillemot, F., and Kageyama, R. Roles of the basic helix-loop-helix genes Hes1 and Hes5 in expansion of neural stem cells of the developing brain. *J Biol Chem* **2001**; *276*: 30467–30474.

39 Ishibashi, M., Ang, S.L., Shiota, K., Nakanishi, S., Kageyama, R., and Guillemot, F. Targeted disruption of mammalian hairy and enhancer of split homolog–1 (HES–1) leads to up-regulation of neural helix-loop-helix factors, premature neurogenesis, and severe neural tube defects. *Genes Dev* **1995**; *9*: 3136–3148.

40 Norton, J.D. ID helix-loop-helix proteins in cell growth, differentiation and tumorigenesis. *J Cell Sci* **2000**; *113*: 3897–3905.

41 Lyden, D., Young, A.Z., Zagzag, D., Wei, Y., Gerald, W., O'Reilly, R., Bader, B.L., Hynes, R.O., Zhuang, Y., Manova, K., and Benezra, R. Id1 and Id3 are required for neurogenesis, angiogenesis and vascularization of tumour xenografts. *Nature* **1999**; *401*: 670–677.

42 Toma, J.G., El-Bizri, H., Barnabe-Heider, F., Aloyz, R., and Miller, F.D. Evidence that helix-loop-helix proteins collaborate with retinoblastoma tumor suppressor protein to regulate cortical neurogenesis. *J Neurosci* **2000**; *20*: 7648–7656.

43 Roztocil, T., Matter-Sadzinski, L., Alliod, C., Ballivet, M., and Matter, J.M. NeuroM, a neural helix-loop-helix transcription factor, defines a new transition stage in neurogenesis. *Development* **1997**; *124*: 3263–3272.

44 Mutoh, H., Naya, F.J., Tsai, M.J., and Leiter, A.B. The basic helix-loop-helix protein BETA2 interacts with p300 to coordinate differentiation of secretin-expressing enteroendocrine cells. *Genes Dev* **1998**; *12*: 820–830.

45 Miyata, T., Maeda, T., and Lee, J.E. NeuroD is required for differentiation of the granule cells in the cerebellum and hippocampus. *Genes Dev* **1999**; *13*: 1647–1652.

46 Olson, J.M., Asakura, A., Snider, L., Hawkes, R., Strand, A., Stoeck, J., Hallahan, A., Pritchard, J., and Tapscott, S.J. NeuroD2 is necessary for development and survival of central nervous system neurons. *Dev Biol* **2001**; *234*: 174–187.

47 Lamar, E., Kintner, C., and Goulding, M. Identification of NKL, a novel Gli-Kruppel zinc-finger protein that promotes neuronal differentiation. *Development* **2001**; *128*: 1335–1346.

48 Garcia-Dominguez, M., Poquet, C., Garel, S., and Charnay, P. Ebf gene function is required for coupling neuronal differentiation and cell cycle exit. *Development* **2003**; *130*: 6013–6025.

49 Wodarz, A., and Nusse, R. Mechanisms of Wnt signaling in development. *Annu Rev Cell Dev Biol* **1998**; *14*: 59–88.

50 Korinek, V., Barker, N., Willert, K., Molenaar, M., Roose, J., Wagenaar, G., Markman, M., Lamers, W., Destree, O., and Clevers, H. Two members of the Tcf family implicated in Wnt/beta-catenin signaling during embryogenesis in the mouse. *Mol Cell Biol* **1998**; *18*: 1248–1256.

51 Schroeder, J.A., Troyer, K.L., and Lee, D.C. Cooperative induction of mammary tumorigenesis by TGFalpha and Wnts. *Oncogene* **2000**; *19*: 3193–3199.

52 Lee, S.M., Tole, S., Grove, E., and McMahon, A.P. A local Wnt–3a signal is required for development of the mammalian hippocampus. *Development* **2000**; *127*: 457–467.

53 Megason, S.G., and McMahon, A.P. A mitogen gradient of dorsal midline Wnts organizes growth in the CNS. *Development* **2002**; *129*: 2087–2098.

54 Nusse, R. An ancient cluster of Wnt paralogues. *Trends Genet* **2001**; *17*: 443.

55 Huelsken, J., and Behrens, J. The Wnt signalling pathway. *J Cell Sci* **2002**; *115*: 3977–3978.

56 Kawano, Y., and Kypta, R. Secreted antagonists of the Wnt signalling pathway. *J Cell Sci* **2003**; *116*: 2627–2634.

57 Gurdon, J.B., and Bourillot, P.Y. Morphogen gradient interpretation. *Nature* **2001**; *413*: 797–803.

58 Zecca, M., Basler, K., and Struhl, G. Direct and long-range action of a winglesss morphogen gradient. *Cell* **1996**; *87*: 833–844.

59 Cadigan, K.M., Fish, M.P., Rulifson, E.J., and Nusse, R. Wingless repression of Drosophila frizzled 2 expression shapes the Wingless morphogen gradient in the wing. *Cell* **1998**; *93*: 767–777.

60 Arias, A.M. Wnts as morphogens? The view from the wing of Drosophila. *Nat Rev Mol Cell Biol* **2003**; *4*: 321–325.

61 Burrus, L.W. Wnt–1 as a short-range signaling molecule. *Bioessays* **1994**; *16*: 155–157.

62 Howes, R., and Bray, S. Pattern formation: Wingless on the move. *Curr Biol* **2000**; *10*: R222–226.

63 Pfeiffer, S., Alexandre, C., Calleja, M., and Vincent, J.P. The progeny of wingless-expressing cells deliver the signal at a distance in Drosophila embryos. *Curr Biol* **2000**; *10*: 321–324.

64 Moline, M.M., Southern, C., and Bejsovec, A. Directionality of wingless protein transport influences epidermal patterning in the Drosophila embryo. *Development* **1999**; *126*: 4375–4384.

65 Strigini, M., and Cohen, S.M. Wingless gradient formation in the Drosophila wing. *Curr Biol* **2000**; *10*: 293–300.

66 Gavin, B.J., McMahon, J.A., and McMahon, A.P. Expression of multiple novel Wnt–1/int–1–related genes during fetal and adult mouse development. *Genes Dev* **1990**; *4*: 2319–2332.

67 Grove, E.A., Tole, S., Limon, J., Yip, L., and Ragsdale, C.W. The hem of the embryonic cerebral cortex is defined by the expression of multiple Wnt genes and is compromised in Gli3–deficient mice. *Development* **1998**; *125*: 2315–2325.

68 Kim, A.S., Anderson, S.A., Rubenstein, J.L., Lowenstein, D.H., and Pleasure, S.J. Pax–6 regulates expression of SFRP–2 and Wnt–7b in the developing CNS. *J Neurosci* **2001**; *21*: RC132.

69 Kim, A.S., Lowenstein, D.H., and Pleasure, S.J. Wnt receptors and Wnt inhibitors are expressed in gradients in the developing telencephalon. *Mech Dev* **2001**; *103*: 167–172.

70 Rubenstein, J.L., Anderson, S., Shi, L., Miyashita-Lin, E., Bulfone, A., and Hevner, R. Genetic control of cortical regionalization and connectivity. *Cereb Cortex* **1999**; *9*: 524–532.

71 McMahon, A.P., and Bradley, A. The Wnt–1 (int–1) protooncogene is required for development of a large region of the mouse brain. *Cell* **1990**; *62*: 1073–1085.

72 Thomas, K.R., and Capecchi, M.R. Targeted disruption of the murine int–1 proto-oncogene resulting in severe abnormalities in midbrain and cerebellar development. *Nature* **1990**; *346*: 847–850.

73 Galceran, J., Miyashita-Lin, E.M., Devaney, E., Rubenstein, J.L., and Grosschedl, R. Hippocampus development and generation of dentate gyrus granule cells is regulated by LEF1. *Development* **2000**; *127*: 469–482.

74 Ikeya, M., Lee, S.M., Johnson, J.E., McMahon, A.P., and Takada, S. Wnt signalling required for expansion of neural crest and CNS progenitors. *Nature* **1997**; *389*: 966–970.

75 Patapoutian, A., and Reichardt, L.F. Roles of Wnt proteins in neural development and maintenance. *Curr Opin Neurobiol* **2000**; *10*: 392–399.

76 Dickinson, M.E., Krumlauf, R., and McMahon, A.P. Evidence for a mitogenic effect of Wnt–1 in the developing mammalian central nervous system. *Development* **1994**; *120*: 1453–1471.

77 Liu, P., Wakamiya, M., Shea, M.J., Albrecht, U., Behringer, R.R., and Bradley, A. Requirement for Wnt3 in vertebrate axis formation. *Nat Genet* **1999**; *22*: 361–365.

78 Chenn, A., and Walsh, C.A. Regulation of cerebral cortical size by control of cell cycle exit in neural precursors. *Science* **2002**; *297*: 365–369.

79 Chenn, A., and Walsh, C.A. Increased neuronal production, enlarged forebrains and cytoarchitectural distortions in beta-catenin overexpressing transgenic mice. *Cereb Cortex* **2003**; *13*: 599–606.

80 Zechner, D., Fujita, Y., Hulsken, J., Muller, T., Walther, I., Taketo, M.M., Crenshaw, E.B., 3rd, Birchmeier, W., and Birchmeier, C. beta-Catenin signals regulate cell growth and the balance between progenitor cell expansion and differentiation in the nervous system. *Dev Biol* **2003**; *258*: 406–418.

81 Theil, T., Aydin, S., Koch, S., Grotewold, L., and Ruther, U. Wnt and Bmp signalling cooperatively regulate graded Emx2 expression in the dorsal telencephalon. *Development* **2002**; *129*: 3045–3054.

82 Tole, S., Ragsdale, C.W., and Grove, E.A. Dorsoventral patterning of the telencephalon is disrupted in the mouse mutant extra-toes(J). *Dev Biol* **2000**; *217*: 254–265.

83 Heins, N., Cremisi, F., Malatesta, P., Gangemi, R.M., Corte, G., Price, J., Goudreau, G., Gruss, P., and Gotz, M. Emx2 promotes symmetric cell divisions and a multipotential fate in precursors from the cerebral cortex. *Mol Cell Neurosci* **2001**; *18*: 485–502.

84 Aberle, H., Schwartz, H., and Kemler, R. Cadherin-catenin complex: protein interactions and their implications for cadherin function. *J Cell Biochem* **1996**; *61*: 514–523.

85 Van Aken, E., De Wever, O., Correia da Rocha, A.S., and Mareel, M. Defective E-cadherin/catenin complexes in human cancer. *Virch Arch* **2001**; *439*: 725–751.

86 Chenn, A., Zhang, Y.A., Chang, B.T., and McConnell, S.K. Intrinsic polarity of mammalian neuroepithelial cells. *Mol Cell Neurosci* **1998**; *11*: 183–193.

87 Ganzler-Odenthal, S.I., and Redies, C. Blocking N-cadherin function disrupts the epithelial structure of differentiating neural tissue in the embryonic chicken brain. *J Neurosci* **1998**; *18*: 5415–5425.

88 Lele, Z., Folchert, A., Concha, M., Rauch, G.J., Geisler, R., Rosa, F., Wilson, S.W., Hammerschmidt, M., and Bally-Cuif, L. parachute/n-cadherin is required for morphogenesis and maintained integrity of the zebrafish neural tube. *Development* **2002**; *129*: 3281–3294.

89 Machon, O., van den Bout, C.J., Backman, M., Kemler, R., and Krauss, S. Role of beta-catenin in the developing cortical and hippocampal neuroepithelium. *NeuroScience* **2003**; *122*: 129–143.

90 Viti, J., Gulacsi, A., and Lillien, L. Wnt regulation of progenitor maturation in the cortex depends on Shh or fibroblast growth factor 2. *J Neurosci* **2003**; *23*: 5919–5927.

91 Behrens, J. Cross-regulation of the Wnt signalling pathway: a role of MAP kinases. *J Cell Sci* **2000**; *113*(Pt 6): 911–919.

92 Penton, A., and Hoffmann, F.M. Decapentaplegic restricts the domain of wingless during Drosophila limb patterning. *Nature* **1996**; *382*: 162–164.

93 Marti, E., and Bovolenta, P. Sonic hedgehog in CNS development: one signal, multiple outputs. *Trends Neurosci* **2002**; *25*: 89–96.

94 Ingham, P.W., and McMahon, A.P. Hedgehog signaling in animal development: paradigms and principles. *Genes Dev* **2001**; *15*: 3059–3087.

95 Goetz, J.A., Suber, L.M., Zeng, X., and Robbins, D.J. Sonic Hedgehog as a mediator of long-range signaling. *Bioessays* **2002**; *24*: 157–165.

96 McCarthy, R.A., Barth, J.L., Chintalapudi, M.R., Knaak, C., and Argraves, W.S. Megalin functions as an endocytic sonic hedgehog receptor. *J Biol Chem* **2002**; *277*: 25660–25667.

97 McCarthy, R.A., and Argraves, W.S. Megalin and the neurodevelopmental biology of sonic hedgehog and retinol. *J Cell Sci* **2003**; *116*: 955–960.

98 Marti, E., Takada, R., Bumcrot, D.A., Sasaki, H., and McMahon, A.P. Distribution of Sonic hedgehog peptides in the developing chick and mouse embryo. *Development* **1995**; *121*: 2537–2547.

99 Traiffort, E., Charytoniuk, D., Watroba, L., Faure, H., Sales, N., and Ruat, M. Discrete localizations of hedgehog signalling components in the developing and adult rat nervous system. *Eur J Neurosci* **1999**; *11*: 3199–3214.

100 Traiffort, E., Moya, K.L., Faure, H., Hassig, R., and Ruat, M. High expression and anterograde axonal transport of amino-terminal sonic hedgehog in the adult hamster brain. *Eur J Neurosci* 2001; *14*: 839–850.

101 Dahmane, N., Sanchez, P., Gitton, Y., Palma, V., Sun, T., Beyna, M., Weiner, H., and Ruiz i Altaba, A. The Sonic Hedgehog-Gli pathway regulates dorsal brain growth and tumorigenesis. *Development* 2001; *128*: 5201–5212.

102 Hui, C.C., Slusarski, D., Platt, K.A., Holmgren, R., and Joyner, A.L. Expression of three mouse homologs of the Drosophila segment polarity gene cubitus interruptus, Gli, Gli–2, and Gli–3, in ectoderm- and mesoderm-derived tissues suggests multiple roles during postimplantation development. *Dev Biol* 1994; *162*: 402–413.

103 Dahmane, N., and Ruiz-i-Altaba, A. Sonic hedgehog regulates the growth and patterning of the cerebellum. *Development* 1999; *126*: 3089–3100.

104 Wallace, V.A. Purkinje-cell-derived Sonic hedgehog regulates granule neuron precursor cell proliferation in the developing mouse cerebellum. *Curr Biol* 1999; *9*: 445–448.

105 Wechsler-Reya, R.J., and Scott, M.P. Control of neuronal precursor proliferation in the cerebellum by Sonic Hedgehog. *Neuron* 1999; *22*: 103–114.

106 Pons, S., Trejo, J.L., Martinez-Morales, J.R., and Marti, E. Vitronectin regulates Sonic hedgehog activity during cerebellum development through CREB phosphorylation. *Development* 2001; *128*: 1481–1492.

107 Wechsler-Reya, R.J. Caught in the matrix: how vitronectin controls neuronal differentiation. *Trends Neurosci* 2001; *24*: 680–682.

108 Rowitch, D.H., S-Jacques, B., Lee, S.M., Flax, J.D., Snyder, E.Y., and McMahon, A.P. Sonic hedgehog regulates proliferation and inhibits differentiation of CNS precursor cells. *J Neurosci* 1999; *19*: 8954–8965.

109 Gunhaga, L., Jessell, T.M., and Edlund, T. Sonic hedgehog signaling at gastrula stages specifies ventral telencephalic cells in the chick embryo. *Development* 2000; *127*: 3283–3293.

110 Nery, S., Wichterle, H., and Fishell, G. Sonic hedgehog contributes to oligodendrocyte specification in the mammalian forebrain. *Development* 2001; *128*: 527–540.

111 Gaiano, N., Kohtz, J.D., Turnbull, D.H., and Fishell, G. A method for rapid gain-of-function studies in the mouse embryonic nervous system. *Nat Neurosci* 1999; *2*: 812–819.

112 Machold, R., Hayashi, S., Rutlin, M., Muzumdar, M.D., Nery, S., Corbin, J.G., Gritli-Linde, A., Dellovade, T., Porter, J.A., Rubin, L.L., Dudek, H., McMahon, A.P., and Fishell, G. Sonic hedgehog is required for progenitor cell maintenance in telencephalic stem cell niches. *Neuron* 2003; *39*: 937–950.

113 Chiang, C., Litingtung, Y., Lee, E., Young, K.E., Corden, J.L., Westphal, H., and Beachy, P.A. Cyclopia and defective axial patterning in mice lacking Sonic hedgehog gene function. *Nature* 1996; *383*: 407–413.

114 Rallu, M., Machold, R., Gaiano, N., Corbin, J.G., McMahon, A.P., and Fishell, G. Dorsoventral patterning is established in

the telencephalon of mutants lacking both Gli3 and Hedgehog signaling. *Development* **2002**; *129*: 4963–4974.

115 Lu, Q.R., Yuk, D., Alberta, J.A., Zhu, Z., Pawlitzky, I., Chan, J., McMahon, A.P., Stiles, C.D., and Rowitch, D.H. Sonic hedgehog-regulated oligodendrocyte lineage genes encoding bHLH proteins in the mammalian central nervous system. *Neuron* **2000**; *25*: 317–329.

116 Lai, K., Kaspar, B.K., Gage, F.H., and Schaffer, D.V. Sonic hedgehog regulates adult neural progenitor proliferation *in vitro* and *in vivo*. *Nat Neurosci* **2003**; *6*: 21–27.

117 Ruiz i Altaba, A., Sanchez, P., and Dahmane, N. Gli and hedgehog in cancer: tumours, embryos and stem cells. *Nat Rev Cancer* **2002**; *2*: 361–372.

118 Ruiz, I.A.A., Palma, V., and Dahmane, N. Hedgehog-Gli signalling and the growth of the brain. *Nat Rev Neurosci* **2002**; *3*: 24–33.

119 Harris, W.A. Cellular diversification in the vertebrate retina. *Curr Opin Genet Dev* **1997**; *7*: 651–658.

120 McCabe, K.L., Gunther, E.C., and Reh, T.A. The development of the pattern of retinal ganglion cells in the chick retina: mechanisms that control differentiation. *Development* **1999**; *126*: 5713–5724.

121 Levine, E.M., Roelink, H., Turner, J., and Reh, T.A. Sonic hedgehog promotes rod photoreceptor differentiation in mammalian retinal cells *in vitro*. *J Neurosci* **1997**; *17*: 6277–6288.

122 Jensen, A.M., and Wallace, V.A. Expression of Sonic hedgehog and its putative role as a precursor cell mitogen in the developing mouse retina. *Development* **1997**; *124*: 363–371.

123 Neumann, C.J., and Nuesslein-Volhard, C. Patterning of the zebrafish retina by a wave of sonic hedgehog activity. *Science* **2000**; *289*: 2137–2139.

124 Laessing, U., and Stuermer, C.A. Spatiotemporal pattern of retinal ganglion cell differentiation revealed by the expression of neurolin in embryonic zebrafish. *J Neurobiol* **1996**; *29*: 65–74.

125 Raymond, P.A., Barthel, L.K., and Curran, G.A. Developmental patterning of rod and cone photoreceptors in embryonic zebrafish. *J Comp Neurol* **1995**; *359*: 537–550.

126 Schmitt, E.A., and Dowling, J.E. Comparison of topographical patterns of ganglion and photoreceptor cell differentiation in the retina of the zebrafish, *Danio rerio*. *J Comp Neurol* **1996**; *371*: 222–234.

127 Hu, M., and Easter, S.S. Retinal neurogenesis: the formation of the initial central patch of postmitotic cells. *Dev Biol* **1999**; *207*: 309–321.

128 Heberlein, U., and Moses, K. Mechanisms of Drosophila retinal morphogenesis: the virtues of being progressive. *Cell* **1995**; *81*: 987–990.

129 Dominguez, M., and Hafen, E. Hedgehog directly controls initiation and propagation of retinal differentiation in the Drosophila eye. *Genes Dev* **1997**; *11*: 3254–3264.

130 Masai, I., Stemple, D.L., Okamoto, H., and Wilson, S.W. Midline signals regulate retinal neurogenesis in zebrafish. *Neuron* **2000**; *27*: 251–263.

131 Yung, S.Y., Gokhan, S., Jurcsak, J., Molero, A.E., Abrajano, J.J., and Mehler, M.F. Differential modulation of BMP signaling promotes the elaboration of cerebral cortical GABAergic neurons or oligodendrocytes from a common sonic hedgehog-responsive ventral forebrain progenitor species. *Proc Natl Acad Sci USA* **2002**; *99*: 16273–16278.

132 Mehler, M.F., Mabie, P.C., Zhu, G., Gokhan, S., and Kessler, J.A. Developmental changes in progenitor cell responsiveness to bone morphogenetic proteins differentially modulate progressive CNS lineage fate. *Dev Neurosci* **2000**; *22*: 74–85.

133 Palma, V., and Ruiz i Altaba, A. Hedgehog-GLI signaling regulates the behavior of cells with stem cell properties in the developing neocortex. *Development* **2004**; *131*: 337–345.

134 Liu, F., Massague, J., and Ruiz i Altaba, A. Carboxy-terminally truncated Gli3 proteins associate with Smads. *Nat Genet* **1998**; *20*: 325–326.

135 Jacob, J., and Briscoe, J. Gli proteins and the control of spinal-cord patterning. *EMBO Rep* **2003**; *4*: 761–765.

136 Ornitz, D.M. FGFs, heparan sulfate and FGFRs: complex interactions essential for development. *Bioessays* **2000**; *22*: 108–112.

137 Reuss, B., von Bohlen und Halbach, O. Fibroblast growth factors and their receptors in the central nervous system. *Cell Tissue Res* **2003**; *313*: 139–157.

138 Bartlett, P.F., Brooker, G.J., Faux, C.H., Dutton, R., Murphy, M., Turnley, A., and Kilpatrick, T.J. Regulation of neural stem cell differentiation in the forebrain. *Immunol Cell Biol* **1998**; *76*: 414–418.

139 Cameron, H.A., Hazel, T.G., and McKay, R.D. Regulation of neurogenesis by growth factors and neurotransmitters. *J Neurobiol* **1998**; *36*: 287–306.

140 Qian, X., Davis, A.A., Goderie, S.K., and Temple, S. FGF2 concentration regulates the generation of neurons and glia from multipotent cortical stem cells. *Neuron* **1997**; *18*: 81–93.

141 Ghosh, A., and Greenberg, M. E. Distinct roles for bFGF and NT–3 in the regulation of cortical neurogenesis. *Neuron* **1995**; *15*: 89–103.

142 Ray, J., and Gage, F.H. Spinal cord neuroblasts proliferate in response to basic fibroblast growth factor. *J Neurosci* **1994**; *14*: 3548–3564.

143 Menard, C., Hein, P., Paquin, A., Savelson, A., Yang, X.M., Lederfein, D., Barnabe-Heider, F., Mir, A.A., Sterneck, E., Peterson, A.C., Johnson, P.F., Vinson, C., and Miller, F.D. An essential role for a MEK-C/EBP pathway during growth factor-regulated cortical neurogenesis. *Neuron* **2002**; *36*: 597–610.

144 Ortega, S., Ittmann, M., Tsang, S.H., Ehrlich, M., and Basilico, C. Neuronal defects and delayed wound healing in mice lacking fibroblast growth factor 2. *Proc Natl Acad Sci USA* **1998**; *95*: 5672–5677.

145 Vaccarino, F.M., Schwartz, M.L., Raballo, R., Nilsen, J., Rhee, J., Zhou, M., Doetschman, T., Coffin, J.D., Wyland, J.J., and Hung, Y.T. Changes in cerebral cortex size are governed by fibroblast growth factor during embryogenesis. *Nat Neurosci* **1999**; *2*: 246–253.

146 Lee, S.M., Danielian, P.S., Fritzsch, B., and McMahon, A.P. Evidence that FGF8 signalling from the midbrain-hindbrain junction regulates growth and polarity in the developing midbrain. *Development* **1997**; *124*: 959–969.

147 Xu, J., Liu, Z., and Ornitz, D.M. Temporal and spatial gradients of Fgf8 and Fgf17 regulate proliferation and differentiation of midline cerebellar structures. *Development* **2000**; *127*: 1833–1843.

148 Faux, C.H., Turnley, A.M., Epa, R., Cappai, R., and Bartlett, P.F. Interactions between Fibroblast Growth Factors and Notch regulate neuronal differentiation. *J Neurosci* **2001**; *21*: 5587–5596.

149 Diez del Corral, R., Breitkreutz, D.N., and Sato, K.G. Onset of neuronal differentiation is regulated by paraxial mesoderm and requires attenuation of FGF signalling. *Development* **2002**; *129*: 1681–1691.

150 Diez del Corral, R., Olivera-Maritnez, I., Goriely, A., Gale, E., Maden, M., and Storey, K. Opposing FGF and retinoid pathways control ventral neural pattern, neuronal differentiation, and segmentation during body axis extension. *Neuron* **2003**; *40*: 65–79.

151 Alroy, I., and Yarden, Y. The ErbB signaling network in embryogenesis and oncogenesis: signal diversification through combinatorial ligand-receptor interactions. *FEBS Lett* **1997**; *410*: 83–86.

152 Kornblum, H.I., Hussain, R.J., Bronstein, J.M., Gall, C.M., Lee, D.C., and Seroogy, K.B. Prenatal ontogeny of the epidermal growth factor receptor and its ligand, transforming growth factor alpha, in the rat brain. *J Comp Neurol* **1997**; *380*: 243–261.

153 Seroogy, K.B., Gall, C.M., Lee, D.C., and Kornblum, H.I. Proliferative zones of postnatal rat brain express epidermal growth factor receptor mRNA. *Brain Res* **1995**; *670*: 157–164.

154 Okano, H.J., Pfaff, D.W., and Gibbs, R.B. Expression of EGFR-, p75NGFR-, and PSTAIR (cdc2)-like immunoreactivity by proliferating cells in the adult rat hippocampal formation and forebrain. *Dev Neurosci* **1996**; *18*: 199–209.

155 Reynolds, B.A., and Weiss, S. Clonal and population analyses demonstrate that an EGF-responsive mammalian embryonic CNS precursor is a stem cell. *Dev Biol* **1996**; *175*: 1–13.

156 Reynolds, B.A., Tetzlaff, W., and Weiss, S. A multipotent EGF-responsive striatal embryonic progenitor cell produces neurons and astrocytes. *J Neurosci* **1992**; *12*: 4565–4574.

157 Gritti, A., Frolichsthal-Schoeller, P., Galli, R., Parati, E.A., Cova, L., Pagano, S.F., Bjornson, C.R., and Vescovi, A.L. Epidermal and fibroblast growth factors behave as mitogenic regulators for a single multipotent stem cell-like population from the subventricular region of the adult mouse forebrain. *J Neurosci* **1999**; *19*: 3287–3297.

158 Whittemore, S.R., Morassutti, D.J., Walters, W.M., Liu, R.H., and Magnuson, D.S. Mitogen and substrate differentially affect the lineage restriction of adult rat subventricular zone neural precursor cell populations. *Exp Cell Res* **1999**; *252*: 75–95.

159 Kuhn, H.G., Winkler, J., Kempermann, G., Thal, L.J., and Gage, F.H. Epidermal growth factor and fibroblast growth factor-2 have different effects on neural progenitors in the adult rat brain. *J Neurosci* **1997**; *17*: 5820–5829.

160 Doetsch, F., Petreanu, L., Caille, I., Garcia-Verdugo, J.M., and Alvarez-Buylla, A. EGF converts transit-amplifying neurogenic precursors in the adult brain into multipotent stem cells. *Neuron* **2002**; *36*: 1021–1034.

161 Lillien, L., and Cepko, C. Control of proliferation in the retina: temporal changes in responsiveness to FGF and TGF alpha. *Development* **1992**; *115*: 253–266.

162 Lillien, L. Changes in retinal cell fate induced by overexpression of EGF receptor. *Nature* **1995**; *377*: 158–162.

163 Mahanthappa, N.K., and Schwarting, G.A. Peptide growth factor control of olfactory neurogenesis and neuron survival *in vitro*: roles of EGF and TGF-beta s. *Neuron* **1993**; *10*: 293–305.

164 Farbman, A.I., and Buchholz, J.A. Transforming growth factor-alpha and other growth factors stimulate cell division in olfactory epithelium *in vitro*. *J Neurobiol* **1996**; *30*: 267–280.

165 Reynolds, B.A., and Weiss, S. Generation of neurons and astrocytes from isolated cells of the adult mammalian central nervous system. *Science* **1992**; *255*: 1707–1709.

166 Burrows, R.C., Wancio, D., Levitt, P., and Lillien, L. Response diversity and the timing of progenitor cell maturation are regulated by developmental changes in EGFR expression in the cortex. *Neuron* **1997**; *19*: 251–267.

167 Lillien, L., and Raphael, H. BMP and FGF regulate the development of EGF-responsive neural progenitor cells. *Development* **2000**; *127*: 4993–5005.

168 Tropepe, V., Craig, C.G., Morshead, C.M., and van der Kooy, D. Transforming growth factor-alpha null and senescent mice show decreased neural progenitor cell proliferation in the forebrain subependyma. *J Neurosci* **1997**; *17*: 7850–7859.

169 Threadgill, D.W., Dlugosz, A.A., Hansen, L.A., Tennenbaum, T., Lichti, U., Yee, D., LaMantia, C., Mourton, T., Herrup, K., Harris, R.C., et al. Targeted disruption of mouse EGF receptor: effect of genetic background on mutant phenotype. *Science* **1995**; *269*: 230–234.

170 Meyer, D., and Birchmeier, C. Multiple essential functions of neuregulin in development. *Nature* **1995**; *378*: 386–390.

171 Crone, S.A., and Lee, K.F. Gene targeting reveals multiple essential functions of the neuregulin signaling system during development of the neuroendocrine and nervous systems. *Ann NY Acad Sci* **2002**; *971*: 547–553.

172 Gerlai, R., Pisacane, P., and Erickson, S. Heregulin, but not ErbB2 or ErbB3, heterozygous mutant mice exhibit hyperactivity in multiple behavioral tasks. *Behav Brain Res* **2000**; *109*: 219–227.

173 Lee, K.F., Simon, H., Chen, H., Bates, B., Hung, M.C., and Hauser, C. Requirement for neuregulin receptor erbB2 in neural and cardiac development. *Nature* **1995**; *378*: 394–398.

174 Riethmacher, D., Sonnenberg-Riethmacher, E., Brinkmann, V., Yamaai, T., Lewin, G.R., and Birchmeier, C. Severe neu-

ropathies in mice with targeted mutations in the ErbB3 receptor. *Nature* **1997**; *389*: 725–730.
175 Gassmann, M., Casagranda, F., Orioli, D., Simon, H., Lai, C., Klein, R., and Lemke, G. Aberrant neural and cardiac development in mice lacking the ErbB4 neuregulin receptor. *Nature* **1995**; *378*: 390–394.
176 Massague, J. TGFbeta signaling: receptors, transducers, and Mad proteins. *Cell* **1996**; *85*: 947–950.
177 Flanders, K.C., Ludecke, G., Engels, S., Cissel, D.S., Roberts, A.B., Kondaiah, P., Lafyatis, R., Sporn, M.B., and Unsicker, K. Localization and actions of transforming growth factor-beta s in the embryonic nervous system. *Development* **1991**; *113*: 183–191.
178 Schmid, P., Cox, D., Bilbe, G., Maier, R., and McMaster, G.K. Differential expression of TGF beta 1, beta 2 and beta 3 genes during mouse embryogenesis. *Development* **1991**; *111*: 117–130.
179 Millan, F.A., Denhez, F., Kondaiah, P., and Akhurst, R.J. Embryonic gene expression patterns of TGF beta 1, beta 2 and beta 3 suggest different developmental functions *in vivo*. *Development* **1991**; *111*: 131–143.
180 Unsicker, K., Meier, C., Krieglstein, K., Sartor, B.M., and Flanders, K.C. Expression, localization, and function of transforming growth factor-beta s in embryonic chick spinal cord, hindbrain, and dorsal root ganglia. *J Neurobiol* **1996**; *29*: 262–276.
181 Constam, D.B., Schmid, P., Aguzzi, A., Schachner, M., and Fontana, A. Transient production of TGF-beta 2 by postnatal cerebellar neurons and its effect on neuroblast proliferation. *Eur J Neurosci* **1994**; *6*: 766–778.
182 Anchan, R.M., and Reh, T.A. Transforming growth factor-β–3 is mitogenic for rat retinal progenitor cells *in vitro*. *J Neurobiol* **1995**; *28*: 133–145.
183 Kane, C.J., Brown, G.J., and Phelan, K.D. Transforming growth factor-beta 2 both stimulates and inhibits neurogenesis of rat cerebellar granule cells in culture. *Brain Res Dev Brain Res* **1996**; *96*: 46–51.
184 Zhang, J.M., Hoffmann, R., and Sieber-Blum, M. Mitogenic and anti-proliferative signals for neural crest cells and the neurogenic action of TGF-beta1. *Dev Dyn* **1997**; *208*: 375–386.
185 Newman, M.P., Feron, F., and Mackay-Sim, A. Growth factor regulation of neurogenesis in adult olfactory epithelium. *NeuroScience* **2000**; *99*: 343–350.
186 Getchell, M.L., Boggess, M.A., Pruden, S.J., 2nd, Little, S.S., Buch, S., and Getchell, T.V. Expression of TGF-beta type II receptors in the olfactory epithelium and their regulation in TGF-alpha transgenic mice. *Brain Res* **2002**; *945*: 232–241.
187 Briscoe, J., and Ericson, J. Specification of neuronal fates in the ventral neural tube. *Curr Opin Neurobiol* **2001**; *11*: 43–49.
188 Hall, A.K., and Miller, R.H. Emerging roles for bone morphogenetic proteins in central nervous system glial biology. *J Neurosci Res* **2004**; *76*: 1–8.
189 Liem, K.F., Jr., Tremml, G., Roelink, H., and Jessell, T.M. Dorsal differentiation of neural plate cells induced by BMP-mediated signals from epidermal ectoderm. *Cell* **1995**; *82*: 969–979.

190 Lopez-Coviella, I., Berse, B., Krauss, R., Thies, R.S., and Blusztajn, J.K. Induction and maintenance of the neuronal cholinergic phenotype in the central nervous system by BMP–9. *Science* **2000**; *289*: 313–316.

191 Li, W., Cogswell, C.A., and LoTurco, J.J. Neuronal differentiation of precursors in the neocortical ventricular zone is triggered by BMP. *J Neurosci* **1998**; *18*: 8853–8862.

192 Mabie, P.C., Mehler, M.F., and Kessler, J.A. Multiple roles of bone morphogenetic protein signaling in the regulation of cortical cell number and phenotype. *J Neurosci* **1999**; *19*: 7077–7088.

193 Li, W., and LoTurco, J.J. Noggin is a negative regulator of neuronal differentiation in developing neocortex. *Dev Neurosci* **2000**; *22*: 68–73.

194 Lim, D.A., Tramontin, A.D., Trevejo, J.M., Herrera, D.G., Garcia-Verdugo, J.M., and Alvarez-Buylla, A. Noggin antagonizes BMP signaling to create a niche for adult neurogenesis. *Neuron* **2000**; *28*: 713–726.

195 Okano-Uchida, T., Himi, T., Komiya, Y., and Ishizaki, Y. Cerebellar granule cell precursors can differentiate into astroglial cells. *Proc Natl Acad Sci USA* **2004**; *101*: 1211–1216.

196 Mekki-Dauriac, S., Agius, E., Kan, P., and Cochard, P. Bone morphogenetic proteins negatively control oligodendrocyte precursor specification in the chick spinal cord. *Development* **2002**; *129*: 5117–5130.

197 Gross, R.E., Mehler, M.F., Mabie, P.C., Zang, Z., Santschi, L., and Kessler, J.A. Bone morphogenetic proteins promote astroglial lineage commitment by mammalian subventricular zone progenitor cells. *Neuron* **1996**; *17*: 595–606.

198 Nakashima, K., Takizawa, T., Ochiai, W., Yanagisawa, M., Hisatsune, T., Nakafuku, M., Miyazono, K., Kishimoto, T., Kageyama, R., and Taga, T. BMP2–mediated alteration in the developmental pathway of fetal mouse brain cells from neurogenesis to astrocytogenesis. *Proc Natl Acad Sci USA* **2001**; *98*: 5868–5873.

199 Gomes, W.A., Mehler, M.F., and Kessler, J.A. Transgenic overexpression of BMP4 increases astroglial and decreases oligodendroglial lineage commitment. *Dev Biol* **2003**; *255*: 164–177.

200 Winnier, G., Blessing, M., Labosky, P.A., and Hogan, B.L. Bone morphogenetic protein–4 is required for mesoderm formation and patterning in the mouse. *Genes Dev* **1995**; *9*: 2105–2116.

201 Mishina, Y., Suzuki, A., Ueno, N., and Behringer, R.R. Bmpr encodes a type I bone morphogenetic protein receptor that is essential for gastrulation during mouse embryogenesis. *Genes Dev* **1995**; *9*: 3027–3037.

202 Beppu, H., Kawabata, M., Hamamoto, T., Chytil, A., Minowa, O., Noda, T., and Miyazono, K. BMP type II receptor is required for gastrulation and early development of mouse embryos. *Dev Biol* **2000**; *221*: 249–258.

203 Israsena, N., and Kessler, J.A. Msx2 and p21(CIP1/WAF1) mediate the pro-apoptotic effects of bone morphogenetic protein–4 on ventricular zone progenitor cells. *J Neurosci Res* **2002**; *69*: 803–809.

204 Kretzschmar, M., Doody, J., and Massague, J. Opposing BMP and EGF signalling pathways converge on the TGF-beta family mediator Smad1. *Nature* **1997**; *389*: 618–622.

205 Moustakas, A., Souchelnytskyi, S., and Heldin, C.H. Smad regulation in TGF-beta signal transduction. *J Cell Sci* **2001**; *114*: 4359–4369.

206 Neubuser, A., Peters, H., Balling, R., and Martin, G.R. Antagonistic interactions between FGF and BMP signaling pathways: a mechanism for positioning the sites of tooth formation. *Cell* **1997**; *90*: 247–255.

207 Henderson, C.E. Role of neurotrophic factors in neuronal development. *Curr Opin Neurobiol* **1996**; *6*: 64–70.

208 Maisonpierre, P.C., Belluscio, L., Friedman, B., Alderson, R.F., Wiegand, S.J., Furth, M.E., Lindsay, R.M., and Yancopoulos, G.D. NT-3, BDNF, and NGF in the developing rat nervous system: parallel as well as reciprocal patterns of expression. *Neuron* **1990**; *5*: 501–509.

209 Barnabe-Heider, F., and Miller, F.D. Endogenously produced neurotrophins regulate survival and differentiation of cortical progenitors via distinct signaling pathways. *J Neurosci* **2003**; *23*: 5149–5160.

210 Vicario-Abejon, C., Johe, K.K., Hazel, T.G., Collazo, D., and McKay, R.D. Functions of basic fibroblast growth factor and neurotrophins in the differentiation of hippocampal neurons. *Neuron* **1995**; *15*: 105–114.

211 Cattaneo, E., and McKay, R. Proliferation and differentiation of neuronal stem cells regulated by nerve growth factor. *Nature* **1990**; *347*: 762–765.

212 Liu, Z.Z., Zhu, L.Q., and Eide, F.F. Critical role of TrkB and brain-derived neurotrophic factor in the differentiation and survival of retinal pigment epithelium. *J Neurosci* **1997**; *17*: 8749–8755.

213 Friedman, W.J., and Greene, L.A. Neurotrophin signaling via Trks and p75. *Exp Cell Res* **1999**; *253*: 131–142.

214 Strom, A., Castella, P., Rockwood, J., Wagner, J., and Caudy, M. Mediation of NGF signaling by post-translational inhibition of HES–1, a basic helix-loop-helix repressor of neuronal differentiation. *Genes Dev* **1997**; *11*: 3168–3181.

215 Ito, H., Nakajima, A., Nomoto, H., and Furukawa, S. Neurotrophins facilitate neuronal differentiation of cultured neural stem cells via induction of mRNA expression of basic helix-loop-helix transcription factors Mash1 and Math1. *J Neurosci Res* **2003**; *71*: 648–658.

216 Chojnacki, A., Shimazaki, T., Gregg, C., Weinmaster, G., and Weiss, S. Glycoprotein 130 signaling regulates Notch1 expression and activation in the self-renewal of mammalian forebrain neural stem cells. *J Neurosci* **2003**; *23*: 1730–1741.

217 Shimazaki, T., Shingo, T., and Weiss, S. The ciliary neurotrophic factor/leukemia inhibitory factor/gp130 receptor complex operates in the maintenance of mammalian forebrain neural stem cells. *J Neurosci* **2001**; *21*: 7642–7653.

218 Williams, B.P., Park, J.K., Alberta, J.A., Muhlebach, S.G., Hwang, G.Y., Roberts, T.M., and Stiles, C.D. A PDGF-regu-

lated immediate early gene response initiates neuronal differentiation in ventricular zone progenitor cells. *Neuron* **1997**; *18*: 553–562.

219 Turnley, A.M., Faux, C.H., Rietze, R.L., Coonan, J.R., and Bartlett, P.F. Suppressor of cytokine signalling 2 regulates neuronal differentiation by inhibiting growth hormone signaling. *Nat Neurosci* **2002**; *5*: 1155–1162.

220 Sauer, F.C. Mitosis in the neural tube. *J Comp Neurol* **1935**; *62*: 377–405.

221 Luskin, M.B., Pearlman, A.L., and Sanes, J.R. Cell lineage in the cerebral cortex of the mouse studied *in-vivo* and *in-vitro* with a recombinant retrovirus. *Neuron* **1988**; *1*: 635–647.

222 Rakic, P. A small step for the cell, a giant leap for mankind: a hypothesis of neocortical expansion during evolution. *Trends Neurosci* **1995**; *18*: 383–388.

223 McConnell, S.K. Constructing the cerebral cortex: neurogenesis and fate determination. *Neuron* **1995**; *15*: 761–768.

224 Mione, M.C., Cavanagh, J.F., Harris, B., and Parnavelas, J.G. Cell fate specification and symmetrical/asymmetrical divisions in the developing cerebral cortex. *J Neurosci* **1997**; *17*: 2018–2029.

225 Noctor, S.C., Martinez-Cerdeno, V., Ivic, L., and Kriegstein, A.R. Cortical neurons arise in symmetric and asymmetric division zones and migrate through specific phases. *Nat Neurosci* **2004**; *7*: 136–144.

226 Haubensak, W., Attardo, A., Denk, W., and Huttner, W.B. Neurons arise in the basal neuroepithelium of the early mammalian telencephalon: A major site of neurogenesis. *Proc Natl Acad Sci USA* **2004**; *101*: 3196–3201.

227 Takahashi, T., Nowakowski, R.S., and Caviness, V.S., Jr. The cell cycle of the pseudostratified ventricular epithelium of the embryonic murine cerebral wall. *J Neurosci* **1995**; *15*: 6046–6057.

228 Caviness, V.S., Jr., Takahashi, T., and Nowakowski, R.S. Numbers, time and neocortical neuronogenesis: a general developmental and evolutionary model. Trends *Neurosci* **1995**; *18*: 379–383.

229 Cai, L., Hayes, N.L., and Nowakowski, R.S. Local homogeneity of cell cycle length in developing mouse cortex. *J Neurosci* **1997**; *17*: 2079–2087.

230 Hartfuss, E., Galli, R., Heins, N., and Gotz, M. Characterization of CNS precursor subtypes and radial glia. *Dev Biol* **2001**; *229*: 15–30.

231 Calegari, F., Haubensak, W., Haffner, C., and Huttner, W.B. Selective lengthening of the cell cycle in the neurogenic subpopulation of neural progenitor cells during mouse brain development. *J. Neurosci* **2005**; *25*: in press.

232 Cremisi, F., Philpott, A., and Ohnuma, S. Cell cycle and cell fate interactions in neural development. *Curr Opin Neurobiol* **2003**; *13*: 26–33.

233 Bally-Cuif, L., and Hammerschmidt, M. Induction and patterning of neuronal development, and its connection to cell cycle control. *Curr Opin Neurobiol* **2003**; *13*: 16–25.

234 Ohnuma, S., and Harris, W.A. Neurogenesis and the cell cycle. *Neuron* **2003**; *40*: 199–208.

235 Oliver, T.G., Grasfeder, L.L., Carroll, A.L., Kaiser, C., Gillingham, C.L., Lin, S.M., Wickramasinghe, R., Scott, M.P., and Wechsler-Reya, R.J. Transcriptional profiling of the Sonic hedgehog response: a critical role for N-myc in proliferation of neuronal precursors. *Proc Natl Acad Sci USA* **2003**; *100*: 7331–7336.

236 Henrique, D., Hirsinger, E., Adam, J., Le Roux, I., Pourquie, O., Ish-Horowicz, D., and Lewis, J. Maintenance of neuroepithelial progenitor cells by Delta-Notch signalling in the embryonic chick retina. *Curr Biol* **1997**; *7*,: 661–670.

237 Bao, Z.Z., and Cepko, C.L. The expression and function of Notch pathway genes in the developing rat eye. *J Neurosci* **1997**; *17*: 1425–1434.

238 Jang, M.S., Miao, H., Carlesso, N., Shelly, L., Zlobin, A., Darack, N., Qin, J.Z., Nickoloff, B.J., and Miele, L. Notch–1 regulates cell death independently of differentiation in murine erythroleukemia cells through multiple apoptosis and cell cycle pathways. *J Cell Physiol* **2004**; *199*: 418–433.

239 Lukaszewicz, A., Savatier, P., Cortay, V., Kennedy, H., and Dehay, C. Contrasting effects of basic fibroblast growth factor and neurotrophin 3 on cell cycle kinetics of mouse cortical stem cells. *J Neurosci* **2002**; *22*: 6610–6622.

240 Dubreuil, V., Hirsch, M., Pattyn, A., Brunet, J., and Goridis, C. The Phox2b transcription factor coordinately regulates neuronal cell cycle exit and identity. *Development* **2000**; *127*: 5191–5201.

241 Dubreuil, V., Hirsch, M.R., Jouve, C., Brunet, J.F., and Goridis, C. The role of Phox2b in synchronizing pan-neuronal and type-specific aspects of neurogenesis. *Development* **2002**; *129*: 5241–5253.

242 Ohnuma, S., Philpott, A., and Harris, W.A. Cell cycle and cell fate in the nervous system. *Curr Opin Neurobiol* **2001**; *11*: 66–73.

243 Morgan, D.O. Principles of CDK regulation. *Nature* **1995**; *374*: 131–134.

244 Obaya, A.J., and Sedivy, J.M. Regulation of cyclin-Cdk activity in mammalian cells. *Cell Mol Life Sci* **2002**; *59*: 126–142.

245 Dobashi, Y., Shoji, M., Kitagawa, M., Noguchi, T., and Kameya, T. Simultaneous suppression of cdc2 and cdk2 activities induces neuronal differentiation of PC12 cells. *J Biol Chem* **2000**; *275*: 12572–12580.

246 De Laurenzi, V., Raschella, G., Barcaroli, D., Annicchiarico-Petruzzelli, M., Ranalli, M., Catani, M.V., Tanno, B., Costanzo, A., Levrero, M., and Melino, G. Induction of neuronal differentiation by p73 in a neuroblastoma cell line. *J Biol Chem* **2000**; *275*: 15226–15231.

247 Vernon, A.E., Devine, C., and Philpott, A. The cdk inhibitor p27Xic1 is required for differentiation of primary neurones in Xenopus. *Development* **2003**; *130*: 85–92.

248 Durand, B., and Raff, M. A cell-intrinsic timer that operates during oligodendrocyte development. *Bioessays* **2000**; *22*: 64–71.

249 Ohnuma, S., Philpott, A., Wang, K., Holt, C.E., and Harris, W.A. p27Xic1, a Cdk inhibitor, promotes the determination of glial cells in Xenopus retina. *Cell* **1999**; *99*: 499–510.

250 Vernon, A.E., and Philpott, A. A single cdk inhibitor, p27Xic1, functions beyond cell cycle regulation to promote muscle differentiation in Xenopus. *Development* **2003**; *130*: 71–83.

251 Tarui, T., Takahashi, T., Nowakowski, R.S., Hayes, N.L., Bhide, P.G., and Caviness, V.S. Overexpression of p27Kip1, probability of cell cycle exit, and laminar destination of neocortical neurons. *Cereb Cortex* **2005**; E pub ahead of print. PMID 15647527.

252 Matsuda, S., Rouault, J., Magaud, J., and Berthet, C. In search of a function for the TIS21/PC3/BTG1/TOB family. *FEBS Lett* **2001**; *497*: 67–72.

253 Tirone, F. The gene PC3(TIS21/BTG2), prototype member of the PC3/BTG/TOB family: regulator in control of cell growth, differentiation, and DNA repair? *J Cell Physiol* **2001**; *187*: 155–165.

254 Iacopetti, P., Michelini, M., Stuckmann, I., Oback, B., Aaku-Saraste, E., and Huttner, W.B. Expression of the antiproliferative gene TIS21 at the onset of neurogenesis identifies single neuroepithelial cells that switch from proliferative to neuron-generating division. *Proc Natl Acad Sci USA* **1999**; *96*: 4639–4644.

255 Canzoniere, D., Farioli-Vecchioli, S., Conti, F., Ciotti, M.T., Tata, A.M., Augusti-Tocco, G., Mattei, E., Lakshmana, M.K., Krizhanovsky, V., Reeves, S.A., Giovannoni, R., Castano, F., Servadio, A., Ben-Arie, N., and Tirone, F. Dual control of neurogenesis by PC3 through cell cycle inhibition and induction of Math1. *J Neurosci* **2004**; *24*: 3355–3369.

256 Calegari, F., and Huttner, W.B. An inhibition of cyclin-dependent kinases that lengthens, but does not arrest, neuroepithelial cell cycle induces premature neurogenesis. *J Cell Sci* **2003**; *116,*: 4947–4955.

257 Takahashi, T., Nowakowski, R.S., and Caviness, V.S., Jr. The leaving or Q fraction of the murine cerebral proliferative epithelium: a general model of neocortical neuronogenesis. *J Neurosci* **1996**; *16*: 6183–6196.

258 Doe, C.Q. Asymmetric cell division and neurogenesis. *Curr. Opin. Genet. & Development* **1996**; *6*: 562–566.

259 Campos-Ortega, J.A. Early neurogenesis in *Drosophila melanogaster*. In *The Development of* Drosophila melanogaster, Vol. 2, C. M. Bate and A. Martinez-Arias (Eds). Cold Spring Harbor: CSH Press, **1993**; pp. 1091–1129.

260 Goodman, C.S., and Doe, C.Q. Embryonic development of the *Drosophila* central nervous system. In *Development of* Drosophila melanogaster, Vol. 2, C. M. Bate and A. Martinez-Arias (Eds). Cold Spring Harbor: CSH Press, **1993**; pp. 1131–1206.

261 Bossing, T., Udolph, G., Doe, C.Q., and Technau, G.M. The embryonic central nervous system lineages of *Drosophila melanogaster*. I. Neuroblast lineages derived from the ventral half of the neuroectoderm. *Dev Biol* **1996**; *179*, 41–64.

262 Schmidt, H., Rickert, C., Bossing, T., Vef, O., Urban, J., and Technau, G.M. The embryonic central nervous system line-

ages of Drosophila melanogaster. II. Neuroblast lineages derived from the dorsal part of the neuroectoderm. *Dev Biol* **1997**; *189*: 186–204.

263 Schmid, A., Chiba, A., and Doe, C.Q. Clonal analysis of Drosophila embryonic neuroblasts: neural cell types, axon projections and muscle targets. *Development* **1999**; *126*: 4653–4689.

264 Wodarz, A., and Huttner, W.B. Asymmetric cell division during neurogenesis in *Drosophila* and vertebrates. *Mech Dev* **2003**; *120*: 1297–1309.

265 Roegiers, F., and Jan, Y.N. **2004**; Asymmetric cell division. *Curr Opin Cell Biol 16*, 195–205.

266 Matsuzaki, F. Asymmetric division of Drosophila neural stem cells: a basis for neural diversity. *Curr Opin Neurobiol* **2000**; *10*: 38–44.

267 Cowan, C.R., and Hyman, A.A. Asymmetric cell division in C. elegans: cortical polarity and spindle positioning. *Annu Rev Cell Dev Biol* **2004**; *20*: 427–453.

268 Knoblich, J.A. Asymmetric cell division during animal development. *Nat Rev Mol Cell Biol* **2001**; *2*: 11–20.

269 Betschinger, J., and Knoblich, J.A. Dare to be different: asymmetric cell division in Drosophila, *C. elegans* and vertebrates. *Curr Biol* **2004**; *14*: R674–685.

270 Kraut, R., Chia, W., Jan, L.Y., Jan, Y.N., and Knoblich, J.A. Role of inscuteable in orienting asymmetric cell divisions in Drosophila. *Nature* **1996**; *383*: 50–55.

271 Kaltschmidt, J.A., Davidson, C.M., Brown, N.H., and Brand, A.H. Rotation and asymmetry of the mitotic spindle direct asymmetric cell division in the developing central nervous system. *Nat Cell Biol* **2000**; *2*: 7–12.

272 Schober, M., Schaefer, M., and Knoblich, J.A. Bazooka recruits Inscuteable to orient asymmetric cell divisions in Drosophila neuroblasts. *Nature* **1999**; *402*: 548–551.

273 Wodarz, A., Ramrath, A., Kuchinke, U., and Knust, E. Bazooka provides an apical cue for Inscuteable localization in Drosophila neuroblasts. *Nature* **1999**; *402*: 544–547.

274 Shen, C.P., Jan, L.Y., and Jan, Y.N. Miranda is required for the asymmetric localization of Prospero during mitosis in Drosophila. *Cell* **1997**; *90*: 449–458.

275 Ikeshima-Kataoka, H., Skeath, J.B., Nabeshima, Y., Doe, C.Q., and Matsuzaki, F. Miranda directs Prospero to a daughter cell during Drosophila asymmetric divisions. *Nature* **1997**; *390*: 625–629.

276 Li, P., Yang, X., Wasser, M., Cai, Y., and Chia, W. Inscuteable and Staufen mediate asymmetric localization and segregation of prospero RNA during Drosophila neuroblast cell divisions. *Cell* **1997**; *90*: 437–447.

277 Broadus, J., Fuerstenberg, S., and Doe, C.Q. Staufen-dependent localization of prospero mRNA contributes to neuroblast daughter-cell fate. *Nature* **1998**; *391*: 792–795.

278 Hirata, J., Nakagoshi, H., Nabeshima, Y., and Matsuzaki, F. Asymmetric segregation of a homeoprotein, *prospero*, during cell divisions in neural and endodermal development. *Nature* **1995**; *377*: 627–630.

279 Knoblich, J.A., Jan, L.Y., and Jan, Y.N. Asymmetric segregation of numb and prospero during cell division. *Nature* **1995**; *377*: 624–627.
280 Lu, B., Rothenberg, M., Jan, L.Y., and Jan, Y.N. Partner of Numb colocalizes with Numb during mitosis and directs Numb asymmetric localization in Drosophila neural and muscle progenitors. *Cell* **1998**; *95*: 225–235.
281 Guo, M., Jan, L.Y., and Jan, Y.N. Control of daughter cell fates during asymmetric division: interaction of numb and Notch. *Neuron* **1996**; *17*: 27–41.
282 Ohshiro, T., Yagami, T., Zhang, C., and Matsuzaki, F. Role of cortical tumour-suppressor proteins in asymmetric division of Drosophila neuroblast. *Nature* **2000**; *408*: 593–596.
283 Peng, C.Y., Manning, L., Albertson, R., and Doe, C.Q. The tumour-suppressor genes lgl and dlg regulate basal protein targeting in Drosophila neuroblasts. *Nature* **2000**; *408*: 596–600.
284 Betschinger, J., Mechtler, K., and Knoblich, J.A. The Par complex directs asymmetric cell division by phosphorylating the cytoskeletal protein Lgl. *Nature* **2003**; *422*: 326–330.
285 Kornack, D.R., and Rakic, P. Radial and horizontal deployment of clonally related cells in the primate neocortex: relationship to distinct mitotic lineages. *Neuron* **1995**; *15*: 311–321.
286 Reid, C.B., Tavazoie, S.F., and Walsh, C.A. Clonal dispersion and evidence for asymmetric cell division in ferret cortex. *Development* **1997**; *124*: 2441–2450.
287 Fishell, G., and Kriegstein, A.R. Neurons from radial glia: the consequences of asymmetric inheritance. *Curr Opin Neurobiol* **2003**; *13*: 34–41.
288 Noctor, S.C., Flint, A.C., Weissman, T.A., Dammerman, R.S., and Kriegstein, A.R. Neurons derived from radial glial cells establish radial units in neocortex. *Nature* **2001**; *409*: 714–720.
289 Miyata, T., Kawaguchi, A., Okano, H., and Ogawa, M. Asymmetric inheritance of radial glial fibers by cortical neurons. *Neuron* **2001**; *31*: 727–741.
290 Aaku-Saraste, E., Oback, B., Hellwig, A., and Huttner, W.B. Neuroepithelial cells downregulate their plasma membrane polarity prior to neural tube closure and neurogenesis. *Mech Dev* **1997**; *69*: 71–81.
291 Aaku-Saraste, E., Hellwig, A., and Huttner, W.B. Loss of occludin and functional tight junctions, but not ZO-1, during neural tube closure – remodeling of the neuroepithelium prior to neurogenesis. *Dev Biol* **1996**; *180*: 664–679.
292 Wilson, D. Tissue interactions in basal regions of the cranial neuroepithelium in the C57BL mouse. *J Craniofac Genet Dev Biol* **1983**; *3*: 269–279.
293 Nose, A., and Takeichi, M. A novel cadherin cell adhesion molecule: its expression patterns associated with implantation and organogenesis of mouse embryos. *J Cell Biol* **1986**; *103*: 2649–2658.
294 Chenn, A., and McConnell, S.K. Cleavage orientation and the asymmetric inheritance of Notch1 immunoreactivity in mammalian neurogenesis. *Cell* **1995**; *82*: 631–641.

295 Cayouette, M., and Raff, M. Asymmetric segregation of Numb: a mechanism for neural specification from Drosophila to mammals. *Nat Neurosci* **2002**; *5*: 1265–1269.

296 Geldmacher-Voss, B., Reugels, A.M., Pauls, S., and Campos-Ortega, J.A. A 90–degree rotation of the mitotic spindle changes the orientation of mitoses of zebrafish neuroepithelial cells. *Development* **2003**; *130*: 3767–3780.

297 Smart, I.H.M. Proliferative characteristics of the ependymal layer during the early development of the mouse neocortex: a pilot study based on recording the number, location and plane of cleavage of mitotic figures. *J Anat* **1973**; *116*: 67–91.

298 Landrieu, P., and Goffinet, A. Mitotic spindle fiber orientation in relation to cell migration in the neo-cortex of normal and reeler mouse. *Neurosci Lett* **1979**; *13*: 69–72.

299 Huttner, W.B., and Brand, M. Asymmetric division and polarity of neuroepithelial cells. *Curr Opin Neurobiol* **1997**; *7*: 29–39.

300 Kosodo, Y., Röper, K., Haubensak, W., Marzesco, A.-M., Corbeil, D., and Huttner, W.B. Asymmetric distribution of the apical plasma membrane during neurogenic divisions of mammalian neuroepithelial cells. *EMBO J* **2004**; *23*: 2314–2324.

301 Manabe, N., Hirai, S., Imai, F., Nakanishi, H., Takai, Y., and Ohno, S. Association of ASIP/mPAR–3 with adherens junctions of mouse neuroepithelial cells. *Dev Dyn* **2002**; *225*: 61–69.

302 Cayouette, M., Whitmore, A.V., Jeffery, G., and Raff, M. Asymmetric segregation of Numb in retinal development and the influence of the pigmented epithelium. *J Neurosci* **2001**; *21*: 5643–5651.

303 Cayouette, M., and Raff, M. The orientation of cell division influences cell-fate choice in the developing mammalian retina. *Development* **2003**; *130*: 2329–2339.

304 Zhong, W. Diversifying neural cells through order of birth and asymmetry of division. *Neuron* **2003**; *37*: 11–14.

8
Generating Cell Diversity

Ajay Chitnis

8.1
Introduction

How an embryo generates a wide diversity of cells in the correct number and location during development is an enduring question that has captured the imagination of developmental biologists. One of the most striking examples of cell diversity is seen in the nervous system where a wide range of neurons and non-neuronal cells acquire a unique distribution in discrete compartments of the neural tissue. Specification of neuronal identity determines stereotyped patterns of axon outgrowth and synaptic connectivity, which eventually leads to the self-organization of complex neural networks. In this chapter we will review how pattern emerges during vertebrate development and use examples from the nervous system to discuss how cell diversity is generated in the developing embryo.

The generation of cell diversity is a process that unfolds from early steps in development when the three germ layers, ectoderm, mesoderm and endoderm are defined. At about the same time, signaling centers are established that promote dorsal fate and divide the ectoderm into prospective neural and epidermal compartments. Dorsalizing signals induce neural tissue that, by default, has a rostral or "forebrain-like" fate and differential exposure to posteriorizing signals from the blastoderm margin helps determine progressively more caudal fates, eventually defining prospective forebrain, midbrain, hindbrain and spinal cord compartments in the neural plate.

The prospective compartments of the neural plate are established as gastrulation begins. However, convergence and extension movements during gastrulation and the process of neurulation transform the flat sheet of neuroepithelial cells into the neural tube. As development proceeds cells come under the influence of additional inductive signals: new signaling centers are established within the neural tissue and inductive signals from the adjacent axial and paraxial mesoderm influence the fate of cells in the neural tube. Together, these signals regulate neurogenesis and determine the identity of neurons as they are generated in the neural tube. During neurogenesis global patterning mechanisms define broad domains where cells acquire the potential to become neurons, while local interactions determine which

Cell Signaling and Growth Factors in Development. Edited by K. Unsicker and K. Krieglstein
Copyright © 2006 WILEY-VCH Verlag GmbH & Co. KGaA, Weinheim
ISBN 3-527-31034-7

cells are permitted to differentiate at a given time. As progenitors withdraw from the cell cycle and begin differentiating, the combinatorial expression of specific transcription factors determines cell identity and specifies appropriate patterns of axon outgrowth.

A number of experimental vertebrate model systems have contributed to our broader understanding of vertebrate neural development. Although each model system provides a unique window on patterning mechanisms, research from phylogenetically diverse organisms demonstrates that very similar mechanisms operate in a wide of range of metazoan systems. The first part of this chapter describes early patterning mechanisms that define germ layers and initial dorsal-ventral and rostral-caudal compartments of the neural plate in zebrafish and Xenopus. In the second part the emphasis shifts to descriptions of patterning events in the spinal cord of chick and mouse where elegant studies have described how cell diversity is generated. Throughout, comparisons emphasize the similarities in these systems and consider the general relevance of conserved mechanisms to our understanding of the generation of cell diversity.

8.2
Establishing Early Compartments of the Embryo

8.2.1
Establishing the Three Germ Layers

8.2.1.1 Nodals Function as Morphogens to Define Germ Layers

The embryo is partitioned along the animal-vegetal axis into three germ layers: the ectoderm, which forms epidermis, neural crest and the nervous system, the mesoderm, which forms the notochord, muscle, mesenchyme and blood, and the endoderm, which forms the lining of the gut and organs derived from it. Though a number of signaling systems are involved in this process, the Nodals, which belong to the TGFβ family, have a prominent role [1]. In zebrafish two nodals, Cyclops and Squint, enhance their own expression and drive the expression of rapidly diffusing antagonists, Lefty1/2. Together the Nodals and the Leftys form a reaction-diffusion system that establishes a source of diffusible Nodal signals around the blastoderm margin [2]. Blastoderm cells at the margin, exposed to the highest levels of Nodal signaling, acquire an endoderm fate while blastoderm cells nearer the animal pole, exposed to relatively low levels of nodal signaling, acquire an ectodermal fate. Adjacent to the margin, cells are exposed to intermediate levels of Nodal signaling where they acquire a mesodermal fate (Fig. 8.1A). Cells in each germ layer are not only distinguished by expression of distinct transcription factors but also by distinctive cell surface properties and cellular organization. For example, in zebrafish ectodermal cells adopt an epithelial morphology while the nascent endodermal cells adopt a mesenchymal morphology: they delaminate from the margin and spread out as individual cells over the surface of the yolk cell.

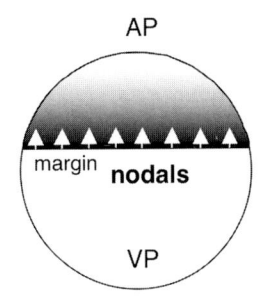

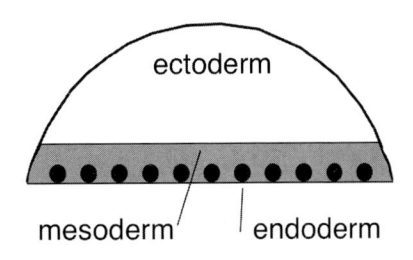

A. DEFINING THE GERM LAYERS

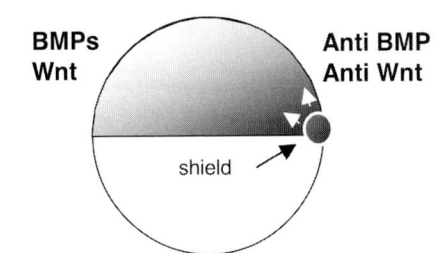

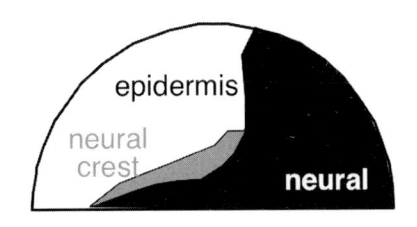

B. DEFINING THE DORSAL-VENTRAL AXIS

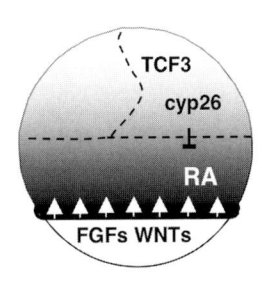

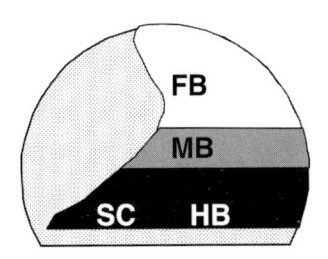

C. DEFINING THE ROSTRAL-CAUDAL AXIS

Fig. 8.1
Establishing the early compartments of the embryo. A, B and C show schematics of the zebrafish embryo at 30, 50 and 80 % epiboly, respectively, showing prominent sources of signals that are responsible for defining germ layers, the dorsal-ventral axis and the rostral-caudal axis. On the right are details of prospective compartments defined in the blastoderm at these stages. AP, animal pole; VP, vegetal pole; RA, retinoic acid; FB, forebrain; MB, midbrain; HB, hindbrain; SC, spinal cord. See text for details.

8.2.1.2 Maternal Factors Regulate Nodal Signaling

In zebrafish and Xenopus, maternal stores of mRNA and protein laid down in the developing oocyte have an essential role in early patterning. Reception of nodal signals is dependent on maternal and zygotic expression of *one eyed pinhead* (*oep*), an EGF-CFC family co-receptor. Loss of nodal signaling in Maternal Zygotic (MZ) *oep* zebrafish mutants leads to a loss of mesendodermal tissue and expansion of ectodermal tissue. In Xenopus, the vegetally localized maternal T-box transcription factor, VegT, has an essential role in promoting nodal signaling [3] while the secreted TGFβ antagonist Coco, localized at the animal pole, inhibits nodal signaling [4]. B1–type Sox transcription factors localized at the animal pole also inhibit nodal signaling and promote an ectodermal fate at the animal pole in Xenopus and zebrafish embryos [5]. Together, these observations illustrate how the maternal distribution of factors with opposing effects on nodal signaling concentrated at the vegetal and animal poles help to establish Nodal signaling at the margin where its function as a morphogen and defines the prospective germ layers.

8.2.2
Establishing the Dorsal-Ventral Axis

8.2.2.1 Formation of the Vertebrate Organizer

Although cells all around the zebrafish blastoderm margin initially secrete nodals their expression is eventually restricted to a morphologically distinct thickening of cells in the margin called the shield. Mesodermal cells adjacent to the blastoderm margin converge toward the shield (Fig. 8.1B). The thickening marks the future dorsal side of the embryo and the site where axial mesodermal cells involute and extend under the prospective neural plate to form the prechordal plate and the notochord. The zebrafish shield corresponds to the dorsal blastopore lip of the amphibian gastrula, which is referred to as the Spemann-Mangold organizer after its co-discoverers, has the ability to induce a secondary axis when transplanted into the belly of a host embryo early in development [6].

The shield functions as an organizer because it is a critical source of a number of secreted BMP antagonists such as Chordin (Chd), Noggin and Follistatin, and Wnt antagonists such as Dickkopf1 (Dkk1) and Frzb1. In the absence of BMP antagonists, high levels of BMP signaling are established in the embryo as a consequence of auto-activation by this signaling pathway. BMP antagonists secreted by the organizer help establish a dorsal-ventral (DV) gradient of BMP signaling in the early embryo: low BMP signaling in proximity of the organizer tissue determines dorsal fates and ventral fates are determined at a distance where BMP signaling remains relatively high. In the ventral ectoderm high levels of BMP signaling induce an epidermal fate while BMP antagonists secreted by the organizer permit the dorsal ectoderm to adopt a neural fate [7].

8.2.2.2 β-Catenin-dependent Dorsalizing Factors Position the Organizer

Early patterning by Nodals defines a ring of mesoderm at the blastoderm margin that is competent to form the organizer. Within this domain the location of the organizer is determined by early β-catenin-mediated transcriptional activation. Microtubule-driven translocation of β-catenin from the vegetal pole to the prospective dorsal side following fertilization establishes a signaling center that facilitates expression of dorsal organizer genes. In Xenopus, β-catenin induces *siamois* expression to establish the Nieuwkoop center, while in zebrafish β-catenin induces the expression of *bozozok (boz,* also called *dharma* or *nieuwkoid)* on the prospective dorsal side of the blastoderm margin in the Yolk Syncytial Layer [8].

β-Catenin functions as a transactivator through interactions with members of the LEF/TCF family of transcription factors that have an HMG DNA binding domain. However, members of the TCF family have binding domains for both activators such as β-catenin, and repressors such as Groucho and CtBP. In the absence of β-catenin some TCF family members, e.g. TCF3, bind co-repressors and have an essential role in repressing target genes (see later Section 8.2.3.2 on rostral-caudal patterning). Consistent with its role as an essential repressor, TCF3 helps maintain repression of organizer genes on the ventral side where there is no β-catenin signaling [9].

Boz is a homeodomain (HD) transcriptional repressor that facilitates organizer formation by two important mechanisms: it directly inhibits dorsal BMP transcription prior to establishment of the organizer [10] and it inhibits expression of Vox/Vega1, Vent/Vega2 and Ved, transcriptional repressors that in turn inhibit expression of organizer-specific genes and the expression of *boz* [8]. As soon as the zygotic genome is activated there is broad expression of BMPs in the zebrafish embryo. By 30 % epiboly, before establishment of the organizer as a source of BMP antagonists, there is already some dorsal reduction of BMP transcripts. Although repression by Boz may account for some of this reduction, the extent of dorsal BMP reduction extends beyond the domain of *boz* expression, suggesting that an additional mechanism must be present. Analysis of the early role of FGF signaling in zebrafish shows that a center of FGF expression is established in the prospective dorsal blastula before 30 % epiboly and activation of FGF signaling pathway has an early role in repressing BMP transcription dorsally [11]. The establishment of this dorsal center of *FGF* expression, like that of *boz*, is also dependent on early dorsal β-catenin-mediated transcription [12]. The discovery of an early role for FGF signaling in zebrafish complements earlier studies in the chick that also showed FGFs to have an essential early role in induction of dorsal fates [13].

8.2.2.3 The Vertebrate Dorsal Organizer is a Self-organizing Tissue

It is important to note that while studies in Xenopus and zebrafish define an early sequence of events, dependent on the polarized distribution of maternal determinants that leads to formation of the organizer, this tissue has a remarkable ability to self-regulate. For example, unless a large domain surrounding the shield is removed, shield ablations in zebrafish are often followed by re-establishment of the organizer [14]. Furthermore, the organizer regulates its own size and actively prevents the establishment of an ectopic organizer in its neighborhood through the secretion of Anti-Dorsalizing Morphogenetic Protein (ADMP), a rapidly diffusing antagonist of organizer formation whose expression is driven by Nodals and Boz [15, 16]. Indeed, secreted BMP antagonists like Chd and ADMP, and the long-range inhibitor of organizer formation, most likely are essential components of a reaction-diffusion system that accounts for the remarkable self-organizing properties of the vertebrate dorsal organizer [17]. The self-organizing properties of the vertebrate organizer have best been described in chick where the organizer is called Henson's node. Lineage labeling experiments show that the node is not a fixed group of cells, but rather its inductive properties are associated with a group of cells whose population is constantly changing as cells move in and out during gastrulation [18]. Amniotes like chick and mouse are less dependent on maternal determinants and their embryos therefore show a much greater ability to regulate: new organizers are formed in individual fragments obtained by cutting the early blastula and these fragments can form complete embryos [19].

8.2.2.4 The BMP Gradient Determines Compartments of the Ectoderm

Classical studies suggested an early response to dorsalizing signals in the ectoderm was the induction of a dorsal neural compartment and a ventral epidermal compartment. Induction of the neural crest at the boundary between epidermal and neural compartments was thought to be due to subsequent interactions between these compartments. However, studies in both zebrafish and Xenopus have suggested that cells interpret an early gradient of BMP signaling to define three prospective ectodermal compartments [20]. A comparison of the relative size of the neural, neural crest and epidermal compartments in *somitabun (sbn)/smad5, snailhouse/bmp7* and *swirl/bmp2b* mutants with progressively greater loss of BMP signaling, reveals a progressive increase in the size of the relatively dorsal compartments and corresponding reduction in the size of ventral compartments [21]. Expansion of the relatively dorsal neural crest domain accompanied by the absence of the ventral epidermal compartment in some *sbn* mutants is a strong argument against an absolute requirement for interactions between epidermal and neural compartments for the initial induction of the intermediate neural crest compartment. Furthermore, the systematic increase in the size of the dorsal compartments accompanied by reduction in size of ventral compartments in mutants with progressive decrease in BMP signaling can most easily be explained by a gradient model that suggests that differential exposure to BMP signaling is interpreted by cells in the ectoderm to define epidermal, neural crest and neural compartments. This model is supported by additional studies that show that intermediate levels of BMP signaling can optimally induce expression of transcription factors responsible for determination of neural crest fate [22–24].

Although a BMP gradient may have an early role in determining the size of the three compartments of the ectoderm, it is likely that at later stages local interactions mediated by BMPs and Wnts between neural, epidermal and underlying mesodermal compartments contribute to specification of neural crest fate. During early somitogenesis, when markers for the neural crest are first seen, BMPs are no longer expressed in a dorsal-ventral gradient, instead they are expressed at relatively high levels near the margin of the neural plate, and after formation of the neural tube they are expressed dorsally in the roof plate. Consequently, although an early dorsal-ventral gradient of BMP signaling in the ectoderm sets the stage for defining at least three prospective compartments of the ectoderm, the establishment of local sources of BMPs and Wnt signals is eventually likely to play a critical role in establishment and maintenance of crest progenitors in the dorsal spinal cord [20, 25].

8.2.2.5 Neural Genes Prevent Differentiation and Maintain a Neural State

Neural inducers like Chordin and Noggin promote neural fate in a dorsal compartment of the ectoderm by permitting expression of a number of genes that would otherwise be inhibited by early BMP signaling. Some of these genes, first identified in Xenopus, include *SoxD*, Sox–2 class genes (*Sox1, Sox2, Sox3*), *zic1, zic3* and *pou2* domain genes [26]. *SoxD* is broadly expressed in the ectoderm at the late blastula stage. BMP antagonists maintain *SoxD* expression in a dorsal compartment as its expression is inhibited ventrally by BMP signaling by mid-gastrula stages. *SoxD* can induce neural tissue and formation of neurons in Xenopus animal caps, while dominant-negative forms of *SoxD* inhibit formation of anterior neural tissue [27]. Sox–2 class transcription factors have an essential role in keeping cells in a neural stem cell-like state and in preventing them from differentiating as neurons [28]. *Zic1* may have a similar role in maintaining a neural state and in preventing premature differentiation [29]. Together it appears that many of the genes that are an early response to neural induction promote neural fate and cooperate to maintain growth and plasticity of the neural tissue by maintaining an undifferentiated neural state.

8.2.3 Establishing the Rostral-Caudal (Anterior-Posterior) Axis

8.2.3.1 The Blastoderm Margin is an Early Source of Posteriorizing Factors

BMP antagonists initially induce neuroectoderm with anterior fate within a defined radius of the dorsal organizer. Its fate can be recognized by its early expression of *Otx* genes [30]. By the beginning of gastrulation FGFs and Wnts are expressed around the blastoderm margin and differential exposure to these factors posteriorizes the neuroectoderm (Fig. 8.1C). In zebrafish, the induction of an initial caudal neural compartment can be recognized by the expression of *gbx1* and *hoxb1b*, which marks the prospective hindbrain during early gastrulation [31]. This prospective caudal compartment, which is initially adjacent to the margin, moves progressively toward the midline and extends rostrally because of convergence and extension movements during gastrulation. As a consequence by the end of gastrulation the prospective hindbrain is caudal to the forebrain and midbrain along the rostral-caudal axis.

The division of the neuroectoderm into a broad anterior domain defined by *otx* expression and a caudal domain defined by *hoxb1b* at the end of gastrulation, is accomplished by the synergistic function of FGFs and WNTS in the margin and Retinoic acid (RA) signals produced in the underlying paraxial mesoderm [31]. FGFs and WNTs have distinct roles in suppressing expression of rostral genes and in the activation of genes that define caudal domains. While their ability to inhibit expression of rostral genes is not dependent on RA, RA is essential for activation of caudal genes by FGFs and Wnts. Determination of the rostral *otx*-positive domain requires mechanisms that actively protect the prospective anterior neuroectoderm from the posteriorizing influence of FGFs and WNTs. One important mechanism is the expression of CYP26 in the prospective anterior neuroectoderm, which encodes an RA-degrading enzyme that prevents RA from effectively activating caudal genes in this domain. Another mechanism is the broad expression of TCF3 in the prospective anterior neuroectoderm, which in this context provides basal repression of caudal genes that would otherwise be activated by posteriorizing factors [32].

8.2.3.2 Rostral-Caudal Patterning of the Anterior Neuroectoderm

Differential exposure to posteriorizing factors is responsible for the further division of a broad *otx2* domain into sub-compartments of the forebrain and midbrain. By the end of gastrulation the expression of transcription factors *emx* and *pax6* defines the prospective telencephalic and diencephalic compartments of the forebrain [33] while *mbx* defines prospective eye and tectal domains [34]. More caudally *pax2* is an early marker for a domain adjacent to the midbrain-hindbrain boundary (MHB domain) where the isthmic organizer will form (see below) [35].

Manipulations that increase the efficacy of Wnt/β-catenin signaling reduce the size of rostral compartments and expand the size of caudal compartments. For example, the *masterblind* (*mbl*) mutant has a loss of function of *axin1*, a gene that promotes destruction of β-catenin. Exaggerated β-catenin signaling in *mbl* mutants, where β-catenin is more stable, leads to a loss of the telencephalon, and eyes expansion of the diencephalon to the front of the brain [36]. *hdl* encodes a TCF3 homolog and loss of essential basal repression mediated by this DNA binding factor with co-repressors makes it easier for posteriorizing factors to promote expression of caudal genes in more rostral domains where they would not normally be as effective. MZ *headless* (*hdl*) mutants typically are slightly more posteriorized compared to *mbl* and in these mutants loss of the eyes and forebrain is accompanied by expansion of the relatively caudal MHB domain but there is little change in the hindbrain domain [32]. Reduced function of an additional TCF3 homolog, *tcf3b* in a MZ *hdl* background leads to even more severe posteriorization where reduction or complete loss of rostral domains is now accompanied by more obvious expansion of hindbrain domains

[37]. The systematic changes in the relative sizes of rostral-caudal compartments seen in these mutants and complementary changes observed when Wnt signals are blocked, support the idea that cells interpret a gradient of posteriorizing activity to define discrete compartments in the anterior neuroectoderm (reviewed in [38]). Wnt/β-catenin signaling has a primary role in de-repressing genes whose expression would otherwise be prevented by TCF3 acting as a repressor. In the absence of basal TCF3–mediated repression, the additional loss of Wnt signals leads to almost no reduction in the dramatic expansion of caudal gene expression. This underscores the fact that expression of caudal genes is most likely actively promoted by additional posteriorizing factors such as FGFs [37].

8.2.3.3 Secondary Organizers and Compartment Boundaries

By the end of gastrulation new sources of inductive signals, called secondary organizers, embellish the pattern established in the neuroectoderm by early organizers. As axial mesoderm involutes at the shield it first forms the prechordal plate, a source of a number of secreted factors including the Wnt antagonist, Dkk1, Hedgehogs and BMP antagonists [39–41]. The prechordal plate moves rostrally and passes under the prospective anterior neuroectoderm and, as a source of Wnt antagonists, it serves as a "sink" to sharpen the posteriorizing gradient established by caudal sources of Wnt signals. A source of an additional Wnt antagonist, Tlc, is also established at the zebrafish Anterior Neural Ridge by the end of gastrulation. By antagonizing Wnt signaling it ensures the formation of the telencephalon whose development would otherwise be prevented by Wnts secreted from the posterior diencephalon [33].

Many secondary organizers are established at compartment boundaries. For example, the juxtaposition of an anterior *OTX* expression domain and a caudal *GBX* expression domain in the hindbrain helps to position the isthmic organizer at the midbrain-hindbrain boundary (MHB) [42]. It becomes a new source of Wnts and FGFs at the end of gastrulation and helps pattern the adjacent midbrain and hindbrain. Later, the juxtaposition of a rostral *six3* domain with a caudal *Irx3* domain helps the establishment of the Zona Limitans Intrathalamica (ZLI) at the future boundary of the dorsal thalamus (dT) and ventral thalamus (vT). Establishment of the ZLI defines a boundary in the diencephalon with low Wnt signaling rostrally and high Wnt signaling caudally. High Wnt signaling facilitates expression of Gbx2, which specifies dT fate, while low Wnt signaling in the diencephalon facilitates Dlx2 expression, which specifies vT fate [43]. While low Wnt signaling is required for establishment of rostral compartments and relatively high Wnt signaling is required to establish caudal compartments, sharp boundaries between the OTX and GBX domains or the six3 and Irx3 domains is dependent on a mutually antagonist relationship between these pairs of transcription factors.

8.2.3.4 Rostral-Caudal Patterning of the Hindbrain

Increasing exposure of prospective caudal neuroectoderm to posteriorizing factors induces progressively more caudal domains of the hindbrain and eventually the spinal cord. The hindbrain or rhombencephalon is divided into seven morphological compartments called rhombomeres. The establishment of these discrete compartments is a key step in the functional organization of the hindbrain. Individually identifiable reticulospinal neurons in the dorsal part of each rhombomere are organized to form a ladder-like array in the hindbrain, and the branchiomotor neurons of the cranial nerves have a rhombomere-specific distribution in the ventral part of the rhombomeres. Even- and odd-numbered rhombomeres share specific features: Branchiomotor axons exit the CNS from even-numbered rhombomeres, as do neural crest cells that migrate out from the dorsal surface of the neural tube, and neurogenesis is delayed in odd-numbered rhombomeres. Even- and odd-numbered rhombomeres, respectively, share cell surface properties that contribute to sharp segregation of cells between these compartments: odd-numbered rhombomeres express the Eph receptor such as EphA4a, while even-numbered rhombomeres express cognate signaling partners like ephrin-B2a.

A complex genetic network establishes the segmental identity of rhombomeres. Discrete rostral-caudal sections of the vertebrate hindbrain and spinal cord are defined by the expression of Hox gene Paralogs. A defining feature of the Hox genes includes their conserved Antenappedia-class homeobox sequences and their organization into gene clusters. There are four clusters in most tetrapods, however, the teleost fishes have undergone an extra duplication and as a result there are seven Hox gene clusters in zebrafish. The genes fall into 13 paralog groups (PGs) although no individual cluster has representatives of all 13 paralogs. The 3′ to 5′ arrangement of Hox genes in each cluster predicts the relative rostral limit and temporal sequence of expression in a tissue: 3′ genes in the cluster have the most rostral and earliest expression in the CNS. Consistent with this observation PGs 1–4 are responsible for patterning the hindbrain, while, PGs 5–9 define brachial, thoracic and lumbar columns in the spinal cord. Details of the genetic network that defines rhombomere identity are reviewed elsewhere and will not be discussed in this chapter due to limitations of space (reviewed in [44, 45].

8.2.3.5 Establishing a Rostral-Caudal FGF Gradient

Rostral-caudal patterning events in the anterior neuroectoderm and the hindbrain begin at the late blastula stage and continue through gastrulation into the beginning of somitogenesis. Patterning in the caudal prospective spinal cord domain becomes much more apparent during somitogenesis. In the chick embryo, as the node recedes, a rostral-caudal gradient of FGF transcripts becomes apparent in the neuroectoderm and presomitic mesoderm. In the chick, FGF transcription is actively maintained in a progenitor population adjacent to the primitive streak (equivalent to the blastoderm margin in zebrafish). As cells leave the zone where active FGF transcription is maintained to become part of the neuroectoderm or presomitic mesoderm, FGF transcription ceases and decay of FGF transcripts results in progressively lower levels of FGF. Rostral cells leave the zone of active FGF transcription earlier than caudal cells and a gradient of FGF signaling is established by coupling the caudal movement of the zone of active FGF transcription with the progressive decay of FGF transcripts in rostral cells [46]. This mechanism is attractive because the shape of the FGF activity gradient is not as constrained by the diffusion properties of the FGF molecule. Instead the shape of the slope is now determined by the speed at which the FGF source recedes and by the rate of FGF transcript decay in the presomitic mesoderm, parameters that can be regulated by the cells.

It has become increasingly clear that in many biological situations, where secreted factors establish morphogen gradients, the shape of the gradient is actively determined by a number of mechanisms that regulate import and export of proteins rather than by passive diffusion in the extracellular space. Most prominent among these is the process of ligand-induced receptor endocytosis [47]. Studies in Drosophila have shown clearly that a gradient of Dpp (related to the TGFβ factor BMP) is established in the Wing imaginal disc by a mechanism where Dpp is internalized with its receptor and then released by exocytosis. In this situation the receptors are not only involved in signal transduction but in active mechanisms that regulate movement of the ligand within a field of cells. In this context when a patch of endocytosis mutant cells lies in front of a source of Dpp, failure to effectively transport Dpp by transendocytosis leads to a "shadow" of low Dpp distribution on the side away from the source. In zebrafish, recent studies have revealed a very different role for endocytosis in establishing an FGF gradient. These studies suggest that endocytosis is essential to *limit* the range of FGF signals. At the shield stage the blastoderm margin is the source of FGF signals, and *sef*, which is a target of FGF signaling, is normally expressed in a zone that is relatively close to the margin. However manipulations that reduce endocytosis lead to a dramatic broadening of the *sef* expression domain consistent with a significant increase in the range of FGF signals. These observations suggest that, rather than facilitating the spread of FGF, endocytosis limits the range of FGF signaling in the zebrafish blastoderm.

8.2.3.6 The FGF Gradient Regulates Mesoderm and Neuroectoderm Fate

Cells in the caudal part of the neuroectoderm and mesoderm respond to the FGF gradient in distinct ways. A caudally receding gradient of FGF activity regulates somitogenesis and Hox gene expression in the paraxial mesoderm: as the FGF activity drops below a defined threshold, Notch signaling-dependent gene oscillations stop in the presomitic mesoderm and somite formation is initiated. As the source of FGF signaling recedes caudally, FGF activity needed for maintaining Notch-dependent oscillations drops below the threshold in a progressively more caudal region of the presomitic mesoderm. When the oscillations cease cells that are at a particular phase of their oscillation, initiate the process for defining a somite boundary. In this manner, during each oscillation, as the FGF gradient recedes, a new somite boundary is established caudal to the boundary formed in the previous oscillation, and a new prospective somite is defined (reviewed in [48]). At the same time that FGF signaling is regulating somitogenesis in the mesoderm it is regulating expression of Hox genes. In this context, progressively longer exposure to FGFs signals induces expression of genes located at progressively more 5′ positions in the Hox clusters. The earlier a progenitor leaves the caudal zone of high FGF signaling the shorter the time that it is exposed to FGF signals, and the more 3′ the position of the Hox gene that is expressed in the cell. Activation of Hox genes is not only linked to the length of exposure to FGF signals but also to Notch signaling and each oscillation facilitates activation of progressively more caudal Hox genes. By simultaneously regulating both somite number and Hox gene expression, Notch signaling-dependent oscillations ensure that the rostral limit of each Hox gene expression domain consistently correlates with the position defined by somite number [49].

Differentiation and patterning is also regulated in the spinal cord by a gradient or differential exposure over time to FGF signaling. In this caudal compartment of the neuroectoderm FGF signals progressively induce Hox genes. Unlike expression of PG 1–4 Hox genes in the hindbrain, the coordinated expression of PG 5 and greater along the rostral-caudal axis of the spinal cord is not as directly dependent on RA. Instead, RA signals have a key role in regulating the time at which neurons differentiate. As long as neural cells remain exposed to relatively high levels of FGF signaling a neural progenitor state is maintained and neurogenesis is prevented in the neuroectoderm. However, when the FGF level drops, somites begin to differentiate and they become the source of RA signals that promote neurogenesis in the adjacent neural plate/neural tube [50]. The coordination of neurogenesis with RA signals and somitogenesis is discussed later in Section 8.3 on neurogenesis.

8.3
Early Neurogenesis in the Neural Plate

8.3.1
Proneural Genes Define Neurogenic Domains

In Xenopus and zebrafish embryos neurogenesis begins relatively early in the neural plate and early neurons are distributed in a relatively simple pattern in the caudal neural plate: Rohon-Beard sensory neurons are distributed in a lateral column, interneurons are in an intermediate column and motor neurons are in a medial column (Fig. 8.2). The identity of neurons along the lateral-medial axis of the neural plate (dorsal-ventral axis of the spinal cord) is determined by opposing BMP and hedgehog signaling systems. The BMP signals promote dorsal fates while hedgehog signals promote ventral fates. Independent of the axial patterning mechanisms that determine neuron identity, however, the potential to be become a neuron is determined by the early expression of basic helix-loop-helix (bHLH) transcription factors such as *neurogenin1* (*ngn1*) that are related to atonal in Drosophila (reviewed in [51]). The early expression of these genes defines three pairs of longitudinal "neurogenic" domains in the caudal neural plate where primarily sensory neurons, interneurons, and motor neurons will differentiate. Ngn1 has a partially redundant role in neurogenesis as additional *atonal* homologs, such as *zath1* and *zath3 (neuroM)*, and *achaete scute* homologs such as *zash1a* and *zash1b* are also expressed. Knock-down of *ngn1* function leads to a specific loss of sensory neurons without affecting interneurons and motor neurons, however, simultaneous knock-down of both *ngn1* and *zath3* leads to a broader reduction of motor and interneurons. These bHLH factors are often collectively referred to as "proneural" factors, a term that is used to refer to their similarity to proneural genes in Drosophila. However, it is important to note that in this context it is their role in defining the potential to become *neurons* that has been studied and not their role in promoting *neural* fate. In the CNS most of these bHLH factors operate within the neuroepithelium that has already been defined as neural at an early stage of development.

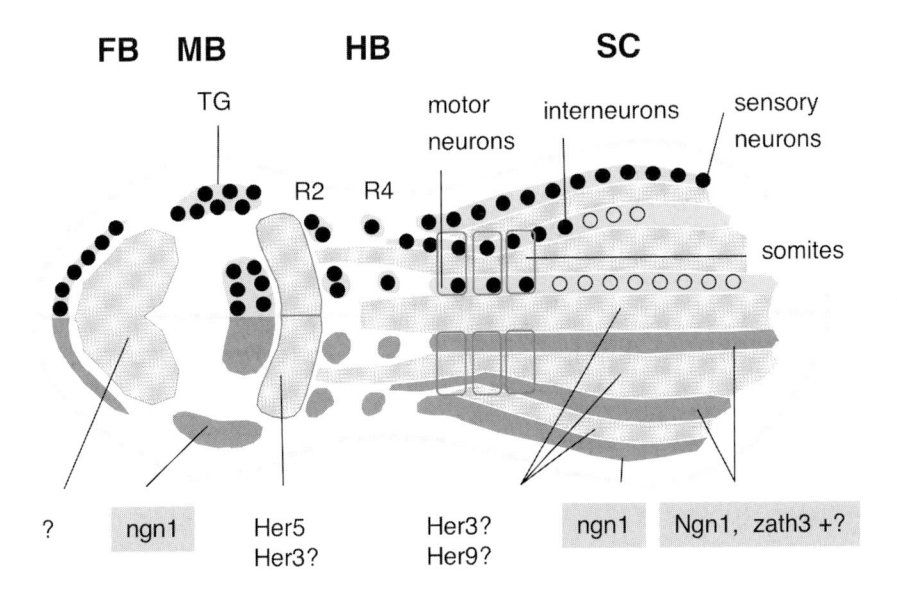

EARLY NEUROGENESIS IN THE ZEBRAFISH NEURAL PLATE

Fig. 8.2
Early neurogenesis in the zebrafish neural plate. A schematic of neurogenic (gray) and non-neurogenic (stippled) domains in the zebrafish neural plate at the 3–4–somite stage. FB, forebrain; MB, midbrain; HB, hindbrain; SC, spinal cord; TG, trigeminal sensory ganglion; R2, rhombomere 2; R4, rhombomere 4. See text for details.

8.3.2
Notch Signaling Regulates Differentiation within Neurogenic Domains

Expression of proneural genes defines neurogenic domains. However, within these domains only a subset of cells is permitted to become neurons. This is because as cells begin expressing progressively higher levels of *ngn1* they drive the expression of the Notch ligand Delta (DeltaA and DeltaD in zebrafish), which in turn activates Notch in the neighboring cells. Activation of Notch leads to the expression of Enhancer of split (E(spl))-related genes (called HES or HER genes for Hairy E(spl) Related in vertebrates). E(spl) genes are activated by Notch signaling in Drosophila where they function as repressors to inhibit transcription of proneural genes. In a similar manner HER genes that are activated by Notch inhibit expression of proneural genes like *ngn1* and prevent cells from becoming neurons in vertebrates. As a result, by driving Delta expression and activating Notch in neighboring cells, each *ngn1*–expressing cell in a neurogenic domain tries to inhibit its neighboring cell from becoming a neuron. As a consequence of this process, called lateral inhibition, only a subset of cells at any time is allowed to become a neuron. While proneural genes drive expression of Notch ligands, they also promote their own expression and in some cells proneural auto-activation overcomes lateral inhibition by neighboring cells. These cells acquire progressively higher levels of *ngn1* and effectively prevent their neighbors from differentiating [51].

Once sufficiently high levels of *ngn1* expression are achieved additional factors like Myt1 are expressed which make cells less sensitive to lateral inhibition. Escape from lateral inhibition allows additional bHLH transcription factors, such as neuroD whose expression is not as sensitive to lateral inhibition [52] but may be regulated by GSK3β-mediated phosphorylation [53], to be expressed further along in the neurogenesis cascade. These genes in turn initiate expression of neuron-specific genes that are required for differentiation as a neuron. A key step in the differentiation of a neuron is the initiation of genes that facilitate exit from the cell cycle. One of these genes encodes the cdk inhibitor p27 that also promotes formation of neurons by stabilizing the Ngn1 protein thereby making cells refractory to the effects of Notch activation [54]. Once a cell is selected to become a neuron it undergoes a number of characteristic morphological changes. It loses its connection with other neuroepithelial cells at the ventricular or apical surface and delaminates to occupy a position closer to the pial surface. After this, depending on the type of neuron it may migrate to an alternate position before it begins extending axons in a cell-specific manner.

8.3.3
Cell Fate Decisions Regulated by Notch

Although the role of Notch signaling has been discussed in the context of whether a cell becomes a neuron, Notch signaling plays a much broader role in generating cell diversity through local interactions in all tissues. In the nervous system it can influence the fate of both neuronal and non-neuronal progenitors. In the zebrafish lateral neural plate Notch activation promotes a neural crest progenitor fate and prevents cells from becoming Rohon-Beard sensory neurons [55]. In the ventral neural plate hedgehog signals promote expression of a bHLH transcription factor, Olig2, which defines a domain where both motor neuron progenitor cells (MNPC) and oligodendrocyte progenitor cells (OPC) are determined. In this domain expression of *olig2* and *ngn1* promotes MNPC while expression of olig2 alone promotes OPC fate. Activation of Notch early in development in a Motor Neuron Precursor Cell (MNPC) can prevent a terminal division to form primary motor neurons and keep the MNPC dividing as a progenitor to produce secondary motor neurons later in development. Notch activation can also promote Oligodendrocyte Precursor Cell (OPC) fate by preventing *ngn1* expression in a cell that initially expresses both *olig2* and *ngn1*. Later, activation of Notch in an *olig2*–expressing OPC can prevent the cell from differentiating as an oligodendrocyte and maintain it in the progenitor state [56]. Zebrafish *mind bomb* (*mib*) mutants have a broad reduction in Notch signaling [57, 58]. In the CNS the deficit in Notch signaling leads typically to an excess of early or primary neurons, a reduction of many late-developing neurons, and a reduction of some glia. However, consistent with a role for Notch in regulating differentiation in both neuron and glial progenitors there are examples of some early glial cells being produced in excess in *mib* mutants. Furthermore, at 1 day of development there is a dramatic reduction of neural progenitors in the CNS, consistent with an essential role for Notch activation in inhibiting differentiation and in maintaining the neural progenitor population.

The examples above illustrate how the consequence of Notch activation depends on the context in which this signaling pathway is activated. In many such situations it appears that Notch plays a permissive role in determining whether a cell will respond to local signals that would otherwise lead it to adopt a specific fate. By preventing a cell from responding to differentiation signals at a given time Notch activation provides the cell with the opportunity to adopt an alternate fate later in development. However, Notch has an instructive role in determining glial fate in specific situations. In one such context it is particularly interesting to note that axons are the source of a novel Notch ligand called F3 or Contactin, a GPI-linked neural cell adhesion molecule. During development F3 is clustered at paranodal regions that are critical for axoglial interactions. In this context F3–mediated, Su(H)-independent, Notch activation promotes oligodendrocyte maturation and myelination [59].

8.3.4
A Brief Review of Notch Signaling

Notch is a transmembrane receptor and during its maturation the full-length peptide undergoes an initial (S1) Furin convertase-mediated cleavage to create two peptides: one containing almost the whole extracellular domain with 29–36 EGF repeats and the other containing the remaining extracellular stub, the transmembrane domain and a long cytoplasmic tail that acts as a transcriptional activator when released from the surface membrane. The two fragments are held together as heterodimers in a calcium-dependent linkage. Removal of the extracellular fragment is a key step in Notch signaling and EDTA-mediated chelation of calcium can dissociate the two fragments causing Notch activation. The extracellular domain mediates interactions with membrane-bound DSL ligands (Delta, Serrate (Drosophila)/Jagged (vertebrates), Lag–2 (*C. elegans*)) that have a conserved DSL domain and a variable number of EGF repeats. In a poorly-understood step, the DSL ligand-Notch interaction facilitates removal of the Notch extracellular domain (NECD) and makes Notch susceptible to the action of TACE metalloproteases that cleave it at a site (S2) outside the Notch transmembrane domain. The membrane-bound Notch fragment that remains (NEXT, Notch extracellular truncation) after the cleavage by TACE metalloproteases is a substrate for γ-secretases which cleave it at a third site (S3) within the membrane to release an intracellular fragment (NICD). NICD functions in a transcriptional activator complex with Su(H)/CBF1/ RBP-J$\varkappa$ [60] to activate Notch target genes related to E(spl) as discussed earlier. In the absence of Notch signaling Su(H) associates with co-repressors and maintains many Notch target genes in a repressed state (for a review see [61]).

8.3.5
Factors that Influence the Outcome of Notch Signaling

When a group of cells competes to adopt a particular fate through the process of lateral inhibition the cell that eventually escapes lateral inhibition could be determined by chance or by biasing mechanisms that can influence the outcome of Notch signaling. Notch activation is under the exquisite control of factors that are responsible for post-translational modification of Notch or its DSL ligands. Some of these factors provide spatial regulation while other factors bias the outcome of Notch signaling.

8.3.5.1 Fringe Glycosyl Transferases
Fringe glycosyl transferases are essential modulators of the Notch EGF repeats [62]. In the Drosophila wing imaginal disc Fringe modification of Notch prevents Serrate from activating Notch in a dorsal compartment and restricts Notch activation to the boundary between dorsal and ventral compartments where Serrate and Delta cells abut each other. Notch activation at the dorsal-ventral compartment boundary establishes a Wingless signaling center that regulates proliferation and patterning of the Wing imaginal disc. Notch is expressed in a similar manner with Fringe, Serrate/jagged and Delta at the boundary of dorsal and ventral compartments of the spinal cord and the limb bud in vertebrates, however, their potential role in setting up signaling centers in this context has not yet been fully investigated.

8.3.5.2 E3 Ubiquitin Ligases

Notch signaling is also regulated by E3 ubiquitin ligases that ubiquitylate Notch signaling components to regulate their efficacy [57, 58, 63]. Ubiquitylation is a modification where one or more molecules of the polypeptide Ubiquitin are covalently linked to a substrate protein recognized by an E3 ubiquitin ligase. Addition of a polyubiquitin chain targets a protein for destruction in the lysosome or proteosome while mono-ubiquitylation targets substrates for internalization at the cell surface. Sel–10 is an E3 ligase that increases the turnover of the activated form of Notch in the nucleus. Itch is an E3 that interacts with Notch at the cell surface, where association with a cytoplasmic factor, Numb, promotes turnover of the intracellular Notch fragment, preventing it from getting to the nucleus. In this manner Numb too acts as a negative regulator of Notch signaling. LYNX is an E3 ligase for which Numb is a substrate and by destroying Numb it promotes Notch signaling.

Neuralized and Mib (mentioned earlier) are E3 ligases that have an unusual role as they *facilitate* Notch signaling by promoting endocytosis of Delta. Ubiquitylation of Delta, after it interacts with Notch in the neighboring cell, promotes internalization of the Notch extracellular domain together with Delta. Removal and trans-endocytosis of the Notch extracellular domain is thought to facilitate S2 and S3 cleavage and release of the active intracellular Notch fragment in the neighboring cell. While Mib appears to be a core component of the Notch signaling pathway, the role of vertebrate Neuralized is less clear.

8.3.5.3 Numb and Biasing Notch Signaling in the SOP

In Drosophila, Numb and Neuralized function synergistically to bias the outcome of Notch signaling between daughter cells of Sensory Organ Precursors (SOP, pI) which divide asymmetrically to form a posterior daughter (pIIa) and an anterior daughter (pIIb) cell [64]. The pIIa and pIIb cells divide asymmetrically to form a bristle and socket cell, and a sensory neuron and sheath cell, respectively. pIIa fate is determined by relatively high Notch activation and pIIb fate is determined by low Notch activation. The asymmetric activation of Notch is determined by asymmetric segregation of both Numb and Neuralized in the anterior pIIb cell by mechanisms linked to planar and cell polarity. Numb ensures low Notch signaling in the pIIb cell by mechanisms that are not yet well understood, while Neuralized promotes endocytosis of Delta to ensure Notch is efficiently activated in the neighboring pIIa cell. By linking the outcome of Notch signaling to planar polarity the embryo coordinates the orientation of Drosophila sensory organs.

The role of a *numb* and a *numb-like* gene has been studied in mouse and in mammalian cell culture; however, their role remains unclear in this system. Some studies have shown that sensitivity to Delta stimulation is inversely proportional to the level of Numb protein in neural crest stem cells and that as neural crest stem cells become progressively more gliogenic and less neurogenic there is a 20–30–fold increase in the ratio of Notch to Numb [65]. This is consistent with Notch signaling inhibiting neurogenesis and promoting gliogenesis. Other studies have shown that differentiation of Oligodendrocyte Progenitor Cells correlates with an increase in Numb and that loss of *numb* delays the differentiation of a specific neuronal population [66]. These studies are consistent with a role for Notch activation in delaying differentiation, and for Numb, as a Notch antagonist, in facilitating differentiation. Double knockouts of Numb and Numblike in mouse, however, result in a depletion of neural progenitors and in this context it has been suggested that Numb is required for self-renewing asymmetric divisions of neural progenitor cells [67].

In Drosophila, the selection of the SOP at the center of a proneural cluster, defined by the expression of proneural genes, is also determined by a biasing mechanism. A central cell is least likely to be selected by lateral inhibition alone because, unlike cells at the edge that are not inhibited by neighbors outside the proneural cluster, a central cell receives lateral inhibition from all its neighbors. Cells in a proneural cluster secrete a diffusible ligand for the EGF receptor. Activation of the EGF receptor promotes proneural gene function and inhibits Notch activation [68]. In this context the central cell has an advantage as it receives the most stimulation from its neighbors. Notch signaling acts in the context of this biasing system to select a central cell for SOP fate. Such central biasing systems have not yet been identified in the context of vertebrate neurogenic domains.

8.3.6
Determination of Neurogenic and Non-neurogenic Domains in the Neuroectoderm

The establishment of the early neurogenic domains is linked to early dorsal-ventral patterning mechanisms. Just as a gradient of BMP signaling helps divide the ectoderm into a prospective neural, neural crest and epidermal compartment it also divides the neuroectoderm into a lateral compartment with Rohon-Beard sensory neurons and an intermediate domain with interneurons [69, 70]. The relatively dorsal sensory neuron and interneuron domains, that are dependent on different levels of BMP signaling, are antagonized by ventral hedgehog signals that are responsible for induction of motor neurons in a ventral domain.

Mechanisms that determine where proneural genes are expressed in the zebrafish neural plate are not yet well defined; however, a few key elements have recently been identified. Early expression of *ngn1* in the caudal neural plate in longitudinal domains is dependent on an Nkx1 family homeodomain transcription factor Pnx (posterior neuron-specific homeobox) [71]. As in earlier studies that showed that the size of individual medio-lateral *ngn1* expression domains is determined by a gradient of BMP signaling, the longitudinal *pnx* expression domains are altered in zebrafish *swl/BMP2b* and chordin mutants. Pnx expression is promoted by posteriorizing factors secreted by the non-axial mesendoderm including Squint, FGF8 and retinoic acid, while it is inhibited by factors that promote a rostral fate including the Wnt antagonist Dkk1 and the nodal antagonist antivin/lefty. *pnx* expression, like *ngn1* is inhibited by Notch signaling. Analysis of the *ngn1* promoter shows that its expression in the anterior neuroectoderm is determined by *pax6* expression although its role in determining *ngn1* expression in the zebrafish neural plate is not as well characterized [72].

Between each of the longitudinal neurogenic domains is a complementary region where there is no early neurogenesis. In Xenopus it has been suggested that *zic2* has a role in defining these non-neurogenic domains, however, it is not clear that *zic2* homologs have a similar role in zebrafish [73]. Another domain where early neurogenesis is actively prevented is the region near the MHB where the isthmic organizer forms and neurogenesis is delayed in additional domains within the anterior neuroectoderm. In zebrafish these non-neurogenic domains are associated with the expression of a subclass of HER genes, *her3*, *her5* and *her9* [74–76]. Functional analysis of *her3* and *her5* has shown that they inhibit neurogenesis. These genes belong to the bHLH-O superfamily because of the tandem arrangement of their bHLH domain and an adjacent sequence known as the Orange domain. Despite sequence similarity to other E(spl) related subfamily members, their expression is not stimulated by Notch, instead their expression is inhibited by Notch activation. In this context their role is more similar to Hairy, which in Drosophila functions as a pre-pattern gene to define non-neurogenic domains.

8.4
Determination of Neuronal Identities in the Spinal Cord

After neurulation the neural plate forms the neural tube and the caudal compartment forms the spinal cord. The preceding sections have discussed early steps in patterning that define a neural compartment in the ectoderm and that divide the neural plate into sequentially smaller rostral-caudal and dorsal-ventral compartments. The emphasis in those sections was on studies in zebrafish and Xenopus that have a relatively simple nervous system. This section will discuss how the identity of neurons is established in the spinal cord with emphasis on what has been learned form the chick and mouse. It will describe the cellular organization of the spinal cord and then discuss how the expression of transcription factors in discrete dorsal-ventral compartments is established by opposing BMP and hedgehog systems that define dorsal and ventral neurons. Finally the organization of motor neurons will be discussed with emphasis on how their individual identity and pattern of axon outgrowth is specified by Hox and LIM domain transcription factors (reviewed in [45, 77, 78]) (Figs 8.3 and 8.4).

8.4.1
Cellular Organization of the Spinal Cord

The mature spinal cord in vertebrates is characterized by two prominent structures, a dorsal horn and a ventral horn. As discussed in the context of early patterning, neurons in the dorsal compartment process sensory input while neurons in the ventral compartment are responsible for motor output. Interneurons that relay cutaneous sensory information to the brain are in the dorsal half, while interneurons

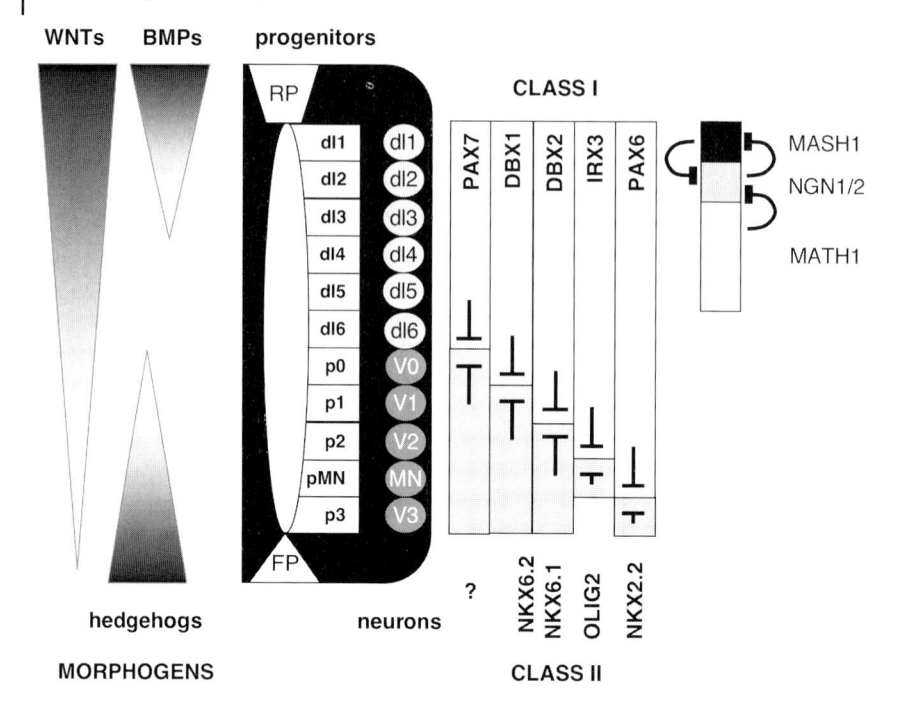

Fig. 8.3
Dorsal-ventral patterning of the chick and mouse spinal cord. Dorsal sources of BMPs and WNTs and ventral sources of hedgehog signals form opposing signaling gradients that define six types of dorsal progenitors (dl1–dl6) and five types of ventral progenitors (p0, p1, p2, pMN and p3). See text for details.

that integrate proprioceptive input with motor output are in the ventral half. Unlike the situation in zebrafish and Xenopus, there are no early sensory neurons like the Rohon-Beard cells in the chick or mouse dorsal spinal cord. Instead the sensory input comes entirely from late-forming sensory neurons in the dorsal root ganglia. These neurons are derived from neural crest progenitors that migrate out of the dorsal neural tube in all vertebrates. Specialized non-neuronal cells are located at the dorsal and ventral boundaries of spinal cord and they are the source of secreted signals that help to pattern the spinal cord.

8.4.2
Specification of Dorsal Neurons

Specification of neurons in the dorsal compartment has not been characterized as well as for the ventral neurons; nevertheless there are some parallels (reviewed in [77]). The dorsal epidermis is initially the source of BMP signals, and these cells may

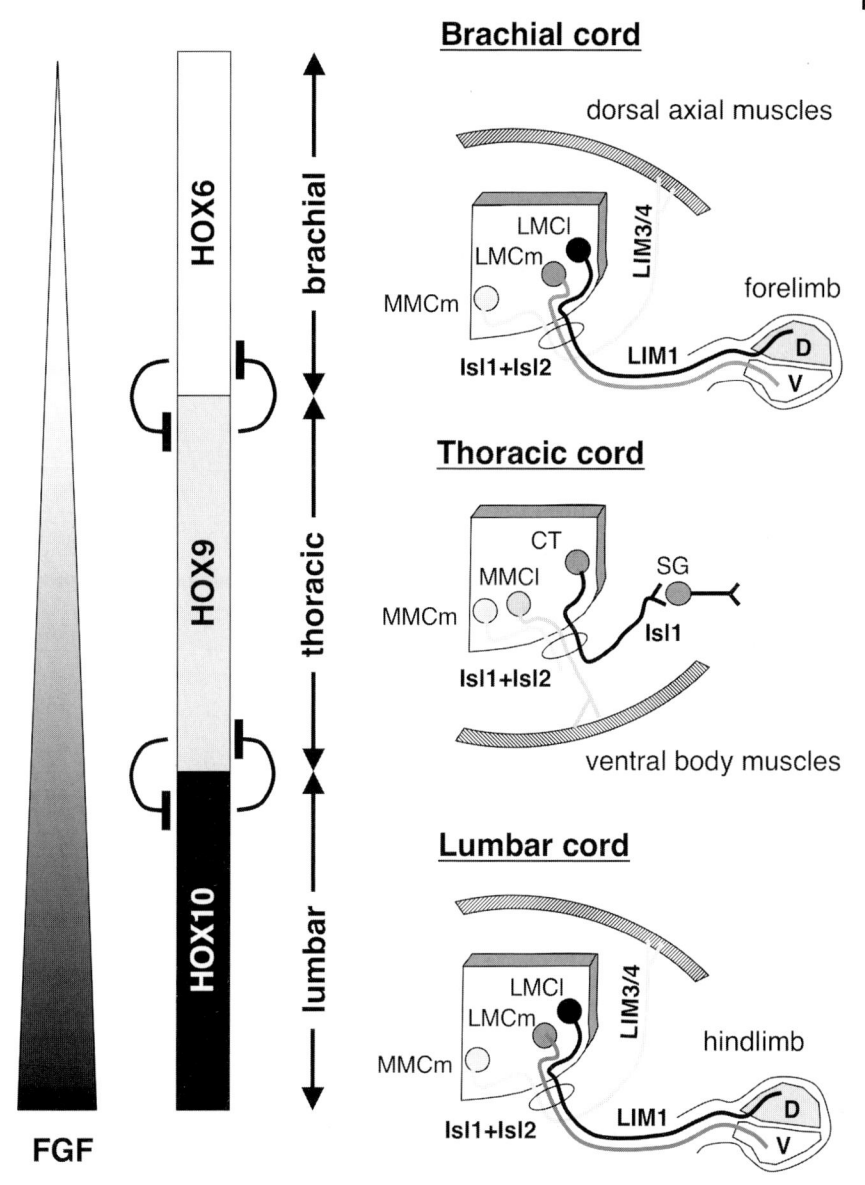

Fig. 8.4
Establishment of motor neuron (MN) identity by Hox and LIM domain genes. Differential exposure to FGF signals helps divide the neural tube along the rostral-caudal axis into brachial, thoracic and lumbar compartments where neurons express Hox6, Hox9 and Hox10, respectively. The MNs are organized into Medial Motor Columns (MMC) and Lateral Motor Columns (LMC) in the brachial and lumbar cord where MNs innervate the limbs. Lateral LMC neurons innervate dorsal (D) muscles in the limb, while medial LMC (LMCm) neurons innervate ventral (V) limb muscles. In the thoracic cord there is no LMC but instead there is a Column of Terni (CT) from which pre-ganglionic neurons innervate the sympathetic ganglia (SG). In the thoracic cord lateral MMC (MMCl) neurons innervate the ventral body muscles. Medial MMC (MMCm) neurons, distributed along the entire spinal cord, innervate dorsal axial muscles. Combinatorial expression of LIM domain genes defines cell-specific axon trajectories. See text for details.

have a role in inducing roof plate fate in dorsal midline cells in the spinal cord, which eventually becomes a source of BMPs and another TGFβ family member, Gdf7. These signals help to establish a dorsal-ventral gradient of BMP signals in the dorsal cord and they help to specify six classes of neurons, dl1–dl6. Loss of BMP signaling leads to a loss of the roof plate and the most dorsal dl1–dl3 neurons. Roof plate cells are also a source of Wnts (Wnt1 and Wnt3a), thus establishing a dorsal-ventral gradient of cell proliferation as Wnt signals have a mitogenic effect. Moreover, Wnts also are likely to contribute to specification of dorsal fates.

Differential activation by the BMP gradient and cross-repressive interactions define three discrete non-overlapping domains of proneural expression in the dorsal spinal cord; the proneural gene *Math1* is expressed in the most dorsal dl1 cells, *Ngn1* and *Ngn2* are expressed in progenitors that form dl2 neurons, and *Mash1* is expressed in a broad region just ventral to the dl2 cells (Fig. 8.3). Many of the *Math1* cells are proliferative due to their proximity to the source of Wnts, however, some post-mitotic cells become dorsal commissural interneurons. dl2 cells in the *ngn1/ngn2* expression domains become lim1/lim2–expressing interneurons. The relatively broad *Mash1* domain specifies dl3–dl5 interneurons. The mechanisms that define dl3–dl6 are complex and are likely to be determined both by interpretation of signaling gradients and by numerous local interactions. It is interesting to note that in this context expression of *Math1*, *ngn1/ngn2* and *Mash1* defines discrete populations of neurons and these proneural genes do not just have a general role in promoting neurogenesis. Eventually, the interneurons occupy five laminae in the dorsal horn. The most dorsal laminae (I and II) receive nociceptive input, laminae II-IV receive input from mechanoreceptors and the most ventral lamina (V) receives proprioceptive input. The specification of neurons in each of these laminae is determined not only by the dorsal-ventral location of the progenitors but also by the time of birth. While all the neurons from the *Math1* domain eventually occupy ventral laminae, early-born interneurons from the *Mash1* domain occupy ventral laminae, while late-born interneurons occupy dorsal laminae.

Homeodomain (HD) genes also determine dorsal fates. *Irx3* and the paired homeodomain transcription factors *Pax3* and *Pax7* are broadly expressed in the dorsal ventricular domain. In mouse, Zic1, Zic2 and Zic3 also have a role in determining dorsal fates and may also antagonize the effects of hedgehog signals in the dorsal spinal cord. However, mechanisms for defining fates in the dorsal cord will not be discussed in detail; instead, the mechanisms that are used in the ventral cord will be used to illustrate how different classes of interneurons and motor neurons are generated in the ventral cord.

8.4.3
The Notochord and Floor Plate Establish a Hedgehog Gradient

During gastrulation the dorsal organizer extends along the midline to form the axial mesoderm. Its derivative, the notochord, comes to lie under the neural plate and is an important source of hedgehog signals. Hedgehog signals from the notochord help to induce and maintain a specialized non-neuronal population called the floor plate in the ventral midline of the neural plate/neural tube. Loss of the notochord in the chick and mouse is associated with loss of the floor plate and ventral neurons. In zebrafish, the axial mesoderm becomes the source of hedgehog signals very early in development prior to overt differentiation of the notochord and floor plate. As a result some zebrafish mutants without a notochord or floor plate lose secondary neurons that are induced relatively late but retain primary motor neurons that are specified by early hedgehog signals from the undifferentiated axial mesoderm (reviewed in [79]). Many studies emphasize the role of Hedgehog signals in specification of the floor plate. However, in zebrafish, where the floor plate consists of three rows of midline cells, Hedgehog signals are only essential for determination of the lateral floor plate cells and not for central floor plate cells. Specification of central floor plate cells is dependent on nodal signals consistent with their origin from the organizer and the role of nodal signaling in formation and function of the organizer. Eventually the notochord and the floor plate become the source of a Hedgehog gradient in the spinal cord, with highest levels ventrally and progressively lower levels dorsally. In the chick and mouse the hedgehog gradient specifies four classes of interneurons and the motor neurons (reviewed in [78]). The V0 interneurons are most dorsal and they are ventral to the dI6 neurons of the dorsal compartment. These are followed ventrally by V1, V2, the motor neurons Vmn, and finally most ventrally, V3 (Fig. 8.3).

8.4.4
Class I and Class II Transcription Factors Respond to a Hedgehog Gradient to Define Discrete Compartments of Neuron Progenitors in the Ventral Cord

The ventral spinal cord is divided into discrete dorsal-ventral compartments based on the paired expression of Class I and Class II transcription factors (see Fig. 8.2). Class I transcription factors are expressed *dorsally* in the spinal cord. The ventral limit of their expression is determined by dose-dependent *repression* by hedgehog signals and by the mutually antagonistic effect of ventrally-expressed Class II transcription factors to which their expression is juxtaposed. The dorsal limit of Class II factors is determined by dose-dependent *activation* by hedgehog signals and by the antagonistic relationship with Class I transcription factors. Class I transcription factors Dbx1, Dbx2, Irx3, and Pax6 have progressively more ventral limits of expression and they are juxtaposed with Class II factors Nkx6.2, Nkx6.1, Olig2, Nkx2.2/Nkx2.9, respectively, with whom they have a mutually antagonistic relationship. The most dorsally expressed Class I factor, Pax7, is predicted to be juxtaposed with expression of a Class II transcription factor, however, this factor has not yet been identified. The juxtaposed Class I and Class II HD expression boundaries define dorsal-ventral compartments of overlapping transcription factor expression that specify the fate of p0, p1, p2, pMN (motor neurons), and p3 progenitors. The progenitors differentiate to form V0, V1, V2, MN, and V3 neurons respectively. Based on this scheme the pMNs are specified in a domain where the *nkx6.1/nkx6.2*, *pax6* and *olig2* expression domains overlap.

8.4.5
Specification of Motor Neuron Fate

The juxtaposed Class I and Class II HD factors define a ventral compartment of Olig2 expression where motor neuron progenitors are specified. By repressing the expression of the ventrally-expressed Class II Olig2, Class I Irx3 defines the dorsal limit of Olig2 expression. Conversely, by repressing ventrally-expressed Class II Nkx2.2/Nkx2.9 HD factors, Class I Pax6 defines the ventral limit of the Olig2 compartment. Olig2 expression in turn regulates the expression of motor neuron specific HD genes including Mnx (Mnr2 and Hb9) and LIM homeodomain factors (LIM HD) Lim3/Lhx3 and Isl1/Isl2. At the same time Olig2 facilitates Ngn2 expression, which has a general role in promoting neurogenesis and, in this context, a specific role in determining expression of motor neuron-specific HD factors such as Hb9. Pax6, which is expressed more broadly in the spinal cord, also has a role in directly promoting Ngn2 expression in the spinal cord.

By promoting expression of both MN-specific HD factors and proneural genes Olig2 coordinates dorsal-ventral patterning mechanisms as well as neurogenesis. However, as discussed earlier, Ngn is sensitive to lateral inhibition mediated by Notch signaling. So although cells may begin by expressing both MN-specific HD factors and proneural genes such as *ngn2*, a subset will eventually not maintain high levels of proneural expression due to lateral inhibition. Analysis of a motor neuron enhancer (MN^E) from the *Hb9* gene reveals that synergistic interactions between LIM HD factors and proneural factors like Ngn1/2 and neuroM are required for activating the *Hb9* gene effectively [80]. Isl1 and Lhx3 binding elements (IL-Es) flank an E box for binding of bHLH neuroM and E47. Isl1 and Lhx3 bind as dimers to the IL-Es and they in turn are linked by dimerization of two Nuclear Lim Interacting (NLI) or LIM-domain binding (Ldb) proteins. Dimerization of NLIs bound to Isl1 and Lhx3 on adjacent IL-Es is essential for effective activation of Hb9 by these LIM HDs. Furthermore, binding of neuroM and E47 or Ngn1/2 and E47 to the intervening E box is thought to create a complex that synergistically drives Hb9 expression and ensures that only cells expressing both the LIM HD factors and proneural factors will become motor neurons. Interestingly, Mash1 which can also bind this E box with E47, does not have this synergistic function, and in a context where it competes with NeuroM for binding to this site it would inhibit activation of Hb9. Olig2 has been shown to function as a repressor and it facilitates function of proneural genes like NeuroM and Ngn1/2 by repressing expression of Mash1.

Generation of spatially-restricted progenitor populations in the spinal cord is to a large extent determined by HD factors that function as repressors by recruiting Groucho/TLE co-repressors. Analysis of the regulatory elements upstream of Hb9 shows how broadly expressed general activators function in the context of spatially regulated de-repression to specify MN fate in a unique dorsal-ventral compartment. A 2.5–kb distal regulatory element in the *Hb9* gene binds Irx3 and Nkx2.2 both of which function as repressors to prevent activation of *Hb9*. Olig2 and Pax6 function as repressors to set the ventral limit of the Irx3 expression and the dorsal limit of Nkx2.2, respectively, to prevent repression of Hb9 in the MN progenitor domain. In the resulting spatially restricted de-repressed domain multiple activator elements cooperate to drive robust expression of Hb9. First a proximal promoter element with Sp1 and E2F binding domains utilizes these general activators to drive Hb9 expression. However, these elements can only drive modest expression on their own and they work synergistically with the IL-Es and E box elements in the MN^E, described above, to drive robust expression [81].

In addition to the activation mechanisms described above, retinoic acid signals have a critical role in activation of target genes at multiple stages during the specification of the neurons in the spinal cord [82]. During early development, as discussed earlier, the caudal neural plate is exposed to relatively high levels of FGF signaling from the presomitic mesoderm. At this time not only is neurogenesis inhibited but expression of Class I and Class II HD factors is inhibited as well. As FGF levels fall and somites begin to differentiate they begin to express RALDH2, an enzyme that converts retinaldehyde to RA. The RA signals promote expression of Class I HD factors in the dorsal cord. At the same time the notochord becomes a source of hedgehog signals and promotes expression of Class II HD factors in a dose-dependent manner. As discussed earlier antagonistic interactions between Class I and Class II HD factors defines a de-repressed ventral domain where Olig2 can be expressed. RA signals subsequently facilitate Olig2 expression in this de-repressed ventral compartment and later RA signals also facilitate expression of proneural genes. In particular, RA signals from the somites promote expression of NeuroM and this provides a spatiotemporal link between somitogenesis and neurogenesis in the adjacent neural tube.

8.4.6
Organization of the Motor Neurons and the LIM Code

Motor neurons are responsible for the somatic and visceral output to targets in the muscles and visceral organs, respectively. Motor neurons that innervate a particular muscle form a pool. These pools are further organized into discrete columns based on which targets they innervate (see Fig. 8.3). Somatic motor neurons that innervate the trunk muscles are in the Medial Motor Column (MMC), along the rostral-caudal axis. Motor neurons from the MMC that innervate the relatively dorsal axial muscles form a continuous medial MMCm group, while motor neurons that innervate muscles in the ventral body wall form a lateral MMCl group which exists only at thoracic levels. Somatic motor neurons that innervate the forelimb and hindlimb muscles are in brachial and lumbar Lateral Motor columns (LMC), respectively. These columns are also divided into a medial LMCm group that innervates ventral limb muscles and a lateral LMCl group that innervates dorsal limb muscles. At thoracic levels, visceral motor neurons that innervate neurons in the sympathetic ganglia form a specialized lateral column called the preganglionic motor Column of Terni (CT).

The specification of motor neuron fate and axonal outgrowth is determined by a remarkable LIM HD code [78]. These genes regulate where the cell bodies of the neurons settle within the cord to form discrete columns and they determine how growth cones will behave at various choice points in their trajectory. To begin, all motor neurons express Islet1 after their final division. Motor neurons that will exit the spinal cord via the ventral root express Lhx3/4 (LIM 3/4). To send a process out of the spinal cord they require the additional function of Islet 2. As they exit the ventral root all the somatic motor neurons express Islet1, Islet2 and Lhx3/4. At this point most of these motor neurons stop expressing Lhx3/4. However, the motor neurons in the MMCm are an exception and continued expression of Lhx3/4 directs their growth cones to dorsal axial muscles. The LMC motor column which innervates the limb muscles forms characteristic bulges at the brachial and lumbar levels of the spinal cord. Of these, the lateral LMC motor neurons are distinguished by the expression of Lim1, which allows their growth cones to respond to cues that direct them to dorsal muscles once they enter the limb. Finally, visceral motor neurons in the thoracic cord that become pre-ganglionic cells in the CT are distinguished from somatic motor neurons that contribute to lateral MMC by the relatively late loss of Islet2 expression [83].

The organization of motor neurons into distinct columns in the brachial, thoracic and lumbar regions is determined by Hox HD factors [84]. These genes, as discussed earlier, help define discrete rostral-caudal compartments of the caudal neural plate and mesoderm in response to progressively higher exposure to FGF signals. As in the case of Class I and Class II HD factors, a mutually antagonistic relationship between pairs of Hox genes defines discrete compartment boundaries along the rostral-caudal axis of the spinal cord. In contrast to dorsal-ventral compartments that are specified in progenitor cells along the ventricular surface of the spinal cord, compartments that distinguish motor neuron identities are determined by expression of Hox genes post-mitotically in neurons. The juxtaposition of a Hox6 expression domain with a Hox9 expression domain distinguishes the relatively rostral brachial compartment where LMC neurons differentiate from the relatively caudal thoracic compartment where the CT neurons differentiate. Similarly, it is thought that Hox10, which is expressed in an even more caudal compartment juxtaposed with Hox9 cells, distinguishes the fate of CT neurons from the lumbar neurons in the LMC. Brachial and lumbar LMC neurons are distinguished by their expression of RALDH2, which makes them a source of RA signals. These RA signals have a local inductive role in determining the fate of lateral MMC neurons that express Lim1 and innervate dorsal limb muscles. The CT neurons are distinguished by the expression of BMP5. It is thought these signals may have a role that is analogous to RALDH2 in LMC neurons and they may have a critical role in further differentiation of CT neurons. RA signals also have an early role in determination of motor neurons that innervate the forelimbs. RA functions synergistically with Hox6 in induction of brachial LMC fate. RA signals do not, however, induce lumbar LMC fate.

Once motor neurons extend axons to target muscles they become dependent on trophic signals like GDNF (glial-cell line-derived neurotrophic factor) for survival [85–88]. These signals promote survival of motor neurons that successfully innervate the muscle. Motor neurons that do not make appropriate connection are eliminated by programmed cell death. In the chick almost half the motor neurons are eliminated. Beyond selecting cells for survival the signals from target muscles also ensure that neurons which innervate the same muscle express the same ETS transcription factors and cell surface cadherins, which allows them to sort and form a common motor pool. Like motor neurons, cranial neural crest cells that contribute to the formation of the head cartilage also receive trophic signals when they reach their final destination in the branchial arches, in this case FGF signals [89]. Such trophic signals add to the robustness of developmental programs by completing the differentiation program only when cells have successfully reached or sent processes to their final destination.

8.5
Concluding Remarks

This chapter illustrates how cell diversity is progressively determined by morphogen gradients that are established and interpreted by cells at multiple stages of development to define progressively smaller compartments. Local inductive and inhibitory interactions within these compartments lead to the specification and selection of subsets of cells with unique identity, cellular organization and behavior. While an elaborate set of signaling events is responsible for generation of cellular diversity, the fate of individual cells is ultimately determined by additional mechanisms that maintain specific cell populations only when the cells are generated in the correct number and in their appropriate environment. This chapter illustrates how generation of cell diversity is a dynamic process that involves robust self-regulating genetic networks with multiple partially redundant molecular components and elaborate positive and negative feedback mechanisms.

8.5.1
Gradients, Compartments, Boundaries and Local Interactions

From the earliest steps in development localized sources of signaling factors are established and they are interpreted to define discrete domains of gene expression. The signaling factors do not act in isolation as morphogens but rather are often components of a robust self-regulating reaction-diffusion system. Although in many of these situations it has been suggested that a *spatial* morphogen gradient is interpreted to define different fates, it is exposure for different amounts of *time* to the morphogen that is often the critical factor that determines distinct fates. The definition of discrete compartments of gene expression is accomplished in a stepwise manner. Cells respond to morphogens to define paired domains of juxtaposed HD factor expression and mutually antagonistic relationships between these factors define compartment boundaries. Eventually, unique cell surface properties and the sorting of cells that belong to different compartments contributes to the sharpening of boundaries. The definition of compartment boundaries defines tissue domains with distinct fate. Juxtaposition of specific HD factor expression domains is also often the first step in establishing secondary organizers that are the source of additional morphogens responsible for division of the tissue into smaller compartments.

8.5.2
Three Principles of Transcription Factor Regulation

At each stage of development many of the same signaling pathways are used in an iterative manner in different combinations to define cell fate. Analysis of their role in driving transcription factor expression reveals some broad principles about how they are able to drive transcriptional activation of different targets in spatially-restricted domains [90]. The receptors for individual signaling systems lead to the production of transcription activators in response to ligand activation. Typically, receptor-activated transcriptional factors cannot effectively drive expression of target genes on their own *in vivo*. They function synergistically with one or more local activators whose function is signal independent. A key element in the specificity of local activator mechanisms is that their function is regulated by default repressor mechanisms and de-repression is essential for their function.

8.5.3
Temporal Regulation of Cell Diversity

An import goal for the future will be to understand how cells progressively acquire the potential to adopt distinct cell fates over the course of development. Although their contribution has not been discussed here, chromatin remodeling factors play an important role in determining the developmental potential of cells. Furthermore, recent studies have shown that while chromatin remodeling confers transcriptional competence to the Hox gene cluster, sequential 3′ to 5′ extrusion of genes from their chromosome territory may determine the coordinated expression of paralogs over time [91]. This chapter has emphasized how spatial patterning mechanisms determine cell identity and less has been discussed about how the potential of a progenitor in defined location changes over time. This question has been studied in Drosophila neuroblasts and in the vertebrate retina and cerebral cortex (reviewed in [92]). In each of these situations, the time of birth has an important role in determining the identity of the cells produced by progenitor cells. In Drosophila the progressive expression of Hunchback, Krupple, Pdm and Castor determines the temporal identity of neuroblasts and the lineage of the Ganglion Mother Cell it will produce. This system promises to serve as an excellent model for understanding specification of temporal identity. An integration of mechanisms that determine spatial and temporal identity and their regulation by cell-cell interactions promises to provide key insights into self-organization of cellular diversity in the growing embryo.

References

1. A. F. Schier. Nodal signaling in vertebrate development. *Annu Rev Cell Dev Biol* **2003**; *19*: 589–621.
2. Y. Chen, A. F. Schier. The zebrafish Nodal signal Squint functions as a morphogen. *Nature* **2001**; *411*: 607–610.
3. M. Rex, E. Hilton, R. Old. Multiple interactions between maternally-activated signalling pathways control Xenopus nodal-related genes. *Int J Dev Biol* **2002**; *46*: 217–226.
4. E. Bell, I. Munoz-Sanjuan, C. R. Altmann, A. Vonica, A. H. Brivanlou. Cell fate specification and competence by Coco, a maternal BMP, TGFbeta and Wnt inhibitor. *Development* **2003**; *130*: 1381–1389.
5. C. Zhang, T. Basta, L. Hernandez-Lagunas, P. Simpson, D. L. Stemple, K. B. Artinger, M. W. Klymkowsky. Repression of nodal expression by maternal B1–type SOXs regulates germ layer formation in Xenopus and zebrafish. *Dev Biol* **2004**; *273*: 23–37.
6. A. F. Schier, W. S. Talbot. Nodal signaling and the zebrafish organizer *Int J Dev Biol* **2001**; *45*: 289–297.
7. E. M. De Robertis, J. Larrain, M. Oelgeschlager, O. Wessely. The establishment of Spemann's organizer and patterning of the vertebrate embryo. *Nat Rev Genet* **2000**, *1*: 171–181.
8. L. Solnica-Krezel, W. Driever. The role of the homeodomain protein Bozozok in zebrafish axis formation. *Int J Dev Biol* **2001**; *45*: 299–310.
9. D. W. Houston, M. Kofron, E. Resnik, R. Langland, O. Destree, C. Wylie, J. Heasman. Repression of organizer genes in dorsal and ventral Xenopus cells mediated by maternal XTcf3. *Development* **2002**; *129*: 4015–4025.
10. T. Leung, J. Bischof, I. Soll, D. Niessing, D. Zhang, J. Ma, H. Jackle, W. Driever. bozozok directly represses bmp2b transcription and mediates the earliest dorsoventral asymmetry of bmp2b expression in zebrafish. *Development* **2003**; *130*: 3639–3649.
11. M. Furthauer, J. Van Celst, C. Thisse, B. Thisse. Fgf signalling controls the dorsoventral patterning of the zebrafish embryo. *Development* **2004**; *131*: 2853–2864.
12. M. Tsang, S. Maegawa, A. Kiang, R. Habas, E. Weinberg, I. B. Dawid. A role for MKP3 in axial patterning of the zebrafish embryo. *Development* **2004**; *131*: 2769–2779.
13. A. Streit, A. J. Berliner, C. Papanayotou, A. Sirulnik, C. D. Stern. Initiation of neural induction by FGF signalling before gastrulation. *Nature* **2000**; *406*: 74–78.
14. L. Saude, K. Woolley, P. Martin, W. Driever, D. L. Stemple. Axis-inducing activities and cell fates of the zebrafish organizer. *Development* **2000**; *127*: 3407–3417.
15. V. Willot, J. Mathieu, Y. Lu, B. Schmid, S. Sidi, Y. L. Yan, J. H. Postlethwait, M. Mullins, F. Rosa, N. Peyrieras. Cooperative action of ADMP- and BMP-mediated pathways in regulating cell fates in the zebrafish gastrula. *Dev Biol* **2002**; *241*: 59–78.
16. Z. Lele, M. Nowak, M. Hammerschmidt. Zebrafish admp is required to restrict the size of the organizer and to promote

posterior and ventral development. *Dev Dyn* **2001**; *222*: 681–687.

17 H. Meinhardt. Organizer and axes formation as a self-organizing process. *Int J Dev Biol* **2001**; *45*: 177–188.

18 K. Joubin, C. D. Stern. Molecular interactions continuously define the organizer during the cell movements of gastrulation. *Cell* **1999**; *98*: 559–571.

19 F. Bertocchini, I. Skromne, L. Wolpert, C. D. Stern. Determination of embryonic polarity in a regulative system: evidence for endogenous inhibitors acting sequentially during primitive streak formation in the chick embryo. *Development* **2004**; *131*: 3381–3390.

20 R. Mayor, M. J. Aybar. Induction and development of neural crest in *Xenopus laevis*. *Cell Tissue Res* **2001**; *305*: 203–209.

21 V. H. Nguyen, B. Schmid, J. Trout, S. A. Connors, M. Ekker, M. C. Mullins. Ventral and lateral regions of the zebrafish gastrula, including the neural crest progenitors, are established by a bmp2b/swirl pathway of genes. *Dev Biol* **1998**; *199*: 93–110.

22 C. Tribulo, M. J. Aybar, V. H. Nguyen, M. C. Mullins, R. Mayor. Regulation of Msx genes by a Bmp gradient is essential for neural crest specification. *Development* **2003**; *130*: 6441–6452.

23 L. Marchant, C. Linker, P. Ruiz, N. Guerrero, R. Mayor. The inductive properties of mesoderm suggest that the neural crest cells are specified by a BMP gradient. *Dev Biol* **1998**; *198*: 319–329.

24 T. Luo, M. Matsuo-Takasaki, J. H. Lim, T. D. Sargent. Differential regulation of Dlx gene expression by a BMP morphogenetic gradient. *Int J Dev Biol* **2001**; *45*: 681–684.

25 A. K. Knecht, M. Bronner-Fraser. Induction of the neural crest: a multigene process. *Nat Rev Genet* **2002**; *3*: 453–461.

26 Y. Sasai. Regulation of neural determination by evolutionarily conserved signals: anti-BMP factors and what next? *Curr Opin Neurobiol* **2001**; *11*: 22–26.

27 K. Mizuseki, M. Kishi, K. Shiota, S. Nakanishi, Y. Sasai. SoxD: an essential mediator of induction of anterior neural tissues in Xenopus embryos. *Neuron* **1998**; *21*: 77–85.

28 V. Graham, J. Khudyakov, P. Ellis, L. Pevny. SOX2 functions to maintain neural progenitor identity. *Neuron* **2003**; *39*: 749–765.

29 P. J. Ebert, J. R. Timmer, Y. Nakada, A. W. Helms, P. B. Parab, Y. Liu, T. L. Hunsaker, J. E. Johnson. Zic1 represses Math1 expression via interactions with the Math1 enhancer and modulation of Math1 autoregulation. *Development* **2003**; *130*: 1949–1959.

30 P. P. Boyl, M. Signore, A. Annino, J. P. Barbera, D. Acampora, A. Simeone. Otx genes in the development and evolution of the vertebrate brain. *Int J Dev Neurosci* **2001**; *19*: 353–363.

31 T. Kudoh, S. W. Wilson, I. B. Dawid. Distinct roles for Fgf, Wnt and retinoic acid in posteriorizing the neural ectoderm. *Development* **2002**; *129*: 4335–4346.

32 C. H. Kim, T. Oda, M. Itoh, D. Jiang, K. B. Artinger, S. C. Chandrasekharappa, W. Driever, A. B. Chitnis. Repressor activity of Headless/Tcf3 is essential for vertebrate head formation. *Nature* **2000**; *407*: 913–916.

33 C. Houart, L. Caneparo, C. Heisenberg, K. Barth, M. Take-Uchi, S. Wilson. Establishment of the telencephalon during gastrulation by local antagonism of Wnt signaling. *Neuron* **2002**; *35*: 255–265.

34 A. Kawahara, C. B. Chien, I. B. Dawid. The homeobox gene mbx is involved in eye and tectum development. *Dev Biol* **2002**; *248*: 107–117.

35 K. Lun, M. Brand. A series of no isthmus (noi) alleles of the zebrafish pax2.1 gene reveals multiple signaling events in development of the midbrain-hindbrain boundary. *Development* **1998**; *125*: 3049–3062.

36 C. P. Heisenberg, C. Houart, M. Take-Uchi, G. J. Rauch, N. Young, P. Coutinho, I. Masai, L. Caneparo, M. L. Concha, R. Geisler, et al. A mutation in the Gsk3–binding domain of zebrafish Masterblind/Axin1 leads to a fate transformation of telencephalon and eyes to diencephalons. *Genes Dev* **2001**; *15*: 1427–1434.

37 R. I. Dorsky, M. Itoh, R. T. Moon, A. Chitnis. Two tcf3 genes cooperate to pattern the zebrafish brain. *Development* **2003**; *130*: 1937–1947.

38 S. W. Wilson, C. Houart. Early steps in the development of the forebrain. *Dev Cell* **2004**; *6*: 167–181.

39 M. Shinya, C. Eschbach, M. Clark, H. Lehrach, M. Furutani-Seiki. Zebrafish Dkk1, induced by the pre-MBT Wnt signaling, is secreted from the prechordal plate and patterns the anterior neural plate. *Mech Dev* **2000**; *98*: 3–17.

40 H. Hashimoto, M. Itoh, Y. Yamanaka, S. Yamashita, T. Shimizu, L. Solnica-Krezel, M. Hibi, T. Hirano. Zebrafish Dkk1 functions in forebrain specification and axial mesendoderm formation. *Dev Biol* **2000**; *217*: 138–152.

41 O. Kazanskaya, A. Glinka, C. Niehrs. The role of Xenopus dickkopf1 in prechordal plate specification and neural patterning. *Development* **2000**; *127*: 4981–4992.

42 W. Wurst, L. Bally-Cuif. Neural plate patterning: upstream and downstream of the isthmic organizer. *Nat Rev Neurosci* **2001**; *2*: 99–108.

43 M. M. Braun, A. Etheridge, A. Bernard, C. P. Robertson, H. Roelink. Wnt signaling is required at distinct stages of development for the induction of the posterior forebrain. *Development* **2003**; *130*: 5579–5587.

44 C. B. Moens, V. E. Prince. Constructing the hindbrain: insights from the zebrafish. *Dev Dyn* **2002**; *224*: 1–17.

45 K. R. Melton, A. Iulianella, P. A. Trainor. Gene expression and regulation of hindbrain and spinal cord development. *Front Biosci* **2004**; *9*: 117–138.

46 J. Dubrulle, O. Pourquie. fgf8 mRNA decay establishes a gradient that couples axial elongation to patterning in the vertebrate embryo. *Nature* **2004**; *427*: 419–422.

47 M. Gonzalez-Gaitan, H. Stenmark. Endocytosis and signaling: a relationship under development. *Cell* **2003**; *115*: 513–521.

48 O. Pourquie. Vertebrate somitogenesis: a novel paradigm for animal segmentation? *Int J Dev Biol* **2003**; *47*: 597–603.

49 J. Dubrulle, M. J. McGrew, O. Pourquie. FGF signaling controls somite boundary position and regulates segmentation

clock control of spatiotemporal Hox gene activation. *Cell* **2001**; *106*: 219–232.

50 R. Diez del Corral, K. G. Storey. Opposing FGF and retinoid pathways: a signalling switch that controls differentiation and patterning onset in the extending vertebrate body axis. *Bioessays* **2004**; *26*: 857–869.

51 A. B. Chitnis. Control of neurogenesis – lessons from frogs, fish and flies. *Curr Opin Neurobiol* **1999**; *9*: 18–25.

52 A. Chitnis, C. Kintner. Sensitivity of proneural genes to lateral inhibition affects the pattern of primary neurons in Xenopus embryos. *Development* **1996**; *122*: 2295–2301.

53 E. A. Marcus, C. Kintner, W. Harris. The role of GSK3beta in regulating neuronal differentiation in *Xenopus laevis*. *Mol Cell Neurosci* **1998**; *12*: 269–280.

54 A. E. Vernon, C. Devine, A. Philpott. The cdk inhibitor p27Xic1 is required for differentiation of primary neurones in Xenopus. *Development* **2003**; *130*: 85–92.

55 R. A. Cornell, J. S. Eisen. Delta/Notch signaling promotes formation of zebrafish neural crest by repressing Neurogenin 1 function. *Development* **2002**; *129*: 2639–2648.

56 H. C. Park, B. Appel. Delta-Notch signaling regulates oligodendrocyte specification. *Development* **2003**; *130*: 3747–3755.

57 M. Itoh, C. H. Kim, G. Palardy, T. Oda, Y. J. Jiang, D. Maust, S. Y. Yeo, K. Lorick, G. J. Wright, L. Ariza-McNaughton, et al. Mind bomb is a ubiquitin ligase that is essential for efficient activation of Notch signaling by Delta. *Dev Cell* **2003**; *4*: 67–82.

58 W. Chen, D. Casey Corliss. Three modules of zebrafish Mind bomb work cooperatively to promote Delta ubiquitination and endocytosis. *Dev Biol* **2004**; *267*: 361–373.

59 Q. D. Hu, B. T. Ang, M. Karsak, W. P. Hu, X. Y. Cui, T. Duka, Y. Takeda, W. Chia, N. Sankar, Y. K. Ng, et al. F3/contactin acts as a functional ligand for Notch during oligodendrocyte maturation. *Cell* **2003**; *115*: 163–175.

60 B. De Strooper, W. Annaert, P. Cupers, P. Saftig, K. Craessaerts, J. S. Mumm, E. H. Schroeter, V. Schrijvers, M. S. Wolfe, W. J. Ray, et al. A presenilin-1–dependent gamma-secretase-like protease mediates release of Notch intracellular domain. *Nature* **1999**; *398*: 518–522.

61 J. S. Mumm, R. Kopan. Notch signaling: from the outside in. *Dev Biol* **2000**; *228*: 151–165.

62 R. S. Haltiwanger, P. Stanley. Modulation of receptor signaling by glycosylation: fringe is an O-fucose-beta1,3–N-acetylglucosaminyltransferase. *Biochim Biophys Acta* **2002**; *1573*: 328–335.

63 E. C. Lai. Protein degradation: four E3s for the notch pathway. *Curr Biol* **2002**; *12*: R74–R78.

64 R. Le Borgne, F. Schweisguth. Unequal segregation of Neuralized biases Notch activation during asymmetric cell division. *Dev Cell* **2003**; *5*: 139–148.

65 C. J. Kubu, K. Orimoto, S. J. Morrison, G. Weinmaster, D. J. Anderson, J. M. Verdi. Developmental changes in Notch1 and numb expression mediated by local cell-cell interactions underlie progressively increasing delta sensitivity in neural crest stem cells. *Dev Biol* **2002**; *244*: 199–214.

66 M. I. Givogri, V. Schonmann, R. Cole, J. De Vellis, E. R. Bongarzone. Notch1 and Numb genes are inversely expressed as oligodendrocytes differentiate. *Dev Neurosci* **2003**; *25*: 50–64.
67 P. H. Petersen, K. Zou, S. Krauss, W. Zhong. Continuing role for mouse Numb and Numbl in maintaining progenitor cells during cortical neurogenesis. *Nat Neurosci* **2004**; *7*: 803–811.
68 J. Culi, E. Martin-Blanco, J. Modolell. The EGF receptor and N signalling pathways act antagonistically in Drosophila mesothorax bristle patterning. *Development* **2001**; *128*: 299–308.
69 K. A. Barth, Y. Kishimoto, K. B. Rohr, C. Seydler, S. Schulte-Merker, S. W. Wilson. Bmp activity establishes a gradient of positional information throughout the entire neural plate. *Development* **1999**; *126*: 4977–4987.
70 V. H. Nguyen, J. Trout, S. A. Connors, P. Andermann, E. Weinberg, M. C. Mullins. Dorsal and intermediate neuronal cell types of the spinal cord are established by a BMP signaling pathway. *Development* **2000**; *127*: 1209–1220.
71 Y. K. Bae, T. Shimizu, T. Yabe, C. H. Kim, T. Hirata, H. Nojima, O. Muraoka, T. Hirano, M. Hibi. A homeobox gene, pnx, is involved in the formation of posterior neurons in zebrafish. *Development* **2003**; *130*: 1853–1865.
72 R. Scardigli, N. Baumer, P. Gruss, F. Guillemot, I. Le Roux. Direct and concentration-dependent regulation of the proneural gene Neurogenin2 by Pax6. *Development* **2003**; *130*: 3269–3281.
73 R. Brewster, J. Lee, A. Ruiz i Altaba. Gli/Zic factors pattern the neural plate by defining domains of cell differentiation. *Nature* **1998**; *393*: 579–583.
74 A. Geling, C. Plessy, S. Rastegar, U. Strahle, L. Bally-Cuif. Her5 acts as a prepattern factor that blocks neurogenin1 and coe2 expression upstream of Notch to inhibit neurogenesis at the midbrain-hindbrain boundary. *Development* **2004**; *131*: 1993–2006.
75 A. Geling, M. Itoh, A. Tallafuss, P. Chapouton, B. Tannhauser, J. Y. Kuwada, A. B. Chitnis, L. Bally-Cuif. bHLH transcription factor Her5 links patterning to regional inhibition of neurogenesis at the midbrain-hindbrain boundary. *Development* **2003**; *130*: 1591–1604.
76 C. Leve, M. Gajewski, K. B. Rohr, D. Tautz. Homologues of c-hairy1 (her9) and lunatic fringe in zebrafish are expressed in the developing central nervous system, but not in the presomitic mesoderm. *Dev Genes Evol* **2001**; *211*: 493–500.
77 T. Caspary, K. V. Anderson. Patterning cell types in the dorsal spinal cord: what the mouse mutants say. *Nat Rev Neurosci* **2003**; *4*: 289–297.
78 R. Shirasaki, S. L. Pfaff. Transcriptional codes and the control of neuronal identity. *Annu Rev Neurosci* **2002**; *25*: 251–281.
79 K. E. Lewis, J. S. Eisen. From cells to circuits: development of the zebrafish spinal cord. *Prog Neurobiol* **2003**; *69*: 419–449.
80 S. K. Lee, S. L. Pfaff. Synchronization of neurogenesis and motor neuron specification by direct coupling of bHLH and homeodomain transcription factors. *Neuron* **2003**; *38*: 731–745.
81 S. K. Lee, L. W. Jurata, J. Funahashi, E. C. Ruiz, S. L. Pfaff. Analysis of embryonic motoneuron gene regulation: derepres-

sion of general activators function in concert with enhancer factors. *Development* **2004**; *131*: 3295–3306.

82 B. Appel, J. S. Eisen. Retinoids run rampant: multiple roles during spinal cord and motor neuron development. *Neuron* **2003**; *40*: 461–464.

83 J. P. Thaler, S. J. Koo, A. Kania, K. Lettieri, S. Andrews, C. Cox, T. M. Jessell, S. L. Pfaff. A postmitotic role for Isl-class LIM homeodomain proteins in the assignment of visceral spinal motor neuron identity. *Neuron* **2004**; *41*: 337–350.

84 J. S. Dasen, J. P. Liu, T. M. Jessell. Motor neuron columnar fate imposed by sequential phases of Hox-c activity. *Nature* **2003**; *425*: 926–933.

85 S. Arber, D. R. Ladle, J. H. Lin, E. Frank, T. M. Jessell. ETS gene Er81 controls the formation of functional connections between group Ia sensory afferents and motor neurons. *Cell* **2000**; *101*: 485–498.

86 J. Livet, M. Sigrist, S. Stroebel, V. De Paola, S. R. Price, C. E. Henderson, T. M. Jessell, S. Arber. ETS gene Pea3 controls the central position and terminal arborization of specific motor neuron pools. *Neuron* **2002**; *35*: 877–892.

87 G. Haase, E. Dessaud, A. Garces, B. de Bovis, M. Birling, P. Filippi, H. Schmalbruch, S. Arber, O. deLapeyriere. GDNF acts through PEA3 to regulate cell body positioning and muscle innervation of specific motor neuron pools. *Neuron* **2002**; *35*: 893–905.

88 D. R. Ladle, E. Frank. The role of the ETS gene PEA3 in the development of motor and sensory neurons. *Physiol Behav* **2002**; *77*: 571–576.

89 N. B. David, L. Saint-Etienne, M. Tsang, T. F. Schilling, F. M. Rosa. Requirement for endoderm and FGF3 in ventral head skeleton formation. *Development* **2002**; *129*: 4457–4468.

90 S. Barolo, J. W. Posakony. Three habits of highly effective signaling pathways: principles of transcriptional control by developmental cell signaling. *Genes Dev* **2002**; *16*: 1167–1181.

91 S. Chambeyron, W. A. Bickmore. Chromatin decondensation and nuclear reorganization of the HoxB locus upon induction of transcription. *Genes Dev* **2004**; *18*: 1119–1130.

92 B. J. Pearson, C. Q. Doe. Specification of temporal identity in the developing nervous system. *Annu Rev Cell Dev Biol* **2004**; *20*: 619–647.

9
The Molecular Basis of Directional Cell Migration

Hans Georg Mannherz

9.1
Introduction

Locomotion is one of the main attributes of animal life. Multicellular animals have developed special tissues – the muscle tissues – that are activated during locomotion and provide the necessary force. Normally parenchymal cells of most organs have differentiated in order to perform particular tasks at defined locations therefore these cells have become sessile and do not move. But also inside a highly developed multicellular organism there exists certain cells that almost permanently exhibit migratory activity or are stimulated to do so. Examples of such cells are white blood cells such as the polymorphonuclear leukocytes or neutrophils and macrophages, which actively migrate through the capillary wall towards a chemotactic signal, or fibroblasts that after wounding differentiate to myofibroblasts and crawl along the surface of for instance a fibrin clot. During embryogenesis many more cell types exhibit migratory activity in order to reach their destined location, thus complete epithelial sheets move, neural crest cells migrate long distances to reach their final destinations, and neuronal cells send out long extensions that are guided from a highly motile region at their tips – the growth cones. Many of these migratory activities are stimulated and regulated by extracellular cues such as growth factors that act locally or systemically. Thus extracellular factors influence the organization and activity of the cellular motile apparatus. The migratory activity of these cells is highly regulated by for instance bacterial chemotactic signals which activate neutrophils and macrophages. In contrast, under adverse pathological conditions transformed parenchymal cells can exhibit migratory activity. Metastatic tumor cell dissemination largely depends on the regained ability to migrate through vessel walls and connective tissue.

Cell Signaling and Growth Factors in Development. Edited by K. Unsicker and K. Krieglstein
Copyright © 2006 WILEY-VCH Verlag GmbH & Co. KGaA, Weinheim
ISBN 3-527-31034-7

9.2
How and What Moves Cells

Motile events can be observed in any living cell. Even sessile cells constantly exhibit within their cytoplasm motile processes such as the transport of cargo-carrying vesicles or the relocation or reshaping of intracellular organelles. Besides these intracellular motile events, some cells also migrate, i.e. they alter their location. Only sperm cells of multicellular organisms swim by being propelled by the beating movement of their tails. All other mobile cells move by crawling. In order to do this they have to be free i.e. not restrained by gluey cell-cell contacts (tight and adherens junctions or desmosomes) to neighboring cells. They engage however in transient contacts to the extracellular matrix (ECM) as they move forward and they exhibit a remarkable ability for flexibility, and shape changes. A fibroblastic cell migrating on a flat substratum is shown in Fig. 9.1. Fig. 9.1A-C shows a migrating cell fluorescently stained for actin (red), tubulin (green) and the merged image with chromatin stain (blue), respectively. It has a triangular form with a broad leading edge (bottom) that determines the direction of movement and a small trailing end. The leading edge extends into a thin broad veil-like plasma membrane extension termed lamellipodium as seen best in the scanning electron micrograph (Fig. 9.1D) which represents one type of transient membrane specialization generating the protrusive force necessary for cell locomotion. In connective tissues migrating cells often have to penetrate a dense meshwork of extracellular fibrils. Under these circumstances migratory cells often secrete proteases like the MMPs (matrix metalloproteases) in order to generate a path for forward movement. Incidentally, MMPs have been shown to play an essential role during the invasive migration of tumor cells.

During locomotion the lamellipodium extends and by subsequent formation of adherence points to the substratum at the lamellipodial ventral face, generates new points of attachment for the cell which is subsequently pulled forward. The extension of the lamellipodium at the leading edge is solely achieved by increased the polymerization and depolymerization activity of the cytoskeletal component actin alone which becomes organized into branched microfilaments (see Section 9.3.2). During the second phase of movement, the attachment phase, integrin-based points of attachment are formed that connect the lamellipodial plasma membrane to components of the ECM. These attachment contacts serve as points of fixation for the third phase, the traction phase, during which the remaining cell body is pulled or pushed forward by forces generated by acto-myosin interactions. Finally the points of attachment at the trailing end of the cell are weakened finally leading to de-adhesion (see scheme in Fig. 9.2).

The process of lamellipodial membrane protrusion is powered by actin dynamics and recently has been intensively investigated, therefore its basic mechanism and the role of the main key players are fairly well understood. Similarly, much has been learned about the formation of the establishment of cell contacts to the ECM, but relatively little is known about the mechanisms and structural prerequisites underlying the forward movement of the cell body.

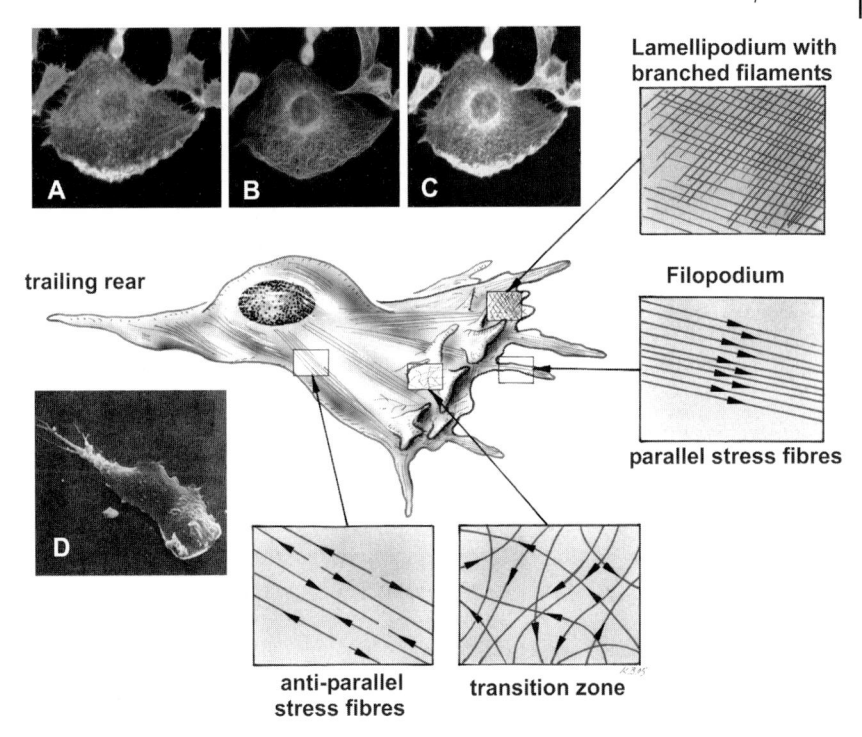

Fig. 9.1
Shape and F-actin organization in migrating cells. (A) Fibroblastic cell fluorescently stained for F-actin, (B) for tubulin and (C) merger of A and B with additional chromatin stain (blue). Note in (A) the different forms of F-actin organization: a homogenously stained rim in the region of the lamellipodium (bottom of the cell) and straight bundled stress fibers in the cell body. (Center) Schematic diagram of a migrating fibroblastic cell and its different forms of F-actin organization. (D) Scanning electron micrograph of a migrating fibroblast. Note the triangular shape with a small lamellipodium on the right lower side. The different organizational forms of actin and their location are depicted in the schematic diagram of a migrating cell. Special regions are magnified in the four insets representing the lamellipodial region containing a branched F-actin network, a filopodium with parallel stress fibers, the transition zone with an open network, and bundles of anti-parallel orientated actin filaments forming the stress fibers crossing the cell body from front to rear. (This figure also appears with the color plates.)

9.3
The Cytoskeleton

The cytoskeleton and specific motor proteins are the prerequisites for cell movement. Three types of cytoskeletal filaments can be distinguished: the actin-containing microfilaments, the microtubules built from the heterodimeric tubulin (α- and

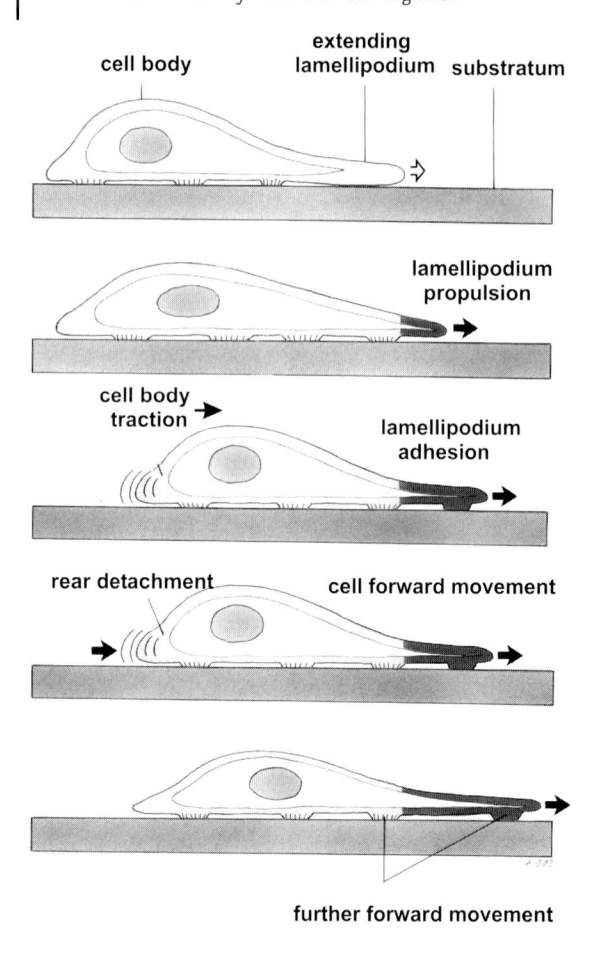

Fig. 9.2
Basic steps of cell forward movement. Schematic representation of a migrating cell exhibiting on its right a lamellipodial plasma membrane extension (blue): different phases of migration are depicted from top to bottom.

β-tubulin), and the intermediate filaments. Only the microfilaments and microtubules together with specific motor proteins generate the force necessary for motility or mobility, whereas the intermediate filaments perform only cytoskeletal functions, although the functional organization of the microfilaments and microtubules is interconnected with the organization of the intermediate filament (for reviews see [1–4]).

9.3.1
Microtubules

Cytoplasmic microtubules are filamentous structures of about 25 nm diameter; 13 tubulin-protofilaments (linear aggregates of tubulin α, β heterodimers) form one microtubule. Microtubules (MT) are polarized and possess two different ends with different affinities for hetero-tubulin. Tubulin molecules preferentially associate to the so-called plus-ends, whereas there is net dissociation from the minus-ends. The tubulin subunits possess a molecular mass of about 55 kDa and bind one molecule of GTP. During the polymerization process, the GTP of the β-subunit is hydrolyzed to GDP. Within the cell, the MTs originate from the microtubule organization centre (MTOC) close to the cell nucleus and extend to the cell periphery. Their minus-ends are tagged into the MTOC probably by association with a particular tubulin molecule; γ-tubulin. Their plus-ends are located peripherally where they can be blocked by capping-proteins (see also Fig. 9.1B). MTs are responsible for the intracellular transport of vesicles, which are moved by specific motor proteins. The main representatives of MT-associated motor proteins are the members of the kinesin and dynein family. These motor proteins associate with the vesicular membrane and move these in a processive manner along the MTs. Most kinesins translocate from the minus- towards the plus-end, whereas the dyneins move in the opposite direction. MTs together with their motor proteins also determine the organization of the intracellular organelles (see also [5]).

Furthermore, the orientation of the MTOC and consequently of the MTs, helps to determine the cell polarity. Thus in polarized cells like enterocytes, the MTOC is always located on the upper side of the nucleus towards the apical face, and in mobile fibroblasts it is located towards the lamellipodial region, i.e. pointing in the direction of migration (see also Fig. 9.1B).

9.3.2
Microfilaments

The microfilaments (MF) are built from actin molecules (molecular mass 42 kDa). Thin filaments or microfilaments of actin dominate in the highly organized sarcomeres of striated muscle cells. However, actin filaments are also present in many other non-muscle cells (see Fig. 9.1A). The structure of a microfilament can be described in two ways: (i) the actins form two strings of monomers that are twisted around each other with a helical repeat of 2×380 nm or (ii) the actins form a genetic helix in which each monomer is rotated by $-166°$ and displaced by 2.74 nm relative to the previous monomer (Fig. 9.3A). Like MTs they also posses a fast growing barbed- or plus-end and a pointed- or minus-end from which actin monomers preferentially dissociate. The actin molecule is firmly bound to one molecule of ATP that is hydrolyzed to ADP during the polymerization process which is normally initiated by an increase in salt concentration. Within the cytoplasm an equilibrium is established between monomeric (also termed G-) and filamentous (F-) actin. In most non-muscle cells, the content of monomeric actin is about equal to that of filamentous actin although the high intracellular salt concentration that would normally favor actin polymerization. This equilibrium is maintained by a number of actin-binding proteins that bind to G-actin and sequester it from the depolymerization and polymerization cycles. The main G-actin-stabilizing proteins are profilin (about 17 kDa) and the small peptide thymosin β4 (5.3 kDa). These proteins form 1 : 1 stoichiometric complexes with monomeric actin. The profilin : actin complex is able to re-associate to the plus-ends of the microfilaments while in contrast, binding of thymosin β4 to G-actin inhibits its re-association.

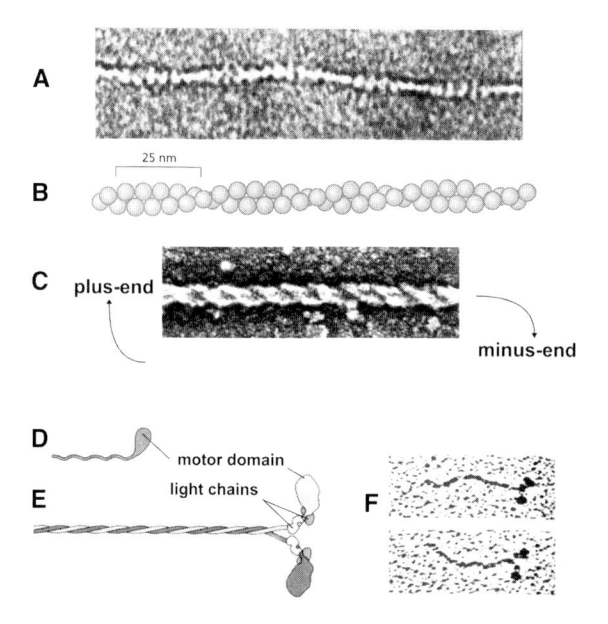

Fig. 9.3
Polarized structure of the actin filament. (a) EM picture of a single actin filament after negative staining. (B) Schematic representation of (A), each actin molecule is represented as a sphere. (C) EM picture of F-actin after binding of myosin heads (decorated F-actin) in order to identify the pointed- or minus-end (right) and barbed- or plus-end by the resulting spearhead appearance. (D) Schematic representation of a myosin I molecule; the globular head on the right contains the motor domain. The length of the α-helical shaft varies according to the type of myosin I and may contain varying numbers of bound calmodulin molecules. (E) Schematic representation of a myosin II molecule composed of two heavy chains that form a long α-helical part towards their C-terminus. Both heavy chains form the long coiled-coil shaft. Towards the N-terminus each heavy chain forms a globular head that comprises the motor domain. The α-helical linker region between shaft and head is stabilized by two light chains. (F) EM of single myosin II molecule after rotary shadowing.

The motor proteins associated with F-actin are members of the myosin family. Muscle myosin is a long filamentous protein composed of two heavy chains forming the coiled-coil tail region and two globular heads each possessing an actin-binding and an ATPase center. It contains two globular heads each able to interact with F-actin and to hydrolyze ATP and represents the prototype for myosin II molecules (Fig. 9.3B). Associated with each head region are two light chains. During the production of force the myosin heads interact cyclically with F-actin by the consumption of ATP. In non-muscle cells, many different types of myosin variants co-exist such as cytoplasmic two-headed myosins (cytoplasmic myosin II), in addition to many variants of single-headed myosins (myosin I) that can form different types of aggregates or even be associated with membranes depending on the chemistry of their tail domains.

9.3.3
Intermediate Filaments

The intermediate filaments (IF) represent the third component of the cytoskeleton. In contrast to microfilaments and microtubules, there are no motor proteins that move along IFs. They are therefore not directly involved in active motile processes. IFs have cytoskeletal functions as they stabilize the cell shape by providing tensile strength. IFs are composed of intermediate filament proteins that exist in many variants and are expressed in a tissue-specific manner. IFs of epithelial cells are composed of members of the large family of keratin proteins; muscle and glial cells and cells of mesenchymal origin contain the single IF-proteins desmin, acid glial fibrillar protein and vimentin, respectively. Neuronal cells contain three neurofilament proteins of different molecular masses (see also [5]). Generally, IF-proteins are rod-shaped with a long central, predominantly α-helical domain and globular N- and C-terminal domains of varying but type-specific sizes. They form coiled-coil dimers, which aggregate to staggered anti-parallel tetramers. End-to-end aggregation of these tetramers leads to protofilaments, eight of which form an intermediate filament of 10 nm in diameter. Due to the anti-parallel orientation of the dimers within the tetramers IFs are not polarized, their ends are of identical chemical composition and properties. They disassemble upon phosphorylation and reassemble after dephosphorylation. A number of reports have indicated that they are functionally interconnected with the other cytoskeletal filament systems. Indeed, it is obvious that during cell migration they also have to undergo cycles of de-and re-polymerization most probably regulated by changes in the extent of phosphorylation.

9.4
Motile Membrane Extensions containing Microfilaments

Motile cells transiently form specific plasma membrane protrusions that are essential for cell migration (see Figs 9.1 and 9.2). Apart from these transient structures, many cells posses stable actin-containing membrane specializations such as microvilli, stereocilia or microrafts, which primarily serve to increase the surface area but have no motile function. The protrusions responsible for cell movement are collectively called pseudopods and comprise the thin sheet-like lamellipodia, which originate from the leading lamella and represent the plasma membrane organelle mainly responsible for cell migration. Additional structures are the filopodia, which are long thin finger-like extensions originating from lamellipodia. Filopodia contain a parallel bundle of actin filaments that continues into the lamellipodium as a so-called microspine (see schematic representation in Fig. 9.1 with insets giving the locally different organizations of the microfilaments). Phagocytotic cells form rim-like extensions form their lamellipodia that enclose the material to be endocytosed. Filopodia extending from lamellipodia as thin protrusions are believed to explore the surroundings for signaling cues or points of attachment. The sheet-like lamellipodia are newly formed after cell stimulation and push the leading edge (lamella) of the cell further outward. Lamellipodia are obvious in migrating cells like myoblasts or fibroblasts during wound healing, in polymorphonuclear leukocytes after receiving a chemotactic signal, or on growth cones of extending neurites. Membrane ruffles represent lamellipodia that have folded back on the dorsal surface of a migrating cell which has failed to attach to the substratum (see Fig. 9.1D).

All these transient membrane extensions contain actin filaments in different forms of organization, either as parallel bundles (filopodia or phagocytotic cups) or in sheet-like networks of intermediate length (lamellipodia). The actin filaments are orientated with their fast growing plus-ends towards the tips of filopodia or the leading membrane zone of lamellipodia. Thus the addition of new actin monomers occurs at the peripherally-located plus-ends of microfilaments and is believed to generate the protrusive force necessary for the membrane extension of the leading zone (see insets in Fig. 9.1D).

9.4.1
The Cytoskeletal Components of Transient Actin-Powered Protrusive Organelles

In protrusive membrane specializations actin-containing microfilaments constitute the sole cytoskeletal component. In lamellipodia they form a branched filamentous network within the flat plasma membrane-enclosed extension. Lamellipodial activity and dynamics have been studied in fast moving fish keratocytes where they form a particularly broad protrusive organelle [6]. During locomotion their width does not change indicating that the actin monomer addition at the leading plus-ends of the microfilaments must be balanced by an equal rate of actin dissociation, preferentially at their minus-ends located in the lamellipodial zone adjacent to the lamella and cytoplasm of these cells. Thus their protrusive activity is powered by a high rate of actin turnover also termed treadmilling. Actin monomers dissociate at the minus-ends and associate at their distal plus-ends maintaining the overall width of the lamellipodium, although the length of individual microfilaments will vary (Fig. 9.4).

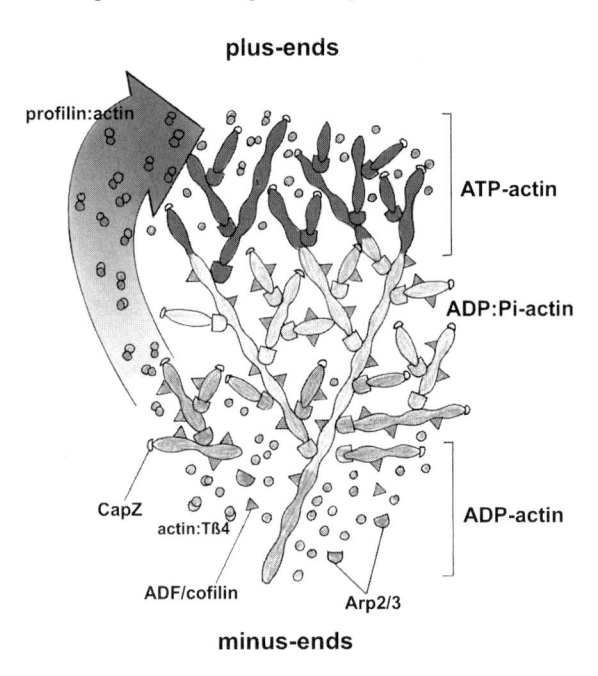

Fig. 9.4
Schematic representation of the branched F-actin network within the lamellipodium. Plus-ends on top are located at the lamellipodial periphery. Here new actin subunits are added to existing filaments or new filaments are nucleated. Within the newly added filaments actin contains bound ATP which is subsequently hydrolyzed to ADP and inorganic phosphate (Pi) both of which remain bound to actin. Towards the minus-ends the Pi is released and the actin contains bound ADP. Minus-end dissociation is stimulated by ADF/cofilin (green triangles), thereafter ADF/cofilin is replaced by either profilin (circles) or thymosin β4 (half moons). The profilin : actin complex is reshuffled to the plus-ends for further outgrowth; the actin : thymosin β4 complex represents the storage form of G-actin. (This figure also appears with the color plates.)

Indeed, isolated F-actin possesses properties that make it well suited for this kind of behavior. During polymerization of G- to F-actin, nuclei composed of three actin monomers must first be formed onto which further monomers are added. Further addition of monomers or polymerization continues until the generated F-actin is in equilibrium with G-actin maintained at the critical concentration of polymerization. In the presence of ATP, actin monomers constantly dissociate from the minus-ends and re-associate at the plus-ends. This mechanism ensures a constant unidirectional flow of actin molecules through the filament and has been termed treadmilling, which can be arrested by selective binding of so-called capping proteins to the filament ends. CapZ is a plus-end capping protein that stops addition of actin monomers to this end. Conversely, treadmilling can be increased by proteins which stimulate the rate of actin monomer dissociation from the minus-ends, since this step is normally rate limiting (Fig. 9.4).

A model system used to study the protrusive ability of polymerizing or treadmilling F-actin is the intracellular movement of certain pathogenic bacteria such as *Listeria monocytogenes* [7]. These bacteria move through the cytoplasm of their hosts by recruiting F-actin together with a few actin-binding proteins to one pole of their surface. Special bacterial surface proteins such as AktA induce the formation of long F-actin tails – also called comet tails – that by continuous unidirectional addition of actin subunits propel these bacteria forward (Fig. 9.5). The set of cytoskeletal proteins recruited by these bacteria appears to be similar to the proteins present in lamellipodia. The minimal protein requirements for *Listeria* movement have been determined by using an *in vitro* system. In addition to actin, only three (out of the large number of) actin-binding proteins are necessary to produce bacterial propulsion in a synthetic medium containing ATP and salt: ADF/cofilin, the Arp2/3 complex (see Section 9.4.2), and an F-actin plus-end capping protein. Bacterial surface proteins like AktA induce the polymerization of cellular actin. The actin filaments of *Listeria* comet tails are short and their barbed-ends are oriented towards the bacterial surface and exhibit a branched network-like organization. Further addition of profilin increased the propulsive speed, whereas the F-actin cross-linking protein α-actinin straightens the F-actin bundle of the comet tail, but does not increase the speed of movement. Thus F-actin treadmilling alone can generate the force necessary for bacterial movement. Interestingly, it has been shown that a number of intracellular particles and endosomal vesicles use similar mechanisms for their movement through the cytoplasm.

9.4.2
Lamellipodial Protrusion

A mechanism similar to bacterial propulsion and also based on the interaction of actin with a few actin-binding proteins operates during the protrusive activity of lamellipodia. The fast growing or plus-ends of F-actin are oriented towards the plasma membrane, most of them being blocked by capping proteins inhibiting addition of new monomers to the plus-ends. The hetero-heptameric Arp2/3 complex containing the actin-related proteins 2 and 3 is located more proximally close to the minus-ends where when activated, it nucleates new filament growth. Nucleation of new filaments is possible because a considerable fraction of the cellular actin is maintained in the monomeric state by the sequestering peptide thymosin β4 or by binding to profilin. Actin molecules from this monomeric pool are reused in filament growth. Other actin-binding proteins such as the actin depolymerizing factor (ADF) and its homolog cofilin, are located more proximally within the region of the

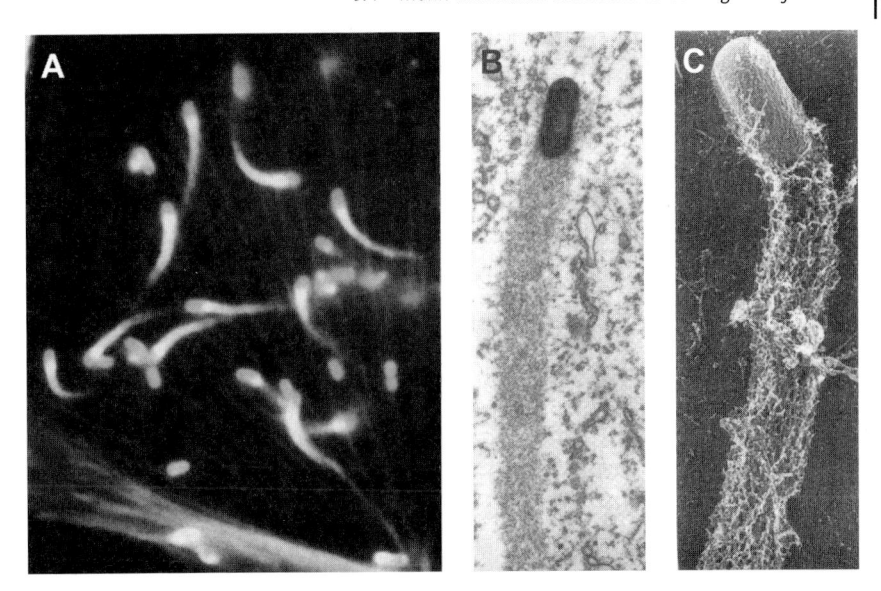

Fig. 9.5
Comet tails induced by the intracellular bacteria *Listeria monocytogenes*. (A) Fluorescent staining of a listeria-infected cell. The bacteria are stained red and the F-actin comet tail in green. Points of F-actin attachment to the bacteria are yellow due to overlap of both stains (from [8]). Different electron micrographs of listeria comet tails: (B) transmission EM at lower magnification [9] and (C) visualized by the deep-etch technique [10]. (This figure also appears with the color plates.)

minus-ends where they stimulate the dissociation of actin monomers. Activation of the Arp2/3 complex and ADF/cofilin is under the control of signaling proteins that themselves are regulated by external cues. The energy necessary for this actin-based motility is provided by ATP hydrolysis produced by actin itself during treadmilling (see Fig. 9.4).

The rate-determining step of treadmilling is the rate of monomer dissociation from the pointed-ends. During treadmilling of pure F-actin, the rate of filament growth at the plus-end is about 0.2/s corresponding to a change in length of 0.04 μm/min, whereas fish keratocytes migrate at a speed of 10 μm/min. Therefore, the rate of actin treadmilling must be enhanced considerably in lamellipodia of migrating cells. This is achieved by the activity of a number of actin-binding proteins. Proteins of the ADF/cofilin family (Mr = 18 kDa) stimulate the rate of monomer dissociation from the pointed ends. The Arp2/3 complex generates new barbed ends by nucleating actin polymerization from the pointed-ends. The presence of both ADF/cofilin and Arp2/3 in lamellipodia is well established. An active Arp2/3 complex also introduces F-actin branches with an angle of about 70° towards the periphery (Fig. 9.4). These branches are added on to either existing or newly nucleated filaments thus generating the dendritic network of lamellipodial microfilaments.

The mechanism by which this system uses actin treadmilling to produce a propulsive force is also illustrated in Fig. 9.4. The active Arp2/3 complex nucleates actin polymerization and by binding to the minus-ends produces free plus-ends, which rapidly grow by addition of new actin monomers. This leads to filament growth immediately underneath the plasma membrane and to its outward extension. Subsequently, the plus-ends of these filaments are capped and the actin-bound ATP is hydrolyzed to ADP. Binding of capping proteins facilitates the release of the pointed-end capper, the Arp2/3 complex. These filaments are de-branched and their ends are then amenable to ADF/cofilin for rapid monomer dissociation. Parallel to the disassembly of old actin filaments new ones are formed either at new branch points further out initiated by Arp2/3 or nucleated from the plus-ends by proteins of the formin family such as p140mDia (see below). Thus the process can restart from the distal end of the lamellipodia.

The branched network organization of F-actin within lamellipodia is generated by activated Arp2/3 complexes that bind sidewise to existing actin filaments and induce the outgrowth of new F-actin with their pointed ends bound to the Arp2/3 complex and their barbed-ends free to extend quickly by the addition of actin monomers. This process has been termed dendritic actin nucleation and results in the formation of actin filaments characteristically branched at an angle of 70°. This mechanism is responsible for the production of the thin lattice-like meshwork of branched F-actin, typical of lamellipodia, which by growth at the plus-ends and shrinkage at the minus-ends (treadmilling) pushes the lamellipodial rim outwards. Disassembly of old filaments after capping by CapZ occurs from their minus-ends where it is favored by the preferential binding of ADF/cofilin to ADP-containing actin monomers. Due to the equilibrium of formation and disassembly of actin filaments, the width of this F-actin meshwork remains constant, although individual actin filaments grow and shrink. Thus the polarity of the actin filament together with its property of treadmilling subunits from the minus- towards the plus-end produces a propulsive force at the lamellipodial tips by extension through addition of monomers to the plus-ends, which after treadmilling through the filament are recycled at the minus-end located close to the lamellipodial-lamellar junction.

The filaments within lamellipodia are shorter than cytoplasmic microfilaments. Binding of plus-end capping proteins restricts filament length. The main intracellular capper is CapZ, a heterodimeric protein present in high concentration in many eukaryotic cells. Capping of growing filaments within the lamellipodia might appear counterproductive, but serves to restrict filament length. Long actin filaments are flexible, shorter ones posses more tensile strength and are better suited for pushing forward the peripheral membrane. Furthermore, capping of lamellipodial F-actin at their plus-ends is necessary to funnel actin monomers into propulsive filaments in regions of directional migration.

The heptameric Arp2/3 complex is composed of two actin-related proteins (Arp 2 and 3) and five different additional proteins called ARPC1–5 [3]. The Arp2/3 complex is activated to stimulate actin nucleation by proteins of the WASP family (mutated in Wiskott-Aldrich syndrome, a bleeding disorder), the SCAR/WAVE family and/or by cortactin. The ADF/cofilin proteins are inactivated by phosphorylation of

Ser3 by LIM-kinases and activated by the calcium/calmodulin-dependent phosphatase 2B (PP2B) and PP1 or PPA2 or by a more specific phosphatase that has been identified from a loss-of-function mutant in *Drosophila*. According to the locus of the gene in *Drospohila* this phosphatase has been named Slingshot.

Apart from the listed minimal number of components, a number of additional proteins have been shown to be present in protrusive lamellipodia, although their role is thought to be mainly supportive. Thus biochemical and genetic evidence has demonstrated the presence of a number of additional proteins such as non-filament-forming or single-headed myosins, profilin and filamin, and of the bundling proteins, α-actinin, fimbrin and fascin among others. A number of these proteins might be involved in either the formation of F-actin bundles within lamellipodia (microspikes) which can grow out from the lamellipodium as filipodia, or the formation of adhesive contacts to the substratum. Cells with a defect in filamin have an interesting phenotype; they migrate poorly and exhibit numerous membrane protrusions (blebs) on their surface. Myosins are motor proteins that normally move towards the plus-ends of F-actin, therefore the presence of non-filament-forming myosin I found in lamellipodia, may indicate that they are involved in the transport of cargo to the lamellipodial tips carrying components needed for their protrusion.

Lamellipodial tips or regions predestined to form filopodia must have the capability to nucleate new filaments. The class of proteins that fulfills these requirements are the formins or p140mDia (named after the Diaphanos locus in *Drosophila*). These dimeric proteins have been shown to nucleate F-actin elongation at their plus-ends by alternate addition of actin monomers to both ends of the two long-pitched strands of F-actin. It was recently demonstrated that the formins are particularly specialized to elongate F-actin by transferring actin monomers from the profilin : actin complex [11]. Profilins are small G-actin-binding proteins ($M_r = 18$ kDa) that are involved in maintaining a high pool of monomeric actin.

9.5
Modulation of the Structure of Transient Protusive Membrane Extensions

The presence of additional actin binding proteins appears to determine the overall shape of these membrane extensions. Filopodia are long thin protrusive extensions that contain parallel bundles of F-actin and often form from existing lamellipodia (see Figs 9.1 and 9.2). They are believed to act as environmental explorers and contain on their surface membrane receptors for extracellular signaling molecules. Their propulsive activity can best be illustrated by the activity of a particular intracellular pathogen – *Rickettsia*. In contrast to *Listeria*, the comet-tail structure of *Rickettsia* is characterized by long parallel actin filaments whose plus-ends are oriented towards the bacterial wall. Actin assembly in these tails is promoted by VASP (vasodilator stimulated phosphoprotein), which binds to their plus-ends and suppresses binding of the branch-inducing Arp2/3 complex. Furthermore, the parallel actin filaments are bundled by fascin, which is also present in filopodia together with fimbrin. The propulsive force for the intracellular movement of these bacteria is also generated by the rapid treadmilling of the actin subunits through the filaments of the comet tails. It is supposed that filipodial growth and extension occurs in a similar fashion. The bundling of the core actin filaments permits localized plasma membrane protrusion. Nucleators of actin polymerization such as the formins, are suspected to reside at the filopodial tips and to induce the F-actin outgrowth.

9.6
Formation of Substratum Contacts

Lamellipodia and filopodia are not only protrusive organelles, but also initiate and form contacts with the substratum. *In vivo* the substratum is the extracellular matrix (ECM), a complicated meshwork of extracellular collagen fibrils and glycosaminoglucans with associated glycoproteins such as the laminins and fibronectin. Cell adhesion to the ECM is achieved by integral membrane receptor proteins called integrins that interact with their large extracellular domains via particular components of the ECM. The heterodimeric integrins are composed of one relatively variable α- and one more conserved β-chain both of about 95 kDa possessing a single transmembrane domain. In humans 24 α- and nine β-chains have been identified that can associate with each other and form heterodimeric functional receptors. The heterodimeric association is not absolutely free, for instance the $β_1$-chain can associate with only 12 different α-chains; nevertheless, a large number of different heterodimers can form whose extracellular domains have highly selective binding properties. Their binding specificity to components of the ECM is mainly determined by the α-subunit, although both chains bind to the ECM. The minimal recognition sequence is RGD (arginine-gylcine-aspartic acid). Binding is of low affinity, and firm association only occurs when a minimal number of integrin receptors have clustered and participate in the attachment process. After binding to ECM, the β-subunits undergo a conformational change at their cytoplasmic domains which initiates binding of cytoskeletal components.

For binding to components of the ECM, the integrin receptors have to be in an active state. Integrin activity can be regulated by external signals and also by signals from inside the cell. In general, integrins may function as signaling molecules in both directions: outside-in or inside-out signaling. When cells are migrating, they form so-called close contacts with the substratum underneath or just behind the lamellipodium (see Fig. 9.2). Interference reflection microscopy has revealed that by these contacts the cell or its lamellipodium remains separated from the substratum by about 30 nm. When cells cease to migrate, the close contacts are transformed into so-called focal adhesions, which are characterized by a separation distance between the plasma membrane and the substratum of 10–15 nm. Furthermore, stress fibers terminate in focal adhesions. The appearance of focal adhesions with terminating stress fibers is correlated with decreased migratory activity.

Contacts to the ECM or substratum are often located underneath the so-called microspikes that represent bundles of F-actin within the lamellipodium. The proximal part of these microspikes is fixed to a small area of the plasma membrane containing an integrin cluster and forming the initial contact, whereas the distal part can further extend as the lamellipodium pushes forward. The attached F-actin bundles of these initial contacts are stabilized by the signaling activity of the ECM-engaged integrins which leads to the recruitment of additional actin-binding proteins such as talin. Talin accumulation is followed by α-actinin and filamin accumulation. These cytoskeletal proteins bind to the cytoplasmic domains of the activated integrins, after which binding of F-actin is induced possibly via vinculin (Fig. 9.6). After the initial contact, substrate adhesion contacts alter their adhesive strength

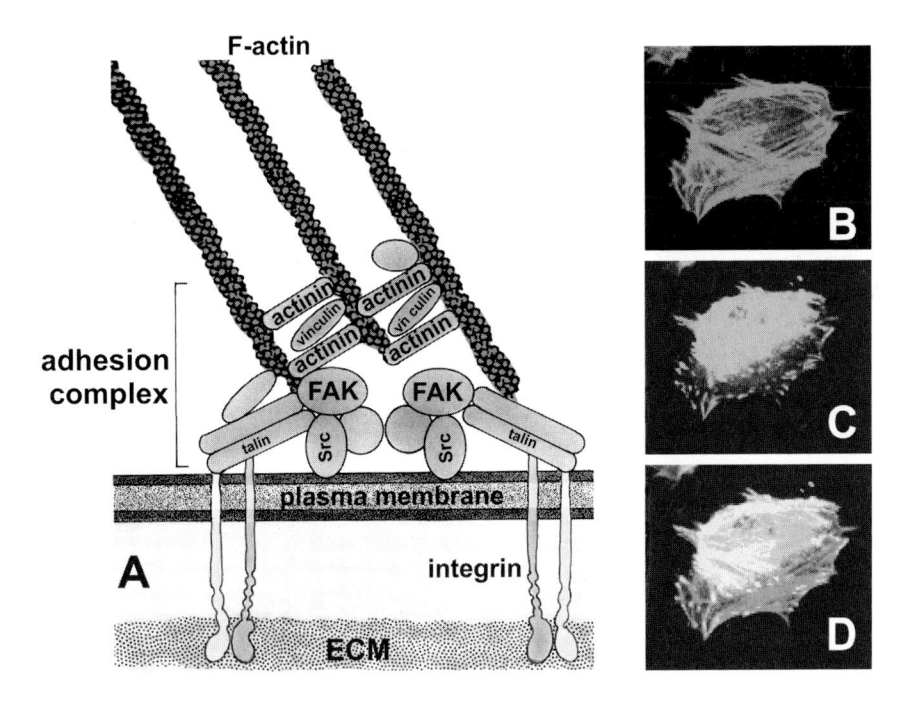

Fig. 9.6
Organization and composition of focal contacts underneath the plasma membrane. (A) Diagram of the organization of a focal contact to the extracellular matrix (ECM). Heterodimeric integrin as transmembrane ECM-receptors and the organization of a parallel bundle of F-actin with a number of associated proteins essential for the attachment of the actin filaments to the plasma membrane (see text for details). Open circles represent various components depending on the type of contact. (B) Fibroblastic cell stained for F-actin (red), (C) for vinculin (green); note the punctate or streaky pattern marking focal contacts from which F-actin bundles radiate (yellow in (D) by the superposition of both stains). (This figure also appears with the color plates.)

and size. Initially, small adhesion complexes are formed underneath the lamellipodia, which subsequently ripen into larger focal adhesions after becoming translocated to the cell edges and the more posterior parts of the cell (for review see [12]).

The formation of substrate contacts is regulated by external and intracellular signals such as small GTPases which also influence to the actin cytoskeleton (see Section 9.11). Although integrins do not posses enzymatic activity they can induce signaling processes by binding and activation of small GTPases (see Section 9.11), focal adhesion kinases (FAK) and members of the Src family of kinases [13]. The action of these kinases can lead to the stimulation of several downstream signaling pathways.

9.7
Cell Body Traction

The third step in cell migration after the establishment of points of contact is the forward traction of the main cell body. From genetic analyses in the slime mould *Dictyostelium discoideum* it has become clear that this process depends on the formation of force-generating complexes between F-actin and a cytoplasmic two-headed myosin (myosin II). Myosin II null cells hardly move, although they are able to extend lamellipodial structures. The force for cell body traction has to be generated behind the lamellipodium. Immediately behind the lamellipodium, the F-actin meshwork loosens into a wider network within a transitional zone or lamella and thereafter is abruptly replaced by a number of different organizational arrays of F-actin underneath the main cell body mass (schematic drawings in Fig. 9.1). Arcs of thick bundles of F-actin may underlie the lamella; these arcs are in either convex or concave orientation towards the lamellipodium. Long stress fibers which transverse the cell parallel to the direction of migration originate more centrally. The stress fibers contain actin filaments in repetitive alternating anti-parallel orientation together with short bipolar filaments of myosin II.

The dense F-actin network of the lamellipodia widens in the transitional zone and the concentration of F-actin decreases. Within the transitional zone the rate of actin treadmilling is considerably lower probably due to the presence of tropomyosin and myosin II; both are only present in the rear part of the lamellipodium. Concomitant with the decrease in actin towards the lamella there is an increase in the number of short bipolar filaments of myosin II which gradually increase in length within the lamella and rear parts of the cell.

It is still not yet quite clear where and how the force required for the forward movement of the cell body is generated. A detailed morphological investigation of myosin II distribution and organization in fish keratocytes revealed that within the transitional region F-actin is organized into an open net-like configuration – similar to a folding grille. A similar F-actin configuration was also found in migrating mammalian fibroblasts. The heads of bipolar filaments of myosin II interspersed within this system are able to reach the boundary of the F-actin net. It is assumed that by this activity the folds of the F-actin network are pulled together leading to a traction force on the cell body [6]. This traction force can be used for forward movement of the cell, because focal adhesions generated underneath the lamellipodium and/or lamella fix the anterior part of the cell to the substratum. A recent speckle microscopic analysis has corroborated the different organizational forms of F-actin between the lamellipodium and lamella and furthermore demonstrated that the force for cell body traction is most likely generated in this transitional area [14].

Stress fibers may originate from behind the lamella and extend into the rear part of the cell (see also schematic drawings in Fig. 9.1). Although these have been shown to possess contractile activity, it appears unlikely that they are responsible for the forward movement of the cell body. The stress fibers are connected to focal contacts, either to newly formed substrate contacts in the anterior of the cell or to older focal adhesions in the trailing posterior of the cell. They are therefore well positioned to transport the contractile force generated in the transversely-orientated lamellar network to posterior parts of the cell. It has also been suggested that the traction on older focal contacts leads to a weakening of their adhesive strength.

Immunostaining with antibodies against myosin I and II has demonstrated that myosin II is concentrated behind the lamellipodia and underneath the plasma membrane of the trailing posterior of the cell, especially the in highly mobile *Dictyostelium* cells, whereas myosin I is close to or within the lamellipodium as detailed above. Derived from these localization studies an alternative or supplementary mechanism for cell body movement in *Dictyostelium* was suggested which proposed that contraction of the trailing sides and posterior of the cell forced the cytoplasm to move in a forward direction. Such a mechanism might occur simultaneously with or in support of the contraction of the lamellar actin network. It is also possible that different cell types have evolved slight variations of the mechanism for forward cell body traction.

9.8
Rear Detachment

After completion of the forward movement of the cell body, the final process for a migratory cell must be the detachment of the posterior of the cell from the substratum. Since migratory cells continuously migrate forward, the process of rear detachment must occur in a manner co-ordinated to forward cell traction. Thus it has been shown that the migratory speed can be modulated by the rate of rear detachment. The exact mechanism by which the adhesive strength of focal contacts at the cell posterior is weakened is not yet clear. It has been proposed that intracellular calcium-dependent kinases such as calcineurin and proteases such as the calpains modify or degrade particular components of the focal contacts. It has also been reported that talin is the first component of the adhesion complex to be degraded by calpain subsequently leading to its disassembly by dissociation of other components. However, it has not yet been determined what makes an adhesion complex prone to calpain targeting, although it might be plausible that changes in the state of phosphorylation of particular cytoplasmic adaptors render them vulnerable to proteolytic attack. It has also been observed that components or even rear cellular fragments containing membrane-enclosed macro-aggregates of integrins are lost and left behind as migration proceeds during cell movement (for review see [15]).

9.9
The Role of Microtubules in Cell Migration

Microtubules do not seem to play a role in fast-moving cells like neutrophils and fish keratocytes. Inhibition of microtubule cycling by nocadazole does not affect the rate of their migration. However, in slowly crawling cells such as fibroblasts, inhibition of MT dynamics leads to inhibition of migratory activity. This effect has been correlated with the strength and number of focal contacts of these cells. MTs appear essential for tail retraction. In this context it seems relevant to highlight the observation that microtubules appear to target and physically contact adhesion complexes and regulate their substrate adhesion dynamics. It has been shown that the larger focal complexes present at the posterior of the cell and at its edges disassemble after MT contact. These data and additional evidence seem to suggest that molecular motor proteins associated with MTs deliver "relaxing" signals or substances to the adhesion complexes rendering them prone to dissociation in order to allow forward movement [16].

9.10
The Role of Intermediate Filaments in Cell Migration

The effect of intermediate filaments also depends on the cell type. It has been shown that fibroblasts derived from vimentin-deficient mice exhibit impaired migratory activity. The reason for this behavior is presently unclear. Conversely, it has been shown that Rho-GTPase-dependent kinases such as ROCK phosphorylate intermediate filament proteins which then depolymerize and are subsequently reorganized. Most probably it is the ability of the cell to reorganize its IFs that is essential for efficient migratory behavior.

9.11
Modulation of the Cytoskeletal Organization by External Signals

The organization of the cytoskeleton, in particular of the microfilament system, is influenced by external signaling molecules which bind to plasma membrane receptors which modulate via protein or phospho-inositide (PIKs) kinases to monomeric GTPases. The small monomeric GTPases of the Ras family are present in all mammalian cells and regulate a wide variety of cellular responses. The small GTPases are active in the so-called on-state which is characterized by the presence of firmly bound GTP; hydrolysis of bound GTP to GDP leads to their inactivation or the off-state. When in the on-state conformation, they bind to target proteins whose activity is subsequently altered. There are more than 60 different Ras proteins in mammals, which fall into five groups: Ras, Rho, Ran, Arf, and Rab. Guanine nucleotide exchange factors (GEFs) stimulate the exchange of GDP to GTP and therefore switch these proteins from the off-state to the on-state. GTPase-activating proteins (GAPs) stimulate GTP hydrolysis and support their transition into the off-state. Thus the small GTPase proteins cycle between the on- and off-state conformations according to external signals acting on GEFs and GAPs.

Of particular interest for the regulation of the supramolecular organization of the actin cytoskeleton in mammalian cells is the class of the Rho proteins; the most important members being RhoA, Rac and Cdc42 (for reviews see also [17–19]). There are over 60 GEFs and 70 GAPs for the Rho-family of small GTPases indicating intricate regulatory possibilities. The members of the Rho-family are activated by different external cues, which stimulate the appropriate GEFs. Rho proteins can be post-translationally modified at their C-termini by prenylation, which allows their close apposition to the plasma or other membrane systems.

Transfection of fibroblasts with constitutively active RhoA or activation of endogenous RhoA by addition of extracellular ligands like lysophosphatidic acid induces the formation of long stress fibers and of associated adhesion complexes. Activation of Rac by for instance, PDGF or EGF induces the translocation of actin to the cell periphery and formation of lamellipodia. The third member of the Rho-subfamily, Cdc42, induces the formation of filopodia after stimulation by for instance, bradykinin (Fig. 9.7). Moreover, there is considerable cross-talk between the different GTPases, in some cell fibroblastic cell lines they appear ordered in a hierarchical manner in that Cdc42 stimulates Rac which then activates Rho. Furthermore, activated integrins have been shown to activate Rac and Rho and thereby influence the organization of the actin cytoskeleton after their engagement with ECM components. The GTPase-specific effects are elicited in a number of different cell types such as fibroblasts, neutrophils, macrophages, and epithelial cells [14]. They also regulate the migratory behaviour of axonal growth cones although particular GTPases induce opposing effects leading to their extension or retraction (see below) [20].

Thus different supramolecular organization of the actin cytoskeleton can be controlled by external factors that signal to different Rho-family members. A large number of receptors are able to signal to Rho-GTPases by an increasing number of different intermediate signaling proteins or phospho-inositides (for a review see [17]). The signaling pathways converging towards the Rho-GTPases will not be described here. Instead this chapter will introduce a few downstream target proteins of the small GTPases that have been identified recently. The previously-mentioned WASP proteins have been identified as target proteins for Rac. Binding of active Rac to WASP activates it to induce the minus-end nucleation of the Arp2/3 complex and

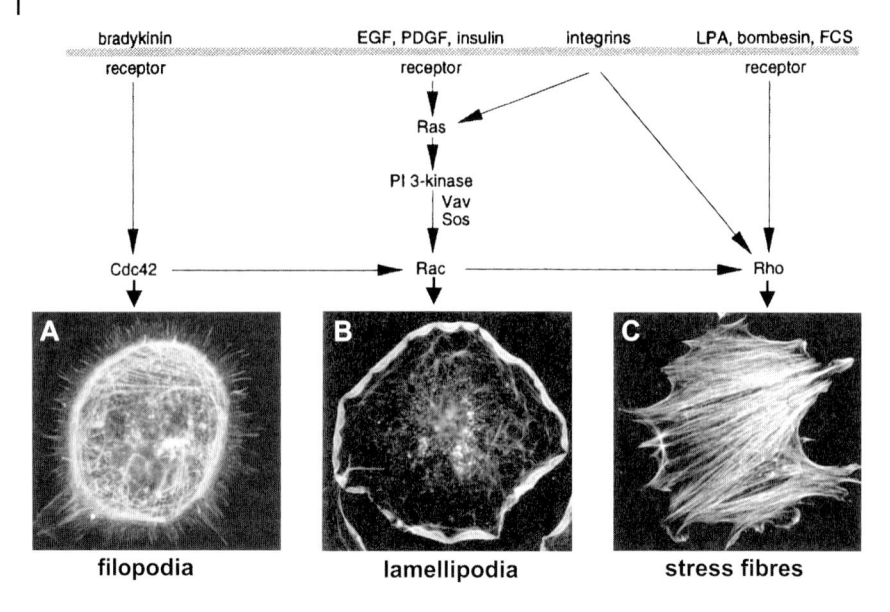

Fig. 9.7
Effect of the small GTPases on the organization of the microfilament system. Diagrammatic representation of the signaling cascades from the extracellular space through the plasma membrane to the small GTPases, Cdc42, Rac, and Rho whose effect on the organization of the fluorescently-stained actin cytoskeleton is shown in (A), (B), and (C), respectively.

subsequent formation of the branched F-actin network of lamellipodia. Binding of active Cdc42 to formins is believed to stimulate their plus-end nucleating activity and the formation of long actin bundles as observed in filopodia.

The mechanism of activation of actin-modulating proteins by small GTPases can be illustrated by the stimulation of formins by Rho [21]. Like many other Rho-GTPase dependent proteins, the formins like p140mDia contain an N-terminal GTPase binding domain (GBD). Unless occupied by an active Rho-GTPase, this GBD folds back to a short auto-inhibitory C-terminal sequence located within the formin homology domain 2 (FH2) (see also [11, 22]). Binding of the specific small GTPase results in dissociation of the GBD from the auto-inhibitory domain and frees the FH2 domain which is then able to induce plus-end actin elongation [19].

Among the targets of active RhoA are also signaling kinases such as ROCK (Rho-activated kinase). ROCK phosphorylates and thereby activates the LIM kinases, which then inactivate ADF/cofilin by phosphorylation at serine–3 supporting the nucleation of long actin cables by formins as found in stress fibers. Furthermore, ROCK phosphorylates and thus activates MLCK (myosin light chain kinase) subsequently leading to MLC phosphorylation and myosin II stimulation, a prerequisite for forward traction of the cell body [17].

Directional cell migration requires cell polarization into a leading edge and a trailing rear part and also requires the polarized distribution of the small GTPases signaling to the cytoskeleton. Unidirectional migration of macrophages has been shown to depend on intact Cdc42. Inhibition of Cdc42 abolishes correct interpretation of external chemotactic concentration gradients and macrophage movement becomes random. It has been suggested that Cdc42 directs the intracellular localization of Rac to the leading edge. Signaling via Rac will then induce the characteristic lamellipodial branched F-actin organization. How the strategic Rac localization is achieved is presently unclear. It is known that the organization of the microtubular system supports polarization, furthermore the MTOC (microtubular organization centre) re-orientates towards the anterior of the cell and MTs stabilized by de-tyrosinated tubulins terminate in the vicinity of the leading edge.

The small GTPases not only influence the organization of actin, but also have a profound effect on adhesion complexes. It has been shown that active Rho is essential for the maintenance of focal adhesions although the mechanism of this effect is unclear. Inactivation of Rho induces disassembly of focal adhesions with subsequent cell detachment. Normally focal adhesions are stable cell-ECM contacts that signal cell cycle progression and survival. In migrating cells, Rac and Cdc42 signaling not only induces the characteristic different supramolecular organizations of F-actin, but also leads to their attachment to focal complexes whose composition is similar to focal contacts, albeit of different morphology. Conversely integrins are reported to signal to small GTPases. Thus there exists a complicated and intertwined network of signaling pathways.

Furthermore, it has been shown that the Rho-GTPases not only have profound effects on the microfilament system, but also influence MT and intermediate filament organization. The mutual interaction of MTs with the microfilament system is presently being intensively investigated.

9.12
The Cytoskeletal Organization of Neuronal Growth Cones

The neural growth cone represents a particularly obvious and unique *in vivo* demonstration of transient protrusive membrane specialization [20]. During neurogenesis the neuroblast is first surrounded by a continuous lamellipodium that fragments into neurite processes after the in-growth of microtubules. The neurite tips enlarge into so-called growth cones that are remnants of the original lamellipodium. For axonogenesis one of the neurites becomes the axon, and the others differentiate into dendrites. The growth cone of the axon enlarges by further accumulation of cytoskeletal components. Its enlarged distal end represents a lamellipodium with extending filopodia. The growth cone moves away from its perikaryon and tries to reach its target cell that may be located a considerable distance away. The outgrowth of the axon is not accompanied by cell body traction, the axon elongates with the perikaryon remaining fixed at its original place.

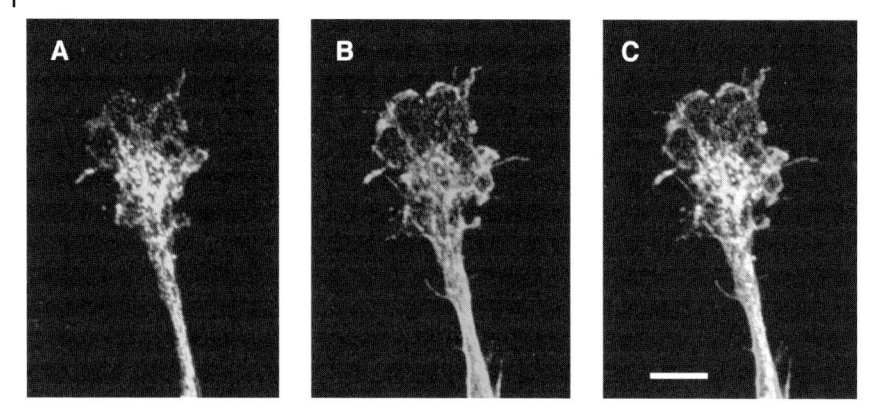

Fig. 9.8
Growth cone of explanted embryonic chicken (E 10) dorsal root ganglia cells fluorescently stained with anti-tubulin (A), anti-actin (B), and overlay of both stains (C). Bar corresponds to 5 μm (kindly provided by Dr. Carsten Theiss, Department of Cytology, Bochum). (This figure also appears with the color plates.)

A growth cone can be divided in two regions: the P-(peripheral) and the C-(central) region. MTs terminate within the C-region (Fig. 9.8A and C), which contains vesicles and organelles, whereas the P-region is flat, devoid of organelles and rich in actin (Fig. 9.8B and C). The growth cone is the pathfinder to guide the axon to the appropriate target. It acts almost independently of the rest of the cell as has been demonstrated after its separation from the rest of the axon. Isolated growth cones are able to move almost like single migratory cells.

The peripheral zone of the growth cone resembles lamellipodia and filopodia with a similar supramolecular organization of F-actin as described above (see Fig. 9.8B). The plus-ends of the actin filaments are located at the distal end, whereas their minus-ends are located in a transition zone (T-domain) between the P- and C-region. Dissociation of actin subunits from actin filaments occurs within the T-zone. Treadmilling, i.e. the retrograde flow of newly added actin molecules, has been directly observed in large growth cones. Propulsion of the leading edge of the growth cone is achieved by addition of actin monomers to the distal plus-ends. The speed of propulsion has been shown to be highest when dissociation from the minus-ends is lowest. The mechanism and protein components responsible for growth cone propulsion are identical or at least very similar to those of migrating cells as described above.

Exposure to external signaling cues can lead to rapid lamellipodial and filopodial re-organization including extension on exposure to positive signaling molecules or retraction and even complete collapse when exposed to repulsive molecules. The filopodia extend from the lamellipodial zone and explore the surroundings with membrane receptors that are concentrated on their surface membrane. Binding of external cues to these receptors induces a signaling cascade which also leads to small GTPases that transmit the signals to the cytoskeleton, in particular to the microfilament system, and induce the assembly of adhesive focal complexes. Positive cues are elicited by neurotrophins, netrins and laminin which induce actin assembly and lamellipodial extension. Repulsive signals are elicited by semaphorins and ephrins, which lead to inhibition of growth cone extension and F-actin disassembly. Attractive and repulsive signaling molecules bind to receptors located on the plasma membrane of filopodia. Their signals presumably lead to differential changes in the cytoplasmic Ca^{2+} concentration that are finally relayed to the cytoskeleton by small GTPases [23]. Growth cone protrusion is stimulated by Rac and Cdc42, whereas Rho inhibits its extension [20]. Growth cone extension is governed by the same mechanisms that lead to lamellipodial protrusion. The mechanism of growth cone retraction can be explained by complete capping of the distal plus-ends of F-actin leading to the inhibition of treadmilling and the predominance of actin subunit dissociation from the minus-ends. However, the role of the microtubules is not yet quite clear. They extend together with growth cone outgrowth and have been shown to shrink under the influence of Rho which directly stimulates their dynamic behavior.

9.13
Conclusions

Eukaryotic cells have evolved mechanisms for locomotion. The ability to move is inherent in single cell organisms. In multicellular organisms the ability to move is important during embryonic development but later becomes confined to a few specialized cell types. Migratory cells crawl along a fixed substratum, which in most cases is composed of components of the extracellular matrix. The mechanisms of locomotion appear to be similar for all migratory cells. For directed locomotion the migratory cells first have to polarize, i.e. to organize a particular area of their plasma membrane for the directional forward movement. The process of movement can be divided into several phases: an initial exploring phase during forward extension of the leading plasma membrane, then the establishment of new adhesive contacts with the substratum followed by forward traction of the cell body and finally the release of the trailing rear part of the cell. The cytoskeletal components involved in active movement are the actin filament system and the myosins as associated motor proteins. Forward extension of the leading plasma membrane is achieved solely by the directed polymerization and depolymerization of actin. The forward movement of the cell body depends on interactions between actin and myosin II. The crawling movement also necessitates reorganization of the microtubular and intermediate filament systems. Cell migration is strictly controlled by external signals. Metastatic tumor cells regain the ability to migrate and escape the external control of migratory regulation. Since more than 50% of all tumor patients succumb to tumor cell dissemination, a complete understanding of the molecular basis of cell locomotion will also be of utmost clinical relevance.

Acknowledgement

The author thanks the Deutsche Forschungsgemeinschaft, Bonn, Germany, for continuous support and is grateful to Mr. K. Barteczko (Bochum) for preparing the artwork. Particular thanks go to Dr. Carsten Theiss (Department of Cytology, Bochum) for providing Fig. 9.8.

References

1 B. Alberts, A. Johnson, J. Lewis, M. Raff, K. Roberts, P. Walter. *The Molecular Biology of the Cell*, 4th edn. New York, USA: Garland Science, **2002**.

2 K. Burridge, K. Fath, T. Kelly, G. Nuckolls, C. Turner. Focal adhesions: transmembrane junctions between the extracellular matrix and the cytoskeleton. *Ann Rev Cell Biol* **1988**; *4*: 487–525.

3 S.S. Brown. Cooperation between microtubule- and actin-based motor proteins. *Ann Rev Cell Dev Biol* **1999**; *15*: 63–80.

4 M. Endo, K. Ohashi, Y. Sasaki, Y. Goshima, R. Niwa, T. Uemura, K. Mizuno. Control of growth cone motility and morphology by LIM kinase and Slingshot via phosphorylation and dephosphorylation of cofilin. *J Neurosci* **2003**; *23*: 2527–2537.

5 S. Etienne-Manneville, A. Hall. Rho GTPases in cell biology. *Nature* **2002**; *420*: 629–635.

6 E. Fuchs, K. Weber. Intermediate filaments: structure, dynamics, function, and disease. *Annu Rev Biochem* **1994**; *63*: 345–382.

7 M.A. Goldberg. Actin based motility of intracellular microbial pathogens. *Microbiol Mol Biol Rev* **2001**; *65*: 595–626.

8 B. Alberts, A. Johnson, J. Lewis, M. Raff, K. Roberts, P. Walter. *The Molecular Biology of the Cell*, 4th edn. New York, USA: Garland Science, **2002**.

9 C. Kocks, E. Gouin, M. Tabouret, B. Berche, H. Ohayon, P. Cossart. Listeria monocytogeneses-induced actin assembly requires the act A gene product, a surface protein, *Cell* **1992**; *68*: 521–531.

10 J. E. Heuser, R. Cooke. Actin-myosin interactions visualized by the quick-freeze, deep-etch replica technique. *J. Mol. Biol.* **1983**; *169*: 97–122.

11 J. Henley, M. Poo. Guiding neuronal growth cone using Ca^{2+} signals. *TICB* **2004**; *14*: 320–330.

12 H.N. Higgs, T.D. Pollard. Regulation of actin filament network formation through Arp2/3 complex: activation by a diverse array of proteins. *Annu Rev Biochem* **2001**; *70*: 649–676.

13 I. Kaverina, O. Krylyshkina, J.V. Small. Microtubule targeting of substrate contacts promotes their relaxation and dissociation. *J Cell Biol* **1999**; *146*: 1033–1043.

14 G. Kirfel, A. Rigort, B. Born, V. Herzog. Cell migration: mechanism of rear detachment and the formation of migration tracks. *Eur J Cell Biol* **2004** *83*: 717–724.

15 L. Luo. Actin cytoskeleton regulation in neuronal morphogenesis and structural plasticity. *Ann Rev Cell Dev Biol* **2002**; *18*: 601–635.
16 S.K, Mitra, D.A. Hanson, D.D. Schlaepfer. Focal adhesion kinase: in command and control of cell motility. *Nat Rev Mol Cell Biol* **2005**; *6*: 56–68.
17 T.D. Pollard, G.G. Borisy. Cellular motility by assembly and disassembly of actin filaments. *Cell* **2003** 112 : 453–465.
18 A. Ponti, M. Machaceck, S.L. Gupton, C.M. Waterman-Storer, G. Danuser. Two distinct actin networks drive protrusion of migrating cells. *Science* **2004**; *305*: 1782–1786.
19 A.J, Ridley. Rho GTPases and cell migration. *J Cell Sci* **2001**; *114*: 2713–2722.
20 O.C. Rodriguez, A.W. Schaefer, C.A. Mandato, P. Forscher, W.M. Bement, C.M. Waterman-Storer. Conserved microtubule-actin interactions in cell movement and morphogenesis. *Nat Cell Biol* **2003**; *5*: 599–609.
21 S. Romero, C. Le Clainche, D. Didry, C. Egile, D. Pantaloni, M.-F. Carlier. Formin is a processive motor that requires profilin to accelerate actin assembly and associated ATP hydrolysis. *Cell* **2004**; *119*: 419–429.
22 A. Shimada, M. Nyitrai, I.R. Vetter, D. Kühlmann, B. Bugyi, S. Narumiya, M.A. Geeves, A. Wittinghofer. The core FH2 domain of Diaphanus-related formins is an elongated actin binding protein that inhibits polymerisation. *Mol Cell* **2004**; *13*: 511–522.
23 A. Schmidt, M.N. Hall. Signalling to actin cytoskeleton. *Ann Rev Cell Dev Biol* **1998**; *14*: 305–338.

10
Cell Death in Organ Development

Kerstin Krieglstein

10.1
Introduction

Programmed cell death is a fundamental and essential process operating in development and assuring tissue homeostasis of multicellular organisms. About half of all cells produced during development die before maturation. Failure to kill appropriate cells during development can lead to severe defects or malformations. In adulthood, cancer and degenerative diseases may result from disturbing the balance of cell proliferation, survival, and death (cf. [1, 2]). We find it appropriate in the context of this book to focus on cell death in developing organs and tissues rather than presenting another extensive review on the molecular signaling cascades of programmed cell death ("apoptosis"). The latter issue has been covered by numerous excellent reviews (e.g. [3]) and will therefore not be extensively dealt with here.

10.2
Programmed Cell Death

The term "programmed cell death" was coined to describe cell death during the embryonic development of multicellular organisms. Cells appear to be "programmed" to die (cf. [4, 5]). Since developmental cell death seems to follow an intracellular "program", programmed cell death is also used as a synonym for apoptosis.

Apoptosis is a genetically-defined, cell-autonomous process, in which cells respond to internal or external signals by actively participating in their suicide and organization of their disposal [5]. "Apoptosis" was initially defined by a characteristic pattern of morphological changes [6, 7], but is now increasingly being used to describe the underlying molecular mechanisms [8]. Forms of cell death lacking the features indicative of active cell death are referred to as necrosis. Necrosis represents a passive form of cell death, where disintegration of cells follows initial destruction of the cell membrane [9]. Furthermore, there is also evidence available suggesting

Cell Signaling and Growth Factors in Development. Edited by K. Unsicker and K. Krieglstein
Copyright © 2006 WILEY-VCH Verlag GmbH & Co. KGaA, Weinheim
ISBN 3-527-31034-7

the existing of evolutionarily conserved mechanistic pathways of necrosis (for a review, see [10]).

The morphological changes, by which apoptotic cell death can be characterized and identified, occur in a consecutive, highly conserved sequence [6, 7]. Dying cells begin to detach from neighboring cells and extracellular matrix and round up. Next, the cells start to show protrusions from the plasma membrane, referred to as blebs. Many dying cells show nuclear condensation and disintegration of the nucleus into several fragments. Importantly and differently from necrosis, where organelles are affected early, organelles of apoptotic cells are still intact at this stage, but may be affected at later stages. Mitochondria have been described to either swell or condense. Additional features include dilatation of the ER, release and aggregation of ribosomes and cytoplasmic vacuoles. Whole cells condense and reorganize into so-called apoptotic bodies [6], which are membrane-bound vesicles containing cytosolic elements, organelles, and parts of the condensed nuclei in various combinations. Apoptotic cells are rapidly engulfed and digested by neighboring cells [6, 11], which makes it quite difficult to study morphological changes *in vivo*.

10.3
Genetics of Programmed Cell Death

Most of what we know about the genetics of programmed cell death in development comes from the analysis of the nematode *C. elegans* as a model system (cf. [5] for a review and references therein). In *C. elegans* the development of each individual cell seems to be programmed into their fate, as development is precisely repeated in each worm. A total of 131 cells of the 1090 somatic cells generated during development of the *C. elegans* hermaphrodite (most of them being neurons), die during *C. elegans* development. Genetic analysis of *C. elegans* by Horvitz and collaborators revealed that cell death is executed by CED–3 (vertebrate homologs are the caspases), which needs to be activated by CED–4 (corresponding to Apaf–1 in vertebrates). The activity of CED–4 is usually blocked by CED–9 (corresponding to the anti-apoptotic Bcl–2 family members in vertebrates), unless CED–9 is counteracted by EGL–1 (BH3–only Bcl–2 proteins in vertebrates). EGL–1 has been shown to be induced in cells destined to die. Thus, the "central dogma of programmed cell death" holds true for programmed cell death from worms to vertebrates, which all share prototypic signaling pathways including a death signal ("thanatin"), an inhibitor, an adaptor, and an executioner step. However, vertebrates have been shown to rely on a much wider set of molecules than *C. elegans* (see [3] for a review). In vertebrates, additional molecular steps have been added to this cascade, such as release of cytochrome c from mitochondria and its molecular control. While in *C. elegans* the entire regulation of programmed cell death seems to be cell-intrinsic, there is ample evidence for an additional cell-extrinsic control in vertebrates. In vertebrates apoptosis may be actively induced via two cell-extrinsic mechanisms: (1) signaling via cell death receptors, and (2) lack of signaling as a consequence of removal of growth factors and cytokines.

10.4
Cell Death Receptors

The cell death receptors belong to the tumor necrosis factor receptor (TNFR, Tab. 10.1) gene family and are defined by a "death domain" or TRAF binding site. CD95L is an example, where ligation of the death receptor results in the clustering of the receptor's death domains [36]. A Fas-associated death domain (FADD) then serves as an adaptor and binds to the death domains of the receptor. In turn, FADD binds through its "death effector domain" the zymogen form of caspase 8 [37, 38]. The "death effector domain" functions as a caspase recruitment domain (CARD), which is found in several caspases (e.g. caspase $-2, -8, -9, -10$) [39]. Following its interactions with FADD, procaspase 8 oligomerizes, self-cleaves, and activates the apoptotic pathway by cleaving appropriate effector caspases [40].

Tab. 10.1
Cell death factors and receptors.

Death factor	Death receptor	References (for example)
CD95L (Apo–1L, FasL)	CD95 (Fas, Apo1)	12–14
TNFα, lymphotoxin α	TNFR1 (p55, CD120a)	15, 16
Unknown	CAR1 (avian)	17
Apo3L (TWEAK)	Death receptor 3 (DR3, Apo3, WSL–1, TRAMP, LARD)	18–21
Apo2L (TRAIL)	DR4 (TNF-related apoptosis-inducing ligand receptor–1	22–25
Apo2L (TRAIL)	DR5 (Apo2, TRAIL-R2, TRICK 2, KILLER)	26–30
Unknown	DR6	31
(pro-)NGF	p75 – sortilin	32–35

10.5
Growth Factor Deprivation

A central problem in developmental biology is the understanding of the regulation of cell survival and death. Exploration of molecular mechanisms underlying cell survival and death is relatively far advanced in developmental neurobiology. The "neurotrophic theory" has provided a hypothesis for experimentally testing and possibly understanding key features of neuron survival and death. According to the "neurotrophic theory" the well-known naturally occurring death of approximately 50 % of all post-mitotic neurons may, at least in part, be explained by a non-saturating supply of soluble neurotrophic factors. Nerve growth factor is the paradigmatic protein growth factor, which is synthesized by the target cells of select sensory and sympathetic neurons, binds to specific receptors, is internalized into axon terminals, and in a retrograde manner is transported to the neuronal cell body. It is speculated but not proven that by these mechanisms neuronal numbers are adjusted to the size of their targets providing adequate rather than supernumerary innervation. Growing evidence, however, suggests that other mechanisms may be involved in regulating cell death, including active signaling of neuron death [41]. Thus, active triggering of neuronal cell death by extrinsic factors is now emerging as a mechanism operating in addition to trophic factor deprivation. For example, BMPs have been shown to signal death in the early development of the nervous system, as e.g. in the dorsal telencephalon and neural crest delaminating from the hindbrain [42, 43]. Even the prototypic neurotrophic factor NGF has been shown to cause cell death by activating the neurotrophin receptor p75 [44, 45] (for reviews see [32, 33]). p75 contains a death domain and a neurotrophin receptor-interacting factor (NRIF) serves as an intracellular p75 binding protein mediating the transmission of the death signal [34].

10.6
Programmed Cell Death in Developing Organs

Programmed cell death during development has been suggested to serve a variety of functions. These include the shaping of developing tissues and organs, removal of transient structures and supernumerary cells, controlling cell numbers, tissue homeostasis [2, 46], and last but not least, disposal of mal-developed or mal-instructed cells (for a review see [1]). In the context of this chapter, we will present selected examples rather than giving a full account of programmed cell death in organ development.

10.6.1
Cavitation of the Early Embryo during Implantation

Cavitation represents a process by which hollow structures arise from solid primordia. This morphogenetic transformation occurs for example twice in early embryogenesis, i.e. in the pre-implantation and post-implantation mouse embryo (cf. [47, 48]). Coucouvanis and Martin have provided evidence that cavitation is a two-step process, whereby extrinsic signals, such as bone morphogenetic proteins (BMP), regulate both the formation of the definite epithelium as well as the induction of apoptosis of cells belonging to the inner cell mass to generate the cavity.

10.6.2
Cardiovascular Development

A complete overview of 31 locations of cell death in the embryonic chicken heart, aorta, and pulmonary trunk during days 2 to 8 of development has been published by Pexieder [49]. These data have been supplemented by descriptions of the rat and human heart. In the developing heart, apoptosis occurs mainly in non-myocardial tissues. Cells undergoing apoptosis belong to the endo- and pericardium as well as to the neural crest (cf. [50]). An apoptotic cell population has also been identified surrounding the conduction system. Following their migration to the endocardial cushions, crest cell become apoptotic, possibly due to interactive signaling of BMP2/4 and Msx1/2, which are expressed in the adjacent endocardial and myocardial compartments. Concerning the relevant apoptotic cascades, caspase–8 knockout mice are lethal at E11 due to impaired cardiac muscle formation [51]. Activation of latent TGF-β seems to be necessary for cardiomyocyte migration and myocardialization of the endocardial cushions. Involvement of TGF-β2 in cardiac apoptotic events has been suggested by the abnormal apoptotic pattern in the outflow tract seen in TGF-β2 knockout mice [52]. Aberrant apoptosis has been shown to accompany cardiac abnormalities. (For a comprehensive review of apoptosis in cardiac development, see [50]).

10.6.3
Renal Development

Apoptosis occurs in an orchestrated fashion in kidney development (cf. [53] for a review). At any time during kidney development, 3–5 % of the cells have been reported to be apoptotic. They seem to be mainly located in islands between developing nephrons. Apoptotic cells apparently include non-induced metanephric blastema [54, 55]. At E19 to P20 in rat kidney, apoptosis occurs in both the nephrogenic zone and in the developing papilla. Approximately 40 % of the apoptotic cells are located in the tubules.

10.6.4
Nervous System Development

Many of the first reports on developmental programmed cell death (PCD) concerned the nervous system [4, 56]. A great majority, if not all neuron populations, are produced in numbers greater than those in the adult, and are subject to developmental death. Neither the biological significance nor the molecular pathways leading to neuron death or survival are fully understood. Whether the best popularized idea about PCD in neurons is correct, i.e. their production is in excess in order to compete for their targets to adjust neuron numbers to the size of the target, is entirely unclear.

Intrinsic "predisposition" and extrinsic factors have been identified as mechanisms leading to developmental neuron death. In addition, deprivation of trophic factors has been proposed as an explanation for PCD in the nervous system. There is evidence to suggest that these three mechanisms operate in concert rather than being completely separated. Molecular pathways underlying intrinsic influences, extrinsic factors, and trophic factor deprivation have been briefly outlined above.

10.6.5
Inner Ear Development

The inner ear is an extremely complex sensory organ responsible for equilibrium and sound detection in vertebrates (cf. [57] for a review). The inner ear develops from the otic placode, which invaginates to give rise to the otic vesicle. Canals and utricle develop from the dorsal portion of the otic vesicle, whereas saccule and cochlea (or lagena) differentiate from the ventral area. In addition, neurons of the 8th cranial ganglion develop from the otic vesicle. The cristae in the semicircular canals, the maculae found in the saccule, utricle, and lagena, and the organ of Corti are the sensory structures. These sensory structures contain the hair cells, i.e. the mechanosensory transducers, and support cells.

Morphogenesis of the inner ear is accompanied by apoptosis, which occurs in distinct spatio-temporal patterns. The earliest apoptotic cells are found during invagination and detachment of the otic vesicle and delamination of the epithelial cells that form the primordium in the 8th cranial ganglion. As development progresses distinct differences in apoptotic areas are found in different vertebrate species probably reflecting the main morphological differences in the inner ear and vestibular apparatus between lower and higher vertebrates. Cell death also accompanies the fusion process of the semicircular canals, and hair cell and ganglionic development.

10.6.6
Removal of Interdigital Tissue and Separation of Digits

The process of digit formation is an excellent example of programmed cell death leading to tissue modeling, which is highly conserved among amniotes [58–60]. Massive cell death in the mesenchyme of the interdigital spaces accompanies the formation of free digits in reptiles [61], birds [62], and mammals [63, 64]. Mouse mutants lacking the essential protein machinery which triggers cell death, i.e. Apaf1 [65] have consistently revealed persistence of interdigital webs. Mice lacking genes of the pro-apoptotic bcl–2 family, such as $bax^{-/-}bak^{-/-}$ double knockouts, fail to lose the interdigital webs [66]. Current knowledge suggests two growth-factor families as master molecules in the extrinsic regulation of interdigital cell death, TGF-βs and BMPs. Dünker and co-workers [67] have shown that interdigital spaces in developing limbs are significantly reduced in $Tgf\beta2^{-/-}Tgf\beta3^{-/-}$ double knockout mice. In the final stages of limb morphogenesis, autopodial cells leaving the progress zone differentiate into cartilage or undergo apoptosis, depending on whether they are incorporated into the digital rays or interdigital spaces. Substantial evidence suggests that these two opposite fates are controlled by BMP signaling (cf. [68]).

10.6.7
Removal of Mal-instructed Cells of the Immune System

During development and maturation lymphocytes undergo cell death if they are not selected or their antigen receptors are not properly arranged, or potentially self-reactive. These criteria define several life/death check points that guide the differentiation of both B-cell and T-cell precursors to develop an immune system that contains large numbers of potentially functional B and T cells and only limited numbers of non-responsive or even self-reactive lymphocytes (for review see [69–71]). Death of immune cells is dictated by at least two distinct mechanisms: (1) growth factor deprivation, and (2) activation-induced cell death. Growth factor withdrawal is regulated by the balance of pro- and anti-apoptotic members of the B-cell lymphoma 2 (Bcl–2) family of proteins (for review, see [72]). Activation-induced cell death is mediated via cell-extrinsic death factors and their corresponding receptors. Identified death receptors are members of the tumor necrosis factor (TNF) receptor gene family (Tab. 10.1) containing a so-called death domain in their intracellular domain [12, 73].

10.7
Concluding Remarks

Cell death is a general phenomenon which accompanies tissue and organ development in order to control morphogenesis, cell number, and removal of malfunctional cells. Cell death is a physiological process required for normal development. The cell-intrinsic mechanisms mediating and executing apoptosis are beginning to be understood at the molecular level. The requirements of future work are to focus on regulatory processes controlling extrinsic and contextual cues governing developmental cell death.

Acknowledgments

The author's work is supported by grants from the Deutsche Forschungsgemeinschaft.

References

1 Vaux DL, Korsmeyer SJ. Cell death in development. *Cell* **1999**; *96*: 245–254.
2 Abrams JM. Competition and compensation: coupled to death in development and cancer. *Cell* **2002**; *110*: 403–406.

3 Putcha GV, Johnson EM Jr. Men are but worms: neuronal cell death in *C. elegans* and vertebrates. *Cell Death Differ* **2004**; *11*: 38–48.

4 Oppenheim RW. Cell death during development of the nervous system. *Annu Rev Neurosci* **1991**; *14*: 453–501.

5 Horvitz HR. Genetic control of programmed cell death in the nematode *Caenorhabditis elegans*. *Cancer Res* **1999**; *59*: 1701s–1706s.

6 Kerr JF, Wyllie AH, Currie AR. Apoptosis: a basic biological phenomenon with wide-ranging implications in tissue kinetics. *Br J Cancer* **1972**; *26*: 239–257.

7 Häcker G. The morphology of apoptosis. *Cell Tiss Res* **2000**; *301*: 5–17.

8 Shi Y. Mechanisms of caspases activation and inhibition during apoptotsis. *Mol Cell* **2002**; *9*: 459–470.

9 Majino G, Joris I. Apoptosis, oncosis, and necrosis. An overview of cell death. *Am J Pathol* **1995**; *146*: 3–15.

10 Waymire KG, Mahar P, Frauwirth K, Chen Y, Wei M, Eng VM, Adelman DM, Simon MC, Yuan J, Lipinski M, Degterev A. Diversity in the mechanisms of neuronal cell death. *Neuron* **2003**; *40*: 401–413.

11 Ellis RE, Yuan JY, Horvitz HR. Mechanisms and functions of cell death. *Annu Rev Cell Biol* **1991**; *7*: 663–698.

12 Krammer PH. CD95(APO–1/Fas)-mediated apoptosis: live and let die. *Adv Immunol* **1999**; *71*: 163–210.

13 Oehm A, Behrmann I, Falk W, Pawlita M, Maier G, Klas C, Li-Weber M, Richards S, Dhein J, Trauth BC. Purification and molecular cloning of the APO–1 cell surface antigen, a member of the tumor necrosis factor/nerve growth factor receptor superfamily. Sequence identity with Fas antigen. *J Biol Chem* **1992**; *267*: 10709–10715.

14 Nagata S, Golstein P. The Fas death receptor. *Science* **1995**; *267*: 1449–1456.

15 Nagata S. Apoptosis by death factor. *Cell* **1997**; *88*: 355–365.

16 Baker SJ, Reddy EP. Modulation of life and death by the TNF receptor superfamily. *Oncogene* **1998**; *17*: 3261–3270.

17 Brojatsch J, Naughton J, Rolls MM. CAR1, a TNFR-related protein, is a cellular receptor for cytopathic avian leucosis-sarcoma viruses and mediates apoptosis. *Cell* **1996**; *87*: 845–855.

18 Chinnaiyan AM, O'Rourke K, Yu GL, et al. Signal transduction by DR3, a death domain-containing receptor related to TNFR–1 and CD95. *Science* **1996**; *274*: 990–992.

19 Kitson J, Raven T, Jiang YP, et al. A death-domain-containing receptor that mediates apoptosis. *Nature* **1996**; *384*: 371–375.

20 Chicheportiche Y, Bourdon PR, Xu H, et al. TWEAK, a new secreted ligand in the tumor necrosis factor family that weakly induces apoptosis. *J Biol Chem* **1997**; *272*: 32401–32410.

21 Marsters SA, Sheridan JP, Pitti RM, et al. Identification of a ligand for the death-domain-containing receptor Apo3. *Curr Biol* **1998**; *8*: 525–528.

22 Pan G, Ni J, Wei YF, et al. An antagonist decoy receptor and a death domain-containing receptor for TRAIL. *Science* **1997**; *277*: 815–818.

23 Wiley SR, Schooley K, Smolak PJ, et al. Identification and characterization of a new member of the TNF family that induces apoptosis. *Immunity* **1995**; *3*: 673–682.
24 Pitti R;M, Marsters SA, Ruppert S, et al. Induction of apoptosis by Apo–2 ligand, a new member of the tumor necrosis factor cytokine family. *J Biol Chem* **1996**; *271*: 12687–12690.
25 Marsters SA, Pitti RM, Donahue CJ, et al. Activation of apoptosis by Apo–2 ligand is independent of FADD but blocked by CrmA. *Curr Biol* **1996**; *6*:750–752.
26 Pan G, O'Rourke K, Chinnaiyan AM. The receptor for the cytotoxic ligand TRAIL. *Science* **1997**; *276*: 111–113.
27 Sheridan JP, Marsters SA, Pitti RM, et al. Control of TRAIL-induced apoptosis by a family of signaling and decoy receptors. *Science* **1997**; *277*: 818–821.
28 Walczak H, Degli-Esposti MA, Johnson RS, et al. TRAIL-Ra: a novel apoptosis-mediating receptor for TRAIL. *EMBO J* **1997**; *16*: 5386–5397.
29 Wu GS, Burns TF, McDonald ER 3rd, et al. KILLER/DR5 is a DNA damage-inducible p53–regulated death receptor gene. *Nat Genet* **1997**; *17*: 141–143.
30 Chaudhary PM, Eby M, Jasmin A, et al. Death receptor 5, a new member of the TNFR family, and DR4 induce FADD-dependent apoptosis and activate the NF-kappaB pathway. *Immunity* **1997**; *7*: 821–830.
31 Pan G, Bauer JH, Harides V, Wang S, Liu D, Yu G, Vincenz C, Aggarwall BB, Ni J, Dixit VM. Identification and functional characterization of DR6, a novel death domain-containing TNF receptor. *FEBS Lett* **1998**; *431*: 351–356.
32 Miller FD, Kaplan DR. Neurotrophin signaling pathways regulating neuronal apoptosis. *CMLS* **2001**; *58*: 1045–1053.
33 Chao MV, Bothwell M. Neurotrophins: to cleave or not to cleave. *Neuron* **2002**; *33*: 9–12.
34 Casademunt E, Carter BD, Benzel I, Frade JM, Dechant G, Barde YA. The zinc finger protein NRIF interacts with the neurotrophin receptor p75NTR and participates in programmed cell death. *EMBO J* **1999**; *21*: 6050–6061.
35 Nykjaer N, Lee R, Teng KK, Jansen P, Madsen P, Nielsen MS, Jacobsen C, Kliemannel M, Schwarz E, Willnow TE, Hemstead BL, Peterson CM. Sortilin is essential fpr proNGF.inced neuronal cell death. *Nature* **2004**; *427*: 843–848.
36 Huang B, Eberstadt M, Olejniczak ET, Meadows RP, Fesik SW. NMR structure and mutagenesis of the Fas (APO–1/CD95) death domain. *Nature* **1996**; *384*: 638–641.
37 Boldin MP, Goncharov TM, Goltsev YV, Wallach D. Involvement of MACH, a novel MORT1/FADD-interacting protease, in Fas/APO–1– and TNF receptor-induced cell death. *Cell* **1996**; *85*: 803–815.
38 Muzio M, Chinnaiyan AM, Kischkel FC, O'Rourke K, Shevchenko A, Ni J, Scaffidi C, Bretz JD, Zhang M, Gentz R, Mann M, Krammer PH, Peter ME, Dixit VM. FLICE, a novel FADD-homolgous ICE/CED–3–like protease, is recruited to the CD95 (Fas/APO–1) death-inducing signaling complex. *Cell* **1996**; *85*: 817–827.

39 Hofmann K, Bucher P, Tschopp J. The CARD domain – a new apoptotic signaling motif. *Trends Biochem Sci* **1997**; *22*: 155–156.

40 Muzio M, Stockwell BR, Stennicke HR, Salvesen GS, Dixit VM. An induced proximity model for caspases–8 activation. *J Biol Chem* **1998**; *273*: 2926–2930.

41 Pettmann B, Henderson CE. Neuron cell death. *Neuron* **1998**; *20*: 633–647.

42 Furuta Y, Piston DW, Hogan BL. Bone morphogenetic proteins (BMPs) as regulators of dorsal forebrain development. *Development* **1997**; *124*: 2203–2212.

43 Graham A, Koentges G, Lumsden A. Neural crest apoptosis and the establishment of craniofacial pattern: an honorable death. *Mol Cell Neurosci* **1996**; *8*: 76–83.

44 Frade JM, Rodriguez-Tebar A, Barde YA. Induction of cell death by endogenous nerve growth factor through its p75 receptor. *Nature* **1996**; *383*: 166–168.

45 Naumann T, Casademunt E, Hollerbach E, Hofmann J, Dechant G, Frotscher M, Barde YA. Complete deletion of the neurtrophin receptor p75NTR leads to long-lasting increases in the number of basal forebrain cholinergic neurons. *J Neurosci* **2002**; *22*: 2409–2418.

46 Abud HE. Shaping developing tissues by apoptosis. *Cell Death Differ* **2004**; *11*: 797–799.

47 Coucouvanis E, Martin GR. Signals for death and survival: A two-step mechanism for cavitation in the vertebrate embryo. *Cell* **1995**; *83*: 279–287.

48 Coucouvanis E, Martin GR. BMP signaling plays a role in visceral endoderm differentiation and cavitation in the early mouse embryo. *Development* **1999**; *126*: 535–546.

49 Pexieder T. Cell death in the morphogenesis and teratogenesis of the heart. *Adv Anat Embryol Cell Biol* **1975**; *51*: 3–99.

50 Poelmann RE, Molin D, Wisse LJ, Gittenberger-de Groot AC. Apoptosis in cardiac development. *Cell Tiss Res* **2000**; *301*: 43–52.

51 Varfolomeev EE, Schuchmann M, Luria V, Chiannilkulchai N, Beckmann JS, Mett IL, Rebrikov D, Brodianski VM, Kemper OC, Kollet O, Lapidot T, Soffer D, Sobe T, Avraham KB, Goncharov T, Holtmann H, Lonai P, Wallach D. Targeted disruption of the mouse Caspase 8 gene ablates cell death induction by the TNF receptors, Fas/Apo1, and DR3 and is lethal prenatally. *Immunity* **1998**; *9*: 267–276.

52 Sanford LP, Ormsby I, Gittenberger-de Groot AC, Sariola H, Friedman R, Boivin GP, Cardell EL, Doetschman T. TGFbeta2 knockout mice have multiple developmental defects that are non-overlapping with other TGFbeta knockout phenotypes. *Development* **1997**; *124*: 2659–2670.

53 Hammerman MR. Regulation of cell survival during renal development. *Pediatr Nephrol* **1998**; *12*: 596–602.

54 Coles HS, Burne JF, Raff MC. Large-scale normal cell death in the developing rat kidney and its reduction by epidermal growth factor. *Development* **1993**; *118*: 777–784.

55 Koseki C. Cell death programmed in uninduced metanephric mesenchymal cells. *Pediatr Nephrol* **1993**; *7*: 609–711.

56 Henderson CE. Programmed cell death in the developing nervous system. *Neuron* **1996**; *17*: 579–585.
57 Leon Y, Sanchez-Galiano S, Gorospe I. Programmed cell death in the development of the vertebrate inner ear. *Apoptosis* **2004**; *9*: 255–264.
58 Saunder JW Jr. Death in embryonic systems. *Science* **1966**; *154*: 604–612.
59 Gilbert. S.F. Developmental Biology. (6th ed.) Sinauer Associates Inc. (Pub.), Sunderland, MA **1997**
60 Jacobson MD, Weil M, Raff MC. Programmed cell death in animal development. *Cell* **1997**; *88*: 347–354.
61 Fallon JF, Cameron J. Interdigital cell death during limb development of the turtle and lizard with an interpretation of evolutionary significance. *J Embryol Exp Morphol* **1977**; *40*: 285–289.
62 Hinchliffe JR, Ede DA. Cell death and the development of limb form and skeletal pattern in normal and wingless (ws) chick embryos. *J Embryol Exp Morphol* **1973**; *30*: 753–772.
63 Ballard KJ, Holt SJ. Cytological and cytochemical studies on cell death and digestion in the foetal rat foot: the role of macrophages and hydrolytic enzymes. *J Cell Sci* **1968**; *3*: 245–262.
64 Zakeri Z, Quaglino D, Ahuja HS. Apoptotic cell death in the mouse limb and its suppression in the hammer-toe mutant. *Dev Biol* **1994**; *165*: 294–297.
65 Cecconi F, Alvarez-Bolado G, Meyer BI, Roth KA, Gruss P. Apaf1 (CED–4 homolog) regulates programmed cell death in mammalian development. *Cell* **1998**; *94*: 727–737.
66 Lindsten T, Ross AJ, King A, Zong WX, Rathmell JC, Shiels HA, Ulrich E, Ma A, Golden JA, Evan G, Korsmeyer SJ, MacGregor GR, Thompson CB. The combined functions of pro-apoptotic Bcl–2 family members bak and bax are essential for normal development of multiple tissues. *Mol Cell* **2000**; *6*: 1389–1399.
67 Dünker N, Schmitt K, Krieglstein K. TGF-beta is required for programmed cell death in interdigital webs of the developing mouse limb. *Mech Dev* **2002**; *113*: 111–120.
68 Merino R, Ganan Y, Macias D, Rodriguez-Leon J, Hurle JM. Bone morphogenetic proteins regulate interdigital cell death in the avian embryo. *Ann NY Acad Sci* **1999**; *887*: 120–132.
69 Daniel NN, Korsmeyer SJ. Cell death: critical control points. *Cell* **2004**; *116*: 205–219.
70 Marsden V, Strasser A. Control of apoptosis in the immune system: Bcl–2, BH3–only proteins and more. *Annu Rev Immunol* **2003**; *21*: 71–105.
71 Jannssen O, Sanzenbacher R, Kabelitz D. Regulation of activation induced cell death of mature T-lymphocyte populations. *Cell Tiss Res* **2000**; *301*: 85–99.
72 Strasser A. The role of BH3–only proteins in the immune system. *Nat Rev Immunol* **2005**; *5*: 189–200.
73 Tartaglia LA, Ayres TM, Wong GH, Goeddel DV. A novel domain within the 55 kd TNF receptor signals cell death. *Cell* **1993**; *74*: 845–853.

**Cell Signaling and
Growth Factors in Development**

Edited by
K. Unsicker and
K. Krieglstein

Further Titles of Interest

R.A. Meyers
Encyclopedia of Molecular Cell Biology and Molecular Medicine
2005, second edition
3-527-30542-4

D. Wedlich
Cell Migration in Development and Disease
2005
ISBN 3-527-30587-4

A. Ridley
Cell Motility
2004
ISBN 0-470-84872-3

G.S. Stein, A.B. Pardee
Cell Cycle and Growth Control
2004, second edition
ISBN 0-471-25071-6

W.W. Minuth, R. Strehl, K. Schumacher
Tissue Engineering
2005
ISBN 3-527-31186-6

E.A. Nigg
Centrosomes in Development and Disease
2004
ISBN 3-527-30980-2

M. Schliwa
Molecular Motors
2003
ISBN 3-527-30594-7

Cell Signaling and Growth Factors in Development

From Molecules to Organogenesis

Edited by
Klaus Unsicker and
Kerstin Krieglstein

WILEY-VCH

WILEY-VCH Verlag GmbH & Co. KGaA

Editor

Prof. Dr. Klaus Unsicker
University of Heidelberg
Interdisciplinary Center for Neurosciences (IZN)
Department of Neuroanatomy
Im Neuenheimer Feld 307
69120 Heidelberg
Germany

Prof. Dr. Kerstin Krieglstein
University of Göttingen
Medical School
Department of Neuroanatomy
Kreuzbergring 36
37075 Göttingen
Germany

Coverillustration
The simplified signaling network (see Fig. 15.5 from Susan Mackem) and the mouse embryo section represent the starting point and goal of the fascinating way from molecules to organo-genesis (mouse embryo section reprinted from »The Atlas of Mouse Development« edited by M.H. Kaufmann, cover page, 1992 with permission from Elsevier).

■ This book was carefully produced. Nevertheless, authors, editor and publisher do not warrant the information contained therein to be free of errors. Readers are advised to keep in mind that statements, data, illustrations, procedural details or other items may inadvertently be inaccurate.

Library of Congress Card No.: applied for
A catalogue record for this book is available from the British Library

Bibliographic information published by Die Deutsche Bibliothek

Die Deutsche Bibliothek lists this publication in the Deutsche Nationalbibliografie; detailed bibliographic data is available in the Internet at http://dnb.dbb.de.

© 2006 WILEY-VCH Verlag GmbH & Co. KGaA, Weinheim
All rights reserved (including those of translation in other languages). No part of this book may be reproduced in any form – by photoprinting, microfilm, or any other means – nor transmitted or translated into machine language without written permission from the publishers. Registered names, trademarks, etc. used in this book, even when not specifically marked as such, are not to be considered unprotected by law.

Printed in the Federal Republic of Germany.
Printed on acid-free paper.
Typesetting pagina GmbH, Tübingen
Printing betz-druck GmbH, Darmstadt
Bookbinding J. Schäffer GmbH i. G., Grünstadt
ISBN–13: 978-3-527-31034-0
ISBN–10: 3-527-31034-7

Contents

Volume 1

Preface   *XXVII*

List of Contributors   *XXIX*

Color Plates   *XXXVII*

I	**Cell Signaling and Growth Factors in Development**
1	**Stem Cells**
	Christian Paratore and Lukas Sommer
1.1	Introduction 3
1.2	Maintenance of Stemness in Balance with Stem Cell Differentiation 5
1.2.1	Wnt Signaling 7
1.2.2	Wnt Signaling Regulates ‚Stemness' in ESCs 8
1.2.3	Wnt Signaling in Hematopoietic Stem Cells 8
1.2.4	Wnt Signal Activation in the Skin 10
1.2.5	Multiple Roles of Canonical Wnt Signaling in Neural Stem Cells 11
1.2.6	Aberrant Wnt signal activation in carcinogenesis 13
1.3	The Notch Signaling Pathway 13
1.3.1	Notch Signaling During Hematopoiesis 14
1.3.2	Notch1 Functions as a Tumor Suppressor in Mouse Skin 15
1.3.3	Notch Signaling in the Nervous System and its Role in Neural Differentiation and Stem Cell Maintenance 15
1.3.4	Aberrant Notch Signaling 18
1.4	Signaling Pathway of the TGFβ Family Members 18
1.4.1	BMP Signaling in ESCs 20

1.4.2	The Influence of TGFβ Family Members on MSC Differentiation	*20*
1.4.3	Tgfβ Factors Act Instructively on NCSC Differentiation	*21*
1.4.4	Aberrant Growth Regulation by Mutations in the Tgfβ Signaling Pathway	*22*
1.5	Shh Signaling	*23*
1.5.1	Hematopoiesis and T-cell Maturation	*24*
1.5.2	The Role of Shh in the Nervous System	*24*
1.5.3	Shh Signaling in Tumorigenesis	*25*
1.6	Conclusions	*26*

2 Germ Cells
Pellegrino Rossi, Susanna Dolci, Donatella Farini, and Massimo De Felici

2.1	Introduction	*39*
2.2	Primordial Germ Cells	*41*
2.2.1	Growth Factors in PGC Commitment	*41*
2.2.2	Growth Factors in PGC Proliferation	*42*
2.2.2.1	The KL/KIT System	*42*
2.2.2.2	LIF/OSM/IL6 Superfamily	*43*
2.2.2.3	TGFβ Superfamily and Neuropeptides	*44*
2.3	Oocytes	*45*
2.3.1	Growth Factors for Fetal Oocytes	*45*
2.3.2	Growth Factors for Growing and Mature Oocytes	*46*
2.3.3	Growth Factors in the Initiation of Follicle and Oocyte Growth	*48*
2.3.4	From Pre-antral to Ovulatory Follicles: Oocytes take Control	*50*
2.4	Male Germ Cells	*52*
2.4.1	GDNF	*53*
2.4.2	BMPs	*55*
2.4.3	KL/Kit System	*56*
2.4.4	Control of Meiotic Progression of Spermatocytes	*58*

3 Implantation and Placentation
Susana M. Chuva de Sousa Lopes, Christine L. Mummery, and Sólveig Thorsteinsdóttir

3.1	Introduction	*73*
3.1.1	Formation of the Trophectoderm, the Precursor of the Invading Trophoblast	*74*
3.1.2	Initial Blastocyst-Uterine Interaction	*74*
3.1.3	Differentiation of Extra-embryonic Lineages During Implantation and Gastrulation	*76*

3.1.4	Steps towards a Functional Placenta 78
3.1.5	Comparison of Mouse and Human Implantation and Placentation 79
3.2	Molecular Mechanisms and Biological Effects 80
3.2.1	Preparation of the Blastocyst for Implantation (E3.5–E4.5) 80
3.2.2	Uterine-Embryonic Signaling and Adhesion (E4.5) 81
3.2.3	The Invasion of the Uterus (E4.5–E7.5) 85
3.2.3.1	Decidualization 85
3.2.3.2	Anchoring of Trophoblast Cells and Differentiation during Peri-implantation 86
3.2.4	The Formation and Development of the Chorio-allantoic Placenta (E8.5–E16.5) 88
3.3	Clinical Relevance in Humans 90
3.4	Summary 92
	Acknowledgments 93

4	**Cell Movements during Early Vertebrate Morphogenesis**
	Andrea Münsterberg and Grant Wheeler
4.1	Introduction 107
4.2	History: Classic Experiments in the Study of Gastrulation 108
4.3	Gastrulation and Neurulation in Different Vertebrate Species 109
4.3.1	Amphibians: *Xenopus laevis* 109
4.3.2	The Teleost, Zebrafish 111
4.3.3	Amniotes: Chick and Mouse 112
4.4	Mechanistic Aspects, Molecules and Molecular Networks 114
4.4.1	Wnt Signaling 115
4.4.2	Convergent Extension and the Regulation of Cytoskeletal Dynamics 118
4.4.3	Fibroblast Growth Factors (FGF) 119
4.4.4	Regulation of Epithelial Mesenchymal Transition (EMT) 122
4.4.5	Cadherins and Protocadherins 123
4.4.6	Extracellular Matrix and Integrin Receptors 124
4.4.7	Platelet Derived Growth Factor (PDGF) 125
4.5	Conclusion and Outlook 126

5	**Head Induction**
	Clemens Kiecker
5.1	Introduction 141
5.2	Classical Concepts for Head Induction 141
5.3	Tissues Involved in Anterior Specification 145
5.3.1	The Gastrula Organizer 145

5.3.2	Primitive and Definitive Anterior Endoderm	146
5.3.3	Ectodermal Signaling Centers	149
5.3.4	Summary	149
5.4	Signaling Pathways Involved in Anterior Specification	149
5.4.1	BMP Signaling	150
5.4.2	FGF Signaling	153
5.4.3	IGF Signaling	154
5.4.4	Nodal Signaling	155
5.4.5	Retinoid Signaling	157
5.4.6	Wnt/β-Catenin Signaling	158
5.4.7	Noncanonical Wnt Signaling	160
5.4.8	Summary	161
5.5	Mechanistic Models for Head Induction	162
5.5.1	The Two- and Three-Inhibitor Models	162
5.5.2	The Organizer-Gradient Model Dualism	162
5.6	Transcriptional Regulators in Anterior Specification	163
5.6.1	ANF/Hesx1	164
5.6.2	Blimp1	164
5.6.3	FoxA2/HNF3β	164
5.6.4	Goosecoid	165
5.6.5	Hex	165
5.6.6	Lim1	165
5.6.7	Otx2	166
5.6.8	Six3	167
5.6.9	Transcriptional Repression in Anterior Specification	167
5.6.10	Summary	167
5.7	Conclusions and Outlook	168
	Abbreviations	169
	Acknowledgments	170
	Notes added in proof	170

6 Anterior-Posterior Patterning of the Hindbrain: Integrating Boundaries and Cell Segregation with Segment Formation and Identity
Angelo Iulianella and Paul A. Trainor

6.1	Introduction	189
6.2	Hindbrain Development	190
6.2.1	Segmentation into Rhombomeres and a Blueprint for Craniofacial Development	190
6.2.2	Segment- and Boundary-restricted *Hox* Gene Expression	193
6.2.3	Altering the *Hox* code: *Hox* Gene Loss- and Gain-of-Function	194
6.2.4	Initiating *Hox* Gene Expression in the Hindbrain	196

6.2.5	Interactions between FGF and Retinoid Signaling	199
6.2.6	*Krox20* and *Kreisler* Regulate Paralogous Groups 2 and 3 in the Vertebrate Hindbrain	201
6.2.7	Establishing the *Hox* Code in the Vertebrate Hindbrain: the Role of Auto- and Cross-regulation	204
6.2.8	A Mechanism for Establishing and Maintaining Hindbrain Segmentation	206
6.2.9	Coupling Cell Lineage Restriction with Discrete Domains of *Hox* Gene Expression	209
6.3	Conclusions	211
	Acknowledgments	211

7 Neurogenesis in the Central Nervous System
Véronique Dubreuil, Lilla Farkas, Federico Calegari, Yoichi Kosodo, and Wieland B. Huttner

7.1	Introduction	229
7.2	Intrinsic Regulation of Neurogenesis	231
7.2.1	Neuronal Determination	231
7.2.1.1	Proneural Genes	231
7.2.1.2	Regulation of Proneural Activity	232
7.2.1.3	Proliferative Determinants	236
7.2.2	Neuronal Differentiation	236
7.3	Extrinsic Regulation of Neurogenesis	237
7.3.1	Wnt Factors	237
7.3.1.1	Wnt Pathway	238
7.3.1.2	Propagation of the Wnt Signal	238
7.3.1.3	Expression and Function during Neurogenesis	239
7.3.1.4	Effectors of the Wnt Pathway	240
7.3.1.5	Cross-talk between Signaling Pathways	241
7.3.2	Hedgehog Factors	241
7.3.2.1	Shh Pathway	241
7.3.2.2	Expression and Function during Neurogenesis	242
7.3.2.3	Eye Differentiation as a Model for Shh Activity	243
7.3.2.4	Cross-talk between Signaling Pathways	244
7.3.3	Fibroblast Growth Factors	244
7.3.3.1	Ligands, Receptors and Expression during Neurogenesis	244
7.3.3.2	Function during Neurogenesis	245
7.3.3.3	Cross-talk between Signaling Pathways	246
7.3.4	Transforming Growth Factors-α, Neuregulins and Epidermal Growth Factors	246
7.3.4.1	Ligands, Receptors and Expression during Neurogenesis	247
7.3.4.2	Function during Neurogenesis	247
7.3.5	Transforming Growth Factors-β	248

7.3.5.1	Ligands, Receptors and Expression during Neurogenesis	248
7.3.5.2	Function in Neurogenesis	249
7.3.5.3	Cross-talk between Signaling Pathways	249
7.3.5.4	BMP Family	250
7.3.6	Other Factors	251
7.3.6.1	Neurotrophins	252
7.3.6.2	Neurokines: LIF, CNTF	253
7.3.6.3	PDGF	253
7.3.6.4	GH	254
7.4	Cell Cycle Regulation and Neuronal Fate Determination	254
7.4.1	Cell Cycle of Neuroepithelial Cells	255
7.4.1.1	Interkinetic Nuclear Migration and Cell Division	255
7.4.1.2	Length of Cell Cycle of Neural Stem Cells	255
7.4.2	Cell-fate Determinants Influencing the Cell Cycle	256
7.4.2.1	Extrinsic Cell-fate Determinants Regulate the Cell Cycle	256
7.4.2.2	Intrinsic Cell-fate Determinants Regulate the Cell Cycle	257
7.4.3	Cell Cycle Regulators Influencing Cell Fate	258
7.4.3.1	Cell Cycle Regulators	258
7.4.3.2	Cell Cycle Regulators Regulate Cell-fate Determination	258
7.4.4	Model of Cell Cycle Lengthening	259
7.5	Neuron-generating Asymmetric Cell Division	260
7.5.1	Background	260
7.5.2	Asymmetric Cell Divisions of Neuroblasts in the *Drosophila* CNS	260
7.5.2.1	Cell Lineage	260
7.5.2.2	Apical-Basal NB Polarity	261
7.5.2.3	Control of Spindle Orientation	262
7.5.2.4	Cell-fate Determinants	262
7.5.3	The Neuronal Cell Lineage in the Mammalian CNS	263
7.5.3.1	Neuron-generating Divisions and Cell-fate Asymmetry	263
7.5.3.2	Cell Polarity in the Mammalian Neuroepithelium	264
7.5.3.3	Neuron-generating Division and Asymmetric Cell Division in Mammals	264
7.5.3.4	Asymmetric Distribution of Cell-fate Determinants in Mammals	266
7.6	Conclusions	266

8	**Generating Cell Diversity**	
	Ajay Chitnis	
8.1	Introduction	287
8.2	Establishing Early Compartments of the Embryo	288
8.2.1	Establishing the Three Germ Layers	288
8.2.1.1	Nodals Function as Morphogens to Define Germ Layers	288

8.2.1.2	Maternal Factors Regulate Nodal Signaling	290
8.2.2	Establishing the Dorsal-Ventral Axis	290
8.2.2.1	Formation of the Vertebrate Organizer	290
8.2.2.2	β-Catenin-dependent Dorsalizing Factors Position the Organizer	291
8.2.2.3	The Vertebrate Dorsal Organizer is a Self-organizing Tissue	291
8.2.2.4	The BMP Gradient Determines Compartments of the Ectoderm	292
8.2.2.5	Neural Genes Prevent Differentiation and Maintain a Neural State	293
8.2.3	Establishing the Rostral-Caudal (Anterior-Posterior) Axis	293
8.2.3.1	The Blastoderm Margin is an Early Source of Posteriorizing Factors	293
8.2.3.2	Rostral-Caudal Patterning of the Anterior Neuroectoderm	294
8.2.3.3	Secondary Organizers and Compartment Boundaries	295
8.2.3.4	Rostral-Caudal Patterning of the Hindbrain	296
8.2.3.5	Establishing a Rostral-Caudal FGF Gradient	296
8.2.3.6	The FGF Gradient Regulates Mesoderm and Neuroectoderm Fate	297
8.3	Early Neurogenesis in the Neural Plate	298
8.3.1	Proneural Genes Define Neurogenic Domains	298
8.3.2	Notch Signaling Regulates Differentiation within Neurogenic Domains	300
8.3.3	Cell Fate Decisions Regulated by Notch	300
8.3.4	A Brief Review of Notch Signaling	301
8.3.5	Factors that Influence the Outcome of Notch Signaling	302
8.3.5.1	Fringe Glycosyl Transferases	302
8.3.5.2	E3 Ubiquitin Ligases	303
8.3.5.3	Numb and Biasing Notch Signaling in the SOP	303
8.3.6	Determination of Neurogenic and Non-neurogenic Domains in the Neuroectoderm	304
8.4	Determination of Neuronal Identities in the Spinal Cord	305
8.4.1	Cellular Organization of the Spinal Cord	305
8.4.2	Specification of Dorsal Neurons	306
8.4.3	The Notochord and Floor Plate Establish a Hedgehog Gradient	308
8.4.4	Class I and Class II Transcription Factors Respond to a Hedgehog Gradient to Define Discrete Compartments of Neuron Progenitors in the Ventral Cord	309
8.4.5	Specification of Motor Neuron Fate	310
8.4.6	Organization of the Motor Neurons and the LIM Code	311
8.5	Concluding Remarks	313

8.5.1	Gradients, Compartments, Boundaries and Local Interactions *313*	
8.5.2	Three Principles of Transcription Factor Regulation *314*	
8.5.3	Temporal Regulation of Cell Diversity *314*	

9 The Molecular Basis of Directional Cell Migration
Hans Georg Mannherz

9.1	Introduction *321*	
9.2	How and What Moves Cells *322*	
9.3	The Cytoskeleton *323*	
9.3.1	Microtubules *325*	
9.3.2	Microfilaments *325*	
9.3.3	Intermediate Filaments *327*	
9.4	Motile Membrane Extensions containing Microfilaments *328*	
9.4.1	The Cytoskeletal Components of Transient Actin-Powered Protrusive Organelles *328*	
9.4.2	Lamellipodial Protrusion *330*	
9.5	Modulation of the Structure of Transient Protusive Membrane Extensions *333*	
9.6	Formation of Substratum Contacts *334*	
9.7	Cell Body Traction *336*	
9.8	Rear Detachment *337*	
9.9	The Role of Microtubules in Cell Migration *338*	
9.10	The Role of Intermediate Filaments in Cell Migration *338*	
9.11	Modulation of the Cytoskeletal Organization by External Signals *338*	
9.12	The Cytoskeletal Organization of Neuronal Growth Cones *341*	
9.13	Conclusions *343*	
	Acknowledgement *344*	

10 Cell Death in Organ Development
Kerstin Krieglstein

10.1	Introduction *347*	
10.2	Programmed Cell Death *347*	
10.3	Genetics of Programmed Cell Death *348*	
10.4	Cell Death Receptors *349*	
10.5	Growth Factor Deprivation *349*	
10.6	Programmed Cell Death in Developing Organs *350*	
10.6.1	Cavitation of the Early Embryo during Implantation *350*	
10.6.2	Cardiovascular Development *350*	
10.6.3	Renal Development *351*	

10.6.4 Nervous System Development *351*
10.6.5 Inner Ear Development *352*
10.6.6 Removal of Interdigital Tissue and Separation of Digits *352*
10.6.7 Removal of Mal-instructed Cells of the Immune System *353*
10.7 Concluding Remarks *353*
Acknowledgments *353*

Volume 2

II Cell Signaling and Growth Factors in Organogenesis

11 Dorso-Ventral Patterning of the Vertebrate Central Nervous System
Elisa Martí, Lidia García-Campmany, and Paola Bovolenta
11.1 Introduction *361*
11.2 Generating Cell Diversity in the Dorsal Neural Tube *363*
11.2.1 The Neural Crest *363*
11.2.1.1 Cellular and Molecular Inducers of Neural Crest *364*
11.2.1.2 Molecular Identity of Newly-induced Neural Crest Cells *366*
11.2.2 The Roof Plate *367*
11.2.3 Signals from the Roof Plate Pattern the Dorsal Spinal Cord *369*
11.2.3.1 Transcriptional Code Defining Dorsal Spinal Cord Interneurons *370*
11.3 Generating Cell Diversity in the Ventral Neural Tube *371*
11.3.1 The Floor Plate *371*
11.3.2 Sonic Hedgehog Secreted from the Notochord and the Floor Plate Patterns the Ventral Spinal Cord *373*
11.4 Generation of D-V Patterning in the Anterior Neural Tube Follows Rules Similar to those Used in the Spinal Cord *374*
11.4.1 The Eye *375*
11.4.1.1 D-V Patterning of the Optic Vesicle *376*
11.4.1.2 D-V Patterning of the Optic Cup *379*
11.4.2 D-V Patterning of the Telencephalon *381*
11.5 Conclusions and Perspectives *384*

12 Novel Perspectives in Research on the Neural Crest and its Derivatives
Chaya Kalcheim, Matthias Stanke, Hermann Rohrer, Kristjan Jessen, and Rhona Mirsky

12.1 General Introduction 395
12.2 Novel Techniques to Investigate NC Development and Fate 396
12.2.1 Cellular Tracers 396
12.2.2 Molecules Expressed by NC Cells 397
12.2.3 Investigating the Function of Molecules Expressed by NC Cells 397
12.3 Mechanisms of Specification and Emigration of NC Cells 399
12.3.1 Specification of the NC 399
12.3.1.1 Transcription Factors in NC Formation 400
12.3.2 The Delamination of NC Progenitors from the Neural Tube 401
12.3.2.1 A Balance between BMP and its Inhibitor Noggin Regulates NC Delamination in the Trunk 402
12.3.2.2 BMP-dependent Genes and NC Delamination 402
12.3.2.3 The Role of the Cell Cycle in NC Delamination 403
12.4 Peripheral Neuronal Lineages: Pluripotentiality and Early Restrictions of Migratory NC 404
12.4.1 Evidence from Cell Lineage Analysis *In Vivo* and *In Vitro* 404
12.4.2 NC Stem Cells and the PNS Lineage 406
12.5 Molecular Control of Neuron Development in Peripheral Ganglia 408
12.5.1 Extrinsic Signals in Sensory, Sympathetic, Parasympathetic and Enteric Neuron Development 408
12.5.1.1 Sensory Ganglia 408
12.5.1.2 Sympathetic Ganglia 408
12.5.1.3 Parasympathetic Ganglia 410
12.5.1.4 Enteric Ganglia 410
12.5.2 Transcriptional Control of Peripheral Neuron Development 411
12.5.2.1 Sensory Neurons 411
12.5.2.2 Sympathetic and Parasympathetic Neurons 413
12.5.2.3 Enteric Neurons 414
12.6 Growth Cone Navigation and Innervation of Peripheral Targets 414
12.6.1 Semaphorins 415
12.6.2 GDNF-family Ligands 415
12.6.3 Hepatocyte Growth Factor (HG) 416
12.6.4 Macrophage Stimulating Protein (MSP) 416
12.6.5 TGF-β/BMP Family 416

12.7	Factors Controlling Neuronal Survival in the PNS	417
12.8	Glial Cell Development from the NC-Peripheral Glial Lineages: Fate Choice and Early Developmental Events	417
12.8.1	The Main Cell Types	417
12.8.2	Molecular Markers of Early Glial Development	419
12.8.3	Transcription Factors that Control the Emergence of Glia	421
12.8.4	Inductive Signals Involved in Glial Specification	422
12.8.5	Developmental Plasticity of Early Glia	423
12.9	Signals that Control Schwann Cell-precursor Survival and Schwann Cell Generation	423
12.9.1	Schwann Cell Precursors	423
12.9.2	Schwann Cell Generation	425
	Acknowledgments	426

13 Eye Development
Filippo Del Bene and Joachim Wittbrodt

13.1	Vertebrate Eye Development: An Overview	449
13.2	Patterning of the Anterior Neural Plate	449
13.2.1	Posteriorizing Factors	450
13.2.2	Repression of Posteriorizing Factors is required for Head Formation	451
13.2.3	Subdivision of the Anterior Neural Plate	451
13.3	Transcription Factors Function in the Establishment of Retinal Identity	453
13.4	Patterning of the Eye Field and its Medio-lateral Separation	456
13.5	Differentiation of Neuroretina and Pigmented Retinal Epithlium	458
13.5.1	Trigger from the Ventral Midline	458
13.5.2	Retinal Polarity	458
13.5.3	Intrinsic Signals in the Neuroretina	460
13.5.4	Six3 and Geminin	461
13.5.5	Chx10	462
13.6	Differentiation of Retinal Cell Types	463
13.6.1	Retinal Fate Determination by Basic Helix-Loop-Helix (bHLH) and Homeobox Transcription Factors	465
13.6.2	Retinal Ganglion Cells	466
13.6.3	Amacrine Cells	467
13.6.4	Bipolar Cells	467
13.6.5	Horizontal Cells	468
13.6.6	Photoreceptor Cells	468
13.6.7	Maintenance of Retinal Stem Cells and Generation of Müller Glia	468

14 Mammalian Inner Ear Development: Of Mice and Man
Bernd Fritzsch and Kirk Beisel

- 14.1 Introduction *487*
- 14.1.1 An Outline of Mammalian Ear Evolution *487*
- 14.1.2 Human Deafness Related Mutations *488*
- 14.2 Morphological and Cellular Events in Mammalian Ear Development *491*
- 14.2.1 Molecular Basis of Ear Placode Induction *493*
- 14.2.2 Molecular Basis of Ear Morphogenesis *494*
- 14.2.2.1 FGFs *495*
- 14.2.2.2 The EYA/SIX/DACH Complex *497*
- 14.2.2.3 GATA3 *499*
- 14.2.3 Molecular Biology of Otoconia, Cupula and Tectorial Membrane Formation and Maintenance *500*
- 14.2.4 Molecular Basis of Mammalian Ear Histogenesis *503*
- 14.2.4.1 Sensory Neuron Development *504*
- 14.2.4.2 Hair Cell Development *507*
- 14.3 Postnatal Maturation *510*
- 14.4 Aging, Hair Cell Loss and the Molecular Biology of Hair Cell Regeneration *510*

15 Limb Development
Susan Mackem

- 15.1 Introduction *523*
- 15.1.1 Morphological Landmarks of the Limb *524*
- 15.1.2 Embryological Elucidation of Inductive Interactions in Limb Development *524*
- 15.2 The Major Signaling Pathways in Limb Development *528*
- 15.2.1 Retinoids *528*
- 15.2.2 Fgfs *530*
- 15.2.3 Wnts *535*
- 15.2.4 Bmps *541*
- 15.2.5 Shh *547*
- 15.2.6 Other Players *551*
- 15.3 Limb Initiation and Formation of the Limb "Organizers" *553*
- 15.3.1 Axial Cues for Limb Initiation and AER Induction *553*
- 15.3.2 Establishment of Early DV Polarity and AER Formation in Limb Bud Ectoderm *556*
- 15.3.3 AER Maturation and Maintenance *558*
- 15.3.4 ZPA Formation *560*
- 15.4 Function of Organizers in Regulating Pattern and Growth along the Limb Axes *562*
- 15.4.1 The AER-Fgfs, RA, and the PD Axis *562*

15.4.2	Dorsal Ectodermal Wnt7a and the DV Axis *566*	
15.4.3	ZPA-Shh and the AP Axis *568*	
15.5	Coordination of Patterning, Outgrowth and Differentiation: Positive Co-regulation and Feedback Loops Synchronize Output from Limb Organizers, and Antagonism between Signaling Pathways Contributes to Polarity *571*	
15.6	Ongoing Late Regulation and Realization of Pattern *573*	
15.6.1	Condensation and Segmentation of Skeletal Elements *573*	
15.6.2	Apoptosis and Sculpting the Final Limb Form *576*	
15.7	Potential Biomedical Applications and Future Directions *578*	

16	**Skeletal Development**	
	William A. Horton	
16.1	Introduction *619*	
16.2	Skeletal Morphogenesis *620*	
16.2.1	Pre-condensation *620*	
16.2.2	Membranous Skeletal Development *620*	
16.2.3	Endochondral Skeletal Development *621*	
16.3	Skeletal Growth *624*	
16.3.1	Growth Plate Structure and Organization *624*	
16.3.2	Regulation of Linear Growth *626*	
16.3.2.1	Ihh-PTHrP Circuit *626*	
16.3.2.2	FGF-FGFR3 Circuit *627*	
16.3.2.3	CNP-GC-B Circuit *628*	
16.3.2.4	VEGF-MMP9 Circuit *628*	
16.4	Therapeutic Considerations *629*	
16.5	Conclusions *631*	

17	**Musculature and Growth Factors**	
	Petra Neuhaus, Herbert Neuhaus, and Thomas Braun	
17.1	Introduction *641*	
17.2	Induction of Primary Myogenesis by Cell Signaling and Growth Factors *642*	
17.2.1	Wnt Molecules *643*	
17.2.2	Shh, Ihh *646*	
17.2.3	BMPs *647*	
17.2.4	FGFs *649*	
17.3	Tour Guides: How Growth Factors Guide Migration of Muscle Precursor Cells During Embryonic Development *651*	
17.3.1	HGF *651*	
17.3.2	FGFs *652*	
17.4	Control of Muscle Size and Muscle Fiber Diversity by Local and Circulating Growth Factors *653*	

17.4.1	Control of Muscle Size	653
17.4.1.1	Myostatin	653
17.4.1.2	IGFs	654
17.4.2	Muscle Fiber Diversity	656
17.4.2.1	Shh	656
17.4.2.2	Wnt	657
17.5	Growth Factors in Muscle Wasting and Skeletal Muscle Regeneration	659
17.5.1	Mediation of Muscle Wasting by Cell Signaling Molecules	659
17.5.2	Activation of Muscle Satellite Cells by Growth Factors	660
17.5.2.1	HGF	662
17.5.2.2	FGF	662
17.6	Does Growth Factor Mediated Recruitment of Uncommitted Stem Cells Contribute to Skeletal Muscle Cell Regeneration?	664

18 Skin Development
Lydia Sorokin and Leena Bruckner-Tuderman

	Abbreviations	679
18.1	Introduction	679
18.2	Morphology of the Skin and Development of Hair Follicles and Other Adnexal Structures	681
18.2.1	Epidermis	681
18.2.2	Non-Keratinocyte Cell Types in the Epidermis	683
18.2.3	Hair Follicle	686
18.2.4	Sebaceous Glands	687
18.2.5	Nails	688
18.2.6	Eccrine and Apocrine Sweat Glands	688
18.3	Cell Adhesion and the Role of Adhesion Molecules in Development	689
18.3.1	Cell-Cell Adhesion	689
18.3.2	Cell-Extracellular Matrix Adhesion in the Epidermis	690
18.3.2.1	Hemidesmosomes	690
18.3.2.2	Inter-Hemidesmosomal Spaces	693
18.3.3	Role of Laminins in Epidermal Basement Membrane Development	695
18.4	Development of the Dermal Matrix	698
18.5	Epithelial-Mesenchymal Interactions and Signaling	700
18.6	Epidermal Stem Cells	703
18.7	Pathological Skin Conditions Caused by Developmental Abnormalities	705
18.8	Future Considerations	706
	Acknowledgments	707

19 Tooth Development
Xiu-Ping Wang and Irma Thesleff

19.1 Introduction 719
19.2 Signaling Pathways and their Networks Regulating Tooth Development 722
19.2.1 Fibroblast Growth Factors 722
19.2.2 TGFβ Family 725
19.2.3 Sonic Hedgehog 729
19.2.4 Wnt Family 730
19.2.5 Tumor Necrosis Factors 732
19.2.6 Other Growth Factors 733
19.2.7 Integrations between Growth Factor Signal Pathways 734
19.3 Biological Effects 735
19.3.1 Formation of Dental Placodes 736
19.3.2 Epithelial Cell Proliferation 736
19.3.3 Patterning of the Tooth Crown 738
19.3.4 Cell Differentiation 738
19.4 Biomedical Application 739
19.5 Summary 742

20 Gastrointestinal Tract
Daniel Ménard, Jean-François Beaulieu, François Boudreau, Nathalie Perreault, Nathalie Rivard, and Pierre H. Vachon

20.1 Introduction 755
20.2 Development of the Gastrointestinal Mucosa 756
20.2.1 Specialization of Epithelium and Functional Units 756
20.2.1.1 Esophagus 756
20.2.1.2 Stomach 756
20.2.1.3 Small Intestine 756
20.2.1.4 Colon 757
20.2.2 Mesenchymal-Epithelial Interactions 757
20.2.2.1 Growth Factors and Morphogens 758
20.2.2.2 Hox Genes 759
20.2.2.3 Epimorphin/Syntaxin 2 760
20.2.2.4 Basement Membrane Proteins 760
20.2.2.5 Mesodermal Transcription Factors 761
20.3 Gastric Cell Proliferation and Differentiation 761
20.3.1 Functional Compartmentalization of Gastric Glands 761
20.3.2 Hormones and Growth Factors 762
20.3.3 ECM and Integrins 763
20.4 Intestinal Cell Proliferation and Differentiation 765
20.4.1 Hormones and Growth Factors 765
20.4.2 ECM and Integrins 766

20.4.2.1	Development	768
20.4.2.2	Anteroposterior (AP) Axis	769
20.4.2.3	Crypt-Villus Axis	769
20.4.3	Cell Signaling Pathways	770
20.3.4.1	ERK-MAP Kinase Cascade	770
20.4.3.2	p38–MAP Kinase Cascade	771
20.4.3.3	Wnt Pathway	772
20.4.3.4	Phosphatidylinositol 3–Kinase Signaling Pathway	772
20.4.4	Transcription Factors	774
20.4.4.1	Transcription Factors Involved in the Determination of Intestinal Epithelial Cell Lineage	775
20.4.4.2	Transcriptional Regulators of Intestinal-specific Genes	776
20.4.5	Cell Survival, Apoptosis and Anoikis	778
20.4.5.1	Crypt-Villus Axis Distinctions	779
20.4.5.2	Proximal-Distal Axis Distinctions	780
20.4.5.3	Intestinal Cell Survival and Death: Differences and Differentiation	781
20.5	Biomedical Applications	782

21 **Cell Signaling and Growth Factors in Lung Development**
David Warburton, Saverio Bellusci, Pierre-Marie Del Moral, Stijn DeLanghe, Vesa Kaartinen, Matt Lee, Denise Tefft, and Wei Shi

21.1	Introduction	791
21.1.1	The Stereotypic Branch Pattern of Respiratory Organs	791
21.1.2	Transduction of Candidate Growth Factor Peptide Ligand Signals	792
21.1.3	Examples of Peptide Growth Factor Signaling Pathways	792
21.2	Growth Factors and Lung Development	792
21.2.1	Candidate Growth Factors in Lung Development	792
21.2.2	Growth Factor-mediated Epithelial-Mesenchymal Interactions and Lung Development	793
21.2.3	FGF10	793
21.2.4	The Role of BMP4	794
21.2.5	The Role of the Vasculature and VEGF Signaling	794
21.2.6	Postnatal Lung Development	795
21.2.7	The Influence of Peptide Growth Factor Signaling on the Correct Organization of the Matrix	795
21.3	Growth Factor Signaling Pathways in Lung Development	795
21.3.1	Critical Signaling Pathways in Lung Development	795
21.3.1.1	BMPs	795
21.3.1.2	Activin Receptor-like Kinases	796
21.3.1.3	ALKs in Pulmonary Development	796
21.3.1.4	ALKs and the Pulmonary Vasculature	797

21.3.1.5   ALKs, Pulmonary Fibrosis and Inflammation   797
21.3.2   FGF Signaling Promotes Outgrowth of Lung Epithelium   798
21.3.3   FGF2b   798
21.3.4   FGF10   799
21.3.5.   FGF10 Activity   799
21.3.5.1. Regulation of FGF Signaling   800
21.3.5.2. Mice and Humans Possess Several *Spry* Genes (*Mspry1–4* and *Hspry1–4*)   800
21.3.5.3. Sprouty Binds Downstream Effector Complexes   800
21.3.5.4. FGFs as Tyrosine Kinase Receptors   801
21.3.5.5. Relationship between FGF Signaling and Spry during Development   801
21.3.5.6. Spry2 and Spry4 Share a Common Inhibitory Mechanism   801
21.3.4   Sonic Hedgehog, Patched and Hip   802
21.3.4.1 Expression and Activation of TGFβ Family of Peptides   802
21.3.4.2. Developmental Specificity of the TGFβ1 Overexpression Phenotype   803
21.3.4.3. TGFβ Signaling   804
21.3.4.4. The Bleomycin-induced Model of Lung Fibrosis   804
21.3.5.   VEGF Isoform and Cognate Receptor Signaling in Lung Development   805
21.3.5.1. Both Humans and Mice have Three Different VEGF Isoforms   805
21.3.5.2. *Vegf* Knockout Mice have a Lethal Phenotype within the Early Stages of Embryonic Development (E8.5–E9)   805
21.3.5.3. The Role of VEGF in Maintaining Alveolar Structure   806
21.3.5.4. VEGF-C and VEGF-D   806
21.3.5.5. VEGF Isoforms Induce Vasculogenesis, Angiogenesis and Lymphoangiogenesis   806
21.3.6.   Wnt Signaling   806
21.3.7.   Inactivation of the β-Catenin Gene   808
21.3.8.   Dickkopf Regulates Matrix Function   808
21.4   Regulation of Signaling Networks   809
21.4.1.   Growth Factor Tyrosine Kinase and TGF-β Pathways   809
21.4.2.   Mutual Regulation of Intracellular Signaling Networks   809
21.4.3.   Regulation of TGF-β signaling by EGF Signaling   810
21.4.4.   Calcium Signaling and the Mitochondrial Apoptosis Pathway   810
21.5   Developmental Modulation of Growth Factor Signaling by Adapter Proteins   810
21.6   Morphogens and Morphogenetic Gradients   811
21.6.1.   Coordination of Growth Factor Morphogenetic Signals to Determine Lung Development   811

21.6.2.	Action of Morphogens	*811*
21.6.3.	Other Morphogen Gradient Systems	*812*
21.6.4.	The APR Model	*812*
21.6.5.	The Modified Turing Model	*813*
21.6.6.	Expression of *Fgf10*	*813*
21.6.7.	Tip-splitting Event	*813*
21.6.8.	The Value of Hypothetical Models	*814*
21.6.9.	Retinoic Acid Receptors	*815*
21.7.	Conclusion	*815*

22 Molecular Genetics of Liver and Pancreas Development
Tomas Pieler, Fong Cheng Pan, Solomon Afelik, and Yonglong Chen

22.1	Introduction	823
22.2	From the Fertilized Egg to Primitive Endodermal Precursor Cells	825
22.3	Commitment to Pancreas and Liver Fates in Xenopus	826
22.4	Liver and Pancreas Specification in Mouse, Chicken and Zebrafish	826
22.5	Proliferation and Differentiation of Functionally Distinct Pancreatic and Hepatic Cell Populations	829
22.6	Transdifferentiation of Pancreas and Liver	830
22.7	Pancreas and Liver Regeneration	830
22.8	Generation of Pancreatic and Hepatic Cells from Pluripotent Embryonic Precursor Cells	832

23 Molecular Networks in Cardiac Development
Thomas Brand

23.1	Introduction	841
23.2	Implications of Studying Heart Development for Adult Cardiology	841
23.3	Model Organisms	842
23.3.1	The Mouse Embryo (*Mus musculus*)	842
23.3.2	The Chick Embryo (*Gallus gallus*)	843
23.3.3	The Frog Embryo (*Xenopus laevis*)	843
23.3.4	The Zebrafish Embryo (*Danio rerio*)	843
23.3.5	Lower Chordates	844
23.3.6	*Drosophila melanogaster*	844
23.3.7	*Homo sapiens*	844
23.4	Anatomical Description of Heart Development	845
23.5	Cardiac Induction	846
23.5.1	The Role of BMP2 in Heart Induction	847
23.5.2	Canonical Wnt Signaling Interferes with Heart Formation in Vertebrates	849

23.5.3	FGF Cooperates with BMP2	850
23.5.4	Cripto	851
23.5.5	Shh	852
23.5.6	Notch Signaling Interferes with Myocardial Specification	852
23.6	Transcription Factor Families Involved in Early Heart Induction	853
23.6.1	The NK Family of Homeobox Genes	853
23.6.2	The GATA Family of Zinc-finger Transcription Factors	854
23.6.3	Serum Response Factor	855
23.6.4	Synergistic Interaction of Cardiac Transcription Factors	855
23.7	Tubular Heart Formation	856
23.8	Left-Right Axis Development	857
23.8.1	Looping Morphogenesis	857
23.8.1.1	Mechanisms of L-R Axis Determination	857
23.8.1.2	Generation of Initial Asymmetry	858
23.8.1.3	Transfer of L-R Asymmetry to the Organizer Tissue	858
23.8.1.4	The Nodal Flow Model	859
23.8.1.5	Transfer of L-R Asymmetry to the Lateral Plate	861
23.8.1.6	Asymmetric Organ Morphogenesis	862
23.8.2	A-P Axis Formation in the Heart	864
23.8.3	Dorso-ventral Polarity of the Heart Tube	864
23.9	Chamber Formation	865
23.9.1	Analysis of Growth Patterns in the Heart	865
23.9.2	Reprogramming of Gene Expression at the Onset of Chamber Development	865
23.9.3	The Secondary or Anterior Heart Field (AHF)	866
23.9.4	The Right Ventricle is a Derivative of the Anterior Heart Field	868
23.9.5	T-box Genes Pattern the Cardiac Chambers and are Involved in Septum Formation	869
23.9.6	Cell-Cell Interaction in Chamber Formation	869
23.9.6.1	Formation of Compact and Trabecular Layer	869
23.9.6.2	The Epicardium Controls Ventricular Compact Layer Formation	870
23.9.6.3	Epicardial Cells Form the Coronary Vasculature after Epithelial Mesenchymal Transition	871
23.10	Outflow Tract Patterning and the Role of the Neural Crest	872
23.10.1	Tbx1 is Mutated in the DiGeorge Syndrome	873
23.10.2	The Neural Crest Cells have an Early Role in the Heart Tube	873
23.11	Signals Governing Valve Formation	874
23.11.1	The Tgfβ Superfamily and Valve Formation	874
23.11.2	AV Cushion Formation Requires Hyaluronic Acid	875
23.11.3	NFAT2 probably Mediates VEGF Signaling during Cushion Formation	876

23.11.4	Wnt and Notch/Delta Signaling Pathways and Cardiac Valve Formation 876
23.12	Epigenetic Factors 877
23.13	Outlook 877
	Acknowledgments 879

24 Vasculogenesis
Georg Breier

24.1	Introduction 909
24.2	Modes of Blood Vessel Morphogenesis: Vasculogenesis, Angiogenesis and Arteriogenesis 909
24.3	Endothelial Cell Differentiation and Hematopoiesis 910
24.4	Adult Arteriogenesis and Vasculogenesis 911
24.5	Endothelial Cell Growth and Differentiation Factors 912
24.6	VEGF and VEGF Receptors 913
24.7	Other VEGF Family Members: Involvement in Angiogenesis and Lymphangiogenesis 914
24.7.1	Angiopoietins and Ties in Angiogenesis and Lymphangiogenesis 915
24.7.2	Ephrins, Notch, and Arteriovenous Differentiation 916
24.8	Hypoxia-inducible Factors and Other Endothelial Transcriptional Regulators 917
24.9	Outlook 917

25 Inductive Signaling in Kidney Morphogenesis
Hannu Sariola and Kirsi Sainio

25.1	Early Differentiation of the Kidney 925
25.2	Regulation of Ureteric Bud Branching 928
25.3	Genes Affecting Early Nephrogenesis 928
25.4.	Signaling Molecules, their Receptors, and the Integrins 930
25.4.1	Glial Cell Line-derived Neurotrophic Factor 930
25.4.2	Fibroblast Growth Factors 932
25.4.3	Leukemia Inhibitory Factor 932
25.4.4	Wnt Proteins 933
25.4.5.	Bone Morphogenetic Proteins 933
25.4.6.	Sonic Hedgehog 934
25.4.7	Formins 934
25.4.8	Hepatocyte Growth Factor and Met Receptor Tyrosine Kinase 934
25.4.9.	Integrins 935
25.4.10.	Other Molecules 935
25.5	Transcription Factors 935

25.5.1	Wilms' Tumor Gene 1	935
25.5.2	Pax–2	936
25.5.3	Six-Eya Complex	936
25.5.4	Emx–2	937
25.5.5	Lim–1	937
25.5.6	Fox Genes	937
25.6	Future Perspectives	938

26 Molecular and Cellular Pathways for the Morphogenesis of Mouse Sex Organs
Humphrey Hung-Chang Yao

26.1	Introduction	947
26.2	Building the Foundation: Establishment of the Urogenital Ridge	948
26.2.1	Gonadogenesis in Mice	949
26.2.2	Molecular Mechanisms of Gonadogenesis	949
26.3	Parting the Way: Sexually Dimorphic Development of the Gonad	950
26.3.1	Embryonic Testis	951
26.3.1.1	Sry: The Master Switch from the Y Chromosome	951
26.3.1.2	Sox9: The Master Switch Downstream of Sry	953
26.3.1.3	Specification of Sertoli Cell Lineage	954
26.3.1.4	Cellular Events Triggered by *Sry*	955
26.3.2	Embryonic Ovary	958
26.3.2.1	The Quest for the Ovary-determining Gene	958
26.3.2.2	Female Germ Cells: The Key to Femaleness	960
26.4	Dimorphic Development of the Reproductive Tracts	961
26.4.1	Initial Formation of the Wolffian and Müllerian Ducts	961
26.4.2	Sexually Divergent Development of Reproductive Tracts	963
26.4.3	Patterning of the Reproductive Tracts	964
26.5	Morphogenesis of the External Genitalia	966
26.6	Summary	967

Index   979

Preface

Developmental biology investigates and describes, in the broadest sense, changes in organisms from their origins to birth, and on into adulthood, or even to death. Developmental changes are characterized by increasing complexity and specialization of cells and organs. Research in developmental biology has always been a combination of experiment and description. Although mostly morphologically oriented in its beginnings, developmental biology has experienced exponential growth in the 20th century following its fruitful merger with molecular biology and genetics during the 1980s and 1990s.

What we are currently witnessing is the integration of molecular cell biology into developmental biology, with the goal of eventually understanding all developmental events at the level of and as the result of cell biological events. Following this trend, we may ultimately expect to understand development at the resolutional level of molecular machines, individual molecules, and even atoms. The elucidation of signaling between developing cells and the analysis of intracellular signaling networks represent an intermediate but important and indispensable step in the attainment of this goal.

This book addresses representative examples of inter- and intracellular signaling in development involving growth and transcription factors. The various chapters cover general aspects of development such as e.g. stem cells, implantation and placentation, generation of cell diversity, cell migration, cell death, anterior-posterior and dorso-ventral patterning. Chapters concerning the development of specific structures and organs address, *inter alia*, the nervous system, gastrointestinal, respiratory, and reproductive tracts, cardiovascular and excretory systems, sense organs, and limb, skeletal and muscle development.

We thank all the authors for having taken on the burden of writing these extensive reviews and sharing with the community their profound knowledge of their specialist subjects. We are particularly grateful to those authors who submitted their chapters early and who have shown great patience in awaiting the completed volumes.

We also thank all the reviewers, who through their critically assessments have helped the authors to maximize the quality of their chapters.

Special thanks go to Ursel Lindenberger for secretarial assistance in Heidelberg (Neuroanatomy Secretary Heidelberg) and to Dr Andreas Sendtko and the Wiley team for having initiated and promoted the project.

Heidelberg and Göttingen
September 2005   *Klaus Unsicker and Kerstin Krieglstein*

List of Contributors

Solomon Afelik
Georg-August-University of
Goettingen
Goettingen Center for Molecular
Biosciences
Department of Developmental
Biochemistry
Justus-von-Liebig-Weg 11
37077 Göttingen
Germany

Jean-François Beaulieu
CIHR Group on Functional
Development and Physiopathology of
the Digestive Tract
Department of Anatomy and Cell
Biology
Faculté de médicine et des sciences de
la santé
Université de Sherbrooke
Sherbrooke (Québec) J1H 5N4
Canada

Kirk W. Beisel
Creighton University
Department of Biomedical Sciences
723 N 18th Street
Omaha, NE 68178
USA

Saverio Bellusci
University of Southern California Keck
School of Medicine
Developmental Biology Program
Saban Research Institute
Childrens Hospital Los Angeles
4650 Sunset Boulevard
Los Angeles CA 90027
USA

François Boudreau
CIHR Group on Functional
Development and Physiopathology of
the Digestive Tract
Department of Anatomy and Cell
Biology
Faculté de médicine et des sciences de
la santé
Université de Sherbrooke
Sherbrooke (Québec) J1H 5N4
Canada

Paola Bovolenta
Instituto Cajal de Neurobiología
(CSIC)
Av. Doctor Arce 37
28002 Madrid
Spain

Cell Signaling and Growth Factors in Development. Edited by K. Unsicker and K. Krieglstein
Copyright © 2006 WILEY-VCH Verlag GmbH & Co. KGaA, Weinheim
ISBN 3-527-31034-7

Thomas Brand
University of Wuerzburg
Molecular Developmental Biology
Cell and Developmental Biology
(Zoology I)
Am Hubland
97074 Würzburg
Germany

Thomas Braun
Department of Physiological
Chemistry
Martin-Luther-University
Halle-Wittenberg
Hollystr. 1
06114 Halle
Germany

Georg Breier
Medical Faculty Carl Gustav Carus
University of Technology Dresden
Institute of Pathology
Fetscherstr. 74
01307 Dresden
Germany

Leena Bruckner-Tuderman
Department of Dermatology
University of Freiburg
Hauptstr. 7
79104 Freiburg
Germany

Federico Calegari
Max Planck Institute of Molecular Cell
Biology and Genetics
Pfotenhauerstrasse 108
01307 Dresden
Germany

Yonglong Chen
Georg-August-University Goettingen
Goettingen Center for Molecular
Biosciences
Department of Developmental
Biochemistry
Justus-von-Liebig-Weg 11
37077 Göttingen
Germany

Ajay Chitnis
Unit on Vertebrate Neural
Development
Laboratory of Molecular Genetics
National Institute of Child Health and
Human Development
National Institutes of Health
6 Center Drive
Bethesda, MD 20892
USA

Susana M. Chuva de Sousa Lopes
Hubrecht Laboratory
Netherlands Institute for
Developmental Biology
Uppsalalaan 8
3584 CT Utrecht
The Netherlands

Massimo De Felici
Section of Histology and Embryology
Department of Public Health and Cell
Biology
University of Rome "Tor Vergata"
Via Montpellier 1
00133 Rome
Italy

Stijn DeLanghe
Developmental Biology Program
Saban Research Institute
Childrens Hospital Los Angeles
4650 Sunset Boulevard
Los Angeles CA 90027
USA

Filippo Del Bene
European Molecular Biology
Laboratory
EMBL
Developmental Biology Programme
Meyerhofstrasse 1
69117 Heidelberg
Germany

Pierre-Marie Del Moral
University of Southern California
School of Dentistry
Developmental Biology Program
Saban Research Institute
Childrens Hospital Los Angeles
4650 Sunset Boulevard
Los Angeles CA 90027
USA

Susanna Dolci
Section of Human Anatomy
Department of Public Health and Cell
Biology
University of Rome "Tor Vergata"
Via Montpellier 1
00133 Rome
Italy

Véronique Dubreuil
Max Planck Institute of Molecular Cell
Biology and Genetics
Pfotenhauerstrasse 108
01307 Dresden
Germany

Donatella Farini
Section of Histology and Embryology
Department of Public Health and Cell
Biology
University of Rome "Tor Vergata"
Via Montpellier 1
00133 Rome
Italy

Lilla Farkas
Max Planck Institute of Molecular Cell
Biology and Genetics
Pfotenhauerstrasse 108
01307 Dresden
Germany

Bernd Fritzsch
Creighton University
Department of Biomedical Sciences
723 N 18th Street
Omaha, NE 68178
USA

Lidia García-Campmany
Instituto de Biología Molecular de
Barcelona (CSIC)
Parc Cientific de Barcelona
Josef Samitier 1–5
08028 Barcelona
Spain

William A. Horton
Director, Research Center
Shriners Hospital for Children
3101 S. W. Sam Jackson Park Road
Portland, OR 97239
USA

Wieland B. Huttner
Max Planck Institute of Molecular Cell
Biology and Genetics
Pfotenhauerstrasse 108
01307, Dresden
Germany

Angelo Iulianella
Stowers Institute for Medical Research
1000 E. 50th Street,
Kansas City, MO 64110
USA

Kristjan Jessen
Department of Anatomy and
Developmental Biology
University College London
Gower Street
WC1E 6BT London
UK

Vesa Kaartinen
University of Southern California Keck
School of Medicine
Developmental Biology Program
Saban Research Institute
Childrens Hospital Los Angeles
4650 Sunset Boulevard
Los Angeles CA 90027
USA

Chaya Kalcheim
Department of Anatomy and Cell
Biology
Hebrew University-Hadassah Medical
School
P.O. Box 12272
Jerusalem 91120
Israel

Clemens Kiecker
MRC Centre for Developmental
Neurobiology
Guy's Hospital Campus
King's College
University of London
London SE1 1UL
UK

Yoichi Kosodo
Max Planck Institute of Molecular Cell
Biology and Genetics
Pfotenhauerstrasse 108
01307 Dresden
Germany

Kerstin Krieglstein
Department of Neuroanatomy
Medical Faculty
University of Göttingen
Kreuzbergring 36
37075 Göttingen
Germany

Matt Lee
University of Southern California
School of Dentistry
Center for Craniofacial Molecular
Biology
Health Sciences Campus
Los Angeles, CA 90033
USA

Susan Mackem
National Cancer Institute
Laboratory of Pathology
Building 10, Room 2A33
9000 Rockville Pike
Bethesda, MD 20892–1500
USA

Hans Georg Mannherz
Department of Anatomy and Cell
Biology
Ruhr-University
Universitätsstr. 150
44780 Bochum
Germany

Elisa Martí
Instituto de Biología Molecular de
Barcelona (CSIC)
Parc Cientific de Barcelona
Josep Samitier 1–5
08028 Barcelona
Spain

Daniel Ménard
Département d'Anatomie et de
Biologie Cellulaire
Faculté de Médecine
Université de Sherbrooke
Sherbrooke
Québec, J1H 5N4
Canada

Rhona Mirsky
Department of Anatomy and
Developmental Biology
University College London
Gower Street
WC1E 6BT London
UK

Christine L. Mummery
Hubrecht Laboratory
Netherlands Institute for
Developmental Biology
Uppsalalaan 8
3584 CT Utrecht
The Netherlands

Andrea Münsterberg
School of Biological Sciences
Developmental Biology Research
Groups
University of East Anglia
Norwich NR4 7TJ
UK

Herbert Neuhaus
Department of Physiological
Chemistry
Martin-Luther-University
Halle-Wittenberg
Hollystr. 1
06114 Halle
Germany

Petra Neuhaus
Department of Physiological
Chemistry
Martin-Luther-University
Halle-Wittenberg
Hollystr. 1
06114 Halle
Germany

Fong Cheng Pan
Georg-August-University of
Goettingen
Goettingen Center for Molecular
Biosciences
Department of Developmental
Biochemistry
Justus-von-Liebig-Weg 11
37077 Göttingen
Germany

Christian Paratore
Institute of Cell Biology
Swiss Federal Institute of Technology
ETH-Hoenggerberg HPM E47
8093 Zürich
Switzerland

Nathalie Perreault
CIHR Group on Functional
Development and Physiopathology of
the Digestive Tract
Department of Anatomy and Cell
Biology
Faculté de médecine et des sciences de
la santé
Université de Sherbrooke
Sherbrooke (Québec) J1H 5N4
Canada

Tomas Pieler
Georg-August-University of
Goettingen
Goettingen Center for Molecular
Biosciences
Department of Developmental
Biochemistry
Justus-von-Liebig-Weg 11
37077 Göttingen
Germany

Nathalie Rivard
CIHR Group on Functional
Development and Physiopathology of
the Digestive Tract
Department of Anatomy and Cell
Biology
Faculté de médicine et des sciences de
la santé
Université de Sherbrooke
Sherbrooke (Québec) J1H 5N4
Canada

Hermann Rohrer
Max-Planck-Institut für
Hirnforschung
Abteilung Neurochemie
Deutschordenstrasse 46
60528 Frankfurt
Germany

Pellegrino Rossi
Section of Human Anatomy
Department of Public Health and Cell
Biology
University of Rome "Tor Vergata"
Via Montpellier 1
00133 Rome
Italy

Kirsi Sainio
Institute of Biomedicine
University of Helsinki
P.O. Box 63
00014 Helsinki
Finland

Hannu Sariola
Institute of Biomedicine
University of Helsinki
P.O. Box 63
00014 Helsinki
Finland

Wei Shi
University of Southern California Keck
School of Medicine
Developmental Biology Program
Saban Research Institute
Childrens Hospital Los Angeles
4650 Sunset Boulevard
Los Angeles CA 90027
USA

Lukas Sommer
Institute of Cell Biology
Swiss Federal Institute of Technology
ETH-Hoenggerberg HPM E38
8093 Zürich
Switzerland

Lydia Sorokin
Department of Experimental Pathology
Lund University Hospital
Sölvegatan 8B
22362 Lund
Sweden

Matthias Stanke
Max-Planck-Institut für
Hirnforschung
Abteilung Neurochemie
Deutschordenstrasse 46
60528 Frankfurt
Germany

Denise Tefft
University of Southern California Keck
School of Medicine
Developmental Biology Program
Saban Research Institute
Childrens Hospital Los Angeles
4650 Sunset Boulevard
Los Angeles CA 90027
USA

Irma Thesleff
Institute of Biotechnology
University of Helsinki
P.O. Box 63
00014 Helsinki
Finland

Sólveig Thorsteinsdóttir
Department of Animal Biology and
Centre for Environmental Biology
Faculty of Sciences
University of Lisbon
Lisbon
Portugal
and
Gulbenkian Institute of Science
Oeiras
Portugal

Paul A. Trainor
Stowers Institute for Medical Research
1000 E. 50th Street,
Kansas City, MO 64110
USA

Xiu-Ping Wang
Institute of Biotechnology
University of Helsinki
P.O. Box 63
00014 Helsinki
Finland

Grant Wheeler
School of Biological Sciences
University of East Anglia
Norwich NR4 7TJ
UK

Pierre H. Vachon
CIHR Group on Functional
Development and Physiopathology of
the Digestive Tract
Department of Anatomy and Cell
Biology
Faculté de médicine et des sciences de
la santé
Université de Sherbrooke
Sherbrooke (Québec) J1H 5N4
Canada

David Warburton
University of Southern California Keck
School of Dentistry
Vice Chair, Department of Surgery,
Leader, Developmental Biology
Program,
Saban Research Institute
Childrens Hospital Los Angeles
4650 Sunset Boulevard
Los Angeles CA 90027
USA

Joachim Wittbrodt
European Molecular Biology
Laboratory
EMBL
Developmental Biology Programme
Meyerhofstrasse 1
69117 Heidelberg
Germany

Humphrey Hung-Chang Yao
Department of Veterinary Biosciences
University of Illinois
3806 VMBSB
2001 South Lincoln Avenue
Urbana, IL 61802
USA

Color Plates

Cell Signaling and Growth Factors in Development. Edited by K. Unsicker and K. Krieglstein
Copyright © 2006 WILEY-VCH Verlag GmbH & Co. KGaA, Weinheim
ISBN 3-527-31034-7

A. Differentiation of roof plate

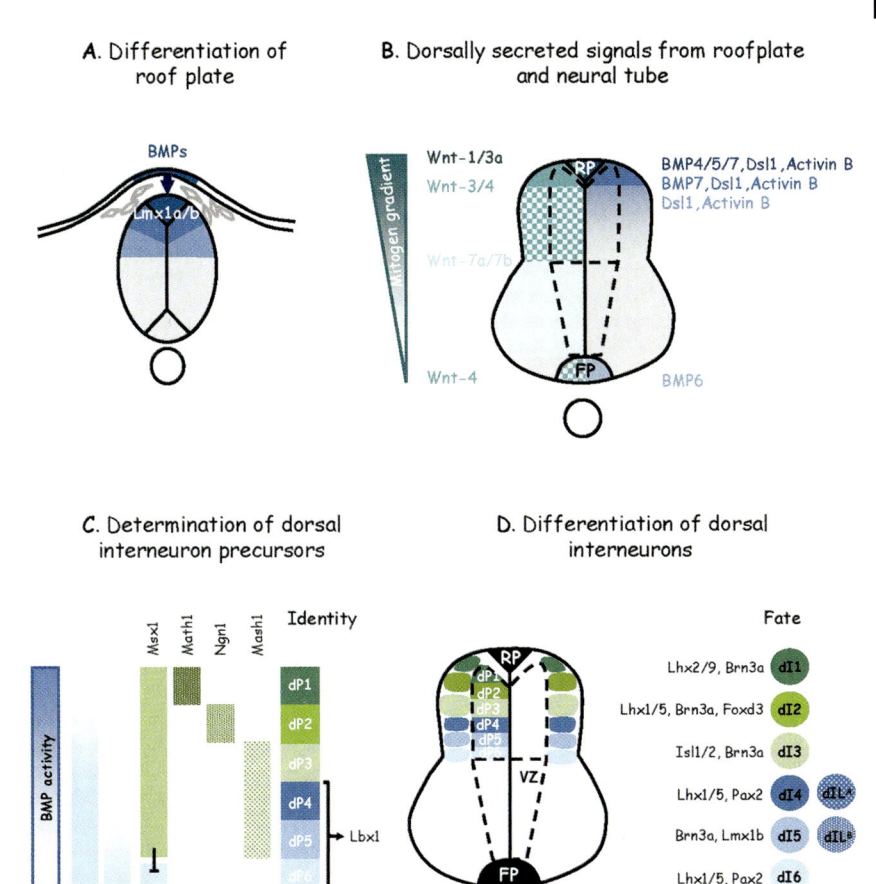

B. Dorsally secreted signals from roofplate and neural tube

C. Determination of dorsal interneuron precursors

D. Differentiation of dorsal interneurons

Fig. 11.2
Extracellular secreted signals and transcriptional codes patterning the dorsal neural tube. (A) Diagram of the neural tube during roof plate specification. BMP-dependent activation of Lmx1a/b specifies the roof plate. (B) Schematic representation of a roof plate-derived gradient of TGFβ/BMP molecules (color coded on the right) and Wnt proteins (color coded on the left) involved in the induction of the progenitor domains within the dorsal neural epithelium. (C) A graded distribution of BMP activity (shaded bar on the left) mediates the expression of different HD and bHLH transcription factors (color-coded bar graphs in the middle) required for the specification of the dorsal progenitor domains (represented on the right with a color-coded bar). (D) Diagram showing the relationship between neural progenitor domains (color coded) and the positions at which post-mitotic neurons (spheres on the right with the same color code) are generated along the dorso-ventral axis of the dorsal spinal cord. This figure also appears on page 368.

XL | Color Plates

A. c*Shh* expression

B. Differentiation of floorplate

C. Ventral gradient of Shh from the floorplate

D. Determination of ventral neuron precursors

E. Differentiation of ventral neurones

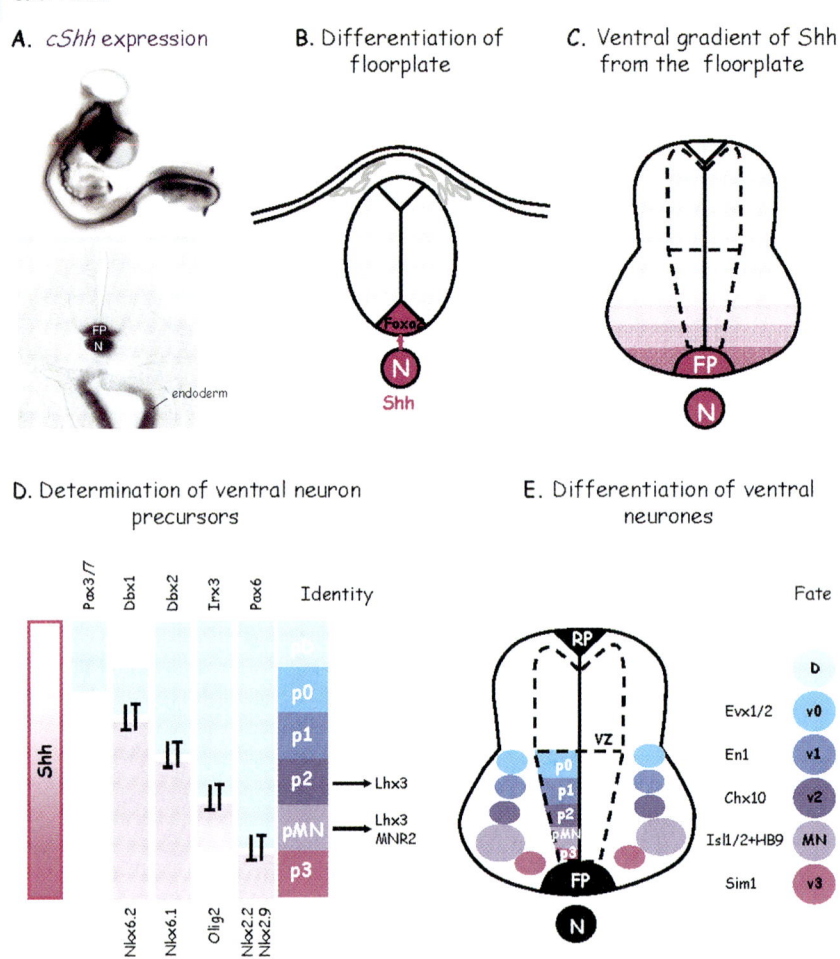

Fig. 11.3
Extracellular secreted signals and transcriptional codes patterning the ventral neural tube. (A) Whole mount *in situ* hybridization in a St 18 embryo showing the expression of *Shh*. Note the specific localization in the notochord, floor plate and endoderm, as shown in the cross-section of the neural tube. (B) Diagram of the neural tube during floor plate (FP) differentiation. A tightly-regulated network arranged in a feedback loop, including *FoxA2* and *Shh*, is required for floor plate specification. (C) Shh secreted from the notochord and the floor plate presumably forms a gradient in the ventral neural tube distributed in a ventral-high, dorsal-low concentration (shown in graded levels of pink) within the ventral neural epithelium. (D) Different levels of Shh (shaded bar on the left) mediate the repression of Class I HD proteins (Pax7, Dbx1, Dbx2, Irx3 and Pax6) at different threshold concentrations and induce the expression of Class II proteins (Nkx6.1/2; Nkx2.2/9 and Olig2), represented in the center by bar graphs. Cross-repressive interactions between Class I and Class II proteins define six different progenitor domains, represented on the right with a color-coded bar. (E) Diagram showing the relationship between neural progenitor domains (color coded) and the positions at which post-mitotic neurons (spheres on the right with the same color code) are generated along the dorso-ventral axis of the ventral spinal cord. This figure also appears on page 372.

A. Unpatterned optic vesicle

- extraocular mesenchyme
- lens placode
- axial midline
- optic vesicle

B. Patterned optic vesicle

- presumptive RPE
- presumptive neural retina
- presumptive optic stalk
- lens pit

C. Optic cup

- RPE
- neural retina
- optic stalk
- lens

A'
D/V patterning of the optic vesicle

Otx2, Otx1, Pax6, Rx, Lhx2, Six3, Pax2

(BMP/activin, Shh, FGF)

B'
Induction of RPE, NR, OS

Dorsal-Distal

- Otx2/1, Mitf, Pax6
- Chx10, Rx, Lhx2, Six3/6, Pax6
- Pax2, Vax1

Ventral-Proximal

C'
D/V patterning of the neural retina

- Otx2/1, Mitf — RPE
- Tbx5 — dorsal Neural Retina
- Vax2 — ventral
- Pax2, Vax1, Vax2 — Optic stalk

(Vnt+Shh+RA?, BMP4)

Fig. 11.4
Molecular network involved in the patterning of the optic vesicle and optic cup. Schematic representation of the transition from unpatterned (A) to patterned optic vesicle (B) and optic cup (C). Signaling molecules and progressive tissue specification are represented in color codes. Color-coded bar graphs represent the molecular components implicated in the transition from transition from unpatterned (A') to patterned optic vesicle (B') and optic cup (C'). In (A, A') all the neuroepithelial cells are indistinguishable (gray color) and express a common set of transcription factors. Shh signaling from the axial midline is required for the specification of the proximo-ventral optic stalk (pink in B, B'). TGFβ-like signals from the extraocular mesenchyme activate cells from the optic vesicle to become RPE (blue in B, B'), whereas FGF signals from the lens placode repress RPE and activate NR identity (green in B, B'). During optic cup formation the graded distribution of BMP4 dorsally and ventroptin Vnt, possibly together with RA and Shh, establishes the D-V polarity of the neural retina (shaded green in C, C'). This figure also appears on page 377.

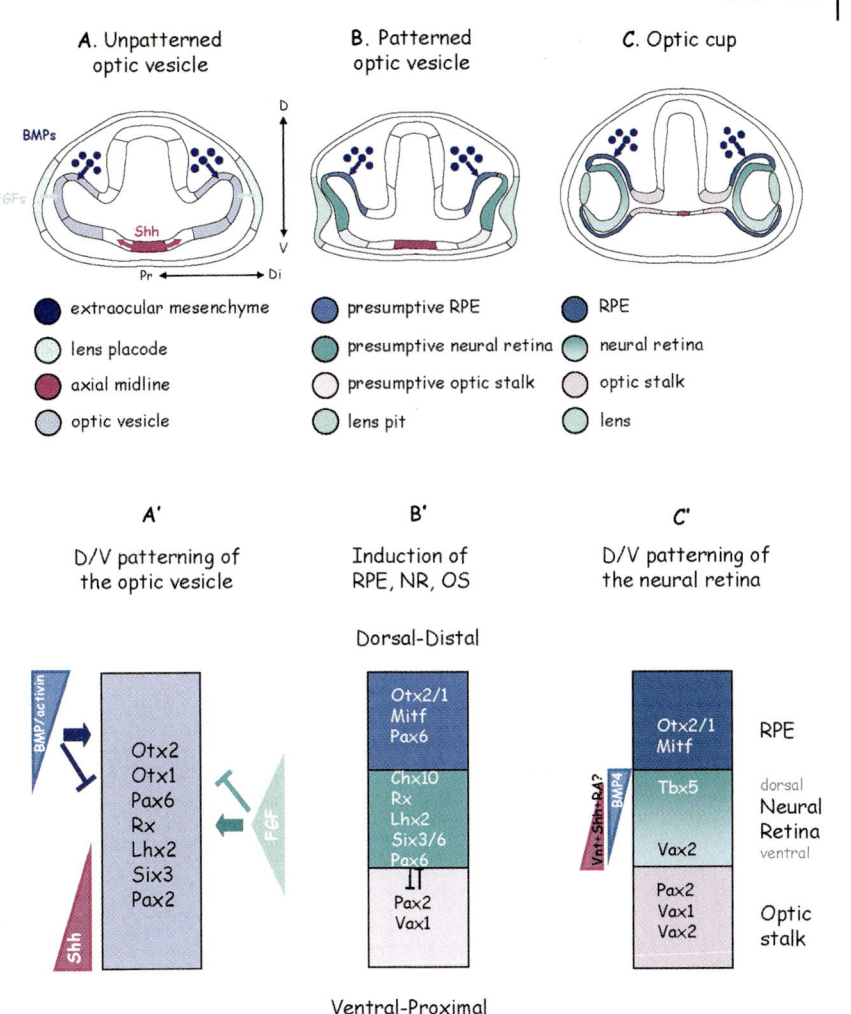

A. D/V patterning of the mouse telencephalic vesicle at E11.5

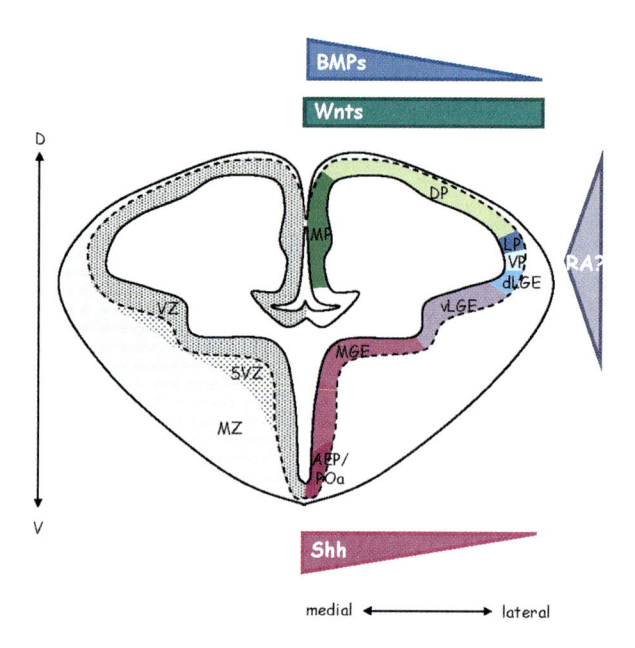

B. Determination of VZ progenitor domains and regionalization of the telencephalon

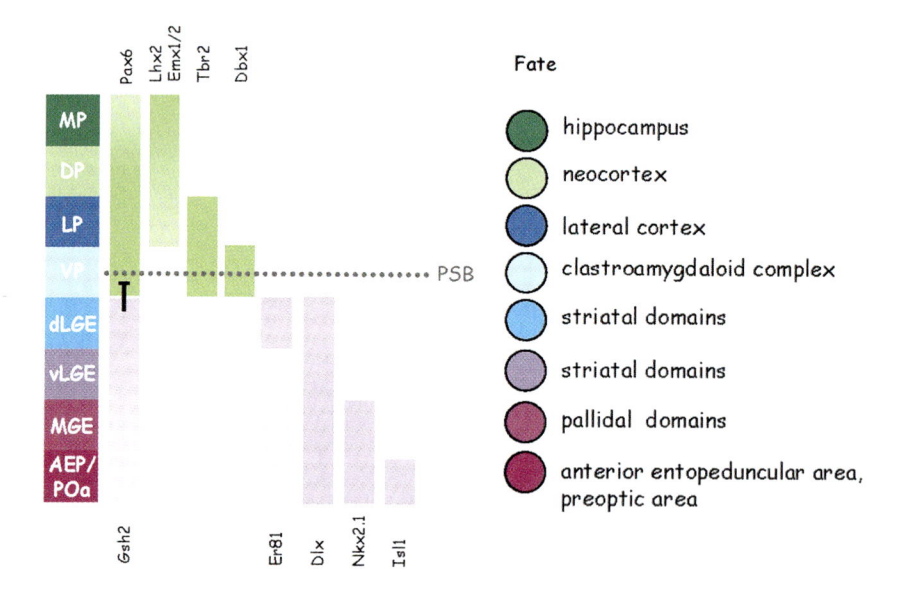

Fate:
- hippocampus
- neocortex
- lateral cortex
- clastroamygdaloid complex
- striatal domains
- striatal domains
- pallidal domains
- anterior entopeduncular area, preoptic area

Fig. 11.5
Extracellular secreted signals and transcriptional codes patterning the telencephalon. (A) Schematic representation of the mouse telencephalic vesicle at E11.5 showing extrinsic signals involved in the establishment of the progenitor domains along the D-V and medio-lateral axis. (B) A color-coded bar represents telencephalic subdivisions on the left. In the middle are bar graphs representing gene expression patterns, as indicated at the top or bottom of each bar. Different levels of expression are represented by the shading. A horizontal broken line represents the PSB. On the right the fates of each telencephalic subdivision are described using the same color code. AEP, anterior entopeduncular area; DP, dorsal pallium; dLGE, dorsal lateral ganglionic eminence; vLGE, ventral ganglionic eminence; LP, lateral pallial mantle; MGE, medial ganglionic eminence; MP, medial pallium; MZ, marginal zone; Pd, pallidum; POa, preoptic area; PSB, pallial-subpallial boundary; St, striatum; SVZ, subventricular zone; VP, ventral pallium; VZ, ventricular zone. This figure also appears on page 382.

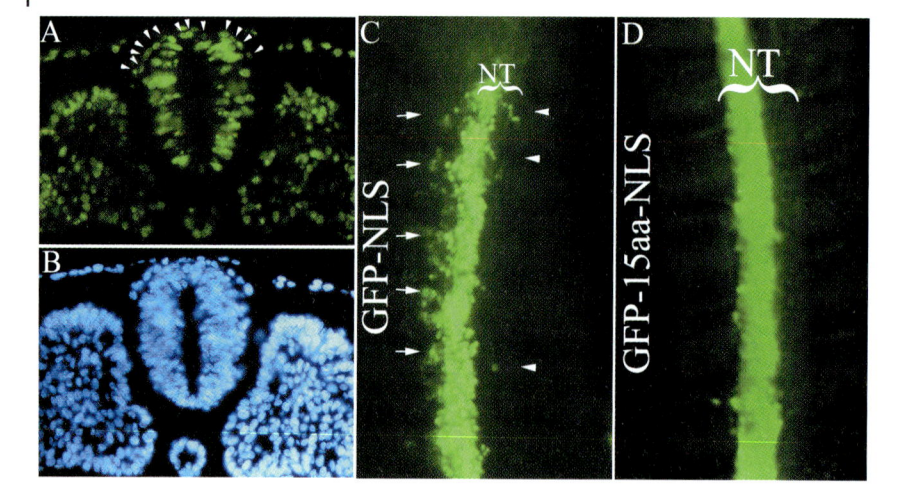

Fig. 12.1
Emigration of NC cells from the dorsal tube depends upon successful G1/S transition. (A, B) Neural crest cells emigrate from the neural tube in the S phase of the cell cycle. Most emigrating NC cells are Brdu+ following a 1–h pulse and immediate fixation (A, arrowheads). Compare with total Hoechst-stained nuclei in B. (C, D) The cdk-binding domain of MyoD and dn E2F–1 inhibits delamination of NC cells *in ovo*. Dorsal views of embryos that received control GFP-NLS vector (C) showing axial level-dependent migration of GFP+ crest cells through intersomites and into the rostral somitic domains (arrows). Some labeled cells also cross to the contralateral side (arrowheads). (D) Electroporation with the 15–amino acid carboxy terminus moiety of MyoD fused to a GFP reporter (GFP–15aa-NLS), resulted in no migration of labeled crest cells. This small sequence of MyoD binds to Cdk4/5 and CyclinD1 forming a complex that ultimately inhibits the G1/S transition specifically. This figure also appears on page 405.

Fig. 12.2
Extrinsic signals and cell-intrinsic factors that are known to be involved in the generation of sympathetic, parasympathetic, enteric and sensory neurons. The specification of sympathetic and parasympathetic neurons is dependent on bone morphogenetic proteins (BMPs), whereas the development of the sensory neuronal lineage depends on Wnt signaling. In zebrafish there is also evidence of a role for sonic hedgehog (shh) in sensory neuron generation. The early pre-specification of sensory and autonomic neural crest precursor cells is indicated by the color code. The molecular basis of this pre-specification, i.e. the difference between a totipotent neural crest stem cell (gray) and pre-specified progenitors (colors) is unknown. In sympathetic precursor cells, BMPs induce the expression of Ascl1, Phox2a/b, dHand and Gata2/3, which, in turn, control the expression of noradrenergic (TH, DBH) and generic neuronal properties. In precursor cells of the parasympathetic ciliary ganglion Ascl1 and Phox2a/b, but not dHand and Gata2/3 are induced by BMPs, and TH/DBH are expressed only transiently. All enteric neurons depend on Phox2b and a subpopulation on Ascl1, it is unclear, however, whether their development is also controlled by BMPs. Wnt signaling is essential and sufficient to induce sensory neuron generation, which involves the neurogenins ngn1 and ngn2, the first members of a cascade of bHLH factors acting in sensory neuron differentiation, i.e. NeuroD, NSCL–1/2 and homeodomain transcription factors Brn3a and DRG11. Class III-semaphorins (Sema3a) play an important role for sensory and sympathetic neuron projections and in the location of sympathetic cell bodies. Members of the GDNF-family ligands (GFLs) are involved in the initial axon projections of sympathetic and parasympathetic neurons. The survival of peripheral neurons depends on neurotrophins at different stages of their development (not shown). This figure also appears on page 409.

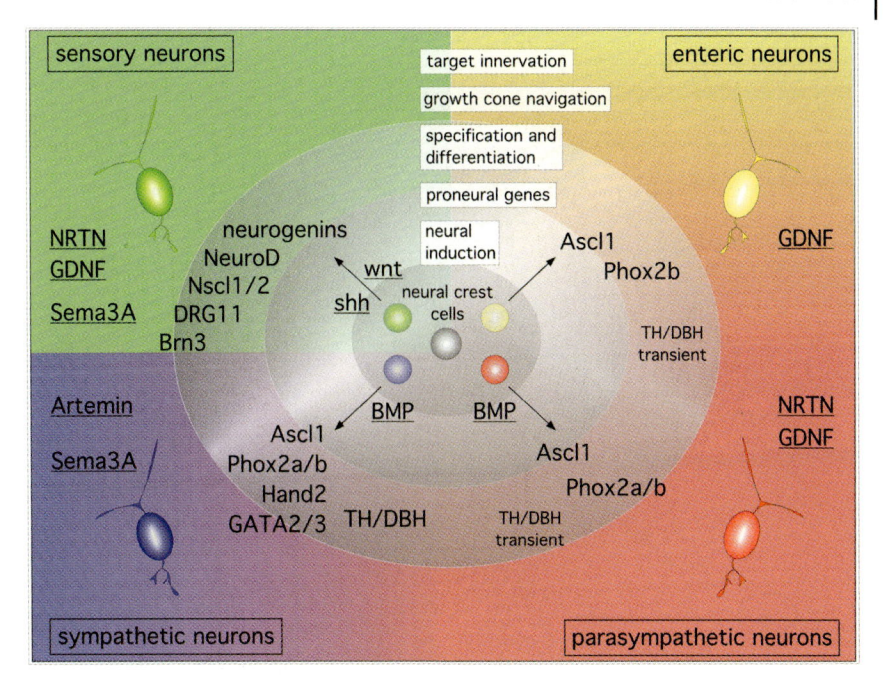

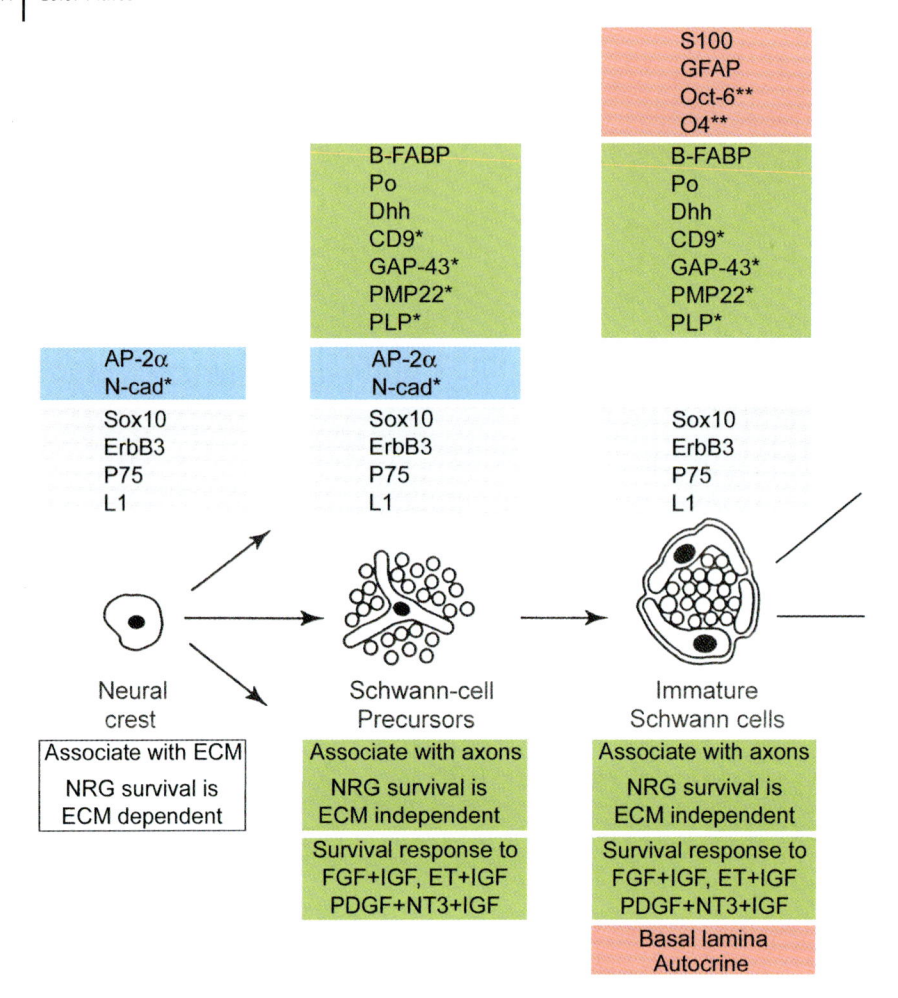

Fig. 12.4

Differentiation markers in embryonic Schwann cell development. Gray: markers shared by migrating, undifferentiated neural crest cells, Schwann cell precursors and immature Schwann cells. Blue: markers present on migrating crest cells and precursors but not (or at very low levels) on immature Schwann cells. Green: markers present on Schwann cell precursors and immature Schwann cells, but absent from migrating undifferentiatedcrest cells. These proteins also appear on neuroblasts/early neurons. Pink: markers expressed by immature Schwann cells but not (or at much lower levels) by Schwann cell precursors. These markers are acutely dependent on axons for expression. In addition Schwann cells have autocrine survival circuits not present in precursors. This figure also appears on page 420.

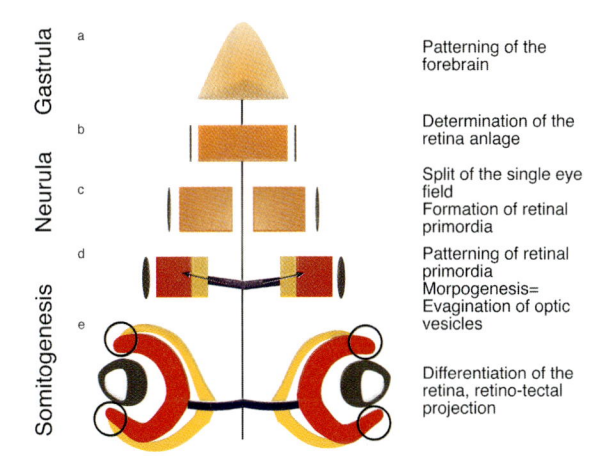

Fig. 13.1
Schematic representation of vertebrate eye development. In the figure, the central line represents both temporal axes and embryonic midline. (A) Late gastrula stages, patterning of the forebrain. (B) End of gastrulation, determination of the eye field. (C) Splitting of the eye field in two by signals emanating from the ventral midline. Initial proximo-distal patterning of the eyes primordia. (D) Optic vesicle evagination as result of cell proliferation and active morphogenesis. Formation of neuroretina (red), pigmented retina epithelium (yellow) and optic stalk (blue). Signals from the optic stalk initiate retina differentiation (arrows). (E) Differentiation of the neuroretina and formation of optic nerve along the optic stalk to target areas in the brain. In anamniotes retinal proliferation continues at the ciliary marginal zone (circles). Adapted from [229]. This figure also appears on page 450.

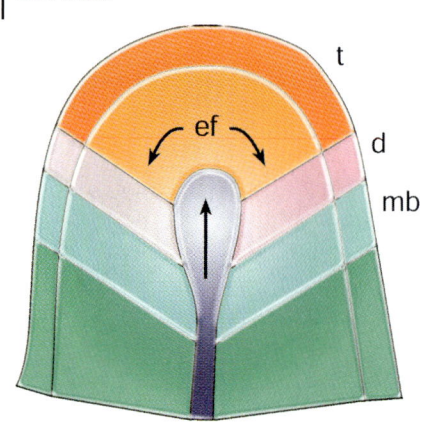

Fig. 13.2
Embryonic origin of the eye field. Cartoon of the anterior neural plate. t, telencephalon (red); ef, eye field (orange); d, diencephalon (pink); mb, mid-brain (turquoise). The migrating axial tissue is shown in blue. Adapted from [84]. This figure also appears on page 452.

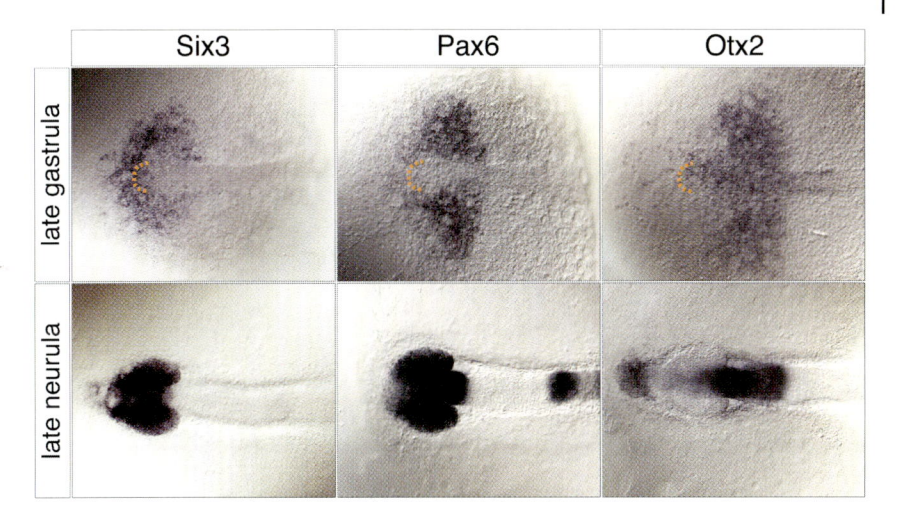

Fig. 13.3
Expression of Six3, Pax6 and Otx2 during early eye development in medaka fish (*Oryzias latipes*). Expression at late gastrula stages and early neurula stages. The orange-hatched line demarcates the anterior end of the embryonic axis. Note the partial overlap of the *Six3* and *Pax6*, *Pax6* and *Otx2* expression domains both at late gastrula and early neurula stages. At the same time *Six3* and *Otx2* are mutually exclusive. Adapted from [46]. This figure also appears on page 453.

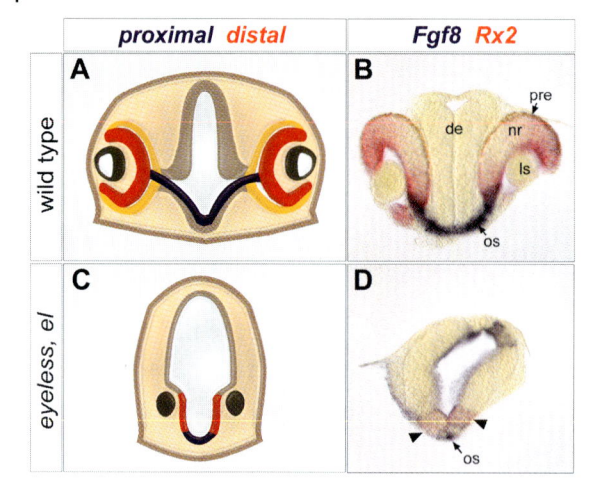

Fig. 13.4
Rx3/eyeless mutation impairs morphogenesis but does not affect the patterning of the retinal primordium. In medaka, a mutation in the *Rx3* gene (*eyeless, el*) impairs optic vesicle evagination and results in a patterned retinal primordium that remains in the ventral diencephalon. (A, C) Schematic comparison of eye structures in *wt* and *el* embryos. Optic stalk (blue), neuroretina (red), pigmented retina epithelium (yellow), lens (green). (B, D) transverse sections of *wt* and *el* embryos, showing *Fgf8* (blue) and *Rx2* (red) expression in the optic stalk and neuroretina respectively. de, diencephalon; ls, lens; nr, neuroretina; os, optic stalk; pre, pigmented retina epithelium. Adapted from [68]. This figure also appears on page 457.

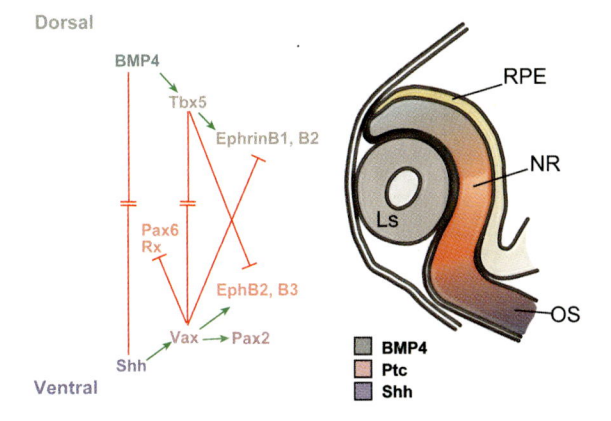

Fig. 13.6
Vax2 inactivation causes eye coloboma. In mouse the targeted inactivation of *Vax2* causes failure in optic fissure closure, causing eye coloboma. (A, B) Frontal view of wild-type and mutant eyes disected from 16.5–day embryos. Adapted from [126]. This figure also appears on page 460.

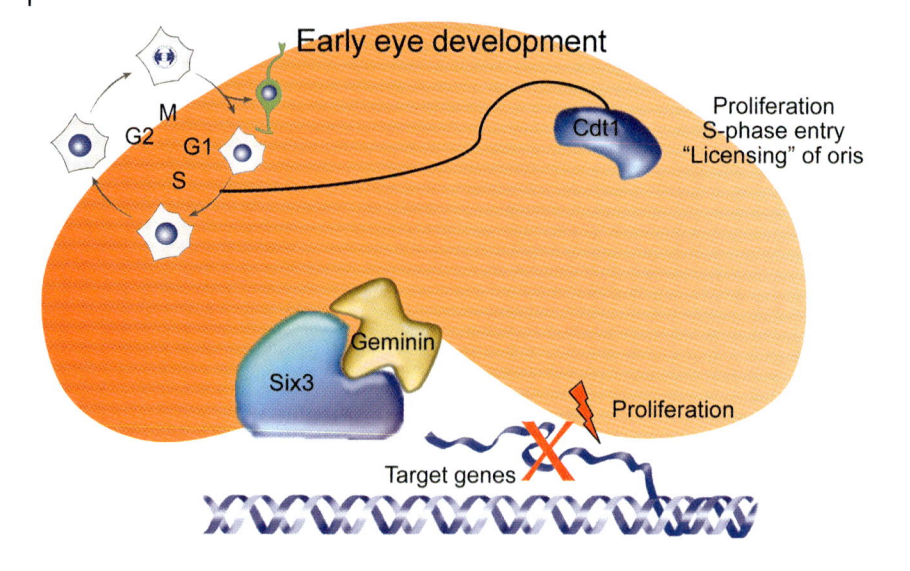

Fig. 13.7
Six3 binding to Geminin stimulates cell proliferation. The cartoon represents the mechanism by which Six3 promotes cell proliferation during early eye development. By sequestering Geminin, Six3 frees Ctd1 to promote cell cycle progression and thus generates conditions that are permissive for cell proliferation. This figure also appears on page 462.

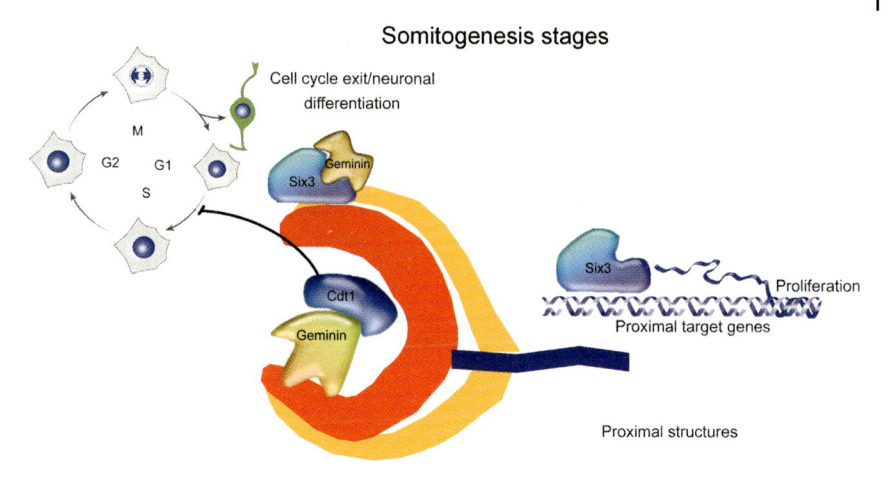

Fig. 13.8
The interaction between Six3 and Geminin controls cell proliferation and differentiation. At somitogenesis stages, *Geminin* and *Six3* are co-expressed in the ciliary marginal zone, where proliferation still occurs. In the more central retina *Six3* transcription is repressed, allowing Geminin to facilitate the cell cycle exit as a prerequisite for neuronal differentiation. In more proximal regions high levels of Six3 activity, without Geminin expression, are required for optic stalk specification. This figure also appears on page 463.

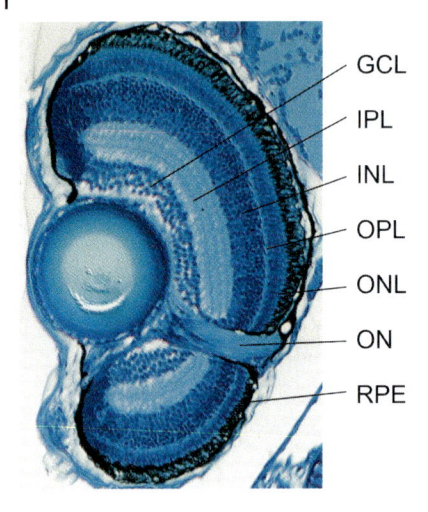

Fig. 13.9
Regular histological structure of the retina. A transverse plastic section through the fish (medaka) retina at the exit point of the optic nerve. The layered arrangement of neurons is clearly visible. GCL, ganglion cell layer; IPL, inner plexiform layer; INL, inner nuclear cell layer; OPL, outer plexiform layer; ONL, outer nuclear layer, ON, optic nerve; RPE, retinal pigmented epithelium. This figure also appears on page 464

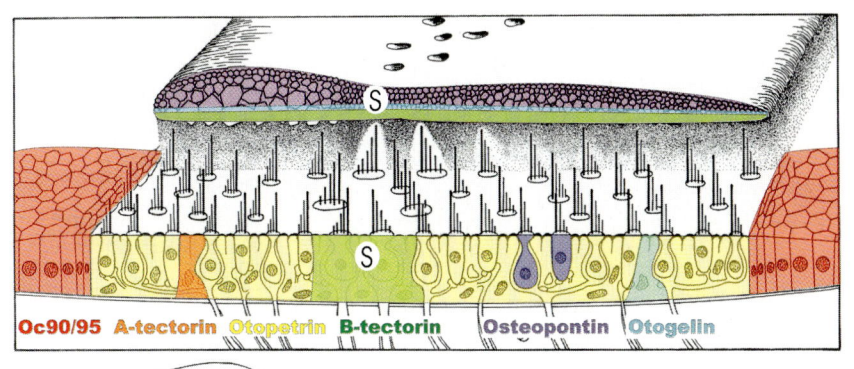

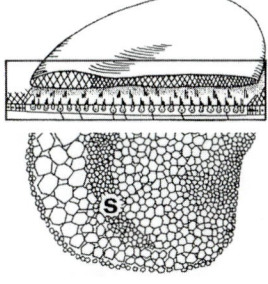

Fig. 14.3
The regional differences of otoconia-related gene expression in the utricle and structural and protein differences in the utricular otoconia are shown in this diagram (adapted from Lindeman [163]). Note the differences in the size of otoconia crystals in striola (S) and extrastriolar regions. Oc90/95 is expressed in the non-sensory area of the ear (red); otopetrin is expressed throughout the sensory epithelium (yellow); osteopontin is present in the hair cells (dark blue); otogelin is present in the supporting cells (cyan); α-tectorin is shown in the supporting cells (orange) and β-tectorin in the striola (green). All proteins are distributed throughout the otoconia (lilac) except for otogelin (cyan) and β-tectorin (green) which occur only in the lower layers. No information on the concentration differences of proteins within otoconia is available. This figure also appears on page 501.

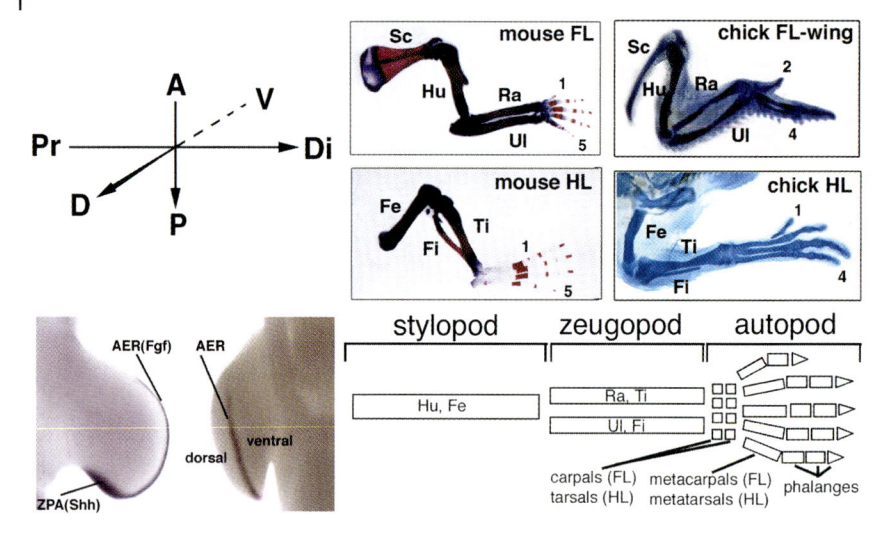

Fig. 15.1
Limb bud signaling centers and the morphological components of the mature limb skeleton. The three orthogonal limb axes (upper left panel) show the orientation below in limb bud stage (left-most panel), and to the right, in skeletons of late stage mouse and chick embryos. Chick limb buds hybridized with *Fgf8* and *Shh* probes to highlight the AER and ZPA, respectively, are shown in the lower left, oriented as in the diagram above (left panel) and rotated 90° counterclockwise to show AER relative to dorsal and ventral limb bud surfaces (right panel). Standard generic names for the different PD skeletal components of the limb are indicated in the diagram on the lower right. Note that the chick wing has only three digits, and the hindlimb four digits. Sc, scapula; Hu, humerus; Ra, radius; Ul, ulna; Fe, femur; Ti, tibia; Fi, fibula; FL, forelimb; HL, hindlimb. This figure also appears on page 525.

Color Plates | LVII

	initiation	patterning: signaling centers	condensation & differentiation	growth, segmentation, late regulation digit identity?
RA	initiation, ZPA induction	proximal fate? Fgf/AER antagonism	← modulation of AER maintenance? →	apoptosis, inhibition of chondrogenesis
Bmp	initiation, AER induction	DV polarity -ventral identity	← modulation of AER maintenance →	AER regression, apoptosis, chondrogenesis
Can. Wnt	initiation, AER induction	DV polarity -ventral identity	← AER maintenance →	inhibition of chondrogenesis, segmentation for joints
non-Can. Wnt		DV polarity* -dorsal identity		chondrogenesis, hypertrophy
non-AER Fgf	initiation, AER induction		← AER maintenance →	inhibition of cartilage proliferation in growth plates

Fig. 15.3
Timeline of different roles of major signaling pathways during progression of limb development. AER-Fgfs and Hedgehogs (Shh, Ihh) are shown in the schematic depicting limb stages at the top, because they play "dedicated" roles in mediating AER function, ZPA function, and in chondrogenic growth, respectively. *The evidence regarding whether Wnt7a, the major signal regulating dorsal polarity in mesodermal limb derivatives, acts through the canonical or a non-canonical pathway is highly conflicted, as discussed in text. This figure also appears on page 528.

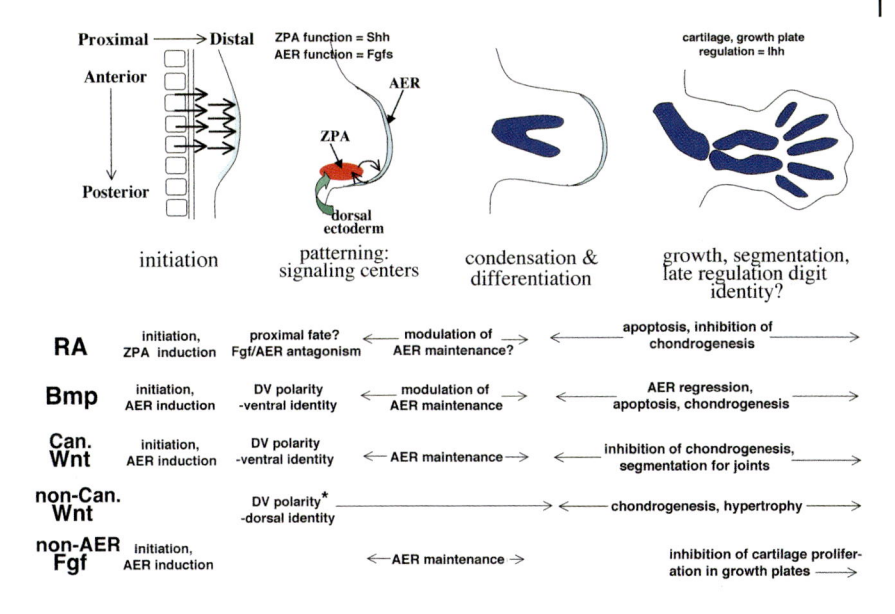

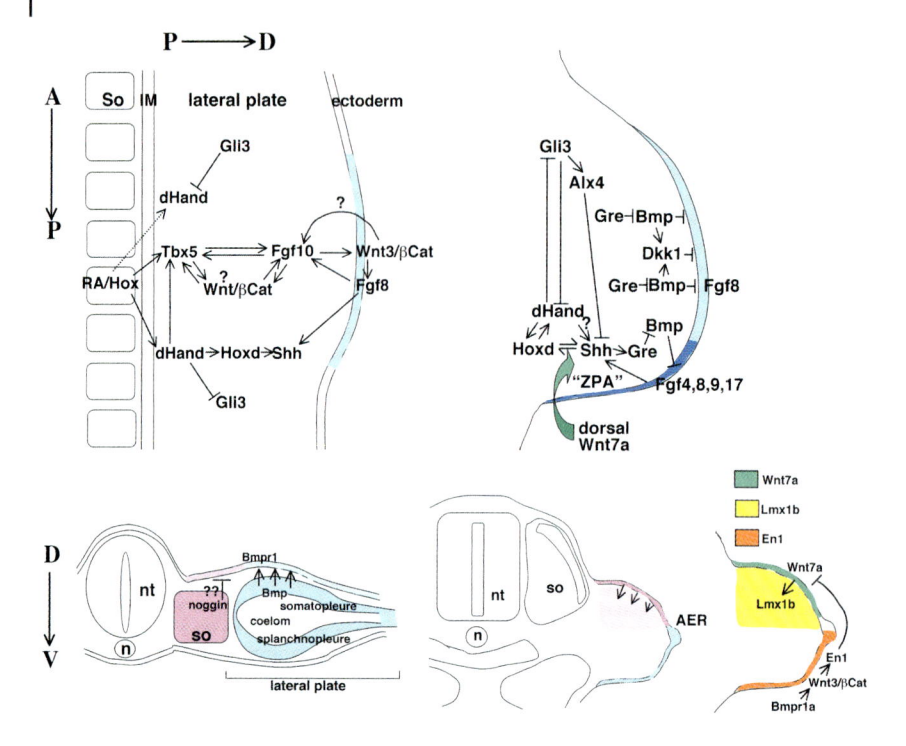

Fig. 15.5
Regulatory cascades leading to formation and maintenance of limb "organizers". As described in text, signaling cascades are shown that lead to limb initiation (forelimb cascade shown) and regulate the formation and maintenance of AER and ZPA (upper panels); and cascades that co-regulate early and later stages in specification of DV polarity and AER formation/positioning from within the ectoderm (lower panels). Arrows and bars are not necessarily meant to signify direct interactions. Where specific signals vary between organisms, only the relevant mammalian (mouse) factors and likely sequences of induction are shown. (For example, several Wnts that regulate chick limb initiation and AER formation do not appear to play the same roles in mouse; see text). n, notochord; nt, neural tube; so, somite; IM, intermediate mesoderm. This figure also appears on page 536.

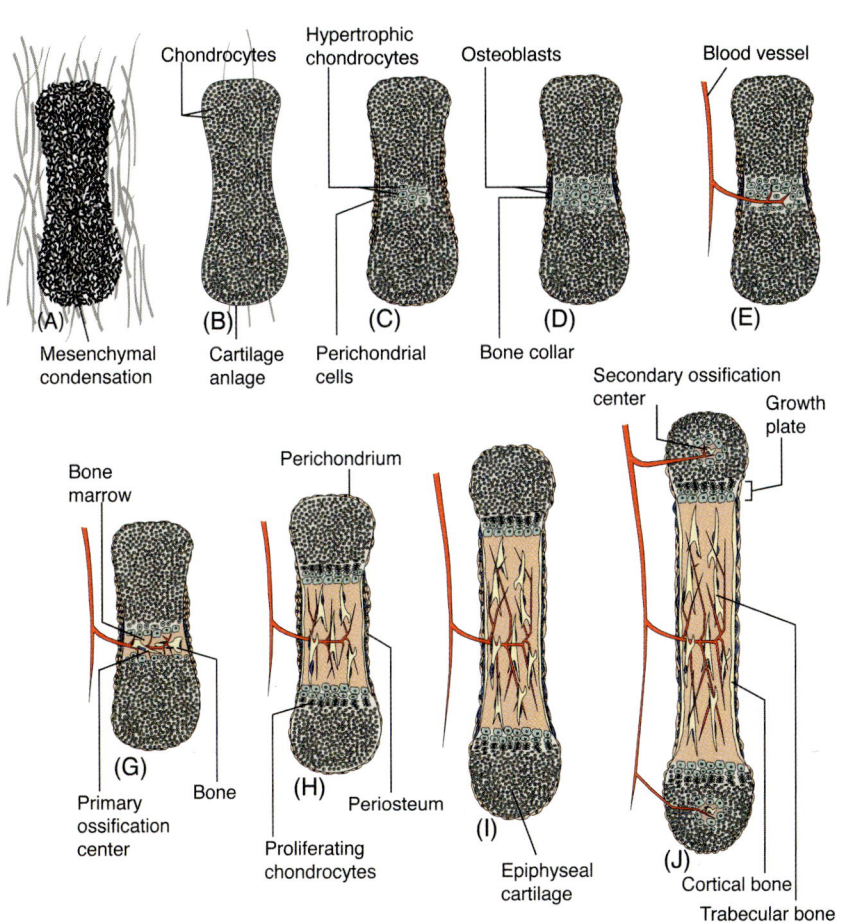

Fig. 16.1
Endochondral bone development from mesenchymal condensation stage to formation of secondary ossification centers (see text for discussion of events). This figure also appears on page 622.

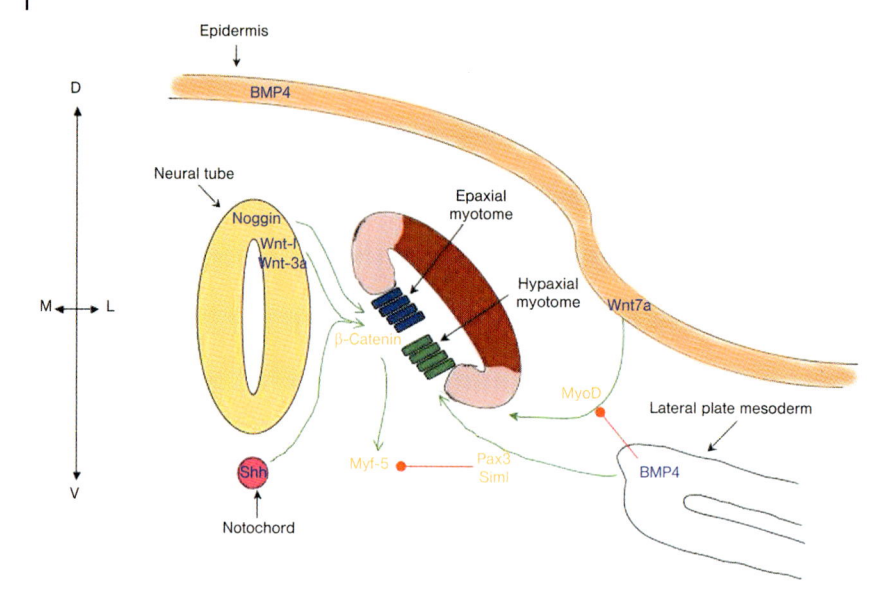

Fig. 17.1
Growth factors determining the dorso-ventral and medial-lateral polarity of the somite are shown (typed in blue). Inductive signals are marked by green arrows and repressive signals are shown in red. Genes induced by the growth factors are indicated in bright orange. The scheme gives only an overview; more details can be found in the text (see Chapter 2). This figure also appears on page 650.

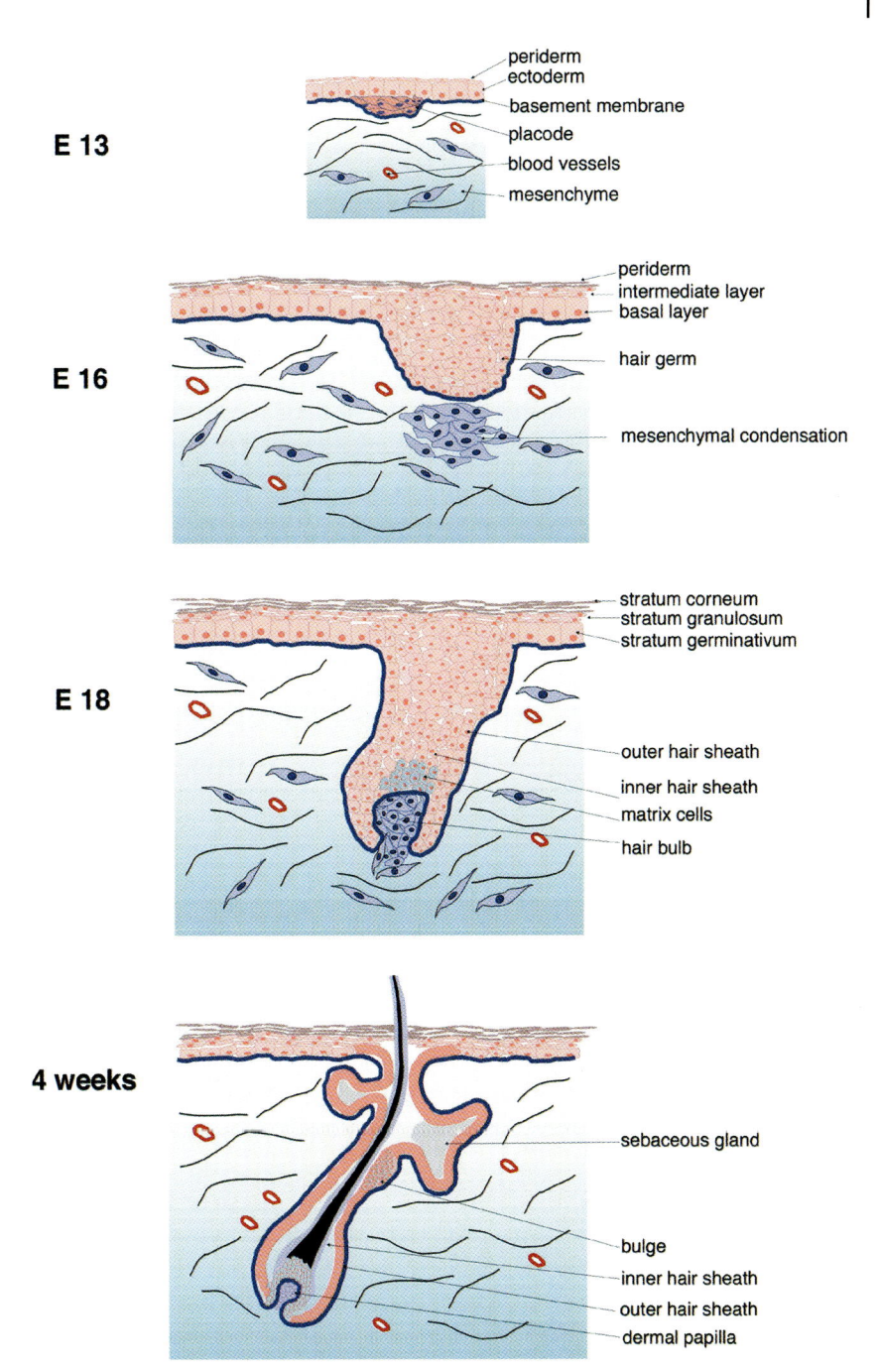

E 13
- periderm
- ectoderm
- basement membrane
- placode
- blood vessels
- mesenchyme

E 16
- periderm
- intermediate layer
- basal layer
- hair germ
- mesenchymal condensation

E 18
- stratum corneum
- stratum granulosum
- stratum germinativum
- outer hair sheath
- inner hair sheath
- matrix cells
- hair bulb

4 weeks
- sebaceous gland
- bulge
- inner hair sheath
- outer hair sheath
- dermal papilla

Fig. 18.1 (see p. LXI)
Schematic representation of skin and hair follicle development in the mouse. The epidermis originates from a single-layered ectoderm, which forms the periderm. By E13, placodes form, marking initiation of hair follicle development. An intermediate proliferating ectodermal layer forms between the periderm and basal layer by E15. Cells of this layer enter terminal differentiation to form a stratified epidermis by E16. Hair follicle development involves initial mesenchymal condensation at E16 (papilla primordium) and subsequent follicular epithelial cell migration into the dermis to form hair bulb, inner and outer hair sheaths, and hair core at E18. The lowermost epithelial part of the hair follicle, the hair bulb, originates from proliferating matrix cells. By 4 weeks postnatally, a population of transient amplifying matrix cells terminally differentiate to generate the different cell layers of the follicle, including the outer root sheath, inner root sheath and the innermost hair shaft. The outer root sheath of a mature hair follicle is contiguous with the basal epidermal layer. The hair bulb surrounds specialized mesenchymal cells of the dermis, the dermal papilla, which develops from the mesenchymal condensate/papilla primordium. Sebaceous glands form in upper portions of the hair follicle and are bordered basally by a population of stem cells in the bulge region. This figure also appears on page 680.

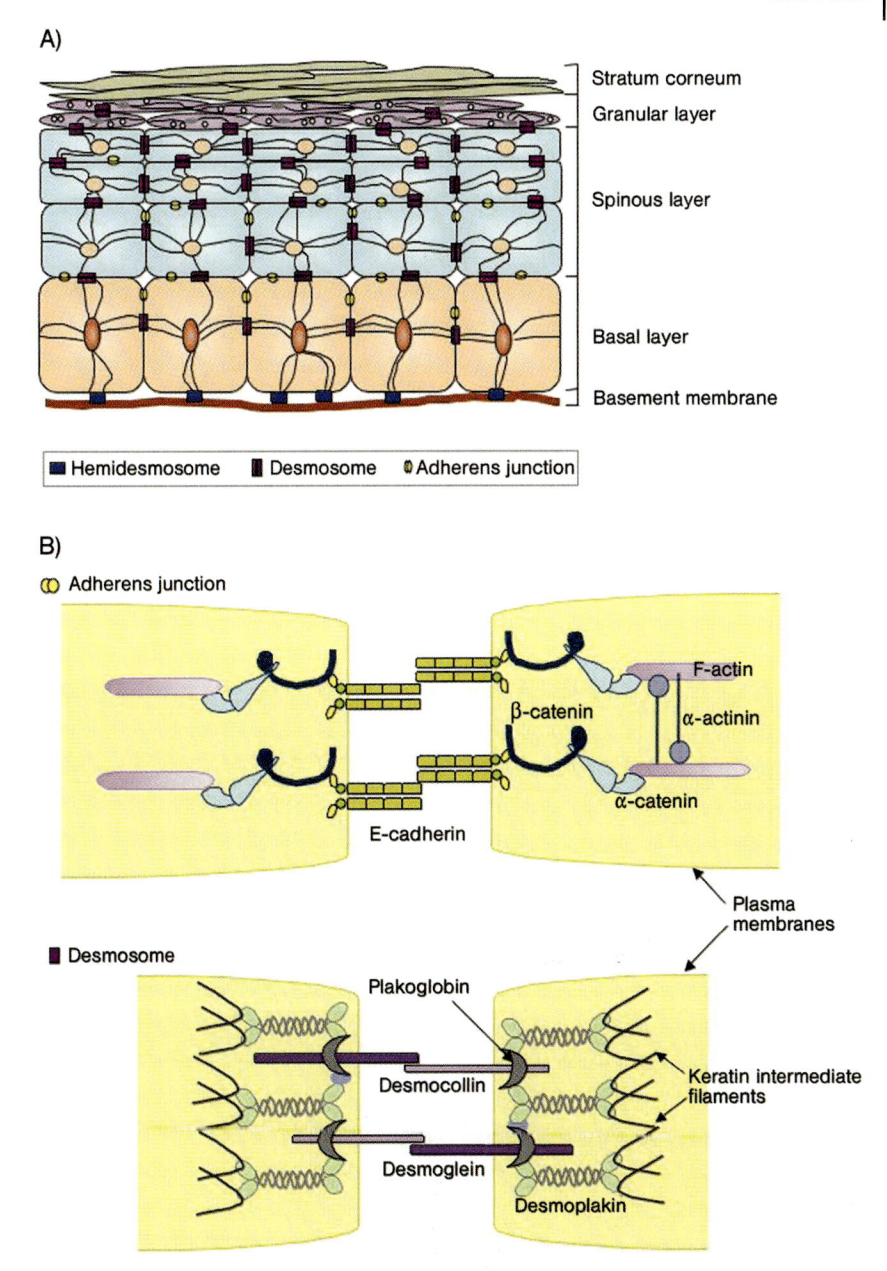

Fig. 18.2
(A) Schematic representation of the cell layers of the mature stratified epithelium of the skin and the cell-cell and cell-extracellular matrix junctions characteristic of these layers. Adherens junctions and desmosomes on the lateral surfaces of the keratinocytes act to seal the epithelial sheet, while hemidesmosomes anchor the basal keratinocytes to the epidermal-dermal basement membrane. (B) Molecular composition of the adherens junctions and the desmosomes. Figure is modified according to [2]. This figure also appears on page 684.

A)

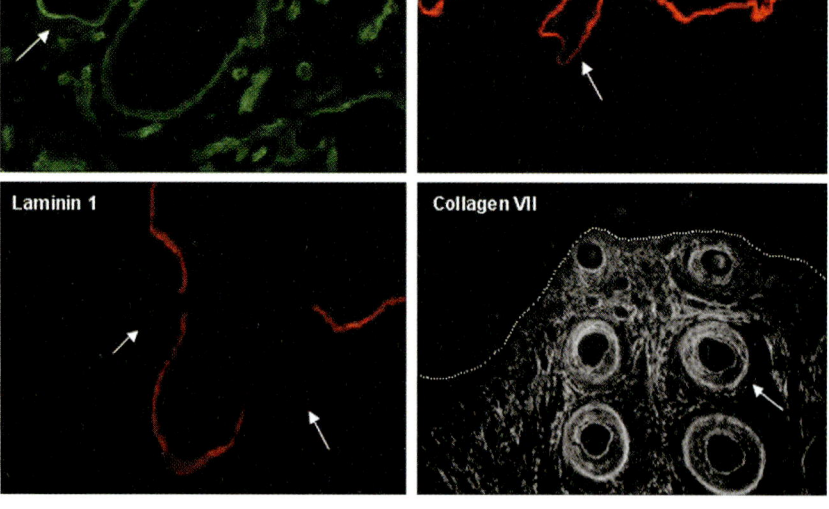

N-terminus

Shortarms { α5, α3

β1 ~~~~~ γ1   β3 ~~~~~ γ2

→ Proteolytic cleavage

Globular domain: G1, G2, G3, G4, G5

C-terminus

→ Proteolytic cleavage

B)

Pan-Laminin | Laminin 5

Laminin 1 | Collagen VII

Fig. 18.4
(A) Structure of laminin 10 (composed of α5, β1 and γ1 chains) as a prototype laminin, as compared to laminin 5 (composed of α3, β3, γ2 chains). Laminin α3 occurs in two spliced variants, one of which lacks the N-terminal domains (boxed area) and represents the form that occurs in the epidermal-dermal basement membrane [149]. Laminin α3 and γ2 chains are further proteolytically cleaved *in vivo* (cleavage sites marked). (B) Comparison of immunofluorescent staining of E17 mouse skin with antibodies to pan-laminin, laminin α3, laminin α1, and collagen VII. Pan-laminin staining reveals all basement membranes, including the epidermal-dermal basement membrane and basement membranes of blood vessels in the dermis. Laminin α3 and α1 are restricted to the epidermal-dermal basement membrane, with laminin α1 being excluded from the basement membrane surrounding the hair follicles. Collagen VII shows extensive fibrillar staining of the dermis, extending from the epidermal-dermal basement membrane (dotted line). Positions of developing hair follicles or in the case of the collagen VII panel, whiskers, are marked by arrows. Laminin images × 200 magnification; collagen VII × 100. This figure also appears on page 694.

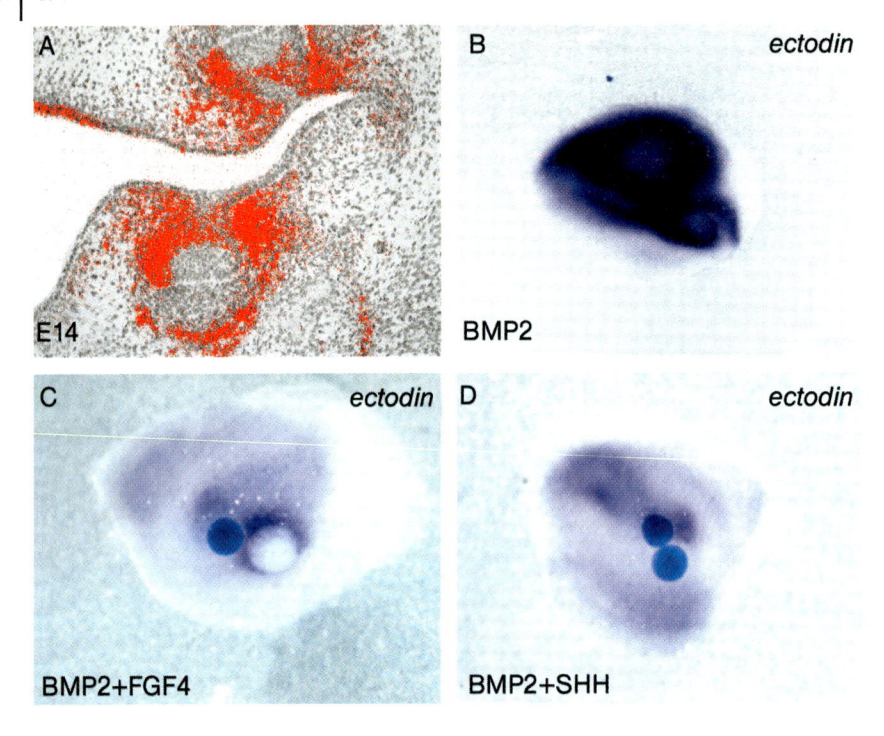

Fig. 19.4
Ectodin is a BMP inhibitor integrating the BMP, FGF and Shh signals from the enamel knot. (A) *In situ* hybridization analysis of *ectodin* expression in a cap stage tooth germ indicates downregulation in the enamel knot and its surrounding tissue. (B) Bead-released BMP2 induces ectodin expression in a dental explant *in vitro*. (C) Bead-released Shh together with the BMP2 bead, inhibits ectodin expression. (D) An FGF-releasing bead also inhibits the ectodin-inducing effect of BMP. (Courtesy of Johanna Laurikkala). This figure also appears on page 728.

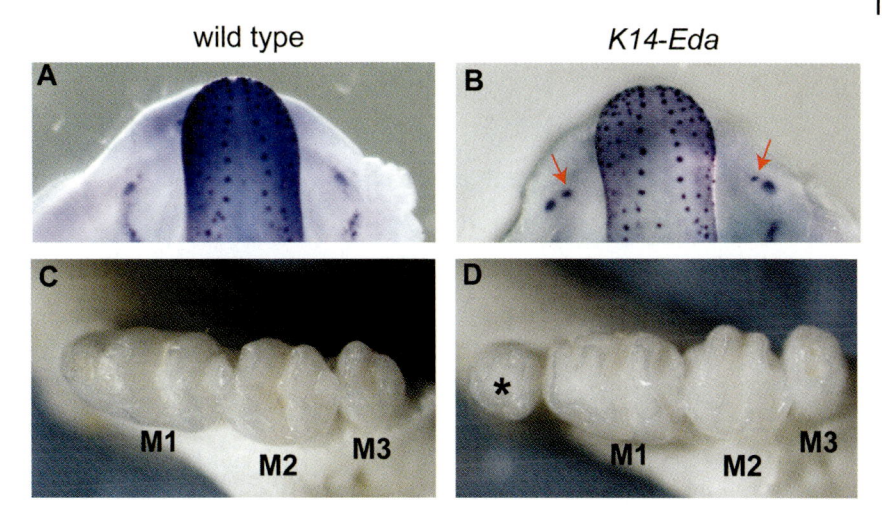

Fig. 19.6
Excess ectodysplasin stimulates dental placode formation and induces the formation of extra teeth. (A, B) When *ectodysplasin* is overexpressed in the dental epithelium (*K14–Eda* transgenic mice), large dental placodes (arrows) are evident in front of the molar placode, as indicated by intense *Shh* expression. (C, D) An extra tooth (*) has formed from this placode in front of the first molar in *K14–Eda* mice. (Courtesy of Aapo T. Kangas). This figure also appears on page 737.

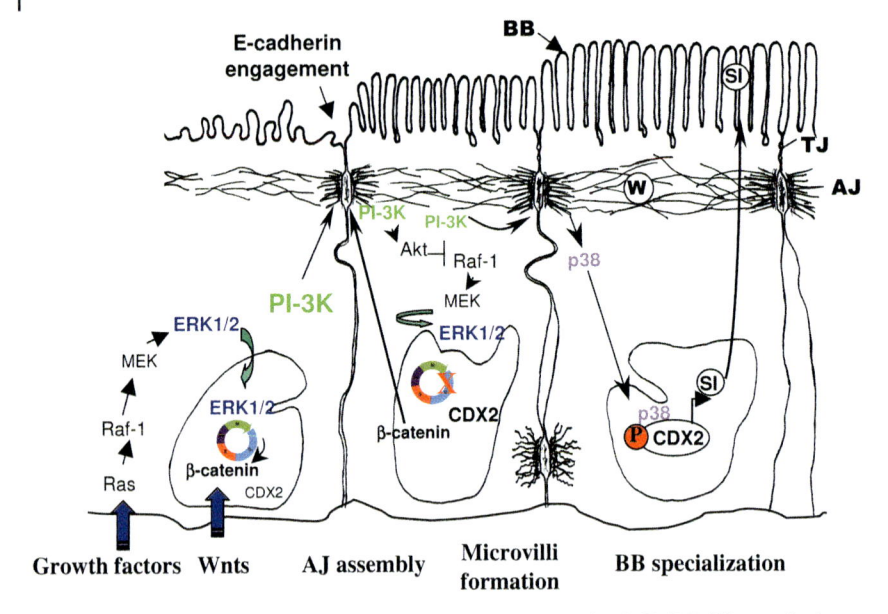

Fig. 20.3.
Schematization of proposed pathways involved in the regulation of proliferation and differentiation of intestinal epithelial cells. Growth factors stimulate proliferation of undifferentiated cells by activating the ERK pathway while Wnt factors stimulate proliferation by inducing accumulation of nuclear β-catenin/TCF–4 target genes. Upon cell-cell contacts, E-cadherin engagement recruits catenin proteins and PI–3K. This recruitment of PI–3K stimulates PI–3K signaling which, by increasing the amount of F-actin at the sites of cell-cell contact (through Rac?), promotes the assembly of adherens junction components with the cytoskeleton and induces the activation of the p38α MAPK cascade which enhances the transactivation capacity of CDX2 toward intestine-specific genes. TJ, tight junctions; AJ, adherent junctions; BB, brush border; W, terminal web of actin; SI, sucrase-isomaltase. This figure also appears on page 774.

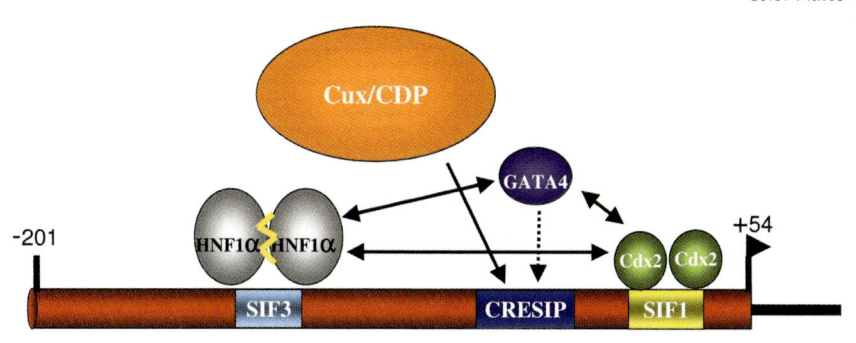

Fig. 20.4
Model of SI transcriptional regulation. HNF–1α, GATA–4 and Cdx2 physically interact with each other as well as with specific elements of the SI promoter to stimulate transcription. Cux/CDP interacts with the CRESIP element of the SI promoter and is involved in the repression of SI colonic transcription. This figure also appears on page 779.

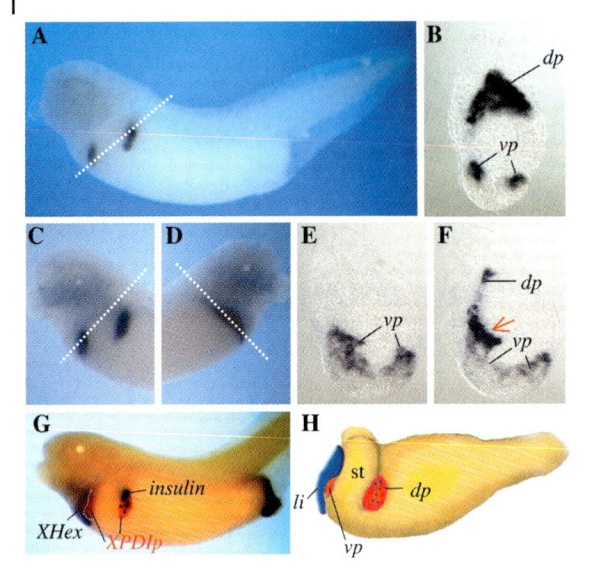

Fig. 22.2
Xenopus embryonic liver and pancreas. (A and B) Whole-mount *in situ* hybridization reveals that exocrine marker gene, XPDIp, is exclusively expressed in the dorsal and two ventral pancreatic anlagen of a stage–39 embryo. Panel B shows a transversal section across the white dashed line, as indicated in panel A. (C-F) During subsequent development (stage 40), XPDIp expression reflects the fusion process of dorsal and ventral pancreatic buds, first at the right side of the embryo (red arrow in F). Panels C and D are left and right side views of the same stage–40 embryo. Panels E and F represent transversal sections across the white dashed lines indicated in panels C and D, respectively. (G) Left side view of a stage–40 Xenopus embryo after whole-mount *in situ* hybridization analysis of liver (XHex) and pancreas (insulin and XPDIp) marker gene expression. Note that the liver (XHex expression in dark blue) is adjacent to the ventral pancreatic buds (XPDIp expression in red). Insulin expression (dark blue) is restricted to the dorsal pancreatic bud at this stage of development. The white dashed line demarcates the boundary between the liver and the ventral pancreatic bud. (H) A schematic drawing reflecting the situation depicted in A (modified from [81]). The blue dots in the dorsal pancreatic bud represent insulin expression. dp, dorsal pancreatic bud; li, liver anlage; st, stomach; vp, ventral pancreatic bud. This figure also appears on page 824.

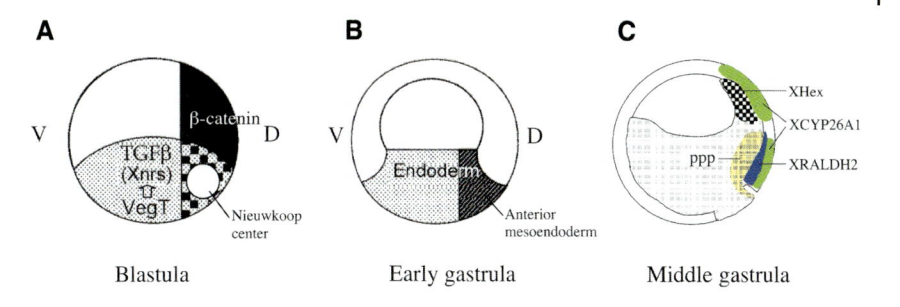

Fig. 22.3
Schematic drawings showing some of the early signaling pathways that are involved in the pre-patterning of the liver and the pancreas precursor cells in Xenopus. The VegT-activated TGFβ (Xnrs) signaling activities and the canonical Wnt pathway mediated by β-catenin are thought to coordinate the formation of an early signaling center, the so-called Nieuwkoop center, in the presumptive dorsal vegetal hemisphere of blastula stage embryos, which will give rise to the anterior mesoendoderm during gastrulation (A and B after [16]). (C) The anterior mesoendodermal cells give rise to liver, as reflected in XHex expression. The putative pancreatic precursor cells (ppp) are likely to be under the influence of the RA signaling center that is created by the RA-generating (XRALDH2) and -degrading (XCYP26A1) enzymes during gastrulation. D, dorsal; V, ventral. This figure also appears on page 825.

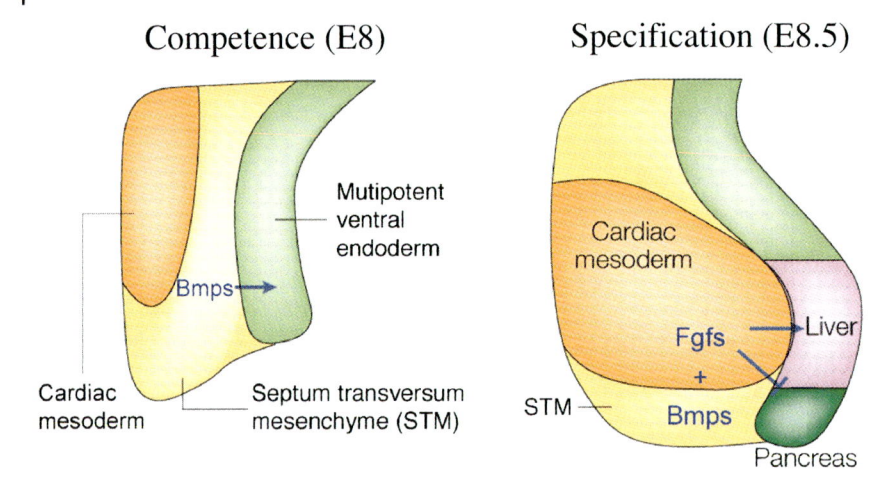

Fig. 22.5
Specification of liver and ventral pancreas in the mouse. The default state of the ventral foregut endoderm is pancreas. Signaling molecules, such as Bmps from septum transversum mesenchyme and Fgfs from cardiac mesoderm, direct subpopulations of the ventral foregut endoderm to form liver (after [2]). This figure also appears on page 828.

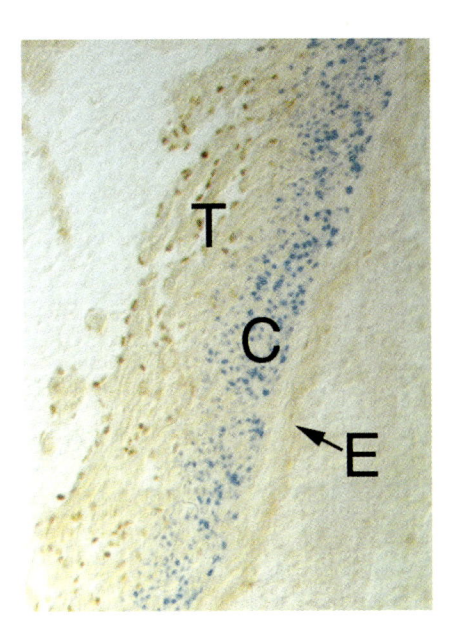

Fig. 23.4
During chamber development the ventricles generate two different compartments, i.e. the compact layer and the trabecular layer myocardium. These two compartments are shown here in a section of a 13.5 Popdc1–Lac/ transgenic mouse heart with an antibody against the cell cyle inhibitor p57Kip2 [91, 404, 405] which labels the trabecular layer and by LacZ staining which visualizes the expression domain of Popdc1 in the compact layer myocardium [352]. T, trabecular layer; C, compact layer; E, epicardium. This figure also appears on page 870.

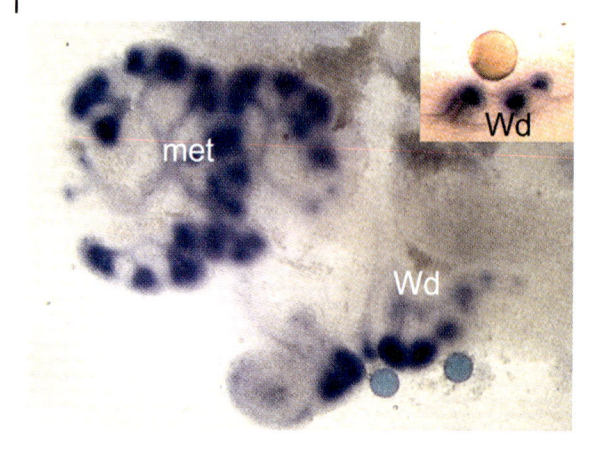

Fig. 25.4
GDNF regulates ureteric budding and branching in the metanephros (Met). Experimentally, GDNF-releasing beads promote supernumerary budding from the Wolffian duct (Wd). The branches of the ureteric buds and the ectopic GNDF-induced buds from the Wd are shown in the whole mount of *in situ* hybridization by GFRα1 mRNA. The inset shows a GDNF-releasing bead and supernumerary budding from Wd at high magnification (courtesy of Marjo Hytönen). This figure also appears on page 931.

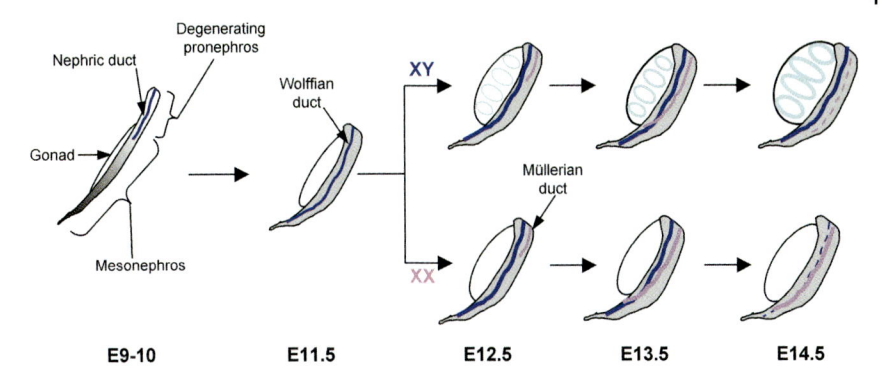

Fig. 26.3
Sexually dimorphic development of the Wolffian and Müllerian ducts in mouse between E9 and E14.5. This figure also appears on page 962.

Part II
Cell Signaling and Growth Factors in Organogenesis

11
Dorso-Ventral Patterning of the Vertebrate Central Nervous System

Elisa Martí, Lidia García-Campmany, and Paola Bovolenta

11.1
Introduction

Pattern formation is the process by which embryonic cells form ordered spatial arrangements in different tissues. Within the neural plate, the wide array of neurones and glial cells must be produced in a highly organized temporal-spatial order to form a functional nervous system.

In recent years considerable progress has been made in determining the cellular and molecular events that control the patterning of the neural plate along its three major axes: anterior-posterior, dorsal-ventral and left-right. The general view that emerged from these studies indicates that acquisition of a specific neural cell fate depends on the initial spatial coordinates of a precursor cell within the neural plate. This initial position defines the exposure of a progenitor cell to specific local environmental signals that progressively restrict its developmental potential. These local signals direct cell fate by activating the expression of transcriptional regulators, which in turn, control the genetic network necessary for the proper specific function of each neuronal cell.

The spinal cord, the anatomically simplest and most conserved region of the vertebrate CNS, has been instrumental in the understanding of the cellular and molecular mechanisms that control the progressive acquisition of neural cell identity along one of the embryonic axes: the dorsal-ventral (D-V) axis. The accumulation of information regarding the patterning of the CNS has been triggered by the identification of the notochord (an axial mesodermal structure underlying the midline of the neural plate) as the source of inductive signals for ventral spinal cord specification. Indeed, the notochord provides local information crucial for the specification of the midline floor plate cells and a longer-range signal necessary for motor neuron generation (reviewed in [1]). The subsequent isolation of Sonic hedgehog [2–4] and the determination of its activity as the main signaling molecule responsible for both inductive processes [4–7] has thereafter allowed an exponential growth of knowledge and understanding of the molecular networks responsible for D-V patterning in the entire neural plate.

Cell Signaling and Growth Factors in Development. Edited by K. Unsicker and K. Krieglstein
Copyright © 2006 WILEY-VCH Verlag GmbH & Co. KGaA, Weinheim
ISBN 3-527-31034-7

A. Setting up the border/ Creating the neural folds
Neural Plate Stage

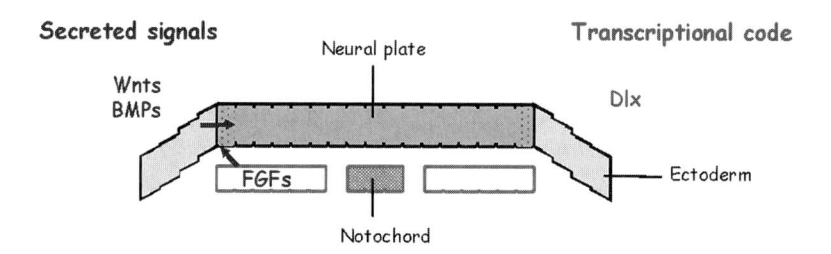

Secreted signals

Wnts
BMPs

Neural plate

Transcriptional code

Dlx

FGFs

Ectoderm

Notochord

B. Determination of neural crest precursors
Neural fold Stage

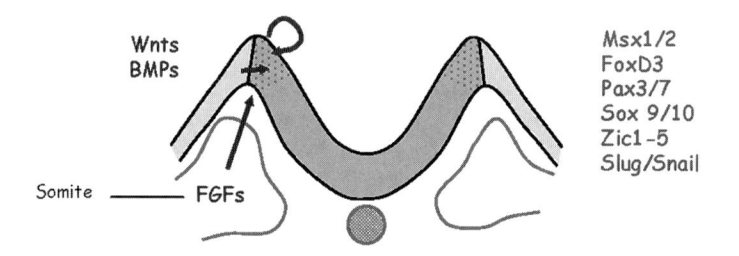

Wnts
BMPs

Msx1/2
FoxD3
Pax3/7
Sox 9/10
Zic1-5
Slug/Snail

Somite — FGFs

C. Differentiation of roof plate/delamination of neural crest cells
Neural tube

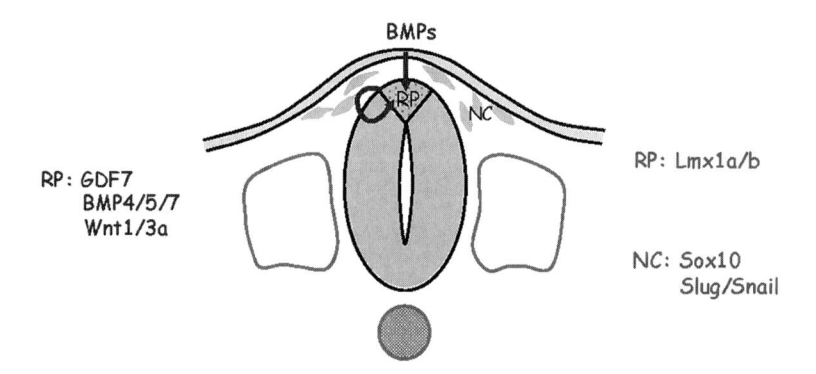

BMPs

RP: GDF7
BMP4/5/7
Wnt1/3a

RP: Lmx1a/b

NC: Sox10
Slug/Snail

Fig. 11.1
Extracellular secreted signals and transcriptional codes determining neural crest precursors. (A) Diagram of a neural plate stage in which Wnt and BMP proteins secreted from the lateral non-neural ectoderm, and FGF proteins secreted from the lateral plate mesoderm are responsible for creating the neural/non-neural border. The border cells are defined by the expression of *Dlx*. (B) Diagram of a neural groove in which the same Wnt, BMP and FGF signals are responsible for induction of neural crest (NC) precursors within the neural folds. NC precursors are defined by a combination of transcription factors including Msx1/2, FoxD3, Pax3/7, Sox 9/10, Zic1–5 and Slug/Snail. (C) Diagram of a neural tube in which BMP signaling from the non-neural dorsal ectoderm induces roof plate (RP) differentiation, identified by the expression of *Lmx1a/b*. Expression of *Lmx1a/b* differentiates roof plate cells from migrating neural crest cells (NC) which express *Sox10* and *Slug/Snail*. A differentiated roof plate becomes an organizing center and secretes GDF7, BMP4/5/7, and Wnt1/3a.

In this chapter, we will review our current understanding of this topic, focusing on those studies that have used the spinal cord as a model. Because it is now clear that the basic strategies used to establish neuronal diversity in the caudal part of the CNS are also used in more anterior regions, as examples, we will also briefly discuss establishment of the D-V patterning in the eye and the telencephalon, structures originating from the anterior-most neural plate.

11.2
Generating Cell Diversity in the Dorsal Neural Tube

Three distinct populations of cells develop from the dorsal neural tube/lateral neural plate in a time-dependent and spatially-organized sequence of events. Neural crest cells are generated from the border region between the neural plate and the adjacent non-neural ectoderm and migrate out of the neural tube before or during neural tube closure (Fig. 11.1). Neural crest cells migrate to diverse locations throughout the embryo and when they reach their final locations differentiate into a remarkable variety of cell types [8]. Roof plate cells develop at the dorsal midline of the neural tube as a group of specialized cells that in turn become a signaling center, which can promote the differentiation of a third population of cells: the dorsal spinal cord interneurons. Dorsal interneurons integrate and relay somatosensory information, conveyed into the spinal cord by periphery sensory neurons to the central targets [9, 10].

11.2.1
The Neural Crest

Neural crest cells are a transient migratory population of multipotent progenitor cells. There are important and remarkable differences in the origins and the descendants of the neural crest in cranial and trunk regions. Cranial neural crest cells originate from distinct groups of precursor cells located in the diencephalon, midbrain and hindbrain. In addition to the cell types originating from neural crest of the trunk (sensory nervous system ganglia, enteric nervous system, satellite and Schwann cells of the ganglia, melanocytes and endocrine and paraendocrine cells), cranial crest cells have the potential to differentiate into cartilage, bone and connective tissue. The important contribution of this cell population to the formation of the head and their differences with their trunk counterparts have been thoroughly and recently reviewed [11].

Evolutionarily, neural crest cells are of outstanding interest since they are a vertebrate invention [8]. This, however, makes them difficult to study since they cannot be followed in yeast, worms or fly, traditional organisms for rapid, large scale genetics. However, classical embryological experiments provide a detailed account of where the neural crest arises, which tissues are needed for neural crest induction, and which cell types they form. More recent work has also identified the molecular nature of the secreted factors that induce crest cells and the transcriptional code needed to confer crest cell identity to their precursor cells.

There seems to be no doubt about the topological origin of the neural crest cells. Ectopic grafts of explanted epidermis or neural plate in avian, amphibian, and zebrafish embryos, showed that neural crest cells are generated wherever new boundaries between neural and non-neural tissues are created [12–15]. The neural folds, i.e. the interphase between neural and non-neural tissues, contain pre-migratory neural crest cells, although only a subset of them will actually migrate. Cell marking experiments have shown that the progeny of individual cells within the neural folds can contribute to neural tube, epidermis and neural crest [16]. Thus, neural crest cells are not a fully defined cell population until the cells begin to migrate. Once they have emerged from the neural tube, neural crest cells will generate their progeny following multiple intrinsic and extrinsic cues (reviewed in [17, 18]). We will now discuss how neural crest precursors are determined.

11.2.1.1 Cellular and Molecular Inducers of Neural Crest

There is a continuing debate about which tissue in the embryo provides the signals that induce neural crest development. Most likely and in line with many other embryonic inductions, the induction of the neural crest is not a single-step process but rather a series of progressive cellular interactions that culminate in the triggering of cell migration. After gastrulation, the region that will form the neural crest is in contact with the neural plate, the epidermis and the paraxial mesoderm. All of these three tissues have been proposed to participate in neural crest induction through the activity of three class of signaling molecules, which are expressed at the proper time and locations: Wnt proteins, a family of highly conserved secreted signaling molecules related to the Drosophila wingless protein; Bone Morphogenetic Proteins (BMPs), multifunctional secreted proteins of the transforming growth factor-β (TGFβ) superfamily and Fibroblast Growth Factors (FGFs), multifunctional factors that regulate growth and morphogenesis of multiple tissues and organs (Fig. 11.1). In various assays, all of these proteins can mimic the tissue interactions that induce the neural crest. Participation of retinoic acid has also been proposed [19].

In avian embryos, BMPs are both necessary and sufficient to induce the neural crest and other dorsal neural phenotypes [20–22]. Consistent with the idea that BMPs induce the neural crest, zebrafish mutants of the BMP pathway (*swirl/bmp2b*, *snailhouse/bmp7* and *somitabun/smad5*) have either reduced or expanded domains of neural crest progenitors, depending on the precise alteration of the BMP signaling levels [23, 24]. Thus, for several years BMPs (in particular BMP2/4 and BMP7) have been considered as the "favorite" signal mediating the neural/epidermal interaction. However, weakening this view is the fact that BMPs appear to be strongly expressed in the neural folds and the dorsal neural tube, but only weakly present in the ectoderm during stages when neural crest induction is ongoing. Thus, the current model suggests that BMPs may act in an autocrine way to maintain the border state [25]. Furthermore, it appears that neural crest formation requires BMP signaling only after the initial induction step, indicating that BMPs might serve as maintenance factors in the induction process [26] or as triggers for the delamination of neural crest cells from the neural tube [27].

Wnt signaling activation may precede the function of BMPs during neural crest induction. Activation of the canonical Wnt pathway results in stabilization and nuclear translocation of β-catenin, which activates transcription of Wnt target genes in cooperation with DNA binding proteins of the LEF/TCF family [28]. During open neural plate stages β-catenin is localized to the nucleus of presumptive neural crest precursors indicating that these cells receive and respond to Wnt signals [29]. Activation of Wnt target genes may include *Slug*, one of the first molecules expressed in response to neural crest-inducing signals, since there are functional LEF/TCF sites in the *Slug* promoter [30]. Furthermore, inhibition of endogenous Wnt/β-catenin signaling at discrete times in zebrafish development interferes with neural crest formation without blocking development of dorsal spinal neurons [31]. Several members of the Wnt family are expressed in and around the tissues where neural crest induction is taking place, and it is unclear which of the multiple Wnt ligands can be considered to be the endogenous crest inducer(s). Wnt6 is an intriguing new candidate. In the chick embryo, it is expressed prior to neural tube closure throughout the non-neural ectoderm, up to but not crossing the neural-epidermal border [29, 32, 33] This pattern, related to that observed in *Xenopus* [34] and mouse embryos [35], is consistent with a role as an endogenous inducer. "Classical" Wnt genes have also been implicated in neural crest formation; Wnt1 and Wnt3a are strongly expressed in the dorsal neural tube and might be important for the proliferation of neural crest progenitors [36, 37]. Wnt7b and Wnt8 are also worth considering as potential neural crest inducers, since they are expressed in early embryos and have neural crest-inducing activity in *Xenopus* [38] and zebrafish [31].

Early FGF signaling is required for initiating and maintaining the border between the neural plate and the adjacent non-neural ectoderm [25, 39, 40]. Slightly after gastrulation, FGF ligands are expressed in the paraxial mesoderm and can induce neural crest precursors by posteriorization of the neural folds [40]. In particular, FGF8 is strongly expressed by the paraxial mesoderm in *Xenopus* embryos and is sufficient to induce early neural crest markers [41].

In summary, neural crest precursors are probably determined by the integration of various signals coming from the developing neural plate, the paraxial mesoderm and the non-neural ectoderm, and involving at least BMPs, Wnts and FGFs (Fig. 11.1).

11.2.1.2 Molecular Identity of Newly-induced Neural Crest Cells

Secreted inductive signals bring about the expression of transcription factors that control the specification, migration and final differentiation of neural crest cells. Neural crest precursors in the neural folds express a large number of genes thought to be necessary and/or sufficient to initiate neural crest development, although their precise function and relationship are still poorly understood (see Table 1 in [18]).

Among these genes are members of the *Msx*, *Sox*, *Fox*, *Zic*, *Pax* and *Slug* families of transcription factors (Fig. 11.1). Prior to the onset of any of these neural crest markers, members of the *Dlx* family of transcription factors repress neural properties in the ectoderm and promote the expression of *Msx1* in the neural folds, thus positioning the border between neural and non-neural ectoderm [42, 43]. The *Msx1* and *Msx2* transcriptional repressors are direct downstream targets of the BMP pathway in several animal models. In the developing neural tube, *Msx1* transcription is activated in the neural folds at the same precise concentration of BMP signaling required for neural crest induction [44, 45]. Further, *Msx1* expression is sufficient to induce other neural crest markers (*Sox9/10*, *FoxD3*, and *Snail/Slug*) and it has been proposed as an upstream regulator of *Snail* and *Slug* in the genetic cascade that specifies the neural crest [45].

Among the remaining neural crest markers, *Sox* and *Fox* genes are expressed within the dorsal neural tube only in the neural folds. *Sox9* and *Sox10* are high mobility group (HMG)-domain transcriptional activators that seem specific to the neural crest in fish, frog, chick and mouse embryos. In chick [46] and mouse embryos [47], *Sox9* is expressed in pre-migratory neural crest, whereas *Sox10* is expressed just as crest emigration is initiated. *Sox9* overexpression in the chick neural tube is sufficient to induce other markers of pre-migratory neural crest (*Sox10*, *FoxD3* and *Slug*), but not to induce markers for delamination such as RhoB [46, 48]. Thus, *Sox9* expression, induced by an as yet unknown extracellular signal, would be sufficient to induce other neural crest markers including *Sox10*, which in turn, is necessary for neural crest migration and later differentiation as shown in mice and zebrafish mutants [49–52]. However, complicating this picture, mice and zebrafish lacking *Sox9* have no defects in neural crest specification and emigration [47, 53], and in *Xenopus* embryos, both *Sox9* and *Sox10* seem to be required for the early specification of neural crests, since loss of function of either of them [54, 55] results in inhibition of neural crest precursors. Similarly to *Sox* genes, the forkhead class transcription factor *FoxD3* is expressed in early neural crest precursors prior to delamination, and forced expression of *FoxD3* induces some aspects of neural crest differentiation, although it is not sufficient to induce cells that exhibit all the characteristics of the neural crest [56, 57]. Thus, *Sox* and *Fox* function seems important for neural crest development. However, their precise position in the neural crest developmental program and their interactions with other early genes, i.e. *Msx1*, remain to be clarified.

Transcription factors of the *Zic* zinc-finger and Paired (*Pax*) families are also expressed in the neural folds and are considered as neural crest markers, although their expression is broader and they extend to the dorsal neural tube. Members of these families (*Zic1, Zic2, Zic3, Zic5* and *Pax3, Pax7*) contribute to the generation of neural crest precursors. The onset of *Zic* gene expression occurs early in the neural ectoderm, likely in response to neural inducers. Overexpression of any *Zic* gene induces neural crest markers [58–61], promotes precursor cell proliferation and inhibits neuronal differentiation by activating Notch signaling [62], functions confirmed by the analysis of mouse mutants for *Zic1* [63], *Zic2* [64] and *Zic5* [65]. The transcriptional activator *Pax3* is expressed earlier than *Pax7* in neural tube development [66], and both genes are independently required for neural crest formation. Mice mutants for *Pax3* or *Pax7* display defects in various neural crest derivatives [67, 68].

Particularly crucial to neural crest development are the *Slug* and *Snail* genes, members of the Snail family of transcription factors of the zinc-finger type. *Slug* or *Snail*, depending on the species considered, are expressed in the pre-migratory neural crest population of all vertebrate species analyzed, and their activity is required for both the formation of neural crest precursors and their subsequent delamination from the neural tube. Thus, in both the chick and *Xenopus* embryos, antisense oligonucleotides to *Slug* impair neural crest migration while *Slug* gain-of-function increases neural crest production in the chick embryo, interestingly only in the head region. As mentioned above, *Slug/Snail* genes are among the first target activated by neural crest inducers and direct or indirect functional interaction between BMP, Wnt and FGF signaling and Slug activity has been demonstrated during neural crest development. In turn, activation of Slug/Snail has the fundamental role of down-regulating cadherin expression in neural crest cells, thus setting up the conditions for their delamination (reviewed in [39, 69]).

11.2.2
The Roof Plate

The second distinct cell population that is generated from the lateral neural plate (neural folds) around the time of neural tube closure is the roof plate. Lineage-tracing experiments in chick [16] and mice [70] have indicated that roof plate and neural crest cells share a common cellular precursor. Moreover, early in neural tube development, molecular markers that later define roof plate and neural crest cell populations, are co-expressed by dorsal midline cells. Therefore the mechanisms of induction of neural crest and roof plate, and the nature of the signals responsible for

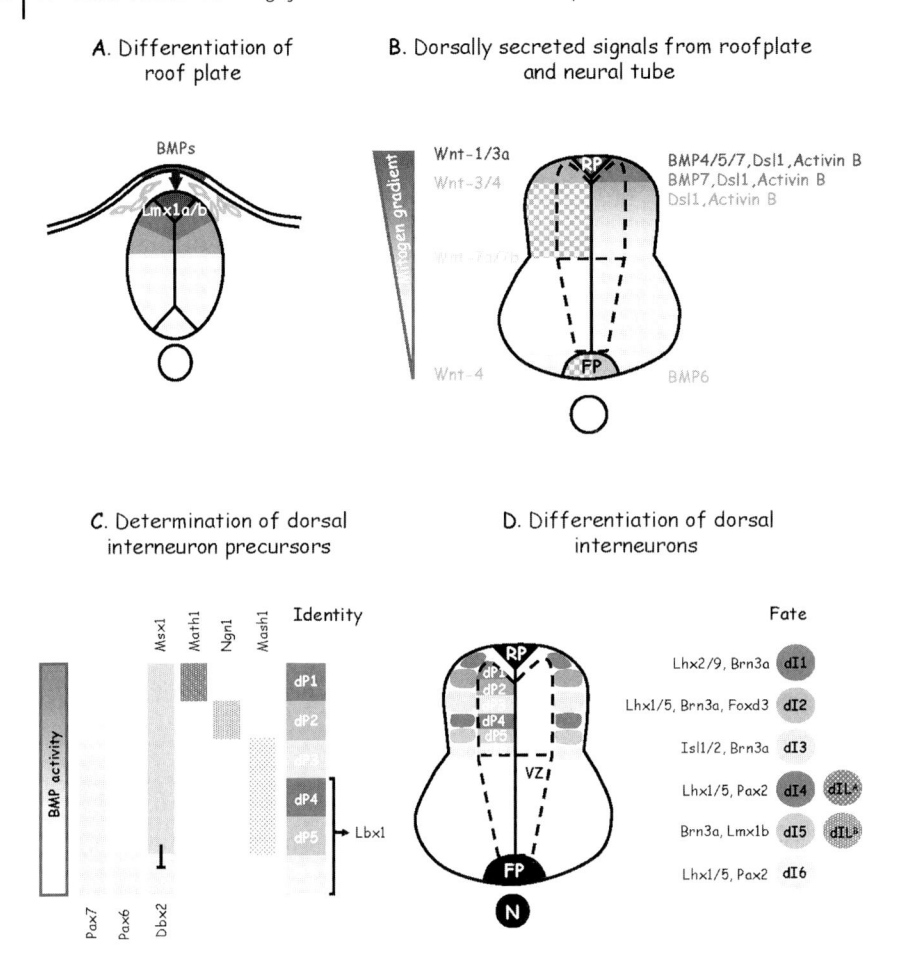

368 | *11 Dorso-Ventral Patterning of the Vertebrate Central Nervous System*

A. Differentiation of roof plate

B. Dorsally secreted signals from roofplate and neural tube

C. Determination of dorsal interneuron precursors

D. Differentiation of dorsal interneurons

Fig. 11.2
Extracellular secreted signals and transcriptional codes patterning the dorsal neural tube. (A) Diagram of the neural tube during roof plate specification. BMP-dependent activation of Lmx1a/b specifies the roof plate. (B) Schematic representation of a roof plate-derived gradient of TGFβ/BMP molecules (color coded on the right) and Wnt proteins (color coded on the left) involved in the induction of the progenitor domains within the dorsal neural epithelium. (C) A graded distribution of BMP activity (shaded bar on the left) mediates the expression of different HD and bHLH transcription factors (color-coded bar graphs in the middle) required for the specification of the dorsal progenitor domains (represented on the right with a color-coded bar). (D) Diagram showing the relationship between neural progenitor domains (color coded) and the positions at which post-mitotic neurons (spheres on the right with the same color code) are generated along the dorso-ventral axis of the dorsal spinal cord. (This figure also appears with the color plates.)

such inductions, should be closely related. In fact, roof plate cells, like neural crest cells, are induced in neural explants by exposure to signals from adjacent epidermal ectoderm [22]. How then do precursor cells within the neural folds know whether to become neural crest cells and emigrate out of the neural tube, or whether to remain within the neural tissue and differentiate into roof plate cells?

The analysis of the spontaneous mutant mouse dreher (*dr*) appeared crucial to resolve this issue since its phenotype results from the failure of the roof plate to develop [71]. Positional cloning of *dr* identified the LIM homeodomain transcription factor *Lmx1a* as the cause of defective roof plate development [72]. Further, analysis of the role of *Lmx1a* and the closely related *Lmx1b* [73, 74] in the roof plate, revealed that they are not directly involved in neural crest development but are sufficient to direct neural tube precursors into functional roof plate cells. Then, if *Lmx1a/b* are definitive roof plate determinants, which signal might mediate *Lmx1a/b* activation and roof plate formation? Wnts and BMP signals have been tested for this role and it appears that *Lmx1a/b* activation is dependent on BMP (Fig. 11.2A) [73, 74].

11.2.3
Signals from the Roof Plate Pattern the Dorsal Spinal Cord

Eight distinct populations of interneurons arise in the developing dorsal neural tube; six of them (dI1–dI6, following a dorsal to ventral order) are generated early in development, whilst the remaining two (dIL^A and dIL^B) arise later (Fig. 11.2C and D).

Once differentiated, the roof plate itself becomes an organizing center required for the determination of the three most dorsal classes of interneurons (dI1–dI3) [9, 10]. The role of the roof plate in patterning the dorsal neural tube was elegantly demonstrated by genetic ablation of the roof plate, which resulted in the severe loss of dorsal interneurons (dI1–dI3) [75]. The next obvious question was which individual secreted factors mediate this process. Again members of the same families of signaling molecules, Wnts and BMPs, were the obvious candidates on the basis of their expression pattern.

Wnt1 and Wnt3a are expressed in the roof plate during dorsal interneuron specification. Wnt1/Wnt3a double knockout mice have reduced populations of dorsal interneurons suggesting a role for these Wnts in proliferation [36], an idea that was supported by Wnt1 and Wnt3a overexpression experiments in the neural tube of chick embryos [37]. However, further analysis of the Wnt1/Wnt3a double mutant revealed a diminished development of dI1 and dI3 and a compensatory increase of dI4, without changes in the expression of BMPs [76], supporting a direct role for Wnt1/Wnt3a in the specification of at least two roof plate-dependent cell types (dI1 and dI3).

The role of roof plate-derived BMP signaling in the specification of dorsal phenotypes has been more extensively addressed. It has been proposed that, *in vivo*, a gradient of BMP extends throughout the entire D-V axis of the neural tube providing positional information and thus controlling neuronal determination [24, 77]. However, multiple BMPs, as well as non-BMP members of the TGFβ superfamily, are expressed in the roof plate and the dorsal neural tube. These proteins can all induce dorsal phenotypes *in vitro* [20, 22], raising the question of whether different BMPs have redundant or distinct functions in the generation of specific dorsal cell types. Initial studies point to the latter possibility since GDF7 (a BMP family member specifically expressed in the roof plate) is required *in vivo* only for the generation of one discrete class of dorsal interneurons, dI1 [78]. However, lack of further experimental data, leaves this issue unresolved. Nevertheless, manipulating the levels of BMP signaling by expressing constitutively active BMP receptors (BMPR-IA and BMPR-IB) resulted in differential expansion and repression of basic-helix-loop-helix (bHLH) factors in progenitor cells and in the specification of dI1–dI3 interneurons [79]. Additional roles for BMPR-IA in promoting proliferation and for BMPR–1B in inducing differentiation have also been shown *in vivo* [80]. Thus, BMP signaling plays a fundamental role in growth, patterning and differentiation of the dorsal spinal cord, although we are still far from understanding how a precursor cell integrates the multiple BMP signals received to generate distinct phenotypes or how different concentrations of the same signal may drive precursor cells toward diverse identity.

11.2.3.1 Transcriptional Code Defining Dorsal Spinal Cord Interneurons

Eight distinct interneuron subtypes have been identified in the dorsal spinal cord. Each of them is defined by its birth date, pattern of migration, dependence of roof plate signals and by the combinatorial expression of homeodomain (HD) and bHLH factors [81, 82] (reviewed in [9]). The expression of specific transcription factors defines six early-born post-mitotic dorsal neuron populations known as dI1–dI6 and two later-born post-mitotic populations known as dIL^A and dIL^B (Fig. 11.2D). These neurons can be further classified on the basis of their specification dependence on roof plate signaling: roof plate signals are required by Class A (dI1–dI3) but not by Class B (dI4–dI6 and dIL$^{A/B}$) [81, 82]. dI1–dI3 interneurons migrate to deep dorsal horn layers and differentiate into relay interneurons that project contralaterally, later-born dIL$^{A/B}$ migrate to more superficial dorsal horn layers and comprise the association neurons.

Prior to neuronal birth, progenitor domains within the dorsal neural tube express discrete members of the bHLH family of transcriptional activators, all of them capable of inducing a generic neuronal fate [83, 84]. However, overexpression experiments demonstrated that each factor has an additional distinct activity that generates a specific type of dorsal interneuron. Progenitor cells expressing *Math1* (*Cath1* in chick embryos) give rise to dI1 cells, those expressing *Neurogenin1* to dI2, and those expressing *Mash1/Cash1* are likely to give rise to dI3–dI5 [84, 85]. In each case, the increase in one precursor population appears to occur at the expense of the other two. Distinct domains within the HLH motif of *Math1* are responsible for driving neural differentiation and cell type specification [84], suggesting the importance of the unique protein-protein interactions involved in these functions. As such, combinations of bHLH and HD factor expression are likely to further define progenitor domains [86, 87].

A similar mechanism could apply for dorso-medial cell fates, since they are determined by the combinatorial expression of bHLH and HD factors; dI4–dI6 and the late-born dIL$^{A/B}$ all express *Lbx1* [81, 82]. However, we are still far from understanding what regulates the precise expression of bHLH and HD in a particular set of precursors, and how the expression of a particular combination of genes is interpreted in the differentiation of a particular neuronal cell type.

In conclusion, the final identification of progenitor domains is determined by their relative position within the developing neural tube, together with the expression of molecular markers (Fig. 11.2D). Each discrete progenitor pool will give rise to a particular dorsal interneuron population, which in turn, can be identified by the expression of a specific set of molecular markers. All these data are just beginning to provide a framework for understanding the patterning of the dorsal neural tube but will eventually permit molecular analyses of the assembly of functional neuronal circuits.

11.3
Generating Cell Diversity in the Ventral Neural Tube

11.3.1
The Floor Plate

Situated at the ventral-most part of the vertebrate neural tube, the floor plate (FP) is an important organizing center that controls two fundamental aspects of CNS development: regionalization of the ventral nervous system and navigation of commissural axons. Interestingly, these two key roles are largely assigned to two key players: Netrin–1 that promotes the outgrowth of spinal commissural axons [88] and the morphogen Sonic hedgehog (Shh, Fig. 11.3A) that controls patterning of the ventral neural tube (discussed below) and also collaborates with Netrin–1 to guide commissural axons [89].

Although the function of the FP is well conserved from fish to humans, small variations do exist with respect to the embryonic origin of the FP in the various vertebrate species studied and along the A-P axis within the same species [90]. Common features are the early determination during gastrulation, the requirement of the winged-helix transcription factor *FoxA2*, and the participation of the Shh and Activin/Nodal signaling pathways. These molecules operate in a tightly-regulated network arranged in a feedback loop: *FoxA2* directly controls *Shh* and *Nodal/Activin* transcription and in turn, is a downstream effector of both signaling pathways (reviewed in [91]). Recent experiments also provide evidence that Nodal signaling can cooperate with Shh to induce the epiblast precursors toward a floor-plate fate [92].

A. cShh expression

B. Differentiation of floorplate

C. Ventral gradient of Shh from the floorplate

D. Determination of ventral neuron precursors

E. Differentiation of ventral neurones

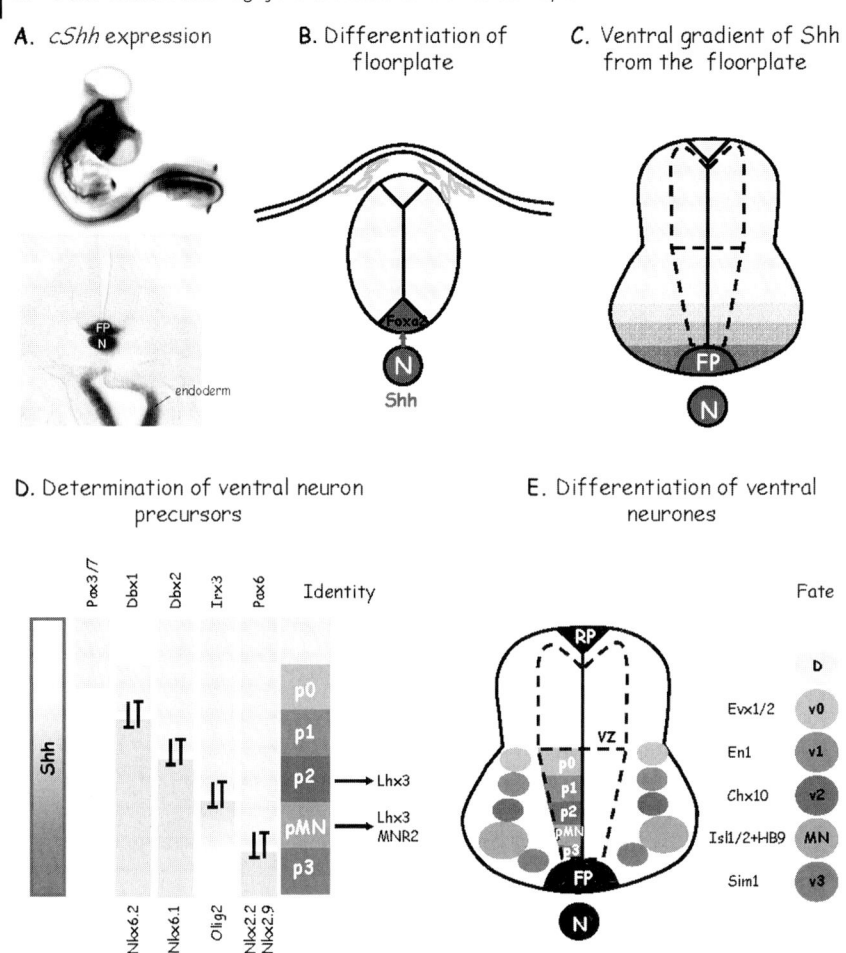

Fig. 11.3
Extracellular secreted signals and transcriptional codes patterning the ventral neural tube. (A) Whole mount *in situ* hybridization in a St 18 embryo showing the expression of *Shh*. Note the specific localization in the notochord, floor plate and endoderm, as shown in the cross-section of the neural tube. (B) Diagram of the neural tube during floor plate (FP) differentiation. A tightly-regulated network arranged in a feedback loop, including *FoxA2* and *Shh*, is required for floor plate specification. (C) Shh secreted from the notochord and the floor plate presumably forms a gradient in the ventral neural tube distributed in a ventral-high, dorsal-low concentration (shown in graded levels of pink) within the ventral neural epithelium. (D) Different levels of Shh (shaded bar on the left) mediate the repression of Class I HD proteins (Pax7, Dbx1, Dbx2, Irx3 and Pax6) at different threshold concentrations and induce the expression of Class II proteins (Nkx6.1/2; Nkx2.2/9 and Olig2), represented in the center by bar graphs. Cross-repressive interactions between Class I and Class II proteins define six different progenitor domains, represented on the right with a color-coded bar. (E) Diagram showing the relationship between neural progenitor domains (color coded) and the positions at which post-mitotic neurons (spheres on the right with the same color code) are generated along the dorso-ventral axis of the ventral spinal cord. (This figure also appears with the color plates.)

11.3.2
Sonic Hedgehog Secreted from the Notochord and the Floor Plate Patterns the Ventral Spinal Cord

A functional floor plate acts as an organizer patterning the ventral CNS. The signaling protein Sonic hedgehog (Shh), secreted from floor plate and notochord cells [93], mimics all floor plate patterning functions. Ectopic expression of Shh *in vivo* and *in vitro* can induce the differentiation of floor plate, motor neurons and ventral interneurons [4, 6, 7, 94]. Conversely, elimination of Shh function *in vitro* [7] or *in vivo* by gene targeting [95] prevented the differentiation of floor plate, motor neurons and most classes of ventral interneurons. Progressive changes in Shh concentration generate five molecularly distinct classes of ventral progenitor cells at defined positions within the ventral ventricular zone (V3, MN, V2, V1, V0, Fig. 11.3E). Each neuronal subtype (p3, pMN, p2, p1, p0; Fig. 11.3D) is generated from a spatially-discrete progenitor domain, and these distinct domains are defined by the expression of a characteristic combination of genes that encode HD and bHLH transcription factors. These HD proteins can be subdivided into two classes on the basis of their pattern of expression and mode of regulation by Shh. Class I proteins are constitutively expressed by neural progenitor cells and their expression is repressed by distinct threshold concentrations of Shh. As a consequence their ventral boundaries of expression delineate progenitor domains. Conversely, the expression of each Class II HD requires Shh signaling and is activated at a distinct Shh threshold concentration. Thus their dorsal boundaries delineate progenitor domains (reviewed in [96, 97]). However, the different transcriptional responses to graded Shh signaling cannot adequately explain the striking sharp boundaries of Class I and Class II protein expression observed *in vivo* [98]. Gain- and loss-of-function experiments suggest that selective cross-repressive interactions between pairs of Class I and Class II proteins that abut the same progenitor domain function to refine such boundaries and to consolidate progenitor domain identity as the neural tube grows in size. As such, Pax6 (Class I) and Nkx2.2 (Class II); Irx3 (Class I) and Olig2 (Class II); Dbx2 (Class I) and Nkx6.1 (Class II) and Dbx1 (Class I) and Nkx6.2 (Class II) are incompatible within the same cell because of their cross-inhibitory activities [98–102] (Fig. 11.3D).

The subdivision of progenitors within the ventricular zone is the initial requirement for the generation of distinct neuronal subtypes. Subsequently, the profile of HD proteins expressed by precursor cells acts to specify the identity of neurons derived from each progenitor domain. The ventral-most progenitor domain p3 generates the ventral-most interneurons type V3, identified by the expression of *Sim1*. Expression of *Nkx2.2* within the p3 domain is necessary for the generation of V3 interneurons [100], conversely ectopic activation of *Nkx2.2* is sufficient for the generation of ectopic V3–Sim1–positive progenitor neurons [98] (Fig. 11.3E).

Dorsal to p3, motor neuron precursors (pMN) are determined by the combined expression of HD factors including Nkx6.1/6.2 and Pax6, together with the expression of the bHLH factors Olig1/2 and Ngn2 [97, 98, 103–105]. The initiation of post-mitotic motor neuron differentiation is accompanied by the downregulation of progenitor cell factors and the selective expression of MNR2 (Motor Neuron Restricted 2) and HB9 (Homoebox 9) proteins in these neurons. Expression of HB9 is tightly regulated by the Nkx2.2 and Irx3 HD factors, which act as potent repressors of HB9 transcription outside the pMN domain [106]. Conversely, a high level of HB9 transcription in post-mitotic motor neurons is activated by neurogenic factors such as Ngn2, NeuroD and NeuroM [105] and by LIM factors such as Lhx3 and islet1 [106]. Further, the combination of LIM proteins forms a high order complex with the LIM co-factor nuclear LIM interactor (NLI) through cell type-specific protein-protein interaction that consolidates the motor neuron fate [107].

Three different types of ventral interneurons are generated dorsal to MN (v2, v1 and v0 respectively, Fig. 11.3E). p2 precursors are determined by the combined expression of *Nkx6.1* and *Irx3*, and generate *Chx10*–expressing ventral interneurons. Misexpression of Nkx6.1 within the Irx3 domain (but not ventral to it) results in the ectopic generation of many *Chx10*–positive v2 neurons in the p1 and p0 domains together with a marked decrease in the number of v1 and v0 subtypes of interneurons. Conversely, misexpression of *Irx3* in regions ventral to p2 is sufficient for the generation of ectopic v2 subtype of neurons within the normal domain of MN generation [98].

v1 and v0 subtypes of interneurons derive from adjacent progenitor domains (p1 and p0) that are distinguished by the expression of the homeodomain proteins Dbx1 and Dbx2. The spatially restricted expression of *Dbx1* has a critical role in establishing the distinction in v0 and v1 neuronal fate. In *Dbx1* mutant mice, neural progenitors fail to generate v0 neurons and instead give rise to interneurons that express many characteristics of v1 neuron [108].

In conclusion, within the ventral spinal cord, a gradient of Shh secreted from the notochord and floor plate helps create domains of expression of HD and bHLH transcription factors. The specification of each neuronal subtype then results from the mutual repression, mediated by the recruitment of transcriptional co-repressors from the *groucho* family [102], which establish sharp domains of identified ventral neural progenitors.

11.4
Generation of D-V Patterning in the Anterior Neural Tube Follows Rules Similar to those Used in the Spinal Cord

The studies described above have provided a framework within which to address the question of how D-V specification of the neural tube is also achieved at more anterior levels, although our current understanding is not as precise.

The tightly controlled and combined activity of different signaling pathways, including Wnts, FGFs, BMPs and Retinoids, allow the regionalization of the neural plate along its anterior-posterior axis (reviewed in [109]). Thus, the hindbrain, midbrain and forebrain are generated anterior to the spinal cord. While the hindbrain forms in a similar manner and maintains a geometrical organization close to that of the spinal cord (for an account of hindbrain development see [110]), the remaining portions of the anterior neural tube result from a more complex series of morphogenetic movements that progressively reposition neural plate precursor cells along the anterior-posterior and dorsal-ventral axes. For instance, in chick and fish the cells of the ventral forebrain (hypothalamus) are located caudal to the telencephalic and eye fields that together constitute the dorsal forebrain precursors. Rostral migration of hypothalamic precursor cells splits the eye field in two separate entities, also displacing the dorsal telecencephalic domain laterally [109]. Furthermore, after closure, the anterior neural tube undergo a series of folding processes that, as in the case of the optic vesicle (see below), progressively re-specifies the medial-lateral and D-V position of precursor cells. On the basis of these observations, it might be expected that the mechanisms underlying the establishment of D-V polarity in the anterior neural tube would be more elaborate than those described for the spinal cord. Although this is partially true, the specification of the D-V axis in both the posterior and anterior neural tube by and large follows similar rules. Thus, inductive signals, derived from a specific group of cells located in critical regions surrounding the developing forebrain, establish restricted domains of progenitor cells identified by the specific expression of transcriptional regulators of the HD and bHLH families. More interestingly, as in the spinal cord, the activation of the Shh signaling pathway governs the generation of ventral forebrain structures while TGF-β/BMPs molecules are responsible for the establishment of dorsal phenotypes.

A brief account of the current view of the molecular mechanisms of D-V patterning in the eye and the telencephalon are outlined below.

11.4.1
The Eye

The eye is a bilateral organ that originates from a single field positioned in the anterior portion of the neural plate. Although still often described as evaginations of the diencephalon [111], one of the most accepted models of CNS regionalization considers the eyes to be structures of the secondary prosencephalon, strictly related to the telencephalic derivatives [112]. In support of this association, studies on anterior neural plate specification have highlighted a close relationship between telencephalic and eye precursor cells [113, 114]. As a result of anterior migration of hypothalamic precursors along the midline, the eye field is divided into two. Proliferation and evagination of these regions give rise to two visible optic primordia, the optic vesicles. In vertebrates, the cells that compose the early optic vesicle are morphologically and molecularly undistinguishable and they all co-express a number of transcription factors, including *Otx2, Pax6, Rx1 Lhx2* and *Six3*, which are required to initiate eye development (Fig. 11.4) [115]. Initially, these cells are all competent to become neural retina (NR), optic stalk (OS) or retina pigmented epithelium (RPE). Nevertheless, optic vesicle progenitors are soon patterned along the D-V axis: the dorsal-most precursor cells will give rise to the presumptive RPE, while the ventral cells will generate the OS. Because the optic vesicles are bulges of the neural tube, their subdivision is often described in relation to the proximal-distal axis. Thus, OS precursors have a proximal location, while the distal cells will produce both the NR and the RPE. According to this view, the distal-most cells (intermediate in a D-V axis) constitute the NR precursors (Fig. 11.4).

11.4.1.1 D-V Patterning of the Optic Vesicle

Proximo-ventral cells of the OS are generated under the influence of Shh emanating from the axial ventral midline (Fig. 11.4A). Targeted disruption of the *Shh* gene in the mouse and mutations in the human *SHH* gene cause severe anterior neural tube defects that include the formation of a cyclopic eye lacking OS tissue. A likely explanation of this phenotype is provided by the observation that SHH activity regulates the spatial expression of *Pax6* and *Pax2*, HD transcription factors of the paired type, which normally demarcate the distal-dorsal and proximal-ventral optic primordium, respectively. Thus, widespread overexpression of SHH (or of the related tiggywinkle hedgehog, *twhh*) causes the expansion of the *Pax2* domain at the expense of the *Pax6*–positive tissue. In contrast, absence of *Shh* and *twhh* expression in cyclopic zebrafish mutants correlates with a clear reduction of the *Pax2* domain compensated by a corresponding increase in *Pax6* across the midline, responsible for the fusion of the two retina fields (reviewed in [111, 116]). A similar dependence on Shh signaling has also been observed for the expression of *Vax* genes, other HD transcription factors involved in OS generation [117].

Targeted inactivation of *Pax2* [118] or *Vax1* [119, 120] results in severe optic nerve abnormalities, consistent with a role for these genes in proximal-ventral eye development. The phenotype of these mice is characterized by the presence of coloboma (failure to close) of the ventral optic fissure, and, in the case of *Pax2*, an extension of the RPE into the optic nerve region. Related but milder defects have also been reported in *Vax2*–deficient mice [121, 122]. Thus, as in the spinal cord, acquisition of a ventral phenotype (the optic stalk in this case) depends on the Shh-dependent expression of HD transcription factors that impose ventral character on optic vesicle precursor cells. In an extended analogy, the establishment of a precise boundary between the presumptive optic stalk and the more dorso-distal optic vesicle (presumptive neural retina and RPE) may result from the reciprocal transcriptional repression between *Pax6* and *Pax2* [123].

Again, resembling the spinal cord the generation of dorsal optic vesicle progenitor cells (presumptive RPE) may depend on the activation of the TGFβ/BMP signaling pathway. With a notable difference however: the acquisition of intermediate (neural retina) identity does not depend on the graded activity of either Shh or TGFβ/BMP signaling pathway, but rather seems to require the presence of FGF signaling, provided by a more closely located tissue, the lens placode (Fig. 11.4). TGFβ/BMP and FGF signaling act antagonistically on the specification of RPE and NR precursors. Removal of the FGF-enriched lens placode results in an impaired segregation of the RPE and NR cells, an effect that can be rescued by exogenous application of FGFs. Conversely, generation of transgenic mice that ectopically express FGF in the pre-

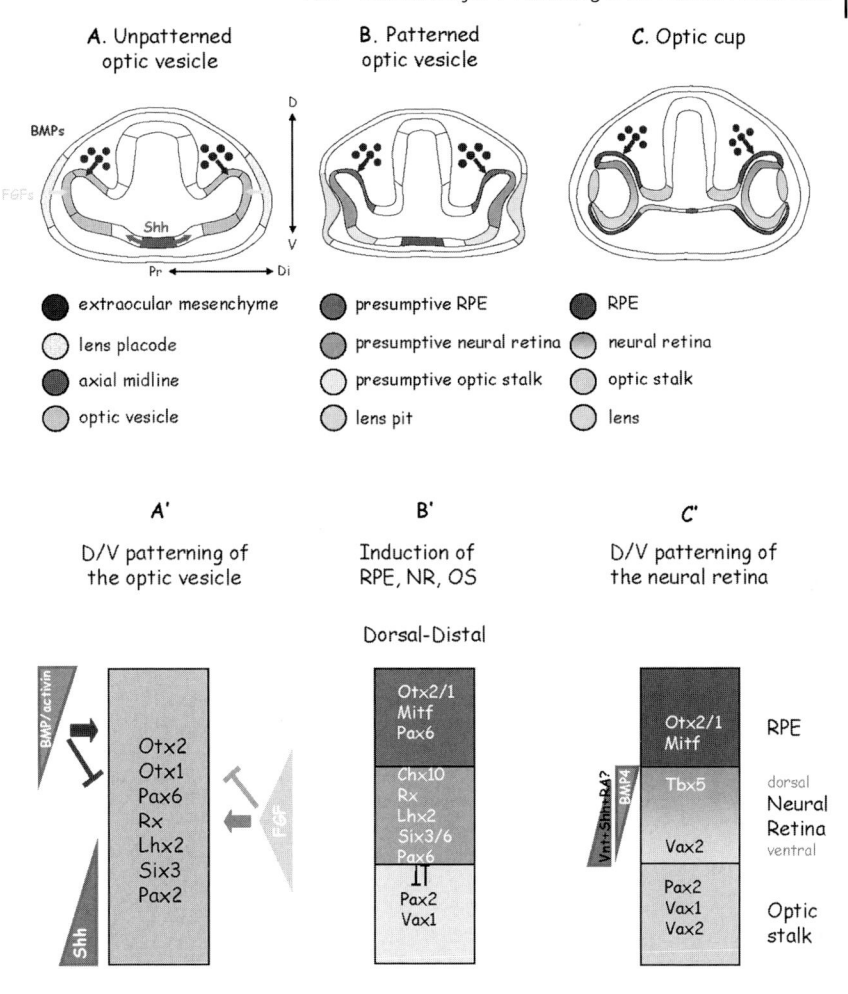

11.4 Generation of D-V Patterning in the Anterior Neural Tube

A. Unpatterned optic vesicle

B. Patterned optic vesicle

C. Optic cup

- extraocular mesenchyme
- lens placode
- axial midline
- optic vesicle
- presumptive RPE
- presumptive neural retina
- presumptive optic stalk
- lens pit
- RPE
- neural retina
- optic stalk
- lens

A' D/V patterning of the optic vesicle

Otx2
Otx1
Pax6
Rx
Lhx2
Six3
Pax2

B' Induction of RPE, NR, OS

Dorsal-Distal

Otx2/1
Mitf
Pax6

Chx10
Rx
Lhx2
Six3/6
Pax6

Pax2
Vax1

Ventral-Proximal

C' D/V patterning of the neural retina

Otx2/1
Mitf — RPE

Tbx5 — dorsal
Vax2 — Neural Retina ventral

Pax2
Vax1
Vax2 — Optic stalk

Fig. 11.4
Molecular network involved in the patterning of the optic vesicle and optic cup. Schematic representation of the transition from unpatterned (A) to patterned optic vesicle (B) and optic cup (C). Signaling molecules and progressive tissue specification are represented in color codes. Color-coded bar graphs represent the molecular components implicated in the transition from transition from unpatterned (A') to patterned optic vesicle (B') and optic cup (C'). In (A, A') all the neuroepithelial cells are indistinguishable (gray color) and express a common set of transcription factors. Shh signaling from the axial midline is required for the specification of the proximo-ventral optic stalk (pink in B, B'). TGFβ-like signals from the extraocular mesenchyme activate cells from the optic vesicle to become RPE (blue in B, B'), whereas FGF signals from the lens placode repress RPE and activate NR identity (green in B, B'). During optic cup formation the graded distribution of BMP4 dorsally and ventroptin Vnt, possibly together with RA and Shh, establishes the D-V polarity of the neural retina (shaded green in C, C'). (This figure also appears with the color plates.)

sumptive RPE converts this tissue into a second neural retina, indicating that FGF signaling normally inhibits RPE formation, whilst it activates NR specification. TGFβ family members, derived from the extraocular mesenchyme that surrounds the dorsal optic vesicle, have the opposite effect. In fact, optic vesicle explants which have been cultured in the absence of the surrounding mesenchyme lose the expression of RPE markers and upregulate the expression of those markers specific for the NR. Although a link has not been firmly establish, these signaling pathway seems to impinge upon the expression of specific transcriptional regulators, the function of which is instrumental to the acquisition of either RPE or NR identities (reviewed in [124]).

Thus, functional inactivation of either *Otx* or *Mitf* genes (HD and bHLH transcription factors, respectively) impairs RPE differentiation [125, 126] and their expression is downregulated upon FGF-mediated trans-differentiation of the RPE into neural retina. Transfection of either *Otx* or *Mitf* genes confers an RPE phenotype to retina neuroepithelial cells by activating the expression of melanogenic genes [127]. This suggests that the cooperation between the two transcription factors is sufficient to impose a dorsal/RPE character to optic vesicle cells. Initial activation of *Mitf* expression may involve the activity of *Pax6* (Fig. 11.4.B-C). Indeed, ectopic expression of *Pax6* in the mouse optic stalk, driven by the *Pax2* promoter, induces the formation of *Mitf*-positive RPE cells, disrupting the normal optic cup/optic stalk boundary [123]. The function of *Pax6* in RPE specification is, however, transient since its expression is rapidly downregulated at optic cup stages (reviewed in [124]).

A number of transcription factors, initially expressed in the unpatterned optic vesicle, are also expressed in the presumptive neural retina as this is specified under the influence of FGF and TGFβ signaling. Currently, it is not clear which transcription factor and how many of them are really needed to impose neural retina identity. Two members of the *Six* family of HD transcription factors, *Six3* and *Six6*, for instance, soon disappear from the dorsal optic vesicle, suggesting that their function is incompatible with the acquisition of RPE character (reviewed in [128] Fig. 11.4). In support of this idea, overexpression of *Six6* or *Six3* in RPE cells leads to the acquisition of a neural retina phenotype [129]. These genes however are also expressed in the optic stalk making their participation in neural retina specification questionable.

The expression of *Lhx2*, a factor also involved in dorsal neural tube specification, becomes restricted to the prospective neural retina. Mice deficient in *Lhx2* function are characterized by anophthalmia [130], as in the case of mice mutants for the *Pax6* gene [131]. Although the two genes seems to act independently, they are both required during the period of close contact between the surface ectoderm and optic vesicle. It remains to be established whether this implies that their activity is necessary for the specification of the intermediate region of the optic vesicle (prospective neural retina).

The *paired*-like homeobox gene *Chx10* is possibly the best candidate to impose a neural retina character to naïve optic vesicle cells under the influence of a lens-derived signal (Fig. 11.4B). The expression of *Chx10* first appears to be restricted to the presumptive neural retina territory, when close contact between the optic vesicle neuroepithelium and the lens placode is established [132, 133]. Explant recombination studies using naïve optic vesicle and lens placode ectoderm also indicate a tight link between the onset of *Chx10* expression and the presence of a lens-derived inductive signal [126]. Furthermore, *Mitf* and *Chx10* expression appears incompatible suggesting that mutual repression between the two genes may also contribute to the definition of the presumptive RPE and NR territories, respectively [126].

11.4.1.2 D-V Patterning of the Optic Cup

Specification of the optic vesicle along its proximal-distal and D-V axis occurs as the optic vesicle folds over, generating the optic cup (Fig. 11.4). During this process, cells of the presumptive RPE spread along the entire cup also covering the dorsal portion of the retina. The OS, initially a ventral structure, shifts towards a central position in most vertebrate species except in birds where it remains in a more ventral location. The presumptive NR occupies the inner layer of the cup and spreads dorso-ventrally. As mentioned above, these movements establish a new position for most of the optic vesicle precursors, thus changing their initial D-V position. Once this rearrangement is accomplished, the optic cup undergoes a new wave of polarization causing cells of the dorsal optic cup to assume different characteristics from those occupying a ventral position.

A gradient of BMP signaling seems to be mainly responsible for this new pattern (Fig. 11.4). BMP4 is expressed in the dorsal retina of most vertebrate species analysed, where it contributes to the control proliferation and apoptosis of dorsal neural retina cells [134]. Misexpression of BMP4 in the ventral portion of the chick optic cup suppress the expression of *Pax2* and *cVax* and activates that of *Tbx5*, a transcription factor of the *T-box* class normally expressed only in the dorsal portion of the optic cup [135]. The counterbalance for this dorsalizing activity is ventroptin, a BMP signaling antagonist expressed in the ventral optic cup. Forced expression of ventroptin (*Vnt*) in the dorsal optic cup decreases BMP4 expression and increases that of *Vax* [136].

From this description, it would seem that the ventral character of the optic cup is a default state depending on the inhibition of the dorsalizing activity of BMP4. However, the story is not that simple, because inhibition of BMP signaling as a result of the overexpression of the BMP antagonist, noggin, in the ventral part of the chick optic cup causes colobomas, pecten agenesis, replacement of the ventral RPE by neuroepithelium-like tissue and ectopic expression of optic stalk markers in the region of the ventral retina and RPE [137]. Furthermore, there are several studies indicating that Retinoic Acid (RA) may contribute to the establishment of the ventral structures of the eye. Thus, enzymes involved in the synthesis and degradation of RA as well as RA receptors are expressed with complex but very polarized patterns within the eye (reviewed in [19]). RA treatment upregulates *Pax2* expression in the eye of zebrafish embryos, while deprivation of vitamin A during embryogenesis, inactivation of RA receptors or inhibition of RA synthesis all lead to embryos with clear defects in the ventral retina [19]. The importance of these findings has been recently challenged by the lack of alteration in D-V eye patterning observed in mice deficient in *Raldh1*, *Raldh2* and *Raldh3*, genes coding for RA synthesis enzymes

[138, 139]. Thus, both RA and BMPs may contribute to ventral optic cup specification but their relative contribution and precise action need further analysis.

Equally unclear is whether and how Shh signaling contributes to the imposition of a ventral character to the optic cup. Although Shh is needed to control the initial expression of *Pax2* and *Vax* genes in the optic vesicle, its forced dorsal expression at slightly later stages of development does not result in the ventralization of the dorsal optic cup, although BMP4 expression is downregulated. Surprisingly, defects in optic disc formation, *Otx2* activation and the appearance of pigmented cells in the neural retina were observed instead. Conversely, Shh blocking antibodies perturbed *Otx2* expression and pigment formation in the ventral RPE, leading to the loss of its characteristics [140]. Similarly, the cyclopamine-based inhibition of the hedgehog pathway in *Xenopus* embryos at the optic cup stage resulted in a lack of pigmentation and a decrease in the expression levels of RPE markers, particularly evident in the RPE periphery, where downstream components of the Hh pathway are expressed [141]. Thus, the Shh signaling pathway may contribute to the specification of the RPE cells in the ventral optic cup and of those that compose the optic fissure. The physiological source of the ligand might be different in the two cases. RPE cells may depend on a self-feeding mechanism since Hh-related proteins have been detected in the RPE of different vertebrate species [141, 142]. Development of optic stalk neuroepithelial cells instead seems to depend on Shh derived from the differentiating retina ganglion cells (RGC). In fact, conditional ablation of Shh in RGC causes a complete loss of optic disc astrocyte precursor cells [143].

Independent of the signaling mechanism, the transcriptional control of D-V polarization of the optic cup depends largely on the activity of *Tbx5* in the dorsal portion of the cup and *Vax* genes in the ventral part. Overexpression of *Tbx5* causes dorsalization of the optic cup [135] while overexpression of *Vax* leads to its ventralization [144]. The significance of D-V polarity in the neural retina at this stage and, thus, of the function of *Tbx5* and *Vax* genes, might be tightly linked to the establishment of the proper spatial order of the retino-tectal projections. In fact, overexpression studies suggest that *Tbx5* and *Vax1/2* directly control the expression of the Eph family of receptor tyrosine kinases and their Ephrin ligands which are axon guidance cues essential for target recognition of RGC axons (see e.g. [145]). Indeed, overexpression of *Tbx5* or *Vax* genes leads to alterations in the expression of EphrinB-EphB in the retina and is associated with misprojection of the RGC axons [135, 144]. Consistent with these data, *Vax2* null mice show converse alterations in the expression of EphrinB-EphB and defects in the retino-tectal projections [121, 122]. Inactivation of *Tbx5* does not provide support for the overexpression studies but the ocular phenotype of $Tbx5^{-/-}$ mice could be obscured by the presence of additional T-box genes such as *Tbx2* or *Tbx3* [146].

In conclusion, D-V patterning of the eye is a multi-step process with different functional purposes. At optic vesicle stage, acquisition of D-V identity is linked to the specification of the different territories that form the mature eye (Fig. 11.4). In a second step, definition of D-V polarity in the retina may be a prerequisite for the establishment of proper retino-tectal projections.

11.4.2
D-V Patterning of the Telencephalon

The two principal structures of the telencephalon are the cerebral cortex and the basal ganglia. Large projection neurons in the cortex derive from the dorsal telencephalon while most of the cortical interneurons are generated in the ventral telencephalon and subsequently migrate tangentially into the cortex. Neurons of the basal ganglia derive only from the ventral telencephalon. Possibly because of the morphological rearrangement during development and the structural complexity of the telencephalon, our understanding of its specification has progressed more slowly. Part of the genetic network required for this process has only recently been identified. Excellent and complete accounts of these findings can be found in [147, 148]. Here, we will only provide an outline.

As in the spinal cord, the main source of signals responsible for dorsal telencephalic patterning is the roof plate. Genetic ablation of the telencephalic dorsal midline results in a severely reduced expression of *Lhx2* [149], one of the transcription factors that together with *Emx1* and *Emx2* has a dorsal to ventral graded expression in the telencephalic primordium. Both *BMPs* and *Wnt* genes are strongly expressed in the dorsal midline of the telencephalon and thus are likely mediators of the roof plate activity (Fig. 11.5). Conditional inactivation of the BMPRIa impairs choroid plexus (a dorsal midline derivative) development but does not otherwise alter the telencephalic D-V patterning [150], suggesting that Wnt2b, whose expression is lost after ablation of the roof plate [149], may cooperate with BMPs to establish dorsal telencephalic identity. This idea is supported by the presence of a consensus binding site for Smads and Tcf proteins (ultimate downstream effectors of the BMP and Wnt signaling cascade respectively) in the promoter region of *Emx2* [151]. Thus, *Emx2* together with *Emx1* and *Lhx2* may act, in response to BMP and Wnt signaling, in the specification of the medial and dorsal pallium. This is supported by *Lhx2* mutants, which lack most of the hippocampus and neocortex [149], but it is less evident in *Emx* mutants, where the dorsal telencephalon is only mildly affected [152]. Gli3, a transcription factor of the zinc-finger family that normally inhibits Shh signaling, may contribute, upstream of *Emx*, to establishing a dorsal character in the telencephalon, since loss of *Gli3* function results in the absence of *Emx* expression and ectopic expression of ventral telencephalic markers including *Gsh2* [153] (Fig. 11.5).

The analysis of *Shh* null mice indicates that SHH is also involved in some aspect of ventral telencephalic patterning (Fig. 11.5) [95]. As for the optic vesicle, where only specification of the most ventro-medial component (optic stalk) is Shh dependent, *Shh* is required for ventro-medial (medial ganglionic eminence) telencephalic development, since the expression of *Nkx2.1*, a marker for the medial ganglionic eminence, is absent in *Shh* null mice while ventro-lateral (lateral ganglionic eminence) specification, determined by the presence of the *Gsh2* and *Dlx2* markers, is largely independent of Shh signaling [153]. Telencephalic ventral specification further requires a mutual antagonistic interaction between *Shh* and *Gli3* function, a mechanism also involved in the generation of the ventral-most cell types in the spinal cord. Thus, in $Shh^{-/-};Gli^{-/-}$ mouse embryos ventral specification in the spinal

A. D/V patterning of the mouse telencephalic vesicle at E11.5

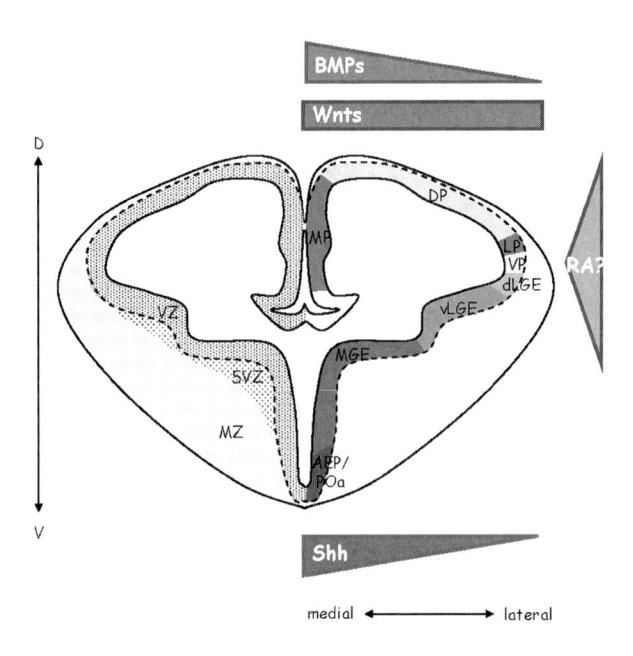

B. Determination of VZ progenitor domains and regionalization of the telencephalon

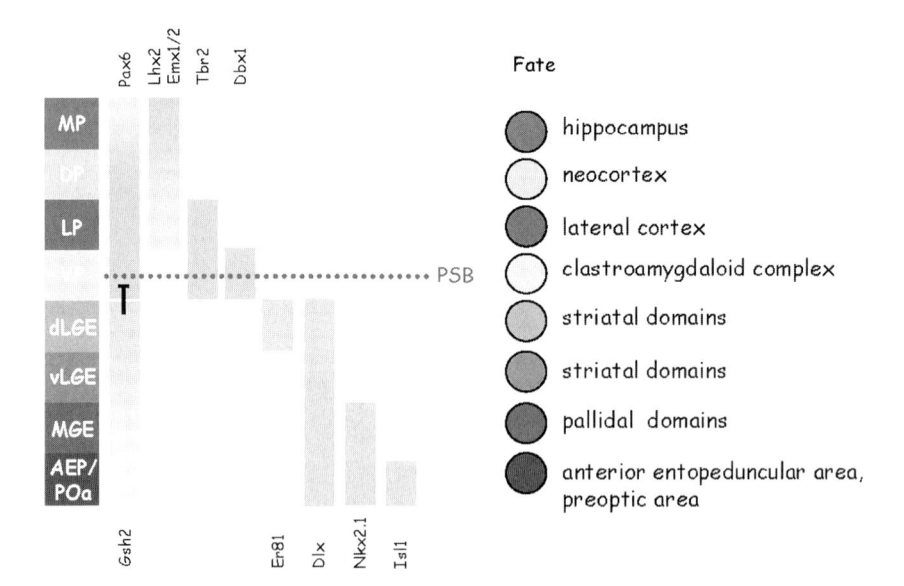

Fig. 11.5
Extracellular secreted signals and transcriptional codes patterning the telencephalon. (A) Schematic representation of the mouse telencephalic vesicle at E11.5 showing extrinsic signals involved in the establishment of the progenitor domains along the D-V and medio-lateral axis. (B) A color-coded bar represents telencephalic subdivisions on the left. In the middle are bar graphs representing gene expression patterns, as indicated at the top or bottom of each bar. Different levels of expression are represented by the shading. A horizontal broken line represents the PSB. On the right the fates of each telencephalic subdivision are described using the same color code. AEP, anterior entopeduncular area; DP, dorsal pallium; dLGE, dorsal lateral ganglionic eminence; vLGE, ventral ganglionic eminence; LP, lateral pallial mantle; MGE, medial ganglionic eminence; MP, medial pallium; MZ, marginal zone; Pd, pallidum; POa, preoptic area; PSB, pallial-subpallial boundary; St, striatum; SVZ, subventricular zone; VP, ventral pallium; VZ, ventricular zone. (This figure also appears with the color plates.)

cord and the telencephalon is partially rescued as compared to the ventral specification in $Shh^{+/-}$ embryos. Furthermore, patterning of the dorsal telencephalon of the double mutants is also improved over that of $Gli3$ single mutants, suggesting that an increase in Shh signaling may be responsible for the $Gli3$ mutant phenotype [153]. An interesting observation can be made from the analysis of $Shh^{+/-};Gli^{+/-}$ double mutants. Because in the absence of both genes the telencephalon is roughly properly patterned, other signals must contribute to the patterning process. Discovering their nature is one of the next challenges in understanding the development of the telencephalon.

A similarly unanswered question is the source of Shh involved in telencephalic specification. A recent study suggests that this might be linked to the node during gastrulation, implying that telencephalic ventral patterning may be initiated earlier than originally envisaged [154].

The intermediate region of the telencephalon includes the lateral ganglionic eminence (LGE), and the ventral and lateral pallium. Signals from both the dorsal and the ventral telencephalon may influence the specification of this region, but lateral-arising signals, including RA derived form mesenchymal cells of the olfactory placodes are likely to be involved (Fig. 11.5, reviewed in [148]). These would explain why the dorsal *Pax6* and ventral *Gsh2* graded expression is maximal in the medial region of the telencephalon at the pallio-subpallial boundary (PBS; Fig. 11.5). Downstream of any possible signal, the mutual repression of *Pax6* and *Gsh2* is required for positioning of the boundary, which, in turn, regulates the specification of the two intermediate telencephalic regions lying at each side of the boundary: the dorso-lateral ganglionic eminence and the ventral pallium. In the absence of *Pax6*, a number of dorso-lateral ganglionic eminence markers become ectopically expressed initially in the ventral pallium and then also in the dorsal pallium. The opposite is true in *Gsh2* mutants, where some ventral pallial markers are ectopically expressed in the lateral ganglionic eminence [155]. These alterations are not complete and other markers of D-V identity are expressed normally in these mutants, indicating that other molecules are involved. Again Gli3 is a likely candidate since *Gli3* mutants show alterations similar to those observed in the pallium of *Pax6* mutants [153]. Furthermore, in *Emx2$^{-/-}$;Pax6$^{-/-}$* double mutants the entire pallium loses its dorsal identity, a phenotype considerably more obvious than that observed in single mutants [156].

11.5
Conclusions and Perspectives

Work carried out to date has provided a substantial understanding of the mechanisms that control cell specification along the D-V axis. Some general principles have been established and a considerable number of key cellular and molecular players have been identified. Do we need to identify more inducers and effectors responsible for cell identity specification? Whilst this may well be useful, it is more essential to elucidate the details concerning the relationship between the molecules which have already been identified and to understand their interaction and regulation.

In this chapter, we have stressed the idea that the same signaling pathways are used along the entire A-P axis to determine dorsal or ventral cell fates. If the inducers are the same, it is clear that in many cases, different effectors are activated in different regions of the CNS according to the A-P axial level. This indicates that information imposed during the establishment of the A-P axis modifies the instructions received by D-V inductive signals in each cell. Possibly one of the most important challenges for the future is to understand how this information is integrated to produce progenitor domains that are unique in relation to their spatial position.

References

1 Tanabe Y, Jessell TM. Diversity and pattern in the developing spinal cord. *Science* **1996**; *274*: 1115–1123.
2 Krauss S, Concordet JP, Ingham PW. A functionally conserved homolog of the Drosophila segment polarity gene hh is expressed in tissues with polarizing activity in zebrafish embryos. *Cell* **1993**; *75*: 1431–1444.
3 Riddle RD, Johnson RL, Laufer E, Tabin C. Sonic hedgehog mediates the polarizing activity of the ZPA. *Cell* **1993**; *75*: 1401–1416.
4 Echelard Y, Epstein DJ, St-Jacques B, Shen L, Mohler J, McMahon JA, McMahon AP. Sonic hedgehog, a member of a family of putative signaling molecules, is implicated in the regulation of CNS polarity. *Cell* **1993**; *75*: 1417–1430.
5 Roelink H, Augsburger A, Heemskerk J, Korzh V, Norlin S, Ruiz i Altaba A, Tanabe Y, Placzek M, Edlund T, Jessell TM, et al. Floor plate and motor neuron induction by vhh–1, a ver-

tebrate homolog of hedgehog expressed by the notochord. *Cell* **1994**; *76*: 761–775.
6. Roelink H, Porter JA, Chiang C, Tanabe Y, Chang DT, Beachy PA, Jessell TM. Floor plate and motor neuron induction by different concentrations of the amino-terminal cleavage product of sonic hedgehog autoproteolysis. *Cell* **1995**; *81*: 445–455.
7. Marti E, Bumcrot DA, Takada R, McMahon AP. Requirement of 19K form of Sonic hedgehog for induction of distinct ventral cell types in CNS explants. *Nature* **1995**; *375*: 322–325.
8. Le Douarin NM, Kalcheim C. *The Neural Crest* (2nd edn). Cambridge, UK: Cambridge University Press, **1999**.
9. Helms AW, Johnson JE. Specification of dorsal spinal cord interneurons. *Curr Opin Neurobiol* **2003**; *13*: 42–49.
10. Lee KJ, Jessell TM. The specification of dorsal cell fates in the vertebrate nervous system. *Annu Rev Neurosci* **1999**; *22*: 261–294.
11. Santagati F, Rijli FM. Cranial neural crest and the building of the vertebrate head. *Nat Rev Neurosci* **2003**; *4*: 806–818.
12. Rollhauser-ter Host J. Artificial neural crest formation in amphibia. *Anat Embryol* **1979**; *157*: 113–120.
13. Moury JD, Jacobson AG. Neural fold formation at newly created boundaries between neural plate and epidermis in the axolot. *Dev Biol* **1989**; *133*: 44–57.
14. Selleck MAJ, Bronner-Fraser M. Origins of the avian neural crest: the role of neural plate-epidermal interactions. *Development* **1995**; *121*: 525–538.
15. Woo K, Fraser S. Specification of the hindbrain fate in the zebrafish. *Dev Biol* **1989**; *197*: 283–296.
16. Bronner-Fraser M, Fraser S. Cell linage analysis shows multipotentiality of some avian neural crest cells. *Nature* **1988**; *335*: 161–164.
17. Le Douarin NM, Dupin E. Multipotentiality of the neural crest. *Curr Opin Genet Dev* **2003**; *13*: 529–536.
18. Gammill LS, Bronner-Fraser M. Neural crest specification: migration into genomics. *Nat Rev Neurosci* **2003**; *4*: 795–805.
19. Ross SA, McCaffery PJ, Drager UC, De Luca LM. Retinoids in embryonal development. *Physiol Rev* **2000**; *80*: 1021–1054.
20. Basler K, Edlund T, Jessell TM, Yamada T. Control of cell pattern in the neural tube: regulation of cell differentiation by dorsalin-1, a novel TGF beta family member. *Cell* **1993**; *73*: 687–702.
21. Liem K, Tremml G, Roelink H, Jessell TM. Dorsal differentiation of neural plate cells induced by BMP4–mediated signals from epidermal ectoderm. *Cell* **1995**; *82*: 969–979.
22. Liem K, Tremml G, Jessell TM. A role for the roof plate and its resident TGFβ-related proteins in neuronal patterning in the dorsal spinal cord. *Cell* **1997**; *91*: 127–138.
23. Nguyen VH, Schmid B, Trout J, Connors SA, Ekker M, Mullins M. Ventral and lateral regions of the zebrafish gastrula, including the neural crest progenitors, are established by a bmp2b/swirl pathway of genes. *Dev Biol* **1998**; *199*: 93–110.
24. Nguyen VH, Trout J, Connors SA, Andermann P, Weinberg E, Mullins M. Dorsal and intermediate neuronal cell types of the

spinal cord are established by a BMP signaling pathway. *Development* **2000**; i: 1209–1220.

25 Streit A, Stern C. Establishment and maintenance of the border of the neural plate in the chick: involvement of FGF and BMP activity. *Mech Dev* **1999**; *82*: 51–66.

26 Selleck M, García-Castro M, Artinger K, Bronner-Fraser M. Effects of Shh and Noggin on neural crest formation demonstrate that BMP is required in the neural tube but not the ectoderm. *Development* **1998**; *125*: 4919–4930.

27 Sela-Donenfeld D, Kalcheim K. Regulation of the onset of neural crest emigration by coordinated activity of BMP4 and Noggin in the dorsal neural tube. *Development* **1999**; *126*: 4749–4762.

28 Wodarz A, Nusse R. Mechanisms of Wnt signaling in development. *Ann Rev Cell Dev Biol* **1998**; *14*: 59–88.

29 García-Castro MI, Marcelle C, Bronner-Fraser M. Ectodermal Wnt functions as a neural crest inducer. *Science* **2002**; *297*: 848–851.

30 Vallin, J., Thuret, R., Giacomello, E., Faraldo, M.M., Thiery, J.P. and Broders, F. Cloning and characterization of three Xenopus slug promoters reveal direct regulation by lef/beta-catenin signaling. *J Biol Chem* **2001**; *276*: 30350–30358.

31 Lewis JL, Bonner J, Modrell M, Ragland JW, Moon RT, Dorsky RI, Raible DW. Reiterated Wnt signaling during zebrafish neural crest development. *Development* **2004**; *131*: 1299–1308.

32 Cauthen CA, Berdougo E, Sandler J, Burrus L. Comparative analysis of the expression patterns of Wnts and Frizzleds during early myogenesis in chick embryos. *Mech Dev* **2001**; *104*: 133–138.

33 Schubert F, Mootoosamy R, Walters E, Graham A, Tumiotto L, Munsterberg A, Lumsden A, Dietrich S. Wnt6 marks sites of epithelial transformations in the chick embryo. *Mech Dev* **2002**; *114*: 143–148.

34 Wolda SL, Moon RT. Cloning and developmental expression in *Xenopus laevis* of seven additional members of the Wnt family. *Oncogene* **1992**; *7*: 1941–1947.

35 Gavin BJ, McMahon JA, McMahon AP. Expression of multiple novel Wnt1/Int–1 related genes during fetal and adult mouse development. *Genes Dev* **1990**; *4*: 2319–2332.

36 Ikeya M, Lee SM, Johnson JE, McMahon AP, Takada S. Wnt signaling required for expansion of neural crest and CNS progenitors. *Nature* **1997**; *389*: 966–970.

37 Megason SG, McMahon AP. A mitogen gradient of dorsal midline Wnts organizes growth in the CNS. *Development* **2002**; *129*: 2087–2098.

38 Yanfeng W, Saint-Jeannet J-P, Klein PS. Wnt-frizzled signaling in the induction and differentiation of neural crest. *BioEssays* **2003**; *25*: 317–325.

39 Nieto MA. The snail superfamily of zinc-finger transcription factors. *Nat Rev Mol Cell Biol* **2002**; *3*: 155–166.

40 Villanueva S, Glavic A, Ruiz P, Mayor R. Posteriorization by FGF, Wnt and retinoic acid is required for neural crest induction. *Dev Biol* **2002**; *241*: 289–301.

41 Monsoro-Burq AH, Fletcher RB, Harland RM. Neural crest induction by paraxial mesoderm in Xenopus embryos requires FGF signals. *Development* **2003**; *130*: 3111–3124.

42 Woda JM, Pastiaga J, Mercola M, Artinger KB. Dlx position the neural plate border and determine adjacent cell fates. *Development* **2003**; *130*: 331–342.

43 McLarren KW, Litsiou A, Streit A. DLX5 positions the neural crest and preplacode region at the border of the neural plate. *Dev Biol* **2003**; *259*: 34–47.

44 Suzuki A, Ueno N, Hemmati-Brivanlou A. Xenopus msx1 mediates epidermal induction and neural inhibition by BMP4. *Development* **1997**; *124*: 3037–3044.

45 Tríbulo C, Aybar MJ, Nguyen VH, Mullins M, Mayor R. Regulation of Msx genes by a Bmp gradient is essential for neural crest specification. *Development* **2003**; *130*: 6441–6452.

46 Cheung M, Briscoe J. Neural crest development is regulated by the transcription factor Sox9. *Development* **2003**; *130*: 5681–5693

47 Mori-Akiyama Y, Akiyama H, Rowitch DH, Crombrugghe B. Sox9 is required for determination of chondrogenic cell lineage in the cranial neural crest. *Proc Natl Acad Sci USA* **2003**; *100*: 9360–9365.

48 Liu JP, Jessell TM. A role for rhoB in the delamination of neural crest cells from the dorsal neural tube. *Development* **1998**; *125*: 5055–5067.

49 Britsch SE, Goerich D, Reithmacher D, Peirano RI, Rossner M, Nave KA, Brichmeier C, Wegner M. The transcription factor Sox10 is a key regulator of peripheral glia development. *Genes Dev* **2001**; *15*: 66–78.

50 Dutton KA, Pauliny A, Lopes SS, Elworth S, Carney TJ, Rauch J, Geisler R, Haffter P, Kelsh RN. Zebrafish colourless encodes Sox10 and specifies non-ectomesemchymal neural crest fates. *Development* **2001**; *128*: 4113–4125.

51 Kim J, Lo L, Dormand E, Anderson DJ. SOX10 maintains multipotency and inhibits neuronal differentiation of neurula crest stem cells. *Neuron* **2003**; *38*: 17–31.

52 Paratore C, Eichenberger C, Suter U, Sommer L. Sox10 haploinsufficiency affects maintenance of progenitor cells in a mouse model of Hirschsprung disease. *Hum Mol Genet* **2002**; *11*: 3075–3085.

53 Yan YL, Miller CT, Nissen RM, Singer A, Liu D, Kirn A, Draper B, Willoughby J, Morcos PA, Amsterdam A, Chung BC, Westerfield M, Haffter P, Hopkins N, Kimmel C, Postlethwait JH, Nissen R. A zebrafish sox9 gene required for cartilage morphogenesis. *Development* **2002**; *129*: 5065–5079.

54 Sopokony RF, Aoki Y, Saint-Germain N, Magner-Fink E, Saint-Jeannet J-P. The transcription factor Sox9 is required for cranial neural crest development. *Development* **2002**; *129*: 421–432.

55 Honore SM, Aybarm MJ, Mayor R. Sox10 is required for the early development of the prospective neural crest in Xenopus embryos. *Dev Biol* **2003**; *260*: 79–96.

56 Dottori M, Gross MK, Labosky P, Goulding M. The winged-helix transcription factor Foxd3 suppresses interneuron differ-

entiation and promotes neural crest cell fate. *Development* **2001**; *128*: 4127–4138.

57 Kos R, Reedy MV, Johnson RL, Erickson CA. The winged-helix transcription factor FoxD3 is important for establishing the neural crest lineage and repressing melanogenesis in avian embryos. *Development* **2001**; *128*: 1467–1479.

58 Brewster R, Lee J, Ruiz i Altaba A. Gli/Zic factors pattern the neural plate by defining domains of cell differentiation. *Nature* **1998**; *393*: 579–583.

59 Nakata K, Nagai T, Aruga J, Mikoshiba K. Xenopus Zic3, a primary regulator both in neural and neural crest development. *Proc Natl Acad Sci USA* **1997**; *94*: 11980–11985.

60 Nakata K, Koyabu Y, Aruga J, Mikoshiba K. A novel member of the Xenopus Zic family, Zic5, mediates neural crest development. *Mech Dev* **2000**; *99*: 83–91.

61 Mizuseki K, Kishi M, Matsui M, Nakanishi S, Sasay Y. Xenopus Zic-related1 and Sox2, two factors induced by chordin, have distinct activity in the initiation of neural induction. *Development* **1998**; *125*: 579–583.

62 Aruga J. The role of Zic genes in neural development. *Mol Cell Neurosci* **2004**; *26(2)*: 205–221.

63 Aruga J, Tohmonda T, Homma S, Mikoshiba K. Zic1 promotes the expansion of dorsal neural progenitors in spinal cord by inhibiting neuronal differentiation. *Dev Biol* **2002**; *244(2)*: 329–341.

64 Aruga J, Inoue T, Hoshino J, Mikoshiba K. Zic2 controls cerebellar development in cooperation with Zic1. *J Neurosci* **2002**; *22*: 218–225.

65 Inoue T, Hatayama M, Tohmonda T, Itohara S, Aruga J, Mikoshiba K. Mouse Zic5 deficiency results in neural tube defects and hypoplasia of cephalic neural crest derivatives. *Dev Biol* **2004**; *270*: 146–162.

66 Mansouri A, Gruss P. Pax3 and Pax7 are expressed in commissural neurons and restrict ventral neuronal identity in the spinal cord. *Mech Dev* **1998**; *78*: 171–178.

67 Epstein DJ, Vekemans M, Gros P. Splotch (Sp2H), a mutation affecting development of the mouse neural tube, shows a deletion within the paired homeodomain of Pax–3. *Cell* **1991**; *67*: 767–774.

68 Mansouri A, Stoykova A, Torres M, Gruss P. Dysgenesis of cephalic neural crest derivatives in Pax7$^{-/-}$ mutant mice. *Development* **1996**; *122*: 831–838.

69 Nieto MA. Early steps in neural crest induction. *Mech Dev* **2001**; *105*: 27–35.

70 Echelard Y, Vassileva G, McMahon AP. Cis-acting regulatory sequences governing Wnt1 expression in the developing mouse CNS. *Development* **1994**; *120*: 2213–2224.

71 Manzanares M, Trainor PA, Ariza-McNaughton L, Nonchev S, Krumlauf R. Dorsal patterning defects in the hindbrain, roof plate and skeleton in the dreher (dr(J)) mouse mutant. *Mech Dev* **2000**; *94*: 147–156.

72 Millonig JH, Millen KJ, Hatten ME. The mouse Dreher gene Lmx1a controls formation of the roof plate in the vertebrate CNS. *Nature* **2000**; *403*: 764–749.

73. Chizhikov VV, Millen KJ. Control of roof plate formation by Lmx1a in the developing spinal cord. *Development* **2004**; *131*: 2693–2705.
74. Chizhikov VV, Millen KJ. Control of roof plate development and signalling by Lmx1b in the caudal vertebrate CNS. *J Neurosci* **2004**; *24*: 5694–5703.
75. Lee KJ, Dietrich P, Jessell TM. Genetic ablation reveals that the roof plate is essential for dorsal interneuron specification. *Nature* **2000**; *403*: 734–740.
76. Muroyama Y, Fujihara M, Ikeya M, Kondoh H, Takada S. Wnt signaling plays an essential role in neuronal specification of the dorsal spinal cord. *Genes Dev* **2002**; *16*: 548–553.
77. Barth KA, Kishimoto Y, Rohr KB, Seydler C, Schulte-Merker S, Wilson SW. Bmp activity establishes a gradient of positional information throughout the entire neural plate. *Development* **1999**; *126*: 4977–4987.
78. Lee KJ, Mendelsohn M, Jessell TM. Neuronal patterning by BMPs: a requirement for GDF7 in the generation of a discrete class of commissural interneurons in the mouse spinal cord. *Genes Dev* **1998**; *12*: 3394–3407.
79. Timmer JR, Wang C, Niswander L. BMP signaling patterns the dorsal and intermediate neural tube via regulation of homeobox and helix-loop-helix transcription factors. *Development* **2002**; *129*: 2459–2472.
80. Panchision DM, Pickel JM, Studer L, Lee SH, Turner PA, Hazel TG, McKay RD. Sequential actions of BMP receptors control neural precursor cell production and fate. *Genes Dev* **2001**; *15*: 2094–2110.
81. Gross MK, Dottori M, Goulding M. Lbx1 specifies somatosensory association interneurons in the dorsal spinal cord. *Neuron* **2002**; *34*: 535–549.
82. Muller T, Brohmann H, Pierani A, Heppenstall PA, Lewin GR, Jessell TM, Birchmeier C. The homeodomain factor Lbx1 distinguishes two major programs of neuronal differentiation in the dorsal spinal cord. *Neuron* **2002**; *34*: 551–562.
83. Bertrand N, Castro DS, Guillemot F. Proneural genes and the specification of neural cell types. *Nat Rev Neurosci* **2002**; *3*: 517–530.
84. Nakada Y, Hunsaker TL, Henke RM, Johnson JE. Distinct domains within Mash1 and Math1 are required for function in neuronal differentiation versus neuronal cell-type specification. *Development* **2004**; *131*: 1319–1330.
85. Gowan K, Helms AW, Hunsaker TL, Collisson T, Ebert PJ, Odom R, Johnson JE. Cross-inhibitory activities of Ngn1 and Math1 allow specification of distinct dorsal interneurons. *Neuron* **2001**; *31*: 219–232.
86. Liu Y, Helms AW, Johnson J. Distinct activities of Msx1 and Msx3 in dorsal neural tube development. *Development* **2004**; *131*: 1017–1028.
87. Scardigli R, Schuurmans C, Gradwohl G, Guillemot F. Cross-regulation between neurogenin2 and pathways specifying neuronal identity in the spinal cord. *Neuron* **2001**; *31*: 203–217.
88. Kennedy TE, Serafini T, de la Torre JR, Tessier-Lavigne M. Netrins are diffusible chemotropic factors for commissural axons in the embryonic spinal cord. *Cell* **1994**; *78*: 425–435.

89 Charron F, Stein E, Jeong J, McMahon AP, Tessier-Lavigne M. The morphogen sonic hedgehog is an axonal chemoattractant that collaborates with netrin–1 in midline axon guidance. *Cell* **2003**; *113*: 11–23.

90 Charrier JB, Lapointe F, Le Douarin NM, Teillet MA. Dual origin of the floor plate in the avian embryo. *Development* **2002**; *129*: 4785–4796.

91 Strahle U, Lam CS, Ertzer R, Rastegar S. Vertebrate floor-plate specification: variations on common themes. *Trends Genet* **2004**; *20*: 155–162.

92 Patten I, Kulesa P, Shen MM, Fraser S, Placzek M. Distinct modes of floor plate induction in the chick embryo. *Development* **2003**; *130*: 4809–4821.

93 Marti E, Takada R, Bumcrot DA, Sasaki H, McMahon AP. Distribution of Sonic hedgehog peptides in the developing chick and mouse embryo. *Development* **1995**; *120*: 2537–2547.

94 Ericson J, Morton S, Kawakami A, Roelink H, Jessell TM. Two critical periods of Sonic Hedgehog signaling required for the specification of motor neuron identity. *Cell* **1996**; *87*: 661–673.

95 Chiang C, Litingtung Y, Lee E, Young KE, Corden JL, Westphal H, Beachy PA. Cyclopia and defective axial patterning in mice lacking Sonic hedgehog gene function. *Nature* **1996**; *383*: 407–413.

96 Jessell TM. Neuronal specification in the spinal cord: inductive signals and transcriptional codes. *Nat Rev Genet* **2000**; *1*: 20–29.

97 Briscoe J, Ericson J. Specification of neuronal fates in the ventral neural tube. *Curr Opin Neurobiol* **2001**; *11*: 43–49.

98 Briscoe J, Pierani A, Jessell TM, Ericson J. A homeodomain protein code specifies progenitor cell identity and neuronal fate in the ventral neural tube. *Cell* **2000**; *101*: 435–445.

99 Ericson J, Rashbass P, Schedl A, Brenner-Morton S, Kawakami A, van Heyningen V, Jessell TM, Briscoe J. Pax6 controls progenitor cell identity and neuronal fate in response to graded Shh signaling. *Cell* **1997**; *90*: 169–180.

100 Briscoe J, Sussel L, Serup P, Hartigan-O'Connor D, Jessell TM, Rubenstein JL, Ericson J. Homeobox gene Nkx2.2 and specification of neuronal identity by graded Sonic hedgehog signalling. *Nature* **1999**; *398*: 622–627.

101 Sander M, Paydar S, Ericson J, Briscoe J, Berber E, German M, Jessell TM, Rubenstein JL. Ventral neural patterning by Nkx homeobox genes: Nkx6.1 controls somatic motor neuron and ventral interneuron fates. *Genes Dev* **2000**; *14*: 2134–2139.

102 Muhr J, Andersson E, Persson M, Jessell TM, Ericson J. Groucho-mediated transcriptional repression establishes progenitor cell pattern and neuronal fate in the ventral neural tube. *Cell* **2001**; *104*: 861–873.

103 Vallstedt A, Muhr J, Pattyn A, Pierani A, Mendelsohn M, Sander M, Jessell TM, Ericson J. Different levels of repressor activity assign redundant and specific roles to Nkx6 genes in motor and interneuron specification. *Neuron* **2001**; *31*: 743–755.

104 Zhou Q, Anderson DJ. The bHLH transcription factors OLIG2 and OLIG1 couple neuroneal and glial subtype specification. *Cell* **2002**; *109*: 61–73.

105 Lee SK, Pfaff SL. Synchronization of neurogenesis and motor neuron specification by direct coupling of bHLH and homeodomain transcription factors. *Neuron* **2003**; *38*: 731–745.
106 Lee SK, Jurata LW, Funahashi J, Ruiz EC, Pfaff SL. Analysis of embryonic motoneuron gene regulation: derepression of general activators function in concert with enhancer factors. *Development* **2004**; *131*: 3295–3306.
107 Thaler JP, Lee SK, Jurata LW, Gill GN, Pfaff SL. LIM factor Lhx3 contributes to the specification of motor neuron and interneuron identity through cell-type-specific protein-protein interactions. *Cell* **2002**; *110*: 237–249.
108 Pierani A, Moran-Rivard L, Sunshine MJ, Littman DR, Goulding M, Jessell TM. Control of interneuron fate in the developing spinal cord by the progenitor homeodomain protein Dbx1. *Neuron* **2001**; *29*: 367–384.
109 Wilson SW, Houart C. Early steps in the development of the forebrain. *Dev Cell* **2004**; *6*: 167–181.
110 Lumsden A. Segmentation and compartition in the early avian hindbrain. *Mech Dev* **2004**; *121*: 1081–1088.
111 Chow RL, Lang RA. Early eye development in vertebrates. *Annu Rev Cell Dev Biol* **2001**; *17*: 255–296.
112 Redies C, Puelles L. Modularity in vertebrate brain development and evolution. *BioEssays* **2001**; *23*: 1100–1111.
113 Houart C, Caneparo L, Heisenberg C, Barth K, Take-Uchi M, Wilson S. Establishment of the telencephalon during gastrulation by local antagonism of Wnt signaling. *Neuron* **2002**; *35*: 255–265.
114 Esteve P, López-Ríos J, Bovolenta, P. Sfrp1 is required for the proper establishment of the eye field in the medaka fish. *Mech Dev* **2004**; *121*: 687–701.
115 Zuber ME, Gestri G, Viczian AS, Barsacchi G, Harris WA. Specification of the vertebrate eye by a network of eye field transcription factors. *Development* **2003**; *130*: 5155–5167.
116 Marti E, Bovolenta P. Sonic hedgehog in CNS development: one signal, multiple outputs. *Trends Neurosci* **2002**; *25*: 89–96.
117 Take-uchi M, Clarke JD, Wilson SW. Hedgehog signaling maintains the optic stalk-retinal interface through the regulation of Vax gene activity. *Development* **2003**; *130*: 955–968.
118 Torres M, Gomez-Pardo E, Gruss P. Pax2 contributes to inner ear patterning and optic nerve trajectory. *Development* **1996**; *122*: 3381–3391.
119 Bertuzzi S, Hindges R, Mui SH, O'Leary DD, Lemke G. The homeodomain protein vax1 is required for axon guidance and major tract formation in the developing forebrain. *Genes Dev* **1999**; *13*: 3092–3105.
120 Hallonet M, Hollemann T, Pieler T, Gruss P. Vax1, a novel homeobox-containing gene, directs development of the basal forebrain and visual system. *Genes Dev* **1999**; *13*: 3106–3114.
121 Barbieri AM, Broccoli V, Bovolenta P, Alfano G, Marchitiello A, Mocchetti C, Crippa L, Bulfone A, Marigo V, Ballabio A, Banfi S. Vax2 inactivation in mouse determines alteration of the eye dorsal-ventral axis, misrouting of the optic fibres and eye coloboma. *Development* **2002**; *129*: 805–813.

122 Mui SH, Hindges R, O'Leary DD, Lemke G, Bertuzzi S. The homeodomain protein Vax2 patterns the dorsoventral and nasotemporal axes of the eye. *Development* **2002**; *129*: 797–804.

123 Schwarz M, Cecconi F, Bernier G, Andrejewski N, Kammandel B, Wagner M, Gruss P. Spatial specification of mammalian eye territories by reciprocal transcriptional repression of Pax2 and Pax6. *Development* **2000**; *127*: 4325–4334.

124 Martínez-Morales JR, Rodrigo I, Bovolenta P. Eye development: a view from the pigmented epithelium. *BioEssays* **2004**; *26*: 766–777.

125 Martinez-Morales JR, Signore M, Acampora D, Simeone A, and Bovolenta P. Otx genes are required for tissue specification in the developing eye. *Development* **2001**; *128*: 2019–2030.

126 Nguyen M, and Arnheiter H. Signaling and transcriptional regulation in early mammalian eye development: a link between FGF and MITF. *Development* **2000**; *127*: 3581–3591.

127 Martínez-Morales JR, Dolez V, Rodrigo I, Zaccarini R, Leconte L, Bovolenta P, Saule S. OTX2 activates the molecular network underlying retina pigment epithelium differentiation. *J Biol Chem* **2003**; *278*: 21721–21731.

128 Rodríguez de Córdoba S, Gallardo ME, Lopez-Rios J, Bovolenta P. The human Six family of homeobox genes. *Curr Genomics* **2001**; *2*: 231–242.

129 Toy J, Yang JM, Leppert GS, Sundin OH. The optx2 homeobox gene is expressed in early precursors of the eye and activates retina-specific genes. *Proc Natl Acad Sci USA* **1998**; *95*: 10643–10648.

130 Porter FD, Drago J, Xu Y, Cheema SS, Wassif C, Huang SP, Lee E, Grinberg A, Massalas JS, Bodine D, Alt F, Westphal H. Lhx2, a LIM homeobox gene, is required for eye, forebrain, and definitive erythrocyte development. *Development* **1997**; *124*: 2935–2944.

131 Grindley JC, Davidson DR, Hill RE. The role of Pax–6 in eye and nasal development. *Development* **1995**; *121*: 1433–1442.

132 Liu IS, Chen JD, Ploder L, Vidgen D, van der Kooy D, Kalnins VI, McInnes RR. Developmental expression of a novel murine homeobox gene (Chx10): evidence for roles in determination of the neuroretina and inner nuclear layer. *Neuron* **1994**; *13*: 377–393.

133 Burmeister M, Novak J, Liang MY, Basu S, Ploder L, Hawes NL, Vidgen D, Hoover F, Goldman D, Kalnins VI, Roderick TH, Taylor BA, Hankin MH, McInnes RR. Ocular retardation mouse caused by Chx10 homeobox null allele: impaired retinal progenitor proliferation and bipolar cell differentiation. *Nat Genet* **1996**; *12*: 376–384.

134 Trousse F, Esteve P, Bovolenta P. BMP4 mediates apoptotic cell death and proliferation in the chick optic cup. *J Neurosci* **2001**; *21*: 1292–1301.

135 Koshiba-Takeuchi K, Takeuchi JK, Matsumoto K, Momose T, Uno K, Hoepker V, Ogura K, Takahashi N, Nakamura H, Yasuda K, et al. Tbx5 and the retinotectum projection. *Science* **2000**; *287*: 134–137.

136 Sakuta H, Suzuki R, Takahashi H, Kato A, Shintani T, Iemura S, Yamamoto TS, Ueno N, Noda M. Ventroptin: a BMP4 an-

tagonist expressed in a double-gradient pattern in the retina. *Science* **2001**; *293*: 111–115.

137 Belecky-Adams TL, Adler R, Beebe DC. Bone morphogenetic protein signaling and the initiation of lens fiber cell differentiation. *Development* **2002**; *129*: 3795–3802.

138 Dupe V, Matt N, Garnier JM, Chambon P, Mark M, Ghyselinck NB. A newborn lethal defect due to inactivation of retinaldehyde dehydrogenase type 3 is prevented by maternal retinoic acid treatment. *Proc Natl Acad Sci USA* **2003**; *100*: 14036–14041.

139 Fan X, Molotkov A, Manabe S, Donmoyer CM, Deltour L, Foglio MH, Cuenca AE, Blaner WS, Lipto SA, Duester G. Targeted disruption of Aldh1a1 (Raldh1) provides evidence for a complex mechanism of retinoic acid synthesis in the developing retina. *Mol Cell Biol* **2003**; *23*: 4637–4648.

140 Zhang XM, Yang XJ. Temporal and spatial effects of Sonic hedgehog signaling in chick eye morphogenesis. *Dev Biol* **2001**; *233*: 271–290.

141 Perron M, Boy S, Amato MA, Viczian A, Koebernick K, Pieler T, Harris WA. A novel function for Hedgehog signaling in retinal pigment epithelium differentiation. *Development* **2003**; *130*: 1565–1577.

142 Stenkamp DL, Frey RA, Prabhudesai SN, Raymond PA. Function for Hedgehog genes in zebrafish retinal development. *Dev Biol* **2000**; *220*: 238–252.

143 Dakubo GD, Wang YP, Mazerolle C, Campsall K, McMahon AP, Wallace VA. Retinal ganglion cell-derived sonic hedgehog signaling is required for optic disc and stalk neuroepithelial cell development. *Development* **2003**; *130*: 2967–2980.

144 Schulte D, Furukawa T, Peters MA, Kozak CA, Cepko CL Misexpression of the *Emx*-related homeobox genes *cVax* and *mVax2* ventralizes the retina and perturbs the retinotectal map. *Neuron* **1999**; *24*: 541–553.

145 Godement P, Mason CA. It's all in the assay: a new model for retinotectal topographic mapping. *Neuron* **2004**; *42*: 697–699.

146 Takabatake Y, Takabatake T, Takeshima K. Conserved and divergent expression of T-box genes Tbx2–Tbx5 in Xenopus. *Mech Dev* **2000**; *91*: 433–437.

147 Wilson SW, Rubinstein, JL. Induction and dorsoventral patterning of the telencephalon. *Neuron* **2000**; *28*: 641–651.

148 Campbell K. Dorsal-ventral patterning in the mammalian telencephalon. *Curr Opin Neurobiol* **2003**; *13*: 50–56.

149 Monuki ES, Porter FD, Walsh CA. Patterning of the dorsal telencephalon and cerebral cortex by a roof plate-Lhx2 pathway. *Neuron* **2001**; *32*: 591–604.

150 Hebert J, Mishina Y, McConnell S. BMP signaling is required locally to pattern the dorsal telencephalic midline. *Neuron* **2002**; *35*: 1029.

151 Theil T, Aydin S, Koch S, Grotewold L, Ruther U. Wnt and Bmp signaling cooperatively regulate graded Emx2 expression in the dorsal telencephalon. *Development* **2002**; *129*: 3045–3054.

152 Yoshida M, Suda Y, Matsuo I, Miyamoto N, Takeda N, Kuratani S, Aizawa S. Emx1 and Emx2 functions in development of dorsal telencephalon. *Development* **1997**; *124*: 101–111.

153 Rallu M, Machold RP, Gaiano N, Corbin JG, McMahon AP, Fishell G. Dorsoventral patterning is established in the telencephalon of mutants lacking both Gli3 and Hedgehog signaling. *Development* **2002**; *129*: 4963–4974.

154 Gunhaga L, Jessell TM, Edlund T Sonic hedgehog signalling at gastrula stages specifies ventral telencephalic cells in the chick embryo. *Development* **2000**; *127*: 3283–3293.

155 Yun K, Potter S, Rubenstein JL. Gsh2 and Pax6 play complementary roles in dorsoventral patterning of the mammalian telencephalon. *Development* **2001**; *128*: 193–205.

156 Muzio L, DiBenedetto B, Stoykova A, Boncinelli E, Gruss P, Mallamaci A. Conversion of cerebral cortex into basal ganglia in Emx2$^{-/-}$ Pax6Sey/Sey double-mutant mice. *Nat Neurosci* **2002**; *5*: 737–745.

12
Novel Perspectives in Research on the Neural Crest and its Derivatives

Chaya Kalcheim, Matthias Stanke, Hermann Rohrer, Kristjan Jessen, and Rhona Mirsky

12.1
General Introduction

The neural crest (NC) is a vertebrate-specific cell population that arises at the border between neural plate and epidermal ectoderm. Although being a discrete structure with transient existence in the early embryo, it gives rise to an extraordinary variety of cell types, from peripheral neurons of many kinds, glial and endocrine cells to bones, tendons, connective, dermal and adipose tissues, and also to melanocytes [1].

The NC of vertebrates probably evolved from cells on either side of the neural plate-epidermal boundary in protochordate ancestors. In amphioxus, for example, homologs of several vertebrate NC marker genes are expressed precisely at these sites. Some of these markers are also similarly expressed in tunicates. In contrast to vertebrates, however, these marker-expressing cells neither migrate nor differentiate into a wide spectrum of cell types, suggesting that only some of the features that characterize the emergence of the NC were initially present [2]. In basal vertebrates such as the lamprey, migration of NC cells takes place and similarities with the behavior of NC cells in gnathostomes was observed. Yet, significant differences in migratory pathways were also unraveled as well as the lack of some known vertebrate derivatives such as cranial sensory ganglia [3].

Since its first description in the avian embryo by His [4, 5] as a *"Zwischenstrang"*, a strip of cells lying between the dorsal ectoderm and the neural tube, the NC has attracted the attention of many generations of researchers. Particular to the NC are its structural and lineage relations with neuroepithelial progenitors that will generate the entire central nervous system, their dramatic conversion from a pseudostratified epithelial structure to become mesenchymal and migratory, their often long and tortuous, yet accurate routes of stereotypic migrations through a diversity of embryonic sites; and finally, their dramatic diversification into many differentiated cell types (reviewed in [1]). The classical contributions resulting from the pioneering work of His and later of Newth were reviewed in a comprehensive monograph by Hörstadius [6]. After these early studies, the ontogeny of the NC remained largely unresolved for many years. In the 1960s Chibon and Weston took advantage of the proliferative property of these cells to directly track their migration by labeling

Cell Signaling and Growth Factors in Development. Edited by K. Unsicker and K. Krieglstein
Copyright © 2006 WILEY-VCH Verlag GmbH & Co. KGaA, Weinheim
ISBN 3-527-31034-7

S-phase nuclei with tritiated thymidine in association with the grafting of neural primordia. This technique enabled only a transient follow-up of NC progenitors, a limitation that was later overcome with the introduction of the quail-chick marker by Le Douarin [7]. The latter method provided a stable means of tracing the migration and fate of NC progenitors throughout embryonic development.

Past years have witnessed a dramatic shift from descriptive and experimental embryology to the elucidation of molecular mechanisms underlying the development of the NC and its derivatives. A wealth of information is now available on genes that control the various stages of NC ontogeny and work in progress aims now at synthesizing cellular and molecular perspectives into an integrated picture of NC development. This chapter attempts to present such an integrated picture.

12.2
Novel Techniques to Investigate NC Development and Fate

12.2.1
Cellular Tracers

Because during their early ontogeny, NC cells are intimately associated with neighboring cells of either the neuroepithelium or their migratory pathways, it was essential to devise techniques that differentially label this cell population. Classical methods were all discussed in Le Douarin and Kalcheim [1] and will therefore be only briefly mentioned. These include direct cell visualization in some amphibia and fish where NC cells are characterized by their comparatively larger size; these studies were complemented by extirpation experiments and by *in situ* destruction of NC cells followed by analysis of the resulting phenotypes.

Vital dye labeling makes it possible to track the development of the NC in normal and experimental embryos. Lineage tracers such as Lysinated rhodamine dextran served to mark the fate of individual cells *in ovo* [8]. Lipophilic dyes such as DiI and DiO permitted a transient follow-up of migratory progenitors in whole mounts and sectioned material. When applied to relatively transparent embryos in association with time lapse cinematography, a dynamic picture of NC behavior was obtained [9–12].

Of all the techniques, the advent of the quail-chick marker system mentioned previously was a landmark in NC research. It set the scene for a wealth of studies that clarified the migratory routes and fates of NC cells along the entire axis of avian embryos. This already classical technique also provided significant insights into the developmental potential of NC cells which was found to be broader than the actually expressed fates. From such studies the widely accepted notion emerged that the NC is a highly multipotent structure at least when considered at the population level. This concept was subsequently challenged by clonal analysis of NC progenitors *in vitro*. Although confirming the multipotency of many NC cells, these experiments also highlighted the existence of restricted progenitors including cells that are specified to unique fates from early stages onwards.

12.2.2
Molecules Expressed by NC Cells

Elucidation of molecules differentially expressed by NC cells constituted a leap forward not only in our ability to identify spatio-temporally distinct cell subsets but also to begin to understand gene function in NC development (see below). The initial contribution to this subject was provided with the advent of monoclonal antibodies. Perhaps the most notable among the many NC-specific markers found in this way is the HNK–1/NC–1 sugar epitope that is carried by proteins and lipids and that labels large sets of migrating and differentiating NC cells.

Presently, NC cell subsets can be defined in different species, axial levels and stages by a variety of specific transcription factors (FoxD3, various Sox genes, AP2, Slug or Snail, Neurogenins, PAX 3 and 7, Mitf, etc.) [13–23], growth factor receptors (neuropilin–2, Trks, p75, c-kit, Robos, erbB, receptors to BMPs, Wnts, etc.; reviewed in [1], see also [24–30]) and by expression of specific secreted factors (BMPs, Wnts, Noelin, etc. [31–37]) or cell adhesion molecules such as N-cadherin, cadherin 6B, 7 and 11 [38–43].

12.2.3
Investigating the Function of Molecules Expressed by NC Cells

Elucidating the functions of these gene products in the development of the NC is clearly a major challenge. To achieve this goal, a diversity of technologies is currently being exploited in various animal species. A large number of NC mutants have been identified in zebrafish screens and analysis of gene function is underway. Research on NC in Xenopus embryos has greatly benefited from antisense RNA technologies. Constitutive and inducible overexpression experiments are routine in this species. Here, only a few approaches applicable to mice and avian embryos will be discussed. Genetic technologies in mice offer the possibility to overexpress or to abrogate gene activity and to determine their function during embryogenesis [44]. For instance, combined inactivation of *Wnt1* and *Wnt3a*, both expressed in the dorsal midline of the neural tube, has revealed a role for these factors in NC expansion [45]. These results were further confirmed in cultures of mammalian NC infected with Wnt1 or β-catenin-expressing RCAS viruses [46]. Inactivation of neurotrophins and their receptors demonstrated their significant function in survival/ differentiation of sensory and/or sympathetic lineages [1]. Knockout of *endothelin 3* and of *endothelin receptor B* revealed their importance in the development of the enteric nervous system and that of NC-derived melanocytes [47, 48]. Embryos lacking *Neurogenin1* function failed to generate the proximal cranial sensory ganglia that comprise the trigeminal, vestibulo-cochlear, accessory, jugular and superior ganglia [22]. Complementary to this phenotype, deletion of *Neurogenin 2* resulted in elimination of the distal cranial ganglia including the geniculate and petrosal with no apparent effect on the proximal ganglia [21]. Notably, the nodose ganglion which expresses both *Neurogenins* was spared in the single mutants of either type, suggesting a mutual compensation of gene activity [21].

The use of transgenic mice expressing Cre recombinase under the control of spatio-temporally regulated promoters is a powerful tool to promote conditional deletion of floxed genes. A growing number of transgenic Cre-lines is becoming available [49], among which some are potentially useful for abolishing NC-specific genes [50–57]. For example, β-catenin was inactivated using Wnt1–Cre-mediated deletion revealing the importance of the former gene in craniofacial, sensory and melanocyte development [58, 59]. An additional application of transgenic lines in which β-galactosidase expression is driven by Cre recombinase under the control of specific NC promoters is their utilization as lineage tracers to follow NC cell fates. Studies of this kind have used promoters such as the *human plasminogen activator* [57], *Wnt1* [50], and *Neurogenin 2* [60].

The avian embryo is an organism well suited for experimental embryology and indeed much of our knowledge on NC development stems from studies in chick and quail embryos. Until recently, however, progress towards understanding molecular mechanisms was hampered in these species because gene misexpression was limited to antibody delivery or to viral infections. The newly introduced technique of *in ovo electroporation*, which uses pulses of electrical current to transfect DNA into cells, is being widely employed already to generate gene expression via constitutive promoters [61, 62]. Foreign genes can thus be transfected into the neural primordium, becoming expressed in pre-migratory NC cells as early as 3 to 5 h following electroporation and their activities can be monitored as a function of time [63, 15–17]. Additional advantages of this technique include the ability to electroporate only a hemi-neural tube therefore using the contralateral tube as a matching control. Electroporation also offers the possibility to transfect genes at discrete time points which is particularly important for factors exerting multiple activities at various stages of NC ontogeny. Recent reports indicated that this technique could also be used to generate restricted patterns of gene expression. Hox gene enhancers were used to drive restricted expression of reporter genes along the rostrocaudal extent of the hindbrain [64]. Similarly, murine enhancers were introduced that restricted expression of reporter genes to the dorsoventral aspect of the tube and enabled rapid evaluation of novel vertebrate enhancer sequences [65].

An additional method with promising prospects for NC research is the use of RNA interference (RNAi) to inhibit gene activity in the avian neural tube. RNAi-mediated gene knock-down is a useful and rapid alternative to gene knockouts in plants, worms and vertebrate systems. RNAi is caused by sequence-specific mRNA degradation and is mediated by small interfering RNAs (siRNAs). siRNAs typically consist of two 21–nt single-stranded RNAs that form a 19–bp duplex with 2–nt 3' overhangs. When introduced into cells, these siRNAs are recruited into yet-to-be-identified proteins of an endonuclease complex (RNA-induced silencing complex), which then guides target mRNA cleavage [66]. Until now, most research was dedicated to studying RNAi in invertebrate and mammalian systems. Development of a convenient system for testing RNAi using electroporation of chick embryos was recently reported. In this research, siRNA against the green fluorescent protein (*gfp*) gene inhibited expression of GFP in the neural tube of embryos. Moreover, RNAi inhibited the replication of a nonpathogenic derivative of Rous sarcoma virus (RSV). These results provide evidence that chick embryos, in particular cells in the developing neural tube/NC, contain the machinery to use siRNAs as a substrate for the targeted degradation of mRNAs [67, 68].

Morpholino antisense oligonucleotides are DNA analogs in which a morpholine ring has been substituted for the riboside moiety. They function by blocking translation, rather than mediating RNAse H-dependent degradation as is the case for DNA oligos. They are long lived and of low toxicity thus opening additional avenues for gene knock-down in embryos. In contrast to other systems such as frog or zebrafish that permit direct microinjection into blastomeres, square pulse electroporation was found to be the best method for efficient delivery into the neural tube of avians [69]. Use of this technique confirmed the negative role of FoxD3 in induction of melanogenesis (compare [17] with [69]). Furthermore, an endogenous effect of metalloproteinase–2 in migration of NC cells was shown both upon use of a specific inhibitor as well as with morpholino inhibition ([69] and references therein). Likewise knock-down of Tenascin-C expression resulted in retention of NC cells in the dorsal tube and its lumen and failure of cell dispersion, thus extending similar results stemming from previous antibody perturbation studies [70].

12.3
Mechanisms of Specification and Emigration of NC Cells

12.3.1
Specification of the NC

During neurulation, the embryonic ectoderm becomes subdivided into the neural plate and prospective epidermis. The boundary region between these tissues becomes the NC, as defined by expression of a variety of specific markers. These include transcription factors such as *Slug* or *Snail, AP–2, Foxd3, PAX3, twist, Sox9, Zic5,* etc. which appear in different species at changing rostrocaudal levels of the axis. Interactions between the epidermal ectoderm and neural plate, and contribution of mesodermal signals were found to underlie early expression of these NC-specific traits. Evidence obtained primarily in Xenopus embryos suggests that two independent signals mediate these interactions at the various phases of the process, a BMP signal which must be modulated by its inhibitors and separate inputs that can be either a canonical Wnt signal, FGF or retinoic acid. The involvement of these factors and factor combinations was documented and recently reviewed [13, 71, 72]. Hence, we shall only consider new data on the possible role of specific transcription factors in NC formation.

12.3.1.1 Transcription Factors in NC Formation

Snail family members that include the *Snail* and *Slug* transcription factors are among the earliest markers of NC development and therefore their expression is widely used as indicative of NC formation. Notably, it was shown that cells in the dorsal neural folds which express *Slug* have the potential to give rise to both neural tube and NC lineages [73, 74], and *Slug* transcription is downregulated in the trunk at levels where pre-migratory NC is still being produced [36], suggesting that this factor is not a bona fide marker for early NC. The involvement of *Slug* in the production of NC cells was first documented in the chick by application of antisense oligonucleotides which prevented cell emigration perhaps due to an earlier defect in their specification [19]. Conversely, overexpression of *Slug* led to an increase in the production of cranial but not of trunk-level NC cells [75]. In the mouse, an exchange between expression of *Slug* and *Snail* in the crest domain took place when compared to avian embryos [76, 20], suggesting that Snail functions in the mouse as Slug does in the chick. Functional evidence for a possible role of Snail in NC formation in mice is, however, still lacking due to early lethality of the embryos by the time of gastrulation [77]. The strongest evidence for an effect of these factors in NC formation stems from recent work in Xenopus. Inhibition of Slug function at early stages was shown to prevent the formation of crest progenitors while inhibition at later stages interfered with cell emigration [78]. The ability to temporally dissociate between these two sequential events makes the argument more compelling. Aybar et al. [79] have further elaborated on this question and recently showed that *Snail* is also required for NC specification in Xenopus and moreover, is able to induce transcription of *Slug* suggesting that *Snail* lies upstream of *Slug* in a genetic cascade leading to the formation of NC cells. The expression of *Slug* and *Snail* at the neural plate border is preceded by that of the proto-oncogene *c-myc*. Morpholino-mediated depletion of c-myc protein led to embryos that failed to express NC-specific markers and developed without a peripheral nervous system and other NC derivatives [80].

In addition to *Slug*, the cascade of *Snail*-induced genes was shown to comprise *Zic5, FoxD3, Twist* and *Ets1* [79]. Consistent with the notion that these genes are also part of the NC-producing genetic repertoire, it was shown that *Zic5* overexpression induces NC markers, including *Slug*, at the expense of epidermal markers [81]. A family of transcription factors that has been shown to have important functions in cell specification and lineage segregation is the winged-helix or forkhead class [82], recently renamed Fox proteins for the forkhead box [83]. One member of the family was cloned in the chick, *FoxD3*, and shown to be expressed in the neural folds and later in early migrating NC [16, 17]. Overexpression in the neural tube consistently led to a widening of the HNK–1–positive domain and concomitant suppression of interneuron development, suggesting that *FoxD3* biases neuroepithelial progenitors towards a NC fate. In Xenopus, *FoxD3* which was suggested to be a downstream signal of both *Snail* and *Slug* [79] was found to act as a transcriptional repressor, downregulating *Slug* and also *Cadherin 11* as well as its own expression. Likewise, NC formation was reduced [84]. These results are difficult to interpret in light of a positive effect of *Snail* and *Slug* on NC production and of the timely expression of *FoxD3* in the presumptive NC territory.

SOX genes encode a family of proteins containing a 79–amino acid High Mobility Group (HMG)-type DNA binding domain [85]. Sox 8, 9 and 10 comprise the subgroup E of Sox proteins and are highly expressed in NC. *Sox9* was shown to affect craniofacial development. Xenopus *Sox9* is expressed maternally and accumulates shortly after gastrulation at the lateral edges of the neural plate in the NC-forming region. At later stages, it persists in migrating NC cells at cranial regions as they populate the pharyngeal arches. Depletion of *Sox9* using antisense morpholinos caused a loss of NC cells, which was also reflected in reduced *Slug* transcription, and a compensatory expansion of the neural plate territory [18]. Conversely, forced expression in the chicken neural tube promoted NC-like properties at the expense of CNS neuronal differentiation [15]. Along with *Sox9*, the transcription factor *AP2α* was also implicated in cranial NC development. Gene targeting of *AP2α* resulted in reduced transcription of both *Slug* and *Sox9* [86] and caused severe craniofacial abnormalities [87, 88]. In chick embryos, *Sox 10* is expressed prior to neural tube closure, but later than *Sox9*, and in contrast to the latter, it continues to be expressed when cells become migratory [89, 15]. The role of *Sox10* in early chick NC development remains unclear but work in Xenopus confirmed that expression of *Sox10* is restricted to the NC [90] suggesting a role for this factor in early NC induction [91].

Taken together, growing evidence points to the involvement of multiple genes in the formation of the NC. Elucidating additional proteins of relevance as well as deciphering the networks that connect their activities will be a major challenge for the future. Notably, such studies will have to take into consideration that NC populations differ along the rostrocaudal axis. Emerging data point to differences between genetic cascades leading to NC specification in the head versus the trunk, an issue of unequivocal evolutionary significance. This is exemplified first, by the restriction of expression of certain early genes to either region (for example *Noelin* [92], *Id2* [93], *Slug* which persists in migratory NC in the head but only transiently marks pre-migratory NC in the trunk [36], etc.); second, by the differential responsiveness of head versus trunk epithelium to *Slug/Snail* overexpression [75]; and third, by differences in the response of cultured NC to various environmental factors such as FGF2/8, BMP2/4 and TGFβ1 [94]. It is thus expected that the progressing analysis at the molecular level clarifies both common mechanisms as well as regional differences.

12.3.2
The Delamination of NC Progenitors from the Neural Tube

The delamination of NC cells from the dorsal midline of the neural tube and their migration through neighboring structures represents a unique feature of the neuroepithelium, as CNS counterparts migrate and differentiate within the confines of the neural tube. This process of epithelial-mesenchymal conversion occurs during development of several embryonic structures and also underlies the formation of metastases during tumor progression (reviewed in [95]), Hence, investigating NC delamination represents a model system for understanding the underlying molecular basis of epithelio-mesenchymal transitions, and for evaluating how conserved this process is at various axial levels, across developmental systems and during tumor spreading.

12.3.2.1 A Balance between BMP and its Inhibitor Noggin Regulates NC Delamination in the Trunk

The onset of NC cell migration is a complex morphogenetic process which involves the coordinated action of several categories of molecules (cell adhesion molecules, cytoskeletal components, extracellular matrix macromolecules and transcription factors) upon which environmental signals were shown to act [96, 13]. The identity of some signals was recently elucidated. Sela-Donenfeld and Kalcheim [36] reported that along the rostrocaudal extent of the neural primordium, a dynamic interplay between two resident molecules, the inhibitor noggin and its ligand BMP4, generates graded concentrations of the latter that in turn trigger delamination of NC progenitors. This occurs without affecting their initial specification, a process which is also dependent upon BMP signaling, yet the two are separable both temporally and molecularly [15, 17, 36, 37]. Furthermore, a role for BMP2 in NC emigration/migration from rhombencephalic areas of mouse embryos was subsequently proposed [97, 98].

As pointed out previously, the activity of BMP4 along the rostrocaudal axis of the neural tube is modulated by changing levels of *noggin* [36]. To understand the etiology underlying the regulation of NC delamination, it was necessary to clarify which signals help to establish this gradient of *noggin* production. The temporal coordination between somite dissociation and the onset of NC migration, suggested that factors produced by the paraxial mesoderm might regulate *noggin* synthesis in the dorsal neural tube which is high opposite the unsegmented mesoderm and declines towards levels of dissociated somites. In line with this suggestion, experimental manipulations of the paraxial somitic mesoderm which consisted of grafting dissociating somites in place of unsegmented mesoderm, resulted in premature downregulation of *noggin* transcription in the dorsal neural tube and consequently triggered precocious emigration of NC progenitors from the caudal tube, an area which never releases mesenchymal cells under normal conditions. Thus, an inhibitory cross-talk exists between the paraxial mesoderm and the neural primordium that is mediated by regulating levels of *noggin* transcription. This interaction controls the timing of NC delamination to match the development of the somites which are the actual substrates for NC migration [99].

12.3.2.2 BMP-dependent Genes and NC Delamination

Genes such as *Slug, FoxD3, PAX3, rhoB, Cad- 6, Msx1* and *2, Wnt 1* and *3a*, etc. are either specifically expressed or become restricted to the dorsal tube from early stages onward, making it in some instances difficult to discriminate between possible roles in specification of the NC, subsequent delamination, or both. Experiments had to be designed to inhibit delamination without affecting initial specification, and therefore BMP activity was abrogated following initial expression of these genes. The inhibition of NC emigration observed *in vivo* following noggin treatment was preceded by a partial or total reduction in the expression of *cadherin 6B, rhoB, PAX3, Msx1,2* and *Wnt1*, but not of *Slug* (see above, [36, 63], and unpublished data). Their local downregulation suggests that these genes may be part of a molecular cascade triggered by BMP4, which leads to the separation of NC cells from the neural tube. This hypothesis requires that the effect of each factor be tested in experimental contexts in which it is possible to dissociate specification and delamination events.

Growing evidence suggests that these two processes can indeed be dissociated from each other. *FoXD3* and *Sox9*, two genes whose transcription depends on BMP signaling, induced NC specification in the neuroepithelium but were not sufficient to trigger ectopic cell delamination [15–17]. Likewise, Slug overexpression did not induce NC emigration at trunk levels; and an effect of the Slug protein on delamination of cranial NC cells is still lacking as neither loss- nor gain-of-function experiments discriminated between specification versus epithelio-mesenchymal conversion [19, 75]. Furthermore, deleting the *zfhx1b* gene, a zinc finger and homeodomain-containing transcription factor that encodes Smad-interacting protein–1, caused arrest of delamination of cranial NC cells without impairing their specification, and yet resulted in a failure of the actual formation of vagal-level progenitors. The latter led to a phenotype partially resembling the aganglionic megacolon syndrome observed in humans carrying a mutation in this gene [100].

Pax–3 is expressed in both the dorsal neural tube and the adjacent somites [101]. The mouse mutation *Splotch* [102] represents a deletion in the gene coding for *Pax–3* [103, 104]. *Splotch* mutants are characterized by defects in neural tube closure and severe reduction or even absence of certain NC derivatives including pigment cells, sympathetic and spinal ganglia, enteric neurons and cardiac structures. These defects were suggested to result from a delay in the onset of NC emigration from the neural tube [105]. Another study found that crest cell emigration (or formation) was severely affected in the vagal and rostral thoracic areas, while virtually no cells emigrated from the tube more caudally, perhaps as a result of aberrant interactions among adjacent neural tube progenitors or between NC and somitic cells [106]. A possible role for *Pax–3* in mediating epithelial-mesenchymal interactions was suggested in other systems [107] as well as the possibility that *Pax–3* triggers a non-canonical Wnt signaling cascade entailing JNK activation and cytoskeletal rearrangements leading to the generation of cell movement [108].

12.3.2.3 The Role of the Cell Cycle in NC Delamination

NC cells are mitotically active progenitors while residing in the dorsal neural tube and throughout migration. This initially discrete population must expand to reach the final number of cells that populates peripheral ganglia and other derivatives. The first post-mitotic cells appear by the time of gangliogenesis [109]. Prior to emigration, prospective NC progenitors are an integral part of the neuroepithelium and, as such, they undergo interkinetic nuclear migration whereby the position of the cell soma with its nucleus changes upon the phase of the cell cycle [110, 111]. Moreover, NC cells reveal similar cell cycle characteristics to laterally-located progenitors [111–113].

When *noggin* is downregulated and consequently BMP activity is augmented, the dorsal area of the neural tube becomes highly distinct from the rest of the neuroepithelium and NC cells begin to delaminate. Surprisingly, it was found that trunk-level NC cells become synchronized to the S-phase of the cell cycle during emigration from the neural tube (Fig. 12.1A and B; see [63]). This observation suggested that the transition from G1 to S is important for generating cell movement. Consistent with this hypothesis, specific inhibition of the G1–S transition prevented initial delamination of NC cells (Fig. 12.1C and D), but arresting the cell cycle during S or at G2 had no effect. These results showed for the first time that the transition between G1 and S is a necessary event for the epithelial-to-mesenchymal conversion of pre-migratory NC cells [63]. Is there a functional relationship between BMP signaling and the cell cycle and if so, what mechanisms mediate this interaction? Clarifying this question will undoubtedly expand our understanding of the molecular regulation of the onset of NC migration.

12.4
Peripheral Neuronal Lineages: Pluripotentiality and Early Restrictions of Migratory NC

12.4.1
Evidence from Cell Lineage Analysis *In Vivo* and *In Vitro*

The NC gives rise to a variety of different cell types, including the different neurons of the sensory, autonomic and enteric ganglia [1]. The NC is composed of uncommitted, pluripotent cells but also contains cells with restricted developmental potential. Classical studies, using heterotopic transplantation of pre-migratory NC along the rostrocaudal axis of chick/quail embryos, demonstrated the pluripotentiality of NC within a given area [114, 115]. However, as discussed in Section 12.3.1.1, the equivalence in developmental potential of NC along the rostrocaudal axis is not complete. Trunk NC cells, unlike cranial crest cells, are unable to form cartilage [116] or to develop into heart mesenchymal cells [117] even when transplanted at the appropriate anteroposterior position. This difference is paralleled by a differential expression of anteroposterior patterning genes, the Hox genes [118, 119], and by the selective expression of *Id2* [93] and *Noelin–1* [92] in cranial NC. Id2 and Noelin–1 are expressed in cranial neural folds and are involved in NC generation [93, 92], but their role in the specification of cranial crest identity is unclear. Hox genes indeed seem to function in the specification of cranial versus trunk NC, since the ectopic expression of the trunk-specific gene *Hoxa10* in cranial crest induces characteristic trunk NC properties [94].

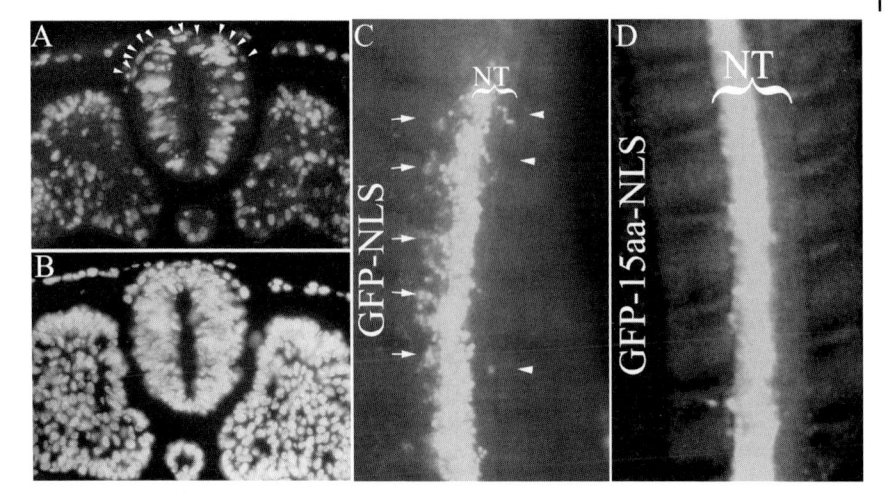

Fig. 12.1
Emigration of NC cells from the dorsal tube depends upon successful G1/S transition. (A, B) Neural crest cells emigrate from the neural tube in the S phase of the cell cycle. Most emigrating NC cells are Brdu+ following a 1-h pulse and immediate fixation (A, arrowheads). Compare with total Hoechst-stained nuclei in B. (C, D) The cdk-binding domain of MyoD and dn E2F–1 inhibits delamination of NC cells *in ovo*. Dorsal views of embryos that received control GFP-NLS vector (C) showing axial level-dependent migration of GFP+ crest cells through intersomites and into the rostral somitic domains (arrows). Some labeled cells also cross to the contralateral side (arrowheads). (D) Electroporation with the 15–amino acid carboxy terminus moiety of MyoD fused to a GFP reporter (GFP–15aa-NLS), resulted in no migration of labeled crest cells. This small sequence of MyoD binds to Cdk4/5 and CyclinD1 forming a complex that ultimately inhibits the G1/S transition specifically. (This figure also appears with the color plates.)

Rhombomere transplantation suggested that NC identities in the hindbrain are determined by the Hox-code of a particular rhombomere [120]. More recent findings have revealed that the Hox-code expression is maintained only in transplanted cell groups, not in single cells [121]. Transposition of single cells resulted in a switch of Hox gene expression and change of cell fate appropriate to the new location [122]. Thus, single cranial NC cells are plastic and can be reprogrammed by environmental cues, whereas community effects maintain a given NC fate [123]. In addition to rostrocaudal restrictions, there is also a restriction with respect to the NC cell populations that leave the dorsal neural tube at different time points. At trunk levels, the first NC cells that delaminate from the neural tube migrate to the sympathetic ganglia, whereas the sensory ganglion precursors leave later and the last emigrating cells migrate along the dorsolateral pathway to become melanocytes [124, 125]. As the late-migrating cells had lost the ability to give rise to noradrenergic neurons upon implantation into young host embryos, it was concluded that the developmental potential of the migrating NC becomes progressively restricted, although the alternative explanation of changing environmental signals was not excluded [125, 126].

Although heterotopic graft studies and mass cultures demonstrate differences and changes in the pluripotentiality of NC cell populations, they do not allow us to draw conclusions about the properties of single cells, as already discussed for A/P patterning of cranial NC. The identification of the fate and developmental potential of individual NC cells requires clonal cultures or lineage tracing by labeling single cells *in vitro* or *in vivo*, with the caveat that the properties of the cells may be altered upon culturing.

The analysis of the developmental potential of NC cells using single cell cultures showed that the majority is pluripotent [127, 128] and subcloning of single cells provided evidence for self-renewal, at least for mammalian NC [129]. However, a considerable proportion of the cells was found to give rise to a single cell type [130, 128]. More recently, the developmental potential of single NC cells was analyzed using a lineage tracer in mass cultures, which is considered to reflect better the *in vivo* situation. Although pluripotent cells were initially present, their number was small and decreased rapidly in culture. Nearly half of the early migrating NC cells gave rise to a single phenotype [131]. *In vivo*, following the fate of single migrating avian NC cells by tracer injection [8, 132] or by infecting them with a replication-deficient retrovirus [133] provided evidence for multipotentiality. In agreement with the in vitro studies, these experiments also showed that a considerable proportion of clones contributed to only one NC derivative, i.e. either sensory ganglia, sympathetic ganglia or peripheral nerve ganglia [132]. In zebrafish, lineage restriction seems to be much greater than that among birds and amphibians, as more than 80 % of tracer-injected cells generated single derivatives [134] (but see [122]).

The generation of restricted progenitors could be explained either by the action of positive or negative instructive signals on a homogenous population of initially pluripotent cells or by stochastic differentiation of these cells. Recent studies provide strong evidence for NC heterogeneity at the earliest stages of crest development, in particular with respect to the sensory ganglion lineage [59, 135]. Taken together, these data suggest that the NC is composed of a mixture of pluripotent cells and cells with restricted potential. A major interest in this field of research is to identify, select and characterize the pluripotent NC stem cells and the progenitor cells of the different neuronal and non-neuronal lineages, resulting in a defined lineage of the peripheral nervous system.

12.4.2
NC Stem Cells and the PNS Lineage

The pluripotentiality of single identified migratory NC cells, demonstrated both *in vitro* and *in vivo*, together with the evidence for self-renewal, was the basis for using the term, NC stem cell (NCSC) for these cells [129]. As stem cells in other tissues are maintained throughout the lifetime of the organism, the issue is raised of the extent to which such pluripotent cells are maintained during development and in the adult

12.4 Peripheral Neuronal Lineages: Pluripotentiality and Early Restrictions of Migratory NC

stage. This question was initially addressed by back-implantation of peripheral ganglia into the NC migration pathway of early avian embryos. These experiments revealed the existence of cells that had retained the ability to migrate along NC migration pathways and subsequently to differentiate into PNS neurons, glia and melanocytes according to their final location (reviewed by [136]). *In vitro* clonal analysis provided evidence for the pluripotentiality of such cells [137, 138]. Interestingly, the cell populations of different ganglia differed in their developmental potential. Sensory neuron precursors were detected only in sensory ganglia and only during neurogenesis, whereas cells with the potential for autonomic neurogenesis were found in both sensory and autonomic ganglia and also after the period of neurogenesis [139, 140]. These experiments also indicated that the PNS lineage includes progenitor cells restricted to sensory or sympathetic phenotypes including sensory versus sympathetic satellite cells [136]. It should be pointed out that in these studies the identification of progenitor cells and their characterization was based exclusively on retrospective evidence. More recent studies succeeded in prospectively identifying specific precursor populations and in investigating their developmental potential. In avian NC cultures progenitor cells, identified by their expression of c-kit or trkC, were shown to be committed to a melanocyte or neuronal fate, respectively [141]. On the other hand, pluripotent cells were prospectively isolated from the peripheral nervous system. A cell population from embryonic rat sciatic nerve using p75 and P0 as positive and negative cell surface selection markers, showed *in vitro* stem cell properties, i.e. self-renewal and pluripotentiality [142]. When injected into the NC migration pathway of chick embryos these cells populated the peripheral nervous system and differentiated into autonomic neurons and glial cells [142, 143]. In view of the gliogenic bias of the cells isolated from the nerve and the possible re-programming of Schwann cell precursors it is presently unclear whether the cells isolated from peripheral nerve represent Schwann cell precursors or a separate cell population of sciatic nerve NCSSs (see also Section 12.8.5). However, they did not give rise to sensory neurons and displayed a restriction towards glial and autonomic neuron fate, although they failed to acquire noradrenergic properties. Such peripheral NCSCs could be demonstrated in peripheral nerve, ganglia and in the gut, in the latter location up to the adult stage [144–146]. NCSCs can be propagated as neurospheres, which may be taken as additional evidence for their stem cell properties [147]. Cells isolated using the same cell surface markers but from different locations in the embryo, or isolated at different developmental time points, had characteristic and different developmental potentials and responsiveness to extrinsic differentiation signals such as BMPs [144]. As NCSCs with different identities were obtained using the same set of surface markers, the early stages of the PNS lineage, i.e. progenitors defined by specific cell surface antigens and specific developmental potential, are still unclear. Using gene expression profiling techniques, these approaches will, however, lead to a molecular signature of different populations [148] which may represent an important step forward in the definition of PNS lineages.

12.5
Molecular Control of Neuron Development in Peripheral Ganglia

12.5.1
Extrinsic Signals in Sensory, Sympathetic, Parasympathetic and Enteric Neuron Development

12.5.1.1 Sensory Ganglia

The earliest marker for the sensory lineage is the bHLH proneural transcription factor ngn2 whose expression starts in the dorsal neural tube and/or in early migrating NC cells [148]. This early wave of sensory neurogenesis, giving rise to early differentiating proprioceptive and mechanosensitive sensory neurons, is followed by the ngn1–dependent generation of nociceptive sensory neurons. Whereas extrinsic signals have been identified that control the development of autonomic neurons, glia and melanocytes (for reviews see [149–151]), factors specifying sensory neurons from NC have been reported only very recently (see Fig. 12.2 for an overview). In zebrafish, sensory DRG neuron development was shown to depend on the morphogen hedgehog, secreted by notochord and floorplate [152]. It is unclear whether this is relevant to the situation in other vertebrate species, as in the chick shh acts to arrest the migration of sensory precursors rather than to induce sensory neurons *de novo* [153]. Wnt molecules acting through β-catenin comprise a second class of extrinsic factors. The selective elimination of β-catenin in NC cells, using Wnt-Cre mice to create a conditional β-catenin knockout, results in a selective loss of sensory neurons and melanocytes, whereas sympathetic and enteric neurons are not affected [59]. Conversely, overexpression of β-catenin in NC results in the massive generation of ectopic sensory neurons at the expense of virtually all other NC derivatives [154]. The multiple effects of Wnt signaling observed in NC development, i.e. NC induction [155], NC proliferation [156], melanocyte specification [45, 157] and sensory neuron generation [59, 154] raises the issue of how these potential roles are coordinated during normal development. BMPs, controlling peripheral autonomic neuron development (see below) also induce the generation of the epibranchial placode-derived cranial sensory neurons, in agreement with their function in Phox2–dependent peripheral autonomic circuits [158].

12.5.1.2 Sympathetic Ganglia

Sympathetic ganglia are formed in the vicinity of the dorsal aorta and notochord by migrating Sox10–, HNK1– and p75–positive NC cells that aggregate at this location to form the primary sympathetic ganglia [151, 159]. The first marker gene which allows us to distinguish autonomic precursors from uncommitted migrating NC cells is *Ascl1*, the vertebrate homolog of Drosophila achaete scute complex proneural genes (also known as *Mash1*) [160–162]. *Ascl1* expression is followed by other transcriptional control genes, marker genes of noradrenergic sympathetic neurons such as tyrosine hydroxylase (TH) and dopamine-β-hydroxylase (DBH) as well as generic

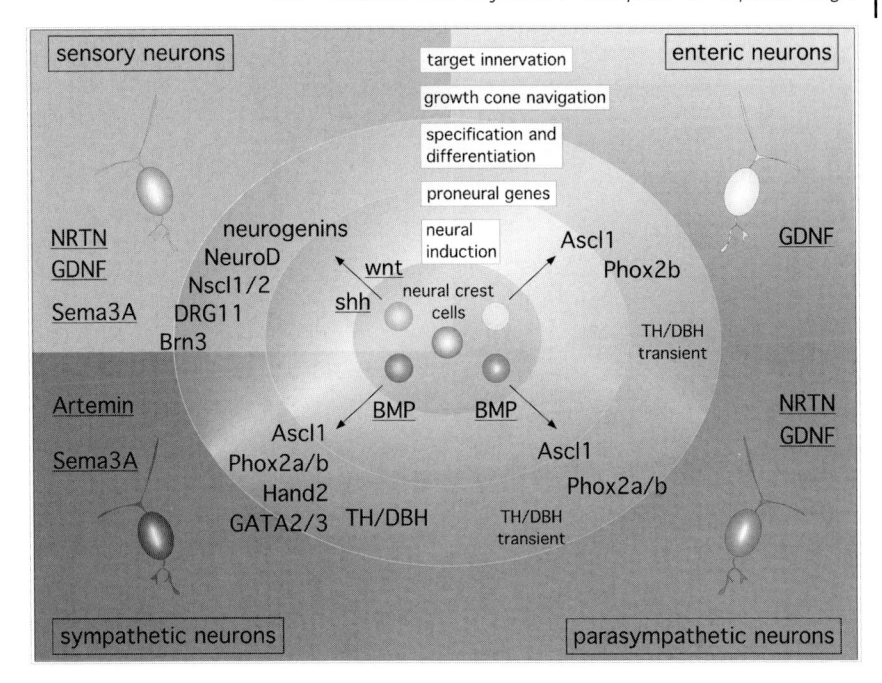

12.5 Molecular Control of Neuron Development in Peripheral Ganglia | 409

Fig. 12.2
Extrinsic signals and cell-intrinsic factors that are known to be involved in the generation of sympathetic, parasympathetic, enteric and sensory neurons. The specification of sympathetic and parasympathetic neurons is dependent on bone morphogenetic proteins (BMPs), whereas the development of the sensory neuronal lineage depends on Wnt signaling. In zebrafish there is also evidence of a role for sonic hedgehog (shh) in sensory neuron generation. The early pre-specification of sensory and autonomic neural crest precursor cells is indicated by the color code. The molecular basis of this pre-specification, i.e. the difference between a totipotent neural crest stem cell (gray) and pre-specified progenitors (colors) is unknown. In sympathetic precursor cells, BMPs induce the expression of Ascl1, Phox2a/b, dHand and Gata2/3, which, in turn, control the expression of noradrenergic (TH, DBH) and generic neuronal properties. In precursor cells of the parasympathetic ciliary ganglion Ascl1 and Phox2a/b, but not dHand and Gata2/3 are induced by BMPs, and TH/DBH are expressed only transiently. All enteric neurons depend on Phox2b and a subpopulation on Ascl1, it is unclear, however, whether their development is also controlled by BMPs. Wnt signaling is essential and sufficient to induce sensory neuron generation, which involves the neurogenins ngn1 and ngn2, the first members of a cascade of bHLH factors acting in sensory neuron differentiation, i.e. NeuroD, NSCL–1/2 and homeodomain transcription factors Brn3a and DRG11. Class III-semaphorins (Sema3a) play an important role for sensory and sympathetic neuron projections and in the location of sympathetic cell bodies. Members of the GDNF-family ligands (GFLs) are involved in the initial axon projections of sympathetic and parasympathetic neurons. The survival of peripheral neurons depends on neurotrophins at different stages of their development (not shown). (This figure also appears with the color plates.)

neuronal genes such as *SCG10* and neurofilament [163–167]. The differentiation of sympathetic neurons depends on environmental signals from the notochord, ventral neural tube and dorsal aorta as suggested by experimental embryology [168–171]. Whereas the identity of the signals from the ventral neural tube/notochord remained unclear, bone morphogenetic proteins (BMPs) have been identified as dorsal aorta-derived factors that are essential for sympathetic neuron development. The dorsal aorta expresses BMP2 and BMP7 in the mouse [172] but BMP4 and BMP7 in the chick [173] during the period when sympathetic neuron development is initiated. NC cells differentiate to sympathetic neurons in response to BMP application both *in vitro* [172–174] and *in vivo* [173] and the TH-inducing effect of co-cultured dorsal aorta can be blocked by the application of the BMP antagonist noggin [175]. Most importantly, inhibition of BMP signaling *in vivo* by the local application of noggin led to an inhibition of sympathetic neuron differentiation, demonstrating the essential role of BMPs in this lineage [175]. Why is autonomic neuron development not already observed during migration, as BMPs are already involved in NC specification and delamination in the dorsal spinal cord? The most straightforward explanation is that the properties of NC cells change during their migration, which may reflect a change in BMP receptor expression. In substantiation of this idea is the fact that BMPReceptor IA is rapidly downregulated following NC emigration from the neural tube [176]. Furthermore, specific signals may also prevent precocious neuronal differentiation in emigrating NC cells (e.g. Wnts, Sox9, Sox10) or in the ganglion environment. Interestingly, Sox10–deficient DOM-mice display premature noradrenergic differentiation during their migration [177], raising the additional issue of whether this is induced by BMPs from the dorsal neural tube, dorsal aorta or unidentified signals from the notochord/ventral neural tube.

12.5.1.3 Parasympathetic Ganglia

BMPs are also involved in the generation of parasympathetic ganglia. BMP expression was observed in the vicinity of avian ciliary ganglion [178] and the pelvic ganglion [143] at the time of ganglion formation. The essential role of BMP signaling was demonstrated by the application of the BMP inhibitor noggin, resulting in the complete lack of ciliary neurons [178]. In noggin-treated embryos the mesencephalic NC cells migrated correctly, aggregated to form ciliary ganglion primordia but the expression of fate-determining transcription factors, generic neuronal and subtype-specific genes was prevented. BMP overexpression led to an increased number of ciliary ganglion neurons from NC precursor cells, which is reflected by enlarged ciliary ganglia [178].

12.5.1.4 Enteric Ganglia

The enteric nervous system is derived from the NC at three levels, i.e. vagal, sacral and from the rostral trunk [1]. Vagal crest cells populate the entire gut, sacral crest cells contribute to the enteric nervous system in the post-umbilical gut, whereas truncal NC colonizes the esophagus and stomach. Interestingly, vagal and sacral crest-derived cells depend on GDNF signaling to form the enteric ganglia [179–182], whereas trunk derivatives develop in a Ret-independent manner [183]. This correlates with mutations in the genes for Ret [184], GDNF [185] and GFRa1 [186] in familial forms of Hirschprung's disease, characterized by an aganglionic terminal colon. Genetic analysis also implicates an essential role for endothelin–3 and endothelin receptor-B in enteric nervous system development [187, 188]. Ret and endothelin receptor-B act synergistically to allow colonization of the distal bowel as demonstrated by genetic analysis [189] and *in vitro* studies [146, 190]. GDNF increases proliferation and survival of neural precursors as well as subsequent neuronal differentiation and axon outgrowth [191, 192]. Recent studies on the mechanism of GDNF and endothelin–3 signaling demonstrate cooperative effects on the proliferation of enteric progenitors [190] and on distal gut colonization [146]. Although BMP4 has been shown to be involved in specification and differentiation of gut mesoderm and smooth muscle [193–195] it is unclear whether BMPs, similar to the situation in sympathetic and parasympathetic precursor cells, initiate enteric neuron development by inducing the expression of Phox2 genes. At a later stage BMPs seem to be involved in the induction of NT–3 responsiveness, and thus in the differentiation and survival of the NT–3–dependent subpopulation of enteric neurons [196].

12.5.2
Transcriptional Control of Peripheral Neuron Development

12.5.2.1 Sensory Neurons

The development of NC-derived peripheral neurons is critically dependent on the vertebrate homologs of Drosophila proneural genes. Similar to the situation in Drosophila, different types of peripheral neurons depend on different classes of proneural genes. Sensory neurons are dependent on the expression of the Drosophila atonal-related bHLH factors *Ngn1* and *Ngn2*, whereas sympathetic, parasympathetic and some enteric neurons depend on the achaete scute homolog Ascl1. The elimination of *Ngn1* and *Ngn2* affects different sensory neuronal populations. In the mouse DRG, Ngn2 is initially required by the early differentiating lineage of proprioceptive and mechanoreceptive neurons, whereas *Ngn1* is required for the later differentiating lineage of nociceptive neurons [197]. Also at cranial levels different types of sensory ganglia are affected in Neurogenin knockouts. *Ngn2* is essential for epibranchial placode-derived sensory neurons [21], whereas NC-derived proximal cranial sensory ganglia are affected in the *Ngn1* knockout [22]. It should be noted, however, that Ngn1 can finally compensate for the lack of *Ngn2* in trunk DRG. How do ngn bHLH proteins function? Do they determine the sensory neuron fate of the cells, or do they act only as proneural genes? In Drosophila, misexpression of proneural genes revealed that they contribute to the specification of neuronal identity as well as generic neuronal fate [198]. Whereas there is strong evidence for an early restriction of sensory precursors, it is less clear whether *Ngn2* is involved in this process. *In vivo* lineage tracing showed that *Ngn2*–expressing cells preferentially contribute to sensory rather than autonomic ganglia however, giving rise to both sensory neurons and sensory glial cells [149]. As some *Ngn2*–expressing cells are able to develop into autonomic neurons, *Ngn2* expression is either not a strong or not the only restrictive signal. In rat NC explants, fate-committed sensory neuron precursors were identified by the property of the cells to induce neurogenin expression and to be unable to respond to factors that induce the generation of autonomic neurons [135]. Further evidence for a fate-determining role is the generation of sensory neurons upon ectopic expression of *Ngns* in the chick embryo [23]. On the other hand, there are data that argue against an essential fate-specification role for *Ngn2*. Using spinal cord cultures it was found that both Ascl1 and Ngn2 can induce autonomic neuron generation, implicating decisive roles for additional lineage-specifying factors [199]. Also *in vivo*, using a knock-in strategy to express *Ngn2* in the locus of *Ascl1* and vice versa, replacement of *Ascl1* by *Ngn2* did not affect noradrenergic differentiation and, conversely, Ascl1 could replace *Ngn2* in sensory neuron development [200].

Neurogenins are only the first members of a cascade of bHLH factors involved in sensory neuron differentiation [201]. *Ngn* expression in sensory precursors is followed by *NeuroD*, *NeuroM*, *NSCL–1* and *NSCL–2*. Accordingly, Neurogenin overexpression induces *NeuroD* and *NSCL* in NC cells [23]. A further downstream target of Neurogenins in sensory precursors is the POU homeodomain transcription factor Brn3a [202]. Brn3a is essential for the development of sensory neurons [203] and seems to be directly or indirectly involved in the control of generic and subtype-specific sensory neuron properties [165, 204]. Interesting direct target genes for *Brn3a* are the *trk* neurotrophin receptor genes that are required for the correct innervation and survival of sensory neurons [165]. With respect to the mechanism of action of Neurogenins, and of bHLH transcription factors in general, it has become clear that in addition to transcriptional activation by binding to E-boxes in the promoter of target genes, they may also act without DNA binding, sequestering or acting in a complex with other DNA-binding factors [205–207].

The paired homeodomain transcription factor *DRG11* is expressed from early stages of development both in sensory DRG neurons and in the dorsal spinal cord. Embryos deficient in DRG11 display abnormalities in the initial ingrowth of sensory afferent fibers and subsequent lack of nociceptive circuits [208]. Interestingly, visceral sensory neurons of the epibranchial placode-derived cranial ganglia, representing the sensory components of autonomic circuits are also dependent on the paired homeodomain *Phox2a*/*Phox2b* transcription factors that control the development of peripheral, central, sensory and motor components of the autonomic nervous system [209–211]. As shown for DRG11 and sensory projections there is also evidence to suggest that *Phox2a/2b* control the development of connections in the autonomic nervous system (see below).

The differentiation of proprioceptive sensory neurons depends on additional transcription factors of the Runt and ETS-domain families of transcription factors (reviewed in [212]). *Runx3* controls axonal projections and/or survival through transcriptional control of trkC expression [213, 214]. Under the influence of NT–3, produced by the innervated muscle, proprioceptive sensory neurons were shown to express ETS-domain transcription factors such as *PEA3* and *ER81* [215, 216]. These transcription factors control in turn the selective connection of Ia afferent fibers with appropriate motoneurons in the spinal cord [213, 217].

12.5.2.2 Sympathetic and Parasympathetic Neurons

The intracellular signal cascade that is induced by BMPs in NC cells involves, in addition to *Ascl1* [161], the paired homeodomain transcription factors *Phox2a* and *Phox2b* [163, 209, 210], the bHLH transcription factor dHand [218, 219] and the Zn-finger transcription factors Gata2/3 [166] (summarized in Fig. 12.2). The induction of *Ascl1* and *Phox2* is (transiently) dependent on the HMG-box factor *Sox10*, which displays a complex function in the maintenance of neuronal and glial potential as well in delaying neuronal differentiation of NC stem cells [177]. The importance of the BMP-induced transcription factors *Ascl1*, *Phox2b* and *Gata3* for sympathetic neuron development was demonstrated by loss-of-function approaches (reviewed in [151]). Overexpression of these BMP-induced transcription factors showed that they are also sufficient to elicit neuron generation from NC precursors *in vitro* and *in vivo* [209, 219, 220]. Interestingly, *Phox2* and *dHand* coordinate the expression of subtype-specific and generic neuronal genes whereas *Ascl1* and *Gata2/3* preferentially induce pan-neuronal gene expression in these overexpression experiments [220] (Tsarovina and Rohrer, in preparation). This may be correlated with the finding that *Phox2a/b* activate transcription of *TH* and *DBH* genes [221–226] and that *dHand* strongly potentiates this effect at the DBH promotor [221, 227, 228]. The *dHand/Phox2a* interaction requires DNA binding of *Phox2a* but not of *dHand*. Promotor studies also implicate *AP–2* [229, 230] and *Creb* [222] in the control of noradrenergic differentiation. Indeed, in zebrafish devoid of *AP2alpha*, neurons are generated in the location of sympathetic ganglia, but they are devoid of TH [231]. This suggests that *AP2alpha* would selectively control noradrenergic gene expression. A similar role has also been implicated for *Gata3* [166]. Thus, although the signal transduction pathway from the activated BMP receptor to the induction of Ascl1 and Phox2b is not fully understood, the direct activation of characteristic, subtype-specific genes by *Phox2a/b*, *dHand* and *Gata3* provides a framework within which to understand the generation of neuron subtypes in the autonomic nervous system.

As parasympathetic neurons in general display a non-noradrenergic phenotype the function of *Phox2a/b* in the expression of the noradrenergic marker genes must be modified in these neurons. Data obtained for the chick ciliary ganglion suggest that noradrenergic gene expression is initiated in these parasympathetic neurons, but not maintained due to the absence of *dHand* [178].

12.5.2.3 Enteric Neurons

NC cells entering the gut at E8 in the mouse critically depend on the transcription factor *Phox2b* which is required for the generation of the entire autonomic nervous system [210]. A subpopulation of enteric neurons controlled by Ascl 1, transiently expresses catecholaminergic properties and subsequently acquires other phenotypes, in particular serotonergic characteristics [232, 233]. The bHLH transcription factor *dHand* (also known as *Hand2*) is expressed on neurons of both the myenteric and submucosal ganglia in all segments of the developing gut [234]. Although BMP4 is able to induce *dHand* expression in gut-derived NC cells, loss-of-function experiments are required to define the action of BMPs in the genesis of enteric neurons.

12.6
Growth Cone Navigation and Innervation of Peripheral Targets

The specification of neuronal identity also includes its specific axonal projection pattern which will lead to the generation of physiologically meaningful circuits. This should be reflected by the expression of subtype-specific receptors that enable the growth cone of these neurons to respond to diffusible and cell- or substrate-bound environmental cues. In agreement with this notion, the specification of different types of spinal cord motoneurons includes a specific projection pattern. For example, medial motoneurons are specified by *Lhx3* and project into axial musculature, whereas the innervation of dorsal axial muscles by LMC [l] neurons is dependent on the Lim1–induced motoneuron specification (reviewed in [235]). The EphA receptor is an important target gene translating and mediating subtype specification into growth cone navigation in the limb [236]. What is known about the transcriptional control of target innervation in NC-derived neurons? The selective expression of *Phox2b* in neurons of the medullary visceral reflexes [237] and the absence of these circuits in the *Phox2b* knockout, implicates *Phox2b* in the control of circuit connectivity. Although there is some evidence for axon navigation effects of the *Phox2* ortholog in C. elegans [238], such potential late effects of *Phox2* genes cannot be analyzed in the conventional knockouts, as affected progenitors do not differentiate in the absence of *Phox2b* and die before projections are established. The *Phox2*-dependent survival can be explained in some neurons by the control of c-ret expression, the signaling receptor subunit of the GDNF-family ligands (GFLs) [209]. Effects on axon navigation as well as later target genes will be accessible either in compound mutants where neuron death is prevented or in conditional knockouts. The former approach has been used to demonstrate a direct regulation of trkA expression and effects on axon outgrowth in developing sensory neurons by the POU domain transcription factor *Brn3a* [164, 239]. In the following, we will shortly summarize the main players involved in the initial axonal projection for sensory and sympathetic neurons.

12.6.1
Semaphorins

An important mechanism that governs peripheral afferent trajectories of spinal nerves, including sensory projections, is provided by the class III-semaphorins. The secreted chemorepellent protein Semaphorin 3A (Sema3A) induces growth cone collapse in cultured sensory and sympathetic neurons by interacting with the plexin1A receptor and the co-receptor neuropilin1 [240, 241]. For sensory efferent projections the action of Sema3A is dependent on L1 [242]. At the time of ganglion formation, *Sema3A* is expressed in the dermamyotome and paraxial mesoderm, limiting and therefore defining the projection routes for sympathetic and sensory efferent fibers. In neuropilin1–mutant mouse embryos, the prominent sensory neuron phenotype shows projectional aberrations including defasciculation, overshooting of their targets and ectopic projections, whereas the positioning of sensory ganglia does not seem to be affected [243]. In contrast, neuropilin-deficient sympathetic neurons not only develop similar abnormal projections, but also show mislocation of sympathetic cell bodies with ectopic neurons in the forelimb along the trajectories of the spinal nerve [244]. Incidentally, outgrowing blood vessels and major motor and sensory nerves share the same Sema3A dependency and a tight congruency between the projection pattern of nerves and vasculature has been observed. As ablation of either nerves or vessels in quail embryo limbs does not affect the patterning of the other, it is concluded that the development of both systems, although dependent on Sema3A signaling, is not causally linked [245, 246].

12.6.2
GDNF-family Ligands

Members of the GDNF-family ligands (GFLs) are particularly important for the initial axon projections of sympathetic and parasympathetic neurons. GFLs bind to GFRalpha receptors and this complex is able to activate the common signal transducing receptor ret. GDNF, Neurturin (NRTN), Artemin (ARTN) and Persephin (PSPN) bind to GFRa1, GFRa2, GFRa3 and GFRa4, respectively [247].

Whereas *in vitro* several family members have been shown to support proliferation of sympathetic precursors, axon growth and survival of embryonic and postnatal sympathetic neurons [248, 249], *in vivo* essential functions were shown only for Artemin. In mice deficient for either *ARTN*, *GFRa3* or *ret*, the trunk sympathetic ganglia are smaller and sometimes missing. Sympathetic neurons show severe defects in axonal outgrowth, aberrant projections and subsequently reduced innervation of target areas [183, 249–252]. Conversely, implantation of Artemin-soaked beads near the sympathetic trunk rescues the outgrowth of sympathetic neurites in Art$^{-/-}$ embryos. *Artemin* is expressed in smooth muscle cells of central and peripheral blood vessels as early as E10.5 in the mouse and its expression proceeds to more distal peripheral vessels in the course of vasculogenesis. This expression pattern, together with the finding that sympathetic projections follow blood vessels, suggested that Artemin might direct sympathetic growth cones along the vascularized tissue towards their innervation targets [250] (but see [245]).

GFLs are also important for the development of parasympathetic neurons. Adult mice deficient in *NRTN* and *GFRa2* display strong defects in the parasympathetic target innervation [253, 254]. Ret signaling is required for the proper development of sphenopalatine, otic and submandibular ganglia. In addition to NRTN, which is late acting, GDNF is an early survival factor for parasympathetic neurons [255, 256]. In chick ciliary ganglion cultures GDNF and NRTN promote neuron survival and axon outgrowth. Interestingly, the neurotrophic response to GDNF but not NRTN is reduced during development, in line with a decline in GFRa1 and maintained expression of *GFRa2* [257, 258]. Blocking GDNF *in vivo* by neutralizing antibodies strongly impaired ciliary axon outgrowth [257]. Although a long line of research has established the importance of GFL and c-ret signaling in the development of the enteric nervous system (see Section 12.5.1) it is not known whether axon outgrowth is controlled by GFLs *in vivo*.

12.6.3
Hepatocyte Growth Factor (HG)

HG has recently been shown to promote axon growth in several different types of neurons and to enhance survival during development [258].

12.6.4
Macrophage Stimulating Protein (MSP)

The HG-related secreted growth factor, MSP, has age-related effects on growth and survival of NGF-dependent sensory and sympathetic neurons and is expressed in the corresponding target tissues [259].

12.6.5
TGF-β/BMP Family

Although BMPs have recently been shown to affect commissural axon trajectories in the spinal cord [260] there is no evidence so far of any effect on sensory or autonomic axon outgrowth. However, dendrite growth of sympathetic neurons seems to be controlled by BMPs [261, 262]. TGF-β increases the number of neurites, as well as neurite length in explant cultures of chick DRG [263]. Activin and BMPs are present in sensory neuron target tissues and, since they induce neuropeptide expression *in vitro*, may also control the development of the transmitter phenotype of sensory neurons *in vivo* [264, 265].

12.7
Factors Controlling Neuronal Survival in the PNS

From the time of birth and throughout their life, peripheral neurons are dependent on survival factors. The neurotrophic theory, i.e. the proposal that neuron survival depends on target-derived factors, was formulated based on the finding that the survival of peripheral neurons depends on the size of their peripheral targets. The prototype of neurotrophic factors, nerve growth factor (NGF), was identified as a survival factor for sensory and sympathetic neurons. The simple view of a single neurotrophic factor maintaining a specific type of neuron has changed in several ways: Various neurotrophic signals are required by an individual peripheral neuron at different stages of its development, including the stage before axon outgrowth starts; these different factors are not only produced by the final target, but also by tissues and cells in the environment of the ganglia and along the axonal projectories; these factors not only affect neuronal survival but in addition, may control growth cone guidance, cell type specification and differentiation. As this broad field of research cannot be covered within the limits of the present overview, we refer the interested reader to the literature [266, 267, 268–270]. A further novel aspect is the cooperative interaction of neurotrophic factors with signals that control the responsiveness of the neurons [271]. A well-understood example represents the action of TGFβs that are essential for the action of neurotrophic factors such as NGF, CNTF, GDNF and FGF2 on different types of peripheral ganglion neurons [272–274]. TGFβ-mediated GDNF responsiveness was shown to be due to the recruitment of GFRa1 to the plasma membrane [275]. A role in the induction of neurotrophic dependence of sympathetic neurons has also been suggested for BMP4 [276]. TGFβs have an additional role in the regulation of ontogenetic neuron death and in neuron death induced upon target removal [277]. Interestingly, members of the neurotrophin and TGFβ growth factor families are also involved in synapse formation during development and synaptic plasticity in the adult [278–281].

12.8
Glial Cell Development from the NC-Peripheral Glial Lineages: Fate Choice and Early Developmental Events

12.8.1
The Main Cell Types

Several glial cell types develop from the NC. These include the myelinating and non-myelinating Schwann cells that ensheath axons in nerve trunks, satellite cells that envelop neuronal cell bodies in sensory, sympathetic and parasympathetic ganglia, which are very similar to non-myelinating Schwann cells and enteric glial cells that are found in the enteric ganglia of the gut and show some intriguing similarities with astrocytes. Lastly, terminal glia or teloglia are associated with axon terminals at the skeletal neuro-muscular junction [282–284]. In this chapter we will deal mainly with Schwann cell development from the trunk NC, because this system represents the best characterized crest-derived glial lineage.

Between migrating crest cells and mature Schwann cells there are three main developmental transitions [285–287] (Fig. 12.3): first, the formation of Schwann cell precursors, second, the formation of immature Schwann cells and, lastly, the reversible generation of myelinating and non-myelinating cells. Since crest cells generate other cell types in addition to glia, and immature Schwann cells will either myelinate or not, the first and last of these transition points involves fate choice. The second step, in contrast, represents lineage transition, since, apart from programmed cell death, presumably the only fate of Schwann cell precursors in normal development is to become Schwann cells.

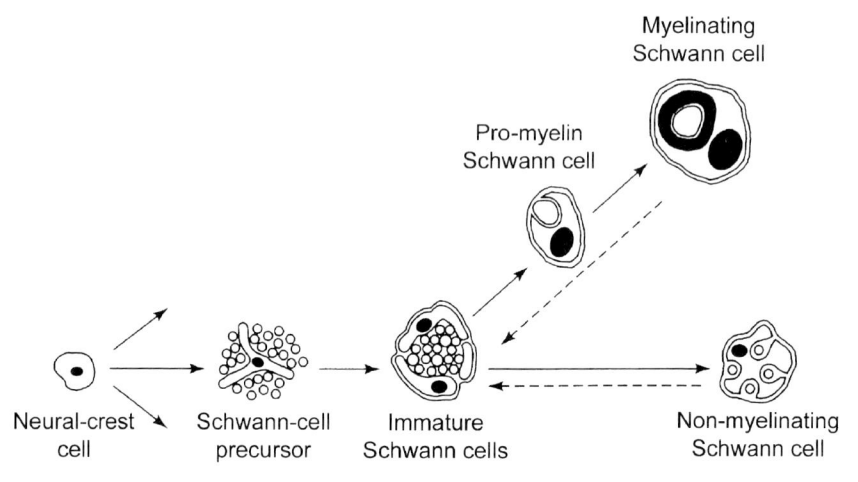

Fig. 12.3
The Schwann cell lineage in rat and mouse. There are three main transitions in the Schwann cell lineage. I. The formation of Schwann cell precursors from undifferentiated migrating neural crest cells. This stage is characterized by the appearance of several differentiation markers (a subclass of which is shared by early neurons), by changes in survival factor responses and by the assumption of an intimate relationship with axons. II. The precursor-Schwann cell transition. This stage is characterized by the appearance of a number of differentiation markers (some of which are acutely dependent on axonal contact and remain readily reversible) downregulation of the transcription factor AP2α, and the establishment of autocrine survival circuits. III. The formation of mature myelinating and non-myelinating Schwann cells. This stage involves radical morphological and molecular changes, particularly in myelinating cells; nevertheless, the transition is readily reversible when axonal contact is lost (dashed arrows) and is re-established in regenerating nerves.

12.8.2
Molecular Markers of Early Glial Development

It has been difficult to study the onset of glial differentiation in NC cells due to a lack of early glial differentiation markers, namely, genes that are activated as soon as cells start to differentiate as glial cells but remain silent in other crest-derived lineages. The activation of such genes indicates that in normal development the cell is fated to become a glial cell, or to die a developmental death. It does not mean that the cell is irreversibly committed to a glial fate and cannot be diverted to other lineages by alternative signals e.g. *in vitro*, since cell differentiation may remain reversible for a considerable time after initial lineage choice is made. Because of this lack of early markers, many studies of PNS glial development have used markers of later stages of development such as S100 or glial fibrillary acidic protein (GFAP; Fig. 12.4 and see below). Recently, however, at least three early glial markers have been identified. These are protein zero (P_0), brain fatty acid binding protein (B-FABP) and desert hedgehog (Dhh).

The myelin gene P_0 is, in the chick, activated in a subpopulation of migrating crest cells in stage 19 embryos and subsequently in the earliest glia associated with newly formed nerves [288]. In rats P_0 is first activated in migrating crest cells located in a region ventral to condensing DRGs at E11 [289]. P_0 remains expressed in the cells that associate with the earliest axons emanating from the ventral spinal cord and is subsequently seen in most/all cells in E14/15 rat nerves, i.e. Schwann cell precursors, and later, in immature Schwann cells. Post-natally, P_0 is strongly upregulated in cells that become myelinated and is suppressed in non-myelinating cells [290]. At no time is a P_0 *in situ* hybridization signal seen in neurons identified by the early marker β-III tubulin (detected by TUJ1 antibodies), indicating that P_0 gene expression is excluded from the neuronal lineage from the onset (E. Calle and K.R. Jessen, unpublished data) [288, 289, 291]. These findings indicate that P_0 gene expression can be used as a very early marker of glial specification in NC development in chick and rat [288, 290].

The fatty acid binding protein B-FABP is expressed in the developing CNS, in particular by radial glia. It is absent from undifferentiated crest cells and crest-derived neurons but appears in early glia within DRGs and associated nerves at embryo day (E) 10–11 in the mouse, although not at this early stage in rat [292, 293].

Desert hedgehog (Dhh) is a member of the Hedgehog family of signaling proteins. Peri- and post-natally, Schwann cell-derived Dhh is necessary for the formation of the perineurial and epineurial sheaths that protect peripheral nerves [294]. A Dhh *in situ* hybridization signal is already present in mouse Schwann cell precursors at E12, although it is absent from the NC and neurons [295] (R. Mirsky and K.R. Jessen, unpublished data). Both P_0 and B-FABP appear earlier or are expressed more strongly, in cells in ventral roots and proximal spinal nerves than inside DRGs. Furthermore, a P_0 *in situ* hybridization signal is first seen in individual crest cells that appear not to be associated with neurons and lie ventrally to condensing DRGs at the level of the ventral part of the neural tube. These observations suggest that glial differentiation of those crest cells destined to become Schwann cells of peripheral nerves

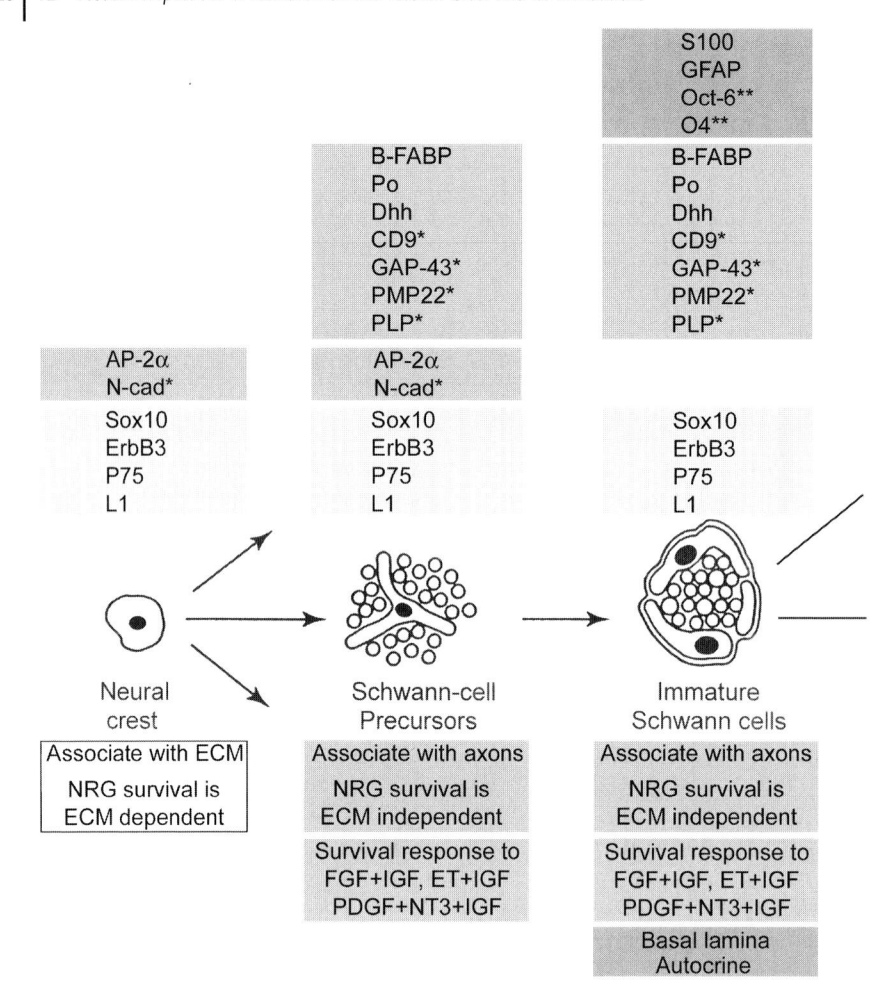

420 | 12 Novel Perspectives in Research on the Neural Crest and its Derivatives

		S100 GFAP Oct-6** O4**
	B-FABP Po Dhh CD9* GAP-43* PMP22* PLP*	B-FABP Po Dhh CD9* GAP-43* PMP22* PLP*
AP-2α N-cad*	AP-2α N-cad*	
Sox10 ErbB3 P75 L1	Sox10 ErbB3 P75 L1	Sox10 ErbB3 P75 L1
Neural crest	Schwann-cell Precursors	Immature Schwann cells
Associate with ECM NRG survival is ECM dependent	Associate with axons NRG survival is ECM independent	Associate with axons NRG survival is ECM independent
	Survival response to FGF+IGF, ET+IGF PDGF+NT3+IGF	Survival response to FGF+IGF, ET+IGF PDGF+NT3+IGF
		Basal lamina Autocrine

Fig. 12.4
Differentiation markers in embryonic Schwann cell development. Gray: markers shared by migrating, undifferentiated neural crest cells, Schwann cell precursors and immature Schwann cells. Blue: markers present on migrating crest cells and precursors but not (or at very low levels) on immature Schwann cells. Green: markers present on Schwann cell precursors and immature Schwann cells, but absent from migrating undifferentiated crest cells. These proteins also appear on neuroblasts/early neurons. Pink: markers expressed by immature Schwann cells but not (or at much lower levels) by Schwann cell precursors. These markers are acutely dependent on axons for expression. In addition Schwann cells have autocrine survival circuits not present in precursors. (This figure also appears with the color plates.)

commences relatively ventrally in structures that are on the level with the ventral regions of the neural tube.

12.8.3
Transcription Factors that Control the Emergence of Glia

Clonal cell culture experiments, *in vivo* cell-fate tracing and studies on the generation of sensory neurons all point to the possibility that a significant number of crest cells may already be strongly lineage biased or even irreversibly committed to particular fates (e.g. [132, 296, 297]). Most crest cells, however, are likely to be faced with lineage choice, as discussed previously. This is generally expected to be controlled, in the main, by inductive signals, although it is also possible that stochastic mechanisms operate in crest development [298] (also see Section 12.4.2).

The best established molecular regulator of crest gliogenesis *in vivo* is the transcription factor *Sox–10* [292, 299–305]. Sox–10 is expressed in migrating crest cells, is downregulated in early neurons but persists in developing glia both in ganglia and along nerve trunks, and in melanocytes [299]. By the time of birth the levels of *Sox–10* mRNA in peripheral nerves are, however, much lower than those in NC cells (U. Lange, K.R. Jessen and R Mirsky, unpublished data). In mice in which *Sox–10* is absent or non-functional, early glial cells fail to form in DRGs and nerve trunks. Nevertheless, there is evidence for the presence of a population of undifferentiated crest cells in these mice, particularly in the DRG [304], which indicates that *Sox–10* is required for the specification of glial cells from crest cells. Lack of *Sox–10* in crest cells leads to reduced levels of the neuregulin (NRG–1) receptor ErbB3 (see below), a receptor that mediates survival and proliferation signals in Schwann cell precursors along peripheral nerves [25, 306, 307]. This could provide an additional reason for the lack of early glia in nerve trunks of these mice, although this mechanism would not contribute to the absence of satellite cells since the development of satellite cells in DRGs does not require NRG–1 signaling (see below). Although B-FABP-positive glial precursors are not present in Sox–10–deficient mice, DRG sensory neurons, identified by expression of β-III tubulin, are initially formed normally, although they die later, underlining the idea that *Sox–10* has a particular function in glial development [304]. A further link between *Sox–10* and glial differentiation is provided by the finding that enforced expression of *Sox–10* induces expression of the endogenous P_0 gene, a marker of glial crest differentiation, in the neuroblastoma cell line N2A [301].

Besides being required for the formation of glia, Sox–10 is also required to maintain a precursor pool that can give rise to autonomic and enteric neurons, melanocytes, smooth muscle and glia [177, 303, 308]. A related transcription factor, *Sox–9*, is expressed earlier than *Sox–10* in pre-migratory NC. When it is expressed ectopically in chick neural cells it induces *Sox–10*, and in NC cells is permissive for glial development, promotes melanocyte development and is incompatible with neuronal differentiation [15].

Pax–3 is involved in the development of NC derivatives including Schwann cells, as seen for instance in mice homozygous for the *splotch* mutation, which is a deletion in the *Pax–3* homeodomain [104]. In these mice, DRGs are absent from lumbar and sacral regions and spinal nerves, and because they are formed from ventral roots only, are devoid of Schwann cell precursors or Schwann cells, perhaps due to failure of NC migration (Section 12.3.2.2).

12.8.4
Inductive Signals Involved in Glial Specification

As yet, inductive cell-cell signals that specify glial development from the crest have not been unambiguously identified in *in vivo* studies. Using crest cell cultures, however, at least three signals have been implicated in gliogenesis, namely beta forms of neuregulin–1 (NRG–1), delta-notch signaling, and bone morphogenetic proteins (BMPs). NRG–1 strongly suppresses the generation of neurons in crest cell cultures. By barring crest cells from the neuronal lineage, NRG–1 might leave crest cells exposed to glial-inducing signals and therefore, indirectly, promote glial development. Nevertheless, GFAP-positive (glial) cells emerge irrespective of the presence or absence of NRG–1 [309] and in mouse mutants in which NRG–1 signaling is inactivated, satellite cell development is normal, confirming that NRG–1 is not required for the general development of glial cells from the crest. Despite this, in NRG mutants, Schwann cell precursors are missing. This is likely to reflect the role of NRG–1, in particular the membrane-associated isoform III, as a Schwann cell precursor survival signal (see below). In addition to the possible role of NRG–1 in early glial development, there is strong evidence for the involvement of this factor in the control of survival, proliferation and differentiation of later cells in the Schwann cell lineage.

Notch activation promotes the appearance of glial cells in the CNS *in vivo* [310, 311], and in crest cultures notch activation leads to an increase in the number of GFAP$^+$ cells [312, 313]. This is consistent with the action of a positive inductive signal promoting gliogenesis. Two other explanations are, however, equally plausible. First, delta-notch signaling, which also strongly suppresses neurogenesis could act indirectly as outlined above for NRG–1. Second, GFAP is a marker for immature Schwann cells and not for Schwann cell precursors. It is therefore possible to obtain an increase in the number of GFAP$^+$ cells in experiments of this kind without instructing crest cells to develop as glia, simply by promoting survival/proliferation of Schwann cell precursors or accelerating their transition to GFAP$^+$ Schwann cells. It remains to be determined, whether notch activation is involved in either of these roles.

BMPs, which are important in the development of sympathetic neurons *in vivo* and *in vitro* [175] (see also Section 12.5.1), inhibit the generation of GFAP$^+$ cells *in vitro* [172]. BMPs may therefore act as negative instructive signals for gliogenesis. Dorsally expressed BMPs might for instance suppress gliogenesis in the dorsal crest and contribute to the pattern of ventral crest gliogenesis. On the other hand, the suppressive effect of BMP on gliogenesis raises the question of how glia emerge in sympathetic ganglia where BMPs are required for neuronal development. One possibility would be that this took place through instructive notch activation, since there is evidence that delta-notch signaling dominates over the anti-gliogenic effect of BMPs [313].

12.8.5
Developmental Plasticity of Early Glia

Although crest-derived cells in early embryonic nerve trunks are likely to be fated to generate only glial cells during development, it has been clear for some time that when such cells are placed in culture, they can generate crest derivatives other than glia. It is possible that some of this plasticity can be accounted for by a small population of cells that, although biased towards glial development, remains essentially similar to migrating NC cells in developmental potential [142, 143, 312]. The main explanation for the generation of neurons and melanocytes from young nerve trunks *in vitro* is however, likely to be that cells that have unambiguously started glial development, i.e. Schwann cell precursors, retain a degree of developmental plasticity. In the chick, for instance, cells expressing the glial marker P_0 [314, 315] or the glial marker SMP [316] can be induced to generate melanocytes *in vitro*. Similarly in the rat, P_0-positive cells can be induced to form neurons [291, 142]. This re-programming of early, specified glial cells is, in principle, in line with observations on developmental plasticity in a number of other systems [317–319].

12.9
Signals that Control Schwann Cell-precursor Survival and Schwann Cell Generation

12.9.1
Schwann Cell Precursors

Schwann cell precursors are the glial cells of E14/15 rat (E12/13 mouse) nerves [285, 286, 320]. By E17/18 these cells have converted into immature Schwann cells. Unlike undifferentiated crest cells, the precursors express B-FABP, Dhh and P_0. They show very low levels of S100 protein, which is found at much higher levels in Schwann cells. Another critical difference between these cells and Schwann cells relates to survival support. When dissociated from neurons and plated *in vitro*, the precursors die rapidly unless they are rescued by specific survival factors. The most potent of these is NRG–1, although precursor survival is also promoted by endothelins, and FGF2 in the presence of IGF–1/2. This acute requirement for extrinsic survival signals is due to the inability of precursors to support their own survival by autocrine or other cell density-dependent mechanisms. On the other hand, Schwann cells from E18 or older nerves survive in comparable cultures without added factors due to their ability to survive by autocrine mechanisms (see below). Other differences between undifferentiated crest cells, Schwann cell precursors and immature Schwann cells are outlined in Fig. 12.4.

In vivo, Schwann cell precursors lie in close apposition with axons that they envelop communally by extension of flattened processes along the surface and in the interior of embryonic nerves. A large body of evidence from *in vivo* and *in vitro* studies shows that precursor survival depends on survival signals from the axons they contact, and that the most important of these by far is NRG–1, in particular isoform III, acting on ErbB2/3 receptors in the precursors. The dependence of precursors on neuron-derived signals has been shown both *in vitro* [306, 320] and *in vivo* [321, 322]. The neuron-derived survival signal *in vitro* has been identified as NRG–1 [306]. This is in line with observations in mouse mutants in which NRG–1 signaling had been inactivated, since the nerves of these animals are essentially devoid of Schwann cell precursors, although glial cells within DRGs are normal [323]. A survival function for NRG–1 is further supported by observations in mutants lacking isoform III of NRG–1 [324]. In these mice, early Schwann cell precursors (at E11) populate nerves, but by E14, a time when normal nerves contain late precursors that are converting rapidly to Schwann cells, the number of cells in the nerves is severely depleted. Neuregulin has also been found to support the survival of early glia in embryonic chick nerves [322]. Neuregulin–1 is expressed *in vivo* in embryonic DRG and motor neurons and accumulates along axonal tracts [325–328]. All of these findings taken together strongly support the notion that a major function of NRG–1 in embryonic nerves is to support the survival of Schwann cell precursors [285, 286, 306].

Although the evidence is not unequivocal, reduction in glial migration from DRGs may also contribute to the lack of glia in nerve trunks. This notion derives mainly from the observation that the failure to form sympathetic ganglia in mice lacking the NRG–1 receptor ErbB3 appears to be due to impaired crest cell migration [25].

The absence of Schwann cell precursors seen in mice lacking isoform III of *NRG–1*, the *ErbB3* receptor, or *Sox–10* as described above, has an unexpected effect on neuronal development. In these animals, DRG neurons and motor neurons, that initially formed normally, later die in large numbers at limb levels of the spinal cord. It has been suggested that this cell death is due to the absence of glial-derived factors that are needed for the survival of sensory and motor neurons [26, 292]. This raises the possibility that a key function of Schwann cell precursors and early Schwann cells is to regulate the survival of discrete pools of embryonic CNS and PNS neurons. In the *NRG–1* mutants, loss of contact between nerve terminals and peripheral targets may also contribute to sensory and motor neuron death [328, 329]. This is less likely in the case of DRG neuron death in Sox–10 mutants since these cells die relatively early and before most of them have had significant interactions with target tissues. This observation points to a direct trophic role not only for developing Schwann cells but also for satellite cells [292]. DRG neuron death occurs earlier and also more extensively in Sox–10 mutants, where satellite cells and Schwann cell precursors are missing, than in Erb B3 mutants which only lack Schwann cell precursors.

12.9.2
Schwann Cell Generation

In the nerves of the rat hind limb, the conversion of precursors to Schwann cells takes place between E14/15 and E17/18. DNA synthesis accelerates during this period and is considerably higher in E18 cells than in the precursors [330]. Coincident with this change in glial phenotype, the entire architecture of peripheral nerves switches from compact to loose: at E14 and 15 nerves contain no significant extracellular space, and collagen fibrils and basal lamina are absent as are blood vessels. By E18, however, nerves contain numerous extracellular spaces that are starting to separate into discrete axon-Schwann cell units. These spaces contain nascent collagen fibrils, basal lamina have started to form and blood vessels are now present [324]. Little is known about the mechanisms that control these changes in cell and tissue relationships.

The precursor/Schwann cell transition can be studied *in vitro*, since in the presence of NRG–1, even in cultures without neurons, E14 cells convert to cells with the phenotype of E18 Schwann cells [306]. The time course of the transition can be retarded by endothelins, acting via endothelin B receptors. Endothelins are expressed in peripheral nerves and in the spotting lethal rat, in which endothelin B receptors are inactive, Schwann cells are generated ahead of schedule [331]. In the presence of endothelin alone *in vitro*, precursors convert very slowly to Schwann cells, but the process is accelerated by the addition of NRG–1. Together these observations show that endothelin acts as a negative regulator of the rate of Schwann cell generation *in vitro* and *in vivo*. The experiments also suggest that NRG–1 accelerates the conversion of precursors to Schwann cells, in addition to supporting precursor survival.

The precursor/Schwann cell conversion requires a coordinated change in features as diverse as molecular expression, survival regulation, response to mitogens, cell shape and migration. The transcriptional control of this program remains to be described. The only transcription factor that has been functionally implicated in this process is AP2α. AP2α expression is sharply downregulated at the precursor/Schwann cell transition in both rat and mouse. This may be important, since if it is prevented by enforcing AP2α expression in precursors *in vitro*, the conversion is delayed [286].

The two strikingly different Schwann cell types that we know from normal adult nerves, the myelinating and non-myelinating cells, develop from immature Schwann cells in a process that starts around birth (E21 in rats), when the first cells fall out of division and start wrapping axons to form myclin sheaths. The discussion of this process lies outside the scope of this chapter (for reviews see [332–334]).

Acknowledgments

This work was supported by grants from the March of Dimes Birth Defects Foundation, the ICRF and the DFG (SFB488) to C. K.; the DFG (SFB 269 and DFG 5301) and EU (grants Bio4–1998–0112 and QLG3–2000–0072) to H.R.; K.R.J. and R. M. were supported in part by a program grant from the Wellcome Trust.

References

1 Le Douarin, N.M. and Kalcheim, C. *The Neural Crest* (2nd edn). New York: Cambridge University Press, **1999**.
2 Holland, L. and Holland, N. Evolution of neural crest and placodes: amphioxus as a model for the ancestral vertebrate? *J Anat* **2001**; *199*: 85–98.
3 McCauley, D. and Bronner Fraser, M. neural crest contributions to the lamprey head. *Development* **2003**; *130*: 2317–2327.
4 His W, Jr. Die Häute und Höhlen des Körpers. *Arch Anat Physiol, Anat Abt* **1903**; 368.
5 His, W. *Untersuchungen über die erste Anlage des Wirbeltierleibes. Die erste Entwicklung des Hühnchens im Ei*. Leipzig: F.C.W. Vogel, **1868**.
6 Hörstadius, S. *The Neural Crest: Its Properties and Derivatives in the Light of Experimental Research*. London: Oxford University Press, **1950**.
7 Le Douarin, N.M. Particularités du noyau interphasique chez la Caille japonaise (*Coturnix coturnix japonica*). Utilisation de ces particularités comme "marquage biologique" dans les recherches sur les interactions tissulaires et les migrations cellulaires au cours de l'ontogenèse. *Bull Biol Fr Bel*. **1969**; *103*: 435–452.
8 Bronner-Fraser, M. and Fraser, S.E. Cell lineage analysis reveals multipotency of some avian neural crest cells. *Nature* **1988**; *335*: 161–164.
9 Krull, C.E., Lansford, R., Gale, N.W., et al. Interactions of Eph-related receptors and ligands confer rostrocaudal pattern to trunk neural crest migration. *Curr Biol* **1997**; *7*: 571–580.
10 Kulesa, P., Bronner Fraser, M. and Fraser, S. *In ovo* time-lapse analysis after dorsal neural tube ablation shows rerouting of chick hindbrain NC. *Development* **2003**; *127*: 2843–2852.
11 Kulesa, P. and Fraser, S. Confocal imaging of living cells in intact embryos. *Methods Mol Biol* **1999**; *122*: 205–222.
12 Kulesa, P. and Fraser, S. *In ovo* time-lapse analysis of chick hindbrain neural crest cell migration shows cell interactions during migration to the branchial arches. *Development* **2003**; *127*: 1161–1172.
13 Kalcheim, C. Mechanisms of early neural crest development: From cell specification to migration. *International Rev Cytol: Survey Cell Biol* **2000**; *200*: 143–195.
14 Gammill, L.S. and Bronner-Fraser, M. Neural crest specification: Migrating into genomics. *Nat Rev Neurosci* **2003**; *4*: 795–805.

15 Cheung, M. and Briscoe, J. Neural crest development is regulated by the transcription factor Sox9. *Development* **2003**; *130*: 5681–5693.
16 Dottori, M., Gross, M.K., Labosky, P. and Goulding, M. The winged-helix transcription factor Foxd3 suppresses interneuron differentiation and promotes neural crest cell fate. *Development* **2001**; *128*: 4127–4138.
17 Kos, R., Reedy, M.V., Johnson, R.L. and Erickson, C.A. The winged-helix transcription factor FoxD3 is important for establishing the neural crest lineage and repressing melanogenesis in avian embryos. *Development* **2001**; *128*: 1467–1479.
18 Spokony, R.F., Aoki, Y., Saint-Germain, N., Magner-Fink, E. and Saint-Jeannet, J.P. The transcription factor Sox9 is required for cranial neural crest development in Xenopus. *Development* **2002**; *129*: 421–432.
19 Nieto, M.A., Sargent, M.G., Wilkinson, D.G. and Cooke, J. Control of cell behavior during vertebrate development by slug, a zinc finger gene. *Science* **1994**; *264*: 835–839.
20 Sefton, M., Sanchez, S. and Nieto, M.A. Conserved and divergent roles for members of the Snail family of transcription factors in the chick and mouse embryo. *Development* **1998**; *125*: 3111–3121.
21 Fode, C., Gradwohl, G., Morin, X., et al. The bHLH protein NEUROGENIN 2 is a determination factor for epibranchial placode-derived sensory neurons. *Neuron* **1998**; *20*: 483–494.
22 Ma, Q.F., Chen, Z.F., Barrantes, I.D., De la Pompa, J.L. and Anderson, D.J. *Neurogenin1* is essential for the determination of neuronal precursors for proximal cranial sensory ganglia. *Neuron* **1998**; *20*: 469–482.
23 Perez, S.E., Rebelo, S. and Anderson, D.J. Early specification of sensory neuron fate revealed by expression and function of neurogenins in the chick embryo. *Development* **1999**; *126*: 1715–1728.
24 Gammill, L. and Bronner Fraser, M. Genomic analysis of neural crest induction. *Development* **2002**; *129*:5731–5741.
25 Britsch, S., Li, L., Kirchhoff, S., et al. The ErbB2 and ErbB3 receptors and their ligand, neuregulin–1, are essential for development of the sympathetic nervous system. *Genes Dev* **1998**; *12*: 1825–1836.
26 Riethmacher, D., Sonnenberg Riethmacher, E., Brinkmann, V., Yamaai, T., Lewin, G. and Birchmeier, C. Severe neuropathies in mice with targeted mutations in the ErbB3 receptor. *Nature* **1997**; *389*: 725–730.
27 Kahane, N. and Kalcheim, C. Expression of trkC receptor mRNA during development of the avian nervous system. *J Neurobiol* **1994**; *25*: 571–584.
28 Rifkin, J., Todd, V., Anderson, L. and Lefcort, F. Dynamic expression of neurotrophin receptors during sensory neuron genesis and differentiation. *Dev Biol* **2003**; *227*: 465–480.
29 De Bellard, M.E., Rao, Y. and Bronner-Fraser, M. Dual function of Slit2 in repulsion and enhanced migration of trunk, but not vagal, neural crest cells. *J Cell Biol* **2003**; *162*: 269–279.
30 Deardorff, M., Tan, C., Saint Jeannet, J. and Klein, P. A role for frizzled 3 in neural crest development. *Development* **2003**; *128*: 3655–3663.

31 Roelink, H., and Nusse, R. Expression of two members of the Wnt family during mouse development-restricted temporal and spatial patterns in the developing neural tube. *Genes Dev* **1991**; *5*: 381–388.

32 McMahon AP, Joyner AL, Bradley A, McMahon JA. The midbrain-hindbrain phenotype of Wnt–1–/Wnt–1– mice results from stepwise deletion of engrailed-expressing cells by 9.5 days postcoitum *Cell* **1992**; *69*, 581–595.

33 Parr, B., Shea, M., Vassileva, G. and McMahon, A. Mouse Wnt genes exhibit discrete domains of expression in the early embryonic CNS and limb buds. *Development* **1993**; *119*: 247–261.

34 Echelard, Y., Vassileva, G. and McMahon, A.P. Cis-acting regulatory sequences governing Wnt–1 expression in the developing mouse CNS. *Development* **1994**; *120*: 2213–2224.

35 Moreno, T. and Bronner Fraser, M. The secreted glycoprotein Noelin–1 promotes neurogenesis in Xenopus. *Dev Biol* **2003**; *240*: 340–360.

36 Sela-Donenfeld, D. and Kalcheim, C. Regulation of the onset of neural crest migration by coordinated activity of BMP4 and noggin in the dorsal neural tube. *Development* **1999**; *126*: 4749–4762.

37 Liem, K.F.Jr., Tremml, G. and Jessell, T.M. A role for the roof plate and its resident TGFβ-related proteins in neuronal patterning in the dorsal spinal cord. *Cell* **1997**; *91*:127–138.

38 Akitaya, T. and Bronner Fraser, M. Expression of cell adhesion molecules during initiation and cessation of neural crest cell migration. *Dev Dyn* **1992**; *194*: 12–20.

39 Duband, J.L., Volberg, T., Sabanay, I., Thiery, J.P. and Geiger, B. Spatial and temporal distribution of the adherens-junction-associated adhesion molecule A-CAM during avian embryogenesis. *Development* **1988**; *103*: 325–344.

40 Nakagawa, S. and Takeichi, M. neural crest cell-cell adhesion controlled by sequential and subpopulation-specific expression of novel cadherins. *Development* **1995**; *121*: 1321–1332.

41 Newgreen, D. and Gooday, D. Control of the onset of migration of neural crest cells in avian embryos. Role of Ca -DEPENDENT CELL ADHESION. *Cell Tissue Res* **1985**; *239*: 329–336.

42 Newgreen, D.F. and Minichiello, J. Control of epitheliomesenchymal transformation. I. Events in the onset of neural crest cell migration are separable and inducible by protein kinase inhibitors. *Dev Biol* **1995**; *170*: 91–101.

43 Vallin, J., Girault, J.M., Thiery, J.P. and Broders, F. Xenopus cadherin–11 is expressed in different populations of migrating neural crest cells. *Mech Dev* **1998**; *75*: 171–174.

44 Lewandoski, M. Conditional control of gene expression in the mouse. *Nat Rev Genet* **2003**; *2*: 743–755.

45 Ikeya, M., Lee, S.M.K., Johnson, J.E., McMahon, A.P. and Takada, S. Wnt signalling required for expansion of neural crest and CNS progenitors. *Nature* **1997**; *389*: 966–970.

46 Dunn, K., Williams, B., Li, Y. and Pavan, W. Neural crest-directed gene transfer demonstrates Wnt1 role in melanocyte expansion and differentiation during mouse development. *Proc Natl Acad Sci USA* **2003**; *97*: 10050–10055.

47 Greenstein-Baynash, A., Hosoda, K., Giaid, A., Richardson, J.A., Emoto, N., Hammer, R.E. and Yanagisawa, M. Interaction of endothelin–3 with endothelin-B receptor is essential for development of neural crest-derived melanocytes and enteric neurons : missense mutation of endothelin–3 gene in *lethal spotting* mice. *Cell* **1994**; *79*: 1277–1285.

48 Hosoda, K., Hammer, R.E., Richardson, J.A., Greenstein-Baynash, A., Cheung, J.C., Giaid, A. and Yanagisawa, M. Targeted and natural (piebald-lethal) mutations of endothelin-B receptor gene produce megacolon associated with spotted coat color in mice. *Cell* **1994**; *79*: 1267–1276.

49 Nagy, A. Cre recombinase: the universal reagent for genome tailoring. *Genesis* **2003**; *26*: 99–109.

50 Danielian, P., Muccino, D., Rowitch, D., Michael, S. and McMahon, A. Modification of gene activity in mouse embryos *in utero* by a tamoxifen-inducible form of Cre recombinase. *Curr Biol* **1998**; *8*: 1323–1326.

51 Jiang, X., Choudhary, B., Merki, E., Chien, K., Maxson, R. and Sucov, H. Normal fate and altered function of the cardiac neural crest cell lineage in retinoic acid receptor mutant embryos. *Mech Dev* **2003**; *117*: 115–122.

52 Jiang, X., Rowitch, D., Soriano, P., McMahon, A. and Sucov, H. Fate of the mammalian cardiac neural crest. *Development* **2003**; *127*: 1607–1616.

53 Li, J., Chen, F. and Epstein, J. Neural crest expression of Cre recombinase directed by the proximal Pax3 promoter in transgenic mice. *Genesis* **2003**; *26*: 162–164.

54 Yamauchi, Y., Abe, K., Mantani, A., et al. A novel transgenic technique that allows specific marking of the neural crest cell lineage in mice. *Dev Biol* **1999**; *212*: 191–203.

55 Feltri, M., D'Antonio, M., Quattrini, A., et al. A novel P0 glycoprotein transgene activates expression of lacZ in myelin-forming Schwann cells. *Eur J Neurosci* **1999**; *11*: 1577–1586.

56 Voiculescu, O., Charnay, P. and Schneider Maunoury, S. Expression pattern of a Krox–20/Cre knock-in allele in the developing hindbrain, bones, and peripheral nervous system. *Genesis* **2003**; *26*: 123–126.

57 Pietri, T., Eder, O., Blanche, M., Thiery, J. and Dufour, S. The human tissue plasminogen activator-Cre mouse: a new tool for targeting specifically neural crest cells and their derivatives *in vivo*. *Dev Biol* **2003**; *259*: 176–187.

58 Brault, V., Moore, R., Kutsch, S., et al. Inactivation of the beta-catenin gene by Wnt1–Cre-mediated deletion results in dramatic brain malformation and failure of craniofacial development. *Development* **2003**; *128*: 1253–1264.

59 Hari, L., Brault, V., Kleber, M., et al. Lineage-specific requirements of beta-catenin in neural crest development. *J Cell Biol* **2003**; *159*: 867–880.

60 Zirlinger, M., Lo, L., McMahon, J., McMahon, A.P. and Anderson, D.J. Transient expression of the bHLH factor neurogenin–2 marks a subpopulation of neural crest cells biased for a sensory but not a neuronal fate. *Proc Natl Acad Sci USA* **2002**; *99*: 8084–809.

61 Momose, T., Tonegawa, A., Takeuchi, J., Ogawa. H., Umesono, K. and Yasuda, K. Efficient targeting of gene expression in chick embryos by microelectroporation. *Dev Growth Differ* **1999**; *41*: 335–344.

62 Muramatsu, T., Mizutani, Y., Ohmori, Y. and Okumura, J. Comparison of three nonviral transfection methods for foreign gene expression in early chicken embryos *in ovo*. *Biochem Biophys Res Commun* **1997**; *230*: 376–380.

63 Burstyn-Cohen, T. and Kalcheim, C. Association between the cell cycle and neural crest delamination through regulation of G1–S transition. *Dev Cell* **2002**; *3*: 383–395.

64 Itasaki, N. Bel-Vialar, S. and Krumlauf, R. "Shocking developments in chick embryology". Electroporation and *in ovo* gene expression. *Nat Cell Biol* **1999**; *1*: E203–E207.

65 Timmer, J., Johnson, J. and Niswander, L. The use of *in ovo* electroporation for the rapid analysis of neural-specific murine enhancers. *Genesis* **2003**; *29*:123–132.

66 Elbashir, S., Lendeckel, W. and Tuschl, T. RNA interference is mediated by 21– and 22–nucleotide RNAs. *Genes Dev* **2003**; *15*: 188–200.

67 Hu, W., Myers, C., Kilzer, J., Pfaff, S. and Bushman, F. Inhibition of retroviral pathogenesis by RNA interference. *Curr Biol* **2003**; *12*: 1301–1311.

68 Pekarik, V., Bourikas, D., Miglino, N., Joset, P., Preiswerk, S. and Stoeckli, E. Screening for gene function in chicken embryo using RNAi and electroporation. *Nat Biotechnol* **2003**; *21*: 93–96.

69 Kos, R., Tucker, R., Hall, R., Duong, T. and Erickson, C. Methods for introducing morpholinos into the chicken embryo. *Dev Dyn* **2003**; *226*: 470–477.

70 Tucker, R. Abnormal neural crest cell migration after the in vivo knockdown of tenascin-C expression with morpholino antisense oligonucleotides. *Dev Dyn* **2003**; *222*: 115–119.

71 Wu, J., Saint Jeannet, J. and Klein, P. Wnt-frizzled signaling in neural crest formation. *Trends Neurosci* **2003**; *26*: 40–45.

72 Knecht, A. and Bronner Fraser, M. Induction of the neural crest: a multigene process. *Nat Rev Genet* **2003**; *3*: 453–461.

73 Selleck, M.A.J. and Bronner-Fraser, M. Origins of the avian neural crest: The role of neural plate-epidermal interactions. *Development* **1995**; *121*: 525–538.

74 Collazo, A., Bronner-Fraser, M. and Fraser, S.E. Vital dye labelling of *Xenopus laevis* trunk neural crest reveals multipotency and novel pathways of migration. *Development* **1993**; *118*: 363–376.

75 del Barrio, M.G. and Nieto, M.A. Overexpression of Snail family members highlights their ability to promote chick neural crest formation. *Development* **2002**; *129*: 1583–193.

76 Jiang, R., Lan, Y., Norton, C.R., Sundberg, J.P. and Gridley, T. The Slug gene is not essential for mesoderm or neural crest development in mice. *Dev Biol* **1998**; *198*: 277–285.

77 Carver, E.A., Jiang, R.L., Lan, Y., Oram, K.F. and Gridley, T. The mouse snail gene encodes a key regulator of the epithelial-mesenchymal transition. *Mol Cell Biol* **2001**; *21*: 8184–8188.

78. LaBonne, C. and Bronner-Fraser, M. Snail-related transcriptional repressors are required in *Xenopus* for both the induction of the neural crest and its subsequent migration. *Dev Biol* **2000**; *221*: 195–205.
79. Aybar, M., Nieto, M. and Mayor, R. Snail precedes slug in the genetic cascade required for the specification and migration of the Xenopus neural crest. *Development* **2003**; *130*: 483–494.
80. Bellmeyer, A., Krase, J., Lindgren, J. and LaBonne, C. The protooncogene c-myc is an essential regulator of neural crest formation in xenopus. *Dev Cell* **2003**; *4*: 827–839.
81. Nakata, K., Koyabu, Y., Aruga, J. and Mikoshiba, K. A novel member of the xenopus zic family, zic5, mediates neural crest development. *Mech Dev* **2000**; *99*: 83–91.
82. Kaufmann, E. and Knochel, W. Five years on the wings of fork head. *Mech Dev* **1996**; *57*: 3–20.
83. Kaestner, K., Knochel, W. and Martinez, D. Unified nomenclature for the winged helix/forkhead transcription factors. *Genes Dev* **2003**; *14*: 142–146.
84. Pohl, B.S. and Knochel, W. Overexpression of the transcriptional repressor FoxD3 prevents neural crest formation in Xenopus embryos. *Mech Dev* **2001**; *103*: 93–106.
85. Wegner, M. From head to toes: the multiple facets of Sox proteins. *Nucleic Acids Res* **1999**; *27*: 1409–1420.
86. Luo, T., Lee, Y., Saint Jeannet, J. and Sargent, T. Induction of neural crest in Xenopus by transcription factor AP2alpha. *Proc Natl Acad Sci USA* **2003**; *100*: 532–537.
87. Schorle, H., Meier, P., Buchert, M., Jaenisch, R. and Mitchell, P.J. Transcription factor AP–2 essential for cranial closure and craniofacial development. *Nature* **1996**; *381*: 235–238.
88. Zhang, J.A., Hagopian-Donaldson, S., Serbedzija, G., et al. Neural tube, skeletal and body wall defects in mice lacking transcription factor AP–2. *Nature* **1996**; *381*: 238–241.
89. Cheng, Y., Cheung, M., Abu-Elmagd, M.M., Orme, A., and Scotting, P.J. Chick sox 10, a transcription factor expressed in both early neural crest cells and central nervous system. *Dev Brain Res* **2000**; *121*: 233–241.
90. Aoki, Y. Sait-Germain, N, Gyda, M., Magner-Fink, E., Lee, Y.H., Credidio, C, Saint-Jeannet, JP. Sox 10 regulates the development of neural crest-derived melanocytes in Xenopus. *Dev Biol* **2003**; *259*: 19–33.
91. Honore, S., Aybar, M. and Mayor, R. Sox10 is required for the early development of the prospective neural crest in Xenopus embryos. *Dev Biol* **2003**; *260*: 79–96.
92. Barembaum, M., Moreno, T., LaBonne, C., Sechrist, J. and Bronner Fraser, M. Noelin–1 is a secreted glycoprotein involved in generation of the neural crest. *Nat Cell Biol* **2003**; *2*: 219–225.
93. Martinsen, B.J. and Bronner-Fraser, M. neural crest specification regulated by the helix-loop-helix repressor Id2. *Science* **1998**; *281*: 988–991.
94. Abzhanov, A., Tzahor, E., Lassar, A.B. and Tabin, C.J. Dissimilar regulation of cell differentiation in mesencephalic (cranial) and sacral (trunk) neural crest cells *in vitro*. *Development* **2003**; *130*: 4567–4579.

95 Thiery, J. Epithelial-mesenchymal transitions in tumour progression. *Nat Rev Cancer* **2003**; *2*:442–454.
96 Christiansen, J.H., Coles, E.G. and Wilkinson, D.G. Molecular control of neural crest formation, migration and differentiation. *Curr Opin Cell Biol* **2000**; *12*: 719–724.
97 Nieto, A.M. The early steps in neural crest development. *Mech Dev* **2001**; *105*: 27–35.
98 Kanzler, B., Foreman, R.K., Labosky, P.A. and Mallo, M. BMP signaling is essential for development of skeletogenic and neurogenic cranial neural crest. *Development* **2000**; *127*: 1095–1104.
99 Sela-Donenfeld, D. and Kalcheim, C. Inhibition of noggin expression in the dorsal neural tube by somitogenesis: a mechanism for coordinating the timing of neural crest emigration. *Development* **2000**; *127*: 4845–4854.
100 Van de Putte, T., Maruhashi, M., Francis, A., et al. Mice lacking ZFHX1B, the gene that codes for Smad-interacting protein–1, reveal a role for multiple neural crest cell defects in the etiology of Hirschsprung disease-mental retardation syndrome. *Am J Hum Genet* **2003**; *72*: 465–470.
101 Goulding, M.D, Chalepakis, G., Deutch, U., Erselius, J.R. and Gruss, P. Pax–3, a novel murine DNA binding protein expressed during early neurogenesis. *EMBO J* **1991**; *10*: 1135–1147.
102 Russell, W.L. Splotch, a new mutation in the house mouse *Mus musculus*. *Genetics* **1947**; *32*: 107–111.
103 Kessel, M. and Gruss, P. Murine developmental control genes. *Science* **1990**; *249*: 374–379.
104 Epstein, D.J., Vekemans, M. and Gros, P. *Splotch* (*Sp2H*), a mutation affecting development of the mouse neural tube, shows a deletion within the paired homeodomain of *Pax–3*. *Cell* **1991**; *67*: 767–774.
105 Moase, C.E. and Trasler, D.G. Delayed neural crest cell emigration from *Sp* and *Spd* mouse neural tube explants. *Teratology* **1990**; *42*: 171–182.
106 Serbedzija, G.N. and McMahon, A.P. Analysis of neural crest cell migration in Splotch mice using a neural crest-specific LacZ reporter. *Dev Biol* **1997**; *185*: 139–148.
107 Wiggan, O., Fadel, M. and Hamel, P. Pax3 induces cell aggregation and regulates phenotypic mesenchymal-epithelial interconversion. *J Cell Sci* **2003**; *115*: 517–529.
108 Wiggan, O. and Hamel, P. Pax3 regulates morphogenetic cell behavior in vitro coincident with activation of a PCP/non-canonical Wnt-signaling cascade. *J Cell Sci* **2003**; *115*: 531–541.
109 Kahane N, and Kalcheim C. Identification of early postmitotic cells in distinct embryonic sites and their possible roles in morphogenesis. *Cell Tissue Res* **1998**; *294*: 297–307.
110 Martin, A. and Langman, J. The development of the spinal cord examined by autoradiography. *J Embryo Exper Morphol* **1965**; *14*: 25–35.
111 Langman J, Guerrant R.L. and Freeman B.G. Behavior of neuro-epithelial cells during closure of the neural tube. *J Comp Neurol* **1966**; *127*: 399–411.
112 Smith, J.L. and Schoenwolf, G.C. Cell cycle and neuroepithelial cell shape during bending of the chick neural plate. *Anat Rec* **1987**; *218*: 196–206.

113 Smith, J.L, and Schoenwolf, G.C. Role of cell-cycle in regulating neuroepithelial cell shape during bending of the chick neural plate. *Cell Tissue Res* **1998**; *252*: 491–500.
114 Le Douarin, N.M. and Teillet, M.-A. Experimental analysis of the migration and differentiation of neuroblasts of the autonomic nervous system and of neuroectodermal mesenchymal derivatives using a biological cell marking technique. *Dev Biol* **1974**; *41*: 162–184.
115 Le Douarin, N.M.. The ontogeny of the neural crest in chick embryo chimaeras. *Nature* **1980**; *286*: 663–669.
116 Nakamura, H. and Ayer-Le Lievre, C. Neural crest and thymic myoid cells. *Curr Top Dev Biol* **1986**; *20*: 111–115.
117 Kirby, M.L. Plasticity and predetermination of mesencephalic and trunk neural crest transplanted into the region of the cardiac neural crest. *Dev Biol* 1989; *134*: 402–412.
118 Rijli, F.M., Mark, M., Lakkaraju, S., Dietrich, A., Dolle, P. and Chambon, P. A homeotic transformation is generated in the rostral branchial region of the head disruption of Hoxa–2, which acts as a selector gene. *Cell* 1993; *75*: 1333–1349.
119 Grammatopoulos, GA., Bell, E., Toole, L., Lumsden, A. and Tucker, A. Homeotic transformation of branchial arch identity after Hoxa2 overexpression. *Development* **2000**; *127*: 5355–5365.
120 Noden, D.M. The role of the neural crest in patterning of avian cranial skeletal, connective, and muscle tissues. *Dev Biol* **1983**; *96*: 144–165.
121 Trainor, P.A. and Krumlauf, R. Plasticity in mouse neural crest cells reveals a new patterning role for cranial mesoderm. *Nat Cell Biol* **2000**; *2*: 96–102.
122 Schilling, T. Plasticity of zebrafish Hox expression in the hindbrain and cranial neural crest hindbrain. *Dev Biol* **2001**; *231*: 201–216.
123 Trainor, P.A. and Krumlauf, R. Hox genes, neural crest cells and branchial arch patterning. *Curr Opin Cell Biol* **2001**; *13*: 698–705.
124 Serbedzida, G.N., Bronner-Fraser, M. and Fraser, S.E. A vital dye analysis of the timing and pathways of avian trunk neural crest cell migration. *Development* **1989**; *106*: 809–816.
125 Raible, D.W. and Eisen, J.S. Regulative interactions in zebrafish neural crest. *Development* **1996**; *122*: 510–507.
126 Artinger, K.B. and Bronner-Fraser, M. Partial restriction in the developmental potential of late emigrating avian neural crest cells. *Dev Biol* **1992**; *149*: 149–157.
127 Sieber-Blum, M. and Cohen, A.M. Clonal analysis of quail neural crest cells: they are pluripotent and differentiate *in vitro* in the absence of noncrest cells. *Dev Biol* **1980**; *80*: 96–106.
128 Baroffio, A., Dupin, E. and Le Douarin, N.M. Clone-forming ability and differentiation potential of migratory neural crest cells. *Proc Natl Acad Sci USA* **1988**; *85*: 5325–5329.
129 Stemple, D.L. and Anderson, D.J. Isolation of a stem cell for neurons and glia from the mammalian neural crest. *Cell* **1992**; *71*: 973–985.
130 Ito, K. and Sieber-Blum, M. In vitro clonal analysis of quail cardiac neural crest development. *Dev Biol* **1991**; *148*: 95–106.

131 Henion, P.D. and Weston, J.A. Timing and pattern of cell fate restrictions in the neural crest lineage. *Development* **1997**; *124*: 4351–4359.

132 Fraser, S.E. and Bronner-Fraser, M. Migrating neural crest cells in the trunk of the avian embryo are multipotent. *Development* **1991**; *112*: 913–920.

133 Frank, E. and Sanes, J.R. Lineage of neurons and glia in chick dorsal root ganglia: Analysis *in vivo* with a recombinant retrovirus. *Development* **1991**; *111*: 895–908.

134 Raible, D.W. and Eisen, J.S. Restriction of neural crest cell fate in the trunk of the embryonic zebrafish. *Development* **1994**; *120*: 495–503.

135 Greenwood, A.L., Turner, E.E. and Anderson, D.J. Identification of dividing, determined sensory neuron precursors in the mammalian neural crest. *Development* **1999**; *126*: 3545–3559.

136 LeDouarin, N.M. Cell line segregation during peripheral nervous system ontogeny. *Science* **1986**; *231*: 1515–1522.

137 Duff, R.S., Langtimm, C.J., Richardson, M.K. and Sieber-Blum, M. *In vitro* clonal analysis of progenitor cell patterns in dorsal root and sympathetic ganglia of the quail. *Dev Biol* **1991**; *147*: 451–459.

138 Sextier-Sainte-Claire Deville, F., Ziller, C. and Le Douarin, N. Developmental potentialities of cells derived from the truncal neural crest in clonal cultures. *Dev Brain Res* **1992**; *66*: 1–10.

139 Schweizer, G., Ayer-Lelievre, C. and LeDouarin, N.M. Restrictions of developmental capacities in the dorsal root ganglia in the course of development. *Cell Different* **1983**; *13*: 191–200.

140 Le Lievre, C.S., Schweizer, G.G., Ziller, C.M. and LeDouarin, N.M. Restrictions of developmental capabilities in neural crest cell derivatives as tested by *in vivo* transplantation experiments. *Dev Biol* **1980**; *77*: 362–378.

141 Luo, R., Gao, J., Wehrle-Haller, B. and Henion, P.D. Molecular identification of distinct neurogenic and melanogenic neural crest sublineages. *Development* **2003**; *130*: 321–330.

142 Morrison, S.J., White, P.M., Zock, C. and Anderson, D.J. Prospective identification, isolation by flow cytometry, and *in vivo* self-renewal of multipotent mammalian neural crest stem cells. *Cell* **1999**; *96*: 737–749.

143 White, P.M., Morrison, S.J., Orimoto, K., Kubu, C.J., Verdi, J.M. and Anderson, D.J. neural crest stem cells undergo cell-intrinsic developmental changes in sensitivity to instructive differentiation signals. *Neuron* **2001**; *29*: 57–71.

144 Bixby, S., Kruger, G.M., Mosher, J.T., Joseph, N.M. and Morrison, S.J. Cell-intrinsic differences between stem cells from different regions of the peripheral nervous system regulate the generation of neural diversity. *Neuron* **2002**; *35*: 643–656.

145 Kruger, G.M., Mosher, J.T., Bixby, S., Joseph, N., Iwashita, T. and Morrison, S.J. neural crest stem cells persist in the adult gut but undergo changes in self-renewal, neuronal subtype potential, and factor responsiveness. *Neuron* **2002**; *35*: 657–669.

146 Kruger, G.M., Mosher, J.T., Tsai, Y.-H., Yeager, K.J., Iwashita, T., Gariepy, C.E. and Morrison, S.J. Temporally distinct requirements for endothelin receptor B in the generation and

migration of gut neural crest stem cells. *Neuron* **2003**; *40*: 917–929.

147 Molofsky, A.V., Pardal, R., Iwashita, T., Park, I.-K., Clarke, M.F. and Morrison, S.J. Bmi–1 dependence distinguishes neural stem cell self-renewal from progenitor proliferation. *Nature* **2003**; *425*: 962–967.

148 Iwashita, T., Kruger, G.M., Pardal, R., Kiel, M.J. and Morrison, S.J. Hirschsprung disease is linked to defects in neural crest stem cell function. *Science* **2003**; *310*: 972–976.

149 Anderson, D.J., Groves, A., Lo, L., Ma, Q., Rao, M., Shah, N.M. and Sommer, L. Cell lineage determination and the control of neuronal identity in the neural crest. *Cold Spring Harbor Symp Quant Biol* **1997**; *62*: 493–504.

150 Sommer, L. Context-dependent regulation of fate decisions in multipotent progenitor cells of the peripheral nervous system. *Cell Tissue Res* **2001**; *305*: 211–216.

151 Goridis, C. and Rohrer, H. Specification of catecholaminergic and serotonergic neurons. *Nat Rev Neurosci* **2002**; *3*: 531–541.

152 Ungos, J.M., Karlstrom, R.O. and Raible, D.W. Hedgehog signaling is directly required for the development of zebrafish dorsal root ganglia neurons. *Development* **2003**; *130*: 5351–5362.

153 Fedtsova, N., Perris, R. and Turner, E.E. Sonic hedgehog regulates the position of the trigeminal ganglia. *Dev Biol* **2003**; *261*: 456–469.

154 Lee, H.-Y., Kléber, M., Hari, L., Brault, V., Suter, U., Taketo, M.M., Kemler, R. and Sommer, L. Instructive role of Wnt/b-catenin in sensory fate specification in neural crest stem cells. *Science* **2004**; *303*, 1020–1023.

155 Garcia-Castro, M.I., Marcelle, C. and Bronner-Fraser, M. Ectodermal Wnt function as a neural crest inducer. *Science* **2002**; *297*: 848–851.

156 Megason, S.G. and McMahon, A.P. A mitogen gradient of dorsal midline Wnts organizes growth in the CNS. *Development* **2002**; *129*: 2087–2098.

157 Dorsky, R.I., Moon, R.T. and Raible, D.W. Control of neural crest cell fate by the Wnt signalling pathway. *Nature* **1998**; *396*: 370–373.

158 Begbie, J., Brunet, J.-F., Rubenstein, J.L. and Graham, A. Induction of the epibranchial placodes. *Development* **1999**; *126*: 895–870.

159 Rohrer, H. The role of bone morphogenetic proteins in sympathetic neuron development. *Drug News Perspective* **2003**; *16*: 589–596.

160 Guillemot, F., Lo, L.-C., Johnson, J.E., Auerbach, A., Anderson, D.J. and Joyner, A.L. Mammalian achaete-scute homolog 1 is required for the early development of olfactory and autonomic neurons. *Cell* **1993**; *75*: 463–476.

161 Guillemot, F. and Joyner, A.L. Dynamic expression of the murine Achaete-Scute homologue Mash–1 in the developing nervous system. *Mech Dev* **1993**; *42*: 171–185.

162 Ernsberger, U., Patzke, H., Tissier-Seta, J.P., Reh, T., Goridis, C. and Rohrer, H. The expression of tyrosine hydroxylase and

the transcription factors cPhox–2 and Cash–1: Evidence for distinct inductive steps in the differentiation of chick sympathetic precursor cells. *Mech Develop* **1995**; *52*: 125–136.

163 Pattyn, A., Morin, X., Cremer, H., Goridis, C. and Brunet, J.F. Expression and interactions of the two closely related homeobox genes Phox2a and Phox2b during neurogenesis. *Development* **1997**; *124*: 4065–4075.

164 Huang, E.J., Zang, K.L., Schmidt, A., Saulys, A., Xiang, M.Q. and Reichardt, L.F. POU domain factor Brn–3a controls the differentiation and survival of trigeminal neurons by regulating Trk receptor expression. *Development* **1999**; *126*: 2869–2882.

165 Valarché, I., Tissier-Seta, J.-P., Hirsch, M.-R., Martinez, S., Goridis, C. and Brunet, J.-F. The mouse homeodomain protein Phox2 regulates Ncam promoter activity in concert with Cux/CDP and is a putative determinant of neurotransmitter phenotype. *Development* **1993**; *119*: 881–896.

166 Lim, K.-C., Lakshmanan, G., Crawford, S.E., Gu, Y., Grosveld, F. and Engel, J.D. Gata3 loss leads to embryonic lethality due to noradrenaline deficiency of the sympathetic nervous system. *Nature Genet* **2000**; *25*: 209–212.

167 Ernsberger, U., Reissmann, E., Mason, I. and Rohrer, H. The expression of dopamine b-hydroxylase, tyrosine hydroxylase, and Phox2 transcription factors in sympathetic neurons: evidence for common regulation during noradrenergic induction and diverging regulation later in development. *Mech Dev* **2000**; *92*: 169–177.

168 Teillet, M.-A. and Le Douarin, N.M. Consequences of neural tube and notochord excision on the development of the peripheral nervous system in the chick embryo. *Dev Biol* **1983**; *98*: 192–211.

169 Howard, M.J. and Bronner-Fraser, M. The influence of neural tube-derived factors on differentiation of neural crest cells *in vitro*. I. Histochemical study on the appearance of adrenergic cells. *J Neurosci* **1985**; *5*: 3302–3309.

170 Stern, C.D., Artinger, K.B. and Bronner-Fraser, M. Tissue interactions affecting the migration and differentiation of neural crest cells in the chick embryo. *Development* **1991**; *113*: 207–216.

171 Groves, A.K., George, K.M., Tissier-Seta, J.-P., Engel, J.D., Brunet, J.-F. and Anderson, D, J. Differential regulation of transcription factor gene expression and phenotypic markers in developing sympathetic neurons. *Development* **1995**; *121*: 887–901.

172 Shah, N.M., Groves, A.K. and Anderson, D.J. Alternative neural crest cell fates are instructively promoted by TGFb superfamily members. *Cell* **1996**; *85*: 331–343.

173 Reissmann, E., Ernsberger, U., Francis-West, P.H., Rueger, D., Brickell, P.M. and Rohrer, H. Involvement of bone morphogenetic proteins–4 and –7 in the specification of the adrenergic phenotype in developing sympathetic neurons. *Development* **1996**; *122*: 2079–2088.

174 Varley, J.E., Wehby, R.G., Rueger, D.C. and Maxwell, G.D. Number of adrenergic and islet–1 immunoreactive cells is in-

creased in avian trunk neural crest cultures in the presence of human recombinant osteogenic protein–1. *Dev Dyn* **1995**; *203*: 434–447.

175 Schneider, C., Wicht, H., Enderich, J., Wegner, M. and Rohrer, H. Bone morphogenetic proteins are required *in vivo* for the generation of sympathetic neurons. *Neuron* **1999**; *24*: 861–870.

176 Serbedzija, G.N., Bronner-Fraser, M. and Fraser, S.E. A vital dye analysis of the timing and pathways of avian trunk neural crest cell migration. *Development* **1989**; *106*: 809–816.

177 Kim, J., Lo, L., Dormand, E. and Anderson, D.J. SOX10 maintains multipotency and inhibits neuronal differentiation of neural crest stem cells. *Neuron* **2003**; *38*: 17–31.

178 Müller, F. and Rohrer, H. Molecular control of ciliary neuron development: BMPs and downstream transcriptional control in the parasympathetic lineage. *Development* **2002**; *129*: 5707–5717.

179 Schuchardt, A., D'Agati, V., Larsson-Blomberg, L., Costantini, F. and Pachnis, V. Defects in the kidney and enteric nervous system of mice lacking the tyrosine kinase receptor Ret. *Nature* **1994**; *367*: 380–383.

180 Moore, M.W., Klein, R.D., Fariñas, I., Sauer, H., Armanini, M., Phillips, H., Reichardt, L.F., Ryan, A.M., Carver-Moore, K. and Rosenthal, A. Renal and neuronal abnormalities in mice lacking GDNF. *Nature* **1996**; *382*: 76–79.

181 Pichel, J.G., Shen, L.Y., Sheng, H.Z., Granholm, A.C., Drago, J., Grinberg, A., Lee, E.J., Huang, S.P., Saarma, M., Hoffer, B.J., Sariola, H. and Westphal, H. Defects in enteric innervation and kidney development in mice lacking GDNF. *Nature* **1996**; *382*: 73–76.

182 Enomoto, H., Araki, T., Jackman, A., Heuckeroth, R.O., Snider, W.D., Johnson, E.M. Jr. and Milbrandt, J. GFRa1–deficient mice have deficits in the enteric nervous system and kidneys. *Neuron* **1998**; *21*: 317–324.

183 Durbec, P.L., Larsson-Blomberg, L.B., Schuchardt, A., Costantini, F. and Pachnis, V. Common origin and developmental dependence on c-ret of subsets of enteric and sympathetic neuroblasts. *Development* **1996**; *122*: 349–358.

184 Edery, P., Lyonnet, S., Mulligan, L.M., Pelet, A., Dow, E., Abel, L., Holder, S., Nihoul-Fékété, C., Ponder, B.A.J. and Munnich, A. Mutations of the RET proto-oncogene in Hirschprung's disease. *Nature* **1994**; *367*: 378–380.

185 Ivanchuk, S.M., Myers, S.M., Eng, C. and Mulligan, L.M. De novo mutation of GDNF, ligand for the RET/GDNF-alpha receptor complex in Hirschsprung's disease. *Hum Mol Genet* **1996**; *5*: 2020–2026.

186 Angrist, M., Jing, S., Bolk, S., Bentley, K., Nallasamy, S., Haluska, M., Fox, G.M. and Chakravarti, A. Human GFRA1: cloning, mapping, genomic structure, and evaluation as a candidate gene for Hirschsprung's disease susceptibility. *Genomics* **1998**; *15*: 354–362.

187 Edery, P., Attie, T., Amiel, J., Pelet, A., Eng, C., Hostra, R.M., Martelli, H., Bidaud, C., Munnich, A. and Lyonnet, S. Mutation of the endothelin–3 gene in the Waardenburg-Hirsch-

prung disease (Shah-Waardenburg syndrome). *Nat Genet* **1996**; *12*: 442–444.

188 Baynash, A.G., Hosada, K., Giaid, A., Richardson, J.A., Emoto, N., Hammer, R.E. and Yanagisawa, M. Interaction of endothelin–3 with endothelin-B receptor is essential for development of epidermal melanocytes and enteric neurons. *Cell* **1994**; *79*: 1277–1285.

189 Carrasquillo, M.M., McCallion, A.S., Puffenberger, E.G., Kashuk, C.S., Nouri, N. and Chakravarti, A. Genome-wide association study and mouse model identify interaction between RET and EDNRB pathways in Hirschsprung disease. *Nat Genet* **2002**; *32*: 237–244.

190 Barlow, A., de Graaf, E. and Pachnis, V. Enteric nervous system progenitors are coordinately controlled by the G protein-coupled receptor EDNRB and the receptor tyrosine kinase RET. *Neuron* **2003**; *40*: 905–916.

191 Chalazonitis, A., Rothman, T.P., Chen, J.X. and Gershon, M.D. Age-dependent differences in the effects of GDNF and NT–3 on the development of neurons and glia from neural crest-derived precursors immunoselected from the fetal rat gut: Expression of GFRa–1 *in vitro* and *in vivo*. *Dev Biol* **1998**; *204*: 385–406.

192 Hearn, C.J., Murphy, M. and Newgreen, D. GDNF and EZ–3 differentially modulate the numbers of avian enteric neural crest cells and enteric neurons *in vitro*. *Dev Biol* **1998**; *197*: 93–105.

193 Apelqvist, A., Ahlgren, U. and Edlund, H. Sonic hedgehog directs specialized mesoderm differentiation in the intestine and pancreas. *Curr Biol* **1997**; *7*: 801–804.

194 Roberts, D.J., Johnson, R.L., Burke, A.C., Nelson, C.E., Morgan, B.A. and Tabin, C. Sonic hedgehog is an endodermal signal inducing Bmp–4 and Hox genes during induction and regionalization of the chick hindgut. *Development* **1995**; *121*: 3163–3174.

195 Smith, D.M., Nielsen, C., Tabin, C.J. and Roberts, D.J. Roles of BMP signaling and Nkx2.5 in patterning at the chick midgut-foregut boundary. *Development* **2000**; *127*: 3671–3681.

196 Chalazonitis, A. Neurotrophin–3 in the development of the enteric nervous system. *Prog Brain Res* **2003**; *146*: 243–263.

197 Ma, Q.F., Fode, C., Guillemot, F. and Anderson, D.J. NEUROGENIN1 and NEUROGENIN2 control two distinct waves of neurogenesis in developing dorsal root ganglia. *Genes Dev* **1999**; *13*: 1717–1728.

198 Jarman, A.P., Grau, Y., Jan, L.Y. and Jan, Y.N. Atonal is a proneural gene that directs chordotonal organ formation in the Drosophila peripheral nervous system. *Cell* **1993**; *73*: 1307–1321.

199 Lo, L., Dormand, E., Greenwood, A. and Anderson, D.J. Comparison of the generic neuronal differentiation and neuron subtype specification functions of mammalian achaete-scute and atonal homologs in cultured neural progenitor cells. *Development* **2002**; *129*: 1553–1567.

200 Parras, C.M., Schuurmans, C., Scardigli, R., Kim, J., Anderson, D.J. and Guillemot, F. Divergent functions of the proneu-

ral genes Mash1 and Ngn2 in the specification of neuronal subtype identity. *Genes Dev* **2002**; *16*: 324–338.
201 Anderson, D.J. Lineages and transcription factors in the specification of vertebrate primary sensory neurons. *Curr Opin Neurobiol* **1999**; *9*: 517–524.
202 Gerrero, M.R., McEvilly, R.J., Turner, E., Lin, C.R., O'Connell, S., Jenne, K.J., Hobbs, M.V. and Rosenfeld, M.G. Brn–3.0: A POU-domain protein expressed in the sensory, immune, and endocrine systems that functions on elements distinct from known octamer motifs. *Proc Natl Acad Sci USA* **1993**; *90*: 10841–10845.
203 McEvilly, R.J., Erkman, L., Luo, l., Sawchenko, P.E., Ryan, A.F. and Rosenfeld, M.G. Requirement for Brn–3.0 in differentiation of sensory and motor-neurons. *Nature* **1996**; *384*: 574–577.
204 Smith, M.D., Dawson, S.J. and Latchman, D.S. The Brn-3a transcription factor induces neuronal process outgrowth and the coordinate expression of genes encoding synaptic proteins. *Mol Cell Biol* **1997**; *17*: 345–354.
205 Sun, Y., Nadal-Vicens, M., Misono, S., Lin, M.Z., Zubiaga, A., Hua, X.X., Fan, G.P. and Greenberg, M.E. Neurogenin promotes neurogenesis and inhibits glial differentiation by independent mechanisms. *Cell* **2001**; *104*: 365–376.
206 McFadden, D.G., McAnally, J., Richardson, J.A., Charité, J. and Olson, E.N. Misexpression of dHAND induces ectopic digits in the developing limb bud in the absence of direct DNA binding. *Development* **2002**; *129*: 3077–3088.
207 Ross, S.E., Greenberg, M.E. and Stiles, C.D. Basic helix-loop-helix factors in cortical development. *Neuron* **2003**; *39*: 13–25.
208 Chen, Z.F., Rebelo, S., White, F., Malmberg, A.B., Baba, H., Lima, D., Woolf, C.J., Basbaum, A.I. and Anderson, D.J. The paired homeodomain protein DRG11 is required for the projection of cutaneous sensory afferent fibers to the dorsal spinal cord. *Neuron* **2001**; *31*: 59–73.
209 Morin, X., Cremer, H., Hirsch, M.-R., Kapur, R.P., Goridis, C. and Brunet, J.-F. Defects in sensory and autonomic ganglia and absence of locus coeruleus in mice deficient for the homeobox gene Phox2. *Neuron* **1997**; *18*: 411–423.
210 Pattyn, A., Morin, X., Cremer, H., Goridis, C. and Brunet, J.-F. The homeobox gene Phox2b is essential for the development of all autonomic derivatives of the neural crest. *Nature* **1999**; *399*: 366–370.
211 Dauger, S., Pattyn, A., Lofaso, F., Gaulthier, C., Goridis, C., Gallego, J. and Brunet, J.-F. Phox2b controls the development of peripheral chemoreceptors and afferent visceral pathways. *Development* **2003**; *130*: 6635–6642.
212 Chen, H.-H., Hippenmeyer, S., Arber, S. and Frank, E. Development of the monosynaptic stretch reflex circuit. *Curr Opin Neurobiol* **2003**; *13*: 96–102.
213 Levanon, D., Brenner, O., Otto, F. and Groner, Y. Runx3 knockouts and stomach cancer. *EMBO Rep* **2003**; *4*: 560–564.
214 Inoue, K., Ozaki, S., Ito, K., Iseda, T., Kaeaguchi, S., Ogawa, M., Bae, S.C., Yamashita, N., Itohara, S., Kudo, N. and Ito, Y. Runx3 is essential for the target-specific axon pathfinding of

trkC-expressing dorsal root ganglion neurons. *Blood Cells Mol Dis* **2003**; *30*: 157–160.

215 Lin, J.H., Saito, T., Anderson, D.J., Lance-Jones, C., Jessell, T.M. and Arber, S. Functionally related motor neuron pool and muscle sensory afferent subtypes defined by coordinate ETS gene expression. *Cell* **1998**; *95*: 393–407.

216 Patel, T.D., Kramer, I., Kucera, J., Niederkofler, V., Jessell, T.M., Arber, S. and Snider, W.D. Peripheral NT3 signaling is required for ETS protein expression and central patterning of proprioceptive sensory afferents. *Neuron* **2003**; *38*: 403–416.

217 Arber, S., Ladle, D.R., Lin, J.H., Frank, E. and Jessell, T.M. ETS gene Er81 controls the formation of functional connections between group Ia sensory afferents and motor neurons. *Cell* **2000**; *101*: 485–498.

218 Howard, M.J., Stanke, M., Schneider, C., Wu, X. and Rohrer, H. The transcription factor dHAND is a downstream effector of BMPs in sympathetic neuron specification. *Development* **2000**; *127*: 4073–4081.

219 Stanke, M., Junghans, D., Geissen, M., Goridis, C., Ernsberger, U. and Rohrer, H. The Phox2 homeodomain proteins are sufficient to promote the development of sympathetic neurons. *Development* **1999**; *126*: 4087–4094.

220 Stanke, M., Stubbusch, J. and Rohrer, H. Interaction of Mash1 and Phox2b in sympathetic neuron development. *Mol Cell Neurosci* **2004** (in press).

221 Swanson, D.J., Zellmer, E. and Lewis, E.J. The homeodomain protein arix interacts synergistically with cyclic AMP to regulate expression of neurotransmitter biosynthetic genes. *J Biol Chem* **1997**; *272*: 27382–27392.

222 Swanson, D.J., Adachi, M. and Lewis, E.J. The homeodomain protein Arix promotes protein kinase A-dependent activation of the dopamine b-hydroxylase promoter through multiple elements and interaction with the coactivator cAMP-response element-binding protein-binding protein. *J Biol Chem* **2000**; *275*: 2911–2923.

223 Adachi, M., Browne, D. and Lewis, E.J. Paired-like homeodomain proteins Phox2a/Arix and Phox2b/NBPhox have similar genetic organization and independently regulate dopamine-b-hydroxylase gene transcription. *DNA Cell Biol.* **2000**; *19*: 539–554.

224 Zellmer, E., Zhang, Z., Greco, D., Rhodes, J., Cassel, S. and Lewis, E.J. A homeodomain protein selectively expressed in noradrenergic tissue regulates transcription of neurotransmitter biosynthetic genes. *J Neurosci* **1995**; *15*: 8109–8120.

225 Yang, C., Kim, H.-S., Seo, H., Kim, C.-H., Brunet, J.-F. and Kim, K-S. Paired-like homeodomain proteins, Phox2a and Phox2b, are responsible for noradrenergic cell-specific transcription of the dopamine b-hydroxylase gene. *J Neurochem* **1998**; *71*: 1813–1826.

226 Kim, H.S., Seo, H., Yang, C.Y., Brunet, J.F. and Kim, K.S. Noradrenergic-specific transcription of the dopamine b-hydroxylase gene requires synergy of multiple cis-acting elements including at least two Phox2a-binding sites. *J Neurosci* **1998**; *18*: 8247–8260.

227 Xu, H., Firulli, A.B., Zhang, X. and Howard, M.J. HAND2 synergistically enhances transcription of dopamine-b-hydroxylase in the presence of Phox2a. *Dev Biol* **2003**; *262*: 183–193.

228 Rychlik, J.L., Gerbasi, V. and Lewis, E.J. The interaction between dHand and Arix at the dopamine b-hydroxylase promotor region is independent of direct dHand binding to DNA. *J Biol Chem* **2003**; *278*: 49652–49660.

229 Kim, H.S., Hong, S.J., LeDoux, M.S. and Kim, K.S. Regulation of the tyrosine hydroxylase and dopamine b-hydroxylase genes by the transcription factor AP–2. *J Neurochem* **2001**: *76*: 280–294.

230 Greco, D., Zellmer, E., Zhang, Z. and Lewis, E. Transcription factor AP–2 regulates expression of the dopamine b-hydroxylase gene. *J Neurochem* **1995**; *65*: 510–516.

231 Holzschuh, J., Barrallo-Gimeno, A., Ettl, A.K., Durr, K., Knapik, E.W. and Driever, W. Noradrenergic neurons in zebrafish hindbrain are induced by retinoic acid and require tfap2a for expression of the neurotransmitter phenotype. *Development* **2003**; *130*: 5741–5754.

232 Lo, L.-C., Johnson, J.E., Wuenschell, C.W., Saito, T. and Anderson, D.J. Mammalian achaete-scute homolog 1 is transiently expressed by spatially restricted subsets of early neuroepithelial and neural crest cells. *Genes Dev* **1991**; *5*: 1524–1537.

233 Blaugrund, E., Pham, T.D., Tennyson, V.M., Lo, L., Sommer, L., Anderson, D.J. and Gershon, M.D. Distinct subpopulations of enteric neruonal progenitors defined by time of development, sympathoadrenel lineage markers and MASH–1–dependence. *Development* **1996**; *122*: 309–320.

234 Wu, X. and Howard, M.J. Transcripts encoding HAND genes are differentially expressed and regulated by BMP4 and GDNF in developing avian gut. *Gene Expression* **2002**; *10*: 279–293.

235 Jessell, T.M. Neuronal specification in the spinal cord: inductive signals and transcriptional codes. *Nat Rev Genet* **2000**; *1*: 20–29.

236 Kania, A. and Jessell, T.M. Topographic motor projections in the limb imposed by LIM homeodomain protein regulation of Ephrin-A : EphA interactions. *Neuron* **2003**; *38*: 581–596.

237 Tiveron, M.C., Hirsch, M.R. and Brunet, J.F. The expression pattern of the transcription factor Phox2 delineates synaptic pathways of the autonomic nervous system. *J Neurosci* **1996**; *16*: 7649–7660.

238 Pujol, N., Torregrossa, P., Ewbank, J.J. and Brunet, J. F. The homeodomain protein CePHOX2/CEH–17 controls anteroposterior axonal growth in C. elegans. *Development* **2000**; *127*: 3361–3371.

239 Ma, L., Lei, L., Eng, S.L., Turner, E. and Parada, L.F. Brn3a regulation of TrkA/NGF receptor expression in developing sensory neurons. *Development* **2003**; *130*: 3525–3534.

240 Pasterkamp, R.J. and Kolodkin, A.L. Semaphorin junction: making tracks toward neural connectivity. *Curr Opin Neurobiol* **2003**; *13*: 79–89.

241 Castellani, V. and Rougon, G. Control of semaphorin signaling. *Curr Opin Neurobiol* **2002**; *12*: 532–541.

242 Castellani, V., Chédotal, A., Schachner, M., Faivre-Sarraih, C. and Rougon, G. Analysis of the L1–deficient mouse phenotype reveals cross-talk between Sema3A and L1 signaling in axonal guidance. *Neuron* **2000**; *27*: 237–249.

243 Taniguchi, M., Yuasa, S., Fujisawa, H., Naruse, I., Saga, S., Mishina, M. and Yagi, T. Disruption of Semaphorin III/D gene causes severe abnormality in peripheral nerve projection. *Neuron* **1997**; *19*: 519–530.

244 Kawasaki, T., Bekku, Y., Suto, F., Kitsukawa, T., Taniguchi, M., Nagatsu, I., Nagatsu, T., Itoh, K., Yagi, T. and Fujisawa, H. Requirement of neuropilin 1–mediated Sema3A signals in patterning of the sympathetic nervous system. *Development* **2002**; *129*: 671–680.

245 Bates, D., Taylor, G.I., Minichiello, J., Farlie, P., Cichowitz, A., Watson, N., Klagsbrun, M., Mamluk, R. and Newgreen, D.F. Neurovascular congruence results from a shared patterning mechanism that utilizes Semaphorin3A and Neuropilin–1. *Dev Biol* **2003**; *255*: 77–98.

246 Carmeliet, P. Blood vessels and nerves: common signals, pathways and diseases. *Nat. Rev Genet* **2003**; *4*: 710–720.

247 Airaksinen, M.S. and Saarma, M. The GDNF family: signalling, biological functions and therapeutic value. *Nat Rev Neurosci* **2002**; *3*: 383–394.

248 Baloh, R.H., Tansey, M.G., Lampe, P.A., Fahrner, T.J., Enomoto, H., Simburger, K.S., Leitner, M.L., Araki, T., Johnson, E.M. Jr. and Milbrandt, J. Artemin, a novel member of the GDNF ligand family, supports peripheral and central neurons and signals through the GFRa3–RET receptor complex. *Neuron* **1998**; *21*: 1291–1302.

249 Andres, R., Forgie, A., Wyatt, S., Chen, Q., De Sauvage, F.J. and Davies, A.M. Multiple effects of artemin on sympathetic neurone generation, survival and growth. *Development* **2001**; *128*: 3685–3695.

250 Honma Y, Araki T, Gianino S, Bruce A, Heuckeroth R, Johnson E, Milbrandt J. Artemin is a vascular-derived neurotropic factor for developing sympathetic neurons. *Neuron* **2002**; *35*: 267–282.

251 Enomoto, H., Crawford, P.A., Gorodinsky, A., Heuckeroth, R.O., Johnson, E.M. Jr. and Milbrandt, J. RET signaling is essential for migration, axonal growth and axon guidance of developing sympathetic neurons. *Development* **2001**; *128*: 3963–3974.

252 Nishino, J., Mochida, K., Ohfuji, Y., Shimazaki, T., Meno, C., Ohishi, S., Matsuda, Y., Fujii, H., Saijoh, Y. and Hamada, H. GFRa3, a component of the artemin receptor, is required for migration and survival of the superior cervical ganglion. *Neuron* **1999**; *23*: 725–736.

253 Heuckeroth, R.O., Enomoto, H., Grider, J.R., Golden, J.P., Hanke, J.A., Jackman, A., Molliver, D.C., Bardgett, M.E., Snider, W.D., Johnson, E.M Jr. and Milbrandt, J. Gene targeting reveals a critical role for neurturin in the development and maintenance of enteric, sensory, and parasympathetic neurons. *Neuron* **1999**; *22*: 253–263.

254 Rossi, J., Luukko, K., Poteryaev, D., Laurikainen, A., Sun, Y.F., Laakso, T., Eerikäinen, S., Tuominen, R., Lakso, M., Rauvala, H., Arumäe, U., Pasternack, M., Saarma, M. and Airaksinen, M.S. Retarded growth and deficits in the enteric and parasympathetic nervous system in mice lacking GFRa2, a functional neurturin receptor. *Neuron* **1999**; *22*: 243–252.

255 Enomoto, H., Heuckeroth, R.O., Golden, J.P., Johnson, E.M. Jr. and Milbrandt, J. Development of cranial parasympathetic ganglia requires sequential actions of GDNF and neurturin. *Development* **2000**; *127*: 4877–4889.

256 Rossi, J., Tomac, A., Saarma, M. and Airaksinen, M.S. Distinct roles for GFRa1 and GFRa2 signalling in different cranial parasympathetic ganglia *in vivo*. *Eur J Neurosci* **2000**; *12*: 3944–3952.

257 Hashino, E., Shero, M., Junghans, D., Rohrer, H., Milbrandt, J. and Johnson, E.M. Jr. GDNF and neurturin are target-derived factors essential for cranial parasympathetic neuron development. *Development* **2001**; *128*: 3773–3782.

258 Maina, F. and Klein, R. Hepatocyte growth factor, a versatile signal for developing neurons. *Nat Neurosci* **1999**; *2*: 213–217.

259 Forgie, A., Wyatt, S., Correll, P.H. and Davies, A.M. Macrophage stimulating protein is a target-derived neurotrophic factor for developing sensory and sympathetic neurons. *Development* **2003**; *130*: 995–1002.

260 Hall, A.K., Burke, R.M., Anand, M. and Dinsio, K.J. Activin and bone morphogenetic proteins are present in perinatal sensory neuron target tissues that induce neuropeptides. *J Neurobiol* **2002**; *52*: 52–60.

261 Lein, P., Johnson, M., Guo, X., Rueger, D. and Higgins, D. Osteogenic protein–1 induces dendritic growth in rat sympathetic neurons. *Neuron* **1995**; *15*: 597–605.

262 Lein, P.J., Beck, H.N., Chandrasekaran, V., Gallagher, P.J., Chen, H.L., Lin, Y., Guo, X., Kaplan, P.L., Tiedge, H. and Higgins, D. Glia induce dendritic growth in cultured sympathetic neurons by modulating the balance between bone morphogenetic proteins (BMPs) and BMP antagonists. *J Neurosci* **2002**; *22*: 10377–10387.

263 Unsicker, K., Meier, C., Krieglstein, K., Sartor, B.M. and Flanders, K.C. Expression, localization, and function of transforming growth factor-bs in embryonic chick spinal cord, hindbrain, and dorsal root ganglia. *J Neurobiol* **1996**; *29*: 262–276.

264 Hall, A.K., Dinisio, K.J. and Cappuzzello, J. Skin cell induction of calcitonin-reated peptide in embryonic sensory neurons *in vitro* involves activin. *Dev Biol* **2001**; *229*, 263–270.

265 Butler, S.J. and Dodd, J. A role for BMP heterodimers in roof plate-mediated repulsion of commissural axons. *Neuron* **2003**; *8*: 389–401.

266 Davies, A.M. Regulation of neuronal survival and death by extracellular signals during development. *EMBO J* **2003**; *22*: 2537–2545.

267 Chao, M.V. Neurotrophins and their receptors: A convergence point for many signalling pathways. *Nat Rev Neurosci* **2003**; *4*: 299–309.

268 Huang, E.J. and Reichardt, L.F. Trk receptors: roles in neuronal signal transduction. *Ann Rev Biochem* **2003**; *72*: 609–642.

269 Huang, E.J. and Reichardt, L.F. Neurotrophins: Roles in neuronal development and function. *Annu Rev Neurosci* **2001**; *24*: 677–736.

270 Sofroniew, M.V., Howe, C.L. and Mobley, W.C. Nerve growth factor signaling, neuroprotection, and neural repair. *Annu Rev Neurosci* **2001**; *24*: 1217–1281.

271 Böttner, M., Krieglstein, K. and Unsicker, K. The transforming growth factor-bs: Structure, signaling, and roles in nervous system development and functions. *J .Neurochem* **2000**; *75*: 2227–2240.

272 Krieglstein, K., Farkas, L. and Unsicker, K. TGF-b regulates the survival of ciliary ganglionic neurons synergistically with ciliary neurotrophic factor and neurotrophins. *J Neurobiol* **1998**; *37*: 563–572

273 Krieglstein, K., Henheik, P., Farkas, L., Jaszai, J., Galter, D., Krohn, K. and Unsicker, K. Glial cell line-derived neurotrophic factor requires transforming growth factor-b for exerting its full neurotrophic potential on peripheral and CNS neurons. *J Neurosci* **1998**; *18*: 9822–9834.

274 Krieglstein, K. and Unsicker, K. Distinct modulatory actions of TGF-b and LIF on neurotrophin-mediated survival of developing sensory neurons. *Neurochem Res* **1996**; *21*: 849–856.

275 Yuan, X.B., Jin, M., Xu, X.H., Song, Y.Q., Wu, C.P., Poo, M.M. and Duan, S.M. Signalling and crosstalk of Rho GTPases in mediating axon guidance. *Nat Cell Biol* **2003**; *5*: 38–45.

276 Gomes, W.A. and Kessler, J.A. Msx-2 and p21 mediate the pro-apoptotic but not the antiproliferative effects of BMP4 on cultured sympathetic neuroblasts. *Dev Biol* **2001**; *237*: 212–221.

277 Krieglstein, K., Richter, S., Farkas, L., Schuster, N., Dünker, N., Oppenheim, R.W. and Unsicker, K. Reduction of endogenous transforming growth factors prevents ontogenetic neuron death. *Nat Neurosci* **2000**; *3*: 1085–1090.

278 Thoenen, H. Neurotrophins and neuronal plasticity. *Science* **1995**; *270*: 593–598.

279 McCabe, B.D., Marqués, G., Haghighi, A.P., Fetter, R.D., Crotty, M.L., Haerry, T.E., Goodman, C.S. and O'Connor, M.B. The BMP homolog Gbb provides a retrograde signal that regulates synaptic growth at the Drosophila neuromuscular junction. *Neuron* **2003**; *39*: 241–254.

280 Miller, F.D. and Kaplan, D.R. Signaling mechanisms underlying dendrite formation. *Curr Opin Neurobiol* **2003**; *13*: 391–398.

281 Kovalchuk, Y., Hanse, E., Kafitz, K.W. and Konnerth, A. Postsynaptic induction of BDNF-mediated long-term potentiation. *Science* **2002**; *295*: 1729–1734.

282 Jessen, K.R. Glial cells. *Int J Biochem Cell Biol* **2004** (in press).

283 Jessen, K.R. and Richardson, W.D. (Eds). *Glial Cell Development; Basic Principles and Clinical Relevance*. Oxford, UK: Oxford University Press, **2001**.

284 Dyck, P.J., Thomas, P.K., Lambert A., et al. (Eds). *Peripheral Neuropathy* (3rd edn). Toronto, Canada: W.B. Saunders Company, **1993**.

285 Jessen, K.R., and Mirsky, R. Schwann cells and their precursors emerge as major regulators of nerve development. *Trends Neurosci* **1999**; *22*: 402–410.

286 Jessen, K.R., and Mirsky, R. Signals that determine Schwann cell identity. *J Anat* **2003**; *200*: 367–376.

287 Lobsiger, C.S., Taylor, V., and Suter, U. The early life of a Schwann cell. *Biol Chem* **2002**; *383*: 245–253.

288 Bhattacharyya, A., Frank, E., Ratner, N., and Brackenbury, R. P_0 is an early marker of the Schwann cell lineage in chickens. *Neuron* **1991**; *7*: 831–844.

289 Lee, M.-J., Calle, E., Brennan, A., Ahmed, S., Sviderskaya, E., Jessen, K.R., and Mirsky, R. In early development of the rat mRNA for the major myelin protein P_0 is expressed in non-sensory areas of the embryonic inner ear, notochord, enteric nervous system, and olfactory ensheathing cells. *Dev Dyn* **2001**; *222*: 40–51.

290 Lee, M.-J., Brennan, A., Blanchard, A., Zoidl, G., Dong, Z., Tabernero, A., Zoidl, C., Dent, M.A.R., Jessen, K.R., and Mirsky, R. P_0 is constitutively expressed in the rat neural crest and embryonic nerves and is negatively and positively regulated by axons to generate non-myelin-forming and myelin-forming Schwann cells, respectively. *Mol Cell Neurosci* **1997**; *8*: 336–350.

291 Hagedorn, L., Suter, U., and Sommer, L. P0 and PMP22 mark a multipotent neural crest-derived cell type that displays community effects in response to TGF-beta family factors. *Development* **1999**; *12*: 3781–3794.

292 Britch, S., Goerich, D. E., Liethmacher, D., Peirano, R. I., Rossner, M., Nave, K. A., Birchmeier, C., and Wegner, M. The transcription factor Sox10 is a key regulator of peripheral glial development. *Genes Dev* **2001**; *15*: 66–78.

293 Woodhoo, A., Dean, C., Droggiti, A., Mirsky, R., Jessen, K.R. The trunk neural crest and its early glial derivatives: a study of survival responses, developmental secludes and autocrine mechanisms. *Mol Cell Neurosci* **2003**; *23*: 13–27.

294 Parmantier, E., Lynn, B., Lawson, D., Turmaine, M., Sharghi Namini, S., Chakrabarti, L., McMahon, A.P., Jessen, K.R., and Mirsky, R. Schwann cell-derived Desert Hedgehog controls the development of peripheral nerve sheaths. *Neuron* **1999**; *23*: 713–724.

295 Bitgood, M.J., and McMahon, A.P. Hedgehog and Bmp genes are coexpressed at many diverse sites of cell-cell interaction in the mouse embryo. *Dev Biol* **1995**, *172*: 126–138.

296 Anderson, D.J. Genes, lineages and the neural crest: a speculative review. *Phil Trans R Soc Lond B Biol Sci* **2000**; *355*: 953–964.

297 Le Douarin, N.M. and Dupin, E. Multipotentiality of the neural crest. *Curr Opin Gen Dev* **2003**; *13*: 529–536.

298 Baroffio, A. and Blot, M. Statistical evidence for a random commitment of pluripotent cephalic neural crest cells. *J Cell Sci* **1992**; *103*: 581–587.

299 Kuhlbrodt, K., Herbarth, B., Sock, E., Hermans-Borgmeyer, I., and Wegner, M. Sox10, a novel transcriptional modulator in glial cells. *J Neurosci* **1998**; *18*: 237–250.

300 Southard-Smith, E.M., Kos, L., and Pavan, W. J. *Sox 10* mutation disrupts neural crest development in *Dom* Hirschsprung mouse model. *Nature Genet* **1998**; *18*: 60–64.

301 Peirano, R.I., Goerich, D.E., Riethmacher, D., and Wegner, M. Protein zero gene expression is regulated by the glial transcription factor Sox10. *Mol Cell Biol* **2000**; *20*: 3198–3209.

302 Wegner, M. Transcriptional control in myelinating glia: the basic recipe. *Glia* **2000**; *29*: 118–123.

303 Paratore, C., Goerich, D.E., Suter, U., Wegner, M. and Sommer, L. Survival and glial fate acquisition of neural crest cells are regulated by an interplay between the transcription factor Sox10 and extrinsic combinatorial signaling. *Development* **2001**; *128*: 3949–3961.

304 Sonnenberg-Riethmacher, E., Miehe, M., Stolt, C.C., Goerich, D.E., Wegner, M. and Riethmacher, D. Development and degeneration of dorsal root ganglia in the absence of the HMG-domain transcription factor Sox10. *Mech Dev* **2001**; *10*: 253–265.

305 Mollaaghababa, R., Pavan, W.J. The importance of having your SOX on: role of Sox–10 in the development of neural crest-derived melanocytes and glia. *Oncogene* **2003**; *22*: 3024–3034.

306 Dong, Z., Brennan, A., Liu, N., Yarden, Y., Lefkowitz, G., Mirsky, R. and Jessen, K. R. NDF is a neuron-glia signal and regulates survival, proliferation, and maturation of rat Schwann cell precursors. *Neuron* **1995**; *15*: 585–596.

307 Dong, Z., Sinanan, A., Parkinson, D., Parmantier, E., Mirsky, R., and Jessen, K.R. Schwann cell development in embryonic mouse nerves. *J Neurosci Res* **1999**; *56*: 334–348.

308 Paratore, C., Eichenberger, C., Suter, U. and Sommer, L. Sox10 haploinsufficiency affects maintenance of progenitor cells in a mouse model of Hirschprung disease. *Hum Mol Genet* **2002**; *11*: 3075–3085.

309 Shah, N.M., Marchionni, M.A., Isaacs, I., Stroobant, P. and Anderson, D.J. Glial growth factor restricts mammalian neural crest stem cells to a glial fate. *Cell* **1994**; *77*: 349–360.

310 Furukawa, T., Mukherjee, S., Bao, Z.Z., Morrow, E., and Cepko, C. L. rax, Hes1, and notch1 promote the formation of Muller glia by postnatal retinal progenitor cells. *Neuron* **2000**; *26*: 383–394.

311 Gaiano, N., Nye, J.S., and Fishell, G. Radial glial identity is promoted by Notch1 signaling in the murine forebrain. *Neuron* **2000**; *26*: 395–404.

312 Kubu, C.J., Orimoto, K., Morrison, S.J., Weinmaster, G., Anderson, D.J., and Verdi, J.M. Developmental changes in Notch1 and numb expression mediated by local cell-cell interactions underlie progressively increasing delta sensitivity in neural crest stem cells. *Dev Biol* **2002**; *244*: 199–214.

313 Morrison, S.J., Perez, S.E., Qiao, Z., Verdi, J.M., Hicks, C., Weinmaster, G., and Anderson, D.J. Transient Notch activation initiates an irreversible switch from neurogenesis to gliogenesis by neural crest stem cells. *Cell* **2000**; *101*: 499–510.

314 Sherman, L., Stocker, K.M., Morrison, R. and Ciment, G. Basic fibroblast growth factor (bFGF) acts intracellularly to cause the

transdifferentiation of avian neural crest-derived Schwann cell precursors into melanocytes. *Development* **1993**; *118*: 1313–1326.

315 Stocker, K.M., Sherman, L., Rees, S. and Ciment, G. Basic FGF and TGF-β1 influence commitment to melanogenesis in neural crest-derived cells of avian embryos. *Development* **1991**; *111*: 635–645.

316 Eguchi, G. and Kodama, R. Transdifferentiation. *Curr Opin Cell Biol* **1993**; *5*: 1023–1028.

317 Dupin, E., Real, C., Galvieux-Pardanaud, C., Vaigot, P. Le Dourain, N.M. Reversal of developmental restrictions in neural crest lineages: Transition from Schwann cells to glial-melanocytic precursors *in vitro*. *Proc Natl Acad Sci USA* **2003**; *100*: 5229–5233.

318 Tosh, D. and Slack, J.M. How cells change their phenotype. *Nat Rev Mol Cell Biol* **2002**; *3*: 187–194.

319 Kondo, T. and Raff, M. Oligodendrocyte precursor cells reprogrammed to become multipotential CNS stem cells. *Science* **2000**; *289*: 1754–1757.

320 Jessen, K.R., Brennan, A., Morgan, L., Mirsky, R., Kent, A., Hashimoto, Y., and Gavrilovic, J. The Schwann cell precursor and its fate: A study of cell death and differentiation during gliogenesis in rat embryonic nerves. *Neuron* **1994**; *12*: 509–527.

321 Ciutat, D., Calderó, J., Oppenheim, R. W. and Esquerda, J.E. Schwann cell apoptosis during normal development and after axonal degeneration induced by neurotoxins in the chick embryo. *J Neurosci* **1996**; *16*: 3979–3990.

322 Winseck, A.K., Calderó, J., Ciutat, D., Prevette, D., Scott, S.A., Wang, G., Esquerda, J.E., and Oppenheim, R. W. *In vivo* analysis of Schwann cell programmed cell death in the embryonic chick: regulation by axons and glial growth factor. *J Neurosci* **2002**; *22*: 4509–4521.

323 Garratt, A.N., Britsch, S. and Birchmeier, C. Neuregulin, a factor with many functions in the life of a Schwann cell. *BioEssays* **2000**; *22*: 987–996.

324 Wanner, I., B., Kumar, A., Wood, P.M., Mirsky, R. and Jessen, K.R. Role of Glial N-cadherin in peripheral nerve development. *Soc Neurosci Abstr* **2002**; *527*: 15.

325 Falls, D.L., Rosen, K.M., Corfas, G., Lane, W. S. and Fischbach, G. D. ARIA, a protein that stimulates acetylcholine receptor synthesis, is a member of the neu ligand family. *Cell* **1993**; *72*: 801–815.

326 Marchionni, M.A., Goodearl, A.D.J., Chen, M.S., Bermingham-McDonogh, O., Kirk, C., Hendricks, M., Danehy, F., Misumi, D., Sudhalter, J., Kobayashi, K. et al. Glial growth factors are alternatively spliced erbB2 ligands expressed in the nervous system. *Nature* **1993**; *362*: 312–318.

327 Orr-Urtreger, A., Trakhtenbrot, L., Ben-Levy, R., Wen, D., Rechavi, G., Lonai, P. and Yarden, Y. Neural expression and chromosomal mapping of Neu differentiation factor to 8p12–p21. *Proc Natl Acad Sci USA* **1993**; *90*: 1867–1871.

328 Loeb, J.A., Khurana, T.S., Robbins, J.T., Yee, A. G. and Fischbach, G.D. Expression patterns of transmembrane and re-

leased forms of neuregulin during spinal cord and neuromuscular synapse development. *Development* **1999**; *126*: 781–791.

329 Wolpowitz, D., Mason, T.B., Dietrich, P., Mendelsohn, M., Talmage, D. A. and Role, L.W. Cysteine-rich domain isoforms of the neuregulin–1 gene are required for maintenance of peripheral synapses. *Neuron* **2000**; *25*: 79–91.

330 Stewart, H.J.S., Morgan, L., Jessen, K.R. and Mirsky, R. Changes in DNA synthesis rate in the Schwann cell lineage *in vivo* are correlated with the precursor-Schwann cell transition and myelination. *Eur J Neurosci* **1993**; *5*: 1136–1144.

331 Brennan, A., Dean, C.H., Zhang, A.L., Cass, D.T., Mirsky, R. and Jessen, K.R. Endothelins control the timing of Schwann cell generation *in vitro* and *in vivo*. *Dev Biol* **2000**; *227*: 545–557.

332 Scherer, S.S. and Arroyo, E.J. Recent progress on the molecular architecture of myelinated axons. *J Periph Nerv Syst* **2002**; *7*: 1–12.

333 Topilko, P. and Meijer, D. Transcription factors that control Schwann cell development and myelination. In *Glial Cell Development: Basic Principles and Clinical Relevance* (2nd edn), K.R. Jessen and W.D. Richardson, (Eds), pp. 223–244. Oxford University Press, **2001**.

334 Jessen, K.R. and Mirsky, R. Schwann cell development. In *Myelin and its Diseases*, R. Lazzarini (Ed.). San Diego CA, USA: Academic Press, **2004**.

13
Eye Development

Filippo Del Bene and Joachim Wittbrodt

13.1
Vertebrate Eye Development: An Overview

At the end of gastrulation vertebrate eye development is initiated with the specification of the eye field in the anterior neural plate [1, 2]. After formation of the neural tube by convergent extension movements the optic vesicles evaginate from the lateral walls of the diencephalon. The optic vesicles remain connected to the brain via the optic stalk and moreover play a crucial role in the induction and differentiation of the lens in the overlaying ectoderm. Following complex morphogenetic movements, the optic vesicle rounds-up around the developing lens to form the optic cup that can be subdivided into the (inner) neuroretina and the (surrounding) retinal pigmented epithelium (RPE). Finally, the proliferating neuroblasts of the neuroretina undergo neuronal differentiation to give rise to the stereotypical laminar pattern found in the mature vertebrate retina. Axons projecting from the retinal ganlion cells exit the retina at the optic disk and form the optic nerve which projects into integration centers within the central nervous system (Fig. 13.1).

13.2
Patterning of the Anterior Neural Plate

In vertebrates the first events in eye development occur at the end of gastrulation and are closely linked to early neural induction and anterior neural plate patterning [1]. The neuro-ectoderm is specified from the embryonic ectoderm in amniotes and anamniotes. Despite apparent differences in *Xenopus* and chick, a common molecular mechanism underlying this process has recently been proposed [3]. This involves an inductive signal by secreted factors of the Fibroblast Growth Factor (Fgf) family that act before the onset of gastrulation [4–8] and induce neural fate when (wingless/int) wnt-signaling is suppressed. Subsequently, expression of antagonists of the Bone Morphogenetic Proteins (BMPs, secreted factors of the Transforming Growth Factor β (TGFβ) superfamily) such as Follistatin, Noggin, Chordin and

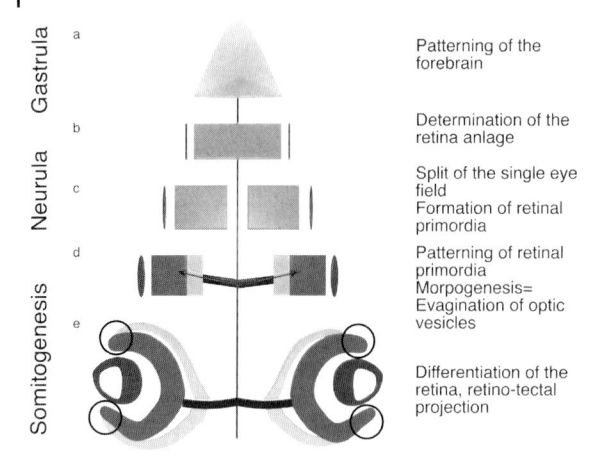

Fig. 13.1
Schematic representation of vertebrate eye development. In the figure, the central line represents both temporal axes and embryonic midline. (A) Late gastrula stages, patterning of the forebrain. (B) End of gastrulation, determination of the eye field. (C) Splitting of the eye field in two by signals emanating from the ventral midline. Initial proximo-distal patterning of the eyes primordia. (D) Optic vesicle evagination as result of cell proliferation and active morphogenesis. Formation of neuroretina (red), pigmented retina epithelium (yellow) and optic stalk (blue). Signals from the optic stalk initiate retina differentiation (arrows). (E) Differentiation of the neuroretina and formation of optic nerve along the optic stalk to target areas in the brain. In anamniotes retinal proliferation continues at the ciliary marginal zone (circles). Adapted from [229]. (This figure also appears with the color plates.)

Cerberus is required to maintain the neural fate [9–12]. BMP antagonists are expressed during gastrulation by structures such as the Spemann organizer (in amphibia), Hensen's node, the equivalent structure in chick or the mouse Node. In this unified model, repression of BMP-signaling is only required to maintain "pre"-neural fate induced by Fgf factors during blastula stages, and thus to prevent epidermal fate [13].

13.2.1
Posteriorizing Factors

The neural plate is subsequently patterned along its anterior-posterior axis and again, through comparative analysis of multiple vertebrate species in recent years, a general model has emerged. Induced neural tissues are committed to develop an anterior character unless exposed to secreted factors with posteriorizing activity. Candidates for such factors include wnts, Fgfs, retinoic acid, Bmps and Nodal family proteins [14, 15]. In this model, that now identifies and names factors that had previously been proposed by Nieuwkoop more than 50 years ago [16], antagonists of posteriorizing factors are required for the induction of the most anterior fates in the neural plate, part of which is the eye Anlage.

Zebrafish mutants affecting Wnt-signaling provide compelling evidence for the involvement and role of the Wnt pathway in anterior-posterior patterning of the neural plate. The *headless* (*hdl*) mutation affects the transcription factor Tcf3 a negative modulator of the Wnt-signaling pathway, resulting therefore in an enhanced Wnt-signaling [17]. Hdl mutant embryos lack any telencephalic structures and are severely reduced in the diencephalon. Consequently they fail do develop optic vesicles and eyes.

Conversely, mutants affected in Nodal signaling like the one-eyed pinhead (*oep*) show an expanded telencephalic territory [18]. *Oep* encodes an essential co-receptor for nodal proteins and its disruption is believed to block nodal signaling completely [19].

13.2.2
Repression of Posteriorizing Factors is required for Head Formation

In addition to posteriorizing factors, a number of antagonistic molecules act anteriorly to refine the anterior-posterior polarity of the central nervous system. Likely candidates for the posteriorizing signals required to induce anterior neural fate are Cerberus [9, 20] a multifunctional antagonist of Wnt, Bmp and Nodal signaling and the Wnt antagonist Dickkopf [21, 22], among others. In the mouse the proposed source of these molecules is the anterior visceral endoderm (AVE), a tissue with an extra-embryonic fate [23, 24]. The AVE is physically separated from the Node and this led to the hypothesis that, at least in mouse, a specific head organizer exists, that is distinct from the node. However induction of the AVE depends on the function of the Node [25] and therefore the relative contribution of these two embryonic structures to anterior neural fate induction is difficult to investigate. Moreover in other vertebrate species, an extra-embryonic structure equivalent to the AVE is not clearly identifiable, although other tissues with embryonic fate have been proposed to play a similar role, e.g. the cells located at the anterior margin of the neural plate (ANB) in zebrafish [26].

13.2.3
Subdivision of the Anterior Neural Plate

The anterior neural plate is then further subdivided rostro-caudally, to define the different regions of the forebrain, namely the telencephalon, the eye field and the diencephalon (Fig. 13.2). The neural plate territory posterior to the eye field expresses Wnt proteins and it is believed that a gradient of Wnt activity is involved in the subdivision within the anterior neural plate [27]. The analysis of the zebrafish mutant *masterblind* (*mbl*) illustrates this concept. *Mbl* encodes for the scaffolding protein Axin1 required for the downregulation of the Wnt-signaling pathway [28]. In *mbl* mutant embryos the local increase in Wnt-signaling disrupts the anterior-posterior subdivision of the anterior neural plate. The telencephalon and the eyes are re-specified to a more posterior diencephalic fate [28, 29]. Wnt antagonists secreted by the ANB help to sharpen the local Wnt activity gradient and thus to generate boundaries rather than gradual transitions. Examples of these proteins are TLC [27] and sFRP1 [30], both members of the secreted Frizzled Related Protein (sFRP) family that antagonize Wnt-signaling by sequestering the secreted Wnt factors [31].

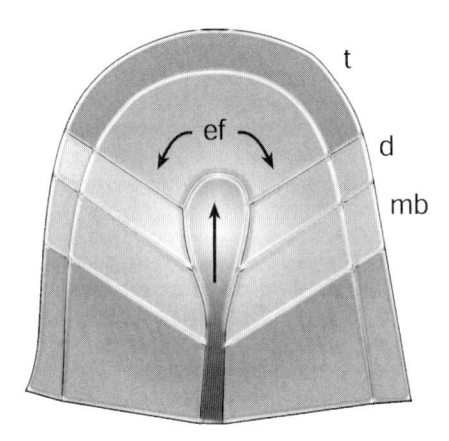

Fig. 13.2
Embryonic origin of the eye field. Cartoon of the anterior neural plate. t, telencephalon (red); ef, eye field (orange); d, diencephalon (pink); mb, mid-brain (turquoise). The migrating axial tissue is shown in blue. Adapted from [84]. (This figure also appears with the color plates.)

In summary, eye field specification involves a series of inductive signals, conventionally subdivided into three major steps: neural induction which occurs in the presumptive ectoderm, the antero-posterior subdivision of the neural plate and finally specification of the eye field between the telencephalic and the diencephalic territories of the anterior neural plate. Although this model is convincingly supported by experimental evidence it does not yet fully explain and represent the continuous series of events occurring *in vivo*.

Since the inhibition of Bmp signaling is crucial for both neural induction and anterior neural plate specification, both these processes are tightly interconnected [32, 33]. Moreover, as mentioned above, the requirement of the Node for the induction of the AVE in mouse indicated inductive cross-talk between two organizing structures. Consequently, the model of a head organizer represents only an approximation to the situation *in vivo*. To overcome this a new model has been proposed in mouse where the Node alone is capable of inducing head formation and the main role of the AVE would be to direct cell movements in the epiblast thus keeping some of the cells of the anterior neural plate at a distance from caudalizing signals [13, 23, 34, 35].

In teleosts the role of the ANB has been proposed to be analogous to the AVE in protecting the anterior neural plate from posteriorizing signals, but it is also an essential signaling center required for the subspecification of the anterior neural plate into restricted fates [26, 27]. These two functions are again difficult to separate and distinguish since inhibition of Wnt-signaling plays a crucial role in both processes. It is clear at the moment that the complex mechanisms playing a role in patterning the anterior neural plate are far from being completely understood and comparative approaches taking advantage of the different model organisms will be required to elucidate the detailed molecular interactions controlling this process.

13.3
Transcription Factors Function in the Establishment of Retinal Identity

The first molecular markers that are expressed and play a role in the eye field belong to the homeobox containing the transcription factor family. Among these, the best characterized and studied in several model organisms are Pax6, Rx and Six3 (Fig. 13.3). Although the expression of these genes temporally follows the inductive events that pattern the anterior neural plate, the molecular nature of their induction is unknown as is their interaction with cues that specify the eye anlage.

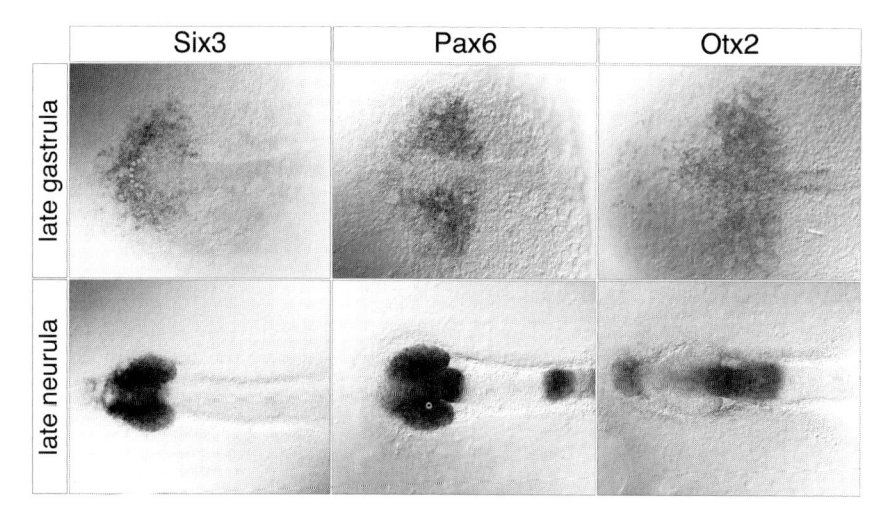

Fig. 13.3
Expression of Six3, Pax6 and Otx2 during early eye development in medaka fish (*Oryzias latipes*). Expression at late gastrula stages and early neurula stages. The orange-hatched line demarcates the anterior end of the embryonic axis. Note the partial overlap of the *Six3* and *Pax6*, *Pax6* and *Otx2* expression domains both at late gastrula and early neurula stages. At the same time *Six3* and *Otx2* are mutually exclusive. Adapted from [46]. (This figure also appears with the color plates.)

Pax6 is a transcription factor containing two DNA binding domains, a paired-type homeodomain and a paired domain. The most C-terminal part of the protein is rich in proline, serine and threonine (PST) and acts as a transcriptional activator [36–39]. *Pax6* is highly conserved between vertebrates and invertebrates. In *Drosophila* two Pax6 homologs, *eyeless* (*ey*) and *twin of eyeless* (*toy*), are expressed in the single ectodermal epithelium of the eye imaginal disk and are required for normal eye development [40, 41]. Moreover misexpression of *ey* or of mouse *Pax6 Drosophila* imaginal disks is sufficient to induce the formation of entire ectopic eyes at the site of misexpression [42]. These results, together with the early expression of vertebrate *Pax6* at the end of gastrulation in the anterior neural plate (Fig. 13.3) [43–48], led to the hypothesis that Pax6 is key regulator of eye development conserved through evolution of Bilateria (master control genes hypothesis) [41, 42, 49]. This revolutionary hypothesis implied that the development of such different structures as the vertebrate camera eye and the arthropods compound eye is under the control of the same gene, suggesting a monophyletic origin for these two structures.

While *Pax6* obviously plays an evolutionarily ancient role it does not function early in the establishment of the retina but rather in the differentiation of the vertebrate retina. Mice and humans carrying mutations in the *Pax6* gene are born with no eyes [50–53], but studies in the mouse suggest that this is a secondary consequence of the loss of Pax6 activity. Mice homozygous for the *Pax6* mutation *Small eye* do not develop eye cups, but they show optic vesicle evagination during embryogenesis [43, 54, 55], indicating that the formation of the eye field and establishment of retinal identity is not impaired. These optic vesicles eventually form optic cups, although lens formation in the abutting head ectoderm (where *Pax6* function is also required) does not occur. Strikingly, eye field specification occurs in these embryos and the expression of early markers such as *Rx*, *Six3* and *Pax6* itself is not affected in the anterior neural plate [43, 56, 57]. Morpholino knock-down experiments in fish (medaka) confirm this scenario. Here *Pax6* loss-of-function results in a small eye phenotype without affecting early eye field markers such as *Rx3*, *Six3* and *Pax6* itself [58]. Thus gross defects in eye development observed in absence of *Pax6* function in vertebrates, are likely due to a dual role of *Pax6*, first in lens formation, which indirectly affects proper optic cup morphogenesis and differentiation [59, 60], and second during the differentiation of retinal progenitor cells. Elegant experiments in mouse, using a conditional knockout approach, did indeed demonstrate an essential role of Pax6 in the maintenance of a pluripotent state of retina progenitor cells [61].

Another homeobox-containing transcription factor of the paired-like family expressed in the early eye field is the retina-specific homeobox gene *Rx* [62–65]. In fish, three paralogs of this gene arose during teleosts evolution, *Rx1*, *Rx2* and *Rx3*, whereas mammals have only a single Rx gene [63–67]). The targeted inactivation of *Rx* in mouse leads to the loss of any morphological sign of eye development, including early optic vesicle evagination [65]. The expression of eye field markers is also missing, indicating an early role for Rx in the specification of retinal precursor cells [57]. In fish, mutations in the *Rx3* gene are reported for both medaka (*eyeless*/*Rx3*) and zebrafish (*chokh*/*Rx3*). They both result in the failure of the optic vesicle to evaginate and consequently in the absence of eye structures [63, 64, 68]. *Rx3* is the first of the three *Rx* paralogs to be expressed during teleosts eye development. Its loss of function does not affect the expression of other early eye field markers such as *Six3* and *Pax6*, but rather the proper evagination of the optic vesicle [63, 64, 68]. This indicates a role for *Rx3* in eye morphogenesis rather than in patterning. These data suggest that although *Rx* genes have essentially conserved functions in vertebrate eye development, the genetic cascade in which they act may have subtle differences in different vertebrate species.

Six3 is a homeobox-containing transcription factor of the *Six/sine oculis* family [69, 70]. Initially identified because of its homology with the *Drosophila* gene *sine oculis* [71], it is now considered to be the ortholog of another *Drosophila* gene *optix* that was discovered after its vertebrate ortholog [72, 73]. Like *Pax6* and *Rx* genes, *Six3* is first expressed at the end of gastrulation in the anterior neural plate in the presumptive eye field in all the vertebrate species analyzed (Fig. 13.3) [46, 70, 74–77]. Gain-of-function studies in medaka first uncovered the role of *Six3* in the establishment of retinal identity. Overexpression of *Six3* leads to retinal hyperplasia and to the induction of ectopic retinal primorida which eventually differentiate to form ectopic eye cups [78]. Epistasis experiments indicate that *Six3* and *Pax6* positively regulate the expression of each other [78, 79], suggesting that both genes cooperate in the establishment of retinal identity.

Loss-of-function data confirm the crucial role of *Six3* in eye field specification and the patterning of the anterior neural plate. Both, in medaka knock-down embryos and in mouse knockout embryos structures anterior to the midbrain are missing in the absence of *Six3* function [58, 80]. No eye-specific markers were detectable in these embryos at any stage of development. In the absence of *Six3* function, cells in the anterior neural plate normally expressing *Six3*, undergo apoptosis [58]. In the mouse knockout model anterior expansion of more posterior markers was observed, raising the possibility that Six3 is directly involved in shaping the Wnt activity gradient that is responsible for patterning the anterior neural plate [80]. More strikingly *Six3* overexpression is able to rescue telencephalic and eye fate in *hdl* mutant embryos which lack the activity of the transcriptional repressor of Wnt target genes, Tcf3 [80]. Conversely increased Wnt-signaling blocks *Six3* expression (and eye field specification as outlined above) [81–83]. Taken together *Six3* is active in regions of low Wnt activity in the anterior neural plate, where it acts as a negative modulator of Wnt-signaling in a self-reinforcing pathway [84]. At the moment the molecular nature of the cross-talk between *Six3* and the Wnt-signaling pathway remains obscure.

Finally it is worth discussing the role of another homeobox-containing transcription factor, namely, Otx2. Its expression in the anterior neuro-ectoderm precedes any other eye field marker [85] and Otx2 function is essential for forebrain and midbrain formation [86, 87]. When the expression of genes such as *Rx* and *Six3* starts to demarcate the eye field, *Otx2* becomes downregulated in this region and its expression never overlaps with that of *Rx* or *Six3* [46, 88]. *Otx2* at this point is excluded from the eye field, but persists in more anterior and posterior regions (Fig. 13.3). *Six3* and *Rx* misexpression is sufficient to restrict the *Otx2* expression domain [88]. Interestingly the potential of Six3 to induce ectopic eyes is limited to the Otx expression domain in the midbrain [78]. Only co-injection of *Otx2* with one or more eye-specification genes in *Xenopus* can force ectopic induction of eye structures outside the endogenous *Otx* expression domain [89]. This observation supports a permissive role for *Otx2* in eye field specification. *Otx2* confers to the neural plate the competence to become the eye field, but must then be downregulated during eye field specification.

13.4
Patterning of the Eye Field and its Medio-lateral Separation

So far we have always referred to the eye field as a continuous domain within the anterior neural plate. Early eye field markers are in fact expressed across the entire medio-lateral width of the anterior neuro-ectoderm. Only in a second step does the initially uniform eye field become separated into two lateral domains, concomitant with proximo-distal patterning, to give rise to two bilateral eyes. Although not immediately obvious, this concept was hypothesized in the 1920s by Adelmann, who studied the eye-forming potential of a region he transplanted from the midline of anterior neural plate in amphibians [90]. Later he postulated the crucial role of axial tissue in separating the eye field to form two bilateral eyes and the optic stalks. He noted the generation of cyclopic embryos by removal of the anterior prechordal plate mesoderm that underlies the eye field [91]. More recently studies in chick and *Xenopus* have confirmed this old concept using molecular markers [92, 93]. Removal of the prechordal plate causes a failure in splitting the *Pax6* expression domain in the overlying neuro-ectoderm. Moreover grafted prechordal plate mesoderm is able to downregulate *Pax6* expression in more lateral neural plate regions. Several reports in zebrafish have identified the Nodal pathway as one of the crucial prechordal plate-derived signals responsible for the splitting of the eye field (for a recent review see [19]). In zebrafish three well-characterized mutants show severe cyclopia and failure of ventral forebrain formation (hypothalamus) among other midline defects [94, 95]. They all encode components of the Nodal signaling pathway, and are *cyclops* (*cyc*) [96–98] and *squint* (*sqt*) [96], nodal-related members of the Tgf-β superfamily. The crucial role of nodal signaling had already been uncovered in the characterization of the mutant *oep* [18, 99] which affects a nodal co-receptor.

Downstream of nodal signaling the *sonic hedgehog* (*shh*) pathway has been implicated in optic stalk specification, and thus proximo-distal patterning of the eye field. Shh is a vertebrate homolog of the *Drosophila* patterning molecule hedgehog, implicated in many aspects of *Drosophila* development (for a recent review see [100]). In mouse and human, mutations in *shh* result in severe cyclopia or holoprosencephaly, a syndrome characterized by reduction in anterior midline structure formation [101–103]. In fish, mutations in *shh* do not give rise to any obvious midline defects possibly due to functional redundancy with other *hedgehog* family members expressed in overlapping domains [104, 105]. Two mechanisms have been proposed for eye field spitting and proximo-distal patterning, and possibly act in a coordinated manner during development. Posteriorly located hypothalamic cells, whose fate is induced by nodal signaling, move anteriorly as a result of the convergence-extension movements occurring in the neural plate. This forces the eye field to move laterally and eventually leads to optic vesicle evagination. Mutations impairing theses movements, such as in mutants where gastrulation movements or nodal signaling has been affected, show cyclopia as eye field cells fail to be displaced from the midline [106, 107]. At the same time, inductive signals like Shh and Bmp7 emanating from the ventral neural tube, specify medially-located cells to adopt an optic stalk fate [104, 108]. In accordance with this model, *shh* has been reported in zebrafish to be directly upstream of the induction of optic stalk and proximo-ventral eye markers such as *Vax1*, *Vax2* and *Pax2* [109]. Fgf signaling, through a mechanism that is not well characterized, has also been implicated in proximo-distal patterning and in inducing optic stalk identity [110–112].

Evidence that optic vesicle evagination and proximo-distal pattering are two genetically separable processes comes from the *Rx3* mutants in medaka and zebrafish [63, 64]. In these embryos optic vesicle evagination does not occur. Nevertheless retinal progenitor cells remaining within the neural tube are patterned normally and express both optic stalk markers in more proximal regions (ventral), and neuro-retinal markers in more distal regions (dorsally) [68] (Fig. 13.4).

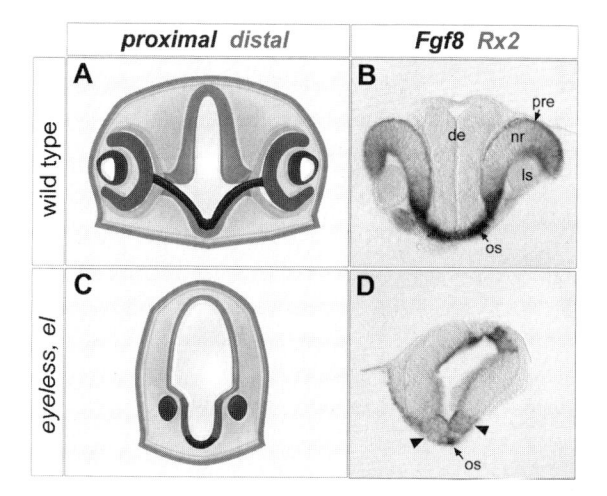

Fig. 13.4
Rx3/eyeless mutation impairs morphogenesis but does not affect the patterning of the retinal primordium. In medaka, a mutation in the *Rx3* gene (*eyeless, el*) impairs optic vesicle evagination and results in a patterned retinal primordium that remains in the ventral diencephalon. (A, C) Schematic comparison of eye structures in *wt* and *el* embryos. Optic stalk (blue), neuroretina (red), pigmented retina epithelium (yellow), lens (green). (B, D) transverse sections of *wt* and *el* embryos, showing *Fgf8* (blue) and *Rx2* (red) expression in the optic stalk and neuroretina respectively. de, diencephalon; ls, lens; nr, neuroretina; os, optic stalk; pre, pigmented retina epithelium. Adapted from [68]. (This figure also appears with the color plates.)

13.5
Differentiation of Neuroretina and Pigmented Retinal Epithlium

13.5.1
Trigger from the Ventral Midline

Under the inductive influence of Shh and Fgf, proximal eye regions adopt optic stalk identity and express a number of homeobox transcription factors such as *Vax1*, *Vax2* and *Pax2* [110, 113–117]. The cells constituting the optic stalk will eventually adopt a non-neural fate and will give rise to the glial cells that together with the axons of the retinal ganglion cells will constitute the optic nerve. Nodal and Shh signaling are responsible for optic stalk formation and result in upregulation of *Pax2* and downregulation of *Pax6* in proximal retinal structures [104, 108]. Fgf signaling seems to establish a competence domain, within which midline-derived Shh can act [110]. At the same time, *Pax6* restriction to the most distal part of the optic vesicle defines the region that will give rise to the neuro-retina and RPE [118]. Mutual repression between Pax6 and Pax2 has been implicated in refining the boundary between these two regions, so that at a later stage the formation of the so-called optic disk is positioned between the optic nerve and the optic cup [118]. Graded *Six3* loss-of-function has revealed a surprisingly new role for this gene in proximo-distal patterning. While severe loss of *Six3* activity causes the complete loss of telencephalic and eye structures, moderate knock-down of *Six3* by morpholino oligos affects only the formation of proximal eye structures and causes cyclopia [58]. *Vax1* expression in particular is extremely sensitive to *Six3* inactivation and is lost at doses of the morpholino that do not block the expression of more distal markers such as *Pax6* and *Rx2*. These observations are in accordance with observed cyclopia and holoprosencephy in humans when only one copy of the *Six3* gene is mutated [119].

In the mouse a *Vax1* null model has been created to investigate the proximal-distal patterning of the eye. *Vax1* knockout embryos show severe coloboma, i.e. the choroid fissure formed by the ventral extension of the optic cup fails to close [120, 121]. In addition they have major defects in optic nerve formation and loss of the eye/nerve boundary with the expansion of the neuroretina along the optic nerve. A very similar phenotype is observed in mice when *Pax2* is inactivated [122] and in human patients harboring a frame-shift mutation in the *Pax2* locus [123].

13.5.2
Retinal Polarity

Formation of the optic cup, and thus differentiation of optic stalk, neural retina and RPE, involves both patterning of the optic rudiment and its simultaneous morphogenetic transformation. This complex process requires the activity of soluble molecules, secreted either by the optic primordia itself or by neighboring structures (ventral midline, mesenchyme, and lens). For instance, dorsoventral patterning of the eye is controlled by *Shh* signaling derived from the ventral midline. In addition to

13.5 Differentiation of Neuroretina and Pigmented Retinal Epithlium

Vax1 and Pax2 Shh activates another Vax gene in the optic stalk, Vax2 [109]. Later *Vax2* expression is also detected in the ventral portion of the optic cup [113, 117]. In contrast dorsally-restricted Bmp signaling acts antagonistically to Shh and activates the expression of factors specific for the dorsal retina such as the T-box family transcription factor, Tbx5 [124]. As in the cases of *Six3/Irx3* and *Pax6/Pax2*, *Vax2* and *Tbx5* negatively regulate each other and define the dorsal and ventral identity of the retina [125]. Downstream effectors of Vax2 and Tbx5 in this patterning process are the EphB receptor tyrosine kinase and ephrin-B ligands, respectively [125] (Fig. 13.5). *Tbx5* misexpression in the ventral retina or *Vax2* in the dorsal retina leads to the ectopic expression of *ephrin-B* or *Eph-B* respectively [117, 124]. Consistent with these data the mouse knockout model for *Vax2* shows dorsalization of the optic cup with ventral expansion of *ephrin-B* ligands, in addition to the mild coloboma and optic nerve defects similar to those observed for the targeted inactivation of *Pax2* and *Vax1* [126, 127] (Fig. 13.6).

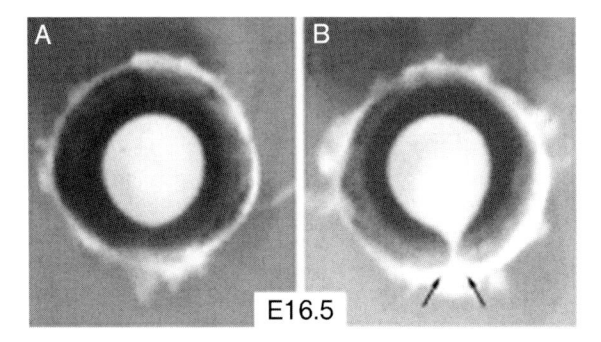

Fig. 13.5
Roles of Shh and Bmp4 in D-V pattern formation of the vertebrate eye. The schematic drawing shows a section through the plane of the optic fissure of an early optic cup. During the transition from the optic vesicle to the optic cup, Shh signals emanating from the ventral forebrain and Bmp4 signals expressed in the dorsal retina act antagonistically to establish D-V properties. Adapted from [228].

A detailed description of dorso-ventral patterning is presented by Marti et al. in Chapter 11.

Concomitant with these patterning events, dramatic morphogenetic changes occur that eventually lead to the formation of the optic cup. The optic vesicle folds around the lens and the anterior and posterior ends of the optic vesicle fuse to form the optic fissure. Failure of this process leads to the formation of eye coloboma, one of the most common genetic diseases affecting the human eye [128]. At the end of this process the initially dorsal-most area of the optic vesicle is surrounding the optic cup, and will give rise to the RPE, whereas the ventral part of the optic vesicle remains confined internally and will eventually differentiate into the neuroretina. Close cross-talk between the lens placode and the optic vesicle is essential for proper

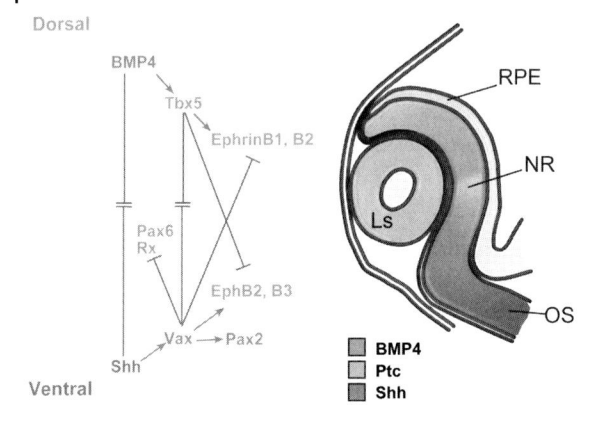

Fig. 13.6
Vax2 inactivation causes eye coloboma. In mouse the targeted inactivation of
Vax2 causes failure in optic fissure closure, causing eye coloboma. (A, B) Frontal
view of wild-type and mutant eyes disected from 16.5-day embryos. Adapted
from [126]. (This figure also appears with the color plates.)

optic cup and neuroretina formation as shown in classical embryological experiments in amphibians. Surprisingly, comparably little is known about the genetic factors regulating this process.

13.5.3
Intrinsic Signals in the Neuroretina

Cell proliferation must be tightly regulated during the morphogenetic movements and inductive events occurring during eye development in order to establish the development of an eye of the proper shape and size. Later in development the proliferative potential of retinal progenitor cells must be downregulated to permit the initiation of neuronal differentiation, and consequently the formation of a mature and functional neuroretina. Control of cell proliferation must ultimately act on cell cycle progression. Several components of the cell cycle machinery show ubiquitous expression during embryogenesis, but others are expressed only in very restricted areas and control regional growth. For instance *cyclin D1* (*cycD1*) is the predominant cyclin of the D-type expressed in retinal progenitor cells (RPCs). *CycD1* null mice are born with hypocellular retinas due to reduced proliferation [129–131]. Other components of the cell cycle seem to have the ability to bias the cell fate of RPCs. The CDK inhibitor p27^{Xic1} promotes Muller glia differentiation when overexpressed in the *Xenopus* retina [132], while p57^{kip2} is required for the maturation of a subset of amacrine cells in the mouse retina [133].

Several transcription factors have been proposed to control cell proliferation during eye development. Overexpression of *Rx* in *Xenopus*, medaka and zebrafish results in retinal hyperplasia indicating a positive role for these genes during eye development [64, 88, 134]. Six3 overexpression also results in hyper-proliferation of the endogenous retinal tissue, in addition to the induction of ectopic retinal primordia, hinting at a role for *Six3* in promoting proliferation during the establishment of retinal identity [78]. Experiments in *Xenopus* indicated a very similar role for *Six6*, a very close homolog of *Six3* [135]. However, *Six6* expression is initiated only after optic vesicle evagination indicating a later role for this gene in relation to *Six3* [56, 73, 136, 137]. Because of their structural and biochemical similarity [138–140], the phenotype observed in response to the overexpression of *Six6* may be due to the induction of endogenous *Six3* that is known to feedback positively on its own expression. *Six6* knockout mice show retinal hypoplasia, detectable at a stage compatible with its later expression [141].

13.5.4
Six3 and Geminin

Transcription factors can alter cell proliferation by acting on the regulation of transcription of key components of the cell cycle. Both *Six3* and *Six6* act mainly in association with co-repressors belonging to the Groucho family to repress the transcription of target genes [138–140]. *Six6* has been shown to bind and repress the promoter of the cell cycle inhibitor $p27^{kip1}$ providing evidence for a direct influence of *Six6* on cell cycle progression [141].

In addition to direct transcriptional regulation a novel interaction has been elucidated for Six3 and Six6. Both transcription factors can influence the proliferation of RPC by directly interfering with the cell cycle machinery by binding to and inhibiting the replication initiation inhibitor Geminin [142]. Geminin is a protein of about 30 kDa identified in two independent and parallel screens, which is on one hand able to arrest cell cycle progression by inhibiting S phase initiation [143], and on the other hand induces premature neuronal differentiation [144]. Intracellular Geminin levels oscillate during the cell cycle and its degradation is required to initiate DNA replication in the S phase. At the molecular level this is due to the ability of Geminin to directly bind and sequester Cdt1 [145, 146], a key component of the pre-replicative complex. It thereby prevents Cdt1 being loaded onto the replication origins to prime them for a new round of DNA replication [147].

In a yeast two-hybrid screen for interactors of Six3 [148], Geminin was identified to physically interact with Six3. Experiments in medaka showed that Geminin and Six3 act at early stages of eye development in an antagonistic manner to regulate cell proliferation. In particular, *Geminin* overexpression strikingly recapitulates *Six3* loss-of-function phenotypes and *Geminin* inactivation promotes RPC proliferation similar to *Six3* gain-of-function experiments [142]. These data indicate that the binding of Six3 to Geminin established conditions that are permissive for cell cycle progression while the eye field and retinal identity is being established. Conversely Geminin can influence Six3 activity and since cell cycle exit is a prerequisite for neuronal differentiation, Geminin interacting with Six3 is able to combine these two distinct, but related functions in a single molecule. At early stages of eye development high levels of Six3 promote cell proliferation by sequestering Geminin and blocking its function (Fig. 13.7), while later Geminin promotes cell cycle exit and thus establishes conditions that are permissive for neuronal differentiation (Fig. 13.8). Consistent with this model, *Geminin* and *Six3* expression overlaps in the early eye field. Later *Geminin* is maintained in the central part of the optic cup where the first mature neurons differentiate, while high levels of *Six3* and *Geminin* co-localize to the most peripheral part of the developing retina where cell proliferation is ongoing [142].

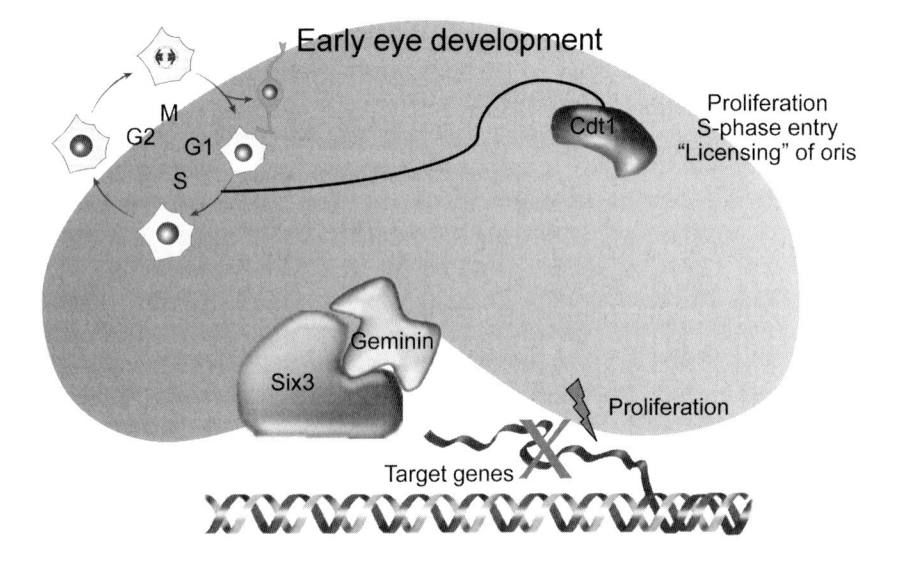

Fig. 13.7
Six3 binding to Geminin stimulates cell proliferation. The cartoon represents the mechanism by which Six3 promotes cell proliferation during early eye development. By sequestering Geminin, Six3 frees Ctd1 to promote cell cycle progression and thus generates conditions that are permissive for cell proliferation. (This figure also appears with the color plates.)

13.5.5
Chx10

Chx10 (also known as Vsx2 or Alx1 in different vertebrate species) is a homeobox-containing transcription factor expressed during eye development exclusively in the neurorctina [149–152]. The transcripts can be detected at early stages of optic cup development in proliferating neuroblasts, but at later stages are absent in postmitotic retinal cells with the exception of bipolar cells, where *Chx10* seems to have an additional function. This expression pattern suggests a specific role in RPCs

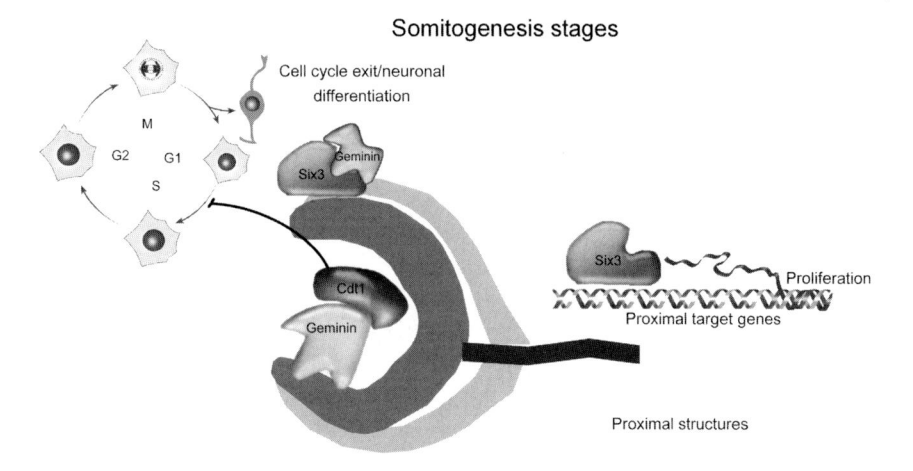

Fig. 13.8
The interaction between Six3 and Geminin controls cell proliferation and differentiation. At somitogenesis stages, *Geminin* and *Six3* are co-expressed in the ciliary marginal zone, where proliferation still occurs. In the more central retina *Six3* transcription is repressed, allowing Geminin to facilitate the cell cycle exit as a prerequisite for neuronal differentiation. In more proximal regions high levels of Six3 activity, without Geminin expression, are required for optic stalk specification. (This figure also appears with the color plates.)

proliferation. In mouse the mutation *ocular retardation* affects Chx10 function [153]. These mice show reduced cell numbers in the neuroretina, but all the neuronal cell types as well as the progenitor cells are present with the single exception of bipolar cells. A similar phenotype is observed in zebrafish after *Chx10* inactivation by antisense oligonucleotides [154]. This indicates a specific role for Chx10 in the regulation of the proliferation rate of RPCs, without affecting their multipotential state. As in the case of Six6, Chx10 seems to regulate the levels of $p27^{kip1}$ and $Chx10/p27^{kip1}$ double mutants show a partial rescue of the retinal phenotype [155]. In contrast, the conditional inactivation of *Pax6* in the retina leads to both the reduction of proliferation of retinal progenitor cells, accompanied by the restriction of their developmental potential and the formation of only one type of retinal neuron [61].

13.6
Differentiation of Retinal Cell Types

The mature vertebrate retina comprises six major classes of neurons and one type of glial cell all originating from a retinal progenitor cell (RPC). As revealed by histological analysis, retinal cells are arranged in layers that arch around the vitreal cavity where the lens is located. Three main layers are easily distinguishable from inside out: the ganglion cell layer (GCL, closest to the lens), the inner nuclear layer (INL) and the photoreceptor layer or outer nuclear layer (ONL). The GCL consists of ganglion cells that project their axons into the brain by a limited number of interneurons usually referred to as displaced amacrine cells. In the INL three classes of interneurons are found which connect the two other cellular layers. These include the majority of amacrine cells, bipolar cells and horizontal cells. The photoreceptor layer is exclusively populated by the radially disposed, elongated bodies of two types of photoreceptors, cones and rods. These three main layers are separated by the so-called inner and outer plexiform layers, which are mainly formed by the cellular projections from the nuclear layers. In addition to neuronal cells, the vertebrate retina contains also Müller glial cells, whose cell-bodies lie within the INL while their processes span the retina apico-basally (Fig. 13.9).

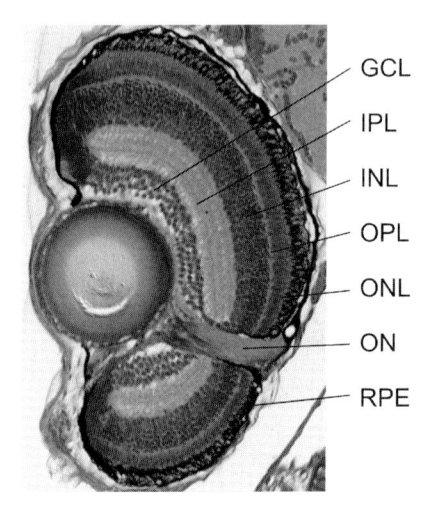

Fig. 13.9
Regular histological structure of the retina. A transverse plastic section through the fish (medaka) retina at the exit point of the optic nerve. The layered arrangement of neurons is clearly visible. GCL, ganglion cell layer; IPL, inner plexiform layer; INL, inner nuclear cell layer; OPL, outer plexiform layer; ONL, outer nuclear layer, ON, optic nerve; RPE, retinal pigmented epithelium. (This figure also appears with the color plates.)

This well-organized architecture is established in a stepwise process in which different cells types are generated in a stereotypic temporal order that is conserved throughout all vertebrates. In all vertebrate species examined so far, ganglion cells are generated first, followed by horizontal cells, cone photoreceptors, amacrine cells, rod photoreceptors, bipolar cells and, finally, Müller glial cells [156, 157]. The timing of cell cycle exit and cell fate specification is therefore closely linked. Cells exiting the cell cycle earlier are likely to adopt an early cell fate and vice versa [158–160].

Neurogenesis in the retina is also tightly regulated spatially. It is always initiated centrally and spreads from the center like a wave to the more peripheral tissue [161–163]. This neurogenic wave resembles the morphogenetic furrow that progresses through the developing eye imaginal disk of *Drosophila* [164]. In zebrafish embryos a small group of cells located in a ventro-nasal position are the first to exit the cell cycle and give rise to the first patch of post-mitotic ganglion cell neurons [165]. *Shh* and *Atonal 5 (Ath5)* are among the first molecular markers expressed in this group of cells. Shh has been implicated in the spreading of this neurogenic wave [166]. Blocking *hedgehog* signaling either with the drug cyclopamine or by mutation of the *shh* gene *(sonic you, syu)*, results in the neurogenic wave failing to spread. Nevertheless, in both situations, initiation of retinal differentiation occurs normally in the first clutch of cells [166]. The ability of shh to induce its own expression in neighboring cells tentatively explains its wave-like spreading over the retina. The expression of the basic helix-loop-helix (bHLH) transcription factor *Ath5*, a marker for differentiating retinal ganglion cells as the first read-out of differentiation, also spreads across the retina in a wave-like manner [167]. Although shh is required for the progression of the differentiation wave, it is clear that the molecular nature of the signal that initiates neurogenesis in vertebrates does not depend on hedgehog signaling and remains unknown. Data in zebrafish indicate that this signal derives from the optic stalk territory [167]. In *oep* mutant embryos affected in the nodal pathway and consequently lacking axial structures, the optic stalk territory does not differentiate and retinal differentiation fails to be initiated. This led to the hypothesis of an optic stalk-derived trigger for retinal differentiation that is further supported by the fact that in several vertebrate species, including mouse and zebrafish, the differentiation of RGCs is stereotypically initiated in proximity to the optic stalk [165, 168]. In transplantation experiments the tissue at the boundary between optic stalk and neuroretina is sufficient to activate *ath5* expression [167].

13.6.1
Retinal Fate Determination by Basic Helix-Loop-Helix (bHLH) and Homeobox Transcription Factors

During retinogenesis, multipotent RPCs pass through a series of competence states during each of which the progenitors are able to generate a subset of retinal cell types. Heterochronic transplantation experiments mainly undertaken in rodents and chick, demonstrated that the competence state is an intrinsic cellular condition and that the external environmental signals can alter the proportion of each cell type at a given time, but cannot force progenitors to become temporally inappropriate cell types [169–171].

Thus, the competence state must be determined by an intrinsic timer that may be set through the number of cell cycles that a single progenitor passes through and the repertoire of transcription factors and signal transduction components expressed at a given time.

Very little is known about the intrinsic changes in RPCs over time. It has been proposed that levels of p27^{Xic1} increase over time and accumulation of p27^{Xic1} above a certain threshold promotes the specification of Müller glial cells, the last cell type to be generated in the vertebrate retina [132].

Much more is known about the influence of transcription factors on cell fate decision. For each retinal cell type, a particular combination of basic helix-loop-helix (bHLH) and homeobox transcription factors has been described (Fig. 13.10).

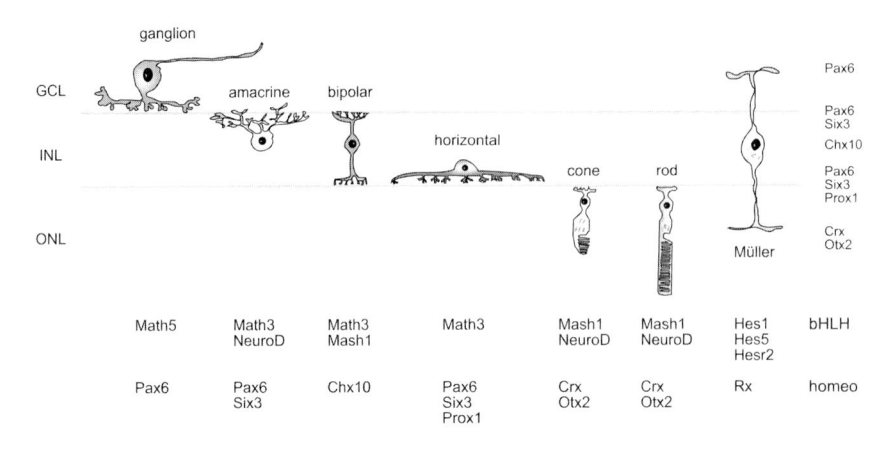

Fig. 13.10
Cooperation of bHLH and homeodomain genes for retinal cell type specification. Hes1 inhibits neuronal differentiation and maintains progenitors. A specific combination of bHLH and homeobox gene expression is characteristic for the different neuronal cell types in the retina. The cells that do not lose *Hes1/Hes5* expression during the stages of neurogenesis adopt the final cell fate, Muller glial cells. Adapted from [186].

13.6.2
Retinal Ganglion Cells

Retinal ganglion cells (RGCs) are among the best-studied retinal cell types. Lineage commitment of the RGCs is marked by the early expression of the bHLH proneural gene *ath5*. Although it seems that not all cells expressing *ath5* will eventually give rise to RGCs [172], *ath5* expression is an absolute requirement for RGC formation. In both zebrafish and mouse, mutations or targeted inactivation of the *ath5* gene results in the complete absence of RGCs with an increased number of other cell types such as amacrine cells and cone photoreceptors [173–175]. Overexpression experiments in chick and *Xenopus* have shown that *ath5* was also able to promote RGC formation at the expense of other cell types [176, 177]. Downstream of *ath5*, a number of transcription factors act to activate RGC-specific genes. Among these the best characterized are the homeobox gene *BarH1* [178] and members of the class IV POU domain transcription factors, *Brn3a*, *Brn3b* and *Brn3c* which share partially redundant functions in directing RGC formation [179–182].

13.6.3
Amacrine Cells

Two bHLH transcription factors are essential for the specification of the amacrine cell fate versus the RGC fate: *NeuroD* and *Math3*. *NeuroD* expression occurs in amacrine cells [183] while *Math3* is only transiently expressed in these cells and persists only in horizontal cells [184]. They share a redundant function and only when both are inactivated in the retina is the formation of amacrine cells severely affected in favor of RGCs [184]. Under those conditions *ath5* is also upregulated indicating that *NeuroD/Math3* and *ath5* can act antagonistically and can negatively regulate each other in a mechanism that remains to be elucidated. Even if *NeuroD* and *Math3* are essential for amacrine cell formation they are not sufficient since the overexpression of either of them favors the generation of photoreceptors cells. The co-expression of the homeobox genes *Pax6* or *Six3* is essential for the differentiation of amacrine cells [184]. Both *Pax6* and *Six3* are transiently expressed in the INL and their overexpression at later stages can increase the formation of undifferentiated INL cells. Interestingly while *Pax6/NeuroD* misexpression increases amacrine cell formation, *Pax6/Math3* misexpression preferentially specifies horizontal cell fate. These data support a model in which the homeobox gene Pax6 specifies the retinal layer (INL) while the different bHLH molecules can direct cell fate choice.

13.6.4
Bipolar Cells

A similar situation is observed during the specification of bipolar cells. *Mash1* and *Math3* are two bHLH transcription factors expressed in bipolar cells. In mouse, the targeted inactivation of both genes completely abolishes bipolar cell formation and instead Müller glial cells are formed. The single knock-out only mildly affects bipolar cell differentiation [185]. The ectopic expression of either *Mash1* or *Math3* interestingly generates photoreceptors cells, indicating that although *Mash1* and *Math3* can induce neuronal differentiation versus glial cell fate, none of them is sufficient for bipolar cell specification [186]. Previous studies had indicated that the homeobox gene *Chx10* is expressed in differentiating bipolar cells, as well in proliferating RPCs [151]. As mentioned above, a mutation in this gene causes the complete loss of bipolar cells (*ocular retardation* mouse mutant) [153]. *Chx10* overexpression alone increases the number of cells in the INL, but these cells are either undifferentiated or Müller glial cells. Instead, when *Chx10* is misexpressed together with either *Mash1* or *Math3* bipolar cell formation is significantly promoted [186]. These results demonstrate again how the combination of homeobox and bHLH transcription factors determines retinal cell fate specification.

13.6.5
Horizontal Cells

In addition to *Math3* and *Pax6* another transcription factor is important for the specification of horizontal cells, the homeobox-containing transcription factor Prox1. *Prox1* is the vertebrate ortholog of the *Drosophila* gene *prospero*, which has an important role in asymmetric cell division, and cell cycle exit [187, 188]. *Prox1* is expressed in RPCs and horizontal cells in a variety of vertebrate species [189–191]. In mouse, *Prox1* inactivation abolishes horizontal cell differentiation and misexpression of *Prox1* in the retina of postnatal progenitor cells by retroviral infection, promotes horizontal cell formation [192].

13.6.6
Photoreceptor Cells

Photoreceptor development has also been intensively studied since they are affected in many human retinal degenerative diseases. The bHLH transcription factors *Mash1* and *NeuroD* are both expressed in photoreceptor cells [183, 193]. When they are inactivated in the retina, reduced or delayed photoreceptor differentiation is observed [194]. Several homeobox transcription factors are expressed in photoreceptor cells. Among these are *Crx*, *Rx* (also known as *Rax* in chick) and *Otx2*. In *Crx* null mice photoreceptors are formed, but they do not differentiate properly and lack the outer segment [195]. Because of their essential function during early eye development, it is more difficult to address the function of *Rx* genes in photoreceptor development. Overexpression of a putative dominant negative form of this gene in chick supported its role in the early steps of photoreceptor differentiation [196]. *Otx2* expression has been described to occur transiently, in precursors of photoreceptor cells in mouse, but not in *Xenopus* [197–200]. In the mouse model using conditional gene inactivation, it has been shown that *Otx2* is essential for cell fate determination of the photoreceptor cells [201]. In *Xenopus* a similar function has been attributed to a related gene known as *Otx5b* which has no clear ortholog in other vertebrate species and which shares a closer sequence similarity to *Crx* [200, 202].

13.6.7
Maintenance of Retinal Stem Cells and Generation of Müller Glia

The minimal definition of a stem cell includes the ability to self-renew and the competence for behaving as a multipotent precursor cell. In the developing vertebrate eye several cell populations can fulfill these criteria. Lineage analysis has shown that RPCs in the developing retina are multipotent [156, 203–205]. In culture in medium supplemented with EGF and FGF2 these cells can generate neurospheres consisting mainly of proliferating cells expressing the neuroectodermal stem cell marker, *nestin* [206, 207]. In presence of complete fetal calf serum these cells generate differentiated neurons and glial cells. The eyes of fish and amphibians continue to grow in size during the entire life of the animal and retinogenesis continues in the adult although at a low rate. A population of cells at the periphery of the neuroretina maintains embryonic retina stem cell characteristics. These cells, found in the ciliary marginal zone (CMZ), continue to proliferate during the entire life of the animal and allow the retina to grow by adding cells of all types at its periphery [208–210]. At the very periphery of the CMZ cells proliferate very slowly and express genes such as *Six3*, *Six6*, *Rx1*, *Rx2*, and *Pax6*. Moving more to the central retina cells start to proliferate faster and express proneural genes of the bHLH family such as *ath5*, *NeuroD*, *Ath3* and *Ash1* [211]. They progressively lose their multipotency, become post-mitotic and finally generate mature neurons which are added to the retina. The sequence of gene expression in the CMZ recapitulates the genetic sequence observed during retinal development [212]. The Notch pathway is involved in the maintenance of progenitor cells at the CMZ as well in the selection of which RPCs will differentiate at a given time [169, 213, 214]. Activation of Notch signaling by Notch ligands expressed by differentiating neighboring cells, leads to the expression of the bHLH transcription factors *Hes1* and *Hes5* [215, 216]. Hes1 and Hes5 are transcriptional repressors that recruit the co-repressor Groucho and block the transcription of proneural genes such as *ath5* and *Ash1* [217, 218]. In mouse, where retinogenesis is completed shortly after birth and CMZ stem cells do not exist, Hes1 is involved during embryogenesis and *Hes1* null mice develop small eyes due to premature differentiation [217, 219]. In amniotes like mouse and chick, neural stem cells/progenitors are not reported to exist in the adult neural retina. However a few years ago several groups detected a small population of mitotically quiescent cells that proliferated when removed from their niche and cultured *in vitro* [220–222]. These cells located in the ciliary epithelium (CE) can self-renew since they clonally generate neurospheres. The CE is a pigmented tissue located between the boundary of neuroretina, RPE, and the iris. In presence of the correct inductive environment, they can generate retinal neurons. Their niche and their pigmented nature lead to the hypothesis that they represent an intermediate state of differentiation between RPE and neuroretina, and that their stem cell potential is the result of genetic reprogramming *in vitro*. Nevertheless, it is interesting to observe that the CE can proliferate *in vivo* after injury or growth factor stimulation [206]. It remains to be investigated whether these cells can migrate and integrate in the neuroretina. Several signaling pathways have been implicated in the maintenance of the multipotent and mitotically-inactive state of CE cells. Notch signaling is likely to play a similar role here as it does in CMZ specification. Perturbation of the Notch pathways significantly affects the ability of CE cells to generate neurospheres *in vitro*. A similar situation is observed for the Wnt-and shh pathways and these observations are supported by expression data that show components of both expressed in the CE [206]. Recently it has been reported that mice having only one copy of *Patched*, a negative modulator of the shh pathway, show ectopic proliferation in the CE [223]. Finally, *in vitro* data support the role of scatter factor (SCF) and its receptor c-Kit in the maintenance of stem cell potential in the CE but no *in vivo* evidence has so far been reported [206].

Recently Müller glial cells have attracted much attention for their neuronal progenitor properties. Müller glial cells are the last type of retinal cells to differentiate. Several lines of evidence indicated the Müller glial fate to be a default fate, when other neuron-specific fates are not adopted. According to this Müller glial cells are the cell type that undergoes fewer changes from the RPC state. Morphologically Müller glial cells are similar to RPCs and like them maintain connections to both the basal and the apical retinal surfaces. They are the only cell type whose processes span all three cellular layers in the mature retina. Misexpression of genes promoting RPC proliferation or maintenance in a variety of vertebrate model systems, such as *Rx*, *Hes1*, *Hes5* and *Notch*, promotes Müller glial cell formation at the expense of other neuronal fates [213, 224, 225]. It is likely that cells that go through several competence states without adopting any neuronal fate and without losing the expression of RPC genes such as *Hes1* and *Hes5* eventually adopt this last available cell fate, becoming mitotically quiescent. In support of this model, Müller glia cells have recently been reported to posses the potential to become neurogenic RPCs [226, 227]. These studies performed mainly in the chick model, demonstrated that in response to retinal damage by toxin injection or exogenous growth factors like Fgf2, Müller glial cells could de-differentiate and re-enter the cell cycle. These cells can give rise to new neurons and glia sequentially expressing all the transcription factors which normally act during embryonic retinal development.

References

1 Chow, R.L., and Lang, R.A. Early eye development in vertebrates. *Annu Rev Cell Dev Biol* **2001**; *17*: 255–296.

2 Chuang, J.C., and Raymond, P.A. Embryonic origin of the eyes in teleost fish. *Bioessays* **2002**; *24*: 519–529.

3 Wilson, S.I., and Edlund, T. Neural induction: toward a unifying mechanism. *Nat Neurosci* **2001**; *4*(Suppl): 1161–1168.

4 Gamse, J.T., and Sive, H. Early anteroposterior division of the presumptive neurectoderm in Xenopus. *Mech Dev* **2001**; *104*: 21–36.

5 Itoh, K., Tang, T.L., Neel, B.G., and Sokol, S.Y. Specific modulation of ectodermal cell fates in Xenopus embryos by glycogen synthase kinase. *Development* **1995**; *121*: 3979–3988.

6 Streit, A., Berliner, A.J., Papanayotou, C., Sirulnik, A., and Stern, C.D. Initiation of neural induction by FGF signalling before gastrulation. *Nature* **2000**; *406*: 74–78.

7 Wessely, O., Agius, E., Oelgeschlager, M., Pera, E.M., and De Robertis, E.M. Neural induction in the absence of mesoderm: beta-catenin-dependent expression of secreted BMP antagonists at the blastula stage in Xenopus. *Dev Biol* **2001**; *234*: 161–173.

8 Wilson, S.I., Graziano, E., Harland, R., Jessell, T.M., and Edlund, T. An early requirement for FGF signalling in the acquisition of neural cell fate in the chick embryo. *Curr Biol* **2000**; *10*: 421–429.

9. Bouwmeester, T., Kim, S., Sasai, Y., Lu, B., and De Robertis, E.M. Cerberus is a head-inducing secreted factor expressed in the anterior endoderm of Spemann's organizer. *Nature* **1996**; *382*: 595–601.
10. Hemmati-Brivanlou, A., Kelly, O.G., and Melton, D.A. Follistatin, an antagonist of activin, is expressed in the Spemann organizer and displays direct neuralizing activity. *Cell* **1994**; *77*: 283–295.
11. Lamb, T.M., Knecht, A.K., Smith, W.C., Stachel, S.E., Economides, A.N., Stahl, N., Yancopolous, G.D., and Harland, R.M. Neural induction by the secreted polypeptide noggin. *Science* **1993**; *262*: 713–718.
12. Sasai, Y., Lu, B., Steinbeisser, H., Geissert, D., Gont, L.K., and De Robertis, E.M. Xenopus chordin: a novel dorsalizing factor activated by organizer-specific homeobox genes. *Cell* **1994**; *79*: 779–790.
13. Stern, C.D. Induction and initial patterning of the nervous system – the chick embryo enters the scene. *Curr Opin Genet Dev* **2002**; *12*: 447–451.
14. Agathon, A., Thisse, C., and Thisse, B. The molecular nature of the zebrafish tail organizer. *Nature* **2003**; *424*: 448–452.
15. Munoz-Sanjuan, I., and A, H.B. Early posterior/ventral fate specification in the vertebrate embryo. *Dev Biol* **2001**; *237*: 1–17.
16. Nieuwkoop, P.D. Activation and organisation of the central nervous system in amphibians. II Differentiation and organisation. *J Exp Zool* **1952**; *120*: 33–81.
17. Kim, C.H., Oda, T., Itoh, M., Jiang, D., Artinger, K.B., Chandrasekharappa, S.C., Driever, W., and Chitnis, A.B. Repressor activity of Headless/Tcf3 is essential for vertebrate head formation. *Nature* **2000**; *407*: 913–916.
18. Gritsman, K., Zhang, J.J., Cheng, S., Heckscher, E., Talbot, W.S., and Schier, A.F. The EGF-CFC protein one-eyed pinhead is essential for nodal signaling. *Cell* **1999**; *97*: 121–132.
19. Schier, A.F. Nodal signaling in vertebrate development. *Annu Rev Cell Dev Biol* **2003**; *19*: 589–621.
20. Piccolo, S., Agius, E., Leyns, L., Bhattacharyya, S., Grunz, H., Bouwmeester, T., and De Robertis, E.M. The head inducer Cerberus is a multifunctional antagonist of Nodal, BMP and Wnt signals. *Nature* **1999**; *397*: 707–710.
21. Glinka, A., Wu, W., Delius, H., Monaghan, A.P., Blumenstock, C., and Niehrs, C. Dickkopf–1 is a member of a new family of secreted proteins and functions in head induction. *Nature* **1998**; *391*: 357–362.
22. Niehrs, C. Head in the WNT. *Trends Genet* **1999**; *15*: 314–319.
23. Kimura, C., Yoshinaga, K., Tian, E., Suzuki, M., Aizawa, S., and Matsuo, I. Visceral endoderm mediates forebrain development by suppressing posteriorizing signals. *Dev Biol* **2000**; *225*: 304–321.
24. Perea-Gomez, A., Rhinn, M., and Ang, S.L. Role of the anterior visceral endoderm in restricting posterior signals in the mouse embryo. *Int J Dev Biol* **2001**; *45*: 311–320.
25. Bachiller, D., Klingensmith, J., Kemp, C., Belo, J.A., Anderson, R.M., May, S.R., McMahon, J.A., McMahon, A.P., Harland,

R.M., Rossant, J., and De Robertis, E.M. The organizer factors Chordin and Noggin are required for mouse forebrain development. *Nature* **2000**; *403*: 658–661.

26 Houart, C., Westerfield, M., and Wilson, S.W. A small population of anterior cells patterns for forebrain during zebrafish gastrulation. *Nature* **1998**; *391*: 788–792.

27 Houart, C., Caneparo, L., Heisenberg, C., Barth, K., Take-Uchi, M., and Wilson, S. Establishment of the telencephalon during gastrulation by local antagonism of Wnt signaling. *Neuron* **2002**; *35*: 255–265.

28 Heisenberg, C.P., Houart, C., Take-uchi, M., Rauch, G.J., Young, N., Coutinho, P., Masai, I., Caneparo, L., Concha, M.L., Geisler, R., et al. A mutation in the Gsk3–binding domain of zebrafish Masterblind/Axin1 leads to a fate transformation of telencephalon and eyes to diencephalon. *Genes Dev* **2001**; *15*: 1427–1434.

29 van de Water, S., van de Wetering, M., Joore, J., Esseling, J., Bink, R., Clevers, H., and Zivkovic, D. Ectopic Wnt signal determines the eyeless phenotype of zebrafish masterblind mutant. *Development* **2001**; *128*: 3877–3888.

30 Esteve, P., Lopez-Rios, J., and Bovolenta, P. SFRP1 is required for the proper establishment of the eye field in the medaka fish. *Mech Dev* **2004**; *121*: 687–701.

31 Uren, A., Reichsman, F., Anest, V., Taylor, W.G., Muraiso, K., Bottaro, D.P., Cumberledge, S., and Rubin, J.S. Secreted frizzled-related protein–1 binds directly to Wingless and is a biphasic modulator of Wnt signaling. *J Biol Chem* **2000**; *275*: 4374–4382.

32 De Robertis, E.M., Larrain, J., Oelgeschlager, M., and Wessely, O. The establishment of Spemann's organizer and patterning of the vertebrate embryo. *Nat Rev Genet* **2000**; *1*: 171–181.

33 Stern, C.D. Initial patterning of the central nervous system: how many organizers? *Nat Rev Neurosci* **2001**; *2*: 92–98.

34 Foley, A.C., Skromne, I., and Stern, C.D. Reconciling different models of forebrain induction and patterning: a dual role for the hypoblast. *Development* **2000**; *127*: 3839–3854.

35 Perea-Gomez, A., Lawson, K.A., Rhinn, M., Zakin, L., Brulet, P., Mazan, S., and Ang, S.L. Otx2 is required for visceral endoderm movement and for the restriction of posterior signals in the epiblast of the mouse embryo. *Development* **2001**; *128*: 753–765.

36 Czerny, T., and Busslinger, M. DNA-binding and transactivation properties of Pax–6: three amino acids in the paired domain are responsible for the different sequence recognition of Pax–6 and BSAP (Pax–5). *Mol Cell Biol* **1995**; *15*: 2858–2871.

37 Glaser, T., Jepal, L., Edwards, J.G., Young, S.R., Favor, J., and Maas, R. L. PAX6 gene dosage effect in a family with congenital catatacts, aniridia, anophthalmia and central nervous system defects. *Nature Genet* **1994**; *7*: 463–471.

38 Mikkola, I., Bruun, J.A., Bjorkoy, G., Holm, T., and Johansen, T. Phosphorylation of the transactivation domain of Pax6 by extracellular signal-regulated kinase and p38 mitogen-activated protein kinase. *J Biol Chem* **1999**; *274*: 15115–15126.

39 Tang, H.K., Singh, S., and Saunders, G.F. Dissection of the transactivation function of the transcription factor encoded by the eye developmental gene PAX6. *J Biol Chem* **1998**; *273*: 7210–7221.

40 Czerny, T., Halder, G., Kloter, U., Souabni, A., Gehring, W.J., and Busslinger, M. Twin of eyeless, a second Pax–6 gene of Drosophila, acts upstream of eyeless in the control of eye development. *Mol Cell* **1999**; *3*: 297–307.

41 Quiring, R., Walldorf, U., Kloter, U., and Gehring, W.J. Homology of the eyeless gene of drosophila to the small eye gene in mice and aniridia in humans. *Science* **1994**; *265*: 785–789.

42 Halder, G., Callaerts, P., and Gehring, W.J. Induction of ectopic eyes by targeted expression of the eyeless gene in Drosophila. *Science* **1995**; *267*: 1788–1792.

43 Grindley, J.C., Davidson, D.R., and Hill, R.E. The role of PAX6 in eye and nasal development. *Development* **1995**; *121*: 1433–1442.

44 Hirsch, N., and Harris, W.A. Xenopus Pax–6 and retinal development. *J Neurobiol* **1997**; *32*: 45–61.

45 Li, Y., Allende, M.L., Finkelstein, R., and Weinberg, E.S. Expression of two zebrafish orthodenticle-related genes in the embryonic brain. *Mech Dev* **1994**; *48*: 229–244.

46 Loosli, F., Koster, R.W., Carl, M., Krone, A., and Wittbrodt, J. Six3, a medaka homologue of the Drosophila homeobox gene sine oculis is expressed in the anterior embryonic shield and the developing eye. *Mech Dev* **1998**; *74*: 159–164.

47 Püschel, A.W., Gruss, P., and Westerfield, M. Sequence and expression pattern of pax–6 are highly conserved between zebrafish and mice. *Development* **1992**; *114*: 643–651.

48 Walther, C., and Gruss, P. Pax–6, a murine paired box gene, is expressed in the developing CNS. *Development* **1991**; *113*: 1435–1449.

49 Gehring, W.J., and Ikeo, K. Pax 6: mastering eye morphogenesis and eye evolution. *Trends Genet* **1999**; *15*: 371–377.

50 Hanson, I.M., Seawright, A., Hardman, K., Hodgson, S., Zaletayev, D., Fekete, G., and van Heyningen, V. PAX6 mutations in aniridia. *Hum Mol Genet* **1993**; *2*: 915–920.

51 Hill, R.E., Favor, J., Hogan, B.L.M., Ton, C.C.T., Saunders, G.F., Hanson, I.M., Prosser, J., Jordan, T., Hastie, N.D., and van Heyningen, V. Mouse small eye results from mutations in a paired-like homeobox containing gene. *Nature* **1991**; *354*: 522–525.

52 Jordan, T., Hanson, I., Zaletayev, D., Hodgson, S., Prosser, J., Seawright, A., Hastie, N., and van Heyningen, V. The human PAX6 gene is mutated in two patients with aniridia. *Nat Genet* **1992**; *1*: 328–332.

53 Ton, C.C., Hirvonen, H., Miwa, H., Weil, M.M., Monaghan, P., Jordan, T., van Heyningen, V., Hastie, N.D., Meijers-Heijboer, H., Drechsler, M., et al. Positional cloning and characterization of a paired box- and homeobox-containing gene from the aniridia region. *Cell* **1991**; *67*: 1059–1074.

54 Hogan, B.L.M., Horsburgh, G., Cohen, J., Hetherington, C.M., Fisher, G., and Lyon, M.F. Small eyes (Sey): a homozy-

gous lethal mutation on chromosome 2 which affects the differentiation of both lens and nasal placodes in the mouse. *J Embryol Exp Morphol* **1986**; *97*: 95–110.

55 Matsuo, T., Osumi-Yamashita, N., Noji, S., Ohuchi, H., Koyama, E., Myokai, F., Matsuo, N., Toniguchi, S., Dari, H., Jseki, S., et al. A mutation at the *Pax–6* gene in rat small eye is associated with impaired migration of midbrain crest cells. *Nat Genet* **1993**; *3*: 299–304.

56 Jean, D., Bernier, G., and Gruss, P. Six6 (Optx2) is a novel murine Six3–related homeobox gene that demarcates the presumptive pituitary/hypothalamic axis and the ventral optic stalk. *Mech Dev* **1999**; *84*: 31–40.

57 Zhang, L., Mathers, P.H., and Jamrich, M. Function of Rx, but not Pax6, is essential for the formation of retinal progenitor cells in mice. *Genesis* **2000**; *28*: 135–142.

58 Carl, M., Loosli, F., and Wittbrodt, J. Six3 inactivation reveals its essential role for the formation and patterning of the vertebrate eye. *Development* **2002**; *129*: 4057–4063.

59 Hyer, J., Mima, T., and Mikawa, T. FGF1 patterns the optic vesicle by directing the placement of the neural retina domain. *Development* **1998**; *125*: 869–877.

60 Nguyen, M.T., and Arnheiter, H. Signaling and transcriptional regulation in early mammalian eye development: a link between FGF and MITF. *Development* **2000**; *127*: 3581–3591.

61 Marquardt, T., Ashery-Padan, R., Andrejewski, N., Scardigli, R., Guillemot, F., and Gruss, P. Pax6 is required for the multipotent state of retinal progenitor cells. *Cell* **2001**; *105*: 43–55.

62 Furukawa, T., Kozak, C.A., and Cepko, C.L. Rax, a Novel Paired-Type Homeobox Gene, Shows Expression in the Anterior Neural Fold and Developing Retina. *Proc Natl Acad Sci USA* **1997**; *94*: 3088–3093.

63 Loosli, F., Staub, W., Finger-Baier, K., Ober, E., Verkade, H., Wittbrodt, J., and Baier, H. Loss of eyes in zebrafish caused by mutation of chokh/rx3. *EMBO Rep* **2003**; *4*: 894–899.

64 Loosli, F., Winkler, S., Burgtorf, C., Wurmbach, E., Ansorge, W., Henrich, T., Grabher, C., Arendt, D., Carl, M., Krone, A., et al. Medaka *eyeless* is the key factor linking retinal determination and eye growth. *Development* **2001**; *128*: 4035–4044.

65 Mathers, P.H., Grinberg, A., Mahon, K.A., and Jamrich, M. The Rx homeobox gene is essential for vertebrate eye development. *Nature* **1997**; *387*: 603–607.

66 Chuang, J.C., Mathers, P.H., and Raymond, P.A. Expression of three *Rx* homeobox genes in embryonic and adult zebrafish. *Mech Dev* **1999**; *84*: 195–198.

67 Deschet, K., Bourrat, F., Ristoratore, F., Chourrout, D., and Joly, J.-S. Expression of the medaka (*Oryzias latipes*) Ol-Rx3 *paired*-like gene in two diencephalic derivatives, the eye and the hypothalamus. *Mech Dev* **1999**; *83*: 179–182.

68 Winkler, S., Loosli, F., Henrich, T., Wakamatsu, Y., and Wittbrodt, J. The conditional medaka mutation eyeless uncouples patterning and morphogenesis of the eye. *Development* **2000**; *127*: 1911–1919.

69 Kawakami, K., Sato, S., Ozaki, H., and Ikeda, K. *Six* family genes-structure and function as transcription factors and their roles in development. *Bioessays* **2000**; *22*: 616–626.

70 Oliver, G., Mailhos, A., Wehr, R., Copeland, N.G., Jenkins, N.A., and Gruss, P. Six3, a murine homologue of the sine oculis gene, demarcates the most anterior border of the developing neural plate and is expressed during eye development. *Development* **1995**; *121*: 4045–4055.

71 Cheyette, B.N.R., Green, P.J., Martin, K., Garren, H., Hartenstein, V., and Zipursky, S.L. The Drosophila sine oculis locus encodes a homeodomain-containing protein required for the development of the entire visual system. *Neuron* **1994**; *12*: 977–996.

72 Seimiya, M., and Gehring, W.J. The Drosophila homeobox gene optix is capable of inducing ectopic eyes by an eyeless-independent mechanism. *Development* **2000**; *127*: 1879–1886.

73 Toy, J., Yang, J.-M., Leppert, G., and Sundin, O.H. The Optx2 homeobox gene is expressed in early precursors of the eye and activates retina-specific genes. *Proc Natl Acad Sci USA* **1998**; *95*: 10643–10648.

74 Bovolenta, P., Mallamaci, A., Puelles, L., and Boncinelli, E. Expression pattern of *cSix3*, a member of the Six/sine oculis family of transcription factors. *Mech Dev* **1998**; *70*: 201–203.

75 Kobayashi, M., Toyama, R., Takeda, H., Dawid, I.B., and Kawakami, K. Overexpression of the forebrain-specific homeobox gene six3 induces rostral forebrain enlargement in zebrafish. *Development* **1998**; *125*: 2973–2982.

76 Seo, H.C., Drivenes, O., Ellingsen, S., and Fjose, A. Expression of two zebrafish homologues of the murine Six3 gene demarcates the initial eye primordia. *Mech Dev* **1998**; *73*: 45–57.

77 Zhou, X., Hollemann, T., Pieler, T., and Gruss, P. Cloning and expression of *xSix3*, the *Xenopus* homologue of murine *Six3*. *Mech Dev* **2000**; *91*: 327–330.

78 Loosli, F., Winkler, S., and Wittbrodt, J. Six3 overexpression initiates the formation of ectopic retina. *Genes Dev* **1999**; *13*: 649–654.

79 Goudreau, G., Petrou, P., Reneker, L.W., Graw, J., Loster, J., and Gruss, P. Mutually regulated expression of Pax6 and Six3 and its implications for the Pax6 haploinsufficient lens phenotype. *Proc Natl Acad Sci USA* **2002**; *99*: 8719–8724.

80 Lagutin, O.V., Zhu, C.C., Kobayashi, D., Topczewski, J., Shimamura, K., Puelles, L., Russell, H.R.C., McKinnon, P.J., Solnica-Krezel, L., and Oliver, G. Six3 repression of Wnt signaling in the anterior neuroectoderm is essential for vertebrate forebrain development. *Genes Dev* **2003**; *17*: 368–379.

81 Braun, M.M., Etheridge, A., Bernard, A., Robertson, C.P., and Roelink, H. Wnt signaling is required at distinct stages of development for the induction of the posterior forebrain. *Development* **2003**; *130*: 5579–5587.

82 Kiecker, C., and Niehrs, C. A morphogen gradient of Wnt/beta-catenin signalling regulates anteroposterior neural patterning in Xenopus. *Development* **2001**; *128*: 4189–4201.

83 Nordstrom, U., Jessell, T.M., and Edlund, T. Progressive induction of caudal neural character by graded Wnt signaling. *Nat Neurosci* **2002**; *5*: 525–532.

84 Wilson, S.W., and Houart, C. Early steps in the development of the forebrain. *Dev Cell* **2004**; *6*: 167–181.

85 Simeone, A., Acampora, D., Mallamaci, A., Stornaiuolo, A., Dapice, M.R., Nigro, V., and Boncinelli, E. A vertebrate gene related to orthodenticle contains a homeodomain of the bicoid class and demarcates anterior neuroectoderm in the gastrulating mouse embryo. *EMBO J* **1993**; *12*: 2735–2747.

86 Matsuo, I., Kuratani, S., Kimura, C., Takeda, N., and Aizawa, S. Mouse Otx2 functions in the formation and patterning of rostral head. *Genes Dev* **1995**; *9*: 2646–2658.

87 Pannese, M., Polo, C., Andreazzoli, M., Vignali, R., Kablar, B., Barsacchi, G., and Boncinelli, E. The Xenopus homologue of Otx2 is a maternal homeobox gene that demarcates and specifies anterior body regions. *Development* **1995**; *121*: 707–720.

88 Andreazzoli, M., Gestri, G., Angeloni, D., Menna, E., and Barsacchi, G. Role of *Xrx1* in *Xenopus* eye and anterior brain development. *Development* **1999**; *126*: 2451–2460.

89 Zuber, M.E., Gestri, G., Viczian, A.S., Barsacchi, G., and Harris, W.A. Specification of the vertebrate eye by a network of eye field transcription factors. *Development* **2003**; *130*: 5155–5167.

90 Adelmann, H.B. Experimental studies on the development of the eye. II. The eye-forming potencies of the median portions of the urodelan neural plate (*Triton teniatus* and *Amblystoma punctatum*). *J Exp Zool* **1929**; *54*: 291–317.100.

91 Adelmann, H.B. The problem of cyclopia. Part I. *Quart Rev Biol* **1936**; *11*: 161–182.

92 Li, H., Tierney, C., Wen, L., Wu, J.Y., and Rao, Y. A single morphogenetic field gives rise to two retina primordia under the influence of the prechordal plate. *Development* **1997**; *124*: 603–615.

93 Pera, E.M., and Kessel, M. Patterning of the chick forebrain anlage by the prechordal plate. *Development* **1997**; *124*: 4153–4162.

94 Hammerschmidt, M., Pelegri, F., Mullins, M.C., Kane, D.A., Brand, M., van Eeden, F.J., Furutani-Seiki, M., Granato, M., Haffter, P., Heisenberg, C.P., et al. Mutations affecting morphogenesis during gastrulation and tail formation in the zebrafish, Danio rerio. *Development* **1996**; *123*: 143–151.

95 Hatta, K., Kimmel, C.B., Ho, R.K., and Walker, C. The cyclops mutation blocks specification of the floor plate of the zebrafish central nervous system. *Nature* **1991**; *350*: 339–341.

96 Feldman, B., Gates, M.A., Egan, E.S., Dougan, S.T., Rennebeck, G., Sirotkin, H.I., Schier, A.F., and Talbot, W.S. Zebrafish organizer development and germ-layer formation require nodal-related signals. *Nature* **1998**; *395*: 181–185.

97 Rebagliati, M.R., Toyama, R., Haffter, P., and Dawid, I.B. Cyclops encodes a nodal-related factor involved in midline signaling. *Proc Natl Acad Sci USA* **1998**; *95*: 9932–9937.

98 Sampath, K., Rubinstein, A.L., Cheng, A.M., Liang, J.O., Fekany, K., Solnica-Krezel, L., Korzh, V., Halpern, M.E., and Wright, C.V. Induction of the zebrafish ventral brain and floorplate requires cyclops/nodal signalling. *Nature* **1998**; *395*: 185–189.

99 Zhang, J.J., Talbot, W.S., and Schier, A.F. Positional cloning identifies zebrafish one-eyed pinhead as a permissive Egf-re-

lated ligand required during gastrulation. *Cell* **1998**; *92*: 241–251.

100 McMahon, A.P., Ingham, P.W., and Tabin, C.J. Developmental roles and clinical significance of hedgehog signaling. *Curr Top Dev Biol* **2003**; *53*: 1–114.

101 Belloni, E., Muenke, M., Roessler, E., Traverso, G., Siegel-Bartelt, J., Frumkin, A., Mitchell, H.F., Donis-Keller, H., Helms, C., Hing, A.V., et al. Identification of Sonic hedgehog as a candidate gene responsible for holoprosencephaly. *Nat Genet* **1996**; *14*: 353–356.

102 Chiang, C., Litingtung, Y., Lee, E., Young, K.E., Corden, J.L., Westphal, H., and Beachy, P.A. Cyclopia and defective axial patterning in mice lacking Sonic hedgehog gene function. *Nature* **1996**; *383*: 407–413.

103 Roessler, E., Belloni, E., Gaudenz, K., Jay, P., Berta, P., Scherer, S.W., Tsui, L.C., and Muenke, M. Mutations in the human Sonic Hedgehog gene cause holoprosencephaly. *Nat Genet* **1996**; *14*: 357–360.

104 Ekker, S.C., Ungar, A.R., Greenstein, P., v. Kessler, D., Porter, J.A., Moon, R.T., and Beachy, P.A. Patterning activities of vertebrate hedgehog proteins in the developing eye and brain. *Curr Biol* **1995**; *5*: 944–955.

105 Schauerte, H.E., van Eeden, F.J., Fricke, C., Odenthal, J., Strahle, U., and Haffter, P. Sonic hedgehog is not required for the induction of medial floor plate cells in the zebrafish. *Development* **1998**; *125*: 2983–2993.

106 Heisenberg, C.P., Tada, M., Rauch, G.J., Saude, L., Concha, M.L., Geisler, R., Stemple, D.L., Smith, J.C., and Wilson, S.W. Silberblick/Wnt11 mediates convergent extension movements during zebrafish gastrulation. *Nature* **2000**; *405*: 76–81.

107 Varga, Z.M., Wegner, J., and Westerfield, M. Anterior movement of ventral diencephalic precursors separates the primordial eye field in the neural plate and requires *cyclops*. *Development* **1999**; *126*: 5533–5546.

108 Macdonald, R., Barth, K.A., Xu, Q., Holder, N., Mikkola, I., and Wilson, S.W. Midline signalling is required for Pax gene regulation and patterning of the eyes. *Development* **1995**; *121*: 3267–3278.

109 Take-uchi, M., Clarke, J.D., and Wilson, S.W. Hedgehog signalling maintains the optic stalk-retinal interface through the regulation of Vax gene activity. *Development* **2003**; *130*: 955–968.

110 Carl, M., and Wittbrodt, J. Graded interference with FGF-signaling reveals its dorso-ventral asymmetry at the mid-hindbrain boundary. *Development* **1999**; *126*: 5659–5667.

111 Shanmugalingam, S., Houart, C., Picker, A., Reifers, F., Macdonald, R., Barth, A., Griffin, K., Brand, M., and Wilson, S.W. Ace/Fgf8 is required for forebrain commissure formation and patterning of the telencephalon. *Development* **2000**; *127*: 2549–2561.

112 Walshe, J., and Mason, I. Unique and combinatorial functions of Fgf3 and Fgf8 during zebrafish forebrain development. *Development* **2003**; *130*: 4337–4349.

113 Barbieri, A.M., Lupo, G., Bulfone, A., Andreazzoli, M., Mariani, M., Fougerousse, F., Consalez, G.G., Borsani, G., Be-

ckmann, J.S., Barsacchi, G., et al. A homeobox gene, vax2, controls the patterning of the eye dorsoventral axis. *Proc Natl Acad Sci USA* **1999**; *96*: 10729–10734.

114 Dressler, G.R., Deutsch, U., Chowdhury, K., Nornes, H.O., and Gruss, P. Pax2, a new murine paired-box-containing gene and its expression in the developing excretory system. *Development* **1990**; *109*: 787–795.

115 Hallonet, M., Hollemann, T., Wehr, R., Jenkins, N.A., Copeland, N.G., Pieler, T., and Gruss, P. Vax1 is a novel homeobox-containing gene expressed in the developing anterior ventral forebrain. *Development* **1998**; *125*: 2599–2610.

116 Nornes, H.O., Dressler, G.R., Knapik, E.W., Deutsch, U., and Gruss, P. Spatially and temporally restricted expression of Pax2 during murine neurogenesis. *Development* **1990**; *109*: 797–809.

117 Schulte, D., Furukawa, T., Peters, M.A., Kozak, C.A., and Cepko, C.L. Misexpression of the Emx-related homeobox genes cVax and mVax2 ventralizes the retina and perturbs the retinotectal map. *Neuron* **1999**; *24*: 541–553.

118 Schwarz, M., Cecconi, F., Bernier, G., Andrejewski, N., Kammandel, B., Wagner, M., and Gruss, P. Spatial specification of mammalian eye territories by reciprocal transcriptional repression of Pax2 and Pax6. *Development* **2000**; *127*: 4325–4334.

119 Wallis, D.E., Roessler, E., Hehr, U., Nanni, L., Wiltshire, T., Richieri-Costa, A., Gillessen-Kaesbach, G., Zackai, E.H., Rommens, J., and Muenke, M. Mutations in the homeodomain of the human SIX3 gene cause holoprosencephaly. *Nat Genet* **1999**; *22*: 196–198.

120 Bertuzzi, S., Hindges, R., Mui, S.H., O'Leary, D.D., and Lemke, G. The homeodomain protein vax1 is required for axon guidance and major tract formation in the developing forebrain. *Genes Dev* **1999**; *13*: 3092–3105.

121 Hallonet, M., Hollemann, T., Pieler, T., and Gruss, P. Vax1, a novel homeobox-containing gene, directs development of the basal forebrain and visual system. *Genes Dev* **1999**; *13*: 3106–3114.

122 Torres, M., Gómez-Pardo, E., and Gruss, P. *Pax2* contributes to inner ear patterning and optic nerve trajectory. *Development* **1996**; *122*: 3381–3391.

123 Sanyanusin, P., Schimmenti, L.A., McNoe, L.A., Ward, T.A., Pierpont, M.E., Sullivan, M.J., Dobyns, W.B., and Eccles, M.R. Mutation of the PAX2 gene in a family with optic nerve colobomas, renal anomalies and vesicoureteral reflux. *Nat Genet* **1995**; *9*: 358–364.

124 Koshiba-Takeuchi, K., Takeuchi, J.K., Matsumoto, K., Momose, T., Uno, K., Hoepker, V., Ogura, K., Takahashi, N., Nakamura, H., Yasuda, K., and Ogura, T. Tbx5 and the retinotectum projection. *Science* **2000**; *287*: 134–137.

125 Peters, M.A. Patterning the neural retina. *Curr Opin Neurobiol* **2002**; *12*: 43–48.

126 Barbieri, A.M., Broccoli, V., Bovolenta, P., Alfano, G., Marchitiello, A., Mocchetti, C., Crippa, L., Bulfone, A., Marigo, V., Ballabio, A., and Banfi, S. Vax2 inactivation in mouse deter-

mines alteration of the eye dorsal-ventral axis, misrouting of the optic fibres and eye coloboma. *Development* **2002**; *129*: 805–813.

127 Mui, S.H., Hindges, R., O'Leary, D.D., Lemke, G., and Bertuzzi, S. The homeodomain protein Vax2 patterns the dorsoventral and nasotemporal axes of the eye. *Development* **2002**; *129*: 797–804.

128 Onwochei, B.C., Simon, J.W., Bateman, J.B., Couture, K.C., and Mir, E. Ocular colobomata. *Surv Ophthalmol* **2000**; *45*: 175–194.

129 Fantl, V., Stamp, G., Andrews, A., Rosewell, I., and Dickson, C. Mice lacking cyclin D1 are small and show defects in eye and mammary gland development. *Genes Dev* **1995**; *9*: 2364–2372.

130 Ma, C., Papermaster, D., and Cepko, C. L. A unique pattern of photoreceptor degeneration in cyclin D1 mutant mice. *Proc Natl Acad Sci USA* **1998**; *95*: 9938–9943.

131 Sicinski, P., Donaher, J.L., Parker, S.B., Li, T., Fazeli, A., Gardner, H., Haslam, S.Z., Bronson, R.T., Elledge, S.J., and Weinberg, R.A. Cyclin D1 provides a link between development and oncogenesis in the retina and breast. *Cell* **1995**; *82*: 621–630.

132 Ohnuma, S., Philpott, A., Wang, K., Holt, C.E., and Harris, W. A. p27Xic1, a Cdk inhibitor, promotes the determination of glial cells in Xenopus retina. *Cell* **1999**; *99*: 499–510.

133 Dyer, M.A., and Cepko, C. L. p57(Kip2) regulates progenitor cell proliferation and amacrine interneuron development in the mouse retina. *Development* **2000**; *127*: 3593–3605.

134 Chuang, J.C., and Raymond, P.A. Zebrafish genes rx1 and rx2 help define the region of forebrain that gives rise to retina. *Dev Biol* **2001**; *231*: 13–30.

135 Zuber, M.E., Perron, M., Philpott, A., Bang, A., and Harris, W.A. Giant eyes in *Xenopus laevis* by overexpression of XOptx2. *Cell* **1999**; *98*: 341–352.

136 Ghanbari, H., Seo, H.C., Fjose, A., and Brandli, A.W. Molecular cloning and embryonic expression of Xenopus Six homeobox genes. *Mech Dev* **2001**; *101*: 271–277.

137 Lopez-Rios, J., Gallardo, M.E., Rodriguez de Cordoba, S., and Bovolenta, P. Six9 (Optx2), a new member of the six gene family of transcription factors, is expressed at early stages of vertebrate ocular and pituitary development. *Mech Dev* **1999**; *83*: 155–159.

138 Kobayashi, M., Nishikawa, K., Suzuki, T., and Yamamoto, M. The homeobox protein Six3 interacts with the Groucho corepressor and acts as a transcriptional repressor in eye and forebrain formation. *Dev Biol* **2001**; *232*: 315–326.

139 Lopez-Rios, J., Tessmar, K., Loosli, F., Wittbrodt, J., and Bovolenta, P. Six3 and Six6 activity is modulated by members of the groucho family. *Development* **2003**; *130*: 185–195.

140 Zhu, C.C., Dyer, M.A., Uchikawa, M., Kondoh, H., Lagutin, O.V., and Oliver, G. Six3–mediated auto repression and eye development requires its interaction with members of the Groucho-related family of co-repressors. *Development: Suppl* **2002**; *129*: 2835–2849.

141 Li, X., Perissi, V., Liu, F., Rose, D.W., and Rosenfeld, M.G. Tissue-specific regulation of retinal and pituitary precursor cell proliferation. *Science* **2002**; *297*: 1180–1183.

142 Del Bene, F., Tessmar-Raible, K., and Wittbrodt, J. Direct interaction of geminin and Six3 in eye development. *Nature* **2004**; *427*: 745–749.

143 McGarry, T.J., and Kirschner, M.W. Geminin, an inhibitor of DNA replication, is degraded during mitosis. *Cell* **1998**; *93*: 1043–1053.

144 Kroll, K.L., Salic, A.N., Evans, L.M., and Kirschner, M.W. Geminin, a neuralizing molecule that demarcates the future neural plate at the onset of gastrulation. *Development: Suppl* **1998**; *125*: 3247–3258.

145 Tada, S., Li, A., Maiorano, D., Mechali, M., and Blow, J.J. Repression of origin assembly in metaphase depends on inhibition of RLF-B/Cdt1 by geminin. *Nat Cell Biol* **2001**; *3*: 107–113.

146 Wohlschlegel, J.A., Dwyer, B.T., Dhar, S.K., Cvetic, C., Walter, J.C., and Dutta, A. Inhibition of eukaryotic DNA replication by geminin binding to Cdt1. *Science* **2000**; *290*: 2309–2312.

147 Maiorano, D., Moreau, J., and Mechali, M. XCDT1 is required for the assembly of pre-replicative complexes in *Xenopus laevis*. *Nature* **2000**; *404*: 622–625.

148 Tessmar, K., Loosli, F., and Wittbrodt, J. A screen for co-factors of Six3. *Mech Dev* **2002**; *117*: 103–113.

149 Chen, C.M., and Cepko, C.L. Expression of Chx10 and Chx10–1 in the developing chicken retina. *Mech Dev* **2000**; *90*: 293–297.

150 Levine, E.M., Passini, M., Hitchcock, P.F., Glasgow, E., and Schechter, N. Vsx–1 and Vsx–2: two Chx10–like homeobox genes expressed in overlapping domains in the adult goldfish retina. *J Comp Neurol* **1997**; *387*: 439–448.

151 Liu, I.S.C., Chen, J.D., Ploder, L., Vidgen, D., Vanderkooy, D., Kalnins, V.I., and Mcinnes, R.R. Developmental expression of a novel murine homeobox gene (Chx10): Evidence for roles in determination of the neuroretina and inner nuclear layer. *Neuron* **1994**; *13*: 377–393.

152 Passini, M.A., Levine, E.M., Canger, A.K., Raymond, P.A., and Schechter, N. *Vsx–1* and *Vsx–2*: Differential expression of two *paired*-like homeobox genes during zebrafish and goldfish retinogenesis. *J Comp Neurol* **1997**; *388*: 495–505.

153 Burmeister, M., Novak, J., Liang, M.-Y., Basu, S., Ploder, L., Hawes, N.L., Vidgen, D., Hoover, F., Goldman, D., Kalnins, V.I., et al. Ocular retardation in mouse caused by *Chx10* homeobox null allele: impaired retinal progenitor proliferation and bipolar cell differentiation. *Nat Genet* **1996**; *12*: 376–384.

154 Barabino, S. M.L., Spada, F., Cotelli, F., and Boncinelli, E. Inactivation of the zebrafish homologue of Chx10 by antisense oligonucleotides causes eye malformations similar to the ocular retardation phenotype. *Mech Dev* **1997**; *63*: 133–143.

155 Green, E.S., Stubbs, J.L., and Levine, E.M. Genetic rescue of cell number in a mouse model of microphthalmia: interactions between Chx10 and G1–phase cell cycle regulators. *Development* **2003**; *130*: 539–552.

156 Cepko, C.L., Austin, C.P., Yang, X.J., and Alexiades, M. Cell fate determination in the vertebrate retina. *Proc Natl Acad Sci USA* **1996**; *93*: 589–595.

157 Stiemke, M.M., and Hollyfield, J.G. Cell birthdays in Xenopus laevis retina. *Differentiation* **1995**; *58*: 189–193.

158 Cremisi, F., Philpott, A., and Ohnuma, S. Cell cycle and cell fate interactions in neural development. *Curr Opin Neurobiol* **2003**; *13*: 26–33.

159 Livesey, F.J., and Cepko, C.L. Vertebrate neural cell-fate determination: lessons from the retina. *Nat Rev Neurosci* **2001**; *2*: 109–118.

160 Marquardt, T., and Gruss, P. Generating neuronal diversity in the retina: one for nearly all. *Trends Neurosci* **2002**; *25*: 32–38.

161 Laessing, U., and Stuermer, C.A. Spatiotemporal pattern of retinal ganglion cell differentiation revealed by the expression of neurolin in embryonic zebrafish. *J Neurobiol* **1996**; *29*: 65–74.

162 Prada, C., Puga, J., Perez-Mendez, L., Lopez, and Ramirez, G. Spatial and temporal patterns of neurogenesis in the chick retina. *Eur J Neurosci* **1991**; *3*: 1187.

163 Young, R.W. Cell differentiation in the retina of the mouse. *Anat Rec* **1985**; *212*: 199–205.

164 Bonini, N.M., and Choi, K.W. Early decisions in Drosophila eye morphogenesis. *Curr Opin Genet Dev* **1995**; *5*: 507–515.

165 Hu, M., and Easter, S.S. Retinal neurogenesis: the formation of the initial central patch of postmitotic cells. *Dev Biol* **1999**; *207*: 309–321.

166 Neumann, C.J., and Nüsslein-Volhard, C. Patterning of the zebrafish retina by a wave of sonic hedgehog activity. *Science* **2000**; *289*: 2137–2139.

167 Masai, I., Stemple, D.L., Okamoto, H., and Wilson, S.W. Midline signals regulate retinal neurogenesis in zebrafish. *Neuron* **2000**; *27*: 251–263.

168 Xiang, M. Requirement for Brn–3b in early differentiation of postmitotic retinal ganglion cell precursors. *Dev Biol* **1998**; *197*: 155–169.

169 Austin, C.P., Feldman, D.E., Ida, J.A., and Cepko, C.L. Vertebrate retinal ganglion cells are selected from competent progenitors by the action of Notch. *Development* **1995**; *121*: 3637–3650.

170 Belliveau, M.J., and Cepko, C.L. Extrinsic and intrinsic factors control the genesis of amacrine and cone cells in the rat retina. *Development: Suppl* **1999**; *126*: 555–566.

171 Belliveau, M.J., Young, T.L., and Cepko, C.L. Late retinal progenitor cells show intrinsic limitations in the production of cell types and the kinetics of opsin synthesis. *J Neurosci* **2000**; *20*: 2247–2254.

172 Yang, Z., Ding, K., Pan, L., Deng, M., and Gan, L. Math5 determines the competence state of retinal ganglion cell progenitors. *Dev Biol* **2003**; *264*: 240–254.

173 Brown, N.L., Patel, S., Brzezinski, J., and Glaser, T. Math5 is required for retinal ganglion cell and optic nerve formation. *Development* **2001**; *128*: 2497–2508.

174 Kay, J.N., Finger-Baier, K.C., Roeser, T., Staub, W., and Baier, H. Retinal Ganglion Cell Genesis Requires lakritz, a Zebrafish atonal Homolog. *Neuron* **2001**; *30*: 725–736.

175 Wang, S.W., Kim, B.S., Ding, K., Wang, H., Sun, D., Johnson, R.L., Klein, W.H., and Gan, L. (**2001**). Requirement for math5 in the development of retinal ganglion cells. *Genes Dev* 15, 24–29.

176 Kanekar, S., Perron, M., Dorsky, R., Harris, W.A., Jan, L.Y., Jan, Y.N., and Vetter, M.L. Xath5 participates in a network of bHLH genes in the developing Xenopus retina. *Neuron* **1997**; *19*: 981–994.

177 Liu, W., Mo, Z., and Xiang, M. The Ath5 proneural genes function upstream of Brn3 POU domain transcription factor genes to promote retinal ganglion cell development. *Proc Natl Acad Sci USA* **2001**; *98*: 1649–1654.

178 Poggi, L., Vottari, T., Barsacchi, G., Wittbrodt, J., and Vignali, R. The homeobox gene Xbh1 cooperates with proneural genes to specify ganglion cell fate in the Xenopus neural retina. *Development* **2004**; *131*: 2305–2315.

179 Erkman, L., McEvilly, R.J., Luo, L., Ryan, A.K., Hooshmand, F., O'Connell, S.M., Keithley, E.M., Rapaport, D.H., Ryan, A.F., and Rosenfeld, M.G. Role of transcription factors Brn-3.1 and Brn-3.2 in auditory and visual system development. *Nature* **1996**; *381*: 603–606.

180 Gan, L., Xiang, M., Zhou, L., Wagner, D.S., Klein, W.H., and Nathans, J. POU domain factor Brn-3b is required for the development of a large set of retinal ganglion cells. *Proc Natl Acad Sci USA* **1996**; *93*: 3920–3925.

181 Liu, W., Khare, S.L., Liang, X., Peters, M.A., Liu, X., Cepko, C.L., and Xiang, M. All Brn3 genes can promote retinal ganglion cell differentiation in the chick. *Development: Suppl* **2000**; *127*: 3237–3247.

182 Wang, S.W., Mu, X., Bowers, W.J., Kim, D.S., Plas, D.J., Crair, M.C., Federoff, H.J., Gan, L., and Klein, W.H. Brn3b/Brn3c double knockout mice reveal an unsuspected role for Brn3c in retinal ganglion cell axon outgrowth. *Development* **2002**; *129*: 467–477.

183 Morrow, E.M., Furukawa, T., Lee, J.E., and Cepko, C.L. NeuroD regulates multiple functions in the developing neural retina in rodent. *Development: Suppl* **1999**; *126*: 23–36.

184 Inoue, T., Hojo, M., Bessho, Y., Tano, Y., Lee, J.E., and Kageyama, R. Math3 and NeuroD regulate amacrine cell fate specification in the retina. *Development* **2002**; *129*: 831–842.

185 Tomita, K., Moriyoshi, K., Nakanishi, S., Guillemot, F., and Kageyama, R. Mammalian achaete-scute and atonal homologs regulate neuronal versus glial fate determination in the central nervous system. *EMBO J* **2000**; *19*: 5460–5472.

186 Hatakeyama, J., Tomita, K., Inoue, T., and Kageyama, R. Roles of homeobox and bHLH genes in specification of a retinal cell type. *Development: Suppl* **2001**; *128*: 1313–1322.

187 Jan, Y.N., and Jan, L.Y. Asymmetric cell division in the Drosophila nervous system. *Nat Rev Neurosci* **2001**; *2*: 772–779.

188 Li, L., and Vaessin, H. Pan-neural Prospero terminates cell proliferation during Drosophila neurogenesis. *Genes Dev* **2000**; *14*: 147–151.

189 Glasgow, E., and Tomarev, S.I. Restricted expression of the homeobox gene prox 1 in developing zebrafish. *Mech Dev* **1998**; *76*: 175–178.

190 Oliver, G., Sosapineda, B., Geisendorf, S., Spana, E.P., Doe, C.Q., and Gruss, P. Prox 1, a Prospero-related homeobox gene

expressed during mouse development. *Mech Dev* **1993**; *44*: 3–16.

191 Tomarev, S.I., Sundin, O., Banerjee-Basu, S., Duncan, M.K., Yang, J.M., and Piatigorsky, J. Chicken homeobox gene Prox 1 related to Drosophila prospero is expressed in the developing lens and retina. *Dev Dyn* **1996**; *206*: 354–367.

192 Dyer, M.A., Livesey, F.J., Cepko, C.L., and Oliver, G. Prox1 function controls progenitor cell proliferation and horizontal cell genesis in the mammalian retina. *Nat Genet* **2003**; *34*: 53–58.

193 Ahmad, I. Mash–1 is expressed during ROD photoreceptor differentiation and binds an E-box, E(opsin)–1 in the rat opsin gene. *Brain Res Dev Brain Res* **1995**; *90*: 184–189.

194 Tomita, K., Nakanishi, S., Guillemot, F., and Kageyama, R. Mash1 promotes neuronal differentiation in the retina. *Genes Cells* **1996**; *1*: 765–774.

195 Furukawa, T., Morrow, E.M., Li, T., Davis, F.C., and Cepko, C.L. Retinopathy and attenuated circadian entrainment in Crx-deficient mice. *Nat Genet* **1999**; *23*: 466–470.

196 Chen, C.M., and Cepko, C.L. The chicken RaxL gene plays a role in the initiation of photoreceptor differentiation. *Development* **2002**; *129*: 5363–5375.

197 Baas, D., Bumsted, K.M., Martinez, J.A., Vaccarino, F.M., Wikler, K.C., and Barnstable, C.J. The subcellular localization of Otx2 is cell-type specific and developmentally regulated in the mouse retina. *Brain Res Mol Brain Res* **2000**; *78*: 26–37.

198 Bovolenta, P., Mallamaci, A., Briata, P., Corte, G., and Boncinelli, E. Implication of *Otx2* in pigmented epithelium determination and neural retina differentiation. *J Neurosci* **1997**; *17*: 4243–4252.

199 Martinez-Morales, J.R., Signore, M., Acampora, D., Simeone, A., and Bovolenta, P. Otx genes are required for tissue specification in the developing eye. *Development* **2001**; *128*: 2019–2030.

200 Viczian, A.S., Vignali, R., Zuber, M.E., Barsacchi, G., and Harris, W. A. XOtx5b and XOtx2 regulate photoreceptor and bipolar fates in the Xenopus retina. *Development* **2003**; *130*: 1281–1294.

201 Nishida, A., Furukawa, A., Koike, C., Tano, Y., Aizawa, S., Matsuo, I., and Furukawa, T. Otx2 homeobox gene controls retinal photoreceptor cell fate and pineal gland development. *Nat Neurosci* **2003**; *6*: 1255–1263.

202 Vignali, R., Colombetti, S., Lupo, G., Zhang, W., Stachel, S., Harland, R.M., and Barsacchi, G. Xotx5b, a new member of the Otx gene family, may be involved in anterior and eye development in Xenopus laevis. *Mech Dev* **2000**; *96*: 3–13.

203 Holt, C.E., Bertsch, T.W., Ellis, H.M., and Harris, W.A. Cellular determination in the Xenopus retina is independent of lineage and birth date. *Neuron* **1988**; *1*: 15–26.

204 Turner, D.L., Snyder, E.Y., and Cepko, C.L. Lineage-independent determination of cell type in the embryonic mouse retina. *Neuron* **1990**; *4*: 833–845.

205 Wetts, R., and Fraser, S.E. Multipotent precursors can give rise to all major cell types of the frog retina. *Science* **1988**; *239*: 1142–1145.

206 Ahmad, I., Das, A.V., James, J., Bhattacharya, S., and Zhao, X. Neural stem cells in the mammalian eye: types and regulation. *Semin Cell Dev Biol* **2004**; *15*: 53–62.

207 Ahmad, I., Dooley, C.M., Thoreson, W.B., Rogers, J.A., and Afiat, S. In vitro analysis of a mammalian retinal progenitor that gives rise to neurons and glia. *Brain Res* **1999**; *831*: 1–10.

208 Johns, P.R., and Easter, S.S., Jr. Growth of the adult goldfish eye. II. Increase in retinal cell number. *J Comp Neurol* **1977**; *176*: 331–341.

209 Straznicky, K., and Gaze, R.M. The growth of the retina in Xenopus laevis: an autoradiographic study. *J Embryol Exp Morphol* **1971**; *26*: 67–79.

210 Wetts, R., Serbedzija, G.N., and Fraser, S.E. Cell lineage analysis reveals multipotent precursors in the ciliary margin of the frog retina. *Dev Biol* **1989**; *136*: 254–263.

211 Perron, M., Kanekar, S., Vetter, M.L., and Harris, W.A. The genetic sequence of retinal development in the ciliary margin of the Xenopus eye. *Dev Biol* **1998**; *199*: 185–200.

212 Harris, W.A., and Perron, M. Molecular recapitulation – the growth of the vertebrate retina. *Int J Dev Biol* **1998**; *42*: 299–304.

213 Dorsky, R.I., Rapaport, D.H., and Harris, W.A. Xotch inhibits cell differentiation in the Xenopus retina. *Neuron* **1995**; *14*: 487–496.

214 Henrique, D., Hirsinger, E., Adam, J., Le Roux, I., Pourquie, O., Ish-Horowicz, D., and Lewis, J. Maintenance of neuroepithelial progenitor cells by Delta-Notch signaling in the embryonic chick retina. *Curr Biol* **1997**; *7*: 661–670.

215 Artavanis-Tsakonas, S., Rand, M.D., and Lake, R.J. Notch signaling: cell fate control and signal integration in development. *Science* **1999**; *284*: 770–776.

216 Honjo, T. The shortest path from the surface to the nucleus: RBP-J kappa/Su(H) transcription factor. *Genes Cells* **1996**; *1*: 1–9.

217 Ishibashi, M., Ang, S.L., Shiota, K., Nakanishi, S., Kageyama, R., and Guillemot, F. Targeted disruption of mammalian hairy and Enhancer of split homolog–1 (HES–1) leads to up-regulation of neural helix-loop-helix factors, premature neurogenesis, and severe neural tube defects. *Genes Dev* **1995**; *9*: 3136–3148.

218 Ohtsuka, T., Ishibashi, M., Gradwohl, G., Nakanishi, S., Guillemot, F., and Kageyama, R. Hes1 and Hes5 as notch effectors in mammalian neuronal differentiation. *EMBO J* **1999**; *18*: 2196–2207.

219 Tomita, K., Ishibashi, M., Nakahara, K., Ang, S.L., Nakanishi, S., Guillemot, F., and Kageyama, R. Mammalian hairy and Enhancer of split homolog 1 regulates differentiation of retinal neurons and is essential for eye morphogenesis. *Neuron* **1996**; *16*: 723–734.

220 Ahmad, I., Tang, L., and Pham, H. Identification of neural progenitors in the adult mammalian eye. *Biochem Biophys Res Commun* **2000**; *270*: 517–521.

221 Fischer, A.J., and Reh, T.A. Identification of a proliferating marginal zone of retinal progenitors in postnatal chickens. *Dev Biol* **2000**; *220*: 197–210.

222 Tropepe, V., Coles, B.L., Chiasson, B.J., Horsford, D.J., Elia, A.J., McInnes, R.R., and van der Kooy, D. Retinal stem cells in the adult mammalian eye. *Science* **2000**; *287*: 2032–2036.

223 Moshiri, A., and Reh, T.A. Persistent progenitors at the retinal margin of ptc+/− mice. *J Neurosci* **2004**; *24*: 229–237.

224 Furukawa, T., Mukherjee, S., Bao, Z.Z., Morrow, E.M., and Cepko, C. L. rax, Hes1, and notch1 promote the formation of Muller glia by postnatal retinal progenitor cells. *Neuron* **2000**; *26*: 383–394.

225 Hojo, M., Ohtsuka, T., Hashimoto, N., Gradwohl, G., Guillemot, F., and Kageyama, R. Glial cell fate specification modulated by the bHLH gene Hes5 in mouse retina. *Development* **2000**; *127*: 2515–2522.

226 Dyer, M.A., and Cepko, C.L. Control of Muller glial cell proliferation and activation following retinal injury. *Nat Neurosci* **2000**; *3*: 873–880.

227 Fischer, A.J., and Reh, T.A. Muller glia are a potential source of neural regeneration in the postnatal chicken retina. *Nat Neurosci* **2001**; *4*: 247–252.

228 Yang, X.J. Roles of cell-extrinsic growth factors in vertebrate eye pattern formation and retinogenesis. *Semin Cell Dev Biol* **2004**; *15*: 91–103.

229 Wittbrodt, J., Shima, A., and Schartl, M. Medaka – A model organism from the Far East. *Nat Rev Genet* **2002**; *3*: 53–64.

14
Mammalian Inner Ear Development: Of Mice and Man

Bernd Fritzsch and Kirk Beisel

14.1
Introduction

14.1.1
An Outline of Mammalian Ear Evolution

The mammalian ear evolved out of the ancestral tetrapod ear which already consisted of a three-dimensional labyrinth of ducts and recesses with a total of eight sensory epithelia arranged to extract specific aspects of angular acceleration, linear acceleration, and sound [1, 2]. These ancestral endorgans are the three semicircular canal cristae for perception of angular acceleration, three otoconia-bearing organs for perception of linear acceleration (saccule, utricle and lagena), the papilla neglecta, an epithelium of variable size and uncertain function [2, 3], and the basilar papilla, an organ that evolved into the mammalian cochlea [4–6]. Compared to this ancestral organization of the inner ear as presented by the monotremes [7], the eutherian and marsupial ears show certain losses, gains and reorganizations (Fig. 14.1). The cochlea is much longer and is composed of fewer rows (three rows of outer hair cells and one row of inner hair cells) than the basilar papilla of other amniotic vertebrates. The mammalian cochlea also shows a high frequency extension beyond the 10–kHz limit typically found in other vertebrates. However, the eutherian and marsupial inner ears are uniquely derived among amniotes by having lost the lagena, an ancestral sarcopterygian organ involved in the perception of linear acceleration as well as of sound [1].

Understanding the evolution of the mammalian (and human) ear therefore requires the dissection of the molecular basis of the ontogenetic re-patterning which led to these gains and losses within the context of the development of a vestibular system that remained qualitatively unchanged for hundreds of millions of years. Overall, the evolutionary changes in the vertebrate ear seem to follow on the one hand a highly conserved pattern of gene expression and their function in cell fate assignment which is embedded into less conserved morphogenetic pathways that seem to implement several molecular developmental modules also used in the development of several other sensory and non-sensory systems [8]. One feature unique

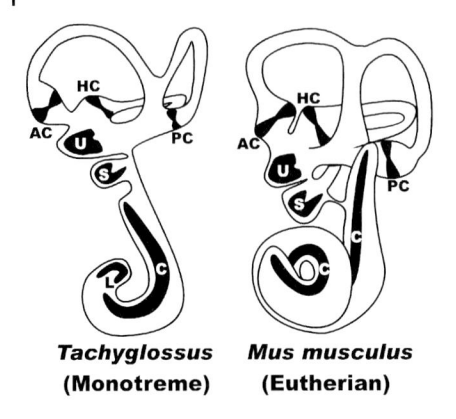

Tachyglossus **Mus musculus**
(Monotreme) (Eutherian)

Fig. 14.1
This comparison shows the differences in ear organization in a monotreme and a eutherian mammal. Note that the vestibular system is largely conserved and shows the canonical five vestibular sensory organs, the gravistatic organs of utricle (U) and saccule (S) and the angular acceleration sensors, the anterior, horizontal and posterior crisata (AC, HC, PC). Note that only monotremes have a six-vestibular organ, the lagena (L). The eutherian cochlea (C) is curved with two or more turns whereas the monotreme cochlea is comma shaped and tipped by the lagena. It therefore appears that loss of the lagena and gain of extra apical elongation of the cochlea are somehow related, but the details are unknown. Modified after Jorgensen and Locket [7].

to the eutherian and marsupial cochlea is coiling, not found in any other vertebrate. Obviously, understanding the process of coiling and whether or not it is linked with or coincidental to the loss of the lagena present at the tip of the cochlea in monotremes, remains a major question for the cochlea development and will be in part addressed here.

14.1.2
Human Deafness Related Mutations

Hearing loss and vestibular disorders, both congenital and acquired, are among the most frequent and most costly diseases of humans. If not treated and recognized in the early stages, congenitally deaf children will have problems becoming fully integrated into average social activities. Hearing-related disorders fall into four major categories based on their mode of inheritance and their exclusivity to the ear. These are non-syndromic dominant (DFNA), non-syndromic recessive (DFNB), non-syndromic X-linked (DFN) and syndromic deafness. There are approximately 54 DFNA, 51 DFNB, 6 DFN, and 32 syndromic loci which have been described to date with over 50% having an identified corresponding gene (see Tab. 14.1). In general, these genes can be placed into three major categories based on their function, which are:

(1) stereocilia-based mechanoelectrical transduction, (2) K$^+$ recirculation and conductances, and (3) the compositional integrity and function of basement membranes. A summary of the deafness loci, the corresponding gene, if known, and the mouse equivalent genes can be found on several websites, for example: http://webhost.ua.ac.be/hhh/ and in Beisel et al 2004 [9] (see supplemental Table 4 therein). Recent advances in mouse genetics have helped outline the molecular basis of ear development, and molecular "engineering" has generated several "humanized" mouse models which have been used to investigate, at the molecular level, the interactions of mutated proteins of some of the 50 characterized human hearing-related genes [10]. In addition, there are many more deafness and vestibular genes known in the mouse. Many of these are associated with spontaneous mutation, but there are a growing number of genetically-engineered mouse lines with inner ear defects.

Numerous genes are known to affect ear development in a syndromic manner and likely lead to non-viable offspring. For example, null mutations of the bHLH gene *neurogenin 1* (*Neurog1*) blocks formation of any and all sensory neuron formation connecting the ear to the brain, but also causes loss of all proximal cranial ganglia and a large fraction of spinal dorsal root ganglia [11–13]. Logically, human mutations in that gene will be lethal as early as the equivalent mutations in the mouse and thus their effect on human hearing, assuming that these genes have been mutated in humans, likely went unnoticed. Likewise, homozygous mutations of the *Pax2* gene in both mouse and humans are lethal at early stages of development [14], but a profound phenotype of variable penetrance exists among heterozygotic animals. Similar problems will plague the large-scale mutagenic screening now underway in several laboratories to identify more hearing-related genes. It is possible that a certain fraction of mutations in developmental transcription factors might not be found in these screens due to embryonic lethality and the lack of a phenotype in heterozygotes. Even so, such screens will reveal numerous new genes and will thus enhance our capacity to eventually generate a functional molecular network of interactions in ear development and adult maintenance of function. Understanding such interactions will be essential for the molecular treatment of hearing loss to regenerate, for example, lost hair cells.

Unfortunately, the analysis of genetic networks for the ear is still in its infancy [15]. Excellent prototypical examples of what needs to be understood are provided by a recent review highlighting the interactions of 7000 proteins [16] in the fruit fly proteome and the *C. elegans* interactome network of 5500 interactions [17]. These data not only confirm the short-range interactions which had previously been noted, but also indicate long-range interactions, likely mediated by the influences of different multiprotein complexes on one another. Another excellent example is to be found in developmental molecular networks such as that recently established for gastrulation in sea urchins [18] and the single cell transcript analysis of pancreas development [19]. Insights comparable to this are emerging in the development and evolution of the chordate central nervous system [20, 21]. Such data is needed to obtain a more logical basis for protein-protein interactions and transcription factor-promoter actions in ear development in order to understand the "second code" of the

Tab. 14.1 Cloned genes for non-syndromic loci.

Locus (OMIM link)	Gene	Reference
Dominant loci		
DFNA1	DIAPH1	Lynch et al., 1997
DFNA2	GJB3 (Cx31)	Xia et al., 1998
DFNA2	KCNQ4	Kubisch et al., 1999
DFNA3	GJB2 (Cx26)	Kelsell et al., 1997
DFNA3	GJB6 (Cx30)	Grifa et al., 1999
DFNA5	DFNA5	Van Laer et al., 1998
DFNA6/DFNA14	WFS1	Bespalova et al., 2001; Young et al., 2001
DFNA8/DFNA12	TECTA	Verhoeven et al., 1998
DFNA9	COCH	Robertson et al., 1998
DFNA10	EYA4	Wayne et al., 2001
DFNA11	MYO7A	Liu et al., 1997
DFNA13	COL11A2	McGuirt et al., 1999
DFNA15	POU4F3	Vahava et al., 1998
DFNA17	MYH9	Lalwani et al., 2000
DFNA20/DFNA26	ACTG1	Zhu et al, 2003
DFNA22	MYO6	Melchionda et al., 2001
DFNA28	TFCP2L3	Peters et al, 2002
DFNA36	TMC1	Kurima et al, 2002
DFNA48	MYO1A	Donaudy et al, 2003
	CRYM	Abe et al., 2003
Recessive loci		
DFNB1	GJB2 (Cx26)	Kelsell et al., 1997
DFNB1	GJB6 (Cx30)	Del Castillo et al, 2002
DFNB2	MYO7A	Liu et al., 1997, Weil et al., 1997
DFNB3	MYO15	Wang et al., 1998
DFNB4	SLC26A4	Li et al., 1998
DFNB6	TMIE	Naz et al, 2002
DFNB7/DFNB11	TMC1	Kurima et al, 2002
DFNB8/DFNB10	TMPRSS3	Scott et al., 2001
DFNB9	OTOF	Yasunaga et al., 1999
DFNB12	CDH23	Bork et al., 2001
DFNB16	STRC	Verpy et al., 2001
DFNB18	USH1C	Ouyang et al, 2002, Ahmed et al, 2002
DFNB21	TECTA	Mustapha et al., 1999
DFNB22	OTOA	Zwaenepoel et al., 2002
DFNB23	PCDH15	Ahmed et al, 2003
DFNB29	CLDN14	Wilcox et al., 2001
DFNB30	MYO3A	Walsh et al., 2002
DFNB31	WHRN	Mburu et al, 2003
DFNB36	ESPN	Naz et al, 2004
DFNB37	MYO6	Ahmed et al, 2003
	GJA1 (Cx43)	Liu et al., 2001
	PRES (Prestin)	Liu et al, 2003
X-linked loci		
DFN3	POU3F4	De Kok et al., 1995

Table after http://webhost.ua.ac.be/hhh/ with permission of Dr. Smith. Citations provided at website.

DNA that is relevant for the gene regulation network [22]. Below we summarize some of the progress made in recent years toward such an understanding to facilitate future analysis of mutational defects in ear development.

14.2
Morphological and Cellular Events in Mammalian Ear Development

The first detailed description of the 3–D development of the human ear was provided through a serial section reconstruction of ears at various stages of development [23] and can be found in textbooks on human development (Fig. 14.2). This analysis has been more recently confirmed using paint fillings of fixed ears [24]. The ear originates as an epithelial thickening, coined over 100 years ago [25] to otic placode. This area is lateral to rhombomeres 5 and 6 of the hindbrain and next to the forming pharyngeal arches. Placodes are thickenings of the ectoderm that likely come about by an altered rate of proliferation compared to the surrounding ectoderm, to generate a multilayered ectoderm. Whether this assumption of early proliferation effects is true and if so, what causes these altered rates of proliferation is still unknown and has not been formally investigated using pulse chase experiments or other techniques to evaluate proliferation. Like the nearby central nervous system, the otic placode undergoes eventually invagination to form the otic cup. At this stage there is an apparent switch in the topology of proliferation: all ectodermal proliferation is in the cellular layer residing on top of the basement membrane. In contrast, proliferation in the otocyst, as well as proliferation in the neural tube, is at the forming lumen. While this change in proliferation position has long been known to occur at later stages, there is no analysis to date that shows exactly at which point in otocyst formation this change in the topology of proliferation occurs. It is also unknown how this change relates to the possibly upregulated overall proliferation rate of the forming otocyst. Interestingly, factors involved in other systems in proliferation regulation are among the earliest expressed in the otic placode: fibroblast growth factor (*Fgf*) [26, 27] and forkhead (*Fox*) genes [28, 29]. Direct assessment of proliferation alteration in the placodal tissue of these mutants needs to be undertaken.

The otocyst will form between the first somite and the more rostral somitomeres. It is therefore likely that the developing otocyst will be exposed to some of the same factors emanating from the notochord and spinal cord such as *Shh*, which upregulate genes specific for muscle and tendon formation such as the *bHLH* genes *MyoD*, *Myf5* and *Scx*, as well as *Pax1* and *Fgf8* in somites [30]. In the ear, in a somewhat similar topology, genes of the same families appear to be equally regulated by sonic hedgehog (*Shh*), the *bHLH* genes, *Neurog1* and *Neurod1*, and Pax family member, *Pax2* [31]. It has been shown that *Fgf10* is expressed early in the mesenchyme under the forming otic placode [32] but it is unknown if this expression is regulated by *Shh*. If that were the case, the major players of early otic placode formation would all be under molecular governance comparable to the nearby somites.

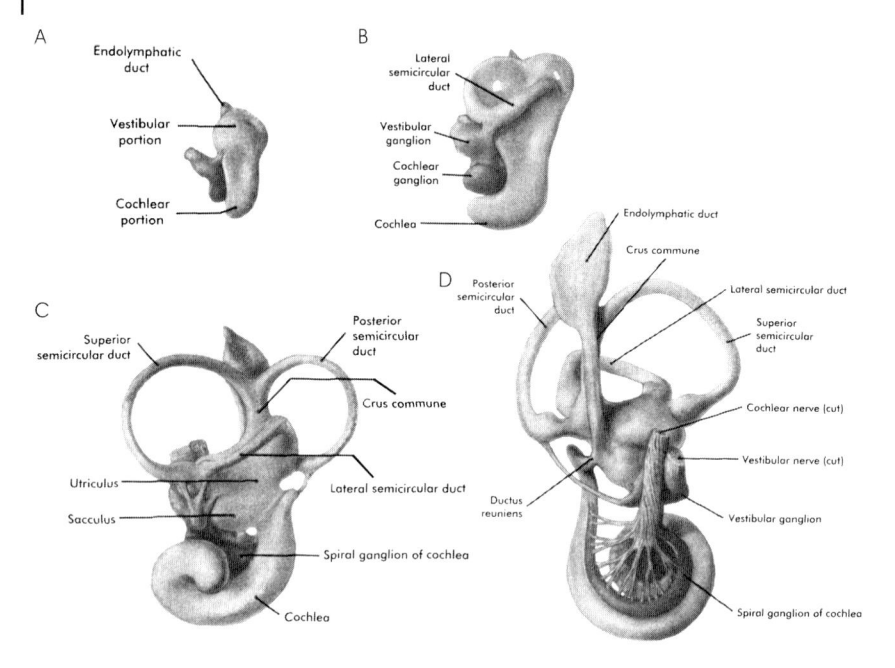

Fig. 14.2
The formation of the human ear as revealed with a wax model. Note the gradual modification of the simple elongated otocyst into a three-dimensional mass of ducts, including the coiled cochlear duct. As the shape of the labyrinth is established the sensory neurons are also formed; these will eventually innervate the vestibular and cochlea (or auditory) areas of the ear individually. A) The otocyst of a 6.6 mm long human fetus of about 30 days shows already a division into the superior (vestibular), inferior (cochlear) and ganglion portion. B) At 11 mm length, human embryos of 40 days show an ear that consists of partially developed canals, a cochlea that has completed a half turn, and a distinct vestibular and cochlear ganglion. C) A 13 mm long human embryo of about 45 days has all three canals fully differentiated, has a distinct utricular and saccular recess, has the cochlear ganglion anlage now forming a distinct spiral ganglion and shows about one entire turn of the cochlear duct. D) A 30 mm, approximately 60 day old human embryo has a fully differentiated inner ear with all six sensory epithelia and their innervation fully established. Modified after [23].

In this context it is important to realize that a population of facial branchial motoneurons becomes uniquely associated with the ear as efferents [33, 34] and that this association may have occurred at the time an otocyst evolved and physically eliminated the formation of a somite [35]. It is noticeable that hair cells express an $\alpha 9$ nicotinic acetylcholine receptor (*Chrna9*) [36, 37] which is considered to be ancestral to the $\alpha 1$ (CHRNA1) receptor found in muscle [38].

Between 24 and 48 h after the placode first becomes apparent as an ectodermal thickening, an otocyst is formed that is detached from the ectoderm (in the mouse this occurs between embryonic days 8.5 and 9.5). The otocyst grows in size over the next 24 h and subsequently undergoes morphological changes. One of the first changes is the outgrowth of the endolymphatic duct. A second change is the growth of a plate that will form the anterior and posterior vertical canal by embryonic day 12.5 [24, 39]. Through adhesion and either dislocation or even cell death of the central parts, the remaining ducts are formed [40, 41]. Partially overlapping with this is the growth of another plate that will eventually form the horizontal or lateral canal (Fig. 14.2). At this stage, another separation has started that eventually separates the superior part of the ear from the inferior area, the utriculo-saccular foramen. Like an iris, this foramen progressively closes, so much so that eventually almost no paint filling of the two compartments can be achieved with a single injection [41]. Approximately 5 days after the first appearance of the placodal thickening, the superior division of the ear is formed with the three semicircular canals and the ampullary enlargements harboring the three semicircular crista organs.

The differentiation of the inferior division is somewhat delayed. In the mouse, around embryonic day 11 the first ventral outgrowth of the cochlea begins to form, reaching about one complete turn 2 days later and showing completeness of all 2.5 turns at approximately embryonic day 15. The saccule, the second component of the inferior division of the otocyst, becomes apparent as a recess at approximately embryonic day 12.5. Around embryonic day 13.5, the segregation of the saccule from the cochlea is completed and the ductus reuniens starts to form. Through this process, the basal part of the cochlea moves from an anterior position in the developing ear to a posterior position, while the saccule retains its anterior position. Superimposed and interdigitating with the morphogenetic process is a histogenetic process that gives rise to the sensory neurons which delaminate and eventually project into distinct sensory patches formed through generation of hair cells and the formation of the acellular covering structures providing the ability to access specific mechanical energies.

This outline of morphological and histogenetic changes that transforms a simple layer of ectoderm into a complex three-dimensional structure with a sophisticated degree of cell differentiation raises a number of questions for which partial answers exist:

1. What defines the boundaries of a placode and how is the transformation of ectoderm into otic ectoderm achieved at the cellular and molecular level?
2. What defines the compartments of the ear and at which stage in development is the complexity of the future 3D space already pre-patterned in the otocyst?
3. What defines spatial origin and directs the cellular differentiation in the otocyst?

Below we will provide some information regarding these questions.

14.2.1
Molecular Basis of Ear Placode Induction

The problems regarding the definition of the "onset of placode induction" revolve around basic issues of developmental biology: classically, morphological criteria defined a starting point. For example, if the thickness of the otic placode exceeds that of the non-otic ectoderm by, say, 30%, it would be labeled as otic placode. More recent work has shown that upregulation of placode-specific marker genes predate any morphological distinction [42] and hence should lead to a new genetic-based definition, subject to changes once another even earlier expressed placode-specific markers is discovered. Beyond the function as a marker, an experimental embryological definition requires the assessment of the function of the marker. Gene knockout studies are the modern replacement for the classical transplantation analysis of achievement of competence of the otic placode. Now, single or several genes suspected of being involved in otic placode induction and/or early ear formation and their alleged function can be critically tested by monitoring the lack of other marker gene expression in the null mutants of such genes. While keeping these caveats in mind, below we will use a working definition for the otic placode that is by necessity open ended. This is in part related to the fact that some genes critical for ear development, as shown by targeted deletion, are not restricted to the otic placode, and that adjacent ectodermal areas expressing such a gene may not be invaginated to become the otocyst. Such data imply that the otic placode should be viewed as a piece of a larger otic morphogenetic field or area of potential transformation that is set up through the overlapping expression of several transcription factors which together establish the otic signature of ectodermal transformation. Precisely how large and how long lasting such a preplacodal area remains, is, at present, speculative [43].

Expression of early otic markers in mammals is shown below (Tab. 14.2) [43, 44]. Note that the techniques used to identify the onset of expression vary (*in situ* hybridization, lacZ marker). No matter which technique has been used, the true onset of a particular expression in terms of the formation of the first transcript is likely to have occurred much earlier. Likewise, there is a gray area with respect to the delay of functionally relevant amounts of transcription factor proteins. Importantly, genes that will regulate the expression of those ear-specific markers have to be specifically expressed even earlier in order to achieve an ectodermal/mesodermal expression mosaic that leads to the local upregulation of otic placode-specific genes. Dissecting ectoderm of the appropriate region and comparing the expression profile of this area with nearby non-placodal ectoderm using quantitative PCR (qPCR) and microarray could help to establish the early onset of a distinct expression profile. At the moment, the role of some of the earliest genes expressed at or around the otic placode is unknown. Moreover, the data across vertebrates suggest that even orthologs of genes such as *Foxi1* [29] and *Fgf10* [43] may not be conserved in their function for placode induction. All of this demonstrates that placodes are an embryological adaptation and as such subject to alterations to fit into the different developmental pathways of various vertebrates and may carry mammalian-specific adaptations.

14.2.2
Molecular Basis of Ear Morphogenesis

An increasing number of genes have been identified through targeted mutations that play major roles in ear morphogenesis. It seems most logical to group those

Table 14.2 Molecular markers of otic placode genes.

Stage	Gene	Reference
E.8 (TS12)	Pax8	153, 156
	Pax2	153
	Foxg1 (BF1)	28
	Dlx5	152
	Tbx1	113
	Gata3	89
	Fgf3	50
	Bmp7	159
E8.5 (TS13)	Sox2	161
	Six1	66
	Hmx3 (Nkx5.1)	158
	c-kit	155
	Notch1	157
	Mab21l2	160
E9.0 (TS14)	Neurog1	11
	Neurod1	11
	Bmp4	24
	Delta1	11
	Acvr1	162
	Epha4	154
	Lfng	24

genes in order of their relative importance as judged by the overall reduction of ear formation in the respective null mutants. A second way of looking at these genes is by analyzing their known or suspected interaction by establishing alterations of gene expression in those null mutants. However, this second aspect is clearly hampered by our limited understanding of some of those interactions and by a generally incomplete analysis of other genes in some of the analyzed mutants. The ever-growing list of morphogenetically important genes makes such analyses, by definition, more and more incomplete. Other measures have to be taken such as microarray combined with qPCR analysis of mutant otocysts to obtain a clearer picture of the quantitative changes in expression and the temporal progression of such expression alterations resulting from specific mutations. Overall, the genes we chose to present in detail here belong to several classes of factors known to play some role in other systems: FGF factors, EYA/SIX/DACH cassette and Zinc finger proteins.

14.2.2.1 FGFs

FGFs are a large family of 22 predominantly secreted ligands (except for FGF11–14) that diffuse up to eight cell diameters (about 60 μm) to interact with their seven specific receptors (differently spliced out of four *Fgfr* genes) in a paracrine or autocrine fashion. FGFRs are composed of an extracellular moiety consisting of three immunoglobin-like domains for ligand binding and an intracellular tyrosine kinase domain for signaling. FGFs need to bind to heparan sulfate proteoglycan for FGFR binding and eliminating synthesis of heparan sulfate abrogates all FGF signaling [45]. Upon ligand binding FGFRs homodimerize and activate various intracellular pathways to ultimately change gene expression [46]. FGFs are mitogenic but also play roles in cellular differentiation, motility and survival.

The most important FGFs for early ear development are FGF10, FGF3 and their common receptor, FGFR2b. *Fgf10* is expressed at the placodal stage in the mesoderm underlying the otic placode. As the placode thickens, *Fgf10* expression moves into the otic ectoderm. In the otocyst, *Fgf10* expression is in the ventral half and the delaminating sensory neurons, but eventually becomes restricted to the developing sensory epithelia [32, 47, 48]. *Fgf3* is initially only found in the nearby rhombomeres 5 and 6 of the hindbrain. Eventually, *Fgf3* is expressed in the antero-ventral aspect of the otocyst, overlapping with *Fgf10* [32]. Expression of *Fgfr1* and *Fgfr2b* initially overlaps in the otic placode. During morphogenesis of the ear, *Fgfr2b* is predominantly expressed in the dorsal half of the otocyst, expanding caudally into the cochlea [47]. Expression of *Fgf10* and *Fgfr2b* suggests both a paracrine and an autocrine interaction, whereas *Fgf3* signals predominantly through a paracrine pathway with *Fgfr2b*.

FGF factors play a pivotal role in mammalian ear formation. A double knockout of both *Fgf3* and *Fgf10* show only limited formation of occasional microvesicles, if any vesicles form at all [32, 49]. An ear forms in single *Fgf3* null mutants [50] but the size reduction of the ear is apparently different in the two *Fgf3* null mutant lines thus far generated, ranging from vesicle-like [50] to apparently normal [49]. A smaller ear also forms in single *Fgf10* null mutants [51] and *Fgf10* null ears do not undergo normal morphogenesis [48]. Clearly, FGF3 and FGF10 interact in a dose-dependent fashion to reduce ear vesicle size which suggests that signaling occurs through identical receptors [32]. In *Fgfr2b* null mutants there is a loss of almost all ear morphogenesis [47]. This reduction in ear development is less severe in *Fgfr2b* than in *Fgf3/10* double null mutations [32, 49] suggesting the existence of yet another receptor, likely *Fgfr1* [32, 47, 52]. Consistent with the signaling redundancy of FGFs through FGFRs, there appears to be a less severe effect of *Fgf10* null mutation on ear morphogenesis: the ear does not develop certain epithelia and shows altered morphogenesis, in particular loss of semicircular canals, which is consistent with the role FGFs in branching morphogenesis [47, 48]. How FGFs achieve this growth of canals in the ear is presently unknown. But, other factors such as *BMP4* [53–55], *Nkx5.1* [56, 57], *Dlx* [58, 59], *Gata3* [33] and *Otx1* [60, 61] must interact in an as yet unknown way in concert with FGF signaling to form canals. Modeling functional molecular interactions between FGF-mediated FGFR signaling and other factors targeting similar promoter regions requires a much deeper knowledge of the genes involved, their qualitative and quantitative expression patterns, and their function in ear development than is currently available. Some data on gene expression alteration exists [27], but it is still unclear whether those genes (*Pax2*, *Pax8*, *Gbx2* and *Dlx5*) are directly regulated by FGF3 and FGF10 or are affected as a consequence of an overall alteration in ear size. Expression changes of other important factors such as *Nkx5.1*, *Gata3*, *Eya1* and *Six1* have not been investigated in these null mutants. In addition, other FGFs appear to be important for ear morphogenesis such as FGF8 and FGF9. At the very least, these data indicate that complete expression profiles as well as mutational analysis of the function of all FGFs is likely needed before the full significance of this admittedly large family is fully characterized.

Nevertheless, these data show that FGFs and their receptors are crucial for otocyst formation and ear morphogenesis. In addition, FGFs also play a role in histogenesis [48, 52, 62]. *Fgfr1c* reduction causes loss of cochlear hair cell formation in a dose-dependent fashion [52]. However, some hair cells form, suggesting that *Fgfr1c* may increase the proliferation of a precursor population specified by other means. Release of *Fgf8* leads to the formation of inner hair cells and may also determine the cell fate of pillar cells which express *Fgfr3* [47, 62]. *Fgf10* null mutant mice not only show a reduction of canal sensory epithelia but also incomplete separation [48]. Other than the size reduction, no direct evaluation of mitogenic activity of FGFs has been performed in the ear, also the reduction of hair cell formation in FGFR1c hypomorphs clearly suggest the involvement of FGFs in that characteristic function. Numerous suggestions about the function of FGFs in neuronal differentiation and survival require conditional mutations to isolate these effects from the overall effects. It is notable that *Fgf10* null mutants show rather normal early development of sensory neurons [48] which suggests compensation of FGF10 as FGFs are known to be involved in neuronalization of the ectoderm [63], which likely includes the otic ectoderm. The loss of neurons to the posterior crista coincides with neurotrophin-mediated cell death and is likely mediated by the absence of neurotrophin expression in that area of the ear normally becoming the posterior crista.

14.2.2.2 The EYA/SIX/DACH Complex

At the early stages in the formation of the embryonic ear, the involvement of four genes is observed in the otic vesicle [64–66]. The interaction of these gene has been best documented in their role in *Drosophila* eye development and represent an evolutionary conserved gene network consisting of paired-box (*Pax*)– eye absent (*Eya*) – sine oculis (*Six*) – dachshund (*Dach*) genes [67]. EYA, SIX and DACH proteins directly interact to form a functional transcription factor. The DNA binding site is contained within the SIX proteins, while EYA mediates transcriptional transactivation and contains SIX and DACH binding domains. Additional regulatory complexity is provided by DACH, which appears to function as a co-factor by directly interacting with EYA. This network of interacting proteins can be considered a developmental module that is used in a variety of organs and across phyla to regulate gene promoter regions. Its fundamental role may be in association with the process of invagination. It has been suggested that these genes permit cells to migrate without altering their cell-fate commitment [68]; they also appear to have a function in the regulation of proliferation and cell death [66].

A conserved expression pattern is found in the otic vesicle where similar isoforms representing these four gene families are found. In the mouse these genes are *Pax2*, *Eya1* and *Eya4*, *Six1* and *Six4*, and both *Dach* isoforms [65, 66, 69]. *Eya1* is upstream of *Six1* and is regulated independently of *Six1* [64, 66]. Starting at the otic cup stage, *Six1* and *Eya1* are therefore expressed in an overlapping manner in the ventral otocyst and later on in the delaminating sensory neurons as well as in the sensory epithelia.

Eya1 and *Six1* are crucial for any development of the ear past the otocyst stage as only microvesicles form in *Eya1/Six1* double null mutants [66]. This conclusion is supported by the single null mutations. They show that *Eya1* and *Six1* appear to be critical for otocyst formation since null mutations in either *Eya1* or *Six1* do not affect the formation of the otic placode and vesicle but there is no further morphogenesis [64, 66, 70]. Biological redundancy is suggested by the formation of the ear in altered *Pax2* gene expression in spontaneous and null mutations [64, 66, 71] and by the absence of any phenotypic change associated with the loss of the *Dach1* gene [65]. These data show that this gene network is crucial for the early stages of ear morphogenesis. Interestingly, the haploid insufficiency observed in human and mouse *Eya1* mutants suggests that alterations in gene dosage, protein levels, or function can affect morphogenesis [66, 72–75].

Interestingly, while *Eya1* null mutation abrogates *Six1* expression, *Six1* null mutation has no effect on either *Eya1* or *Pax2* expression [64, 66]. *Six1* affects the expression of genes known to be important in ear morphogenesis such as *Bmp4*, *Fgf10* and *Fgf3* [32, 48, 55], which are downregulated in *Six1* null mutants. *Six1* also alters the expression pattern of two genes known to be involved in ear morphogenesis, *Nkx5.1* and *Gata3* [33, 57] suggesting the direct involvement of *Six1* in maintaining their correct expression [66]. Interestingly, compound heterozygotic mice show a variable phenotype. The absence of the posterior crista is consistent with the massive downregulation of *Fgf10*. There is also a reduction in growth of the cochlea which might be mediated by changes in *Gata3* expression [33] as *Pax2* is unaffected by *Six1* [66]. Cochlear defects are more profound in *Eya1/Pax2* compound heterozygotic animals but compound *Eya1/Gata3* heterozygotes have not yet been tested.

Similar to other genes involved in ear morphogenesis, these regulatory elements also play a role in histogenesis as indicated by their expression patterns in the developing ear. One example is the *Eya4* gene which is initially expressed in the otic vesicle [76, 77]. This isoform is present primarily in the upper epithelium of the cochlear duct, in a region where Reissner's membrane and the stria vascularis eventually develop. At E18.5 *Eya4* is expressed in areas of the cochlear duct destined to become the spiral limbus, organ of Corti, and spiral prominence with the highest level of expression occurring in the basal turn and in the early external auditory meatus. Diminishing levels of expression are found at later ages in these tissues and in the developing cochlear capsule during the period of ossification after birth to P14. In the vestibular system *Eya4* is observed in the developing sensory epithelia. Interestingly, mutant *Eya4* results in late-onset deafness at the DFNA10 locus [77, 78]. Clearly, *Eya1* and *Six1* are the best understood genes in terms of their effect on other important genes in ear morphogenesis, however, explaining their function still remains tentative in the absence of any information about most of their downstream genes [79]. In addition, nothing is as yet known about the regulation of *Eya1* and neither *Six1* nor *Eya1* have been extensively tested in other mutants such as

Gata3 null mutants or *Fgf10* null mutants. Further dissection of their role in ear development must be approached by using conditional mutant mouse lines to understand the contextual role that these regulatory genes play without compromising the viability of these mutants owing to their systemic effect on other vital organs [80].

14.2.2.3 GATA3

The comparatively few GATA factors of chordates show a highly conserved zinc finger DNA interaction domain. GATA factors have been found to participate directly in several signal-transduction pathways in both invertebrates and vertebrates that are involved in proliferation, morphogenesis and cell-fate assignment [81]. Most interesting in relation to the ear is the apparent interaction of GATA factors with BMP, FGF, FOX, NKX and T-BOX signaling. A GATA factor plays a major role in the gastrulation of echinoderms were it interacts with several of the 50 characterized proteins, in particular with FOX and T-BOX proteins [18]. Interaction between GATA and FGFs to neuralize ectoderm may be a feature of ancestry among chordates as this interaction has already been established in Ascidians [82]. Also, the insect GATA factor pannier mediates as well as maintains the cardiogenic decapentaplegic signal, the insect ortholog of *Bmp4* [83] and plays a role in the formation of the sensory organs in insects [81, 84]. Pannier can regulate the activation of bHLH factors as a possible molecular basis for its neuralizing effect [85]. Consistent with these findings in insects, in the vertebrate heart, *Gata6* is required to maintain *Bmp4* and *Nkx2* upregulation [86]. In other systems, such as the liver, it has been shown that GATA4 is elevated by BMP and FGF signaling [81]. In several tissues, the pattern of expression of GATA4 and GATA6 closely parallels that of BMP4, and the *Bmp4* gene was identified as a direct downstream target for GATA4 and 6 [87]. Arguably, the best characterized action of GATA3 at a molecular level is its role in the formation of Th1 and Th2 thymocytes [88].

Like some *Fox* genes, GATA3 is expressed at the level of the otic placode [89]. Targeted disruption of GATA3 leads to the arrest of ear development at the level of the otocyst [33] with little morphogenesis beyond a sac. Interestingly, *Gata3* null mutants appear to invaginate only a smaller part of the placodal tissue that expresses *Gata3* as shown by LacZ histochemistry. GATA3 is the only early marker that identifies delaminating spiral sensory neurons [33, 89] but its role in the development of the spiral ganglion is unclear. Based on path-finding errors in inner ear efferents that also express this gene [33], it is possible that GATA3 is involved in the path-finding of these sensory neurons. Recent analysis indicates a critical role for sonic hedgehog (SHH) in sensory neuron formation [31], but no analysis of GATA3 was performed in these mutants leaving open the question of whether or not a specific effect on spiral neurons occurs with SHH.

How GATA3 interacts with FGFs [47, 48], NKX 5.1 [56], DLX [58, 59], forkheads [29] and other genes involved in ear morphogenesis [64, 90] is still uncertain. However, multiple binding sites for GATA factors, Otx and E-boxes in the promoter regions of forkhead genes have been described [91] and interaction of GATA factors with FGFs and OTX may be important for neural tube induction in chordates [82]. Given the early expression of GATA3 it is not at all surprising to see influences on transcription factors regulating morphogenesis and histogenesis of the ear [89]. As with FGFs, it remains unclear how GATA3 interacts with BMP4, a likely essential molecule for ear morphogenesis [55]. Based on interactions of related genes in insect development (pannier, decapentaplegic) such interactions with BMP4 are likely [84]. Both BMP-mediated and direct signaling by GATA3 [85] as well as GATA3 interaction with FOX and T-BOX proteins [28, 81] could affect the formation of sensory neurons. Our laboratory is currently exploring the expression of several of these genes in *Gata3* null mutant mice.

Other major factors in early ear morphogenesis such as retinoids, NKX and other homeobox factors have been reviewed recently [15, 57] and will not be dealt with here. It suffices to say that studies of the genes and their effects on ear morphogenesis are no further along in explaining the molecular causality of developmental defects than the three examples given above.

14.2.3
Molecular Biology of Otoconia, Cupula and Tectorial Membrane Formation and Maintenance

As already stated in the Introduction, the vertebrate ear is a complex structure housing three sensory systems which perceive linear acceleration (gravity), angular acceleration, and sound [1, 92]. Specificity of stimulus acquisition is achieved in the ear by the association of particular sensory epithelia such as the utricle, saccule, canal cristae and cochlea with the acellular matrix proteins inside the inner ear space which mediates access to unique mechanical stimuli. Otoconia provide the mass and inertia to generate shearing forces for the utricle and saccule when subjected to linear acceleration such as gravity. The otoconia are typically composed of calcium carbonate crystals embedded in a glycoprotein matrix that is tethered by proteins to the underlying sensory epithelium (Fig. 14.3). The crystals form inside the endolymphatic space of the inner ear which contains a fluid composed of high levels of potassium with very little calcium, through unknown control mechanisms that are subject to disassembly under certain conditions, including age and sex-dependent alterations [93, 94]. In contrast, the cupulae of the semicircular canal cristae and the tectorial membrane of the cochlea do not contain calcium carbonate as this would likely alter the rigidity of these structures and change their physical properties regarding the perception of angular acceleration and sound. Such data are supported by the human condition of otolithiasis where dislocated otoconia are associated with the posterior crista cupula and modify its physical responses such that the patients suffer from dizziness.

In recent years the biochemical composition of the otoconia, tectorial membrane and cupulae has been analyzed and several genes essential for their formation and attachment to the underlying sensory epithelia of the ear have been characterized

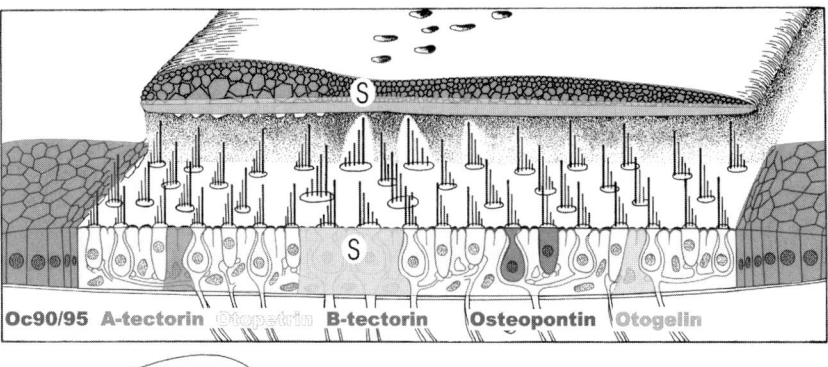

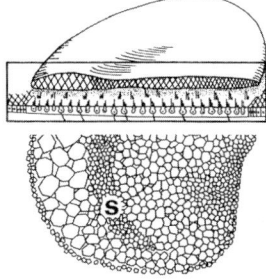

Fig. 14.3
The regional differences of otoconia-related gene expression in the utricle and structural and protein differences in the utricular otoconia are shown in this diagram (adapted from Lindeman [163]). Note the differences in the size of otoconia crystals in striola (S) and extrastriolar regions. Oc90/95 is expressed in the non-sensory area of the ear (red); otopetrin is expressed throughout the sensory epithelium (yellow); osteopontin is present in the hair cells (dark blue); otogelin is present in the supporting cells (cyan); α-tectorin is shown in the supporting cells (orange) and β-tectorin in the striola (green). All proteins are distributed throughout the otoconia (lilac) except for otogelin (cyan) and β-tectorin (green) which occur only in the lower layers. No information on the concentration differences of proteins within otoconia is available. (This figure also appears with the color plates.)

[94–99]. These data suggest specific compositional differences between all three acellular covering structures of the ear. Eight major proteins of the extracellular matrices have been identified in the mammalian ear thus far [94, 98]. Further data on another protein that links the stereocilia of the hair cells to the otoconia are emerging [100].

Six proteins have thus far been identified at the molecular level in otoconia (Fig. 14.3). The most abundant protein fraction in otoconia of mammals is otoconin 90/95 [101, 102]. Otoconin 90 is a modified *phospholipase-a* and comprises the major protein fraction. The phospholipase molecules of the otoconia have a higher content of anionic residues than their non-otic homologs, and thus generate a highly acidic pI. In addition, these proteins have incorporated several chains of sulfated polysaccharides while conventional phospholipases are not glycosylated. mRNA expression of otoconin 90 (*Oc90*) starts early in the development of the ear in mice (around embryonic day 9.5) and expression is restricted to non-sensory areas of the ear. In essence, *Oc90* expression negatively defines sensory epithelia in developing and adult mice [94, 101]. OC90 could thus present an early marker for non-sensory parts of the ear and could help elucidate changes in gene composition between non-sensory and sensory areas of the ear using qPCR and microarrays.

Another important protein that attaches the acellular matrix structures of the ear to the sensory epithelia is otogelin (*otog*). In *Otog* mutant animals, otoconia, cupulae and the tectorial membrane detach from the sensory epithelia [97] rendering the mutants unresponsive to sound and vestibular stimuli. Like otoconin 90, otogelin is expressed early (embryonic day 10.5) and is apparently specifically expressed in the differentiating supporting cells [103]. This early expression thus suggests that some supporting cell commitment may predate proliferation of hair cells [104].

A crucial protein for otoconia formation is the recently identified otopetrin [99]. If this protein is mutated, as in the *tilt* mutation, no otoconia will form [95, 99]. *In situ* hybridization profiles of this gene suggest expression only in the otoconia-bearing sensory maculae, a feature which distinguishes it from other sensory organs of the ear. In contrast to this restricted expression, immunocytochemistry of the protein shows a widespread diffusion of the secreted protein in the ear. It remains unclear how this transmembrane protein may be involved in otoconia formation and digestion of the extracellular loop which contains the mutation and against which the antibody is directed is a logical assumption.

Using immunocytochemistry, unequal distribution profiles have been reported for two of the other proteins common to all acellular covering structures, α-tectorin and β-tectorin [98]. Interestingly, β-tectorin is only found in the inner layers of mammalian otoconia and in birds is restricted to the striola region thus indicating an extraordinary specification in its distribution. Despite that, its function remains unclear as residual capacity to sense linear acceleration has not been tested in either *Tecta* or *Tectb* null mutant mice [96, 98, 105]. Expression data [106] show that in the saccule and utricle, *Tecta* mRNA is detected at E12.5, but *Tectb* mRNA is not observed until E14.5. Expression of α-tectorin mRNA ceases after P15, whereas β-tectorin mRNA expression continues within the striolar region of the utricle until at least P150. The results show *Tecta* and *Tectb* mRNAs are expressed during the early stages of inner ear development, prior to or concomitant with hair-cell differentiation, and before the appearance of hair bundles. Although *Tecta* is only expressed transiently during otoconia development, *Tectb* is continuously expressed within the striolar region of the utricle.

The last protein known to be present in otoconia is osteopontin. Osteopontin is a non-collagenous bone matrix protein that is involved in ossification processes. Osteopontin mRNA has been identified in marginal cells of the stria vascularis, spiral ganglion cells, vestibular sensory hair cells and vestibular dark cells [107, 108]. In the utricle and saccule, osteopontin was detected by immunocytochemistry in the otoconia, suggesting that this protein is one of the matrix proteins of mammalian otoconia. In contrast to all other acellular matrix proteins, the mRNA of osteopontin is expressed exclusively in vestibular hair cells [109].

While the molecular biology of the composition of the acellular covering structure has thus advanced, basic questions related to the development and maintenance of the otoconia, cupula and tectorial membrane remain open. For example, it is still not known how the proteins that form the otoconia assemble in the calcium-poor endolymphatic fluid and how this extracellular protein assembly guides the crystallization of the otoconia, but not the calcification of the other acellular matrices. Of particular importance is the fact that certain otoconia proteins are expressed outside the sensory epithelia [101]. In contrast, others are expressed only in the epithelia and in particular otopetrin, α-tectorin and β-tectorin are apparently exclusively expressed in otoconia-bearing sensory epithelia [99, 106]. However, it remains unclear how this expression relates to the other identified proteins of otoconia and how otoconia formation relates to the detailed spatio-temporal expression of all the proteins outlined above as this has not been consistently analyzed. Moreover, how all these different proteins relate to the distinctive fine structure observed in acellular matrices of the ear [94, 98] as well as to the different crystal structures of otoconia [110, 111] and the different striolar/extrastriolar organization of otoconia [1] remains unclear.

Most important here is that the three different acellular covering structures of the mammalian ear, the cupulae of the cristae, the otoconia of the linear acceleration system and the tectorial membrane represent variations of a general molecular theme with a number of shared proteins being present in all three of these structures [98]. Transforming an otoconia-bearing organ like the lagena or the saccule into the tectorial membrane covering of the mammalian cochlea might be accomplished with very few changes, for example simply by suppression of otopetrin (*Otop1*) [99]. More information about the developmental expression of these proteins is needed to reveal the ontogenetic differences between the different epithelia and to understand the developmental distinctions between these acellular covering structures. Likewise, a correlation needs to be established between the maturation of stereocilia on the hair cells and the development of the extracellular matrix covering structures.

14.2.4
Molecular Basis of Mammalian Ear Histogenesis

Histogenesis of the ear produces four distinct groups of cells with several subtypes. These types are the sensory neurons and the hair cells, which are intimately connected on a functional basis to extract mechanosensory stimuli and transmit them to the brain. The supporting cells, clonally related to hair cells [90], are involved in physically supporting the hair cells as well as facilitating their ionic homeostatic function. They also have certain glia cell-like features [112]. The least well-characterized cell type is the non-sensory cells of the remaining otic labyrinth. However, certain areas involved in potassium secretion (stria vascularis, vestibular dark cells) have been extensively characterized in the adult system, but have been the subject of much less developmental work. We will focus here on the first two cell types, the sensory neurons and hair cells, as most available developmental data concern those two cell types.

14.2.4.1 Sensory Neuron Development

Sensorineural development requires precise spatial and temporal regulation of many genes. Those genes are responsible for: (1) specifying areas of neurosensory formation; (2) organizing cell-fate specification; (3) achieving cell-fate commitment; (4) governing fiber growth to distinct peripheral and central targets; and (5) mediating cell survival and death.

Numerous candidate genes exist that might be, at least in part, responsible for the precise localization of the delaminating sensory neurons. It is entirely possible that the entire invaginating otocyst might have the capacity to form neuroblasts and that several genes are needed to reduce this capacity. The reduction in expression of FGF10 might relate to this as does the change in expression of TBX1 [113]. Also, other genes such as GATA3, EYA1, SIX1 and FGFs may be important for the maintenance of neurogenic capacity. Recent work on EYA1 and SIX1 showed reduction and even complete loss of sensory neuron formation [66]. Likewise, *Gata3* null mutants [33] and *Shh* null mutants [31] show reduced formation of sensory neurons. In both cases it is unclear whether this relates directly to the regulation of neuronal capacity or to the reduction in size of the otocyst.

Loss-of-function (targeted null mutations of the respective genes) experiments have clarified some of the proneural genes crucial for inner ear primary sensory neuron development (Fig. 14.4). The work of Ma et al. [11] showed that inner ear primary sensory neuron formation requires the vertebrate bHLH gene, neurogenin 1 (*Neurog1*) [11]. Indeed, a follow-up study showed that no primary sensory neurons ever form in these mutants [13]. As a consequence of the absence of primary sensory neuron formation in *Neurog1* null mutants, the ear develops in complete isolation without direct connections to the brainstem as afferents do not form and neither efferents nor autonomic fibers appear to reach the ear in these animals [13]. Nevertheless, such ears develop fairly normally in their overall histology. This suggests that ear formation and development, even of many hair cells, is largely autonomous of innervation. Whilst those hair cells that do form are morphologically normal despite the absence of innervation (except for some minor disorientation), the numbers of hair cell are reduced to various degrees in Neurog1 null mutant mice. Most interestingly, the cochlea is shortened and the saccule almost completely lost. In addition, extra rows of hair cells form in the shortened cochlea. These data suggest a significant interaction between the progenitor cells that form primary neurons and progenitor cells that give rise to hair cells, supporting cells and other inner ear epithelial cells. The simplest explanation would be a clonal relationship between primary sensory clones and the hair cell/supporting cell clones [90, 92]. Other possible interactions cannot be excluded however [114].

A gene that is immediately downstream of and mostly regulated by *Neurog1* is *Neurod1*, another *bHLH* gene [11]. Null mutations of this gene, also known as *Beta*

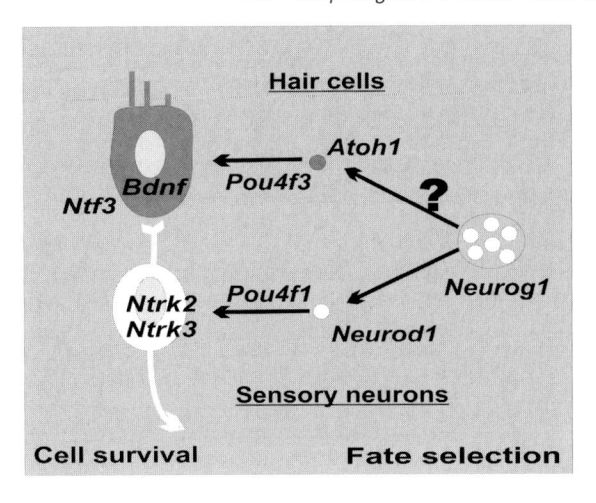

14.2 Morphological and Cellular Events in Mammalian Ear Development

Fig. 14.4
Diagram of the sequence of gene expression and their likely upstream regulation which relates causally to the formation of hair cells and sensory neurons in the ear and to neurotrophin-mediated survival. Note that precursors will be selected by pro-neuronal activation genes which will lead to the expression of bHLH genes in the neurosensory precursors. Some of these precursors may be common to both hair cells and sensory neurons. Activation of *Neurod1* occurs downstream of *Neurog1* and will eventually lead to the upregulation of the Pou domain factor *Pou4f1*. *Pou4f1* is responsible for the upregulation of neurotrophin receptors in hair cells (*Ntrk2*, *Ntrk3*). In the hair cell precursors there is an upregulation of *Atoh1* followed by an upregulation of another Pou domain factor, *Pou4f3*. Whether mediated by these genes or independently regulated, the hair cells will express the neurotrophin *Bdnf* whereas the supporting cells around them will express *Ntf3*. Ultimately, *Ntf3* expression will also be upregulated in the inner hair cells.

or *NeuroD*, have been analyzed and show severe reduction [115] or even complete loss of sensory neurons [116]. It appears that the surviving vestibular and cochlear sensory neurons may be in an unusual position and may project in an aberrant pattern into only parts of the cochlea and some vestibular sensory epithelia [8, 115]. Judging from these data, *Neurod1* appears to play a role in neuronal differentiation and survival, migration to appropriate areas, and target selection of peripheral neurites. Consistent with the effect on survival is the reduction and/or absence of certain neurotrophin receptor genes known to be essential for neuronal survival [115, 116]. Moreover it is possible that the effects of sensory neurons on survival in *Neurod1* null mutants are in part mediated by the reduced expression of a Pou domain factor *Pou4f1* (alias *Brn3a*), a factor that appears to have somewhat similar phenotypes with respect to the lack of innervation of certain sensory epithelia [117]. As with *Neurod1*, *Pou4f1* affects upregulation of certain neurotrophin receptors and thus might be only indirectly achieving its effect. While no promoter analysis of these genes has been carried out, their delayed expression compared to *Neurog1* and the less severe effects in the respective null mutants suggest that *Neurod1* and *Pou4f1* might affect survival to some extent via regulation of neurotrophin receptors.

Recent molecular data suggest some candidate genes that regulate peripheral process development. These candidate genes belong to the family of ephrin receptor and ligand complexes and semaphorin ligand and receptor systems which are known to be important in other developing systems [118]. More simplistically, one might assume that hair cells simply attract growing neurites. If so, the molecular principle of neurite attraction should be abolished by ablation of hair cells. A recent study on *Pou4f3* (alias *Brn3c*) null mutant mice which do not develop differentiated hair cells past early neonatal stages, reported the absence of growth of afferents and the disappearance of hair cells in addition to the expression of neurotrophic factors in these mutants [119]. Recent data on *Atoh1* (alias *Math1*) null mutant mice that never even form differentiated hair cells also suggest rather normal fiber growth [164]. Moreover, even if the entire formation of sensory epithelia is abrogated in specific null mutant mice, the initial growth of afferents is targeted instead [48] suggesting that at least some of the path-finding properties must reside in the interaction between growing neurites and the otocyst substrate.

Null mutants for the ephrin receptor, *Ephb2*, show circling behavior and altered axonal guidance of midline crossing efferents, but no data on the alteration of peripheral innervation have been provided [120]. Other ephrin mutants exist but their phenotype has not yet been characterized for the ear and likely requires more detailed analysis than simple uniform fiber labeling as detailed errors might be missed using non-specific techniques. Recently data has been presented indicating the role of one semaphorin, *Sema3a*, in providing a stop signal for growing afferents. Specifically, in this mutant growing neurites do not stop at the vestibular sensory epithelia but rather continue to grow and may extend outside the ear as far as the skin above the ear [121]. While these data indicate some progress in the molecular guidance of peripheral neurite guidance, they also highlight how rudimentary this insight is [122]. Moreover, virtually nothing is known about the molecular guidance of developing vestibular and cochlear afferents into the brain [123, 124].

In contrast to this apparent paucity of data on molecular guidance, extensive knowledge exists on the molecular basis of sensory neuron survival [125]. Briefly, two neurotrophins (BDNF, NTF3) and their receptors (NTRK2, NTRK3) as well as the so-called low affinity receptor p75 have been identified in the developing ear using various techniques. Double null mutations of either the two neurotrophins or the two neurotrophin receptors have shown complete loss of all sensory neurons, demonstrating that these ligands and receptors are crucial for the survival of inner ear sensory neurons. Remaining issues center on the differential function of each neurotrophin receptor combination. Since most other vertebrates express only one neurotrophin in the ear, it also remains unclear why mammals have two. In addition, the expression patterns described show a highly dynamic change along the cochlea as well as alterations in the vestibular system that beg the question of the functional significance of these changes [126, 127]. This problem moved further into the forefront when recent data on transgenic knock-in animals showed that each neurotrophin can be functionally replaced by the other neurotrophin, essentially suggesting that there is a topological difference in expression rather than any specific molecular effect [128, 129]. This apparent capacity of functional replacement of one neurotrophin by the other reinvigorated the debate concerning the functional significance of the presence of two neurotrophins as well as their differential expression.

An answer to this puzzle was recently provided using transgenic knock-in animals in which *Bdnf* replaced *Ntf3*. Selective tracer injection in these animals showed reorganization of the peripheral projection pattern of vestibular neurons into the basal turn of the cochlea [130]. These projections developed at the time when the neurons are known to become susceptible to the neurotrophins for their survival, suggesting that they become dependent on neurotrophins both neurotrophically as well as neurotropically at the same time. These data also suggest that the differential expression of neurotrophins in the ear plays a significant role in patterning the ear innervation. Interestingly, the central projection remained unchanged in these animals suggesting little if any influence of neurotrophins on the patterning of the central projections. Overall, these data suggest that the molecular processes that guide peripheral and central projection are distinct.

In summary, certain molecular aspects of sensory neuron formation, guidance and survival have been clarified in recent years. Still, numerous open issues remain in this fast moving field and virtually no molecular data exist for the basic patterning of central projections. Elucidation of the apparent relationship between sensory neurons and hair cells at a molecular level will provide an interesting topic for future work.

14.2.4.2 Hair Cell Development

As with sensory neuron development, hair cell development requires precise spatial and temporal regulation of many genes. Those genes are responsible for: (1) specifying areas of hair cell formation (future sensory patches); (2) selecting the topology of hair cell precursors inside those areas; (3) upregulating genes necessary for hair cell differentiation; (4) specifying different types of hair cells inside the sensory epithelia; (5) organizing the polarity of hair cells, differentiating the asymmetric distribution of the stereocilia and regulating the precise localization of the mechanoelectric transduction system; and (6) specifying and organizing the ribbon synapse formation of the afferents as well as the subsynaptic cisternae for the efferent synapse.

There is no good molecular data that indicates specificity of sensory patch formation but it is clear that the topology of these patches is dependent on the acquisition of the proper polarity by the developing otocyst [90]. A candidate for the establishment of dorso-ventral and possibly anterior-posterior axis formation is sonic hedgehog [31]. As already described some 100 years ago, sensory patches seem not to form as discrete entities but rather a single patch forms that divides into the final number of sensory patches. Certain markers for sensory patch formation and their patterns have been described [24, 127]. Specifically, it appears that lunatic fringe and

NTF3 expression highlights the cochlea, utricle and saccular sensory patch whereas BMP4, BDNF and FGF10 highlight the canal cristae [24, 48, 127]. Null mutants of *Fgf10* strongly support the involvement of FGF10 in sensory patch and hair cell formation, but the role of BMP4 in this context is less clear.

Unknown factors initiate the proliferation of hair cell precursors inside the growing and dividing sensory patches. Interestingly, at least the formation of some hair cells is linked to the sensory neuron formation as *Neurog1* null mutants lack some hair cells and fail to develop certain sensory patches such as the saccule [13]. It seems reasonable to assume that growth of epithelia depends on the proliferation of hair cell precursors which may accumulate as early as E10.5 in the mouse. How all these early patterning processes ultimately relate to the upregulation of the genes relevant for hair cell differentiation remains unclear.

It is clear that hair cell differentiation requires a single *bHLH* gene, atonal homolog 1 (*Atoh1*), both during development [131] and during regeneration [132, 133]. However, *Atoh1*, the bHLH gene responsible for hair cell differentiation, may not function as a true proneural gene to establish a neural lineage in the mammalian brain [85]. In contrast to most vertebrates, in mammals *Atoh1* functions only to select progenitor cells from a pool of already specified neuroepithelial stem cells [85]. In this context recent data suggest that the gene that establishes the hair cell lineage in the cochlea is still unknown [134]. Confirming and extending this conclusion are recent data on organ of Corti tissue culture showing that *Atoh1* is not required for the initial expression of MYOVIIa, MYOVI, nicotinic acetylcholine receptor α9, fimbrin, POU4F3 and other hair cell markers [135]. This conclusion is also supported by the precocious expression of the neurotrophin BDNF in the cochlea, prior to the upregulation of *Atoh1*, in what appears to be hair cell progenitors [127, 134]. Recent data from our group (Fritzsch et al. 2005) suggest that BDNF is even expressed in some hair cell progenitors in *Atoh1* null mutants. In contrast, in other hair cells *Atoh1* is essential for BDNF expression. This suggests that other genes also regulate BDNF expression in hair cell precursors and supports the conclusion of the *Atoh1–LacZ* analysis that hair cell precursor formation does not require *Atoh1*. Nevertheless, the possible involvement of *Atoh1* in the selection of proliferating hair cell precursors cannot be ruled out in the case of vestibular sensory epithelia. If proven, this would suggest that hair cell formation with respect to early specifying genes is not uniform throughout the ear.

Several crucial genes for hair cell differentiation and maintenance have now been characterized: *Atoh1*, *Pou4f3*, *Gfi1* and *Barhl1* [119, 131, 136–138]. More recent work indicates that the Pou domain factor (*Pou4f3*) regulates the zinc finger protein, GFI1 [139]. How these genes in turn tie in with the regulation of *Barhl1* as well as the differentiation of hair cell remains unclear. It appears, however, that major aspects of hair cell differentiation can be accomplished in the absence of either *Pou4f3* or *Gfi1* [139]. Thus *Atoh1* is certainly able to promote hair cell differentiation to the extent of stereocilia formation even without these genes.

The molecular nature of the hair cell competence area (the greater and lesser epithelial ridge of the cochlea) is unknown, but transfection experiments using a virus with *Atoh1* gene insertions, have shown the competence of these cells to transform into hair cells in neonates and adults [132, 133]. How *Atoh1* is initially upregulated in the cells that will become hair cells and how early in development of hair cells this occurs (i.e. in proliferating precursors or only in postmitotic undifferentiated hair cells) is a not fully established and may vary according to the type of sensory epithelia, the species used and the technical limitations of the specific techniques employed [134, 140]. While the role of p27 in terminal mitosis of hair cell precursors is now clear [141], other genes relevant for mitosis such as blastoma-related genes, need to be studied as well. Interestingly, mutating retinoblastoma for example, results in the overproduction of hair cells (Mantela J., Jiang Z., Ylikoski J., Fritzsch B., Zacksenhaus E., Pirvola U. 2005. The retinoblastoma gene pathway regulates the postmitotic state of hair cells of the mouse inner ear. *Development* 132: 2377–2388). When Math1 is upregulated there is also an upregulation of related *bHLH* genes (*Hes1* and *Hes5*) which regulate the mosaic of hair cell and supporting cell formation [132, 142]. Mutation in the *Hes* genes result in extra inner (*Hes1*) or outer (*Hes5*) hair cell formation, supporting the notion that these genes regulate the hair cell/supporting cell mosaic and have limited effected on the overall topology and extent of hair cell progenitor formation which is regulated by as yet unknown genes.

How all these cell-fate determination processes are molecularly integrated into the process of cell contact formation that uses a variety of different cadherins and related molecules [143] which appear to play crucial roles in hair cell polarity formation [144, 145], remains unknown as no direct connection with the differential distribution of actin and espin to form the stereociliary staircase has yet been shown [146]. It is conceivable, but unknown, that differential distribution of espin that regulates the assembly of actin in a quantitative fashion may be at the basis of the differential growth of stereocilia. Thus the effects described at the moment for various transmembrane and secreted signals in the establishment of hair cell polarity require a more direct link with known molecules that are quantitatively differentially distributed in various hair cells.

In summary, while our molecular understanding of hair cell formation and differentiation is by far the most advanced of all ear morphogenetic and histogenetic processes, there are also blatant holes of ignorance that require further work so that logical and causal sequences can be established which can then be used in a controlled fashion to overcome a major condition in sensorineural hearing loss: replacement of lost hair cells. Moreover, while significant efforts have been focused on the formation of the mechanoelectric transduction system, the equally important process of afferent and efferent synapse formation which is relevant to the flow of information to the central nervous system has attracted much less attention. Notable exceptions are recent attempts to relate the functional maturation of hair cells [146] to the onset of transmitter release mechanisms [147]; these studies which use appropriate mutational analysis of conditional mutants with specific deletions of genes relevant to the ear only, remain to be completed.

14.3
Postnatal Maturation

An excellent overview of the structural and functional maturation of the mammalian ear was recently published and the reader is referred to this paper for details [146]. Briefly, the sequence of activation of the various conductance-mediating channels as well as the final maturation of the mechanoelectric transduction system and the afferent and efferent synaptic complexes are active and fast growing areas of research that have, however, benefited much less from the mutational analysis of critical proteins since mutations in such proteins appear to be lethal early in development. Therefore, it is necessary to generate conditional mutants with mutations specific for the inner ear to facilitate the study of more of these relevant genes.

14.4
Aging, Hair Cell Loss and the Molecular Biology of Hair Cell Regeneration

One in three Americans over the age of 65 years will suffer some degree of hearing loss with an age-dependent increase in sensorineural hearing loss. Thus, by the age of 70 years the decline in the numbers of vestibular and cochlear hair cells may have depleted the capacity of the system to compensate and hearing and vestibular disorders become apparent. While numerous devices such as cochlear and cochlear nucleus implants, have been designed to restore hearing in particular, these devices still suffer from a lack of signal-to-noise discrimination thus preventing a focused conversation from taking place in the middle of a party. Ultimately it thus appears most appropriate to use the insights into hair cell and sensory neuron formation to regenerate hair cells and neurons and to connect them properly so that hearing and eventually vestibular sensation is fully restored in the elderly.

Significant progress toward these goals has been achieved in the last year. Gene therapy using *Atoh1*–loaded adeno viruses has shown that transformation of various cells in the larger epithelium of the cochlea into hair cells can be achieved [133]. More recently, these authors reported that numerous newly-formed hair cells can be found in freshly deafened cochleas using this technique [148]. Unfortunately, this remarkable structural recovery is not accompanied by an equally impressive physiological recovery. In part this may relate to the changed structure of the cochlea as these transfection experiments convert supporting cells into hair cells, thus changing the mechanical properties of the organ of Corti. However, it is also possible that the innervation process does not develop normally as some afferent fibers may have degenerated following trauma to the hair cells.

Other avenues in the search to achieve functional recovery are the regeneration of hair cells and sensory neurons from embryonic or adult stem cells. Regeneration of cells expressing numerous hair cell markers in tissue culture and upon implantation into the inner ear have recently been reported [149, 150] and isolation of proliferating ganglion cell precursors from the human cochlea appears to be possible

[151]. Implanting such cells into ears depleted of either hair cells or sensory neurons or both might be a future possibility in the quest to restore hearing. While important, these findings are still a long way off being therapeutically relevant. Therefore in the meantime it appears prudent to improve the already existing techniques such as cochlear implants by improving their near range excitation. It may be possible to use the recent insights into the directional growth-promoting role of BDNF [130] to direct growing spiral ganglion neurites specifically towards the stimulating electrodes of the implants.

Overall, these data provide significant advances in support of the notion that curing hearing loss might become a possibility in the near future using either gene therapy or cell transplantation strategies. Clearly, more work is needed before any of these techniques can be used for clinical trials, but the rapid progress achieved in recent years through the translation of developmental insights into regeneration tools should be beneficial for the future of this research.

References

1 Lewis ER, Leverenz EL, Bialek WS. *The Vertebrate Inner Ear.* Boca Raton: CRC Press, **1985**.
2 Fritzsch B, Wake MH. The inner ear of gymnophione amphibians and its nerve supply: a comparative study of regressive events in a complex sensory system. *Zoomorphol* **1988**; *108*: 210–217.
3 Brichta AM, Goldberg JM. The papilla neglecta of turtles: a detector of head rotations with unique sensory coding properties. *J Neurosci* **1998**; *18*: 4314–4324.
4 Retzius G. Das Gehörorgan der Wirbeltiere: II. Das Gehörorgan der Amnioten. Stockholm: Samson und Wallin, **1884**.
5 de Burlet HM. Vergleichende Anatomie des statoakustischen Organs. (a) Die innere Ohrsphäre. In *Handbuch der Vergleichenden Anatomie der Wirbeltiere*, Bolk L, Göppert E, Kallius E, Lubosch W, (Eds). Berlin: Urban and Schwarzenberg, **1934**; pp. 1293–1432.
6 Fritzsch B. The ear of Latimeria chalumnae revisited. *Zoology* **2003**; *106*: 243–248.
7 Jorgensen JM, Locket NA. The inner ear of the echidna Tachyglossus aculeatus: the vestibular sensory organs. *Proc R Soc Lond B Biol Sci* **1995**; *260*: 183–189.
8 Fritzsch B, Beisel KW Molecular conservation and novelties in vertebrate ear development. *Curr Top Dev Biol* **2003**; *57*: 1–44.
9 Beisel KW, Shiraki T, Morris KA, Pompeia C, Kachar B, Arakawa T, Bono H, Kawai J, Hayashizaki Y, Carninci P. Identification of unique transcripts from a mouse full-length, subtracted inner ear cDNA library. *Genomics* **2004**; *83*: 1012–1023.
10 Quint E, Steel KP. Use of mouse genetics for studying inner ear development. *Curr Top Dev Biol* **2003**; *57*: 45–83.
11 Ma Q, Chen Z, del Barco Barrantes I, de la Pompa JL, Anderson DJ. neurogenin1 is essential for the determination of neu-

ronal precursors for proximal cranial sensory ganglia. *Neuron* **1998**; *20*: 469–482.

12 Ma Q, Fode C, Guillemot F, Anderson DJ. Neurogenin1 and neurogenin2 control two distinct waves of neurogenesis in developing dorsal root ganglia. *Genes Dev* **1999**; *13*: 1717–1728.

13 Ma Q, Anderson DJ, Fritzsch B. Neurogenin 1 null mutant ears develop fewer, morphologically normal hair cells in smaller sensory epithelia devoid of innervation. *J Assoc Res Otolaryngol* **2000**; *1*: 129–143.

14 Favor J, Sandulache R, Neuhauser-Klaus A, Pretsch W, Chatterjee B, Senft E, Wurst W, Blanquet V, Grimes P, Sporle R, Schughart K. The mouse Pax2(1Neu) mutation is identical to a human PAX2 mutation in a family with renal-coloboma syndrome and results in developmental defects of the brain, ear, eye, and kidney. *Proc Natl Acad Sci USA* **1996**; *93*: 13870–13875.

15 Romand R, Varela-Nieto I. Development of auditory and vestibular systems – 3. Molecular development of the inner ear. *Curr Top Dev Biol* **2003**; *57*: 1–481.

16 Giot L, Bader JS, Brouwer C, Chaudhuri A, Kuang B, Li Y, Hao YL, Ooi CE, Godwin B, Vitols E, Vijayadamodar G, Pochart P, Machineni H, Welsh M, Kong Y, Zerhusen B, Malcolm R, Varrone Z, Collis A, Minto M, Burgess S, McDaniel L, Stimpson E, Spriggs F, Williams J, Neurath K, Ioime N, Agee M, Voss E, Furtak K, Renzulli R, Aanensen N, Carrolla S, Bickelhaupt E, Lazovatsky Y, DaSilva A, Zhong J, Stanyon CA, Finley RL, Jr., White KP, Braverman M, Jarvie T, Gold S, Leach M, Knight J, Shimkets RA, McKenna MP, Chant J, Rothberg JM. A protein interaction map of *Drosophila melanogaster*. *Science* **2003**; *302*: 1727–1736.

17 Li S, Armstrong CM, Bertin N, Ge H, Milstein S, Boxem M, Vidalain P-O, Han J-DJ, Chesneau A, Hao T, Goldberg DS, Li N, Martinez M, Rual J-F, Lamesch P, Xu L, Tewari M, Wong SL, Zhang LV, Berriz GF, Jacotot L, Vaglio P, Reboul J, Hirozane-Kishikawa T, Li Q, Gabel HW, Elewa A, Baumgartner B, Rose DJ, Yu H, Bosak S, Sequerra R, Fraser A, Mango SE, Saxton WM, Strome S, van den Heuvel S, Piano F, Vandenhaute J, Sardet C, Gerstein M, Doucette-Stamm L, Gunsalus KC, Harper JW, Cusick ME, Roth FP, Hill DE, Vidal M. A Map of the interactome network of the metazoan *C. elegans*. *Science* **2004**; *303*: 540–543.

18 Davidson EH, Rast JP, Oliveri P, Ransick A, Calestani C, Yuh CH, Minokawa T, Amore G, Hinman V, Arenas-Mena C, Otim O, Brown CT, Livi CB, Lee PY, Revilla R, Rust AG, Pan Z, Schilstra MJ, Clarke PJ, Arnone MI, Rowen L, Cameron RA, McClay DR, Hood L, Bolouri H. A genomic regulatory network for development. *Science* **2002**; *295*: 1669–1678.

19 Chiang MK, Melton DA. Single-cell transcript analysis of pancreas development. *Dev Cell* **2003**; *4*: 383–393.

20 Fritzsch B. Evolution of the ancestral vertebrate brain. In *The Handbook of Brain Theory and Neural Networks*, Arbib MA (Ed.). Cambridge: MIT Press **2002**; 426–430.

21 Lowe CJ, Wu M, Salic A, Evans L, Lander E, Stange-Thomann N, Gruber CE, Gerhart J, Kirschner M Anteroposterior patter-

ning in hemichordates and the origins of the chordate nervous system. *Cell* **2003**; *113*: 853–865.
22. Pennisi E. Searching for the genome's second code. *Science* **2004**; *306*: 632–635.
23. Streeter GL. On the development of the membraneous labyrinth and the acoustic and facila nerves in the human embryo. *Amer J Anat* **1907**; *6*: 6.
24. Morsli H, Choo D, Ryan A, Johnson R, Wu DK. Development of the mouse inner ear and origin of its sensory organs. *J Neurosci* **1998**; *18*: 3327–3335.
25. von Kupffer C. The development of the cranial nerves of vertebrates. *J Comp Neurol* **1891**; *1*: 246–264, 315–332.
26. Wright TJ, Hatch EP, Karabagli H, Karabagli P, Schoenwolf GC, Mansour SL. Expression of mouse fibroblast growth factor and fibroblast growth factor receptor genes during early inner ear development. *Dev Dyn* **2003**; *228*: 267–272.
27. Wright TJ, Mansour SL. FGF signaling in ear development and innervation. *Curr Top Dev Biol* **2003**; *57*: 225–259.
28. Hatini V, Ye X, Balas G, Lai E. Dynamics of placodal lineage development revealed by targeted transgene expression. *Dev Dyn* **1999**; *215*: 332–343.
29. Solomon KS, Kudoh T, Dawid IB, Fritz A. Zebrafish foxi1 mediates otic placode formation and jaw development. *Development* **2003**; *130*: 929–940.
30. Brent AE, Schweitzer R, Tabin CJ. A somitic compartment of tendon progenitors. *Cell* **2003**; *113*: 235–248.
31. Riccomagno MM, Martinu L, Mulheisen M, Wu DK, Epstein DJ. Specification of the mammalian cochlea is dependent on Sonic hedgehog. *Genes Dev* **2002**; *16*: 2365–2378.
32. Wright TJ, Mansour SL. Fgf3 and Fgf10 are required for mouse otic placode induction. *Development* **2003**; *130*: 3379–3390.
33. Karis A, Pata I, van Doorninck JH, Grosveld F, de Zeeuw CI, de Caprona D, Fritzsch B Transcription factor GATA–3 alters pathway selection of olivocochlear neurons and affects morphogenesis of the ear. *J Comp Neurol* **2001**; *429*: 615–630.
34. Tiveron MC, Pattyn A, Hirsch MR, Brunet JF. Role of Phox2b and Mash1 in the generation of the vestibular efferent nucleus. *Dev Biol* **2003**; *260*: 46–57.
35. Fritzsch B. Ontogenetic and evolutionary evidence for the motoneuron nature of vestibular and cochlear efferents. In *The Efferent Auditory System: Basic Science and Clinical Applications*, Berlin CI (Ed.). San Diego: Singular Publishing, **1999**; 31–59.
36. Vetter DE, Liberman MC, Mann J, Barhanin J, Boulter J, Brown MC, Saffiote-Kolman J, Heinemann SF, Elgoyhen AB. Role of alpha9 nicotinic ACh receptor subunits in the development and function of cochlear efferent innervation. *Neuron* **1999**; *23*: 93–103.
37. Zuo J, Treadaway J, Buckner TW, Fritzsch B. Visualization of alpha9 acetylcholine receptor expression in hair cells of transgenic mice containing a modified bacterial artificial chromosome. *Proc Natl Acad Sci USA* **1999**; *96*: 14100–14105.
38. Fritzsch B. Evolution and desensitization of LGIC receptors. *Trends Neurosci* **1995**; *18*: 297; discussion 298–299.

39. Martin P, Swanson GJ. Descriptive and experimental analysis of the epithelial remodellings that control semicircular canal formation in the developing mouse inner ear. *Dev Biol* 1993; *159*: 549–558.
40. Fekete DM, Homburger SA, Waring MT, Riedl AE, Garcia LF. Involvement of programmed cell death in morphogenesis of the vertebrate inner ear. *Development* 1997; *124*: 2451–2461.
41. Cantos R, Cole LK, Acampora D, Simeone A, Wu DK. Patterning of the mammalian cochlea. *Proc Natl Acad Sci USA* 2000; *97*: 11707–11713.
42. Baker CV, Bronner-Fraser M. Vertebrate cranial placodes I. Embryonic induction. *Dev Biol* 2001; *232*: 1–61.
43. Brown ST, Martin K, Groves AK. Molecular basis of inner ear induction. *Curr Top Dev Biol* 2003; *57*: 115–149.
44. Noramly S, Grainger RM. Determination of the embryonic inner ear. *J Neurobiol* 2002; *53*: 100–128.
45. Inatani M, Irie F, Plump AS, Tessier-Lavigne M, Yamaguchi Y. Mammalian brain morphogenesis and midline axon guidance require heparan sulfate. *Science* 2003; *302*: 1044–1046.
46. Ornitz DM, Marie PJ. FGF signalling pathways in enchondral and intramembranous bone development and human genetic disease. *Genes Dev* 2002; *16*: 1446–1465.
47. Pirvola U, Spencer-Dene B, Xing-Qun L, Kettunen P, Thesleff I, Fritzsch B, Dickson C, Ylikoski J. FGF/FGFR-2(IIIb) signaling is essential for inner ear morphogenesis. *J Neurosci* 2000; *20*: 6125–6134.
48. Pauley S, Wright TJ, Pirvola U, Ornitz D, Beisel K, Fritzsch B. Expression and function of FGF10 in mammalian inner ear development. *Dev Dyn* 2003; *227*: 203–215.
49. Alvarez Y, Alonso MT, Vendrell V, Zelarayan LC, Chamero P, Theil T, Bosl MR, Kato S, Maconochie M, Riethmacher D, Schimmang T. Requirements for FGF3 and FGF10 during inner ear formation. *Development* 2003; *130*: 6329–6338.
50. Mansour SL. Targeted disruption of int–2 (fgf–3) causes developmental defects in the tail and inner ear. *Mol Reprod Dev* 1994; *39*: 62–67; discussion 67–68.
51. Ohuchi H, Hori Y, Yamasaki M, Harada H, Sekine K, Kato S, Itoh N. FGF10 acts as a major ligand for FGF receptor 2 IIIb in mouse multi-organ development. *Biochem Biophys Res Commun* 2000; *277*:643–649.
52. Pirvola U, Ylikoski J, Trokovic R, Hebert J, McConnell S, Partanen J. FGFR1 is required for the development of the auditory sensory epithelium. *Neuron* 2002; *35*: 671.
53. Chang W, Nunes FD, De Jesus-Escobar JM, Harland R, Wu DK. Ectopic noggin blocks sensory and nonsensory organ morphogenesis in the chicken inner ear. *Dev Biol* 1999; *216*: 369–381.
54. Gerlach LM, Hutson MR, Germiller JA, Nguyen-Luu D, Victor JC, Barald KF. Addition of the BMP4 antagonist, noggin, disrupts avian inner ear development. *Development* 2000; *127*: 45–54.
55. Chang W, ten Dijke P, Wu DK. BMP pathways are involved in otic capsule formation and epithelial-mesenchymal signaling

in the developing chicken inner ear. *Dev Biol* **2002**; *251*: 380–394.
56. Hadrys T, Braun T, Rinkwitz-Brandt S, Arnold HH, Bober E. Nkx5–1 controls semicircular canal formation in the mouse inner ear. *Development* **1998**; *125*: 33–39.
57. Bober E, Rinkwitz S, Herbrand H. Molecular basis of otic commitment and morphogenesis: a role for homeodomain-containing transcription factors and signaling molecules. *Curr Top Dev Biol* **2003**; *57*: 151–175.
58. Merlo GR, Paleari L, Mantero S, Zerega B, Adamska M, Rinkwitz S, Bober E, Levi G. The Dlx5 homeobox gene is essential for vestibular morphogenesis in the mouse embryo through a BMP4–mediated pathway. *Dev Biol* **2002**; *248*: 157–169.
59. Solomon KS, Fritz A. Concerted action of two dlx paralogs in sensory placode formation. *Development* **2002**; *129*: 3127–3136.
60. Morsli H, Tuorto F, Choo D, Postiglione MP, Simeone A, Wu DK. Otx1 and Otx2 activities are required for the normal development of the mouse inner ear. *Development* **1999**; *126*: 2335–2343.
61. Fritzsch B, Signore M, Simeone A Otx1 null mutant mice show partial segregation of sensory epithelia comparable to lamprey ears. *Dev Genes Evol* **2001**; *211*: 388–396.
62. Colvin JS, Bohne BA, Harding GW, McEwen DG, Ornitz DM. Skeletal overgrowth and deafness in mice lacking fibroblast growth factor receptor 3. *Nat Genet* **1996**; *12*: 390–397.
63. Sheng G, dos Reis M, Stern CD. 2003. Churchill, a zinc finger transcriptional activator, regulates the transition between gastrulation and neurulation, *Cell* 115: 603–613.
64. Xu PX, Adams J, Peters H, Brown MC, Heaney S, Maas R. Eya1–deficient mice lack ears and kidneys and show abnormal apoptosis of organ primordia. *Nat Genet* **1999**; *23*: 113–117.
65. Davis RJ, Shen W, Sandler YI, Heanue TA, Mardon G. Characterization of mouse Dach2, a homologue of Drosophila dachshund. *Mech Dev* **2001**; *102*: 169–179.
66. Zheng W, Huang L, Wei ZB, Silvius D, Tang B, Xu PX. The role of Six1 in mammalian auditory system development. *Development* **2003**; *130*: 3989–4000.
67. Hanson IM. Mammalian homologues of the Drosophila eye specification genes. *Semin Cell Dev Biol* **2001**; *12*: 475–484.
68. Streit A. Extensive cell movements accompany formation of the otic placode. *Dev Biol* **2002**; *249*: 237–254.
69. Davis RJ, Shen W, Heanue TA, Mardon G. Mouse Dach, a homologue of Drosophila dachshund, is expressed in the developing retina, brain and limbs. *Dev Genes Evol* **1999**; *209*: 526–536.
70. Xu PX, Woo I, Her H, Beier DR, Maas RL. Mouse Eya homologues of the Drosophila eyes absent gene require Pax6 for expression in lens and nasal placode. *Development* **1997**; *124*: 219–231.
71. Torres M, Giraldez F. The development of the vertebrate inner ear. *Mech Dev* **1998**; *71*: 5–21.
72. Abdelhak S, Kalatzis V, Heilig R, Compain S, Samson D, Vincent C, Levi-Acobas F, Cruaud C, Le Merrer M, Mathieu M,

Konig R, Vigneron J, Weissenbach J, Petit C, Weil D. Clustering of mutations responsible for branchio-oto-renal (BOR) syndrome in the eyes absent homologous region (eyaHR) of EYA1. *Hum Mol Genet* **1997**; *6*: 2247–2255.

73 Abdelhak S, Kalatzis V, Heilig R, Compain S, Samson D, Vincent C, Weil D, Cruaud C, Sahly I, Leibovici M, Bitner-Glindzicz M, Francis M, Lacombe D, Vigneron J, Charachon R, Boven K, Bedbeder P, Van Regemorter N, Weissenbach J, Petit C. A human homologue of the Drosophila eyes absent gene underlies branchio-oto-renal (BOR) syndrome and identifies a novel gene family. *Nat Genet* **1997**; *15*: 157–164.

74 Vincent C, Kalatzis V, Abdelhak S, Chaib H, Compain S, Helias J, Vaneecloo FM, Petit C BOR and BO syndromes are allelic defects of EYA1. *Eur J Hum Genet* **1997**; *5*: 242–246.

75 Johnson KR, Cook SA, Erway LC, Matthews AN, Sanford LP, Paradies NE, Friedman RA. Inner ear and kidney anomalies caused by IAP insertion in an intron of the Eya1 gene in a mouse model of BOR syndrome. *Hum Mol Genet* **1999**; *8*: 645–653.

76 Borsani G, DeGrandi A, Ballabio A, Bulfone A, Bernard L, Banfi S, Gattuso C, Mariani M, Dixon M, Donnai D, Metcalfe K, Winter R, Robertson M, Axton R, Brown A, van Heyningen V, Hanson I. EYA4, a novel vertebrate gene related to Drosophila eyes absent. *Hum Mol Genet* **1999**; *8*: 11–23.

77 Wayne S, Robertson NG, DeClau F, Chen N, Verhoeven K, Prasad S, Tranebjarg L, Morton CC, Ryan AF, Van Camp G, Smith RJ. Mutations in the transcriptional activator EYA4 cause late-onset deafness at the DFNA10 locus. *Hum Mol Genet* **2001**; *10*: 195–200.

78 Pfister M, Toth T, Thiele H, Haack B, Blin N, Zenner HP, Sziklai I, Murnberg P, Kupka S. A 4–bp insertion in the eya-homologous region (eyaHR) of EYA4 causes hearing impairment in a Hungarian family linked to DFNA10. *Mol Med* **2002**; *8*: 607–611.

79 Zou D, Silvius D, Fritzsch B, Xu PX. Eya1 and Six1 are essential for early steps of sensory neurogenesis in mammalian cranial placodes. *Development* **2004**; *131*: 5561–5572.

80 Xu PX, Zheng W, Laclef C, Maire P, Maas RL, Peters H, Xu X. Eya1 is required for the morphogenesis of mammalian thymus, parathyroid and thyroid. *Development* **2002**; *129*: 3033–3044.

81 Patient RK, McGhee JD. The GATA family (vertebrates and invertebrates). *Curr Opin Genet Dev* **2002**; *12*: 416–422.

82 Bertrand V, Hudson C, Caillol D, Popovici C, Lemaire P. Neural tissue in ascidian embryos is induced by FGF9/16/20, acting via a combination of maternal GATA and Ets transcription factors. *Cell* **2003**; *115*: 615–627.

83 Klinedinst SL, Bodmer R. Gata factor Pannier is required to establish competence for heart progenitor formation. *Development* **2003**; *130*: 3027–3038.

84 Sato M, Saigo K. Involvement of pannier and u-shaped in regulation of decapentaplegic-dependent wingless expression in developing Drosophila notum. *Mech Dev* **2000**; *93*: 127–138.

85. Bertrand N, Castro DS, Guillemot F. Proneural genes and the specification of neural cell types. *Nat Rev Neurosci* **2002**; *3*: 517–530.
86. Peterkin T, Gibson A, Patient R. GATA-6 maintains BMP-4 and Nkx2 expression during cardiomyocyte precursor maturation. *EMBO J* **2003**; *22*: 4260–4273.
87. Nemer G, Nemer M. Transcriptional activation of BMP-4 and regulation of mammalian organogenesis by GATA-4 and -6. *Dev Biol* **2003**; *254*: 131–148.
88. Zhou M, Ouyang W. The function role of GATA-3 in Th1 and Th2 differentiation. *Immunol Res* **2003**; *28*: 25–37.
89. Lawoko-Kerali G, Rivolta MN, Holley M. Expression of the transcription factors GATA3 and Pax2 during development of the mammalian inner ear. *J Comp Neurol* **2002**; *442*: 378–391.
90. Fekete DM, Wu DK. Revisiting cell fate specification in the inner ear. *Curr Opin Neurobiol* **2002**; *12*: 35–42.
91. David ES, Luke NH, Livingston BT. Characterization of a gene encoding a developmentally regulated winged helix transcription factor of the sea urchin *Strongylocentrotus purpuratus*. *Gene* **1999**; *236*: 97–105.
92. Fritzsch B, Beisel KW. Evolution and development of the vertebrate ear. *Brain Res Bull* **2001**; *55*: 711–721.
93. Ross MD, Johnsson L-G, Peacor D. Observations on normal and degenerating human otoconia. *Ann Otol* **1976**; *85*: 310–326.
94. Thalmann R, Ignatova E, Kachar B, Ornitz DM, Thalmann I. Development and maintenance of otoconia: biochemical considerations. *Ann NY Acad Sci* **2001**; *942*: 162–178.
95. Ornitz DM, Bohne BA, Thalmann I, Harding GW, Thalmann R. Otoconial agenesis in tilted mutant mice. *Hear Res* **1998**; *122*: 60–70.
96. Legan PK, Lukashkina VA, Goodyear RJ, Kossi M, Russell IJ, Richardson GP. A targeted deletion in alpha-tectorin reveals that the tectorial membrane is required for the gain and timing of cochlear feedback. *Neuron* **2000**; *28*: 273–285.
97. Simmler MC, Cohen-Salmon M, El-Amraoui A, Guillaud L, Benichou JC, Petit C, Panthier JJ. Targeted disruption of otog results in deafness and severe imbalance. *Nat Genet* **2000**; *24*: 139–143.
98. Goodyear RJ, Richardson GP. Extracellular matrices associated with the apical surfaces of sensory epithelia in the inner ear: molecular and structural diversity. *J Neurobiol* **2002**; *53*: 212–227.
99. Hurle B, Ignatova E, Massironi SM, Mashimo T, Rios X, Thalmann I, Thalmann R, Ornitz DM. Non-syndromic vestibular disorder with otoconial agenesis in tilted/mergulhador mice caused by mutations in otopetrin 1. *Hum Mol Genet* **2003**; *12*: 777–789.
100. Zwaenepoel I, Mustapha M, Leibovici M, Verpy E, Goodyear R, Liu XZ, Nouaille S, Nance WE, Kanaan M, Avraham KB, Tekaia F, Loiselet J, Lathrop M, Richardson G, Petit C. Otoancorin, an inner ear protein restricted to the interface between the apical surface of sensory epithelia and their overly-

ing acellular gels, is defective in autosomal recessive deafness DFNB22. *Proc Natl Acad Sci USA* **2002**; *99*: 6240–6245.

101 Wang Y, Kowalski PE, Thalmann I, Ornitz DM, Mager DL, Thalmann R. Otoconin–90, the mammalian otoconial matrix protein, contains two domains of homology to secretory phospholipase A2. *Proc Natl Acad Sci USA* **1998**; *95*: 15345–15350.

102 Verpy E, Leibovici M, Petit C. Characterization of otoconin–95, the major protein of murine otoconia, provides insights into the formation of these inner ear biominerals. *Proc Natl Acad Sci USA* **1999**; *96*: 529–534.

103 El-Amraoui A, Cohen-Salmon M, Petit C, Simmler MC. Spatiotemporal expression of otogelin in the developing and adult mouse inner ear. *Hear Res* **2001**; *158*: 151–159.

104 Ruben RJ. Development of the inner ear of the mouse: a radioautographic study of terminal mitoses. *Acta Otolaryngol:Suppl* **1967**; *220*: 221–244.

105 Verhoeven K, Van Laer L, Kirschhofer K, Legan PK, Hughes DC, Schatteman I, Verstreken M, Van Hauwe P, Coucke P, Chen A, Smith RJ, Somers T, Offeciers FE, Van de Heyning P, Richardson GP, Wachtler F, Kimberling WJ, Willems PJ, Govaerts PJ, Van Camp G. Mutations in the human alpha-tectorin gene cause autosomal dominant non-syndromic hearing impairment. *Nat Genet* **1998**; *19*: 60–62.

106 Rau A, Legan PK, Richardson GP. Tectorin mRNA expression is spatially and temporally restricted during mouse inner ear development. *J Comp Neurol* **1999**; *405*: 271–280.

107 Takemura T, Sakagami M, Nakase T, Kubo T, Kitamura Y, Nomura S. Localization of osteopontin in the otoconial organs of adult rats. *Hear Res* **1994**; *79*: 99–104.

108 Sakagami M. Role of osteopontin in the rodent inner ear as revealed by in situ hybridization. *Med Electron Microsc* **2000**; *33*: 3–10.

109 Uno Y, Horii A, Umemoto M, Hasegawa T, Doi K, Uno A, Takemura T, Kubo T. Effects of hypergravity on morphology and osteopontin expression in the rat otolith organs. *J Vestib Res* **2000**; *10*: 283–289.

110 Hallworth R, Wiederhold ML, Campbell JB, Steyger PS. Atomic force microscope observations of otoconia in the newt. *Hear Res* **1995**; *85*: 115–121.

111 Lins U, Farina M, Kurc M, Riordan G, Thalmann R, Thalmann I, Kachar B. The otoconia of the guinea pig utricle: internal structure, surface exposure, and interactions with the filament matrix. *J Struct Biol* **2000**; *131*: 67–78.

112 Rio C, Dikkes P, Liberman MC, Corfas G. Glial fibrillary acidic protein expression and promoter activity in the inner ear of developing and adult mice. *J Comp Neurol* **2002**; *442*: 156–162.

113 Raft S, Nowotschin S, Liao J, Morrow BE. Suppression of neural fate and control of inner ear morphogenesis by Tbx1. *Development* **2004**; *131*: 1801–1812.

114 Fritzsch B, Beisel KW, Jones K, Farinas I, Maklad A, Lee J, Reichardt LF. Development and evolution of inner ear sensory epithelia and their innervation. *J Neurobiol* **2002**; *53*:143–156.

115 Kim WY, Fritzsch B, Serls A, Bakel LA, Huang EJ, Reichardt LF, Barth DS, Lee JE. NeuroD-null mice are deaf due to a

severe loss of the inner ear sensory neurons during development. *Development* **2001**; *128*: 417–426.
116 Liu M, Pereira FA, Price SD, Chu MJ, Shope C, Himes D, Eatock RA, Brownell WE, Lysakowski A, Tsai MJ. Essential role of BETA2/NeuroD1 in development of the vestibular and auditory systems. *Genes Dev* **2000**; *14*: 2839–2854.
117 Huang EJ, Liu W, Fritzsch B, Bianchi LM, Reichardt LF, Xiang M. Brn3a is a transcriptional regulator of soma size, target field innervation and axon pathfinding of inner ear sensory neurons. *Development* **2001**; *128*: 2421–2432.
118 Tessier-Lavigne M, Goodman CS. The molecular biology of axon guidance. *Science* **1996**; *274*: 1123–1133.
119 Xiang M, Maklad A, Pirvola U, Fritzsch B. Brn3c null mutant mice show long-term, incomplete retention of some afferent inner ear innervation. *BMC Neurosci* **2003**; *4*: 2.
120 Cowan CA, Yokoyama N, Bianchi LM, Henkemeyer M, Fritzsch B. EphB2 guides axons at the midline and is necessary for normal vestibular function. *Neuron* **2000**; *26*: 417–430.
121 Gu C, Rodriguez ER, Reimert DV, Shu T, Fritzsch B, Richards LJ, Kolodkin AL, Ginty DD. Neuropilin–1 conveys semaphorin and VEGF signaling during neural and cardiovascular development. *Dev Cell* **2003**; *5*: 45–57.
122 Fritzsch B. Development of inner ear afferent connections: forming primary neurons and connecting them to the developing sensory epithelia. *Brain Res Bull* **2003**; *60*: 423–433.
123 Rubel EW, Fritzsch B. Auditory system development: primary auditory neurons and their targets. *Annu Rev Neurosci* **2002**; *25*: 51–101.
124 Maklad A, Fritzsch B. Development of vestibular afferent projections into the hindbrain and their central targets. *Brain Res Bull* **2003**; *60*: 497–510.
125 Fritzsch B, Coppola V, Tessarollo L, Reichardt LF. Neurotrophins in the ear: their roles in sensory neuron survival and fiber guidance. *Prog Brain Res* **2004**; *146*: 265–278.
126 Pirvola U, Ylikoski J, Palgi J, Lehtonen E, Arumae U, Saarma M. Brain-derived neurotrophic factor and neurotrophin 3 mRNAs in the peripheral target fields of developing inner ear ganglia. *Proc Natl Acad Sci USA* **1992**; *89*: 9915–9919.
127 Farinas I, Jones KR, Tessarollo L, Vigers AJ, Huang E, Kirstein M, de Caprona DC, Coppola V, Backus C, Reichardt LF, Fritzsch B. Spatial shaping of cochlear innervation by temporally regulated neurotrophin expression. *J Neurosci* **2001**; *21*: 6170–6180.
128 Coppola V, Kucera J, Palko ME, Martinez-De Velasco J, Lyons WE, Fritzsch B, Tessarollo L. Dissection of NT3 functions *in vivo* by gene replacement strategy. *Development* **2001**; *128*: 4315–4327.
129 Agerman K, Hjerling-Leffler J, Blanchard MP, Scarfone E, Canlon B, Nosrat C, Ernfors P. BDNF gene replacement reveals multiple mechanisms for establishing neurotrophin specificity during sensory nervous system development. *Development* **2003**; *130*: 1479–1491.
130 Tessarollo L, Coppola V, Fritzsch B. 2004. NT–3 replacement with brain-derived neurotrophic factor redirects vestibular nerve fibers to the cochlea. *J Neurosci* 24: 2575–2584.

131 Bermingham NA, Hassan BA, Price SD, Vollrath MA, Ben-Arie N, Eatock RA, Bellen HJ, Lysakowski A, Zoghbi HY. Math1: an essential gene for the generation of inner ear hair cells. *Science* **1999**; *284*: 1837–1841.

132 Gao WQ. Hair cell development in higher vertebrates. *Curr Top Dev Biol* **2003**; *57*: 293–319.

133 Kawamoto K, Ishimoto S, Minoda R, Brough DE, Raphael Y. Math1 gene transfer generates new cochlear hair cells in mature guinea pigs *in vivo*. *J Neurosci* **2003**; *23*:4395–4400.

134 Chen P, Johnson JE, Zoghbi HY, Segil N. The role of Math1 in inner ear development: Uncoupling the establishment of the sensory primordium from hair cell fate determination. *Development* **2002**; *129*: 2495–2505.

135 Rivolta MN, Halsall A, Johnson CM, Tones MA, Holley MC. Transcript profiling of functionally related groups of genes during conditional differentiation of a mammalian cochlear hair cell line. *Genome Res* **2002**; *12*: 1091–1099.

136 Erkman L, McEvilly RJ, Luo L, Ryan AK, Hooshmand F, O'Connell SM, Keithley EM, Rapaport DH, Ryan AF, Rosenfeld MG. Role of transcription factors Brn–3.1 and Brn–3.2 in auditory and visual system development. *Nature* **1996**; *381*: 603–606.

137 Li S, Price SM, Cahill H, Ryugo DK, Shen MM, Xiang M. Hearing loss caused by progressive degeneration of cochlear hair cells in mice deficient for the Barhl1 homeobox gene. *Development* **2002**; *129*: 3523–3532.

138 Wallis D, Hamblen M, Zhou Y, Venken KJ, Schumacher A, Grimes HL, Zoghbi HY, Orkin SH, Bellen HJ. The zinc finger transcription factor Gfi1, implicated in lymphomagenesis, is required for inner ear hair cell differentiation and survival. *Development* **2003**; *130*: 221–232.

139 Hertzano R, Montcouquiol M, Rashi-Elkeles S, Elkon R, Yucel R, Frankel WN, Rechavi G, Moroy T, Friedman TB, Kelley MW, Avraham KB. Transcription profiling of inner ears from Pou4f3ddl/ddl identifies Gfi1 as a target of the Pou4f3 deafness gene. *Hum Mol Genet* **2004**; *13*: 2143–2153.

140 Riley BB. Genes controlling the development of the zebrafish inner ear and hair cells. *Curr Top Dev Biol* **2003**; *57*: 357–388.

141 Chen P, Segil N. p27(Kip1) links cell proliferation to morphogenesis in the developing organ of Corti. *Development* **1999**; *126*: 1581–1590.

142 Zine A, Aubert A, Qiu J, Therianos S, Guillemot F, Kageyama R, de Ribaupierre F. Hes1 and Hes5 activities are required for the normal development of the hair cells in the mammalian inner ear. *J Neurosci* **2001**; *21*: 4712–4720.

143 Kelley MW. Cell adhesion molecules during inner ear and hair cell development, including notch and its ligands. *Curr Top Dev Biol* **2003**; *57*: 321–356.

144 Dabdoub A, Donohue MJ, Brennan A, Wolf V, Montcouquiol M, Sassoon DA, Hseih JC, Rubin JS, Salinas PC, Kelley MW. Wnt signaling mediates reorientation of outer hair cell stereociliary bundles in the mammalian cochlea. *Development* **2003**; *130*: 2375–2384.

145 Montcouquiol M, Kelley MW. Planar and vertical signals control cellular differentiation and patterning in the mammalian cochlea. *J Neurosci* **2003**; *23*: 9469–9478.
146 Eatock RA, Hurley KM. Functional development of hair cells. *Curr Top Dev Biol* **2003**; *57*: 389–448.
147 Schimmang T, Tan J, Muller M, Zimmermann U, Rohbock K, Kopschall I, Limberger A, Minichiello L, Knipper M. Lack of Bdnf and TrkB signalling in the postnatal cochlea leads to a spatial reshaping of innervation along the tonotopic axis and hearing loss. *Development* **2003**; *130*: 4741–4750.
148 Izumikawa M, Minoda R, Kawamoto K, Abrashkin KA, Swiderski DL, et al. 2005. Auditory hair cell replacement and hearing improvement by Atoh1 gene therapy in deaf mammals. *Nat Med* 11: 271–276.
149 Li H, Liu H, Heller S. Pluripotent stem cells from the adult mouse inner ear. *Nat Med* **2003**; *9*: 1293–1299.
150 Li H, Roblin G, Liu H, Heller S. Generation of hair cells by stepwise differentiation of embryonic stem cells. *Proc Natl Acad Sci USA* **2003**; *100*: 13495–13500.
151 Rask-Andersen H, Bostrom M, Gerdin B, Kinnefors A, Nyberg G, et al. 2005. Regeneration of human auditory nerve. In vitro/in video demonstration of neural progenitor cells in adult human and guinea pig spiral ganglion. *Hear Res* 203: 180–91.
152 Depew MJ, Liu JK, Long JE, Presley R, Meneses JJ, Pedersen RA, Rubenstein JL. Dlx5 regulates regional development of the branchial arches and sensory capsules. *Development* **1999**; *126*: 3831–3846.
153 Hans S, Liu D, Westerfield M. Pax8 and Pax2a function synergistically in otic specification, downstream of the Foxi1 and Dlx3b transcription factors. *Development* **2004**; *131*: 5091–5102.
154 Nieto MA, Gilardi-Hebenstreit P, Charnay P, Wilkinson DG. A receptor protein tyrosine kinase implicated in the segmental patterning of the hindbrain and mesoderm. *Development* **1992**; *116*: 1137–1150.
155 Orr-Urtreger A, Avivi A, Zimmer Y, Givol D, Yarden Y, Lonai P. Developmental expression of c-kit, a proto-oncogene encoded by the W locus. *Development* **1990**; *109*: 911–923.
156 Pfeffer PL, Gerster T, Lun K, Brand M, Busslinger M. Characterization of three novel members of the zebrafish Pax2/5/8 family: dependency of Pax5 and Pax8 expression on the Pax2.1 (noi) function. *Development* **1998**; *125*: 3063–3074.
157 Reaume AG, Conlon RA, Zirngibl R, Yamaguchi TP, Rossant J. Expression analysis of a Notch homologue in the mouse embryo. *Dev Biol* **1992**; *154*: 377–387.
158 Rinkwitz-Brandt S, Justus M, Oldenettel I, Arnold HH, Bober E. Distinct temporal expression of mouse Nkx–5.1 and Nkx–5.2 homeobox genes during brain and ear development. *Mech Dev* **1995**; *52*: 371–381.
159 Solloway MJ, Robertson EJ. Early embryonic lethality in Bmp5;Bmp7 double mutant mice suggests functional redundancy within the 60A subgroup. *Development* **1999**; *126*: 1753–1768.
160 Wong RL, Chan KK, Chow KL. Developmental expression of Mab21l2 during mouse embryogenesis. *Mech Dev* **1999**; *87*: 185–188.

161 Wood HB, Episkopou V. Comparative expression of the mouse Sox1, Sox2 and Sox3 genes from pre-gastrulation to early somite stages. *Mech Dev* **1999**; *86*: 197–201.

162 Yoshikawa SI, Aota S, Shirayoshi Y, Okazaki K. The ActR-I activin receptor protein is expressed in notochord, lens placode and pituitary primordium cells in the mouse embryo. *Mech Dev* **2000**; *91*: 439–444.

163 Lindeman HH. 1969. Studies on the morphology of the sensory regions of the vestibula apparatus with 45 figures. *Ergeb Anat Entwicklungsgesch* 42: 1–113.

164 Fritzsch B, Matei VA, Nichols DH, Bermingham N, Jones K, et al. 2005. Atoh1 null mice show directed afferent fiber growth to undifferentiated ear sensory epithelia followed by incomplete fiber retention. *Dev Dyn* 233: 570–583.

15
Limb Development

Susan Mackem

15.1
Introduction

Now well more than 50 years since it first became a focus for experimental study, the developing vertebrate limb has endured as one of the premiere systems for unraveling the regulatory signaling cascades directing pattern formation and the mechanics of morphogenesis. Early studies of embryonic limb development in chick embryos (and the related process of regeneration in urodele amphibians) focused on these systems because of accessibility and ease of experimental manipulation, which led to the identification of several signaling centers, or organizers, that regulate and coordinate limb patterning and outgrowth (e.g. see [1–4]). This early elucidation of key inductive interactions provided a rich biological framework and laid the foundation for later studies using molecular-genetic approaches to characterize the signaling components regulating limb development. The latter have been propelled by the development of powerful genetic approaches in mouse and in zebrafish; early events in teleost fin bud formation closely parallel early limb bud development [5]. As a structure that is strongly driven by evolutionary adaptation in which modest anomalies and diversity in form may be tolerated without loss of viability, and which also leaves a rich fossil record due to the bony nature of its final morphogenetic product, the limb has served as a central model in evolutionary biology [6–10] and in classical genetic studies of mutants in experimental organisms [1, 11–15], both of which have also enriched our understanding of this highly regulated and robust developmental process. What has emerged is a landscape of rich and highly redundant regulation with many levels of cross-talk, feedback and feed-forward circuits. Although a real mechanistic understanding of how early patterns are translated into precise morphologies remains as yet elusive, the developing limb promises to be one of the first systems in which this ultimate goal will be achieved.

This review will focus mainly on the regulatory centers that are set up and function in the early limb bud, the major signaling pathways involved, and how they regulate the pattern of the "intrinsic" structure formed from limb bud mesenchyme, the limb skeleton. The limb mesenchyme gives rise to the entire limb skeleton as a cartilage model, later replaced by bone, as well as all of the associated tendons [16–19]. The patterning and differentiation of "extrinsic" structures (muscle, vasculature, nerves), that form from progenitors entering the limb bud from medial trunk tissues such as the somites [18, 20, 21], are also highly regulated and take many cues from the same organizer signals that pattern the primary cartilage skeleton. While the cartilage pattern appears to be regulated primarily by signaling centers arising from within the early limb bud, the tendons are also patterned by extrinsically-derived tissues (e.g. muscle masses) [22, 23]. Development of the extrinsic components [21, 24–28], as well as tendon differentiation and patterning, and late events such as cartilage maturation, enchondral ossification and formation of cartilage growth plates for elongation of bones [29–31], have all been the subject of intensive analysis, but lie beyond the scope of this chapter.

15.1.1
Morphological Landmarks of the Limb

The limb elements form in a proximal to distal (hip to toe) order, usually as contiguous branching condensations of the mesenchyme that then differentiate to cartilage and segment to form joints, thereby segregating individual elements [32, 33]. These cartilage models are subsequently replaced by bone. The resulting fore and hindlimbs are composed of homologous skeletal elements with distinctions in morphology that give each their unique "identity". Since the early patterning of these structures in both limb types is similar, common terminology is often used to refer to them (Fig. 15.1): the proximal long bone (humerus/femur) is referred to as the stylopod, the anterior (radius/tibia) and posterior (ulna/fibula) paired long bones constitute the zeugopod, and the distal hand/foot bones are collectively referred to as the autopod (these include the carpals/tarsals of the wrist/ankle, the metacarpals/metatarsals of the palmar/plantar region, and the phalanges of the digits). In different amniote vertebrates, both digit number and identity (the number and size of phalanges within a given digit) are varied along the autopod AP axis. The pattern of different digits is highly regulated and particularly subject to evolutionary variation [1, 10, 32]. For example, the anterior-most digit 1 in mice and humans is biphalangeal, while the rest are triphalangeal. In contrast, the chick hindlimb has a different number of phalanges in each of its four hindlimb digits, increasing from two to five in digit 1–4. Views are conflicting with regard to whether the proximal limb girdles ought to be considered part of the limb proper in terms of their developmental regulation. The pelvis forms from lateral plate mesoderm [17] in contiguity with more distal condensations (discussed in [34]), while the scapula is largely of somitic origin (the entire blade), except for the articulating head/neck component [17, 35]. Both appear to be regulated by some signals conjointly with more distal limb elements (e.g. [36–38]), but not by others (e.g. [39–41]), perhaps because the girdle elements develop very early, preceding the appearance of a morphologic limb "bud".

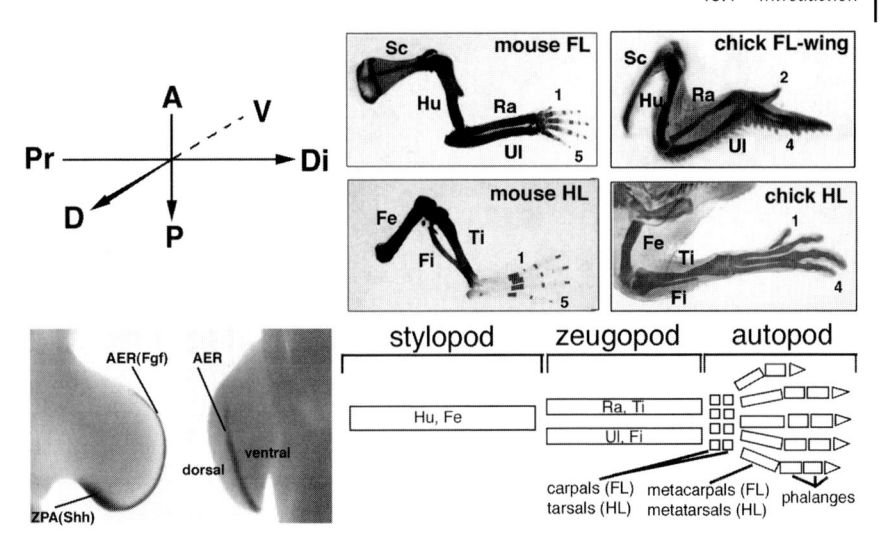

Fig. 15.1
Limb bud signaling centers and the morphological components of the mature limb skeleton. The three orthogonal limb axes (upper left panel) show the orientation below in limb bud stage (left-most panel), and to the right, in skeletons of late stage mouse and chick embryos. Chick limb buds hybridized with *Fgf8* and *Shh* probes to highlight the AER and ZPA, respectively, are shown in the lower left, oriented as in the diagram above (left panel) and rotated 90 ° counterclockwise to show AER relative to dorsal and ventral limb bud surfaces (right panel). Standard generic names for the different PD skeletal components of the limb are indicated in the diagram on the lower right. Note that the chick wing has only three digits, and the hindlimb four digits. Sc, scapula; Hu, humerus; Ra, radius; Ul, ulna; Fe, femur; Ti, tibia; Fi, fibula; FL, forelimb; HL, hindlimb. (This figure also appears with the color plates.)

15.1.2
Embryological Elucidation of Inductive Interactions in Limb Development

Early work in chick embryos using tissue grafts, ablation, and barrier inserts to study tissue interactions (see below and [1]), as well as various cell-fate tracers recently including use of dyes [42, 43], chick-quail chimeras [44], and genetic marking techniques in mouse [45–47], have illuminated early inductive events leading to limb bud formation and the establishment of signaling centers that function as "organizers" for regulating polarity, pattern, and growth roughly along the three orthogonal limb axes: proximodistal (PD), dorsoventral (DV) and anteroposterior (AP).

The limb bud initiates as a swelling of increased cell density in the lateral plate mesoderm (somatopleure) at precise positions along the body axis [48–51], in response to signals from more medial tissues, possibly including axial, paraxial, and/or intermediate (nephrogenic) mesoderm [52–58]. Which of these several more midline tissues is essential for induction and whether there is a sequential medial-to-lateral cascade of signals is still unresolved. Early limb forming capacity is broadly distributed along the trunk and the flank can respond to ectopic signals and initiate limb budding, even for a short while after the normal limb buds have formed [48, 59, 60]. It also remains unclear whether limb initiation becomes constrained to the future forelimb and hindlimb domains simply due to spatial restriction of inductive signals or the presence of other inhibitory factors.

Subsequent signaling from the incipient limb bud mesoderm induces formation and maintenance of a specialized ectodermal structure along the limb bud's DV rim or apex, the apical ectodermal ridge (AER, Fig. 15.1) [61, 62]. Prospective limb mesoderm induces AER formation, even when grafted heterotopically beneath non-limb ectoderm [62–64]. Cell marking studies show that limb ectoderm already has a defined DV polarity and discrete dorsal and ventral compartments prior to the emergence of a limb bud, and the mature AER forms from broadly distributed progenitors in the ventral ectoderm that later become positioned along the limb bud DV border [42, 44, 45]. Signals from the AER maintain the survival and proximodistal outgrowth of the limb bud mesoderm leading to the sequential formation of skeletal elements [65]. However, limb type and the nature of limb structures formed at different stages are determined intrinsically by the mesoderm [16, 66], while AER signals play a permissive rather than instructive role in outgrowth, since AER of different stages function similarly when exchanged [67]. AER removal causes limb truncations at progressively more distal positions after later stages of ridge removal (Fig. 15.2). This led to the concept of the "progress zone" [68, 69], a distal zone of sub-AER undifferentiated mesenchyme that progressively gives rise to different skeletal elements depending on the length of time cells are exposed to AER signals before growth displaces them out of this zone. Recent fate mapping of the "progress zone" region [43] and genetic studies of AER function [40] have prompted a re-evaluation of this model (discussed in [70–74] and in Section 15.4.1).

The establishment of DV polarity in the future limb ectoderm depends on early signals from underlying mesodermal tissues before limb bud formation [54, 63, 75, 76]. More recent experiments indicate that signals from paraxial mesoderm (somites) and from lateral plate mesoderm, respectively, regulate dorsal and ventral ectodermal fates prior to limb bud initiation in chick embryos [44], resulting in discrete dorsal and ventral ectodermal compartments (see Fig. 15.5). Regulation of DV polarity is subsequently transferred back to the underlying limb bud mesoderm by ectodermal signals shortly after the limb bud forms. Thus, reversing the DV orientation of the limb bud ectodermal jacket relative to the bud mesoderm results in a global reversed DV polarity of underlying limb structures that later form distally [76–79].

The third organizing center, regulating AP polarity in the early limb bud, and particularly the pattern of different digits (e.g. from thumb to pinky) that form, was operationally named the "zone of polarizing activity" (ZPA, Fig. 15.1) because of its ability to induce duplications of limb structures with AP mirror image symmetry (Fig. 15.2) when grafted to the anterior border of the limb bud in the chick [80]. Like limb-forming capacity, ZPA activity is also first broadly distributed in the lateral plate at early stages in chick and mouse prior to bud formation [59, 81], and then

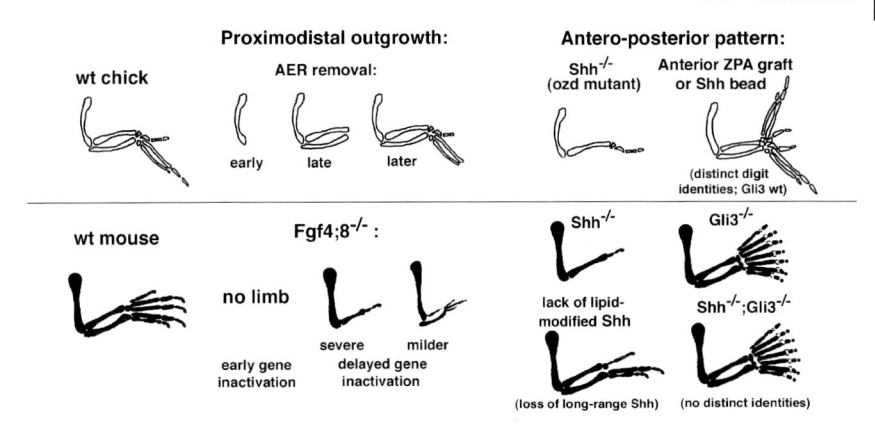

Fig. 15.2
Schematic summary of phenotypes following embryologic manipulations or due to loss-of-function mutations that affect AER and ZPA in chick and mouse (forelimbs shown). As discussed in text, AER removal causes successive loss of skeletal elements in a distal to proximal order with earlier time of removal; these results led to development of the "progress zone" model. In contrast, AER-*Fgf* gene inactivation after transient *Fgf* expression (a situation similar to later AER removal), results in reduced size of several elements rather than complete loss of distal structures. The major function of *Shh* in the limb is to prevent formation of Gli3R; genetic removal of *Gli3* in *Shh* mutants extensively rescues limb formation. Ectopic Shh signaling leads to patterned digit duplications, whereas *Gli3* mutants have unpatterned digits that lack normal or distinct digit identities, indicating an essential role for *Gli3* in specifying digit identity. Mutants that cannot lipid-modify Shh show digit patterning defects due to loss of long-range signaling and lack digits 2,3 (remaining digit identities are preserved). ozd, oligozeugodactyly mutation in chick.

becomes restricted to the posterior margin of the early limb bud. Recent work indicates that ZAP induction is also regulated by midline cues; classical experimental embryology has been less revealing in this regard, since limb-forming ability and AP polarity are both established very early and coordinately [82, 83].

Development along the three limb axes is coordinated by several cross-regulatory interactions between the signaling centers (Figs 15.3 and 15.5), some of which were revealed by early embryological experiments [67, 84], and some have been additionally uncovered with molecular approaches. Most importantly, the AER, which promotes mesodermal survival and outgrowth, is reciprocally maintained in a functional state by signals from the underlying mesoderm, especially from the posterior ZPA [85, 86], and conversely, ZPA graft experiments suggested that the ZPA may depend on AER signals [87, 88]. With the advent of molecular tools, later studies established that the ZPA is maintained by continued signaling from both the AER [40, 89, 90] and the dorsal ectoderm [91].

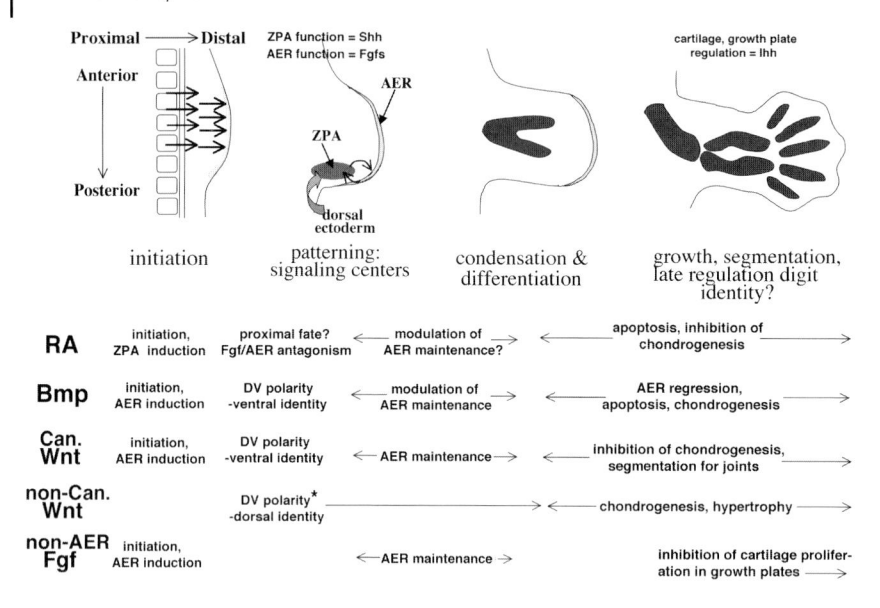

Fig. 15.3
Timeline of different roles of major signaling pathways during progression of limb development. AER-Fgfs and Hedgehogs (Shh, Ihh) are shown in the schematic depicting limb stages at the top, because they play "dedicated" roles in mediating AER function, ZPA function, and in chondrogenic growth, respectively. *The evidence regarding whether Wnt7a, the major signal regulating dorsal polarity in mesodermal limb derivatives, acts through the canonical or a non-canonical pathway is highly conflicted, as discussed in text. (This figure also appears with the color plates.)

15.2
The Major Signaling Pathways in Limb Development

15.2.1
Retinoids

Retinoic acid (RA) is involved in several aspects of limb development and, like most signals regulating the limb, plays different roles in different contexts and at different stages (see Fig. 15.3), including regulation of limb initiation and ZPA induction [92–96], proximodistal outgrowth [97, 98], and apoptosis to sculpt limb shape [99, 100], as well as later events during chondrogenesis [98, 99, 101]. RA signals are directly transferred to the nucleus to impact gene expression by specific binding to nuclear receptors of the steroid receptor superfamily that can function either as transcriptional activators or repressors depending on whether or not they are ligand bound [102–106]. There are three RA receptor genes, $RAR\alpha$, $RAR\beta$, and $RAR\gamma$, each capable of generating multiple different isoforms through alternate splicing and promoter usage [107], as well as three *RXRs* (RAR-related receptors) with a similar complexity of isoforms. RARs heterodimerize with RXRs to regulate transcription of target promoters via binding to specific elements (RAREs) [108], while RXRs can alternatively function as homodimers or can heterodimerize with other nuclear receptors [104, 107, 109]. *RARα* and *RARγ* are expressed broadly in the early limb bud and later in association with prechondrogenic condensations [110, 111], and compound null mutants have some defects in limb skeletal elements including reduced size and variable digit loss [101]. *RARβ* is proximally restricted and also expressed in the flank at early stages, and later is expressed in areas associated with apoptosis, such as interdigital mesenchyme [110]. *RARβ,γ* compound null mutants display syndactyly [112]. *RXRα,β* are expressed ubiquitously during embryogenesis [113, 114] and are functionally highly redundant [115, 116], while *RXRγ* is completely absent from the early limb bud. In addition, CRABPs (cytosolic RA binding proteins) can bind RA and may further modulate the ultimate accessibility of cellular RA to the nuclear receptors as well as chaperone access of RA to metabolic enzymes, although genetic functional analyses have not revealed a critical role in embryonic development [117–119].

The bewildering complexity of co-expressed functionally redundant receptors and receptor isoforms has complicated genetic dissection of their roles, particularly at early stages in limb development, and suggests that differential receptor responses may be regulated largely by ligand availability. Consequently, recent analyses have focused on the metabolic pathways for retinoid biosynthesis and degradation and the evaluation of the consequences of inactivating key synthetic and degradative enzymes. Conversion of dietary retinol to retinaldehyde occurs via enzymes that are ubiquitously expressed during development [120]. All-trans RA appears to play the most critical role in early development, although 9–cis RA (more specific for RXRs) is also present, and chick embryos have 3,4–didehydro RA as well [118, 121, 122]. The critical step in the synthesis of all these retinoids is the conversion of their aldehyde precursor to the acid by retinaldehyde dehydrogenases (Raldhs), which are encoded by three genes differentially expressed in vertebrate embryos [92, 123, 124]. *Raldh2* plays the most important role in early stages of embryogenesis, including forelimb initiation, which fails to occur in null mouse mutants [92–94] (*Raldh3* may compensate in the hindlimb region), and *Raldh2* is also essential for pectoral fin bud initiation in zebrafish embryos [125]. Raldh inhibitors applied to early chick embryos likewise prevent limb initiation [96]. It has long been known that RA applied to the early limb bud can mimic the ZPA [126, 127] and can also induce an ectopic ZPA in the anterior limb bud, although it is not the endogenous ZPA signal ([128, 129], see also [130]). Partial rescue of the *Raldh2* mouse mutant by RA further reveals an involvement in ZPA induction [93, 94]. Degradation of RA is achieved by the action of cytochrome P450 family members [131], for which there are three genes, *Cyp26a1,b1,c1* [132–135]. *Cyp26b1* is expressed in the distal limb bud mesenchyme. In contrast to the *Raldh2* knockout, inactivation of the *Cyp26b1* gene phenocopies the effects of excess RA on the established limb bud, including distal truncations, excessive apoptosis and inhibition of chondrogenesis [98].

Consistent with their opposing null phenotypes and proposed roles in generating gradient RA distributions through active synthesis and active removal, the expression of *Raldh* and *Cyp26* family members is orchestrated to be largely complementary in early embryos [132–134, 136–139]. During gastrulation, *Cyp26* members are expressed rostrally and *Raldh* caudally. Later, *Raldh2* is expressed in paraxial, intermediate, and lateral plate mesoderm in the trunk and in the proximal limb bud, while *Cyp26a1* is expressed caudally in extending tail-bud tissues and *Cyp26b1* is expressed distally in the limb bud. The availability of an RA-responsive LacZ reporter [140] has aided mutant analyses by allowing visualization of the effective RA "activity" gradients generated and reveals a graded PD distribution of RA in the wild-type limb bud [94] consistent with the proposed role for RA in antagonizing distal Fgfs in later proximodistal limb development [97]. Trunk retinoid levels that may be critical for regulating limb initiation and axial positioning appear to be uniformly high at early stages as assayed by such reporters [140]. Newer, more rapid real-time *in vivo* reporters of RA activity may enable detection of transient, dynamic fluxes in RA at these early stages [141] and allow future re-evaluation of whether selective RA spatial distributions are related to sites of limb initiation.

15.2.2
Fgfs

The *Fgfs* constitute a large family of growth factor genes with at least 23 different members identified to date [58, 142, 143], which is further diversified by the production of several different alternately spliced forms for certain family members [144]. A number of different *Fgfs* are expressed selectively either in early limb bud mesoderm or in the ectoderm, usually becoming localized specifically to the AER [58]. Later, other *Fgfs* are expressed during chondrogenic differentiation and growth [145–147]. There are four Fgf receptor genes (*Fgfrs*) encoding receptor tyrosine kinases (RTKs) that homodimerize and undergo trans-phosphoryation upon ligand binding [148]. Ligand binding and receptor dimerization occurs in cooperation with heparan sulfate proteoglycans (HSPGs, Fig. 15.4), which strongly promote the ligand-receptor interaction [142, 149]. The Fgfrs contain two or three extracellular Ig domains (IgI, II and III) and each of the four Fgfrs occurs as multiple different isoforms generated by alternative splicing and polyA addition site selection [150–153]. Most notably, alternative splicing of IgIII, which is involved in ligand recognition, alters the domain structure and relative ligand affinities of different Fgfr isoforms. Consequently, particular Fgfr isoforms have strong preferences for interaction with certain Fgf ligands [153, 154]. *Fgfr1* is expressed exclusively in the mesoderm of the early limb bud [155], while different *Fgfr2* isoforms are expressed selectively in either mesoderm or in AER ectoderm [156]. *Fgfr3* is expressed and functions mainly later in regulating growth plates of long bones [157], whereas *Fgfr4* function is not required in limb [158]. The distribution of *Fgf* and *Fgfr* expression in the early limb bud facilitates reciprocal epithelial-mesenchymal signaling that induces AER formation and limb mesoderm outgrowth. Thus, mesodermal Fgf10 preferentially binds to the Fgfr2–IIIb isoform, which is expressed selectively in ectoderm, while the ectodermal Fgf4 and Fgf8 bind to the mesodermally expressed Fgfr2–IIIc, as well as to Fgfr1–IIIc isoforms [153, 154].

Fgf10, expressed in lateral plate mesoderm prior to limb initiation, is essential for AER induction and null mutants fail to sustain limb bud outgrowth beyond the initial formation of a small mesenchymal bud that has no evidence of an AER or of Fgf8 expression in progenitors [36, 37]. Likewise, Fgf10 misexpression in chick can induce ectopic limb bud and AER formation in the flank [159]. Inactivation of Fgfr2 results in very similar phenotypes [156, 160]. Isoform selective knockouts have revealed that Fgfr2IIIb is an essential receptor in the ectoderm for tranducing Fgf10 signals [161], although in this case transient Fgf8 expression was initiated, but not maintained. Based on these results and analysis of Fgf10 function in chick, Fgf10 has also been proposed to play a role in AER maintenance [159, 161]. A role in AER maintenance is also supported by the results of transgenic misexpression of a dominant-interfering secreted Fgfr2–IIIb that sequesters Fgf10 ligand and causes progressive PD limb truncations dependent on the timing of onset of transgene expression, phenocopying the results of chick AER removal at different times [162], including distal limb mesodermal apoptosis. However, in the mesoderm, Fgfr2IIIc is dispensable and its function is apparently redundant with other receptors in responding to Fgf signals from the AER [163]. The AER Fgfs can substitute for all AER function and rescue limb development in chick embryos following ridge removal [164, 165]. Like Fgf10, they can also induce ectopic limb buds in flank, probably by first inducing mesodermal Fgf10 expression via the reciprocal signaling loop [60]. The Fgfs expressed in the AER have some functional overlap, but Fgf8 is expressed the earliest in AER progenitors and plays the most important role in promoting mesoderm survival and proximodistal outgrowth, and Fgf8 inactivation in pre-limb ectoderm results in the loss or severe reduction of multiple limb skeletal elements [166, 167]. Fgf4, expressed later together with Fgf9 and Fgf17 in the posterior half of the AER, is dispensable by itself [168, 169], but becomes upregulated and can partly compensate for lack of Fgf8. Genetic removal of Fgf4 together with Fgf8 results in a much more severe phenotype than loss of Fgf8 alone; although a morphological AER initially forms, the lack of virtually all AER function is evident by the complete absence of all limb elements (except a rudiment of the limb girdle) when gene removal is carried out early enough [40, 41]. Fgf9 and Fgf17 appear to be largely nonessential for AER function (discussed in [40]), although compound knockout analyses have yet to be reported.

The critical mesodermal Fgfrs responsive to AER Fgf signals have yet to be definitively determined. Since the isoform-specific knockout of mesodermal Fgfr2IIIc alone does not affect early limb development, other receptors must also be involved and several lines of evidence suggest that Fgfr1, particularly the -IIIc isoform, plays a key role in the response to ectodermal Fgfs. Definitive analysis of Fgfr1 function in limb awaits analysis of the conditional null allele because of early embryonic lethality of Fgfr1 null mutants during gastrulation (see [170]). However, chimeric embryos aggregated from admixed wild-type and Fgfr1 null cells show a complete exclusion of the mutant cells from the distal limb "progress zone" mesenchyme and chimeras

Fig. 15.4

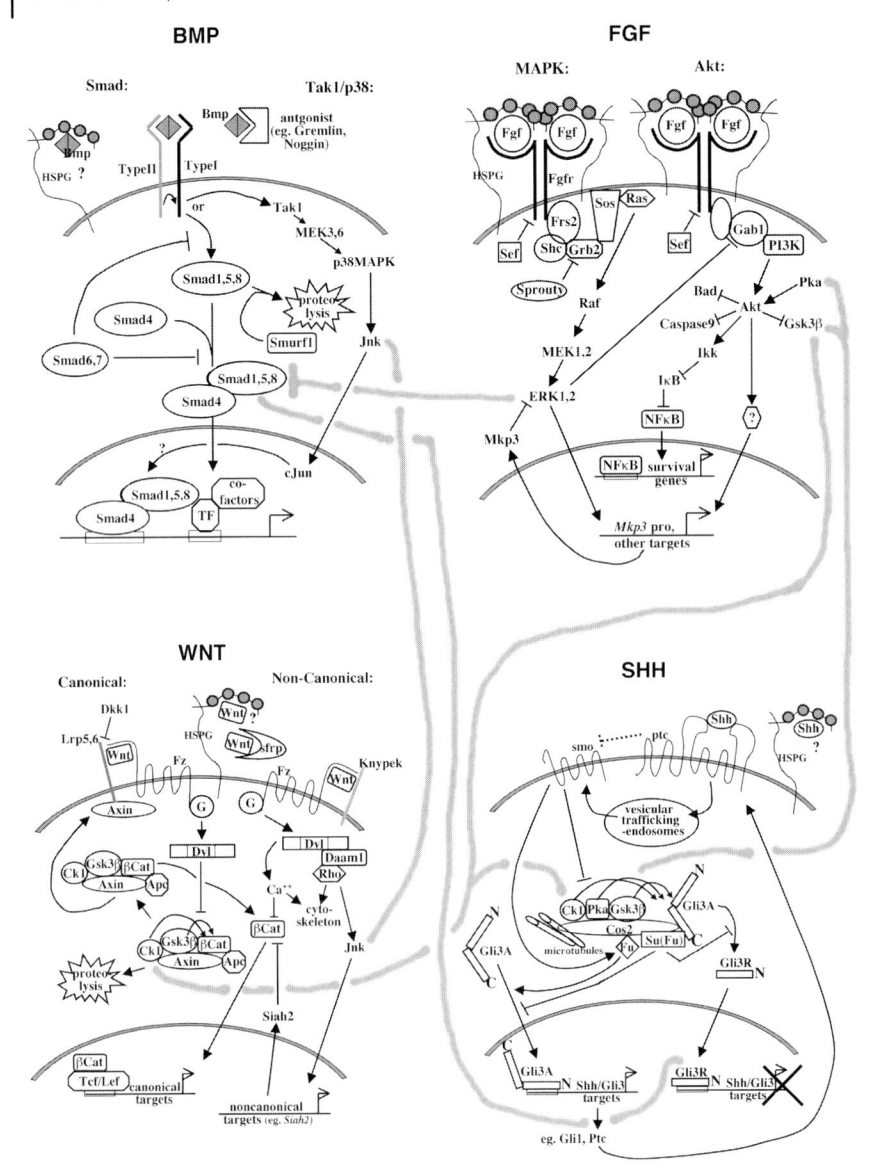

Major signal transduction pathways regulating limb development and hypothetical points of crosstalk. Schematics showing Bmp, Fgf, Wnt, and Shh signal transduction pathways, as discussed in text. For Bmp, Fgf, and Wnt there are several alternate intracellular pathways, some of which are shown (p38 pathway not shown for Fgf; atypical RTK pathways, Adenyl cyclase/PKA pathway and cadherins not shown for Wnt). Not all components for each pathway are shown, and certain components may not necessarily be operative in the context of the developing limb. Arrows and bars do not necessarily signify direct relationships, but may be indirect. Several points of potential cross-talk between pathways are shown as fuzzy gray lines connecting components in the different pathways. Note the conservation of many components between the canonical Wnt and Shh pathways, suggesting that these pathways are evolutionarily related (discussed in [216, 243, 530]).

with high percentage mutant cells display distal limb bud truncations, strongly suggesting a requirement for this receptor in survival and/or proliferation of the skeletal element progenitor population in the distal limb mesoderm [171–173]. Furthermore, hypomorphic *Fgfr1* alleles [155] and transgenic misexpression of *Fgfr1* using BACs [174] produce opposing skeletal phenotypes displaying patterning and growth changes consistent with an important role for this receptor in the distal limb mesenchyme receiving AER signals. *Fgfr3* function is essential later in cartilage growth plates of long bones, as a key negative regulator of proliferation and growth; null mutants display elongated long bones [157]. Of several *Fgfs* expressed at these late growth stages, knockout analyses have implicated *Fgf18* as an essential ligand [146, 147].

The intracellular kinase signal transduction cascades operating downstream of the Fgf receptors are highly varied and each pathway can ultimately lead to phosphorylation of multiple different substrates and activation of a number of different transducing transcription factors (Fig. 15.4). Receptor dimerization and trans-autophosphorylation is first relayed through several docking proteins, such as Frs2 (Fgfr stimulated Grb2 binding protein) and Grb2 (growth factor receptor-bound protein 2) [172, 175–177], to form membrane-associated complexes that can recruit either Ras or PI3K (phosphatidyl-inositol 3'OH kinase) to the cell membrane [178], leading either to the activation of one of the several MAPK (mitogen activated protein kinase) pathways (which include ERK, JNK or p38 branches [179]), or to the activation of the Akt pathway, respectively [180]. Alternatively, Fgfr transduction can also lead to phosphorylation and activation of JAK/STATs [181], as well as PLC-γ1 [182]. In the early limb bud, the MAPK pathway involving phospho-ERK and the Akt pathway appear to play the most important roles [183–185]. The MAPK/p38 pathway may play a role in interdigital apoptosis, but this more likely occurs downstream of Bmp/TAK1, rather than Fgf signaling [186]. During regulation of growth plates by Fgfr3 activity, both the phosphorylation of STATs and p21/Inks and the activation of the MAPK/ERK pathway are important for inhibiting chondrocyte proliferation and hypertrophic differentiation respectively [157, 187, 188]. The Rb relatives p107 and p130 are key targets of this pathway and are required to arrest chondrocyte proliferation [189].

Recent studies examining the phosphorylation status of ERK and Akt in chick and mouse detected active phospho-ERK (pERK) at high levels in the ectoderm/AER and active phospho-Akt (pAkt) mainly in the mesoderm [183–185]. This complementary distribution appears to arise by mutual antagonism between the pathways (Fig. 15.4) and may relate to antagonistic roles in cell survival. pERK can promote dissociation of PI3K from GAB1 docking protein, leading to decreased pAkt [177]. Conversely, Akt activity induces the expression of *Mkp3/Pyst1* (MAPK-phosphatase-3 [190]), which acts as a feedback inhibitor that inactivates the MAPK pathway in limb bud mesoderm [185] by dephosphorylating ERK1,2. *Mkp3* expression has also been reported to be induced as a MAPK target involved in an auto-negative feedback loop in the limb [184]. Akt activity promotes cell survival in many systems, inhibiting pro-apoptotic and activating survival factors at both the protein and gene expression levels [180]. Thus, either forced expression of a constitutively active MEK1 or removal of Mkp3 by siRNA caused limb bud mesodermal apoptosis [185]. Conversely, the AER normally displays foci of ongoing apoptosis that regulate its extent [100], and normally has high levels of pERK. However, this view of MAPK/ERK and Akt roles may be overly simplistic. pERK is also present in the distal mesoderm [183, 184], and likely plays a role there as well. In fact, overexpression of *Mkp3* throughout the limb bud actually inhibited limb outgrowth in association with AER defects [185], which could be interpreted either as showing a requirement for ERK activity within the AER, or alternatively that mesodermal ERK plays some role required for AER maintenance by mesoderm. Furthermore, Shp2 is a relay protein phosphatase that potentiates the MAPK/ERK pathway, and *Shp2* null mutant cells are completely excluded from the distal limb mesoderm in aggregation chimeras, suggesting an essential function of Shp2 and MAPK/ERK in distal mesoderm [172]. Finally, it is also surprising that combined inactivation of *Fgf4* and *Fgf8*, which should eliminate Fgf signaling to the mesoderm and hence eliminate mesodermal Akt/Mkp3 activity, does not lead to distal mesodermal apoptosis [40]. Clearly, given the conflicting results, much remains to be resolved in understanding the roles played by the MAPK and Akt pathways in the early limb bud, as well as whether and how receptor-isoform specificity dictates which pathway is utilized.

In addition to *Mkp3/Pyst1*, Fgf signaling induces the expression of several other targets that act as feedback inhibitors of the Fgf transduction cascade. Sef (similar-expression-to-fgf) prevents activation of both MAPK and PI3K pathways through a direct interaction with Fgfrs [191, 192]. *Sprouty* genes are targets that have been postulated to act at the level of interfering with adaptor/relay proteins such as Grb2, although other mechanisms have also been proposed [193, 194]. Several *Sprouty* family members are expressed in distal limb bud mesoderm and overexpression in chick suggests they may play a role in modulating early Fgf function, as well as later roles in chondrogenic differentiation [195]; confirmation awaits loss-of-function analyses.

There are also cross-regulatory interactions of Fgf with other signaling pathways that can be antagonistic or in some cases, cooperative. Most notably, Bmps and Fgfs have antagonistic effects at many stages in limb development (including AER maintenance, cell survival, and chondrocyte proliferation). This occurs partially through opposing functions of their downstream targets, but cross-talk between signal transduction components may contribute as well (Fig. 15.4). Although its relevance to the limb remains to be determined, pERK can phosphorylate Smad1, a nuclear transducer of Bmp signals, and thereby interfere with its transcriptional function [196–198]. Recent evidence using MAPK inhibitors indicates that Bmps may promote interdigital apoptosis via activation of p38MAPK, which lies downstream of the Bmp/Tak1 pathway, as well as the Fgf/p38MAPK pathway [199, 200]. Of note in this regard, it has been proposed that Fgf and Bmp pathways may synergize in the mesoderm to regulate cell death in interdigits since Bmp requires prior Fgf activity to promote apoptosis [201]. However this cooperative effect of Fgfs may be very indirect (occurring after 24 h), making activation of p38 a less likely mechanism. Regulation of Akt activation by PKA and the ability of pAkt to phosphorylate and inhibit Gsk–3β function [180], also raise intriguing possibilities for cross-talk with the canonical Wnt and Shh pathways, whose nuclear effectors are regulated by phosphorylation by one or both of these kinases (see Fig. 15.4 and Sections 15.2.3 and 15.2.5), but its relevance to the limb remains to be determined.

Fgfs and RA also have antagonistic effects during proximodistal limb outgrowth [97, 98]. In the early embryo similar Fgf-RA antagonism is seen during patterning of the neural tube [138] and segmentation of paraxial mesoderm [139]. In this context, Fgfs regulate RA metabolic enzymes, inducing expression of *Cyp26* members and repressing *Raldh2*. Conversely, RA has also been reported to induce *Mkp3* expression in the early embryo. The most notable cooperative cross-regulatory interaction of Fgfs in the limb occurs between AER-Fgfs and Shh (Fig. 15.5). In this case, AER-Fgfs and Shh each maintain one another's expression via a positive feedback loop [89, 90, 169], and Fgfs and Shh may also cooperate by virtue of their roles in maintaining mesodermal cell survival [40, 202], although this has not been formally demonstrated.

In the developing limb, the transcription factors that are activated to transduce Fgf signals have not been delineated. One possible candidate is Rel/NF$\varkappa$B, which is regulated downstream of RTKs in other systems and can act as a survival factor (Fig. 15.4). Akt activates IKK (IK Kinase) which phosphorylates and thereby promotes degradation of Inhibitor of **NF$\varkappa$B** (I$\varkappa$B); consequently Rel/NF$\varkappa$B is released to the nucleus to regulate genes for cell survival [180]. *Rel/NF$\varkappa$B* family members are expressed in early distal limb bud mesoderm (where pAkt is high) and misexpression of *I$\varkappa$B* in chick has been used to demonstrate a role for Rel members in limb outgrowth [203, 204]. However, there is no clear evidence of a similar role for Rel family members in mouse, possibly reflecting redundancy. Previous reports have demonstrated limb skeletal defects in *IKK* null mutants due to impaired AER function [205, 206], consistent with a possible role for Rel factors in AER maintenance and limb outgrowth. However, recent analyses using selective targeting of *IKK* inactivation to ectoderm indicate that these phenotypes may in fact be due to an IKK function that is unrelated to its role in NF$\varkappa$B regulation [207]. Although no compelling candidate transducing transcription factors for Fgfs acting within the ectoderm have yet been identified, several factors may be key targets of Fgf10 signaling from the mesoderm. These include *p63*, a *p53* relative, and the *Sp1* relatives *Sp8* and *Sp9*, both of which are expressed in limb ectoderm and which have been implicated in AER maturation and maintenance from gene inactivation studies [208–211]. These factors all appear to function downstream of Fgf10, early Bmp, and canonical Wnt signaling that are involved in initiating AER formation and it is not yet established for which pathways they are direct or indirect targets, although in zebrafish fin buds, *p63* appears to be a direct target of the Bmp pathway [212].

15.2.3
Wnts

Wnt signaling plays complex roles in virtually every aspect of early limb development as well as later cartilage differentiation, maturation and joint formation (Fig. 15.3).

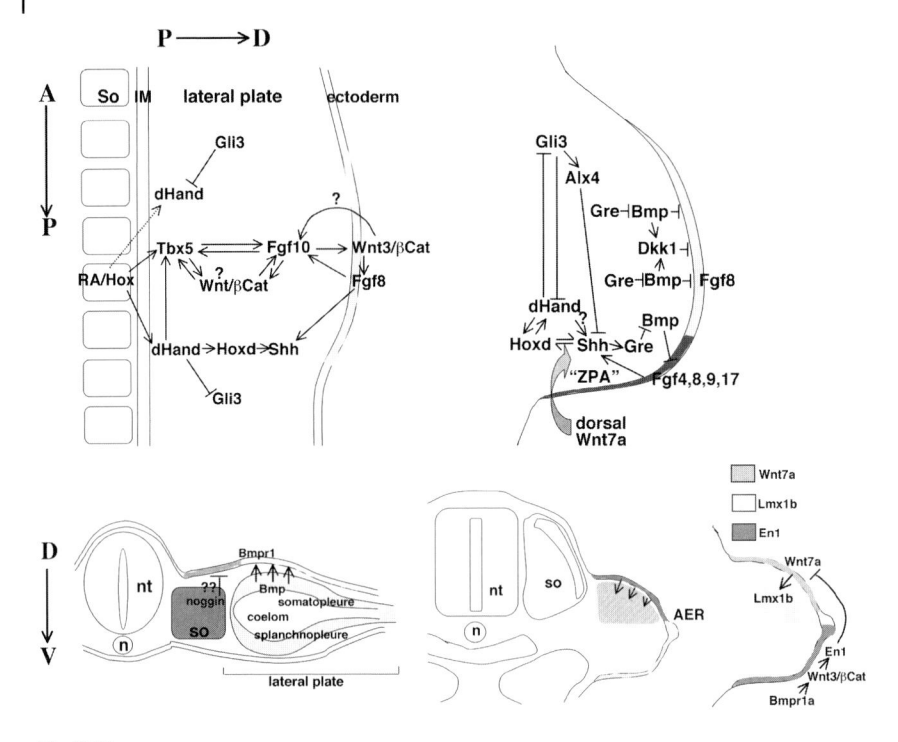

Fig. 15.5
Regulatory cascades leading to formation and maintenance of limb "organizers". As described in text, signaling cascades are shown that lead to limb initiation (forelimb cascade shown) and regulate the formation and maintenance of AER and ZPA (upper panels); and cascades that co-regulate early and later stages in specification of DV polarity and AER formation/positioning from within the ectoderm (lower panels). Arrows and bars are not necessarily meant to signify direct interactions. Where specific signals vary between organisms, only the relevant mammalian (mouse) factors and likely sequences of induction are shown. (For example, several Wnts that regulate chick limb initiation and AER formation do not appear to play the same roles in mouse; see text). n, notochord; nt, neural tube; so, somite; IM, intermediate mesoderm. (This figure also appears with the color plates.)

The vertebrate Wnt ligands, homologs of *Drosophila wingless* (*wg*), constitute a large family including at least 19 members to date [213], a number of which are expressed in the developing limb in restricted domains at different times [214, 215]. Wnt proteins are lipid modified (N-acylated) and are not freely released or diffusible in tissues [216]. In vertebrates they are thought to signal over a short range, although the details of their transport, secretion and modulation of range have as yet been mainly worked out in *Drosophila* [149, 216, 217]. Besides binding to specific receptors, Wnts also bind to HSPGs on glypican core proteins, which may serve accessory roles such as sequestering and/or concentrating ligand at the cell surface to enhance interaction with specific receptors; consequently, in *Drosophila*, mutations in certain

HSPGs and in their biosynthetic enzymes have *wg*-like phenotypes [149, 217]. HSPG involvement in Wnt signaling has as yet been less extensively characterized in vertebrates, but will likely play similar roles, as exemplified by a sulfatase implicated in heparan-dependent Wnt signaling that regulates *MyoD* [218].

Wnt signals are transduced by the Frizzled (Fz) family of seven-pass transmembrane receptors (Fig. 15.4) that all have conserved cysteine-rich domains (Fz domains) involved in ligand binding [219]. The *Fz* family includes at least 10 members, many of which have complex and overlapping expression domains in the limb [215], and that can lead to activation of several different intracellular cascades referred to as "canonical" and non-canonical pathways. Canonical signaling requires cooperative binding of Fz with single-pass transmembrane co-receptors of the LDL-family, Lrp5,6 (homologs of *arrow* in *Drosophila*) [220, 221]. Lrp6 plays the main role in canonical signaling during early limb development [222], while *Lrp5* function is more important later during osteogenesis [223], but has a partly overlapping function in the limb, as revealed in the context of compound null mutants affecting other Wnt pathway regulators [224]. Non-canonical signaling employs Fz receptors in conjunction with a different co-receptor, Knypek [225]. Additionally, Wnts may also signal by binding certain receptor tyrosine kinase (RTK) orphan receptors (Ror) that also have an extracellular Fz-like domain. Rors may employ some of the same intracellular components as does Fz, and appear to use a non-canonical pathway [226]. Another type of orphan RTK, Ryk, also transduces Wnt signals [227]. *Ror2* is expressed in chondrogenic condensations and growth plates and gene inactivation in mice results in brachydactyly (biphalangeal digits) ands also later cartilage growth defects [228], and mutations in *Ror2* are responsible for the related Robinow syndrome in humans [229, 230]. *Ror1,2* are partly redundant and compound mutants have more severe phenotypes [226]. *Ryk* is involved mainly in late aspects of limb skeletal development; mutants have delayed cartilage growth [231].

Overall, the vertebrate *Fz* genes have broad and complex expression patterns, suggesting that at least partial functional redundancy will likely be found. Further, determinants of Wnt-Fz signaling specificity in vertebrates have not been fully elucidated. One Wnt can bind to any of a number of Fz receptors, although evaluation of affinity among different ligand-receptor pairs has been hampered by the general difficulty in preparing soluble, active Wnt proteins [213]. While certain Wnts tend to transduce through the canonical, and others through one or more non-canonical pathways, signaling is highly context dependent and in some cases seems to depend primarily on which receptors and intracellular components are present [232], while in other cases, different Wnts behave differently in an apparently identical cellular environment (see e.g. [233], discussion in [215], and discussion of Wnt3a and Wnt7a function below, and in Section 15.4.2).

Both canonical and non-canonical signaling through Fz (see Fig. 15.4), like other seven-pass transmembrane receptors, likely involves heterotrimeric G proteins [219, 234, 235] and results in the activation of Dishevelled proteins (Dsh in *Drosophila*, Dvl in mammals), but how the Dvl/Dsh and G protein activities are coordinated is not understood. The three vertebrate *Dvl* genes are broadly expressed and likely have considerable functional overlap (see e.g. [236]). Dvl/Dsh links the Fz receptors to the several divergent downstream pathways (canonical and several non-canonical) through the use of different interaction domains in the Dvl protein that are specific to these pathways, dependent on the receptor/co-receptor complex involved [237–239].

In the canonical pathway (Fig. 15.4), a complex between the docking protein Axin (*Drosophila* Arrow), APC (adenomatous polyposis coli), and Gsk–3β (Glycogen Synthetase Kinase–3β) phosphorylates β-Catenin, which is then degraded via a ubiquitin-proteosome pathway [240–243]. Upon Wnt signaling, activated Dvl/Dsh, through a poorly-understood mechanism, inhibits the activity of this complex, allowing free β-Catenin to translocate to the nucleus and regulate target promoters as part of a complex with the sequence-specific HMG transcription factors Tcf/Lef [244–246]. The Lrp5,6 co-receptors bind to Axin, thus bringing the Axin/ β-Catenin complex to active Dvl in the canonical Fz activation pathway [247]. Since β-Catenin also functions in cytoskeletal complexes with Cadherins, this cytoplasmic association may also indirectly regulate Wnt-responsiveness through competition [248], but remains to be rigorously proven experimentally.

In contrast to the canonical pathway, the non-canonical pathways are less well understood in vertebrates, and which of several alternative non-canonical cascades are employed, particularly in the context of limb development, remain largely unexamined. The canonical pathway culminates in activation of a single transcriptional complex of β-Catenin with Tcf1/Lef1that regulates target genes involved in proliferation and cell fate in a context-dependent manner, and inactivation of these final common mediators unambiguously reveals the requirement of the canonical pathway in particular processes (e.g. see [249–251]). Activity levels of this pathway in specific contexts can also be directly visualized in embryos using transgenic LacZ reporters driven by β-Catenin-responsive cis elements [252]. In contrast, the non-canonical pathways do not lead to activation of a single nuclear mediator/transducer and in fact often regulate complex cell behaviors, such as migration and adhesion, via cytoskeletal changes that do not easily lend themselves to simple reporter assays [238].

The *Drosophila* non-canonical pathway, the PCP (planar cell polarity) pathway, transduces Wnt/Wg signals via Dsh, employing different Dsh domains than used in the canonical pathway, and recruits Dsh to the cell membrane to activate small GTPases, such as Rho, that regulate cytoskeletal changes [253], as well as JNK (June activated kinase) [254]. This pathway also operates similarly during Wnt signaling in vertebrate gastrulation and employs the Formin-homology protein Daam1 to mediate interaction between Dvl and Rho [255–259] (Fig. 15.4). Alternatively, in vertebrates, non-canonical Dvl activation can also stimulate calcium flux and activity of calcium-responsive kinases (Protein kinase C, PKC, and calcium/calmodulin-dependent kinase II CaMKII) [260]. The involvement of Dvl and certain other "PCP" components in the Wnt/calcium pathway suggests this may be a vertebrate branch of the PCP pathway [238]. For example, the same Wnts, Wnt5a and Wnt11, which are proposed to be involved in PCP-like signaling in vertebrates also activate calcium signaling [256, 257, 261, 262].

In addition, there is potential antagonism between the canonical and non-canonical pathways, not only by competing for shared components, but also via cross-regulation, which has been observed in several systems [238]. For example, one target of non-canonical Wnt5a signaling, *Siah2*, acts to increase degradation of β-Catenin [263]; hence *Wnt5a* mutant embryos exhibit increased β-Catenin activity at sites where *Wnt5a* would normally be expressed, such as the distal limb bud.

Recently, yet another non-canonical Wnt signaling route has been demonstrated in vertebrates, in which the Adenyl Cyclase/PKA cascade is employed downstream of Wnts to activate the transcription factor CREB via cAMP binding [264]. This pathway was shown to control myogenesis in somites, but it remains to be seen whether it also plays roles elsewhere, and what components are deployed for transduction. Since PKA regulates Hedgehog signal transduction [243, 265, 266], this new pathway reveals potential avenues for Wnt-Hh cross-talk.

Wnt signaling is also further modulated by different types of secreted and/or cell surface antagonists. Several of these, including Wif, frizzled-related proteins (Frp or Frzb) and Cerberus, bind directly to Wnts to compete with receptor binding [215, 216]. Others, including Dickkopf (Dkk) and Wnt-modulator-in-surface-ectoderm (Wise), block canonical Wnt signaling by interacting with Lrp5,6 co-receptors [216, 267]. Dkk binding, together with another cell surface protein, Kremen, promotes endocytosis of Lrp co-receptors [268]. Both *Frps* and *Dkk* members are expressed in the developing limb [215]. *Dkk1* function has been analyzed both in chick and in mouse, where it has been found to play both a negative role in AER maintenance and a positive role in interdigital apoptosis [269–272]. However, certain Frp and Dkk members can act to potentiate, rather than inhibit, Wnt signaling ([273], see also discussion in [215]), perhaps by increasing ligand concentration in the case of Frps, which are membrane tethered, or by dominant interfering effects and competition with other members in the case of Dkks.

Wnt signaling plays roles in various stages of limb development (Fig. 15.3). A number of different Wnt signals regulate both early and late aspects of limb development including the early cascade leading to limb initiation, AER formation and maintenance, DV polarity of the limb, and several aspects of chondrogenesis as well as joint formation [215, 274]. Compared to the other signaling pathways playing major roles in limb development, the Wnt pathway displays more apparent distinctions among different vertebrates, although overall functions are highly conserved. Some of these roles are discussed in detail in later sections; a brief overview summary is given here.

Canonical Wnt signaling has been implicated in limb initiation in both chick and mouse, operating upstream of mesodermal *Fgf10*. In chick, *Wnt2b* and *Wnt8c*, expressed in future forelimb and hindlimb fields of the lateral plate respectively, are good candidates for inducing *Fgf10*, as is *Wnt2b* in zebrafish pectoral fin field [275, 276]. In mouse, no candidate mesodermal Wnt ligands that might act during limb initiation have as yet been identified, but the double knockout of *Tcf1/Lef1*, which effectively abrogates canonical signal transduction, arrests limb development at a very early bud stage with no evidence of AER formation [249]. Yet, a small limb bud does still form in the mouse *Tcf1/Lef1* mutants and these genes seem to function downstream of Tbox genes (e.g. *Tbx5*) which regulate limb initiation in mouse [277], in contrast to *Wnt2b* in zebrafish, which acts upstream of *Tbx5* [276]. Furthermore, the phenotype following conditional knockout of *β–Catenin* in limb ectoderm is very similar to that of the *Tcf1/Lef1* null mutant [250, 251], suggesting the possibility that Wnt pathway activation may be most important for the ectodermal response to mesodermal signals for limb initiation in mammals. Alternatively, another as yet unidentified Tcf/Lef member could compensate for the loss of *Tcf1/Lef1* function in mesoderm.

In the chick, *Wnt3a* (and also *Wnt10a*) has been implicated to act downstream of *Fgf10* in the limb ectoderm and regulate AER formation through the canonical pathway [233, 275, 278]. In mouse, neither *Wnt3a* nor other candidate *Wnts* are expressed in limb AER progenitors, but the related *Wnt3* is expressed ubiquitously in limb ectoderm and carries out an analogous function in AER formation [250]. The mouse *Wnt3*, *Tcf1/Lef1*, and ectodermal *β–Catenin* mutants all have very similar phenotypes [249–251], indicating that Wnt3 signals to the canonical pathway in the ectoderm and again suggesting that the canonical pathway may function mainly via autocrine ectodermal signaling to regulate AER formation.

Wnts play critical roles in regulating early DV polarity of the limb bud, apparently involving both canonical and non-canonical pathways, although the evidence for involvement of a non-canonical pathway is conflicting and remains to be resolved (see also Section 15.4.2). In both chick and mouse, *Wnt7a* expression is restricted to the dorsal ectoderm, where it regulates dorsal limb fates in the underlying limb mesoderm, reportedly through a non-canonical pathway [233, 279, 280]. In mouse, analyses of *Wnt3*, *Tcf1/Lef1*, and ectodermal *β–Catenin* mutants indicate that the canonical pathway also plays a role in DV polarity, regulating normal ventral ectodermal fate. Consistent with this view, loss of dorsal *Wnt7a* results in a biventral limb phenotype, whereas loss of any one of the canonical components gives a bidorsal limb bud phenotype. Likewise, in the chick enforced expression of *Wnt3a* and of *Wnt7a* results in distinct phenotypes that do not mimic each other, and constitutively active *β–Catenin* misexpression in ectoderm mimics only *Wnt3a* effects. However, mouse *Lrp6* mutations, which should affect only the canonical pathway, instead produce a biventral phenotype opposite to that of the canonical pathway components and similar to *Wnt7a* mutants [220, 222]. Furthermore, at least in cell culture, there is evidence that Wnt7a can activate canonical signaling via a Fz-Lrp6 receptor complex [281]. Possibly both Wnt3/Wnt3a and Wnt7a employ the canonical pathway, but interact very selectively with distinct ectodermal and mesodermal Fz receptors. The phenotypes of compound *Tcf1/Lef1* compared to *Lrp6* null mutants remain difficult to explain. In both cases, despite global gene inactivation in both ectoderm and mesoderm, the limb phenotypes of these cell-autonomous "canonical" pathway components display opposite effects on DV polarity [222, 249], suggesting either some undiscovered Tcf/Lef redundancy or that Lrp6 may act via a very atypical route.

At later stages, both canonical and non-canonical Wnt pathways play roles in condensation, segmentation to produce joints, cartilage proliferation and maturation, and growth plate regulation [215, 274, 282]. At these later stages, non-canonical pathway(s) appear to play a major role in promoting chondrogenesis and regulate the pacing of chondrocyte proliferation and hypertrophy in growth plates (via *Wnt 5a,5b*), while the canonical pathway (via *Wnt4, 14, 16*) promotes joint formation by antagonizing the chondrogenic fate regulator, Sox9 [283–286]. Canonical Wnt signaling represses *Sox9* expression [286] and β-Catenin also potentially inhibits Sox9 functionally, through a direct binding interaction that promotes degradation of both transcription factors in the complex [287]. Conversely *N-cadherin*, which is necessary for cartilage condensation [248, 288], may act partly through competitive antagonism of canonical signaling. *Wnt5a* is also required for formation of condensations in the distal limb bud. Since canonical signaling induces joint formation by reversing and repressing the chondrogenic differentiation program, non-canonical signaling has been proposed to play a role in the condensation process by antagonizing the canonical pathway, which is highly active in the early limb bud. Wnt5a signaling activates at least one target, *Siah2*, involved in β-Catenin degradation [263]. *Wnt5a* is also expressed early in the distal mesenchyme as well as in the ectoderm of the limb bud, and *Wnt5a* null mutants display progressive shortening of more distal skeletal elements [289]. Based on this phenotype, *Wnt5a* was suggested to act by affecting proliferation of distal mesenchyme, but mutant limb buds do also display increased canonical signaling activity, as well as altered proliferation, which could inhibit chondrogenesis. Interestingly, the non-canonical (RTK) *Ror2* receptor mutant has some similarity in phenotypes; however, it is expressed later in cartilage condensations [228]. The timing of non-canonical Wnt signaling that is critical for regulating the condensation process is thus still not entirely defined.

15.2.4
Bmps

Bmps (bone morphogenetic proteins; so named for their bone inducing properties) belong to the TGFβ superfamily of signaling factors that, together with the related GDF (growth-differentiation factor) family members, play a number of different roles throughout early and late phases of limb development [290–292]. A number of different *Bmps* are expressed in the developing limb, including *Bmps 2,4,7* in early limb bud mesoderm and ectoderm [290, 293, 294], and later, together with *Bmps5,6* and *Gdfs5,6,11*, during stages of mesenchymal condensation, cartilage differentiation and skeletal growth [186, 292, 295–297]. Bmps, as well as the secreted antagonists with which they interact, contain cysteine-knot motifs involved in folding and dimerization, and are secreted and bind to their receptors as dimers [298]. Although they formally have the potential to heterodimerize, as reported in other systems, the *in vivo* relevance of this has yet to be established in the limb and remains doubtful [290, 299, 300]. Bmps also bind to HSPGs, but the binding properties and crystal structures of the BMP-receptor complex suggest HSPG binding does not play a direct role in ligand-receptor interaction, as it does for FGFs [149]. In *Drosophila*, genetic studies of HSPG biosynthetic mutants have revealed that the tissue signaling-range of the Bmp homolog Dpp (decapentaplegic) and morphogen gradient formation by Dpp are strongly influenced by ligand interaction with cell surface HSPGs [149]; presumably HSPGs will be found to play similar roles in vertebrate Bmp signaling.

TGFβ signals are transduced by heterodimers of Type I and Type II receptors (Fig. 15.4). In the limb, the high affinity receptors for Bmps and/or Gdfs include the Type I receptors [301, 302], *Bmpr1a* (*Alk3*) which is expressed widely, *Bmpr1b* (*Alk6*) which is expressed more selectively in condensing mesenchyme and later perichondrium [303, 304], and low levels of *ActRI* (*Alk2*) [291], and the Type II receptors *BmpRII* and *ActRII* [305–307], which are also very broadly expressed [291]. Both Type I and II receptors are transmembrane receptor Ser/Thr kinases and, upon ligand binding, activated Type II receptors phosphorylate Type I receptors, which in turn phosphorylate receptor-activated Smads (Smad 1,5, and 8 are activated by Bmps) [308–310]. Smad proteins all contain Mad (for *Drosophila* homolog "mothers against dpp") homology domains MH1,2 which mediate protein-protein and protein-DNA interactions [311, 312]. Receptor-activated phospho-Smads recruit and heterodimerize with a common "partner" Smad (usually Smad4) and translocate to the nucleus to bind to any of a number of transcription factor partners and regulate target gene expression in a context- and tissue-specific manner [312–314]. Nuclear Smads can interact with a large number of different sequence-specific transcription factors, and in addition also regulate transcription by binding general co-activators (such as CBP, p300, MSG) that act as and/or recruit histone acetyltransferases, and co-repressors (such as TGIF or Ski) that recruit histone deacetylases. Specific factor interactions that are relevant in the context of the developing limb have not been extensively analyzed. Smad1 has been reported to bind Gli3 [315], which acts as a major antagonist in transducing Shh signaling (see Section 15.2.5). The potential for nuclear interaction of Bmp and Shh pathways is interesting in light of evidence suggesting that Bmps may cooperate with Shh in AP patterning of digits [316], but the relevance of this observed interaction requires further follow up.

The Bmp signal transduction cascade is modulated in several ways (Fig. 15.4). Inhibitory Smads (Smad 6,7) can compete with receptor interaction and possibly also Smad4 binding of the receptor-specific Smads [312], and *Smad6,7* expression are also regulated as TGFβ targets, providing a negative feedback mechanism [317]. Smurf1binds to and promotes the ubiquitin-mediated degradation of Smad1,5 [318]. Phosphorylation of the Smad MH1–2 linker region by MAPKs (downstream of Fgf signaling) antagonizes the Bmp pathway by interfering with Smad nuclear translocation [196, 198, 319] and serves as a point of cross-talk with Fgf signaling. This may serve as one basis for the frequent antagonism between Bmp and Fgf pathways in several aspects of limb development, but its relevance remains to be evaluated in the context of the limb.

In an alternative signal transduction pathway (Fig. 15.4), Bmp-receptor interaction can also lead to the activation of p38MAPK via a cascade involving the phosphorylation of Tak1 (TGFβ activated kinase 1) by activated BMP receptors, and resulting in downstream activation of cJun by Jnk (Jun NH2–terminal kinase) [320]. This pathway may play a role in Bmp-regulated apoptosis in the limb interdigits and in other sites of programmed cell death [186]. Smad6 has been reported to prevent

Bmp/Tak1–stimulated apoptosis, suggesting that Smads may also function as downstream transducers of the Tak1 pathway as well [199]. In this regard, cJun can also bind to Smads, providing a possible point of cross-talk [312].

Bmps are also highly regulated at the level of ligand availability, which is limited by several secreted antagonists that bind to Bmps directly, preventing receptor interaction [321–323]. In the limb, these include Noggin [324, 325], Chordin [326], and the DAN family members Gremlin and DAN [327–330]. All of these antagonists, like the Bmp ligands that they bind, belong to a superfamily of cysteine knot-containing proteins [298]. *Gremlin* and *DAN* are expressed in the early limb bud mesenchyme with *Gremlin* continuing later in interdigit regions, while *Noggin* and *Chordin* become more highly expressed at the onset of differentiation in early condensations and in prospective joint regions. The roles of *Gremlin* and of *Noggin* have been most clearly delineated from gene inactivation studies in mice as well as misexpression experiments in chick (also discussed further below). *Gremlin* acts as a key intermediate in the Shh-AER/Fgf regulatory loop that maintains the AER and limb outgrowth [327, 329, 331, 332]. *Noggin* plays an essential role in regulating chondrogenesis for joint formation [324].

In addition to their key roles in limb development, Bmp antagonists have served as a useful experimental tool, since they can be misexpressed or applied focally using loaded beads to abolish endogenous Bmp signaling without the potential artifacts of using dominant negative constructs, and consequently have also been instrumental for unraveling the many different context-dependent Bmp functions at different stages in limb development in the chick (see Fig. 15.3). This has served as an important complement to gene inactivation studies in mouse, since some of the Bmps and receptors display partial or considerable redundancy necessitating analysis of compound mutants, while several other Bmp members, along with some receptors and Smad transducers, are essential at early gastrulation stages, requiring the use of conditional null alleles to evaluate their later roles [291].

Bmps2,4,7 are expressed in early limb region ventral ectoderm and underlying mesoderm just prior to limb bud formation. Manipulation of Bmp signaling using Noggin beads or misexpression of constitutively active *Bmprs1a,1b*, reveals a positive role for the Bmp pathway at very early stages in AER formation including the ability of Bmp activity to induce ectopic AERs, as well as a role in the establishment of limb DV polarity by regulating ventral ectoderm fate [294]. Likewise, conditional inactivation of *Bmpr1a* in mice reveals that Bmp signaling to the ventral limb ectoderm is required both to specify ventral ectoderm fate and to promote AER formation from ventral progenitors; ectodermal *Bmpr1a* inactivation caused failure of AER formation and *Fgf8* expression and also defective DV patterning leading to formation of a bidorsal limb bud [293].

After the AER has formed, Bmps negatively regulate AER maintenance and eventually promote AER regression. Thus Noggin treatment after the AER has formed leads to a prolonged and hyperfunctional AER and upregulated *Fgf4* expression in the AER [333]. Shh promotes AER maintenance and maintains *Fgf4* expression by upregulating *Gremlin* expression in the mesoderm and thereby antagonizing Bmps [327–329]. *Gremlin* inactivation in mice results in failure to establish a mature AER and also causes increased apoptosis in limb core mesenchyme, leading to skeletal reduction defects [331, 332]. Gremlin may promote survival both by antagonizing the pro-apoptotic Bmp function in the mesoderm directly (see below), and/or indirectly by antagonizing the negative effect of Bmps on AER maintenance and expression of AER-*Fgfs* that act as survival factors for the mesoderm. *Bmp2* and *Bmp4* knockouts die during gastrulation [291], but recently conditional inactivation of *Bmp4* in early limb field mesoderm has confirmed these roles; early gene inactivation led to defects in AER formation and DV fate specification, while in embryos where the timing of gene inactivation was presumably delayed, prolonged AER maintenance, reduced mesenchymal apoptosis and subsequent polydactyly were observed [334]. Consistent with this, the non-conditional *Bmp4* knockout is also mildly haploinsufficient, and heterozygotes display a low penetrance polydactyly phenotype [335, 336].

At later stages, as condensations form, Bmps also regulate normal cell death in adjacent soft tissues such as interdigit mesenchyme. Misexpression of either a dominant negative *Bmpr1b* [337], or of *Noggin* [328, 333] in chick inhibits apoptosis in regions normally destined to regress and results in soft-tissue overgrowth. Conversely, expression of a constitutively active form of *Bmpr1b*, although not *Bmpr1a*, enhances soft-tissue cell death in both chick and mouse [304, 338], implicating *Bmpr1b* in this process. However, as *Bmpr1b* is not expressed appreciably in interdigit regions, it may be mimicking the action of another receptor in this context and dominant-negative forms may act by simply sequestering pro-apoptotic Bmps [100]. The role of the Bmp pathway in apoptosis in mouse is not readily evident from analysis of a number of mutants (see summary in [291]), being apparently obscured by extensive functional redundancy of Bmps that mediate apoptosis. Indeed, *Noggin* has also been misexpressed in the limb ectoderm of transgenic mouse embryos and results in severe syndactyly as well as some polydactyly due to delayed AER regression, consistent with the role for Bmps in promoting apoptosis [339]. Bmps regulate programmed cell death, at least in part, by activating expression of the Wnt antagonist *Dkk1* that also acts to promote apoptosis in the limb [269, 270].

Work in the chick has pointed to a critical role for Bmp signaling in the formation and differentiation of chondrogenic condensations, as well as interdigital apoptosis (which overlap, again indicating context dependence). This has been demonstrated by manipulating the Bmp pathway both *in vivo* and *in vitro*, in micromass cultures (high density culture of limb mesenchyme cells which closely mimic the events in normal condensation and chondrogenesis *in vivo*). In fact, extensive *Noggin* misexpression in chick causes a complete failure to initiate chondrogenic differentiation and form condensations [340, 341]. Interestingly, transient apoptosis in the limb core regions where condensations are forming was also observed after *Noggin* misexpression, suggesting that Bmps may promote cell survival in chondrocytes. Noggin treatment also interferes with the maintenance of the chondrogenic differentiation state after it has been initiated [341].

Analysis of mouse mutants indicates that Bmp function in chondrogenesis is not uniformly associated with any single ligand or receptor, probably reflecting extensive functional overlaps between multiple Bmps, and possibly also other Tgfβs (see

[325]). *Bmps2–5* and *Bmp7* are all expressed around the periphery of early condensations and future joint segmentations and later in the perichondrium, which regulates rates of cartilage proliferation and maturation [292]. *Bmps6,7* are also expressed in subsets of maturing or proliferating chondrocytes at later growth stages. At late skeletal growth stages, Bmps play roles in growth plate regulation in complex interactions with both Ihh and Fgf signaling pathways [31, 342–344]. *Bmp5* inactivation in mice affects formation of some condensations, but not in the limb [345, 346]. *Bmp6* mutants likewise do not have limb skeletal abnormalities [295]. *Bmp7* mutation causes mainly polydactyly similar to *Bmp4* mutant heterozygotes, with which it synergizes [336, 347, 348]; these phenotypes are probably related more to early effects of Bmp function on AER than on chondrogenesis *per se*. In addition, *Gdfs5–7* are expressed in early condensed cartilage, localized to sites of future segmentation well before joint formation begins, and *Gdf5* and *Gdf6* inactivation affects the formation of particular phalangeal and wrist/ankle joints [297, 349–351]. *Gdf11* is expressed in distal limb mesenchyme and peripherally around condensations and misexpression in chick limb bud inhibits chondrogenesis [352], but mouse mutants do not display skeletal defects in the limb [353].

Manipulating *Bmpr1* levels in the chick also points to the critical role of the Bmp pathway during chondrogenesis. Misexpression of either constitutively active *Bmpr1a* or *Bmpr1b* causes cartilage overgrowth, but only a dominant negative form of *Bmpr1b* is able to completely block chondrogenic differentiation [304, 354], suggesting that *Bmpr1b* normally functions early in formation of condensations, while *Bmpr1a* may function in later steps of chondrogenic differentiation. However, interpretation of these studies is complicated by the fact that the endogenous receptors are also still present and competition or potency of the dominant-negative forms may vary. Analysis of mouse mutants in the *Bmpr1b* receptor suggests a greater degree of functional overlap with other receptors, since several compound null mutants of *Bmpr1b* and other pathway components (*Gdf5*, *Bmp7*, conditional *Bmpr1a* allele) display stronger phenotypes [292, 355, 356]. Moreover, in both chick and mouse, interfering with *Bmpr1b* function causes only selective loss of certain phalanges, rather than a complete absence of cartilage, and some of the affected condensations still form but do not proliferate normally [304, 356].

Gdf5 and *Bmpr1b* null mutants have similar phenotypes, consisting of small biphalangeal digits (brachydactyly) [355, 356], but *Gdf5*, as well as *Gdf6*, are expressed very selectively in prospective future joint regions and have been proposed to regulate joint formation [297, 349–351], whereas the receptor is more broadly expressed in condensations prior to joint formation [304, 354, 356]. Yet, *Gdf5* misexpression in chick or in mouse does not induce ectopic joint formation, but seems to cause cartilage overgrowth instead [351, 357–359]. Alternatively, it has been suggested that Gdf5 may serve as a feedback modulator of joint formation, inhibiting further joint formation too near to an incipient joint and so ensuring proper spacing of joints [285, 286]. In this regard, Gdf5 can bind to Noggin [358] and may possibly function to sequester Noggin, which appears to be required for all normal joint formation *in vivo* [324]. Interestingly, either Noggin treatment [341] or loss of *Bmpr1b* [356] cause increased *Gdf5* expression; this could be interpreted either as reflecting an expansion of joint progenitors and/or conversion of chondrocytes to joint fate, or alternatively as part of a negative feedback mechanism attempting to compensate for the excess joint specification.

Signaling from the different interdigital mesenchyme zones between digit primordia, before becoming apoptotic, has also been shown to differentially regulate the identities of adjacent digital rays at late stages, just before segmentation of the digital condensations to form joints begins [360]. The nature of these signals remains unknown, but the ability to modulate interdigit effects with Bmp antagonists suggests that Bmps may act as at least one of the interdigital signals. *Bmp2,4,7* are all expressed in interdigital mesenchyme and hence are candidates for such signals; perhaps other Bmp members, such as *Gdfs* expressed around future segmentation sites, also play a role in mediating these late "interdigit" effects.

Finally, Bmps (particularly Bmp2) have also been proposed to play a role as a downstream relay signal of Shh in AP patterning of digits to regulate more posterior digit identities, suggesting a conserved regulatory cascade with *Drosophila*, where Dpp acts as a downstream morphogen for Hh [361]. Supporting such a relay mechanism, the identity of duplicated digits induced by Shh treatment in chick is dosage dependent (more posterior digits with a greater number of phalanges require progressively higher dosage) and correlates with levels of *Bmp* induction by Shh [362]. Furthermore, interfering with Bmp signaling shortly after Shh treatment prevents formation of more posterior-type digits that are dependent on high Shh dosage [316]. Also consistently in mouse, transgenic misexpression of a constitutively active form of *BmprIb* in early limb mesoderm causes polydactyly consisting of posterior (triphalangeal) digit types [338]. Yet, in light of the multiple, complex roles played by Bmps, particularly in the formation of condensations and in joint formation, which ultimately contribute to final digit identity (morphology), it may be difficult to distinguish with certainty between a bona fide Shh relay mechanism and more indirect downstream regulation of digit identities by the effect of Bmps on growth and segmentation.

The *Msx1,2* homeobox genes are very often expressed at sites of Bmp activity and have been proposed to function as key direct downstream targets of Bmp signaling [290]. *Msx* genes act as repressors and often play roles in maintenance of an undifferentiated state, particularly during myogenesis [363, 364]. Their expression in distal limb mesenchyme suggests a possible similar role in this context. In misexpression studies, *Msx* genes have also been implicated as acting downstream of Bmp signals to regulate both AER induction and later mesenchymal apoptosis, and thus must likely engage in very context-specific interactions ([294, 337, 365] but see also [339]). However, the genetic inactivation of *Msx2*, alone or in combination with *Msx1*, results in only mild limb phenotypes with some long bone shortening related to later roles in chondrogenesis and growth plate regulation [366, 367]. Autopod phenotypes of compound mutants have not yet been described in detail, but do not reveal an absolute requirement for AER formation or early outgrowth, perhaps due to functional overlap with, or the parallel actions of other regulators in AER formation and maturation (see Sections 15.3.2 and 15.3.3 below).

Another possible direct target of Bmp signaling in ventral ectoderm is the *p53* relative *p63*, which is directly regulated by Smads4,5 in zebrafish fin bud [212]. *p63* is thought to function as an AER maintenance factor in the ectoderm. In zebrafish, the N-deleted dominant repressor isoform of p63 is required to antagonize p53 and promote epithelial proliferation and in p63 null mutants AER maturation and fin bud outgrowth fail [212, 368]. *p63* is highly expressed in mouse AER ectoderm and null mutants have skeletal truncations due to severe AER defects as well as evidence of limb bud dorsalization [208, 209]. Likewise, semi-dominant human mutations in *p63* cause ECC syndromes (ectrodactyly-ectodermal dysplasia) consistent with AER defects [369].

15.2.5
Shh

Much of the Hedgehog (Hh) signaling pathway in vertebrates has been delineated using cues taken from work in *Drosophila*, given the high degree of conservation [370], although a few components of the pathway may be unique to insects or vertebrates. There are three *Hh* genes in amniote vertebrates (teleosts have more), *Desert, Indian* and *Sonic Hedgehog* (*Dhh, Ihh, Shh*) that are functionally largely interchangeable [371] and play different *in vivo* roles mainly by virtue of differing expression domains [130, 372–375] and operating in different developmental contexts [370]. *Dhh* is not expressed in the developing limb, while *Ihh* is expressed during chondrogenesis and plays key roles in cartilage proliferation, maturation and growth plate regulation [376, 377]. *Shh* is expressed in the posterior limb bud mesoderm; misexpression and mutational analyses have unequivocally established *Shh* as the signaling factor responsible for ZPA activity which regulates limb AP polarity. Ectopic *Shh* mimics RA or ZPA effects on digit pattern and like an ectopic ZPA is rapidly induced by RA [130, 374]. *Shh* null mutants both in mouse and chick (*ozd*) lack a detectable ZPA and have limb phenotypes very similar to those seen after ZPA removal [378–381]. *Shh* function is particularly critical for autopod development; null mutants form anterior long bones but lack the ulna/fibula, and form only a single dysmorphic digit of digit1–like morphology (Fig. 15.2). Like ZPA grafts [88], Shh regulates both digit number and pattern in a dose-dependent fashion; low doses specify more anterior, and high doses more posterior digit identities [362, 382]. Many lines of evidence suggest that Shh acts as a true morphogen (dose-dependent responses occur directly via a long-range spatial gradient), particularly with respect to patterning of the neural tube [370], and movement of Shh protein out of the ZPA has been directly demonstrated in the limb bud by immunodetection [382]. However, other models, such as relays, have not been definitively disproven.

Hedgehog ligands undergo lipid modifications that affect their signaling range [216, 383], surprisingly promoting their long-range action. Auto-catalytic cleavage of Hh protein in conjunction with nucleophilic attack by cholesterol generates an active N-terminal peptide that is cholesterol-modified at its C-terminus [384]. Hh proteins are also palmitoylated near their N-terminus by the "skinny hedgehog" acyltransferase (*Ski* in *drosophila, Skn* in mouse, [385]). The unmodified Hh N-peptide is also active, but with lower potency and shorter signaling range, at least in animal tissues [382, 386, 387]. In vertebrate embryos, both of these lipid modifications appear to promote higher level bio-activity and enhanced solubility in culture [386, 387] and also facilitate long range signaling *in vivo* [382, 387], possibly via formation of lipid rafts or micelle-like structures that present focal high ligand concentrations to the cell surface. In mouse, genetic removal of *Skn* causes phenotypes consistent with impairment of both Shh and Ihh signaling range, and engineered *Shh* mutations that prevent cholesterol addition or palmitoylation likewise decrease the effective range of Shh activity [382, 387]. Loss of lipid modification causes failure to specify digits 2 and 3 (Fig. 15.2), revealing a requirement for lipid-modified Shh in long-range signaling. Based on lineage analysis of Shh-producing ZPA cells, digits 2 and 3 depend on paracrine Shh, whereas digits 4,5 arise entirely from ZPA cells through autocrine Shh signaling [47], and formation of digit 1 is Shh independent [378, 379, 382]. These results are compatible with Shh acting as a direct long-range morphogen in digit patterning. Alternatively, a "temporal" model of varying duration of exposure to short-range Shh signals followed by proliferation and exit from the ZPA has also been proposed [47], while other work has also suggested a relay mechanism involving downstream Bmp signals [316]; (see also Section 15.4.3).

The long-range action of the lipid-modified Hh proteins is also selectively regulated by the transmembrane protein Dispatched (Disp), which is functionally required only in the Hh secreting cells [388–390]. Genetic inactivation of mouse *Disp1* only in Shh-producing cells gives phenotypes similar to a global *Disp1* knockout and is partially rescued by non-lipid modified Shh N-peptide. However, *Disp1* may not act by simply facilitating release of the lipid-modified Hh proteins from cells [391], perhaps also by promoting formation of lipid rafts in some manner. HSPGs also play roles in regulating the range of Hh signaling. *Drosophila tout velu* (*Ext* genes, for exostosis, in mammals) is involved in heparan sulfate polymerization and is necessary for long range Hh signaling [392], perhaps by promoting ligand transport or stability [149, 370]. However, vertebrate *Ext* mutants may function more to restrict the range of Hh signaling, at least in some contexts, as demonstrated for Ihh signaling in growth plates [393]. Other transmembrane proteins, such as Hip (Hedgehog interacting protein), that bind Hh ligands specifically and function to sequester ligands, also play important roles in modulating Hh signals; but *Hip* is not essential in the limb [394, 395].

Hh ligands all signal by binding to Patched (Ptc) surface receptors (Fig. 15.4) that, along with Disp, are 12–pass transmembrane proteins having sterol sensor domains and structural similarities to bacterial membrane transporter proteins [383, 391, 396]. Unlike Disp, the Ptc receptor functions in responding cells and plays a negative role in the Hh pathway that is relieved by ligand binding; thus high levels of surface Ptc also sequester and restrict the activity of Hh proteins [386, 397–399]. Consequently, *Ptc* mutant mouse embryos display effects of excess Hh signal transduction such as polydactyly [398]. *Ptc* expression is also induced as a target of Hh signaling, providing negative feedback regulation [400]. Hh ligand binding to Ptc relieves the tonic inhibition of Smoothened (Smo), a seven-pass transmembrane protein related to G-protein-coupled receptors, thereby enabling signal transduction by Smo [370]. Ptc and Smo do not interact physically and the inhibitory effect of Ptc on Smo function probably occurs indirectly and not at the cell surface, possibly via a catalytic effect on trafficking [401–403]. Through a mechanism not yet fully understood, Hh ligand binding promotes the internalization and degradation of Ptc, enabling the recruitment of Smo away from vesicular compartments to the cell membrane [404]. In *Drosophila*, phosphorylation of Smo by PKA positively regulates Smo stability and membrane recruitment [266, 405]. It is not yet clear whether phosphorylation plays a similar role in activating vertebrate Smo, but consensus PKA sites are not conserved. Hh proteins function in early neural development and left/right asymmetry and the *Smo* null mutant is a very early embryonic lethal, prior to organogenesis stages [406]. Indeed, because the Hh pathway plays so many crucial roles is early development, evaluation of the roles of many of the pathway components and modulators in limb development will require analysis of conditional knockout alleles.

In *Drosophila*, activation of Smo prevents the phosphorylation and cleavage of the zinc-finger transcription factor Ci (Cubitus Interruptus) into a repressor form, enables the nuclear translocation of full-length Ci activator form, and also enhances Ci activator function via some modification [370]. Vertebrates have three *Ci* homologs, *Gli1–3* (for glioma expression), that have diverged to carry out different *Ci* functions, and each of which play roles of varying importance in different contexts. Gli1 is not regulated by cleavage in the absence of Hh signals, but functions more as a dedicated activator and *Gli1* is a target, rather than a transducer, of Hh signals [407–411]. As elsewhere, *Gli1* expression is induced by Shh in the limb, but is functionally dispensable for limb development [412, 413]. Gli2 and Gli3 are both regulated by cleavage to generate repressor forms in the absence of Hh signals and in many contexts Gli3 functions as the major nuclear transducer of Hh signals [407–409, 414, 415]. In the developing limb, *Gli3* function is most critical, while *Gli2* plays a very modest and partly redundant role [416–419]. In contrast to some other developmental contexts, Gli3 repressor (Gli3R) and its relative levels at different positions along the limb bud AP axis plays the main role downstream of Shh, while activator (Gli3A) function may be dispensable. Genetic removal of *Gli3* causes polydactyly, whereas *Shh* inactivation results in the loss of most of the autopod with a single dysmorphic digit remaining. Inactivation of both *Gli3* and *Shh* rescues autopod formation (Fig. 15.2), indicating that *Gli3* functions downstream of *Shh* and that the major role of *Shh* in enabling digit formation is to prevent the formation of Gli3R [418, 419]. However, both digit number and morphologies are abnormal in the absence of *Gli3*; constraint to form 5 digits is lost and the digits are dysmorphic, often appearing biphalangeal, and lacking distinct identities [418, 420]. Since Gli3A is also absent in *Gli3* null mutants, it has been proposed that Gli3A is necessary for specification of normal digit identities and pattern. Alternatively, the anterior-to-posterior Gli3R gradient generated by Shh activity in the limb bud has been proposed to regulate digit pattern, a model that may also explain human mutations expected to generate a constitutive Gli3R form, yet causing polydactyly [46, 415]. Genetic marking in mouse embryos for cells that have responded to Shh signals (an assay for Gli-activator activity), in wild-type and several Gli-mutant contexts, showed that digit identity phenotypes did not correlate in any consistent fashion with either Gli2A and/or Gli3A levels, but did correlate with the Gli3R gradient along the limb AP axis [46]. The dependence of normal limb patterning primarily on graded levels of Gli3R also explains the sometimes paradoxical phenotypes of Shh pathway mutants in the limb as compared to other organ systems where Gli activator forms are also critical (e.g. [421]; see also discussion of IFT mutants below). Furthermore, Gli3 can bind at least several other transcription factors, which may modify its function further and impact on target gene regulation by Gli3R gradients [315, 420, 422]. In the limb bud, stoichiometric interaction of Gli3R with Hoxd proteins that also regulate digit identity may convert Gli3R to an activator as total Gli3R levels decline relative to Hoxd levels posteriorly.

Much of the intracellular signal transduction cascade operating between Smo activity at the cell membrane and nuclear Ci/Gli levels serves to regulate cleavage and also nuclear translocation of Ci/GliA forms (Fig. 15.4). In the absence of Hh signals, Ci/GliA is sequestered on cytoplasmic microtubules with Costal2 (Cos2) and other cytoskeletal docking proteins that recruit several kinases and promote cleavage to the repressor form [423, 424], which is released and translocates to the nucleus to repress Hh target promoters. Proteolysis requires sequential phosphorylation by CkI, Pka, and Gsk3β, followed by Slimb recognition of phosphorylated Ci/GliA and recruitment of proteases [243, 265, 370]. Although most of this regulatory cascade has been demonstrated in *Drosophila*, Pka regulation of Gli3 processing has been shown in the vertebrate limb [415], and vertebrate Cos2 homologs have recently been identified and functionally characterized in zebrafish [425]. Suppressor of fused (SuFu) is recruited to Ci/Cos2 complexes upon low level Hh signaling, which releases the complex from micotubules and suppresses the proteolysis of Ci, but bound SuFu also sequesters Ci in the cytoplasm. Maximal Hh signaling activates the Fused kinase which somehow promotes the release of Ci to allow nuclear translocation [370]. Unlike *Drosophila*, where SuFu and Fused are non-essential and seem to play an accessory, modulatory role, inactivation of the mouse *SuFu* causes early embryonic lethality with phenotypes indicating Hh pathway hyperactivity, consistent with a more central role for the vertebrate homolog [426] and a role in limiting nuclear import of Gli activator forms. Conversely, a vertebrate Fused homolog appears to stimulate nuclear import of Gli2 [427].

Several other recently identified factors, such as Iguana in zebrafish, and Rab23, FKBP8, and several intraflagellar transport proteins (IFTs) in mouse, appear to function uniquely in vertebrates to regulate Hh signal transduction, and all appear to play roles in regulating the relative balance between GliA and GliR levels. The zebrafish Iguana zinc-finger protein appears to regulate nuclear translocation of Gli proteins and *Iguana* mutants display a low constitutive level of Hh signaling that is refractory to ligand stimulation [428, 429]. The *Rab23* gene, which appears to play a role in vesicle trafficking, functions as a negative regulator of the Hh pathway [430], and hypomorphic *Rab23* mutants (*open brain* mutation) display polydactyly consistent with excess Hh activity in the limb [431]. FKBP8 is associated with regulatory subunits of Pka and loss-of-function mutants display phenotypes indicating excessive Hh signal transduction [432]. Another class of proteins uniquely involved in Hh signal transduction in vertebrates include the intraflagellar transport proteins (IFTs) that have been identified through genetic analyses of several mouse mutants, and IFTs function genetically downstream of trafficking factors such as *Rab23*, but upstream of the transducing *Gli* transcription factors [433]. IFT proteins are all part of the transport motors that control anterograde (kinesin) and retrograde (dynein) flow in non-motile cilia. These cilia, which are widespread throughout many tissues [434], are either lost or abnormal in the mutants. But it is not yet clear whether function of IFTs in Hh signal transduction is related to the cilia *per se*, or in some other way to their cargo transport roles. Interestingly, although genetic analyses reveal an essential role for IFTs in Hh signaling in the neural tube, mutants in one of the IFTs, (*polaris* or *flexo*), cause polydactyly in the limb [435]. This may indicate a requirement of the Polaris/Flexo IFT protein for both Gli2,3A and Gli2,3R function in some manner; the differing phenotypes in limb and CNS would then reflect the relative importance of Gli activator versus repressor forms in particular developmental contexts.

15.2.6
Other Players

There are of course a number of other signaling factors that have been implicated in the developing limb, for example Insulin-like and Epidermal growth factors (Igfs, Egfs) and Notch pathway in early inductive interactions and Eph receptor-Ephrin pairs, as well as certain cell adhesion molecules, in regulating selective cell-cell interactions during the condensation process. The proposed roles of some of these are briefly summarized below and in subsequent sections; they are not covered in any detail in this chapter because they have not been extensively analyzed in relation to the major regulatory interactions governing limb development *in vivo*, and relatively less is known about how they are integrated into these regulatory cascades.

Igf-Igfr signaling employs an RTK pathway and generally regulates proliferation and cell survival (reviewed in [436]). *Igfs* are expressed in both ectoderm and mesoderm of the early limb, but are also broadly expressed in the embryo, and have been found to induce ectopic limb bud formation in the flank and can rescue AER formation in certain chick mutants lacking an AER [437]. How their activity relates to other signals mediating limb initiation and AER induction remains unclear, and Igf pathway-mutant mice show late growth defects in limb skeleton but do not display phenotypes suggesting an early role for Igfs in limb initiation and outgrowth (discussed in [436]), although this might reflect functional redundancy between different ligands and receptors (*Igf1* and *Insulin*; *IgfrI* and *IgfrII*).

Recently Egf-Egfr signaling, which also uses an RTK pathway, has been shown to affect AER formation in chick, which is intriguing since the Egf pathway regulates formation of more distal leg structures in *Drosophila* [438]. *Egfr* misexpression induces ectopic AERs and dorsalizes the limb bud in chick [439], possibly indicating a complementary role to Bmp signaling in regulating dorsal (as opposed to ventral) ectodermal identity during AER induction, but this remains to be determined. Once again, mouse mutants have not as yet supported an essential role for the Egf pathway in limb outgrowth and patterning, perhaps reflecting functional redundancy of both ligands and receptors.

In the Notch pathway (reviewed in [440, 441]), membrane-bound ligand-Notch receptor interactions promote the proteolytic release of the Notch intracellular domain, which translocates to the nucleus to regulate target gene expression, often regulating cell fate and formation of compartments. The Fringe family members are glycosyltransferases that modify and thereby alter the affinity between Notch and its ligands. A few years ago, the discovery that chick *Radical fringe* is expressed in dorsal ectoderm and AER and acts to position AER formation along the edge of its expression border, generated considerable excitement, given the very analagous role of *Fringe* and the Notch pathway in *Drosophila*, in regulating the position of the DV compartment boundary in the wing disc [442, 443]. However, subsequent mutational analysis of the mouse *Radical Fringe* homolog failed to reveal an essential role in the developing limb [444], and a mouse mutant in the Serrate ligand (*Jagged2*) has only subtle AER abnormalities that lead to interdigital soft tissue retention, syndactyly [445, 446], raising speculation about whether this pathway curiously only plays a major role in AER formation in some vertebrates. The role of Notch pathway in vertebrate limb development remains enigmatic; thus far analyses of additional pathway components are more suggestive of roles during the later differentiation of specific tissues in the limb (e.g. [447]).

Among cell adhesion factors, members of the immunoglobulin supergene family of membrane glycoprotein cell adhesion molecules (CAMs), and the calcium-dependent transmembrane glycoproteins (Cadherins) have been implicated in the pre-cartilaginous condensation process [448]. CAMs and Cadherins each engage in homophilic binding interactions to mediate adhesion between expressing cells. Cadherins interact via cyotplasmic domains with Catenins to modify the cytoskeletal machinery. Both N-cadherin and N-CAM are expressed in the limb in early condensing mesenchyme and appear to be required for initial condensation formation [288, 449]. N-Cadherin binding interactions may also affect gene expression and cartilage fate commitment by sequestering β-Catenin from the nucleus [248] thereby directly preventing canonical Wnt target gene expression, and also the inhibitory effects of β-Catenin on *Sox9* expression and function, a master regulator of chondrogenic fate [286, 287].

Eph receptors are RTKs and interact with the membrane-associated, glycosyl-phosphatidylinositol-anchored Ephrin ligands. Eph/Ephrin signaling is bi-directional (there is both forward and reverse signaling, although the mechanisms of Ephrin signal transduction are poorly understood), and interactions of these molecules affect cell-cell behavior mediated by cytoskeletal regulation [450, 451]. In particular, interactions induce cell-cell repulsive movements in various developmental processes and additionally, interactions can also mediate cell adhesion in certain contexts. The Eph family is divided into two subclasses, EphA and EphB. EphA receptors generally bind to Ephrin-A ligands, whereas EphB receptors bind to Ephrin-B transmembrane proteins, although interactions across classes may occur.

Members of both A and B subclasses are expressed in the limb and chick misexpression as well as mouse knockout analyses have indicated their involvement in subdividing forming condensations, regulating the selective sorting of mesenchymal progenitors for different skeletal elements [452–455].

15.3
Limb Initiation and Formation of the Limb "Organizers"

15.3.1
Axial Cues for Limb Initiation and AER Induction

Limb bud formation is initiated at very specific sites in the lateral plate mesoderm along the embryo AP (rostro-caudal) axis [48–51] in response to signals that are relayed from more axial/midline embryonic tissues to the periphery [52–58] (see also Section 15.1.2). A swelling of increased cell density is first induced in the lateral mesoderm, which then induces AER formation in the overlying ectoderm. The capacity to form limbs is initially broadly distributed in both chick and mouse lateral plate [48, 59], becoming progressively restricted to the limb fields. Recent work has led to the identification of several canonical pathway Wnts and Fgfs that may act in a cascade to regulate limb initiation [275], and all of which can induce ectopic limb buds in early chick embryo flank region [60, 159, 275], reflecting the fact that these factors also cross-regulate each other in positive feedback loops (see Fig. 15.5, and discussed below). In chick, *Wnt2b* and *Wnt8c* are expressed in broad rostral and caudal (pre-limb) regions of the early lateral plate mesoderm while *Fgf10* is initially expressed more widely, subsequently becoming restricted to the limb fields coincident with restricted limb-forming capacity. *Wnt2b* or *Wnt8c* misexpression rapidly induce *Fgf10* and it has been proposed that these Wnt signals maintain *Fgf10* expression in the rostral and caudal limb fields [275], but how latent limb-forming capacity becomes restricted and localized precisely to certain sites is still unclear. In mouse, Wnts that might function similarly in lateral plate at these stages have not yet been identified, but *Tcf1/Lef1* compound mutants that interrupt all canonical Wnt signaling cause early arrest of limb buds similar to that seen in *Fgf10* or *Fgfr2* mutants [36, 37, 156, 160, 249]. Fgf10 signaling to ectoderm then induces AER formation by activating and maintaining ectodermal *Wnt3a* (and *Wnt10a*) expression, which in turn induce *Fgf8* expression in AER progenitor cells, and initiate a positive feedback loop in which Fgf8 signals maintain continued *Fgf10* expression in mesoderm [159, 233, 275, 278]. In mouse, mutant analysis has revealed that *Wnt3* carries out the function of *Wnt3a,10a* in chick [250].

Although both Fgf and canonical Wnt signaling can induce limb bud formation, neither pathway may operate at the earliest signaling steps medial to the lateral plate. *Fgf8* is expressed in intermediate (mesonephric) mesoderm (one of the proposed relay sites), but selective *Fgf8* inactivation in this tissue using a conditional knockout has no effect on limb bud formation [41]. Furthermore, limb buds still form, although they become arrested prior to AER formation, in null mutant embryos for any of the relevant factors capable of inducing ectopic limbs, including either *Fgf4,8* (ectodermal), *Fgf10* (mesodermal), *Fgfr2* (transduces ectodermal response to Fgf10) or *Tcf1/Lef1* (transduce canonical Wnt signals) [36, 37, 40, 156, 160, 249]. Hence, all of these factors probably function mainly in the lateral plate mesoderm and/or ectoderm (not as early midline signals), and possibly in redundant or partly parallel cascades.

In contrast, another candidate midline signaling factor, RA, probably acts upstream of this lateral plate mesoderm cascade. Although RA does not induce ectopic limb buds in flank, (probably because it acts at very early stages more medially), RA is clearly critical for limb initiation. RA synthetic inhibitors can block limb bud formation in chick [96] and zebrafish *Raldh2* mutants fail to form a pectoral fin bud [125]. Likewise, *Raldh2* null mutant mouse embryos form no forelimb bud at all [92, 93]. Partial RA rescue of caudal axis truncations in these mutants shows that the hindlimb bud can still form, perhaps because of compensation by overlapping expression of *Raldh3*. Notably, RA administration in mouse embryos prior to gastrulation can induce the development of supernumerary hind limbs [456, 457], consistent with a very early role in limb induction although it remains possible that this represents a form of caudal axis (tail bud) duplication. In mouse, the timing of early events in limb field formation are uncertain, but in chick prospective limb fields can already be detected shortly after gastrulation [82].

RA may specify limb position by regulating expression of the *Hox* clusters in the trunk and may also regulate subsequent limb outgrowth by inducing *Tbox* transcription factor expression (*Tbx5* in forelimb and *Tbx4* in hind limb; see Fig. 15.5 for forelimb cascade). It remains to be seen whether the *Hox* genes are in fact involved in regulating expression of these *Tbox* genes in the early limb field. *Hox* genes are expressed along the trunk rostrocaudal axis in nested overlapping domains that are collinear with their genomic order. Normal limb bud position is related to the expression limits of certain *Hox* genes, both in paraxial mesoderm [49] and in the lateral plate [50]. Formation of both endogenous and ectopic flank limb buds, as well as their absence in snakes, is correlated with specific shifts in the lateral plate expression limits of *Hox9* paralogs [50, 458]. Furthermore, mutations in *Hoxb5*, *Hoxd10* and *Hoxd11* cause shifts in limb position [459, 460]. The relation between *Hox* expression domains and limb position may be difficult to establish definitively, because cooperative effects of multiple *Hox* genes are probably involved. Retinoic acid activity levels are high in early posterior mesoderm and later in the trunk in mouse embryos [140] and RA has been shown to regulate patterns of *Hox* cluster expression both *in vitro* and *in vivo*. Intriguingly, RA activates *Hox* cluster members sequentially *in vitro* in a dose- and/or duration-dependent manner according to their collinear expression order *in vivo* [461–463]. RA administration *in vivo* causes vertebral transformations that coincide with changes in the nested collinear *Hox* expression pattern along the trunk rostrocaudal axis [464].

The T-box transcription factors, *Tbx5* and *Tbx4*, are selectively expressed in presumptive forelimb and hind limb regions, respectively [465] and misexpression of active or dominant negative forms in chick has implicated them as regulators of both

15.3 Limb Initiation and Formation of the Limb "Organizers"

outgrowth and limb type identity [276, 466–468], being both necessary and sufficient to induce specific limb type formation endogenously and ectopically in the flank. Although their role as master regulators of limb identity has recently been called into question based on the ability of *Tbx4* to rescue normal forelimb formation in *Tbx5* mutant mice [469], their essential role in regulating limb initiation and outgrowth has been confirmed by mutational analyses in mouse [38, 277, 469, 470], and in zebrafish where the *Tbx5* ortholog is required for fin bud formation [276, 471]. Although *Tbx4* can rescue limb formation in *Tbx5* mutants, in the normal hind limb *Tbx4* probably also cooperates with the hind limb-specific paired-homoebox genes *Pitx1* and *Pitx2*, to induce limb bud formation since mouse *Tbx4* null mutants form a hindlimb bud that arrests very early [470], whereas compound null mutants of both of the *Pitx1,2* upstream regulators of *Tbx4* have a more complete arrest of limb bud formation [472]. In the forelimb, genetic removal of *Tbx5*, like *Raldh2*, results in a total absence of limb bud formation [38, 92, 277]. The *Raldh2* null mutant fails to express *Tbx5* in the forelimb field [93, 94], implicating RA as an early signal for limb initiation, operating upstream of *Tbx5*. Notably, several early "limb field" markers, including *dHand*, are still expressed in *Tbx5* mutant lateral plate, suggesting *Tbx5* is necessary for limb bud initiation, but not for limb field specification [277]. In contrast, *dHand* expression is not activated in the *Raldh2* mutant, also suggesting that RA operates at an earlier step, but other early limb field markers still need to be evaluated [94].

Tbx5 likely functions upstream of the Wnt/Fgf signaling loops in the lateral plate since *Tbx5* is still expressed in Fgf and Wnt pathway mutants that arrest at early limb bud stages, whereas *Fgf10* and the *Lef1/Tcf1* mediators of canonical Wnt signaling fail to be expressed in the lateral plate of *Tbx5* null mutant embryos [37, 38, 277]. Subsequently, several of these factors clearly cross-regulate each other in positive feedback loops and possible parallel pathways, thereby also complicating the interpretation of some epistasis experiments (see Fig. 15.5). Possibly, some of the epistatic relationships may vary in different organisms as well. In chick, either ectopic Fgf or Wnt activity induce sustained *Tbx5,4* expression in the flank and inhibition of these pathways results in failure to maintain *Tbx* expression; conversely, *Tbx5* or *Tbx4* induce both *Fgf10* and canonical *Wnt* expression, and are also required for their maintenance [276, 468]. Mainly kinetic arguments have been used to order the induction cascade and suggest that Wnts may initially function upstream of *Fgf10* and that subsequently both cooperate to maintain limb outgrowth. In zebrafish, *Wnt2b* is expressed in more medial trunk tissues and seems to act genetically upstream of *Tbx5* based on gene inactivation studies with morpholino oligonucleotides [276]. In mouse, *Tcf1/Lef1* mutant embryos, unlike *Tbx5* mutants, do still initiate but later fail to maintain *Fgf10* expression, placing *Fgf10* upstream of Wnt signaling [277]. Interestingly, *Wnt2b,8c* equivalents have not been identified in mouse and the ectodermal *Wnt3* knockout displays phenotypes quite similar to the *Tcf1/Lef1* knockout [249, 250], raising the possibility that signaling from the ectoderm in mouse may be primarily responsible for Wnt pathway activation in the lateral plate mesoderm. In this case, ectodermal Wnt3 could contribute directly to the Wnt/Fgf regulatory loop that maintains limb outgrowth (rather than indirectly by regulating *Fgf8* expression). With these several caveats in mind, most of the data from applying factors or antagonists, and various manipulations of gene expression in several organisms supports a sequential cascade in which Tbx5 initially operates upstream of both Fgf and canonical Wnt pathways in the lateral plate to ultimately regulate ectodermal Wnt/Fgf signals. These ectodermal signals subsequently maintain expression of the mesodermal components, including *Tbx5*, thereby completing a positive feedback loop (Fig. 15.5). Continued mesodermal signaling maintains, as well as induces, a structural and functional AER at early stages. AER induction and maturation is also regulated together with the establishment of DV polarity in the limb as discussed further below.

15.3.2
Establishment of Early DV Polarity and AER Formation in Limb Bud Ectoderm

The AER forms from movement and compaction of progenitors that arise broadly in the ventral limb ectoderm (Fig. 15.5), and factors that regulate limb DV polarity also regulate AER positioning and morphogenesis along the DV rim of the limb bud. Either mosaic misexpression of regulators of ectodermal DV identity [294, 442, 443, 473], or the ectopic juxtaposition of dorsal and ventral limb field ectoderm by grafting [474], can induce AER formation at the new "boundaries" in chick. Consequently, the establishment of DV polarity and AER formation are normally tightly interlinked, although they can be uncoupled in some instances [440, 441]. The DV polarity of the ectoderm is initially conferred by early mesodermal signals [75, 79], resulting in the formation of discrete ventral and dorsal ectodermal compartments [44]. The homeodomain repressor *Engrailed–1* (*En1*) becomes selectively expressed in the ventral ectoderm where it represses *Wnt7a* expression, thereby restricting *Wnt7a* to the dorsal ectoderm [473, 475]. Misexpression experiments in chick and mutant analyses in mouse have shown that Wnt7a is a major regulator of dorsal fate and signals to the underlying dorsal mesoderm to activate expression of the LIM-homeobox gene *Lmx1b*, which in turn specifies dorsal fates [279, 280, 476, 477]. The *En1* knockout has bidorsal limbs associated with unrestrained *Wnt7a* expression throughout the limb bud, whereas *Wnt7a* mutants have biventral limbs. *En1*, *Wnt7a* compound mutants display a biventral limb phenotype very similar to that of *Wnt7a* mutants alone, indicating that the main role of *En1* in regulating DV polarity is to prevent ventral *Wnt7a* expression [478]. However, a ventral compartment boundary is already established at very early stages in the ectoderm prior to, and independent of, *En1* expression (see [42, 44, 45, 294, 479]).

Both Bmps and canonical Wnts act upstream of *En1* to regulate ventral ectodermal identity and the specification of AER progenitors, and may also be responsible for initial ventral compartment formation [250, 251, 293, 294]. Manipulation of Bmp signaling using Noggin or a constitutively active *Bmpr1* in chick showed that Bmps regulate both DV polarity and AER formation at early stages, and activate expression of *En1*, as well as *Msx1,2* (frequent Bmp targets) [294]. These experiments also demonstrated that ectopic AERs are induced wherever a boundary between Bmp "active" and non-infected ectoderm occurs. In mouse, knockout of either *Wnt3*, *Bmpr1a* or *β–catenin* in the limb ectoderm all disrupt both AER formation and DV polarity, causing loss of ventral ectoderm identity [250, 251, 293]. Rescue of *Bmpr1a* mutants with a constitutively active *β–catenin* knock-in mutant indicates that *Bmpr1a* acts upstream of the canonical Wnt pathway in AER formation in the ventral ectoderm, but may operate in parallel or downstream to regulate DV polarity [251]. Bmp activity may act upstream of Wnts partly by regulating specific *Fz* receptor expression in ectoderm, but activated β-Catenin can also induce ectodermal *Bmp* expression in an apparent feedback loop. Thus the epistatic relations are not entirely clear and *Wnt3* has also been proposed to act upstream of Bmp signaling in AER formation as well as regulating DV polarity [250]. In any case, both ventral Bmp and canonical Wnt signaling, via Wnt3, are required to activate expression of *En1*, which then represses *Wnt7a* expression, thereby restricting it to dorsal ectoderm (Fig. 15.5). Since Wnt3 inactivation and ectodermal *β–Catenin* inactivation have very similar phenotypes [250], in this instance it appears that canonical Wnt signaling operates within the ventral ectoderm in an autocrine manner (through Wnt3 in mouse, or Wnt3a and Wnt10a in chick) to regulate ventral fate, whereas Wnt7a regulates dorsal fate (see also discussion in Section 15.3.1).

Whether Bmp signals arising from mesoderm, or within ectoderm, or both normally bind and activate ectodermal Bmpr1a is not yet established; several *Bmps* are expressed in both sites [293, 294]. A recent conditional knockout of *Bmp4*, removed selectively in limb mesoderm, had modest effects on AER maturation, which was delayed, as well as mild DV patterning defects, implicating mesodermal Bmps in regulating early patterning of limb field ectoderm [334]. Dorsal and ventral limb ectodermal polarity is initially transferred from the underlying somitic and lateral plate mesoderm, respectively, to specify DV polarity in the overlying ectoderm [44, 75] (see Fig. 15.5). Based on the timing of gene inactivation required to obtain different phenotypes in the *Bmpr1a* conditional knockout, it has been proposed that mesodermal Bmps in the ventral lateral plate together with dorsal Bmp antagonists, such as Noggin, in somitic mesoderm constitute the early mesodermal signals that transfer DV polarity to the limb ectoderm [293]. Whether Wnt signals may also play a role in transferring DV polarity from mesoderm to ectoderm, and the epistatic relations between the canonical Wnt pathway and Bmps remain to be clarified.

Both *En1* and *Msx1,2* are downstream targets of Bmp signaling in the ventral ectoderm [293, 294]. *Msx1* misexpression in chick can induce ectopic AER formation at early stages [294], although loss-of-function analyses in mouse have not revealed an essential role for *Msx* genes in AER formation [366, 367]. *En1* regulates both AER positioning and DV polarity and can also induce ectopic AERs due to abnormal maturation, but is not required for AER formation *per se*, since *En1* null mutants still form an AER [442, 443, 475, 480]. Besides *En1* and *Msx* targets, a possible direct target of Bmp signaling in ventral ectoderm is the *p53* relative *p63*, which is regulated downstream of Bmps by Smad4 and Smad5 in the zebrafish fin bud [212]. *p63* regulates both AER formation and DV polarity in the zebrafish fin bud and in the mouse limb bud [208, 209, 212, 368]. Although described as an AER maintenance factor, inactivation of *p63* in fact already affects ventral ectoderm and AER formation at early stages. Whether *p63* is a direct target of Bmp signals in the amniote limb bud remains to be demonstrated.

Relatives of the *Sp1* transcription factor, *Sp8* and *Sp9*, are expressed in pre-AER ectodermal progenitors and also regulate early steps in AER formation and maturation, as revealed by inactivation of *Sp8* in mouse [210], and inactivation of *Sp8* or *Sp9* in the zebrafish pectoral fin bud [211]. Both *Sp8* and *Sp9*, function downstream of Fgf10 signaling to the ectoderm. In zebrafish, *Sp8* is also a target of β-Catenin, but *Sp9*, which has later expression onset, is not. The epistatic relations remain to be established, but *Sp1* family members appear to function downstream of earlier AER initiating signals (e.g. Bmp and Wnt pathways) and also act within the ectodermal progenitors to regulate AER formation. With respect to AER defects, the *Sp8* null mutant has similar but somewhat less severe phenotypes than *p63*. Unlike *p63*, *Sp8* mutants appear to initiate DV polarity normally, as judged by *En1* expression, although this is subsequently not maintained [210]. This suggests that *Sp1* relatives may act downstream of *p63*.

15.3.3
AER Maturation and Maintenance

There are two distinct events involved in normal AER formation: the expression of appropriate *Fgfs*, especially *Fgf8*, which is the most critical for carrying out AER function; and formation of a morphological AER, which occurs through movement and compaction of ventral precursors to form a ridge along the limb bud apex. AER morphogenesis and functional AER integrity are to some extent separable. The mouse *Fgf4,8* compound null mutant initially forms a morphological AER that is not maintained at later stages due to the loss or reduced expression of mesodermal AER maintenance factors regulated by AER-Fgfs [40]. However, at present it remains formally possible that residual expression of other AER-Fgfs (*Fgf 9,17*) preserves AER morphogenesis. In contrast, null mutations affecting several pathways involved in early steps of AER formation, such as the canonical Wnt pathway [249–251], early Bmp signaling [251, 293], or Fgf10 signaling [36, 37, 156, 160, 161] all result in failure to form a mature, morphologically distinct AER. Notably, these signals also need to be sustained over time during AER maturation for AER formation to be maintained stably.

The murine AER forms from a broad zone of *Fgf8*–expressing progenitors in the ventral ectoderm of the prospective limb region, both by movement of pre-AER cells to the DV border and by loss of AER gene expression in remaining ventral cells, to form a sharply demarcated ridge of tall columnar cells during maturation [45, 441]. The chick differs somewhat in that AER progenitors are already more restricted from the onset of *Fgf8* expression [481]. However, the ventral origin of AER progenitors and compaction to form a pseudo-stratified, columnar AER along the limb bud DV apex from flat ectodermal precursors occurs similarly [42, 44], and the movement and compaction process is regulated similarly by *En1* in ventral ectoderm in both organisms [45, 473, 480].

In addition to loss of ventral polarity, the limb buds of *En1* mutants also exhibit abnormal AER morphogenesis [475, 480]. The initial induction and expression of AER markers, particularly *Fgf8*, occurs normally but a mature, compact AER fails to form. Genetic marking experiments have shown that *En1* regulates dorsal movement of AER progenitors and that the limits of *En1* expression appear to set a border determining the position of AER formation and maturation [45]. Loss of this "border" perturbs movements and leads to formation of secondary ventral AERs with subsequent digit duplications in *En1* mutants [480]. Misexpression experiments show that *En1* also affects AER formation in the chick, again apparently setting a border along which the AER forms [442, 443, 473]. In chick, this involves the Notch-pathway modifier, *Radical Fringe*, which is repressed by En1 and hence dorsally restricted; however, there is no evidence for involvement of mouse *Fringe* homologs in this process [444].

In contrast to the effect of *En1* inactivation, AER formation is unperturbed in *Wnt7a* as well as in *Wnt7a,En1* compound null mutants [279, 478]. Thus, the ectopic expression of *Wnt7a* in *En1* mutants is necessary to lead to abnormal AER formation, possibly being mediated by movement of AER progenitors towards new ventral domains of *Wnt7a* expression, which would be absent when *Wnt7a* is also inactivated [45, 480]. The lack of AER defects in single *Wnt7a* mutants would then require redundancy of other signals, which may indeed exist. Genetic removal of the canonical Wnt antagonist *Dkk1* from either *Wnt7a* or from *Wnt7a,En1* compound null mutants reveals AER abnormalities, including ectopic ventral AERs, and suggests that other Wnts can partly compensate for *Wnt7a* function in regulating AER positioning [272].

Once AER formation is initiated, its maintenance requires the ongoing input of several signals involved in AER induction. Positive feedback loops operate between mesoderm and AER-ectoderm via reciprocal mesodermal Fgf10 and ectodermal Wnt3 signaling, which may involve Fgf8 as an intermediate (Fig. 15.5). Wnt3 (Wnt3a,10a in chick) regulates downstream ectodermal *Fgf8* [233, 250, 278] and Fgf8 signaling maintains *Fgf10* expression [156, 159, 275]. Wnt3 may also signal to the mesoderm to regulate *Fgf10* since mesodermal canonical pathway activation also regulates *Fgf10* ([275]; see also Section 15.3.1). Varied timing of gene inactivation in conditional mouse mutants indicates that both continued Wnt3/β-Catenin activity in the ectoderm [250], and continued Fgf10 signaling from the mesoderm [161], are necessary for AER maintenance. *Sp8* and *p63* play downstream roles in maintaining stable AER induction and maturation [208–210]. In addition, several other regulators, such as homeobox genes of the *Dlx* family [482], the *Lhx* (LIM) family [483], and the *Cux* family [484], and the prototype T-box gene *Brachyury* [485], as well as *Gas1* (a membrane protein of uncertain function that maintains distal *Fgf10* expression [486]), act either within the ectoderm or from the mesoderm to regulate AER maturation and/or maintenance, based on gene inactivation or misexpression phenotypes; but these have not yet been extensively characterized with respect to their relationship to other signaling and transcriptional components in the AER regulatory cascade.

By contrast, after the mature AER has formed, Bmps have a negative effect on later AER maintenance, and actually promote AER regression at late stages. Thus late stage Noggin application in the chick leads to AER persistence and extended *Fgf4* expression beyond the normal time of regression [333] and late inactivation of mesodermal *Bmp4* in mouse also prolongs AER function [334]. Gremlin promotes AER maintenance by directly antagonizing Bmps, which also results in upregulation of posterior AER-Fgf expression [327, 329]. Indeed, genetic inactivation of *Gremlin* interferes with normal AER maturation and maintenance after *Fgf8* expression has been initiated, and results in skeletal phenotypes due to severe AER defects [331, 332]. The feedback loop by which Shh and ectodermal Fgfs maintain one another's expression [90, 487] is mediated by mesodermal *Gremlin* downstream of *Shh* [327, 329]. Shh may contribute to the "posterior AER maintenance factor", previously identified embryologically (see [333]), by maintaining *Gremlin* expression and hence posterior AER-*Fgfs4,9,17*. However, mesodermal *Gremlin* is regulated by other factors as well and is expressed prior to *Shh* and along the entire extent of the sub-AER mesoderm, consistent with the much more severe AER phenotypes observed in *Gremlin* as compared to *Shh* null mutants [331, 332, 378, 379]. Recently, restricted upregulation of *Gremlin* expression by paracrine Shh has also been proposed to play a role in the timing of AER regression, because cells descended from the ZPA (*Shh*-expressing zone) cannot upregulate *Gremlin* expression and these descendants expand across the distal limb bud over time [488].

Bmp signaling may act directly on the AER to negatively regulate maintenance and promote regression, since expression of a dominant negative *Bmpr1b* in limb ectoderm causes AER persistence and extended *Fgf4* expression [333]. In addition, Bmps may also act indirectly, by upregulating the expression of the Wnt antagonist *Dkk1* in the distal mesoderm and AER [269, 270]. *Dkk1* misexpression disrupts AER integrity and maintenance, whereas *Dkk1* inactivation prolongs AER persistence and also alters AER compaction to produce widened and sometimes ectopic ventral AERs [269, 271, 272]. These effects may be a consequence of Dkk1 antagonism of Wnt7a, Wnt3 (Wnt3a,10a in chick), or both. AER phenotypes due to loss of *Dkk1* are at least partly rescued when *Wnt7a* is also inactivated [272]. Wnt3 regulates AER formation and maintenance [250], and would also seem a likely target of Dkk1 in modulating AER maintenance, but this remains to be evaluated.

15.3.4
ZPA Formation

Like early limb forming capacity, early ZPA activity is widespread in both chick and mouse lateral plate [59, 81] and may correlate with certain early *Hox* gene expression patterns along the embryo axis, although unlike limb bud induction, the ZPA is not reprogrammed later by Fgf beads; ectopic flank buds have an inverted ZPA [50, 60, 458]. As in the case of limb initiation, several lines of evidence point to an early role for RA signaling from more midline tissues in ZPA induction. RA was the first factor identified that could mimic the effects of ZPA grafts and induce digit duplications in chick [126, 127, 489]. In this context, RA acts by inducing *de novo* *Shh* expression in the anterior limb bud [130], perhaps recapitulating an earlier role. More importantly, early inhibition of RA synthesis or application of RA antagonists can also prevent normal *Shh* induction in the chick limb [95, 96]. Moreover, *Raldh2* mutant mouse embryos fail to express *dHand* in the forelimb field (an upstream factor that regulates *Shh*, as discussed below). Following partial rescue with RA administration, *dHand* expression is restored in *Raldh2* mutants, but is not polarized normally and extends across the limb bud AP axis, presumably because the normal spatial RA distribution is not reconstituted [93, 94].

As mentioned earlier in the context of limb initiation, a number of (par)axially expressed *Hox* genes are regulated by RA both *in vitro* and *in vivo* [461–464]. ZPA induction in the forelimb has been linked to altered *Hoxb8* expression limits in the trunk based on transgenic misexpression studies in mouse [490], and inhibition of *Shh* expression by RA antagonists or metabolic inhibitors is associated with the loss of lateral plate *Hoxb8* expression in the forelimb field [491, 492]. However, if *Hoxb8* plays a role in ZPA induction normally, it must be redundant with other trunk *Hox* gene groups, since genetic inactivation of either *Hoxb8* alone, or of all *Hox8* group paralogs together, has no effect on ZPA formation or limb pattern [493, 494]. More recently, in both loss-of-function and misexpression studies, the basic helix-loop-helix transcription factor *dHand* was shown to be essential for initiation of limb *Shh* expression; furthermore, *dHand* is expressed in a manner correlating with ZPA extent in the early embryo and is regulated by RA [495, 496]. Interestingly, *dHand* also regulates *Tbx5* expression; in zebrafish *dHand* mutants, *Tbx5* expression in the pectoral fin field is poorly maintained [497]. Co-regulation of *Tbx5* and *dHand* by RA also serves to link and thereby coordinate the regulation of limb outgrowth with ZPA formation, in a manner analogous to the linkage of DV axis and AER formation through common upstream regulation by Bmp/Wnt signals.

As the early ZPA activity becomes restricted to the posterior limb field, *Shh* is not the first polarized gene to be expressed selectively in the posterior limb bud, but is induced through a set of regulatory interactions between *dHand* and other upstream factors in the anterior and posterior limb bud (Fig. 15.5). Early antagonism between *dHand* and *Gli3* appears to be critical in setting up polarized gene expression in the limb. *Gli3* acts as both a transducer and antagonist of the Shh pathway [370, 407]; Shh signals prevent formation of Gli3R [415], which represses Shh target genes [407, 408, 414] and also expression of *Shh* itself [498]. Analyses of mouse mutants reveal that *dHand* and *Gli3* antagonize each other functionally in the regulation of downstream target genes and cross-repress each other's expression in the limb [499]. For example, *Gli3* upregulates, whereas *dHand* downregulates, the expression of the paired-homeobox gene, *Alx4*, which itself functions primarily as a repressor of *Shh* expression in the anterior limb bud [500, 501]. These antagonistic interactions then effectively divide the limb bud into posterior (*dHand* expressing) and anterior (*Gli3* expressing) domains and regulate the polarized expression of downstream genes that further regulate restricted *Shh* induction (Fig. 15.5). Subsequently, Shh also maintains *dHand* expression, establishing yet another positive feedback circuit [495, 496]. 5′*Hoxd* genes, although targets of Shh regulation once the ZPA has been established [130], are also initially expressed in a polarized, nested order prior to the onset of *Shh* expression [39], and require *dHand* for their initial induction in zebrafish [497]. Both misexpressed and endogenously expressed *Hoxd* genes have been shown to induce *Shh* and are likely responsible, as a group, for initiating *Shh* expression downstream of *dHand* [34, 502]. When the posterior restriction of endogenous *Hoxd* expression domains is perturbed, expansion of *dHand* and loss of *Gli3* expression also occur [502]. Epistasis is again difficult to assess unequivocally, given the extensive cross-regulation and feedback between different factors, but plausibly, *Hoxd* genes may regulate *dHand* by inducing *Shh* expression and closing the feedback loop, considering that normal *dHand* expression and function precede that of the *Hoxd* genes. Finally, Fgf8 signals from the forming AER also cooperate with axial cues to induce, and later on maintain, *Shh* expression [40, 481, 503] (Fig. 15.5). In the context of ectopic ZPA experiments, it has long been known that proximity to the AER was a crucial factor for ZPA maintenance [87, 88] and AER-*Fgf* inactivation studies have confirmed the essential role of Fgfs in activating and maintaining *Shh* expression [40].

15.4
Function of Organizers in Regulating Pattern and Growth along the Limb Axes

15.4.1
The AER-Fgfs, RA, and the PD Axis

AER removal interrupts further distal limb outgrowth, resulting in truncation of all distal elements not yet specified; both survival and outgrowth are completely rescued by distal Fgf application [164, 165]. While these chick experiments showed that Fgfs are both necessary and sufficient to carry out all AER function, the analysis of engineered mouse mutants has shed considerable new light on how Fgfs act mechanistically to maintain outgrowth. At least four *Fgfs* are selectively expressed in the AER including *Fgf8*, which is expressed earlier than the others throughout AER progenitors [144, 481, 503]. Subsequently *Fgf4, 9,* and *17* are expressed in the posterior AER and their expression is maintained through a positive feedback loop with Shh from the ZPA [89, 90, 169]. Since *Fgf 4* and *Fgf8* null mutants are early embryonic lethals, analyzing their role in AER function has required analysis of conditional knockouts. Analysis of single mouse mutants for all the AER-*Fgfs* as well compound *Fgf4,8* mutants has revealed both partly overlapping and unique roles in recruitment of mesoderm to form a limb bud, in cell survival and expansion of skeletal element progenitor pools, and in initiating and maintaining Shh expression.

Genetic inactivation of either *Fgf4, Fgf9,* or *fgf17* alone causes no AER-related phenotype ([168, 169] and discussion in [40]), calling into question the functional importance of the Fgf-Shh feedback loop. However, the several posterior-AER Fgfs are likely all redundant in this *Shh* maintenance function. In contrast, inactivation of

Fgf8 alone results in delayed onset of *Shh* expression and in the loss of certain skeletal elements, particularly the earliest forming proximal stylopod, but with more distal elements relatively preserved, limited to loss of a single digit [166, 167]. Although *Fgf4* function is dispensable in the AER, *Fgf4* expression becomes upregulated in the *Fgf8* knockout and partly compensates the loss of *Fgf8*, as confirmed by removal of both *Fgf4,8* together [40, 41]. Upregulation of *Fgf4* in *Fgf8* null mutants rescues later forming distal elements and also *Shh* activation; suggesting that these Fgfs behave interchangeably and functional differences reflect different onset of their normal expression. Both the *Fgf8* and the *Fgf4,8* compound knockouts have progressively more severe phenotypes with earlier removal of the conditional alleles (varied in timing of Cre-mediated recombination before or after onset of first *Fgf* expression). Gene inactivation prior to onset of any *Fgf4,8* expression results in complete agenesis of the limb skeleton (except limb girdle) [40, 41]. Under these conditions, *Shh* expression is never initiated either, confirming the essential role of AER-Fgfs in *Shh* regulation. Even when *Fgf4,8* are expressed very transiently due to slightly delayed gene inactivation, *Shh* expression is not maintained, and the resulting limb phenotype also resembles that of the Shh null mutant (Fig. 15.2).

The initial size of the limb bud is notably smaller than normal in *Fgf4,8* compound mutants, despite a lack of any change in proliferation index or cell survival at this stage [40], suggesting a need for early Fgf signals to either alter adhesive behavior or movement of lateral plate mesoderm cells for their recruitment into the initial limb bud swelling. The ability of Fgfs to stimulate cell migration even within the established limb bud lends support to this view [504]. Although a role for cell migration has not been established in chick or mouse limb bud formation, active cell movement into the limb field has been directly demonstrated to give rise to the pectoral fin bud in zebrafish [471]. The role of ectodermal Fgfs in regulating initial limb bud size is important because the initial mesodermal cell mass of the limb bud ultimately impacts on the cell number available to form condensations, and hence on the final size and number of skeletal elements.

Fgf function appears to be unnecessary for the formation of a normal morphological AER structure, which is still present in *Fgf4,8* compound null mutants even after early gene inactivation [40]. However, it remains possible that residual low levels of Fgf from *Fgf9,17* expression are sufficient to form a mature AER structure. Massive cell death in the distal "progress zone" mesenchyme, as seen upon AER removal, is also notably absent in *Fgf4,8* compound mutants, perhaps reflecting *Fgf9,17* activity, or because of some other signal that is preserved along with AER integrity. The massive apoptosis following AER removal in chick [43] has raised questions regarding the existence of the Fgf-Shh feedback loop, since apoptosis by itself would interrupt *Shh* expression. However, the loss of *Shh* expression in the *Fgf4,8* compound mutant, which lacks distal mesoderm apoptosis, confirms the validity of the feedback loop model. There is however a high level of proximal cell death in the compound mutant limb buds, the mechanism of which is unclear (discussed further below). Signaling of AER-Fgfs to the mesoderm may employ predominantly the Akt pathway [185] (see also Section 15.2.2), which would be consistent with their role in maintaining mesoderm survival since Akt pathway activation promotes survival through several mechanisms [180]. However, the remoteness of the observed limb bud apoptosis from the site of likely Fgf action (distal mesoderm) suggests a much more indirect mechanism.

The intermediate limb phenotypes resulting from slightly delayed gene inactivation and transient *Fgf4,8* expression reveal a dual role for Fgfs in limb outgrowth. In addition to determining initial limb bud size, they act as survival factors and promote expansion of progenitor pools for different skeletal elements sequentially over time. Thus, high proximal apoptosis together with a lack of very early Fgf signals is thought to lead to selective loss of the proximal stylopod when *Fgf8* is inactivated; the later upregulation of *Fgf4* rescues formation of subsequent distal elements in the *Fgf8* conditional knockout [166, 167]. In the context of later onset gene inactivation in the *Fgf4,8* compound null mutant, (with transient early activity of both *Fgf4,8*), proximal limb elements form but distal elements are progressively reduced in size, apparently due to rapidly declining Fgf levels at later times [40]. In this case, although many of the distal elements are severely hypoplastic, they are nevertheless still present in a minute form (Fig. 15.2).

These results have led to the proposal that differential timing of Fgf exposure promotes the sequential expansion of different PD progenitor pools that are already specified in the very early limb bud. This model is partly also based on results of recent fate-mapping studies, using fluorescent dyes in chick, that revealed the presence of small groups of progenitor cells for different proximal and distal elements already present as distinct, distinguishable populations subjacent to the AER in the very early distal limb bud [43]. This pre-specification and expansion model differs significantly from, and has prompted a reconsideration of, the "progress zone" model for AER function. The progress zone model is based on the loss of limb elements in a distal to proximal sequence following AER removal at progressively earlier times in the chick and the fact that AER signals are permissive rather than instructive for outgrowth; according to this model uncommitted stem cell-like progenitors in the distal sub-AER become programmed to form more distal elements as they are exposed to AER signals for longer periods of time before exiting this "progress" zone to grow and differentiate [68, 69]. The progress zone model would predict that distal structures should be more sensitive than proximal elements to decreased total duration of AER signals (in the case of delayed *Fgf* expression in the conditional *Fgf8* knockout), and that elements would be absent when not specified, rather than hypoplastic. The progress zone model would also predict cell marking studies to demonstrate cells capable of contributing to all elements at early times, rather than only to specific PD elements. Advocates of the newer "pre-specification" or "allocation" model argue that the AER removal phenotypes were artifactually caused by the high level of associated distal apoptosis occurring with AER excision, which would destroy all of the progenitors in distal mesoderm and thereby produce complete truncation of distal structures, rather than small elements that have failed to undergo expansion.

Some experimental data can be used in support of either model and there is still an ongoing debate regarding the validity of each model (see e.g. discussions in [70–74]). Several facts are clear and will need to be accounted for as these models evolve. First, there are progenitor populations discernable by early fate mapping. Secondly, when the temporal duration of signaling is reduced, the size, rather than the presence/specification of elements is usually affected. On the other hand, although the data support early specification of elements, there is still a high degree of regulation in the limb bud and these populations are not irreversibly determined until much later. Thus, even proximal tissue after distal amputation at early limb stages can reconstitute all the limb elements if re-provided with Fgf and Shh signals [505]. Furthermore, if limb mesoderm is dissociated into single cells, then reaggregated, placed into an ectodermal jacket and re-supplied with Shh or ZPA signals, all the limb elements can re-form in correct PD order and this does not occur simply by sorting of different progenitor cells, but by re-specification [43, 506]. Thus, limb mesoderm is very plastic and can easily be reprogrammed to form different structures even after an established limb bud with mature AER and ZPA has formed.

In both the progress zone and pre-specification models, Fgf signals are viewed as being permissive rather than instructive; in the former, Fgfs are thought to somehow allow the mesoderm to monitor time and become more "distalized" when exposed for a longer period before exiting the distal zone, while in the latter, they serve to regulate proximodistal pattern mainly by sequentially promoting the orderly expansion and growth of different element progenitors that have already been specified at a very early stage through as yet unknown mechanisms. Some evidence for a distalizing role of Fgfs during outgrowth comes from studies examining the relation between Fgfs and RA/Meis, which have been proposed to function as "proximalizing" factors during limb outgrowth [97, 327, 507]. An RA responsive LacZ reporter has demonstrated a PD gradient of RA in the limb bud, which is generated by complementary *Raldh2* expression proximally and *Cyp26b1* expression distally [94, 98]. Functional antagonism between opposing RA and Fgf "gradients" are known to regulate the periodic segmentation of paraxial mesoderm to produce somites [139], and the AP patterning of the neural tube [138]. Furthermore, in these systems cross-regulatory interactions steepen the opposing gradients; Fgfs induce *Cyp26* and repress *Raldh2* expression, while RA represses *Fgf8* expression. RA has also been reported to induce *Mkp3* expression in paraxial mesoderm [139], which, if operational in the limb, could serve to inactivate high pERK that is normally associated with the AER and distal mesoderm [183]. However, this interpretation would not fit well with the proposed model that Akt promotes mesoderm cell survival in the limb by inducing *Mkp3* expression and thus inactivating pERK [185].

In the limb, RA regulates the proximal expression of the tale-homeobox gene *Meis/Meis2* [94, 97], noted for its role in specifying "proximal" segment identity in the *Drosophila* limb (*Hth* homolog), and proposed to function similarly in vertebrates [327, 507]. Misexpression of either *Meis1* or *Meis2* causes reduction or truncation of distal structures and apparent shifts to proximal-type gene expression in the limb. Like RA treatment, *Meis* inhibits expression of a number of distal genes, including 5'*Hoxd* genes, which in turn also repress *Meis2* expression. Meis can also bind directly to Hoxd proteins [508], which may also antagonize normal Hoxd function and provide a possible mechanism for interfering with distal fate specification. Conversely, Fgfs inhibit *Meis* expression in the limb [97], as well as *Raldh2* expression in early limb lateral plate. The limb phenotypes of *Raldh2* and of *Cyp26b1* mutants can also be viewed as consistent with a "proximalizing" role for RA. Interestingly, when the mouse *Raldh2* mutant is rescued with increasing doses of RA, there is a trend for rescue of distal structures before proximal, consistent with a need for higher RA levels to specify proximal fates [93]. *Cyp26b1* gene inactivation, like *Meis* misexpression or RA treatment at limb bud stages, causes reductions of distal limb elements and proximalized gene expression patterns including elevated *Meis* expression [98]. However, *Fgf8* expression is notably preserved and hence the distal skeletal reductions are not merely due to a loss of AER-Fgfs.

On the other hand, the distal hypoplasia phenotypes due to excessive RA or Meis in the developing amniote limb do not really recapitulate the proximalizing role of RA during amphibian limb regeneration, where RA treatment clearly leads to a re-establishment of proximal fates with ensuing outgrowth of a complete set of duplicated PD structures [509]. Alternatively, rather than directly altering PD fate specification, functional antagonism between Fgfs and RA might also act to affect cell survival, with an RA-Fgf balance regulating the amount of sequential expansion of different segment progenitor populations. This view would be more in line with the pre-specification/expansion model. RA can have a pro-apoptotic function, as demonstrated in late interdigital mesenchyme [99], and it is possible that the extensive proximal cell death seen in *Fgf4,8* compound null mutants relates to high unopposed RA activity in the proximal limb; but since Fgfs are thought not to diffuse very far in tissues, their effect would have to be mediated indirectly. Apoptosis does also occur in the chondrogenic core regions in *Cyp26b1* mutant limb buds, but at a much later time than in the *Fgf* mutants, possibly because AER-Fgfs are still preserved in this case [98], and could counteract the high RA levels. Overall, how RA and Fgf-RA antagonism function to affect PD pattern and outgrowth still remains open to interpretation and merits further evaluation.

Finally, it is also worth noting that Fgfs antagonize Bmps to maintain the mesenchyme in an undifferentiated state and prevent premature chondrogenesis from occurring. Excess Fgf signaling at later stages inhibits chondrogenesis, both *in vivo* [503], as well as *in vitro* in micromass cultures [189]. Thus, extinction of *Fgf* expression is also important to allow chondrogenic differentiation to proceed.

15.4.2
Dorsal Ectodermal Wnt7a and the DV Axis

The DV polarity of the developing limb bud is regulated primarily by *Wnt7a* expressed from the dorsal ectoderm (Fig. 15.5). The loss of ventral identity in either Bmp pathway, canonical Wnt pathway, or *En1* mutants are all associated with downstream ectopic *Wnt7a* expression in ventral ectoderm [249–251, 293, 475, 478]. The major function of *Wnt7a* is to specify dorsal fates of mesodermal structures by regulating the expression of the LIM domain-homeobox gene, *Lmx1b*, in the underlying dorsal limb mesoderm [279, 280]. However, there may be other as yet unidentified factors that also regulate polarized *Lmx1b* expression in mesoderm since

Wnt7a maintains Lmx1b expression mainly in the distal limb mesoderm; proximal limb *Lmx1b* expression is preserved in *Wnt7a* null mutant mouse embryos [478]. Also consistent with the existence of other early dorsal regulatory signals, the *Lmx1b* null mutant has similar but more extensive phenotypes than does the *Wnt7a* knockout, which mainly affects DV polarity of distal but not of proximal limb structures [477, 478]. *Lmx1b* function is both necessary and sufficient to specify dorsal limb pattern. Both misexpression analyses in chick [280, 476], and gene inactivation studies in mice [477] reveal global effects on the DV pattern, including extrinsically derived components such as muscle masses and integumental appendages (hair, feathers, nails), as well as the intrinsic skeletal elements and tendons derived from the limb mesenchyme. These structures all become symmetrically bi-dorsal in their morphologic pattern upon *Lmx1b* misexpression, or bi-ventral following *Lmx1b* inactivation. Autosomal dominant *Lmx1b* mutations in humans are responsible for nail-patella syndrome [510, 511], and cause loss of, or defects in, certain dorsal structures including the patella, elbow, and nail beds.

In addition to, and independent of regulating *Lmx1b* expression, Wnt7a also upregulates and maintains *Shh* expression in the ZPA [91], and *Wnt7a* gene inactivation leads to posterior defects in digit pattern associated with the loss of *Shh* expression [279]. This role of Wnt7a is not mediated through regulating *Lmx1b* expression, since *Shh* expression remains normal in *Lmx1b* mutants, as does *Wnt7a* expression, which is not itself subject to feedback regulation by Lmx1b [76, 477]. The involvement of Wnt7a in maintaining *Shh* expression serves to link and coordinate the regulation of DV polarity with AP patterning (Fig. 15.5).

The experimental evidence addressing whether Wnt7a actually functions using a canonical or non-canonical Wnt pathway remains highly conflicted. Several lines of evidence indicate that the canonical Wnt pathway regulates ventral ectodermal identity and that *Wnt7a* must act via a non-canonical pathway to specify dorsal fate (see also Section 15.2.3) including the fact that global misexpression of *Wnt7a* or of *Wnt3a* in the chick limb bud produces different phenotypes [233], and gene inactivation of nuclear mediators of the canonical pathway (*Tcf1/Lef1*) in mouse phenocopies *Wnt3* rather than *Wnt7a* mutants [249]. However, genetic removal of *Lrp6*, which acts as a co-receptor in the canonical pathway [220, 222], results in phenotypes very similar to loss of *Wnt7a*, including both loss of dorsal identities and loss of posterior digits secondary to failure to maintain *Shh* expression [222]. Perhaps Lrp6 can also act in non-canonical pathways in certain contexts, but there is no experimental evidence to date that supports this. *Wnt7a* also interacts genetically with another canonical Wnt component, the antagonist *Dkk1*; phenotypes due to loss of *Dkk1* are at least partly rescued when *Wnt7a* is also inactivated [272]. Furthermore, at least in cell culture, there is evidence that Wnt7a can activate canonical signaling via a Fz-Lrp6 receptor complex [281]. An alternative possibility is that Wnt7a and Wnt3 both act through the canonical pathway but produce distinct, opposite effects because their target cells are spatially distinct; Wnt7a signals primarily to cells in the dorsal mesoderm whereas Wnt3 signals mainly to ventral ectodermal cells [250, 279]. Since these Wnts cannot mimic each other functionally upon broad misexpression, their receptor affinities must also be very different and selective receptor use may also explain how Wnt7a and Wnt3 might act differentially in the canonical pathway. Even so, it remains difficult to reconcile such models with the phenotypes of compound *Tcf1/Lef1* compared to *Lrp6* null mutants; in both cases, despite global gene inactivation in all tissues, the limb phenotypes of these "canonical" pathway components display opposite effects on DV polarity [222, 249].

15.4.3
ZPA-Shh and the AP Axis

A large body of evidence confirms that Shh mediates ZPA function. Ectopic Shh mimics the mirror image duplications and dose-dependency of ZPA grafts or RA treatment, is rapidly induced by RA like ectopic ZPA, *Shh* is expressed in various tissues having ZPA-like activity, and *Shh* null mutants lack detectable ZPA activity [130, 362, 374, 378]. Shh regulates both digit-forming capacity of the mesoderm and AP patterning of digits. Shh treatment in chick promotes the formation of more posterior type digits with increasing dose [316, 362] and genetic modification in mouse reveals that lipid modification of Shh is needed for long-range effects on digit formation and pattern [382, 387]. Conversely, *Shh* null mutants form primarily anterior long bones (e.g. humerus, radius) and a single dysmorphic digit 1–like structure [378, 379]. Thus, *Shh* function is most critical for autopod development, regulating both digit number and identities. How this is accomplished mechanistically is still far from resolved. Regulation of digit number must involve an expansion of the early limb mesoderm, which determines the capacity to form condensations, whereas certain or many aspects of digit patterning to produce distinct identities may entail ongoing and/or later regulation [360, 512]. The phenotype of *Shh,Gli3* compound null mutants highlights these two distinct roles of *Shh*. Gli3 acts as both a transducer and antagonist of the Shh pathway since Shh activity prevents conversion of Gli3A to Gli3R [407, 415]. Removal of Gli3R by *Gli3* inactivation in the compound mutants restores digit-forming capacity, which is in fact de-regulated, and polydactyly ensues [418, 419]. In contrast, the polydactylous digits are not patterned, but are very dysmorphic and uniform in morphology, suggesting a distinct Shh role that depends on some other factor absent in the compound null mutants, variously proposed to be Gli3A, graded levels of Gli3R, or possibly Gli3R-Hoxd complexes [46, 418–420].

Shh is mitogenic in many contexts (for example in various adult organs where deregulated Hh signaling in humans is responsible for cancers, such as skin, CNS, gastrointestinal tract, and prostate [370, 513–516]), but does not seem to be a major feature of Shh action in the limb. Shh may however, act partly as a survival factor. Shh beads in the chick limb bud can rescue anterior necrotic zones and interdigital mesenchyme from apoptosis. This effect is context dependent; Shh also appears to be pro-apoptotic autonomously within the ZPA, which has been suggested to serve as a feedback mechanism that modulates the extent of the ZPA [202]. Increased anterior mesodermal survival may serve as one mechanism by which Shh regulates digit-forming capacity, thereby increasing the total number of condensations that are able to form.

However, at least part of the Shh effect on digit-forming capacity may occur indirectly, through its role in regulating AER-*Fgf* expression via the Shh-Fgf feedback loop first identified in chick [89, 90], but also operative in mouse [40, 41, 169]. *Shh* signals are relayed by *Gremlin* to activate posterior AER-*Fgfs4,9,17*, which in turn signal to the posterior mesoderm to activate *Shh* expression and complete the loop [327, 329, 332]. Shh is not required for the initiation of AER-*Fgf* expression, but is important for its maintenance [89, 90, 378, 379], whereas AER-Fgfs (normally Fgf8) are required to first activate *Shh* expression [40, 166, 481, 503]. Likewise, expression of the Bmp antagonist *Gremlin* is also initiated prior to *Shh*, but subsequently requires Shh for maintenance of high-level expression in the distal mesoderm [329, 418, 419]. Gremlin relays the Shh signal to the AER by antagonizing Bmps, which negatively regulate maintenance of the established AER [333], and *Gremlin* thereby maintains a functional AER and *Fgf* expression, [327, 329]. *Shh* regulation of *Gremlin* was thought to occur indirectly through regulation of *Formin*, a gene first identified through mutants causing a "limb deformity" phenotype (known as *Ld*). However, it was subsequently uncovered that the *Formin* gene is linked to, and co-regulated with *Gremlin*; *Formin* mutants alter a *Gremlin* cis-regulatory element, resulting in a hypomorphic *Gremlin* allele and consequent limb phenotypes with reduced zeugopod and digit number due to AER-Fgf defects [331, 332, 517].

Although upregulation of posterior AER-Fgfs *Fgf4,9,17* by the Shh feedback loop may be dispensable in that inactivation of any one of these *Fgfs* alone does not affect the limb, a more complex feedback regulation involving Shh and probably other factors also sustains later *Fgf8* expression and AER integrity. Gremlin expression is not maintained in *Shh* null mutant limb buds [418, 419] and in fact, AER regression and the loss of *Fgf8* expression are accelerated in *Shh* null mutant embryos and may contribute to the phenotype [378, 379]. This regulation is more complex than the simple Shh-Fgf loop since *Fgf8* declines in response to loss of *Shh* more gradually than the posterior AER-Fgfs, and *Fgf8* is also expressed throughout the AER, not just adjacent to the ZPA. In any case, part of the effect of Shh on outgrowth and digit capacity could likely be related to AER-*Fgf* maintenance through Gremlin and their function as survival factors. However, although extensive cell death is also present in *Shh* null mutant limb mesoderm, the regions of mesodermal cell death seen in *Shh*, *Gremlin* and *Fgf4,8* mutant limb buds are all somewhat different [40, 332, 419].

Shh appears to regulate the timing of AER regression as well as its maintenance, and this also affects the extent of distal limb bud expansion and ultimate digit-forming capacity. Although Shh is required to maintain *Gremlin* expression, *Gremlin* is not expressed appreciably within the ZPA region itself and over time becomes progressively more restricted, allowing AER regression to occur. Misexpression experiments at these later times in chick have revealed that *Shh*-expressing cells and their descendants in the limb bud cannot upregulate *Gremlin* expression autonomously in an autocrine manner; *Gremlin* is activated only in adjacent cells that are not progeny of the ZPA [488]. Thus, at late stages, the Shh-Fgf feedback loop can be re-instated by enforced *Gremlin* misexpression, but not by either Shh or Fgf. Recent fate-mapping studies, using genetic marking in mice, show that Shh-descendants (prior ZPA cells) actually expand during limb bud growth to include an increasing proportion of the distal mesenchyme (see also below) [47]. Since these descendants are themselves refractory to *Gremlin* induction, they have been proposed to generate an insulating zone that eventually disrupts the Shh-Gremlin-Fgf loop, leading to AER regression after a certain degree of autopod expansion has occurred.

Regulation of digit pattern by Shh is less well understood, probably because it is ultimately a complex process entailing regulation and coordination of a number of morphogenetic events (e.g. differential growth, segmentation, cell death to shape morphologies). Since these events are ongoing at later stages when *Shh* expression in the limb has ceased, their regulation by Shh is necessarily at least partly indirect (see also Section 15.7.1). At a molecular level, Shh is known to regulate expression of several downstream targets including *Bmps* that may function as downstream relay signals in patterning [316, 362], and 5′*Hoxd* gene members that are also major downstream effectors of Shh in regulating digit identities [130, 518, 519]. *Hoxd* genes have been implicated in regulating digit pattern and identity from both misexpression and mutant analyses [34, 520–523]. How they accomplish this is still unclear, although several *Hoxd* and *Hoxa* genes appear to cooperate and function in a partly redundant and quantitative, dose-dependent manner, rather than by playing qualitatively different roles in digit patterning [524, 525]. Comparison of the effects of misexpression at different stages in the chick limb and the progression of mouse mutant phenotypes during development, indicate that *Hoxd* genes act at both early and later stages of limb development and regulate both the initial formation of prechondrogenic condensations and their subsequent growth [521, 526, 527]. Very little is as yet known regarding the Hoxd targets that are critical for carrying out these functions. For some of the *Hoxa* gene paralogs with functional overlap, roles in regulating cell adhesion and particularly the expression of *Ephrins-Eph* receptors have been demonstrated [528, 529]; these targets directly affect the condensation process. Whether *Hoxd* genes regulate similar targets remains to be established.

Bmp2 is regulated by Shh and has been proposed to serve as a relay signal through which Shh regulates digit patterning, based on chick experiments with antagonists showing that active Bmp signaling is required to obtain high-dosage Shh effects on digit identity [316]. Furthermore, engineered tethering of Shh does not seem to alter its dosage-dependent effects on the types of digits formed [362]. However, whether lipid anchorage actually prevents all movement through tissues remains unclear, and Bmps can easily be envisioned to cooperate with Shh signals through mechanisms other than a relay. Regardless of whether *Bmps* serve in a relay capacity, they are indeed important *Shh* targets that play essential roles in both the formation and differentiation of condensations, in soft tissue apoptosis, and potentially in the regulation of late aspects of digit identity, all of which may be at least partly under the sphere of influence of Shh signals (see Sections 15.2.4 and 15.2.5 for discussion and references).

But does Shh long-range signaling occur directly, via a morphogen gradient, or through relays? Movement of endogenous lipid-modified Shh protein out of the ZPA has been directly demonstrated in the limb bud by immunodetection [382]. Mutations preventing lipid modification of Shh specifically affect only the formation of digits 2 and 3 (dependent on paracrine signaling, see below) and are compatible with lipid modifications facilitating Shh transport through tissue [382, 387]. In contrast to the lipid-facilitated diffusion and relay models, it was recently proposed that Shh may act as a morphogen through a temporal gradient, whereby digit type is specified based on cumulative Shh dosage integrated over time, rather than a spatial gradient; such a model would also obviate the need for relay signals. This "temporal" model is based on fate mapping results using an inducible Cre knock-in expressed from the *Shh* promoter to genetically mark cells that have expressed *Shh* at particular times [47]. The results showed that, of the Shh-dependent digits (2–5), digits 4 and 5 are derived completely, and digit 3 partly from cells that have previously expressed, and later shut-off *Shh* upon exiting the ZPA at different times. Only digit 2 derives entirely from non-expressing cells and hence depends entirely on paracrine (rather than autocrine) Shh signals for its specification. It must be noted, however, that both in the case of analyses of *Shh* mutants with altered lipid modification and fate mapping experiments, as well as ectopic Shh application, the effects of Shh on digit formation *per se* cannot be separated from effects on digit identity, complicating simple interpretation of how Shh acts over a distance to regulate different identities.

15.5
Coordination of Patterning, Outgrowth and Differentiation: Positive Co-regulation and Feedback Loops Synchronize Output from Limb Organizers, and Antagonism between Signaling Pathways Contributes to Polarity

To form a morphologically normal, proportioned structure, there must be mechanisms of communication between signaling centers that regulate different aspects of growth and patterning along the PD, DV and AP limb axes to ensure synchronized development. Common upstream regulation serves to initially coordinate the induction of these several signaling centers (Fig. 15.6). The early determination of DV polarity and AER induction are both regulated jointly by upstream Bmp and canonical Wnt signals; these regulatory cascades subsequently branch to regulate distinct downstream targets involved in determining the DV polarity of the ectoderm and inducing the formation of the AER [250, 251, 293, 294]. The early PD and AP axes appear to be likewise linked by common upstream regulators; RA synthesized by Raldh2 in early axial tissues regulates both axes by inducing *Tbx5* in the lateral plate to initiate limb budding and cascades for AER induction, and *dHand* in the posterior limb field, to initiate the cascade leading to ZPA formation [92–94]. Later linkage and coordination of signaling along the PD and AP axes is sustained by Fgf-Shh cross-regulation; Fgfs are required to activate *Shh* and establish a positive feedback loop between Shh-Fgf whereby Shh and Fgfs in posterior AER subsequently maintain one anothers' expression [40, 89, 90, 169]. The activity of this ongoing cross-regulation is evident from the phenotypes of AER-*Fgf* and *Shh* loss-of-function mutants, which share certain features: *Fgf* mutants display patterning defects of digits [40, 41, 166–169] and *Shh* mutants also have an outgrowth defect [378, 379]. Signaling along the developing AP and DV axes is also coordinated by the activity of

Wnt7a, regulating both dorsal mesodermal fate and maintenance of *Shh* expression [91, 279]. Epithelial-mesenchymal interaction between AER and subjacent limb mesoderm to maintain both tissues in a functional state and sustain outgrowth occurs via reciprocal positive feedback loops between Wnt/Fgf signals in both tissues [156, 159, 233, 275, 468].

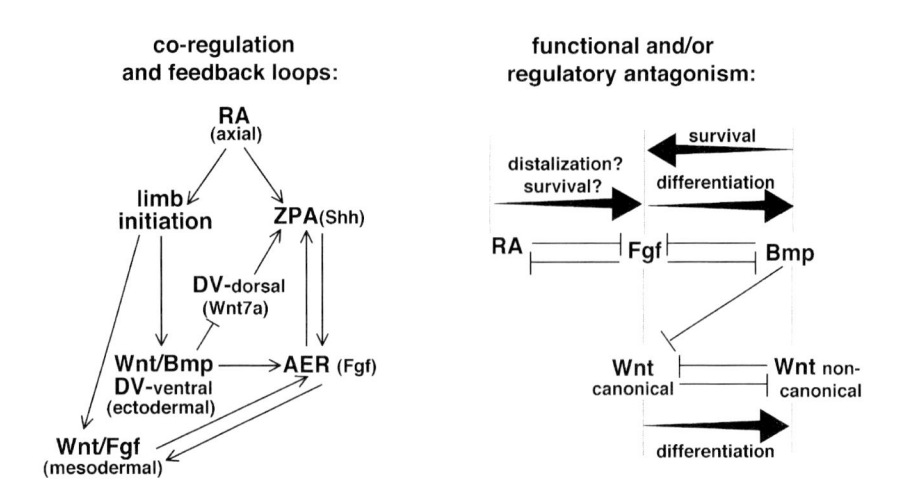

Fig. 15.6
Regulatory interactions coordinate outgrowth, patterning and differentiation in the developing limb. As discussed in the text, co-regulation links the formation of several different signaling centers in the limb and positive feedback loops serve to maintain these links (left panel). Functional and/or regulatory antagonistic interactions (right panel) contribute to polarity in patterning, survival and differentiation status. The regulatory relationships shown are generally not direct, but often involve multiple intermediates and are simplified to highlight the overall cooperative and antagonistic relationships.

Interestingly, while some positive regulatory interactions are sustained, other relationships reverse and mutual regulatory and functional antagonism between signaling pathways also serves to regulate polarity along the limb axes, as well as influencing the pacing of fate specification and/or differentiation status (Fig. 15.6). Thus after limb initiation, RA antagonizes the output of the forming AER; both Fgfs and RA negatively cross-regulate each others' effective levels and also the expression of downstream targets, such as proximal *Meis* and distal *Hox* genes [97–99, 327]. Functional and regulatory antagonism between RA and Fgfs is conserved in several other developmental processes [138, 139]. Following early events in AER formation, Bmps are also functionally antagonistic with Fgfs (the output of the AER); a relationship that is also conserved in other developmental processes [198]. Bmps negatively regulate AER maintenance and positively regulate programmed cell death in certain contexts [333], as well as promoting differentiation, particularly chondrogenesis, and cartilage growth at later stages [292]. In contrast, Fgfs promote cell survival and the maintenance of an undifferentiated, proliferative state in early limb mesoderm [40, 503] and, in the context of the later stage growth plate, negatively regulate cartilage growth [157]. In addition, canonical and non-canonical Wnt pathways often play antagonistic roles and also modulate one anothers' activity, particularly during chondrogenesis, both early in maintaining an undifferentiated state versus promoting differentiation, and later in joint formation versus differentiation [263, 274, 283, 285, 286]. In this regard, canonical Wnt signals and Bmps also behave antagonistically, both in regulating cell survival and in chondrogenic differentiation.

There are also many potential points of cross-talk between the signal transduction pathways, some of which have been highlighted in this chapter, based on work in other systems (Fig. 15.4). As of yet, the function of these cross-talk junctions in the limb remains hypothetical; both the extent to which they are used and whether they contribute to cooperative and/or antagonistic effects between the signaling pathways is largely unknown, but will likely be relevant. To recap, some of these include: inhibitory phosphorylation of Bmp-activated Smads by Fgf-activated ERK that may partly account for Bmp-Fgf antagonism [196]; the common regulation of β-Catenin and Gli2,3 processing by Ck1 and Gsk–3β [243] that links regulation of canonical Wnt and Hh pathways to each other and potentially also to Fgf, via the inhibitory phosphorylation of Gsk–3β by Akt [180]; the involvement of PKA in multiple pathways including transduction of Wnt signals [264], Akt activation downstream of Fgf [180] and regulation of Gli2,3 and possibly of Smo phosphorylation [265, 266]; interaction between Bmp-activated Smads and Gli3 [315] that may relate to cooperative effects between Hh and Bmp signals in digit patterning and in growth plate regulation [316, 342]; activation of Jnk/cJun by both Bmp/Tak1 [320] and non-canonical Wnt pathways [254, 274] and Jun-Smad interaction [312] all of which may perhaps link Bmp and non-canonical Wnt effects during chondrogenesis. Notably, the Wnt and Hh pathways share many similarities suggesting their evolutionary relatedness (including specific lipid modifications, features of transport, signal reception and regulation of transduction by similar phosphorylation-processing events) [216, 243, 530], which also raises the potential for numerous points of co-regulation of these pathways by cross-talk from other signaling pathways.

15.6
Ongoing Late Regulation and Realization of Pattern

15.6.1
Condensation and Segmentation of Skeletal Elements

Although the regulation of skeletal pattern is initiated very early in the limb bud, the final morphogenetic outcome is not realized until much later and involves formation and elongation of mesenchymal condensations, commitment to chondrogenesis and differentiation, and segmentation to produce joints. The steps connecting early global patterning with these late morphogenetic realizations are still poorly understood, but do require ongoing regulation. Subsequent enchondral ossification and bone elongation are similar throughout the skeletal system (see recent reviews [29–31]) and are remote from early "patterning" signals in that overall size may still be modified, but the "identity" of skeletal elements is largely determined. However, experiments in chick indicate that, at least in the autopod, the earlier processes of condensation and segmentation are still subject to ongoing regulation by patterning signals likely set up in the early limb bud; thus, even though digit number is already determined, the final morphologic identity of digits remains plastic until very late [360]. At this stage, when the digit primordia have condensed as rays but have not yet segmented to give distinctive phalangeal patterns, signals from interdigital mesenchyme (or possibly adjacent perichondrial precursors) are able to differentially modify digit identity. "Swaps" between different digital ray condensations, that expose them either to more anterior or more posterior interdigit tissues, leads to either anterior or posterior transformations of identity respectively [360]. Since each digit in the manipulated chick hind limb has a different number of phalanges (from two to five segments) an unequivocal assignment of identity changes is possible. The signals involved in regulating final phalangeal pattern have not been unequivocally determined, but there are several possible candidates: Bmps in the interdigit mesoderm [316, 360]; Fgfs in the overlying ectoderm [512]; and Gdfs [297, 326], Bmp antagonists [324] and canonical Wnts [285, 286] in relation to sites of future segmentation (see also discussion in [10, 531]). Several of these factors may also act in concert, and some may operate upstream and/or downstream of 5'Hoxd and Hoxa13 genes. Several 5'Hoxd genes and Hoxa13 are also expressed in interdigits and later in prospective joint regions and together they regulate both formation of condensations and segmentation to give different phalangeal identities [521, 523–526, 528].

A considerable body of evidence points to an essential role for Bmp signaling in both the formation and differentiation of prechondrogenic mesenchymal condensations. Both genetic removal and misexpression of constitutively active Bmp receptors, or of Bmp antagonists has revealed an essential role for Bmp signaling in formation of condensations and in particular, in regulating expression of the HMG-box transcription factor *Sox9* [292, 304, 324, 340, 341, 354–356] (see also Section 15.2.4). *Sox9* is a master regulator essential for specifying chondrogenic fate and differentiation [532]. Cadherins regulate the cell-cell adhesion required in forming condensations and may function downstream of Bmp signaling [248, 292]. Through their binding to β-Catenin, N-Cadherins may promote chondrogenesis by modulating canonical Wnt activity, which antagonizes Sox9 [248, 285–287] (see also below).

A number of different *Ephrins* and *Eph* receptors (of both A and B subclasses) are expressed in the limb, often in complementary position-specific domains [452–454]. Ephrin-Eph receptor interactions regulate selective cell adhesion by inducing repulsive cell sorting [450, 451]. Both misexpression experiments in chick and mosaic gene inactivation in mouse lead to abnormal condensation patterns in the limb, such as polydactyly, by causing cell sorting and splitting of condensations where abnormal Eph-Ephrin interfaces occur [452–455]. Eph-Ephrin interactions appear to function in parallel with Bmp signaling and perhaps cooperate with Bmps in regulating the condensation process, but the relationship between these pathways has not yet been extensively analyzed. Several *Hox* genes play late roles in regulating the adhesion, size and growth of condensations [521, 526, 528, 529]. For *Hoxa13*, there is evidence that this involves regulation of *EphA7* receptor expression [529], as well as regulation of *Bmp* expression [533] by *Hoxa13*. Interestingly, the paralogs of different *Hox* gene groups function most prominently (but not exclusively) in the formation of particular PD components of the limb (i.e. paralogs of the *Hox9,10* groups in stylopod, of the *Hox11* group in zeugopod, and of the *Hox12,13* groups in autopod) [524, 534, 535]. Perhaps these selective roles of different *Hox* groups in formation of specific condensations may be mediated in part by their regulation of different *Ephs* and *Ephrins* to generate condensation patterns of mesenchymal progenitors for different skeletal elements along the limb PD axis. The evidence is suggestive, but the actual extent to which selective *Eph-Ephrin* expression and interaction of different Eph-Ephrin pairs regulates fine skeletal pattern, particularly the normal splitting of condensations to give rise to the digital pattern in the autopod, remains to be established.

Lengthening of phalanges, especially the distal tip, may also be regulated by late Fgf signaling from the AER overlying digit primordia. Manipulations of late chick limb bud to sustain *Fgf8* expression for longer times can generate additional phalanges through lengthening and segmentation of the most distal element, and it has been suggested that cessation of Fgf8 signaling normally serves to "cap" the digit distal tip [512]. This finding raises the possibility, requiring further investigation, that Shh and downstream mesenchymal signals such as Gremlin, may "pattern" the AER by maintaining differential levels of Fgf activity along the AP extent of the AER, which in turn regulates final phalangeal number and length in the underlying digital ray.

Finally, recent work has also revealed insights into the mechanisms regulating segmentation and joint formation. Misexpression studies in chick and mutant and transgenic analyses in mouse indicate that the canonical Wnt pathway (via several *Wnts4,14,16*) positively regulates joint formation by locally reversing the cartilage differentiation program [285, 286]. These Wnts appear to act by repressing *Sox9* expression to reverse the cartilage differentiation program. In addition, *Sox9* and *β–Catenin* appear to be mutually antagonistic functionally; Sox9 interferes with Tcf/Lef interaction and target gene activation by β-Catenin by competitively binding to β-Catenin, which also results in degradation of Sox9–β-Catenin complexes, and hence decreased Sox9 activity as well [287]. The Bmp family member *Gdf5* and the Bmp antagonist *Noggin* also play essential roles in the segmentation process and genetic removal of either causes defects in joint formation; in the case of *Noggin*, there is a total absence of joints in mutants [324], and in the case of *Gdf5* mutants or inactivation of its receptor *Bmpr1b*, a loss of particular phalangeal segments [349, 355, 356]. However, unlike the canonical Wnts, neither misexpressed *Gdf5* nor *Noggin* is able to induce ectopic joints. Excess Noggin prevents chondrogenesis altogether [340, 341], while excess Gdf5 causes cartilage overgrowth [351, 357–359]. One model proposes that these factors may be regulated downstream of canonical Wnts to create a refractory zone and inhibit nearby joint formation [285, 286]. Wnts repress *Noggin* (and *Chordin*) expression and induce *Gdf5*, which binds to, and could sequester *Noggin* as well. This local modulation of Bmp antagonists would promote chondrogenesis via unrestrained Bmp activity and suppress further joint formation until Noggin activity recovers at a distance from the segmentation site.

However, much remains to be elucidated in understanding why joints form where they do, and how size of different phalangeal segments is regulated. Joint formation is not merely a periodic segmentation process occurring at regular intervals or after a certain amount of growth. In a given species, there is no strict correlation between digit length and the frequency of segments (phalange number) within that digit. Phalange number and length, and hence spacing of segmentation sites, vary dramatically both among different species and within the autopod of a given species [1]. For example, bats have remarkably long wing digits, mostly with only two phalanges each [536], while marine mammals display hyperphalange of their flipper digits [10, 531]. In the chick hindlimb, digit 4 has the greatest number of segments, but is not the longest digit. In the bat, the hindlimb digits are all of identical length but differ in phalange number. Thus, some higher order regulation must be imposed on a periodic tendency to generate joints and segment phalanges, perhaps operating from the interdigit mesenchyme, but still enigmatic.

15.6.2
Apoptosis and Sculpting the Final Limb Form

The limb bud begins as a hemispherically shaped structure that becomes elongated and modeled both through directed elongation of its chondrogenic condensations, through poorly understood mechanical forces [1], and through selective loss of mesenchyme via apoptosis [100]. Loss of excess anterior and posterior soft tissue may be more extensive in the chick wing which has readily evident anterior and posterior necrotic zones, than in mouse (see [43]), but in all vertebrates without webbed feet, the interdigital mesenchyme must be removed by cell death in order to free the digits.

Several pathways regulate apoptosis and may act partly in a cascade, including RA [99, 112, 537], Bmp [337, 339], and Wnt via antagonism by Dkk1 [269, 270]. Involvement of each of these pathways has been demonstrated both by misexpression and gene inactivation analyses that alter their activity. RA can induce expression of *Bmps*, several of which may function redundantly in the interdigit mesenchyme to regulate apoptosis [99, 112, 337, 339], and Bmps in turn induce expression of *Dkk1* [270]. Excess RA antagonizes the function of Bmps in chondrogenic differentiation and inhibits chondrogenesis, both in chick, and in mice lacking *Cyp26b1* [98, 99]. Conversely, RA antagonists alone can induce chondrogenesis in interdigit regions, suggesting RA may normally interfere with the chondrogenic function of Bmps. Finally RA also regulates extracellular proteases that degrade extracellular matrix, as well as caspases, and probably facilitates removal of dead tissue from interdigit regions [112].

Bmps have long been analyzed for pro-apoptotic functions and implicated as major regulators of apoptosis, partly because of the strong correlation between their expression and sites of cell death (including anterior and posterior necrotic zones and interdigital mesenchyme). *Bmp* expression, although complex, is clearly regulated by patterning signals from the organizing centers in the early limb bud. Recent work in chick has demonstrated that Gli3 repressor in the early limb bud is a positive regulator of apoptosis, and acts via inducing *Bmp4* expression anteriorly [538]. Consistent with this, compound heterozygous mutants for *Gli3* and *Bmp4* show genetic interaction with reduced apoptosis resulting in extensive syndactyly [335]. Shh-regulated expression of *Bmps* in posterior limb bud might also contribute to its observed pro-apoptotic action in posterior limb mesoderm [202]. Bmps promote both cell death in interdigit regions and chondrogenesis of digital rays in the autopod, highlighting the fact that their function is clearly context dependent [100, 304, 325, 337, 340, 341]. Misexpressed *Bmpr1b* behaves similarly to Bmps in both roles [304, 337, 354], but since *Bmpr1b* is not expressed appreciably in interdigit regions, it may be mimicking the action of another as yet unidentified receptor in the context of apoptosis [100]. Several mechanisms have been proposed to explain the striking context-dependent behavior of Bmps, including the possible use of distinct receptors as suggested above, the formation of functionally different Bmp heterodimers (which remains largely hypothetical; see [290, 299, 300]), and the presence of distinct intracellular interacting components, for example through the synergistic action of other pathways, such as RA (see above) which also promotes apoptosis and inhibits chondrogenesis. Any or all of these mechanisms may contribute to some extent, but require further analysis to be verified.

A recent study of the role of Bmp5 in interdigital apoptosis revealed involvement of the p38 MAPK pathway [186], known to be pro-apoptotic in other contexts, and which is probably activated via Tak1 downstream of Bmp5 signaling. Bmp5 signals activate expression of both *Msx* target genes, which can act as pro-apoptotic regulators [365], and *Dkk1* (discussed below). However, use of p38MAPK inhibitors suggested that *Dkk1*, but not *Msx2*, induction by Bmp5 employs the Tak1 pathway [186], and hence, both Tak1 and Smad pathways downstream of Bmps may cooperate to mediate pro-apoptotic effects.

The Fgf pathway has surprisingly also been implicated in interdigit cell death [201], notwithstanding its survival role in undifferentiated limb mesoderm [58, 73, 74]. Bmps appear to require Fgfr activity in order to promote apoptosis, based on the use of chemical inhibitors of Fgfr function [201]. Fgf signals can also activate p38MAPK [200]; however unlike their survival effect, a pro-apoptotic effect of Fgfs was very indirect, requiring 24 h after Fgf treatment to become evident. More likely, this Fgf effect results from induction of target genes that act in apoptotic cascades. *Msx2* is regulated by Fgfs and its induction in interdigits has been shown to require both Fgf and Bmp signaling, by combined use of ligands for one pathway with inhibitors of the other [201]. Fgf signaling also upregulates expression of *Dkk1* [270] (discussed below), which likely contributes to the observed cooperative effects between Bmp and Fgf pathways in promoting interdigital cell death. These selective effects of Fgfs on late interdigit mesenchyme may be mediated through *Fgfr3*, which becomes upregulated in interdigital mesenchyme [186].

The canonical Wnt antagonist *Dkk1* is co-regulated by both Fgf and Bmp pathways and plays a key role downstream of Bmp signals in regulating apoptosis [269, 270]. *Dkk1* mutants display polysyndactyly, while misexpression of *Dkk1* promotes excess apoptosis. *Dkk1* expression also depends on cJun activity, which is itself activated by p38MAPK/Jnk signaling [179]. This is again consistent with the observed activation of *Dkk1* by Bmp-Tak1/p38 signaling in interdigits [186]. Smad6 has also been reported to prevent Bmp/Tak1–stimulated apoptosis [199], which may reflect cooperation of both pathways in promoting apoptosis downstream of Bmp signals, or could also be due to cross-talk involving cJun, which can bind to Smads to regulate target promoters [312]. The canonical Wnt pathway targets regulating cell survival that are affected downstream by Dkk1–mediated antagonism remain to be elucidated. Furthermore, much of how these several signaling cascades ultimately funnel to regulate survival effectors, such as *Bcl2* and *Bag1* (which is downregulated immediately preceding apoptotic commitment in interdigit mesenchyme), and to regulate death genes, such as *Apaf1*, *Bax*, and caspases (all of which are upregulated and/or activated during programmed cell death), remains to be unraveled [100].

15.7
Potential Biomedical Applications and Future Directions

The developing limb has served as an excellent system for illuminating the varied biological functions of signal transduction cascades in different contexts as well as for deciphering the regulatory inputs to these pathways. A broad and in-depth understanding of limb development has benefited biomedical disciplines in several ways.

First, much of what has been learned has strong relevance for understanding the basis of genetic diseases affecting the skeleton, as well as the pathogenesis of some skeletal degenerative diseases, and also figures prominently in the pathogenesis of many cancers. Although, in general, mutations in the key signaling pathways would be expected to cause early embryonic lethality when homozygous, hypomorphic, activating, or dominant-interfering mutations are often associated with genetic disease syndromes that affect the limb as well as other developing organs. Mutations in many signaling components and their associated transcriptional regulators surprisingly often result in haploinsufficiency phenotypes as well. For example, mutations affecting the Hh pathway are responsible for several polydactyly syndromes (Grieg's Cephalopolysyndactyly, Postaxial polydactylys [416, 539]), and Gorlin's (Basal cell-nevus) syndrome [513]. Wnt pathway mutations have been implicated in nail-patella syndrome [510, 511], Robinow (brachydactyly) syndrome [229, 230], as well as an osteoporosis syndrome [223]. Abnormal Wnt activity also plays a pathogenetic role in degenerative inflammatory joint diseases (see discussion in [215, 274]). Bmp pathway mutations are responsible for several skeletal developmental disorders including chondrodysplasias and brachydactyly syndromes [359, 540, 541] and sclerosteosis (a Bmp antagonist mutation causing bone overgrowth) [542, 543]. Fgf receptor mutations are involved in a number of different chondrodysplastic syndromes [143]. Although genetic diseases related to the RA pathway are less evident, maternal vitamin A deficiency or exposure to teratogenic (elevated) RA levels *in utero* have long been known to be responsible for a broad range of developmental syndromes [119]. Furthermore, several of the pathways, particularly Hh, Wnt, and Fgf promote cell proliferation and survival, and these pathways have been strongly implicated in neoplasia as well. In addition to Gorlin's syndrome, in which excess Hh signaling due to Ptc mutations causes basal cell tumors of skin, as well as medulloblastomas and rhabdomyosarcomas [513], unrestrained Hh signaling has recently been demonstrated as a major causative factor in the genesis of a number of adult sporadic cancers of the GI tract (especially esophageal, gastric, pancreatic) and prostate [514–516]. Canonical Wnt pathway mutations leading to unrestrained activity have also been identified in certain GI tumor types, the classic example being APC in colon cancers [241, 544]. Fgfs have been implicated in adult CNS gliomas as well as other tumors [545].

Our knowledge of these signaling pathways and the developmental consequences of their alteration have also provided numerous insights regarding chemical/environmental teratogenesis, which also finds utility in intelligent drug discovery that may have therapeutic potential for some of the signaling pathway diseases briefly summarized above. A striking example of this is the discovery of the mechanism of action of cyclopamine, a natural plant product teratogen which causes holoprosencephaly and also limb abnormalities by interfering with Hh signal reception [546, 547]. This compound may find use as an anti-tumor therapeutic agent for certain of the GI cancers that involve excess Hh (Ihh) production, and is currently under evaluation in clinical trials [514]. It is not difficult to imagine that a more systematic evaluation of compounds with highly distinct actions that phenocopy the developmental effects of signaling pathway mutations will lead to the identification of additional drugs with efficacy in various diseases related to these pathways, and could even serve as a large-scale screening tool.

Another potential biomedical application holding promise for the future is the use of tissue bioengineering approaches to repair tissue damage from degenerative inflammatory and traumatic causes. Bmps were named based on their surprising (at the time of discovery) ability to induce and recapitulate the entire sequence of bone formation from condensation through enchondral ossification when applied to adult soft tissue *in vivo*, and this property first raised the exciting possibility of developing tissue bioengineering as a practical application [548]. With what has been learned since then about the activation and modulation of chondrogenesis and osteogenesis by other signaling pathways, especially Wnts (discussed in [274]), and their interactions with Bmp signaling, efforts are underway to develop practical methods for stimulating tissue re-growth and differentiation to repair degenerated and diseased joint structures [549, 550].

Urodele (tailed) amphibians retain the ability to regenerate an entire amputated limb as adults through a process of tissue de-differentiation that creates a pluripotent stump blastema which undergoes outgrowth and patterning in response to signaling that in many respects, though not all, recapitulates the limb development process [4]. Notably, the study of amphibian limb regeneration, which is now providing many new insights into mechanisms of de-differentiation and stem cell pluripotency [551], has also relied substantially on an understanding of the roles of different signaling pathways gleaned from work on the developing amniote embryonic limb, which is in many respects much more amenable to molecular genetic analyses. Mammals have largely lost the ability to regenerate limb structures, although they retain a very limited capability for re-growth of the distal digit tip [552]. Recent work suggests some similarity to events in amphibian regeneration; both appear to involve differentiation reversal regulated by *Msx1* [553]. *Msx* genes play a role in inhibiting differentiation in several systems and particularly in myogenic progenitors [363, 364], partly through increased *Cyclin D1* expression. Together, these results suggest a potential role for *Msx* genes in the maintenance of stem cell populations. Although it is questionable that it will ever be possible to promote extensive limb regeneration in mammals (because of inherent limitations of the system, which would also require the coordinated regeneration of extrinsic tissues such as nerves, muscles and vessels), the studies of regeneration in both amphibian and mammalian systems promise to shed new light on regulatory pathways that will likely find adaptation in tissue bioengineering efforts and applications of stem cell research as well.

Finally, although much has been learned about the signaling networks that are set up very early to regulate pattern formation, the final morphogenetic outcome is not realized until considerably later and also requires ongoing regulation. The steps connecting the early global patterning with the late morphogenetic realization of the limb skeleton are as yet poorly understood. But a clear understanding of these steps will feature importantly in developing bioengineering tools to re-program tissues and in novel approaches for the treatment of diseases associated with signaling pathway abnormalities, both of which will ultimately require the ability to (re)create appropriately differentiated tissues. Furthermore, the rich cross-regulatory and homeostatic mechanisms in play in the developing limb are clearly too complex for intuitive predictions based on simple linear models to adequately anticipate the downstream biological consequences of perturbing the process by altering one or another of the regulatory components. Thus, a major challenge for the future, both to fully understand the regulatory landscape generated by signaling interactions and for the design of practical biomedical applications, will be to develop robust models for the processes governing limb development that have strong predictive value. To achieve these goals, it will be essential to adapt quantitative approaches from Systems Biology to develop strong predictive paradigms; the ongoing intense focus and analysis of mechanism in the context of limb development makes this a promising area for employing "Systems" approaches very fruitfully in the future.

References

1. Hinchliffe, J.R. and Johnson, D. R. *The Development of the Vertebrate Limb: An Approach through Experiment, Genetics and Evolution.* Clarendon Press: Oxford, **1980**.
2. Tickle, C. The contribution of chicken embryology to the understanding of vertebrate limb development. *Mech Dev* **2004**; *121*: 1019–1029.
3. Stern, C.D. The chick: A great model system becomes even greater. *Dev Cell* **2005**; *8*: 9–17.
4. Gardiner, D.M., Endo, T., and Bryant, S.V. The molecular basis of amphibian limb regeneration: integrating the old with the new. *Sem Cell Dev Biol* **2002**; *13*: 345–352.
5. Grandel, H., and Schulte-Merker, S. The development of the paired fins in the Zebrafish (*Danio rerio*). *Mech Dev* **1998**; *79*: 99–120.
6. Coates, M.I. The origin of vertebrate limbs. *Development* **1994**; Supplement 169–180.
7. Shubin, N., Tabin, C., and Carroll, S. Fossils, genes and the evolution of animal limbs. *Nature* **1997**; *388*: 639–648.
8. Coates, M.I., and Cohn, M.J. Vertebrate axial and appendicular patterning: The early development of paired appendages. *Am Zoologist* **1999**; *39*: 676–685.
9. Shubin, N.H. Origin of evolutionary novelty: Examples from limbs. *J Morphol* **2002**; *252*: 15–28.
10. Richardson, M.K., Jeffery, J.E., and Tabin, C.J. Proximodistal patterning of the limb: insights from evolutionary morphology. *Evol Dev* **2004**; *6*: 1–5.
11. Johnson, D.R. *The Genetics of the Skeleton.* Clarendon Press: Oxford, **1986**.
12. Haffter, P., Granato, M., Brand, M., Mullins, M.C., Hammerschmidt, M., Kane, D.A., Odenthal, J., vanEeden, F.J.M., Jiang, Y.J., Heisenberg, C.P., Kelsh, R.N., FurutaniSeiki, M., Vogelsang, E., Beuchle, D., Schach, U., Fabian, C., and Nusslein-Volhard, C. The identification of genes with unique and essential functions in the development of the zebrafish, *Danio rerio*. *Development* **1996**; *123*: 1–36.
13. vanEeden, F.J.M., Granato, M., Schach, U., Brand, M., FurutaniSeiki, M., Haffter, P., Hammerschmidt, M., Heisenberg, C.P., Jiang, Y.J., Kane, D.A., Kelsh, R.N., Mullins, M.C., Odenthal, J., Warga, R.M., and NussleinVolhard, C. Genetic analysis of fin formation in the zebrafish, *Danio rerio*. *Development* **1996**; *123*: 255–262.
14. Innis, J.W., and Mortlock, D.P. Limb development: molecular dysmorphology is at hand! *Clin Genet* **1998**; *53*: 337–348.
15. Cohn, M.J., and Bright, P.E. Molecular control of vertebrate limb development, evolution and congenital malformations. *Cell Tissue Res* **1999**; *296*: 3–17.
16. Zwilling, E. Ectoderm-mesoderm relationship in the development of the chick embryo limb bud. *J Exp Zool* **1955**; *128*: 423–441.
17. Chevallier, A. Origin of scapular and pelvic girdles of bird embryo. *J Embryol Exp Morphol* **1977**; *42*: 275–292.

18 Christ, B., Jacob, H.J., and Jacob, M. Experimental-analysis of origin of wing musculature in avian embryos. *Anat Embryol* **1977**; *150*: 171–186.

19 Kieny, M., and Chevallier, A. Autonomy of tendon development in the embryonic chick wing. *J Embryol Exp Morphol* **1979**; *49*: 153–165.

20 Chevallier, A., Kieny, M., and Mauger, A. Limb-somite relationship – origin of limb musculature. *J Embryol Exp Morphol* **1977**; *41*: 245–258.

21 Kardon, G., Campbell, J.K., and Tabin, C.J. Local extrinsic signals determine muscle and endothelial cell fate and patterning in the vertebrate limb. *Dev Cell* **2002**; *3*: 533–545.

22 Brent, A.E., Braun, T., and Tabin, C.J. Genetic analysis of interactions between the somitic muscle, cartilage and tendon cell lineages during mouse development. *Development* **2005**; *132*: 515–528.

23 Edom-Vovard, F., and Duprez, D. Signals regulating tendon formation during chick embryonic development. *Dev Dyn* **2004**; *229*: 449–457.

24 Berg, J.S., and Farel, P.B. Developmental regulation of sensory neuron number and limb innervation in the mouse. *Dev Brain Res* **2000**; *125*: 21–30.

25 Duprez, D. Signals regulating muscle formation in the limb during embryonic development. *Int J Dev Biol* **2002**; *46*: 915–925.

26 Kardon, G., Harfe, B.D., and Tabin, C.J. A Tcf4–positive mesodermal population provides a prepattern for vertebrate limb muscle patterning. *Dev Cell* **2003**; *5*: 937–944.

27 Wang, G., and Scott, S.A. Independent development of sensory and motor innervation patterns in embryonic chick hindlimbs. *Dev Biol* **1999**; *208*: 324–336.

28 Wilting, J., Brandsaberi, B., Huang, R.J., Zhi, Q.X., Kontges, G., Ordahl, C.P., and Christ, B. Angiogenic potential of the avian somite. *Dev Dyn* **1995**; *202*: 165–171.

29 Karsenty, G., and Wagner, E.F. Reaching a genetic and molecular understanding of skeletal development. *Dev Cell* **2002**; *2*: 389–406.

30 Karsenty, G. The complexities of skeletal biology. *Nature* **2003**; *423*: 316–318.

31 Kronenberg, H.M. Developmental regulation of the growth plate. *Nature* **2003**; *423*: 332–336.

32 Shubin, N.H., and Alberch, P. A morphogenetic approach to the origin and basic organization of the tetrapod limb. *Evol Biol* **1986**; *20*: 319–387.

33 Cohn, M.J., Lovejoy, C.O., Wolpert, L., and Coates, M.I. Branching, segmentation and the metapterygial axis: pattern versus process in the vertebrate limb. *Bioessays* **2002**; *24*: 460–465.

34 Knezevic, V., De Santo, R., Schughart, K., Huffstadt, U., Chiang, C., Mahon, K.A., and Mackem, S. Hoxd–12 differentially affects preaxial and postaxial chondrogenic branches in the limb and regulates Sonic hedgehog in a positive feedback loop. *Development* **1997**; *124*: 4523–4536.

35 Huang, R.J., Zhi, Q.X., Patel, K., Wilting, J., and Christ, B. Dual origin and segmental organisation of the avian scapula. *Development* **2000**; *127*: 3789–3794.
36 Min, H.S., Danilenko, D.M., Scully, S.A., Bolon, B., Ring, B.D., Tarpley, J.E., DeRose, M., and Simonet, W.S. Fgf-10 is required for both limb and lung development and exhibits striking functional similarity to Drosophila branchless. *Genes Dev* **1998**; *12*: 3156–3161.
37 Sekine, K., Ohuchi, H., Fujiwara, M., Yamasaki, M., Yoshizawa, T., Sato, T., Yagishita, N., Matsui, D., Koga, Y., Itoh, N., and Kato, S. Fgf10 is essential for limb and lung formation. *Nat Genet* **1999**; *21*: 138–141.
38 Rallis, C., Bruneau, B.G., Del Buono, J., Seidman, C.E., Seidman, J.G., Nissim, S., Tabin, C.J., and Logan, M.P.O. Tbx5 is required for forelimb bud formation and continued outgrowth. *Development* **2003**; *130*: 2741–2751.
39 Ros, M.A., LopezMartinez, A., Simandl, B.K., Rodriguez, C., Belmonte, J.C.I., Dahn, R., and Fallon, J.F. The limb field mesoderm determines initial limb bud anteroposterior asymmetry and budding independent of sonic hedgehog or apical ectodermal gene expressions. *Development* **1996**; *122*: 2319–2330.
40 Sun, X., Mariani, F.V., and Martin, G.R. Functions of FGF signalling from the apical ectodermal ridge in limb development. *Nature* **2002**; *418*: 501–508.
41 Boulet, A.M., Moon, A.M., Arenkiel, B.R., and Capecchi, M.R. The roles of Fgf4 and Fgf8 in limb bud initiation and outgrowth. *Dev Biol* **2004**; *273*: 361–372.
42 Altabef, M., Clarke, J.D.W., and Tickle, C. Dorso-ventral ectodermal compartments and origin of apical ectodermal ridge in developing chick limb. *Development* **1997**; *124*: 4547–4556.
43 Dudley, A.T., Ros, M.A., and Tabin, C.J. A re-examination of proximodistal patterning during vertebrate limb development. *Nature* **2002**; *418*: 539–544.
44 Michaud, J.L., Lapointe, F., and LeDouarin, N.M. The dorsoventral polarity of the presumptive limb is determined by signals produced by the somites and by the lateral somatopleure. *Development* **1997**; *124*: 1453–1463.
45 Kimmel, R.A., Turnbull, D.H., Blanquet, V., Wurst, W., Loomis, C., and Joyner, A.L. Two lineage boundaries coordinate vertebrate apical ectodermal ridge formation. *Genes Dev* **2000**; *14*: 1377–1389.
46 Ahn, S., and Joyner, A.L. Dynamic changes in the response of cells to positive hedgehog signaling during mouse limb patterning. *Cell* **2004**; *118*: 505–516.
47 Harfe, B.D., Scherz, P.J., Nissim, S., Tian, F., McMahon, A.P., and Tabin, C.J. Evidence for an expansion-based temporal Shh gradient in specifying vertebrate digit identities. *Cell* **2004**; *118*: 517–528.
48 Stephens, T.D., Beier, R.L.W., Bringhurst, D.C., Hiatt, S.R., Prestridge, M., Pugmire, D.E., and Willis, H.J. Limbness in the early chick-embryo lateral plate. *Dev Biol* **1989**; *133*: 1–7.
49 Burke, A.C., Nelson, C.E., Morgan, B.A., and Tabin, C. Hox genes and the evolution of vertebrate axial morphology. *Development* **1995**; *121*: 333–346.

50 Cohn, M.J., Patel, K., Krumlauf, R., Wilkinson, D.G., Clarke, J.D.W., and Tickle, C. Hox9 genes and vertebrate limb specification. *Nature* **1997**; *387*: 97–101.

51 Coates, M.I., and Cohn, M.J. Fins, limbs, and tails: outgrowths and axial patterning in vertebrate evolution. *Bioessays* **1998**; *20*: 371–381.

52 Kieny, M. (On the relations between somatic and somatopleural mesoderm before and during primary induction of chick embryo limbs). *C R Acad Sci Hebd Seances Acad Sci D* **1969**; *268*: 3183–3186.

53 Pinot, M. (The role of the somitic mesoderm in the early morphogenesis of the limbs in the fowl embryo). *J Embryol Exp Morphol* **1970**; *23*: 109–151.

54 Kieny, M. Les phases d'activité morphogene du mésoderme somatopleural pendant le développement précoce du membre chez l'embryon de poulet. *Ann Embryol Morphog* **1971**; *4*: 281–298.

55 Stephens, T.D., Spall, R., Baker, W.C., Hiatt, S.R., Pugmire, D.E., Shaker, M.R., Willis, H.J., and Winger, K.P. Axial and paraxial influences on limb morphogenesis. *J Morphol* **1991**; *208*: 367–379.

56 Dealy, C.N. Hensen's node provides an endogenous limb-forming signal. *Dev Biol* **1997**; *188*: 216–223.

57 FernandezTeran, M., Piedra, M.E., Simandl, B.K., Fallon, J.F., and Ros, M.A. Limb initiation and development is normal in the absence of the mesonephros. *Dev Biol* **1997**; *189*: 246–255.

58 Martin, G.R. The roles of FGFs in the early development of vertebrate limbs. *Genes Dev* **1998**; *12*: 1571–1586.

59 Tanaka, M., Cohn, M.J., Ashby, P., Davey, M., Martin, P., and Tickle, C. Distribution of polarizing activity and potential for limb formation in mouse and chick embryos and possible relationships to polydactyly. *Development* **2000**; *127*: 4011–4021.

60 Cohn, M.J., Izpisuabelmonte, J.C., Abud, H., Heath, J.K., and Tickle, C. Fibroblast growth-factors induce additional limb development from the flank of chick-embryos. *Cell* **1995**; *80*: 739–746.

61 Kieny, M. (Inductive role of the mesoderm in the early differentiation of the limb bud in the chick embryo). *J Embryol Exp Morphol* **1960**; *8*: 457–467.

62 Kieny, M. (Variation in the inductive capacity of mesoderm and the competence of ectoderm during primary induction in the chick embryo limb bud). *Arch Anat Microsc Morphol Exp* **1968**; *57*: 401–418.

63 Saunders, J.W., and Reuss, C. Inductive and axial properties of prospective wing-bud mesoderm in chick-embryo. *Dev Biol* **1974**; *38*: 41–50.

64 Carrington, J.L., and Fallon, J.F. The stages of flank ectoderm capable of responding to ridge induction in the chick-embryo. *J Embryol Exp Morphol* **1984**; *84*: 19–34.

65 Saunders, J.W. The proximo-distal sequence of origin of parts of the chick wing and the role of the ectoderm. *J Exp Zool* **1948**; *108*: 363–404.

66 Pautou, M.P. (Determining role of the mesoderm in the specific differentiation of the leg in birds). *Arch Anat Microsc Morphol Exp* **1968**; *57*: 311–328.

67 Rubin, L., and Saunders, J.W., Jr. Ectodermal-mesodermal interactions in the growth of limb buds in the chick embryo: constancy and temporal limits of the ectodermal induction. *Dev Biol* **1972**; *28*: 94–112.

68 Summerbell, D., Lewis, J.H., and Wolpert, L. Positional information in chick limb morphogenesis. *Nature* **1973**; *244*: 492–496.

69 Summerbell, D., and Lewis, J.H. Time, place and positional value in chick limb-bud. *J Embryol Exp Morphol* **1975**; *33*: 621–643.

70 Saunders, J.W. Is the progress zone model a victim of progress? *Cell* **2002**; *110*: 541–543.

71 Tickle, C., and Wolpert, L. The progress zone – alive or dead? *Nat Cell Biol* **2002**; *4*: E216–E217.

72 Wolpert, L. Limb patterning: Reports of model's death exaggerated. *Curr Biol* **2002**; *12*: R628–R630.

73 Mariani, F.V., and Martin, G.R. Deciphering skeletal patterning: clues from the limb. *Nature* **2003**; *423*: 319–325.

74 Niswander, L. Pattern formation: Old models out on a limb. *Nat Rev Genet* **2003**; *4*: 133–143.

75 Geduspan, J.S., and Maccabe, J.A. Transfer of dorsoventral information from mesoderm to ectoderm at the onset of limb development. *Anat Rec* **1989**; *224*: 79–87.

76 Chen, H.X., and Johnson, R.L. Dorsoventral patterning of the vertebrate limb: a process governed by multiple events. *Cell Tissue Res* **1999**; *296*: 67–73.

77 Maccabe, J.A., Errick, J., and Saunders, J.W. Ectodermal control of dorsoventral axis in leg bud of chick-embryo. *Dev Biol* **1974**; *39*: 69–82.

78 Pautou, M.P. Establishment of dorso-ventral axis in foot of chick-embryo. *J Embryol Exp Morphol* **1977**; *42*: 177–194.

79 Geduspan, J.S., and Maccabe, J.A. The ectodermal control of mesodermal patterns of differentiation in the developing chick wing. *Dev Biol* **1987**; *124*: 398–408.

80 Saunders, J.W., and Gasseling. M.T. Ectodermal-mesenchymal interactions in the origin of limb symmetry. In *Epithelial-Mesenchymal Interactions*, R.F.a.R.E. Billingham, (Ed.). Williams and Wilkins: Baltimore, **1968**; pp. 78–97.

81 Hornbruch, A., and Wolpert, L. The Spatial and temporal distribution of polarizing activity in the flank of the pre-limb-bud stages in the chick-embryo. *Development* **1991**; *111*: 725–731.

82 Rudnick, D. Limb-forming potencies of the chick blastoderm: Including notes on associated trunk structures. *Trans Conn Acad Arts Sci* **1945**; *36*: 353–377.

83 Chaube, S. On axiation and symmetry in transplanted wing of the chick. *J Exp Zool* **1959**; *140*: 29–78.

84 Zwilling, E. Limb morphogenesis. *Adv Morphog* **1961**; *1*: 301–330.

85 Zwilling, E., and Hansborough, L. Interaction between limb bud ectoderm and mesoderm in the chick embryo. III Experiments with polydactylous limbs. *J Exp Zool* **1956**; *132*: 219–239.

86 Saunders, J.W., Jr., and Gasseling, M.T. Trans-filter propagation of apical ectoderm maintenance factor in the chick embryo wing bud. *Dev Biol* **1963**; *7*: 64–78.

87 Saunders, J.W., Jr. The experimental analysis of chick limb bud development. In *Vertebrate Limb and Somite Morphogenesis*, J.R.H. D.A. Ede, and M.J. Balls, (Eds). Cambridge University Press: Cambridge, **1977**; pp. 1–24.

88 Tickle, C. The number of polarizing region cells required to specify additional digits in the developing chick wing. *Nature* **1981**; *289*: 295–298.

89 Laufer, E., Nelson, C.E., Johnson, R.L., Morgan, B.A., and Tabin, C. Sonic Hedgehog and Fgf–4 act through a signaling cascade and feedback loop to integrate growth and patterning of the developing limb bud. *Cell* **1994**; *79*: 993–1003.

90 Niswander, L., Jeffrey, S., Martin, G.R., and Tickle, C. A positive feedback loop coordinates growth and patterning in the vertebrate limb. *Nature* **1994**; *371*: 609–612.

91 Yang, Y.Z., and Niswander, L. Interaction between the signaling molecules wnt7a and shh during vertebrate limb development – dorsal signals regulate anteroposterior patterning. *Cell* **1995**; *80*: 939–947.

92 Niederreither, K., Subbarayan, V., Dolle, P., and Chambon, P. Embryonic retinoic acid synthesis is essential for early mouse post-implantation development. *Nat Genet* **1999**; *21*: 444–448.

93 Niederreither, K., Vermot, J., Schuhbaur, B., Chambon, P., and Dolle, P. Embryonic retinoic acid synthesis is required for forelimb growth and anteroposterior patterning in the mouse. *Development* **2002**; *129*: 3563–3574.

94 Mic, F.A., Sirbu, I.O., and Duester, G. Retinoic acid synthesis controlled by Raldh2 is required early for limb bud initiation and then later as a proximodistal signal during apical ectodermal ridge formation. *J Biol Chem* **2004**; *279*: 26698–26706.

95 Helms, J.A., Kim, C.H., Eichele, G., and Thaller, C. Retinoic acid signaling is required during early chick limb development. *Development* **1996**; *122*: 1385–1394.

96 Stratford, T., Horton, C., and Maden, M. Retinoic acid is required for the initiation of outgrowth in the chick limb bud. *Curr Biol* **1996**; *6*: 1124–1133.

97 Mercader, N., Leonardo, E., Piedra, M.E., Martinez-A, C., Ros, M.A., and Torres, M. Opposing RA and FGF signals control proximodistal vertebrate limb development through regulation of Meis genes. *Development* **2000**; *127*: 3961–3970.

98 Yashiro, K., Zhao, X.L., Uehara, M., Yamashita, K., Nishijima, M., Nishino, J., Saijoh, Y., Sakai, Y., and Hamada, H. Regulation of retinoic acid distribution is required for proximodistal patterning and outgrowth of the developing mouse limb. *Dev Cell* **2004**; *6*: 411–422.

99 Rodriguez-Leon, J., Merino, R., Macias, D., Ganan, Y., Santesteban, E., and Hurle, J.M. Retinoic acid regulates programmed cell death through BMP signalling. *Nat Cell Biol* **1999**; *1*: 125–126.

100 Zuzarte-Luis, V., and Hurle, J.M. Programmed cell death in the developing limb. *Int J Dev Biol* **2002**; *46*: 871–876.

101 Lohnes, D., Mark, M., Mendelsohn, C., Dolle, P., Dierich, A., Gorry, P., Gansmuller, A., and Chambon, P. Function of the Retinoic acid receptors (rars) during development .1. cranio-

facial and skeletal abnormalities in rar double mutants. *Development* **1994**; *120*: 2723–2748.
102. Giguere, V., Ong, E.S., Segui, P., and Evans, R.M. Identification of a receptor for the morphogen retinoic acid. *Nature* **1987**; *330*: 624–629.
103. Petkovich, M., Brand, N.J., Krust, A., and Chambon, P. A human retinoic acid receptor which belongs to the family of nuclear receptors. *Nature* **1987**; *330*: 444–450.
104. Chambon, P. A decade of molecular biology of retinoic acid receptors. *FASEB J* **1996**; *10*: 940–954.
105. Wei, L.N. Retinoid receptors and their coregulators. *Annu Rev Pharmacol Toxicol* **2003**; *43*: 47–72.
106. Weston, A.D., Blumberg, B., and Underhill, T.M. Active repression by unliganded retinoid receptors in development: less is sometimes more. *J Cell Biol* **2003**; *161*: 223–228.
107. Sucov, H.M., and Evans, R.M. Retinoic acid and retinoic acid receptors in development. *Mol Neurobiol* **1995**; *10*: 169–184.
108. Kastner, P., Mark, M., Ghyselinck, N., Krezel, W., Dupe, V., Grondona, J.M., and Chambon, P. Genetic evidence that the retinoid signal is transduced by heterodimeric RXR/RAR functional units during mouse development. *Development* **1997**; *124*: 313–326.
109. Mangelsdorf, D.J. Vitamin-a receptors. *Nutr Rev* **1994**; *52*: 32–44.
110. Dolle, P., Ruberte, E., Kastner, P., Petkovich, M., Stoner, C.M., Gudas, L.J., and Chambon, P. Differential expression of genes encoding alpha,beta and gamma-retinoic acid receptors and crabp in the developing limbs of the mouse. *Nature* **1989**; *342*: 702–705.
111. Ruberte, E., Dolle, P., Krust, A., Zelent, A., Morrisskay, G., and Chambon, P. Specific spatial and temporal distribution of retinoic acid receptor-gamma transcripts during mouse embryogenesis. *Development* **1990**; *108*: 213–222.
112. Dupe, V., Ghyselinck, N.B., Thomazy, V., Nagy, L., Davies, P.J.A., Chambon, P., and Mark, M. Essential roles of retinoic acid signaling in interdigital apoptosis and control of BMP-7 expression in mouse autopods. *Dev Biol* **1999**; *208*: 30–43.
113. Mangelsdorf, D.J., Borgmeyer, U., Heyman, R.A., Zhou, J.Y., Ong, E.S., Oro, A.E., Kakizuka, A., and Evans, R.M. Characterization of 3 Rxr genes that mediate the action of 9–cis retinoic acid. *Genes Dev* **1992**; *6*: 329–344.
114. Dolle, P., Fraulob, V., Kastner, P., and Chambon, P. Developmental expression of murine retinoid-X-receptor (RXR) genes. *Mech Dev* **1994**; *45*: 91–104.
115. Sucov, H.M., IzpisuaBelmonte, J.C., Ganan, Y., and Evans, R.M. Mouse embryos lacking RXR alpha are resistant to retinoic-acid-induced limb defects. *Development* **1995**; *121*: 3997–4003.
116. Krezel, W., Dupe, V., Mark, M., Dierich, A., Kastner, P., and Chambon, P. RXR gamma null mice are apparently normal and compound RXR alpha(+/−)/RXR beta(−/−)/RXR gamma(−/−) mutant mice are viable. *Proc Natl Acad Sci USA* **1996**; *93*: 9010–9014.

117 Lampron, C., Rochetteegly, C., Gorry, P., Dolle, P., Mark, M., Lufkin, T., Lemeur, M., and Chambon, P. Mice deficient in cellular retinoic acid-binding protein-II (CRABPII) or in both CRABPI and CRABPII are essentially normal. *Development* **1995**; *121*: 539–548.

118 Napoli, J.L. Retinoid binding-proteins redirect retinoid metabolism: biosynthesis and metabolism of retinoic acid. *Sem Cell Dev Biol* **1997**; *8*: 403–415.

119 Lee, G.S., Kochhar, D.M., and Collins, M.D. Retinoid-induced limb malformations. *Curr Pharm Design* **2004**; *10*: 2657–2699.

120 Molotkov, A., Fan, X.H., Deltour, L., Foglio, M.H., Martras, S., Farres, J., Pares, X., and Duester, G. Stimulation of retinoic acid production and growth by ubiquitously expressed alcohol dehydrogenase Adh3. *Proc Natl Acad Sci USA* **2002**; *99*: 5337–5342.

121 Mic, F.A., Molotkov, A., Benbrook, D.M., and Duester, G. Retinoid activation of retinoic acid receptor but not retinoid X receptor is sufficient to rescue lethal defect in retinoic acid synthesis. *Proc Natl Acad Sci USA* **2003**; *100*: 7135–7140.

122 Scott, W.J., Walter, R., Tzimas, G., Sass, J.O., Nau, H., and Collins, M.D. Endogenous status of retinoids and their cytosolic binding-proteins in limb buds of chick vs mouse embryos. *Dev Biol* **1994**; *165*: 397–409.

123 Dupe, V., Matt, N., Garnier, J.M., Chambon, P., Mark, M., and Ghyselinck, N.B. A newborn lethal defect due to inactivation of retinaldehyde dehydrogenase type 3 is prevented by maternal retinoic acid treatment. *Proc Natl Acad Sci USA* **2003**; *100*, 14036–14041.

124 Fan, X.H., Molotkov, A., Manabe, S.I., Donmoyer, C.M., Deltour, L., Foglio, M.H., Cuenca, A.E., Blaner, W.S., Lipton, S.A., and Duester, G. Targeted disruption of Aldh1a1 (Raldh1) provides evidence for a complex mechanism of retinoic acid synthesis in the developing retina. *Mol Cell Biol* **2003**; *23*: 4637–4648.

125 Grandel, H., Lun, K., Rauch, G.J., Rhinn, M., Piotrowski, T., Houart, C., Sordino, P., Kuchler, A.M., Schulte-Merker, S., Geisler, R., Holder, N., Wilson, S.W., and Brand, M. Retinoic acid signalling in the zebrafish embryo is necessary during pre-segmentation stages to pattern the anterior-posterior axis of the CNS and to induce a pectoral fin bud. *Development* **2002**; *129*: 2851–2865.

126 Tickle, C., Alberts, B., Wolpert, L., and Lee, J. Local application of retinoic acid to the limb bond mimics the action of the polarizing region. *Nature* **1982**; *296*: 564–566.

127 Tickle, C., Lee, J., and Eichele, G. A quantitative-analysis of the effect of all-trans-retinoic acid on the pattern of chick wing development. *Dev Biol* **1985**; *109*: 82–95.

128 Noji, S., Nohno, T., Koyama, E., Muto, K., Ohyama, K., Aoki, Y., Tamura, K., Ohsugi, K., Ide, H., Taniguchi, S., and Saito, T. Retinoic acid induces polarizing activity but is unlikely to be a morphogen in the chick limb bud. *Nature* **1991**; *350*: 83–86.

129 Wanek, N., Gardiner, D.M., Muneoka, K., and Bryant, S.V. Conversion by retinoic acid of anterior cells into Zpa cells in the chick wing bud. *Nature* **1991**; *350*: 81–83.

130 Riddle, R.D., Johnson, R.L., Laufer, E., and Tabin, C. Sonic-Hedgehog mediates the polarizing activity of the Zpa. *Cell* **1993**; *75*: 1401–1416.

131 Fujii, H., Sato, T., Kaneko, S., Gotoh, O., FujiiKuriyama, Y., Osawa, K., Kato, S., and Hamada, H. Metabolic inactivation of retinoic acid by a novel P450 differentially expressed in developing mouse embryos. *EMBO J* **1997**; *16*: 4163–4173.

132 MacLean, G., Abu-Abed, S., Dolle, P., Tahayato, A., Chambon, P., and Petkovich, M. Cloning of a novel retinoic-acid metabolizing cytochrome P450, Cyp26B1, and comparative expression analysis with Cyp26A1 during early murine development. *Mech Dev* **2001**; *107*: 195–201.

133 Reijntjes, S., Gale, E., and Maden, M. Expression of the retinoic acid catabolising enzyme CYP26B1 in the chick embryo and its regulation by retinoic acid. *Gene Expression Patt* **2003**; *3*: 621–627.

134 Reijntjes, S., Gale, E., and Maden, M. Generating gradients of retinoic acid in the chick embryo: Cyp26C1 expression and a comparative analysis of the Cyp26 enzymes. *Dev Dyn* **2004**; *230*: 509–517.

135 Tahayato, A., Dolle, P., and Petkovich, M. Cyp26C1 encodes a novel retinoic acid-metabolizing enzyme expressed in the hindbrain, inner ear, first branchial arch and tooth buds during murine development. *Gene Expression Patt* **2003**; *3*: 449–454.

136 Swindell, E.C., Thaller, C., Sockanathan, S., Petkovich, M., Jessell, T.M., and Eichele, G. Complementary domains of retinoic acid production and degradation in the early chick embryo. *Dev Biol* **1999**; *216*: 282–296.

137 Blentic, A., Gale, E., and Maden, M. Retinoic acid signalling centres in the avian embryo identified by sites of expression of synthesising and catabolising enzymes. *Dev Dyn* **2003**; *227*: 114–127.

138 del Corral, R.D., Olivera-Martinez, I., Goriely, A., Gale, E., Maden, M., and Storey, K. Opposing FGF and retinoid pathways control ventral neural pattern, neuronal differentiation, and segmentation during body axis extension. *Neuron* **2003**; *40*: 65–79.

139 Moreno, T.A., and Kintner, C. Regulation of segmental patterning by retinoic acid signaling during Xenopus somitogenesis. *Dev Cell* **2004**; *6*: 205–218.

140 Rossant, J., Zirngibl, R., Cado, D., Shago, M., and Giguere, V. Expression of a retinoic acid response element-hsplacz transgene defines specific domains of transcriptional activity during mouse embryogenesis. *Genes Dev* **1991**; *5*: 1333–1344.

141 Mackem, S., Baumann, C.T., and Hager, G.L. A glucocorticoid/retinoic acid receptor chimera that displays cytoplasmic/nuclear translocation in response to retinoic acid – A real time sensing assay for nuclear receptor ligands. *J Biol Chem* **2001**; *276*: 45501–45504.

142 Ornitz, D.M. FGFs, heparan sulfate and FGFRs: complex interactions essential for development. *Bioessays* **2000**; *22*: 108–112.

143 Ornitz, D.M., and Marie, P.J. FGF signaling pathways in endochondral and intramembranous bone development and human genetic disease. *Genes Dev* **2002**; *16*: 1446–1465.

144 Crossley, P.H., and Martin, G.R. The mouse Fgf8 gene encodes a family of polypeptides and is expressed in regions that direct outgrowth and patterning in the developing embryo. *Development* **1995**; *121*: 439–451.

145 Maruoka, Y., Ohbayashi, N., Hoshikawa, M., Itoh, N., Hogan, B.L.M., and Furuta, Y. Comparison of the expression of three highly related genes, Fgf8, Fgf17 and Fgf18, in the mouse embryo. *Mech Dev* **1998**; *74*: 175–177.

146 Ohbayashi, N., Shibayama, M., Kurotaki, Y., Imanishi, M., Fujimori, T., Itoh, N., and Takada, S. FGF18 is required for normal cell proliferation and differentiation during osteogenesis and chondrogenesis. *Genes Dev* **2002**; *16*: 870–879.

147 Liu, Z.H., Xu, J.S., Colvin, J.S., and Ornitz, D.M. Coordination of chondrogenesis and osteogenesis by fibroblast growth factor 18. *Genes Dev* **2002**; *16*: 859–869.

148 Stauber, D.J., DiGabriele, A.D., and Hendrickson, W.A. Structural interactions of fibroblast growth factor receptor with its ligands. *Proc Natl Acad Sci USA* **2000**; *97*: 49–54.

149 Lin, X.H. Functions of heparan sulfate proteoglycans in cell signaling during development. *Development* **2004**; *131*: 6009–6021.

150 Hou, J.Z., Kan, M., McKeehan, K., McBride, G., Adams, P., and McKeehan, W.L. Fibroblast growth-factor receptors from liver vary in 3 structural domains. *Science* **1991**; *251*: 665–668.

151 Werner, S., Duan, D.S.R., Devries, C., Peters, K.G., Johnson, D.E., and Williams, L.T. Differential splicing in the extracellular region of fibroblast Growth-Factor Receptor–1 generates receptor variants with different ligand-binding specificities. *Mol Cell Biol* **1992**; *12*, 82–88.

152 Orr-Urtreger, A., Bedford, M.T., Burakova, T., Arman, E., Zimmer, Y., Yayon, A., Givol, D., and Lonai, P. Developmental localization of the splicing alternatives of fibroblast growth factor receptor–2 (FGFR2). *Dev Biol* **1993**; *158*: 475–486.

153 Ornitz, D.M., Xu, J.S., Colvin, J.S., McEwen, D.G., MacArthur, C.A., Coulier, F., Gao, G.X., and Goldfarb, M. Receptor specificity of the fibroblast growth factor family. *J Biol Chem* **1996**; *271*: 15292–15297.

154 Igarashi, M., Finch, P.W., and Aaronson, S.A. Characterization of recombinant human fibroblast growth factor (FGF)–10 reveals functional similarities with keratinocyte growth factor (FGF–7). *J Biol Chem* **1998**; *273*: 13230–13235.

155 Partanen, J., Schwartz, L., and Rossant, J. Opposite phenotypes of hypomorphic and Y766 phosphorylation site mutations reveal a function for Fgfr1 in anteroposterior patterning of mouse embryos. *Genes Dev* **1998**; *12*: 2332–2344.

156 Xu, X.L., Weinstein, M., Li, C.L., Naski, M., Cohen, R.I., Ornitz, D.M., Leder, P., and Deng, C.X. Fibroblast growth factor receptor 2 (FGFR2)-mediated reciprocal regulation loop between FGF8 and FGF10 is essential for limb induction. *Development* **1998**; *125*: 753–765.

157 Deng, C.X., WynshawBoris, A., Zhou, F., Kuo, A., and Leder, P. Fibroblast growth factor receptor 3 is a negative regulator of bone growth. *Cell* **1996**; *84*: 911–921.

158 Weinstein, M., Xu, X.L., Ohyama, K., and Deng, C.X. FGFR–3 and FGFR–4 function cooperatively to direct alveogenesis in the murine lung. *Development* **1998**; *125*: 3615–3623.

159 Ohuchi, H., Nakagawa, T., Yamamoto, A., Araga, A., Ohata, T., Ishimaru, Y., Yoshioka, H., Kuwana, T., Nohno, T., Yamasaki, M., Itoh, N., and Noji, S. The mesenchymal factor, FGF10, initiates and maintains the outgrowth of the chick limb bud through interaction with FGF8, an apical ectodermal factor. *Development* **1997**; *124*: 2235–2244.

160 Arman, E., Haffner-Krausz, R., Gorivodsky, M., and Lonai, P. Fgfr2 is required for limb outgrowth and lung-branching morphogenesis. *Proc Natl Acad Sci USA* **1999**; *96*: 11895–11899.

161 Revest, J.M., Spencer-Dene, B., Kerr, K., De Moerlooze, L., Rosewell, I., and Dickson, C. Fibroblast growth factor receptor 2–IIIb acts upstream of Shh and Fgf4 and is required for limb bud maintenance but not for the induction of Fgf8, Fgf10, Msx1, or Bmp4. *Dev Biol* **2001**; *231*: 47–62.

162 Celli, G., LaRochelle, W.J., Mackem, S., Sharp, R., and Merlino, G. Soluble dominant-negative receptor uncovers essential roles for fibroblast growth factors in multi-organ induction and patterning. *EMBO J* **1998**; *17*: 1642–1655.

163 Eswarakumar, V.P., Monsonego-Ornan, E., Pines, M., Antonopoulou, I., Morriss-Kay, G.M., and Lonai, P. The IIIc alternative of Fgfr2 is a positive regulator of bone formation. *Development* **2002**; *129*: 3783–3793.

164 Niswander, L., Tickle, C., Vogel, A., Booth, I., and Martin, G.R. Fgf–4 Replaces the Apical Ectodermal Ridge and Directs Outgrowth and Patterning of the Limb. *Cell* **1993**; *75*: 579–587.

165 Fallon, J.F., Lopez, A., Ros, M.A., Savage, M.P., Olwin, B.B., and Simandl, B.K. Fgf–2 – apical ectodermal ridge growth signal for chick limb development. *Science* **1994**; *264*: 104–107.

166 Lewandoski, M., Sun, X., and Martin, G.R. Fgf8 signalling from the AER is essential for normal limb development. *Nat Genet* **2000**; *26*: 460–463.

167 Moon, A.M., and Capecchi, M.R. Fgf8 is required for outgrowth and patterning of the limbs. *Nat Genet* **2000**; *26*: 455–459.

168 Moon, A.M., Boulet, A.M., and Capecchi, M.R. Normal limb development in conditional mutants of Fgf4. *Development* **2000**; *127*: 989–996.

169 Sun, X., Lewandoski, M., Meyers, E.N., Liu, Y.H., Maxson, R.E., and Martin, G.R. Conditional inactivation of Fgf4 reveals complexity of signalling during limb bud development. *Nat Genet* **2000**; *25*: 83–86.

170 Xu, X.L., Qiao, W.H., Li, C.L., and Deng, C.X. Generation of Fgfr1 conditional knockout mice. *Genesis* **2002**; *32*: 85–86.

171 Deng, C.X., Bedford, M., Li, C.L., Xu, X.L., Yang, X., Dunmore, J., and Leder, P. Fibroblast growth factor receptor–1 (FGFR–1) is essential for normal neural tube and limb development. *Dev Biol* **1997**; *185*: 42–54.

172 Saxton, T.M., Ciruna, B.G., Holmyard, D., Kulkarni, S., Harpal, K., Rossant, J., and Pawson, T. The SH2 tyrosine phosphatase Shp2 is required for mammalian limb development. *Nat Genet* **2000**; *24*: 420–423.

173 Xu, X.L., Weinstein, M., Li, C.L., and Deng, C.X. Fibroblast growth factor receptors (FGFRs) and their roles in limb development. *Cell Tissue Res* **1999**; *296*: 33–43.

174 Hajihosseini, M.K., Lalioti, M.D., Arthaud, S., Burgar, H.R., Brown, J.M., Twigg, S.R.F., Wilkie, A.O.M., and Heath, J.K. Skeletal development is regulated by fibroblast growth factor receptor 1 signalling dynamics. *Development* **2004**; *131*; 325–335.

175 Kouhara, H., Hadari, Y.R., SpivakKroizman, T., Schilling, J., BarSagi, D., Lax, I., and Schlessinger, J. A lipid-anchored Grb2–binding protein that links FGF-receptor activation to the Ras/MAPK signaling pathway. *Cell* **1997**; *89*: 693–702.

176 Blaikie, P., Immanuel, D., Wu, J., Li, N.X., Yajnik, V., and Margolis, B. A Region in Shc distinct from the Sh2 domain can bind tyrosine-phosphorylated growth-factor receptors. *J Biol Chem* **1994**; *269*: 32031–32034.

177 Yu, C.F., Liu, Z.X., and Cantley, L.G. ERK negatively regulates the epidermal growth factor-mediated interaction of Gab1 and the phosphatidylinositol 3–kinase. *J Biol Chem* **2002**; *277*: 19382 *J Biol Chem* **1994**;19388.

178 Ong, S.H., Hadari, Y.R., Gotoh, N., Guy, G.R., Schlessinger, J., and Lax, I. Stimulation of phosphatidylinositol 3–kinase by fibroblast growth factor receptors is mediated by coordinated recruitment of multiple docking proteins. *Proc Natl Acad Sci USA* **2001**; *98*: 6074–6079.

179 Roux, P.P., and Blenis, J. ERK and p38 MAPK-activated protein kinases: a family of protein kinases with diverse biological functions. *Microbiol Mol Biol Rev* **2004**; *68*: 320.

180 Datta, S.R., Brunet, A., and Greenberg, M.E. Cellular survival: a play in three Akts. *Genes Dev* **1999**; *13*: 2905–2927.

181 Deo, D.D., Axelrad, T.W., Robert, E.G., Marcheselli, V., Bazan, N.G., and Hunt, J.D. Phosphorylation of STAT–3 in response to basic fibroblast growth factor occurs through a mechanism involving platelet-activating factor, JAK–2, and Src in human umbilical vein endothelial cells – Evidence for a dual kinase mechanism. *J Biol Chem* **2002**; *277*: 21237–21245.

182 Mohammadi, M., Honegger, A.M., Rotin, D., Fischer, R., Bellot, F., Li, W., Dionne, C.A., Jaye, M., Rubinstein, M., and Schlessinger, J. A tyrosine-phosphorylated carboxy-terminal peptide of the fibroblast growth-factor receptor (Flg) is a binding-site for the Sh2 domain of phospholipase C-gamma–1. *Mol Cell Biol* **1991**; *11*: 5068–5078.

183 Corson, L.B., Yamanaka, Y., Lai, K.M.V., and Rossant, J. Spatial and temporal patterns of ERK signaling during mouse embryogenesis. *Development* **2003**; *130*: 4527–4537.

184 Eblaghie, M.C., Lunn, J.S., Dickinson, R.J., Munsterberg, A.E., Sanz-Ezquerro, J.J., Farrell, E.R., Mathers, J., Keyse, S.M., Storey, K., and Tickle, C. Negative feedback regulation of FGF signaling levels by Pyst1/MKP3 in chick embryos. *Curr Biol* **2003**; *13*: 1009–1018.

185 Kawakami, Y., Rodriguez-Leon, J., Koth, C.M., Buscher, D., Itoh, T., Raya, A., Ng, J.K., Esteban, C.R., Takahashi, S., Henrique, D., Schwarz, M.F., Asahara, H., and Belmonte, J.C.I. MKP3 mediates the cellular response to FGF8 signalling in the vertebrate limb. *Nat Cell Biol* **2003**; *5*: 513–519.
186 Zuzarte-Luis, V., Montero, J.A., Rodriguez-Leon, J., Merino, R., Rodriguez-Rey, J.C., and Hurle, J.M. A new role for BMP5 during limb development acting through the synergic activation of Smad and MAPK pathways. *Dev Biol* **2004**; *272*: 39–52.
187 Li, C.L., Chen, L., Iwata, T., Kitagawa, M., Fu, X.Y., and Deng, C.X. A Lys644Glu substitution in fibroblast growth factor receptor 3 (FGFR3) causes dwarfism in mice by activation of STATs and ink4 cell cycle inhibitors. *Hum Mol Genet* **1999**; *8*: 35–44.
188 Sahni, M., Ambrosetti, D.C., Mansukhani, A., Gertner, R., Levy, D., and Basilico, C. FGF signaling inhibits chondrocyte proliferation and regulates bone development through the STAT–1 pathway. *Genes Dev* **1999**; *13*: 1361–1366.
189 Laplantine, E., Rossi, F., Sahni, M., Basilico, C., and Cobrinik, D. FGF signaling targets the pRb-related p107 and p130 proteins to induce chondrocyte growth arrest. *J Cell Biol* **2002**; *158*: 741–750.
190 Groom, L.A., Sneddon, A.A., Alessi, D.R., Dowd, S., and Keyse, S.M. Differential regulation of the MAP, SAP and RK/p38 kinases by Pyst1, a novel cytosolic dual-specificity phosphatase. *EMBO J* **1996**; *15*: 3621–3632.
191 Tsang, M., Friesel, R., Kudoh, T., and Dawid, I.B. Identification of Sef, a novel modulator of FGF signalling. *Nat Cell Biol* **2002**; *4*: 165–169.
192 Kovalenko, D., Yang, X.H., Nadeau, R.J., Harkins, L.K., and Friesel, R. Sef inhibits fibroblast growth factor signaling by inhibiting FGFR1 tyrosine phosphorylation and subsequent ERK activation. *J Biol Chem* **2003**; *278*: 14087–14091.
193 Yusoff, P., Lao, D.H., Ong, S.H., Wong, E.S.M., Lim, J., Lo, T.L., Leong, H.F., Fong, C.W., and Guy, G.R. Sprouty2 inhibits the Ras/MAP kinase pathway by inhibiting the activation of raf. *J Biol Chem* **2002**; *277*: 3195–3201.
194 Fong, C.W., Leong, H.F., Wong, E.S.M., Lim, J., Yusoff, P., and Guy, G.R. Tyrosine phosphorylation of Sprouty2 enhances its interaction with c-Cbl and is crucial for its function. *J Biol Chem* **2003**; *278*: 33456–33464.
195 Minowada, G., Jarvis, L.A., Chi, C.L., Neubuser, A., Sun, X., Hacohen, N., Krasnow, M.A., and Martin, G.R. Vertebrate Sprouty genes are induced by FGF signaling and can cause chondrodysplasia when overexpressed. *Development* **1999**; *126*: 4465–4475.
196 Massague, J. Integration of Smad and MAPK pathways: a link and a linker revisited. *Genes Dev* **2003**; *17*: 2993–2997.
197 Kretzschmar, M., Doody, J., and Massague, J. Opposing BMP and EGF signalling pathways converge on the TGF-beta family mediator Smad1. *Nature* **1997**; *389*: 618–622.
198 Pera, E.M., Ikeda, A., Eivers, E., and De Robertis, E.M. Integration of IGF, FGF, and anti-BMP signals via Smad1 phosphorylation in neural induction. *Genes Dev* **2003**; *17*: 3023–3028.

199 Kimura, N., Matsuo, R., Shibuya, H., Nakashima, K., and Taga, T. BMP2–induced apoptosis is mediated by activation of the TAK1–p38 kinase pathway that is negatively regulated by Smad6. *J Biol Chem* **2000**; *275*: 17647–17652.

200 Boilly, B., Vercoutter-Edouart, A.S., Hondermarck, H., Nurcombe, V., and Le Bourhis, X. FGF signals for cell proliferation and migration through different pathways. *Cytokine Growth Factor Rev* **2000**; *11*: 295–302.

201 Montero, J.A., Ganan, Y., Macias, D., Rodriguez-Leon, J., Sanz-Ezquerro, J.J., Merino, R., Chimal-Monroy, J., Nieto, M.A., and Hurle, J.M. Role of FGFs in the control of programmed cell death during limb development. *Development* **2001**; *128*: 2075–2084.

202 Sanz-Ezquerro, J.J., and Tickle, C. Autoregulation of Shh expression and Shh induction of cell death suggest a mechanism for modulating polarising activity during chick limb development. *Development* **2000**; *127*: 4811–4823.

203 Bushdid, P.B., Brantley, D.M., Yull, F.E., Blaeuer, G.L., Hoffman, L.H., Niswander, L., and Kerr, L.D. Inhibition of NF-kappa B activity results in disruption of the apical ectodermal ridge and aberrant limb morphogenesis. *Nature* **1998**; *392*: 615–618.

204 Kanegae, Y., Tavares, A.T., Belmonte, J.C.I., and Verma, I.M. Role of Rel/NF-kappa B transcription factors during the outgrowth of the vertebrate limb. *Nature* **1998**; *392*: 611–614.

205 Li, Q.T., Lu, Q.X., Hwang, J.Y., Buscher, D., Lee, K.F., Izpisua-Belmonte, J.C., and Verma, I.M. IKK1–deficient mice exhibit abnormal development of skin and skeleton. *Genes Dev* **1999**; *13*: 1322–1328.

206 Takeda, K., Takeuchi, O., Tsujimura, T., Itami, S., Adachi, O., Kawai, T., Sanjo, H., Yoshikawa, K., Terada, N., and Akira, S. Limb and skin abnormalities in mice lacking IKK alpha. *Science* **1999**; *284*: 313–316.

207 Sil, A.K., Maeda, S., Sano, Y., Roop, D.R., and Karin, M. IkB kinase-alpha acts in the epidermis to control skeletal and craniofacial morphogenesis. *Nature* **2004**; *428*: 660–664.

208 Mills, A.A., Zheng, B.H., Wang, X.J., Vogel, H., Roop, D.R., and Bradley, A. p63 is a p53 homologue required for limb and epidermal morphogenesis. *Nature* **1999**; *398*: 708–713.

209 Yang, A., Schweitzer, R., Sun, D.Q., Kaghad, M., Walker, N., Bronson, R.T., Tabin, C., Sharpe, A., Caput, D., Crum, C., and McKeon, F. p63 is essential for regenerative proliferation in limb, craniofacial and epithelial development. *Nature* **1999**; *398*: 714–718.

210 Bell, S.M., Schreiner, C.M., Waclaw, R.R., Campbell, K., Potter, S.S., and Scott, W.J. Sp8 is crucial for limb outgrowth and neuropore closure. *Proc Natl Acad Sci USA* **2003**; *100*: 12195–12200.

211 Kawakami, Y., Esteban, C.R., Matsui, T., Rodriguez-Leon, J., Kato, S., and Belmonte, J.C.I. Sp8 and Sp9, two closely related buttonhead-like transcription factors, regulate Fgf8 expression and limb outgrowth in vertebrate embryos. *Development* **2004**; *131*: 4763–4774.

212 Bakkers, J., Hild, M., Kramer, C., Furutani-Seiki, M., and Hammerschmidt, M. Zebrafish Delta Np63 is a direct target of bmp signaling and encodes a transcriptional repressor blocking neural specification in the ventral ectoderm. *Dev Cell* **2002**; *2*: 617–627.

213 Hsieh, J.C. Specificity of WNT-receptor interactions. *Front Biosci* **2004**; *9*: 1333–1338.

214 Parr, B.A., Shea, M.J., Vassileva, G., and McMahon, A.P. Mouse Wnt genes exhibit discrete domains of expression in the early embryonic Cns and limb buds. *Development* **1993**; *119*: 247–261.

215 Church, V.L., and Francis-West, P. Wnt signalling during limb development. *Int J Dev Biol* **2002**; *46*: 927–936.

216 Nusse, R. Wnts and Hedgehogs: lipid-modified proteins and similarities in signaling mechanisms at the cell surface. *Development* **2003**; *130*: 5297–5305.

217 Lin, X.H., and Perrimon, N. Developmental roles of heparan sulfate proteoglycans in Drosophila. *Glycoconjug J* **2003**; *19*: 363–368.

218 Dhoot, G.K., Gustafsson, M.K., Ai, X.B., Sun, W.T., Standiford, D.M., and Emerson, C.P. Regulation of Wnt signaling and embryo patterning by an extracellular sulfatase. *Science* **2001**; *293*: 1663–1666.

219 Malbon, C.C. Frizzleds: New members of the superfamily of G-protein-coupled receptors. *Front Biosci* **2004**; *9*: 1048–1058.

220 Tamai, K., Semenov, M., Kato, Y., Spokony, R., Liu, C.M., Katsuyama, Y., Hess, F., Saint-Jeannet, J.P., and He, X. LDL-receptor-related proteins in Wnt signal transduction. *Nature* **2000**; *407*: 530–535.

221 Wehrli, M., Dougan, S.T., Caldwell, K., O'Keefe, L., Schwartz, S., Vaizel-Ohayon, D., Schejter, E., Tomlinson, A., and DiNardo, S. Arrow encodes an LDL-receptor-related protein essential for Wingless signalling. *Nature* **2000**; *407*: 527–530.

222 Pinson, K.I., Brennan, J., Monkley, S., Avery, B.J., and Skarnes, W.C. An LDL-receptor-related protein mediates Wnt signalling in mice. *Nature* **2000**; *407*: 535–538.

223 Gong, Y.Q., Slee, R.B., Fukai, N., Rawadi, G., Roman-Roman, S., Reginato, A.M., Wang, H.W., Cundy, T., Glorieux, F.H., Lev, D., Zacharin, M., Oexle, K., Marcelino, J., Suwairi, W., Heeger, S., Sabatakos, G., Apte, S., Adkins, W.N., Allgrove, J., Arslan-Kirchner, M., Batch, J.A., Beighton, P., Black, G.C.M., Boles, R.G., Boon, L.M., Borrone, C., Brunner, H.G., Carle, G.F., Dallapiccola, B., De Paepe, A., Floege, B., Halfhide, M.I., Hall, B., Hennekam, R.C., Hirose, T., Jans, A., Juppner, H., Kim, C.A., Keppler-Noreuil, K., Kohlschuetter, A., LaCombe, D., Lambert, M., Lemyre, E., Letteboer, T., Peltonen, L., Ramesar, R.S., Romanengo, M., Somer, H., Steichen-Gersdorf, E., Steinmann, B., Sullivan, B., Superti-Furga, A., Swoboda, W., van den Boogaard, M.J., Van Hul, V., Vikkula, M., Votruba, M., Zabel, B., Garcia, T., Baron, R., Olsen, B.R., and Warman, M.L. LDL receptor-related protein 5 (LRP5) affects bone accrual and eye development. *Cell* **2001**; *107*: 513–523.

224 MacDonald, B.T., Adamska, M., and Meisler, M.H. Hypomorphic expression of Dkk1 in the doubleridge mouse: dose de-

pendence and compensatory interactions with Lrp6. *Development* **2004**; *131*: 2543–2552.
225 Topczewski, J., Sepich, D.S., Myers, D.C., Walker, C., Amores, A., Lele, Z., Hammerschmidt, M., Postlethwait, J., and Solnica-Krezel, L. The zebrafish glypican knypek controls cell polarity during gastrulation movements of convergent extension. *Dev Cell* **2001**; *1*: 251–264.
226 Yoda, A., Oishi, I., and Minami, Y. Expression and function of the Ror-family receptor tyrosine kinases during development: Lessons from genetic analyses of nematodes, mice, and humans. *J Recept Signal Transd* **2003**; *23*: 1–15.
227 Lu, W.G., Yamamoto, V., Ortega, B., and Baltimore, D. Mammalian Ryk is a Wnt coreceptor required for stimulation of neurite outgrowth. *Cell* **2004**; *119*: 97–108.
228 DeChiara, T.M., Kimble, R.B., Poueymirou, W.T., Rojas, J., Masiakowski, P., Valenzuela, D.M., and Yancopoulos, G.D. Ror2, encoding a receptor-like tyrosine kinase, is required for cartilage and growth plate development. *Nat Genet* **2000**; *24*: 271–274.
229 Oldridge, M., Fortuna, A.M., Maringa, M., Propping, P., Mansour, S., Pollitt, C., DeChiara, T.M., Kimble, R.B., Valenzuela, D.M., Yancopoulos, G.D., and Wilkie, A.O.M. Dominant mutations in ROR2, encoding an orphan receptor tyrosine kinase, cause brachydactyly type B. *Nat Genet* **2000**; *24*: 275–278.
230 Afzal, A.R., Rajab, A., Fenske, C.D., Oldridge, M., Elanko, N., Ternes-Pereira, E., Tuysuz, B., Murday, V.A., Patton, M.A., Wilkie, A.O.M., and Jeffery, S. Recessive Robinow syndrome, allelic to dominant brachydactyly type B, is caused by mutation of ROR2. *Nat Genet* **2000**; *25*: 419–422.
231 Halford, M.M., Armes, J., Buchert, M., Meskenaite, V., Grail, D., Hibbs, M.L., Wilks, A.F., Farlie, P.G., Newgreen, D.F., Hovens, C.M., and Stacker, S.A. Ryk-deficient mice exhibit craniofacial defects associated with perturbed Eph receptor crosstalk. *Nat Genet* **2000**; *25*: 414–418.
232 Tao, Q.H., Yokota, C., Puck, H., Kofron, M., Birsoy, B., Yan, D., Asashima, M., Wylie, C.C., Lin, X.H., and Heasman, J. Maternal Wnt11 activates the canonical wnt signaling pathway required for axis formation in Xenopus embryos. *Cell* **2005**; *120*: 857–871.
233 Kengaku, M., Capdevila, J., Rodriguez-Esteban, C., De La Pena, J., Johnson, R.L., Belmonte, J.C.I., and Tabin, C.J. Distinct WNT pathways regulating AER formation and dorsoventral polarity in the chick limb bud. *Science* **1998**; *280*: 1274–1277.
234 Liu, T., DeCostanzo, A.J., Liu, X.X., Wang, H.Y., Hallagan, S., Moon, R.T., and Malbon, C.C. G protein signaling from activated rat Frizzled–1 to the beta-catenin-Lef-Tcf pathway. *Science* **2001**; *292*: 1718–1722.
235 Katanaev, V.L., Ponzielli, R., Semeriva, M., and Tomlinson, A. Trimeric G protein-dependent frizzled signaling in Drosophila. *Cell* **2005**; *120*: 111–122.
236 Hamblet, N.S., Lijam, N., Ruiz-Lozano, P., Wang, J.B., Yang, Y.S., Luo, Z.G., Mei, L., Chien, K.R., Sussman, D.J., and Wyns-

haw-Boris, A. Dishevelled 2 is essential for cardiac outflow tract development, somite segmentation and neural tube closure. *Development* **2002**; *129*: 5827–5838.
237 Boutros, M., and Mlodzik, M. Dishevelled: at the crossroads of divergent intracellular signaling pathways. *Mech Dev* **1999**; *83*: 27–37.
238 Veeman, M.T., Axelrod, J.D., and Moon, R.T. A second canon: Functions and mechanisms of beta-catenin-independent wnt signaling. *Dev Cell* **2003**; *5*: 367–377.
239 Wharton, K.A. Runnin' with the Dvl: Proteins that associate with Dsh/Dvl and their significance to Wnt signal transduction. *Dev Biol* **2003**; *253*: 1–17.
240 Zeng, L., Fagotto, F., Zhang, T., Hsu, W., Vasicek, T.J., Perry, W.L., Lee, J.J., Tilghman, S.M., Gumbiner, B.M., and Costantini, F. The mouse fused locus encodes Axin, an inhibitor of the Wnt signaling pathway that regulates embryonic axis formation. *Cell* **1997**; *90*: 181–192.
241 Korinek, V., Barker, N., Morin, P.J., vanWichen, D., deWeger, R., Kinzler, K.W., Vogelstein, B., and Clevers, H. Constitutive transcriptional activation by a beta-catenin-Tcf complex in APC(–/–) colon carcinoma. *Science* **1997**; *275*: 1784–1787.
242 Liu, C.M., Li, Y.M., Semenov, M., Han, C., Baeg, G.H., Tan, Y., Zhang, Z.H., Lin, X.H., and He, X. Control of beta-catenin phosphorylation/degradation by a dual-kinase mechanism. *Cell* **2002**; *108*: 837–847.
243 Harwood, A.J. Signal transduction in development: Holding the key. *Dev Cell* **2002**; *2*: 384–385.
244 Clevers, H., and Van de Wetcring, M. TCF/LEF factors earn their wings. *Trends Genet* **1997**; *13*: 485–489.
245 Hurlstone, A., and Clevers, H. T-cell factors: turn-ons and turn-offs. *EMBO J* **2002**; *21*: 2303–2311.
246 Bienz, M., and Clevers, H. Armadillo/beta-catenin signals in the nucleus – proof beyond a reasonable doubt? *Nat Cell Biol* **2003**; *5*: 179–182.
247 Mao, J.H., Wang, J.Y., Liu, B., Pan, W.J., Farr, G.H., Flynn, C., Yuan, H.D., Takada, S., Kimelman, D., Li, L., and Wu, D.Q. Low-density lipoprotein receptor-related protein–5 binds to Axin and regulates the canonical Wnt signaling pathway. *Mol Cell* **2001**; *7*: 801–809.
248 DeLise, A.M., and Tuan, R.S. Alterations in the spatiotemporal expression pattern and function of N-cadherin inhibit cellular condensation and chondrogenesis of limb mesenchymal cells in vitro. *J Cell Biochem* **2002**; *87*: 342–359.
249 Galceran, J., Farinas, I., Depew, M.J., Clevers, H., and Grosschedl, R. Wnt3a(–/–)-like phenotype and limb deficiency in Lef1(–/–)Tcf1(–/–) mice. *Genes Dev* **1999**; *13*: 709–717.
250 Barrow, J.R., Thomas, K.R., Boussadia-Zahui, O., Moore, R., Kemler, R., Capecchi, M.R., and McMahon, A.P. Ectodermal Wnt3/beta-catenin signaling is required for the establishment and maintenance of the apical ectodermal ridge. *Genes Dev* **2003**; *17*: 394–409.
251 Soshnikova, N., Zechner, D., Huelsken, J., Mishina, Y., Behringer, R.R., Taketo, M.M., Crenshaw, E.B., and Birchmeier,

W. Genetic interaction between Wnt/beta-catenin and BMP receptor signaling during formation of the AER and the dorsal-ventral axis in the limb. *Genes Dev* **2003**; *17*: 1963–1968.

252 Maretto, S., Cordenonsi, M., Dupont, S., Braghetta, P., Broccoli, V., Hassan, A.B., Volpin, D., Bressan, G.M., and Piccolo, S. Mapping Wnt/beta-catenin signaling during mouse development and in colorectal tumors. *Proc Natl Acad Sci USA* **2003**; *100*: 3299–3304.

253 Strutt, D.I., Weber, U., and Mlodzik, M. The role of RhoA in tissue polarity and Frizzled signalling. *Nature* **1997**; *387*: 292–295.

254 Boutros, M., Paricio, N., Strutt, D.I., and Mlodzik, M. Dishevelled activates JNK and discriminates between JNK pathways in planar polarity and wingless signaling. *Cell* **1998**; *94*: 109–118.

255 Li, L., Yuan, H.D., Xie, W., Mao, J.H., Caruso, A.M., McMahon, A., Sussman, D.J., and Wu, D.Q. Dishevelled proteins lead to two signaling pathways – Regulation of LEF-1 and c-Jun N-terminal kinase in mammalian cells. *J Biol Chem* **1999**; *274*: 129–134.

256 Heisenberg, C.P., Tada, M., Rauch, G.J., Saude, L., Concha, M.L., Geisler, R., Stemple, D.L., Smith, J.C., and Wilson, S.W. Silberblick/Wnt11 mediates convergent extension movements during zebrafish gastrulation. *Nature* **2000**; *405*: 76–81.

257 Tada, M., and Smith, J.C. Xwnt11 is a target of Xenopus Brachyury: regulation of gastrulation movements via Dishevelled, but not through the canonical Wnt pathway. *Development* **2000**; *127*: 2227–238.

258 Habas, R., Dawid, I.B., and He, X. Coactivation of Rac and Rho by Wnt/Frizzled signaling is required for vertebrate gastrulation. *Genes Dev* **2003**; *17*: 295–309.

259 Habas, R., Kato, Y., and He, X. Wnt/Frizzled activation of Rho regulates vertebrate gastrulation and requires a novel formin homology protein Daam1. *Cell* **2001**; *107*: 843–854.

260 Sheldahl, L.C., Slusarski, D.C., Pandur, P., Miller, J.R., Kuhl, M., and Moon, R.T. Dishevelled activates Ca2+ flux, PKC, and CamKII in vertebrate embryos. *J Cell Biol* **2003**; *161*: 769–777.

261 Slusarski, D.C., YangSnyder, J., Busa, W.B., and Moon, R.T. Modulation of embryonic intracellular Ca2+ signaling by Wnt–5A. *Dev Biol* **1997**; *182*: 114–120.

262 Kilian, B., Mansukoski, H., Barbosa, F.C., Ulrich, F., Tada, M., and Heisenberg, C.P. The role of Ppt/Wnt5 in regulating cell shape and movement during zebrafish gastrulation. *Mech Dev* **2003**; *120*: 467–476.

263 Topol, L., Jiang, X.Y., Choi, H., Garrett-Beal, L., Carolan, P.J., and Yang, Y.Z. (2003). Wnt–5a inhibits the canonical Wnt pathway by promoting GSK–3–independent beta-catenin degradation. *J Cell Biol* **2003**; *162*: 899–908.

264 Chen, A.E., Ginty, D.D., and Fan, C.M. Protein kinase A signalling via CREB controls myogenesis induced by Wnt proteins. *Nature* **2005**; *433*: 317–322.

265 Price, M.A., and Kalderon, D. Proteolysis of the Hedgehog signaling effector Cubitus interruptus requires phosphoryla-

tion by glycogen synthase kinase 3 and casein kinase 1. *Cell* **2002**; *108*: 823–835.

266 Apionishev, S., Katanayeva, N.M., Marks, S.A., Kalderon, D., and Tomlinson, A. Drosophila Smoothened phosphorylation sites essential for Hedgehog signal transduction. *Nat Cell Biol* **2005**; *7*: 86.

267 Mao, B.Y., Wu, W., Li, Y., Hoppe, D., Stannek, P., Glinka, A., and Niehrs, C. LDL-receptor-related protein 6 is a receptor for Dickkopf proteins. *Nature* **2001**; *411*: 321–325.

268 Mao, B.Y., Wu, W., Davidson, G., Marhold, J., Li, M.F., Mechler, B.M., Delius, H., Hoppe, D., Stannek, P., Walter, C., Glinka, A., and Niehrs, C. Kremen proteins are Dickkopf receptors that regulate Wnt/beta-catenin signalling. *Nature* **2002**; *417*: 664–667.

269 Mukhopadhyay, M., Shtrom, S., Rodriguez-Esteban, C., Chen, L., Tsukui, T., Gomer, L., Dorward, D.W., Glinka, A., Grinberg, A., Huang, S.P., Niehrs, C., Belmonte, J.C.I., and Westphal, H. Dickkopf1 is required for embryonic head induction and limb morphogenesis in the mouse. *Dev Cell* **2001**; *1*: 423–434.

270 Grotewold, L., and Ruther, U. The Wnt antagonist Dickkopf–1 is regulated by Bmp signaling and c-Jun and modulates programmed cell death. *EMBO J* **2002**; *21*: 966–975.

271 Adamska, M., MacDonald, B.T., and Meisler, M.H. doubleridge, a mouse mutant with defective compaction of the apical ectodermal ridge and normal dorsal-ventral patterning of the limb. *Dev Biol* **2003**; *255*: 350–362.

272 Adamska, M., MacDonald, B.T., Sarmast, Z.H., Oliver, E.R., and Meisler, M.H. En1 and Wnt7a interact with Dkk1 during limb development in the mouse. *Dev Biol* **2004**; *272*: 134–144.

273 Wu, W., Glinka, A., Delius, H., and Niehrs, C. Mutual antagonism between dickkopf1 and dickkopf2 regulates Wnt/beta-catenin signalling. *Curr Biol* **2000**; *10*: 1611–1614.

274 Yang, Y. Wnts and wing: Wnt signaling in vertebrate limb development and musculoskeletal morphogenesis. *Birth Defects Res C Embryo Today* **2003**; *69*: 305–317.

275 Kawakami, Y., Capdevila, J., Buscher, D., Itoh, T., Esteban, C.R., and Belmonte, J.C.I. WNT signals control FGF-dependent limb initiation and AER induction in the chick embryo. *Cell* **2001**; *104*: 891–900.

276 Ng, J.K., Kawakami, Y., Buscher, D., Raya, A., Itoh, T., Koth, C.M., Esteban, C.R., Rodriguez-Leon, J., Garrity, D.M., Fishman, M.C., and Belmonte, J.C.I. The limb identity gene Tbx5 promotes limb initiation by interacting with Wnt2b and Fgf10. *Development* **2002**; *129*: 5161–5170.

277 Agarwal, P., Wylie, J.N., Galceran, J., Arkhitko, O., Li, C.L., Deng, C.X., Grosschedl, R., and Bruneau, B.G. Tbx5 is essential for forelimb bud initiation following patterning of the limb field in the mouse embryo. *Development* **2003**; *130*: 623–633.

278 Narita, T., Sasaoka, S., Udagawa, K., Ohyama, T., Wada, N., Nishimatsu, S.I., Takada, S., and Nohno, T. Wnt10a is involved in AER formation during chick limb development. *Dev Dyn* **2005** (in press).

279 Parr, B.A., and McMahon, A.P. Dorsalizing signal Wnt–7a required for normal polarity of D-V and A-P axes of mouse limb. *Nature* **1995**; *374*: 350–353.

280 Riddle, R.D., Ensini, M., Nelson, C., Tsuchida, T., Jessell, T.M., and Tabin, C. Induction of the Lim Homeobox Gene Lmx1 by Wnt7a Establishes Dorsoventral Pattern in the Vertebrate Limb. *Cell* **1995**; *83*: 631–640.

281 Caricasole, A., Ferraro, T., Iacovelli, L., Barletta, E., Caruso, A., Melchiorri, D., Terstappen, G.C., and Nicoletti, F. Functional characterization of WNT7A signaling in PC12 cells – Interaction with a FZD5 center dot LRP6 receptor complex and modulation by dickkopf proteins. *J Biol Chem* **2003**; *278*: 37024–37031.

282 Church, V., Nohno, T., Linker, C., Marcelle, C., and Francis-West, P. Wnt regulation of chondrocyte differentiation. *J Cell Sci* **2002**; *115*: 4809–4818.

283 Hartmann, C., and Tabin, C.J. Dual roles of Wnt signaling during chondrogenesis in the chicken limb. *Development* **2000**; *127*: 3141–3159.

284 Yang, Y.Z., Topol, L., Lee, H., and Wu, J.L. Wnt5a and Wnt5b exhibit distinct activities in coordinating chondrocyte proliferation and differentiation. *Development* **2003**; *130*: 1003–1015.

285 Hartmann, C., and Tabin, C.J. Wnt–14 plays a pivotal role in inducing synovial joint formation in the developing appendicular skeleton. *Cell* **2001**; *104*: 341–351.

286 Guo, X.Z., Day, T.F., Jiang, X.Y., Garrett-Beal, L., Topol, L., and Yang, Y.Z. Wnt/beta-catenin signaling is sufficient and necessary for synovial joint formation. *Genes Dev* **2004**; *18*: 2404–2417.

287 Akiyama, H., Lyons, J.P., Mori-Akiyama, Y., Yang, X.H., Zhang, R., Zhang, Z.P., Deng, J.M., Taketo, M.M., Nakamura, T., Behringer, R.R., McCrea, P.D., and de Crombrugghe, B. Interactions between Sox9 and beta-catenin control chondrocyte differentiation. *Genes Dev* **2004**; *18*: 1072–1087.

288 Oberlender, S.A., and Tuan, R.S. Expression and functional involvement of N-cadherin in embryonic limb chondrogenesis. *Development* **1994**; *120*: 177–187.

289 Yamaguchi, T.P., Bradley, A., McMahon, A.P., and Jones, S. A Wnt5a pathway underlies outgrowth of multiple structures in the vertebrate embryo. *Development* **1999**; *126*: 1211–1223.

290 Hogan, B.L.M. Bone morphogenetic proteins: Multifunctional regulators of vertebrate development. *Genes Dev* **1996**; *10*: 1580–1594.

291 Zhao, G.Q. Consequences of knocking out BMP signaling in the mouse. *Genesis* **2003**; *35*: 43–56.

292 Yoon, B.S., and Lyons, K.M. Multiple functions of BMPs in chondrogenesis. *J Cell Biochem* **2004**; *93*: 93–103.

293 Ahn, K., Mishina, Y., Hanks, M.C., Behringer, R.R., and Crenshaw, E.B. BMPR-IA signaling is required for the formation of the apical ectodermal ridge and dorsal-ventral patterning of the limb. *Development* **2001**; *128*: 4449–4461.

294 Pizette, S., Abate-Shen, C., and Niswander, L. BMP controls proximodistal outgrowth, via induction of the apical ectodermal ridge, and dorsoventral patterning in the vertebrate limb. *Development* **2001**; *128*: 4463–4474.

295 Solloway, M.J., Dudley, A.T., Bikoff, E.K., Lyons, K.M., Hogan, B.L.M., and Robertson, E.J. Mice lacking Bmp6 function. *Dev Genet* **1998**; *22*: 321–339.

296 Nakashima, M., Toyono, T., Akamine, A., and Joyner, A. Expression of growth/differentiation factor 11, a new member of the BMP/TGF beta superfamily during mouse embryogenesis. *Mech Dev* **1999**; *80*: 185–189.

297 Settle, S.H., Rountree, R.B., Sinha, A., Thacker, A., Higgins, K., and Kingsley, D.M. Multiple joint and skeletal patterning defects caused by single and double mutations in the mouse Gdf6 and Gdf5 genes. *Dev Biol* **2003**; *254*: 116–130.

298 Avsian-Kretchmer, O., and Hsueh, A.J.W. Comparative genomic analysis of the eight-membered ring cystine knot-containing bone morphogenetic protein antagonists. *Mol Endocrinol* **2004**; *18*: 1–12.

299 Suzuki, A., Kaneko, E., Maeda, J., and Ueno, N. Mesoderm induction by BMP–4 and –7 heterodimers. *Biochem Biophys Res Commun* **1997**; *232*: 153–156.

300 Nishimatsu, S., and Thomsen, G.H. Ventral mesoderm induction and patterning by bone morphogenetic protein heterodimers in Xenopus embryos. *Mech Dev* **1998**; *74*: 75–88.

301 Tendijke, P., Yamashita, H., Sampath, T.K., Reddi, A.H., Estevez, M., Riddle, D.L., Ichijo, H., Heldin, C.H., and Miyazono, K. Identification of Type-I receptors for Osteogenic Protein–1 and Bone Morphogenetic Protein–4. *J Biol Chem* **1994**; *269*: 16985–16988.

302 Nishitoh, H., Ichijo, H., Kimura, M., Matsumoto, T., Makishima, F., Yamaguchi, A., Yamashita, H., Enomoto, S., and Miyazono, K. (1996). Identification of type I and type II serine/threonine kinase receptors for growth/differentiation factor–5. *J Biol Chem* **1996**; *271*: 21345–21352.

303 Dewulf, N., Verschueren, K., Lonnoy, O., Moren, A., Grimsby, S., Vandespiegle, K., Miyazono, K., Huylebroeck, D., and Tendijke, P. Distinct spatial and temporal expression patterns of 2 Type-I receptors for bone morphogenetic proteins during mouse embryogenesis. *Endocrinology* **1995**; *136*: 2652–2663.

304 Zou, H.Y., Wieser, R., Massague, J., and Niswander, L. Distinct roles of type I bone morphogenetic protein receptors in the formation and differentiation of cartilage. *Genes Dev* **1997**; *11*: 2191–2203.

305 Liu, F., Ventura, F., Doody, J., and Massague, J. Human Type-II receptor for Bone Morphogenic Proteins (Bmps) – Extension of the 2–kinase receptor model to the Bmps. *Mol Cell Biol* **1995**; *15*: 3479–3486.

306 Nohno, T., Ishikawa, T., Saito, T., Hosokawa, K., Noji, S., Wolsing, D.H., and Rosenbaum, J.S. Identification of a Human Type-II Receptor for Bone Morphogenetic Protein–4 That Forms Differential Heteromeric Complexes with Bone Morphogenetic Protein Type-I Receptors. *J Biol Chem* **1995**; *270*: 22522–22526.

307 Rosenzweig, B.L., Imamura, T., Okadome, T., Cox, G.N., Yamashita, H., Tendijke, P., Heldin, C.H., and Miyazono, K. Cloning and characterization of a Human Type-II receptor for Bone Morphogenetic Proteins. *Proc Hatl Acad Sci USA* **1995**; *92*: 7632–7636.

308 Hoodless, P.A., Haerry, T., Abdollah, S., Stapleton, M., Oconnor, M.B., Attisano, L., and Wrana, J.L. MADR1, a MAD-re-

lated protein that functions in BMP2 signaling pathways. *Cell* **1996**; *85*: 489–500.

309 Kretzschmar, M., Liu, F., Hata, A., Doody, J., and Massague, J. The TGF-β family mediator Smad1 is phosphorylated directly and activated functionally by the BMP receptor kinase. *Genes Dev* **1997**; *11*: 984–995.

310 Massague, J. TGF signal transduction. *Annu Rev Biochem* **1998**; *67*, 753–791.

311 Derynck, R., Zhang, Y., and Feng, X.H. Smads: Transcriptional activators of TGF-beta responses. *Cell* **1998**; *95*: 737–740.

312 Wrana, J.L. Crossing Smads. Science's STKE, (www.stke.org/cgi/content/full/OC_sig-trans;2000/2023/re2001), **2000**.

313 Attisano, L., and Wrana, J.L. Smads as transcriptional co-modulators. *Curr Opin Cell Biol* **2000**; *12*: 235–243.

314 Massague, J., and Wotton, D. Transcriptional control by the TGF-beta/Smad signaling system. *EMBO J* **2000**; *19*: 1745–1754.

315 Liu, F., Massague, J., and Altaba, A.R.I. Carboxy-terminally truncated Gli3 proteins associate with Smads. *Nat Genet* **1998**; *20*: 325–326.

316 Drossopoulou, G., Lewis, K.E., Sanz-Ezquerro, J.J., Nikbakht, N., McMahon, A.P., Hofmann, C., and Tickle, C. A model for anteroposterior patterning of the vertebrate limb based on sequential long- and short-range Shh signalling and Bmp signalling. *Development* **2000**; *127*: 1337–1348.

317 Afrakhte, M., Moren, A., Jossan, S., Itoh, S., Westermark, B., Heldin, C.H., Heldin, N.E., and ten Dijke, P. Induction of inhibitory Smad6 and Smad7 mRNA by TGF-beta family members. *Biochem Biophys Res Commun* **1998**; *249*: 505–511.

318 Zhu, H.T., Kavsak, P., Abdollah, S., Wrana, J.L., and Thomsen, G.H. A SMAD ubiquitin ligase targets the BMP pathway and affects embryonic pattern formation. *Nature* **1999**; *400*: 687–693.

319 Kretzschmar, M., Doody, J., Timokhina, I., and Massague, J. A mechanism of repression of TGF beta/Smad signaling by oncogenic Ras. *Genes Dev* **1999**; *13*: 804–816.

320 Derynck, R., and Zhang, Y.E. Smad-dependent and Smad-independent pathways in TGF-beta family signalling. *Nature* **2003**; *425*: 577–584.

321 Piccolo, S., Sasai, Y., Lu, B., and DeRobertis, E.M. Dorsoventral patterning in xenopus: Inhibition of ventral signals by direct binding of Chordin to BMP–4. *Cell* **1996**; *86*: 589–598.

322 Zimmerman, L.B., DeJesusEscobar, J.M., and Harland, R.M. The Spemann organizer signal noggin binds and inactivates bone morphogenetic protein 4. *Cell* **1996**; *86*: 599–606.

323 Hsu, D.R., Economides, A.N., Wang, X.R., Eimon, P.M., and Harland, R.M. The Xenopus dorsalizing factor gremlin identifies a novel family of secreted proteins that antagonize BMP activities. *Mol Cell* **1998**; *1*: 673–683.

324 Brunet, L.J., McMahon, J.A., McMahon, A.P., and Harland, R.M. Noggin, cartilage morphogenesis, and joint formation in the mammalian skeleton. *Science* **1998**; *280*: 1455–1457.

325 Merino, R., Ganan, Y., Macias, D., Economides, A.N., Sampath, K.T., and Hurle, J.M. Morphogenesis of digits in the avian limb is controlled by FGFs, TGF beta s, and noggin through BMP signaling. *Dev Biol* **1998**; *200*: 35–45.

326 Francis-West, P.H., Parish, J., Lee, K., and Archer, C.W. BMP/GDF-signalling interactions during synovial joint development. *Cell Tissue Res* **1999**; *296*: 111–119.

327 Capdevila, J., Tsukui, T., Esteban, C.R., Zappavigna, V., and Belmonte, J.C.I. Control of vertebrate limb outgrowth by the proximal factor Meis2 and distal antagonism of BMPs by Gremlin. *Mol Cell* **1999**; *4*: 839–849.

328 Merino, R., Rodriguez-Leon, J., Macias, D., Ganan, Y., Economides, A.N., and Hurle, J.M. The BMP antagonist Gremlin regulates outgrowth, chondrogenesis and programmed cell death in the developing limb. *Development* **1999**; *126*: 5515–5522.

329 Zuniga, A., Haramis, A.P.G., McMahon, A.P., and Zeller, R. Signal relay by BMP antagonism controls the SHH/FGF4 feedback loop in vertebrate limb buds. *Nature* **1999**; *401*: 598–602.

330 Pearce, J.J.H., Penny, G., and Rossant, J. A mouse cerberus/Dan-related gene family. *Dev Biol* **1999**; *209*: 98–110.

331 Khokha, M.K., Hsu, D., Brunet, L.J., Dionne, M.S., and Harland, R.M. Gremlin is the BMP antagonist required for maintenance of Shh and Fgf signals during limb patterning. *Nat Genet* **2003**; *34*: 303–307.

332 Michos, O., Panman, L., Vintersten, K., Beier, K., Zeller, R., and Zuniga, A. Gremlin-mediated BMP antagonism induces the epithelial-mesenchymal feedback signaling controlling metanephric kidney and limb organogenesis. *Development* **2004**; *131*: 3401–3410.

333 Pizette, S., and Niswander, L. BMPs negatively regulate structure and function of the limb apical ectodermal ridge. *Development* **1999**; *126*: 883–894.

334 Selever, J., Liu, W., Lu, M.F., Behringer, R.R., and Martin, J.F. Bmp4 in limb bud mesoderm regulates digit pattern by controlling AER development. *Dev Biol* **2004**; *276*: 268–279.

335 Dunn, N.R., Winnier, G.E., Hargett, L.K., Schrick, J.J., Fogo, A.B., and Hogan, B.L.M. Haploinsufficient phenotypes in Bmp4 heterozygous null mice and modification by mutations in Gli3 and Alx4. *Dev Biol* **1997**; *188*: 235–247.

336 Katagiri, T., Boorla, S., Frendo, J.L., Hogan, B.L.M., and Karsenty, G. Skeletal abnormalities in doubly heterozygous Bmp4 and Bmp7 mice. *Dev Genet* **1998**; *22*: 340–348.

337 Zou, H.Y., and Niswander, L. Requirement for BMP signaling in interdigital apoptosis and scale formation. *Science* **1996**; *272*: 738–741.

338 Zhang, Z.Y., Yu, X.Y., Zhang, Y.D., Geronimo, B., Lovlie, A., Fromm, S.H., and Chen, Y.P. Targeted misexpression of constitutively active BMP receptor-IB causes bifurcation, duplication, and posterior transformation of digit in mouse limb. *Dev Biol* **2000**; *220*: 154–167.

339 Guha, U., Gomes, W.A., Kobayashi, T., Pestell, R.G., and Kessler, J.A. *In vivo* evidence that BMP signaling is necessary for apoptosis in the mouse limb. *Dev Biol* **2002**; *249*: 108–120.

340 Capdevila, J., and Johnson, R.L. Endogenous and ectopic expression of noggin suggests a conserved mechanism for regulation of BMP function during limb and somite patterning. *Dev Biol* **1998**; *197*: 205–217.

341 Pizette, S., and Niswander, L. BMPs are required at two steps of limb chondrogenesis: Formation of prechondrogenic condensations and their differentiation into chondrocytes. *Dev Biol* **2000**; *219*: 237–249.

342 Minina, E., Wenzel, H.M., Kreschel, C., Karp, S., Gaffield, W., McMahon, A.P., and Vortkamp, A. BMP and Ihh/PTHrP signaling interact to coordinate chondrocyte proliferation and differentiation. *Development* **2001**; *128*: 4523–4534.

343 Minina, E., Kreschel, C., Naski, M.C., Ornitz, D.M., and Vortkamp, A. Interaction of FGF, Ihh/Pthlh, and BMP signaling integrates chondrocyte proliferation and hypertrophic differentiation. *Dev Cell* **2002**; *3*: 439–449.

344 Long, F.X., Chung, U.I., Ohba, S., McMahon, J., Kronenberg, H.M., and McMahon, A.P. Ihh signaling is directly required for the osteoblast lineage in the endochondral skeleton. *Development* **2004**; *131*: 1309–1318.

345 King, J.A., Marker, P.C., Seung, K.J., and Kingsley, D.M. Bmp5 and the molecular, skeletal, and soft-tissue alterations in short ear mice. *Dev Biol* **1994**; *166*: 112–122.

346 Kingsley, D.M. What do Bmps do in mammals – clues from the mouse short-ear mutation. Trends *Genet* **1994**; *10*: 16–21.

347 Dudley, A.T., Lyons, K.M., and Robertson, E.J. A requirement for Bone Morphogenetic Protein–7 during development of the mammalian kidney and eye. *Genes Dev* **1995**; *9*: 2795–2807.

348 Luo, G., Hofmann, C., Bronckers, A., Sohocki, M., Bradley, A., and Karsenty, G. Bmp–7 is an inducer of nephrogenesis, and is also required for eye development and skeletal patterning. *Genes Dev* **1995**; *9*, 2808–2820.

349 Storm, E.E., Huynh, T.V., Copeland, N.G., Jenkins, N.A., Kingsley, D.M., and Lee, S.J. Limb alterations in brachypodism mice due to mutations in a new member of the Tgf-beta-superfamily. *Nature* **1994**; *368*: 639–643.

350 Storm, E.E., and Kingsley, D.M. Joint patterning defects caused by single and double mutations in members of the bone morphogenetic protein (BMP) family. *Development* **1996**; *122*: 3969–3979.

351 Storm, E.E., and Kingsley, D.M. GDF5 coordinates bone and joint formation during digit development. *Dev Biol* **1999**; *209*: 11–27.

352 Gamer, L.W., Cox, K.A., Small, C., and Rosen, V. Gdf11 is a negative regulator of chondrogenesis and myogenesis in the developing chick limb. *Dev Biol* **2001**; *229*, 407–420.

353 McPherron, A.C., Lawler, A.M., and Lee, S.J. Regulation of anterior posterior patterning of the axial skeleton by growth differentiation factor 11. *Nat Genet* **1999**; *22*: 260–264.

354 Kawakami, Y., Ishikawa, T., Shimabara, M., Tanda, N., EnomotoIwamoto, M., Iwamoto, M., Kuwana, T., Ueki, A., Noji, S., and Nohno, T. BMP signaling during bone pattern determination in the developing limb. *Development* **1996**; *122*: 3557–3566.

355 Baur, S.T., Mai, J.J., and Dymecki, S.M. Combinatorial signaling through BMP receptor IB and GDF5: shaping of the distal mouse limb and the genetics of distal limb diversity. *Development* **2000**; *127*: 605–619.

356 Yi, S.E., Daluiski, A., Pederson, R., Rosen, V., and Lyons, K.M. The type IBMP receptor BMPRIB is required for chondrogenesis in the mouse limb. *Development* **2000**; *127*: 621–630.

357 Francis-West, P.H., Abdelfattah, A., Chen, P., Allen, C., Parish, J., Ladher, R., Allen, S., MacPherson, S., Luyten, F.P., and Archer, C.W. Mechanisms of GDF–5 action during skeletal development. *Development* **1999**; *126*: 1305–1315.

358 Merino, R., Macias, D., Ganan, Y., Economides, A.N., Wang, X., Wu, Q., Stahl, N., Sampath, K.T., Varona, P., and Hurle, J.M. Expression and function of Gdf–5 during digit skeletogenesis in the embryonic chick leg bud. *Dev Biol* **1999**; *206*, 33–45.

359 Tsumaki, N., Tanaka, K., Arikawa-Hirasawa, E., Nakase, T., Kimura, T., Thomas, J.T., Ochi, T., Luyten, F.P., and Yamada, Y. Role of CDMP–1 in skeletal morphogenesis: Promotion of mesenchymal cell recruitment and chondrocyte differentiation. *J Cell Biol* **1999**; *144*, 161–173.

360 Dahn, R.D., and Fallon, J.F. Interdigital regulation of digit identity and homeotic transformation by modulated BMP signaling. Science **2000**; *289*: 438–441.

361 Basler, K., and Struhl, G. Compartment boundaries and the control of drosophila limb pattern by Hedgehog protein. *Nature* **1994**; *368*: 208–214.

362 Yang, Y., Drossopoulou, G., Chuang, P.T., Duprez, D., Marti, E., Bumcrot, D., Vargesson, N., Clarke, J., Niswander, L., McMahon, A., and Tickle, C. Relationship between dose, distance and time in Sonic Hedgehog-mediated regulation of anteroposterior polarity in the chick limb. *Development* **1997**; *124*: 4393–4404.

363 Woloshin, P., Song, K.N., Degnin, C., Killary, A.M., Goldhamer, D.J., Sassoon, D., and Thayer, M.J. Msx1 inhibits Myod expression in fibroblast X 10t1/2 cell hybrids. *Cell* **1995**; *82*: 611–620.

364 Hu, G.Z., Lee, H., Price, S.M., Shen, M.M., and Abate-Shen, C. Msx homeobox genes inhibit differentiation through upregulation of cyclin D1. *Development* **2001**; *128*:2373–2384.

365 Marazzi, G., Wang, Y.Q., and Sassoon, D. Msx2 is a transcriptional regulator in the BMP4–mediated programmed cell death pathway. *Dev Biol* **1997**; *186*:127–138.

366 Satokata, I., and Maas, R. Msx1 deficient mice exhibit cleft-palate and abnormalities of craniofacial and tooth development. *Nat Genet* **1994**; *6*: 348–356

367 Satokata, I., Ma, L., Ohshima, H., Bei, M., Woo, I., Nishizawa, K., Maeda, T., Takano, Y., Uchiyama, M., Heaney, S., Peters, H., Tang, Z.Q., Maxson, R., and Maas, R. Msx2 deficiency in mice causes pleiotropic defects in bone growth and ectodermal organ formation. *Nat Genet* **2000**; *24*: 391–395.

368 Lee, H., and Kimelman, D. A dominant-negative form of p63 is required for epidermal proliferation in zebrafish. *Dev Cell* **2002**; *2*: 607–616.

369 Celli, J., Duijf, P., Hamel, B.C.J., Bamshad, M., Kramer, B., Smits, A.P.T., Newbury-Ecob, R., Hennekam, R.C.M., Van Buggenhout, G., van Haeringen, B., Woods, C.G., van Essen, A.J., de Waal, R., Vriend, G., Haber, D.A., Yang, A., McKeon, F., Brunner, H.G., and van Bokhoven, H. Heterozygous germline mutations in the p53 homolog p63 are the cause of EEC syndrome. *Cell* **1999**; *99*: 143–153.

370 Ingham, P.W., and McMahon, A.P. Hedgehog signaling in animal development: paradigms and principles. *Genes Dev* **2001**; *15*: 3059–3087.

371 Pathi, S., Pagan-Westphal, S., Baker, D.P., Garber, E.A., Rayhorn, P., Bumcrot, D., Tabin, C.J., Pepinsky, R.B., and Williams, K.P. Comparative biological responses to human Sonic, Indian, and Desert hedgehog. *Mech Dev* **2001**; *106*: 107–117.

372 Echelard, Y., Epstein, D.J., Stjacques, B., Shen, L., Mohler, J., McMahon, J.A., and McMahon, A.P. Sonic-Hedgehog, a member of a family of putative signaling molecules, is implicated in the regulation of CNS polarity. *Cell* **1993**; *75*: 1417–1430.

373 Krauss, S., Concordet, J.P., and Ingham, P.W. A functionally conserved homology of the Drosophila segment polarity gene-hh is expressed in tissues with polarizing activity in zebrafish embryos. *Cell* **1993**; *75*: 1431–1444.

374 Chang, D.T., Lopez, A., Vonkessler, D.P., Chiang, C., Simandl, B.K., Zhao, R.B., Seldin, M.F., Fallon, J.F., and Beachy, P.A. Products, genetic-linkage and limb patterning activity of a murine Hedgehog gene. *Development* **1994**; *120*: 3339–3353.

375 Roelink, H., Augsburger, A., Heemskerk, J., Korzh, V., Norlin, S., Altaba, A.R.I., Tanabe, Y., Placzek, M., Edlund, T., Jessell, T.M., and Dodd, J. Floor plate and motor-neuron induction by Vhh–1, a vertebrate homolog of Hedgehog expressed by the notochord. *Cell* **1994**; *76*: 761–775.

376 Vortkamp, A., Lee, K., Lanske, B., Segre, G.V., Kronenberg, H.M., and Tabin, C.J. Regulation of rate of cartilage differentiation by Indian hedgehog and PTH-related protein. *Science* **1996**; *273*: 613–622.

377 St-Jacques, B., Hammerschmidt, M., and McMahon, A.P. Indian hedgehog signaling regulates proliferation and differentiation of chondrocytes and is essential for bone formation. *Genes Dev* **1999**; *13*: 2072–2086.

378 Chiang, C., Litingtung, Y., Harris, M.P., Simandl, B.K., Li, Y., Beachy, P.A., and Fallon, J.F. Manifestation of the limb prepattern: Limb development in the absence of sonic hedgehog function. *Dev Biol* **2001**; *236*: 421–435.

379 Kraus, P., Fraidenraich, D., and Loomis, C.A. Some distal limb structures develop in mice lacking Sonic hedgehog signaling. *Mech Dev* **2001**; *100*: 45–58.

380 Ros, M.A., Dahn, R.D., Fernandez-Teran, M., Rashka, K., Caruccio, N.C., Hasso, S.M., Bitgood, J.J., Lancman, J.J., and Fallon, J.F. The chick oligozeugodactyly (ozd) mutant lacks sonic hedgehog function in the limb. *Development* **2003**; *130*: 527–537.

381 Pagan, S.M., Ros, M.A., Tabin, C., and Fallon, J.F. Surgical removal of limb bud Sonic hedgehog results in posterior skeletal defects. *Dev Biol* **1996**; *180*: 35–40.

382 Lewis, P.M., Dunn, M.P., McMahon, J.A., Logan, M., Martin, J.F., St-Jacques, B., and McMahon, A.P. Cholesterol modification of sonic hedgehog is required for long-range signaling activity and effective modulation of signaling by Ptc1. *Cell* **2001**; *105*: 599–612.

383 Mann, R.K., and Beachy, P.A. Novel lipid modifications of secreted protein signals. *Annu Rev Biochem* **2004**; *73*: 891–923.

384 Lee, J.J., Ekker, S.C., Vonkessler, D.P., Porter, J.A., Sun, B.I., and Beachy, P.A. Autoproteolysis in Hedgehog protein biogenesis. *Science* **1994**; *266*: 1528–1537.

385 Chamoun, Z., Mann, R.K., Nellen, D., von Kessler, D.P., Bellotto, M., Beachy, P.A., and Basler, K. Skinny Hedgehog, an acyltransferase required for palmitoylation and activity of the Hedgehog signal. *Science* **2001**); *293*: 2080–2084.

386 Zeng, X., Goetz, J.A., Suber, L.M., Scott, W.J., Schreiner, C.M., and Robbins, D.J. A freely diffusible form of Sonic hedgehog mediates long-range signalling. *Nature* **2001**; *411*: 716–720.

387 Chen, M.H., Li, Y.J., Kawakami, T., Xu, S.M., and Chuang, P.T. Palmitoylation is required for the production of a soluble multimeric Hedgehog protein complex and long-range signaling in vertebrates. *Genes Dev* **2004**; *18*: 641–659.

388 Caspary, T., Garcia-Garcia, M.J., Huangfu, D.W., Eggenschwiler, J.T., Wyler, M.R., Rakeman, A.S., Alcorn, H.L., and Anderson, K.V. Mouse dispatched homolog1 is required for long-range, but not juxtacrine, Hh signaling. *Curr Biol* **2002**; *12*: 1628–1632.

389 Kawakami, T., Kawcak, T., Li, Y.J., Zhang, W.H., Hu, Y.M., and Chuang, P.T. Mouse dispatched mutants fail to distribute hedgehog proteins and are defective in hedgehog signaling. *Development* **2002**; *129*: 5753–5765.

390 Tian, H., Jeong, J.H., Harfe, B.D., Tabin, C.J., and McMahon, A.P. Mouse Disp1 is required in sonic hedgehog-expressing cells for paracrine activity of the cholesterol-modified ligand. *Development* **2005**; *132*: 133–142.

391 Ma, Y., Erkner, A., Gong, R.Y., Yao, S.Q., Taipale, J., Basler, K., and Beachy, P.A. Hedgehog-mediated patterning of the mammalian embryo requires transporter-like function of dispatched. *Cell* **2002**; *111*: 63–75.

392 Bellaiche, Y., The, I., and Perrimon, N. Tout-velu is a Drosophila homologue of the putative tumour suppressor EXT–1 and is needed for Hh diffusion. *Nature* **1998**; *394*: 85–88.

393 Koziel, L., Kunath, M., Kelly, O.G., and Vortkamp, A. Ext1-dependent heparan sulfate regulates the range of Ihh signaling during endochondral ossification. *Dev Cell* **2004**; *6*: 801–813.

394 Chuang, P.T., and McMahon, A.P. Vertebrate Hedgehog signalling modulated by induction of a Hedgehog-binding protein. *Nature* **1999**; *397*: 617–621.

395 Chuang, P.T., Kawcak, T., and McMahon, A.P. Feedback control of mammalian hedgehog signaling by the hedgehog-binding protein, Hip1, modulates Fgf signaling during branching morphogenesis of the lung. *Genes Dev* **2003**; *17*: 342–347.

396 Marigo, V., Davey, R.A., Zuo, Y., Cunningham, J.M., and Tabin, C.J. Biochemical evidence that Patched is the Hedgehog receptor. *Nature* **1996**; *384*: 176–179.

397 Chen, Y., and Struhl, G. Dual roles for patched in sequestering and transducing hedgehog. *Cell* **1996**; *87*: 553–563.

398 Goodrich, L.V., Milenkovic, L., Higgins, K.M., and Scott, M.P. Altered neural cell fates and medulloblastoma in mouse patched mutants. *Science* **1997**; *277*: 1109–1113.

399 Briscoe, J., Chen, Y., Jessell, T.M., and Struhl, G. A hedgehog-insensitive form of patched provides evidence for direct long-range morphogen activity of Sonic hedgehog in the neural tube. *Mol Cell* **2001**); *7*: 1279–1291.

400 Goodrich, L.V., Johnson, R.L., Milenkovic, L., McMahon, J.A., and Scott, M.P. Conservation of the hedgehog/patched signaling pathway from flies to mice: Induction of a mouse patched gene by Hedgehog. *Genes Dev* **1996**; *10*: 301–312.

401 Denef, N., Neubuser, D., Perez, L., and Cohen, S.M. Hedgehog induces opposite changes in turnover and subcellular localization of patched and smoothened. *Cell* **2000**; *102*: 521–531.

402 Ingham, P.W., Nystedt, S., Nakano, Y., Brown, W., Stark, D., van den Heuvel, M., and Taylor, A.M. Patched represses the Hedgehog signalling pathway by promoting modification of the Smoothened protein. *Curr Biol* **2000**; *10*: 1315–1318.

403 Taipale, J., Cooper, M.K., Maiti, T., and Beachy, P.A. Patched acts catalytically to suppress the activity of Smoothened. *Nature* **2002**; *418*: 892–897.

404 Incardona, J.P., Gruenberg, J., and Roelink, H. Sonic hedgehog induces the segregation of patched and smoothened in endosomes. *Curr Biol* **2002**; *12*: 983–995.

405 Jia, J.H., Tong, C., Wang, B., Luo, L.P., and Jiang, J. Hedgehog signalling activity of Smoothened requires phosphorylation by protein kinase A and casein kinase I. *Nature* **2004**; *432*: 1045–1050.

406 Zhang, X.M., Ramalho-Santos, M., and McMahon, A.P. Smoothened mutants reveal redundant roles for Shh and Ihh signaling including regulation of L/R asymmetry by the mouse node. *Cell* **2001**; *105*: 781–792.

407 Dai, P., Akimaru, H., Tanaka, Y., Maekawa, T., Nakafuku, M., and Ishii, S. Sonic hedgehog-induced activation of the Gli1 promoter is mediated by GLI3. *J Biol Chem* **1999**; *274*: 8143–8152.

408 Altaba, A.R.I. Gli proteins encode context-dependent positive and negative functions: implications for development and disease. *Development* **1999**; *126*: 3205–3216.

409 Sasaki, H., Nishizaki, Y., Hui, C.C., Nakafuku, M., and Kondoh, H. Regulation of Gli2 and Gli3 activities by an amino-terminal repression domain: implication of Gli2 and Gli3 as primary mediators of Shh signaling. *Development* **1999**; *126*: 3915–3924.

410 Lee, J., Platt, K.A., Censullo, P., and Altaba, A.R.I. Gli1 is a target of Sonic hedgehog that induces ventral neural tube development. *Development* **1997**; *124*: 2537–2552.

411 Bai, C.B., Auerbach, W., Lee, J.S., Stephen, D., and Joyner, A.L. Gli2, but not Gli1, is required for initial Shh signaling and ectopic activation of the Shh pathway. *Development* **2002**; *129*: 4753–4761.

412 Marigo, V., Johnson, R.L., Vortkamp, A., and Tabin, C.J. Sonic hedgehog differentially regulates expression of GLI and GLI3 during limb development. *Dev Biol* **1996**; *180*: 273–283.

413 Park, H.L., Bai, C., Platt, K.A., Matise, M.P., Beeghly, A., Hui, C.C., Nakashima, M., and Joyner, A.L. Mouse Gli1 mutants are viable but have defects in SHH signaling in combination with a Gli2 mutation. *Development* **2000**; *127*: 1593–1605.

414 Shin, S.H., Kogerman, P., Lindstrom, E., Toftgard, R., and Biesecker, L.G. GLI3 mutations in human disorders mimic Drosophila Cubitus interruptus protein functions and localization. *Proc Natl Acad Sci USA* **1999**; *96*: 2880–2884.

415 Wang, B.L., Fallon, J.F., and Beachy, P.A. Hedgehog-regulated processing of Gli3 produces an anterior/posterior repressor gradient in the developing vertebrate limb. *Cell* **2000**; *100*: 423–434.

416 Hui, C.C., and Joyner, A.L. A mouse model of Greig Cephalopolysyndactyly syndrome – the extra-toes(J) mutation contains an intragenic deletion of the Gli3 gene. *Nat Genet* **1993**; *3*: 241–246.

417 Mo, R., Freer, A.M., Zinyk, D.L., Crackower, M.A., Michaud, J., Heng, H.H.Q., Chik, K.W., Shi, X.M., Tsui, L.C., Cheng, S.H., Joyner, A.L., and Hui, C.C. Specific and redundant functions of Gli2 and Gli3 zinc finger genes in skeletal patterning and development. *Development* **1997**; *124*: 113–123.

418 Litingtung, Y., Dahn, R.D., Li, Y.N., Fallon, J.F., and Chiang, C. Shh and Gli3 are dispensable for limb skeleton formation but regulate digit number and identity. *Nature* **2002**; *418*: 979–983.

419 Welscher, P.T., Zuniga, A., Kuijper, S., Drenth, T., Goedemans, H.J., Meijlink, F., and Zeller, R. Progression of vertebrate limb development through SHH-mediated counteraction of GLI3. *Science* **2002**; *298*: 827–830.

420 Chen, Y.T., Knezevic, V., Ervin, V., Hutson, R., Ward, Y., and Mackem, S. Direct interaction with Hoxd proteins reverses Gli3-repressor function to promote digit formation downstream of Shh. *Development* **2004**; *131*: 2339–2347.

421 Stamataki, D., Ulloa, F., Tsoni, S.V., Mynett, A., and Briscoe, J. A gradient of Gli activity mediates graded Sonic Hedgehog signaling in the neural tube. *Genes Dev* **2005**; *19*: 626–641.

422 Callahan, C.A., Ofstad, T., Horng, L., Wang, J.K., Zhen, H.H., Coulombe, P.A., and Oro, A.E. MIM/BEG4, a Sonic hedgehog-responsive gene that potentiates Gli-dependent transcription. *Genes Dev* **2004**; *18*: 2724–2729.

423 Lum, L., Zhang, C., Oh, S., Mann, R.K., von Kessler, D.P., Taipale, J., Weis-Garcia, F., Gong, R.Y., Wang, B.L., and Beachy, P.A. Hedgehog signal transduction via smoothened association with a cytoplasmic complex scaffolded by the atypical kinesin, Costal-2. *Mol Cell* **2003**; *12*: 1261–1274.

424 Zhang, W.S., Zhao, Y., Tong, C., Wang, G.L., Wang, B., Jia, J.H., and Jiang, J. Hedgehog-regulated Costal2–kinase complexes control phosphorylation and proteolytic processing of Cubitus interruptus. *Dev Cell* **2005**; *8*: 267–278.

425 Tay, S.Y., Ingham, P.W., and Roy, S. A homologue of the Drosophila kinesin-like protein Costal2 regulates Hedgehog

signal transduction in the vertebrate embryo. *Development* **2005**; *132*: 625–634.

426 Kogerman, P., Grimm, T., Kogerman, L., Krause, D., Unden, A.B., Sandstedt, B., Toftgard, R., and Zaphiropoulos, P.G. Mammalian Suppressor-of-Fused modulates nuclear-cytoplasmic shuttling of GLI–1. *Nat Cell Biol* **1999**; *1*: 312–319.

427 Murone, M., Luoh, S.M., Stone, D., Li, W.L., Gurney, A., Armanini, M., Grey, C., Rosenthal, A., and de Sauvage, F.J. Gli regulation by the opposing activities of Fused and Suppressor of Fused. *Nat Cell Biol* **2000**; *2*: 310–312.

428 Sekimizu, K., Nishioka, N., Sasaki, H., Takeda, H., Karlstrom, R.O., and Kawakami, A. The zebrafish iguana locus encodes Dzip1, a novel zinc-finger protein required for proper regulation of Hedgehog signaling. *Development* **2004**; *131*: 2521–2532.

429 Wolff, C., Roy, S., Lewis, K.E., Schauerte, H., Joerg-Rauch, G., Kirn, A., Weiler, C., Geisler, R., Haffter, P., and Ingham, P.W. Iguana encodes a novel zinc-finger protein with coiled-coil domains essential for Hedgehog signal transduction in the zebrafish embryo. *Genes Dev* **2004**; *18*: 1565–1576.

430 Eggenschwiler, J.T., Espinoza, E., and Anderson, K.V. Rab23 is an essential negative regulator of the mouse Sonic hedgehog signalling pathway. *Nature* **2001**; *412*: 194–198.

431 Gunther, T., Struwe, M., Aguzzi, A., and Schughart, K. Open brain, a new mouse mutant with severe neural-tube defects, shows altered gene-expression patterns in the developing spinal-cord. *Development* **1994**; *120*: 3119–3130.

432 Bulgakov, O.V., Eggenschwiler, J.T., Hong, D.H., Anderson, K.V., and Li, T.S. FKBP8 is a negative regulator of mouse sonic hedgehog signaling in neural tissues. *Development* **2004**; *131*: 2149–2159.

433 Huangfu, D.W., Liu, A.M., Rakeman, A.S., Murcia, N.S., Niswander, L., and Anderson, K.V. Hedgehog signalling in the mouse requires intraflagellar transport proteins. *Nature* **2003**; *426*: 83–87.

434 Wheatley, D.N., Wang, A.M., and Strugnell, G.E. Expression of primary cilia in mammalian cells. *Cell Biol Int* **1996**; *20*: 73–81.

435 Zhang, Q.H., Murcia, N.S., Chittenden, L.R., Richards, W.G., Michaud, E.J., Woychik, R.P., and Yoder, B.K. Loss of the Tg737 protein results in skeletal patterning defects. *Dev Dyn* **2003**; *227*: 78–90.

436 McQueeney, K., and Dealy, C.N. Roles of insulin-like growth factor-I (IGF-I) and IGF-I binding protein–2 (IGFBP2) and–5 (IGFBP5) in developing chick limbs. *Growth Horm Igf Res* **2001**; *11*: 346–363.

437 Dealy, C.N., and Kosher, R.A. IGF-I, insulin and FGFs induce outgrowth of the limb buds of amelic mutant chick embryos. *Development* **1996**; *122*: 1323–1330.

438 Galindo, M.I., Bishop, S.A., Greig, S., and Couso, J.P. Leg patterning driven by proximal-distal interactions and EGFR signaling. *Science* **2002**; *297*: 256–259.

439 Omi, M., Fisher, M., Maihle, N.J., and Dealy, C.N. Studies on epidermal growth factor receptor signaling in vertebrate limb patterning. *Dev Dyn* **2005** (in press).

440 Zeller, R., and Duboule, D. Dorso-ventral limb polarity and origin of the ridge: On the fringe of independence? *Bioessays* **1997**; *19*: 541–546.

441 Tickle, C., and Munsterberg, A. Vertebrate limb development – the early stages in chick and mouse. *Curr Opin Genet Dev* **2001**; *11*: 476–481.

442 Laufer, E., Dahn, R., Orozco, O.E., Yeo, C.Y., Pisenti, J., Henrique, D., Abbott, U.K., Fallon, J.F., and Tabin, C. Expression of radical fringe in limb-bud ectoderm regulates apical ectodermal ridge formation. *Nature* **1997**; *386*: 366–373.

443 RodriguezEsteban, C., Schwabe, J.W.R., DeLaPena, J., Foys, B., Eshelman, B., and Belmonte, J.C.I. Radical fringe positions the apical ectodermal ridge at the dorsoventral boundary of the vertebrate limb. *Nature* **1997**; *386*: 360–366.

444 Moran, J.L., Levorse, J.M., and Vogt, T.F. Limbs move beyond the Radical fringe. *Nature* **1999**; *399*: 742–743.

445 Sidow, A., Bulotsky, M.S., Kerrebrock, A.W., Bronson, R.T., Daly, M.J., Reeve, M.P., Hawkins, T.L., Birren, B.W., Jaenisch, R., and Lander, E.S. Serrate2 is disrupted in the mouse limb-development mutant syndactylism. *Nature* **1997**; *389*: 722–725.

446 Jiang, R.L., Lan, Y., Chapman, H.D., Shawber, C., Norton, C.R., Serreze, D.V., Weinmaster, G., and Gridley, T. Defects in limb, craniofacial, and thymic development in Jagged2 mutant mice. *Genes Dev* **1998**; *12*: 1046–1057.

447 Vasiliauskas, D., Laufer, E., and Stern, C.D. A role for hairy1 in regulating chick limb bud growth. *Dev Biol* **2003**; *262*: 94–106.

448 DeLise, A.M., Fischer, L., and Tuan, R.S. Cellular interactions and signaling in cartilage development. *Osteoarthr Cartil* **2000**; *8*: 309–334.

449 Widelitz, R.B., Jiang, T.X., Murray, B.A., and Chuong, C.M. Adhesion molecules in skeletogenesis .2. Neural Cell-adhesion molecules mediate precartilaginous mesenchymal condensations and enhance chondrogenesis. *J Cell Physiol* **1993**; *156*: 399–411.

450 Poliakov, A., Cotrina, M., and Wilkinson, D.G. Diverse roles of Eph receptors and ephrins in the regulation of cell migration and tissue assembly. *Dev Cell* **2004**; *7*: 465–480.

451 Davy, A., and Soriano, P. Ephrin signaling *in vivo*: Look both ways. *Dev Dyn* **2005**; *232*: 1–10.

452 Wada, N., Kimura, I., Tanaka, H., Ide, H., and Nohno, T. Glycosylphosphatidylinositol-anchored cell surface proteins regulate position-specific cell affinity in the limb bud. *Dev Biol* **1998**; *202*: 244–252.

453 Wada, N., Tanaka, H., Ide, H., and Nohno, T. Ephrin-A2 regulates position-specific cell affinity and is involved in cartilage morphogenesis in the chick limb bud. *Dev Biol* **2003**; *264*: 550–563.

454 Compagni, A., Logan, M., Klein, R., and Adams, R.H. Control of skeletal patterning by EphrinB1–EphB interactions. *Dev Cell* **2003**; *5*: 217–230.

455 Davy, A., Aubin, J., and Soriano, P. Ephrin-B1 forward and reverse signaling are required during mouse development. *Genes Dev* **2004**); *18*: 572–583.

456 Rutledge, J.C., Shourbaji, A.G., Hughes, L.A., Polifka, J.E., Cruz, Y.P., Bishop, J.B., and Generoso, W.M. Limb and lower-body duplications induced by retinoic acid in mice. *Proc Natl Acad Sci USA* **1994**; *91*: 5436–5440.

457 Niederreither, K., Ward, S.J., Dolle, P., and Chambon, P. Morphological and molecular characterization of retinoic acid-induced limb duplications in mice. *Dev Biol* **1996**; *176*: 185–198.

458 Cohn, M.J., and Tickle, C. Developmental basis of limblessness and axial patterning in snakes. *Nature* (1999). *399*, 474–479.

459 Rancourt, D.E., Tsuzuki, T., and Capecchi, M.R. Genetic interaction between Hoxb–5 and Hoxb–6 is revealed by nonallelic noncomplementation. *Genes Dev* **1995**; *9*: 108–122.

460 Gerard, M., Chen, J.Y., Gronemeyer, H., Chambon, P., Duboule, D., and Zakany, J. In vivo targeted mutagenesis of a regulatory element required for positioning the Hoxd–11 and Hoxd–10 expression boundaries. *Genes Dev* **1996**; *10*: 2326–2334.

461 Simeone, A., Acampora, D., Arcioni, L., Andrews, P.W., Boncinelli, E., and Mavilio, F. Sequential activation of Hox2 homeobox genes by retinoic acid in human embryonal carcinoma-cells. *Nature* **1990**; *346*: 763–766.

462 Papalopulu, N., Lovellbadge, R., and Krumlauf, R. The expression of murine Hox–2 genes is dependent on the differentiation pathway and displays a collinear sensitivity to retinoic acid in F9 cells and Xenopus embryos. *Nucleic Acids Res* **1991**; *19*: 5497–5506.

463 Simeone, A., Acampora, D., Nigro, V., Faiella, A., Desposito, M., Stornaiuolo, A., Mavilio, F., and Boncinelli, E. Differential regulation by retinoic acid of the homeobox genes of the 4 Hox loci in human embryonal carcinoma-cells. *Mech Dev* **1991**; *33*: 215–228.

464 Kessel, M. Respecification of vertebral identities by retinoic acid. *Development* **1992**; *115*: 487–501.

465 GibsonBrown, J.J., Agulnik, S.I., Chapman, D.L., Alexiou, M., Garvey, N., Silver, L.M., and Papaioannou, V.E. Evidence of a role for T-box genes in the evolution of limb morphogenesis and the specification of forelimb/hindlimb identity. *Mech Dev* **1996**; *56*: 93–101.

466 Rodriguez-Esteban, C., Tsukui, T., Yonei, S., Magallon, J., Tamura, K., and Belmonte, J.C.I. The T-box genes Tbx4 and Tbx5 regulate limb outgrowth and identity. *Nature* **1999**; *398*: 814–818.

467 Takeuchi, J.K., Koshiba-Takeuchi, K., Matsumoto, K., Vogel-Hopker, A., Naitoh-Matsuo, M., Ogura, K., Takahashi, N., Yasuda, K., and Ogura, T. Tbx5 and Tbx4 genes determine the wing/leg identity of limb buds. *Nature* **1999**; *398*: 810–814.

468 Takeuchi, J.K., Koshiba-Takeuchi, K., Suzuki, T., Kamimura, M., Ogura, K., and Ogura, T. Tbx5 and Tbx4 trigger limb initiation through activation of the Wnt/Fgf signaling cascade. *Development* **2003**; *130*: 2729–2739.

469 Minguillon, C., Del Buono, J., and Logan, M.P. Tbx5 and Tbx4 are not sufficient to determine limb-specific morphologies but

have common roles in initiating limb outgrowth. *Dev Cell* 2005; *8*: 75–84.

470 Naiche, L.A., and Papaioannou, V.E. Loss of Tbx4 blocks hindlimb development and affects vascularization and fusion of the allantois. *Development* 2003; *130*: 2681–2693.

471 Ahn, D.G., Kourakis, M.J., Rohde, L.A., Silver, L.M., and Ho, R.K. T-box gene tbx5 is essential for formation of the pectoral limb bud. *Nature* 2002; *417*: 754–758.

472 Marcil, A., Dumontier, E., Chamberland, M., Camper, S.A., and Drouin, J. Pitx1 and Pitx2 are required for development of hindlimb buds. *Development* 2003; *130*: 45–55.

473 Logan, C., Hornbruch, A., Campbell, I., and Lumsden, A. The role of Engrailed in establishing the dorsoventral axis of the chick limb. *Development* 1997; *124*: 2317–2324.

474 Tanaka, M., Tamura, K., Noji, S., Nohno, T., and Ide, H. Induction of additional limb at the dorsal-ventral boundary of a chick embryo. *Dev Biol* 1997; *182*: 191–203.

475 Loomis, C.A., Harris, E., Michaud, J., Wurst, W., Hanks, M., and Joyner, A.L. The mouse Engrailed–1 gene and ventral limb patterning. *Nature* 1996; *382*: 360–363.

476 Vogel, A., Rodriguez, C., Warnken, W., and Belmonte, J.C.I. Dorsal cell fate specified by chick Lmx1 during vertebrate limb development. *Nature* 1995; *378*: 716–720.

477 Chen, H., Lun, Y., Ovchinnikov, D., Kokubo, H., Oberg, K.C., Pepicelli, C.V., Gan, L., Lee, B., and Johnson, R.L. Limb and kidney defects in Lmx1b mutant mice suggest an involvement of LMX1B in human nail patella syndrome. *Nat Genet* 1998; *19*: 51–55.

478 Cygan, J.A., Johnson, R.L., and McMahon, A.P. Novel regulatory interactions revealed by studies of murine limb pattern in Wnt–7a and En–1 mutants. *Development* 1997; *124*: 5021–5032.

479 Altabef, M., Logan, C., Tickle, C., and Lumsden, A. Engrailed–1 misexpression in chick embryos prevents apical ridge formation but preserves segregation of dorsal and ventral ectodermal compartments. *Dev Biol* 2000; *222*: 307–316.

480 Loomis, C.A., Kimmel, R.A., Tong, C.X., Michaud, J., and Joyner, A.L. Analysis of the genetic pathway leading to formation of ectopic apical ectodermal ridges in mouse Engrailed–1 mutant limbs. *Development* 1998; *125*: 1137–1148.

481 Crossley, P.H., Minowada, G., MacArthur, C.A., and Martin, G.R. Roles for FGF8 in the induction, initiation, and maintenance of chick limb development. *Cell* 1996; *84*: 127–136.

482 Robledo, R.F., Rajan, L., Li, X., and Lufkin, T. The Dlx5 and Dlx6 homeobox genes are essential for craniofacial, axial, and appendicular skeletal development. *Genes Dev* 2002; *16*: 1089–1101.

483 Rodriguez-Esteban, C., Schwabe, J.W.R., De la Pena, J., Rincon-Limas, D.E., Magallon, J., Botas, J., and Belmonte, J.C.I. Lhx2, a vertebrate homologue of apterous, regulates vertebrate limb outgrowth. *Development* 1998; *125*: 3925–3934.

484 Tavares, A.T., Tsukui, T., and Belmonte, C.I. Evidence that members of the Cut/Cux/CDP family may be involved in AER positioning and polarizing activity during chick limb development. *Development* 2000; *127*: 5133–5144.

485 Liu, C.Q., Nakamura, E., Knezevic, V., Hunter, S., Thompson, K., and Mackem, S. A role for the mesenchymal T-box gene Brachyury in AER formation during limb development. *Development* **2003**; *130*: 1327–1337.

486 Liu, Y., Liu, C.Q., Yamada, Y., and Fan, C.M. Growth arrest specific gene 1 acts as a region-specific mediator of the Fgf10/Fgf8 regulatory loop in the limb. *Development* **2002**; *129*: 5289–5300.

487 Laufer, E., Marigo, V., Nelson, C., Johnson, R.L., Scott, M.P., and Tabin, C. Coordination of Limb Patterning by Sonic Hedgehog and Fgf-4. *Dev Biol* **1995**; *170*: 739–739.

488 Scherz, P.J., Harfe, B.D., McMahon, A.P., and Tabin, C.J. The limb bud Shh-Fgf feedback loop is terminated by expansion of former ZPA cells. *Science* **2004**; *305*: 396–399.

489 Tickle, C., Crawley, A., and Farrar, J. Retinoic acid application to chick wing buds leads to a dose-dependent reorganization of the apical ectodermal ridge that is mediated by the mesenchyme. *Development* **1989**; *106*: 691–705.

490 Charite, J., Degraaff, W., Shen, S.B., and Deschamps, J. Ectopic expression of Hoxb–8 causes duplication of the zpa in the forelimb and homeotic transformation of axial structures. *Cell* **1994**; *78*: 589–601.

491 Lu, H.C., Revelli, J.P., Goering, L., Thaller, C., and Eichele, G. Retinoid signaling is required for the establishment of a ZPA and for the expression of Hoxb–8, a mediator of ZPA formation. *Development* **1997**; *124*: 1643–1651.

492 Stratford, T.H., Kostakopoulou, K., and Maden, M. Hoxb–8 has a role in establishing early anterior-posterior polarity in chick forelimb but not hindlimb. *Development* **1997**; *124*: 4225–4234.

493 van den Akker, E., Reijnen, M., Korving, J., Brouwer, A., Meijlink, F., and Deschamps, J. Targeted inactivation of Hoxb8 affects survival of a spinal ganglion and causes aberrant limb reflexes. *Mech Dev* **1999**; *89*: 103–114.

494 van den Akker, E., Fromental-Ramain, C., de Graaff, W., Le Mouellic, H., Brulet, P., Chambon, P., and Deschamps, J. Axial skeletal patterning in mice lacking all paralogous group 8 Hox genes. *Development* **2001**; *128*: 1911–1921.

495 Charite, J., McFadden, D.G., and Olson, E.N. The bHLH transcription factor dHAND controls Sonic hedgehog expression and establishment of the zone of polarizing activity during limb development. *Development* **2000**; *127*: 2461–2470.

496 Fernandez-Teran, M., Piedra, M.E., Kathiriya, I.S., Srivastava, D., Rodriguez-Rey, J.C., and Ros, M.A. Role of dHAND in the anterior-posterior polarization of the limb bud: implications for the Sonic hedgehog pathway. *Development* **2000**; *127*: 2133–2142.

497 Yelon, D., Ticho, B., Halpern, M.E., Ruvinsky, I., Ho, R.K., Silver, L.M., and Stainier, D.Y.R. The bHLH transcription factor Hand2 plays parallel roles in zebrafish heart and pectoral fin development. *Development* **2000**; *127*: 2573–2582.

498 Masuya, H., Sagai, T., Wakana, S., Moriwaki, K., and Shiroishi, T. A duplicated zone of polarizing activity in polydactylous mouse mutants. *Genes Dev* **1995**; *9*: 1645–1653.

499 Welscher, P.T., Fernandez-Teran, M., Ros, M.A., and Zeller, R. Mutual genetic antagonism involving GLI3 and dHAND pre-patterns the vertebrate limb bud mesenchyme prior to SHH signaling. *Genes Dev* **2002**; *16*: 421–426.

500 Qu, S.M., Niswender, K.D., Ji, Q.S., vanderMeer, R., Keeney, D., Magnuson, M.A., and Wisdom, R. Polydactyly and ectopic ZPA formation in Alx–4 mutant mice. *Development* **1997**; *124*: 3999–4008.

501 Takahashi, M., Tamura, K., Buscher, D., Masuya, H., Yonei-Tamura, S., Matsumoto, K., Naitoh-Matsuo, M., Takeuchi, J., Ogura, K., Shiroishi, T., Ogura, T., and Belmonte, J.C.I. The role of Alx–4 in the establishment of anteroposterior polarity during vertebrate limb development. *Development* **1998**; *125*: 4417–4425.

502 Zakany, J., Kmita, M., and Duboule, D. A dual role for Hox genes in limb anterior-posterior asymmetry. *Science* **2004**; *304*: 1669–1672.

503 Vogel, A., Rodriguez, C., and IzpisuaBelmonte, J.C. Involvement of FGF–8 in initiation, outgrowth and patterning of the vertebrate limb. *Development* **1996**; *122*: 1737–1750.

504 Li, S.G., and Muneoka, K. Cell migration and chick limb development: Chemotactic action of FGF–4 and the AER. *Dev Biol* **1999**; *211*: 335–347.

505 Kostakopoulou, K., Vogel, A., Brickell, P., and Tickle, C. 'Regeneration' of wing bud stumps of chick embryos and reactivation of Msx–1 and Shh expression in response to FGF–4 and ridge signals. *Mech Dev* **1996**; *55*: 119–131.

506 Ros, M.A., Lyons, G.E., Mackem, S., and Fallon, J.F. Recombinant limbs as a model to study homeobox gene-regulation during limb development. *Dev Biol* **1994**; *166*: 59–72.

507 Mercader, N., Leonardo, E., Azpiazu, N., Serrano, A., Morata, G., Martinez, C., and Torres, M. Conserved regulation of proximodistal limb axis development by Meis1/Hth. *Nature* **1999**; *402*: 425–429.

508 Shen, W.F., Montgomery, J.C., Rozenfeld, S., Moskow, J.J., Lawrence, H.J., Buchberg, A.M., and Largman, C. AbdB-like Hox proteins stabilize DNA binding by the Meis1 homeodomain proteins. *Mol Cell Biol* **1997**; *17*: 6448–6458.

509 Maden, M. Vitamin-A and pattern-formation in the regenerating limb. *Nature* **1982**; *295*: 672–675.

510 Dreyer, S.D., Zhou, G., Baldini, A., Winterpacht, A., Zabel, B., Cole, W., Johnson, R.L., and Lee, B. Mutations in LMX1B cause abnormal skeletal patterning and renal dysplasia in nail patella syndrome. *Nat Genet* **1998**; *19*: 47–50.

511 Vollrath, D., Jaramillo-Babb, V.L., Clough, M.V., McIntosh, I., Scott, K.M., Lichter, P.R., and Richards, J.E. Loss-of-function mutations in the LIM-homeodomain gene, LMX1B, in nail-patella syndrome. *Hum Mol Genet* **1998**; *7*: 1091–1098.

512 Sanz-Ezquerro, J.J., and Tickle, C. Fgf signaling controls the number of phalanges and tip formation in developing digits. *Curr Biol* **2003**; *13*, 1830–1836.

513 Altaba, A.R., Sanchez, P., and Dahmane, N. Gli and hedgehog in cancer: Tumours, embryos and stem cells. *Nat Reviews Cancer* **2002**; *2*: 361–372.

514 Berman, D.M., Karhadkar, S.S., Maitra, A., de Oca, R.M., Gerstenblith, M.R., Briggs, K., Parker, A.R., Shimada, Y., Eshleman, J.R., Watkins, D.N., and Beachy, P.A. Widespread requirement for Hedgehog ligand stimulation in growth of digestive tract tumours. *Nature* **2003**; *425*: 846–851.

515 Beachy, P.A., Karhadkar, S.S., and Berman, D.M. Tissue repair and stem cell renewal in carcinogenesis. *Nature* **2004**; *432*: 324–331.

516 Karhadkar, S.S., Bova, G.S., Abdallah, N., Dhara, S., Gardner, D., Maitra, A., Isaacs, J.T., Berman, D.M., and Beachy, P.A. Hedgehog signalling in prostate regeneration, neoplasia and metastasis. *Nature* **2004**; *431*: 707–712.

517 Zuniga, A.E., Michos, O., Spitz, F., Haramis, A.P.G., Panman, L., Galli, A., Vintersten, K., Klasen, C., Mansfield, W., Kuc, S., Duboule, D., Dono, R., and Zeller, R. Mouse limb deformity mutations disrupt a global control region within the large regulatory landscape required for Gremlin expression. *Genes Dev* **2004**; *18*: 1553–1564.

518 Izpisuabelmonte, J.C., Tickle, C., Dolle, P., Wolpert, L., and Duboule, D. Expression of the homeobox Hox–4 genes and the specification of position in chick wing development. *Nature* **1991**; *350*: 585–589.

519 Nohno, T., Noji, S., Koyama, E., Ohyama, K., Myokai, F., Kuroiwa, A., Saito, T., and Taniguchi, S. Involvement of the Chox–4 chicken homeobox genes in determination of anteroposterior axial polarity during limb development. *Cell* **1991**; *64*: 1197–1205.

520 Morgan, B.A., Izpisuabelmonte, J.C., Duboule, D., and Tabin, C.J. Targeted misexpression of Hox–4.6 in the avian limb bud causes apparent homeotic transformations. *Nature* **1992**; *358*: 236–239.

521 Dolle, P., Dierich, A., Lemeur, M., Schimmang, T., Schuhbaur, B., Chambon, P., and Duboule, D. Disruption of the Hoxd–13 gene induces localized heterochrony leading to mice with neotenic limbs. *Cell* **1993**; *75*: 431–441.

522 Davis, A.P., and Capecchi, M.R. A mutational analysis of the 5′ HoxD genes: Dissection of genetic interactions during limb development in the mouse. *Development* **1996**; *122*: 1175–1185.

523 Zakany, J., and Duboule, D. Synpolydactyly in mice with a targeted deficiency in the HoxD complex. *Nature* **1996**; *384*: 69–71.

524 FromentalRamain, C., Warot, X., Messadecq, N., LeMeur, M., Dolle, P., and Chambon, P. Hoxa–13 and Hoxd–13 play a crucial role in the patterning of the limb autopod. *Development* **1996**; *122*: 2997–3011.

525 Zakany, J., FromentalRamain, C., Warot, X., and Duboule, D. Regulation of number and size of digits by posterior Hox genes: A dose-dependent mechanism with potential evolutionary implications. *Proc Natl Acad Sci USA* **1997**; *94*: 13695–13700.

526 Goff, D.J., and Tabin, C.J. Analysis of Hoxd–13 and Hoxd–11 misexpression in chick limb buds reveals that Hox genes affect both bone condensation and growth. *Development* **1997**; *124*: 627–636.

527 Boulet, A.M., and Capecchi, M.R. Multiple roles of Hoxa11 and Hoxd11 in the formation of the mammalian forelimb zeugopod. *Development* **2004**; *131*, 299–309.

528 Yokouchi, Y., Nakazato, S., Yamamoto, M., Goto, Y., Kameda, T., Iba, H., and Kuroiwa, A. Misexpression of Hoxa–13 induces cartilage homeotic transformation and changes cell adhesiveness in chick limb buds. *Genes Dev* **1995**; *9*: 2509–2522.

529 Stadler, H.S., Higgins, K.M., and Capecchi, M.R. Loss of Eph-receptor expression correlates with loss of cell adhesion and chondrogenic capacity in Hoxa13 mutant limbs. *Development* **2001**; *128*: 4177–4188.

530 Kalderon, D. Similarities between the Hedgehog and Wnt signaling pathways. *Trends Cell Biol* **2002**; *12*: 523–531.

531 Fedak, T.J., and Hall, B.K. Perspectives on hyperphalangy: patterns and processes. *J Anat* **2004**; *204*: 151–163.

532 Bi, W.M., Deng, J.M., Zhang, Z.P., Behringer, R.R., and de Crombrugghe, B. Sox9 is required for cartilage formation. *Nat Genet* **1999**; *22*: 85–89.

533 Knosp, W.M., Scott, V., Bachinger, H.P., and Stadler, H.S. HOXA13 regulates the expression of bone morphogenetic proteins 2 and 7 to control distal limb morphogenesis. *Development* **2004**; *131*: 4581–4592.

534 Davis, A.P., Witte, D.P., Hsiehli, H.M., Potter, S.S., and Capecchi, M.R. Absence of Radius and Ulna in Mice Lacking Hoxa–11 and Hoxd–11. *Nature* **1995**; *375*: 791–795.

535 Wellik, D.M., and Capecchi, M.R. Hox10 and Hox11 genes are required to globally pattern the mammalian skeleton. *Science* **2003**; *301*: 363–367.

536 Chen, C.H., Cretekos, C.J., Rasweiler, J.J., and Behringer, R.R. Hoxd13 expression in the developing limbs of the short-tailed fruit bat, Carollia perspicillata. *Evol Dev* **2005**; *7*: 130–141.

537 Alles, A.J., and Sulik, K.K. Retinoic-acid-induced limb-reduction defects – perturbation of zones of programmed cell-death as a pathogenetic mechanism. *Teratology* **1989**; *40*: 163–171.

538 Bastida, M.F., Delgado, M.D., Wang, B.L., Fallon, J.F., Fernandez-Teran, M., and Ros, M.A. Levels of Gli3 repressor correlate with Bmp4 expression and apoptosis during limb development. *Dev Dyn* **2004**; *231*: 148–160.

539 Biesecker, L.G.. Strike three for GLI3. *Nat Genet* **1997**; *17*: 259–260.

540 Polinkovsky, A., Robin, N.H., Thomas, J.T., Irons, M., Lynn, A., Goodman, F.R., Reardon, W., Kant, S.G., Brunner, H.G., vanderBurgt, I., Chitayat, D., McGaughran, J., Donnai, D., Luyten, F.P., and Warman, M.L. Mutations in CDMP1 cause autosomal dominant brachydactyly type C. *Nat Genet* **1997**; *17*: 18–19.

541 Thomas, J.T., Kilpatrick, M.W., Lin, K., Erlacher, L., Lembessis, P., Costa, T., Tsipouras, P., and Luyten, F.P. Disruption of human limb morphogenesis by a dominant negative mutation in CDMP1. *Nat Genet* **1997**; *17*: 58–64.

542 Brunkow, M.E., Gardner, J.C., Van Ness, J., Paeper, B.W., Kovacevich, B.R., Proll, S., Skonier, J.E., Zhao, L., Sabo, P.J., Fu, Y.H., Alisch, R.S., Gillett, L., Colbert, T., Tacconi, P., Galas, D.,

Hamersma, H., Beighton, P., and Mulligan, J.T. Bone dysplasia sclerosteosis results from loss of the SOST gene product, a novel cystine knot-containing protein. *Am J Hum Genet* **2001**; *68*: 577–589.

543 Kusu, N., Laurikkala, J., Imanishi, M., Usui, H., Konishi, M., Miyake, A., Thesleff, I., and Itoh, N. Sclerostin is a novel secreted osteoclast-derived bone morphogenetic protein antagonist with unique ligand specificity. *J Biol Chem* **2003**; *278*: 24113–24117.

544 Reya, T., and Clevers, H. Wnt signalling in stem cells and cancer. *Nature* **2005**; *434*: 843–850.

545 Kaur, B., Tan, C., Brat, D.J., Post, D.E., and Van Meir, E.G. Genetic and hypoxic regulation of angiogenesis in gliomas. *J Neuro-Oncol* **2004**; *70*: 229–243.

546 Cooper, M.K., Porter, J.A., Young, K.E., and Beachy, P.A. Teratogen-mediated inhibition of target tissue response to Shh signaling. *Science* **1998**; *280*: 1603–1607.

547 Chen, J.K., Taipale, J., Cooper, M.K., and Beachy, P.A. Inhibition of Hedgehog signaling by direct binding of cyclopamine to Smoothened. *Genes Dev* **2002**; *16*: 2743–2748.

548 Hoffmann, A., Weich, H.A., Gross, G., and Hillmann, G. Perspectives in the biological function, the technical and therapeutic application of bone morphogenetic proteins. *Appl Microbiol Biotechnol* **2001**; *57*: 294–308.

549 Kuo, C.K., and Tuan, R.S. Tissue engineering with mesenchymal stem cells. *IEEE Eng Med Biol Mag* **2003**; *22*: 51–56.

550 Tuan, R.S., Boland, G., and Tuli, R. Adult mesenchymal stem cells and cell-based tissue engineering. *Arth Res Ther* **2003**; *5*: 32–45.

551 Tanaka, E.M. Regeneration: If they can do it, why can't we? *Cell* **2003**); *113*: 559–562.

552 Muller, T.L., Ngo-Muller, V., Reginelli, A., Taylor, G., Anderson, R., and Muneoka, K. Regeneration in higher vertebrates: Limb buds and digit tips. *Sem Cell Dev Biol* **1999**; *10*: 405–413.

553 Han, M.J., Yang, X.D., Farrington, J.E., and Muneoka, K. Digit regeneration is regulated by Msx1 and BMP4 in fetal mice. *Development* **2003**; *130*: 5123–5132.

16
Skeletal Development
William A. Horton

16.1
Introduction

Skeletal development in humans is an extremely complex process. It can be divided into a morphogenesis phase in which the embryonic skeleton is formed and a growth phase during which the final adult skeletal form is reached. A prolonged growth phase accounts for the vast majority of adult size. Although the general scheme responsible for skeletal development has been known for decades, many of the details of how it comes about and how it is regulated have only emerged in recent years.

Several factors have contributed to the recent advances. One is the Human Genome Project, which facilitated the discovery of genes that harbor mutations that cause disturbances in skeletal development in humans [1, 2]. This led to the identification of a number of what might be called skeletal dysplasia genes [3]. Analysis of tissues from patients with skeletal dysplasias has provided insight into how these genes contribute to skeletal development [4]. Identifying genes responsible for defective skeletal development in animals such as the mouse has also contributed to the recent advances [5]. Another factor is refinement in techniques to manipulate the genomes of animal models, especially the mouse. Being able to genetically increase and decrease specific gene functions, often in selected tissues and at different times in gestation, has allowed the dissection of the complex events and regulatory networks that orchestrate normal skeletal development and has also facilitated modeling of human disorders of skeletal development.

This chapter will address current concepts in mammalian skeletal development. The reader should recognize that most of these concepts are based on what has been learned from animal studies, especially experiments performed in the chick limb bud and in recent years, in genetically-engineered mice. The paradigms seem to hold for humans even though the timing of events and size of the developing skeleton are different. Of note, the regulatory networks that control these events appear to be remarkably similar between humans and mice.

Cell Signaling and Growth Factors in Development. Edited by K. Unsicker and K. Krieglstein
Copyright © 2006 WILEY-VCH Verlag GmbH & Co. KGaA, Weinheim
ISBN 3-527-31034-7

16.2
Skeletal Morphogenesis

16.2.1
Pre-condensation

Formation or morphogenesis of the vertebrate skeleton is carried out by cells from three separate embryonic lineages [6–9]. The craniofacial skeleton derives from neural crest cells, cells from the paraxial mesoderm (somites) give rise to the axial skeleton and the limb skeleton arises from cells of the lateral plate mesoderm. The earliest event in skeletal morphogenesis is migration of mesenchymal cells to these locations where they form densely-packed cell condensations [10]. Patterning genes are mainly responsible for directing cells to the proper locations. Most of these genes encode transcription factors such as PAX and HOX genes as well as proteins involved in cell-to-cell communication and intracellular signaling, i.e. growth factors, receptors and signaling molecules [7, 10, 11]. Regulatory networks are established that impart information about identity and spatial position to the migrating cells and control events such as segmentation which generate embryonic structures.

Condensation of mesenchymal cells is a critical stage in skeletal morphogenesis [10–12]. Condensations are either osteogenic or chondrogenic depending on the skeletal elements they initiate. Both types of condensation exhibit a high density of cells and expression of genes encoding a variety of molecules, including receptors, adhesion proteins, matrix proteins and transcription factors that promote adhesion, set boundaries and control the size of the condensations [12, 13]. BMPs and BMP antagonists such as Noggin play important roles in recruiting neighboring mesenchymal precursor cells into condensations and modulating their ultimate size and shape [14].

16.2.2
Membranous Skeletal Development

Osteogenic condensations give rise directly to bone tissue in the cranial bones and in parts of the clavicle [15, 16]. The process is called membranous ossification. It begins as mesenchymal cells differentiate into osteoblasts and produce bone matrix to form ossification centers. The earliest stages of osteoblastic differentiation may precede condensation [14]. In the skull, ossification centers form within the primitive membrane that covers the cranial vault. The ossification centers spread to form bony plates that are separated by unossified sutures.

An essential regulator of osteoblastic differentiation is Cbfa1 (also designated Runx2), which belongs to the *Runt* family of transcription factors [17–19]. It induces expression of a number of bone matrix proteins including type I collagen, osteocalcin, bone sialoprotein and alkaline phosphatase. Mice null for Cbfa1/Runx2 lack bone tissue, and haploinsufficiency for Cbfa1/Runx2 causes cleidocranial dysplasia in both humans and mice [17, 20, 21]. Recent studies in mice suggest that a family member of Cbfa1/Runx2 known as Cbf-beta (Cbfb) or Runx1 forms heterodimers with Cbfa1/Runx2 and is required for its function and for complete osteoblast differentiation and membranous ossification [22, 23]. Osterix (Osx) is another transcription factor that is required for osteoblast differentiation and bone formation [24]. It is thought to act downstream of Cbfa1/Runx2.

Although many local growth factors have been implicated in promoting osteoblast differentiation, Indian hedgehog (Ihh) and FGF18 are the only ones for which there is genetic evidence from mouse mutants for being essential [19, 25–27]. For instance, mice lacking Ihh have no osteoblasts in skeletal elements that normally ossify through endochondral ossification (see below). BMPs 2, 4, 6 and 7, which are often considered to be "bone-inducers" because of their extraordinary bone-forming properties, may act through inducing skeletal mesoderm rather than bone *per se* [28, 29].

Growth of the membranous skeleton occurs through proliferation of precursor cells that undergo osteoblastic differentiation and deposit bone at the outer border of the ossification centers early in development and at the edge of bones later in development [15]. This is sometimes referred to as appositional growth. This process occurs at the sutures for the flat bones of the cranium. Craniosynostosis syndromes, such as Pfeiffer and Apert syndromes, which are due to activating mutations of genes encoding FGF receptors 1 and 2 (FGFR1 and FGFR2) respectively, illustrate disturbances of growth of the membranous calvarium that lead to premature closure of sutures [30–32].

16.2.3
Endochondral Skeletal Development

Mesenchymal cells that reside in chondrogenic condensations give rise to the axial and limb skeletons as well as to the base of the skull [10, 15]. The scheme is depicted in Fig. 16.1. These cells differentiate into chondrocytes and deposit an extracellular matrix typical of cartilage that serves as a template for future bones. This matrix is rich in types II, IX and XI collagen, proteoglycans such as aggrecan and other matrix proteins such as cartilage oligomeric matrix protein (COMP) and matrilins [19, 33]. Chondrocytes in these templates, which are often called anlagen, proliferate in a randomly-oriented fashion. Condensations can be found in humans at 6.5 weeks of gestation, and cartilage anlagen have been detected at 16 weeks gestation [34, 35]. Comparable structures are identified in mice at 10.5 and 11.5 days of the 18.5-day gestation [36]. The timing varies by location in both species.

Members of the Sox family of transcription factors are required for chondrocytic differentiation [9, 37]. Sox9 acts very early in the differentiation scheme and may activate other transcription factors that promote overt chondrocyte differentiation including Sox5 and Sox6 [38, 39]. Genes for a number of cartilage matrix proteins including types II and XI collagen chains contain enhancers that are activated by Sox9 [37]. Haploinsufficiency for Sox9 results in campomelic dysplasia (congenital bowing and angulation of limbs) in humans and defective skeletal development in

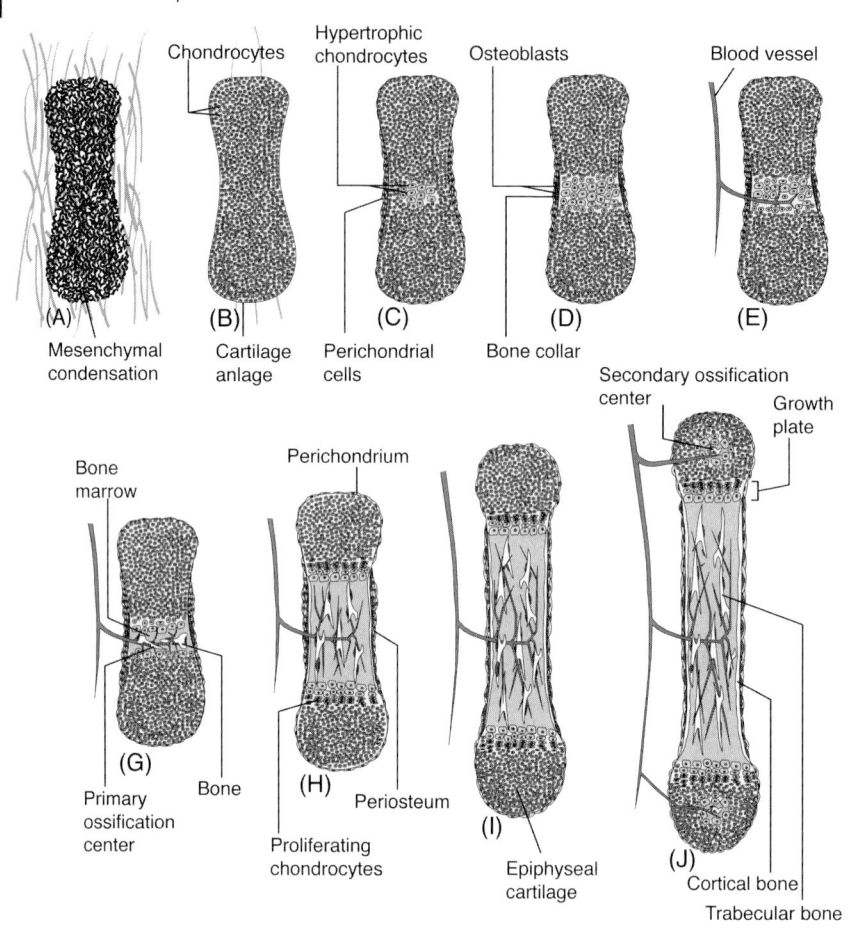

Fig. 16.1
Endochondral bone development from mesenchymal condensation stage to formation of secondary ossification centers (see text for discussion of events). (This figure also appears with the color plates.)

mice [40, 41]. Chondrogenic condensations fail to develop in mouse embryos in whom Sox9 is knocked out [38]. Thus, Sox9 may be equivalent to Cbfa1/Runx2 in osteoblastic differentiation. The two are sometimes referred to as "master genes" for chondrogenesis and osteogenesis, respectively. Several other transcription factors including ATF–2, Pbx1, P105 and P130, have important roles in chondrocyte differentiation [9]. Growth factors including BMPs, FGFs and Hedgehogs influence chondrocyte differentiation and proliferation and the initial stages of cartilage formation, but the precise relationships between these transcription factors and the growth factors is still uncertain [37].

The transition from chondrogenic condensations to cartilage templates of future bones occurs in a proximal to distal fashion and earlier in anterior compared to posterior structures. The shift is accompanied by modifications, such as bifurcation and segmentation, of the forming structures that sculpt the future bones and joints [10, 14].

Soon after cartilage anlagen are formed, centrally-located chondrocytes begin to terminally differentiate. They stop expressing many chondrocyte-specific genes such as Sox9, and begin to express genes characteristic of the hypertrophic chondrocyte cellular phenotype including genes encoding type X collagen, VEGF and alkaline phosphatase [6, 19, 42, 43]. Cbfa1/Runx2 and Cbfb/Runx1 are required for terminal chondrocyte differentiation in some locations [22, 23, 44]. Sox 5 and Sox 6 may help to coordinate the transition from post-mitotic (non-hypertrophic) to hypertrophic chondrocytes [45]. As the chondrocytes exit the cell cycle and begin to terminally differentiate they go through a pre-hypertrophic stage, which is marked by the expression of Ihh [46, 47]. These biosynthetic alterations are accompanied by substantial enlargement of cells and mineralization of the hypertrophic cartilage matrix [33, 48, 49].

Concurrent with these changes, mesenchymal cells that surround the cartilage anlagen differentiate as osteoblasts forming a perichondrial collar of membranous bone around the center of the anlagen [50]. This collar appears to act as a staging area for the subsequent invasion of the cartilage anlagen. VEGF and possibly other angiogenic factors secreted from hypertrophic chondrocytes in the central anlagen induce sprouting angiogenesis from the perichondrium [7, 43]. This invasion brings osteoclasts, osteoblasts and hematopoetic cells along with blood vessels into the anlagen as the most terminally differentiated chondrocytes die by apoptosis. The osteoclasts degrade most of the hypertrophic cartilage matrix leaving fragments that serve as scaffolding for deposition of bone matrix (osteoid) by the osteoblasts. Bone marrow becomes established in spaces between the bony trabeculae. The end result of these events is the formation of primary ossification centers.

Once the centers are established, ossification spreads centripetally as a front within the anlagen [33, 51]. Chondrocytes adjacent to the front terminally differentiate to hypertrophic chondrocytes, which facilitate the hypertrophy, degradation, and replacement of the cartilage. Much of the cartilage anlage is converted to bone by this process. However, as this front nears the ends of the bone, the epiphyses, chondrocytes adjacent to the hypertrophic cells proliferate as stacks of flattened newborn cells that synthesize abundant cartilage matrix before they proceed through the pre-hypertrophic and hypertrophic phases of terminal differentiation [33, 51]. Thus, the source of template cartilage that is replaced by bone changes from cartilage originally derived from mesenchyme to cartilage produced by newly proliferated chondrocytes. This switch allows the epiphyseal cartilages at the ends of the bones, to be pushed apart and the bones to grow in length.

Joints develop through a series of processes that begin with the formation of specialized regions of high cell density, so called interzones, that extend across developing cartilage elements at sites of future joints [7, 52]. After cells in the center of the interzone die by apoptosis, articular chondrocytes differentiate along its cartilaginous edges followed by the formation of fluid-filled spaces that coalesce to establish the joint space. BMPs, especially GFD5 and 6, BMP antagonists such as Noggin and Wnts4 and 14 have been implicated in joint formation [7, 9, 52, 53].

16.3
Skeletal Growth

16.3.1
Growth Plate Structure and Organization

The process of chondrocyte proliferation followed by hypertrophy and replacement becomes structurally organized into the growth plate or physis that occupies a narrow space between the epiphyseal cartilage and the growing bone proper. The mature growth plate can be viewed functionally as a dynamic structure with a leading edge where cells are born that synthesize new cartilage template and a trailing edge where this template is replaced by bone (Figs 16.2 and 16.3).

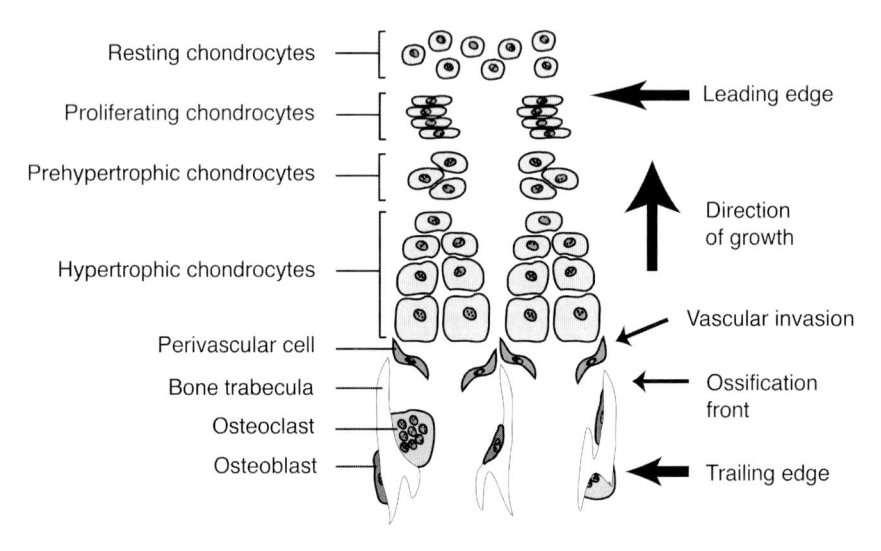

Fig. 16.2
Schematic diagram of the growth plate. Cells and structures are identified on the left; dynamic aspects on the right.

There are a number of human chondrodysplasias in which the cartilage template generated by post-mitotic growth plate chondrocytes is defective. Most notable are members of the spondyloepiphyseal dysplasia (SED) family of disorders, which result from mutations of type II collagen, the principal protein of cartilage matrix [3, 54, 55]. A number of human SEDs have been modeled in transgenic mice, and mice null for type II collagen have severe skeletal defects [56–61]. Other human chondrodysplasias resulting from cartilage matrix template disturbances include pseudoachondroplasia, which is due to mutations of COMP and multiple epiphyseal dysplasia (MED) which results from mutations of COMP, the alpha 2 and alpha 3 chains of type IX collagen and matrilin-3 [55]. The developmental manifestations of Schmid metaphyseal chondrodysplasia and osteogenesis imperfecta, which result

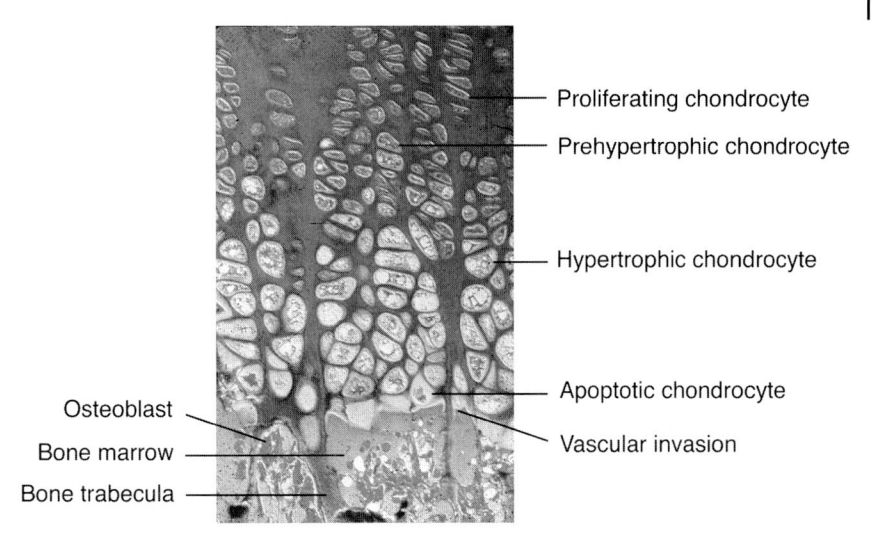

Fig. 16.3
Micrograph of human growth plate showing cells and other anatomical landmarks.

from mutations of types X and I collagen, respectively, reflect disturbances in replacing the cartilage template by bone matrix during endochondral ossification [55, 62].

Growth plates are found in all bones formed by endochondral ossification. They are established near the end of the first trimester in humans and at around 15 days of gestation in mice. However, as with other developmental events, their formation occurs earliest in the proximal skeleton. Growth plates account for essentially all linear skeletal growth from the time they form until the adult size is reached at the end of puberty in both species. However, because of the much prolonged growth phase, they are responsible for a much greater proportion of adult skeletal size in humans than in mice and other small and short-lived animals.

A unique feature of a growing bone is that temporal events that occur within a growth plate are represented spatially [33, 48, 51, 63]. Chondrocytes at the earliest stages of the scheme are furthest from, and those at the latest stages are closest to, the ossification front. Since neighboring chondrocytes are at approximately the same stage of the scheme, they tend to be the same distance from the front. This arrangement creates the appearance of columns or elongated clusters of cells aligned parallel to the direction of growth and zones of cells with similar characteristics perpendicular to this axis. The names of the zones reflect their predominant activity, i.e. reserve, proliferative, pre-hypertrophic and hypertrophic, as schematically represented in Figs 16.2 and 16.3.

Another component of endochondral skeletal development is the formation of secondary ossification centers in most bones [64]. These develop in the centers of the epiphyseal cartilage that remains after growth plates are formed. The timing of their appearance varies widely and they may even appear after birth in some bones in humans. Their appearance is typically delayed in disorders of cartilage matrix proteins, i.e. the SEDs. The formation and expansion of secondary ossification centers within the epiphyses resembles the primary ossification process involving chondrocyte hypertrophy, matrix mineralization, vascular invasion and deposition of bone matrix. However, these centers have much less of the chondrocyte proliferation characteristic of the growth plate and consequently, secondary ossification serves mainly to convert epiphyseal cartilage template into bone.

16.3.2
Regulation of Linear Growth

The regulation of the growth plate, which is responsible for the growth phase of skeletal development, is complex [63, 65]. A number of systemic factors, such as growth hormone, thyroid hormone and steroid hormones clearly influence growth plate function. However, they appear to act through local regulatory circuits that coordinate and couple chondrocyte proliferation, terminal differentiation, vascular invasion and bone formation. Several of these circuits have been defined in the last several years, primarily through identification of mutations in human skeletal dysplasias and through studies of transgenic and gene targeted mice (Fig. 16.4).

16.3.2.1 Ihh-PTHrP Circuit

The best-defined regulatory circuit involves Ihh and parathyroid hormone-related protein (PTHrP) [47, 63, 66]. Secreted by pre-hypertrophic chondrocytes, Ihh promotes chondrocyte proliferation by inducing expression of cyclin D1 [67]. Ihh also activates a negative feedback loop mediated by PTHrP [25, 47, 63]. From analyses of mice carrying different combinations of Ihh, PTHrP and PTHrP receptor (PTHR1) mutations, it appears that Ihh binds its receptor Patched (Ptc), which leads to the activation of members of the Gli family of transcription factors in perichondrial cells adjacent to the pre-hypertrophic chondrocytes. A signal is then transmitted by these cells to perichondrial cells at the ends of the bones where articular surfaces form. These cells secrete PTHrP that diffuses through the growth plate to cells at the border between proliferating and pre-hypertrophic chondrocytes where it binds and activates PTHR1 which blocks chondrocyte terminal differentiation [66, 68].

The Ihh-PTHrP feedback loop acts like a ratchet to control chondrocyte terminal differentiation. As the number of pre-hypertrophic chondrocytes increases, Ihh and PTHrP increase, which feed back via the loop to prevent terminal differentiation while allowing chondrocyte proliferation induced by Ihh. This results in a decreased number of pre-hypertrophic cells which leads to a decrease in Ihh and Ihh-induced proliferation and subsequently in PTHrP, which in turn releases the brake on terminal differentiation. Activating mutations of PTHR1, a G protein-coupled receptor for both PTH and PTHrP in Jansen metaphyseal dysplasia, appear to enhance the braking effect of this feedback loop slowing bone growth [69]. Loss-of-function mutations of PTHR1 are responsible for Bloomstrand chondrodysplasia [70].

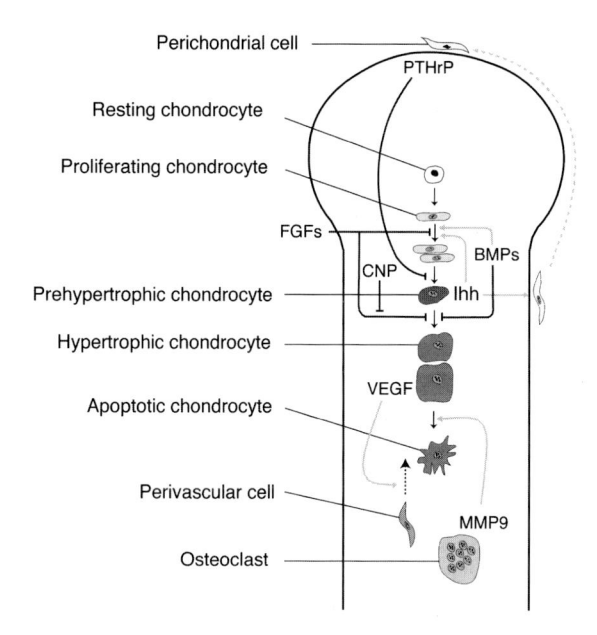

Fig. 16.4
Regulatory networks relevant to linear skeletal growth (see text for discussion). BMPs, bone morphogenetic proteins; CNP, C-type natiuretic peptide; FGFs, fibroblast growth factors; Ihh, Indian hedgehog; MMP9, matrix metalloproteinase; VEGF, vascular endothelial growth factor.

The Ihh-PTHrP feedback loop is modulated by other factors. For instance, hedgehog-interacting protein (HIP) binds to and may attenuate Ihh signals [71]. Similarly, BMPs may act in parallel with Ihh and PTHrP to coordinate chondrocyte proliferation and terminal differentiation [72, 73]. Indeed, BMPs 2, 3, 4, 5 and 6 are expressed in the perichondrium; BMP2 and 6 are expressed in hypertrophic chondrocytes and BMP7 is expressed in proliferating chondrocytes [63]. There is also evidence that Ihh is induced by BMPs [74].

As the skeleton grows and secondary ossification centers form between the articular perichondrial source of PTHrP and the PTHrP-responsive cells in the growth plate, the anatomy of the circuit must also change. One strong possibility is that cells in the upper proliferative zone of the growth plate assume the role of the perichondrial cells responding to Ihh signals and producing PTHrP thus initiating a new feedback loop between proliferating and pre-hypertrophic cells within the growth plate proper.

16.3.2.2 FGF-FGFR3 Circuit

FGF receptor 3 (FGFR3) is a key negative regulator of linear skeletal growth based on observations from both humans and mice. Mice null for FGFR3 have longer than normal bones [75, 76]. In contrast, patients with achondroplasia and related disorders and mouse models of these conditions, all of which involve activation of FGFR3, have short bones [77–83].

FGFR3 is expressed in the developing skeleton primarily in proliferating and prehypertrophic growth plate cells. Although it transmits signals through several pathways, the two most relevant to skeletal development appear to be the STAT1–p21 and the MAP kinase-ERK pathways [84, 85]. Induction of p21, a cell-cycle inhibitor, by STAT1 activation inhibits chondrocyte proliferation [86]. Activation of the MAP kinase-ERK pathway reduces the synthesis of cartilage matrix by post-mitotic growth plate chondrocytes and interferes with terminal chondrocyte differentiation [87, 88]. Several FGF ligands are expressed in the vicinity of the relevant cells, but the physiologic ligand(s) for FGFR3 is (are) not known. The best candidates to date are FGF9 and FGF18 [26, 82, 89].

It is clear that FGFR3 and Ihh signals are integrated in the growth plate although the precise mechanisms are incompletely understood. For instance, FGFR3 signals counter pro-mitotic Ihh signals in the proliferative zone [80]. However, they act in concert with the PTHrP-mediated effects of Ihh to decrease the number of prehypertrophic chondrocytes and secondarily the local abundance of Ihh. It has also been suggested that FGFR3 may act through the MAP kinase pathway to induce expression of Sox9 in post-mitotic growth plate chondrocytes and that the induced Sox9 is activated by phosphorylation by PKA activated by PTHrP [88, 90, 91].

16.3.2.3 CNP-GC-B Circuit

A recently described growth plate regulatory network involves C-type natiuretic peptide (CNP) and its guanylyl cyclase B-coupled receptor (GC-B) also referred to as Npr2 [92, 93]. Both ligand and receptor are expressed in the proliferative and prehypertrophic zones of the growth plate [94]. Mice null for CNP exhibit a reduction in the number of hypertrophic chondrocytes, which is restored to normal when CNP expression is targeted to growth plate cartilage. The two observations suggest that the CNP-GC-B circuit functions as a local positive regulator of terminal differentiation. There is evidence that downstream effects of CNP-GC-B block MAP kinase-ERK signals originating from FGFR3 and thereby counter the growth inhibitory effects of FGFR3 that are mediated by this pathway [87]. In fact, overexpression of CNP in the growth plate of dwarf mice harboring an achondroplasia FGFR3 transgene, restores linear bone growth to normal [87].

16.3.2.4 VEGF-MMP9 Circuit

The regulatory circuit that controls vascular invasion and couples chondrocyte differentiation to ossification in the growth plate, results from an interplay between the terminally differentiated chondrocytes and osteoclastic cells in the bone marrow. The former cells secrete VEGF and other angiogenic factors into the hypertrophic cartilage matrix [43]. Proteases such as matrix metalloproteinases (MMPs) produced by osteoclasts are thought to release these factors that are chemotactic for osteoclasts from the hypertrophic cartilage matrix initiating the invasion process. Vascular invasion involves a combination of apoptosis of the most terminally hypertrophic chondrocytes, degradation of hypertrophic matrix and sprouting angiogenesis from marrow vascular elements. It is severely disturbed in mice null for MMP–9 (Gelatinase B) or in whom VEGF signals are blocked with an antibody [43, 95]. Moreover, chondrocyte apoptosis is blocked in the MMP–9 knockout mice further illustrating the interrelationships between these processes.

VEGF expression appears to be regulated by Cbfa1/Runx2 as part of the terminal chondrocyte differentiation genetic program [96]. There is also evidence that hypoxia-inducing factor (HIF) also upregulates VEGF expression, perhaps as part of a response to tissue hypoxia in the growth plate [97].

Connective tissue growth factor (CTGF) influences vascular invasion of the growth plate [98]. The growth plates of mice null for CTGF display an expanded zone of hypertrophic chondrocytes consistent with defective vascular invasion. This defect is accompanied by reduced expression of VEGF and a paucity of MMP–9–expressing osteoclasts which probably reflects the reduction in VEGF. The mechanism responsible for the VEGF reduction is not known.

Activation of latent regulatory molecules in the matrix by enzyme cleavage may be a common regulatory mechanism in the growth plate. For example, TGF-β, which is abundant in the growth plate, is secreted and stored in the matrix as a latent complex in association with latent TGF-β-binding proteins (LTBPs) [99, 100]. There is evidence that MMP13 (collagenase 3), which is expressed during chondrocyte hypertrophy, cleaves and thereby activates the growth factor during terminal differentiation [101].

The deposition of bone matrix by osteoblasts (osteogenesis) at the ossification front is coupled to chondrogenesis in the growth plate, but the mechanism is not well understood. Similarly, osteoblasts and osteoclasts are thought to be functionally linked during the bone remodeling that occurs during skeletal growth and afterwards.

16.4
Therapeutic Considerations

Defects of early skeletal morphogenesis are typically not recognized until it is too late to act therapeutically. No treatments currently exist for these conditions and it is unlikely that they will be developed in the near future. In contrast, disturbances of linear skeletal growth, i.e. disturbances of growth plate function, are potentially amenable to treatment, especially since most linear growth occurs after birth. Because it is by far the most common as well as the prototype of the human dwarfing conditions, achondroplasia has received much more attention than other constitutional disorders of skeletal growth. A variety of therapeutic strategies have been considered and used in some cases.

There have been a number of short-term trials of human growth hormone (GH) for achondroplasia and to a lesser extent for hypochondroplasia as recently reviewed by Hagenäs and Hertel [102]. GH is known to influence proliferation and differentiation of growth plate chondrocytes through both IGF-I-dependent and IGF-I-independent mechanisms. However it is not known if the latter effects are mediated by IGF-II or reflect direct effects of GH on growth plate cells [103]. GH doses used in the trials have been similar to those used for conditions such as Turner and Noonan syndromes, and results have shown small to modest increases in short-term growth rates. Despite this limited success, growth hormone therapy is not widely accepted as treatment for growth plate disorders.

Surgical limb lengthening, sometimes called distraction osteogenesis, has been used with moderate success in achondroplasia [104–106]. Techniques have varied substantially, but all involve creating a fracture and mechanically stretching it as it heals. Depending on the technique, growth plate-like tissue may form during the healing process and to some extent lengthening probably involves the same cellular and molecular events that occur during normal bone growth with the major exception that it is driven by mechanical force. In other cases, there is little if any cartilage formed and healing is more like prolonged normal fracture healing. When distraction osteogenesis is used to treat achondroplasia, both the femur and the tibia as well as the humeri must be lengthened. Surgical lengthening may increase height by as much as 15–30 cm [105, 106]. However, for a variety of reasons including the time required for healing, complications and the fact that it does not address the non-limb clinical problems in achondroplasia, surgical limb lengthening remains controversial.

Three molecular strategies have been considered for achondroplasia and related disorders, all of which are directed at the excessive FGFR3 signals that reduce chondrocyte proliferation and terminal differentiation in the growth plate [107]. They are based on what is currently known about the biology of this receptor [108–110]. FGFR3 is a tyrosine kinase-coupled receptor. The first step in signal transduction is ligand-induced dimerization of receptor monomers that activates their intrinsic kinase activity which transphosphorylates key tyrosine residues in the cytoplasmic domain of the receptor. These phosphorylated residues serve as docking sites for adaptor proteins and signal effectors that are recruited to the activated receptor and propagate their signals through STAT1, MAP kinase and other pathways. Achondroplasia mutations amplify receptor signaling by stabilizing ligand-induced dimers, introducing free cysteines that induce spontaneous dimerization, interfering with kinase autoinhibition and slowing degradation of activated receptors [111, 112]. Therapies that interfere with activation events and thereby reduce FGFR3 signals would potentially be effective in achondroplasia.

The first strategy is to block the initiation of the signals by inhibiting the intrinsic tyrosine kinase activity of FGFR3. Chemical compounds capable of selectively inhibiting this activity have been identified and their ability to increase bone growth in cultured bones from genetically engineered achondroplastic mice has been documented [107]. However, it is not known if they are selective or specific enough to be used therapeutically in humans.

The second strategy involves inhibiting activation of FGFR3 by interfering with FGF ligand-FGFR3 interactions. Most promising has been the development of a high affinity Fab antibody fragment specific for human FGFR3 that has been shown to inhibit FGFR3 effects on proliferation in cell culture assays [113].

The third strategy is based on observations from mouse cross-breeding experiments in which targeting of CNP expression to growth plate cartilage corrected the bone growth deficiency that had resulted from expression of an achondroplasia FGFR3 transgene in growth plate cartilage [87]. The "therapeutic" effect involved reducing MAP kinase-ERK signals downstream of FGFR3 that restrained chondrocyte terminal differentiation. From observations of skeletal overgrowth in mice with high levels of circulating brain natiuretic peptide (BNP), a closely related peptide that also binds to the CNP receptor, it has been suggested that delivery of CNP through the circulation could potentially promote bone growth in achondroplasia [87, 114].

16.5
Conclusions

Skeletal development in mammals is very complicated. However, as a result of advances in studying human genetic disease and in experimentally manipulating the genomes of mice and other experimental animals, many of the genetic pathways responsible for skeletal development and the mechanisms through which they act have been delineated. These pathways contribute to the sequential phases of skeletal morphogenesis that include patterning, condensation and overt organogenesis of the membranous and endochondral embryonic skeletons. The growth required to reach final adult form comes primarily from cartilage-mediated endochondral ossification in the growth plate. The extent to which growth contributes to final form varies more between species such as man and mouse than do the underlying biologic mechanisms responsible for skeletal development.

Several recurrent themes can be identified. One is the establishment of regulatory networks involving growth factors, receptors, signaling pathways and transcription factors that control cell fates such as migration, adhesion, proliferation, differentiation and apoptosis. Another is the interlocking of the regulatory circuits with one another which allows cellular programs within single cell lineages and even between cells of different lineages to be coupled. For example, the Ihh-PTHrP circuit interacts with the FGFR3 circuit to integrate proliferation and terminal differentiation in the growth plate. Another theme is the use of common molecules for different purposes such as Cbfa1/Runx2 to drive both osteoblastic and terminal chondrocyte differentiation or of BMPs to modulate several regulatory circuits.

Therapies based on understanding the cellular and molecular events that occur in the growth plate are beginning be contemplated. To date they have focused on achondroplasia since it is the most common of the growth plate disorders and because its molecular pathogenesis is somewhat well defined. However, it seems plausible that treatments that stimulate growth plate function in achondroplasia may also provide benefits for bone growth disturbances unrelated to FGFR3 mutations.

References

1 Lander, E.S., Linton, L.M., Birren, B., Nusbaum, C., Zody, M.C., et al., Initial sequencing and analysis of the human genome. *Nature* **2001**; *409*: 860–921.

2 Venter, J.C., Adams, M.D., Myers, E.W., Li, P.W., Mural, R.J., Sutton, G.G., et al., The sequence of the human genome. *Science* **2001**; *291*: 1304–1351.

3 Horton, W.A., Molecular genetic basis of the human chondrodysplasias. *Endocrinol Metab Clin North Am* **1996**; *25*: 683–697.

4 Rimoin, D.L., Lachman, R. S. and Unger, S., Chondrodysplasias. In *Emery and Remoin's Principles and Practice of Medical Genetics* (4th edn), Rimoin, D.L., Connor, J.M., Pyeritz, R. E. and Korf, B. R. (Eds). Churchill Livingstone: New York, **2002**, pp. 4071–4115.

5 Zelzer, E. and Olsen, B.R., The genetic basis for skeletal diseases. *Nature* **2003**; *423*: 343–348.

6 Erlebacher, A., Filvaroff, E.H., Gitelman, S. E. and Derynck, R., Toward a molecular understanding of skeletal development. *Cell* **1995**; *80*: 371–378.

7 Olsen, B.R., Reginato, A. M. and Wang, W., Bone development. *Annu Rev Cell Dev Biol* **2000**; *16*: 191–220.

8 Helms, J. A. and Schneider, R.A., Cranial skeletal biology. *Nature* **2003**; *423*: 326–331.

9 Murakami, S., Akiyama, H. and de Crombrugghe, B., Development of bone and cartilage. In *Inborn Errors of Development*, Epstein, C.J., Erickson, R. P. and Winshaw-Boris, A. (Eds). Oxford University Press: New York, **2004**, pp. 133–147.

10 Hall, B. K. and Miyake, T., All for one and one for all: condensations and the initiation of skeletal development. *Bioessays* **2000**; *22*: 138–147.

11 Mariani, F. V. and Martin, G.R., Deciphering skeletal patterning: clues from the limb. *Nature* **2003**; *423*: 319–325.

12 DeLise, A.M., Fischer, L. and Tuan, R.S., Cellular interactions and signaling in cartilage development. *Osteoarthritis Cartilage* **2000**; *8*: 309–334.

13 Hall, B. K. and Miyake, T., The membranous skeleton: the role of cell condensations in vertebrate skeletogenesis. *Anat Embryol (Berl)* **1992**; *186*: 107–124.

14 Hall, B. K. and Miyake, T., Divide, accumulate, differentiate: cell condensation in skeletal development revisited. *Int J Dev Biol* **1995**; *39*: 881–893.

15 Morriss-Kay, G.M., Derivation of the mammalian skull vault. *J Anat* **2001**; *199*: 143–151.

16 Hall, B.K., Development of the clavicles in birds and mammals. *J Exp Zool* **2001**; *289*: 153–161.
17 Komori, T., Yagi, H., Nomura, S., Yamaguchi, A., Sasaki, K., Deguchi, K., Shimizu, Y., Bronson, R.T., Gao, Y.H., Inada, M., Sato, M., Okamoto, R., Kitamura, Y., Yoshiki, S. and Kishimoto, T., Targeted disruption of Cbfa1 results in a complete lack of bone formation owing to maturational arrest of osteoblasts. *Cell* **1997**; *89*: 755–764.
18 Ducy, P., Zhang, R., Geoffroy, V., Ridall, A. L. and Karsenty, G., Osf2/Cbfa1: a transcriptional activator of osteoblast differentiation. *Cell* **1997**; *89*: 747–754.
19 Karsenty, G. and Wagner, E.F., Reaching a genetic and molecular understanding of skeletal development. *Dev Cell* **2002**; *2*: 389–406.
20 Otto, F., Thornell, A.P., Crompton, T., Denzel, A., Gilmour, K.C., Rosewell, I.R., Stamp, G.W., Beddington, R.S., Mundlos, S., Olsen, B.R., Selby, P. B. and Owen, M.J., Cbfa1, a candidate gene for cleidocranial dysplasia syndrome, is essential for osteoblast differentiation and bone development. *Cell* **1997**; *89*: 765–771.
21 Mundlos, S., Otto, F., Mundlos, C., Mulliken, J.B., Aylsworth, A.S., Albright, S., Lindhout, D., Cole, W.G., Henn, W., Knoll, J.H., Owen, M.J., Mertelsmann, R., Zabel, B. U. and Olsen, B.R., Mutations involving the transcription factor CBFA1 cause cleidocranial dysplasia. *Cell* **1997**; *89*: 773–779.
22 Kundu, M., Javed, A., Jeon, J.P., Horner, A., Shum, L., Eckhaus, M., Muenke, M., Lian, J.B., Yang, Y., Nuckolls, G.H., Stein, G. S. and Liu, P.P., Cbfbeta interacts with Runx2 and has a critical role in bone development. *Nat Genet* **2002**; *32*: 639–644.
23 Yoshida, C.A., Furuichi, T., Fujita, T., Fukuyama, R., Kanatani, N., Kobayashi, S., Satake, M., Takada, K. and Komori, T., Core-binding factor beta interacts with Runx2 and is required for skeletal development. *Nat Genet* **2002**; *32*: 633–638.
24 Nakashima, K., Zhou, X., Kunkel, G., Zhang, Z., Deng, J.M., Behringer, R. R. and de Crombrugghe, B., The novel zinc finger-containing transcription factor osterix is required for osteoblast differentiation and bone formation. *Cell* **2002**; *108*: 17–29.
25 St-Jacques, B., Hammerschmidt, M. and McMahon, A.P., Indian hedgehog signaling regulates proliferation and differentiation of chondrocytes and is essential for bone formation. *Genes Dev* **1999**; *13*: 2072–2086.
26 Ohbayashi, N., Shibayama, M., Kurotaki, Y., Imanishi, M., Fujimori, T., Itoh, N. and Takada, S., FGF18 is required for normal cell proliferation and differentiation during osteogenesis and chondrogenesis. *Genes Dev* **2002**; *16*: 870–879.
27 Long, F., Chung, U.I., Ohba, S., McMahon, J., Kronenberg, H. M. and McMahon, A.P., Ihh signaling is directly required for the osteoblast lineage in the endochondral skeleton. *Development* **2004**; *131*: 1309–1318.
28 Reddi, A.H., Bone morphogenetic proteins: an unconventional approach to isolation of first mammalian morphogens. *Cytokine Growth Factor Rev* **1997**; *8*: 11–20.

29 Capdevila, J. and Izpisua Belmonte, J.C., Patterning mechanisms controlling vertebrate limb development. *Annu Rev Cell Dev Biol* **2001**; *17*: 87–132.

30 Wilkie, A. O. and Morriss-Kay, G.M., Genetics of craniofacial development and malformation. *Nat Rev Genet* **2001**; *2*: 458–468.

31 Muenke, M., Schell, U., Hehr, A., Robin, N.H., Losken, H.W., Schinzel, A., Pulleyn, L.J., Rutland, P., Reardon, W., Malcolm, S. and et al., A common mutation in the fibroblast growth factor receptor 1 gene in Pfeiffer syndrome. *Nat Genet* **1994**; *8*: 269–274.

32 Lomri, A., Lemonnier, J., Hott, M., de Parseval, N., Lajeunie, E., Munnich, A., Renier, D. and Marie, P.J., Increased calvaria cell differentiation and bone matrix formation induced by fibroblast growth factor receptor 2 mutations in Apert syndrome. *J Clin Invest* **1998**; *101*: 1310–1317.

33 Morris, N.P., Keene, D. R. and Horton, W.A., Morphology and chemical composition of connective tissue: cartilage. In *Connective Tissue and its Heritable Disorders* (2nd edn), Royce, P. M. and Steinmann, B. (Eds). Wiley: New York, **2002**, pp. 41–66.

34 Uhthoff, H.K., Development of the growth plate during intrauterine life. In *Behavior of the Growth Plate*, Uhthoff, H. K. and Wiley, J. J. (Eds). Raven Press: New York, **1988**, pp. 17–24.

35 Ogden, J. A. and Rosenberg, L.C., Defining the growth plate. In *Behavior of the Growth Plate*, Uhthoff, H. K. and Wiley, J. J. (Eds). Raven Press: New York, **1988**, pp. 1–16.

36 Kaufman, M.H., *The Atlas of Mouse Development*. Academic Press: London, **1992**.

37 de Crombrugghe, B., Lefebvre, V. and Nakashima, K., Regulatory mechanisms in the pathways of cartilage and bone formation. *Curr Opin Cell Biol* **2001**; *13*: 721–727.

38 Bi, W., Deng, J.M., Zhang, Z., Behringer, R. R. and de Crombrugghe, B., Sox9 is required for cartilage formation. *Nat Genet* **1999**; *22*: 85–89.

39 Lefebvre, V., Behringer, R. R. and de Crombrugghe, B., L-Sox5, Sox6 and Sox9 control essential steps of the chondrocyte differentiation pathway. *Osteoarthritis Cartilage* **2001**; *9*(Suppl. A): S69–75.

40 Wagner, T., Wirth, J., Meyer, J., Zabel, B., Held, M., Zimmer, J., Pasantes, J., Bricarelli, F.D., Keutel, J., Hustert, E. and et al., Autosomal sex reversal and campomelic dysplasia are caused by mutations in and around the SRY-related gene SOX9. *Cell* **1994**; *79*: 1111–1120.

41 Foster, J.W., Dominguez-Steglich, M.A., Guioli, S., Kowk, G., Weller, P.A., Stevanovic, M., Weissenbach, J., Mansour, S., Young, I.D., Goodfellow, P. N. and et al., Campomelic dysplasia and autosomal sex reversal caused by mutations in an SRY-related gene. *Nature* **1994**; *372*: 525–530.

42 Iyama, K., Ninomiya, Y., Olsen, B.R., Linsenmayer, T.F., Trelstad, R. L. and Hayashi, M., Spatiotemporal pattern of type X collagen gene expression and collagen deposition in embryonic chick vertebrae undergoing endochondral ossification. *Anat Rec* **1991**; *229*: 462–472.

43 Gerber, H.P., Vu, T.H., Ryan, A.M., Kowalski, J., Werb, Z. and Ferrara, N., VEGF couples hypertrophic cartilage remodeling, ossification and angiogenesis during endochondral bone formation. *Nat Med* **1999**; *5*: 623–628.

44 Inada, M., Yasui, T., Nomura, S., Miyake, S., Deguchi, K., Himeno, M., Sato, M., Yamagiwa, H., Kimura, T., Yasui, N., Ochi, T., Endo, N., Kitamura, Y., Kishimoto, T. and Komori, T., Maturational disturbance of chondrocytes in Cbfa1-deficient mice. *Dev Dyn* **1999**; *214*: 279–290.

45 Smits, P., Dy, P., Mitra, S. and Lefebvre, V., Sox5 and Sox6 are needed to develop and maintain source, columnar, and hypertrophic chondrocytes in the cartilage growth plate. *J Cell Biol* **2004**; *164*: 747–758.

46 Bitgood, M. J. and McMahon, A.P., Hedgehog and Bmp genes are coexpressed at many diverse sites of cell-cell interaction in the mouse embryo. *Dev Biol* **1995**; *172*: 126–138.

47 Vortkamp, A., Lee, K., Lanske, B., Segre, G.V., Kronenberg, H. M. and Tabin, C.J., Regulation of rate of cartilage differentiation by Indian hedgehog and PTH-related protein. *Science* **1996**; *273*: 613–622.

48 Brighton, C.T., Morphology and biochemistry of the growth plate. *Rheum Dis Clin North Am* **1987**; *13*: 75–100.

49 Buckwalter, J.A., Mower, D., Ungar, R., Schaeffer, J. and Ginsberg, B., Morphometric analysis of chondrocyte hypertrophy. *J Bone Joint Surg Am* **1986**; *68*: 243–255.

50 Bianco, P., Cancedda, F.D., Riminucci, M. and Cancedda, R., Bone formation via cartilage models: the »borderline« chondrocyte. *Matrix Biol* **1998**; *17*: 185–192.

51 Hunziker, E.B., Mechanism of longitudinal bone growth and its regulation by growth plate chondrocytes. *Microsc Res Tech* **1994**; *28*: 505–519.

52 Storm, E. E. and Kingsley, D.M., Joint patterning defects caused by single and double mutations in members of the bone morphogenetic protein (BMP) family. *Development* **1996**; *122*: 3969–3979.

53 King, J.A., Storm, E.E., Marker, P.C., Dileone, R. J. and Kingsley, D.M., The role of BMPs and GDFs in development of region-specific skeletal structures. *Ann NY Acad Sci* **1996**; *785*: 70–79.

54 Horton, W. A. and Hecht, J.T., The skeletal dysplasias: disorders involving cartilage matrix proteins. In *Nelson Textbook of Pediatrics* (16th edn), Behrman, R.E., Kliegman, R. M. and Jenson, H. B. (Eds). W.B. Saunders: Philadelphia, **2000**, pp. 2116–2120.

55 Horton, W. A. and Hecht, J.A., Chondrodysplasias, part II. Disorders involving cartilage matrix proteins. In *Extracellular Matrix and Heritable Disorders of Connective Tissue* (2nd edn), Royce, P. M. and Steinmann, B. (Eds). Alan R. Liss: New York, **2002**.

56 Vandenberg, P., Khillan, J.S., Prockop, D.J., Helminen, H., Kontusaari, S. and Ala-Kokko, L., Expression of a partially deleted gene of human type II procollagen (COL2A1) in transgenic mice produces a chondrodysplasia. *Proc Natl Acad Sci USA* **1991**; *88*: 7640–7644.

57 Garofalo, S., Vuorio, E., Metsaranta, M., Rosati, R., Toman, D., Vaughan, J., Lozano, G., Mayne, R., Ellard, J., Horton, W. and et al., Reduced amounts of cartilage collagen fibrils and growth plate anomalies in transgenic mice harboring a glycine-to-cysteine mutation in the mouse type II procollagen alpha 1–chain gene. *Proc Natl Acad Sci USA* **1991**; *88*: 9648–9652.

58 Metsaranta, M., Garofalo, S., Decker, G., Rintala, M., de Crombrugghe, B. and Vuorio, E., Chondrodysplasia in transgenic mice harboring a 15–amino acid deletion in the triple helical domain of pro alpha 1(II) collagen chain. *J Cell Biol* **1992**; *118*: 203–212.

59 Maddox, B.K., Garofalo, S., Smith, C., Keene, D. R. and Horton, W.A., Skeletal development in transgenic mice expressing a mutation at Gly574Ser of type II collagen. *Dev Dyn* **1997**; *208*: 170–177.

60 Li, S.W., Prockop, D.J., Helminen, H., Fassler, R., Lapvetelainen, T., Kiraly, K., Peltarri, A., Arokoski, J., Lui, H., Arita, M. and et al., Transgenic mice with targeted inactivation of the Col2 alpha 1 gene for collagen II develop a skeleton with membranous and periosteal bone but no endochondral bone. *Genes Dev* **1995**; *9*: 2821–2830.

61 Aszodi, A., Chan, D., Hunziker, E., Bateman, J. F. and Fassler, R., Collagen II is essential for the removal of the notochord and the formation of intervertebral discs. *J Cell Biol* **1998**; *143*: 1399–1412.

62 Myllyharju, J. and Kivirikko, K.I., Collagens and collagen-related diseases. *Ann Med* **2001**; *33*: 7–21.

63 Kronenberg, H.M., Developmental regulation of the growth plate. *Nature* **2003**; *423*: 332–336.

64 Floyd, W.E., 3rd, Zaleske, D.J., Schiller, A.L., Trahan, C. and Mankin, H.J., Vascular events associated with the appearance of the secondary center of ossification in the murine distal femoral epiphysis. *J Bone Joint Surg Am* **1987**; *69*: 185–190.

65 Horton, W.A., Skeletal development: insights from targeting the mouse genome. *Lancet* **2003**; *362*: 560–569.

66 Chung, U.I., Schipani, E., McMahon, A. P. and Kronenberg, H.M., Indian hedgehog couples chondrogenesis to osteogenesis in endochondral bone development. *J Clin Invest* **2001**; *107*: 295–304.

67 Long, F., Zhang, X.M., Karp, S., Yang, Y. and McMahon, A.P., Genetic manipulation of hedgehog signaling in the endochondral skeleton reveals a direct role in the regulation of chondrocyte proliferation. *Development* **2001**; *128*: 5099–5108.

68 Karp, S.J., Schipani, E., St-Jacques, B., Hunzelman, J., Kronenberg, H. and McMahon, A.P., Indian hedgehog coordinates endochondral bone growth and morphogenesis via parathyroid hormone related-protein-dependent and -independent pathways. *Development* **2000**; *127*: 543–548.

69 Schipani, E., Kruse, K. and Juppner, H., A constitutively active mutant PTH-PTHrP receptor in Jansen-type metaphyseal chondrodysplasia. *Science* **1995**; *268*: 98–100.

70 Jobert, A.S., Zhang, P., Couvineau, A., Bonaventure, J., Roume, J., Le Merrer, M. and Silve, C., Absence of functional

receptors for parathyroid hormone and parathyroid hormone-related peptide in Blomstrand chondrodysplasia. *J Clin Invest* **1998**; *102*: 34–40.

71 Chuang, P. T. and McMahon, A.P., Vertebrate Hedgehog signalling modulated by induction of a Hedgehog-binding protein. *Nature* **1999**; *397*: 617–621.

72 Minina, E., Wenzel, H.M., Kreschel, C., Karp, S., Gaffield, W., McMahon, A. P. and Vortkamp, A., BMP and Ihh/PTHrP signaling interact to coordinate chondrocyte proliferation and differentiation. *Development* **2001**; *128*: 4523–4534.

73 Minina, E., Kreschel, C., Naski, M.C., Ornitz, D. M. and Vortkamp, A., Interaction of FGF, Ihh/Pthlh, and BMP signaling integrates chondrocyte proliferation and hypertrophic differentiation. *Dev Cell* **2002**; *3*: 439–449.

74 Seki, K. and Hata, A., Indian hedgehog gene is a target of the bone morphogenetic protein signaling pathway. *J Biol Chem* **2004**; *279*: 18544–18549.

75 Colvin, J.S., Bohne, B.A., Harding, G.W., McEwen, D. G. and Ornitz, D.M., Skeletal overgrowth and deafness in mice lacking fibroblast growth factor receptor 3. *Nat Genet* **1996**; *12*: 390–397.

76 Deng, C., Wynshaw-Boris, A., Zhou, F., Kuo, A. and Leder, P., Fibroblast growth factor receptor 3 is a negative regulator of bone growth. *Cell* **1996**; *84*: 911–921.

77 Shiang, R., Thompson, L.M., Zhu, Y.Z., Church, D.M., Fielder, T.J., Bocian, M., Winokur, S. T. and Wasmuth, J.J., Mutations in the transmembrane domain of FGFR3 cause the most common genetic form of dwarfism, achondroplasia. *Cell* **1994**; *78*: 335–342.

78 Tavormina, P.L., Shiang, R., Thompson, L.M., Zhu, Y.Z., Wilkin, D.J., Lachman, R.S., Wilcox, W.R., Rimoin, D.L., Cohn, D. H. and Wasmuth, J.J., Thanatophoric dysplasia (types I and II) caused by distinct mutations in fibroblast growth factor receptor 3. *Nat Genet* **1995**; *9*: 321–328.

79 Bellus, G.A., McIntosh, I., Smith, E.A., Aylsworth, A.S., Kaitila, I., Horton, W.A., Greenhaw, G.A., Hecht, J. T. and Francomano, C.A., A recurrent mutation in the tyrosine kinase domain of fibroblast growth factor receptor 3 causes hypochondroplasia. *Nat Genet* **1995**; *10*: 357–359.

80 Naski, M.C., Colvin, J.S., Coffin, J. D. and Ornitz, D.M., Repression of hedgehog signaling and BMP4 expression in growth plate cartilage by fibroblast growth factor receptor 3. *Development* **1998**; *125*: 4977–4988.

81 Chen, L., Adar, R., Yang, X., Monsonego, E.O., Li, C., Hauschka, P.V., Yayon, A. and Deng, C.X., Gly369Cys mutation in mouse FGFR3 causes achondroplasia by affecting both chondrogenesis and osteogenesis. *J Clin Invest* **1999**; *104*: 1517–1525.

82 Garofalo, S., Kliger-Spatz, M., Cooke, J.L., Wolstin, O., Lunstrum, G.P., Moshkovitz, S.M., Horton, W. A. and Yayon, A., Skeletal dysplasia and defective chondrocyte differentiation by targeted overexpression of fibroblast growth factor 9 in transgenic mice. *J Bone Miner Res* **1999**; *14*: 1909–1915.

83 Iwata, T., Chen, L., Li, C., Ovchinnikov, D.A., Behringer, R.R., Francomano, C. A. and Deng, C.X., A neonatal lethal mutation in FGFR3 uncouples proliferation and differentiation of growth plate chondrocytes in embryos. *Hum Mol Genet* **2000**; *9*: 1603–1613.

84 Kanai, M., Goke, M., Tsunekawa, S. and Podolsky, D.K., Signal transduction pathway of human fibroblast growth factor receptor 3. Identification of a novel 66–kDa phosphoprotein. *J Biol Chem* **1997**; *272*: 6621–6628.

85 Sahni, M., Ambrosetti, D.C., Mansukhani, A., Gertner, R., Levy, D. and Basilico, C., FGF signaling inhibits chondrocyte proliferation and regulates bone development through the STAT–1 pathway. *Genes Dev* **1999**; *13*: 1361–1366.

86 Su, W.C., Kitagawa, M., Xue, N., Xie, B., Garofalo, S., Cho, J., Deng, C., Horton, W. A. and Fu, X.Y., Activation of Stat1 by mutant fibroblast growth-factor receptor in thanatophoric dysplasia type II dwarfism. *Nature* **1997**; *386*: 288–292.

87 Yasoda, A., Komatsu, Y., Chusho, H., Miyazawa, T., Ozasa, A., Miura, M., Kurihara, T., Rogi, T., Tanaka, S., Suda, M., Tamura, N., Ogawa, Y. and Nakao, K., Overexpression of CNP in chondrocytes rescues achondroplasia through a MAPK-dependent pathway. *Nat Med* **2004**; 10: 80–86.

88 Murakami, S., Balmes, G., McKinney, S., Zhang, Z., Givol, D. and de Crombrugghe, B., Constitutive activation of MEK1 in chondrocytes causes Stat1–independent achondroplasia-like dwarfism and rescues the Fgfr3–deficient mouse phenotype. *Genes Dev* **2004**; *18*: 290–305.

89 Liu, Z., Xu, J., Colvin, J. S. and Ornitz, D.M., Coordination of chondrogenesis and osteogenesis by fibroblast growth factor 18. *Genes Dev* **2002**; *16*: 859–869.

90 Murakami, S., Kan, M., McKeehan, W. L. and de Crombrugghe, B., Up-regulation of the chondrogenic Sox9 gene by fibroblast growth factors is mediated by the mitogen-activated protein kinase pathway. *Proc Natl Acad Sci USA* **2000**; *97*: 1113–1118.

91 Huang, W., Zhou, X., Lefebvre, V. and de Crombrugghe, B., Phosphorylation of SOX9 by cyclic AMP-dependent protein kinase A enhances SOX9's ability to transactivate a Col2a1 chondrocyte-specific enhancer. *Mol Cell Biol* **2000**; *20*: 4149–4158.

92 Nakao, K., Ogawa, Y., Suga, S. and Imura, H., Molecular biology and biochemistry of the natriuretic peptide system. II: Natriuretic peptide receptors. *J Hypertens* **1992**; *10*: 1111–1114.

93 Nakao, K., Ogawa, Y., Suga, S. and Imura, H., Molecular biology and biochemistry of the natriuretic peptide system. I: Natriuretic peptides. *J Hypertens* **1992**; *10*: 907–912.

94 Chusho, H., Tamura, N., Ogawa, Y., Yasoda, A., Suda, M., Miyazawa, T., Nakamura, K., Nakao, K., Kurihara, T., Komatsu, Y., Itoh, H., Tanaka, K., Saito, Y. and Katsuki, M., Dwarfism and early death in mice lacking C-type natriuretic peptide. *Proc Natl Acad Sci USA* **2001**; *98*: 4016–4021.

95 Vu, T.H., Shipley, J.M., Bergers, G., Berger, J.E., Helms, J.A., Hanahan, D., Shapiro, S.D., Senior, R. M. and Werb, Z., MMP–

9/gelatinase B is a key regulator of growth plate angiogenesis and apoptosis of hypertrophic chondrocytes. *Cell* **1998**; *93*: 411–422.

96 Zelzer, E., Glotzer, D.J., Hartmann, C., Thomas, D., Fukai, N., Soker, S. and Olsen, B.R., Tissue specific regulation of VEGF expression during bone development requires Cbfa1/Runx2. *Mech Dev* **2001**; *106*: 97–106.

97 Schipani, E., Ryan, H.E., Didrickson, S., Kobayashi, T., Knight, M. and Johnson, R.S., Hypoxia in cartilage: HIF–1alpha is essential for chondrocyte growth arrest and survival. *Genes Dev* **2001**; *15*: 2865–2876.

98 Ivkovic, S., Yoon, B.S., Popoff, S.N., Safadi, F.F., Libuda, D.E., Stephenson, R.C., Daluiski, A. and Lyons, K.M., Connective tissue growth factor coordinates chondrogenesis and angiogenesis during skeletal development. *Development* **2003**; *130*: 2779–2791.

99 Gleizes, P.E., Munger, J.S., Nunes, I., Harpel, J.G., Mazzieri, R., Noguera, I. and Rifkin, D.B., TGF-beta latency: biological significance and mechanisms of activation. *Stem Cells* **1997**; *15*: 190–197.

100 Pedrozo, H.A., Schwartz, Z., Gomez, R., Ornoy, A., Xin-Sheng, W., Dallas, S.L., Bonewald, L.F., Dean, D. D. and Boyan, B.D., Growth plate chondrocytes store latent transforming growth factor (TGF)-beta 1 in their matrix through latent TGF-beta 1 binding protein–1. *J Cell Physiol* **1998**; *177*: 343–354.

101 D'Angelo, M., Sarment, D.P., Billings, P. C. and Pacifici, M., Activation of transforming growth factor beta in chondrocytes undergoing endochondral ossification. *J Bone Miner Res* **2001**; *16*: 2339–2347.

102 Hagenas, L. and Hertel, T., Skeletal dysplasia, growth hormone treatment and body proportion: comparison with other syndromic and non-syndromic short children. *Horm Res* **2003**; *60*(Suppl. 3): 65–70.

103 Wang, J., Zhou, J., Cheng, C.M., Kopchick, J. J. and Bondy, C.A., Evidence supporting dual, IGF-I-independent and IGF-I-dependent, roles for GH in promoting longitudinal bone growth. *J Endocrinol* **2004**; *180*: 247–255.

104 Ilizarov, G.A., Clinical application of the tension-stress effect for limb lengthening. *Clin Orthop* **1990**; *250*: 8–26.

105 Vilarrubias, J.M., Ginebreda, I. and Fernandez-Fairen, M., (500 cases of lower limb lengthening using a personal technique in achondroplasia) *Acta Orthop Belg* **1988**; *54*: 384–390.

106 Paley, D., Current techniques of limb lengthening. *J Pediatr Orthop* **1988**; *8*: 73–92.

107 Aviezer, D., Golembo, M. and Yayon, A., Fibroblast growth factor receptor–3 as a therapeutic target for Achondroplasia – genetic short limbed dwarfism. *Curr Drug Targets* **2003**; *4*: 353–365.

108 Plotnikov, A.N., Schlessinger, J., Hubbard, S. R. and Mohammadi, M., Structural basis for FGF receptor dimerization and activation. *Cell* **1999**; *98*: 641–650.

109 Hart, K.C., Robertson, S. C. and Donoghue, D.J., Identification of tyrosine residues in constitutively activated fibroblast

growth factor receptor 3 involved in mitogenesis, Stat activation, and phosphatidylinositol 3–kinase activation. *Mol Biol Cell* **2001**; *12*: 931–942.

110 Ornitz, D. M. and Marie, P.J., FGF signaling pathways in endochondral and intramembranous bone development and human genetic disease. *Genes Dev* **2002**; *16*: 1446–1465.

111 Horton, W.A., Fibroblast growth factor receptor 3 and the human chondrodysplasias. *Curr Opin Pediatr* **1997**; *9*: 437–442.

112 Cho, J.Y., Guo, C., Torello, M., Lunstrum, G.P., Iwata, T., Deng, C. and Horton, W.A., Defective lysosomal targeting of activated fibroblast growth factor receptor 3 in achondroplasia. *Proc Natl Acad Sci USA* **2004**; *101*: 609–614.

113 Rauchenberger, R., Borges, E., Thomassen-Wolf, E., Rom, E., Adar, R., Yaniv, Y., Malka, M., Chumakov, I., Kotzer, S., Resnitzky, D., Knappik, A., Reiffert, S., Prassler, J., Jury, K., Waldherr, D., Bauer, S., Kretzschmar, T., Yayon, A. and Rothe, C., Human combinatorial Fab library yielding specific and functional antibodies against the human fibroblast growth factor receptor 3. *J Biol Chem* **2003**; *278*: 38194–38205.

114 Suda, M., Ogawa, Y., Tanaka, K., Tamura, N., Yasoda, A., Takigawa, T., Uehira, M., Nishimoto, H., Itoh, H., Saito, Y., Shiota, K. and Nakao, K., Skeletal overgrowth in transgenic mice that overexpress brain natriuretic peptide. *Proc Natl Acad Sci USA* **1998**; *95*: 2337–2342.

17
Musculature and Growth Factors

Petra Neuhaus, Herbert Neuhaus, and Thomas Braun

17.1
Introduction

Myogenic development begins with the specification of mesodermal cells to become myogenic precursors that will eventually give rise to terminally differentiated myocytes resulting in the formation of myotubes. While most of the head muscles are formed from cephalic paraxial and prechordal mesoderm [1, 2], all skeletal muscle of the body is formed from the somites, which are transient metameric structures laid down in a rostral to caudal direction on each side of the neural tube and notochord during embryogenesis [3]. Initially somites bud off the unsegmented paraxial mesoderm and form epithelial, ball-shaped structures with a centrally located somitocoel. During further development the somites differentiate and eventually give rise to the sclerotome and the dermomyotome [3, 4]. Due to inductive and repressive instructions from surrounding tissues the somite acquires a dorsal-ventral, anterior-posterior and medial-lateral polarity [5]. Cells in the ventral part of the somite de-epithelialize and become mesenchymal cells of the sclerotome, forming the axial skeleton. Cells in the dorsal portion of the somite retain their epithelial character and form the dermomyotome. The dermomyotome is further separated in a medial and a lateral population of cells, giving rise to the dermis, epaxial muscles (deep muscles of the back) and the hypaxial muscles (appendicular musculature, abdominal muscles, diaphragma, hypoglossal chord and intercostal muscles) [3, 6].

The muscle precursor cells (mpcs) in the somite undergo proliferation and differentiation, before they give rise to functional muscle tissue in the adult organism. The differentiation program of mpcs is regulated by the muscle regulatory factors (MRFs), which are basic helix-loop-helix (bHLH) transcription factors, comprising Myf5, MyoD, MRF4 and myogenin. The onset of MRF expression is the first specific indicator of myogenesis. Myf5 and MyoD are determination genes and the first two members of the MRF family to be switched on in the developing dermomyotome, while the differentiation factors MRF4 and myogenin are activated later during myogenesis [7].

Ablation studies in chicken embryos indicated that the onset of myogenesis is dependent on surrounding axial and lateral tissues like the neural tube, notochord

Cell Signaling and Growth Factors in Development. Edited by K. Unsicker and K. Krieglstein
Copyright © 2006 WILEY-VCH Verlag GmbH & Co. KGaA, Weinheim
ISBN 3-527-31034-7

and dorsal ectoderm [8, 9]. Later studies identified different members of the Wnt family secreted from the dorsal neural tube or the surface ectoderm [10], sonic hedgehog (Shh) secreted from the neural tube and the notochord [11] and members of the bone morphogenetic protein (BMP) family secreted from the lateral mesoderm [12, 13] as essential growth factors involved in the induction of myogenesis and patterning of the early somite.

However, none of these factors can be viewed independently, they rather form an interacting network which will finally lead to the initiation of the myogenic program in mesodermal precursor cells and to the differentiation of mpcs resulting in the formation of the different muscle cell lineages such as epaxial and hypaxial muscles [13, 14] as well as the different muscle fiber types [15]. Besides their function in embryogenesis, several of these growth factors are also key regulators of muscle regeneration in adult life and control stem cell differentiation towards the myogenic lineage.

In this chapter we will first focus on growth factors controlling embryonic myogenesis. Later we will discuss the growth factors that may direct the development of specific subsets of skeletal muscle and regeneration processes in adult life.

17.2
Induction of Primary Myogenesis by Cell Signaling and Growth Factors

Dorsoventral patterning of the somite and thus the decision between sclerotome and dermomyotome development is mediated by the dorsal neural tube and the surface ectoderm, which secretes members of the Wnt family as dorsalizing signals [10, 13, 16]. Ventralizing signals are Noggin and Shh, released from the notochord and ventral neural tube [11, 17, 18]. Since Shh and Wnts are able to act over some distance [9, 17, 19, 20], the dorso-ventral patterning process is believed to be established by the balanced activities of these two signals, which presumably work by establishing opposing concentration gradients across the somite field [16]. However, several recent reports have shown direct interaction between the signaling pathways utilized by Shh and Wnts, suggesting that the concentration gradient is not the only cause of the opposing effects of Shh and Wnts [21, 22].

Further players in the patterning of the somite and specification of different myogenic lineages are Noggin and transforming growth factor β (TGFβ) family members, which are secreted from the notochord and the lateral plate mesoderm, respectively [13]. BMPs are members of the TGFβ family and counteract Wnt signaling, keeping migrating muscle precursor cells in an undifferentiated state until they reach their target destination [12]. As with Shh and Wnt, TGFβ and Wnt signaling cascades interact during the establishment of different myogenic lineages in the somite [23]. Noggin antagonizes BMP-signaling by specifically binding the ligand, thereby blocking ligand-receptor interaction and interfering with signal transduction [24]. Together with BMP4, noggin was shown to be essential for the correct patterning of the somite and myotome development [18].

To add even more complexity to the induction of myogenesis, the family of fibroblast growth factors (FGFs) is also involved in the balance between proliferation and differentiation of mpcs, and FGF signaling was shown to be essential for limb bud myogenesis in chicken embryos [25–27].

17.2.1
Wnt Molecules

The Wnt (wingless and integrated) genes encode a large family of secreted, cysteine-rich growth factors that play key roles in intercellular signaling during development. Numerous studies have been carried out *in vitro* and in different model organisms such as Drosophila, *Caenorhabditis elegans* (*C. elegans*), Xenopus, chicken and mouse establishing the role of Wnt family members in various developmental processes including segmentation [21, 28], CNS patterning [29], dorso-ventral patterning of the embryo [23, 30] and myogenesis [9, 31, 32]. Wnt molecules are post-translationally modified, and exert their function through a family of seven pass transmembrane receptors called frizzled (Fz). Fz genes were initially isolated as tissue polarity factors in Drosophila development [33, 34] and later shown to represent a large gene family present in all mammalians species [35]. Intracellular Wnt signaling is achieved either through the so-called canonical signaling pathway by inhibition of glycogen synthase kinase 3 (GSK3) stabilization of β-catenin, translocation of β-catenin into the nucleus and interaction with lymphoid enhancer binding factor/T-cell transcription factor1 (LEF/TCF–1) to induce target gene transcription, or by means of non-canonical pathways via Jun-kinases (JNK) or calcium dependent protein kinase C (PKC) signaling [36–38].

During early myogenesis in somites Wnts, together with Shh are essential for dorso-ventral patterning and hence for sclerotome versus dermomyotome induction [11, 30]. In explanted somites, several Wnts (Wnt–1, Wnt–3, Wnt–4, Wnt–5a, Wnt–6 and Wnt–7a) were shown to induce myogenesis, consistent with the expression of Wnts in the dorsal neural tube and surface ectoderm [9, 39, 40].

With respect to the induction of primary myogenesis and myotome formation Wnt–1 and Wnt–3a are the best characterized members of the Wnt family. Wnt–1 (int–1), originally identified as a proto-oncogene [41], induces muscle cell differentiation by activating the myogenic program in somite explants *in vitro* [39]. In addition, Wnt–1 (and to a lesser extent Wnt–4), secreted from the neural tube, has been implicated in the onset of Myf–5 expression, whereas Wnt–7a, secreted from the dorsal ectoderm, has been proposed to switch on MyoD expression thus inducing myogenesis in the epaxial and hypaxial compartment of the dermomyotome respectively [40]. However, a null mutation of Wnt–1 in mice did not result in altered myogensis [42], suggesting that it might be dispensable for the normal onset of muscle development *in vivo*. The expression pattern of Wnt–3a in the dorsal neural tube overlaps with the expression of Wnt–1 [10] and both genes have been associated with the correct onset of patterning and myogenesis in the somite. As in the case of

Wnt–1, a null mutation of Wnt–3a did not result in a muscle phenotype [43]. Due to the overlapping expression patterns of the two genes, normal muscle development in single mutants might allow compensation of Wnt–1 by Wnt–3a and vice versa. Analysis of double mutant mice revealed that the medial compartment of the dermomyotome (giving rise to the myotome and subsequently the epaxial muscles) is missing during the early stages of development while the epithelial dermomyotome is formed readily. Interestingly, although the myotome will eventually form, it shows a reduction in cell number and poor organization [28]. Both effects could probably be due to the delayed onset of Myf–5 expression, which might be normally activated by Wnt–1 and/or Wnt–3a [28].

Wnt–11 expression corresponds to the onset of myotome formation and is specific for the medial lip of the dermomyotome. Interestingly, Wnt–11 is induced by BMP signaling from the dorsal neural tube, which acts indirectly via Wnt–1 and Wnt–3a and can be counteracted by Noggin and Shh signaling from the notochord/ventral neural tube [8, 13]. In accordance with the results obtained with compound Wnt–1;Wnt–3 mutants, inhibition of Wnt–11 signaling results in a very similar defect, with the delayed formation of the medial compartment of the dermomyotome and epaxial muscles.

Experiments in chicken embryo demonstrated that Wnt–1 and Wnt–3a induce myogenesis via Frizzled1 (Fz1) and β-catenin/Lef1/TCF signaling [44]. The onset of myogenic induction is marked by the expression of β-catenin, which becomes apparent before MyoD expression is detectable. However, in this system, Wnt signaling is necessary but not sufficient for the onset of myogenesis in the early somites, since neither Wnt–1 nor Wnt–3a alone are able to induce β-catenin expression, but require Shh secreted from the notochord. In contrast to the data obtained *in vivo*, experiments *in vitro* using P19 cells (embryonic carcinoma cells) suggest that β-catenin expression is both, necessary and sufficient for MRF expression, leading to skeletal muscle development [45]. Similar results have been obtained using mesenchymal stem cells derived from several adult tissues (Belaga-Begana et al., in preparation).

In accordance with the data provided by Petropoulus et al. [45], indicating that β-catenin expression and thus early myogenesis is induced by Wnt signaling independent of other growth factors, there have been recent reports that the initiation of myogenesis *in vivo* is Shh independent. Wnt–5b, produced in the presomitic mesoderm, has been shown to induce MyoD expression without external signaling molecules and Myf–5 expression in the presomitic mesoderm might be activated in a mesoderm-autonomous fashion without activation by additional factors outside the paraxial mesoderm [46]. These data are supported by the normal expression of Myf–5 during early somitic mesoderm development in Shh mutants indicating that Shh is dispensable for the induction of the early myogenic program [47, 48].

Although the onset of myogenesis might occur as a mesoderm-autonomous process, maintenance and correct patterning of myogenic gene expression in the developing somites and subsequent muscle development is dependent on environmental factors [46, 47].

The distinction between hypaxial and epaxial myogenesis is achieved by combined action of Wnts, Shh, Noggin and BMPs [13]. While Shh induces a medial, epaxial fate in the dorsomedial compartment of the dermomyotome, BMP–4 was shown to inhibit Wnt signaling and β-catenin expression in the lateral lip of the dermomyotome, thus keeping epaxial precursor cells in an undifferentiated state [12, 44, 49]. These inductive and repressive signaling pathways were also analyzed *in vitro* in P19 cells. As shown in these reports, Myf5 is preferentially activated by Wnts secreted from the dorsal neural tube (Wnt–1 and Wnt–3) and acts via the canonical pathway, and MyoD expression is regulated by Wnt signals derived from the dorsal ectoderm (Wnt–7a) using the non-canonical PKC signaling cascade [31]. The combination of BMP inhibition of Wnt signaling derived from the neural tube and the differential activation of Myf5 and MyoD by different members of the Wnt family utilizing different signaling cascades might also explain earlier observations, showing that Myf5 and MyoD differentially regulate the development of limb versus trunk skeletal muscle [14]. Nevertheless, the specificity of activation of individual MRFs by distinct signaling molecules is still not fully resolved.

The interaction of the different signaling pathways (Wnt, Shh and TGFβ) leading to the specification of different muscle lineages is well established at the level of inhibitors/effectors of the different growth factors. A family of Wnt-signaling inhibitors, called secreted frizzled-related proteins (Sfrps), have been isolated and shown to inhibit Wnt signaling *in vivo*. There are four known members of the Sfrp family: Sfrp1 (also known as hFRP, FrzA, SARP2, Frzb1), Sfrp2 (SARP1, SDF–5), Sfrp3 (Frzb, Fritz) and Sfrp4 [50–53]. Sfrp proteins act as dominant negative inhibitors of Wnt signaling by keeping the Wnts away from their receptors. They contain the cysteine-rich Wnt binding region homologous to the ligand-binding domain of frizzled genes, yet lack the transmembrane domain of frizzled proteins and thus signaling-capacity. In Xenopus and mouse, Sfrp1/Frzb1 was shown to antagonize Wnt signaling during the onset of myogenesis by inhibiting MyoD expression [52, 54]. In chicken, Sfrp–2 is expressed in regions linked to Wnt-mediated myogenesis, and Sfrp2 has been shown to act as a Wnt antagonist in animal cap assays [32]. Sfrp2 is expressed in migrating hypaxial mpcs presumably preventing precocious differentiation of developing myoblasts until they have reached their destination. It is also possible that Sfrp2 selectively antagonizes one group of Wnts, thereby promoting a single myogenic pathway [22, 53]. Interestingly, Shh has been shown to induce Sfrp2 expression in the sclerotome, thus preventing Wnt signals from inducing dermomyotome formation in the medial compartment of the somite. In this context, Sfrp2 might specifically inhibit Wnt–1 and Wnt–4, but not Wnt–3a [22]. In contrast, Wnt signaling might also inhibit Shh by induction of the growth arrest specific gene1 (GAS1), a Shh antagonist leading to a ventral specification of the dermomyotome [55]. Taken together, the currently available data suggest that dorsalizing and ventralizing signals not only form rival concentration gradients across the somitic field, but might also communicate with each other by both inducing opposing signal transduction pathways and by direct competition within an individual pathway.

17.2.2
Shh, Ihh

Shh is a growth factor secreted from the floor plate and notochord during early development, [56] and mediates its signal through the cell surface receptor patched (ptc), a 12–pass transmembrane protein [57, 58]. Ptc does not directly transmit the Shh-induced signal, but interacts with Smoothened (Smo), which is the signal-transducing unit of the Shh signaling pathway. Smo is a seven-pass transmembrane protein associated with G-proteins [59]. In the absence of Shh signaling, Ptc and Smo are associated at the cell membrane, resulting in the block of Shh signal transduction. Upon binding of Shh to Ptc, Smo is released and represses downstream signaling molecules such as protein kinase A (PKA), GSK3 and Casein kinase 1 (CK1) which inactivate Gli (glioblastoma derived) transcription factors [60]. In mammalian embryos three Gli proteins are known, which can exist in two different forms: a short, repressive form and a longer activating form. Proteolysis leading to cleavage of the full-length form is regulated by phosphorylation. Since Shh inhibits these phosphorylation events Gli proteins are activated by Shh. Interestingly, target genes of Shh signaling via Gli proteins include members of the Wnt, TGFβ and BMP families as well as Ptc, the Shh receptor itself [38, 61–63]. Thus, enhanced transcription of Ptc initiated by its own ligand will create a negative feedback loop and silence Shh signaling [64].

In chicken embryos, the notochord and ventral neural tube were shown to be essential for epaxial muscle development, since ablation of the notochord and ventral neural tube led to an absence of axial structures including epaxial muscle [65]. Because of its expression in the ablated tissues, Shh was proposed to be essential for the development of the missing axial structures. Experiments *in vitro* and *in vivo* revealed that Shh is able to rescue the phenotype of notochord- and neural tube-ectomized chicken embryos, resulting in sclerotome and myotome formation [9, 11]. These data suggest that Shh is essential for cell survival [66] as well as for the expansion of premyogenic cells by enhancing cell proliferation of committed skeletal muscle cells [67]. Both groups report that early sclerotomal (Pax–1) or myotomal markers (Pax–3, MyoD) are expressed in notochord- and neural tube-ablated embryos, indicating that the induction of skeletogenesis and myogenesis did occur, but the cells were not able to expand and survive. This is in accordance with data obtained from Shh mutant mice, which lack axial structures and muscles while early hypaxial muscle development occurs normally. Later during development, hypaxial muscles in Shh mutants disappear [68], which was initially explained by a lack of innervation of muscle fibers leading to apoptosis [69]. In Shh mutant mice the onset of myogenesis, as judged by Myf5 expression in the early somites occurs normally but Myf5 expression is not maintained and somites in the mutants appear smaller and disorganized [47]. These data suggest that the absence of limb muscles at later developmental stages is not due to a lack of innervation, but to an absence of Shh-mediated survival and proliferation of hypaxial muscle cells [47].

There still is an ongoing debate on the question of whether Shh is a trophic factor necessary for myoblast survival and proliferation [47, 67], or an inducer of the epaxial myogenic lineage [69–72]. Some experiments analyzing the promoter region of Myf5 and Myf5 reporter gene constructs in wild-type and Shh mutant mice seemed to indicate an inductive effect of Shh on the establishment of the myotome rather than a trophic role [69]. It was proposed that the weak Myf5 expression detected in the dorsomedial region of somites in Shh mutants by whole mount *in situ* hybridization reflects a lateralization of the somite and expression of hypaxial Myf5 in more medial regions of the somite [72]. Gustafsson et al. also claimed that the activation of Myf5 by Shh is a direct effect of Shh signaling mediated by Gli proteins [72]. More recent data, however, has enforced the original view that Shh does not activate the Myf–5 promoter in early epaxial cells of the somite but is necessary for the maintenance of Myf–5 expression [48]. These authors demonstrated that activation of Myf5 expression depends on neither Shh function nor an intact Gli binding site, although the Gli site is necessary for continuation of expression. It seems likely that this report [72] led to incorrect conclusions due to the existence of specific interactions between the enhancer and the Myf5 promoter.

An interesting aspect of the Shh signaling pathway is the activation of Gli proteins via PKA and GSK3/CK1 [60]. As mentioned above, Wnt proteins also use GSK3 in their signaling pathway, which is in several respects similar to the Shh signaling pathway. Frz and Smo belong to the same family of seven-pass transmembrane molecules associated with G-proteins, and both act via inactivation of GSK3 leading to a less phosphorylated downstream signaling effector (β-catenin in Wnt signaling; Glis in Shh signaling) rendering them more stable against proteolysis. The fact that the activity of GSK3 regulated by Wnt or Shh does not result in an activation of β-catenin or Glis by Shh and Wnt respectively was explained by the need for Glis to be phosphorylated by PKA before GSK3 phosphorylation and inhibition of proteolysis can occur. This is not necessary for β-catenin which can be directly phosphorylated by GSK3 [60]. Nevertheless, Shh and Wnt signaling pathways do converge at the level of Gli proteins, since Wnt molecules were shown to regulate Gli2 and Gli3 in the segmental plate mesoderm (Gli3 repression) and in the somite (Gli2 and Gli3 activation). The activation of Gli 2 and 3 by Wnt in combination with Shh-mediated activation of Sfrp2 (a Wnt–1 and Wnt–4 antagonist, see above) in the ventral somite [22] might be a possible mechanism by which Shh directs Gli2 and 3 dorsalization during somite maturation and thus leads to the activation of the myogenic program in the dorsomedial somite and myotome [21].

17.2.3
BMPs

As discussed in the sections above, Shh and Wnt signaling mainly establish the dorso-ventral patterning of the somite. However, two other key players in mediating the correct dorso-ventral and medio-lateral patterning of the somites are BMP4 and noggin, two opposing signals secreted from the neural tube and lateral mesoderm [18, 49, 73].

Bone morphogenetic proteins (BMPs) belong to the family of TGFβ signaling molecules and have been involved in a number of developmental processes including gastrulation [74, 75], cardiogenesis [76–78], skeletogenesis [79, 80], neurogenesis [81] and myogenesis [12, 49, 73]. By embryonic manipulation of chicken embryos, the lateral plate mesoderm was identified as the source of a signal maintaining Pax3 expression in myoblasts [82], which serves as a hallmark of undifferentiated, proliferating limb muscle precursor cells lacking Myf5 and MyoD expression [83–86]. Later on, BMP4 was identified as the lateralizing signal secreted from the lateral plate mesoderm [12]. In chicken embryos, BMP4 is able to induce Single minded 1 (Sim1), a basic helix-loop-helix transcription factor and marker of lateral lip dermomyotomal cells and Pax3 expression. In addition, BMP4 inhibits expression of MyoD in the hypaxial muscle precursor cells [12]. The effect of BMP4 in the induction of lateral plate gene expression, in contrast to myogenic differentiation in the myotome, was shown to be dose dependent. High levels of BMP4 induce lateral mesoderm markers such as Sim1 and maintain Pax3 expression in the lateral lip of the dermomyotome, whereas low levels of BMP4 control the activity of Pax3 in the medial somite to induce MyoD and Myf5 expression [49].

Based on transplantation and ablation experiments in chicken embryos it was proposed that a signal secreted from the neural tube antagonizes BMP4 activity in the medial somite, preventing medial somitic cells from acquiring lateral somitic cell identity [12]. This molecule was shown to be noggin, a BMP antagonist able to bind BMP2, BMP4 and BMP7 and thereby preventing interaction of the ligands with their receptors [24]. In the chicken embryo, it was demonstrated that BMP4 and noggin signaling cascades interact and regulate their own expression in maturing somites. While BMP4 induces noggin, noggin signals back and inhibits BMP4 expression. This interaction leads to a dynamic pattern in which noggin expression is directed from more lateral to more dorsomedial domains of the somite during somite maturation [73]. Noggin is expressed in the dorsomedial lip of the dermomyotome where Pax3–expressing cells first activate the expression of MyoD and Myf5. Overexpression of noggin more laterally, leads to a suppression of Pax3 expression in the lateral dermomyotome and lateral expansion of MyoD expression, indicating its role in the specification of epaxial versus hypaxial muscle development [49, 87]. This lateralization can be overcome by high doses of BMP4 fostering the idea that noggin inhibits BMP4 signaling from the neural tube and thereby neutralizes the block in differentiation of cells of the dorsomedial lip of the somite [49]. In mice, a targeted mutation of noggin leads to defects in neural tube and somite development. The somites formed in noggin mutants are smaller than in their wild-type littermates and the epithelial to mesenchymal transition of dermomyotomal cells leading to myotome formation does not take place. The medial dermomyotome maintains Pax3 expression and the onset of Myf5 and MyoD expression does not take place, indicating a lateralization of the dorsomedial lip of the somites in noggin mutants [18], which is in agreement with the data obtained in chicken embryos.

Since Pax3 has been shown to activate myogenesis in the absence of inducing tissues [88, 89] and to be essential for the activation of MyoD in the trunk of Myf5–deficient mice [90], it is surprising that Pax3 expression marks undifferentiated myoblasts in the hypaxial lineage [86]. However, this can be explained by the antagonizing effects of noggin and BMP4. BMP4 acts as an inhibitor of myogenic differentiation by inhibiting Myf5/MyoD induction downstream of Pax3, and this inhibition is lifted by noggin which is itself induced by Wnt–1 [49] and acts synergistically with Shh [18] to promote myogenesis in the dorsomedial lip of the somite [49]. Such a view is also compatible with the direct inhibition of myogenesis by BMP4 via the Wnt–3a induced canonical β-catenin pathway [31]. Since TGFβ and Wnt signaling pathways have been shown to interact directly, leading to the inhibition of Wnt signaling via the canonical β-catenin pathway [23], it seems reasonable to assume that BMPs and Wnts might anatogonize each other, both through the regulation of noggin expression and by β-catenin signaling. Nishita et al. showed that Lef1/TCF (a component of the canonical Wnt signal transduction pathway) forms a complex with Smad4 (synonym: Madh4, MAD homolog 4, Drosophila mad – mothers against decapentaplegic), an essential mediator of signals initiated by members of the TGFβ superfamily. Blocking the activity of Smad4 in animal cap assays with a dominant negative Smad4 inhibits Wnt signaling as shown by the lack of Xtwn induction (the Xenopus homeobox gene twin and direct target of Wnt signaling in Spemann's organizer). In addition, the authors were able to detect a specific interaction between Smad4 and TGFβ and Smad4 and Lef1/TCF binding sites in the promoter of Xtwn, indicating that these signaling molecules interact to regulate Xtwn expression in the embryo [23].

A simplified scheme representing the growth factors influencing dorso-ventral and lateral-medial polarity of the somites is given in Fig. 17.1.

17.2.4
FGFs

In contrast to the growth factors discussed so far, members of the fibroblast growth factor family (Fgfs) seem to be involved in the regulation of proliferation and/or differentiation of myogenic precursor cells, rather than in establishing somitic polarities and early cell specification within the somite. FGFs represent a large family of growth factors involved in various biological processes and comprise at least 20 members, which signal through four structurally-related high-affinity receptor tyrosine kinases [91, 92].

In vitro, FGFs were shown in several assays to be potent inhibitors of myogenic differentiation [93]. Hence, successful differentiation is usually accompanied by loss of FGFR expression in the cells [94, 95]. However, under certain conditions, FGFs were also shown to stimulate myogenesis *in vivo* and the decision between induction and inhibition of myogenesis seems to be dose dependent in the case of FGF6 [96]. At present it is not clear whether FGFs are necessary for myoblast proliferation and/or myoblast differentiation, since current *in vitro* data support both views.

Since FGFR1 and FGFR4 are expressed in the myogenic lineage during embryogenesis [97, 98] and FGFR4 is expressed in adult satellite cells, representing the

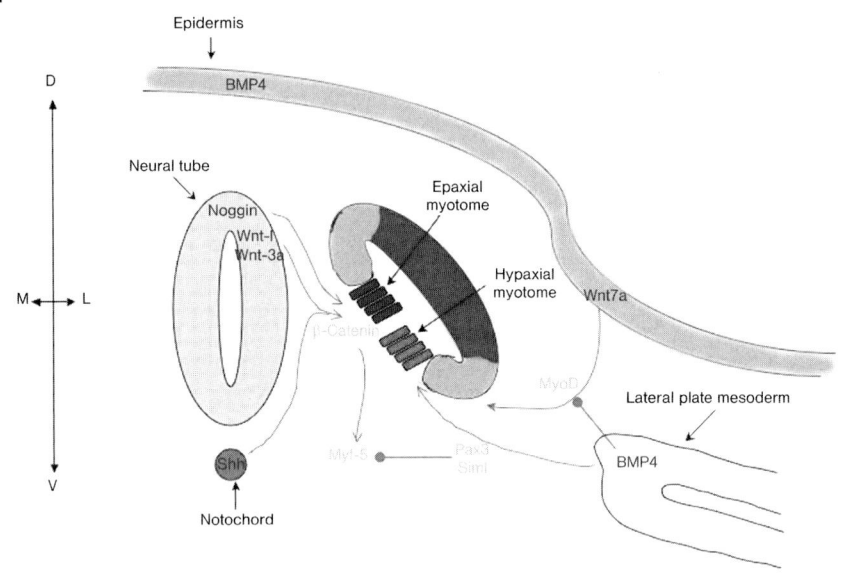

Fig. 17.1
Growth factors determining the dorso-ventral and medial-lateral polarity of the somite are shown (typed in blue). Inductive signals are marked by green arrows and repressive signals are shown in red. Genes induced by the growth factors are indicated in bright orange. The scheme gives only an overview; more details can be found in the text (see Chapter 2). (This figure also appears with the color plates.)

only adult muscle cells capable of proliferation and differentiation in the process of muscle regeneration [97, 99], both receptors have been implicated in mediating FGF function during myogenesis *in vivo*. The targeted mutations of FGFR1 and FGFR4 in mice did not result in the identification of specific functions of the two receptors during myogenesis, since FGFR1 mutant mice die at gastrulation [100] and FGFR4 mutants do not display a muscle phenotype [101]. Because of the overlapping expression patterns of FGFR1 and 4, the lack of a muscle phenotype in FGFR4 mutant mice is probably caused by compensatory effects mediated by FGFR1. Experiments in chicken embryos suggested a role for both receptors in myogenic differentiation. Inhibition of FGFR1 signaling in the developing limb bud of chicken embryos results in a premature terminal differentiation of myoblasts, accompanied by a reduction in the cell number of muscle cell progenitors [26]. In contrast, inhibition of FGFR4 signaling in the chicken embryonic limb bud results in a loss of differentiation of myoblasts, and the cells are kept in a Pax3–expressing, undifferentiated state [27]. No influence of FGFR4 signaling on cell proliferation or apoptosis of myoblasts was observed in these experiments [27]. Taken together FGFR1 and FGFR4 seem to have opposing effects on proliferation and differentiation of myoblasts. The same divergent effects on myogenesis were shown for FGF8 versus FGF4. Ectopically expressed FGF8 stimulates myoblast differentiation via MyoD activation, which is most likely mediated by FGFR4 [27] whereas overexpression of

FGF4 in somites leads to an inhibition of myogenesis and a downregulation of FGFR4, while the FGFR1 is upregulated [102]. Therefore, the combined action of FGF4 and FGF8 might lead to a delicate balance between differentiation and proliferation of myocytes *in vivo*, which determines the correct spatial and temporal patterning of muscle development. A similar model was proposed for development of the mouse skull vault where FGF2 seems to regulate FGFR1 and FGFR2 expression, which is thought to regulate proliferation (FGFR2 mediated) versus differentiation (FGFR1 mediated) of skeletal precursor cells [103].

17.3
Tour Guides: How Growth Factors Guide Migration of Muscle Precursor Cells During Embryonic Development

Migration of mpcs is required during embryonic development as well as in regeneration processes in adult life. Hypaxial muscle precursor cells, including appendicular muscle precursors and precursors of tongue and diaphragm muscles have to migrate long distances from somites to their target region during embryogenesis [15, 86]. In regenerating muscle, activated muscle stem cells have to migrate to the damaged area and participate in wound repair (see Section 17.5.2).

The correct patterning of muscles derived from migratory myocytes depends on several steps. Initially the precursor cell pool has to be specified before the cells are activated (satellite cells) or delaminated from the somite (hypaxial mpcs) and migrate to their target area. During their migration the cells maintain a proliferating, undifferentiated status and differentiate only after reaching their target destination [7, 15, 86]. The balance between proliferation and differentiation, initially leading to the expansion of the premyogenic pool and later to the formation of terminally differentiated myoblasts, is essential for the correct patterning of the developing or regenerating muscle.

17.3.1
HGF

HGF is a heparin-binding growth factor with mitogenic, motogenic and morphogenic properties and its pleiotropic actions are mediated by c-met, a receptor tyrosine kinase [104–107]. Initially, HGF was shown to induce scattering of hepatocytes *in vitro*, an activity which also occurs in muscle precursor cells *in vivo*. Ectopically-expressed HGF, implanted into the flank region of chicken embryos results in delamination of myocytes from the somite and migration of mpcs. Since the cells do not directly migrate towards the signaling source, but rather scatter randomly, HGF seems to act not only as a chemoattractant, as initially hypothesized, but as factor responsible for delamination of epithelial cells from somites [108, 109]. In addition to the scattering effect, HGF renders the cells capable of reacting with as yet unknown factors resulting in their targeted migration and keeps the cells in an undifferentiated state, which is essential for their migratory capacity [110].

In wild-type mice, only medial mpcs at cervical or appendicular axial levels delaminate and start to migrate, although c-met is expressed throughout the dermomyotome at all axial levels of the developing embryo [111, 112]. The restricted action of c-met results from the localized expression of HGF, which is only expressed at the level of the branchial arches and limbs [108, 111–113]. The expression of HGF is controlled by signals from the limb bud, including FGFs, Shh and BMPs [108, 110]. During migration, c-met expression is maintained in mpcs, ensuring the dissociation and motility of the mpcs during migration. The function of HGF/c-met signaling as a trigger for delamination was confirmed by targeted mutations of HGF and c-met, which lead to a loss of hypaxial muscles in the mutants, although the somitic cells are specified normally [111, 114].

The migratory signal emitted by HGF is dependent on the activation of the small GTPases Ras and Ral, since the use of dominant negative isoforms of hRas or RalA *in vitro* and *in vivo* results in the reduced motility of cells normally activated by HGF [115–117]. The function of HGF with respect to satellite cell activation and muscle regeneration in the adult organism will be discussed in Section 17.5.2.1.

17.3.2
FGFs

As mentioned before, myocytes leaving the dermomyotome and migrating into the limb bud have to be maintained in a proliferating, undifferentiated state. This is probably achieved by the interaction of different growth factors similar to the initial determination of mpcs in the somite.

FGFs were shown to initiate limb bud development, since ectopic expression of FGFs leads to formation of additional limb buds at the interlimb level of chicken embryos [108, 118, 119]. FGF-dependent limb patterning, which establishes expression patterns and other regulatory signaling loops of distinct growth factors, like HGF (see above) might affect mpc development indirectly. The environmental signals induced by FGF signaling might cause induction of migratory, inductive or repressive signaling on mpcs migrating into the limb buds. In addition, FGFs directly stimulate proliferation and differentiation of myocytes (as already described in Section 17.2.4).

In vitro, FGFs also act as chemoattractants in myogenic cells, and the small GTPases hRas and RalA have been shown to mediate this action of FGFs in migration assays using C2C12 cells as well as primary muscle satellite cells [116]. For a further discussion of the function of FGFs in muscle regeneration and satellite cell migration see Section 17.5.2.2.

In summary, mpcs that migrate to their target destination depend on different processes and their regulation: I, they leave their tissue of origin, which is (with respect to hypaxial mpcs) regulated by HGF through the induction of the epithelial to mesenchymal transition followed by the scattering of the cells; II; – they remain in an undifferentiated, migrating state, which is regulated by HGF, FGFs and BMP–4; and III, upon reaching their target area they expand by proliferation, which is mainly regulated by FGFs [85].

17.4
Control of Muscle Size and Muscle Fiber Diversity by Local and Circulating Growth Factors

The correct patterning of muscles during myogenesis depends on two simultaneous processes. First, myoblast precursors initiate muscle-specific gene expression (MRFs) leading to the onset of differentiation, and second, myoblasts permanently withdraw from the cell cycle, which leads to an accumulation of myocytes in the G1 phase and ultimately to terminal differentiation of the cells. Disturbing either of these two processes may lead to incorrect patterning of muscles with respect to muscle size and/or formation of the correct fiber type.

17.4.1
Control of Muscle Size

17.4.1.1 Myostatin
Myostatin (Msn, also known as growth and differentiation factor 8, GDF8) belongs to the family of TGFβ growth factors, and was initially thought to be expressed specifically in developing and adult skeletal muscle [120]. Later it was also shown to be expressed in the heart [121] and mammary gland [122]. However, targeted mutation of Msn in mice [120], as well as naturally-occurring mutations in cattle [122], resulted in an obvious muscle overgrowth without another apparent phenotype. The increased muscle mass in mutants results from muscle fiber hyperplasia (increased number of fibers) as well as muscle fiber hypertrophy (larger muscle fiber diameter) [120].

The targeted mutation of Msn established it as a negative regulator of muscle growth, but its mode of action *in vivo* has not yet been determined. *In vitro* experiments using C2C12 cells, a cell line derived from satellite cells with high myogenic potential [124], showed that Msn regulates muscle growth by inhibiting myoblast proliferation and myoblast differentiation.

The withdrawal of myoblasts from the cell cycle, paralleled by the onset of MRF expression, is regulated by different cyclin-dependent kinases (Cdks), a family of kinases involved in cell cycle regulation. During myogenesis, the principal players in the G_1 to S transition are the cyclinD-Cdk4 and cyclinE-Cdk2 complexes, which in turn regulate the retinoblastoma gene (Rb) by phosphorylation [125]. Activity of the Cdks themselves is regulated by cyclin-dependent kinase inhibitors (CKIs), with the p21 family of CKIs being specifically important for the regulation of the G_1/S transition [126].

Overexpression of recombinant Msn protein in C2C12 cells decreases the number of myoblasts in the S-phase and increases the number of myoblasts in G_1 and in the G_2/M phase of the cell cycle resulting in a lower proliferation rate of myogenic cells. This Msn-induced cell cycle arrest is mediated by a rise in p21 CKI activity followed by a reduction in protein levels and activity of Cdk2 and thus hypophosphorylation of Rb, which leads to the concurrent cell cycle arrest of myoblasts in the G_1 phase [127, 128]. In addition to the inhibition of the cell cycle, Msn also inhibits apoptosis in myoblasts, thereby increasing cell numbers and stimulating muscle hypertrophy in Msn mutant animals [129].

Besides inhibiting proliferation, Msn might indirectly affect myoblast differentiation by removing myoblasts from the cell cycle, which is essential for terminal differentiation of muscle cell progenitors [130, 131]. At present it is not clear whether Msn induces exit from the cell cycle by acting directly on MyoD or Myogenin or by different means. While it has been reported that Msn downregulates MyoD expression *in vitro*, leading to a p21–dependent exit from the cell cycle [130], another report stated that Msn downregulates Myogenin [131] and thereby inhibits myoblast differentiation. This discrepancy could, in part, be explained by the different techniques used. While one group overexpressed Msn in C2C12 cells [130], the other group interfered with endogenous Msn signaling [131].

An interesting aspect of the regulation of Msn activity is the presence of several E-boxes within the Msn promoter, which are characteristic binding sites for MRFs [132, 133]. One of these E-boxes was shown to be essential for MyoD-induced activation of the Msn promoter *in vitro*, suggesting that Msn could be a downstream target of MyoD *in vivo*. Since MyoD was shown to regulate cell cycle withdrawal of myoblast precursors at the G_0/G_1 transition [134, 135], it was suggested, that MyoD triggers cell cycle exit by regulating Msn expression [136]. Interestingly, the expression levels of MyoD and Msn do correlate in G_1 phase, while no correlation is detected during G_0 and G_1/S which indicates that the Msn promoter might act independently of MyoD at the latter stages of the cell cycle [136].

Taken together the data obtained *in vitro* suggest that Msn regulates muscle size by inhibiting myoblast proliferation as well as myoblast differentiation, and that the Msn gene is regulated differentially in proliferating versus differentiating muscle precursor cells, reflecting its dual role in proliferation and differentiation during myogenesis.

17.4.1.2 IGFs

Insulin-like growth factors I and II (IGF I and IGF II) are two structurally related proteins regulating survival, proliferation and differentiation of different cell types [137], including muscle cells. Both proteins mediate their action through the IGF–1 receptor, a ligand-activated receptor tyrosine kinase [137]. IGF-I and IGF-II were shown *in vivo* and *in vitro* to be important for muscle growth and differentiation. In mice, the targeted mutation of IGF-I and the IGFR1 results in profound muscle hypoplasia [138], whereas overexpression of IGF–1 leads to increased muscle formation due to muscle fiber hypertrophy [139, 140].

The mode of action of IGFs was mainly analyzed *in vitro*, where IGF signaling regulates proliferation, differentiation and apoptosis of cultured myoblasts. IGF-I expression stimulates determination and early differentiation of myoblasts, which results in upregulation of MyoD, IGF-II, IGFR and thymoma viral proto-oncogene 1 (Akt, a serine-threonine-kinase). This further leads to the upregulation of p21, which renders the cells more resistant to apoptosis as they leave the cell cycle [141–145]. The IGF-induced signal transduction cascade amplifies itself in an autocrine manner leading to Myogenin expression and terminal differentiation of myoblasts into myocytes and ultimately myofibers [146, 147].

In accordance with its anti-apoptotic function during myoblast differentiation, C2C12 cells lacking IGF-II do not differentiate, but undergo apoptosis, which can be overcome by the addition of either IGF-I or platelet derived growth factor (PDGF). Initially this led to the assumption that PDGF and IGF signaling pathways might converge to inhibit apoptosis [143], but later it became clear that IGF promotes cells survival via PI3K/Akt signaling cascades whereas PDGF acts via mitogen activated protein kinase – extracellular signal-regulated kinase (MEK-ERK) signaling [144].

In contrast to the data described above, which seem to indicate that IGF signaling stimulates differentiation and inhibits apoptosis in differentiating myoblasts, there are more recent data suggesting that IGFs are also capable of inhibiting or at least delaying terminal differentiation of myoblasts *in vitro* [148].

In addition to the anabolic effect exerted by IGFs on muscle formation these growth factors can also induce apoptosis in C2 cells, if they are expressed together with TNFα [149]. In this report the authors speculate that progression through the cell cycle, which is stimulated by TNFα and the IGFs, is essential for induction of apoptosis, resulting in a pro-apoptotic action of IGFs in C2 cells. They propose a model, in which balanced actions of TNFα and IGFs regulate fiber size in skeletal muscle: in the absence of TNFα IGFs stimulate proliferation and thus enable muscle precursors and muscle stem cells to increase in number before they differentiate, resulting in an optimized growth and/or repair ability of muscle tissue. The combined action of IGFs and TNFα at low concentrations might be one way to convert the anabolic function of IGF signaling into apoptotic muscle loss and potentially muscle wasting. At high concentrations, TNFα function does not need to be enhanced by IGFs, but induces muscle wasting by inhibition of myoblast differentiation and stimulation of apoptosis independently of IGF, which eventually results in a decrease of stem cells, interference with tissue repair and muscle loss (see Section 17.5.1) [150–152].

Since the studies analyzing the anabolic effects of IGFs with respect to muscle development were carried out mostly without the addition of other cytokines, it might be reasonable to assume that the different effects of IGFs on cellular proliferation, differentiation and apoptosis described here are due to the different cytokine environments of the cells in the respective studies.

17.4.2
Muscle Fiber Diversity

During the formation of mature muscle fibers, myoblast precursor cells undergo massive proliferation prior to their terminal differentiation and subsequent fusion into multinucleated myofibers. These myofibers can be grouped into different classes due to their expression of different myosin heavy chain isoforms (fast versus slow), which determines their contraction speed, metabolic activities, morphology and innervation [153]. Each muscle in the adult organism has a specific arrangement of different fiber types and this complex arrangement of has long raised the question of how these patterns are specified. Basically there have been two points of view: either the myoblasts are determined to become one specific fiber type early on while they are still in the somite, or the surrounding tissues at the site of terminal muscle differentiation determines the fate of the myoblasts [15, 153].

There has been experimental support for both theories. After xenotransplantation of quail somites which give rise to the pectoralis muscle, into chicken hosts, the resulting muscle exhibits the fiber type patterning characteristic of quail rather than chicken pectoralis muscle, indicating that the muscle precursor cells in the quail somite were predetermined with respect to slow and fast fiber type formation [154]. In contrast to these data, retroviral labeling studies showed that even cells that are derived from the same progenitor cell can contribute to slow and fast fibers during limb muscle development, which would clearly argue for environmental signaling to determine fiber type specificity [155]. One way to circumvent the discussion would be the assumption that myogenic cells are biased to form a certain fiber type when they leave the somite, but the system displays some plasticity, enabling environmental signals to overrule the initial determination.

17.4.2.1 Shh

Plasticity of fiber type development can be seen in the zebrafish mutation u-boot [156]. U-boot is a gene located downstream of Shh signaling and myogenic somitic cells of u-boot mutant fishes cannot respond appropriately to Shh signaling, leading to a loss of slow muscle fiber formation [157]. Over-inducing Shh signaling in the mutants leads to formation of slow myoblasts, but these trans-differentiate into fast muscle fibers later in development, indicating that competence and/or specification of muscle fiber types might be reversible and dependent on continued reinforcement by environmental signals [157].

Several reports, including the findings from the u-boot mutation in zebrafish, have established Shh signaling as a key component in fiber type determination. Loss-of-function mutations in components of the Shh signaling network result in the reduced or absent development of slow muscle fibers [156, 158–161], whereas overexpression of Hedgehog (Hh) paralogs results in the ectopic development of slow fiber at the expense of fast fiber type formation [70, 71, 162–164]. These data led to the assumption that Shh acts as a kind of binary switch, with fast fiber type development being the default pathway, which can be overridden by Shh signaling [164]. This is supported by the direct effect of Shh specifically on the proliferation of presumptive slow muscle cell precursors as observed in somite explants. In this system, Shh keeps muscle precursor cells in a proliferating, undifferentiated state, which ultimately leads to an expansion of the precursor cell pool and an increase in the number of slow muscle fibers in the limb bud [159, 165].

However, recent data obtained in zebrafish embryos by interfering with Hh signaling at different levels of the signal transduction pathway, show that Hh signaling specifies different types of muscle precursor cells in a dose- and time-dependent manner and that even some fast myogenic precursor cells are specified by Hh signaling, indicating that the interpretation of slow myofiber development being the default pathway is over-simplified [166]. While the slow-twitch differentiation pathway is induced by low-level Hh signaling at an early stage of development, a subset of fast twitch fibers (derived from the medial fast fiber myogenic precursor cells) is specified by high levels of Hh at a later time-point in development and exposure to Hh signaling [166].

Taken together, it is clear that Shh has an essential role in fiber type specification during myogenesis, but the mode of action is not quite clear as yet. One possibility would be that Hh signaling directly promotes certain fiber type specification, as suggested by the studies of Wolff and colleagues [166]. Since these data show that fiber type specification of myogenic precursor cells in the somite is not just Shh concentration dependent, but also time dependent, a model in which Hh signaling creates a favorable environment enabling certain fiber types to develop may be justified. Recent data, showing that Shh does not act as a mitogen, but rather as an inhibitor of differentiation in limb muscle precursor cells would be consistent with the latter model [165]. By either augmenting or inhibiting Shh signaling in limb buds of chicken embryos, Shh was shown to selectively amplify determined myoblast precursors by keeping them in an undifferentiated state that enables them to react to mitogenic signals in the environment, rather than acting as a mitogen itself. Interestingly, this effect was only detected in myogenic precursor cells of the slow twitch type and thus resulted in an expansion or reduction of slow twitch fibers upon augmentation or inhibition of Shh signaling respectively [165].

17.4.2.2 Wnt

Besides Shh, several members of the Wnt family of growth factors were shown to have distinct functions in myogenic determination *in vitro*. Besides influencing cell proliferation and thus the number of terminally differentiated myogenic cells, overexpression of different Wnts also leads to a change in number of slow and/or fast MyHC-expressing cells respectively [32]. While Wnt5a and Wnt6 do not have a significant effect on the number of terminally differentiated myogenic cells *in vitro*, Wnt3a decreased and Wnts4, 7a, 11 and 14 increased the number of myocytes. These changes in myocyte number are accompanied by different changes in the relation of slow to fast myofibers. The Wnt3a-induced loss of myocytes is accompanied by a loss of slow muscle fibers, while Wnt14 overexpression results in a parallel increase of fast and slow MyHC-expressing cells. Interestingly, Wnt5a and Wnt6 stimulate slow muscle fiber type development, accompanied by a loss of fast MyHC-expressing cells, although the overall number of myocytes remains the same. In contrast, by overexpression of Wnt11, the increase in myocyte number is accompanied by an increase in fast fiber type at the cost of slow fiber type development. Thus, *in vitro*, different members of the Wnt family have different effects on myoblast proliferation and/or fiber type induction [32].

Due to their expression patterns in developing limb buds, two members of the Wnt family were analyzed with reference to their effect on myocyte proliferation and determination in chicken embryos. Wnt5a is expressed in the mesenchymal core of the developing limb bud around the area where most of the slow fiber type muscles will develop later. In contrast, Wnt11 is expressed in the subectodermal mesenchyme next to the areas where most of the fast muscle fibers will be situated [167]. Retroviral overexpression in chicken limb buds confirms that Wnt5 induces slow muscle fiber type development at the cost of fast muscle fibers without changing myocyte numbers, and that Wnt11 increases the number of myocytes accompanied by a decrease in slow muscle fiber development and an induction of fast muscle fibers [32].

These data clearly establish Wnts as proliferation and differentiation factors involved in muscle fiber patterning and indicate that fine-tuning the balance between different Wnt family members is essential for the correct muscle patterning. The different effects of the Wnt family members on fast and slow muscle fiber development are summarized in Tab. 17.1.

Tab. 17.1
The arrows in the table indicate the change in total numbers of myocytes (proliferation), the change in numbers of myocytes belonging to the slow twitch class and change in numbers of myocytes belonging to the fast twitch type after infection of micromass cultures with viruses expressing the indicated Wnt proteins. The direction of the arrows relate to micromass cultures infected with an EGFP virus (Chapter 4), see [32].

Protein	Proliferation	Slow fibers	Fast fibers
Wnt3a	↓	↓	→
Wnt4	↑	↑	→
Wnt5a	→	↑	↓
Wnt6	→	↑	↓
Wnt7a	↑	↑	↓
Wnt11	↑	↓	↑
Wnt14	↑	↑	↑
ΔWnt11	↓	↓	↓

17.5
Growth Factors in Muscle Wasting and Skeletal Muscle Regeneration

Postnatal growth and regeneration of adult muscle is achieved by the activation and differentiation of stem cells which are able to proliferate and differentiate into muscle cells. Stem cells are considered to be non-differentiated cells, displaying a capacity for self-renewal throughout the lifespan of an organism and being able to generate a large number of differentiated progeny. Two populations that have been studied in some detail with respect to their regenerative capacity in adult muscle are satellite cells and side population (SP) cells, both present in adult muscle tissue [168]. *In vitro*, primary cultures of these two cell types show dissimilar differentiation properties, with satellite cells giving rise to myocytes and never forming hematopoietic colonies, whereas SP cells readily differentiate into hematopoietic cells but never form myocytes [169]. Despite these differences *in vitro*, both populations were shown to contribute to muscle regeneration *in vivo*. Several reports also state that non-muscle-derived, uncommitted stem cells derived from non-muscle tissue might be able to contribute to muscle regeneration *in vivo* (see Section 17.6).

17.5.1
Mediation of Muscle Wasting by Cell Signaling Molecules

Muscle wasting (cachexia) is a common component of chronic diseases such as acquired immune deficiency syndrome (AIDS), cancer, chronic heart failure or chronic obstructive pulmonary diseases and it is usually accompanied by chronic elevation of inflammatory cytokines and oxidative stress [170]. Possible explanations for the muscle loss are a decrease in the number of muscle fibers due to apoptosis of mature myotubes, a disturbance of the homeostasis between muscle synthesis and muscle repair, potentially by effects on muscle stem cells or from metabolic dysregulation leading to an imbalance between myofibrillar protein synthesis and proteolysis.

One of the cytokines involved in the reduction of muscle mass in different disease models is tumor necrosis factor α (TNFα). If rodents receive TNFα a transient weight loss is induced and in nude mice sustained exposure to TNFα results in a loss of lean body mass [171, 172]. Depending on the general health status of mice, transgenic overexpression of TNFα results in cachexia-like disease states [150, 151]. Since TNFα blocks differentiation and induces apoptosis in myoblasts *in vitro* [152], the loss of muscle mass is attributed to a loss of regenerative capacity in the muscle tissue of TNFα-treated animals due to apoptotic death of stem cells in the muscle tissue of these mice. The apoptotic effect of TNFα has mostly been described *in vitro*, using relatively high, pharmacological doses of TNFα. If low doses of TNFα are used, differentiation is blocked by the stimulation of myoblast proliferation without inducing apoptosis [149]. The TNFα-induced block of differentiation in cultured myocytes is accompanied by the downregulation of the myogenic regulatory factors MyoD and myogenin [173], which may be mediated by the activation of the nuclear transcription factor-κB (NF-κB) (see below).

NF-κB is usually located in the cytoplasm and inactivated by binding to the inhibitory protein κBα (IκBα). Phosphorylation of κBα and subsequent ubiquitination of IκBα results in the production of active NF-κB which is translocated to the nucleus and induces transcription of a number of target genes including inflammatory cytokines [174, 175]. *In vitro* studies have shown that the TNFα-induced block of differentiation of myoblasts results from NF-κB activation, initially leading to inappropriate proliferation of myoblasts and subsequent block of differentiation, accompanied by low level apoptosis [176]. NF-κB probably causes the block of differentiation seen in myocytes after exposure to TNFα by mediating degradation of MyoD transcripts in myogenic cells and subsequent upregulation of CyclinD1 [177, 178].

Interestingly, IL–1β, another cytokine inducing cachexia [179], also induces NF-κB activation [175] and may thus have similar effects on myoblast differentiation [176] as TNFα..

However, the activation of NF-κB cannot be the only way by which TNFα mediates its block of myogenesis, since there are cell lines, which strongly upregulate NF-κB in response to TNFα yet maintain their ability to differentiate into myocytes and myofibers. In addition, inhibition of caspase activity in C2 cells rescues TNFα-induced loss of differentiation although NF-κB is still upregulated [180]. These observations led to the identification of an alternative pathway for the TNFα-induced block of differentiation in cultured myoblasts, which is mediated by paternally expressed 3 (Peg3, synonym: PW1), a zinc finger protein [181], and caspase pathways [180]. Caspases induce cell death in response to p53 or cytokine signaling, usually accompanied by mitochondrial dysfunction, which can also lead to a block of differentiation in muscle cells [182]. Since PW1 was shown to mediate its action in myoblasts by translocating Bcl2–associated X protein (bax) to the mitochondria [183] and causing mitochondrial dysfunction and block of differentiation [179], PW1–dependent signaling might be an alternative explanation for TNFα-induced cachexia.

Besides affecting the proliferative status of myoblasts TNFα was also shown to inhibit morphological myofiber development by inhibiting the fusion of differentiated, elongated myocytes, essential for the formation of myotubes [184]. However, this is not mediated by the activation of NF-κB, but via induction of oxidative stress [184–186]. Taken together, the data suggest a dual role for TNFα in the induction of muscle wasting: first, the inhibition of the biochemical aspects of myogenesis in a PW1– and/or NF-κB-dependent manner, and second the inhibition of morphological aspects of myogenesis by the induction of oxidative stress.

A summary of the different modes of TNFα action with respect to the induction of cachexia is given in Fig. 17.2.

17.5.2
Activation of Muscle Satellite Cells by Growth Factors

Satellite cells represent a small number of cells (5% of adult muscle cells), which are located adjacent to mature myotubes between the basal lamina and sarcolemma of muscle fibers. Besides morphological criteria they can be defined by gene expres-

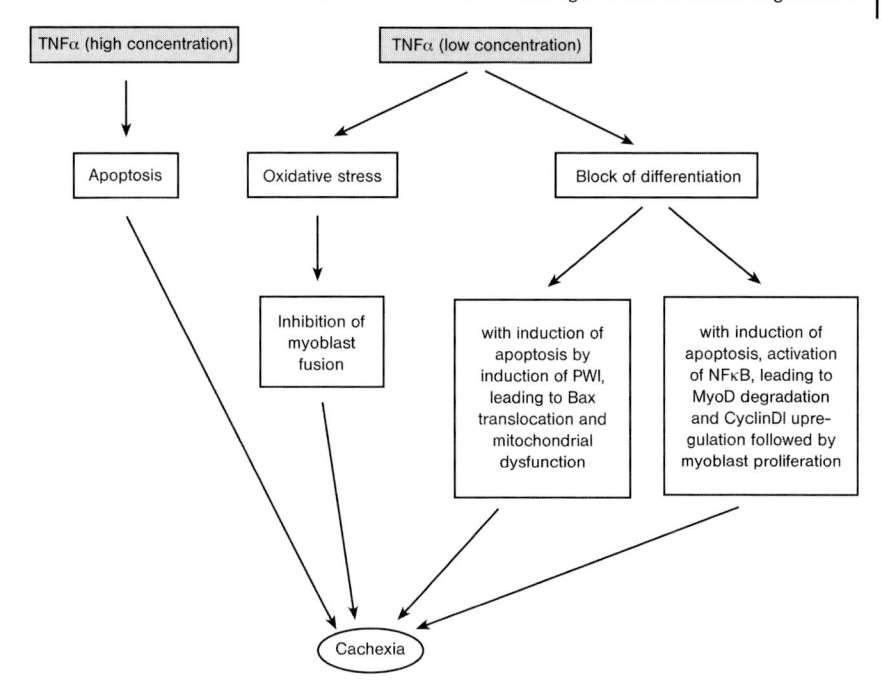

```
┌─────────────────────────┐         ┌─────────────────────────┐
│ TNFα (high concentration)│         │ TNFα (low concentration) │
└─────────────────────────┘         └─────────────────────────┘
           │                              ╱         ╲
           ▼                             ▼           ▼
    ┌───────────┐         ┌────────────────┐    ┌────────────────────┐
    │ Apoptosis │         │ Oxidative stress│    │ Block of differentiation│
    └───────────┘         └────────────────┘    └────────────────────┘
```

Fig. 17.2
This scheme represents the different pathways which are assumed to play a role in TNFα-induced cachexia (see Chapter 5).

Boxes below Block of differentiation: "Inhibition of myoblast fusion"; "with induction of apoptosis by induction of PWI, leading to Bax translocation and mitochondrial dysfunction"; "with induction of apoptosis, activation of NFκB, leading to MyoD degradation and CyclinDl upregulation followed by myoblast proliferation". All pathways lead to **Cachexia**.

sion profiles and *in vivo* function. Satellite cells were shown to be essential for postnatal muscle growth, and upon injury and/or degeneration of mature muscle fibers they give rise to large numbers of daughter myoblasts in addition to new satellite cells [187]. Although satellite cells can be characterized by marker expression (they express m-cadherin, CD34, Msx1, cMet and Pax7 [107, 188, 189, 190]), there are reports questioning whether they represent a unique cell type or rather a heterogenous population of uncommitted and committed muscle precursor cells.

Whilst quiescent satellite cells display very little migration, after muscle trauma activated satellite cells migrate extensively towards the site of injury [191–194] and participate in muscle regeneration. Directed migration of satellite cells was shown to be induced by different growth factors and cytokines released either from inflammatory cells, platelets or the damaged tissue itself. In addition to migration towards the site of injury, satellite cells need to be activated. This means that they need to re-enter the cell cycle in order to replenish the stem cell pool and to give rise to differentiating cells which will form myotubes, thus leading to tissue repair.

17.5.2.1 HGF

Two factors are clearly involved in the activation of quiescent satellite cells. The first is hepatocyte growth factor (HGF, also known as scatter factor, SF) and the second is nitric oxide (NO) [195]. In adult muscle tissue, HGF is localized in the extracellular domain of uninjured skeletal muscle and is released upon muscle injury leading to an association with satellite cells [195, 196]. Although the function of HGF with relation to myoblast migration during embryogenesis *in vivo* is well established (see Section 17.3.1), its function during muscle regeneration has been analyzed mostly *in vitro*.

C2 cells as well as primary satellite cells express HGF and its receptor c-met. This results in an autocrine loop, inducing the phosphorylation and activation of c-met leading to proliferation and blocking of myogenic differentiation of the cells [197]. Sustained activation of c-met signaling leads to apoptosis of myocytes *in vitro*, which led to the hypothesis that other growth factors, potentially induced after satellite cell activation by HGF, must subsequently initiate the differentiation program in the myocytes [197] thereby preventing programmed cell death. One growth factor receptor induced by HGF signaling is FGFR1, and FGFs are able to stimulate proliferation as well as differentiation of mpcs (see Sections 17.2.4 and 17.5.2.2). In addition, both, HGF and FGF are able to act as chemoattractants of primary myocytes *in vitro* indicating additional and potentially co-operative functions of these growth factors in the targeted migration of activated satellite cells towards the site of injury [116, 198, 199].

A potential mechanism for the activation of HGF and thus HGF-mediated satellite cell activation is the induction of oxidative stress by nitric oxide (NO). Stretch injuries in cultured satellite cells lead to the release of NO, which results in the activation of the satellite cells and their subsequent proliferation and differentiation, while the inhibition of the NO synthetase results in the inhibition of myoblast differentiation [200]. Recent experiments in cultured satellite cells show that NO synthesis due to stretch injuries induces the release of HGF from its extracellular binding to heparan sulfate proteoglycans, resulting in its activation and the initiation of satellite cell-mediated tissue repair [201, 202].

17.5.2.2 FGF

Freshly isolated satellite cells show a lag phase during the first day in tissue culture, which represents the quiescent stage and is used for the analysis of genes involved in satellite cell activation. During this phase, the cells are not responsive to FGFs, which cannot activate satellite cells [203]. As soon as the cells are activated (for example by HGF), they start to express FGF family members and FGFRs. Mouse myogenic cell lines and satellite cells express FGFR 1 and/or FGFR 4 [99, 204] displaying an interesting dynamic pattern. The receptor genes are upregulated by FGF or HGF signals in the proliferative phase of muscle precursor cells and become downregulated as differentiation of the myoblasts sets in [204]. *In vitro*, FGFs1, 2, 4, 6, and 9 are able to induce proliferation of primary satellite cells [204], which is in accordance with the known ligand-receptor interactions [205, 206], since FGFR1 is particularly responsive to FGFs1, 2, and 4, whereas FGFR4 is rather specific for FGFs1, 2, 4, 6, and 9. While Sheehan and colleagues [204] used satellite cell cultures, another approach used to determine the function of FGFs *in vitro* is the use of single fiber culture experiments, which resembles myogenesis in an environment in which muscle structure is preserved and thus might actually reflect more subtle injuries, or growth and hypertrophy of muscle, rather than extensive trauma [207, 208]. Using this system, FGF2 and FGF6 were shown to have the most significant effect on satellite cell proliferation, which is most likely transmitted by FGFR4, showing upregulation in muscle fibers after treatment with FGF4 or 6 [99].

Besides stimulation of satellite cell proliferation FGF6 overexpression *in vitro* also results in a delay of differentiation of muscle precursor cells by extending the period during which the cells express myogenin [99], indicating that FGFs might have a dual function in the proliferation and differentiation of satellite cells as well as during embryonic development (see Section 17.2.4).

In vivo, FGF2 and FGF6 have mostly been analyzed with respect to their influence on muscle regeneration. Both single mutations in mice do not reveal any obvious phenotypes with respect to early embryonic myogenesis. During adult muscle regeneration, there are conflicting data for FGF6 *in vivo*, which might result from the different mouse strains used in the two studies. While one report states a reduced regenerative capacity of muscle tissue in FGF6 mutant mice [209], another group does not detect the same phenotype in their mutant mouse strain [210]. However, combined mutation of FGF2 and FGF6 leads to an obvious loss of regenerative capacity in the muscle tissue of compound mutant animals [117], which indicates that FGF2 and FGF6 probably compensate for the loss of each other in the single mutations. The defects seen in these compound mutant animals result from a reduced mobility of satellite cell-derived myoblasts, rather than from a proliferation and/or differentiation defect [117] as might have been expected from *in vitro* experiments on isolated satellite cells or myogenic cell lines (see above).

Both, FGF2 and FGF6 have also been used in gene therapy approaches, to enhance the regeneration of injured muscle tissue. Application of either of the two growth factors enhanced arteriogenesis and myogenesis in skeletal muscle of mice when transferred by an adenovirus embedded in a collagen/gelatin matrix [211] and FGF6 is also able to accelerate regeneration in the soleus muscle in mice when it is injected as recombinant protein [212]. The latter is accompanied by an upregulation of FGFR1 and a downregulation of FGFR4. The authors believe that this reflects the dual action of FGF6 on proliferation, mediated by FGFR1, and differentiation, mediated by FGFR4, of muscle precursor cells during the regeneration process [212]. This would be in accordance with earlier data, showing that FGFR1 is required for proliferation and repression of differentiation of myocytes via an ERK1/2 signaling cascade [213], while FGFR4 signaling results in an arrest of muscle progenitor differentiation [27] (see Section 17.2.4).

Thus far, the data concerning the specific functions and ligand-receptor interactions of FGFs during regeneration processes in muscle tissues are illustrating partly opposing results, but this could well be caused by the different systems and cell lines used. Nevertheless, clear functions concerning FGF2 and FGF6, as well as the FGFR1 and 4 during muscle repair have been established.

17.6
Does Growth Factor Mediated Recruitment of Uncommitted Stem Cells Contribute to Skeletal Muscle Cell Regeneration?

There is an ongoing discussion about whether muscle regeneration is mainly to be attributed to muscle satellite cells, or whether uncommitted stem cells derived from the hematopoietic system, including SP cells, significantly participate in regeneration processes. This is particularly interesting in view of the potential gene therapy applications in patients with muscular dystrophies, since satellite cells cannot colonize muscle tissue if delivered from the circulation and are thus not easily usable in gene therapy approaches.

Two cell types present in the adult organism attracted specific attention with respect to their ability to aid in muscle regeneration after muscle trauma or in dystrophic muscle diseases. These are the so-called side population (SP) cells isolated from muscle (msSP) and bone marrow (bmSP) cells. Generally, SP cells express high levels of the multidrug resistance (mdr) gene, leading to the exclusion of Hoechst-dye, which makes them identifiable by fluorescence-activated cell sorting (FCAS) [214, 215].

In vitro, muscle-derived SP cells are able to differentiate into hematopoietic cells, but never into myogenic cells [168]. However, if these cells are isolated from regenerating muscle or if they are co-cultured with Wnt-secreting cells, they form myocytes and mature myofibers [216]. *In vivo*, Wnt signaling was reported to mediate the regeneration of adult muscle tissue after chemically-induced muscle injury by recruiting SP cells [216]. In addition, inhibiting Wnt signaling in regenerating muscle *in vivo* disturbs muscle regeneration, suggesting that Wnt-mediated recruitment of SP cells in muscle regeneration represents a pathway used *in vivo* [216].

Several other groups showed that bone marrow-derived cells participate in regeneration processes in adult life. Ferrari and colleagues stated that bone marrow-derived stem cells, belonging to the SP population of the bone marrow (BM), are able to participate in muscle regeneration after muscle trauma. Both the injection of the BM cells into the injured muscle or even systemic distribution of BM cells in lethally-irradiated mice through the bloodstream, results in participation of the BM cells in muscle repair [217]. These data were confirmed by other authors who reported the participation of stem cells derived from adult bone marrow in myogenesis after muscle trauma or transfer into mdx mice [215, 218]. Although bone marrow cells were capable of participating in muscle regeneration in these experiments, the efficiency of regeneration was extremely low [217, 219, 220].

The regeneration process which is initiated by putative stem cells from the hematopoietic lineage or muscle tissue presumably always showed the stem cell plasticity of non-committed cells which trans-differentiate into satellite cells and subsequently participate in muscle repair. However, there was some controversy as to whether the observed results really originated from the stem cell plasticity of uncommitted cells, or rather from the fusion of already committed cells with cells of the regenerating tissue which would result in the participation of the "stem cells" in the repair process [221–223].

One argument for stem cell plasticity was that bone marrow-derived, as well as msSP cells were participating in muscle regeneration after trans-differentiating into satellite cells, which then become the actual regenerating cell type [188, 218]. However, these data have been challenged by a new report indicating that bone marrow-derived cells actually form committed myeloid cells which fuse with myocytes during muscle regeneration [220]. That study also attributes the low regenerating capacity after bone marrow transplantation to the rather unlikely, stochastic event, that inflammatory cells (most likely macrophages) are recruited to sites of muscle injury and subsequently trapped in the fusogenic processes during muscle regeneration [220].

A prerequisite for the recruitment of cells for regenerative processes in muscle tissue is that inflammatory and regeneration processes are initiated first. None of the cell types reported so far participates in myogenesis in undamaged muscle [219, 220]. This clearly shows that cell signaling by cytokines and growth factors, besides activating satellite cells (see Section 17.5.2), also mediates the recruitment of cells with regenerative capacity to the injured area. Several approaches were tried to enhance muscle regeneration by locally applying growth factors *in vivo* and showed some positive effects on either recruitment of cells with regenerative capacity or the activation of essential processes such as vascularization and arteriogenesis during wound healing [199, 211, 224, 225].

However, although a number of publications support the idea of stem cell-mediated regeneration processes, to date there are conflicting data in some areas of this research and the exact cell type leading to muscle regeneration cannot be characterized accurately. In addition, several factors have been shown to influence regeneration capacity, yet the mode of action is so far not fully understood.

References

1 Wachtler, F., et al., The extrinsic ocular muscles in birds are derived from the prechordal plate. *Naturwissenschaften* **1984**; *71*: 379–380.
2 Couly, G.F., P.M. Coltey, and N.M. Le Douarin, The developmental fate of the cephalic mesoderm in quail-chick chimeras. *Development* **1992**; *114*: 1–15.
3 Christ, B. and C.P. Ordahl, Early stages of chick somite development. *Anat Embryol (Berl)* **1995**; *191*: 381–396.
4 Brand-Saberi, B. and B. Christ, Evolution and development of distinct cell lineages derived from somites. *Curr Top Dev Biol* **2000**; *48*: 1–42.

5 Brand-Saberi, B., et al., The formation of somite compartments in the avian embryo. *Int J Dev Biol* **1996**; *40*: 411–420.
6 Ordahl, C.P. and N.M. Le Douarin, Two myogenic lineages within the developing somite. *Development* **1992**; *114*: 339–353.
7 Neuhaus, P. and T. Braun, Transcription factors in skeletal myogenesis of vertebrates. *Results Probl Cell Differ* **2002**; *38*: 109–126.
8 Cossu, G., et al., Activation of different myogenic pathways: myf–5 is induced by the neural tube and MyoD by the dorsal ectoderm in mouse paraxial mesoderm. *Development* **1996**; *122*: 429–437.
9 Munsterberg, A.E., et al., Combinatorial signaling by Sonic hedgehog and Wnt family members induces myogenic bHLH gene expression in the somite. *Genes Dev* **1995**; *9*: 2911–2922.
10 Parr, B.A., et al., Mouse Wnt genes exhibit discrete domains of expression in the early embryonic CNS and limb buds. *Development* **1993**; *119*: 247–261.
11 Fan, C.M. and M. Tessier-Lavigne, Patterning of mammalian somites by surface ectoderm and notochord: evidence for sclerotome induction by a hedgehog homolog. *Cell* **1994**; *79*: 1175–1186.
12 Pourquie, O., et al., Lateral and axial signals involved in avian somite patterning: a role for BMP4. *Cell* **1996**; *84*: 461–471.
13 Marcelle, C., M.R. Stark, and M. Bronner-Fraser, Coordinate actions of BMPs, Wnts, Shh and noggin mediate patterning of the dorsal somite. *Development* **1997**; *124*: 3955–3963.
14 Kablar, B., et al., MyoD and Myf–5 differentially regulate the development of limb versus trunk skeletal muscle. *Development* **1997**; *124*: 4729–4738.
15 Francis-West, P.H., L. Antoni, and K. Anakwe, Regulation of myogenic differentiation in the developing limb bud. *J Anat* **2003**; *202*: 69–81.
16 Fan, C.M., C.S. Lee, and M. Tessier-Lavigne, A role for WNT proteins in induction of dermomyotome. *Dev Biol* **1997**; *191*: 160–165.
17 Fan, C.M., et al., Long-range sclerotome induction by sonic hedgehog: direct role of the amino-terminal cleavage product and modulation by the cyclic AMP signaling pathway. *Cell* **1995**; *81*: 457–465.
18 McMahon, J.A., et al., Noggin-mediated antagonism of BMP signaling is required for growth and patterning of the neural tube and somite. *Genes Dev* **1998**; *12*: 1438–1452.
19 Marti, E., et al., Distribution of Sonic hedgehog peptides in the developing chick and mouse embryo. *Development* **1995**; *121*: 2537–2547.
20 Christian, J.L., BMP, Wnt and Hedgehog signals: how far can they go? *Curr Opin Cell Biol* **2000**; *12*: 244–249.
21 Borycki, A., A.M. Brown, and C.P. Emerson, Jr., Shh and Wnt signaling pathways converge to control Gli gene activation in avian somites. *Development* **2000**; *127*: 2075–2087.
22 Lee, C.S., et al., SHH-N upregulates Sfrp2 to mediate its competitive interaction with WNT1 and WNT4 in the somitic mesoderm. *Development* **2000**; *127*: 109–118.

23 Nishita, M., et al., Interaction between Wnt and TGF-beta signalling pathways during formation of Spemann's organizer. *Nature* **2000**; *403*: 781–785.
24 Zimmerman, L.B., J.M. De Jesus-Escobar, and R.M. Harland, The Spemann organizer signal noggin binds and inactivates bone morphogenetic protein 4. *Cell* **1996**; *86*: 599–606.
25 Itoh, N., T. Mima, and T. Mikawa, Loss of fibroblast growth factor receptors is necessary for terminal differentiation of embryonic limb muscle. *Development* **1996**; *122*: 291–300.
26 Flanagan-Steet, H., et al., Loss of FGF receptor 1 signaling reduces skeletal muscle mass and disrupts myofiber organization in the developing limb. *Dev Biol* **2000**; *218*: 21–37.
27 Marics, I., et al., FGFR4 signaling is a necessary step in limb muscle differentiation. *Development* **2002**; *129*: 4559–4569.
28 Ikeya, M. and S. Takada, Wnt signaling from the dorsal neural tube is required for the formation of the medial dermomyotome. *Development* **1998**; *125*: 4969–4976.
29 Muroyama, Y., et al., Wnt signaling plays an essential role in neuronal specification of the dorsal spinal cord: Wnt signaling from the dorsal neural tube is required for the formation of the medial dermomyotome. *Genes Dev* **2002**; *16*: 548–553.
30 Capdevila, J., C. Tabin, and R.L. Johnson, Control of dorsoventral somite patterning by Wnt–1 and beta-catenin. *Dev Biol* **1998**; *193*: 182–194.
31 Ridgeway, A.G., et al., Wnt signaling regulates the function of MyoD and myogenin. *J Biol Chem* **2000**; *275*: 32398–32405.
32 Anakwe, K., et al., Wnt signalling regulates myogenic differentiation in the developing avian wing. *Development* **2003**; *130*: 3503–3514.
33 Bhanot, P., et al., Frizzled and Dfrizzled–2 function as redundant receptors for Wingless during Drosophila embryonic development. *Development* **1999**; *126*: 4175–4186.
34 Bhanot, P., et al., A new member of the frizzled family from Drosophila functions as a Wingless receptor. *Nature* **1996**; *382*: 225–230.
35 Wang, Y., et al., A large family of putative transmembrane receptors homologous to the product of the Drosophila tissue polarity gene frizzled. *J Biol Chem* **1996**; *271*: 4468–4476.
36 Pandur, P., D. Maurus, and M. Kuhl, Increasingly complex: new players enter the Wnt signaling network. *Bioessays* **2002**; *24*: 881–884.
37 Wodarz, A. and R. Nusse, Mechanisms of Wnt signaling in development. Annu Rev *Cell Dev Biol* **1998**; *14*: 59–88.
38 Cohen, M.M., Jr., The hedgehog signaling network. *Am J Med Genet* **2003**; *123A*: 5–28.
39 Stern, H.M., A.M. Brown, and S.D. Hauschka, Myogenesis in paraxial mesoderm: preferential induction by dorsal neural tube and by cells expressing Wnt–1. *Development* **1995**; *121*: 3675–3686.
40 Tajbakhsh, S., et al., Differential activation of Myf5 and MyoD by different Wnts in explants of mouse paraxial mesoderm and the later activation of myogenesis in the absence of Myf5. *Development* **1998**; *125*: 4155–4162.

41 Nusse R., and H.E. Varmus, Many tumors induced by the mouse mammary tumor virus contain a provirus integrated in the same region of the host genome. *Cell* **1982**; *31*: 99–109.

42 McMahon, A.P. and A. Bradley, The Wnt–1 (int–1) proto-oncogene is required for development of a large region of the mouse brain. *Cell* **1990**; *62*: 1073–1085.

43 Takada, S., et al., Wnt–3a regulates somite and tailbud formation in the mouse embryo. *Genes Dev* **1994**; *8*: 174–189.

44 Schmidt, M., M. Tanaka, and A. Munsterberg, Expression of (beta)-catenin in the developing chick myotome is regulated by myogenic signals. *Development* **2000**; *127*: 4105–4113.

45 Petropoulos, H. and I.S. Skerjanc, Beta-catenin is essential and sufficient for skeletal myogenesis in P19 cells. *J Biol Chem* **2002**; *277*: 15393–15399.

46 Linker, C., et al., Intrinsic signals regulate the initial steps of myogenesis in vertebrates. *Development* **2003**; *130*: 4797–4807.

47 Krüger, M., et al., Sonic hedgehog is a survival factor for hypaxial muscles during mouse development. *Development* **2001**; *128*: 743–752.

48 Teboul, L., D. Summerbell, and P.W. Rigby, The initial somitic phase of Myf5 expression requires neither Shh signaling nor Gli regulation. *Genes Dev* **2003**; *17*: 2870–2874.

49 Reshef, R., M. Maroto, and A.B. Lassar, Regulation of dorsal somitic cell fates: BMPs and Noggin control the timing and pattern of myogenic regulator expression. *Genes Dev* **1998**; *12*: 290–303.

50 Wang, S., et al., Frzb, a secreted protein expressed in the Spemann organizer, binds and inhibits Wnt–8. *Cell* **1997**; *88*: 757–766.

51 Leyns, L., et al., Frzb–1 is a secreted antagonist of Wnt signaling expressed in the Spemann organizer. *Cell* **1997**; *88*: 747–756.

52 Borello, U., et al., Transplacental delivery of the Wnt antagonist Frzb1 inhibits development of caudal paraxial mesoderm and skeletal myogenesis in mouse embryos. *Development* **1999**; *126*: 4247–4255.

53 Ladher, R.K., et al., Cloning and expression of the Wnt antagonists Sfrp–2 and Frzb during chick development. *Dev Biol* **2000**; *218*: 183–198.

54 Wang, Y. and R. Jaenisch, Myogenin can substitute for Myf5 in promoting myogenesis but less efficiently. *Development* **1997**; *124*: 2507–2513.

55 Lee, C.S., L. Buttitta, and C.M. Fan, Evidence that the WNT-inducible growth arrest-specific gene 1 encodes an antagonist of sonic hedgehog signaling in the somite. *Proc Natl Acad Sci USA* **2001**; *98*: 11347–11352.

56 Riddle, R.D., et al., Sonic hedgehog mediates the polarizing activity of the ZPA. *Cell* **1993**; *75*: 1401–1416.

57 Marigo, V., et al., Biochemical evidence that patched is the Hedgehog receptor. *Nature* **1996**; *384*: 176–179.

58 Stone, D.M., et al., The tumour-suppressor gene patched encodes a candidate receptor for Sonic hedgehog. *Nature* **1996**; *384*: 129–134.

59 van den Heuvel, M. and P.W. Ingham, Smoothened encodes a receptor-like serpentine protein required for hedgehog signalling. *Nature* **1996**; *382*: 547–551.

60 Price, M.A. and D. Kalderon, Proteolysis of the Hedgehog signaling effector Cubitus interruptus requires phosphorylation by glycogen synthase kinase 3 and casein kinase 1. *Cell* **2002**; *108*: 823–835.

61 Marigo, V. and C.J. Tabin, Regulation of patched by sonic hedgehog in the developing neural tube. *Proc Natl Acad Sci USA* **1996**; *93*: 9346–9351.

62 Marigo, V., et al., Conservation in hedgehog signaling: induction of a chicken patched homolog by Sonic hedgehog in the developing limb. *Development* **1996**; *122*: 1225–1233.

63 Ingham, P.W. and A.P. McMahon, Hedgehog signaling in animal development: paradigms and principles. *Genes Dev* **2001**; *15*: 3059–3087.

64 Ingham, P.W., et al., Patched represses the Hedgehog signalling pathway by promoting modification of the Smoothened protein. *Curr Biol* **2000**; *10*: 1315–1318.

65 Rong, P.M., et al., The neural tube/notochord complex is necessary for vertebral but not limb and body wall striated muscle differentiation. *Development* **1992**; *115*: 657–672.

66 Teillet, M., et al., Sonic hedgehog is required for survival of both myogenic and chondrogenic somitic lineages. *Development* **1998**; *125*: 2019–2030.

67 Duprez, D., C. Fournier-Thibault, and N. Le Douarin, Sonic Hedgehog induces proliferation of committed skeletal muscle cells in the chick limb. *Development* **1998**; *125*: 495–505.

68 Chiang, C., et al., Cyclopia and defective axial patterning in mice lacking Sonic hedgehog gene function. *Nature* **1996**; *383*: 407–413.

69 Borycki, A.G., et al., Sonic hedgehog controls epaxial muscle determination through Myf5 activation. *Development* **1999**; *126*: 4053–4063.

70 Currie, P.D. and P.W. Ingham, Induction of a specific muscle cell type by a hedgehog-like protein in zebrafish. *Nature* **1996**; *382*: 452–455.

71 Blagden, C.S., et al., Notochord induction of zebrafish slow muscle mediated by Sonic hedgehog. *Genes Dev* **1997**; *11*: 2163–2175.

72 Gustafsson, M.K., et al., Myf5 is a direct target of long-range Shh signaling and Gli regulation for muscle specification. *Genes Dev* **2002**; *16*: 114–126.

73 Sela-Donenfeld, D. and C. Kalcheim, Localized BMP4–noggin interactions generate the dynamic patterning of noggin expression in somites. *Dev Biol* **2002**; *246*: 311–328.

74 Kishimoto, Y., et al., The molecular nature of zebrafish swirl: BMP2 function is essential during early dorsoventral patterning. *Development* **1997**; *124*: 4457–4466.

75 Zhao, G.Q., Consequences of knocking out BMP signaling in the mouse. *Genesis* **2003**; *35*: 43–56.

76 Neuhaus, H., V. Rosen, and R.S. Thies, Heart specific expression of mouse BMP-10 a novel member of the TGF-beta superfamily. *Mech Dev* **1999**; *80*: 181–184.

77 Nakajima, Y., et al., Significance of bone morphogenetic protein–4 function in the initial myofibrillogenesis of chick cardiogenesis. *Dev Biol* **2002**; *245*: 291–303.

78 Schneider, M.D., V. Gaussin, and K.M. Lyons, Tempting fate: BMP signals for cardiac morphogenesis. *Cytokine Growth Factor Rev* **2003**; *14*: 1–4.

79 Monsoro-Burq, A.H., et al., The role of bone morphogenetic proteins in vertebral development. *Development* **1996**; *122*: 3607–3616.

80 Pizette, S. and L. Niswander, BMPs are required at two steps of limb chondrogenesis: formation of prechondrogenic condensations and their differentiation into chondrocytes. *Dev Biol* **2000**; *219*: 237–249.

81 Liem, K.F., Jr., G. Tremml, and T.M. Jessell, A role for the roof plate and its resident TGFbeta-related proteins in neuronal patterning in the dorsal spinal cord. *Cell* **1997**; *91*: 127–138.

82 Pourquie, O., et al., Control of somite patterning by signals from the lateral plate. *Proc Natl Acad Sci USA* **1995**; *92*: 3219–3223.

83 Bober, E., et al., Pax–3 is required for the development of limb muscles: a possible role for the migration of dermomyotomal muscle progenitor cells. *Development* **1994**; *120*: 603–612.

84 Goulding, M., A. Lumsden, and A.J. Paquette, Regulation of Pax–3 expression in the dermomyotome and its role in muscle development. *Development* **1994**; *120*: 957–971.

85 Birchmeier, C. and H. Brohmann, Genes that control the development of migrating muscle precursor cells. *Curr Opin Cell Biol* **2000**; *12*: 725–730.

86 Buckingham, M., et al., The formation of skeletal muscle: from somite to limb. *J Anat* **2003**; *202*: 59–68.

87 Capdevila, J. and R.L. Johnson, Endogenous and ectopic expression of noggin suggests a conserved mechanism for regulation of BMP function during limb and somite patterning. *Dev Biol* **1998**; *197*: 205–217.

88 Maroto, M., et al., Ectopic Pax–3 activates MyoD and Myf–5 expression in embryonic mesoderm and neural tissue. *Cell* **1997**; *89*: 139–148.

89 Mennerich, D. and T. Braun, Activation of myogenesis by the homeobox gene Lbx1 requires cell proliferation. *EMBO J* **2001**; *20*: 7174–7183.

90 Tajbakhsh, S., et al., Redefining the genetic hierarchies controlling skeletal myogenesis: Pax–3 and Myf–5 act upstream of MyoD. *Cell* **1997**; *89*: 127–138.

91 Powers, C.J., S.W. McLeskey, and A. Wellstein, Fibroblast growth factors, their receptors and signaling. *Endocr Relat Cancer* **2000**; *7*: 165–197.

92 Xu, X., et al., Fibroblast growth factor receptors (FGFRs) and their roles in limb development. *Cell Tissue Res* **1999**; *296*: 33–43.

93 Olson, E.N., Interplay between proliferation and differentiation within the myogenic lineage. *Dev Biol* **1992**; *154*: 261–272.

94 Moore, J.W., et al., The mRNAs encoding acidic FGF, basic FGF and FGF receptor are coordinately downregulated during myogenic differentiation. *Development* **1991**; *111*: 741–748.

95 Halevy, O., et al., A new avian fibroblast growth factor receptor in myogenic and chondrogenic cell differentiation. *Exp Cell Res* **1994**; *212*: 278–284.
96 Pizette, S., et al., FGF6 modulates the expression of fibroblast growth factor receptors and myogenic genes in muscle cells. *Exp Cell Res* **1996**; *224*: 143–151.
97 Marcelle, C., et al., Distinct developmental expression of a new avian fibroblast growth factor receptor. *Development* **1994**; *120*: 683–694.
98 Marcelle, C., J. Wolf, and M. Bronner-Fraser, The *in vivo* expression of the FGF receptor FREK mRNA in avian myoblasts suggests a role in muscle growth and differentiation. *Dev Biol* **1995**; *172*: 100–114.
99 Kastner, S., et al., Gene expression patterns of the fibroblast growth factors and their receptors during myogenesis of rat satellite cells. *J Histochem Cytochem* **2000**; *48*: 1079–1096.
100 Yamaguchi, T.P., et al., fgfr–1 is required for embryonic growth and mesodermal patterning during mouse gastrulation. *Genes Dev* **1994**; *8*: 3032–3044.
101 Weinstein, M., et al., FGFR-3 and FGFR-4 function cooperatively to direct alveogenesis in the murine lung. *Development* **1998**; *125*: 3615–3623.
102 Edom-Vovard, F., M.A. Bonnin, and D. Duprez, Misexpression of Fgf-4 in the chick limb inhibits myogenesis by down-regulating Frek expression. *Dev Biol* **2001**; *233*: 56–71.
103 Iseki, S., A.O. Wilkie, and G.M. Morriss-Kay, Fgfr1 and Fgfr2 have distinct differentiation- and proliferation-related roles in the developing mouse skull vault. *Development* **1999**; *126*: 5611–5620.
104 Borycki, A.G., L. Mendham, and C.P. Emerson, Jr., Control of somite patterning by Sonic hedgehog and its downstream signal response genes. *Development* **1998**; *125*: 777–790.
105 Naldini, L., et al., Hepatocyte growth factor (HGF) stimulates the tyrosine kinase activity of the receptor encoded by the proto-oncogene c-MET. *Oncogene* **1991**; *6*: 501–504.
106 Weidner, K.M., M. Sachs, and W. Birchmeier, The Met receptor tyrosine kinase transduces motility, proliferation, and morphogenic signals of scatter factor/hepatocyte growth factor in epithelial cells. *J Cell Biol* **1993**; *121*: 145–154.
107 Cornelison, D.D. and B.J. Wold, Single-cell analysis of regulatory gene expression in quiescent and activated mouse skeletal muscle satellite cells. *Dev Biol* **1997**; *191*: 270–283.
108 Heymann, S., et al., Regulation and function of SF/HGF during migration of limb muscle precursor cells in chicken. *Dev Biol* **1996**; *180*: 566–578.
109 Brand-Saberi, B., et al., Scatter factor/hepatocyte growth factor (SF/HGF) induces emigration of myogenic cells at interlimb level *in vivo*. *Dev Biol* **1996**; *179*: 303–308.
110 Scaal, M., et al., SF/HGF is a mediator between limb patterning and muscle development. *Development* **1999**; *126*: 4885–4893.
111 Bladt, F., et al., Essential role for the c-met receptor in the migration of myogenic precursor cells into the limb bud (see comments). *Nature* **1995**; *376*: 768–771.

112 Yang, X.M., et al., Expression of the met receptor tyrosine kinase in muscle progenitor cells in somites and limbs is absent in Splotch mice. *Development* **1996**; *122*: 2163–2171.

113 Dietrich, S., et al., The role of SF/HGF and c-Met in the development of skeletal muscle. *Development* **1999**; *126*: 1621–1629.

114 Maina, F., et al., Uncoupling of Grb2 from the Met receptor *in vivo* reveals complex roles in muscle development. *Cell* **1996**; *87*: 531–542.

115 Hartmann, G., et al., The motility signal of scatter factor/hepatocyte growth factor mediated through the receptor tyrosine kinase met requires intracellular action of Ras. *J Biol Chem* **1994**; *269*: 21936–21939.

116 Suzuki, J., et al., Involvement of Ras and Ral in chemotactic migration of skeletal myoblasts. *Mol Cell Biol* **2000**; *20*: 4658–4665.

117 Neuhaus, P., et al., Reduced mobility of fibroblast growth factor (FGF)-deficient myoblasts might contribute to dystrophic changes in the musculature of FGF2/FGF6/mdx triple-mutant mice. *Mol Cell Biol* **2003**; *23*: 6037–6048.

118 Cohn, M.J., et al., Fibroblast growth factors induce additional limb development from the flank of chick embryos. *Cell* **1995**; *80*: 739–746.

119 Crossley, P.H., et al., Roles for FGF8 in the induction, initiation, and maintenance of chick limb development. *Cell* **1996**; *84*: 127–136.

120 McPherron, A.C., A.M. Lawler, and S.J. Lee, Regulation of skeletal muscle mass in mice by a new TGF-beta superfamily member. *Nature* **1997**; *387*: 83–90.

121 Sharma, M., et al., Myostatin, a transforming growth factor-beta superfamily member, is expressed in heart muscle and is upregulated in cardiomyocytes after infarct. *J Cell Physiol* **1999**; *180*: 1–9.

122 Ji, S., et al., Myostatin expression in porcine tissues: tissue specificity and developmental and postnatal regulation. *Am J Physiol* **1998**; *275*(4 Pt 2): R1265–R1273.

123 Kambadur, R., et al., Mutations in myostatin (GDF8) in double-muscled Belgian Blue and Piedmontese cattle. *Genome Res* **1997**; *7*: 910–916.

124 Yaffe, D. and O. Saxel, Serial passaging and differentiation of myogenic cells isolated from dystrophic mouse muscle. *Nature* **1977**; *270*: 725–727.

125 Guo, K. and K. Walsh, Inhibition of myogenesis by multiple cyclin-Cdk complexes. Coordinate regulation of myogenesis and cell cycle activity at the level of E2F. *J Biol Chem* **1997**; *272*: 791–797.

126 Sherr, C.J. and J.M. Roberts, CDK inhibitors: positive and negative regulators of G1–phase progression. *Genes Dev* **1999**; *13*: 1501–1512.

127 Thomas, M., et al., Myostatin, a negative regulator of muscle growth, functions by inhibiting myoblast proliferation. *J Biol Chem* **2000**; *275*: 40235–40243.

128 Taylor, W.E., et al., Myostatin inhibits cell proliferation and protein synthesis in C2C12 muscle cells. *Am J Physiol Endocrinol Metab* **2001**; *280*: E221–E228.

129 Rios, R., et al., Myostatin regulates cell survival during C2C12 myogenesis. *Biochem Biophys Res Commun* **2001**; *280*: 561–566.
130 Langley, B., et al., Myostatin inhibits myoblast differentiation by down-regulating MyoD expression Myostatin, a negative regulator of muscle growth, functions by inhibiting myoblast proliferation. *J Biol Chem* **2002**; *277*: 49831–49840.
131 Joulia, D., et al., Mechanisms involved in the inhibition of myoblast proliferation and differentiation by myostatin. *Exp Cell Res* **2003**; *286*: 263–275.
132 Lassar, A.B., et al., Functional activity of myogenic HLH proteins requires hetero-oligomerization with E12/E47–like proteins *in vivo*. *Cell* **1991**; *66*: 305–315.
133 Murre, C., et al., Interactions between heterologous helix-loop-helix proteins generate complexes that bind specifically to a common DNA sequence. *Cell* **1989**; *58*: 537–544.
134 Crescenzi, M., et al., MyoD induces growth arrest independent of differentiation in normal and transformed cells. *Proc Natl Acad Sci USA* **1990**; *87*: 8442–8446.
135 Sorrentino, V., et al., Cell proliferation inhibited by MyoD1 independently of myogenic differentiation. *Nature* **1990**; *345*: 813–815.
136 Spiller, M.P., et al., The myostatin gene is a downstream target gene of basic helix-loop-helix transcription factor MyoD. *Mol Cell Biol* **2002**; *22*: 7066–7082.
137 Jones, J.I. and D.R. Clemmons, Insulin-like growth factors and their binding proteins: biological actions. *Endocr Rev* **1995**; *16*: 3–34.
138 Liu, J.P., et al., Mice carrying null mutations of the genes encoding insulin-like growth factor I (Igf–1) and type 1 IGF receptor (Igf1r). *Cell* **1993**; *75*: 59–72.
139 Coleman, M.E., et al., Myogenic vector expression of insulin-like growth factor I stimulates muscle cell differentiation and myofiber hypertrophy in transgenic mice. *J Biol Chem* **1995**; *270*: 12109–12116.
140 Barton-Davis, E.R., et al., Viral mediated expression of insulin-like growth factor I blocks the aging-related loss of skeletal muscle function. *Proc Natl Acad Sci USA* **1998**; *95*: 15603–15607.
141 Florini, J.R., et al., »Spontaneous« differentiation of skeletal myoblasts is dependent upon autocrine secretion of insulin-like growth factor-II. *J Biol Chem* **1991**; *266*: 15917–15923.
142 Stewart, C.E., et al., Overexpression of insulin-like growth factor-II induces accelerated myoblast differentiation. *J Cell Physiol* **1996**; *169*: 23–32.
143 Stewart, C.E. and P. Rotwein, Insulin-like growth factor-II is an autocrine survival factor for differentiating myoblasts. *J Biol Chem* **1996**; *271*: 11330–11338.
144 Lawlor, M.A., et al., Dual control of muscle cell survival by distinct growth factor-regulated signaling pathways. *Mol Cell Biol* **2000**; *20*: 3256–3265.
145 Lawlor, M.A. and P. Rotwein, Insulin-like growth factor-mediated muscle cell survival: central roles for Akt and cyclin-dependent kinase inhibitor p21. *Mol Cell Biol* **2000**; *20*: 8983–8995.

146 Florini, J.R., et al., IGFs and muscle differentiation. *Adv Exp Med Biol* **1993**; *343*: 319–326.

147 Wilson, E.M., M.M. Hsieh, and P. Rotwein, Autocrine growth factor signaling by insulin-like growth factor-II mediates MyoD-stimulated myocyte maturation. *J Biol Chem* **2003**; *278*: 41109–41113. Epub **2003** Aug 25.

148 Stewart, C.E. and P. Rotwein, Growth, differentiation, and survival: multiple physiological functions for insulin-like growth factors. *Physiol Rev* **1996**; *76*: 1005–1026.

149 Foulstone, E.J., et al., Insulin-like growth factors (IGF-I and IGF-II) inhibit C2 skeletal myoblast differentiation and enhance TNF alpha-induced apoptosis. *J Cell Physiol* **2001**; *189*: 207–215.

150 Keffer, J., et al., Transgenic mice expressing human tumour necrosis factor: a predictive genetic model of arthritis. *EMBO J* **1991**; *10*: 4025–4031.

151 Cheng, J., et al., Cachexia and graft-vs.-host-disease-type skin changes in keratin promoter-driven TNF alpha transgenic mice. *Genes Dev* **1992**; *6*: 1444–1456.

152 Meadows, K.A., J.M. Holly, and C.E. Stewart, Tumor necrosis factor-alpha-induced apoptosis is associated with suppression of insulin-like growth factor binding protein–5 secretion in differentiating murine skeletal myoblasts. *J Cell Physiol* **2000**; *183*: 330–337.

153 Stockdale, F.E., Myogenic cell lineages. *Dev Biol* **1992**; *154*: 284–298.

154 Nikovits, W., Jr., et al., Patterning of fast and slow fibers within embryonic muscles is established independently of signals from the surrounding mesenchyme. *Development* **2001**; *128*: 2537–2544.

155 Kardon, G., J.K. Campbell, and C.J. Tabin, Local extrinsic signals determine muscle and endothelial cell fate and patterning in the vertebrate limb. *Dev Cell* **2002**; *3*: 533–545.

156 van Eeden, F.J., et al., Mutations affecting somite formation and patterning in the zebrafish, *Danio rerio*. *Development* **1996**; *123*: 153–164.

157 Roy, S., C. Wolff, and P.W. Ingham, The u-boot mutation identifies a Hedgehog-regulated myogenic switch for fiber-type diversification in the zebrafish embryo. *Genes Dev* **2001**; *15*: 1563–1576.

158 Schauerte, H.E., et al., Sonic hedgehog is not required for the induction of medial floor plate cells in the zebrafish. *Development* **1998**; *125*: 2983–2993.

159 Cann, G.M., J.W. Lee, and F.E. Stockdale, Sonic hedgehog enhances somite cell viability and formation of primary slow muscle fibers in avian segmented mesoderm. *Anat Embryol (Berl)* **1999**; *200*: 239–252.

160 Lewis, K.E., et al., Control of muscle cell-type specification in the zebrafish embryo by Hedgehog signalling. *Dev Biol* **1999**; *216*: 469–480.

161 Barresi, M.J., H.L. Stickney, and S.H. Devoto, The zebrafish slow-muscle-omitted gene product is required for Hedgehog signal transduction and the development of slow muscle identity. *Development* **2000**; *127*: 2189–2199.

162 Hammerschmidt, M., M.J. Bitgood, and A.P. McMahon, Protein kinase A is a common negative regulator of Hedgehog signaling in the vertebrate embryo. *Genes Dev* **1996**; *10*: 647–658.
163 Du, S.J., et al., Positive and negative regulation of muscle cell identity by members of the hedgehog and TGF-beta gene families. *J Cell Biol* **1997**; *139*: 145–156.
164 Norris, W., et al., Slow muscle induction by Hedgehog signalling in vitro. *J Cell Sci* **2000**; *113*(Pt 15): 2695–2703.
165 Bren-Mattison, Y. and B.B. Olwin, Sonic hedgehog inhibits the terminal differentiation of limb myoblasts committed to the slow muscle lineage. *Dev Biol* **2002**; *242*: 130–1148.
166 Wolff, C., S. Roy, and P.W. Ingham, Multiple muscle cell identities induced by distinct levels and timing of hedgehog activity in the zebrafish embryo. *Curr Biol* **2003**; *13*: 1169–1181.
167 Church, V.L. and P. Francis-West, Wnt signalling during limb development. *Int J Dev Biol* **2002**; *46*: 927–936.
168 Seale, P., A. Asakura, and M.A. Rudnicki, The potential of muscle stem cells. *Dev Cell* **2001**; *1*: 333–342.
169 Asakura, A., et al., Myogenic specification of side population cells in skeletal muscle. *J Cell Biol* **2002**; *159*: 123–134.
170 Kotler, D.P., Cachexia. *Ann Intern Med* **2000**; *133*: 622–634.
171 Grunfeld, C., et al., Persistence of the hypertriglyceridemic effect of tumor necrosis factor despite development of tachyphylaxis to its anorectic/cachectic effects in rats. *Cancer Res* **1989**; *49*: 2554–2560.
172 Teng, M.N., et al., Long-term inhibition of tumor growth by tumor necrosis factor in the absence of cachexia or T-cell immunity. *Proc Natl Acad Sci USA* **1991**; *88*: 3535–3539.
173 Szalay, K., Z. Razga, and E. Duda, TNF inhibits myogenesis and downregulates the expression of myogenic regulatory factors myoD and myogenin. *Eur J Cell Biol* **1997**; *74*: 391–398.
174 Karin, M., The beginning of the end: IkappaB kinase (IKK) and NF-kappaB activation. *J Biol Chem* **1999**; *27*: 27339–27342.
175 Pahl, H.L., Activators and target genes of Rel/NF-kappaB transcription factors. *Oncogene* **1999**; *18*: 6853–6866.
176 Langen, R.C., et al., Inflammatory cytokines inhibit myogenic differentiation through activation of nuclear factor-kappaB. *Faseb J* **2001**; *15*: 1169–1180.
177 Guttridge, D.C., et al., NF-kappaB controls cell growth and differentiation through transcriptional regulation of cyclin D1. *Mol Cell Biol* **1999**; *19*: 5785–5799.
178 Guttridge, D.C., et al., NF-kappaB-induced loss of MyoD messenger RNA: possible role in muscle decay and cachexia. *Science* **2000**; *289*: 2363–2366.
179 Fong, Y., et al., Cachectin/TNF or IL–1 alpha induces cachexia with redistribution of body proteins. *Am J Physiol* **1989**; *256*(3 Pt 2): R659–R665.
180 Coletti, D., et al., TNFalpha inhibits skeletal myogenesis through a PW1–dependent pathway by recruitment of caspase pathways. *EMBO J* **2002**; *21*: 631–642.
181 Relaix, F., et al., Pw1, a novel zinc finger gene implicated in the myogenic and neuronal lineages. *Dev Biol* **1996**; *177*: 383–396.

182 Rochard, P., et al., Mitochondrial activity is involved in the regulation of myoblast differentiation through myogenin expression and activity of myogenic factors. *J Biol Chem* **2000**; *275*: 2733–2744.

183 Deng, Y. and X. Wu, Peg3/Pw1 promotes p53–mediated apoptosis by inducing Bax translocation from cytosol to mitochondria. *Proc Natl Acad Sci USA* **2000**; *97*: 12050–12055.

184 Langen, R.C., et al., Tumor necrosis factor-alpha inhibits myogenesis through redox-dependent and -independent pathways. *Am J Physiol Cell Physiol* **2002**; *283*: C714–C721.

185 Li, Y.P., et al., Mitochondria mediate tumor necrosis factor-alpha/NF-kappaB signaling in skeletal muscle myotubes. *Antioxid Redox Signal* **1999**; *1*: 97–104.

186 Williams, G., et al., Cytokine-induced expression of nitric oxide synthase in C2C12 skeletal muscle myocytes. *Am J Physiol* **1994**; *267*(4 Pt 2): R1020–R1025.

187 Seale, P. and M.A. Rudnicki, A new look at the origin, function, and »stem-cell« status of muscle satellite cells. *Dev Biol* **2000**; *218*: 115–124.

188 Seale, P., et al., Pax7 is required for the specification of myogenic satellite cells. *Cell* **2000**; *102*: 777–786.

189 Beauchamp, J.R., et al., Expression of CD34 and Myf5 defines the majority of quiescent adult skeletal muscle satellite cells. *J Cell Biol* **2000**; *151*: 1221–1234.

190 Cornelison, D.D., et al., MyoD(–/–) satellite cells in single-fiber culture are differentiation defective and MRF4 deficient. *Dev Biol* **2000**; *224*: 122–137.

191 Klein-Ogus, C. and J.B. Harris, Preliminary observations of satellite cells in undamaged fibres of the rat soleus muscle assaulted by a snake-venom toxin. *Cell Tissue Res* **1983**; *230*: 671–676.

192 Schultz, E., D.L. Jaryszak, and C.R. Valliere, Response of satellite cells to focal skeletal muscle injury. *Muscle Nerve* **1985**; *8*: 217–222.

193 Schultz, E., et al., Survival of satellite cells in whole muscle transplants. *Anat Rec* **1988**; *222*: 12–17.

194 Phillips, G.D., J.R. Hoffman, and D.R. Knighton, Migration of myogenic cells in the rat extensor digitorum longus muscle studied with a split autograft model. *Cell Tissue Res* **1990**; *262*: 81–88.

195 Anderson, J.E., A role for nitric oxide in muscle repair: nitric oxide-mediated activation of muscle satellite cells. *Mol Biol Cell* **2000**; *11*: 1859–1874.

196 Tatsumi, R., et al., HGF/SF is present in normal adult skeletal muscle and is capable of activating satellite cells. *Dev Biol* **1998**; *194*: 114–128.

197 Anastasi, S., et al., A natural hepatocyte growth factor/scatter factor autocrine loop in myoblast cells and the effect of the constitutive Met kinase activation on myogenic differentiation. *J Cell Biol* **1997**; *137*: 1057–1068.

198 Bischoff, R., Chemotaxis of skeletal muscle satellite cells. *Dev Dyn* **1997**; *208*: 505–515.

199 Villena, J. and E. Brandan, Dermatan sulfate exerts an enhanced growth factor response on skeletal muscle satellite cell proliferation and migration. *J Cell Physiol* **2004**; *198*: 169–178.

200 Lee, K. H., Baek, M. Y., Moon, K. Y., Song, W. K., Chung, C. H., Ha, D. B., and M. S. Kang, Nitric Oxide as a messenger molecule for myoblast fusion. *J. Biol. Chem.* **1994**; *269*: 14371–147374.

201 Tatsumi, R., et al., Release of hepatocyte growth factor from mechanically stretched skeletal muscle satellite cells and role of pH and nitric oxide. *Mol Biol Cell* **2002**; *13*: 2909–2918.

202 Anderson, J. and O. Pilipowicz, Activation of muscle satellite cells in single-fiber cultures. *Nitric Oxide* **2002**; *7*: 36–41.

203 Johnson, S.E. and R.E. Allen, Proliferating cell nuclear antigen (PCNA) is expressed in activated rat skeletal muscle satellite cells. *J Cell Physiol* **1993**; *154*: 39–43.

204 Sheehan, S.M. and R.E. Allen, Skeletal muscle satellite cell proliferation in response to members of the fibroblast growth factor family and hepatocyte growth factor. *J Cell Physiol* **1999**; *181*: 499–506.

205 Ornitz, D.M., et al., Receptor specificity of the fibroblast growth factor family. *J Biol Chem* **1996**; *271*: 15292–15297.

206 Szebenyi, G. and J.F. Fallon, Fibroblast growth factors as multifunctional signaling factors. *Int Rev Cytol* **1999**; *185*: 45–106.

207 Yablonka-Reuveni, Z., et al., The transition from proliferation to differentiation is delayed in satellite cells from mice lacking MyoD. *Dev Biol* **1999**; *210*: 440–455.

208 Yablonka-Reuveni, Z., R. Seger, and A.J. Rivera, Fibroblast growth factor promotes recruitment of skeletal muscle satellite cells in young and old rats. *J Histochem Cytochem* **1999**; *47*: 23–42.

209 Floss, T., H.H. Arnold, and T. Braun, A role for FGF–6 in skeletal muscle regeneration. *Genes Dev*; **1997**; *11*: 2040–2051.

210 Fiore, F., A. Sebille, and D. Birnbaum, Skeletal muscle regeneration is not impaired in Fgf6 −/− mutant mice. *Biochem Biophys Res Commun* **2000**; *272*: 138–143.

211 Doukas, J., et al., Delivery of FGF genes to wound repair cells enhances arteriogenesis and myogenesis in skeletal muscle. *Mol Ther* **2002**; *5*(5 Pt 1): 517–527.

212 Armand, A.S., et al., Injection of FGF6 accelerates regeneration of the soleus muscle in adult mice. *Biochim Biophys Acta* **2003**; *1642*(1–2): 97–105.

213 Jones, N.C., et al., ERK1/2 is required for myoblast proliferation but is dispensable for muscle gene expression and cell fusion. *J Cell Physiol* **2001**; *186*: 104–115.

214 Gussoni, E., et al., Dystrophin expression in the mdx mouse restored by stem cell transplantation. *Nature* **1999**; *401*: 390–394.

215 Jackson, K.A., T. Mi, and M.A. Goodell, Hematopoietic potential of stem cells isolated from murine skeletal muscle. *Proc Natl Acad Sci USA* **1999**; *96*: 14482–14486.

216 Polesskaya, A., P. Seale, and M.A. Rudnicki, Wnt signaling induces the myogenic specification of resident CD45+ adult stem cells during muscle regeneration. *Cell* **2003**; *113*: 841–852.

217 Ferrari, G., et al., Muscle regeneration by bone marrow-derived myogenic progenitors. *Science* **1998**; *279*: 1528–1530.

218 LaBarge, M.A. and H.M. Blau, Biological progression from adult bone marrow to mononucleate muscle stem cell to multinucleate muscle fiber in response to injury. *Cell* **2002**; *111*: 589–601.

219 Ferrari, G. and F. Mavilio, Myogenic stem cells from the bone marrow: a therapeutic alternative for muscular dystrophy? *Neuromuscul Disord* **2002**; *12*(Suppl. 1): S7–S10.

220 Camargo, F.D., et al., Single hematopoietic stem cells generate skeletal muscle through myeloid intermediates. *Nat Med* **2003**; *9*: 1520–1527.

221 Terada, N., et al., Bone marrow cells adopt the phenotype of other cells by spontaneous cell fusion. *Nature* **2002**; *416*: 542–545.

222 Wagers, A.J., et al., Little evidence for developmental plasticity of adult hematopoietic stem cells. *Science* **2002**; *297*: 2256–2259. Epub **2002** Sept 5.

223 Ying, Q.L., et al., Changing potency by spontaneous fusion. *Nature* **2002**; *416*: 545–548.

224 Barbero, A., et al., Growth factor supplemented matrigel improves ectopic skeletal muscle formation – a cell therapy approach. *J Cell Physiol* **2001**; *186*: 183–192.

225 Corti, S., et al., Chemotactic factors enhance myogenic cell migration across an endothelial monolayer. *Exp Cell Res* **2001**; *268*: 36–44.

18
Skin Development

Lydia Sorokin and Leena Bruckner-Tuderman

Abbreviations

Wnt, Wingless-related; Lef, lymphoid enhancer binding factor; Tcf, T cell factor; BMP, bone morphogenic protein; IGF, insulin-like growth factor; FGF, fibroblast growth factor; TGF, transforming growth factor; PDGF, platelet derived growth factor; EGF, epidermal growth factor; Shh, sonic hedgehog; OMIM, online Mendelian inheritance in Man, National Center for Biotechnology Information (http://www.ncbi.nlm.nih.gov:80/entrez/query.fcgi?db=OMIM)

18.1
Introduction

The epidermis of the skin, a stratified layer of epithelial keratinocytes, lies on a highly specialized basement membrane that separates the epidermis from the mesenchymal cells of the dermis. Reciprocal epithelial-mesenchymal interactions between the prospective epidermis and the underlying dermis are the major driving forces in the development of the skin and its appendages. The basement membrane provides specific spatial information that is important for the development of the skin and for its homeostasis, and mediates molecular signals from both skin compartments. This information can be directly conveyed via cellular receptor complexes or, indirectly, via the presentation of growth factors by the basement membrane. The major receptor complex that acts to anchor keratinocytes to the underlying basement membrane is the hemidesmosome. Other integrin and non-integrin receptors occur interdispersed between the hemidesmosomes, which strengthen the interconnection between the extracellular milieu and the actin cytoskeleton, and provide other cellular signals required for the homeostasis of the organ. Further, keratinocytes are interconnected by complex junctions, adherens junctions and desmosomes, all of which are essential for constituting the barrier function of the skin. Analysis of human genetic diseases and of mutant mice has revealed that cell-cell junctional components, and hemidesmosome components and their binding part-

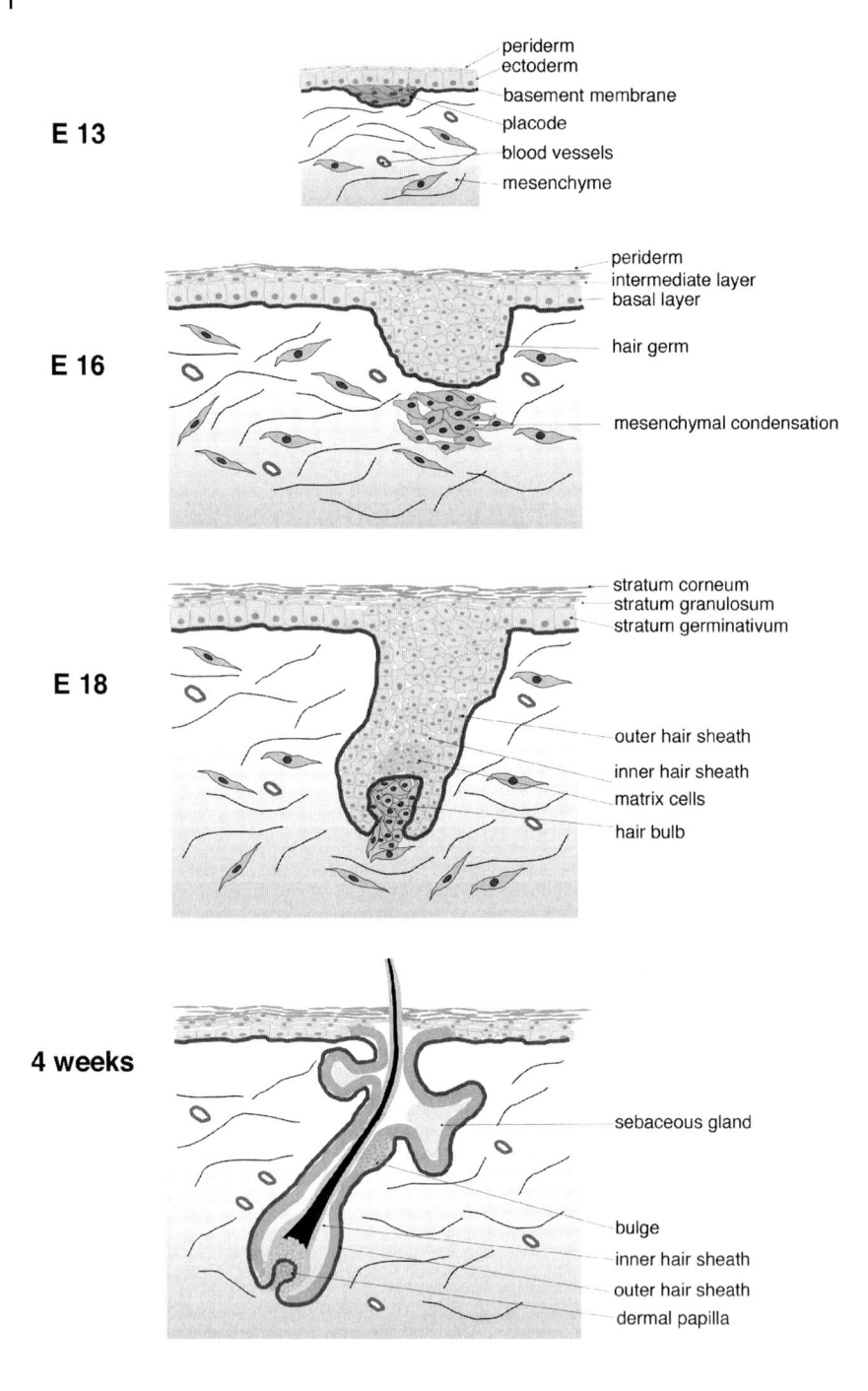

680 | 18 Skin Development

E 13
- periderm
- ectoderm
- basement membrane
- placode
- blood vessels
- mesenchyme

E 16
- periderm
- intermediate layer
- basal layer
- hair germ
- mesenchymal condensation

E 18
- stratum corneum
- stratum granulosum
- stratum germinativum
- outer hair sheath
- inner hair sheath
- matrix cells
- hair bulb

4 weeks
- sebaceous gland
- bulge
- inner hair sheath
- outer hair sheath
- dermal papilla

Fig. 18.1
Schematic representation of skin and hair follicle development in the mouse. The epidermis originates from a single-layered ectoderm, which forms the periderm. By E13, placodes form, marking initiation of hair follicle development. An intermediate proliferating ectodermal layer forms between the periderm and basal layer by E15. Cells of this layer enter terminal differentiation to form a stratified epidermis by E16. Hair follicle development involves initial mesenchymal condensation at E16 (papilla primordium) and subsequent follicular epithelial cell migration into the dermis to form hair bulb, inner and outer hair sheaths, and hair core at E18. The lowermost epithelial part of the hair follicle, the hair bulb, originates from proliferating matrix cells. By 4 weeks postnatally, a population of transient amplifying matrix cells terminally differentiate to generate the different cell layers of the follicle, including the outer root sheath, inner root sheath and the innermost hair shaft. The outer root sheath of a mature hair follicle is contiguous with the basal epidermal layer. The hair bulb surrounds specialized mesenchymal cells of the dermis, the dermal papilla, which develops from the mesenchymal condensate/papilla primordium. Sebaceous glands form in upper portions of the hair follicle and are bordered basally by a population of stem cells in the bulge region. (This figure also appears with the color plates.)

ners in the basement membrane, are essential for integrity of the skin. In addition, there is a growing body of soluble factors, which regulate development and regeneration of the skin and its appendages. Of these, signaling by the Wnt (Wingless-related) pathway is best known. This chapter will focus on the development of the stratified epithelium of the skin, with particular reference to the cell-cell and cell-matrix interactions required for its homeostasis.

18.2
Morphology of the Skin and Development of Hair Follicles and Other Adnexal Structures

18.2.1
Epidermis

Mammalian and non-mammalian epidermis and its appendages are a product of ectodermal organogenesis. As a consequence of their common origin these diverse structures, ranging from hair follicles to feathers, share a common mechanism of development, and all rely upon a series of inductive and reciprocal signals between the epithelium (epidermis) and the mesenchyme (dermis). This chapter will focus on mouse and human skin development, with focal reference to feather development because of its relevance to hair follicle formation.

The epidermis originates from the outer surface cell layer, the ectoderm, of the post-gastrulation embryo between embryonic days 8.5 and 10 (E8.5–E10) in the mouse. Ectoderm consists of a single layer of histologically undifferentiated epithelial cells, which stratifies regionally to form the periderm, a transitory embryonic ectodermal layer that provides the interface between the embryo and the amniotic fluid during the majority of epidermal development *in utero*. In the mouse, the periderm forms at E9 (Fig. 18.1); studies of genetically manipulated mice have revealed that secreted proteins of the Wnt family, fibroblast growth factor (FGF), transforming growth factor (TGF), and bone morphogenic protein (BMP) families are important in these early stages of periderm formation [1, 2]. By E12 an intermediate proliferative ectodermal layer, also known as the stratum intermedium, arises between the basal ectodermal layer and the peridermal sheath, marking the initiation of epidermal stratification. Loose mesenchyme is present below these layers and constitutes the future dermis. By E15 cells from the intermediate ectodermal layer enter terminal differentiation, which involves the formation of the characteristic keratin intermediate filament cytoskeleton of the epidermal cells (also known as keratinization), and markers such as keratins 1 and 10 are induced. By E16 terminal differentiation is essentially completed, marked by the appearance of a specialized outer epidermal layer, the stratum corneum; a multilayered stratified epithelium is also present by this time (Fig. 18.1) [3, 4].

In human skin the periderm forms during the first month of gestation and covers the developing epidermis until keratinization occurs. Morphogenesis of most cutaneous components is evident by the end of the first trimester, and differentiation of the epidermis, the hair follicles and the glands takes place during the second trimester, i.e. between 13th and 26th weeks. As in the mouse, the appendages have both epidermal and dermal components, and their development depends on tightly coordinated epithelial-mesenchymal interactions.

Epidermal stratification begins at 8 weeks of gestation in humans, with the formation of the intermediate ectodermal layer between the basal and periderm layers. The intermediate layer is highly proliferative, and by 24 weeks the epidermis consists of four to five cell layers. The progeny of the intermediate ectodermal cells are similar to later spinous layer cells and, as in the mouse, express keratins 1 and 10, and the desmosomal desmoglein 3. Concomitantly, basal cells become cuboidal in shape and begin to express new keratins, types 6, 8, 16 and 19, which are typical of proliferative stages. The transcription factor p63, a homolog of the tumor suppressor p53, is expressed throughout the basal keratinocyte layer. Mice lacking this transcription factor have a very thin epidermis and exhibit defects in epidermal proliferation, stratification and differentiation [5, 6], indicating that p63 is critical for proliferation and maintenance of the basal cell phenotype. Human mutations that lead to partial loss of p63 function cause genetic syndromes involving bone and skin development (OMIM), e.g. the EEC syndrome (ectrodactyly, ectodermal dysplasia and cleft lip/ palate).

Keratinization also begins at this point, and periderm cells detach from the surface and are sloughed into the amniotic fluid, where they become part of the protective layer of the newly created, vernix caseosa. As keratinization proceeds, the biosynthetic program of keratinocytes is modified, terminal differentiation is initiated and the number of cornified layers increases. Basal keratinocytes express keratins 5 and 14 and basement membrane adhesion molecules, and terminally differentiating cells start expressing components of the cornified envelope, such as filaggrin, loricrin and enzymes required for formation of the skin barrier, e.g. transglutaminase, LEKTI and steroid sulfatase. Although keratinization is mature by about 35 weeks of gestation, the skin barrier function of a newborn is not complete until several weeks after birth. Similarly, in the mouse postnatal maturation of the skin occurs.

At birth a mature stratified epithelium is present which renews itself continuously by cell division in its deepest layer, the stratum germinativum, or basal layer. During self-renewal the basal keratinocytes ascend towards the surface and undergo differentiation and maturation processes, which include keratinization. This differentiation process results in three further layers which, listed in order of their increasing differentiation, are: stratum spinosum (suprabasal layer), stratum granulosum (granular layer), and the stratum corneum (cornified layer) that constitutes the fully keratinized outer layer of the skin which is gradually removed by day to day wear and tear (Fig. 18.2A).

18.2.2
Non-Keratinocyte Cell Types in the Epidermis

Interspersed amongst the basal and suprabasal layers of the epidermis are several non-keratinocyte cell populations, including melanocytes, responsible for melanin pigment production, specialized dendritic cells called Langerhans cells and in the mouse also dendritic epidermal T cells (DETC), and neuroendocrine Merkel cells.

Melanoblasts, the precursors of melanocytes, arise during development from the neural crest and migrate to the developing skin between E10 and E16.5 in the mouse. Key molecules required for their migration, proliferation and survival in developing skin include the transcription factors Sox10, Pax3 and Mitf, the G-coupled receptor, Ednrb and its ligand, endothelin–3, and the tyrosine kinase receptor, c-Kit, and its ligand, stem cell factor [7, 8]. Specification of neural crest cells to the melanocyte lineage requires signaling through the Wnt family (Wnt 1 and Wnt 3) and their downstream effectors, β-catenin and the transcription factors, Tcf/Lef, [8, 9, 10] as well as Sox10 [8]. Melanoblasts enter the dermis between E10 and E12.5 in a process that requires Ednrb activity [11, 12]. Commencing at E13.5, melanoblasts migrate from the dermis to the epidermis through the epidermal-dermal basement membrane and, subsequently, proliferate extensively in the epidermis. At E15.5, melanoblasts undergo a further migration into the forming hair follicles, leaving the epidermis mostly devoid of pigment cells. c-Kit is essential for melanoblast migration and survival in the epidermis, and its expression is dependent upon Mitf/Pax 3 activity [13]. Recent studies suggest that the migration of melanoblasts across the epidermal-dermal basement membrane is precisely regulated, such that only the correct number reaches the epidermis, and that G-protein signaling via Dsk (dark skin) genes regulates this process [14].

A second abundant non-keratinocyte cell type in the epidermis is the Langerhans cells, which originate in the bone marrow and migrate to the epidermis, where they have an antigen-presenting function and, therefore, represent a critical immuno-

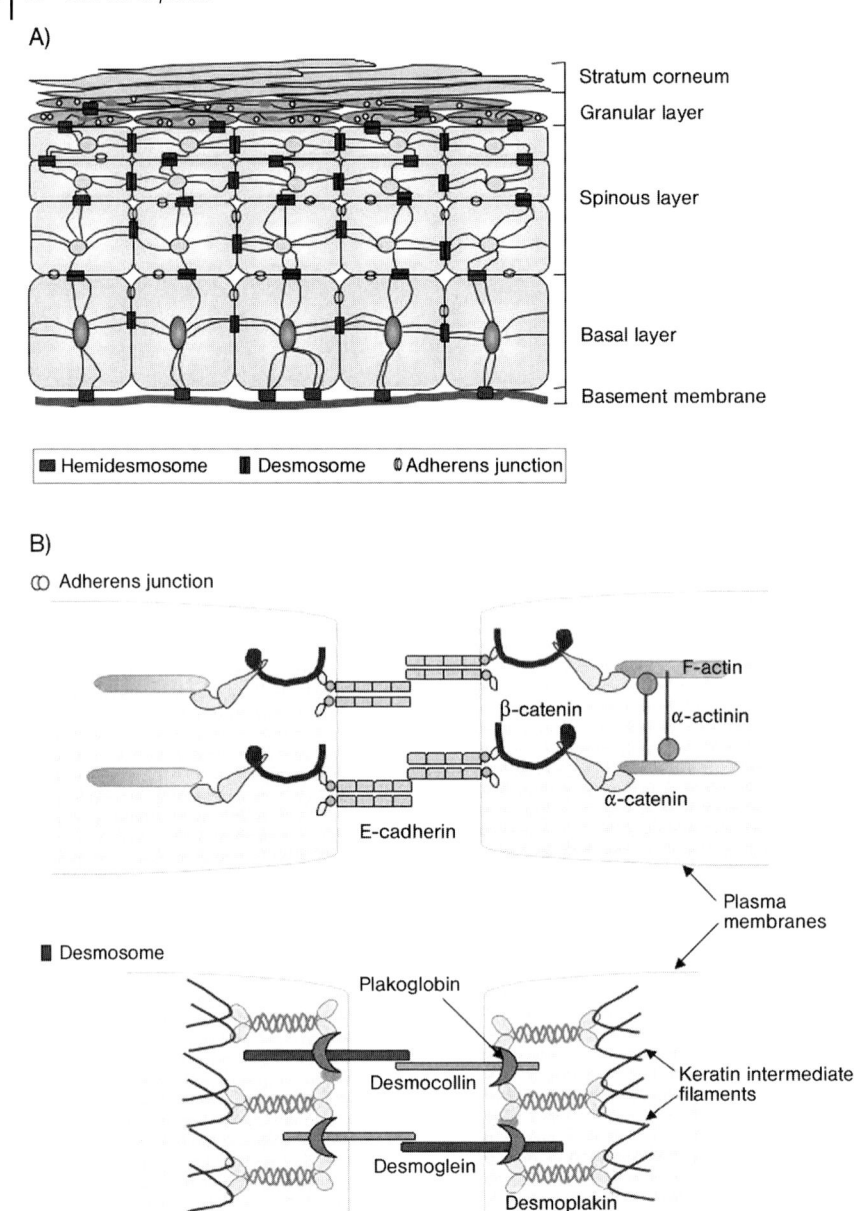

Fig. 18.2
(A) Schematic representation of the cell layers of the mature stratified epithelium of the skin and the cell-cell and cell-extracellular matrix junctions characteristic of these layers. Adherens junctions and desmosomes on the lateral surfaces of the keratinocytes act to seal the epithelial sheet, while hemidesmosomes anchor the basal keratinocytes to the epidermal-dermal basement membrane. (B) Molecular composition of the adherens junctions and the desmosomes. Figure is modified according to [2]. (This figure also appears with the color plates.)

logical barrier to the external environment. In most tissues, dendritic cells constitutively emigrate and are replaced by blood-borne precursors. In contrast, in the skin under steady-state conditions, Langerhans cells are maintained locally in the epidermis for extended periods [15]. Only in the case of an inflammation, do Langerhans cells rapidly emigrate via the lymphatic vessels to the draining lymph nodes [16–18] and are replaced by blood-borne precursors that migrate into the epidermis under the influence of the chemokines, MCP–1 and MCP–3, produced by the keratinocytes [19]. Factors that are responsible for the retention of Langerhans cells in the epidermis under steady-state conditions include high levels of the cell adhesion molecule, E-cadherin, a member of homophilic adhesion molecules that mediate intercellular interactions (see below), and transforming growth factor-β (TGF-β). E-Cadherin is a major component of keratinocyte adherens junctions (see below), suggesting that Langerhans cells may be anchored in the epidermis by forming similar junctions with keratinocytes, while TGF-β knockout mice lack Langerhans cells but not dendritic cells in other organs [20]. The exact mode of action of TGF-β on Langerhans cells is not clear but it has been shown to upregulate E-cadherin on human dendritic cells precursors and to induce morphological changes characteristic of Langerhans cells. In addition, TGF-β has been reported to inhibit the maturation of Langerhans-like dendritic cells induced by proinflammatory cytokines, such as TNF-α and IL–1, which are well documented to induce mobilization of Langerhans cells [21–23]. Finally, TGB-β inhibits TNF-α-induced upregulation of the chemokine receptor CCR7, which is critical for emigration of Langerhan cells from the epidermis [15, 24].

A second immune cell type of the skin is DETC, skin-specific members of the epithelial $\gamma\delta$ T cell family that reside in murine skin [25–27]. They are derived from early fetal thymocytes (E14–E17) and migrate to the epidermis where they undergo maturation in the epidermis during late gestational and post-gestational periods [28–30]. Cytokines secreted by neighboring cells (e.g. IL2–, IL–7, IL–15) promote their residence and regulate their immune function. Conversely, DETC regulate the function of neighboring keratinocytes and Langerhans cells by elaborating other cytokines (e.g. interferon γ, keratinocyte growth factor, and colony-stimulating factors) [27]. This reciprocal interaction represents a unique model of cytokine-mediated intercellular communication by tissue-specific $\gamma\delta$ T cells with nearby epithelial and Langerhans cells, which serves as a modulator of epidermal immunity.

The third major non-keratinocyte cell type of the epidermis is the Merkel cells that are involved in mechano-reception. These cells often occur in clusters in the basal keratinocyte layer in close association with nerve fibers and express nerve growth factor and substance P. They respond to sustained indentation and are considered to be responsible for form and texture perception [31]. The developmental origin of the Merkel cells has not been unambiguously resolved, although good evidence exists for a neural crest origin [32, 33] but there is also evidence that they may be derived from pluripotent keratinocytes [34].

18.2.3
Hair Follicle

In the mouse the first signs of hair follicle development are discernable at E13 with the induction of aggregations in the ectodermal layer, known as follicular *placodes*, by a hitherto unidentified signal from the underlying mesenchyme. The *placode* signals back in turn, to induce mesenchymal condensation in the developing dermis (E15; Fig. 18.1). However, it is not until E18 that the follicle epithelium grows downward through the dermis and into the subcutaneous fat in what is known as the hair germ elongation or anagen phase (Fig. 18.1). The follicle epithelium surrounds the condensate-derived hair follicle *dermal papilla* to form the *hair bulb*. Rapidly proliferating *matrix cells* in the hair bulb give rise to the hair shaft and inner root sheath lineages, which are driven upward towards the skin's surface during hair maturation. Surrounding these cells is the follicle outer root sheath, which is continuous with the interfollicular epithelium. In the following catagen stage, which commences approximately 18 days after birth, hair proliferation ceases and the hair involutes; this is accompanied by apoptosis of the *hair matrix cells* in the hair bulb. This process ends when the *dermal papilla* reaches the stem cell-containing *bulge* region of the hair follicle. After a resting or telogen phase, a new anagen phase is initiated by signaling between the dermal papilla and the overlying hair follicle epithelium (Fig. 18.1) [35, 36].

In human skin, follicular *placodes* form at regularly-spaced intervals in the basal ectodermal cells on the scalp between the 10th and 11th week. They spread ventrally and caudally from the scalp, finally covering the entire skin. The molecular nature of the dermal signal that induces ectodermal aggregation in both mouse and man remains elusive. However, based on its cellular localization, β-catenin is believed to be one of the effectors of this signal. Between weeks 12 and 14 of human development, the placodes induce dermal condensation, a process controlled by a number of cytokines, receptors and signaling molecules, collectively called placode promoters and placode inhibitors [37] (see below).

The mature hair follicle is surrounded by the outer root sheath, an epithelial cell layer that is contiguous with the basal keratinocytes of the interfollicular epidermis. At the base of the hair follicle is the hair bulb composed of proliferating *matrix cells*, which surround specialized mesenchymal cells of the dermis, the *dermal papilla*. A population of *transient-amplifying matrix cells* terminally differentiate to generate the different cell layers of the follicle, including the inner root sheath and the innermost hair shaft (Fig. 18.1).

In postnatal life, the upper portions of the hair follicle and the dermal papilla in the hair bulb are permanent, but the remainder of the hair follicle undergoes cycles of anagen, catagen and telogen. These cycles are dependent on inductive signals between the stem cell-containing bulge region and the base of the follicle for new hair development [36]. A number of genes regulating hair development and cycling have been identified, but we are only beginning to understand the orchestration of the molecular signals and their spatial and temporal, cooperative and antagonizing actions. For example, Sonic hedgehog (Shh), a signaling molecule produced by the epithelial cells of the developing hair follicle, which induces the condensation of dermal papilla cells, also has a role in inducing the progression of the *placodes* and is critical in mediating the transition from telogen to anagen hair cycle phase [38]. Concomitantly, Wnt signaling, i.e. the nuclear β-catenin/lymphoid enhancer binding factor 1 (Lef1)/T cell factor 3 (Tcf3) complex, and bone morphogenic protein (BMP) and fibroblast growth factor (FGF) family members are important regulators of hair development and cycling (reviewed in [36]). One well-studied example is FGF–5, which promotes the transition from the anagen to catagen phase of the hair follicle cycle [37].

Hair follicle development is critical to the maintenance of slow-cycling, epidermal stem cells that reside in the bulge region and regenerate epidermis, hair follicles and sebaceous glands [2]. It is thought that the stem cell progeny exit the bulge, migrate upwards into the epidermis and populate the basal keratinocyte layer. The rate of proliferation and upwards migration is accelerated when the skin is injured and wound healing is induced. A subset of the basal interfollicular keratinocytes is thought to constitute a second population of epidermal stem cells (at least in humans). These cells are capable of proliferation and are multipotent, with the capacity to regenerate epidermis, but it is not yet clear whether they can also renew hair follicles and whether they are identical to the bulge-residing stem-cell population.

18.2.4
Sebaceous Glands

Development of sebaceous glands parallels the formation of hair follicles. The future gland is first visible in weeks 13–16 of human development, and perinatally in mice, in the superficial regions of the maturating hair follicle above the bulge region [35]. The epithelial cells of the superficial hair follicle generate lipogenic cells or *sebocytes*, which produce and accumulate lipids and sebum until they are terminally differentiated and then disintegrate to release the sebum into the upper hair follicle. The growth of sebocytes is under the control of maternal hormones; after birth they become quiescent, to be reactivated in adolescence. The same genes, which orchestrate hair follicle differentiation, regulate sebocyte development. The coordinated action of Wnt signaling and the nuclear β-catenin/Lef1/Tcf3 complex, induces the ectoderm to adopt a hair follicle fate. If the action of this complex is blocked, hair follicle differentiation is impaired, and the cells become sebocytes. Accordingly, skin-specific ablation of the β-catenin gene in mice or transgenic expression of mutant Lef1, which lacks the β-catenin binding site, results in impaired hair follicle formation and differentiation of cells along the epidermal or sebocyte lineages [38–40]. Recently, ectopic Hedgehog expression was shown to induce sebocyte formation, even in regions that normally do not contain hair follicles or associated sebaceous glands [41]. These results raise the possibility that sebocyte cell fate is governed by the relative levels of stimulatory (Hedgehog) and inhibitory (Wnt) signals acting on multipotent progenitors.

18.2.5
Nails

In the mouse, the nail primordium is first seen at E15.5 in all but the anterior-most digit of the forelimb [42]. Nail primordium of this vestigial digit is not formed until E17.5. In man, development of the nails begins at the 8th week of gestation, a little earlier than the initiation of hair follicles. Like the hair follicle, the nail unit in both man and mouse is a tissue composed of keratinized stratified epithelium and underlying mesenchyme. The epithelial component can be divided into several zones that are contiguous with the epidermis [43]. First, the future nail bed is demarcated, then a portion of the ectoderm buds and moves inward to give rise to the future nail fold. The nail fold is a specialized epidermal transition zone, proximal to which is the nail matrix, which contains proliferating cells and will later produce the nail plate. Nail plate formation involves keratinization and flattening of post-mitotic matrix cells, nuclear loss, and cytoplasmic condensation [43]. The plate begins to keratinize at the 11th week of gestation in man and shortly before birth in the mouse, proceeding in a proximal to distal direction. Nail plate cells express a subset of hard keratins expressed in hair and constitute the major structure of the nail unit, which in some species is quite extensive, while in others rudimentary (primates). The first preliminary nail is replaced by a fully-formed and keratinized hard nail by the end of the 5th month in man.

Mutations affecting dorso/ventral patterning perturb nail development. Loss of Wnt 7a, causes ventralization of the limb, resulting in nail truncation as well as formation of the foot pad on the dorsal rather that ventral side of the paw [44, 45]. In contrast, mutations in En1 cause dorsalization of the limb, resulting in circumferential nail structures [44, 46]. In the mesenchyme, LMX1B, a LIM-homeodomain protein acting downstream of Wnt7a and the homeobox gene, MSX–1, are required for nail plate development. Mice and humans lacking LMX1B display dystrophic nail plates, resulting in nail-patella syndrome in humans [47], while mutations in MSX–1 underlie Whitkop syndrome. Developmental nail abnormalities are also found in AEC/EEC syndromes caused by p63 mutations [48]. In contrast to hair follicle development, Shh is not required for nail formation [49, 50].

18.2.6
Eccrine and Apocrine Sweat Glands

Development of sweat glands begins in the first trimester in man and is completed by the end of the second trimester. Initially, mesenchymal pads form on the palms and soles, followed by induction of parallel ectodermal ridges above these pads. The eccrine glands are derived from these ectodermal ridges. Approximately by the 15th week, gland primordia bud from the ridge at regular intervals, elongate as a strand and touch the mesenchymal pad. This induces the formation of glandular structures at the end of the bud, and by the end of week 16, secretory and myoepithelial cells as well as canalization of the eccrine duct are visible. However, the epidermal part of the duct is not fully canalized until week 22. By the time the palmar and plantar sweat glands are already well underway, the apocrine sweat glands and eccrine glands in other skin areas begin to form during the 5th month of gestation. The apocrine glands usually arise from the upper segment of the hair follicles, while the interfollicular eccrine glands arise independently. Apocrine glands are completed by month 7 of gestation and they function temporarily during the third trimester to contribute to the formation of vernix caseosa on the fetal skin surface, but become quiescent in the neonate. In contrast, the eccrine glands begin their secretory activity after birth.

Not much is known about molecular signals guiding the formation and differentiation of sweat glands. However, based on indirect evidence from human diseases, the EDA, EDAR, En1 and Wnt10b genes presumably regulate these events.

18.3
Cell Adhesion and the Role of Adhesion Molecules in Development

18.3.1
Cell-Cell Adhesion

The basal layer of the epidermis is composed of cuboidal cells that are anchored to the epidermal-dermal basement membrane, and interconnected by complex junctions, adherens junctions and desmosomes (Fig. 18.2A), both of which are essential for epithelial sheet formation and, therefore, the establishment of the barrier function of the epidermis [51]. The core component of the adherens junction is E-cadherin, a calcium-dependent transmembrane protein, which undergoes homophilic interactions with other E-cadherin molecules on adjacent cells. The cytoplasmic tail of E-cadherin binds to the actin cytoskeleton via several molecules, including α- and β-catenin (Fig. 18.2B). Desmosomes represent supramolecular aggregates of desmosomal cadherins, desmocollin and desmoglein, which undergo heterophilic interactions. The cytoplasmic tails of the desmosomal cadherins interconnect with the keratin intermediate filament cytoskeleton via desmoglobin and desmoplakin (Fig. 18.2B).

It has recently been shown that the homophilic interaction of E-cadherin on adjacent epidermal cells is the first step required to draw epithelial cells together [52]. This permits interaction between the desmosomal cadherins and results in redirected actin polymerization that can seal the membranes to make adhering sheets of cells that form the epidermal barrier. This process is a basic and essential feature of the skin epidermis and is a highly dynamic process. As the sheets migrate upwards during differentiation, epidermal cells must constantly change their intercellular interactions and, for example, during wound healing, transient downregulation of intercellular adhesion occurs concomitantly with increased cell proliferation. Not much is known about the molecular nature of the regulators, but it has been suggested that the inverse correlation between intercellular adhesion and proliferation is in part controlled by α-catenin and the Ras-MAPK (mitogen-activated protein kinase) pathway, such that modification of phosphorylation of α-catenin or its ligand interactions may result in its functional ablation at the adherens junction and a sustained activation of the Ras-MAPK pathway [53].

E-cadherin also mediates heterotypic cell-cell interactions between different cell types in the epidermis, i.e. between keratinocytes and Langerhans cells, or keratinocytes and immigrating lymphocytes. These non-epithelial cells express low levels of E-cadherin, and homophilic cadherin interactions provide for adhesion [54, 55]. Consequently, these observations imply that E-cadherin expression promotes the persistence of these cells in the epidermis. Recently, adhesive interactions between epithelial cells and intraepithelial lymphocytes were shown to be mediated by ligand interactions between E-cadherin and the αEβ7 integrin. In addition, a new adhesion molecule coined LEEP-CAM, or lymphocyte-endothelial-epithelial-cell adhesion molecule, was identified on non-intestinal epithelial cells [56].

18.3.2
Cell-Extracellular Matrix Adhesion in the Epidermis

18.3.2.1 Hemidesmosomes

The most important receptor complex involved in anchoring the basal keratinocytes, and thus the entire epidermis, to the epidermal-dermal basement membrane is the hemidesmosome [57]. In electron microscopy, hemidesmosomes resemble focal thickenings of the basal plasma membrane of keratinocytes. At a higher magnification, however, they can be seen to have a complicated ultrastructure that resembles half a desmosome. Hemidesmosomes consist of an intracellular adhesion plaque, which is associated with tonofilaments, and an extracellular sub-basal dense plate (Fig. 18.3A). The major components of the hemidesmosome are the transmembrane molecules, integrin α6β4 and collagen XVII (previously known as BP180 or BPAG2 – bullous pemphigoid antigen 2). The integrin β4 chain has an unusually long cytoplasmic tail, which interconnects with the keratin-binding molecule, plectin 1, in the adhesion plaque, but also binds to the cytoplasmic tail of collagen XVII [57]. Collagen XVII is, in turn, interconnected to the intracellular BP230 (or bullous pemphigoid antigen 1), which also binds cytoskeleton keratins (Fig. 18.3A). The major extracellular binding partner of integrin α6β4 is laminin 5, one of the components of the epidermal-dermal basement membrane (see below) [58]. Laminin 5 also binds collagen XVII, and together with the ectodomain of collagen XVII, it is thought to constitute the *anchoring filaments*, electron-dense filaments that extend from the surface of the basal keratinocytes into the basement membrane [58, 59] (Fig. 18.3A). Hence, the hemidesmosome and associated *anchoring filaments* act to bridge the epidermal-dermal basement membrane and the keratin intermediate filament cytoskeleton of the basal keratinocytes.

Analysis of human genetic diseases resulting from null mutations in the genes encoding laminin 5 or integrins α6 and β4 (*Junctional epidermolysis bullosa*) [60–62] (also see OMIM), or mice with ablation of the corresponding genes [63–65], reveals

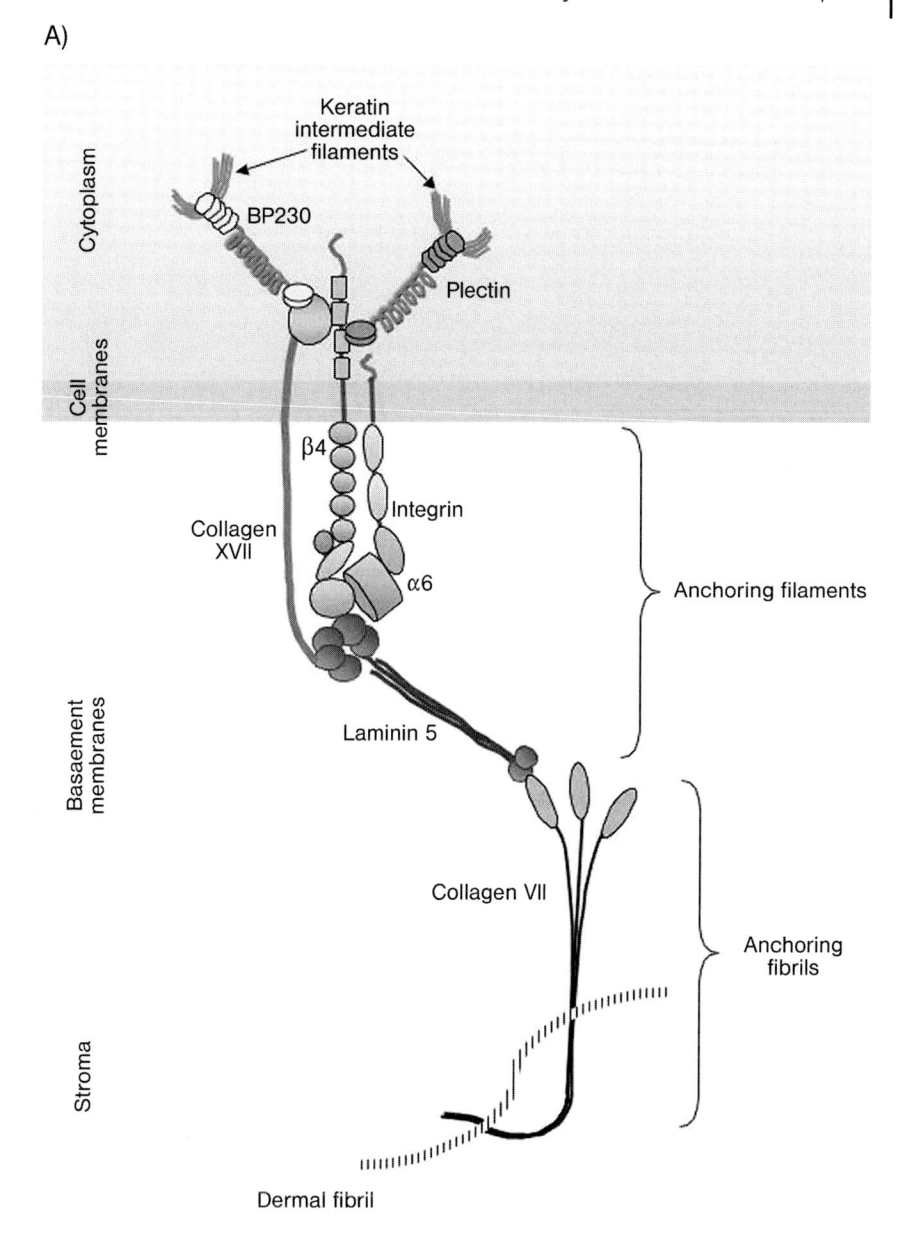

18.3 Cell Adhesion and the Role of Adhesion Molecules in Development

A)

Fig. 18.3
Schematic representation of hemidesmosome (A) and inter-hemidesmosomal (B) receptor complexes in basal keratinocytes, including underlying epidermal-dermal basement membrane components and interconnection with the dermal matrix via collagen VII.

B)

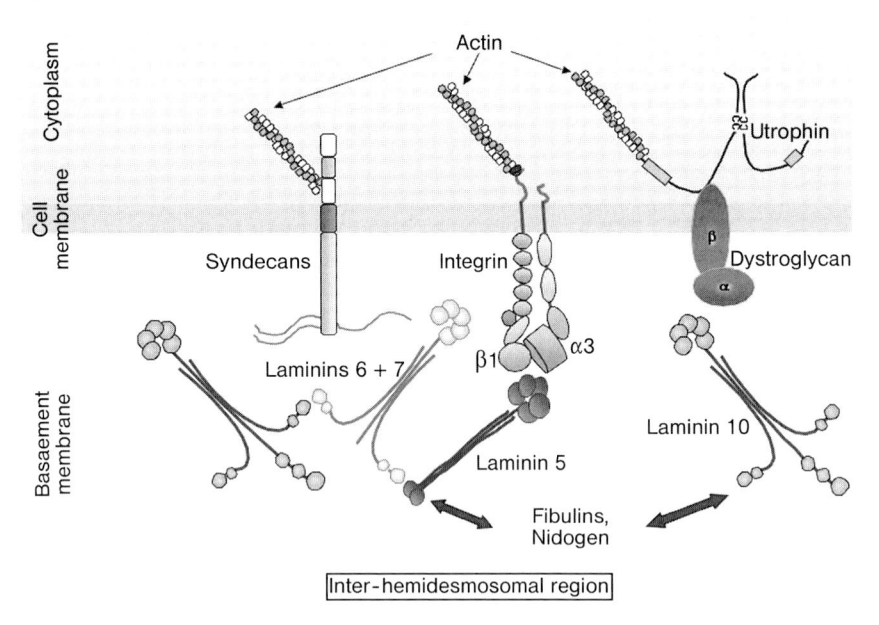

[Figure: Schematic of cell adhesion complexes showing Syndecans binding Laminins 6 + 7, Integrin (α3, β1) binding Laminin 5, and Dystroglycan (α, β) with Utrophin binding Laminin 10, across the cell membrane between cytoplasm (with Actin) and basement membrane. Fibulins and Nidogen are indicated in the Inter-hemidesmosomal region.]

loss of hemidesmosomes and separation of the epidermis from the underlying basement membrane. Although epidermal development is not impaired in these conditions, survival is severely compromised due to blister formation, protein and fluid loss, and susceptibility to infections. Missense mutations in the above genes or in the gene encoding for collagen XVII manifest with milder phenotypes, i.e. trauma-induced dermal-epidermal separation, but not a lethal outcome [66]. By contrast, defects in the adhesion plaque components, BP230, plectin or the endodomain of collagen XVII, cause epidermal cells to become mechanically fragile and to rupture upon physical stress. The phenotype is similar to epidermolysis bullosa simplex [66].

Hemidesmosomes not only maintain firm adhesion of basal keratinocytes to the epidermal-dermal basement membrane, but are also involved in the transduction of cellular signals, via integrin $\alpha 6\beta 4$, that regulate keratinocyte proliferation, differentiation and apoptosis [67]. Integrin $\alpha 6\beta 4$ is expressed only on basal keratinocytes and is downregulated upon upward migration and induction of differentiation, indicating an association with proliferation and an inverse association with migration and differentiation [68]. The cytoplasmic tail of integrin $\beta 4$ is tyrosine phosphorylated upon activation of the $\alpha 6\beta 4$ receptor either by cross-linking with an integrin $\beta 4$ antibody or by addition of laminin 5, resulting in recruitment of Shc adaptor and activation of the Ras-MAP-kinase pathway [69, 70]. Integrin $\beta 4$ is also tyrosine phosphorylated upon treatment of keratinocytes with epidermal growth factor (EGF), which induces keratinocyte differentiation. However, this does not lead to Shc association but rather to hemidesmosome disassembly and induction of cell migration [71, 72]. Differential phosphorylation of the $\beta 4$ subunit may thus lead to the activation of distinct signal transduction pathways, which regulate cell proliferation, hemidesmosome assembly and cell migration.

Much less is known about signaling of the other transmembrane component of hemidesmosomes, collagen XVII. Its intracellular domain contains tyrosine phosphorylation sites, suggesting that this molecule also signals via the MAP kinase pathway, but no detailed studies have been reported yet. A function as a signaling receptor is supported by the fact that the ectodomain of collagen XVII is shed from the cell surface by proteinases of the ADAM (a disintegrin and metalloproteinase) family, in particular by TACE (TNF-α converting enzyme), which releases the soluble ectodomains of a number of other biologically important cell surface receptors and adhesion molecules [73, 74].

18.3.2.2 Inter-Hemidesmosomal Spaces

Integrins α3 and β1 have been shown to be important for basal keratinocyte interactions with the epidermal-dermal basement membrane at sites between hemidesmosomes (Fig. 18.3B) and, in contrast to α6β4, interconnect with the actin cytoskeleton of the basal keratinocytes. Data from knockout mice have shown that integrins β1 and α3 are principally involved in hair follicle development, as ablation of the genes encoding either of these proteins results in hair loss and hair follicle abnormalities in mice [75–77]. In keratinocyte-specific β1 integrin null mice, further defects in the formation of the epidermal-dermal basement membrane at interfollicular sites and reduced numbers of hemidesmosomes are apparent [75, 77], suggesting a role for β1 integrins in the basement membrane formation and possibly also in hemidesmosome formation and/or maintenance. Integrin α3 has been associated with inter-hemidesmosomal interactions, but it seems that in its absence other integrins or non-integrin receptors can compensate functionally [76]. Grafting experiments involving transplantation of integrin α3 null skin onto athymic wild-type mice showed basement membrane assembly defects only surrounding hair follicles, but not in interfollicular regions [78]. Apart from α3β1, basal keratinocytes express several other integrins, including α2β1, α5β1, α9β1 and αvβ5 [68], as well as non-integrin receptors such as α-dystroglycan [79] and syndecans 2 and 4 [80]. The *in vivo* contribution of these receptors to interactions between basal keratinocytes and the basement membrane remains unclear. However, *in vitro* experiments indicated that keratinocytes lacking integrins α6β1, α6β4 and α3β1 can still interact with laminin 5, in part because of additional cell contacts between the αβ2β1 integrin and the short arm of the laminin–5 γ2 chain [80, 81], or between syndecan–2 and syndecan–4 with the heparin binding region in the laminin 5 subdomain LG4 [82, 83]. These data suggest that multiple interactions are involved in anchoring basal keratinocytes at inter-hemidesmosomal sites, possibly varying according to the physiological or pathological situation.

The transmembrane molecule dystroglycan has recently been shown to be expressed by keratinocytes *in vitro* and to be expressed on the basal surface of basal keratinocytes *in vivo* (Fig. 18.3B) [79]. Dystroglycan is post-translationally cleaved

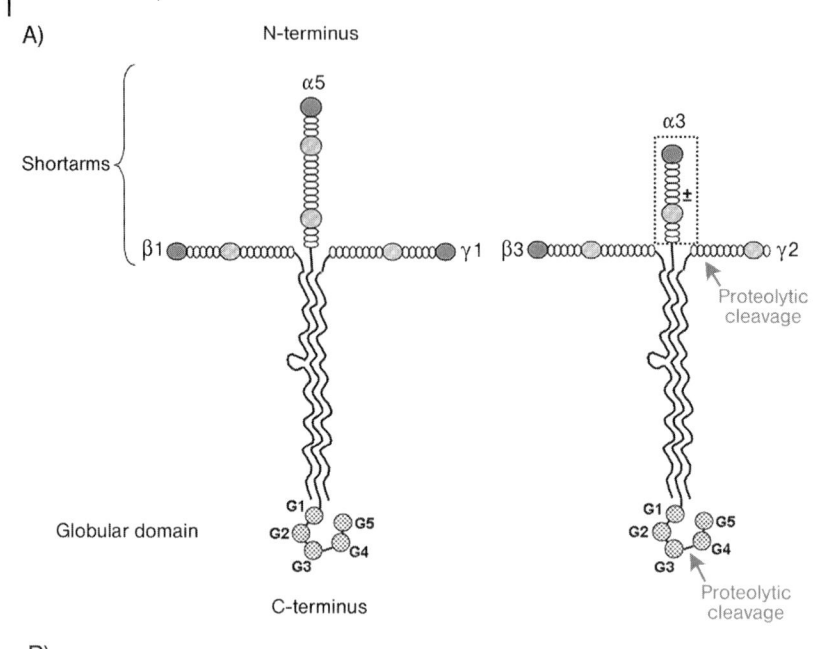

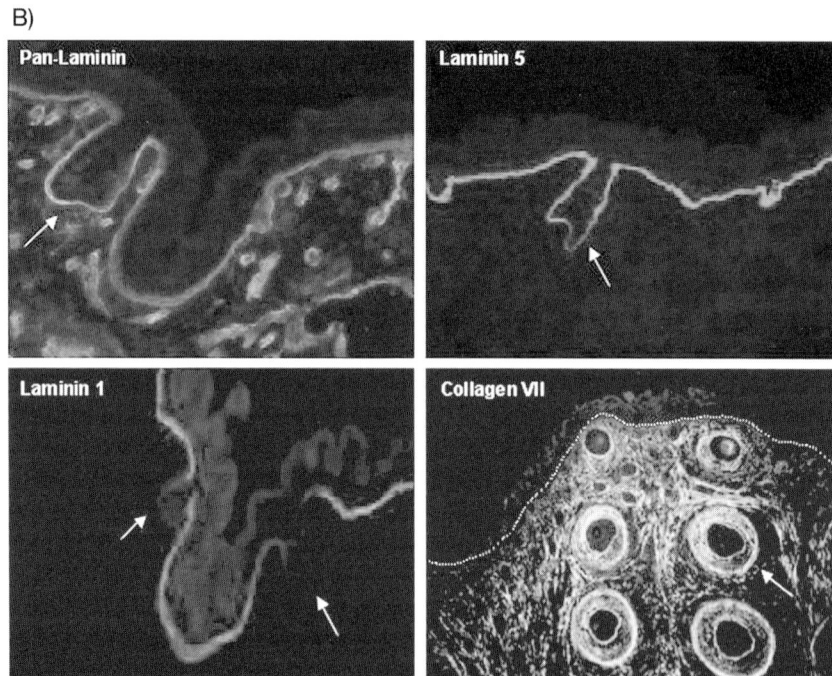

694 | *18 Skin Development*

A)

N-terminus

Shortarms: α5, β1 —— γ1   α3, β3 —— γ2

Proteolytic cleavage

Globular domain: G1, G2, G3, G4, G5

C-terminus

Proteolytic cleavage

B)

Pan-Laminin | Laminin 5

Laminin 1 | Collagen VII

Fig. 18.4
(A) Structure of laminin 10 (composed of α5, β1 and γ1 chains) as a prototype laminin, as compared to laminin 5 (composed of α3, β3, γ2 chains). Laminin α3 occurs in two spliced variants, one of which lacks the N-terminal domains (boxed area) and represents the form that occurs in the epidermal-dermal basement membrane [149]. Laminin α3 and γ2 chains are further proteolytically cleaved *in vivo* (cleavage sites marked). (B) Comparison of immunofluorescent staining of E17 mouse skin with antibodies to pan-laminin, laminin α3, laminin α1, and collagen VII. Pan-laminin staining reveals all basement membranes, including the epidermal-dermal basement membrane and basement membranes of blood vessels in the dermis. Laminin α3 and α1 are restricted to the epidermal-dermal basement membrane, with laminin α1 being excluded from the basement membrane surrounding the hair follicles. Collagen VII shows extensive fibrillar staining of the dermis, extending from the epidermal-dermal basement membrane (dotted line). Positions of developing hair follicles or in the case of the collagen VII panel, whiskers, are marked by arrows. Laminin images × 200 magnification; collagen VII × 100. (This figure also appears with the color plates.)

into a transmembrane portion, β-dystroglycan, and an extracellular portion, α-dystroglycan, the latter of which acts as a high affinity receptor for several basement membrane components, including the heparan sulfate proteoglycans, perlecan and agrin, and laminins 1 and 2 [84]. The cytoplasmic portion of β-dystroglycan interacts via utrophin with the actin cytoskeleton, thereby, acting as a further means of bridging the basement membrane with the actin cytoskeleton (Fig. 18.3B). In skin, the transmembrane β-dystroglycan is constitutively cleaved from the surface of the basal keratinocytes by as yet unidentified sheddases, resulting in release of α-dystroglycan and loss of the connection to the basement membrane. The significance of this physiological cleavage of dystroglycan remains unclear at present, but may represent a further mechanism for regulating basal keratinocyte proliferation and/or induction of upward migration and differentiation processes.

18.3.3
Role of Laminins in Epidermal Basement Membrane Development

The epidermal-dermal basement membrane is an electron microscopically visible bi-layer structure, from which *anchoring fibrils* extend into the superficial dermis (see below) and *anchoring filaments* towards the surface of the basal keratinocytes. The basic components of the basement membrane are the laminins, collagen IV, nidogens and the heparan sulfate proteoglycans. Both the epidermis and the dermis contribute to the formation of the epidermal-dermal basement membrane, but the basal keratinocytes appear to be the principal producers of the laminins.

The laminins are the major functional components of all basement membranes, which play pivotal roles in tissue morphogenesis, including direct conversion of mesenchyme to epithelium and branching morphogenesis of epithelial sheets [85].

They are heterotrimeric glycoproteins, composed of α, β and γ chains (see Fig. 18.4A). To date, five distinct α, four β and three γ laminin chains have been identified, which can combine to form 15 different isoforms [86, 87]. The laminin α-chains are the functionally active portion of the heterotrimers, as they exhibit tissue-specific distribution patterns and contain the major domains that interact with the cellular receptors [86].

Laminin 5 is the major laminin isoform in the epidermal-dermal basement membrane (Fig. 18.4B). Mutations in the genes coding for laminin α3, β3 and γ2 chains, which comprise the laminin 5 isoform, result in a severe skin blistering disease in both man (*junctional epidermolysis bullosa*) and mouse [60, 61, 63, 64, 88, 89], demonstrating that this protein is essential for the attachment of keratinocytes to their basement membrane and for the integrity of the skin, but not for their development. Together with the ectodomain of collagen XVII, laminin 5 forms the anchoring filaments that extend from the basement membrane towards the lower surface of the basal keratinocytes. In turn, laminin 5 interacts with collagen VII, which constitutes the anchoring fibrils in the dermis and, thereby, bridges the surface of the basal keratinocytes with the dermis (Fig. 18.3A).

Most laminins self-aggregate via their N-terminal short arms (Fig. 18.4A) to form organized networks, which are fundamental for basement membrane assembly [90–92]. In addition, these short arms carry important binding sites for other basement membrane components, such as nidogens and collagen IV, which permit incorporation of the laminin network into the basement membrane [87, 93]. Laminin 5 is unusual amongst the laminins in having relatively short N-terminal sequences in all three of its α3, β3 and γ2 chains. In addition, laminin α3 and γ2 chains are further truncated by proteolytic processing, resulting in the almost complete loss of the N-terminal short arms [80, 94]. As a consequence, laminin 5 cannot form a network on its own and has reduced possibilities for incorporation into the epidermal-dermal basement membrane. Several other laminin isoforms exist in the epidermal-dermal basement membrane, including laminin 6 (α3, β1, γ1 chains) and 7 (α3, β2, γ1 chains), which are believed to be covalently linked to laminin 5 and, thereby, to provide the link to the basement membrane [95]. In addition, laminin 5 also links integrin α6β4/ collagen XVII in the hemidesmosome to collagen VII in the dermis, while a laminin 5–6/7 complex is also bound by a β1 integrin (Fig. 18.3B). The recent discovery of further laminin isoforms in the epidermal-dermal basement membrane (see below) suggests that the *in vivo* situation may be more complex.

Laminin 5 is synthesized in a form containing a 200–kDa α3 chain and a 155–kDa γ2 chain. However, in the mature epidermal-dermal basement membrane only proteolytically processed forms of these two chains exist [80, 94]. The α3 chain is processed at the C-terminus, while the γ2 chain can be proteolytically cleaved at two distinct N-terminal sites by independent proteases (BMP–1 or the matrix metalloproteinases, MT-MMP1/MMP2) [96]. *In vitro* studies suggest that BMP–1 cleavage of the γ2 chain results in the exposure of several cryptic binding sites for other basement membrane proteins, such as fibulin–2, which could link laminin 5 to collagen IV and heparan sulfate proteoglycans [96] and, thereby, provide an alternative mode of laminin 5 incorporation into the basement membrane (Fig. 18.3B). A second pathway for laminin γ2 chain processing results in the release of a 70–kDa fragment representing the entire short arm and can be mediated by either MT-MMP1 or MMP2. This mode of proteolytic processing has been suggested to trigger cell migration, at least in pathological situations [97–99].

The C-terminus of laminin α3, like all laminin α chains, consists of five homologous laminin globule domains (LG), of which the last two, LG4–LG5, are cleaved (Fig. 18.4A). The proteases involved in this cleavage *in vivo* remain undefined, although recent *in vitro* experiments have shown that the combined action of MMP and serine proteases is required [100]. The cleavage of the α3 LG4–LG5 domain results in the exposure of a high affinity integrin α3β1 binding site, the function of which is conformation dependent. As with γ2 chain processing, proteolytic cleavage of the α3 LG4–LG5 has also been implicated in keratinocyte migration in wound healing [97]. It has been proposed that after skin injuries, keratinocytes at the wound edge deposit predominantly non-processed laminin 5. Its proteolytic cleavage and loss of LG4–LG5 leads to an increased binding affinity of integrin α3β1, altering integrin signaling and keratinocyte migration into the wound.

Apart from laminin 5, an additional major laminin isoform occurs in epidermal-dermal basement membrane, namely laminin 10, characterized by the α5 chain [101]. During early embryogenesis it is expressed throughout the epidermal-dermal basement membrane but becomes restricted to hair follicles during the elongation phase of the hair germ [102]. Laminin α5–/– mice have reduced numbers of hair germs compared to controls. However, as they die at approximately E16.5, later stages of hair follicle development cannot be analysed directly [103]. Skin grafts from E16.5 laminin α5–/– embryoes have, therefore, been transplanted onto the back of immune deficient mice, revealing failure of hair germ elongation into the dermis and regression of the hair germs. Interestingly, treatment of the laminin α5–/– skin grafts with exogenous purified laminin 10 restores hair follicle development. Similar defects in hair follicle elongation have been reported in the keratinocyte-specific integrin β1 knockout mouse in grafting experiments performed with integrin α3–/– skin [75, 78], and in human scalp xenografts treated with either antibodies to laminin α5 or to the β1 integrin, which result in alopecia. The data provide strong evidence for the role for laminin α5–integrin α3β1 interactions in proliferation of hair follicle keratinocytes, which required for hair germ elongation both during embryogenesis and in mature animals for new hair development. It will be of great interest to define the molecular mechanism of action and the signal transduction pathways involved.

The expression of several other laminin isoforms is temporally and spatially regulated during skin development. For example, laminin α1 is expressed by the early periderm cells (E14); the protein is deposited throughout the basement membrane underlying this cell layer, with the notable exception of the developing hair follicles, up to approximately E16–E17, whereupon it is gradually downregulated (Fig. 18.4B; L. Sorokin, unpublished data). In contrast, laminin α2 occurs in the epidermal-dermal basement membrane specifically surrounding the hair follicles. *In situ* hybridizations have shown that the cellular source of laminin α2 is the dermis, and that its expression becomes increasingly concentrated around the hair follicles with development [104, 105]. Laminin γ3 has a highly restricted distribution in the epidermal-dermal basement membrane, occurring exclusively at points of nerve penetration [106]. The significance of these laminin isoforms in the skin remains to be investigated.

Important components of the epidermal-dermal junction are the anchoring fibrils, which act to reinforce the attachment of the epidermal-dermal basement membrane to the underlying dermis. They consist largely of collagen VII, a ligand of laminin 5 [59]. Collagen VII is first expressed at human epidermal-dermal basement membrane at week 12 of gestation, and mutations of the collagen VII gene lead to a blistering and scarring skin disease (dystrophic epidermolysis bullosa), which is present at birth. Similarly, collagen VII –/– mice die postnatally as a consequence of massive dermal-epidermal separation in the skin and mucous membranes [107].

18.4
Development of the Dermal Matrix

The dermis below the epidermal-dermal basement membrane contains different cell types, predominantly fibroblasts but also endothelial cells and macrophages, embedded in an extensive, hierarchically organized three-dimensional network of extracellular matrix molecules. The developmental characteristics of the dermal mesenchymal cells are incompletely understood. Embryonic fibroblasts, present in the dermis at 6–8 weeks in human skin and up to E13 in the mouse [108], are believed to be pluripotent cells, which can differentiate into adipocytes, fibroblasts or even into cartilage cells. At about 8 weeks in man and E14–E15 in the mouse, the dermal matrix becomes morphologically distinct from the underlying musculo-skeletal condensations, and by 12 to 15 weeks or E16 in the mouse, progressive matrix organization and typical dermal fibroblasts can be distinguished. At this stage collagen containing fibers and elastic fibers can be discerned, and the highly hydrated gel-like, proteoglycan-rich matrix is replaced by more rigid fiber networks containing different fibrillar collagens and their ligands. Elastic fibers are ultrastructurally discernable around 22 to 24 weeks in human skin and in late embryogenesis in the mouse. By the end of the second trimester, the dermis shifts from non-scarring to scarring wound repair, a process regulated by growth factors of the TGF-β family [109].

The proteoglycan-rich matrix of the early dermis imparts massive water-holding capacity which permits rapid diffusion of nutrients, hormones and metabolites and creates turgor pressure to withstand considerable compressive force. Glycosaminoglycans (GAG) linked to protein as structurally diverse proteoglycans, assemble into chains and form gels of varying pore size. These act as sieves to regulate molecular traffic and also play important roles in cellular signaling, binding various secreted molecules, enzymes and inhibitors [110]. Specific proteoglycans also play important roles in regulating dermal architecture; for example, decorin knockout mice exhibit fragile skin due to abnormal dermal fibril cross-linking [111].

Collagens of multiple types, proteoglycans and glycoproteins aggregate to form heterotypic fibers or bundles. Dermal fibers (Fig. 18.3A) comprise collagens I and III as major components, with different minor components such as collagens V, XII, XIV or decorin [112, 113]. Collagen I is considered to be the major form found in the adult dermis, while collagen III is strongly expressed in embryonic skin. However, collagen I and III mRNA are expressed co-ordinately in dermal precursors from very early stages in mouse development (E8.5) [114, 115]. Several dermatological disorders resulting from collagen defects [112, 116, 117], and analysis of genetically manipulated mice suggest that collagen fibrillogenesis in the dermis is a complex process that remains poorly understood but which is crucial for tissue homeostasis. Collagen III is believed to be important for mediating fibrillogenesis during development and homeostasis of the dermis in the adult, as knockout mice develop serious skin lesions [118] and the distribution and size of all types of collagen fibrils is abnormal. Ablation of one of the minor dermal collagens, type V, results in a similar weakened skin and severe scarring [119]. However, these mice specifically display reduced dermal fibril formation, implicating collagen V in the regulation of collagen-containing fibrils in the skin.

Development of the blood and lymphatic vasculature is also important for the homeostasis of the dermis. The basic pattern of dermal vessels is discernable by the end of the first trimester of human development and at approximately E8.5 in the mouse. Extensive remodeling of the vasculature continues as it matures during the entire intra-uterine development. Endothelial cells and simple tube-like vessels are first seen at around 5 weeks of development, and a primitive vascular network between the epidermis and the subcutaneous tissue at around 6 to 8 weeks in man and by E9–E10 in mouse. The molecular control of angiogenesis is complex and the focus of intensive current research. The main factors orchestrating the formation of capillary loops in the skin pre- and postnatally include the vascular endothelial growth factor (VEGF) family, the tyrosine kinase receptors Tie–1 and Tie–2, and their regulators, the angiopoietin stimulatory ligands Ang–1 and Ang–4, as well as the inhibitory ligands Ang–2 and Ang–3 [120].

Dermal vascular endothelium exhibits several tissue-specific characteristics, including abundant anastomosis, constitutive expression of leukocyte adhesion molecules on a subset of venules, and a characteristic response to proinflammatory cytokines. Various proinflammatory stimuli activate endothelial cells in the skin vasculature to express adhesion molecules that lead to intradermal accumulation of specific leukocyte types, e.g. T cells expressing the cutaneous lymphocyte antigen [121, 122]. Further, endothelial cells of the blood vessels of the skin constitutively display the chemokine, CCL17, and the leukocyte adhesion molecule, E-selectin, which have been proposed to be essential for cutaneous immune surveillance [123]. The reason for such tissue-specific characteristics is not clear but implies that the surrounding stroma/dermis plays a role in endothelial maturation and/or physiology.

Lymphatics are believed to originate from endothelial cell buds, which originate in veins at E8.5 in the mouse [124]. Thereafter, the pattern of lymphatic development parallels that of blood vessels. Recent molecular studies into development of lymphatics have identified new markers specific for lymphatic endothelial cells. LYVE–1 and Prox–1 are genes considered to be critical for the earliest lymphatic specification, whereas VEGF-R3 and SLC maybe important in later lymphatic differentiation [124]. The critical role of VEGF-R3 in lymphatic development is indirectly demonstrated by patients with hereditary lymph edema, who have mutations in the VEGF-R3 gene [125, 126]. The significance of the skin lymphatics is best illustrated by its role in providing a route for dendritic cell migration from the skin to the draining lymph nodes [16]. In addition, the afferent lymph transports soluble antigens and chemokines from the periphery to the draining lymph node, thereby, contributing to a rapid immune response [127, 128].

18.5
Epithelial-Mesenchymal Interactions and Signaling

Epithelial-mesenchymal interactions involved in epidermis and hair morphogenesis are complex and, despite recent insight into the roles of different signaling pathways, not yet well understood. The fact that communication between different cell types is critical for skin development points to intercellular signaling molecules as key players in these processes, and many cascades, such as Wnt, Shh, BMP, FGF, PDGF and IGF pathways, appear to be involved in the epithelial-mesenchymal interactions directing morphogenesis and differentiation. Although the most important signals, e.g. the "first dermal signal", still remain unknown, new data on signaling in developing hair follicles [37] have shed light on the molecular components coordinating morphogenesis.

A possible clue to the nature of the "first dermal signal" was derived from the subcellular distribution of β-catenin, which is usually rapidly degraded in the cytoplasm. Under the influence of dermal Wnt intercellular signaling molecules, which diffuse short distances to determine the fate of other cells, degradation of β-catenin is inhibited. As a result, β-catenin accumulates in the cytoplasm, translocates into the nucleus and forms active complexes with members of the Lef/Tcf family of transcription factors [2]. These and other findings suggest that the activation of the Wnt signaling pathway in the dermis is required for establishing "the first dermal signal".

After the first signal, the formation of epithelial placodes is controlled by a balance between several placode promoters and placode repressors. Mouse and chick studies have suggested that members of the Wnt family, Wnt10b and Wnt7a, are expressed uniformly in the epidermis and, in the chick nuclear β-catenin is detected in the epidermis 1 day after the appearance of nuclear β-catenin in the dermis, implicating the activation of the Wnt signaling pathway in the epithelium. In the placode-forming areas, the expression of Wnt10b and Wnt 7a gradually increases, as does the quantity of nuclear β-catenin. Transgenic studies support these assumptions: a reporter gene responsive to Lef/β-catenin signaling was expressed both in dermal condensates and in the placodes [129, 130] and expression of a stable form of β-catenin induced formation of ectopic placodes. Thus, these observations also suggest that the activation of the Wnt signaling pathway is sufficient to induce hair follicle morphogenesis. In concert, genetic models with loss-of-function of Lef/β-catenin are associated with failure of placode development or deficient formation of vibrissae and paucity of body hair [38, 131].

Placode formation is controlled by a number of additional proteins, including FGF/FGFR2–IIIB complex, TGF-β2, Msx–1, Msx–2, EDA/EDAR, Noggin and Delta–1 /Notch1. A loss-of-function mutation in the FGF receptor IIIB results in impaired development of skin and hair follicles. Later in life, FGFs are involved in hair follicle cycling, e.g. FGF–5 promotes transition of the follicle from the anagen to the catagen phase. Growth factors of the TGF-β family, known to regulate a vast number of cellular processes and the production of the extracellular matrix, are involved in the formation of both the placode and the dermal condensate. TGF-β2 can induce hair follicles in mouse embryo skin explants, and transgenic mice lacking functional TGF-β2 exhibit paucity of hair follicles and delayed morphogenesis. Mouse models have also suggested that the homeobox-containing genes Msx–1 and Msx–2 are involved in placode induction, since their ablation results in the same phenotype as the elimination of TGF-β2 expression.

Genetic observations gave the first hint that ectodysplasin A, EDA, a keratinocyte transmembrane protein related to TNF-α, and its receptor, EDAR, are essential for initiating hair follicle formation. Defects in the corresponding human and murine genes result in decreased number of hair follicles, dental defects and absent or underdeveloped sweat glands. EDA mutations cause human X-linked anhidrotic ectodermal dysplasia and, in the mouse, the *Tabby* phenotype, while EDAR mutations underlie human hypohidrotic ectodermal dysplasia, and the mouse *downless* phenotype. The extracellular domain of EDA is shed from the epithelial cell surface [132] and can diffuse to sites of EDAR expression, thus acting as a juxtacrine factor. EDAR expression seems to be important very early in hair follicle morphogenesis, and is a prerequisite for BMP4 and *Sonic hedgehog* (Ssh) expression.

Members of the BMP family of secreted proteins act as inhibitors of placode formation. BMP2 and BMP7 are expressed in the pre-placode ectoderm or the placode, while BMP4 is synthesized in the pre-follicle mesenchyme. Ectopic expression of BMP2 and BMP4 inhibits formation of feather buds in chick, and in mouse skin the placode fate is inhibited by BMP2 and BMP4 in cells surrounding the placode. Many positive and negative regulators of BMPs in placode formation and hair follicle morphogenesis are known, such as Noggin, Follistatin, Gremlin, and Notch1 and its ligand Delta–1. Based on data derived from transgenic and knockout mouse models, these factors are believed to counteract the activity of BMPs in the follicle, but they cannot diffuse into the interfollicular regions. Delta–1 is normally expressed in the mesenchyme, but becomes concentrated in the pre-follicle areas under the influence of FGF [133].

Induction of the dermal condensate is directed by epithelial signals following initiation of placode formation. Wnt signaling is also believed to be required in this process, since the dermal condensate fails to form in the absence of epithelial β-catenin. PDGF-A is likely to be involved in the cross-talk between the epithelium and the dermal condensate, since it is expressed in the placode, and its receptor in the dermal condensate. Consistently, PDGF-A-deficient mice have abnormally small dermal papillae and thin hair.

Another major carrier of epithelial information into the dermis is Sonic hedgehog (Shh), a secreted protein, the expression of which is regulated by epithelial Wnt signaling. It is believed to play a role in a second wave of signaling, since in Shh knockout mice hair follicle formation is initiated and the dermal condensate is formed, but follicle development does not progress. After the finding that mutations of Patched1, an Shh-receptor, and Gli1, a transcriptional effector of Shh signaling, cause human basal cell carcinomas, identification of downstream targets of Shh have been the subject of intense interest due to their potential for molecular therapies. Wnt5a, TGF-β2 and neurotrophin receptors TrkC and p75 have been suggested to be such targets [134].

Taken together, the current understanding is that formation of epithelial placodes in response to the "first dermal signal" includes activation of EDA/EDAR signaling in the epithelium, followed by epithelial Wnt signaling and subsequent activation of BMPs to repress placode formation in the interfollicular epidermis in the adjacent areas (Fig. 18.5). Subsequently, the formation of the dermal condensate is induced by Wnt molecules and PDGF-A from the epithelium. Shh acts later under the control of Wnt signaling to regulate the formation and proliferation of hair follicles and development of the dermal condensate into the dermal papilla of the follicle (Fig. 18.5).

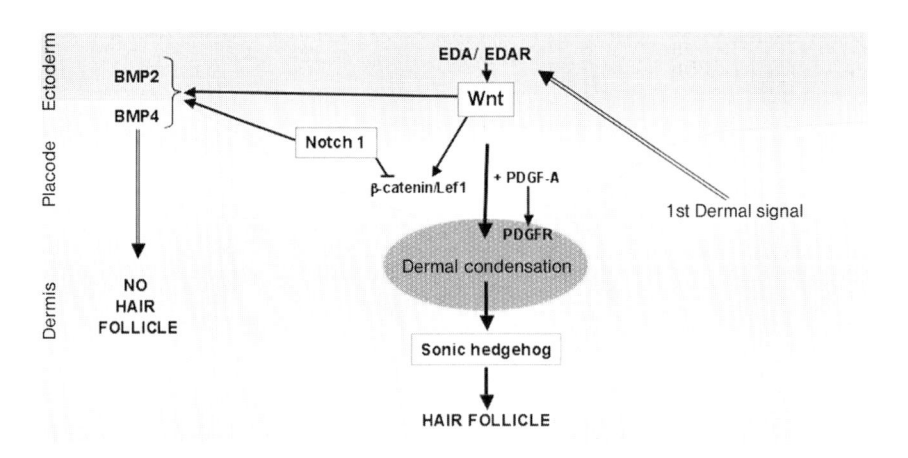

Fig. 18.5
Summary of the major positive and negative regulators of hair follicle development that occur in the ectoderm of the placode and the underlying mesenchyme of the developing dermis. Major positive regulators of hair follicle formation are the Wnt7/β-catenin/Lef I signaling pathways, while Notch 1/BMP2/BMP4 are negative regulators.

Although much is known about expression patterns of structural genes in the epidermis, the appendages and the extracellular matrix, much less is known about how these patterns are established during development, how differentiation programs are orchestrated at the transcriptional level and how the extracellular matrix, especially the proteoglycans, mediate growth factor signals. In the literature, there exists a vast amount of further data, partly conflicting, on additional genes and molecules possibly involved in skin morphogenesis and hair follicle development, differentiation and cycling. For more detailed information, the reader is referred to several recent reviews on these topics [2, 135, 136].

18.6
Epidermal Stem Cells

The identification of epidermal stem cells has been hampered by lack of specific molecular markers, and not much is known about these cells in skin development. One population of adult epidermal stem cells reside in the bulge of the hair follicle, they are slow cycling and multipotent and can give rise to epidermis, hair follicles and sebaceous glands in a wounded epidermis [137]. The stem cell progeny migrate from the bulge into the interfollicular epidermis, populate the basal layer in the epidermis and give rise to the *transient-amplifying cells*, which can divide for a certain number of times before they become committed to terminal differentiation. However, the contribution of these bulge-associated stem cells to epidermal renewal under normal conditions is highly debated [138]. For a long time the basal cell layer in the epidermis has been known to harbor a population of highly proliferative, self-renewing cells, but it is still not clear, whether they represent the transient-amplifying cells and are thus derived from the stem cells in the bulge, or whether they represent a second population of stem cells. Further, it remains unknown, whether they can differentiate into hair follicles and sebaceous glands.

Epidermal stem cells are usually identified on the basis of clonogenicity, growth potential and ability to reconstitute an epidermis. Recently, certain molecular markers have been reported to be characteristic of the epidermal stem cells, such as selective downregulation of the transferrin receptor (CD71) and upregulation of integrin $\alpha 6\beta 1$ [139–142]. Transient amplifying cells, in contrast, express high levels of $\alpha 6$ integrin and of CD71 [142]. As stem cells are restricted to the basal layer of either the hair follicle bulge or the interfollicular epidermis, upregulation of molecules that act to anchor cells to the epidermal-dermal basement membrane is consistent with their role as a stem cell population, as is the downregulation of the transferrin receptor resulting in reduced iron uptake due to slower cell cycling. However, neither of these markers are exclusive to epidermal stem cells and although other molecular events have been correlated with epidermal stem cells *in vitro*, including downregulation of the nuclear receptor protein 14–3–3ø (a nuclear export protein) and cytokeratins K15 and K19 [137, 143], these molecules are also expressed by transient amplifying cells. The question, therefore, arises as to whether such markers function to maintain stem cell character or proliferative capacity.

The data suggest that maintenance of the stem cell phenotype requires an intricate balance between positive and negative regulators. One molecule involved in this process is the p63 protein, which is expressed in both stem cells and in transient amplifying cells. Ablation of the gene encoding p63 results in a thin epidermis and the absence of epidermal proliferation, differentiation and stratification, suggesting that p63 is likely to be involved in the maintenance of a transient amplifying cell population rather than an epidermal stem cell population [2]. In contrast, elevated expression of the cell cycle regulator, c-myc, in stem and transient amplifying cells of transgenic mice leads to epidermal hyperproliferation without loss of differentiation. Additionally, these mice lose their hair and have impaired wound healing, indicating depletion of stem cell populations and/or accelerated use of these cells [144–146].

Wnt signaling has been shown to be linked to hair follicle development (as discussed above), where Lef1 is preferentially expressed in the precortex of the hair follicle and in conjunction with β-catenin activates hair-specific keratin gene expression. In contrast, Tcf3 expression occurs principally in the bulge region where its action is to repress c-myc or cyclin in the basal keratinocytes. Overexpression of this transcription factor in the skin causes repression of epidermal differentiation and generation of cells with outer root sheath and stem cell characteristics. Moreover, when the actions of β-catenin/Tcf3/Lef1 are blocked by elimination of β-catenin expression, hair follicle differentiation is impaired and cells differentiate along the epidermal or sebocyte lineage [38–40]. This suggests that β-catenin/Lef1 mediate hair follicle development and that interference with this pathway and/or relief of the Tcf3–mediated repression might be required for the specification of basal epidermal cells.

Another signaling pathway that is believed to stimulate stem cells to adopt an epidermal fate is the Notch signaling pathway. Cell surface receptors of the Notch family use different ligands to elicit positive or negative signals in various developmental programs in non-mammalians. In mice, Notch1 is expressed throughout the adult epidermis and in the outer root sheath of the hair follicle, but it is not clear whether Notch signaling has a role in maintaining stem cells and specifying epidermal cell fate in mammals. The fact that the Notch ligand Delta–1 is confined to the basal cell layer and is most highly expressed where stem cells are likely to reside, would support this assumption. However, ample evidence also exists for a different role for Notch, namely in epidermal differentiation [147]. For example, ablation of *Notch1* in mouse epithelia leads to epidermal hyperproliferation and reduced differentiation.

It has been suggested that the highly regulated stem cell proliferation and differentiation, required to avoid depletion or amplification of the stem cell pool, can also be achieved by symmetrical or asymmetrical cell divisions of stem cells [148]. Asymmetrical cell division results in two daughter cells one of which is identical to the dividing stem cells while the other is different. This can result from unequal segregation of cell-fate determinants (proteins or RNA) in the daughter cells or from selective responses of the each daughter cell to a gradient of signaling molecules present in the immediate environment. Data suggests that stem cells in the bulge divide symmetrically and that half of the new generation of stem cells remains in the bulge to maintain the pool constant, while the other half leaves. According to this hypothesis the stem cells excluded from the bulge are fully multipotent and can divide symmetrically or asymmetrically to generate committed progenitors depending upon the signals that they receive on their way.

Clearly, additional studies are required on the signals involved in defining the epidermal stem cell phenotype in development and its postnatal maintenance, as well as on the clarification of apparent antagonisms between different signaling pathways, before we understand the molecular mechanisms governing development and maintenance of epidermal stem cells.

18.7
Pathological Skin Conditions Caused by Developmental Abnormalities

Both human and mouse studies have revealed a number of pathological skin conditions to be associated with developmental abnormalities (Tab. 18.1; for references see [117]). These range from genetic diseases to abnormal wound-healing and skin cancer. For example, defective signaling by p63 results in abnormal epidermal development and delayed wound-healing, i.e. the EEC syndrome. Abnormalities of α-catenin-mediated cell-cell adhesion, or of Patched–1– and Gli–1–mediated hair follicle formation and cycling, can be associated with skin cancer. Mutations in the genes encoding proteins in the cornified layer of the epidermis manifest as severe defects of the barrier function of the skin, such as the Netherton syndrome which occurs as a consequence of SPINK–5 gene mutations, leading to various forms of ichthyosis as a result of transglutaminase–2 and steroid sulfatase mutations, or the Vohwinkel syndrome which occurs as a consequence of loricrin abnormalities. Epidermal fragility can be caused by a number of defects, ranging from plakophilin mutations, which are associated with reduced cell-cell adhesion in the ectodermal dysplasia/epidermal fragility syndrome, to mutations in the genes for hemidesmosomal and basement membrane zone proteins, which underlie the skin blistering disorders of the epidermolysis bullosa group. Presumably, not all causative genes for skin disorders have been identified as yet. For example, mouse models with ablation of integrin α3 or a deleted form of BP230 exhibited epidermal-dermal separation and friction-induced skin blistering. However, so far no human mutations in these genes have been reported. In any case, the studies on human genetic disorders and mouse models have shed ample light on physiological processes occurring during development.

Tab. 18.1 Synopsis of pathological conditions associated with abnormalities of skin morphogenesis and differentiation.

Protein	Functional defect	Human disease/Animal model
p63	Abnormal epidermal stratification	AEC, EEC syndromes
	Impaired wound healing	Mouse model
c-Myc	Impaired wound healing	Mouse model
	Loss of hair	
β-Catenin	Impaired hair follicle development	Mouse model
	Abnormal epidermal differentiation	
	Stabilizing mutations	Pilomatricoma
α-Catenin	Defective cell-cell adhesion	Squamous cell carcinoma
Patched–1	Impaired hair follicle development	Basal cell carcinoma
Gli1	Abnormal Shh signaling	Basal cell carcinoma
LEKTI	Barrier defect	Netherton syndrome
Transglutaminase–2	Barrier defect	Lamellar ichthyosis
Steroid sulfatase	Barrier defect	X-linked ichthyosis
Aryl sulfatase	Barrier defect	X-linked ichthyosis
Loricrin	Barrier defect	Ichthyosiform erythroderma
		Vohwinkel's disease
Cadherins	Defective cell-cell adhesion and signaling	Squamous cell carcinoma
Plakophilin	Defective cell-cell adhesion and signaling	Ectodermal dysplasis/ skin fragility syndrome
Plakoglobin	Defective cell-cell adhesion	Naxos disease
		Ectodermal dysplasia
Keratins 1, 10	Barrier defect	Bullous ichthyosis
Keratins 5,14	Defective epidermal adhesion	EB* simplex
EDA, EDAR	Abnormal epidermal development	Anhydrotic or hypohydrotic
XEDAR	Abnormal sweat glands	ectodermal dysplasia
Integrin α3	Abnormal basement membrane, blistering	Mouse model
Integrin α6	Defective epidermal adhesion	Junctional EB
Integrin β1	Excessive scarring, abnormal hair follicles	Mouse model
Integrin β4	Defective epidermal adhesion	Junctional EB
Laminin 5	Defective epidermal adhesion	Junctional EB
Collagen XVII	Defective epidermal adhesion	Junctional EB
Collagen VII	Defective epidermal adhesion	Dystrophic EB
Collagen I	Skin fragility, hyperextensibility, abnormal wound healing and scarring	Ehlers-Danlos syndrome VII
Collagen III	Skin fragility, hyperextensibility, abnormal wound healing and scarring	Ehlers-Danlos syndrome IV
Collagen V	Skin fragility, hyperextensibility, abnormal wound healing and scarring	Ehlers-Danlos syndrome I

18.8
Future Considerations

In compiling this overview it has become evident that considerable progress has been made over the last decade on two major areas of skin research: (1) defining the cell-cell and cell-matrix interactions which are important for skin homeostasis, and (2) growth factor signaling pathways, in particular the Wnt pathway, involved in hair follicle development. Although a wealth of information is now available on growth factors and transcription factors associated with defined stages of skin development, the complex interplay between these different factors still remains to be defined. In this context, potential interactions between the molecules of the epidermal-dermal basement membrane or components of the interstitial matrix of the dermis and secreted growth factors of the Wnt and Hedgehog families will be of great interest. It will be intriguing to elucidate the mechanisms of storage and/or presentation of soluble signaling molecules by matrix components, e.g. highly charged proteoglycans, since such interactions are likely to exert a high degree of control of the inductive and subsequent developmental events during morphogenesis of the skin and its appendages.

Acknowledgments

The authors thank Rupert Hallmann for his tireless help in production of the figures and critical revision of the manuscript.

References

1 Barker, N., and H. Clevers. Catenins, Wnt signaling and cancer. *Bioessays* **2000**; *22*: 961–965.
2 Fuchs, E., and S. Raghavan. Getting under the skin of epidermal morphogenesis. *Nat Rev Genet* **2002**; *3*: 199–209.
3 Byrne, C., and M. Hardman. Integumentary structures. In *Mouse Development. Patterning, Morphogenesis, and Organogenesis*, J.a.T. Rossant, P. (Ed.). Academic Press: San Diego, **2002**; 567–589.
4 Rugh, R. *The Mouse: Its Reproduction and Development.* Burgess: Minneapolis, **1968**.
5 Mills, A.A., B. Zheng, X.J. Wang, H. Vogel, D.R. Roop, and A. Bradley. p63 is a p53 homologue required for limb and epidermal morphogenesis. *Nature* **1999**; *398*: 708–713.
6 Yang, A., R. Schweitzer, D. Sun, M. Kaghad, N. Walker, R.T. Bronson, C. Tabin, A. Sharpe, D. Caput, C. Crum, and F. McKeon. p63 is essential for regenerative proliferation in limb, craniofacial and epithelial development. *Nature* **1999**; *398*: 714–718.
7 Okura, M., H. Maeda, S. Nishikawa, and M. Mizoguchi. Effects of monoclonal anti-c-kit antibody (ACK2) on melanocytes in newborn mice. *J Invest Dermatol* **1995**; *105*: 322–328.
8 Rawls, J.F., E.M. Mellgren, and S.L. Johnson. How the zebrafish gets its stripes. *Dev Biol* **2001**; *240*: 301–314.
9 Christiansen, J.H., E.G. Coles, and D.G. Wilkinson. Molecular control of neural crest formation, migration and differentiation. *Curr Opin Cell Biol* **2000**; *12*: 719–724.

10 Dorsky, R.I., D.W. Raible, and R.T. Moon. Direct regulation of nacre, a zebrafish MITF homolog required for pigment cell formation, by the Wnt pathway. *Genes Dev* **2000**; *14*: 158–162.

11 Dupin, E., Le Douarin, N. M., Development of melanocyte precursors from the vertebrate neural crest. *Oncogene* **2003**; *22*: 3016–3023.

12 Shin, M.K., J.M. Levorse, R.S. Ingram, and S.M. Tilghman. The temporal requirement for endothelin receptor-B signaling during neural crest development. *Nature* **1999**; *402*: 496–501.

13 Wehrle-Haller, B. The role of Kit-ligand in melanocyte development and epidermal homeostasis. *Pigment Cell Res* **2003**; *16*: 287–296.

14 Van Raamsdonk, C.D., K.R. Fitch, H. Fuchs, M.H. de Angelis, and G.S. Barsh. Effects of G-protein mutations on skin color. *Nat Genet* **2004**; *36*: 961–968.

15 Jakob, T., J. Ring, and M.C. Udey. Multistep navigation of Langerhans/dendritic cells in and out of the skin. *J Allergy Clin Immunol* **2001**; *108*: 688–696.

16 Romani, N., G. Ratzinger, K. Pfaller, W. Salvenmoser, H. Stossel, F. Koch, and P. Stoitzner. Migration of dendritic cells into lymphatics-the Langerhans cell example: routes, regulation, and relevance. *Int Rev Cytol* **2001**; *207*: 237–270.

17 Stoitzner, P., S. Holzmann, A.D. McLellan, L. Ivarsson, H. Stossel, M. Kapp, U. Kammerer, P. Douillard, E. Kampgen, F. Koch, S. Saeland, and N. Romani. Visualization and characterization of migratory Langerhans cells in murine skin and lymph nodes by antibodies against Langerin/CD207. *J Invest Dermatol* **2003**; *120*: 266–274.

18 Stoitzner, P., K. Pfaller, H. Stossel, and N. Romani. A close-up view of migrating Langerhans cells in the skin. *J Invest Dermatol* **2002**; *118*: 117–125.

19 Merad, M., M.G. Manz, H. Karsunky, A. Wagers, W. Peters, I. Charo, I.L. Weissman, J.G. Cyster, and E.G. Engleman. Langerhans cells renew in the skin throughout life under steady-state conditions. *Nat Immunol* **2002**; *3*: 1135–1141.

20 Borkowski, T.A., J.J. Letterio, A.G. Farr, and M.C. Udey. A role for endogenous transforming growth factor beta 1 in Langerhans cell biology: the skin of transforming growth factor beta 1 null mice is devoid of epidermal Langerhans cells. *J Exp Med* **1996**; *184*: 2417–2422.

21 Geissmann, F., C. Prost, J.P. Monnet, M. Dy, N. Brousse, and O. Hermine. Transforming growth factor beta1, in the presence of granulocyte/macrophage colony-stimulating factor and interleukin 4, induces differentiation of human peripheral blood monocytes into dendritic Langerhans cells. *J Exp Med* **1998**; *187*: 961–966.

22 Geissmann, F., P. Revy, A. Regnault, Y. Lepelletier, M. Dy, N. Brousse, S. Amigorena, O. Hermine, and A. Durandy. TGF-beta 1 prevents the noncognate maturation of human dendritic Langerhans cells. *J Immunol* **1999**; *162*: 4567–4575.

23 Riedl, E., J. Stockl, O. Majdic, C. Scheinecker, K. Rappersberger, W. Knapp, and H. Strobl. Functional involvement of E-cadherin in TGF-beta 1–induced cell cluster formation of *in*

vitro developing human Langerhans-type dendritic cells. *J Immunol*. **2000**; *165*: 1381–1386.

24. Sato, K., H. Kawasaki, H. Nagayama, M. Enomoto, C. Morimoto, K. Tadokoro, T. Juji, and T.A. Takahashi. TGF-beta 1 reciprocally controls chemotaxis of human peripheral blood monocyte-derived dendritic cells via chemokine receptors. *J Immunol* **2000**; *164*: 2285–2295.
25. Asarnow, D.M., T. Goodman, L. LeFrancois, and J.P. Allison. Distinct antigen receptor repertoires of two classes of murine epithelium-associated T cells. *Nature* **1989**; *341*: 60–62.
26. Stingl, G., and P.R. Bergstresser. Dendritic cells: a major story unfolds. *Immunol Today* **1995**; *16*: 330–333.
27. Takashima, A., and P.R. Bergstresser. Cytokine-mediated communication by keratinocytes and Langerhans cells with dendritic epidermal T cells. *Semin Immunol* **1996**; *8*: 333–339.
28. Elbe, A., O. Kilgus, R. Strohal, E. Payer, S. Schreiber, and G. Stingl. Fetal skin: a site of dendritic epidermal T cell development. *J Immunol* **1992**; *149*: 1694–701.
29. Elbe, A., E. Tschachler, G. Steiner, A. Binder, K. Wolff, and G. Stingl. Maturational steps of bone marrow-derived dendritic murine epidermal cells. Phenotypic and functional studies on Langerhans cells and Thy-1+ dendritic epidermal cells in the perinatal period. *J Immunol* **1989**; *143*: 2431–2438.
30. Stingl, G., A. Elbe, E. Paer, O. Kilgus, R. Strohal, and S. Schreiber. The role of fetal epithelial tissues in the maturation/differentiation of bone marrow-derived precursors into dendritic epidermal T cells (DETC) of the mouse. *Curr Top Microbiol Immunol* **1991**; *173*: 269–277.
31. Johnson, K.O. The roles and functions of cutaneous mechanoreceptors. *Curr Opin Neurobiol* **2001**; *11*: 455–461.
32. Halata, Z., M. Grim, and K.I. Bauman. Friedrich Sigmund Merkel and his »Merkel cell«, morphology, development, and physiology: review and new results. *Anat Rec* **2003**; *271A*: 225–239.
33. Le Douarin, N. Migration and differentiation of neural crest cells. *Curr Top Dev Biol* **1980**; *16*: 31–85.
34. Loomis, C.A., and T. Koss. Embryology. In *Dermatology*, Bologna, J.L., Rapini, R.P. (Eds). Mosby: London, **2003**; 39–48.
35. Davidson, P., and M.H. Hardy. The development of mouse vibrissae *in vivo* and *in vitro*. *J Anat* **1952**; *86*: 342–360.
36. Fuchs, E., B.J. Merrill, C. Jamora, and R. DasGupta. At the roots of a never-ending cycle. *Dev Cell* **2001**; *1*: 13–25.
37. Millar, S.E. Molecular mechanisms regulating hair follicle development. *J Invest Dermatol* **2002**; *118*: 216–225.
38. Huelsken, J., R. Vogel, B. Erdmann, G. Cotsarelis, and W. Birchmeier. beta-Catenin controls hair follicle morphogenesis and stem cell differentiation in the skin. *Cell* **2001**; *105*: 533–545.
39. Merrill, B.J., U. Gat, R. DasGupta, and E. Fuchs. Tcf3 and Lef1 regulate lineage differentiation of multipotent stem cells in skin. *Genes Dev* **2001**; *15*: 1688–1705.
40. Niemann, C., D.M. Owens, J. Hulsken, W. Birchmeier, and F.M. Watt. Expression of DeltaNLef1 in mouse epidermis re-

sults in differentiation of hair follicles into squamous epidermal cysts and formation of skin tumours. *Development* 2002; *129*: 95–109.

41. Allen, M., M. Grachtchouk, H. Sheng, V. Gratchouk, A. Wang, L. Wei, J. Liu, A. Ramirez, D. Metzger, P. Chambon, J.L. Jorcano, and A. Dlugosz. Hedgehog signaling regulates sebaceous gland development. *Am J Pathol* 2004; *163*: 2173–2178.

42. Kaufman, M.H. *The Atlas of Mouse Development*. Academic Press, Inc.: San Diego, CA, USA, 1992.

43. Baden, H.P., and J.C. Kvedar. Epithelial cornified envelope precursors are in the hair follicle and nail. *J Invest Dermatol* 1993; *101*:72S–74S.

44. Cygan, J.A., R.L. Johnson, and A.P. McMahon. Novel regulatory interactions revealed by studies of murine limb pattern in Wnt–7a and En–1 mutants. *Development* 1997; *124*: 5021–5032.

45. Kawakami, Y., N. Wada, S. Nishimatsu, and T. Nohno. Involvement of frizzled–10 in Wnt–7a signaling during chick limb development. *Dev Growth Differ* 2000; *42*: 561–569.

46. Loomis, C.A., R.A. Kimmel, C.X. Tong, J. Michaud, and A.L. Joyner. Analysis of the genetic pathway leading to formation of ectopic apical ectodermal ridges in mouse Engrailed–1 mutant limbs. *Development* 1998; *125*: 1137–1148.

47. Chen, H., Y. Lun, D. Ovchinnikov, H. Kokubo, K.C. Oberg, C.V. Pepicelli, L. Gan, B. Lee, and R.L. Johnson. Limb and kidney defects in Lmx1b mutant mice suggest an involvement of LMX1B in human nail patella syndrome. *Nat Genet* 1998; *19*: 51–55.

48. Brunner, H.G., B.C. Hamel, and H. Van Bokhoven. The p63 gene in EEC and other syndromes. *J Med Genet* 2002; *39*: 377–381.

49. Chiang, C., Y. Litingtung, M.P. Harris, B.K. Simandl, Y. Li, P.A. Beachy, and J.F. Fallon. Manifestation of the limb prepattern: limb development in the absence of sonic hedgehog function. *Dev Biol* 2001; *236*: 421–435.

50. Kraus, P., D. Fraidenraich, and C.A. Loomis. Some distal limb structures develop in mice lacking Sonic hedgehog signaling. *Mech Dev* 2001; *100*: 45–58.

51. Green, K.J., and C.A. Gaudry. Are desmosomes more than tethers for intermediate filaments? *Nat Rev Mol Cell Biol* 2000; *1*:208–216.

52. Vasioukhin, V., C. Bauer, M. Yin, and E. Fuchs. Directed actin polymerization is the driving force for epithelial cell-cell adhesion. *Cell* 2000; *100*: 209–219.

53. Vasioukhin, V., and E. Fuchs. Actin dynamics and cell-cell adhesion in epithelia. *Curr Opin Cell Biol* 2001; *13*: 76–84.

54. Cepek, K.L., S.K. Shaw, C.M. Parker, G.J. Russell, J.S. Morrow, D.L. Rimm, and M.B. Brenner. Adhesion between epithelial cells and T lymphocytes mediated by E-cadherin and the alpha E beta 7 integrin. *Nature* 1994; *372*: 190–193.

55. Tang, A., M. Amagai, L.G. Granger, J.R. Stanley, and M.C. Udey. Adhesion of epidermal Langerhans cells to keratinocytes mediated by E-cadherin. *Nature* 1993; *361*: 82–85.

56. Agace, W.W., J.M. Higgins, B. Sadasivan, M.B. Brenner, and C.M. Parker. T lymphocyte-epithelial-cell interactions: in-

tegrin alpha(E)(CD103)beta(7), LEEP-CAM and chemokines. *Curr Opin Cell Biol* **2000**; *12*: 563–568.
57. Koster, J., L. Borradori, and A. Sonnenberg. *Hemidesmosomes: Molecular Organisation and their Importance for Cell Adhesion and Disease*. Springer Verlag: Heidelberg, **2004**; 243–280.
58. Rouselle, P., G.P. Lunstrum, D.R. Keene, and R.E. Burgeson. Kalinin: an epithelium-specific basement membrane adhesion molecule that is a component of anchoring filaments. *J Cell Biol* **1991**; *114*: 567–576.
59. Roussel, E., and M.-C. Gingras. Transendothelial migration induces rapid expression on neutrophils of granule-release VLA6 used for tissue infiltration. *J Leukoc Biol* **1997**; *62*: 356–362.
60. Aberdam, D., M.-F. Galliano, J. Vailly, L. Pulkkinen, J. Bonifas, A.M. Christiano, K. Tryggvason, J. Uitto, E.H. Epstein, J.-P. Ortonne, and G. Meneguzzi. Herlitz's junctional epidermolysis bullosa is linked to mutations in the gene (LamC2) for the γ2 subunit of nicein/kalinin (laminin-5). *Nat Genet* **1994**; *6*: 299–304.
61. Pulkkinen, L., A.M. Christiano, T. Airenne, H. Haakana, K. Tryggvason, and J. Uitto. Mutations in the γ2 chain gene (LAMC2) of kalinin/laminin 5 in the junctional forms of epidermolysis bullosa. *Nature Genet* **1994**; 6:293–297.
62. Vidal, F., C. Baudoin, C. Miquel, M.F. Galliano, A.M. Christiano, J. Uitto, J.P. Ortonne, and G. Meneguzzi. Cloning of the laminin α3 chain gene (LAMA3) and identification of a homozygous deletion in a patient with Herlitz junctional epidermolysis bullosa. *Genomics* **1995**; *30*: 273–280.
63. Dowling, J., Q.C. Yu, and E. Fuchs. Beta4 integrin is required for hemidesmosome formation, cell adhesion and cell survival. *J Cell Biol* **1996**; *134*: 559–572.
64. Georges-Labouesse, E., N. Messaddeq, G. Yehia, L. Cadalbert, A. Dierich, and M. Le Meur. Absence of integrin α6 leads to epidermolysis bullosa and neonatal death in mice. *Nat Genet* **1996**; *13*: 370–373.
65. Ryan, M., K. Lee, Y. Miyashita, and W. Carter. Targeted disruption of the LAMA3 gene in mice reveals abnormalities in survival and late stage differentiation of epithelial cells. *J Cell Biol* **1999**; *145*: 1309–1323.
66. Christiano, A.M., and J. Uitto. Molecular complexity of the cutaneous basement membrane zone. Revelations from the paradigms of epidermolysis bullosa. *Exp Dermatol* **1996**; *5*: 1–11.
67. Giancotti, F.G. Signal transduction by the alpha 6 beta 4 integrin: charting the path between laminin binding and nuclear events. *J Cell Sci* **1996**; *109*: 1165–1172.
68. Watt, F.M. Role of integrins in regulating epidermal adhesion, growth and differentiation. *EMBO J* **2002**; *21*: 3919–3926.
69. Mainiero, F., A. Pepe, K.K. Wary, L. Spinardi, M. Mohammadi, J. Schlessinger, and F.G. Giancotti. Signal transduction by the alpha 6 beta 4 integrin: distinct beta 4 subunit sites mediate recruitment of Shc/Grb2 and association with the cytoskeleton of hemidesmosomes. *EMBO J* **1995**; *14*: 4470–4481.

70 Nievers, M.G., R.Q. Schaapveld, and A. Sonnenberg. Biology and function of hemidesmosomes. *Matrix Biol* **1999**; *18*: 5–17.
71 Mainiero, F., C. Murgia, K.K. Wary, A.M. Curatola, A. Pepe, M. Blumemberg, J.K. Westwick, C.J. Der, and F.G. Giancotti. The coupling of alpha6beta4 integrin to Ras-MAP kinase pathways mediated by Shc controls keratinocyte proliferation. *EMBO J* **1997**; *16*: 2365–2375.
72 Mainiero, F., A. Pepe, M. Yeon, Y. Ren, and F.G. Giancotti. The intracellular functions of alpha6beta4 integrin are regulated by EGF. *J Cell Biol* **1996**; *134*: 241–253.
73 Franzke, C.W., K. Tasanen, L. Borradori, V. Huotari, and L. Bruckner-Tuderman. Shedding of collagen XVII/BP180: structural motifs influence cleavage from cell surface. *J Biol Chem* **2004**; *279*: 24521–24529.
74 Franzke, C.W., K. Tasanen, H. Schacke, Z. Zhou, K. Tryggvason, C. Mauch, P. Zigrino, S. Sunnarborg, D.C. Lee, F. Fahrenholz, and L. Bruckner-Tuderman. Transmembrane collagen XVII, an epithelial adhesion protein, is shed from the cell surface by ADAMs. *EMBO J* **2002**; *21*: 5026–5035.
75 Brakebusch, C., R. Grose, F. Quondamatteo, A. Ramirez, J.L. Jorcano, A. Pirro, M. Svensson, R. Herken, T. Sasaki, R. Timpl, S. Werner, and R. Fässler. Skin and hair follicle integrity is crucially dependent on beta 1 integrin expression on keratinocytes. *EMBO J* **2000**. *19*: 3990–4003.
76 DiPersio, C.M., K.M. Hodivala-Dike, R. Jaenisch, J.A. Kreidberg, and R. Hynes. α3β1 integrin is required for normal development of the epidermal basement membrane. *J Cell Biol* **1997**; *137*: 729–742.
77 Raghavan, S., C. Bauer, G. Mundschau, Q. Li, and E. Fuchs. Conditional ablation of beta1 integrin in skin. Severe defects in epidermal proliferation, basement membrane formation, and hair follicle invagination. *J Cell Biol* **2000**; *150*: 1149–1160.
78 Conti, F.J., R.J. Rudling, A. Robson, and K.M. Hodivala-Dilke. Alpha3beta1–integrin regulates hair follicle but not interfollicular morphogenesis in adult epidermis. *J Cell Sci* **2003**; *116*: 2737–2747.
79 Herzog, C., C. Has, C.W. Franzke, F.G. Echtermeyer, U. Schlötzer-Schrehardt, S. Kröger, E. Gustafsson, R. Fässler, and L. Bruckner-Tuderman. Dystroglycan in skin and cutaneous cells: beta-subunit is shed from the cell surface. *J Invest Dermatol* **2004**; *122*: 1372–1380.
80 Aumailley, M., A. El Khal, N. Knoss, and L. Tunggal. Laminin 5 processing and its integration into the ECM. *Matrix Biol* **2003**; *22*: 49–54.
81 Decline, F., and P. Rousselle. Keratinocyte migration requires alpha2beta1 integrin-mediated interaction with the laminin 5 gamma2 chain. *J Cell Sci* **2001**; *114*: 811–823.
82 Utani, A., Y. Momota, H. Endo, Y. Kasuya, K. Beck, N. Suzuki, M. Nomizu, and H. Shinkai. Laminin alpha 3 LG4 module induces matrix metalloproteinase–1 through mitogen-activated protein kinase signaling. *J Biol Chem* **2003**; *278*: 34483–34490.
83 Utani, A., M. Nomizu, H. Matsuura, K. Kato, T. Kobayashi, U. Takeda, S. Aota, P.K. Nielsen, and H. Shinkai. A unique se-

quence of the laminin alpha 3 G domain binds to heparin and promotes cell adhesion through syndecan–2 and –4. *J Biol Chem* 2001. *276*: 28779–28788.

84. Michele, D.E., and K.P. Campbell. Dystrophin-glycoprotein complex: post-translational processing and dystroglycan function. *J Biol Chem* 2003; *278*: 15457–15460.
85. Ekblom, M., M. Falk, K. Salmivirta, M. Durbeej, and P. Ekblom. Laminin isoforms and epithelial development. *Ann NY Acad Sci* 1998; *857*: 194–211.
86. Delwel, G.O., and A. Sonnenberg. Laminin isoforms and their integrin receptors. In *Adhesion Receptors as Therapeutic Targets*, M.A. Horton (Ed.). CRC Press: London, 1996; 9–36.
87. Sasaki, T., R. Fässler, and E. Hohenester. Laminin: the crux of basement membrane assembly. *J Cell Biol* 2004; *164*: 959–963.
88. Kivirikko, S., J.A. McGrath, C. Baudoin, D. Aberdam, S. Ciatti, M.G. Dunnill, J.R. McMillan, R.A. Eady, J.P. Ortonne, and G. Meneguzzi. A homozygous nonsense mutation in the α3 chain gene of laminin 5 (LAMA3) in lethal (Herlitz) junctional epidermolysis bullosa. *Hum Mol Genet* 1995; *4*: 959–962.
89. Pulkkinen, L., and J. Uitto. Mutation analysis and molecular genetics of epidermolysis bullosa. *Matrix Biol* 1999; *18*: 29–42.
90. Cheng, Y.S., M.-F. Champliaud, R.E. Burgeson, M.P. Marinkovich, and P.D. Yurchenco. Self-assembly of laminin isoforms. *J Biol Chem* 1997; *272*: 31525–31532.
91. Colognato, H., D.A. Winkelmann, and P.D. Yurchenco. Laminin polymerization induces a receptor-cytoskeletal network. *J Cell Biol* 1999; *145*: 619–631.
92. Yurchenco, P., and Y.S. Cheng. Self-assembly and calcium-binding sites in laminin. *J Biol Chem* 1993; *268*: 17286–17299.
93. Timpl, R., and J. Brown. The laminins. *Matrix Biol* 1994; *14*: 275–281.
94. Marinkovich, M.P., G.P. Lunstrum, D.R. Keene, and R.E. Burgeson. The dermal-epidermal junction of human skin contains a novel laminin variant. *J Cell Biol* 1992; *119*: 695–703.
95. Champliaud, M.F., G.P. Lunstrum, P. Rousselle, T. Nishiyama, D.R. Keene, and R.E. Burgeson. Human amnion contains a novel laminin variant, laminin 7, which like laminin 6, covalently associates with laminin 5 to promote stable epithelial-stromal attachment. *J Cell Biol* 1996; *132*: 1189–1198.
96. Sasaki, T., W. Göhring, K. Mann, C. Brakebusch, Y. Yamada, R. Fässler, and R. Timpl. Short arm region of laminin–5 gamma2 chain: structure, mechanism of processing and binding to heparin and proteins. *J Mol Biol* 2001; *314*: 751–763.
97. Gianelli, G., J. Falk-Marzillier, O. Schiraldi, W.G. Stetler-Stevenson, and V. Quaranta. Induction of cell migration by matrix metalloprotease–2 cleavage of laminin–5. *Science* 1997; *277*: 225–228.
98. Gilles, C., M. Polette, C. Coraux, J.M. Tournier, G. Meneguzzi, C. Munaut, L. Volders, P. Rousselle, P. Birembaut, and J.M. Foidart. Contribution of MT1–MMP and of human laminin–5 gamma2 chain degradation to mammary epithelial cell migration. *J Cell Sci* 2001; *114*: 2967–2976.
99. Goldfinger, L.E., M.S. Stack, and J.C.R. Jones. Processing of laminin–5 and its functional consequences: role of plasmin

and tissue-type plasminogen activator. *J Cell Biol* **1998**; *141*: 255–265.

100. Kunneken, K., G. Pohlentz, A. Schmidt-Hederich, U. Odenthal, N. Smyth, J. Peter-Katalinic, P. Bruckner, and J.A. Eble. Recombinant human laminin–5 domains. Effects of heterotrimerization, proteolytic processing, and N-glycosylation on alpha3beta1 integrin binding. *J Biol Chem* **2004**; *279*: 5184–5193.
101. Sorokin, L.M., M. Frieser, F. Pausch, S. Kröger, E. Ohage, and R. Deutzmann. Developmental regulation of laminin alpha 5 suggests a role in epithelial and endothelial cell maturation. *Dev Biol* **1997**; *189*: 285–300.
102. Li, J., J. Tzu, Y. Chen, Y.P. Zhang, N.T. Nguyen, J. Gao, M. Bradley, D.R. Keene, A.E. Oro, J.H. Miner, and M.P. Marinkovich. Laminin–10 is crucial for hair morphogenesis. *EMBO J* **2003**; *22*: 2400–2410.
103. Miner, J.H., J. Cunningham, and J.R. Sanes. Roles for laminin in embryogenesis: exencephaly, syndactyly, and placentopathy in mice lacking the laminin alpha5 chain. *J Cell Biol.* **1998**; *143*: 1713–1723.
104. Schuler, F., and L.M. Sorokin. Expression of laminin isoforms in mouse myogenic cells *in vitro* and *in vivo*. *J Cell Sci* **1995**; *108*: 3795–3805.
105. Sewry, C.A., M. D'Alessandro, L.A. Wilson, L.M. Sorokin, I. Naom, S. Bruno, A. Ferlini, V. Dubowitz, and F. Muntoni. Expression of laminin chains in skin in merosin-deficient congenital muscular dystrophy. *Neuropediatrics* **1997**; *28*: 217–222.
106. Koch, M., P. Olson, A. Albus, W. Jin, D.D. Hunter, W. Brunken, R.E. Burgeson, and M.-F. Champliaud. Characterization and expression of the laminin gamma 3 chain: a novel non-basement membrane-associated, laminin chain. *J Cell Biol* **1999**; *145*: 605–617.
107. Heinonen, S., M. Mannikko, J.F. Klement, D. Whitaker-Menezes, G.F. Murphy, and J. Uitto. Targeted inactivation of the type VII collagen gene (Col7α1) in mice results in severe blistering phenotype: a model for recessive dystrophic epidermolysis bullosa. *J Cell Sci* **1999**; *112*: 3641–3648.
108. Van Exan, R.J., and M.H. Hardy. The differentiation of the dermis in the laboratory mouse. *Am J Anat* **1984**; 169: 149–164.
109. Cowin, A.J., T.M. Holmes, P. Brosnan, and M.W. Ferguson. Expression of TGF-beta and its receptors in murine fetal and adult dermal wounds. *Eur J Dermatol* **2001**; *11*: 424–431.
110. Keene, D.R., M.P. Marinkovich, and L.Y. Sakai. Immunodissection of the connective tissue matrix in human skin. *Microsc Res Tech* **1997**; *38*: 394–406.
111. Danielson, K.G., H. Baribault, D.F. Holmes, H. Graham, K.E. Kadler, and R.V. Iozzo. Targeted disruption of decorin leads to abnormal collagen fibril morphology and skin fragility. *J Cell Biol* **1997**; *136*: 729–743.
112. Bruckner-Tuderman, L., and P. Bruckner. Genetic diseases of the extracellular matrix: more than just connective tissue disorders. *J Mol Med* **1998**; *76*: 226–237.
113. Mauger, A., H. Emonard, D. Hartmann, J. Foidart, and P. Sengel. Immunofluorescent localization of collagen types I, III,

114 Niederreither, K., R. D'Souza, M. Metsaranta, H. Eberspaecher, P.D. Toman, E. Vuorio, and B. De Crombrugghe. Coordinate patterns of expression of type I and III collagens during mouse development. *Matrix Biol* **1995**; *14*: 705–713.

and IV, fibronectin, laminin, and basement membrane proteoglycan in developing mouse skin. *Roux's Arch Dev Biol* **1987**; *196*: 295–302.

115 Theiler, J. *The House Mouse: Atlas of Embryonic Development.* Springer Verlag: New York, **1989**.

116 Uitto, J. Heritable connective tissue disorders. *Adv Exp Med Biol* **1999**; *455*: 15–21.

117 OMIM. Online Mendelian inheritance in Man, National Center for Biotechnology Information (http://www.ncbi.nlm.nih.gov:80/entrez/query.fcgi?db=OMIM).

118 Liu, X., H. Wu, M. Byrne, S. Krane, and R. Jaenisch. Type III collagen is crucial for collagen I fibrillogenesis and for normal cardiovascular development. *Proc Natl Acad Sci USA* **1997**; *94*: 1852–1856.

119 Andrikopoulos, K., X. Liu, D.R. Keene, R. Jaenisch, and F. Ramirez. Targeted mutation in the col5a2 gene reveals a regulatory role for type V collagen during matrix assembly. *Nat Genet* **1995**; *9*: 31–36.

120 Yancopoulos, G.D., S. Davis, N.W. Gale, J.S. Rudge, S.J. Wiegand, and J. Holash. Vascular-specific growth factors and blood vessel formation. *Nature* **2000**; *407*: 242–248.

121 Reiss, Y., A.E. Proudfoot, C.A. Power, J.J. Campbell, and E.C. Butcher. CC chemokine receptor (CCR)4 and the CCR10 ligand cutaneous T cell-attracting chemokine (CTACK) in lymphocyte trafficking to inflamed skin. *J Exp Med* **2001**; *194*: 1541–1547.

122 Teraki, Y., and L.J. Picker. Independent regulation of cutaneous lymphocyte-associated antigen expression and cytokine synthesis phenotype during human CD4+ memory T cell differentiation. *J Immunol* **1997**; *159*: 6018–6029.

123 Chong, B.F., J.E. Murphy, T.S. Kupper, and R.C. Fuhlbrigge. E-selectin, thymus- and activation-regulated chemokine/CCL17, and intercellular adhesion molecule–1 are constitutively coexpressed in dermal microvessels: a foundation for a cutaneous immunosurveillance system. *J Immunol* **2004**; *172*: 1575–1581.

124 Oliver, G. Lymphatic vasculature development. *Nat Rev Immunol* **2004**; *4*: 35–45.

125 Karkkainen, M.J., A. Saaristo, L. Jussila, K.A. Karila, E.C. Lawrence, K. Pajusola, H. Bueler, A. Eichmann, R. Kauppinen, M.I. Kettunen, S. Yla-Herttuala, D.N. Finegold, R.E. Ferrell, and K. Alitalo. A model for gene therapy of human hereditary lymphedema. *Proc Natl Acad Sci USA* **2001**; *98*: 12677–12682.

126 Makinen, T., L. Jussila, T. Veikkola, T. Karpanen, M.I. Kettunen, K.J. Pulkkanen, R. Kauppinen, D.G. Jackson, H. Kubo, S. Nishikawa, S. Yla-Herttuala, and K. Alitalo. Inhibition of lymphangiogenesis with resulting lymphedema in transgenic mice expressing soluble VEGF receptor–3. *Nat Med* **2001**; *7*: 199–205.

127 Gretz, J.E., C.C. Norbury, A.O. Anderson, A.E. Proudfoot, and S. Shaw. Lymph-borne chemokines and other low molecular weight molecules reach high endothelial venules via specialized conduits while a functional barrier limits access to the lymphocyte microenvironments in lymph node cortex. *J Exp Med* **2000**; *192*: 1425–1440.

128 Sixt, M., Kanazawa, N., M. Selg, T. Samson, G. Roos, D. Reinhardt, R. Papst, and M.S. Lutz, L.M. Sorokin. The conduit system transports soluble antigens from the afferent lymph to residential dendritic cells in the T cell area of the lymph node. *Immunity* **2005**; *22*: 19–30.

129 DasGupta, R., and E. Fuchs. Multiple roles for activated LEF/TCF transcription complexes during hair follicle development and differentiation. *Development* **1999**; *126*: 4557–4568.

130 DasGupta, R., H. Rhee, and E. Fuchs. A developmental conundrum: a stabilized form of beta-catenin lacking the transcriptional activation domain triggers features of hair cell fate in epidermal cells and epidermal cell fate in hair follicle cells. *J Cell Biol* **2002**; *158*: 331–344.

131 Kratochwil, K., M. Dull, I. Farinas, J. Galceran, and R. Grosschedl. Lef1 expression is activated by BMP-4 and regulates inductive tissue interactions in tooth and hair development. *Genes Dev* **1996**; *10*: 1382–1394.

132 Kere, J., A.K. Srivastava, O. Montonen, J. Zonana, N. Thomas, B. Ferguson, F. Munoz, D. Morgan, A. Clarke, P. Baybayan, E.Y. Chen, S. Ezer, U. Saarialho-Kere, A. de la Chapelle, and D. Schlessinger. X-linked anhidrotic (hypohidrotic) ectodermal dysplasia is caused by mutation in a novel transmembrane protein. *Nat Genet* **1996**; *13*: 409–416.

133 Viallet, J.P., F. Prin, I. Olivera-Martinez, E. Hirsinger, O. Pourquie, and D. Dhouailly. Chick Delta–1 gene expression and the formation of the feather primordia. *Mech Dev* **1998**; *72*: 159–168.

134 Ikram, M.S., G.W. Neill, G. Regl, T. Eichberger, A.M. Frischauf, F. Aberger, A. Quinn, and M. Philpott. GLI2 is expressed in normal human epidermis and BCC and induces GLI1 expression by binding to its promoter. *J Invest Dermatol* **2002**; *122*: 1502–1509.

135 Botchkarev, V.A., N.V. Botchkareva, A.A. Sharov, K. Funa, O. Huber, and B.A. Gilchrest. Modulation of BMP signaling by noggin is required for induction of the secondary (nontylotrich) hair follicles. *J Invest Dermatol* **2002**; *118*: 3–10.

136 Cotsarelis, G., and S.E. Millar. Towards a molecular understanding of hair loss and its treatment. *Trends Mol Med* **2001**; *7*: 293–301.

137 Oshima, H., A. Rochat, C. Kedzia, K. Kobayashi, and Y. Barrandon. Morphogenesis and renewal of hair follicles from adult multipotent stem cells. *Cell* **2001**; *104*: 233–245.

138 Ghazizadeh, S., and L.B. Taichman. Multiple classes of stem cells in cutaneous epithelium: a lineage analysis of adult mouse skin. *EMBO J* **2001**; *20*: 1215–1222.

139 Cotsarelis, G., P. Kaur, D. Dhouailly, U. Hengge, and J. Bickenbach. Epithelial stem cells in the skin: definition, markers, localization and functions. *Exp Dermatol* **1999**; *8*: 80–88.

140 Jones, P.H., and F.M. Watt. Separation of human epidermal stem cells from transient amplifying cells on the basis of differences in integrin function and expression. *Cell* **1993**; *73*: 713–724.
141 Li, A., N. Pouliot, R. Redvers, and P. Kaur. Extensive tissue-regenerative capacity of neonatal human keratinocyte stem cells and their progeny. *J Clin Invest* **2004**; *113*: 390–400.
142 Tani, H., R.J. Morris, and P. Kaur. Enrichment for murine keratinocyte stem cells based on cell surface phenotype. *Proc Natl Acad Sci USA* **2000**; *97*: 10960–10965.
143 Taylor, G., M.S. Lehrer, P.J. Jensen, T.T. Sun, and R.M. Lavker. Involvement of follicular stem cells in forming not only the follicle but also the epidermis. *Cell* **2000**; *102*: 451–461.
144 Arnold, I., and F.M. Watt. c-Myc activation in transgenic mouse epidermis results in mobilization of stem cells and differentiation of their progeny. *Curr Biol* **2001**; *11*: 558–568.
145 Frye, M., C. Gardner, E.R. Li, I. Arnold, and F.M. Watt. Evidence that Myc activation depletes the epidermal stem cell compartment by modulating adhesive interactions with the local microenvironment. *Development* **2003**; *130*: 2793–2808.
146 Waikel, R.L., Y. Kawachi, P.A. Waikel, X.J. Wang, and D.R. Roop. Deregulated expression of c-Myc depletes epidermal stem cells. *Nat Genet* **2001**; *28*: 165–168.
147 Savill, N.J., and J.A. Sherratt. Control of epidermal stem cell clusters by Notch-mediated lateral induction. *Dev Biol* **2003**; *258*: 141–153.
148 Gambardella, L., and Y. Barrandon. The multifaceted adult epidermal stem cell. *Curr Opin Cell Biol* **2003**; *15*: 771–777.
149 Galliano, M.-F., D. Aberdam, A. Aguzzi, J.-P. Ortonne, and G. Meneguzzi. Cloning and complete primary structure of the mouse laminin α3 chain. *J Biol Chem* **1995**; *270*: 21820–21826.

19
Tooth Development

Xiu-Ping Wang and Irma Thesleff

19.1
Introduction

Teeth form as ectodermal appendages and their development is regulated by sequential and reciprocal interactions between oral epithelium and the underlying neural crest-derived mesenchyme. The early stages of tooth development resemble morphologically and molecularly other ectodermal derivatives, such as hair, feathers, and various glands [1]. Since the mouse tooth germs are easily accessed and can be experimentally manipulated *in vitro*, they have been used for a long time as models for studying epithelial-mesenchymal interactions and the molecular mechanisms underlying organogenesis [2]. During the last 15 years, there has been rapid progress in the understanding of the genetic regulation of tooth development. The signaling pathways and networks involved in tooth development have begun to be elucidated, and the roles of growth factors in several families including fibroblast growth factor (FGF), transforming growth factor β (TGF-β), hedgehog (Hh), Wnt, and tumor necrosis factor (TNF) are now known in some detail [3, 4].

At the initiation stage, oral epithelium thickens and forms an ectodermal ridge, the dental lamina, marking the future dental arches in the frontonasal, maxillary, and mandibular primordia (Fig. 19.1). Local proliferation of the dental lamina generates multilayered epithelial condensations at the sites of individual teeth. Dental placodes form at the tip of the thickened epithelia and express a number of signal molecules including *Fgf8, Fgf20, Bmp2, Shh, Wnt10a,* and *Wnt10b* [5–8]. The dental placodes resemble the placodes of other ectodermal organs both morphologically and functionally and they act as transient signaling centers during early tooth morphogenesis [1]. Further development of the placodal epithelium leads to the tooth bud invaginating into the underlying mesenchyme and stimulating mesenchymal condensation around the bud. As the dental epithelium grows and folds, it becomes cap-shaped with distinct cell layers forming the enamel organ. The mesenchymal cells enveloped by the enamel organ form the dental papilla, and the surrounding mesenchymal cells form the dental follicle, respectively. A cluster of condensed, non-proliferating epithelial cells at the tip of the enamel organ forms the primary enamel knot. It is also a transient signaling center expressing more than 10 con-

---

Cell Signaling and Growth Factors in Development. Edited by K. Unsicker and K. Krieglstein
Copyright © 2006 WILEY-VCH Verlag GmbH & Co. KGaA, Weinheim
ISBN 3-527-31034-7

served signal molecules including *Fgf3, Fgf4, Fgf9, Fgf20, Bmp2, Bmp4, Bmp7, Shh*, as well as *Wnt3, Wnt10a,* and *Wnt10b* [5,6,8–11].

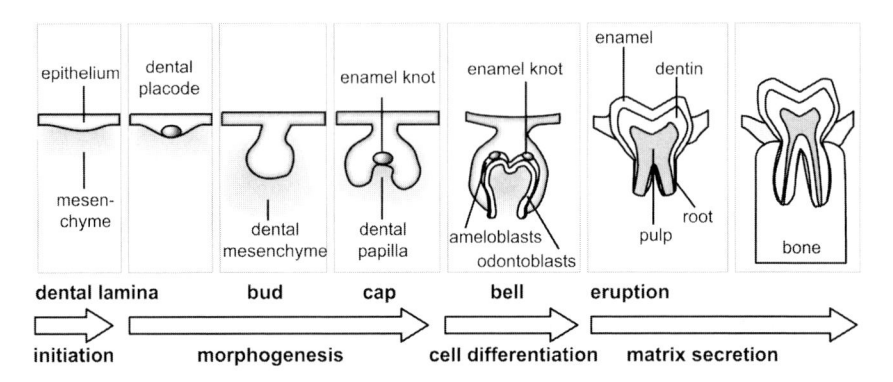

Fig. 19.1
Schematic representation of tooth development. Interactions between the oral epithelium and neural crest-derived mesenchyme direct tooth morphogenesis and cell differentiation. The growth and folding of the dental epithelium determines the shape of the tooth. The dental placode as well as the enamel knots are signaling centers and key regulators of morphogenesis.

During the following bell stage, distinct layers of the enamel organ become more evident with the inner and outer dental epithelia surrounding the stratum intermedium and stellate reticulum cells. The epithelial cervical loops grow and fold rapidly outlining the specific shape of the tooth crown. The primary enamel knot is mostly removed by apoptosis and secondary enamel knots form sequentially at the tips of future cusps within the tooth crown base and determine the locations of tooth cusps (Fig. 19.1). Secondary enamel knots are also composed of packed and non-proliferative cells expressing similar genes as the primary enamel knots and they too are later removed by apoptosis [3, 12]. It has been proposed that the number, size, and position of secondary enamel knots are determined by the primary enamel knot, and thus slight changes in the signals from the primary enamel knot may affect the location and height of the tooth cusps and may even account for the different dental morphologies during vertebrate evolution [13].

At the late bell stage, the mesenchymal cells underlying the dental epithelium at the periphery of the dental papilla differentiate into odontoblasts secreting dentin matrix. The remaining dental papilla cells give rise to the dental pulp (Fig. 19.1). The juxtaposed inner dental epithelial cells differentiate into ameloblasts depositing enamel matrix. The differentiation of odontoblasts and ameloblasts is coordinated and starts at the tips of the tooth cusps, and then gradually sweeps down towards the base of the tooth crown. Once crown formation is completed, the roots start to form. Root development is guided by the continued growth of the cervical part of the dental epithelium, the Hertwig's epithelial root sheath (HERS). The HERS does not differentiate into ameloblasts, but induces dentin formation by differentiation of the odontoblasts from the dental papilla mesenchyme and cementoblasts from the surrounding dental follicle. Cementoblasts lay down a thin layer of bone-like cementum lining the root surface. The dental follicle cells also generate the periodontal ligament connecting the tooth to alveolar bone. During advancing root development, the HERS disintegrates into Malassez epithelial cells forming a network on the root surface, and finally the tooth erupts into the oral cavity (Fig. 19.1) [14].

Evolutionary changes have produced various tooth types among different species, and this has been used as an important tool in studies of fossil records. Most lower vertebrates possess multiple dentitions and homodont teeth with similar shapes, for example the teeth of fish and reptiles are mostly conical and are being replaced throughout life. Mammalian teeth, including the human teeth, are usually diphyodont (consisting of two sets of teeth) and heterodont (teeth with various shapes). According to their shapes, these teeth fall into three groups: incisiform, caniniform, and molariform, which are localized from the anterior to posterior region of the oral cavity. The permanent (secondary) teeth arise from the extension of dental lamina on the lingual aspect of the deciduous (primary) tooth germs (e.g. incisors, canines, and premolars), or from the posterior growth of the dental lamina (e.g. molars) [14]. The molecular mechanisms regulating the formation of successional teeth remain poorly understood. Most rodents, such as mice, have only one dentition with one incisor in the front and three molars in the back of each quadrant of the jaw. Between the incisors and molars is a toothless diastema region containing some rudimentary tooth germs which are degenerated by apoptosis [15, 16].

During embryogenesis, the tooth fate is determined at a very early stage in oral tissue. This was well demonstrated in classical embryological experiments using epithelial-mesenchymal tissue recombinations and transplantations. When tissue from the mouse molar region was dissected before the first molar bud had formed and transplanted to the anterior chamber of the mouse eye, all three molars formed [17]. It was also shown that even disaggregated cells from the incisor cervical loop region can form tooth germs [18, 19]. In the mouse embryos, the early stage (E9–E11) mandibular arch oral epithelium is able to induce tooth formation when recombined with neural crest-derived mesenchyme from the second branchial arch, or even with the trunk region neural crest cells, but not with non-neural crest-derived mesenchyme such as limb mesenchyme. Reverse recombinations between mandibular mesenchyme and non-dental epithelium do not form teeth, indicating that the early-stage oral epithelium possesses odontogenic potential [17, 20]. Recent evidence indicates that early mandibular epithelium can induce tooth formation even in embryonic stem cells [21]. After the initiation stage, around E12 in mice, the odontogenic potential shifts from the dental epithelium to the mesenchyme, which subsequently guides tooth formation and also determines the shape of the tooth [17,19,20].

19.2
Signaling Pathways and their Networks Regulating Tooth Development

The reciprocal epithelial-mesenchymal interactions are mediated by growth factors. To date, more than 350 genes have been demonstrated to be differentially expressed in the developing teeth, and most of them are part of the growth factor signaling networks (see tooth database http://bite-it.helsinki.fi) [4]. Most information on the molecular regulation of tooth development has come from the analysis of gene expression patterns in the developing teeth, as well as from examination of tooth phenotypes in knockout and transgenic mice. The effects of growth factors have also been addressed in organ cultures in which they are introduced either by beads or by aggregates of transfected cells expressing the signals. Growth factor-containing beads are placed in contact with the dissected dental tissue and the response of the tissue is analyzed by using *in situ* hybridization or immunohistochemistry [4, Fig. 19.2]. In the following, we will review the roles of growth factors in the major families, their receptors, and inhibitors, as well as target genes during tooth development.

19.2.1
Fibroblast Growth Factors

FGFs are a large family of intercellular signaling molecules that participate in many developmental and physiological processes of virtually all mammalian tissues. During embryonic development, FGFs have diverse roles in regulating cell survival, proliferation, adhesion, migration, and differentiation [22]. Among the 23 FGFs in vertebrates, seven have been identified in the tooth. *Fgf3*, *Fgf7*, and *Fgf10* are mainly expressed in the dental mesenchyme, except transient weak expression of *Fgf10* at an early stage in the presumptive dental epithelium and *Fgf3* in the primary enamel knot at cap stage. *Fgf3* and *Fgf10* are co-expressed in the dental mesenchyme from the late bud stage until the late bell stage [23]. They are also expressed intensely in the dental mesenchyme around the cervical loop epithelium of the continuously erupting mouse incisors and may be involved in maintaining and/or stimulating the proliferation of epithelial stem cells in the incisors [24]. This idea was supported by the finding that the cervical loop of *Fgf10* knockout mouse incisors is hypoplastic and the incisors failed to grow continuously *in vitro* [25]. *Fgf7* is weakly expressed in the dental mesenchyme around the cervical loop region in the incisors, but is not detected in molars [23].

The expression of *Fgf4*, *Fgf8*, *Fgf9*, and *Fgf20* is confined to the dental epithelium with overlapping patterns in the signaling centers. Before the initiation of tooth development, *Fgf8* transcripts are already present in the oral epithelium. When the teeth start to form, intense expression of *Fgf8* becomes more restricted to the presumptive dental epithelium until the bud stage. This is accompanied by a weak expression of *Fgf9* [11]. *Fgf20* expression appears reiteratively in all the signaling centers, starting from the dental placodes at the early stage to the primary and secondary enamel knots during advancing morphogenesis (Aberg et al., unpublished data). *Fgf4* transcripts are first detected in the primary enamel knot at the late bud stage. Subsequently *Fgf4* and *Fgf20* are co-localized in the secondary enamel knots [9]. *Fgf9* is also expressed in the primary and secondary enamel knots, and more widely in the inner dental epithelium at the late bell stage [11].

FGF signaling is mediated through a family of tyrosine kinase transmembrane receptors. So far, four different FGF receptor genes have been identified in mammals [22]. *Fgfr1, –2, and –3*, but not *Fgfr4*, are expressed in the dental tissues [26]. The IIIb isoforms of *Fgfr1* and *Fgfr2* are expressed mainly in the dental epithelium, whereas the IIIc isoforms are located primarily in the dental mesenchyme. *Fgfr1c* is also expressed in the epithelium. It is noteworthy that *Fgfr2b* is exclusively expressed in the dental epithelium. In line with this, FGF10–releasing beads are able to stimulate cell proliferation only in the dental epithelium, but not in the mesenchyme [26]. It thus appears that FGF10 function is restricted to the dental epithelium, at early stages as an autocrine signal within the epithelium and later as a mediator of signals from dental mesenchyme to the epithelium. Similar actions have also been demonstrated during lung development in which FGF10 from the lung tip mesenchyme binds to FGFR2b in the respiratory epithelium and induces lung branching morphogenesis [27]. Hence, tissue-specific expression of alternatively spliced FGF receptors may direct reciprocal epithelial-mesenchymal interactions. In fact, FGFR2b can bind to all the mesenchymal FGFs detected so far in the tooth: FGF3, FGF7, and FGF10. In *Fgfr2b* knockout mice and in transgenic mice overexpressing a soluble dominant negative form of *Fgfr2b*, most organs developing via epithelial-mesenchymal interactions failed to form (mutant teeth arrest at the bud stage), indicating an absolute requirement for mesenchymal FGF signals for epithelial morphogenesis [28, 29]. Within the enamel organ, there is intense expression of *Fgf* receptors (*Fgfr1b, –1c, –2b*) in the epithelial cervical loops. No FGF receptors are expressed in the non-proliferative enamel knots, although the enamel knot itself expresses several Fgf ligands (*Fgf3, –4, –9, –20*). The FGF proteins secreted by the enamel knot conceivably act on the surrounding cervical loops and the underlying dental mesenchyme, thereby regulating tooth morphogenesis [26]. *Fgfr1b* and *–2b* are also expressed in the epithelial cervical loop of mouse incisors where they may mediate the effects of FGF3 and FGF10 from the underlying mesenchyme for the maintenance of the stem cell niche [23,24,30]. At late bell stage, *Fgfr1* and *Fgfr2* are expressed intensely in the ameloblasts and Fgfr1 is also expressed in the odontoblasts, suggesting that FGFs are also involved in the regulation of tooth-specific cell differentiation. The IIIb and IIIc isoforms of *Fgfr3* are restricted to the dental mesenchyme at the late bell stage [26].

Ligand binding of FGF receptors depends on the presence of heparan sulfate proteoglycans (HSPG), which act as low affinity FGF co-receptors regulating the diffusion of FGF proteins and is essential for the formation of active FGF/FGF receptor signaling complex [31]. Syndecan–1 is a cell surface proteoglycan expressed intensely in the dental mesenchyme during bud and cap stages. It can be induced by epithelial signals and by FGF proteins, and was proposed to be involved in cell proliferation and rapid tooth morphogenesis [32–34]. However, syndecan–1 knockout mice do not have obvious developmental defects, probably due to redundancy of other HSPGs in the teeth [35].

Sprouty genes encode inducible intracellular inhibitors of FGF signaling [36, 37]. In the developing mouse tooth, *Sprouty 1* is expressed very weakly in the dental epithelium and mesenchyme at the bud and cap stages. *Sprouty 2* is expressed intensely in the primary enamel knot at the cap stage, and *Sprouty 4* in the dental mesenchyme and in the epithelial cervical loops [38]. Coordinated and overlapping expression of *Sprouties* and *Fgfs* indicate that the level of FGF signaling is tightly regulated during tooth development.

The roles of FGFs have been examined by targeted gene disruption in mice. *Fgf4* and *Fgf8* knockouts are early embryonic lethal [39–41], but conditional inactivation of *Fgf8* in the ectoderm of the first branchial arch resulted in the absence of molar teeth [42]. Rudimentary lower incisors were present and it was suggested that the function of FGF8 was compensated by FGF9 in the anterior region. In general, redundancy between different FGF family members is common during development [43]. Detailed analysis of the mutants demonstrated that FGF8 is required for the expression of *Barx1, Lhx6,* and *Pax9* in the molar region. *In vitro* bead assays have shown that FGF8 is able to induce in the early jaw mesenchyme the expression of a number of genes, such as *Msx1, Pax9, Dlx2,* and *activin βA*. These genes are critical for tooth morphogenesis since ablation of their function in transgenic mice leads to arrested tooth development [44–47]. Mutations of *MSX1* and *PAX9* cause in humans tooth agenesis [48, 49].

Fgf4, Fgf9, and Fgf20 are reiteratively expressed in the epithelial signaling centers. Beads releasing FGF4 and FGF9 proteins stimulate cell proliferation in both dental epithelium and mesenchyme [26]. Notably, FGF4 and FGF8 proteins can induce *Fgf3*, but not *Fgf10* in the dental mesenchyme [23, 34]. Recent studies suggest that the induction of *Fgf3* by epithelial FGF signals is mediated by the transcription factors Msx1 and Runx2. *Msx1* and *Runx2* are both expressed in the dental mesenchyme and their expression is induced by epithelial signals that can be mimicked by FGF4–releasing beads. FGF4 cannot induce *Fgf3* expression in *Msx1* or *Runx2* mutant dental mesenchyme (Fig. 19.2). In addition, *Runx2* expression is downregulated in *Msx1* mutant tooth germs. These results have placed epithelial FGF signals and mesenchymal Msx1, Runx2, and Fgf3 in the same signaling pathway mediating epithelial-mesenchymal interactions [34, 38]. No tooth phenotype was reported in *Fgf3* knockout mice, which may be due to the redundancy of Fgf10 which is co-expressed in the dental mesenchyme [50]. *Fgf10* knockout mice die at birth and have slightly hypoplastic molars and incisors [25]. Double knockouts of *Fgf3* and *Fgf10* may help to clarify their roles during tooth morphogenesis.

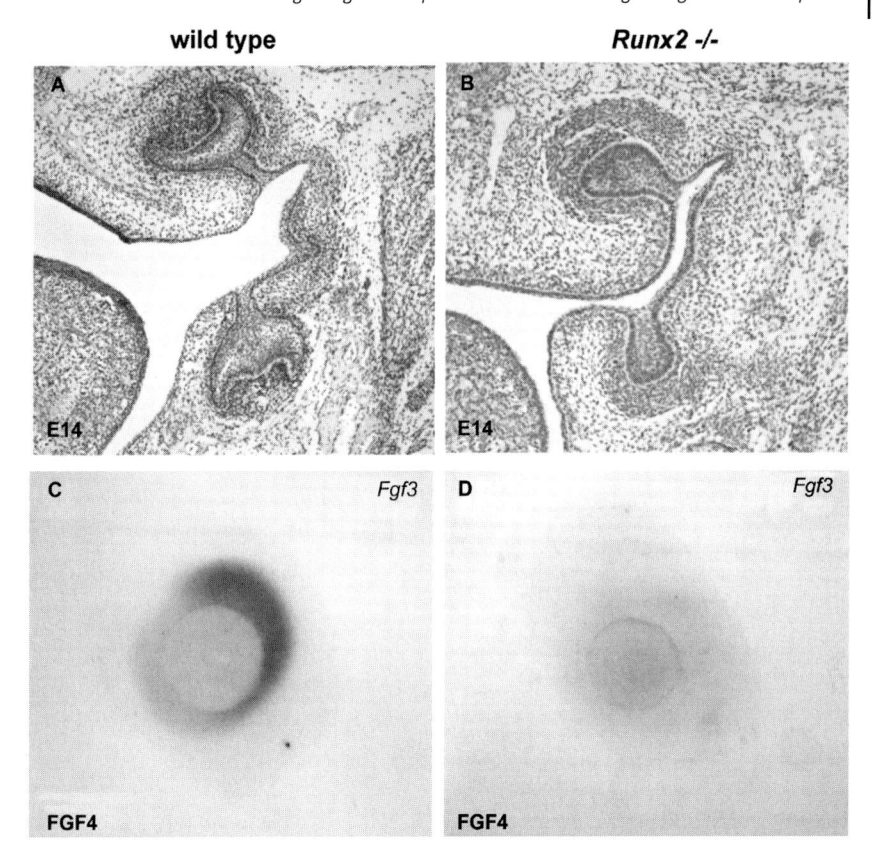

Fig. 19.2
Runx2 is required for tooth development and for the mediation of FGF signaling between epithelium and mesenchyme. (A) At E14, wild-type mouse teeth have developed into the cap stage. (B) *Runx2* mutant teeth are arrested at the bud stage. (C, D) FGF4–releasing beads induce the expression of *Fgf3* in isolated dental mesenchyme in a wild-type mouse embryo, but no induction occurs in the *Runx2* mutant dental mesenchyme.

19.2.2
TGFβ Family

TGFβ superfamily growth factors contain over 50 structurally-related proteins, which play diverse roles during embryogenesis and in the regulation of homeostasis in adult tissues. The TGFβ superfamily is divided into several subfamilies including TGFβs, bone morphogenetic proteins (BMPs), and activin/inhibin family. They bind to the cell surface type II and type I serine/threonine kinase receptors. Upon ligand binding, the type II receptors phosphorylate type I receptors, which then bind and phosphorylate cytoplasmic Smad proteins mediating the signals into the nucleus for the activation of target genes [51, 52]. TGFβ and activin signals are mediated by Smad2 and Smad3, and BMP signals by Smad1, –5, and –8, respectively.

In the tooth bud, *Tgfβ1* is intensely expressed in the epithelial cervical loops and also in the dental mesenchyme. Tissue recombination experiments showed that epithelial *Tgfβ1* expression is induced/maintained by dental mesenchymal signals [53]. During the bell stage, *Tgfβ1, –2*, and *–3* are all expressed intensely in the differentiating odontoblasts and ameloblasts [53, 54].

The expression patterns of several Bmps have been analysed by *in situ* hybridization in the developing teeth of the mouse [5]. Of particular interest is the dynamic expression pattern of *Bmp4*. It is initially expressed in the dental epithelium, but later shifts to the mesenchyme at the time corresponding to the shift of odontogenic potential [55]. During early tooth development, *Bmp2* is expressed intensely in the dental placode. During the bud and cap stages, *Bmp2, Bmp4,* and *Bmp7* transcripts are all localized in the primary enamel knots, and *Bmp4* is expressed intensely in the dental mesenchyme also. During bell stage, *Bmp2, –4,* and *–7* are expressed in the odontoblasts and ameloblasts. *Bmp3* is mainly expressed in the dental follicle mesenchyme. *Bmp5* transcripts are not detected in the tooth region until late bell stage when they are intensely upregulated in the pre-ameloblasts and secretory ameloblasts. *Bmp6* is expressed weakly in the dental mesenchyme during bud and cap stages [5].

Activin βA is expressed in the dental mesenchyme starting from the initiation of tooth development. At the bell stage, intense *activin* transcripts are restricted to the cusp region of the dental papilla mesenchyme [47,56,57]. *In vitro* studies have shown that epithelial FGF8 induces *activin* expression in the mesenchyme. Activin acts reciprocally on the dental epithelium and induces the expression of its own inhibitor *follistatin* [47]. Follistatin is an extracellular inhibitor of several members of TGFβ superfamily proteins, including activin and BMP2, –4, and –7 [43,58–60]. In the developing mouse molars, *follistatin* is expressed in the enamel knots. The strong and overlapping expression of *follistatin* and *Bmps* in the primary and secondary enamel knots and *activin βA* in the underlying dental mesenchyme suggests that they are involved in the regulation of tooth morphogenesis. Indeed, increasing follistatin levels in the dental epithelium in transgenic mice results in misshapen molars with irregular cusp patterning (Fig. 19.3A, B). On the other hand, ablation of follistatin activity also disturbs the formation of secondary enamel knots leading to aberrant folding of the inner dental epithelium and shallow tooth cusps [61]. These observations indicate that the levels of TGFβ family signals must be tightly regulated, and therefore finely tuned antagonistic effects between TGFβ superfamily signals and their inhibitors are critical for normal tooth morphogenesis.

Ectodin is also a secreted BMP inhibitor [62]. It is intensely expressed in developing ectodermal organs, including teeth and hair follicles. However, it is strikingly absent from the hair placodes and enamel knots and, interestingly, several cell layers surrounding the enamel knots are also ectodin-negative and form a sharp boundary with outer ectodin-positive cells (Fig. 19.4). *In vitro* bead experiments showed that BMP2 and BMP7 stimulated *ectodin* expression in tooth explants, but Shh and FGF4, which are co-expressed with *Bmp2* and *–7* in the enamel knot, counteracted this induction. It was therefore proposed that the sharp boundary of *ectodin* expression around the enamel knot may represent the extent of BMP target tissue and may

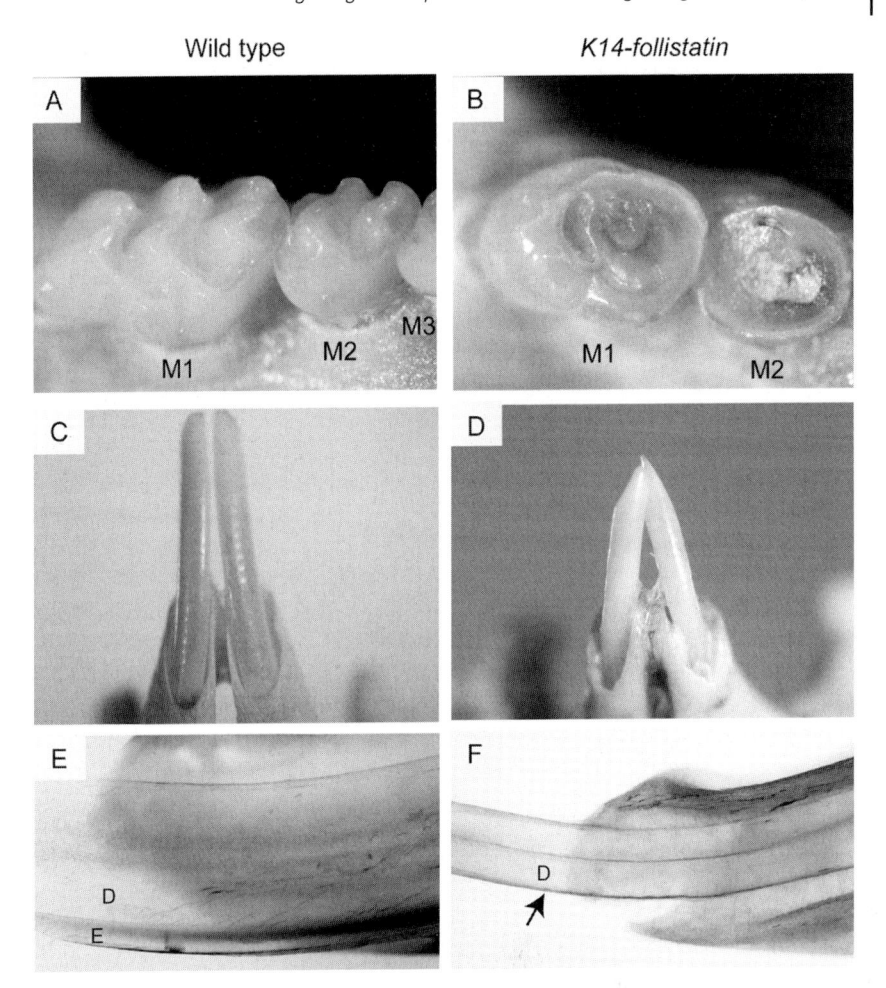

19.2 Signaling Pathways and their Networks Regulating Tooth Development | 727

Fig. 19.3
TGFβ superfamily signals regulate the number, size, shape, and structure of teeth. (A, B) In the mouse overexpressing the TGFβ family inhibitor *follistatin* (*K14–follistatin*), the third molars are missing. The cusps of the first and second molars are aberrant, and the enamel is prematurely worn. (C, D) The lower incisors of transgenic mice are thin, short, and white. (E, F) Sections of the incisors show that no enamel is present in the transgenic mouse (arrow). M1, first molar; M2, second molar; M3, third molar; D, dentin; E, enamel.

also determine the sites of secondary enamel knots [62]. In line with this suggestion, the morphology and cusp patterning of molars in *ectodin* null mutant mice are severely disturbed [63]. This further emphasizes the importance of fine-tuning of BMP activity during tooth morphogenesis.

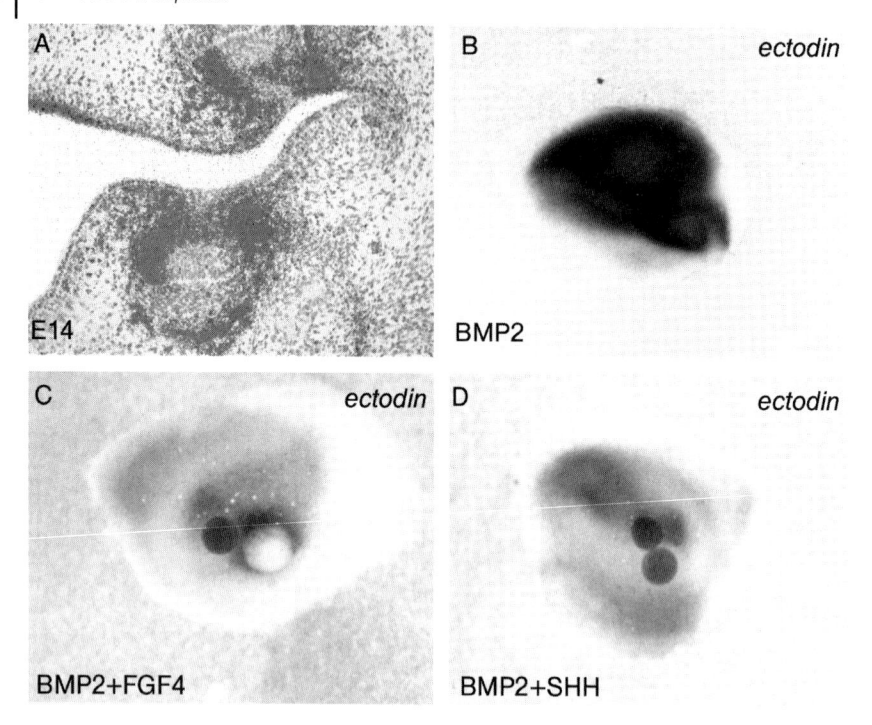

Fig. 19.4
Ectodin is a BMP inhibitor integrating the BMP, FGF and Shh signals from the enamel knot. (A) *In situ* hybridization analysis of *ectodin* expression in a cap stage tooth germ indicates downregulation in the enamel knot and its surrounding tissue. (B) Bead-released BMP2 induces ectodin expression in a dental explant *in vitro*. (C) Bead-released Shh together with the BMP2 bead, inhibits ectodin expression. (D) An FGF-releasing bead also inhibits the ectodin-inducing effect of BMP. (Courtesy of Johanna Laurikkala). (This figure also appears with the color plates.)

In general, it has been difficult to examine the precise functions of individual BMPs *in vivo* since the same signals are used reiteratively at many different developmental stages, in particular during tooth initiation and the transition from bud to cap stage [3,64,65]. *Bmp2* and *Bmp4* knockout mice are early embryonic lethal whereas *Bmp7* knockout mice do not have any tooth phenotype, probably due to the co-expression and redundancy of BMP and BMP4 [66, 67]. So far, the roles of individual BMPs in the developing teeth have mainly been studied *in vitro* by using bead implantation assays and by analysis of the knockouts of BMP signaling pathway genes. During tooth initiation BMP4 stimulates *Msx1* and *Msx2* expression, as well as its own expression in the dental mesenchyme, thus mimicking the inductive effect of the dental epithelium [34, 55]. At later stages, mesenchymal BMP4 has been proposed to be an inducer of the enamel knot signaling center, since BMP-releasing beads induce the expression of enamel knot marker genes *p21* and *Msx2* in the dental epithelium [68]. In *Msx1/Msx2* double knockout mice, tooth development is arrested at the initiation stage, while *Msx1* mutant teeth are arrested at the bud stage without obvious mesenchymal condensation [34, 69]. *Bmp4* expression is downregulated in *Msx1* mutant dental mesenchyme. Strikingly, exogenously added BMP4 protein is able to rescue the *Msx1* mutant tooth germs to develop beyond the cap stage and even start cell differentiation, indicating that BMP4 is necessary for the transition from bud to cap stage [70]. Conditional inactivation of BMP receptor 1A (Bmpr1a) in dental epithelium, results in an arrest of tooth development at the bud stage, further supporting the notion that BMPs are required for epithelial morphogenesis [71].

The deletion of activin βA function results in an arrest of incisors and lower molars at the bud stage, whereas the development of upper molars is not affected [47]. Activin is required in the dental mesenchyme prior to E11.5, but so far no target genes in the mesenchyme have been identified. Since activin receptor IIA+/–, IIB-/– and Smad 2+/– mouse tooth phenotypes are similar to the *activin βA* knockout mice, it was suggested that the maxillary molars do not use the activin signaling pathway during morphogenesis [72]. Activin A-releasing beads stimulate the expression of the TNF receptor *edar* in the dental epithelium, but this effect requires the presence of mesenchymal tissue and may thus be indirect [73].

19.2.3
Sonic Hedgehog

Hedgehog proteins are of pivotal importance for the development of multiple organs [74, 75]. There are three hedgehog genes in mammals – Sonic hedgehog (Shh), Desert hedgehog, and Indian hedgehog – of which only Shh is expressed during tooth development. Shh transmits its signal through a multipass transmembrane receptor complex Patched (Ptc) and Smoothened (Smo). Binding of Shh to Ptc releases its repression of Smo, thereby allowing Smo to activate its intracellular targets including Gli family zinc finger transcription factors. Ptc1 and Gli1 are also downstream target genes of Shh, and so their expression usually indicates active hedgehog signaling. Application of Shh protein to mandibular mesenchyme induces *Ptc1* expression [76]. Ptc1 as well as the hedgehog-interacting protein (Hip) also restrict Shh signaling by binding and sequestering Shh protein. Hip is a membrane-bound protein that binds all three hedgehogs [77]. As Hip can also be induced by Shh in the first branchial arch mesenchyme, it has been proposed to restrict Shh signaling within the tooth-forming region [78].

The expression of *Shh* in the developing tooth is restricted to the dental epithelium. It is initially expressed in the dental lamina before any morphological sign of tooth development, and then appears reiteratively in the epithelial signaling centers during successive stages of tooth morphogenesis. Based on the intense and restricted expression of Shh in the enamel knots, it has been speculated that Shh may be involved in the patterning of tooth cusps similar to the Shh in the zone of polarizing activity for digit patterning in the limb buds [10, 79]. Shh protein, as well as its signaling pathway genes, *Ptc1*, *Smo*, *Gli1*, *Gli2*, and *Gli3* have been localized in both dental epithelium and mesenchyme, indicating that Shh acts as a long-range signal affecting both dental epithelium and mesenchyme [80, 81]. *Ptc2* is expressed only in the epithelial enamel knot region [82]. In *Gli2* and *Gli3* double knockout mice, tooth development is arrested prior to the bud stage, indicating that Shh signaling is essential for bud formation [76]. The introduction of Shh protein with beads on oral epithelium stimulates epithelial proliferation and bud formation, and deletion of Shh function in teeth results in small tooth size (see below) [76, 81].

During the bell stage, *Shh* expression spreads from the secondary enamel knots to the inner dental epithelium. It is downregulated in the secretory ameloblasts, but continues in the adjacent stratum intermedium layer. Interestingly, *Ptc2* and *Gli1*, which are downstream target genes of Shh, exhibit polarized localization in the secretory ameloblasts with enriched expression in the basal and perinuclear compartment. When Shh signaling was deleted only in the dental epithelium (keratin14–smoothened knockout mice), odontoblasts and ameloblasts differentiated, but the dentin and enamel matrices were hypoplastic. It was therefore suggested that Shh is not required for cell differentiation, but may regulate the size, polarization, and/or secretion of ameloblasts [82].

19.2.4
Wnt Family

The vertebrate Wnt family contains at least 21 secreted glycoproteins that are involved in a variety of developmental processes, such as cell fate decision, cell proliferation, differentiation, and migration [83, 84]. Wnts signal by binding to the Frizzled family receptors. In the canonical β-catenin pathway, Wnts have an essential co-receptor, the low-density lipoprotein receptor-related protein (LRP), activating Disheveled in the cytoplasm and leading to the inactivation of the Axin-APC-GSK3 complex. This results in the stabilization of β-catenin, which translocates into the nucleus where it interacts with members of the Lef/TCF transcription factors and activates Wnt-responsive genes [85]. Notably, β-catenin is also a structural adaptor protein linking cadherins to the actin cytoskeleton during cell-cell adhesion. Cadherin may sequester β-catenin from the nucleus by binding it at the cell surface and thus Wnt signaling has been closely linked with cell adhesion [86]. Wnts also signal through alternative pathways, such as the planar cell polarity pathway where binding of Wnt to the Frizzled receptor stimulates intracellular Ca^{2+} release and activates PKC [87].

To date, seven Wnt genes have been reported in developing teeth including *Wnt3*, *–4*, *–5a*, *–6*, *–7b*, *–10a*, and *–10b*. Most of them are expressed in the epithelium, only *Wnt5a* is also localized in the mesenchyme [6, 8]. During the initiation of tooth development, *Wnt10a* and *Wnt10b* transcripts are restricted to the presumptive dental epithelium, whereas *Wnt3* and *Wnt7b* show complementary patterns and are expressed in the flanking oral epithelium. During morphogenesis, *Wnt10a* and *Wnt10b* are confined to the tip of the tooth bud and to the enamel knots. *Wnt3* is also expressed intensely in the enamel knot at the cap and early bell stages. *Wnt4* and *Wnt7b* are expressed in the whole enamel organ but are absent from the enamel knot region. By the early bell stage, *Wnt4* and *Wnt7b* expression are mostly downregulated, whereas *Wnt3* and *Wnt6* continue to be expressed in the inner dental epithelium. *Wnt1, –2, –8,* and *–11* transcripts were not detected in teeth [6, 8].

Of the 11 Frizzled receptors in mammals, only MFz6, –7, and –8 have been analyzed in the developing tooth. *MFz7* and *MFz8* were not detected whereas *MFz6* was initially expressed throughout the oral epithelium and later in the outer dental epithelium and in the enamel knot. MFrzb1 and Mfrp2 are soluble agonist/antagonists of Wnt receptors expressed mainly in the dental mesenchyme [8].

Although many Wnt knockout mice have been generated (*Wnt1, –2, –3, –3a, –4, –5a, –7a*), no tooth phenotypes have been reported. As *Wnts* are expressed in complementary and overlapping patterns in the dental tissues, they may have redundant functions and thus their precise roles remain unclear. The overall importance of the canonical Wnt signaling pathway for tooth development is implied by the phenotype of *Lef1* knockout mice, in which the development of multiple ectodermal appendages, including teeth, hair, and mammary glands is blocked [88]. *Lef1* mutant teeth are arrested at the late bud stage. Although *Lef1* is expressed in both dental epithelium and mesenchyme throughout tooth development, tissue recombination experiments using *Lef1* mutant and wild-type epithelium and mesenchyme indicated that Lef1 is required only transiently in the dental epithelium [89]. The function of Lef1 in the mesenchyme may be compensated by other LEF/TCF family transcription factors. In *Lef1* mutant teeth, *Fgf4* is absent in the enamel knot. Strikingly, exogenous FGF proteins were able to rescue mutant tooth development, suggesting that Lef1 activates *Fgf4* in the enamel knot and thereby directly connects Wnt and FGF signaling pathways during tooth morphogenesis [90]. Recent evidence has shown that overexpression of Dickkopf 1 (a diffusible inhibitor of Wnt) in the epidermis and dental epithelium results in a complete failure of hair placode formation and also blocks tooth and mammary gland development before the bud stage [91]. In addition, treating developing tooth germs with Mfrzb1 protein (inhibitor of Wnt signaling) results in small teeth with reduced cusps [92]. Furthermore, overexpression of *Lef1* in the epithelium under keratin 14 promoter causes increased oral epithelial invaginations and ectopic tooth-like structures in the oral cavity [93]. Taken together, these studies suggest that Wnt signals are important regulators of the initiation and morphogenesis of teeth.

During initiation of tooth development, *Wnt7b* is expressed throughout the oral epithelium but it is remarkably absent from *Shh*-expressing tooth-forming regions. Since ectopic and overexpression of *Wnt7b* in the dental epithelium represses *Shh* expression and prevents tooth bud formation, it was proposed that Wnt7b acts by restricting *Shh* expression to the tooth-forming regions so that Shh can only stimulate cell proliferation locally for the formation of a tooth bud. Hence, antagonistic interactions between Wnt7b and Shh may define boundaries of developing tooth germs [94]. *In vitro* studies have shown that *Wnt6*-expressing cells are able to stimulate the expression of the TNF family protein *ectodysplasin (Eda)* [73]. *Eda* null mutant (*Tabby*) mice exhibit smaller teeth with reduced cusps [95].

19.2.5
Tumor Necrosis Factors

The TNF superfamily consists of more than 20 members, most of which are important regulators of host defense, immune response, and inflammation [96, 97]. TNFs bind on the cell surface to trimeric TNF receptors, which recruit adapter proteins leading to the activation of NF-ϰB, c-JNK, or p38 pathways in the cytoplasm [98, 99]. TNF signaling has mostly been related to cell survival and apoptosis. The identification of ectodysplasin (Eda) as a novel TNF family member and its receptor Edar as a novel TNF receptor, was the first link between TNFs and the regulation of embryonic organogenesis [100–102].

Mutations in *EDA*, its receptor *EDAR*, and the intracellular adaptor *EDARADD* all cause similar hypohidrotic ectodermal dysplasia syndromes in humans (HED) characterized by impaired development of teeth, hair, and a set of exocrine glands, including sweat glands. The equivalent mutations in mice: *Tabby*, *downless*, and *crinkled* cause similar phenotypes [1, 103]. *Eda* and *edar* are co-expressed throughout the oral ectoderm prior to tooth initiation, but when the dental placode forms, *Edar* becomes restricted to the placode while *Eda* remains expressed in the flanking ectoderm. This complementary pattern continues during advancing tooth morphogenesis with *Eda* localized in the outer dental epithelium and *Edar* in the enamel knot [73]. Another TNF receptor, *TNFRSF19*, is expressed in overlapping domains with *Edar* in the developing teeth, as well as in mammary glands and whiskers, suggesting the redundant roles of these receptors [104]. Ectodysplasin has two splice variants, Eda-A1 and Eda-A2 that differ by only two amino acids but bind to distinct receptors, Edar and Xedar. There is evidence that Eda-A1 is the major regulator of ectodermal organogenesis, since transgenic expression of the Eda-A1 isoform can rescue the hair and sweat gland phenotypes in *Tabby* mice (loss-of-function of Eda, see below) [105, 106]. The roles of Eda-A2 and Xedar are not known yet.

In addition to the epithelial signaling centers in teeth, *Edar* is expressed in ectodermal placodes in other organs forming as ectodermal appendages. It is one of the earliest markers of forming ectodermal placodes and recent data indicate that it is a potent stimulator of placode formation (see below) [107]. The *Tabby* mice lack guard hairs forming from the first wave of hair follicles, and the placodes of these hairs are completely missing in the embryos. Incisors and third molars are sometimes missing in *Tabby* mice indicating a similar role of Eda in dental placode formation. *Tabby* mutant teeth exhibit obvious cusp patterning defects as the size and number of cusps are reduced and the buccal and lingual cusps are joined. This is caused by the small size of primary enamel knots leading to the fusion of secondary enamel knots [95]. Although Edar contains a death domain close to the C-terminus, apoptosis is not affected in the developing *Tabby* molars [108]. Exogenously added FGF10 protein can partially restore *Tabby* tooth morphogenesis and stimulates additional tooth cusp formation, suggesting that Eda may stimulate cell proliferation or cell survival. Overexpression of *Eda-A1* in the epidermis and oral epithelium in transgenic mice induces ectopic teeth and mammary glands, stimulates hair and nail growth, and also increases the activity of sweat glands. These observations suggest that Eda/Edar signaling is critical for the initiation, morphogenesis, as well as differentiation of ectodermal organs [1, 109].

Eda/Edar signaling is mediated by the IKK complex and mutations in *IKKγ* cause incontinentia pigmenti syndrome with similar ectodermal dysplasia phenotype as in HED [110]. Suppression of *NF-κB* in mice also causes ectodermal defects including a similar molar phenotype as in the *Tabby* mice [111]. Moreover, targeted deletion of *Ikkα*, another component of the TNF signaling pathway, results in *Tabby*-like molars; this indicates a potential role for Ikkα in the transmission of Edar signaling. However, their incisor phenotype is different and involves evagination of ectoderm implying other functions of IKKα [112].

19.2.6
Other Growth Factors

In addition to the molecules discussed above, several other growth factors are also important for normal tooth development. Notch signaling has been implicated in cell fate decisions and in the formation of tissue compartments during the development of insects and vertebrates [113–115]. The ligands of Notch receptors, Jagged (Serrate) and Delta-like molecules are also membrane bound, and so Notch signaling requires intimate cell-cell contacts. Binding of the ligand causes the release of the Notch intracellular domain which translocates into the nucleus activating the expression of target genes, including Hes family transcription factors. During tooth development, *Notch1, –2*, and *–3* are all expressed in the early oral epithelium and later within the dental epithelium, but interestingly are absent from the basal layer epithelial cells [116]. The Notch ligand *Jagged1* is expressed similarly to *Notch* in the epithelium, and also in the dental mesenchyme during the bud stage. The other ligand, *Delta1*, is weakly expressed at the early stage. During cytodifferentiation, *Delta1* is intensely upregulated in the odontoblasts and ameloblasts, while *Notch* transcripts are expressed in the immediately adjacent cell layers. It has therefore been speculated that Notch-Delta signaling may be involved in the determination of odontoblasts and ameloblasts [117]. Lunatic fringe (L-fng) is a modulator of Notch signaling and in mouse molar teeth it is intensely expressed in the epithelial cervical loops with sharp boundaries around the enamel knot. However, tooth morphogenesis is not affected in the *L-fng* knockout mice [115]. *Hes1* is expressed in both dental epithelium and mesenchyme, and L-fng in addition to BMP4 and FGF10 proteins, is able to induce its expression in the dental epithelium [115]. Notch signaling has also been implicated in the maintenance of epithelial stem cell niche in the continuously growing mouse incisors and vole molars [24,30].

Epidermal growth factor (EGF) and Nerve growth factor (NGF) were among the first growth factors analyzed in tooth development, but their precise roles remain unclear. The expression of *Egf* and *Ngf*, as well as their receptors has been associated with epithelial-mesenchymal interactions [118–121]. Exogenously-added EGF affects tooth morphogenesis and prevents cell differentiation in early mouse tooth germs [122]. EGF can also help to maintain the dental follicle in cultured tooth germs which may be due to its anti-apoptotic activity [12, 122]. NGF and other neurotrophic factors including BDNF and GDNF have also been suggested to regulate tooth morphogenesis [121].

Insulin-like growth factor–1 (IGF–1) expression has been detected in secretory and maturation-stage ameloblasts. IGF–1 receptor is only expressed in secretory ameloblasts, but downregulated in the transitional zone ameloblasts (between secretory and maturation stages). This downregulation of IGF–1 receptor may be associated with apoptosis of matured ameloblasts [123, 124]. Parathyroid hormone-related peptide (PTHrP) is expressed in the dental epithelium and its receptor in the mesenchyme. Conditional deletion of PTHrP function in dental epithelium in mice does not affect tooth morphogenesis but it prevents tooth eruption into the oral cavity [125]. This is caused by the lack of osteoclasts and failure of alveolar bone resorption. PTHrP stimulates osteoclast differentiation around the developing teeth and thereby protects them from bone invasion and allows tooth eruption [125, 126].

19.2.7
Integrations between Growth Factor Signal Pathways

The growth factor signaling pathways are integrated during tooth development at various levels leading to synergistic or counteractive effects [4]. In many cases, FGF and BMP signaling affect the expression of the same transcription factors, in particular during tooth initiation and early morphogenesis. For example, FGFs and BMPs can both induce *Msx1, Dlx2,* and *Hes1* expression in dental tissues [34, 115]. However, mostly they have antagonistic functions. FGFs stimulate and BMPs inhibit the expression of *Pax9* and *Barx1* in the dental mesenchyme [127, 128]. During initiation of tooth development, *Fgf8* and *Bmp4* are expressed in partially overlapping regions in the facial process. The antagonistic interactions between FGF8 and BMP4 restrict *Barx1* expression to the molar-forming region and may thus determine the identity of incisors versus molars [128]. In the dental epithelium, FGFs induce the expression of *Pitx2* and *L-fng,* and these effects can be counteracted by BMPs [129].

Many other signal pathways also integrate. As mentioned above, antagonistic interactions between Wnt and Shh signaling have been implicated in defining the position of tooth initiation [94], and Wnt regulates *Fgf4* expression in the primary enamel knot through Lef1 [90]. In addition, FGFs induce in the dental epithelium the expression of *L-fng*, which can modulate Notch signaling [24, 115]. TNF signaling is integrated with both Wnt and activin pathways, since Wnts induce *Eda* expression, while activin induces its receptor *Edar* in the dental epithelium (19.2.5) [73]. An example of the integration of three major pathways is the counteractive interactions between FGF, Shh and BMP in regulating the expression of the BMP inhibitor *ectodin* around the enamel knot (Fig. 19.4) [62]. Moreover, the formation and signaling functions of the ectodermal signaling centers, the dental placodes and enamel knots involve simultaneous activities of several stimulatory and inhibitory signals (see below). It appears that a spatially and temporally orchestrated interplay between different signaling pathways is required for successful tooth formation (Fig. 19.5).

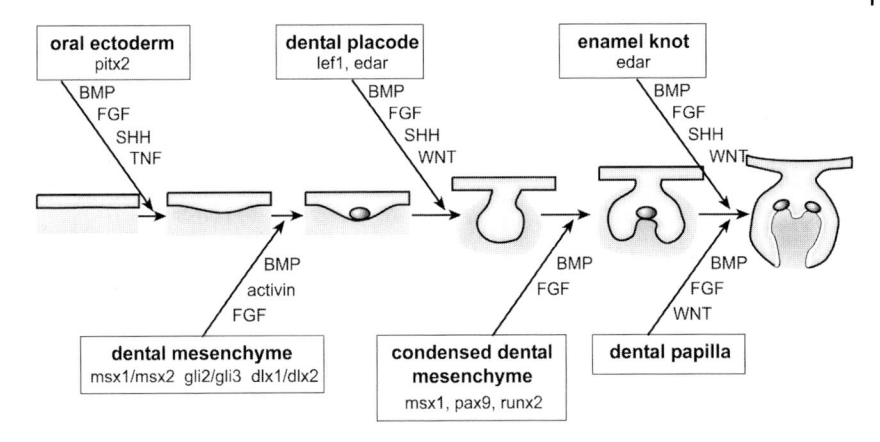

Fig. 19.5
Schematic representation of signaling networks regulating advancing tooth morphogenesis. Arrows indicate signals originating from the dental epithelium (above) or the dental mesenchyme (below). Several signals are used repeatedly at many stages. The genes in the boxes have been shown to be necessary for tooth development as indicated in knockout mice.

19.3
Biological Effects

It is evident that growth factors in different families regulate tooth formation during all stages of development. Most of the individual signals may affect a variety of cellular functions depending on the context and, on the other hand, signals in different families may also affect the same cellular functions. A good example of a signal with many important functions during tooth development is BMP4: its interaction with FGF8 regulates early proximo-distal patterning of the jaws and perhaps tooth position and identity [127, 130]. During early tooth morphogenesis, *Bmp4* expression shifts with the odontogenic potential from the dental epithelium to the mesenchyme. Epithelial BMP4 induces in the mesenchyme the expression of *Msx1* and *Msx2* as well as its own expression, and subsequently BMP4 from the dental mesenchyme induces in the epithelium *p21* expression which is associated with the formation of the enamel knot [55, 68]. Later, *Bmp4* is also expressed intensely in the primary and secondary enamel knots and has been proposed to regulate apoptosis, but its precise role there remains unknown. During cytodifferentiation, *Bmp4* is expressed in the odontoblasts and it is able to induce the expression of *p21* and *ameloblastin* in the overlying dental epithelium, thus triggering terminal differentiation of the ameloblasts (19.3.4) [131].

In the following, we present examples where the biological effects of individual growth factors have been clarified during specific developmental processes of tooth formation.

19.3.1
Formation of Dental Placodes

The molecular mechanisms of ectodermal placode formation have been explored in great detail during feather and hair development, and it is known that the process is regulated by the combined actions of several signals, some of which act as stimulators and some as inhibitors of placode formation [1]. Wnts and FGFs in particular stimulate whereas BMPs inhibit placode development. The molecular regulation of dental placodes has only been studied recently, and it is evident that the same signal pathways as those in the development of other ectodermal organs are involved.

P63 is a transcription factor expressed throughout the surface ectoderm and is required for the formation of placodes of all ectodermal appendages. Mutations of the *p63* gene in humans cause ectodermal dysplasia syndromes [132], and its targeted deletion in mice results in a total failure of ectodermal appendage development together with the absence of dental and hair placodes [133]. *In situ* hybridization analysis of *p63* knockout embryos has demonstrated that *Bmp7, Fgfr2, Jagged1,* and *Notch1* transcripts are absent in the mutant ectoderm, and *β–catenin* and *Edar* expression is dramatically reduced, whereas a large number of other candidate target genes are not affected [134]. p63 is apparently required for the mediation of signals in several different families, and thus placode formation requires coordinated interactions between multiple pathways.

Eda/Edar signaling (see Section 19.2.5, TNF pathway) is the most recent pathway to be analyzed in this context and it is now known to function as a stimulator of ectodermal placode formation. Overexpression of *Eda* in transgenic mice (*K14–Eda*) induces the formation of extra teeth in front of the first molars (Fig. 19.6) [109]. This was shown to result from stimulation of dental placode development in this location. However, a rudimentary placode was also detected in the same location in the wild-type embryo, indicating that Eda does not induce new placode formation and may instead stimulate the growth or survival of already initiated placodes. Careful analysis of these and other placodes in transgenic mice, as well as *in vitro* experiments where recombinant Eda protein was applied on early skin, confirmed that Eda indeed does not regulate placode patterning nor stimulate placodal cell proliferation, but it promotes the placodal cell fate and enhances the incorporation of surrounding cells into the growing placodes [107].

19.3.2
Epithelial Cell Proliferation

Cell proliferation is an integral part of development and practically all growth factors have been shown to have stimulatory effects on cell division. During tooth development such roles have been demonstrated for FGFs and Shh. FGF proteins stimulate proliferation in both dental epithelium and dental mesenchyme when applied with beads to isolated tissue explants [9, 26]. The distribution of FGF receptors correlates with active cell proliferation in the epithelial cervical loop regions, and Fgfs ex-

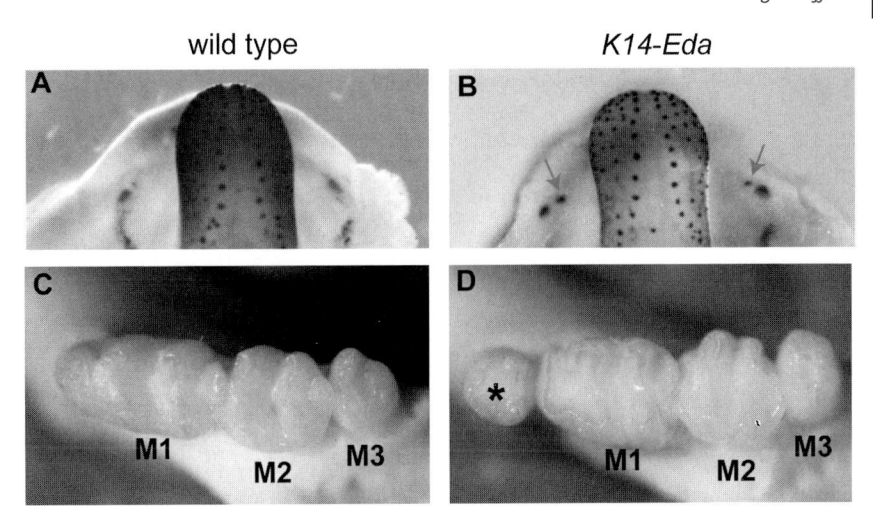

Fig. 19.6
Excess ectodysplasin stimulates dental placode formation and induces the formation of extra teeth. (A, B) When *ectodysplasin* is overexpressed in the dental epithelium (*K14–Eda* transgenic mice), large dental placodes (arrows) are evident in front of the molar placode, as indicated by intense *Shh* expression. (C, D) An extra tooth (*) has formed from this placode in front of the first molar in *K14–Eda* mice. (Courtesy of Aapo T. Kangas). (This figure also appears with the color plates.)

pressed in the epithelial signaling centers as well as in the dental mesenchyme are believed to play pivotal roles in the growth and folding of dental epithelium [23].

Shh has been associated in particular with the proliferation of dental epithelial cells. Local application of Shh with beads to embryonic jaws stimulates epithelial proliferation and induces ectopic epithelial invaginations [76]. Blocking Shh activity with antibodies in the early mouse jaw leads to arrested tooth development [135]. Deletion of *Gli2* and *Gli3* function in double knockout mice arrests tooth development prior to the bud stage [76], and conditional inactivation of Shh function later in tooth development under the keratin 14 promoter results in severe growth retardation [81]. The overall size of the mutant teeth is reduced and in particular the lingual cervical loops fail to grow into the mesenchyme. However, the effect of Shh on the growth of lingual dental epithelium may not be direct, since conditional deletion of Shh signaling only in the dental epithelium (*K14–Smo* knockout mice) does not result in lingual cervical defects [82]. Hence, Shh apparently acts on the underlying dental mesenchyme and regulates the production of reciprocal signals, perhaps FGFs, stimulating epithelial proliferation. The Shh receptors Ptc1 and Ptc2 as well as the Gli signal mediators are expressed in both epithelium and mesenchyme indicating that Shh affects both tissues. It is currently unclear whether Shh affects epithelial cell proliferation directly or indirectly through stimulation of other growth factors.

19.3.3
Patterning of the Tooth Crown

The species-specific cusp patterns in mammalian teeth are thought to be determined by the position, time, and order of appearance of secondary enamel knots [136]. A network model has recently been proposed which is based on activators (such as FGFs and Shh) and inhibitors (such as BMPs) from the primary enamel knot, and it can reproduce the patterns of secondary enamel knots as well as species-specific tooth morphologies when the parameters of the model are changed [3, 13]. This model is supported by the observations that molar morphologies are altered in several mutant mice where specific signal pathways are disturbed. For instance, in the *Tabby* mutants (Eda loss-of-function), molar crowns are small and flattened with fused and fewer cusps. Correspondingly, in *Tabby* embryos, the primary enamel knot is small and most secondary enamel knots are fused. Lack of Eda/Edar signaling appears to lead to the inhibition of the formation of the primary enamel knots in a manner similar to that of the dental placodes, and reduced signaling from the enamel knots apparently results in the aberrant cusp pattern. On the other hand, increased expression of *Eda* in transgenic mice causes increased tooth size and wide cusps (Fig. 19.6). It thus appears that the level of ectodysplasin signaling is closely associated with tooth cusp patterning [95, 137].

The central role of TGFβ/BMP signaling in cusp patterning has also become evident from the aberrant shapes of the molars in *follistatin* transgenic mice. Decreasing or increasing the intensity of activin/BMP signals by overexpression or loss of function of follistatin apparently leads to the imbalance between activators and inhibitors of normal enamel knot formation and results in disturbed patterning of the cusps [61]. The importance of fine-tuning the activity of growth factors for the creation of the correct tooth morphology is also evidenced by the severe defects in cusp patterning when the function of the BMP inhibitor, ectodin, is deleted [63]. However, the effects of ectodin and its ablation may be more complicated since ectodin also functions as a modulator of Wnt signaling and since there are several Wnts expressed in the enamel knot [138].

19.3.4
Cell Differentiation

The differentiation of odontoblasts and ameloblasts is regulated by reciprocal epithelial-mesenchymal interactions, and the same signal molecules that are used for morphogenetic regulation have also been associated with dental cell differentiation. Signals in the TGFβ and FGF families have mostly been implicated in odontoblast differentiation [53,139,140]. *In vitro* experiments have shown that TGFβ1, −3, and BMP2, −4, −6 can induce polarization of preodontoblasts and stimulate dentin matrix secretion [141–144]. FGF1 protein does not stimulate odontoblast differentiation alone, but it can act in a synergistic manner with TGFβ1 in the induction of functional odontoblasts [145]. BMPs induce odontoblast differentiation and dentin matrix production *in vivo* and they have been used for dentin regeneration (see below).

Our recent studies indicate that BMP4 is a major inducer of ameloblast differentiation in mouse incisors. *Bmps*, in particular *Bmp4*, are expressed in the dental papilla mesenchyme and odontoblasts. BMP-releasing beads induced intense expression of the enamel protein *ameloblastin* in the dental epithelium, whereas beads releasing the BMP inhibitor Noggin dramatically downregulated endogenous expression of *ameloblastin* (Fig. 19.7) [63, 131]. Noggin is not expressed in the developing mouse incisors, but another BMP inhibitor, follistatin is expressed in the lingual dental epithelium which does not differentiate into ameloblasts. Follistatin was downregulated in the labial dental epithelium when ameloblasts started to differentiate. We showed that the asymmetrical expression of follistatin and its function as a BMP inhibitor are responsible for the labial-lingual asymmetry of enamel formation in mouse incisors. In transgenic mice which overexpress follistatin throughout the dental epithelium, ameloblast differentiation was completely inhibited and no enamel formed (Fig. 19.3), whereas in follistatin knockout mouse incisors, ameloblasts differentiated ectopically on the lingual surface which is normally enamel free. We also showed that activin, which is mainly expressed in the surrounding dental follicle mesenchyme, induces follistatin expression in the dental epithelium. Hence, ameloblast differentiation is regulated by the antagonistic activities of activin and BMPs emanating from two different mesenchymal cell layers flanking the dental epithelium (Fig. 19.7E) [63, 131].

19.4
Biomedical Application

The new information gathered from studies on the roles of growth factors in tooth development has greatly expanded our understanding of the basic mechanisms of morphogenesis and of the biological functions of various molecules. The goal obviously is to use some of this information in the future for the diagnosis, prevention, and for new therapies in the clinic. The first example of successful translation of genetic and molecular information into clinic comes from treating *Tabby* mice with recombinant EDA-A1 protein. *EDA* was first identified as the gene behind X-linked hypohidrotic ectodermal dysplasia (HED) by positional cloning [100], and the EDA-A1 splice form was found to be the major form regulating ectodermal organogenesis. The spontaneous mouse mutant *Tabby* was shown to be an *Eda* null mutant and was rescued by transgenic expression of Eda-A1 cDNA [146, 147]. Based on this, Gaide and Schneider injected EDA-A1 protein, which had been engineered to cross the placental barrier, intravenously into pregnant *Tabby* mice. The *Tabby* phenotype was rescued resulting in the formation of hairs and sebaceous and sweat glands together with normally shaped teeth. This correction persisted throughout adulthood, indicating that EDA-A1 is only required during embryogenesis for the early development of epidermal structures, but not for maintaining them in adulthood [106]. These studies have opened up new opportunities for the prevention and treatment of HED in human patients.

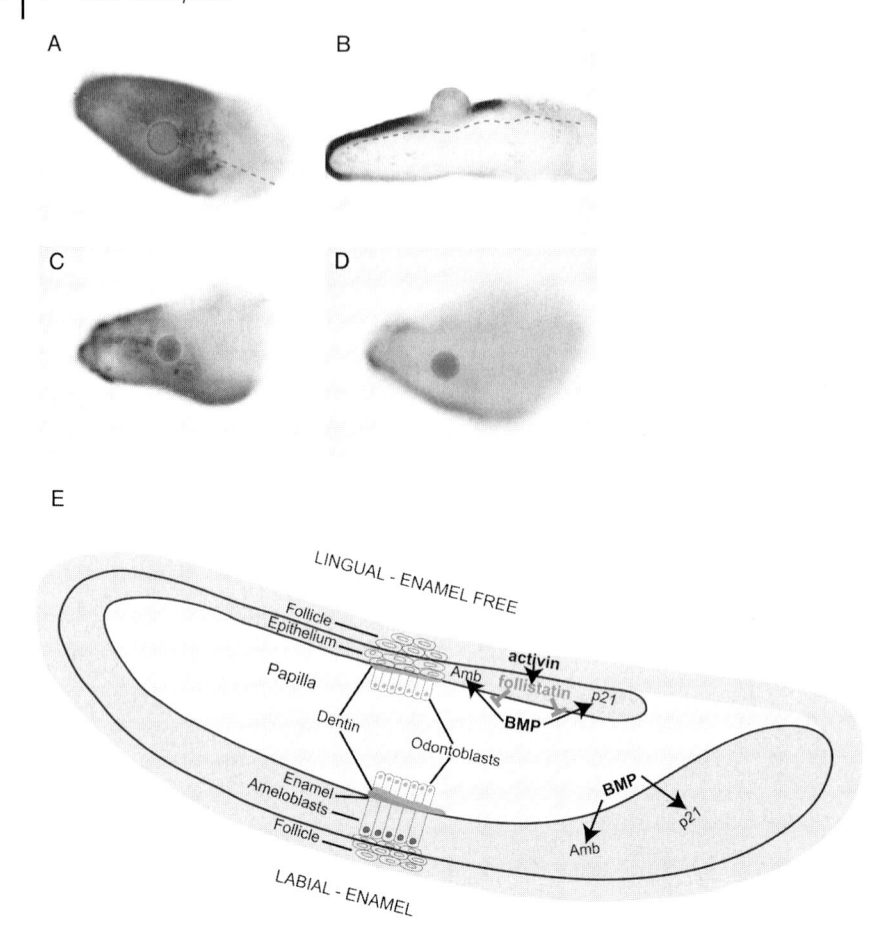

Fig. 19.7
BMP4 is the major inducer of ameloblast differentiation. (A) A BMP4–releasing bead has induced intense *ameloblastin* expression in the mouse incisor around the bead. (B) Corresponding vibratome section showing the induced ameloblastin in the dental epithelium. (C) BSA control bead did not induce ameloblastin (only endogenous expression is seen). (D) The BMP inhibitor noggin has dramatically downregulated endogenous *ameloblastin* expression. (E) Schematic representation of the signaling networks regulating ameloblast differentiation. BMPs are expressed in the odontoblasts and they induce ameloblast differentiation in the overlying dental epithelium, as indicated by the induction of *p21* and ameloblastin. Activin from the dental follicle mesenchyme induces the expression of *follistatin* in dental epithelium, which in turn inhibits the ameloblast-inducing activity of BMPs from the underlying odontoblasts and prevents enamel formation. On the labial surface, follistatin is downregulated in the dental epithelium allowing BMPs from the odontoblasts to induce ameloblast differentiation.

With the recent advances in stem cell biology, tooth bioengineering has also become a topic of interest [21, 148]. However, at present, tooth regeneration is still very challenging. That teeth can form under proper conditions from multipotential stem cells is evidenced by the occurrence of mineralized teeth with specific shapes in teratomas (germ cell tumors). Experiments by embryologists decades ago have also shown that recombination of dissociated cells from embryonic tooth germs can lead to new teeth when cultured *in vitro* or transplanted *in vivo*, suggesting that odontogenic capacity already exists very early in the dental tissues and that tooth development is not dependent on environmental factors, such as jaw bone [17–19]. However, to initiate the tooth program, the correct signals must be present at the right place to create a proper stem cell niche [148]. Although the early embryonic jaw epithelium has the capacity to instruct tooth formation [17,20,21], experimental creation of such a niche in a controlled reproducible manner will be extremely difficult. Further challenges such as the development of correct tooth sizes and shapes, well mineralized and correctly-colored hard tissues, and functioning dental attachment to alveolar bone also need to be addressed before functional teeth can be generated in the oral cavity. More information to solve the problems of tooth regeneration may come from studies on physiological tooth replacement. Interestingly, in the human patients with cleidocranial dysplasia syndrome (CCD, caused by mutations in the *RUNX2* gene), multiple supernumerary teeth form in the jaws and sometimes they even constitute an almost complete third dentition [149–151]. This indicates that humans still have the potential to develop new dentition. The opposite phenotype to CCD is seen in human patients with *AXIN2* mutations. Their deciduous teeth develop normally, but many permanent teeth fail to form indicating a problem in successional tooth formation (Fig. 19.8) [152]. These observations suggest that although Runx2 and Wnt signaling are necessary for tooth morphogenesis (as indicated by the lack of teeth in *Runx2* and *Lef1* null mutant mice) [38, 89], they suppress the formation of successional teeth. Further analysis of the functions of Wnt signals and Runx2 during tooth replacement may shed light on tooth regeneration.

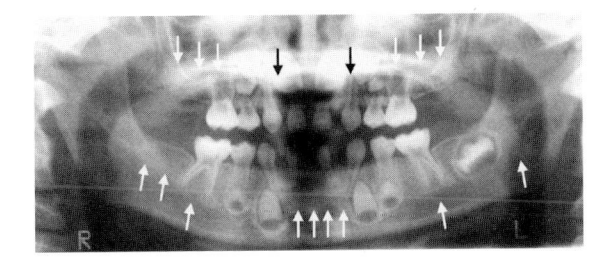

Fig. 19.8
A mutation in the Wnt signal modulator *AXIN2* inhibits the formation of successional teeth. Altogether 17 permanent teeth are missing (arrows) in this patient while the deciduous teeth had developed normally. (Courtesy of Sirpa Arte and Pekka Nieminen).

Experimental generation of whole teeth seems to still have a long way to go, and even if accomplished these treatments would probably be too expensive and take too long a time to replace the already available dental implants and prostethic crowns. On the other hand, the regeneration of parts of teeth may be a more realistic goal for the future. Recombinant human BMP proteins induce new dentin formation in the dental pulp of experimental animals [153, 154], and hence they are also potent stimulators of reparative dentin in human teeth. At present, BMPs are being widely used in clinic to repair bone defects and their potential is being explored in improving the integration of dental implants by stimulating bone growth in and around the implants [155]. An intriguing challenge is the regeneration of periodontal tissues by using stem cells [156] and/or growth factors to improve the treatment of periodontal disease and also to generate a physiological attachment between dental implants and alveolar bone. Looking ahead, the increasing understanding of the biological functions of growth factors in tooth development can be expected to lead to new applications of tissue engineering in the dental clinic.

19.5
Summary

The developing tooth has proven to be a powerful model for studying the roles of growth factors in organogenesis. Since the early 1990s, there have been dramatic advances in our understanding of the genetic control of tooth development and, in particular, the roles of growth factor signaling. Many signals belonging to a number of conserved growth factor families have been found to be reiteratively used during advancing tooth development. The early stages of tooth development resemble morphologically as well as molecularly, those of other organs forming as ectodermal appendages. During the initiation of all ectodermal organs, the same signals such as Wnts, Shh, BMPs, FGFs, Notch and the TNF signal ectodysplasin regulate the formation and function of epithelial placodes. The dental placodes as well as the enamel knots which form in the dental epithelium during advancing morphogenesis are signaling centers expressing multiple growth factors both transiently and locally. Signals in the FGF, Notch ligand and BMP families have been associated with the maintenance of epithelial stem cell niche in mouse incisors, as well as with the differentiation of the cells forming the dental hard tissues. So far, no tooth-specific regulatory genes have been identified, and most of the genes expressed in odontoblasts and ameloblasts have functions in many other tissues. It appears that complex signaling networks involving multiple pathways regulate the correct formation of teeth at the right time and in the right place, although the functions of some genes are redundant. The signals may interact with each other and act synergistically or antagonistically. It has become apparent that the restriction and modulation of the activities of signals by specific inhibitors are also essential for normal tooth development. For example, excessive Wnt signaling as well as loss of Wnt function can both lead to missing teeth [89, 157]. Also, BMP signaling is required for tooth initiation, ameloblast differentiation, and the patterning of the tooth crown, but it must be locally modulated to ascertain correct patterning and morphogenesis [61]. Development of the right number, shape, and structure of teeth thus requires fine-tuning of the effects of the growth factors. At present, systemic approaches and computer modeling are being applied to recapitulate the complex signaling networks accounting for different tooth morphologies [13]. Continuing studies on tooth development will advance our understanding of the regulation of organogenesis and may also lead to translation of the knowledge into dental tissue engineering as well as to the prevention and treatment of dental defects in the clinic.

References

1 Pispa, J. and Thesleff, I. Mechanisms of ectodermal organogenesis. *Dev Biol* **2003**; *262*: 195–205.
2 Thesleff, I. and Nieminen, P. Tooth morphogenesis and cell differentiation. *Curr Opin Cell Biol* **1996**; *8*: 844–850.
3 Jernvall, J. and Thesleff, I. Reiterative signaling and patterning during mammalian tooth morphogenesis. *Mech Dev* **2000**; *92*: 19–29.
4 Thesleff, I. and Mikkola, M.L. The role of growth factors in tooth development. *Int Rev Cytol* **2002**; *217*: 93–135.
5 Åberg, T., Wozney, J. and Thesleff, I. Expression patterns of bone morphogenetic proteins (Bmps) in the developing mouse tooth suggest roles in morphogenesis and cell differentiation. *Dev Dyn* **1997**; *210*: 383–396.
6 Dassule, H. R. and McMahon, A.P. Analysis of epithelial-mesenchymal interactions in the initial morphogenesis of the mammalian tooth. *Dev Biol* **1998**; *202*: 215–227.
7 Keränen, S.V., Åberg, T., Kettunen, P., Thesleff, I. and Jernvall, J. Association of developmental regulatory genes with the development of different molar tooth shapes in two species of rodents. *Dev Genes Evol* **1998**; *208*: 477–486.
8 Sarkar, L. and Sharpe, P.T. Expression of Wnt signalling pathway genes during tooth development. *Mech Dev* **1999**; *85*: 197–200.
9 Jernvall, J., Kettunen, P., Karavanova, I., Martin, L. B. and Thesleff, I. Evidence for the role of the enamel knot as a control center in mammalian tooth cusp formation: non-dividing cells express growth stimulating Fgf-4 gene. *Int J Dev Biol* **1994**; *38*: 463–469.
10 Vaahtokari, A., Åberg, T., Jernvall, J., Keränen, S. and Thesleff, I. The enamel knot as a signaling center in the developing mouse tooth. *Mech Dev* **1996**; *54*: 39–43.
11 Kettunen, P. and Thesleff, I. Expression and function of FGFs–4, –8, and –9 suggest functional redundancy and repetitive use as epithelial signals during tooth morphogenesis. *Dev Dyn* **1998**; *211*: 256–268.
12 Vaahtokari, A., Åberg, T. and Thesleff, I. Apoptosis in the developing tooth: association with an embryonic signaling center

and suppression by EGF and FGF–4. *Development* **1996**; *122*: 121–129.

13 Salazar-Ciudad, I. and Jernvall, J. A gene network model accounting for development and evolution of mammalian teeth. *Proc Natl Acad Sci USA* **2002**; *99*: 8116–8120.

14 Nanci, A. (Ed.), *Ten Cate's Oral Histology: Development, Structure, and Function.* Mosby: Missouri, **2003**.

15 Tureckova, J., Sahlberg, C., Åberg, T., Ruch, J.V., Thesleff, I. and Peterkova, R. Comparison of expression of the msx–1, msx–2, BMP–2 and BMP–4 genes in the mouse upper diastemal and molar tooth primordia. *Int J Dev Biol* **1995**; *39*: 459–468.

16 Keränen, S.V., Kettunen, P., Åberg, T., Thesleff, I. and Jernvall, J. Gene expression patterns associated with suppression of odontogenesis in mouse and vole diastema regions. *Dev Genes Evol* **1999**; *209*: 495–506.

17 Lumsden, A.G. Spatial organization of the epithelium and the role of neural crest cells in the initiation of the mammalian tooth germ. *Development Suppl.* **1988**; *103*: 155–169.

18 Slavkin, H.C., Beierle, J. and Bavetta, L.A. Odontogenesis: cell-cell interactions *in vitro*. *Nature* **1968**; *217*: 269–270.

19 Kollar, E. J. and Baird, G.R. The influence of the dental papilla on the development of tooth shape in embryonic mouse tooth germs. *J Embryol Exp Morph* **1969**; *21*: 131–148.

20 Mina, M. and Kollar, E.J. The induction of odontogenesis in non-dental mesenchyme combined with early murine mandibular arch epithelium. *Arch Oral Biol* **1987**; *32*: 123–127.

21 Ohazama, A., Modino, S.A., Miletich, I. and Sharpe, P.T. Stem-cell-based tissue engineering of murine teeth. *J Dent Res* **2004**; *83*: 518–522.

22 Ornitz, D. and Itoh, N. Fibroblast growth factors. *Genome Biol* **2001**; *2*: 3005.1–3005.12.

23 Kettunen, P., Laurikkala, J., Itäranta, P., Vainio, S., Itoh, N. and Thesleff, I. Associations of FGF–3 and FGF–10 with signaling networks regulating tooth morphogenesis. *Dev Dyn* **2000**; *219*: 322–332.

24 Harada, H., Kettunen, P., Jung, H.S., Mustonen, T., Wang, Y. A. and Thesleff, I. Localization of putative stem cells in dental epithelium and their association with Notch and FGF signaling. *J Cell Biol* **1999**; *147*: 105–120.

25 Harada, H., Toyono, T., Toyoshima, K., Yamasaki, M., Itoh, N., Kato, S., Sekine, K. and Ohuchi, H. FGF10 maintains stem cell compartment in developing mouse incisors. *Development* **2002**; *129*: 1533–1541.

26 Kettunen, P., Karavanova, I. and Thesleff, I. Responsiveness of developing dental tissues to fibroblast growth factors – expression of splicing alternatives of FGFR1, –2, –3, and of FGFR4 – and stimulation of cell proliferation by FGF–2, –4, –8, and –9. *Dev Genet* **1998**; *22*: 374–385.

27 Shannon, J. M. and Hyatt, B.A. Epithelial-mesenchymal interactions in the developing lung. *Annu Rev Physiol* **1904**; *66*: 625–645.

28 Celli, G., Larochelle, W.J., Mackem, S., Sharp, R. and Merlino, G. Soluble dominant-negative receptor uncovers essential

roles for fibroblast growth factors in multi-organ induction and patterning. *EMBO J* **1998**; *17*: 1642–1655.

29. De Moerlooze, L., Spencer-Dene, B., Revest, J., Hajihosseini, M., Rosewell, I. and Dickson, C. An important role for the IIIb isoform of fibroblast growth factor receptor 2 (FGFR2) in mesenchymal-epithelial signalling during mouse organogenesis. *Development* **2000**; *127*: 483–492.

30. Tummers, M. and Thesleff, I. Root or crown: a developmental choice orchestrated by the differential regulation of the epithelial stem cell niche in the tooth of two rodent species. *Development* **2003**; *130*: 1049–1057.

31. Ornitz, D. M. FGFs, heparan sulfate and FGFRs: complex interactions essential for development. *Bioessays* **2000**; *22*: 108–112.

32. Thesleff, I., Jalkanen, M., Vainio, S. and Bernfield, M. Cell surface proteoglycan expression correlates with epithelial-mesenchymal interaction during tooth morphogenesis. *Dev Biol* **1988**; *129*: 565–572.

33. Vainio, S., Jalkanen, M., Vaahtokari, A., Sahlberg, C., Mali, M., Bernfield, M. and Thesleff, I. Expression of syndecan gene is induced early, is transient, and correlates with changes in mesenchymal cell proliferation during tooth organogenesis. *Dev Biol* **1991**; *147*: 322–333.

34. Bei, M. and Maas, R. FGFs and BMP4 induce both Msx1–independent and Msx1–dependent signaling pathways in early tooth development. *Development* **1998**; *125*: 4325–4333.

35. Stepp, M.A., Gibson, H.E., Gala, P.H., Iglesia, D.D., Pajoohesh-Ganji, A., Pal-Ghosh, S., Brown, M., Aquino, C., Schwartz, A.M., Goldberger, O., Hinkes, M. T. and Bernfield, M. Defects in keratinocyte activation during wound healing in the syndecan–1-deficient mouse. *J Cell Sci* **2002**; *115*: 4517–4531.

36. Hacohen, N., Kramer, S., Sutherland, D., Hiromi, Y. and Krasnow, M. A. sprouty encodes a novel antagonist of FGF signaling that patterns apical branching of the Drosophila airways. *Cell* **1998**; *92*: 253–263.

37. Placzek, M. and Skaer, H. Airway patterning: A paradigm for restricted signalling. *Curr Biol* **1999**; *9*: R506–R510.

38. Åberg, T., Wang, X.P., Kim, J.H., Yamashiro, T., Bei, M., Rice, R., Ryoo, H. M. and Thesleff, I. Runx2 mediates FGF signaling from epithelium to mesenchyme during tooth morphogenesis. *Dev Biol* **2004**; *270*: 76–93.

39. Feldman, B., Poueymirou, W., Papaioannou, V.E., DeChiara, T. M. and Goldfarb, M. Requirement of FGF–4 for postimplantation mouse development. *Science* **1995**; *267*: 246–249.

40. Meyers, E.N., Lewandoski, M. and Martin, G.R. An Fgf8 mutant allelic series generated by Cre- and Flp-mediated recombination. *Nat Genet* **1998**; *18*: 136–141.

41. Colvin, J.S., Green, R.P., Schmahl, J., Capel, B. and Ornitz, D.M. Male-to-female sex reversal in mice lacking fibroblast growth factor 9. *Cell* **2001**; *104*: 875–889.

42. Trump, A., Depew, M.J., Rubenstein, J.L., Bishop, J. M. and Martin, G.R. Cre-mediated gene inactivation demonstrates

that FGF8 is required for cell survival and patterning of the first branchial arch. *Genes Dev* **1999**; *13*: 3136–3148.
43 Martin, G. Making a vertebrate limb: new players enter from the wings. *Bioessays* **2001**; *23*: 865–868.
44 Satokata, I. and Maas, R. Msx1 deficient mice exhibit cleft palate and abnormalities of craniofacial and tooth development. *Nat Genet* **1994**; *6*: 348–356.
45 Matzuk, M.M., Kumar, T.R., Vassalli, A., Bickenbach, J.R., Roop, D.R., Jaenisch, R. and Bradley, A. Functional analysis of activins during mammalian development. *Nature* **1995**; *374*: 354–356.
46 Peters, H., Neubüser, A., Kratochwil, K. and Balling, R. Pax9–deficient mice lack pharyngeal pouch derivatives and teeth and exhibit craniofacial and limb abnormalities. *Genes Dev* **1998**; *12*: 2735–2747.
47 Ferguson, C.A., Tucker, A.S., Christensen, L., Lau, A.L., Matzuk, M. M. and Sharpe, P.T. Activin is an essential early mesenchymal signal in tooth development that is required for patterning of the murine dentition. *Genes Dev* **1998**; *12*: 2636–2649.
48 Vastardis, H., Karimbux, N., Guthua, S.W., Seidman, J. G. and Seidman, C. E. A human MSX1 homeodomain missense mutation causes selective tooth agenesis. *Nat Genet* **1996**; *13*: 417–421.
49 Stockton, D.W., Das, P., Goldenberg, M., D'Souza, R. N. and Patel, P.I. Mutation of PAX9 is associated with oligodontia. *Nat Genet* **2000**; *24*: 18–19.
50 Mansour, S.L., Goddard, J. M. and Capecchi, M.R. Mice homozygous for a targeted disruption of the proto-oncogene int–2 have developmental defects in the tail and inner ear. *Development* **1993**; *117*: 13–28.
51 Massague, J. How cells read TGF-beta signals. (Review 109 refs). *Nat Rev Mol Cell Biol* **2000**; *1*: 169–178.
52 Balemans, W. and Van Hul, W. Extracellular Regulation of BMP Signaling in Vertebrates: A cocktail of Modulators. *Dev Biol* **2002**; *250*: 231–250.
53 Vaahtokari, A., Vainio, S. and Thesleff, I. Associations between transforming growth factor beta 1 RNA expression and epithelial-mesenchymal interactions during tooth morphogenesis. *Development* **1991**; *113*: 985–994.
54 Pelton, R.W., Dickinson, M.E., Moses, H. L. and Hogan, B.L. In situ hybridization analysis of TGF beta 3 RNA expression during mouse development: comparative studies with TGF beta 1 and beta 2. *Development* **1990**; *110*: 609–620.
55 Vainio, S., Karavanova, I., Jowett, A. and Thesleff, I. Identification of BMP–4 as a signal mediating secondary induction between epithelial and mesenchymal tissues during early tooth development. *Cell* **1993**; *75*: 45–58.
56 Heikinheimo, K., Bègue-Kirn, C., Ritvos, O., Tuuri, T. and Ruch, J.V. The activin-binding protein follistatin is expressed in developing murine molar and induces odontoblast-like cell differentiation *in vitro*. *J Dent Res* **1997**; *76*: 1625–1636.
57 Åberg, T., Cavender, A., Gaikwad, J.S., Bronckers, A.L., Wang, X., Waltimo-Siren, J., Thesleff, I. and D'Souza, R.N. Phenoty-

pic changes in dentition of Runx2 homozygote-null mutant mice.(erratum appears in *J Histochem Cytochem*). *J Histochem Cytochem* **2004**; *52*: 131–139.

58 Nakamura, T., Takio, K., Eto, Y., Shibai, H., Titani, K. and Sugino, H. Activin-binding protein from rat ovary is follistatin. *Science* **1990**; *247*: 836–838.

59 Yamashita, H., ten Dijke, P., Huylebroeck, D., Sampath, T.K., Andries, M., Smith, J.C., Heldin, C. H. and Miyazono, K. Osteogenic protein–1 binds to activin type II receptors and induces certain activin-like effects. *J Cell Biol* **1995**; *130*: 217–226.

60 Iemura, S., Yamamoto, T.S., Takagi, C., Uchiyama, H., Natsume, T., Shimasaki, S., Sugino, H. and Ueno, N. Direct binding of follistatin to a complex of bone-morphogenetic protein and its receptor inhibits ventral and epidermal cell fates in early Xenopus embryo. *Proc Natl Acad Sci USA* **1998**; *95*: 9337–9342.

61 Wang, X.P., Suomalainen, M., Jorgez C.J., Matzuk, M.M., Wankell, M., Werner, S., and Thesleff, I. *Dev Dyn* **2004**; *270*: 76–93.

62 Laurikkala, J., Kassai, Y., Pakkasjarvi, L., Thesleff, I. and Itoh, N. Identification of a secreted BMP antagonist, ectodin, integrating BMP, FGF, and SHH signals from the tooth enamel knot. *Dev Biol* **2003**; *264*: 91–105.

63 Kassai, Y., Munne, P., Hotta, Y., Penttilä, E., Kavanagh, K., Ohbayashi, N., Takada, S., Tesleff, I., Jernvall, J., Itoh, N. Ectodin integrates induction and inhibition of mammalian tooth cusps (submitted).

64 Winnier, G., Blessing, M., Labosky, P. A. and Hogan, B. L.M. Bone morphogenetic protein–4 is required for mesoderm formation and patterning in the mouse. *Genes Dev* **1995**; *9*: 2105–2116.

65 Zhang, H. and Bradley, A. Mice deficient for BMP2 are nonviable and have defects in amnion/chorion and cardiac development. *Development* **1996**; *122*: 2977–2986.

66 Luo, G., Hofmann, C., Bronckers, A.L., Sohocki, M., Bradley, A. and Karsenty, G. BMP–7 is an inducer of nephrogenesis, and is also required for eye development and skeletal patterning. *Genes Dev* **1995**; *9*: 2808–2820.

67 Dudley, A.T., Lyons, K. M. and Robertson, E. J. A requirement for bone morphogenetic protein–7 during development of the mammalian kidney and eye. *Genes Dev* **1995**; *9*: 2795–2807.

68 Jernvall, J., Åberg, T., Kettunen, P., Keränen, S. and Thesleff, I. The life history of an embryonic signaling center: BMP–4 induces p21 and is associated with apoptosis in the mouse tooth enamel knot. *Development* **1998**; *125*: 161–169.

69 Chen, Y., Bei, M., Woo, I., Satokata, I. and Maas, R. Msx1 controls inductive signaling in mammalian tooth morphogenesis. *Development* **1996**; *122*: 3035–3044.

70 Bei, M., Kratochwil, K. and Maas, R. L. BMP4 rescues a noncell-autonomous function of Msx1 in tooth development. *Development* **2000**; *127*: 4711–4718.

71 Andl, T., Ahn, K., Kairo, A., Chu, E.Y., Wine-Lee, L., Reddy, S.T., Croft, N.J., Cebra-Thomas, J.A., Metzger, D., Chambon,

P., Lyons, K.M., Mishina, Y., Seykora, J.T., Crenshaw, E. B. and Millar, S.E. Epithelial Bmpr1a regulates differentiation and proliferation in postnatal hair follicles and is essential for tooth development. *Development* **2004**; *131*: 2257–2268.

72 Ferguson, C.A., Tucker, A.S., Heikinheimo, K., Nomura, M., Oh, P., Li, E. and Sharpe, P.T. The role of effectors of the activin signalling pathway, activin receptors IIA and IIB, and Smad2, in patterning of tooth development. *Development* **2001**; *128*: 4605–4613.

73 Laurikkala, J., Mikkola, M., Mustonen, T., Åberg, T., Koppinen, P., Pispa, J., Nieminen, P., Galceran, J., Grosschedl, R. and Thesleff, I. TNF signaling via the ligand-receptor pair ectodysplasin and edar controls the function of epithelial signaling centers and is regulated by Wnt and activin during tooth organogenesis. *Dev Biol* **2001**; *229*: 443–455.

74 McMahon, A.P., Ingham, P. W. and Tabin, C.J. Developmental roles and clinical significance of hedgehog signaling. *Curr Top Dev Biol* **2003**; *53*: 1–114.

75 Ingham, P. W. and McMahon, A.P. Hedgehog signaling in animal development: paradigms and principles. *Genes Dev* **2001**; *15*: 3059–3087.

76 Hardcastle, Z., Mo, R., Hui, C. C. and Sharpe, P.T. The Shh signalling pathway in tooth development – defects in Gli2 and Gli3 mutants. *Development* **1998**; *125*: 2803–2811.

77 Chuang, P. T. and McMahon, A.P. Vertebrate Hedgehog signalling modulated by induction of a Hedgehog-binding protein. *Nature* **1999**; *397*: 617–621.

78 Cobourne, M. T. and Sharpe, P.T. Expression and regulation of hedgehog-interacting protein during early tooth development. *Connect Tissue Res* **1904**; *43*: 143–147.

79 Echelard, Y., Epstein, D.J., St-Jacques, B., Shen, L., Mohler, J., McMahon, J. A. and McMahon, A.P. Sonic hedgehog, a member of a family of putative signaling molecules, is implicated in the regulation of CNS polarity. *Cell* **1993**; *75*: 1417–1430.

80 Gritli-Linde, A., Lewis, P., McMahon, A.P., and Linde, A. The whereabouts of a morphogen: direct evidence for short- and graded longrange activity of hedgehog signaling peptides. *Dev Biol* **2001**; *236*: 364–386.

81 Dassule, H.R., Lewis, P., Bei, M., Maas, R. and McMahon, A.P. Sonic hedgehog regulates growth and morphogenesis of the tooth. *Development* **2000**; *127*: 4775–4785.

82 Gritli-Linde, A., Bei, M., Maas, R., Zhang, X.M., Linde, A. and McMahon, A.P. Shh signaling within the dental epithelium is necessary for cell proliferation, growth and polarization. *Development* **2002**; *129*: 5323–5337.

83 Wodarz, A. and Nusse, R. Mechanisms of Wnt signaling in development. *Annu Rev Cell Dev Biol* **1998**; *14*: 59–88.

84 Sharpe, C., Lawrence, N. and Arias, A.M. Wnt signalling: a theme with nuclear variations (Review). *Bioessays* **2001**; *23*: 311–318.

85 Huelsken, J. and Birchmeier, W. New aspects of Wnt signaling pathways in higher vertebrates (Review). *Curr Opin Genet Dev* **2001**; *11*: 547–553.

86 Nelson, W. J. and Nusse, R. Convergence of Wnt, beta-catenin, and cadherin pathways. (Review 66 refs). *Science* **2004**; *303*: 1483–1487.
87 Niehrs, C., Kazanskaya, O., Wu, W. and Glinka, A. Dickkopf1 and the Spemann-Mangold head organizer. *Int J Dev Biol* **1904**; *45*: 237–240.
88 van Genderen, C., Okamura, R.M., Farinas, I., Quo, R.G., Parslow, T.G., Bruhn, L. and Grosschedl, R. Development of several organs that require inductive epithelial-mesenchymal interactions is impaired in LEF–1– deficient mice. *Genes Dev* **1994**; *8*: 2691–2703.
89 Kratochwil, K., Dull, M., Fariñas, I., Galceran, J. and Grosschedl, R. Lef1 expression is activated by BMP–4 and regulates inductive tissue interactions in tooth and hair development. *Genes Dev* **1996**; *10*: 1382–1394.
90 Kratochwil, K., Galceran, J., Tontsch, S., Roth, W. and Grosschedl, R. FGF4, a direct target of LEF1 and Wnt signaling, can rescue the arrest of tooth organogenesis in Lef1(–/–) mice. *Genes Dev* **2002**; *16*: 3173–3185.
91 Andl, T., Reddy, S.T., Gaddapara T and Millar, S. E. WNT signals are required for the initiation of hair follicle development. *Dev Cell* **2002**; *2*: 643–653.
92 Sarkar, L. and Sharpe, P.T. Inhibition of Wnt signaling by exogenous Mfrzb1 protein affects molar tooth size. *J Dent Res* **2000**; *79*: 920–925.
93 Zhou, P., Byrne, C., Jacobs, J. and Fuchs, E. Lymphoid enhancer factor 1 directs hair follicle patterning and epithelial cell fate. *Genes Dev* **1995**; *9*: 700–713.
94 Sarkar, L., Cobourne, M., Naylor, S., Smalley, M., Dale, T. and Sharpe, P.T. Wnt/Shh interactions regulate ectodermal boundary formation during mammalian tooth development. *Proc Natl Acad Sci USA* **2000**; *97*: 4520–4524.
95 Pispa, J., Jung, H.S., Jernvall, J., Kettunen, P., Mustonen, T., Tabata, M.J., Kere, J. and Thesleff, I. Cusp patterning defect in Tabby mouse teeth and its partial rescue by FGF. *Dev Biol* **1999**; *216*: 521–534.
96 Baker, S. J. and Reddy, E.P. Modulation of life and death by the TNF receptor superfamily. (Review 69 refs). *Oncogene* **1998**; *17*: 3261–3270.
97 Locksley, R.M., Killeen, N. and Lenardo, M.J. The TNF and TNF receptor superfamilies: Integrating mammalian biology. *Cell* **2001**; *104*: 487–501.
98 Inoue, J., Ishida, T., Tsukamoto, N., Kobayashi, N., Naito, A., Azuma, S. and Yamamoto, T. Tumor necrosis factor receptor-associated factor (TRAF) family: adapter proteins that mediate cytokine signaling. *Exp Cell Res* **2000**; *254*: 14–24.
99 Thesleff, I. and Mikkola, M. L. 2002. Death receptor signaling giving life to ectodermal organs. *Science's STKE* **2002**; *131*: PE22.
100 Kere, J., Srivastava, A.K., Montonen, O., Zonana, J., Thomas, N., Ferguson, B., Munoz, F., Morgan, D., Clarke, A., Baybayan, P., Chen, E.Y., Ezer, S., Saarialho-Kere, U., de la Chapelle, A. and Schlessinger, D. X-linked anhidrotic (hypohidrotic) ecto-

dermal dysplasia is caused by mutation in a novel transmembrane protein. *Nat Genet* **1996**; *13*: 409–416.

101 Headon, D. J. and Overbeek, P.A. Involvement of a novel TNF receptor homologue in hair follicle induction. *Nat Genet* **1999**; *22*: 370–374.

102 Mikkola, M.L., Pispa, J., Pekkanen, M., Paulin, L., Nieminen, P., Kere, J. and Thesleff, I. Ectodysplasin, a protein required for epithelial morphogenesis, is a novel TNF homologue and promotes cell-matrix adhesion. *Mech Dev* **1999**; *88*: 133–146.

103 Blake, J.A., Richardson, J.E., Bult, C.J., Kadin, J.A., Eppig, J. T. and the mouse database group. The Mouse Genome Database (MGD): the model organism database for the laboratory mouse. *Nucleic Acids Res* **2002**; *30*: 113–115.

104 Pispa, J., Mikkola, M.L., Mustonen, T. and Thesleff, I. Ectodysplasin, Edar and TNFRSF19 are expressed in complementary and overlapping patterns during mouse embryogenesis. *Gene Expr Patterns* **2003**; *3*: 675–679.

105 Schneider, P., Street, S.L., Gaide, O., Hertig, S., Tardivel, A., Tschopp, J., Runkel, L., Alevizopoulos, K., Ferguson, B. M. and Zonana, J. Mutations leading to X-linked hypohidrotic ectodermal dysplasia affect three major functional domains in the tumor necrosis factor family member ectodysplasin-A. *J Biol Chem* **2001**; *276*: 18819–18827.

106 Gaide, O. and Schneider, P. Permanent correction of an inherited ectodermal dysplasia with recombinant EDA. *Nat Med* **2003**; *9*: 614–618.

107 Mustonen, T., Ilmone, M., Pummila, M., Kangas, A.T., Laurikkala, J., Jaatinen, R., Pispa, J., Gaide, O., Schneider, P., Thesleff, I., and Mikkola, M. Ectodysplasin A1 promotes placodal cell fate during early morphogenesis of ectodermal appendages. *Development* **2004**; *131*: 4907–4919.

108 Koppinen, P., Pispa, J., Laurikkala, J., Thesleff, I. and Mikkola, M.L. Signalling and subcellular localization of the TNF receptor Edar. *Exp Cell Res* **2001**; *269*: 180–192.

109 Mustonen, T., Pispa, J., Mikkola, M.L., Pummila, M., Kangas, A.T., Jaatinen, R. and Thesleff, I. Stimulation of ectodermal organ development by ectodysplasin-A1. *Dev Biol* **2003**; *259*: 123–136.

110 Schmidt-Supprian, M., Bloch, W., Courtois, G., Addicks, K., Israel, A., Rajewsky, K. and Pasparakis, M. NEMO/IKK gamma-deficient mice model incontinentia pigmenti. *Mol Cell* **2000**; *5*: 981–992.

111 Schmidt-Ullrich, R., Aebischer, T., Hulsken, J., Birchmeier, W., Klemm, U. and Scheidereit, C. Requirement of NF-kappaB/Rel for the development of hair follicles and other epidermal appendices. *Development* **2001**; *128*: 3843–3853.

112 Ohazama, A., Hu, Y., Schmidt-Ullrich, R., Cao, Y., Scheidereit, C., Karin, M. and Sharpe, P. T. A dual role for Ikk alpha in tooth development. *Dev Cell* **2004**; *6*: 219–227.

113 Artavanis-Tsakonas, S., Rand, M. D. and Lake, R.J. Notch signaling: cell fate control and signal integration in development. *Science* **1999**; *284*: 770–776.

114 Baron, M. An overview of the Notch signalling pathway. *Semin Cell Dev Biol* **2003**; *14*: 113–119.

115 Mustonen, T., Tummers, M., Mikami, T., Itoh, N., Zhang, N., Gridley, T. and Thesleff, I. Lunatic fringe, FGF, and BMP regulate the Notch pathway during epithelial morphogenesis of teeth. *Dev Biol* **2002**; *248*: 281–293.

116 Mitsiadis, T., Lardelli, M., Lendahl, U. and Thesleff, I. Expression of Notch 1, 2 and 3 is regulated by epithelial-mesenchymal interactions and retinoic acid in the developing mouse tooth and associated with determination of ameloblast cell fate. *J Cell Biol* **1995**; *130*: 407–418.

117 Mitsiadis, T.A., Hirsinger, E., Lendahl, U. and Goridis, C. Delta-notch signaling in odontogenesis: correlation with cytodifferentiation and evidence for feedback regulation. *Dev Biol* **1998**; *204*: 420–431.

118 Partanen, A. M. and Thesleff, I. Localization and quantitation of ^{125}I-epidermal growth factor binding in mouse embryonic tooth and other embryonic tissues at different developmental stages. *Dev Biol* **1987**; *120*: 186–197.

119 Snead, M.L., Luo, W., Oliver, P., Nakamura, M., Don-Wheeler, G., Bessem, C., Bell, G.I., Rall, L. B. and Slavkin, H.C. Localization of epidermal growth factor precursor in tooth and lung during embryonic mouse development. *Dev Biol* **1989**; *134*: 420–429.

120 Luukko, K., Arumae, U., Karavanov, A., Moshnyakov, M., Sainio, K., Sariola, H., Saarma, M. and Thesleff, I. Neurotrophin mRNA expression in the developing tooth suggests multiple roles in innervation and organogenesis. *Dev Dyn* **1997**; *210*: 117–129.

121 Luukko, K. Neuronal cells and neurotrophins in odontogenesis. *Eur J Oral Sci* **1998**; *106*: 80–93.

122 Partanen, A.M., Ekblom, P. and Thesleff, I. Epidermal growth factor inhibits morphogenesis and cell differentiation in cultured mouse embryonic teeth. *Dev Biol* **1985**; *111*: 84–94.

123 Joseph, B.K., Savage, N.W., Young, W.G., Gupta, G.S., Breier, B. H. and Waters, M.J. Expression and regulation of insulin-like growth factor-I in the rat incisor. *Growth Factors* **1993**; *8*: 267–275.

124 Joseph, B.K., Harbrow, D.J., Sugerman, P.B., Smid, J.R., Savage, N. W. and Young, W.G. Ameloblast apoptosis and IGF–1 receptor expression in the continuously erupting rat incisor model. *Apoptosis* **1999**; *4*: 441–447.

125 Philbrick, W.M., Dreyer, B.E., Nakchbandi, I. A. and Karaplis, A.C. Parathyroid hormone-related protein is required for tooth eruption. *Proc Natl Acad Sci USA* **1998**; *95*, 11846–11851.

126 Liu, J.G., Tabata, M.J., Fujii, T., Ohmori, T., Abe, M., Ohsaki, Y., Kato, J., Wakisaka, S., Iwamoto, M. and Kurisu, K. Parathyroid hormone-related peptide is involved in protection against invasion of tooth germs by bone via promoting the differentiation of osteoclasts during tooth development. *Mech Dev* **2000**; *95*: 189–200.

127 Neubuser, A., Peters, H., Balling, R. and Martin, G.R. Antagonistic interactions between FGF and BMP signaling pathways: a mechanism for positioning the sites of tooth formation. *Cell* **1997**; *90*: 247–255.

128 Tucker, A.S., Matthews, K. L. and Sharpe, P.T. Transformation of tooth type induced by inhibition of BMP signaling. *Science* **1998**; *282*: 1136–1138.

129 St Amand, T.R., Zhang, Y., Semina, E.V., Zhao, X., Hu, Y., Nguyen, L., Murray, J. C. and Chen, Y. Antagonistic signals between BMP4 and FGF8 define the expression of Pitx1 and Pitx2 in mouse tooth-forming anlage. *Dev Biol* **2000**; *217*: 323–332.

130 Thomas, B.L., Tucker, A.S., Ferguson, C., Qiu, M.S., Rubenstein, J. L. R. and Sharpe, P.T. Molecular control of odontogenic patterning – positional dependent initiation and morphogenesis. *Eur J Oral Sci* **1998**; *106*: 44–47.

131 Wang, X.P., Suomalainen, M., Jorgez C.J., Matzuk, M.M., Werner, S., and Thesleff, I. Follistatin regulates enamel patterning in mouse incisors by asymmetrically inhibiting BMP signaling and ameloblast differentiation. *Dev Cell* **2004**; *7*: 719–730.

132 Celli, J., Duijf, P., Hamel, B. C.J., Bamshad, M., Kramer, B., Smits, A. P.T., Newbury-Ecob, R., Hennekam, R. C.M., Van Buggenhout, G., van Haeringen, B., Woods, CG, van Essen, A.J., de Waal, R., Vriend, G., Haber, D.A., Yang, A., McKeon, F., Brunner, H. G. and van Bokhoven, H. Heterozygous germline mutations in the p53 homolog p63 are the cause of EEC syndrome. *Cell* **1999**; *99*: 143–153.

133 Mills, A.A., Zheng, B.H., Wang, X.J., Vogel, H., Roop, D.R., and Bradley, A. p63 is a p53 homologue required for limb and epidermal morphogenesis. *Nature* **1999**; *398*: 708–713.

134 Laurikkala, J., Mikkola, M. L., James, M., Tummers, M., Mills, A. and Thesleff, I. P63 regulates multiple signalling pathways required for ectodermal organogenesis and differentiation. Submitted.

135 Cobourne, M.T., Hardcastle, Z. and Sharpe, P.T. Sonic hedgehog regulates epithelial proliferation and cell survival in the developing tooth germ. *J Dent Res* **2001**; *80*: 1974–1979.

136 Jernvall, J. Mammalian molar cusp patterns: Developmental mechanisms of diversity. *Acta Zool Fennica* **1995**; *198*: 1–61.

137 Kangas, A.T., Evans, A.R., Thesleff, I., and Jernvall, J. Nonindependence of mammalian dental characters. *Nature* **2004**; *432*: 211–214.

138 Itasaki, N., Jones, C.M., Mercurio, S., Rowe, A., Domingos, P.M., Smith, J. C. and Krumlauf, R. Wise, a context-dependent activator and inhibitor of Wnt signalling. *Development* **2003**; *130*: 4295–4305.

139 Cam, Y., Neumann, M.R., Oliver, L., Raulais, D., Janet, T. and Ruch, J.V. Immunolocalization of acidic and basic fibroblast growth factors during mouse odontogenesis (published erratum appears in *Int J Dev Biol* 1992; 36(4): following 599). *Int J Dev Biol* **1992**; *36*: 381–389.

140 Ruch, J.V. Tooth crown morphogenesis and cytodifferentiations: candid questions and critical comments. *Connect Tissue Res* **1995**; *32*: 1–8.

141 Begue-Kirn, C., Smith, A.J., Ruch, J.V., Wozney, J.M., Purchio, A., Hartmann, D. and Lesot, H. Effects of dentin proteins,

transforming growth factor beta 1 (TGF beta 1) and bone morphogenetic protein 2 (BMP2) on the differentiation of odontoblast *in vitro*. *Int J Dev Biol* **1992**; *36*: 491–503.

142. Begue-Kirn, C., Smith, A.J., Loriot, M., Kupferle, C., Ruch, J. V. and Lesot, H. Comparative analysis of TGF beta s, BMPs, IGF1, msxs, fibronectin, osteonectin and bone sialoprotein gene expression during normal and *in vitro*-induced odontoblast differentiation. *Int J Dev Biol* **1994**; *38*: 405–420.

143. Martin, A., Unda, F.J., Beguekirn, C., Ruch, J. V. and Arechaga, J. Effects of afgf, bfgf, tgf-beta–1 and igf-i on odontoblast differentiation *in vitro*. *Eur J Oral Sci* **1998**; *106*: 117–121.

144. Ruch, J.V. Odontoblast commitment and differentiation. (Review 122 refs). *Biochem Cell Biol* **1998**; *76*: 923–938.

145. Unda, F.J., Martin, A., Hilario, E., Begue-Kirn, C., Ruch, J. V. and Arechaga, J. Dissection of the odontoblast differentiation process in vitro by a combination of FGF1, FGF2, and TGFbeta1. *Dev Dyn* **2000**; *218*: 480–489.

146. Srivastava, A.K., Pispa, J., Hartung, A.J., Du, Y., Ezer, S., Jenks, T., Shimada, T., Pekkanen, M., Mikkola, M.L., Ko, M.S., Thesleff, I., Kere, J. and Schlessinger, D. The Tabby phenotype is caused by mutation in a mouse homologue of the EDA gene that reveals novel mouse and human exons and encodes a protein (ectodysplasin-A) with collagenous domains. *Proc Natl Acad Sci USA* **1997**; *94*: 13069–13074.

147. Srivastava, A.K., Durmowicz, M.C., Hartung, A.J., Hudson, J., Ouzts, L.V., Donovan, D.M., Cui, C. Y. and Schlessinger, D. Ectodysplasin-A1 is sufficient to rescue both hair growth and sweat glands in Tabby mice. *Hum Mol Genet* **2001**; *10*: 2973–2981.

148. Tummers, M. and Thesleff, I. Epithelial stem cell niche in the tooth. In *Handbook of Stem Cells*, Vol. 1, Lanza R., Gearhart J., Hogan B., Melton D., Pedersen R., Thomson J., and West M. (Eds). 265–271. Academic Press: Burlington, MA, USA, **2004**.

149. Jensen, B. L. and Kreiborg, S. Development of the dentition in cleidocranial dysplasia. *J Oral Pathol Med* **1990**; *19*: 89–93.

150. Otto, F., Thornell, A.P., Crompton, T., Denzel, A., Gilmour, K.C., Rosewell, I.R., Stamp, G.W., Beddington, R.S., Mundlos, S., Olsen, B.R., Selby, P. B. and Owen, M.J. Cbfa1, a candidate gene for cleidocranial dysplasia syndrome, is essential for osteoblast differentiation and bone development. *Cell* **1997**; *89*: 765–771.

151. Mundlos, S., Otto, F., Mundlos, C., Mulliken, J.B., Aylsworth, A.S., Albright, S., Lindhout, D., Cole, W.G., Henn, W., Knoll, J.H., Owen, M.J., Mertelsmann, R., Zabel, BU and Olsen, B.R. Mutations involving the transcription factor CBFA1 cause cleidocranial dysplasia. *Cell* **1997**; *89*: 773–779.

152. Lammi, L., Arte, S., Somer, M., Jarvinen, H., Lahermo, P., Thesleff, I., Pirinen, S. and Nieminen, P. Mutations in AXIN2 cause familial tooth agenesis and predispose to colorectal cancer. *Am J Hum Genet* **2004**; *74*: 1043–1050.

153. Rutherford, R.B., Wahle, J., Tucker, M., Rueger, D. and Charette, M. Induction of reparative dentine formation in monkeys by recombinant human osteogenic protein–1. *Arch Oral Biol* **1993**; *38*: 571–576.

154 Nakashima, M. Induction of dentin formation on canine amputated pulp by recombinant human bone morphogenetic proteins (BMP)–2 and –4. *J Dent Res* **1994**; *73*: 1515–1522.

155 Nakashima, M. and Reddi, A.H. The application of bone morphogenetic proteins to dental tissue engineering. (Review 75 refs). *Nat Biotechnol* **2003**; *21*: 1025–1032.

156 Seo, B. M., Miura, M., Granthos, S., Bartold, P. M., Batouli, S., Brahim, J., Young, M., Robey, P. G., Wang, C. Y., Shi, S. Investigation of multipotent postnatal stem cells from human periodontal ligament. Lancet **2004**; *364*: 149–155.

157 Lammi, L., Arte, S., Somer, M., Järvinen, H., Lahermo, P., Thesleff, I., Pirinen, S. and Nieminen, P. Mutations in *AXIN2* cause familial tooth agenesis and predispose to colorectal cancer. *Am J Hum Genet* **2004**; *74*: 1043–1050.

20
Gastrointestinal Tract

Daniel Ménard, Jean-François Beaulieu, François Boudreau, Nathalie Perreault, Nathalie Rivard, and Pierre H. Vachon

20.1
Introduction

The development of specific digestive segments *in utero* and/or after birth occurs at differing rates and involves both morphogenesis and cytodifferentiation. In rodents, the functional changes leading to mature functions in the stomach, pancreas, small and large intestine are characterized by a highly coordinated developmental pattern which occurs at weaning time [1, 2]. For example, the increase in gastric pepsinogen, the appearance of intestinal sucrase activity, the rise of all other brush border enzymes and the fall of lactase activity all occur during the weaning period. In humans, the functional development of the gastrointestinal (GI) tract is much less coordinated chronologically and occurs mainly during the fetal period (Fig. 20.1) [3]. In addition, human digestive functions develop not only at different rates throughout the various GI organs but can also differ within a given organ (Fig. 20.1). Henceforth, the regulatory mechanisms behind human GI tract development are seemingly different and undoubtedly more complex than those in rodents.

It is now well recognized that the final maturation of functional GI epithelial cells, as well as their survival, results from integrated interactions including mesenchymal-epithelial interactions, and interactions between hormones and growth factors, extracellular matrix components and cellular integrins, specific intracellular signaling pathways and selective transcription factors. This chapter focuses on the latest available data concerning our knowledge of cell signaling and involvement of growth factors in the development of the human GI tract.

20.2
Development of the Gastrointestinal Mucosa

20.2.1
Specialization of Epithelium and Functional Units

20.2.1.1 Esophagus

During its development, the human esophageal mucosa exhibits an unusual morphology, i.e. a stratified columnar ciliated epithelium. This particular epithelium appears at 10 weeks, develops and is subsequently replaced by an adult stratified squamous epithelium at about 20–25 weeks. While mostly scattered throughout the stratified epithelium between 10 and 13 weeks, proliferative epithelial cells begin to concentrate in the basal layers as early as 14 weeks of gestation [4]. The regulatory mechanisms governing this particular developmental pattern are as yet unknown.

20.2.1.2 Stomach

The human gastric epithelium is stratified and undifferentiated at 8 to 10 weeks of gestation. By 11–12 weeks, glandular pits are formed, along with the emergence of the first differentiated epithelial cell types. From 11 weeks onward, surface epithelial cells differentiate into columnar mucous cells while gastric glands (functional units) continue to expand with the appearance of endocrine, neck mucous, parietal and chief cells. By 15 weeks of gestation, the fetal gland is very much representative of the adult gastric unit, with all of its various functional compartments having been fully determined: i.e. differentiated epithelial cell types are well in place, all functional markers as well as known hormone and growth factor receptors are expressed and the restricted distribution of extracellular matrix components and integrin-type receptors is also well established [5–8].

20.2.1.3 Small Intestine

The functional unit of the small intestinal mucosa, namely the crypt-villus axis, begins to form in the duodenum and proximal jejunum by 8 to 9 weeks and occurs in a caudal direction [9, 10]. By 11 weeks, the entire small intestinal mucosa is lined by villi most of which are covered by a simple columnar epithelium. Primitive crypts which are epithelial downgrowths into the mesenchyme begin to form between 10 to 12 weeks, also along a craniocaudal axis. It is generally acknowledged that by 14 weeks of gestation, digestive enzymic activities associated with enterocytic brush border such as sucrase, maltase-glucoamylase and trehalase are all present at levels ranging from 70 to 100 % of those present in the mature intestine including a well-established adult craniocaudal gradient of digestive activities. Lactase activity, on the other hand, which is also present at 10 weeks' gestation, remains low until the very latter stages of gestation, when a marked enhancement of activity is observed, elevating term lactase levels two to four times the values found in normal infants of 2 to 11 months of age.

20.2.1.4 Colon

The human colon undergoes several fetal stages before attaining its mature form. One striking event is the formation of villi similar to that seen in the small intestine [9]. The formation of crypt-villus units begins at 11 weeks in the distal portion and proceeds towards the proximal portion of the colon. Between 14 and 16 weeks, the entire colonic mucosa is covered by villus structures which will remain present until 30 weeks of gestation, at which point final maturation of the colonic mucosa begins with the disappearance of villus structures. All brush border digestive enzymes are also present as early as 13 weeks of gestation in the fetal colon, but their activities are significantly and steadily lower than those in the corresponding intestine [11]. Of particular interest is the presence of a sucrase/isomaltase complex, which is the classical marker of enterocytic differentiation and one of the primary intestinal brush border enzymes. This enzymic activity will also disappear between 30 weeks of gestation and birth. As can be surmised, both the onset and differential patterns of development of the various functional units of the human GI tract raise numerous questions with regard to the regulatory signals involved.

20.2.2 Mesenchymal-Epithelial Interactions

The intestinal tube originates embryologically from the ectoderm, endoderm and mesoderm. The epithelium is derived from the ectoderm at the most anterior (mouth) and posterior regions (anus) while the remaining epithelium is formed from the endoderm ultimately giving rise to the digestive organs including the stomach, intestine, liver, and pancreas. The invaginating endoderm recruits the splanchnic mesoderm that will contribute to the musculature and connective tissue of the gastrointestinal tract [12]. The gut tube is subdivided into three regions along the anteroposterior (AP) axis: namely, the foregut, midgut and hindgut. Each of these regions exhibits unique mesodermal and endodermal patterns of differentiation and function. This is achieved by the combined activity of both soluble (growth and transcription factors and Hox genes products) and insoluble (basement membrane proteins and epimorphin) signaling molecules [9, 10, 12]. These molecules are expressed in a spatially restricted pattern, either in the endoderm or mesoderm, or both [13–19]. Thus, interaction between epithelium and mesenchyme is crucial for the regulation of gut morphogenesis and in the normal regulation of proliferation and differentiation [20, 21].

The first evidence of the importance of mesenchymal-epithelial cross-talk in gastrointestinal development stemmed from tissue recombination experiments [22, 23] in which human or rat gut mesenchymal cells combined with rat endoderm or rat

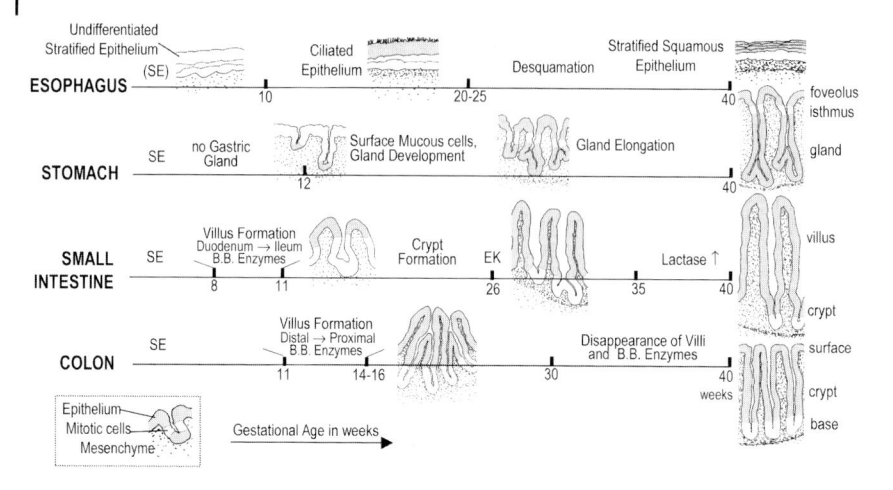

Fig. 20.1
Schematic representation of the key morphological and functional events leading to the development of the human gastrointestinal tract. The onset of morphological and functional changes occurs not only at different gestational periods but can also differ within a given segment of the GI tract. Bb enzymes, intestinal brush border enzymes; distal → proximal, distal to proximal gradient of villus formation; duodenum → ileum, establishment of a proximal-distal gradient of villus formation and brush border digestive enzymic activities; EK, appearance of enteropeptidase activity at 26 weeks of gestation; lactase↑, late gestational rise in lactase activity between 35 and 40 weeks of gestation. Adapted from [3].

intestinal crypt cell lines (IEC–17) were grafted under the kidney capsule of adult rats. It was revealed that the epithelial compartment, in the presence of mesenchymal cells, was able to morphologically (crypt-villus axis) and functionally differentiate within the graft. Interestingly, the mesenchymal counterpart of the graft also underwent gut-specific differentiation to form visceral mesoderm under the influence of the associated epithelial cells [22–24]. These observations strongly suggest that the cross-talk between mesoderm and endoderm is reciprocal and permissive. To date, several factors and signaling events have been identified and shown to be involved in such epithelial-mesenchymal cross-talk in gastrointestinal morphogenesis.

20.2.2.1 Growth Factors and Morphogens

Members of the hedgehog (HH) family are secreted molecules shown to play critical instructional and regulatory roles during the development [25, 26] of numerous tissues and cell types including the intestine. The gut endoderm is known to express sonic hedgehog (SHH) and Indian hedgehog (IHH) but not desert hedgehog (DHH) [27]. The HH family members are thought to be strong candidates for the initial signals from the endoderm to the mesoderm during gut development [18, 28].

Disruption of the SHH and IHH genes has shown that both proteins regulate numerous aspects of gut development. Indeed, *hh* mutant mice have multiple gastrointestinal abnormalities suggesting that multiple tasks are carried out by the HH genes in processes such as AP axis patterning and stem cell proliferation and differentiation [18]. So far, the reciprocal expression pattern of SHH (endoderm) and bone morphogenic protein–4 (BMP–4, mesoderm), a member of the transforming growth factor-β (TGF-β) superfamily, already suggests a potential role for these two morphogens in gut morphogenesis [27]. Indeed, overexpression of SHH in the chick hindgut epithelium induced ectopic expression of BMP–4 and certain Hoxd genes within the underlying mesoderm [28]. Thus, BMPs and Hox genes may act as secondary signals downstream of SHH in early gut development although these two signals could act independently later on. Indeed, in a recent study, Haramis and collaborators [29] using transgenic mice devoid of BMP signaling, demonstrated the importance of these mesodermal factors in the regulation of epithelial architecture of the crypt-villus axis. The absence of BMP signaling in the epithelium induced *de novo* crypt formation in the villus epithelium creating a phenomenon resembling juvenile polyposis syndrome. In addition, various growth factors and hormones are expressed by gut mesenchymal cells but act on the epithelial compartment: for example, hepatocyte growth factor (HGF) is known to influence proliferation [30] and differentiation [19] of intestinal epithelial cells. Neuregulin (NGF), keratinocyte growth factor (KGF), TGF-β and fibroblast growth factors (FGFs) also affect the epithelium. These factors have been extensively reviewed elsewhere [3, 31, 32] and their known effects on the functional development of the intestinal and colonic mucosae are summarized in Tab. 20.1.

20.2.2.2 Hox Genes

The Hox genes encode transcription factors known to be involved in the establishment of the regional identities of the AP axis. The Hox system is composed of four clusters and 13 paralogs; a total of 39 Hox genes have been identified in the mouse. The Hox genes can be expressed solely in the mesoderm or in both cell compartments [17], hence a specific Hox code is found corresponding to each region of the AP axis of the gut [33]. This expression pattern strongly suggests that the major role of the Hox genes is to specify the mesoderm of a region by its position along the AP axis. Thus, this mesoderm will have the capacity to direct its surrounding environment (mainly the epithelial compartments) to its position on the AP axis in order for the epithelial cells to acquire the proper phenotype related to that region. The role of Hox genes in gut specification in the mouse has been reviewed previously [17, 33]. An example of this role in gut specifications was demonstrated by Aubin and coworkers [34] in which the loss of Hoxa5 functions in the mesoderm leads to defects of the gastric epithelium. The mechanism involved appears to be that the loss of Hoxa5 directly affects FGF–10 and TGF-β levels in the mesoderm leading to an anteriorization of IHH expression over SHH expression in the stomach epithelium. This deregulation ultimately results in an altered cellular specification in gastric mucosa. Thus, Hoxa5 is essential for proper morphogenesis and functional specification of the stomach. Other examples of the roles of loss- or gain-of-function of Hox genes in gut morphogenesis can also be found with Hoxa4 [35], Hoxa13 and Hoxd13 [16, 36], and Hoxc8 [37].

20.2.2.3 Epimorphin/Syntaxin 2

Epimorphin was first identified as an extracellular morphogen in lung morphogenesis [38], followed by subsequent identification as syntaxin 2, a member of the syntaxin family of vesicle fusion proteins [39]. Epimorphin/syntaxin 2 has a dual function depending on its location on the membrane. When present on the cytoplasmic side, it plays a role as a fusion protein. When present on the extracellular surface, it acts as a morphogen [40] which is of particular interest in the present case. Epimorphin is expressed by the mesenchymal compartment of many tissues including the gut and has been found to influence the morphogenesis of the epithelial compartment. This protein has a controlled spatio-temporal pattern in the gut mesenchyme where expression is most abundant in the embryonic intestine (E14–E17). This time period corresponds to the initiation of murine villus morphogenesis at day 14–17. Epimorphin expression decreases around E19 (through to birth). A second peak of expression is found during the suckling period, decreasing thereafter to basal levels at weaning time, coinciding with a second transition period in the mouse gut where the epithelium acquires its final functional fate [14]. Moreover, epimorphin expression is at its highest in the ileum on the AP axis [15]. Fritsch and collaborators [41] confirmed the importance of epimorphin in intestinal epithelial morphogenesis with the use of co-cultures of intestinal myofibroblasts (Mic 216) and Caco–2 cells. Mic 216 cells were transfected with sense and antisense epimorphin cDNA. When co-cultures were grafted into the coelomic cavity of chick embryo, the clone with the fibroblasts expressing antisense epimorphin failed to initiate villus morphogenesis. Thus, it will be of major interest to follow further developments in the field and see whether loss- or gain-of-function experiments in the mouse can help elucidate the mechanisms of epimorphin action in gut morphogenesis.

20.2.2.4 Basement Membrane Proteins

The use of tissue recombination to construct chimeric intestines or the dissociation of the epithelial compartment from the mesenchymal compartment has allowed the study of epithelial-mesenchymal cooperation in the formation of the basement membrane (BM) in gastrointestinal development. These experiments have been described elsewhere [20, 21, 42]. Moreover, the role of these proteins in the process of gut morphogenesis and cytodifferentiation will be discussed more extensively in Sections 20.3.3 (stomach) and 20.4.2 (intestine).

20.2.2.5 Mesodermal Transcription Factors

The influence of mesodermal tissues on gastrointestinal development reached new heights in the late 1990s with the emergence of data obtained from the disruption of key mesodermal genes. The intestinal mesenchyme expresses transcription factors that are known today to affect epithelial morphogenesis, including *Foxl1* [13, 43], *Nkx–2.3* [44] and *Foxf1* [45]. Two of these mesenchymal transcription factors are of particular interest and have been shown to alter epithelial architecture during gastrointestinal morphogenesis. The first to be identified was *Foxl1*, a member of the winged helix family, expressed in the mesenchyme of the developing and adult gastrointestinal tract and also found to be in close contact with the gastrointestinal epithelium. The genetic deletion of *Foxl1* revealed that this transcription factor is necessary for the proper development and differentiation of the gastrointestinal epithelium [13]. The reduced levels of BMP–2 and BMP–4 observed in *Foxl1* null mice suggest that these morphogens may be regulated by *Foxl1*. Moreover, Perreault and co-workers [43] have demonstrated that proteoglycan levels were upregulated in null mice. This upregulation was shown to be responsible for the activation of the Wnt pathway in $Foxl1^{-/-}$ mice. The deregulation of the Wnt pathway may explain, in part, the increased proliferation observed in null animals [43]. A second mesenchymal transcription factor, *Nkx–2.3*, was also shown to affect epithelial differentiation in the gastrointestinal tract [44]. Similarly to *Foxl1*, *Nkx–2.3* appears to regulate levels of BMP factors in the mesenchyme although a more complete mode of action for *Nkx–2.3* has yet to be determined. Altogether, these experiments illustrate the importance of transcription factors in the signaling cascade between mesenchyme and epithelial cross-talk in gastrointestinal development.

20.3
Gastric Cell Proliferation and Differentiation

20.3.1
Functional Compartmentalization of Gastric Glands

By 15 to 17 weeks of gestation, fetal gastric glands are representative of the adult digestive glands exhibiting all of the morphological compartments (foveolus, isthmus, neck and base) and their inherent cell lineages with characteristic phenotypes [46]. The proliferation pool responsible for the continuous renewal of all gastric epithelial cells is localized in the isthmus region. Epithelial cells exiting the proliferative zone either differentiate into mucous cells during their upward migration (foveolus and surface epithelium) or into mucous, endocrine, parietal and chief cells during downward migration (neck and glandular epithelium). Interestingly, epithelial cells leaving the proliferative zone differentiate into mucin–5A-expressing mucous cells when migrating upwards and into mucin–6–expressing mucous, endocrine, parietal and chief cells when migrating downwards. The presence of EGF, HGF, IGF and KGF receptors in all glandular compartments clearly suggests the potential role of these growth factors either in the development or in the maintenance of specific gastric epithelial functions [3]. Finally, systemic studies on the presence and distribution of extracellular matrix components (laminin α1, β1 and γ1 chains, type IV collagen, heparin sulfate proteoglycan, fibronectin, tenascin) and integrin-type receptors (α2, α3, α6, α7, β1 and β4 subunits) have clearly established that a restricted distribution of many of these components is already instituted by 15 and 17 weeks [5, 47]. These specific compartmentalizations raise numerous fundamental questions as to the regulation of both development and maintenance of gastric epithelial functions. Until now, research efforts have mainly focused on the modulation of cell proliferation and on zymogenic cell differentiation. An original and important aspect of digestive functions attributed to the human gastric glandular epithelium also involves the hydrolysis of fat, a role which is assumed by a lipase enzyme produced by zymogenic chief cells [46]. Our studies have shown that human gastric lipase (HGL) is expressed early during ontogeny and is uniquely co-localized with fundic-type pepsinogen (Pg5) in fetal chief cells. Thus, it will be essential to clearly understand the molecular mechanisms controlling the acquisition and maintenance of the functions of chief cells when considering the physiological importance of gastric lipolysis and the fact that its role increases in the context of perinatal physiology as well as in pathological conditions associated with pancreatic insufficiency.

20.3.2
Hormones and Growth Factors

Serum-free organ culture has proven to be both an appealing and unique tool for comparative studies of modulators and/or regulators of the developing gastric epithelium [6]. Whereas the addition of hydrocortisone and insulin to cultured gastric explants had no significant effects on [^{3}H]-thymidine incorporation into total DNA, EGF/TGFα and KGF did significantly stimulate epithelial cell proliferation [3]. Furthermore, the potential role of IGFs and their IGF binding proteins (IGFBPs) have also been thoroughly investigated using native forms as well as truncated analogs of IGF-I and IGF-II which do not interact with IGFBPs [8]. The observation that the IGF-I analog was slightly more effective than native IGF-I, suggests that BPs marginally influence the response of proliferative cells. However, while the addition of native IGF-II had no significant effect on gastric cell proliferation, its synthetic analog was not only very potent, but even more so than IGF-I. Thus, IGF-II acts as a growth-promoting agent whose action is tightly regulated by local expression and/or secretion of IGF-II carriers. Interestingly, the relative abundance of IGFBP–2 and IGFBP–6 in the proliferative compartment (visualized by immunofluorescence) may explain why endogenous or added IGF-II does not stimulate mucosal growth in contrast to its truncated analog [8].

The developmental profiles as well as the adult-like regional distribution (fundus → antrum) of HGL and Pg5 activities are achieved at different times between 12 and 20 weeks of gestation, suggesting that both enzymes are under the control of distinct regulatory mechanisms [46]. Experimental data gathered to date using fetal gastric organ cultures definitely support this theory. Indeed, while hydrocortisone stimulates Pg5 activity, it has no effect on HGL. On the other hand EGF/TGFα and IGFs downregulate HGL activity without affecting Pg5 [48].

Correlation between human HGL activity, HGL protein and HGL mRNA signals strongly suggests that HGL expression is most likely regulated at the mRNA level [48]. Initial studies clearly indicate that the downregulation of HGL expression induced by EGF is the result of HGL mRNA downregulation involving, in part, the activation of mitogen-activated protein kinase (MAPK) p42/p44 isoforms [48]. This pathway has been validated using human fetal gastric epithelium in primary culture [49]. Thus, organ and primary culture techniques offer a unique opportunity not only to study the biological effects of given factors but also to identify the molecular mechanisms behind these effects specifically as they relate to humans.

20.3.3
ECM and Integrins

Regulation and maintenance of epithelial differentiation are governed by extracellular signals which are represented not only by growth factors and hormones but also by cell-to-cell interactions and their underlying extracellular matrix (ECM) [20, 21]. ECM provides positional information and cues for cell polarity as well as signals that regulate cell behavior. Laminins (LNs) are among the major constituents of BM, either native or reconstituted (Matrigel) [50], promoting cellular adhesion, migration, and proliferation as well as expression of tissue-specific genes in differentiating epithelial cells [51]. Furthermore, a number of growth factors including transforming growth factor-β1 (TGFβ1) are associated with BM. The latter peptide regulates cell-matrix interactions [52] thus implying that TGF-β1 may function as a regulator of epithelial morphogenesis and expression of tissue specific proteins.

The recent availability of primary cultures of human fetal gastric epithelial cells capable of retaining most of the inherent characteristics found along the foveolus-gland axis, including junctional proteins [49], offers a unique opportunity to address ECM-integrin-growth factor interaction in the regulation of gastric chief cell function. Plating of gastric cell aggregates on Matrigel was shown to result in the presence of three-dimensional clusters lined by well-polarized epithelial cells exhibiting proper intracellular localization of basal nuclei, well-organized junctional complexes at the apical cell margin and abundant secretory organelles [53]. Furthermore, high HGL activity, HGL protein and HGL mRNA levels were also recorded. These experiments clearly illustrate the fundamental role of ECM in the maintenance of the gastric epithelial functions.

In order to further assess the contribution of individual Matrigel components to cell polarity and HGL gene expression, the potential implication of specific laminins (LN–1, LN–2), collagen type I and TGFβ1 was also investigated. Individually, all the above components were unable to maintain epithelial cell polarity or adequate HGL synthesis. However, primary cultures composed of compact organoid structures and well-preserved highly-polarized epithelial cell phenotypes, were observed in the combined presence of LNs and TGFβ1. TGF-β1/LNs even upregulated HGL levels compared to Matrigel. The observed increased expression and redistribution of an α-integrin subunit at the basal membrane of epithelial cells in the inferior portion of the fetal gastric gland is suggestive of a possible key role of α2β1 integrin in the differentiation and/or maintenance of gastric chief cells. The fact that simultaneous addition of a neutralizing antibody against the α2–subunit with LNs/TGFβ1 almost completely abrogated HGL mRNA production, strongly endorses the notion of an important interaction between LNs and α2β1 integrin. Since intracellular signals elicited by hormones, growth factors and integrin ligands may exert their cooperativity on cell responses through a convergent regulation of MAPK cascades, specific inhibitors of these cascades were also tested [53]. The parallel activation of two MAPK cascades (ERK1/2 and p38) was shown to be necessary for optimal regulation of HGL gene expression in chief cells. Thus, a potent synergism between growth factor and basement membrane LNs maintains the functionality of human gastric glandular epithelial cells, including the zymogenic chief cell population, through activation of the α2β1–integrin and effectors of two MAPK pathways. Our current knowledge of the intricate regulatory mechanisms of human gastric chief cells is summarized in Fig. 20.2.

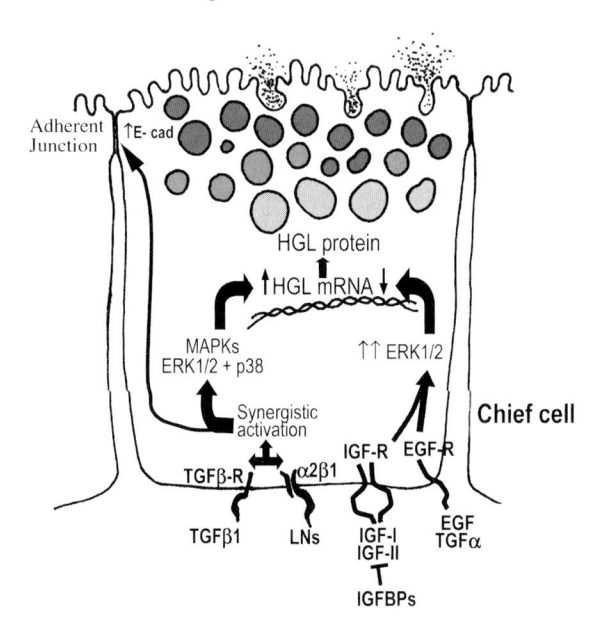

Fig. 20.2
Actual concept for the regulation of gastric lipase and E-cadherin in the human chief cell.

20.4
Intestinal Cell Proliferation and Differentiation

20.4.1
Hormones and Growth Factors

A major issue has always been the precise delineation of specific regulators or modulators involved in functional GI tract development. Since the pioneering work of Moog, glucocorticoid hormones have probably been the most studied modulator of rodent intestinal development [1, 2]. An increase in circulating corticosterone at the onset of the third postnatal week, followed by changes in intestinal enzymic activities, have led to the first suggestion that glucocorticoids could be the foremost important regulator of functional development occurring at weaning. While glucocorticoids have been targeted as having a primary role, over the years many other hormones and growth factors such as insulin, EGF, HGF, IGFs, TGFα, TGFβ, KGF and trefoil peptides acting through endocrine, exocrine, paracrine and autocrine pathways have been proposed as important modulators of functional development leading to what is now a very complex regulatory pattern [1–3]. The extent to which these data can be extrapolated to developing human GI tract remained a debatable issue until the serum-free organ culture technique became available for human fetal tissues [9]. Since human fetal jejunum and colon share the same morphological and functional characteristics, this technique offered the opportunity to determine whether they also share the same regulatory mechanisms. The addition of selected hydrocortisone (HC) levels representative of different gestational periods (15 to 17 weeks, 12.5 ng HC/ml; 35 weeks: 25 ng/ml; 37 to 40 weeks: 50 ng/ml) clearly established the modulatory role of HC on human fetal jejunum development. Indeed, the addition of the highest dosages induced a significant rise in lactase and alkaline phosphatase activities without influencing sucrase, trehalase or glucoamylase activities as well as inducing a significant rise in the epithelial labeling index. It is certainly tempting to correlate these observations with the situation *in utero* since lactase activity remains low during the entire gestational period only to increase during the last weeks of gestation [54], during which period there is a doubling of circulating hydrocortisone levels [9]. By contrast, the addition of HC to human fetal colon had no effect on either brush border digestive enzyme activity or on cell proliferation [55]. While the basic mechanism underlying the differential response of these two intestinal segments remains to be determined, the above observations certainly strengthen the notion that while glucocorticoid hormones are implicated in the regulation of rodent and human GI functional development, both their time of action as well as their physiological parameters which are specifically modulated by HC, are quite dissimilar. Tab. 20.1 illustrates our current knowledge of the specific biological effects of hormones and growth factors on developing human GI. Over the years, it has become obvious that these two GI segments are under very different regulatory mechanisms even though they share similar morphological and functional characteristics. This aspect is further illustrated by the stimulatory effect of KGF on sucrase activity in the small intestine compared to its downregulation effect in the colon [56]. Moreover, a given hormone or growth factor can trigger different and sometimes opposite effects in human fetal GI tract, as opposed to that observed in rodents [1, 2] as exemplified by the comparison of EGF-induced stimulation of intestinal cell proliferation and premature appearance of sucrase activity in the mouse model [57] compared to the effects observed in human fetal intestine [58]. In light of these observations, a great deal of caution is warranted in extrapolating data obtained from rodents directly to human fetal gut. It is also clear that specific cell signals instigated following ligand-receptor binding remain to be fully understood.

Tab. 20.1 Biological effects of growth factors and hormones on the human fetal gut.

	EGF	HC	Insulin	KGF
Jejunum				
Cell proliferation	↓	↑	↑	↑
Sucrase	↓	=	=	↑
Lactase	↑	↑	=	=
Chylomicron	↑	↓	↓	ND
VLDL	↓	↓	=	ND
HDL	↓	↑	=	ND
Apo A-I	↓	↑	=	ND
Apo B–48	↑	↑	=	ND
Apo B–100	↓	↓	=	ND
Colon				
Cell proliferation	↓	=	↑	↑
Sucrase	=	=	=	↓
Lactase	=	=	=	=
Apo A-I	↑	↑	↑	ND
Apo A-IV	↑	↑	=	ND
Apo B–48	↑	↑	=	ND
Apo B–100	↓	=	=	ND

Apo, apolipoprotein; EGF, epidermal growth factor; HC, hydrocortisone; KGF, keratinocyte growth factor; VLDL, very low density lipoprotein; HDL, high density lipoprotein; ↑, increase; ↓, decrease; =, no significant effect; ND, not determined. Results from [56–61].

20.4.2
ECM and Integrins

As in other epithelia, the intestinal epithelium is separated from the underlying interstitial connective tissue by a thin and continuous sheet of specialized matrix: the basement membrane (BM). It is now recognized that BM composition defines the necessary microenvironment required for the regulation of multiple cellular functions during development and at maturity such as proliferation, migration and tissue-specific gene expression [50]. Cell-matrix interactions are mediated by various cell membrane receptors, many of which are members of the integrin superfamily [62].

Analyzing cell-matrix interactions in the intact human intestinal epithelium has been informative in evaluating the role of individual molecules in the regulation of specific cell functions. Indeed, during development, as mentioned in Section 20.2, the process of endodermal differentiation into a functional epithelium is closely related to the sequential morphogenesis of villi and crypts in both proximal and distal intestine leading to the acquisition, in the mature intestine, of specific renewing units consisting of well-separated proliferative and differentiated cell populations [10, 63]. The human intestinal epithelial BM has been found to contain all the major components specific to most BM, as well as some non-exclusive macromolecules also found in the interstitial extracellular matrix. Interestingly, important spatial and temporal differences in BM composition and expression of integrins have been found in relationship to the cell state in the human intestinal epithelium (Tab. 20.2).

Tab. 20.2 BM molecules and integrins identified in association with human epithelial cells.

BM molecules	BM-associated molecules	Integrins
Type IV collagen α1/α2 chains	Fibronectin – D, CV	α1β1 – CV
Type IV collagen α5/α6 chains – D	Tenascin-C – D, PD, CV	α2β1 – D, CV
Laminin–1 – D, CV	SPARC/BM40/Osteonectin – D, PD	αβ1 – D, CV
Laminin–2 – D, CV	Osteopontin – D	α5β1 – CV
Laminin–5 – CV	Decorin – D, CV	α6β1
Laminin–10 – CV		α7Bβ1 – D, PD, CV
Entactin/Nidogen		α9β1 – D, CV
Perlecan		α6β4 – CV

D, differentially expressed during development; PD, Differentially expressed along the proximal-distal axis

CV, Differentially expressed along the crypt-villus axis.

Adapted from [20, 64–67]

Indeed, while some basic BM components such as the classical form of type IV [α1(IV)$_2$ α2(IV)]collagen or perlecan were ubiquitously detected at the base of all intestinal epithelial cells, several others such as laminins, various BM-associated molecules and most integrins, were found to be differentially expressed. The differential expression of these molecules in the human intestine has recently been addressed in detail [20, 64–67]. In this chapter, we will review some of these components in the context of intestinal development, establishment of the proximal-distal axis and organization of the crypt-villus axis.

20.4.2.1 Development

Type IV collagen is a major BM component that forms an intricate framework with which other BM molecules associate [50]. Study of the $\alpha 1$(IV) to $\alpha 6$(IV) collagen chains in the human intestine led to the identification of the $\alpha 5$(IV) and $\alpha 6$(IV) chains in the epithelium [68, 69]. Interestingly, in contrast to the $\alpha 1$(IV) and $\alpha 2$(IV) chains, which exclusively originate from the mesenchymal compartment, the $\alpha 5$(IV) and $\alpha 6$(IV) chains were found to be produced by both epithelial and mesenchymal intestinal compartments [69]. Another peculiarity of these collagen chains is their distinct expression throughout intestinal development, the $\alpha 6$(IV) chain being expressed constitutively at all stages, whereas the $\alpha 5$(IV) chain is subject to a downregulation from the fetus to the adult. The significance of $\alpha 6$(IV) chain expression in the adult epithelial basement membrane without its presumed $\alpha 5$(IV) partner remains to be determined at the functional level but represents a good example of compositional change in the basement membrane throughout development.

Another intriguing example of developmentally-regulated BM molecules is laminin–1. Laminins are a family of $\alpha\beta\gamma$ heterotrimeric molecules that represent the predominant BM glycoprotein, both quantitatively and functionally [70]. In the early developing intestine, laminin–1 is found at the endodermal-mesenchymal interface. Thereafter, a gradual confinement of this laminin to the intervillus area is observed between 8 and 10 weeks of gestation, concomitantly with villus formation. Finally, its progressive disappearance [71] and replacement by laminin–2 occurs a few weeks later in relation with crypt formation [72].

Several integrins were found to be differentially expressed during human intestinal development [67]. In addition to $\alpha 9\beta 1$, a tenascin-C and osteopontin receptor exclusively found in the lower crypt region of developing intestine [73, 74] as well as a number of laminin-binding integrins, such as $\alpha 7B\beta 1$, were also found to be subject to a developmentally-regulated pattern of expression in the human intestinal mucosa. Indeed, the $\alpha 7B$ subunit is first detected at 12 weeks of gestation in cells lining the newly formed villi. From mid-gestation onward, the $\alpha 7B$ subunit is then restricted to the upper crypt and lower villus regions [75]. Another good example is the gradual redistribution of the $\alpha 3\beta 1$ integrin from the entire epithelium to the villus epithelium [72] in parallel with its main ligand, laminin–5 [76].

20.4.2.2 Anteroposterior (AP) Axis

Most epithelial BM molecules studied to date in the developing gut, including type IV collagens, laminins and fibronectin, were found to be expressed in comparable patterns in the small intestine and colon (Tab. 20.2). This was not unexpected in view of the great similarities between these two segments in terms of architecture and functionality during the first two trimesters of gestation [9, 63]. However, some ECM molecules were nevertheless found to be differentially expressed along the length of the developing human intestine, including tenascin-C and SPARC/BM40/osteonectin. Tenascin-C was found to be expressed at the epithelial-mesenchymal interface of the small intestinal mucosa but absent in fetal colonic villi [77]. Conversely, SPARC/BM40/osteonectin was predominantly found at the colonic epithelial-mesenchymal interface at mid-gestation [78]. The complementary patterns of expression of tenascin-C and SPARC/BM40/osteonectin suggest that these two anti-adhesive molecules have a functional relationship in the process of villus formation.

20.4.2.3 Crypt-Villus Axis

A number of BM and BM-associated molecules were also found to be differentially expressed along the crypt-villus axis (Tab. 20.2) suggesting potential interrelationships between compositional changes in BM and modifications in epithelial cell state [20, 67]. Fibronectin expression for instance was found to be closely associated with the proliferative and undifferentiated epithelial cell fractions both in the intact developing and adult intestine [79] as well as in intestinal cell models *in vitro* [80].

It is however the reciprocal expression of laminin–2 and laminin–10 along the crypt-villus axis that are among the better documented BM compositional changes occurring in the human intestinal epithelium [72, 76, 81, 82]. Indeed, the presence of laminin–2 as a lower-crypt form and laminin–10 as an upper-crypt/villus form suggested for the first time a possible correlation between laminin expression and intestinal cell differentiation. Further investigation using experimental cell models confirmed such a correlation between intestinal cell interaction with laminin–10 and the promotion of differentiation-related gene expression [83].

Analysis of the integrin repertoire along the crypt-villus axis in the human intestinal epithelium has enabled the identification of a number of $\beta 1$ and $\beta 4$ integrins exhibiting differential expression patterns. Among these, integrins $\alpha 3\beta 1$, $\alpha 7B\beta 1$ and the functional form of $\alpha 6\beta 4$ appear to be related, in concert with their ligands, laminin–5 and 10, to the process of intestinal cell differentiation. On the other hand, the integrins $\alpha 1\beta 1$, $\alpha 2\beta 1$, $\alpha 5\beta 1$ and the non-functional form of $\alpha 6\beta 4$ appear to be coupled with the undifferentiated and proliferative status of crypt cells [75, 79, 84].

20.4.3
Cell Signaling Pathways

Over the last decade, major attention has been devoted to our understanding of intracellular signaling pathways that transmit extracellular cues for epithelial cell proliferation and differentiation along the intestinal crypt-villus axis. Although many aspects remain to be addressed, strong lines of convergence are emerging in growth and differentiation signaling. At least four signaling networks, not yet fully elucidated, appear to be crucial for cell cycle progression and differentiation of intestinal epithelial cells : (1) the Ras 〉 Raf 〉 MEK 〉 ERK1/2 cascade which is mostly activated by growth factors, resulting in activation of a range of transcription factors such as Elk–1, c-Ets–1 and c-Ets–2; (2) p38 MAP kinase which phosphorylates and activates several tissue-specific transcription factors; (3) the β-catenin signaling pathway which is activated by Wnt factors; and (4) the lipid signaling pathway including phosphatidylinositol 3–kinase and subsequent regulation of AKT. The following discussion will focus on these specific cellular signaling pathways in relation to the control of growth and differentiation of enterocytes.

20.4.3.1 ERK-MAP Kinase Cascade

Mitogen-activated protein kinases, also described as extracellular signal regulatory kinases (ERKs) are termed as such because in a variety of cultured cell lines, they can be activated by many different mitogens, including peptide growth factors such as epidermal growth factor (EGF) and growth-promoting hormones and lipid metabolites such as lysophosphatidic acid (LPA) and prostaglandins. Two closely related mammalian MAP kinases, p44 (ERK1) and p42 (ERK2), have been cloned and were found to be ubiquitously expressed [85]. An interesting feature of this family of kinases is that they require dual phosphorylation of specific threonine and tyrosine residues in order to be activated. ERK1/2 activation is mediated by a dual specificity kinase termed MAP kinase kinase (MEK), which in turn is activated by Raf oncoproteins. Raf proteins appear to be regulated by both the Ras family of oncoproteins and 14–3–3 proteins. Cellular activation of Ras is mediated by the guanine nucleotide exchange factor Sos which, when associated in a complex with the adaptor protein Grb2, binds to activated receptor tyrosine kinases. Hence, the cascade leading from receptors with intrinsic tyrosine kinase activity to ERK activation is relatively complete. Upon stimulation, ERK1/2 translocate to the nucleus where they likely phosphorylate nuclear transcription factors and hence regulate gene expression [85]. There is compelling evidence for the critical involvement of this signaling cascade in the regulation of intestinal epithelial cell proliferation. Firstly, K-ras^{V12G} is the most frequently mutated oncogene in colorectal cancer [86, 87]. It has been demonstrated that targeted expression of K-ras^{V12G} in the intestinal epithelium causes activation of the MAP kinase/ERK cascade and spontaneous tumorigenesis in mice [88]. In addition, blockade of the MEK/ERK pathway suppresses colon tumor growth *in vivo* [89]. Studies on cultured cells have revealed a close correlation between ERK activation and DNA synthesis in which pharmacological and molecular inhibition of cellular ERK activity was shown to block cell cycle progression of IEC–6 and Caco–2/15 cells [90, 91]. Furthermore, the phosphorylated and activated forms of ERK1/2 have mostly been detected in the nucleus of undifferentiated crypt cells in human fetal small intestine [90]. Finally, it has been reported that EGF [91], L-glutamine [92], glucagon-like peptide–2 [93] and vasopressin [94] stimulate proliferation of intestinal crypt cells by activating the ERK pathway.

Several studies have reported that ERKs are selectively inactivated during enterocytic differentiation. Indeed, ERK1/2 activities were markedly reduced in post-confluent differentiating Caco–2/15 cells [90] and in HT29–N2 cells upon changeover to glucose-free growth medium [95]. Hence, inactivation of the ERK pathway may be necessary for cell cycle arrest and induction of terminal differentiation of intestinal epithelial cells. However, low albeit significant levels of activated ERK were still detected in differentiated Caco–2/15 or primary cultures of human differentiated enterocytes, supporting the notion that ERK1/2 are also involved in the regulation of some aspects of enterocyte differentiation. Indeed, pharmacological inhibition of MEK/ERK activation in differentiated enterocytes with PD98059 or U0126, inhibited sucrase-isomaltase expression [90] and altered microvillus architecture [96]. Accordingly, it has been recently reported that ERK1/2 and their upstream modulators Ras and MEK are found in the brush border of differentiated enterocytes. Finally, brush border-associated ERK1/2 are stimulated by both EGF and feeding, suggesting a role in nutrient, ion transport or other brush border-associated functions [96].

20.4.3.2 p38–MAP Kinase Cascade

As discussed above, the founding members of the MAP kinase family, ERK1/2, are predominantly activated by mitogenic stimuli. However, other members of the MAP kinase family have been identified as being preferentially activated by non-mitogenic stimuli such as the c-Jun N-terminal kinase/stress-activated protein kinase (JNK/SAPK) and the p38 MAPK subfamilies, which are activated by most environmental stresses [85]. The p38 MAPKs are also strongly activated by pro-inflammatory cytokines such as interleukin 1 (IL–1) and tumor necrosis factor α (TNF-α), which play an important role in the regulation of the inflammatory response. Four isoforms of p38 MAPKs (p38α, β, γ and δ) have been identified, all of which are phosphorylated and activated by the MAPK kinase (MKK6). Other MKKs exhibit more restricted specificity: for instance, MKK3 activates p38α, p38γ, and p38δ, whereas MKK4 can only activate p38α. Substrates of p38 MAP kinases mostly include other protein kinases and a growing list of transcription factors such as ATF–2, C/EBPβ, CHOP, MEF2C [108]. An important role has recently emerged for p38 MAP kinases in various vertebrate cell differentiation processes, including enterocyte differentiation [97]. In intestinal epithelial cells, p38α was found to be rapidly activated in cells induced to differentiate. Treatment of Caco–2/15 cells or primary cultures of differentiated enterocytes with the specific p38 MAP kinase inhibitor, SB203580, repressed the expression of intestine-specific genes such as sucrase-isomaltase, lactase and villin. The stimulation of intestine-specific gene expression by p38 MAP kinase is apparently mediated by the Cdx2 transcription factor, a p38 substrate also known to be essential in enterocyte differentiation [98]. Indeed, p38 regulates Cdx2 activity via phosphorylation and a rise in the transcriptional activity of Cdx2. Although further studies are needed to pinpoint the upstream pathways activating p38α in committed cells induced to differentiate, this latter study provides novel fundamental insights into the functional role of p38 in the early events of intestinal epithelial differentiation.

20.4.3.3 Wnt Pathway

Activation of canonical Wnt signaling is initiated by binding of secreted Wnt glycoproteins to their transmembrane co-receptors, the Frizzled proteins and the LDL receptor-related proteins 5 and 6. This complex transduces a signal into the cell, resulting in nuclear accumulation of β-catenin and subsequent activation of TCF target genes [99]. Several previous *in vivo* studies have implicated Wnt signals in the regulation of intestinal stem cell proliferation [100–102]. Firstly, proliferative cells at the bottom of small intestinal and colonic crypts accumulate nuclear β-catenin [103, 104]. Secondly, mutations that activate the Wnt/β-catenin pathway can lead to colorectal cancer in humans [100] as well as adenomatous polyp formation in murine intestine [105], whereas mutations of TCF–4 leads to depletion of the intestinal proliferative compartment in fetal mice [106]. Finally, the reduction of intestinal epithelial cell proliferation coincidentally with crypt loss was observed in transgenic mice overexpressing Dickkopf1, a secreted Wnt inhibitor [107]. Together, these studies attest to the critical importance of Wnt signals in the homeostasis of the intestinal epithelium.

20.4.3.4 Phosphatidylinositol 3–Kinase Signaling Pathway

A number of receptor-mediated signals involve phosphatidylinositol 3–kinase (PI3K), including pathways promoting mitogenesis, cell survival, and motility. PI3K catalyzes the production of the potent second messenger phosphatidyl inositol 3,4,5–triphosphate (PIP3), which in turn activates pleckstrin homology domain-containing effectors such as the well characterized serine/threonine kinase PKB (PKB/Akt) isoforms [108]. In intestinal epithelial cells, the PI3K-PKB/Akt pathway has been implicated in EGF- and TGFα-mediated cyclin D1 expression [109] and proliferation as well as in Ras-mediated transformation [110]. However, its role in cellular differentiation is unclear, with conflicting reports as to the promotion or inhibition of differentiation [111, 112]. Wang et al. [111] reported that PI3K inhibition through overexpression of PTEN (which directly dephosphorylates the D3 phosphate group of PI3K lipid products) or wortmannin treatment, results in an increase in Cdx2 gene and protein expression as well as an induction of alkaline phosphatase and sucrase-isomaltase enzymatic activities. By contrast, we have recently demonstrated that PI3K is part of a signaling pathway necessary for functional and morphological differentiation of intestinal epithelial cells [112]. Indeed, inhibition of PI3K decreased the expression of enterocyte markers, namely, sucrase-isomaltase and villin and reduced cell polarization and brush border formation. Furthermore, once Caco–2/15 cells reach confluence, E-cadherin engagement triggered the recruitment of PI3K and activated its signaling at sites of cell-cell contact. This activation appears to be essential for the assembly of adherens junctions and the association of their components with the actin cytoskeleton. Finally, one of the molecular events resulting from E-cadherin-mediated cellular aggregation is the activation of p38 MAPK in a PI3K-dependent manner [112]. Hence, PI3K signaling may play an important role in initiating intestinal epithelial cell differentiation. In addition to these observations, it was recently reported that E-cadherin engagement in intestinal epithelial cells leads to MEK/ERK inhibition in a PI3K/Akt-dependent pathway [113]. Thus, PI3K may play a role in the transition between proliferation and differentiation in the intestinal epithelium, a role also shared by E-cadherin [114, 115]. However, as discussed above, it has been reported that PI3K is also involved in proliferation of intestinal epithelial cells [109, 110]. These roles of PI3K in proliferation versus cell-cell adhesion/differentiation of intestinal epithelial cells could appear contradictory at first, but can be explained by the different subcellular localization of PI3K. Indeed, it has been demonstrated that PI3K exhibited a diffuse cytoplasmic distribution in subconfluent growing intestinal cells whereas a localized distribution pattern at cell-cell interfaces was observed in confluent differentiating cells and in differentiated enterocytes along the intestinal crypt-villus axis [112, 116]. Therefore, a hypothetical model involving the dual role of PI3K according to intestinal epithelial cell status could be proposed whereby PI3K is involved in proliferation in pre-confluent growing cells whereas the same protein controls the onset of differentiation when it is translocated to the E-cadherin adhesion complex upon reaching confluence. Further studies are needed to clarify the mechanisms by which PI3K translocates from cytoplasm to E-cadherin-mediated contacts at this differentiation stage. However, it has been recently demonstrated that the human homolog of Disc-Large (hDlg), a member of the membrane-associated guanylate kinase-like protein family [117], binds to the regulatory subunit of PI3K (p85) with both proteins being part of a common macromolecular complex with E-cadherin at cell-cell contact sites on intestinal epithelial cells. By using an RNA interference approach, hDlg was shown to be essential in recruiting PI3K to cell-cell contact sites, for both adherens junction integrity and for functional differentiation of intestinal epithelial cells [116].

Figure 20.3 summarizes our present understanding of the specific intracellular signaling pathways involved in the regulation of proliferation and differentiation of human intestinal epithelial cells.

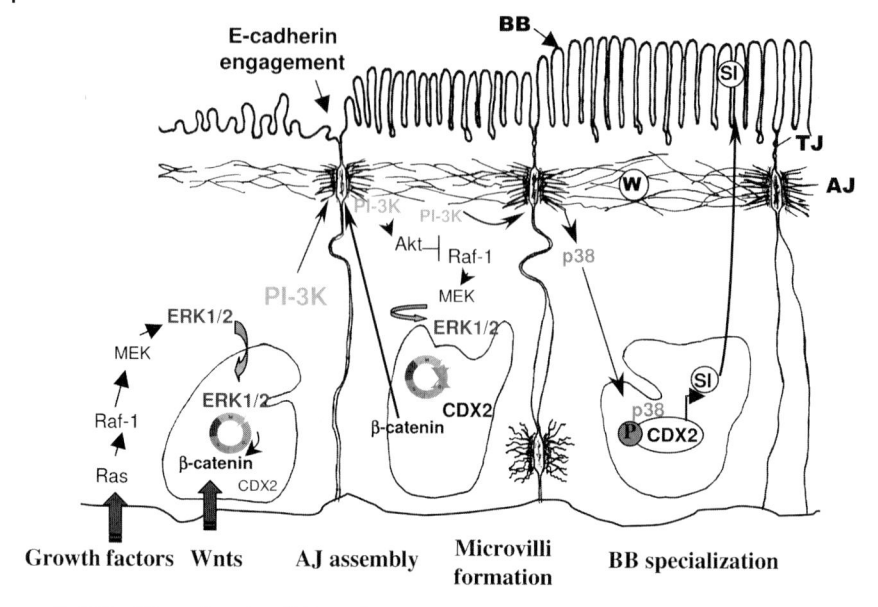

Fig. 20.3.
Schematization of proposed pathways involved in the regulation of proliferation and differentiation of intestinal epithelial cells. Growth factors stimulate proliferation of undifferentiated cells by activating the ERK pathway while Wnt factors stimulate proliferation by inducing accumulation of nuclear β-catenin/TCF–4 target genes. Upon cell-cell contacts, E-cadherin engagement recruits catenin proteins and PI–3K. This recruitment of PI–3K stimulates PI–3K signaling which, by increasing the amount of F-actin at the sites of cell-cell contact (through Rac?), promotes the assembly of adherens junction components with the cytoskeleton and induces the activation of the p38α MAPK cascade which enhances the transactivation capacity of CDX2 toward intestine-specific genes. TJ, tight junctions; AJ, adherent junctions; BB, brush border; W, terminal web of actin; SI, sucrase-isomaltase. (This figure also appears with the color plates.)

20.4.4
Transcription Factors

A much clearer understanding of the molecular mechanisms that govern intestinal epithelial cell proliferation and differentiation has emerged over recent years. Extracellular components that dictate whether epithelial cells should proliferate or differentiate invariably influence a number of intracellular signaling pathways that will ultimately instruct transcription of a specific gene to occur or not to occur. A number of strategies have been successfully utilized to identify transcription factors that specifically regulate intestinal genes. This section will focus on transcription factors that are suspected to be directly involved in the control of intestinal epithelial cell proliferation, determination and differentiation. In a first step, genetic evidence of specific transcription factors involved in the control of intestinal epithelium development and maintenance will be discussed. In a second step, transcriptional components crucial in activating specific intestinal epithelial genes during enterocyte differentiation will be presented.

20.4.4.1 Transcription Factors Involved in the Determination of Intestinal Epithelial Cell Lineage

TCF/LEF transcription factors
The TCF/LEF family of transcription factors comprises several members that share the nature of being the final step in the Wnt cascade. Proteins of this family contain an 80–amino acid high mobility group (HMG) box that binds DNA as monomers. TCF/LEF factors cannot directly activate transcription and thus require the presence of a co-activator, of which the Wnt signaling effector β-catenin is the most prevalent partner (for a review, see [118]). It is well accepted that TCFs act as repressors by complex formation with co-repressors in the absence of nuclear β-catenin. Following the activation of the Wnt signaling pathway, complex formation with β-catenin results in the recruitment of histone acetylases to TCF target genes, thus activating their transcription. Evidence of the importance of this class of transcription factors in the maintenance of intestinal epithelial homeostasis has come from data obtained from the generation of knockout mice. *Tcf–1* mutant mice develop adenomas in the gut [119]. Tcf1, which is the most abundant gut isoform, lacks a β-catenin interaction domain suggesting a natural role for this isoform in a negative feedback loop of Tcf–4/β-catenin signaling in the intestine [120]. In addition, *Tcf–4* mutant mice die shortly after birth and display a complete absence of stem cell compartments in the crypts of the small intestine, therefore lacking actively dividing crypt cells [120]. These observations imply that inappropriate alteration of the TCF genes in intestinal epithelial cells could represent important events in the deregulation of cellular proliferation and initiation of transformation.

Hes1 and Math1 transcription factors
The Notch signaling pathway plays a key role in modulating cell-fate decisions during development. This pathway involves activation of the Notch receptor leading to multiple intracellular proteolytic steps. It is now acknowledged that the Notch pathway is directly involved in the determination of intestinal epithelial cell fate [121]. The basic helix-loop-helix transcription factors Hes1 and Math1 have been identified as downstream targets of this pathway [122, 123]. Deletion of Math1 in mice results in the depletion of the goblet, Paneth and enteroendocrine cell lineages in the small intestine [123]. Hes1 null mice exhibit higher levels of Math1 expression with an increase in the number of enteroendocrine and goblet cells and fewer enterocytes [122]. In crypt progenitor stem cells with high levels of Notch, Hes–1 is switched on to block the expression of Math1, leading to the production of enterocytes. Cells presenting low Notch levels are blocked in the production of Hes–1

leading to the permissive expression of Math1 which in turn allows the cells the choice of becoming either goblet, Paneth or enteroendocrine cells.

Kruppel-like transcription factors
Kruppel-like transcription factors are zinc-finger proteins that have been linked to cell growth regulation and tumorigenesis in a number of systems. KLF4 and KLF5 are the two isoforms that are expressed in the intestinal epithelium. *In vitro* studies have suggested that induction of KLF4 expression is related to differentiation while overexpression of KLF5 likely promotes proliferation [124]. *Klf4* null mice die within 1 day after birth and display abnormal differentiation of colonic goblet cells confirming a crucial role in colonic epithelial cell differentiation *in vivo* [125]. Although some evidence suggests a relationship between KLF5 altered functions and intestinal tumor progression [126], a direct functional analysis has yet to be reported.

20.4.4.2 Transcriptional Regulators of Intestinal-specific Genes

Sucrase-isomaltase (SI), an enzyme expressed in the brush border of mature enterocytes, represents one of the most extensively studied intestinal genes at the transcription level [1–3, 10]. The developmental pattern of SI gene expression coincides with the differentiation and maturation of the intestinal epithelium which explains why this gene has been an excellent model for identifying a number of transcription factors potentially involved in intestinal homeostasis. These factors will be introduced and discussed along with their demonstrated roles in the regulation of proliferation and differentiation.

The caudal-related homeodomain proteins (Cdx)
Cdx1 and *Cdx2* are members of the caudal-related homeobox gene family based on their sequence homology with the *Drosophila melanogaster caudal* gene. These factors were originally demonstrated to both interact with and regulate the SI gene promoter both *in vitro* and *in vivo* [127, 128]. Mutagenesis of the Cdx element of the SI gene impairs promoter activity during intestinal differentiation and postnatal development [129]. The expression of Cdx1 and Cdx2 is restricted to the intestinal epithelium, with the Cdx1 protein mainly located in the crypt compartment whereas the Cdx2 protein is detected in both crypt and villus compartments [130]. Forced expression of either Cdx1 or Cdx2 in the rat intestinal epithelial cell line IEC–6 causes cells to retard proliferation and initiate morphological differentiation with induction of intestinal-specific gene expression such as SI [131, 132]. Similar strategies used in the context of colorectal cell lines have demonstrated a role for both Cdx1 and Cdx2 in cell-cell adhesion and differentiation [133]. However, the function of Cdx1 in cellular proliferation still remains a matter of debate since other groups have reported pro-proliferative properties of this protein in intestinal physiology [134]. *Cdx1* null mice develop an anterior homeotic transformation of the vertebra without any gross abnormalities in the gut [135]. *Cdx2* null mice are developmentally impaired early during embryogenesis with apparent defects in implantation [136].

Heterozygous *Cdx2* null mice develop hamartomas in the proximal colon [136, 137] and show an increased frequency of colonic polyposis when inbred with the APC/Min mouse model [138]. Finally, ectopic expression of Cdx2 in stomach epithelium results in the formation of intestinal metaplasia supporting a role for this factor in intestinal determination [139].

The Hepatocyte Nuclear Factor–1 (HNF–1) family

The HNF–1 family is composed of HNF–1α and HNF–1β isoforms that are distantly related to homeodomain proteins that bind to DNA as either homodimers or heterodimers. These factors were originally identified as liver-enriched transcriptional regulators but are expressed in a variety of tissues including kidney, pancreas, stomach and intestine. HNF–1 interaction is critically important in supporting SI gene transcription during postnatal development and intestinal epithelial differentiation [140]. The HNF–1α homodimer complex promotes activation of SI transcription whereas HNF–1β-containing complexes reduce this transcriptional modulation. The ratio of HNF–1α to HNF–1β increases accordingly during differentiation and postnatal development [150]. It is now evident that several intestinal epithelial genes are positively activated by HNF–1α [141–145]. However, direct implication of HNF–1α in the control of cellular proliferation and/or differentiation awaits further investigation.

The GATA transcription factor family

The GATA transcription family members are characterized by conserved zinc finger DNA-binding domains that recognize the WGATAR consensus sequence. GATA–4, GATA–5 and GATA–6 are the three forms expressed in the intestinal epithelium. These isoforms can interact and regulate intestinal epithelial-specific gene promoters such as SI [129], lactase-phlorizin-hydrolase (LPH) [144], liver fatty acid binding protein [145] and trehalase [146] with varied affinities and magnitude. In general, GATA isoforms influence transcriptional regulation in combination with other tissue-specific transcriptional regulators. For example, the GATA–4 isoform functionally interacts with Cdx–2 and HNF–1α to modulate SI [129] (Figure 20.4) and LPH genes [144]. Recent reports have studied the distribution pattern of GATA members along both the proximal to distal and crypt-villus axes of the intestine. GATA–4 is detectable in the differentiated epithelium of the small intestine but absent in the colon [129] while GATA–6 is detectable in the epithelium of the entire intestine [145]. The presence of GATA–5 protein in adult intestinal epithelium remains speculative [145]. Both GATA–4 and GATA–6 play critical roles in the definitive formation of the endoderm early during embryonic life as demonstrated by the generation of mice with targeted null mutations for each of these genes (for a review, see [147]). Forced expression of GATA–4 and GATA–6 in embryonic stem cells leads to differentiation into extra-embryonic endoderm [148]. Although much evidence has indicated a role for GATA members in the determination and differentiation of the intestinal epithelium, definitive experimental proof is still to come.

The CCAAT-Displacement Protein (CDP) as an intestinal transcriptional repressor

The CDP protein is a transcriptional repressor closely related to the *Drosophila melanogaster cut* protein. CDP homologs structurally share a unique homeodomain and three similar regions designated cut repeats (CR) which function as DNA-binding domains. The presence of multiple DNA-binding modules results in a highly flexible mode of sequence recognition (for a review, see [149]). CDP is believed to repress transcription by two distinct mechanisms: direct competition with transcriptional activators for a given DNA-binding site or active repression with local recruitment of co-repressors such as histone deacetylases (HDAC) [149]. The transcriptional activity of the CDP protein is also controlled by several post-translational modifications such as phosphorylation, acetylation and proteolytic cleavage which produce isoforms with different transcriptional properties [149]. Mutagenesis analysis of the SI promoter has identified a novel element involved in the colonic repression of the gene that interacts with the CDP protein [150]. CDP protein distribution along the cephalo-caudal axis of the intestine supports a role for this protein in the more distal part of the epithelium. Homozygous mutant mice for CDP display colonic re-expression of the SI gene [150], a phenomenon observed in colonic adenomatous polyposis [151]. Several studies have supported a role for CDP in the control of cellular proliferation and/or differentiation. CDP can serve as a cell cycle-dependent transcriptional factor in proliferating cells and form a complex with the retinoblastoma protein-related protein p107 and cyclin A. CDP also controls the transcription of several genes involved in the cell cycle such as p21, thymidine kinase, histone genes and *c-myc* [149]. Whether CDP is directly involved in such phenomena in intestinal epithelial cells still remains unexplored.

In conclusion, delineation of the molecular mechanisms regulating intestinal epithelial-specific gene transcription has been an effective strategy for identifying transcription factors involved in the control of intestinal homeostasis. The question of how particular transcription factors control proliferation and/or differentiation in combinatory processes still represents an interesting experimental challenge. The generation of intestinal conditional knockout and intestinal ectopic expression models for specific transcription factors represents a promising avenue to monitor functions of these entities on epithelial renewal during gastrointestinal development. The remaining challenge will be to reconcile findings obtained from mouse models towards a better comprehension of the mechanisms that govern human gastrointestinal tract development and diseases.

20.4.5
Cell Survival, Apoptosis and Anoikis

Apoptosis, or programmed cell death, is an intricately regulated process which has crucial roles in development, tissue homeostasis and repair, as well as in the pathogenesis of an increasing number of diseases [152–154]. Maintenance of cell survival requires constant regulation through various stimuli originating from growth factors, hormones and cytokines, as well as from cell adhesion [152–155]. To this end,

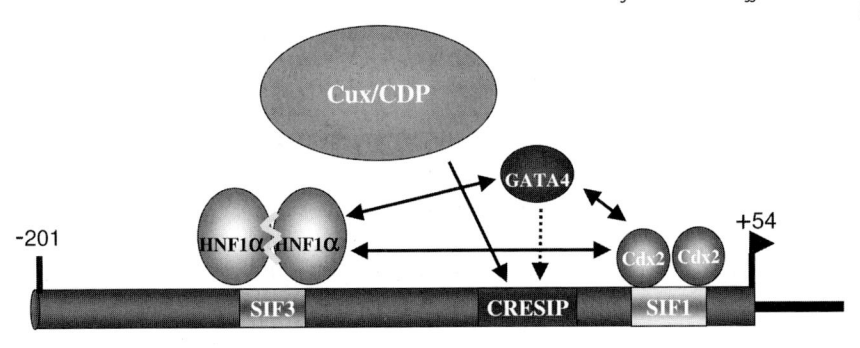

Fig. 20.4
Model of SI transcriptional regulation. HNF-1α, GATA-4 and Cdx2 physically interact with each other as well as with specific elements of the SI promoter to stimulate transcription. Cux/CDP interacts with the CRESIP element of the SI promoter and is involved in the repression of SI colonic transcription. (This figure also appears with the color plates.)

the loss of integrin-ECM interactions results in a form of apoptosis termed "anoikis" [155, 156]. Apoptosis can also be triggered by pro-inflammatory cytokines and various injuries caused by DNA-damaging agents, radiation, free radicals and toxins [152–155]. In any case, the typical death throes of a cell undergoing apoptosis/anoikis include membrane blebbing, nuclear condensation, DNA fragmentation, organelle degradation and cell shrinkage, in order to generate an apoptotic body which is either phagocytosed or released into the lumen for evacuation.

The Bcl–2 homologs are recognized as a critical decisional checkpoint in cell survival and death. More than 15 family members have been identified so far in mammalian cells, functioning either as anti-apoptotic (e.g. Bcl–2, Bcl-X_L, Mcl–1) or pro-apoptotic (e.g. Bax, Bad, Bak) regulators. Bcl–2 homologs are well known to interact among themselves, allowing for the titration of pro- and anti-apoptotic functions. To this end, the suppression or induction of apoptosis will largely depend on a balance of anti- and pro-apoptotic activities from multiple homologs [153, 154]. Such a balance is established primarily through the modulation of Bcl–2 homolog expression; however, post-transcriptional and post-translational modes of regulation of expression and activity can also contribute to this balance [153–155].

20.4.5.1 Crypt-Villus Axis Distinctions

Within the context of continuous cell renewal of the adult intestinal epithelium, apoptosis/anoikis is the normal fate of villus-tip cells in the small intestine, and of surface cells in the colon [157–161]. Spontaneous crypt cell apoptosis, a process of rarer occurrence, serves to remove defective/injured progeny cells, as well as senescent Paneth cells [158, 159, 162]. Such apparent "duality of fate" between undifferentiated and differentiated cells, coupled to distinct patterns of expression of Bcl–2 homologs along the crypt-villus (or gland-surface) axis, has largely contributed to the concept that human adult intestinal epithelial cell survival is regulated distinctively according to its state of differentiation [157–162]. However, little is known of the regulation and roles of apoptosis/anoikis during the development of the human digestive tract. One of the first questions investigated by our group pertained to the establishment of cell survival mechanisms in the developing gut and how these relate to adult intestinal physiology. Between 9 and 17 weeks, epithelial apoptosis is not observed in the human fetal jejunum [163]. Villus-tip apoptosis/anoikis emerges only by the time of the 18-week stage, whereas spontaneous crypt cell apoptosis is not detected [163]. Although Bcl-2, Bcl-X_L, Mcl-1, Bax, Bak and Bad are already expressed at 9 weeks gestation, epithelial localization of the majority of these homologs undergoes a gradual compartmentalization process between 15 and 17 weeks, coinciding with cryptogenesis [163]. Furthermore, such compartmentalization is closely paralleled by changes in expression levels [163]. By 18 weeks, distinct patterns of expression of these Bcl-2 homologs are established along the crypt-villus axis with no further changes in localization or expression occurring thereafter (i.e. 19 and 20 weeks). Interestingly, these established crypt-villus expression patterns in the mid-gestation jejunum correspond to patterns observed in its third-trimester and adult counterparts [164–167]. Consequently, the regulatory mechanisms of human intestinal epithelial cell survival and apoptosis/anoikis along the crypt-villus axis are established by mid-gestation [163], as in the case of other small intestinal architectural and functional features (see Section 20.2).

20.4.5.2 Proximal-Distal Axis Distinctions

Although originating from the same tube-like primitive gut, the adult small and large intestinal mucosae differ in organization and physiological functions. Observed differences in rates of apoptosis, in the susceptibility to apoptosis following insult/injury, and in the expression/localization of Bcl-2 homologs, have suggested that the regulation of epithelial cell survival may also be characterized by distinctions between the adult small and large intestines [157–159, 161, 162]. Physiopathologies involving dysregulation of apoptosis further underlie such proximal-to-distal distinctions. For instance, cancer is extremely rare in the small intestine as opposed to the colon; similarly, the onset of inflammatory bowel disease usually occurs in the proximity of the ileum-colon junction [158, 161, 162, 168]. Therefore, we also looked into whether such apparent segmental specificities resulted from early developmental processes. As in the case of the jejunum, the localization of Bcl-2, Bcl-X_L, Mcl-1, Bax, Bak and Bad in the developing human ileum and colon undergo gradual compartmentalization between 15 and 17 weeks, culminating at 18 weeks and likewise paralleled by developmental changes in their expression levels [169]. However, the resulting crypt-villus patterns and expression profiles of these Bcl-2 homologs differ sharply not only from those of the jejunum, but between the ileum and colon as well [169]. Although the patterns of Bcl-2 homolog expression in the mid-gestation ileum correspond well to those reported in its adult counterpart [157, 167, 170], those observed in the mid-gestation colon correlate poorly with what has been reported in the adult [164–167, 170]. Considering that the colon undergoes its final maturation between 30 weeks and birth in order to acquire its adult morphological and functional features (see Section 20.2), it is therefore likely that additional developmental changes in colonic Bcl–2 homolog expression profiles and apoptotic control mechanisms occur during this period. Nonetheless, proximal-to-distal bowel segment-specific distinctions in the expression profiles of Bcl–2 homologs are already in place in humans by mid-gestation [169].

The next question that presented itself was how proximal-distal-specific patterns of Bcl–2 homolog expression profiles translate into distinctions in the control of cell survival between the small and large intestinal epithelia. Survival signals originating from extracellular cues involve signaling events that affect distinctively the expression and/or functions of the various individual Bcl–2 homologs, depending on the tissue and cell type studied [154–156, 171–173]. Hence, the PI3–K/Akt (see Section 20.4.3.4) and the MEK/ERK (Section 20.4.3.1) pathways, among others, are often involved in the promotion of cell survival [155, 171]. In the case of integrin-mediated cell survival, the focal adhesion kinase (Fak) is also implicated. Conversely, the induction of apoptosis/anoikis can be driven by other signaling pathways, such as the p38 pathway [154–156, 171, 174, 175] (see Section 20.4.3.2). Our *in vitro* analyses, using organotypic culture of explants from 18–20–week-old fetuses, not only demonstrated segment-specific susceptibilities to apoptosis between mid-gestational jejunum and colon epithelia, but also uncovered segment-distinct survival roles for Fak, PI3–K/Akt and the MEK/ERK pathways [176]. Such segment specificities were found in turn to be linked with distinctions in the influence of these pathways, as well as insulin, on the individual expression of Bcl–2 homologs (i.e. Bcl–2, Bcl-X_L, Mcl–1, Bax, Bad and Bak) between the jejunum and colon [176]. Consequently, bowel segment-specific control mechanisms of epithelial cell survival and apoptosis are a fact along the proximal-distal axis of the human gut, and such segmental specificities are already functional by mid-gestation.

20.4.5.3 Intestinal Cell Survival and Death: Differences and Differentiation

The last question to be discussed herein brings us somewhat full circle with regard to the control of intestinal epithelial cell survival, since it deals with how the crypt-to-villus patterns of Bcl–2 homolog expression, in both the mid-gestation and adult human gut, translate into distinctions in the regulation of survival between undifferentiated and differentiated epithelial cells. Our *in vitro* functional analyses, using a combination of human enterocyte-like cell models (e.g. HIEC, PCDE and Caco–2/15; see previous sections), demonstrated that differentiated cells exhibit a susceptibility to apoptosis/anoikis that is distinct from undifferentiated cells through: (1) distinct expression profiles of Bcl–2 homologs that are gradually established during the enterocytic differentiation process [177]; (2) a differential involvement of the PI3–K/Akt and MEK/ERK pathways in survival [177–180]; (3) selective survival roles of Akt isoforms (Akt–1, –2 and –3) [180]; (4) distinct roles enacted by β1, α6β4 integrins and Fak in the suppression of anoikis [177–179]; (5) selective survival/death promoting roles of p38 isoforms (α, β, γ and δ) [178, 179]; and (6) complex, differentiation state-specific roles of cell adhesion, Fak, insulin, PI3–K/Akt, MEK/ERK and p38 pathways in the regulation of the expression of individual Bcl–2 homologs [177, 179]. Therefore, human intestinal epithelial cell survival and apoptosis/anoikis are indeed subject to differentiation state-specific control mechanisms (Fig. 20.5). With regards to anoikis specifically, this in turn fits well with the fact that intestinal epithelial cells, in both the mid-gestational and adult gut, express distinct profiles of integrins and interact with specific ECM components along the crypt-villus axis, depending on their state of differentiation (see Section 20.4.2).

In closing, the present discussion illustrates how a "coming together" of *in vivo* and *in vitro* analyses in the developing human gut can provide new insights into the complex, synchronized regulation of cell survival and apoptosis/anoikis that is required in the process of adult tissue maintenance and renewal. In the particular case of the intestinal epithelium, it is now clear that the survival and death of cells are not only subject to differentiation state-specific control mechanisms, but also to bowel segment-specific distinctions in the aforesaid mechanisms. However, much remains to be understood regarding the regulation and roles of apoptosis/anoikis during development of the human digestive tract, as well as during the normal maintenance of the epithelium throughout adult life. For instance, what is the role of apoptosis during the transformation of the stratified endoderm into a simple columnar epithelium between 7 and 9 weeks of gestation? In addition, the expression, regulation and roles of Bcl–2 homologs and signaling molecules/pathways, other than those mentioned herein, remain to be analyzed *in vivo* and *in vitro* along both the crypt-villus and the proximal-distal axes. Such further studies will most assuredly provide a greater understanding of the roles of apoptosis/anoikis in the homeostasis and repair of the adult human intestinal epithelium, as well as in the pathogenesis of intestinal disorders that involve a dysregulation of apoptosis.

20.5
Biomedical Applications

Over the last decade, our knowledge of the basic cellular and molecular mechanisms governing the morphological and functional development of the GI mucosa has greatly expanded using animal models. Indeed, the generation of intestinal conditional knockout and intestinal ectopic expression models for specific transcription factors, have been and will be instrumental in establishing basic concepts for embryonic and fetal development. The considerable challenge will always be the validation of these concepts in human beings. Indeed, as stated in the Introduction, a great deal of caution is warranted in extrapolating data obtained from rodents directly to the human gut. Our group has been able to develop and characterize several new techniques and appropriate human gastric and intestinal tissue and cell models to specifically address fundamental questions relevant to the physiopathology of the

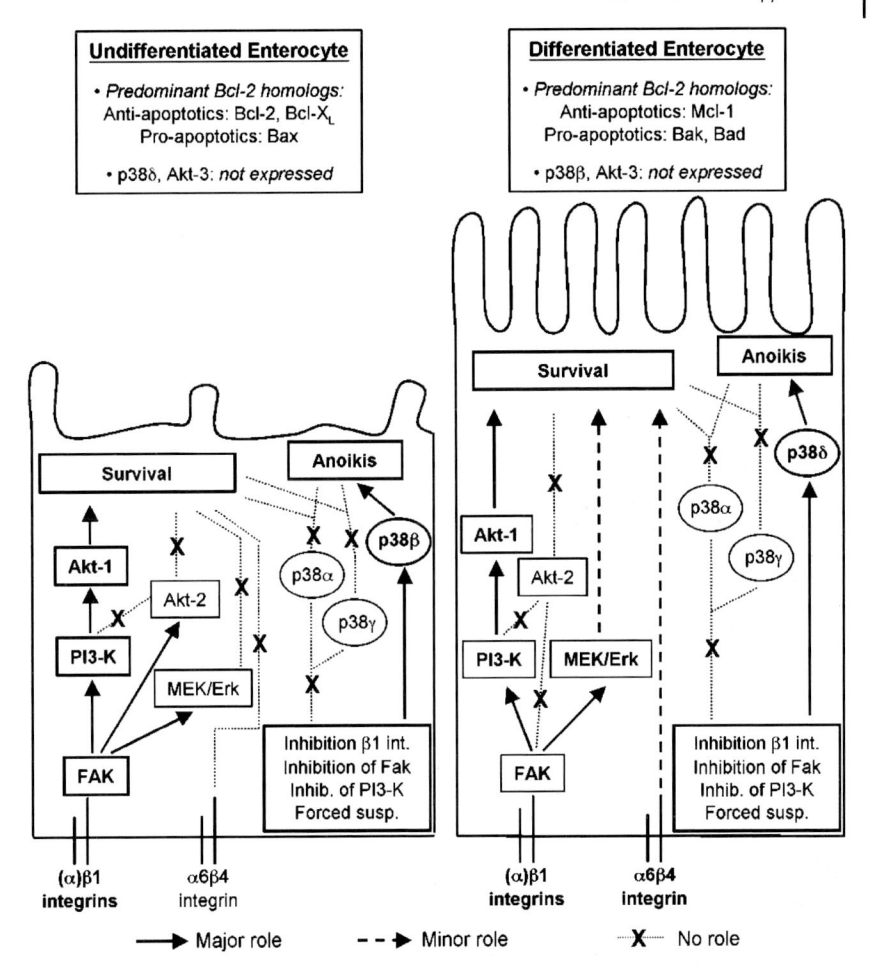

Fig. 20.5
Differentiation state-specific mechanisms in the regulation of human intestinal epithelial cell survival. The expression profiles of Bcl–2 homologs are distinct according to the state of enterocytic differentiation. Cell adhesion and signaling pathways perform distinct roles in cell survival and/or induction of apoptosis/anoikis according to the state of differentiation (as shown here), in addition to having distinct influences on the expression of Bcl–2 homologs (not shown). In *undifferentiated enterocytes*, β1 integrins, Fak, PI3–K and Akt–1 are crucial for survival and suppression of apoptosis/anoikis. However, Akt–2 and the MEK/ERK pathway do not play a role in cell survival or in the induction of apoptosis/anoikis, even though β1 integrins/Fak signaling contributes to their stimulation. Akt–2 is independent of PI3–K, whereas Akt–3 is not expressed. The stress MAP kinase isoforms p38α, p38γ and p38β (but not p38δ) are expressed but only p38β is crucial for the induction of apoptosis. The α6β4 integrin is not involved. In *differentiated enterocytes*, β1 integrins Fak, PI3–K and Akt–1 remain crucial for survival, whereas it is instead p38δ (and not p38β which is no longer expressed) which is required for the induction of apoptosis/anoikis. The MEK/ERK pathway remains dependent on β1 integrins/Fak signaling but this time plays a minor role in the maintenance of cell survival. Furthermore, the α6β4 integrin is likewise a minor contributor to cell survival. On the other hand, Akt–2 remains independent of PI3–K and is now independent of β1 integrins/Fak signaling as well, although it is still not involved in survival or in the induction of apoptosis/anoikis. Akt–3 is still not expressed.

developing gut epithelium. Over and above certain inherent limitations, the integration of the body of data generated by all of these various models will ultimately lead to the emergence of new concepts and therapeutic approaches for many human pathophysiological conditions such as necrotizing enterocolitis, immaturity of the GI tract in premature and newborn infants, pancreatic insufficiency, inflammatory bowel diseases and GI cancers.

References

1 S.J. Henning, *Am J Physiol* **1981**; *241*: G199–G214.
2 D. Ménard, R. Calvert, *Growth of the Gastrointestinal Tract: Gastrointestinal Hormones and Growth Factors*, J. Morisset, T. Solomon (Eds). Boca Raton: CRC Press Inc., **1991**, 147–162.
3 D. Ménard, *Can J Gastroenterol* **2004**; *18*: 39–44.
4 P. Arsenault, D. Ménard, *Anat Rec* **1988**; *220*: 313–317.
5 M. Chénard, J.R. Basque, P. Chailler, J.F. Beaulieu, D. Ménard, *Anat Embryol* **2000**; *202*: 223–233.
6 D. Ménard, P. Arsenault, S. Monfils, *Gastroenterology* **1993**; *104*: 492–501.
7 E. Tremblay, S. Monfils, D. Ménard, *Gastroenterology* **1997**; *112*: 1188–1196.
8 E. Tremblay, P. Chailler, D. Ménard, *Endocrinology* **2001**; *142*: 1795–1803.
9 D. Ménard, Growth-promoting factors and the development of the human gut. In *Human Gastrointestinal Development*, E. Lebenthal (Ed.). New York: Raven Press, **1989**, 123–150.
10 R.K. Montgomery, A.E. Mulberg, R.I. Grand, *Gastroenterology* **1999**; *116*: 702–731.
11 D. Ménard, P. Pothier, *J Pediatr Gastroenterol Nutr* **1987**; *6*: 509–516.
12 D.J. Roberts, *Dev Dyn* **2000**; *219*: 109–120.
13 K.H. Kaestner, D.G. Silberg, P.G. Traber, G. Schutz, *Genes Dev* **1997**; *11*: 1583–1595.
14 A. Goyal, R. Singh, E.A. Swietlicki, M.S. Levin, D.C. Rubin, M. Plateroti, I. Dubuc C. Foltzer-Jourdainne, J.N. Freund, M. Kedinger, *Am J Physiol* **1998**; *275*: G114–G124.
15 M. Plateroti, D.C. Rubin, I. Duluc, R. Singh, C. Foltzer-Jourdainne, J.N. Freund, M. Kedinger, *Am J Physiol* **1998**; *274*: G945–G954.
16 D.J. Roberts, D.M. Smith, D.J. Goff, C.J. Tabin, *Development* **1998**; *125*: 2791–2801.
17 F. Beck, F. Tata, K. Chawengsaksophak, *Bioessays* **2000**; *22*: 431–41.
18 M. Ramalho-Santos, D.A. Melton, A.P. McMahon, *Development* **2000**; *127*: 2763–2772.
19 S. Kermorgant, V. Dessirier, M.J. Lewin, T. Lehy, *Am J Physiol Gastrointest Liver Physiol* **2001**; *281*: G1068–G1080.
20 J.F. Beaulieu, *Prog Histochem Cytochem* **1997**; *31*: 1–78.
21 M. Kedinger, I. Duluc, C. Fritsch, O. Lorentz, M. Plateroti, J.N. Freund, *Ann NY Acad Sci* **1998**; *859*: 1–17.

22 B. Lacroix, M. Kedinger, P.M. Simon-Assmann, K. Haffen, *Differentiation* **1994**; *28*: 129–135.
23 M. Kedinger, P.M. Simon-Assmann, B. Lacroix, A. Marxer, H.P. Hauri, K. Haffen, *Dev Biol* **1986**; *113*: 474–483.
24 M. Kedinger, P. Simon-Assmann, F. Bouziges, C. Arnold, E. Alexandre, K. Haffen, P. Simo, *Differentiation* **1990**; *43*: 87–97.
25 A.P. McMahon, *Cell* **2000**; *100*: 185–188.
26 P.W. Ingham, A.P. McMahon, *Genes Dev* **2001**; *15*: 3059–3087.
27 M.J. Bitgood, A.P. McMahon, *Dev Biol* **1995**; *172*: 126–138.
28 D.J. Roberts, R.L. Johnson, A.C. Burke, C.E. Nelson, B.A. Morgan, C. Tabin, *Development* **1995**; *121*: 3163–3174.
29 A.P. Haramis, H. Begthel, J. van den Born, J. van Es, S. Jonkheer, G.J. Offerhaus, H. Clevers, *Science* **2004**; *303*: 1684–1686.
30 M. Goke, M. Kanai, D.K. Podolsky, *Am J Physiol* **1998**; *274*: G809–G818.
31 C. Birchmeier, D. Meyer, D. Riethmacher, W. Birchmeier, J. Hulsken, J. Behrens, *Int Rev Cytol* **1995**; *160*: 221–266.
32 A.U. Dignass, A. Sturm, *Eur J Gastroenterol Hepatol* **2001**; *13*: 763–770.
33 Y. Kawazoe, T. Sekimoto, M. Araki, K. Takagi, K. Araki, K. Yamamura, *Dev Growth Differ* **2002**; *44*: 77–84.
34 J. Aubin, U. Déry, U. Lemieux, P. Chailler, L. Jeannotte, *Development* **2002**; *129*: 4075–4087.
35 D.J. Wolgemuth, R.R. Behringer, M.P. Mostoller, R.L. Brinster, R.D. Palmiter, *Nature* **1989**; *337*: 464–467.
36 P. de Santa Barbara, D.J. Roberts, *Development* **2002**; *129*: 551–561.
37 H. Le Mouellic, Y. Lallemand, P. Brulet, *Cell* **1992**; *69*: 251–264.
38 Y. Hirai, K. Takebe, M. Takashina, S. Kobayashi, M. Takeichi, *Cell* **1992**; *69*: 471–481.
39 H.R. Pelham, *Cell* **1993**; *73*: 425–426.
40 D.C. Radisky, Y. Hirai, M.J. Bissell, *Trends Cell Biol* **2003**; *13*: 426–434.
41 C. Fritsch, E.A. Swietlicki, O. Lefebvre, M. Kedinger, H. Iordanov, M.S. Levin, D.C. Rubin, A. Goyal, R. Singh, M. Plateroti, I. Duluc, C. Foltzer-Jourdainne, J.N. Freund, *J Clin Invest* **2002**; *110*: 1629–1641.
42 N. Perreault, N., F.E. Herring-Gillam, N. Desloges, I. Bélanger, L.P. Pageot, J.F. Beaulieu, *Biochem Biophys Res Commun* **1998**; *248*: 121–126.
43 N. Perreault, J.P. Katz, S.D. Sackett, K.H. Kaestner, *J Biol Chem* **2001**; *276*: 43328–43333.
44 O. Pabst, R. Zweigerdt, H.H. Arnold, *Development* **1999**; *126*: 2215–2225.
45 M. Mahlapuu, S. Enerback, P. Carlsson, *Development* **2001**; *128*: 2397–2406.
46 D. Ménard, J.R. Basque, Gastrointestinal functions. In *Nestlé Nutrition Workshop Series*, Vol 46, E.E. Delvin, M.J. Lentze (Eds). Philadelphia: Nestec Ltd. Vevey/Lippincott Williams & Wilkins, **2001**, 147–163.
47 E. Tremblay, D. Ménard, *Anat Rec* **1996**; *245*: 668–676.
48 E. Tremblay, J.R. Basque, N. Rivard, D. Ménard, *Gastroenterology* **1999**; *116*: 831–841.

49 J.R. Basque, P. Chailler, N. Perreault, J.F. Beaulieu, D. Ménard, *Exp Cell Res* **1999**; *253*: 493–502.
50 R. Timpl, J.C. Brown, *Bioessays* **1996**; *18*: 123–132.
51 M. Paulsson, *Crit Rev Biochem Mol Biol* **1992**; *27*: 93–127.
52 M.B. Sporn, A.B. Roberts, *J Cell Biol* **1992**; *119*: 1017–1021.
53 J.R. Basque, P. Chailler, D. Ménard, *Am J Cell Physiol* **2002**; *282*: C873–C884.
54 I. Antonowicz, E. Lebenthal, *Gastroenterology* **1977**; *72*: 1299–1303.
55 D. Ménard, L. Corriveau, P. Arsenault, *J Pediatr Gastroenterol Nutr* **1990**; *10*: 13–20.
56 P. Chailler, J.R. Basque, L. Corriveau, D. Ménard, *Pediatr Res* **2000**; *48*: 504–510.
57 C. Malo, D. Ménard, *Gastroenterology* **1982**; *83*: 28–35.
58 D. Ménard, P. Arsenault, P. Pothier, *Gastroenterology* **1988**; *94*: 656–663.
59 P. Arsenault, D. Ménard, *J Pediatr Gastroenterol Nutr* **1985**; *4*: 389–901.
60 D. Ménard, *Acta Pediatr Suppl.* **1994**; *405*: 1–6.
61 E. Lévy, D. Ménard, *Microsc Res Tech* **2000**; *49*: 363–373.
62 F.G. Giancotti, G. Tarone, *Annu Rev Cell Dev Biol* **2003**; *19*: 173–206.
63 D. Ménard, J.F. Beaulieu, Human intestinal brush border membrane hydrolases. In *Membrane Physiopathology*, G. Bkaily (Ed.). Kluwer Norwell: Academic Publisher, **1994**, 319–341.
64 J.F. Beaulieu, *Front Biosci* **1999**; *4*: D310–D321.
65 M. Kédinger, J.N. Freund, J.F. Launay, P. Simon-Assman, Cell interactions through the basement membrane in intestinal development and differentiation. In *Development of the Gastrointestinal Tract*, I.A. Sanderson, W.A. Walker (Eds). Hamilton: B.C. Decker Inc., **1999**, 83–102.
66 C. Lussier, N. Basora, Y. Bouatrouss, J.F. Beaulieu, *Microsc Res Tech* **2000**; *51*: 169–78.
67 I.C. Teller, J.F. Beaulieu, *Expert Rev Mol Med* **2001**; *28 September*, [http://www-ermm.cbcu.cam.ac.uk/01003623h.htm] 1–18.
68 J.F. Beaulieu, P.H. Vachon, F.E. Herring-Gillam, A. Simoneau, N. Perreault, C. Asselin, J. Durand, *Gastroenterology* **1994**; *107*: 957–967.
69 A. Simoneau, F.E. Herring-Gillam, P.H. Vachon, N. Perreault, N. Basora, Y. Bouatrouss, L.P. Pageot, J. Zhou, J.F. Beaulieu, *Dev Dyn* **1998**; *212*: 437–447.
70 H. Colognato, P.D. Yurchenco, *Dev Dyn* **2000**; *218*: 213–234.
71 I. Virtanen, D. Gullberg, J. Rissanen, E. Kivilaakso, T. Kiviluoto, L.A. Laitinen, V.P. Lehto, P. Ekblom, *Exp Cell Res* **2000**; *257*: 298–309.
72 N. Perreault, P.H. Vachon, J.F. Beaulieu, *Anat Rec* **1995**; *242*: 242–250.
73 N. Desloges, N. Basora, N. Perreault, Y. Bouatrouss, D. Sheppard, J.F. Beaulieu, *J Cell Biochem* **1998**; *71*: 536–545.
74 N. Basora, N. Desloges, Q. Chang, Y. Bouatrouss, J. Gosselin, J. Poisson, D. Sheppard, J.F. Beaulieu, *Int J Cancer* **1998**; *75*: 738–743.

75 N. Basora, P.H. Vachon, F.E. Herring-Gillam, N. Perreault, J.F. Beaulieu, *Gastroenterology* **1997**; *113*: 1510–1521.
76 P. Simon-Assmann, B. Duclos, V. Orian-Rousseau, C. Arnold, C. Mathelin, E. Engvall, M. Kedinger, *Dev Dyn* **1994**; *201*: 71–85.
77 I. Bélanger, J.F. Beaulieu, *Histol Histopathol* **2000**; *15*: 577–585.
78 C. Lussier, J. Sodek, J.F. Beaulieu, *J Cell Biochem* **2001**; *81*: 463–476.
79 J.F. Beaulieu, *J Cell Sci* **1992**; *102*: 427–436.
80 P.H. Vachon, A. Simoneau, F.E. Herring-Gillam, J.F. Beaulieu, *Exp Cell Res* **1995**; *216*: 30–34.
81 J.F. Beaulieu, P.H. Vachon, *Gastroenterology* **1994**; *106*: 829–839.
82 Y. Bouatrouss, F.E. Herring-Gillam, J. Gosselin, J. Poisson, J.F. Beaulieu, *Am J Pathol* **2000**; *156*: 45–50.
83 P.H. Vachon, J.F. Beaulieu, *Am J Physiol* **1995**; *268*: G857–867.
84 N. Basora, F.E. Herring-Gillam, F. Boudreau, N. Perreault, L.P. Pageot, A. Simoneau, Y. Bouatrouss, J.F. Beaulieu, *J Biol Chem* **1999**; *274*: 29819–29825.
85 L. Chang, M. Karin, *Nature* **2001**; *410*: 37–40.
86 J.L. Bos, E.R. Fearon, S.R. Hamilton, M. Verlaan-de Vries, J.H. van Boom, A.J. van der Eb, B. Vogelstein, *Nature* **1987**; *327*: 293–297.
87 K. Forrester, C. Almoguera, K. Han, W.E. Grozzle, M. Perucho, *Nature* **1987**; *327*: 298–303.
88 K.P. Janssen, F. El Marjou, D. Pinto, X. Sastre, D. Rouillard, C. Fouquet, T. Soussi, D. Louvard, S. Robine, *Gastroenterology* **2002**; *123*: 492–504.
89 J.S. Sebolt-Leopold, D.T. Dudley, R. Herrera, K. Van Becelaere, A. Wiland, R.C. Gowan, H. Tecle, S.D. Barret, A. Bridges, S. Przybranowski, W.R. Leopold, A.R. Saltiel, *Nature Med* **1999**; *5*: 810–816.
90 J.C. Aliaga, C. Deschênes, J.F. Beaulieu, E.L. Calvo, N. Rivard, *Am J Physiol* **1999**; *277*: G631–G641.
91 N. Rivard, M.J. Boucher, C. Asselin, G. L'Allemain, *Am J Physiol* **1999**; *277*: C652–C664.
92 J.M. Rhoads, R.A. Argenzio, W. Chen, R.A. Rippe, J.K. Westwick, A.D. Cox, H.M. Berschneider, D.A. Brenner, *Am J Physiol* **1997**; *272*: G943–G953.
93 J. Jasleen, N. Shimoda, E.R. Shen, A. Tavakkolizadeh, E.E. Whang, D.O. Jacobs, M.J. Zinner, S.W. Ashley, *J Surg Res* **2000**; *90*: 13–18.
94 T. Chiu, S.S. Wu, C. Santiskulvong, P. Tangkijvanich, H.F.Jr Yee, E. Rozengurt, *Am. J Physiol Cell Physiol* **2002**; *282*: C434–C450.
95 D. Taupin, D.K. Podolsky, *Gastroenterology* **1999**; *116*: 1072–1080.
96 M.J. Boucher, N. Rivard, *Biochem Biophys Res Commun* **2003**; *311*: 121–128.
97 A.R. Nebreda, A.R, Porras, *Trends Biochem Sci* **2000**; *25*: 257–60.
98 M. Houde, P. Laprise, D. Jean, M. Blais, C. Asselin, N. Rivard, *J Biol Chem* **2001**; *276*: 21885–21894.
99 A.M. Zorn, *Curr Biol* **2001**; *11*: R592–R595.

100 K.W. Kinzler, B. Vogelstein, *Cell* **1996**; *87*: 159–170.
101 F.T. Kolligs, G. Bommer, B. Goke, *Digestion* **2002**; *66*: 131–144.
102 C. Booth, G. Brady, C.S. Potten, *Nature Med* **2002**; *8*: 1360–1361.
103 E. Batlle, E., J.T. Henderson, H. Beghtel, M.M. van den Born, E. Sancho, G. Huls, J. Meeldijk, J. Robertson, M. van de Wetering, T. Pawson, et al., *Cell* **2002**; *111*: 251–263.
104 M. van de Wetering, E. Sancho, C. Verweij, W. de Lau, I. Oving, A. Hurlstone, K. van der Horn, E. Batlle, D. Coudreuse, A.P. Haramis, et al., *Cell* **2002**; *111*: 241–250.
105 H. Shibata, K. Toyama, K. Shioya, M. Ito, M. Hirota, S. Hasegawa, H. Matsumoto, H. Takano, T. Akiyama, K. Toyoshima, et al., *Science* **1997**; *278*: 120–123.
106 V. Korinek, N. Barker, P. Moerer, E. van Donselaar, G. Huls, P.J. Peters, H. Clevers, *Nature Genet* **1998**; *19*: 379–383.
107 D. Pinto, A. Gregorieff, H. Begthel, H. Clevers, *Genes Dev* **2003**; *17*: 1709–1713.
108 L.E. Rameh, L.C. Cantley, *J Biol Chem* **1999**; *274*: 8347–8350.
109 H. Sheng, J. Shao, C.M.Jr Townsend, B.M. Evers, *Gut* **2003**; *52*: 1472–1478.
110 H. Sheng, J. Shao, R.N. DuBois, *J Biol Chem* **2001**; *276*: 14498–14504.
111 Q. Wang, X. Wang, A. Hernandez, S. Kim, B.M. Evers, *Gastroenterology* **2001**; *120*: 1381–1392.
112 P. Laprise, P. Chailler, M. Houde, J.F. Beaulieu, M.J. Boucher, N. Rivard, *J Biol Chem* **2002**; *277*: 8226–8234.
113 P. Laprise, M.J. Langlois, M.J. Boucher, C. Jobin, N. Rivard, *J Cell Physiol* **2004**; *199*: 32–39.
114 M.L. Hermiston, J.I. Gordon, *J Cell Biol* **1995**; *129*: 489–506.
115 M.L. Hermiston, M.H. Wong, J.I. Gordon, *Genes Dev* **1996**; *10*: 985–996.
116 P. Laprise, A. Viel, N. Rivard, *J Biol Chem* **2004**; *279*: 10157–10166.
117 G. Caruana, *Int J Dev Biol* **2002**; *46*: 511–518.
118 M. van Noort, H. Clevers, *Dev Biol* **2002**; *244*: 1–8.
119 S. Verbeek, D. Izon, F. Hofhuis, E. Robanus Maandag, H. te Riele, M. van de Wetering, M. Oosterwegel, A. Wilson, H.R. Mac-Donald, H. Clevers, *Nature* **1995**; *374*: 70–74.
120 V. Korinek, N. Barker, P. Moerer, E. van Donselaar, G. Huls, P.J. Peters, H. Clevers, *Naure Genet* **1998**; *19*: 379–383.
121 M. Brittan, N.A. Wright, N.A., *J Pathol* **2002**; *197*: 492–509.
122 J. Jensen, E.E. Pedersen, P. Galante, J. Hald, R.S. Heller, M. Ishibashi, R. Kageyama, F. Guillemot, P. Serup, O.D. Madsen, *Nature Genet* **2000**; 36–44.
123 Q. Yang, N.A. Bermingham, M.J. Finegold, H.Y. Zoghbi, *Science* **2001**; *294*: 2155–2158.
124 R. Sun, X. Chen, V.W. Yang, *J Biol Chem* **2001**; *276*: 6897–6900.
125 J.P. Katz, N. Perreault, B.G. Goldstein, C.S. Lee, P.A. Labosky, V.W. Yang, K.H. Kaestner, *Development* **2002**; *129*: 2619–2628.
126 N.W. Bateman, D. Tan, R.G. Pestell, J.D. Black, A.R. Black, *J Biol Chem* **2004**; *279*: 12093–12101.
127 E. Suh, L. Chen, J. Taylor, P.G. Traber, *Mol Cell Biol* **1994**; *14*: 7340–7351.

128 J.K. Taylor, T. Levy, E.R. Suh, P.G. Traber, *Nucleic Acids Res* **1997**; *25*: 2293–2300.
129 F. Boudreau, E.H. Rings, H.M. van Hering, R.K. Kim, G.P. Swain, S.D. Krasinski, J. Moffett, R.J. Grand, E.R. Suh, P.G.Traber, *J Biol Chem* **2002**; *277*: 31909–31917.
130 D.G. Silberg, G.P. Swain, E.R. Suh, P.G. Traber, *Gastroenterology* **2000**; *119*: 961–971.
131 E.R. Suh, P.G. Traber, *Mol Cell Biol* **1996**; *16*: 619–625.
132 P. Soubeyran, F. Andre, J.C. Lissitzky, G.V. Mallo, V. Moucadel, M. Roccabianca, H. Rechreche, J. Marvaldi, I. Dikic, J.C. Dagorn, J.L. Iovanna, *Gastroenterology* **1999**; *117*: 1326–1338.
133 M.S. Keller, T. Ezaki, R.J. Guo, J.P. Lynch, *Am J Physiol Gastrointest Liver Physiol* **2004**; *287*: G104–G114.
134 R.J. Guo, E.R. Suh, J.P. Lynch, *Cancer Biol Ther* **2004**; *7*: 593–601.
135 V. Subramania, B.I. Meyer, P. Gruss, *Cell* **1995**; *83*: 641–653.
136 Y. Tamai, R. Nakajima, T. Ishikawa, K. Takaku, M.F. Seldin, M.M. Taketo, *Cancer Res* **1999**; *59*: 2965–2970.
137 F. Beck, K. Chawengsaksophak, P. Waring, R.J. Playford, J.B. Furness, *Proc Natl Acad Sci USA* **1999**; *96*: 7318–7323.
138 K. Aoki, Y. Tamai, S. Horiike, M. Oshima, M.M. Taketo, *Nature Genet* **2003**; *35*: 323–330.
139 D.G. Silberg, J. Sullivan, E. Kang, G.P. Swain, J. Moffett, N.J. Sund, S.D. Sackett, K.H. Kaestner, *Gastroenterology* **2002**; *122*: 689–696.
140 F. Boudreau, Y. Zhu, P.G. Traber, *J Biol Chem* **2001**; *276*: 32122–32128.
141 D.B. Rhoads, D.H. Rosenbaum, H. Unsal, K.J. Isselbacher, L.L. Levitsky, *J Biol Chem* **1998**; *273*: 9510–9516.
142 N. Spodsberg, J.T. Troelsen, P. Carlsson, S. Enerback, H. Sjostrom, O. Noren, *Gastroenterology* **1999**; *116*: 842–854.
143 T. Sakaguchi, X. Gu, H.M. Golden, E. Suh, D.B. Rhoads, H.C. Reinecker, *J Biol Chem* **2002**; *277*: 21361–21370.
144 H.M. van Wering, I.L. Huibregtse, S.M. van der Zwan, M.S. de Bie, L.N. Dowl, F. Boudreau, E.H. Rings, R.J. Grand, S.D. Krasinski, *J Biol Chem* **2002**; *277*: 27659–27667.
145 J.K. Divine, L.J. Staloch, H. Haveri, C.M. Jacobsen, D.B. Wilson, M. Heikinheimo, T.C. Simon, *Am J Physiol Gastrointest Liver Physiol* **2004**. Epub ahead of print.
146 T.J. Oesterreicher, S.J. Henning, *Am J Physiol Gastrointest Liver Physiol* **2004**; *286*: G947–G953.
147 J.D. Molkentin, *J Biol Chem* **2000**; *275*: 38949–38952.
148 J. Fujikura, E. Yamato, S. Yonemura, K. Hosoda, S. Masui, K. Nakao, K.I. Miyazaki, H. Niwa, *Genes Dev* **2002**; *16*: 784–789.
149 A. Nepveu, *Gene* **2001**; *270*: 1–15.
150 F. Boudreau, E.H. Rings, G.P. Swain, A.M. Sinclair, E.R. Suh, D.G. Silberg, R.H. Scheuermann, P.G. Traber, *Mol Cell Biol* **2002**; *22*: 5467–5478.
151 J.F. Beaulieu, M.M. Weiser, L. Herrera, A. Quaroni, *Gastroenterology* **1990**; *98*: 1467–1477.
152 M.D. Jacobson, M. Weil, M.C. Raff. *Cell* **1997**; *88*: 347–354.
153 J.C. Reed, *Oncogene* **1998**; *17*: 3225–3236.
154 S. Cory, J.M. Adams, *Nature Rev* **2002**; *2*: 647–656.

155 J. Grossmann, *Apoptosis* **2002**; *7*: 247–260.
156 S.M. Frisch, R.A. Screaton, *Curr Opin Cell Biol* **2001**; *13*: 555–562.
157 S.F. Moss, P.R. Holt, *Gastroenterology* **1996**; *111*: 567–568.
158 C.S. Potten, *Am J Physiol* **1997**; *273*: G253–G257.
159 D.H. Barkla, P.R. Gibson, *Pathology* **1999**; *31*: 230–238.
160 J. Grossmann, K. Walther, M. Artinger, P. Rummele, M. Woenckhaus, J. Schölmerich, *Am J Gastroenterol* **2002**; *97*: 1421–1428.
161 B.A. Jones, G.J. Gores, *Am J Physiol* **1997**; *273*: G1174–G1188.
162 A.B.R. Thomson, L. Drozdowski, C. Iordache, B.K.A. Thomson, S. Vermeire, M.T. Clandinin, G. Wild, *Dig Dis Sci* **2003**; *48*: 1346–1399.
163 P.H. Vachon, É. Cardin, C. Harnois, J.C. Reed, A. Vézina, *Int J Dev Biol* **2000**; *44*: 891–898.
164 S. Krajewski, S. Bodrug, M. Krajewska, A. Shabaik, R. Gascoyne, K. Berean, J.C. Reed, *Am J Pathol* **1995**; *146*: 1309–1319.
165 S. Krajewski, M. Krajewska, J.C. Reed, *Cancer Res* **1996**; *56*: 2849–2855.
166 S. Krajewski, M. Krajewska, A. Shabaik, T. Miyashita, H.-G. Wang, J.C. Reed, *Am J Pathol* **1994**; *145*: 1323–1336.
167 S. Krajewski, M. Krajewska, A. Shabaik, H.-G. Wang, S. Irie, L. Fong, J.C. Reed, *Cancer Res* **1994**; *54*: 5501–5507.
168 A.D. Levine, *Inflamm Bowel Dis* **2000**; *6*: 191–205.
169 P.H. Vachon, É. Cardin, C. Harnois, J.C. Reed, A. Plourde, A. Vézina, *Histol Histopathol* **2001**; *16*: 497–510.
170 S.F. Moss, B. Agarwal, N. Arber, R.J. Guan, M. Krajewska, S. Krajewski, J.C. Reed, P.R. Holt, *Biochem Biophys Res Commun* **1996**; *223*: 199–203.
171 J.M. Kyriakis, J. Avruch, *Physiol Rev* **2001**; *81*: 807–869.
172 M.-J. Boucher, J. Morisset, P.H. Vachon, J.C. Reed, J. Lainé, N. Rivard, *J Cell Biochem* **2000**; *79*: 355–369.
173 F.G. Giancotti, E. Ruoslahti, *Science* **1999**; *285*: 1028–1032.
174 P. Laprise, È.-M. Poirier, A. Vézina, N. Rivard, P.H. Vachon, *J Cell Physiol* **2002**; *191*: 69–81.
175 P. Laprise, K. Vallée, M.-J. Demers, V. Bouchard, È.-M. Poirier, A. Vézina, J.C. Reed, N. Rivard, P.H. Vachon, *J Cell Biochem* **2003**; *89*: 1115–1125.
176 R. Gauthier, P. Laprise, É. Cardin, C. Harnois, A. Plourde, J.C. Reed, A. Vézina, P.H. Vachon, *J Cell Biochem* **2001**; *82*: 339–355.
177 R. Gauthier, C. Harnois, J.-F. Drolet, J.C. Reed, A. Vézina, P.H. Vachon, *Am J Physiol* **2001**; *280*: C1540–C1554.
178 P.H. Vachon, C. Harnois, A. Grenier, G. Dufour, V. Bouchard, J. Han, J. Landry, J.-F. Beaulieu, A. Vézina, A.B. Dydensborg, R. Gauthier, A. Côté, J.-F. Drolet, F. Lareau, *Gastroenterology* **2002**; *123*: 1980–1991.
179 C. Harnois, M.-J. Demers, V. Bouchard, K. Vallée, D. Gagné, N. Fujita, T. Tsuruo, A. Vézina, J.-F. Beaulieu, A. Côté, P.H. Vachon, *J Cell Physiol* **2004**; *198*: 209–222.
180 G. Dufour, M.-J. Demers, D. Gagné, A.B. Dydensborg, I.C. Teller, V. Bouchard, I. Degongre, J.-F. Beaulieu, J.Q. Cheng, N. Fujita, T. Tsuruo, K. Vallée, P.H. Vachon, *J Biol Chem* **2004**; *279*: 44113–44122.

21
Cell Signaling and Growth Factors in Lung Development

David Warburton, Saverio Bellusci, Pierre-Marie Del Moral, Stijn DeLanghe, Vesa Kaartinen, Matt Lee, Denise Tefft, and Wei Shi

21.1
Introduction

Lung development begins with the appearance of the laryngo-tracheal groove as a small diverticulum, arising from the floor of the primitive pharynx at E9 in mouse and 4 weeks in humans. The laryngo-tracheal groove gives rise to the left and right mainstem bronchial buds, which in turn give rise to the left and right lungs. Lobation of the lungs, as well as the first 16 of 23 generations in humans is stereotypic, which implies the presence of a hard-wired, genetic program that controls early embryonic lung branching morphogenesis. The latter phase of lower airway branching and on into the alveolar surface folding and expansion phase is non-stereotypic, but nevertheless follows a recognizable, proximal-distal fractal pattern that is repeated automatically at least 50 million times. The overall lung morphogenetic program drives the formation of an alveolar gas diffusion surface 0.1 μm thick by 70 m^2 in surface area which is perfectly matched to the alveolar capillary and lymphatic vasculature. This huge surface area for gas diffusion is packed into the chest in a highly spatially efficient manner [1].

21.1.1
The Stereotypic Branch Pattern of Respiratory Organs

The stereotypic branch pattern of the respiratory organs is determined by a well-conserved, genetically hard-wired program. Murine and human genetics, organ culture experiments, as well as comparative studies in the fly, have revealed that the stereotypic branch pattern of the respiratory organs is indeed determined by a well-conserved, genetically hard-wired program directed by transcriptional factors, that interact in a coordinated manner with peptide growth factor signaling pathways as well as hypoxia and physical forces [1–4]. Since transposition of early stage distal mesenchyme to the primitive tracheal epithelium can induce supernumerary branching and distal epithelial marker gene expression, epithelial-mesenchymal autocrine and paracrine interactions mediated by soluble factors arising from the mesenchyme have long been implicated in lung branching morphogenesis.

21.1.2
Transduction of Candidate Growth Factor Peptide Ligand Signals

Transduction of candidate growth factor peptide ligand signals can be regulated at many levels. These may include ligand expression, proteolytic activation of latent forms of ligand, ligand binding to matrix bound and/or soluble inhibitors, as well as ligand binding to receptor presentation molecules outside the cell. On the cell surface and within the cell, receptor assembly, kinase activation, and phosphorylation and activation of adapter and messenger protein complexes activate downstream signaling pathways both within and without the nucleus, including the induction of pathway-specific inhibitors. Thus, peptide growth factor signaling is finely coordinated to regulate such essential morphogenetic functions as gene expression, cell cycle progression and cell migration, cytodifferentiation and matrix deposition in the lung.

21.1.3
Examples of Peptide Growth Factor Signaling Pathways

In this chapter we selectively review key examples of peptide growth factor signaling pathways that together mediate lung development. We discuss selected examples of their finely balanced signal transduction mechanisms and propose how they may be functionally coordinated within compound, highly-regulated morphogenetic gradients to drive first stereotypic and then non-stereotypic, automatically repetitive, symmetrical as well as asymmetrical branching events that comprise lung development.

21.2
Growth Factors and Lung Development

21.2.1
Candidate Growth Factors in Lung Development

Epidermal Growth Factor (EGF), Fibroblast Growth Factor (FGF), Hepatocyte Growth Factor (HGF) and Platelet-Derived Growth Factor (PDGF) signal through cognate transmembrane tyrosine kinase receptors to exert a positive effect on lung morphogenesis. In contrast, Transforming Growth Factor (TGF) β and Bone Morphogenetic Protein (BMP) family peptides signal through transmembrane serine-threonine kinase receptors, to exert an inhibitory effect on lung epithelial cell proliferation and hence negatively regulate lung morphogenesis. However, TGFβ isoform-specific null-mutants now show that the latter generalization may not be entirely correct. Moreover, Bone Morphogenetic Protein (BMP) 4 appears to exert a complex negative or positive regulatory influence, depending on whether mesenchymal signaling is intact. Sonic Hedgehog (SHH) family peptide signaling represents another special case. The SHH cognate receptor, patched (PTC), exerts a negative effect on SHH signaling both through the release of the transcriptional repressor Smoothened (SMO) and the induction of the *Hedgehog interacting protein (Hip)*. Recently Wnt signaling has emerged as exerting another critical influence on matrix deposition and hence the formation of branch points. Also the Vascular Endothelial Growth Factor (VEGF) pathway has emerged as playing a critical role in alveolarization.

21.2.2
Growth Factor-mediated Epithelial-Mesenchymal Interactions and Lung Development

BMPs, FGFs, SHH, TGFβ, VEGF and Wnt signaling pathways all play critical roles in driving lung development. These peptide growth factors and their cognate receptors are expressed in repeating patterns within morphogenetic centers that surround and direct each new branch tip in the embryonic lung. Mesenchymally-expressed morphogenetic genes include *Fgf10, Sprouty4 (Spry4), patched, smoothened, Wnt* and *Hox* family members, while *Bmp4, Shh, mSpry2 and Smads 2, 3* and *4* and VEGF are expressed in the adjacent epithelium. The interactions of subsets of these ligand signals, particularly SHH, BMP4 and FGF10 have been extensively reviewed recently and several models have been proposed to explain how they may interact to induce and then regulate epithelial branching morphogenesis [1–3, 5].

21.2.3
FGF10

FGF10 is expressed focally in embryonic lung mesenchyme adjacent to stereotypically-determined branching sites, and acts as a potent chemoattractant to epithelium. Whether this results in a monochotomous or dichotomous branching event, likely depends on additional factors such as the organization of the overlying matrix [6]. However, since FGFR2IIIb, which is the principal and highest affinity FGF10 receptor, is expressed widely throughout the epithelium, the question arises as to how the ligand signal can become stereotypically localized. SHH and BMP4 have been proposed as candidate ligands to play a role in defining the expression and function of FGF10, while Sprouty2 (SPRY2) has been proposed as an inducible negative regulator of FGF signaling.

SHH, which is expressed throughout the epithelium is postulated to suppress *Fgf10* expression and hence prevent branching events at sites where branching is stereotypically determined not to take place. This supposition is based on the finding that *Fgf10* expression is not spatially restricted in the *Shh* null mutant mouse lung.

Moreover, the local suppression of SHH signaling by the induction of *Ptc* and *Hip* at branch tips may serve to facilitate FGF signaling locally where branch outgrowth is stereotypically programmed to take place.

21.2.4
The Role of BMP4

BMP4 is expressed predominantly in the epithelium and is increased at branch tips, and was until recently, postulated to be the localized suppression of epithelial proliferation, thus providing a negative modulatory influence on FGF signaling to mediate the arrest of branch extension and hence to set up branch points. This hypothesis was based upon the hypoplastic phenotype of the epithelium in transgenic misexpression studies of *Bmp4* in the epithelium, as well as upon addition of BMP4 ligand to naked epithelial explants in culture. However, two groups have now shown that BMP4 is actually a potent stimulator of branching in the presence of mesenchyme and at physiologic concentrations in lung explants. Moreover, the effects of BMP4 are in turn negatively modulated by the BMP binding proteins, Gremlin and Noggin. Therefore it seems unlikely that BMP4 signaling merely serves to inhibit epithelial proliferation, particularly since BMP4–specific Smads 1, 5 and 8 are predominantly expressed in the mesenchyme away from the epithelium. BMPs have also been reported to control differentiation of the endoderm along the proximal-distal axis [7]. Inhibition of BMP signaling at the tip of the lung bud by overexpression in the distal epithelium of Noggin (a secreted inhibitor of BMPs) or of a dominant negative form of the Bmp type I receptor, *activin receptor-like kinase 6* (*Alk6*), results in a distal epithelium exhibiting differentiation characteristics, at the molecular and cellular level, of the proximal epithelium.

21.2.5
The Role of the Vasculature and VEGF Signaling

A further puzzle in early lung morphogenesis is how vascularization is perfectly matched to epithelial morphogenesis to ensure optimal gas exchange. Several VEGF isoforms are expressed in the developing epithelium, whereas their cognate receptors are expressed in and direct the emergence of developing vascular and lymphatic capillary networks within the mesenchyme. It is possible that VEGF signaling may lie downstream of FGF signaling, since *in vivo* abrogation of FGF signaling severely affects both epithelial and endothelial morphogenesis.

21.2.6
Postnatal Lung Development

Null mutation studies have revealed essential roles for the PDGF-A chain and for FGFR3 and FGFR4 in the induction of alveolar ridges and the correct orientation of elastic fibers in the postnatal lung. Following delivery, particularly premature delivery, exposure to endotoxin, oxygen and/or barotrauma with the resulting induction of cytokines including excessive amounts of TGFβ, adversely affects alveolarization and can frequently induce interstitial fibrosis, a human pathobiological condition termed bronchopulmonary dysplasia or infantile chronic lung disease.

21.2.7
The Influence of Peptide Growth Factor Signaling on the Correct Organization of the Matrix

This is another critical issue. Recently we have shown that Wnt signaling influences fibronectin deposition at early branch points, while PDGF-A chain and FGFR3 and 4 expression are known to critically regulate elastin expression during alveolarization.

21.3
Growth Factor Signaling Pathways in Lung Development

21.3.1
Critical Signaling Pathways in Lung Development

Bone morphogenetic protein, fibroblast growth factor, sonic hedgehog, transforming growth factorβ, vascular endothelial growth factor and Wnt are critical signaling pathways that drive lung development and each will be considered in turn.

21.3.1.1 BMPs
Several BMPs, including BMP3, 4, 5 and 7, are expressed during embryonic lung development [8–10]. The expression of *Bmp5* and *Bmp7* has been detected in the mesenchyme and endoderm of the developing embryonic lung respectively, while *Bmp4* expression is restricted to the distal epithelial cells and the adjacent mesenchyme [8, 10](King et al., 1994; Bellusci et al., 1996). Most of the BMP signaling pathway components, such as BMP receptors (type II and type I: ALK2, 3, and 6) and BMP-specific receptor-regulated Smads (R-Smads), including Smad1, 5, and 8, are expressed in early mouse embryonic lung [11, 12]. Overexpression of *Bmp4*, driven by the *SP-C* promoter in the distal endoderm of transgenic mice, causes abnormal lung morphogenesis leading to the formation of cystic terminal sacs and the inhi-

bition of epithelial proliferation [8]. In contrast, *SP-C* promoter-driven overexpression of either the BMP antagonist *Xnoggin* or a dominant negative *Alk6* BMP receptor to block BMP signaling, results in severely reduced distal epithelial cell phenotypes and increased proximal cell phenotypes in the lungs of transgenic mice [7]. However, the exact roles of BMP4 in early mouse lung development remain controversial. In isolated E11.5 mouse lung endoderm cultured in Matrigel™ (Collaborative Biomedical Products, Bedford, MA, USA) addition of exogenous BMP4 inhibited epithelial growth induced by the morphogen FGF10 [13]. However, addition of BMP4 to intact embryonic lung explant culture stimulates lung branching morphogenesis [14, 15]. Recently, parallels have been drawn between genetic hard-wiring of tracheal morphogenesis in *Drosophila melanogaster* and mammals [1]. Dpp, the *Drosophila* BMP4 ortholog, has been reported to be essential for the formation of the dorsal and ventral branches of the tracheal system, controlling tracheal branching and outgrowth possibly through induction of the zinc finger proteins *Kni* and *Knrl* [16, 17]. Since conventional murine knockouts for BMP4 and BMP-specific *Smads* cause early embryonic lethality, their functions in lung development *in vivo* still need to be further defined. Interestingly, germ line mutations in BMP type II receptors were recently found in familial primary pulmonary hypertension [18]. Therefore, BMPs may play multiple roles in lung development.

21.3.1.2 Activin Receptor-like Kinases

All TGFβ superfamily members (TGFβs, activins and BMPs) produce their cellular responses through formation of heteromeric complexes of specific type I and type II receptors (reviewed in [19, 20]). The type II receptors are constitutively active kinases, which, upon ligand-mediated complex formation, phosphorylate particular serine and threonine residues in the type I receptor juxtamembrane region. This leads to activation of the type I receptor, which is thereby capable of transducing signals downstream. It has been shown that type I receptors are responsible for determining specificity within the heteromeric signaling complex. Seven type I receptors called activin receptor-like kinases (ALKs) have been discovered in mammals. ALK4 and ALK5 are receptors for activin and TGFβ, respectively, whereas ALK2, ALK3 and ALK6 are receptors for BMPs. Recently, ALK1 was shown to be an endothelial cell-specific TGFβ receptor, while ALK7 has been suggested to mediate signals of another TGFβ-related ligand, nodal. Interestingly, among all TGFβ type I receptors, ALK2 shows the broadest spectrum of specificity. It has been shown to mediate BMP signaling, but it also has been shown to act as a type I receptor for TGFβ, activin and Müllerian inhibitory substance (MIS) [21–25].

21.3.1.3 ALKs in Pulmonary Development

During embryonic days 12–14 (E12–E14), *Alk5* and *Alk4* are expressed predominantly in the lung mesenchyme and the epithelium, respectively [26]. *Alk2* and *Alk6* are expressed in the lung epithelium. However, *Alk6* expression is limited to the lung epithelium [12] (Kaartinen, unpublished data). It was recently suggested that the effect of TGFβ2 on lung branching morphogenesis would be mediated by the TGFβ type II receptor-ALK5 complex. Thus, activins and therefore ALK4 would not have a significant role in this process [26]. The role of ALK6 in pulmonary maturation was recently underscored by Weaver and coworkers, who showed that the BMP signaling mediated by these receptors regulates the proximal-distal differentiation of endoderm in mouse lung development [7]. The role of ALK2 in epithelial differentiation and branching, if any, has yet to be determined.

21.3.1.4 ALKs and the Pulmonary Vasculature

The complex process of vascular development involves vasculogenesis, *de novo* formation of blood vessels through the aggregation of endothelial cells, and angiogenesis, i.e. the growth of new blood vessels from a pre-existing vascular network [27]. Several lines of evidence demonstrate that TGFβ-BMP signaling via ALKs plays a key role in the regulation of angiogenesis. It was recently shown that the TGFβ type I receptor, ALK5, plays a crucial role during vascular development by regulating endothelial cell proliferation, extracellular matrix deposition and migration [28]. Loss-of-function mutations both in the human and mouse genes encoding *Endoglin*, a TGFβ-binding protein, and in *Alk1*, cause hereditary hemorrhagic telangiectasia type 1 (HHT1) and type 2 (HHT2), respectively [29–33]. This disease affects blood vessel integrity and causes arteriovenous malformations of the lung. It has been suggested that ALK1 would function in the establishment of arterial-venous identity, and that the balance between signals mediated by ALK1 and ALK5 is important in determining vascular endothelial properties during angiogenesis [33, 34]. Moreover, recent studies demonstrated that the TGFβ type II receptor, BMPRII, which is one, and maybe the principal binding partner of, ALK2, is mutated in primary pulmonary hypertension (PPH) [18]. Histopathological findings of PPH include intimal fibrosis, *in situ* thrombosis and hypertrophy of smooth muscle cells in walls of pulmonary arteries [35]. Therefore it is evident that TGFβ and particularly BMP signaling, plays a key role in maintaining the normal homeostasis of smooth muscle cells in pulmonary arteries. It will be interesting to see whether Alk signaling plays a role in the remodeling the double alveolar capillary network into a single network during erection of alveolar septae.

21.3.1.5 ALKs, Pulmonary Fibrosis and Inflammation

Several studies have shown that TGFβs are central regulators of pulmonary fibrosis [36, 37]. Interestingly, it has also been shown that TGFβs act as strong anti-inflammatory agents in the lung [38, 39]. Therefore, it is possible that TGFβs contribute to the normal lung repair mechanisms after pulmonary insult, such as inflammation, and that in relatively rare cases this repair process is over-ridden, resulting in life-threatening pulmonary fibrosis. Using the experimental mouse model for allergic airway inflammation, it was recently shown that mRNA levels of *Alk1* and *Alk2* were markedly elevated, while, surprisingly, *Alk5* levels were slightly reduced during allergic airway inflammation [40]. It is expected that the mechanisms used during lung development are similar to those utilized during pulmonary repair, which underscores the importance of understanding complex molecular interactions during lung development *in vivo*.

21.3.2
FGF Signaling Promotes Outgrowth of Lung Epithelium

The mouse embryonic lung represents a uniquely useful system to study the genes involved in bud outgrowth and bud arrest [8, 13, 41–44]. FGF10 promotes directed growth of the lung epithelium and induces both proliferation and chemotaxis of isolated endoderm [13, 42]. The chemotaxis response of the lung endoderm to FGF10 involves the coordinated movement of an entire epithelial tip, containing hundreds of cells, toward an FGF10 source. How this population of cells monitors the FGF gradient and which receptors trigger this effect remains unknown. FGF10 also controls the differentiation of the epithelium by inducing Surfactant Protein C (SP-C) expression and by upregulating the expression of BMP4, a known regulator of lung epithelial differentiation [8, 13, 45, 46]. *In vitro* binding assays have shown that FGF10 acts mostly through FGFR1b and FGFR2b [47]. While there is good evidence that FGF10 acts through FGFR2b *in vivo*, there are as yet no conclusive data involving FGFR1b (or any other receptor) *in vivo*. The biological activities mediated through these two epithelial receptors are likely to be different as FGF7 (acting mostly through FGFR2b) exhibits a different activity compared to FGF10 [42]. This hypothesis is also supported by our recent findings showing that $Fgf10^{-/-}$ lungs exhibit a more severe phenotype than $Fgfr2b^{-/-}$ lungs.

21.3.3
FGF2b

FGFR2b is critical for mesenchymal-epithelial interactions during early lung organogenesis. The mammalian *Fgf* receptor family comprises four genes (*Fgfr1* to *Fgfr4*), which encode at least seven prototype receptors. *Fgfr1, 2* and *3* encode two receptor isoforms (termed IIIb or IIIc) that are generated by alternate splicing, and each binds a specific repertoire of FGF ligands [47]. FGFR2–IIIb (FGFR2b) is found mainly in epithelia and binds four known ligands (FGF1, FGF3, FGF7 and FGF10), which are primarily expressed in mesenchymal cells. Peters et al. reported the first evidence of a key role for *Fgfr2* during lung development [48]. They showed that misexpression of a dominant negative form of *Fgfr2* in the embryonic lung under the *SP-C* promoter led to a severe reduction in branching morphogenesis. Further evidence came from *Fgfr2* inactivation in the embryo. While mice null for the *Fgfr2* gene die early during embryogenesis, those that are null for the *Fgfr2b* isoform, but retain *Fgfr2c*, survive to birth [49–52]. Mice deficient for *Fgfr2b* show agenesis and dysgenesis of multiple organs, including the lungs, indicating that signaling through this receptor is critical for mesenchymal-epithelial interactions during early organogenesis. This idea is supported by the recent finding that prenatally-induced misexpression of a dominant negative FGFR, to abrogate FGF signaling, results in a hypoplastic, emphysematous lung phenotype [53]. In contrast, induced abrogation of FGF signaling postnatally did not produce any recognizable phenotype.

21.3.4
FGF10

FGF10 is a major ligand for FGFR2b during lung organogenesis. The FGF family is comprised of at least 23 members, many of which have been implicated in multiple aspects of vertebrate development (for review see [54]). In particular, FGF10 has been associated with instructive mesenchymal-epithelial interactions, such as those that occur during branching morphogenesis. In the developing lung, *Fgf10* is expressed in the distal mesenchyme at sites where prospective epithelial buds will appear. Moreover, its dynamic pattern of expression and its ability to induce epithelial expansion and budding in organ cultures have led to the hypothesis that FGF10 governs the directional outgrowth of lung buds during branching morphogenesis [42]. Furthermore, FGF10 was shown to induce chemotaxis of the distal lung epithelium [13, 55]. Consistent with these observations, mice deficient for *Fgf10* show multiple organ defects including lung agenesis [56–58]. FGF10 is the main ligand for FGFR2b during the embryonic phase of development as evidenced by the remarkable similarity of phenotypes exhibited by embryos where these genes have been inactivated [44, 51, 58].

21.3.5.
FGF10 Activity

FGF10 activity was initially described as controlling proliferation and chemotaxis of the lung epithelium. The paradigm proposed so far is that FGF10 expressed by the mesenchyme acts on the epithelium (which expresses FGFR1b and 2b). However, a recent report by Sakaue et al. suggests that FGF10 expressed in the fat pad precursor of the developing mammary gland from embryonic day 15.5 (E15.5) onwards, could act in an autocrine fashion to induce the differentiation of adipocytes from the fat pad precursor, but the specific receptors involved are unknown [44, 59]. In *Drosophila melanogaster*, Branchless (bnl), the *Drosophila* counterpart of FGF10, has been involved in the directional growth of the ectoderm-derived cells from the tracheal placode [60]. Bnl expressed by the cells surrounding the placode acts on the ectoderm expressing the *Fgfr2b* ortholog, *breathless* (*btl*). An additional unsuspected function of bnl in the development of the male genital imaginal disc has been recently reported [61]. Here, the FGF signal expressed by ectoderm-derived cells of the male genital disc induces the FGFR-expressing mesodermal cells to migrate into the male disc. These mesodermal cells also undergo a mesenchymal to epithelial transition. The authors suggest that bnl, the FGF10 ortholog, is likely to be involved in this process. Thus, FGF10 is a multifunctional growth factor and additional roles for FGF10 in lung development likely remain to be identified.

21.3.5.1. Regulation of FGF Signaling

Sprouty family members function as inducible negative regulators of FGF signaling in lung development. The role of inhibitory regulators in the formation of FGFR-activated signaling complexes during respiratory organogenesis remains incompletely characterized. The first example of an FGF-inducible signaling antagonist arose from the discovery of the *sprouty* mutant during trachea development in *Drosophila*, in which supernumerary tracheal sprouts arise. In the *Drosophila* tracheae, bnl binds to btl, inducing primary, secondary and terminal branching. The function of bnl is inhibited by Sprouty (Spry), a downstream effector in the bnl pathway [62]. Spry feeds back negatively on bnl, thereby limiting the number of sites at which new secondary tracheal buds form. Spry is not only found downstream in the FGFR pathway, but also appears to be an inhibitor of other tyrosine kinase signaling pathways such as EGF and Torso [63].

21.3.5.2. Mice and Humans Possess Several *Spry* Genes (*Mspry1–4* and *Hspry1–4*)

mSpry2 is the gene that is most closely related to *Drosophila Spry* and is 97% homologous to *hSpry2*. *mSpry2* is localized to the distal tips of the embryonic lung epithelial branches and is downregulated at sites of new bud formation [44]. On the other hand, *mSpry4* is predominantly expressed throughout the distal mesenchyme of the embryonic lung. Abrogation of *mSpry2* expression stimulates murine lung branching morphogenesis and increased expression of specific lung epithelial maturation/differentiation markers [45]. Conversely, overexpression of *mSpry2* under the control of a *SP*-C promoter or by intratracheal microinjection of an adenovirus containing the *mSpry2* cDNA, results in smaller lungs with a particular "moth-eaten" dysplastic appearance along the edges of the lobes, with decreased epithelial cell proliferation [44]. Thus, not only is the function of *Spry* conserved during respiratory organogenesis, but also as seen by loss-of-function and gain-of-function studies, *Spry* plays a vital role in regulating lung branching morphogenesis.

21.3.5.3. Sprouty Binds Downstream Effector Complexes

In *Drosophila, in vitro* co-precipitation studies show that Spry binds to Gap1 and Drk (a Grb2 ortholog), resulting in inhibition of the Ras-MAPK pathway [63]. Upon further investigation of the mechanism by which mSPRY2 negatively regulates FGF10 in mouse lung epithelial cells (MLE15) we recently determined that mSPRY2 differentially binds to FGF downstream effector complexes [64].

21.3.5.4. FGFs as Tyrosine Kinase Receptors

FGFRs are different from other tyrosine kinase receptors in that they require adapter or docking proteins. These include phospholipase Cγ, Shc, FRS2 and various others to recruit the Grb2/Sos complex upon stimulation. Stimulation of FGFR not only results in formation of the FRS2/Grb2/Sos complex, but the binding of a positive tyrosine phosphatase regulator, Shp2, to FRS2 which is required for the full potentiation of MAP-kinase activation [65]. Complex formation leads to catalysis of GDP to GTP on Ras, which is required for Raf (serine/threonine kinase) activation. Raf causes direct activation of ERK, leading to phosphorylation of cytoplasmic proteins followed by cell growth and differentiation [66]. We found that in the native state mSPRY2 associates with Shp2 and Gap, which is a GTPase-activating protein that hydrolyzes GTP to GDP. It is possible that in this state the binding of Shp2 to mSPRY2 regulates mSPRY2 activity. Upon FGFR activation, mSPRY2 disassociates from Shp2 and Gap and translocates to the plasma membrane, where it binds to both FRS2 and Grb2, thus blocking the formation of the FRS2/Grb2/Sos complex, resulting in a net reduction of MAP-kinase activation. Thus, Sprouty would inhibit the formation of specific signaling complexes downstream from the tyrosine-kinase receptors resulting in modulation and coordination of cell growth and development during organogenesis.

21.3.5.5. Relationship between FGF Signaling and Spry during Development

It is interesting to note, that overexpression of *Spry* in chick limb buds results in a reduction in limb bud outgrowth that is consistent with a decrease in FGF signaling [67]. This suggests a possible co-regulatory relationship between FGF signaling and Spry during development. In further support of this model, *Spry4* inhibits branching of endothelial cells as well as sprouting of small vessels in cultured mouse embryos. Endothelial cell proliferation and differentiation in response to FGF and VEGF are also inhibited by *mSpry4*, which acts by repressing ERK activation. Thus, Spry4 may negatively regulate angiogenesis [68].

21.3.5.6. Spry2 and Spry4 Share a Common Inhibitory Mechanism

It has been suggested that both Spry2 and Spry4 share a common inhibitory mechanism. Both Sprouty translocate to membrane ruffles upon EGF stimulation. However, only SPRY2 was shown to associate with microtubules [69]. The C-terminus of hSPRY2 has been shown to be important for modulation of cellular migration, proliferation and membrane co-localization [69, 70]. Interestingly, the C-terminus is the region that is most conserved throughout the Spry family, and contains potential regulatory sites that would modulate Spry activity. Spry has also been shown to interact with c-Cbl resulting in increased EGFR internalization [71]. Although Spry is not a specific inhibitor of either the FGFR signaling pathway or respiratory organogenesis, it appears that Spry plays a vital role in modulating several signaling pathways in order to limit the effects of excessive growth factor receptor tyrosine kinase signaling.

21.3.4
Sonic Hedgehog, Patched and Hip

The role of SHH signaling in lung morphogenesis has recently been reviewed [72]. Hedgehog signaling is essential for lung morphogenesis since *Shh* null produces profound hypoplasia of the lungs and failure of tracheo-esophageal septation [73, 74]. However, proximo-distal differentiation of the endoderm is preserved in the *Shh* null mutant, at least in so far as expression of *surfactant protein-C, SP-C* and *Clara cell protein 10, CC10* genes are concerned. The expression of the SHH receptor, *Patched*, is also decreased in the absence of *Shh* as are the *Gli1* and *Gli3* transcriptional factors. On the other hand lung-specific misexpression of *Shh* results in severe alveolar hypoplasia and a significant increase in interstitial tissue [41]. *Fgf10* expression, which is highly spatially restricted in wild-type, is not spatially restricted and is widespread in the mesenchyme in contact with the epithelium of the *Shh* null mutant mouse lung. Conversely, local suppression of SHH signaling by the induction of *Ptc* and *Hip* at branch tips may serve to facilitate FGF signaling locally, where branch outgrowth is stereotypically programmed to take place [75]. It is interesting to note that the cecum, which forms as a single bud from the mouse midgut and does not branch, also expresses *Fgf10* throughout its mesenchyme (Burns and Bellusci, unpublished data). Thus, temporospatial restriction of *Fgf10* expression by SHH appears to be essential to initiate and maintain branching of lung.

21.3.4.1 Expression and Activation of TGFβ Family of Peptides
Physiological expression and activation of the TGFβ family of peptides is essential for normal lung development. The TGFβ superfamily can be divided into three subfamilies: activin, TGFβ, and BMP [40]. There are three TGFβ isoforms in mammals: TGFβ1, 2, 3. All of them have been detected in murine embryonic lungs [76–79]. In early mouse embryonic lungs (E11.5), TGFβ1 is expressed in the mesenchyme, particularly in the mesenchyme underlying distal epithelial branching points, while TGFβ2 is localized in distal epithelium, and TGFβ3 is expressed in proximal mesenchyme and mesothelium [14]. Mice lacking *Tgfβ1* develop normally but die from aggressive pulmonary inflammation within the first month or two of life. When raised under pulmonary pathogen-free conditions these mice live some-

what longer but die of other forms of inflammation [80]. Thus, physiological concentrations of TGFβ1 appear to suppress the pulmonary inflammation that occurs in response to exogenous factors such as infection end endotoxin. On the other hand *Tgfβ2* null mutants die *in utero* of severe cardiac malformations, while *Tgfβ3* mutants die neonatally of lung dysplasia and cleft palate [81, 82]. Embryonic lung organ and cell cultures reveal that TGFβ2 plays a key role in branching morphogenesis, while TGFβ3 plays a key role in regulating alveolar epithelial cell proliferation during the injury repair response [83, 84]. Thus, finely regulated and correct physiologic concentrations and temporo-spatial distribution of TGFβ1, 2 and 3 are essential for normal lung morphogenesis and defense against lung inflammation. Overexpression in lung epithelium of transgenic mice of *Tgfβ1* initiated by the *SP-C* promoter, causes hypoplastic phenotypes [85]. Similarly, addition of exogenous TGFβ to early embryonic mouse lungs in culture resulted in inhibition of lung branching morphogenesis although each TGFβ isoform has a different IC_{50} (TGFβ2 〉 1 〉 3) [14, 86]. In contrast, abrogation of the TGFβ type II receptor stimulated embryonic lung branching through releasing cell cycle G1 arrest [87]. Moreover, overexpression of constitutively active TGFβ1, but not latent TGFβ1, in airway epithelium, is sufficient to have significant inhibitory effects on lung branching morphogenesis [83]. However, no inhibitory effect on lung branching was observed when TGFβ1 was overexpressed in the pleura and subjacent mesenchymal cells. Furthermore, adenoviral overexpression of a TGFβ inhibitor, *Decorin*, in airway epithelium, completely abrogated exogenous TGFβ1–induced inhibition of embryonic lung growth in culture [87]. On the other hand, reduction of decorin expression by DNA antisense oligonucleotides was able to restore TGFβ1–mediated lung growth inhibition [87]. Therefore, TGFβ signaling in distal airway epithelium seems to be sufficient for its inhibitory function during embryonic lung growth. Interestingly, TGFβ-specific signaling elements, such as Smad2/3/7, are exclusively expressed in distal airway epithelium [88–90]. Attenuation of *Smad2/3* expression by a specific antisense oligonucleotide approach blocked the exogenous TGFβ1–induced inhibitory effects on lung growth. Moreover, expression of *Smad7* in airway epithelium, which was induced by TGFβ, had negative regulatory functions for the TGFβ-Smad pathway in cultured cells, specifically blocking exogenous TGFβ-induced inhibitory effects on lung branching morphogenesis as well as on Smad2 phosphorylation in cultured lung explants. Since blockade of TGFβ signaling not only stimulates lung morphogenesis in culture *per se*, but also potentiates the stimulatory effects of EGF and PDGF-A, it follows that TGFβ signaling functions downstream of or can override tyrosine kinase receptor signaling.

21.3.4.2. Developmental Specificity of the TGFβ1 Overexpression Phenotype

During embryonic and fetal life, epithelial misexpression of TGFβ1 results in hypoplastic branching and decreased epithelial cell proliferation [83]. In contrast, neonatal misexpression of TGFβ1 using an adenoviral vector approach, phenocopies Bronchopulmonary Dysplasia (BPD) with alveolar hypoplasia, some interstitial fibrosis and emphysema [91]. Adult misexpression of TGFβ1, on the other hand, results in a chronic, progressive interstitial pulmonary fibrosis, resulting mainly from increased proliferation and matrix secretion by the mesenchyme; a process that depends on transduction through Smad3 [92, 93]. Thus, the phenotype caused by excessive TGFβ1 production and signaling is always adverse, but the precise effect depends on the developmental stage of the lung: hypoplasia in embryonic, fetal and neonatal lung; fibrosis in premature and adult lung.

21.3.4.3. TGFβ Signaling

TGFβ family peptide signaling is the best studied example of regulation in multiple layers. Selected key aspects of the TGFβ signaling system have recently been reviewed (see [20, 94, 95]). Latent TGFβ ligands require proteolytic activation prior to signal transduction by proteases such as plasmin. Expression of β6 integrin and thrombospondin play key roles in TGFβ ligand activation. Bioavailability of activated TGFβ ligand is further regulated by soluble binding proteins such as Decorin, as well as by binding to matrix proteins such as Fibrillin. Cognate receptor affinity for ligand binding may also be modulated by such factors as betaglycan, Endoglin or Decorin. In the case of TGFβ2 ligand, betaglycan (TGFβ type III receptor) presents activated ligand to the signaling receptor complex and markedly increases ligand-receptor affinity. TGFβ receptors function predominantly as tetrameric transmembrane complexes, comprising pairs of TGFβ type I and II serine-threonine kinase receptors. Following dimeric TGFβ ligand binding, the type I receptor kinase is phosphorylated and activated by the constitutively active TGFβ type II receptor kinase. The activated type I receptor serine/threonine phosphorylates the receptor-activated R-Smads 2 and/or 3. However, this signal transduction step can be negatively modulated by BAMBI, which functions as a dominant negative, kinase dead TGFβ receptor. BAMBI inhibits TGFβ receptor complex signaling to R-Smads. Phosphorylated R-Smads in turn form a complex with the common effector Smad4. This activated complex then becomes rapidly translocated to the nucleus and activates or represses transcription by binding to specific transcriptional complexes on certain gene promoters such as plasminogen activator inhibitor–1 (PAI–1) and cyclin A respectively. Smad complex stability is negatively regulated by Smurf 1, a ubiquitin ligase. Once in the nucleus, Smad complex-mediated gene regulation is antagonized by the transcriptional regulators Sno and Ski.

21.3.4.4. The Bleomycin-induced Model of Lung Fibrosis

The bleomycin induced model of lung fibrosis is mediated by excessive TGFβ production and signaling. The *Smad3* null mutation substantially blocks bleomycin-induced interstitial fibrosis [93]. However, the initial phase of lung inflammation induced by bleomycin is not blocked. Moreover, induction of TGFβ1 expression by bleomycin is not blocked. Rather, the key factor in blockade of bleomycin-induced fibrosis was lack of Smad3 signaling. Thus, Smad3 could act as a final common downstream target in the TGFβ-mediated pathobiologic sequence in the lung. Putative non-Smad signaling pathways provide potential sites for cross-talk with other signaling pathways.

21.3.5.
VEGF Isoform and Cognate Receptor Signaling in Lung Development

Vasculogenesis is initiated as soon as the lung evaginates from the foregut [96]. A critical growth factor during embryonic lung development is VEGF. The loss of even a single allele of *Vegf* leads to embryonic lethality between days E9.5 and E10.5 in the mouse [97]. VEGF is diffusely distributed in pulmonary epithelial and mesenchymal cells and is involved in controlling endothelial proliferation and the maintenance of vascular structure. VEGF is localized in the basement membrane of epithelial cells [98].

21.3.5.1. Both Humans and Mice have Three Different VEGF Isoforms

VEGF–120, VEGF–164 and VEGF–188 are all expressed in mice during development, but VEGF–164 isoform is the most highly expressed and active during embryogenesis. VEGF signals through the cognate receptors FLK–1 (Fetal liver kinase–1) and FLT–1 (Fetal liver tyrosinase–1). VEGF signaling is responsible for the differentiation of embryonic mesenchymal cells into endothelial cells. Interactions between the epithelium and mesenchyme contribute to lung neovascularization, which is crucial in normal lung formation. In fact, epithelial cells of the airways are positive for VEGF and it is expressed to an even greater degree in the budding regions of the distal airway [99]. Also, only lung mesenchyme cultured in the absence of epithelium degenerates significantly and only a few *Flk–1*–positive cells are maintained [96].

21.3.5.2. *Vegf* Knockout Mice have a Lethal Phenotype within the Early Stages of Embryonic Development (E8.5–E9)

Whereas in *Vegf*-misexpressing transgenic mice, where the *Vegf* transgene is under the control of the *SP-C* promoter, gross abnormalities in lung morphogenesis are associated with a decrease in acinar tubules and mesenchyme [97]. VEGF-treated human lung explants show an increase of cellular proliferation in the distal airway epithelial cells with upregulation of the mRNA expression of *Surfactant Protein-A* (*SP-A*) and C (*SP-C*) but not *SP-B* [99].

21.3.5.3. The Role of VEGF in Maintaining Alveolar Structure

VEGF has also been demonstrated to play a role in maintaining alveolar structure [100]. Lungs from newborn mice treated with antibodies to FLT–1 were reduced in size and displayed significant immaturity with a less complex alveolar pattern [101]. In contrast, the accumulation of VEGF in the alveoli appears to make transgenic VEGF mice more resistant to injury by hyperoxia [102, 103].

VEGF is a target of hypoxia-inducible transcription factor–2α (HIF–2α). Hif–2α–deficient newborn mice die from respiratory distress syndrome [103]. In *Hif–2α* null mice the expression of VEGF is dramatically reduced in alveolar epithelial type 2 cells. Additionally we have recently noted that addition of VEGF in early mouse embryonic explants in culture markedly stimulates epithelial as well as vascular morphogenesis (unpublished results). Thus we speculate that VEGF signaling plays an important role in matching the epithelial-capillary interface during lung morphogenesis.

21.3.5.4. VEGF-C and VEGF-D

VEGF-C and VEGF-D are two additional members of the VEGF family. These factors have a restricted expression pattern, with high levels mainly in lung tissues [104]. VEGF-C and -D stimulate lymphoangiogenesis through their cognate receptor VEGFR–3 [105]. Signaling via VEGFR–3 has been shown to be sufficient for lymphoangiogenesis through null mutation [106]. Finally, VEGF-C also interacts with VEGFR–2 and is therefore able to induce angiogenesis *in vivo* [107].

21.3.5.5. VEGF Isoforms Induce Vasculogenesis, Angiogenesis and Lymphoangiogenesis

Thus, VEGF isoforms induce vasculogenesis, angiogenesis and lymphoangiogenesis during lung development and likely play a key role in coordinating epithelial morphogenesis with the developing vascular and lymphatic capillary circulations.

3.6.1.
Wnt Signaling

Wnt signaling controls epithelial and mesenchymal differentiation and plays an important role in lung development. The Wnt growth factor family in the mouse, comprises 19 different secreted ligands that interact with 10 known seven-transmembrane receptors of the Frizzled (Fz) gene family and either one of two single-span transmembrane proteins, low-density-lipoprotein-receptor-related proteins (LRP–5 and LRP–6) [108–110]. Historically, Wnt proteins have been grouped into two classes, canonical and noncanonical. Canonical Wnts bind to frizzled receptors, inhibiting glycogen-synthase kinase–3β (GSK–3β)-mediated phosphorylation of β-catenin. Hypophosphorylated β-catenin accumulates in the cytoplasm, after which it translocates to the nucleus, where it heterodimerizes with members of the TCF/LEF transcription factor family to activate the transcription of TCF/LEF target genes. Noncanonical Wnts activate other Wnt signaling pathways, such as the planar-cell-polarity (PCP)-like pathway that guides cell movements during gastrulation [111]. Activation of the PCP pathway can antagonize the canonical pathway intracellularly [112–114]. Secreted Wnt antagonists can be divided into two functional classes, the secreted Frizzled-related protein (sFRP) class and the Dickkopf class. Members of the sFRP class bind directly to Wnts, thereby altering their ability to bind to the Wnt receptor complex, while members of the Dickkopf class inhibit Wnt signaling by binding to the LRP5/LRP6 component of the Wnt receptor complex. Members of the Dickkopf family are thought to specifically inhibit the canonical Wnt pathway (reviewed in [115]). DKK–1 and –4 act as inhibitors of canonical Wnt signaling, while DKK–2 can function both as an inhibitor in the presence of Kremen–2 or as an activator in its absence. No effect of DKK–3 has been reported on Wnt signaling [116].

Between E10.5 and 17.5, β-catenin is localized in the cytoplasm and often also in the nucleus of the pulmonary epithelium and adjacent mesenchyme [117]. Wnt ligands, Fz receptors and the Tcf/Lef1 transcription factors are expressed during early lung development. In the early mouse lung (at E12.5) *Wnt7b* is expressed in the lung epithelium with a more intense expression distally [118], *Wnt2a* is highly expressed in the distal mesenchyme and *Wnt5a* shows a low expression in the mesenchyme and epithelium and is highest around the trachea and pharynx [119]. *Tcf1* (T-cell factor1), *Lef1* (Lymphoid enhancer factor 1), *sFrp1* (secreted Frizzled-related protein 1) and *sFRP2* are highly expressed in the mesenchyme adjacent to the pulmonary epithelium. *Tcf3* is highly expressed at the apical side of the pulmonary epithelium while *Tcf4* and *sFrp4* are detected in both the epithelium and adjacent mesenchyme [117]. *Fz8* is highly expressed throughout the epithelium while *Fz2, 3, 6* and *7* are expressed both in the epithelium and mesenchyme (De Langhe et al., unpublished data). These expression patterns suggest that Wnt signaling may originate from the epithelium and mesenchyme and may target both tissues in an autocrine and/or paracrine fashion.

TOPGAL or BATGAL mice, which harbor a *β–galactosidase* gene under the control of a LEF/TCF and β-catenin inducible promoter [117], reveal that from E10.5 until E12.5, canonical Wnt signaling occurs throughout the epithelium and in the mesenchyme adjacent to the proximal airways where the bronchial smooth muscle cells (SMC) arise [120] (De Langhe et al., unpublished data). From E13.5 TOPGAL activity is no longer present in the mesenchyme and the activity in the epithelium is reduced distally, concomitant with the onset of expression of *Dkk1* in the distal epithelium (De Langhe et al., unpublished data).

21.3.7.
Inactivation of the β-Catenin Gene

Conditional inactivation of the *β–catenin* gene in the epithelium of the developing mouse lung leads to neonatal death resulting from severe lung defects [121]; branching of secondary bronchi is altered and the number of small peripheral alveolar ducts and terminal saccules is markedly reduced. In addition, the epithelium fails to undergo proper distal differentiation, lacking the expression of pro-*SP-C* protein and *Vascular endothelial growth factor a*, the latter correlating with a reduction in alveolar capillaries. So far, inactivation of only two WNT ligands has resulted in lung defects. *Wnt7b* is normally expressed in epithelial cells of the lung periphery. *Wnt7b$^{-/-}$* mice suffer perinatal death due to respiratory failure. Defects were observed in proliferation of the lung mesenchyme resulting in lung hypoplasia. In addition, *Wnt7b$^{-/-}$* embryos and newborn mice exhibit severe defects in the smooth muscle component of the major pulmonary vessels, with increased apoptosis of the vascular smooth muscle cells (VSMCs), resulting in rupture of the major blood vessels and hemorrhages in the lungs after birth. *Wnt5a* is expressed at high levels in the distal lung mesenchyme. *Wnt5a$^{-/-}$* mice suffer perinatal death as a result of lung defects including truncation of the trachea, overexpansion of the peripheral airways and delayed lung maturation. Absence of WNT5a activity in the mutant lungs leads to increased cell proliferation and upregulation of the expression of *Fgf10*, *Bmp4*, *Shh* and *Ptc*.

21.3.8.
Dickkopf Regulates Matrix Function

Our recent experiments show that early embryonic mouse lung organ cultures treated with Dickkopf (DKK1), a potent and specific diffusible inhibitor of Wnt action which is also endogenously secreted by the distal lung epithelium, display impaired branching, characterized by failed cleft formation and enlarged terminal buds. The DKK1–treated lung explants show reduced expression of α-smooth muscle actin (α-SMA) and defects in the formation of the pulmonary vascular network. These defects coincide with a pattern of decreased fibronectin (FN) deposition and *Platelet derived growth factor-A* (*Pdgf-a*) expression. All of the DKK1–induced morphogenetic defects can be recapitulated by inhibition of FN with an anti-fibronectin antibody and conversely can be rescued by addition of exogenous FN (De Langhe et al., unpublished data). This observation emphasizes the importance of correct orientation of the extracellular matrix in response to growth factor signaling. It also suggests that fibronectin is a downstream target of Wnt signaling.

21.4
Regulation of Signaling Networks

21.4.1.
Growth Factor Tyrosine Kinase and TGF-β Pathways

Growth factor receptor tyrosine kinases and TGF-β pathways interact to regulate MAP kinase activity. Accumulating evidence indicates that signaling pathways interact to regulate cell signaling. In the complex milieu of the developing fetus, these interactions are likely to have important effects on lung development. For example, receptor tyrosine kinases (including those of EGF and FGF) are canonical inducers of MAP kinase activation and cell mitogenesis. In contrast, TGF-β receptors phosphorylate Smad family transcription factors to regulate cell proliferation. Simultaneous stimulation with both EGF and TGF-β results in non-additive regulation of cell proliferation and MAP kinase activation, suggesting mutual regulation of these two signaling pathways [122]. These interactions may be necessary in specification of the normal pulmonary phenotype [123].

Receptor tyrosine kinase and TGF-β signaling pathways appear to interact by multiple mechanisms. First, stimulation of one pathway may regulate the expression of the ligand for the second pathway. In the lung, TGF-β has been reported to regulate expression and activity of EGF and FGF2 receptors [124, 125]. Second, growth factor stimulation may regulate receptor expression. In hepatocytes [126] and cervical epithelial cells [127], TGF-β stimulation decreased EGF receptor expression and activity. Third, EGF receptor phosphorylation [128, 129] or endocytosis [130] may be attenuated by concomitant TGF-β stimulation.

21.4.2.
Mutual Regulation of Intracellular Signaling Networks

The intracellular signaling networks associated with each growth factor may also be mutually regulated. Upon ligand binding, EGF receptors phosphorylate Shc and Grb2 adapter proteins, inducing their association with the GTP exchange protein Sos. The resulting heterotrimer translocates to the plasma membrane where it encounters and activates Ras, leading to sequential activation of Raf and MEK, and the stimulation of Erk1/Erk2 MAP kinases [131, 132] suppresses the EGF signaling pathway, resulting in decreased EGF-induced MEK1 and MAP kinase activation [133]. Numerous reports also document TGF-β-induced activation of p38, p42/p44, and JNK MAP kinases [134–137]. Both Smad-dependent [138] and SMAD-independent mechanisms [134, 139] have been proposed.

21.4.3.
Regulation of TGF-β signaling by EGF Signaling

Conversely, TGF-β signaling is also regulated by EGF signaling. Upon ligand binding, the TGF-β Type I and Type II receptors dimerize and catalyze the phosphorylation of receptor-activated Smad proteins. Smad2 and Smad3 are activated by TGF-β signaling, while Smad1, Smad5, and Smad8 are associated with BMP signaling [121]. These proteins complex with Smad4 and translocate to the nucleus, where they physically associate with DNA to regulate gene transcription. Activated MAP kinases phosphorylate Smad2 and Smad3 [124, 134, 135] to regulate TGF-β signaling, and may be necessary for TGF-β-induced Smad1 phosphorylation [139]. Alternatively, EGF signaling may also phosphorylate and activate the Smad2 repressor TGIF [140], or regulate the transcription of the inhibitory Smad7 protein [126].

21.4.4.
Calcium Signaling and the Mitochondrial Apoptosis Pathway

Calcium signaling and the mitochondrial apoptosis pathway are also subject to interactive regulation by EGF and TGF-β [141, 142]. Taken together, these reports indicate that the interaction between the EGF and TGF-β signaling pathways is extensive and multifactorial. Although the characterization of these interactions is likely to be arduous, the resulting models are more likely to be applicable to the developing lung and, hence, more likely avenues for therapeutic intervention.

21.5
Developmental Modulation of Growth Factor Signaling by Adapter Proteins

The substrates of growth factor receptor kinases are often adapter proteins, which have no intrinsic enzymatic function but combine with other proteins to activate downstream effectors. An important example is that of the Shc protein family, which comprises three isoforms with different functions. All are substrates of receptor tyrosine kinases [143]. The 52–kDa isoform ($p52^{Shc}$) is a mediator of Ras activation. Upon tyrosine phosphorylation, $p52^{Shc}$ forms a heterotrimeric complex with Grb2 and Sos, which then translocates to the plasma membrane where it encounters and activates Ras. Ras activation leads to MAP kinase activation and subsequent induction of cell proliferation. A second isoform of 46 kDa in size is translated from an alternative start site on the $p52^{Shc}$ transcript; the function of this peptide is incompletely understood. A third isoform of 66 kDa ($p66^{Shc}$) in size is transcribed from an alternative splice product of the Shc gene which encodes an additional proline-rich domain to the amino terminus of the $p52^{Shc}$. Unlike $p52^{Shc}$, overexpression of this isoform neither transforms 3T3 fibroblasts nor activates MAP kinases, but appears to antagonize Ras activation, possibly by sequestering Grb2 and making it unavailable for mitogenic signaling [144]. The 66–kDa protein has also been characterized as a mediator of cellular responses to oxidative damage [145]. Cells deficient in p66^{Shc} are resistant to cell death following oxidative damage, and mice deficient in *p66*^{Shc} have a 30 % longer life span. Cellular resistance to oxidation-induced death is reversed by induced expression of the wild-type p66^{Shc}, and this resistance is regulated by serine phosphorylation at amino acid 36 of p66^{Shc} [145]. Induced expression of mutant *p66*^{Shc} in which the Ser^{36} has been ablated does not restore the oxidative response of *p66*^{Shc} null fibroblasts. Phosphorylation of Ser^{36} is induced by a number of cellular stresses including hydrogen peroxide, ultraviolet irradiation, and taxol-induced microtubular disruption [146, 147]. Ser^{36} phosphorylation also occurs in renal mesangial cells following endothelin–1 stimulation, suggesting that the mediated stress response pathway can be induced by intercellular peptide signaling [147]. The p66^{Shc} and 46–kDa isoforms are differentially regulated towards the end of fetal lung development [33].

21.6
Morphogens and Morphogenetic Gradients

21.6.1.
Coordination of Growth Factor Morphogenetic Signals to Determine Lung Development

The terms "morphogen" and "morphogenetic gradient" were coined by Morgan over a century ago to provide a theoretical basis for pattern formation during morphogenesis [148]. The morphogen concept was further advanced by Spemann's classical observation of an "organizer" within the dorsal tip of the blastopore in early *Xenopus* embryos, whose activity is mediated by a diffusible "morphogen" [149]. A mathematical theory explaining how two morphogens might interact to determine form during organ development was proposed by Alan Turing, the World War II Naval Enigma code breaker [150]. His mathematical reaction-diffusion hypothesis states that two homogeneously distributed substances will interact to produce stable patterns during morphogenesis and will thus induce an ordered structure out of a randomly chaotic system [151]. This provides a potential clue to solving the biological enigma of repetitive branching.

21.6.2.
Action of Morphogens

Morphogens, specifically peptide growth factors, can instruct lung morphogenesis through tyrosine kinase signaling, and hence gene induction in target cells. This supports Turing gradients as a possible mechanistic solution to lung morphogenesis [152]. However, while binary reaction-diffusion systems may be adequate to explain relatively simple repeating patterns, it now seems likely that in the lung dual parameter reaction-diffusion is an oversimplification. Instead, in more complex polydimensional biological systems such as the lung, we must consider how the several diffusible ligands mentioned herein may set up repeating morphogenetic fields. Further, as discussed above, the bioactivity of each single morphogen is simultaneously modified by its own system of checks and inducible balances such as the binding proteins and the negative Smads in the BMP signaling pathway and the *Spry* gene family in the FGF pathway. Moreover each of these pathways can respectively positively and/or negatively regulate its fellows through intracellular signaling cross-talk.

21.6.3.
Other Morphogen Gradient Systems

The developing limb bud is another well-studied morphogen gradient system. Certain FGF ligands arising from the distinct apical ectodermal ridge (AER) have been proposed to initiate proximal-distal patterning of the long bones as well as anteroposterior patterning of the digits. Long bone and wrist bones are postulated to arise respectively, based upon the time progenitor cells spend and their orientation within the "progress zone" subjacent to the AER [153]. To the extent that the lung has a recognizable proximal-distal pattern to the airways and alveoli, it is tempting to speculate that inductive and progress zones may also exist in the peripheral lung during morphogenesis. We have recently noted that the peripheral domain of FGF10 expression at the edge of embryonic lung lobes bears a striking resemblance to the domain of FGF expression in the limb AER. Thus it is tempting to speculate that the FGF10 domain in embryonic lung may form an Apical Pulmonary Ridge (APR), which has a pattern-forming function analogous to that of the AER in the limb [44].

The decision of the embryonic airway to branch or not to branch is therefore determined by the integration of multiple peptide growth factor-mediated signals as well as other kinds of signals within automatically repeating morphogenetic signaling centers.

21.6.4.
The APR Model

A noteworthy predictive feature of the APR model is the shape of the FGF10 morphogenetic gradient. The FGF10 morphogenetic gradient decays proximally away from the high level source of FGF10 in the APR. The model predicts that in monopodial lateral branches, such as the lobar bronchial buds at E11, the decaying FGF10 gradient will be symmetrical. In contrast, in the bipodial branch sites at the periphery, the decay of the FGF10 will be asymmetrical. Thus we speculate that the shape of the FGF10 morphogen gradient may play an important role in determining monopodial versus dipodial branching and hence stereotypy of the proximal airway branches.

21.6.5.
The Modified Turing Model

The modified Turing model also predicts that the dynamically changing relative activity of SHH, FGF10 and mSPRY2 may impart automation to the branching process. SHH is high and FGF10 is low where branching is not supposed to take place. In contrast, SHH is suppressed locally by PTC and HIP, so that FGF10 is therefore high where a branch is destined to occur. FGF10 in turn dynamically induces its inhibitor *mSpry2* as branches lengthen. Thus, the net relative activities between SHH, FGF10 and mSPRY2 may determine FGF signal strength in the epithelium and hence the relative rate of bud outgrowth at a given point and hence inter-branch length.

21.6.6.
Expression of *Fgf10*

Fgf10 is expressed locally in the mesenchyme close to a point where a branch will arise from the main epithelial bronchial stem. As the bud begins to elongate, *mSpry2* begins to be expressed in the distal tip. During subsequent elongation, *Fgf10* continues to be expressed in the distal mesenchyme and the level of *mSpry2* gradually increases as the bud lengthens. When the bud finally approaches the pleura, the *Fgf10* expression domain adjacent to the distal tip appears to thin out and some of it appears to be pushed laterally to lie between adjacent branch tips. At the time *mSpry2* expression in the distal tip is at its highest level, perhaps mediating bud outgrowth arrest, a tip-splitting event then occurs. Of note is that *mSpry2* expression is extinguished between the daughter bud tips, but continues to be expressed within the tips of the daughter bud epithelia. This cycle of interaction is then repeated during subsequent branching events.

21.6.7.
Tip-splitting Event

Until recently, the origin of the tip-splitting event was not at all clear. Since tip-splitting events occur widely in nature, where genes are not present, in such processes as viscous fingering, electrolysis of metals, river deltas and oil field deposits, it is not even clear whether physical forces are primary or secondary to growth factors in lung development. The numerous null mutants now available show that growth factors such as FGF10 and SHH are essential. However, exactly what they do remains unclear. FGF10 is clearly capable of inducing a strong chemotactic response in denuded lung epithelial bud tips towards and indeed to engulf an FGF10–soaked bead. Moreover, terminal buds appear to migrate into and towards the FGF10 epitopes within the embryonic lung mesenchyme. However, one of many puzzles has been why *Spry2* expression, which is controlled by FGF10, is focally extinguished in a small stretch of epithelium right in the cleft between two new daughter buds. Another plausible explanation is the "rock in the stream" hypothesis, wherein a solid bar of something in the mesenchyme acts like a rock to divide the flowing stream of epithelium into two as it chemotaxes towards FGF10. Fibronectin and myofibroblasts have been suggested as candidates to play the role of the rock. Certainly our new data on Wnt signaling, using DKK as a means to abrogate the canonical Wnt pathway, suggests that fibronectin deposition is indeed a good candidate to function as the rock in the stream, while Wnt signaling appears to control its deposition. This is interesting because elastin expression by myofibroblasts in response to PDGF and FGF signaling has been adduced as a critical event during the alveolar phase of lung development. Null mutation of *Pdgf-A, Fgfr3/Fgfr4* or of *elastin* also produce a hypoalveolarization phenotype, together with dysplasia of the alveolar extracellular matrix. Since VEGF signaling by the epithelium to the endothelium is also critical for normal alveolarization, it is interesting to speculate that hydraulic force within the capillary vasculature may also be important. Certainly, abrogation of VEGF signaling by inducible misexpression of a dominant negative VEGF receptor under the control of the *Sp-C* promoter abrogates alveolarization as well as peripheral capillary morphogenesis, underscoring the proposal that epithelial to endothelial cross-talk is also an important mechanism in lung morphogenesis.

21.6.8.
The Value of Hypothetical Models

These hypothetical models do go some way to potentially explaining automation and symmetry of early airway branching, but they do not explain stereotypy, antero-posterior orientation or left-right laterality. Nevertheless, we speculate that proximal-distal, antero-posterior and left-right stereotypy must be superimposed on the automatic morphogenetic branching mechanisms just proposed. *Hox* family genes are likely to play a key role in this process, since in *Hoxa–5$^{-/-}$* mice, tracheal occlusion and respiratory distress is associated with a marked decrease in surfactant protein production together with altered gene expression in the pulmonary epithelium [154]. Since *Hoxa–5* expression is restricted to the lung mesenchyme, the null mutant phenotype strongly supports the inference that *Hoxa–5* expression is necessary for induction of epithelial gene expression by the underlying mesenchyme. Could *Fgf10* be a Hox gene target?

21.6.9.
Retinoic Acid Receptors

The retinoic acid receptors, *Gli–2*, *Lefty–1*, *Lefty–2* and *Nodal* are preferentially expressed on the left side of wild-type mouse embryos and are implicated in determination of left-right laterality. *Lefty–1* null mutant mice show a variety of left-right isomerisms in visceral organs, but the most common feature is thoracic left isomerism. The lack of *Lefty–1* expression results in abnormal bilateral expression of *Nodal*, *Lefty–2* and *Pitx2* (a homeobox gene normally expressed on the left side). This suggests that Lefty–1 normally restricts *Lefty–2* and *Nodal* expression to the left side, and that *Lefty–2* or *Nodal* encode a signal for "leftness" in the lung [155].

21.7.
Conclusion

Growth factor signaling through increasingly complex autocrine and paracrine mechanisms is required to mediate the "hard wiring" encoded by the genetic program that determines stereotypy, antero-posterior and lateral symmetry of lung branching. The key growth factor signaling mechanisms best characterized to date as essential or important for normal lung development include BMP, EGF, FGF, HGF, PDGF, SHH, TGFβ, VEGF and Wnt. These pathways are finely regulated and integrated at multiple levels outside as well as inside the cell. Mechanisms of integration between these pathways during signal transduction to the nucleus are emerging. These finely-regulated, interactive growth factor cell signaling mechanisms define such criteria as automation of branching, inter-branch length, stereotypy of branching, left-right asymmetry, and finally gas diffusion surface area. Recently the extracellular matrix has re-emerged as a likely key target for growth factor signaling in lung branching morphogenesis and alveolarization. Classical mesenchymal to epithelial inductive signaling is increasingly realized to collaborate seamlessly with growth factor-mediated mechanisms of epithelial to endothelial cross-talk in coordinating not only branching morphogenesis but also morphogenesis of the alveolar-capillary interface. More complete understanding of these mechanisms and how they are finely and coordinately regulated will be prerequisite to improved prospects for therapeutic applications in lung protection, repair, regeneration and engineering.

References

1 D. Warburton, M. Schwarz, D. Tefft, G. Flores-Delgado, K.D. Anderson, W.V. Cardoso, *Mech Dev* **2000**; *92*: 55–81.
2 M. Affolter, S. Bellusci, B. Itoh, J.P. Thiery, Z. Werb Z, *Dev Cell* **2003**; *4*: 1–20.

3 W.V. Cardoso, *Dev Dyn* **2000**; *219*: 121–130.
4 A.K. Perl, J.A. Whitsett, *Clin Genet* **1999**; *56*: 14–27.
5 B. L.M. Hogan, *Cell* **1999**; *96*: 225–231.
6 V. Fleury, T. Watanabe, *C R Biol* **2002**; *325*: 571–583.
7 M. Weaver, J.M. Yingling, N.R. Dunn, S. Bellusci, B.L. Hogan, *Development* **1999**; *126*: 4005–4015.
8 S. Bellusci, R. Henderson, G. Winnier, T. Oikawa, B.L. Hogan, *Development* **1996**; *122*: 1693–1702.
9 H. Takahashi, T. Ikeda, *Dev Dyn* **1996**; *207*: 439–449.
10 J.A. King, P.C. Marker, K.J. Seung, D.M. Kingsley, *Dev Biol* **1994**; *166*: 112–122.
11 N. Dewulf, K. Verschueren, O. Lonnoy, A. Moren, S. Grimsby, K. Vande Spiegle, K. Miyazono, D. Huylebroeck, P. Ten Dijke, *Endocrinology* **1995**; *136*: 2652–2663.
12 K. Verschueren, N. Dewulf, M.J. Goumans, O. Lonnoy, A. Feijen, S. Grimsby, K. Vandi Spiegle, P. Ten Dijke, A. Moren, P. Vanscheeuwijck, *Mech Dev* **1995**; *52*: 109–123.
13 M. Weaver, N.R. Dunn, B.L. Hogan, *Development* **2000**; *127*: 2695–2704.
14 A.D. Bragg, H.L. Moses, R. Serra, *Mech Dev* **2001**; *109*: 13–26.
15 W. Shi, J. Zhao, K.D. Anderson, D. Warburton, *Am J Physiol Lung Cell Mol Physiol* **2001**; *280*: L1030–L1039.
16 C.K. Chen, R.P. Kuhnlein, K.G. Eulenberg, S. Vincent, M. Affolter, R. Schuh, *Development* **1998**; *125*: 4959–4968.
17 C. Ribeiro, A. Ebner, M. Affolter, *Dev Cell* **2002**; *2*: 677–683.
18 K.B. Lane, R.D. Machado, M.W. Pauciulo, J.R. Thomson, J.A. Phillips, III, J.E. Loyd, W.C. Nichols, R.C. Trembath, *Nat Genet* **2000**; *26*: 81–84.
19 R. Derynck, X.H. Feng, *Biochim Biophys Acta* **1997**; *1333*: F105–F150.
20 J. Massague, Y.G. Chen, *Genes Dev* **2000**; *14*: 627–644.
21 M. Macias-Silva, P.A. Hoodless, S.J. Tang, M. Buchwald, J. L. L. Wrana, *J Biol Chem* **1998**; *273*: 25628–25636.
22 L. Attisano, J. Carcamo, F. Ventura, F.M. Weis, J. Massague, J.L. Wrana, *Cell* **1993**; *75*: 671–680.
23 P.J. Miettinen, R. Ebner, A.R. Lopez, R. Derynck, *J Cell Biol* **1994**; *127*: 2021–2036.
24 J.A. Visser, R. Olaso, M. Verhoef-Post, P. Kramer, A.P. Themmen, H.A. Ingraham, *Mol Endocrinol* **2001**; *15*: 936–945.
25 T.R. Clarke, Y. Hoshiya, S.E. Yi, X. Liu, K.M. Lyons, P.K. Donahoe, *Dev Dyn* **2000**; *217*: 343–360.
26 G.D. Yancopoulos, M. Klagsbrun, J. Folkman, *Cell* **1998**; *93*: 661–664.
27 J. Larsson, M.J. Goumans, L.J. Sjostrand, M. A. van Rooijen, D. Ward, P. Leveen, X. Xu, P. ten Dijke, C.L. Mummery, S. Karlsson, *EMBO J* **2001**; *20*: 1663–1673.
28 K.A. McAllister, K.M. Grogg, D.W. Johnson, C.J. Gallione, M.A. Baldwin, C.E. Jackson, E.A. Helmbold, D.S. Markel, W.C. McKinnon, J. Murrell, *Nat Genet* **1994**; *8*: 345–351.
29 D.W. Johnson, J.N. Berg, M.A. Baldwin, C.J. Gallione, I. Marondel, S.J. Yoon T.T. Stenzel, M. Speer, M.A. Pericak-Vance, A. Diamond, A.E. Guttmacher, C.E. Jackson, L. Attisano, R. Kucherlapati, M.E. Porteous, D.A. Marchuk, *Nat Genet* **1996**; *13*: 189–195.

30 S.P. Oh, E. Li, *Genes Dev* **1997**; *11*: 1812–1826.
31 D.Y. Li, L.K. Sorensen, B.S. Brooke, L.D. Urness, E.C. Davis, D.G. Taylor, B.B. Boak, D.P. Wendel, *Science* **1999**; *284*: 1534–1537.
32 L.D. Urness, L.K. Sorensen, D.Y. Li, *Nat Genet* **2000**; *26*: 328–331.
33 S.P. Oh, T. Seki, K.A. Goss, T. Imamura, Y. Yi, P.K. Donahoe, L. Li, K. Miyazono, P. ten Dijke, S. Kim, E. Li, *Proc Natl Acad Sci USA* **2000**; *97*: 2626–2631.
34 A.J. Peacock, *Thorax* **1999**; *54*: 1107–1118.
35 S.N. Giri, D.M. Hyde, M.A. Hollinger, *Thorax* **1993**; *48*: 959–966.
36 S.H. Phan, M. Gharaee-Kermani, F. Wolber, U.S. Ryan, *J Clin Invest* **1991**; *87*: 148–154.
37 M.M. Shull, I. Ormsby, A.B. Kier, S. Pawlowski, R.J. Diebold, M. Yin, R. Allen, C. Sidman, G. Proetzel, D. Calvin, *Nature* **1992**; *359*: 693–699.
38 J.S. Munger, X. Huang, H. Kawakatsu, M.J. Griffiths, S.L. Dalton, J. Wu, J.F. Pittet, N. Kaminski, C. Garat, M.A. Matthay, D.B. Rifkin, D. Sheppard, *Cell* **1999**; *96*: 319–328.
39 A. Rosendahl, D. Checchin, T.E. Fehniger, P. ten Dijke, C.H. Heldin, P. Sideras P., *Am J Respir Cell Mol Biol* **2001**; *25*: 60–68.
40 J. Massague, *Ann Rev Biochem* **1998**; *67*: 753–791.
41 S. Bellusci, Y. Furuta, M.G. Rush, R. Henderson, G. Winnier, B.L. Hogan, *Development* **1997**; *124*: 53–63.
42 S. Bellusci, J. Grindley, H. Emoto,, N. Itoh, B.L. Hogan, *Development* **1997**; *124*: 4867–4878.
43 J.C. Grindley, S. Bellusci, D. Perkins, B.L. Hogan, *Dev Biol* **1997**; *188*: 337–348.
44 A.A. Mailleux, D. Tefft, D. Ndiaye, N. Itoh, J.P. Thiery, D. Warburton, S. Bellusci, *Mech Dev* **2001**; *102*: 82–94.
45 J.D. Tefft, M. Lee, S. Smith, M. Lienwand, J. Zhao, P. Bringas, Jr., D.L. Crowe, D. Warburton, *Curr Biol* **1999**; *9*: 219–222.
46 D. Lebeche, S. Malpel, W.V. Cardoso, *Mech Dev* **1999**; *86*: 125–136.
47 D.M. Ornitz, J. Xu, J.S. Colvin, D.G. McEwen, C.A. MacArthur, E. Coulier, G. Gao, M. Goldfarb, *J Biol Chem* **1996**; *271*: 15292–15297.
48 K. Peters, S. Werner, X. Liao, S. Wert, J. Whitsett, L. Williams, *EMBO J* **1994**; *13*: 3296–3301.
49 E. Arman, R. Haffner-Krausz, Y. Chen, J.K. Heath, P. Lonai, *Proc Natl Acad Sci USA* **1998**; *95*: 5082–5087.
50 X. Xu, M. Weinstein, C. Li, M. Naski, R.I. Cohen, D.M. Ornitz, P. Leder, C. Deng, *Development* **1998**; *125*: 753–765.
51 L. DeMoerlooze, B. Spencer-Dene, J. Revest, M. Hajihosseini, I. Rosewell, C. Dickson, *Development* **2000**; *127*: 482–492.
52 J.M. Revest, B. Spencer-Dene, K. Kerr, L. DeMoerlooze, I. Rosewell, C. Dickson, *Dev Biol* **2001**; *231*: 47–62.
53 I. Hokuto, A.K. Perl, J.A. Whitsett, *J Biol Chem* **2003**; *278*: 415–421.
54 D.M. Ornitz, N. Itoh, *Genome Biol* **2001**; *2*: REVIEWS3005.
55 W.Y. Park, B. Miranda, D. Lebeche, G. Hashimoto, W.V. Cardoso, *Dev Biol* **1998**; *201*: 125–134.

56 H. Min, D.M. Danilenko, S.A. Scully, B. Bolon, B.D. Ring, J.E. Tarpley, M. DeRose, W.S. Simonet, *Genes Dev* **1998**; *12*: 3156–3161.

57 K. Sekine, H. Ohuchi, M. Fujiwara, M. Yamasaki, T. Yoshizawa, T. Sato, N. Yagishita, D. Matsui, Y. Koga, N. Itoh, S. Kato, *Nat Genet* **1999**; *21*: 138–141.

58 H. Ohuchi, Y. Hori, M. Yamasaki, H. Harada, K. Sekine K, S. Kato, N. Itoh, *Biochem Biophys Res Commun* **2000**; *277*: 643–649.

59 H. Sakaue, M. Konishi, W. Ogawa, T. Asaki, T. Mori, M. Yamasaki, M. Takata, H. Ueno, S. Kato, M. Kasuga, N. Itoh, *Genes Dev* **2002**; *16*: 908–912.

60 D. Sutherland, C. Samakovlis, M.A. Krasnow, *Cell* **1996**; *87*: 1091–1101.

61 S.M. Ahmad, B.S. Baker, *Cell* **2002**; *109*: 651–661.

62 N. Hacohen, S. Kramer, D. Sutherland, Y. Hiromi, M.A. Krasnow, *Cell* **1998**; *92*: 253–263.

63 T. Casci, J. Vinos, M. Freeman, *Cell* **1999**; *96*: 655–665.

64 D. Tefft, M. Lee, S. Smith, D.L. Crowe, S. Bellusci, D. Warburton, *Am J Physiol Lung Cell Mol Physiol* **2002**; *283*: L700–L706.

65 Y.R. Hadari, H. Kouhara, I. Lax, J. Schlessinger, *Mol Cell Biol* **1998**; *18*: 3966–3973.

66 H.J. Shaeffer, W.J. Weber, *Mol Cell Biol* **1999**; *19*: 2435–2444.

67 G. Minewada, L.A. Jarvis, C.L. Chi, A. Neubuser, X. Sun, N. Hacohen, M.A. Krasnow, G.R. Martin, *Development* **1999**; *126*: 4465–4475.

68 M.K. Lee, J. Zhao, S. Smith, J.D. Tefft, P. Bringas, Jr., C. Hwang, D. Warburton, *Pediatr Res* **1998**; *44*: 850–859.

69 J. Lim, E.S. Wong, S.H. Ong, P. Yusoff, B.C. Low, G.R. Guy, *J Biol Chem* **2000**; *275*: 32837–32845.

70 Y. Yigzaw, L. Cartin, S. Pierre, K. Scholich, T.B. Patel, *J Biol Chem* **2001**; *276*: 22742–22747.

71 E.S. Wong, J. Lim, B.C. Low, Q. Chen, G.R. Guy, *J Biol Chem* **2001**; *276*: 5866–5875.

72 M. van Tuyl, M. Post, *Respir Res* **2000**; *1*: 30–35.

73 Y. Litingtung, L. Lei, H. Westphal, C. Chiang, *Nat Genet* **1998**; *20*: 7–8.

74 C.V. Pepicelli, P.M. Lewis, A.P. McMahon, *Curr Biol* **1998**; *8*: 1083–1086.

75 P.T. Chuang, A.P. McMahon, *Nature* **1999**; *397*: 617–621.

76 R.W. Pelton, M.D. Johnson, E.A. Perkett, L.I. Gold, H.L. Moses, *J Cell Biol* **1991**; *115*: 1091–1105.

77 F.A. Millan, F. Denhez, P. Kondaiah, R.J. Akhurst, *Development* **1991**; *111*: 131–143.

78 P. Schmid, D. Cox, G. Bilbe, R. Maier, G.K. McMaster *Development* **1991**; *111*: 117–130.

79 I.S. McLennan, Y. Poussart, K. Koishi, *Dev Dyn* **2000**; *217*: 250–256.

80 U. Bartram, D.G. Molin, L.J. Wisse, A. Mohammad, L.P. Sanford, T. Doetschman, C. Speer, R.F. Poelmann, A.C. Gittenberger-de Groot, *Circulation* **2001**; *103*: 2745–2752.

81 V. Kaartinen, J.W. Voncken, C. Shuler, D. Warburton, D. Bu, N. Heisterkamp, J. Groffen, *Nat Genet* **1995**; *11*: 415–421.

82 S. Buckley, K.C. Bui, M. Hussain, D. Warburton, *Am J Physiol* **1996**; *271* (Suppl.): L54–L60.
83 J. Zhao, P.J. Sime, P. Bringas, Jr., J.D. Tefft, S. Buckley, D. Bu, J. Gauldie, D. Warburton, *Eur J Cell Biol* **1999**; *78*: 715–725.
84 L. Zhou, C.R. Dey, S.E. Wert, J.A. Whitstett, *Dev Biol* **1996**; *175*: 227–228.
85 R. Serra, R.W. Pelton, H.L. Moses, *Development* **1994**; *120*: 2153–2161.
86 J. Zhao, D. Bu, M. Lee, H.C. Slavkin, F.L. Hall, D. Warburton, *Dev Biol* **1996**; *180*: 242–257.
87 J. Zhao, P.J. Sime, P. Bringas, Jr., J. Gauldie, D. Warburton, *Am J Physiol* **1999**; *277*: L412–L422.
88 J. Zhao, D.L. Crowe, C. Castillo, C. Wuenschell, Y. Chai, D. Warburton, *Mech Dev* **2000**; *93*: 71–81.
89 J. Zhao, M. Lee, S. Smith, D. Warburton, *Dev Biol* **1998**; *194*: 182–195.
90 J. Zhao, W. Shi, H. Chen, D. Warburton, *J Biol Chem* **2000**; *275*: 23992–23997.
91 J. Gauldie, T. Galt, P. Bonniaud, C. Robbins, M. Kelly, D. Warburton, *Am J Pathol* **2003**; *163*: 2575–2584.
92 P.J. Sime, Z. Xing, F.L. Graham, K.G. Csaky, J. Gauldie, *J Clin Invest* **1997**; *100*: 768–776.
93 J. Zhao, W. Shi, Y.L. Wang, H. Chen, P. Bringas, Jr., M.B. Datto, J.P. Frederick, X.F. Wang, D. Warburton, *Am J Physiol Lung Cell Mol Physiol* **2002**; *282*: L585–L593.
94 R. Derynck, R.J. Akhurst, A. Balmain, *Nat Genet* **2001**; *2*: 117–129.
95 A.B. Roberts, R. Derynck, *Sci STKE* **2001**; *113*: PE43.
96 S.A. Gebb, J.M. Shannon, *Dev Dyn* **2002**; *217*: 159–169.
97 L. Miquerol, M. Gertsenstein, K. Harpal, J. Rossant, A. Nagy, *Dev Biol* **1999**; *212*: 307–322.
98 M.J. Acarregui, S.C. Penisten, K.L. Goss, K. Ramirez, J.M. Snyder, *Am J Respir Cell Mol Biol* **1999**; *20*: 14–23.
99 K.R. Brown, K.M. England, K.L. Goss, J.M. Snyder, J.M. Acarregui, *Am J Physiol Lung Cell Mol Physiol* **2001**; *281* (Suppl.): L1001–L1010.
100 Y. Kasahara, R.M. Tuder, L. Taraseviciene-Stewart, T.D. Le Cras, S. Abman, P.K. Hirth, J. Waltenberger, N.F. Voelkel, *J Clin Invest* **2000**; *106*: 1311–1319.
101 X. Zeng, S.E. Wert, R. Federici, K.G. Peters, J.A. Whitsett, *Dev Dyn* **1998**; *211*: 215–227.
102 J. Corne, G. Chupp, C.G. Lee, R.J. Homer, Z. Zhu, Q. Chen, B. Ma, Y. Du, F. Roux, J. McArdle, A.B. Waxman, J.A. Elias, *Nat Med* **2002**; *8*: 702–710.
103 V. Compernolle, K. Brusselmans, T. Acker, P. Hoet, M. Tjwa, H. Beck, S. Plaisance, Y. Dor, E. Keshet, F. Lupu, B. Nemery, M. Dewerchin, P. Van Veldhoven, K. Plate, L. Moons, D. Collen, P. Carmeliet, *Nat Med* **2002**; *8*: 702–710.
104 F. Farnebo, F. Piehl, J. Lagercrantz, *Biochem Biophys Res Commun* **1999**; *257*: 891–894.
105 T. Makinen, L. Jussila, T. Veikkola, T. Karpanen, M.I. Kettunen, K.J. Pulkkanen, R. Kaupinen, D.G. Jackson, H. Kubo, S. Nishikawa, S. Yla-Herttuala, K. Alitalo, *Nat Med* **2001**; *7*: 199–205.

106 T. Veikkola, L. Jussila, T. Makinen, T. Karpanen, M. Jeltsch, T.V. Petrova, H. Kubo, G. Thurston, D.M. McDonald, M.G. Achen, S.A. Stacker, K. Alitalo, *EMBO J* **2001**; *20*: 1223–1231.

107 Y. Cao, P. Linden, J. Farnebo, R. Cao, A. Eriksson, V. Kumar, J.H. Qi, L. Claesson-Welsh, K. Alitalo, *Proc Natl Acad Sci USA* **1998**; *95*: 14389–14394.

108 K.I. Pinson, J. Brennan, S. Monkley, B.J. Avery, W.C. Skarnes, *Nature* **2000**; *407*: 535–538.

109 K. Tamai, M. Semenov, Y. Kato, R. Spokony, C. Liu, Y. Katsuyama, F. Hess, J.P. Saint-Jeannet, X. He, *Nature* **2000**; *407*: 530–535.

110 M. Wehrli, S.T. Dougan, K. Caldwell, L. O'Keefe, S. Schwartz, D. Vaizel-Ohayon, E. Schejter, A. Tomlinson, S. DiNardo, *Nature* **2000**; *407*: 527–530.

111 C.P. Heisenberg, M. Tada, G.J. Rauch, L. Saude, M.L. Concha, R. Geisler, D.L. Stemple, J.C. Smith, S.W. Wilson, *Nature* **2000**; *405*: 76–81.

112 M.A. Torres, J.A. Yang Snyder, S.M. Purcell, A.A. DeMarais, L.L. McGrew, R.T. Moon, *J Cell Biol* **1996**; *133*: 1123–1137.

113 M. Kuhl, K. Geis, L.C. Sheldahl, T. Pukrop, R.T. Moon, D. Wedlich, *Mech Dev* **2001**; *106*: 61–76.

114 T. Ishitani, S. Kishida, J. Hyodo-Miura, N. Ueno, J. Yasuda, M. Waterman, H. Shibuya, R.T. Moon, J. Ninomiya-Tsuji, K. Matsumoto, *Mol Cell Biol* **2003**; *23*: 131–139.

115 Y. Kawano, R. Kypta, *J Cell Sci* **2003**; *116*: 2627–2634.

116 B. Mao, C. Niehrs, *Gene* **2003**; *302*: 179–183.

117 M. Tebar, O. Sestree, W. J. de Vree, A.A. Ten Have-Opbroek, *Mech Dev* **2001**; *109*: 437–440.

118 W. Shu, Y.Q. Jiang, M.M. Lu, E.E. Morrisey, *Development* **2002**; *129*: 4831–4842.

119 S. Bellusci, R. Henderson, G. Winnier, T. Oikawa, B.L. Hogan, *Development* **1996**; *122*: 1693–1702.

120 S. Maretto, M. Cordenonsi, S. Dupont, P. Braghetta, V. Broccoli, A.B. Hassan, D. Volpin, G.M. Bressan, S. Piccolo, *Proc Natl Acad Sci USA* **2003**; *100*: 3299–3304.

121 R. Derynck, Y.E. Zhang, *Nature* **2003**; *425*: 577–584.

122 M.K. Lee, C. Hwang, J. Lee, H.C. Slavkin, D. Warburton, *Am J Respir Cell Mol Biol* **1997**; *273*: L374–L381.

123 M.R. Chinoy, S.E. Zgleszewski, R.E. Cilley, C.J. Blewett, T.M. Krummel, S.R. Reisher, S.I. Feinstein, *Pediatr Pulmonol* **1998**; *25*: 244–256.

124 C.M. Li, J. Khosia, P. Hoyle, P.L. Sannes, *Chest* **2001**; *120*: 60S–61S.

125 A. Masuda, M. Kondo, T. Saito, Y. Yatabe, T. Kobayashi, M. Okamoto, M. Suyama, T. Takahashi, *Cancer Res* **1997**; *57*: 4898–4904.

126 M. Afrakhte, A. Moren, S. Jossan, S. Itoh, K. Sampath, B. Westermark, C.H. Heldin, N.E. Heldin, P. ten Dijke, *Biochem Biophys Res Commun* **1998**; *249*: 505–511.

127 J.W. Jacobberger, N. Sizemore, G. Gorodeski, E.A. Rorke, *Exp Cell Res* **1995**; *220*: 390–396.

128 P.R. Chess, R.M. Ryan, J.N. Finkelstein, *Pediatr Res* **1994**; *36*: 481–486.

129 T. Goldkorn, J. Mendelsohn, *Cell Growth Differ* **1992**; *3*: 101–109.
130 G. Baskin, S. Schenker, T. Frosto, G. Henderson, *J Biol Chem* **1991**; *266*: 13238–13242.
131 L.V. Lotti, L. Lanfrancone, E. Migliaccio, C. Zompetta, G. Pelicci, A.E. Salcini, B. Falini, P.G. Pelicci, M.R. Torrisi, *Mol Cell Biol* **1996**; *16*: 1946–1954.
132 G. Pelicci, L. Lanfrancone, F. Grignani, J. McGlade, F. Cavallo, G. Forni, I. Nicoletti, T. Pawson, P.G. Pelicci, *Cell* **1992**; *70*: 93–104.
133 S. Yan, S. Krebs, K.J. Leister, C.E. Wenner. *J Cell Physiol* **2000**; *185*: 107–116.
134 M.E. Engel, M.A. McDonnel, B.K. Law, H.L. Moses, *J Biol Chem* **1999**; *274*: 37413–37420.
135 M. Hannigan, L. Zhan, Y. Ai, C.K. Huang, *Biochem Biophys Res Commun* **1998**; *246*: 55–58.
136 A. Huwiler, J. Pfeilschifter, *FEBS Lett* **1994**; *354*: 255–258.
137 M. Kretzschmar, J. Doody, I. Timokhina, J. Massague, *Genes Dev* **1999**; *13*: 804–816.
138 D.M. Simeone, L. Zhang, K. Graziano, B. Nicke, T. Pham, C. Schaefer, C.D. Logsdon, *Am J Physiol – Cell Physiol* **2001**; *281*: C311–C319.
139 L. Yu, M.C. Hebert, Y.E. Zhang, *EMBO J* **2002**; *21*: 3749–3759.
140 R.S. Lo, D. Wotton, J. Massague, *EMBO J* **2001**; *20*: 128–136.
141 I. Fabregat, B. Herrera, M. Fernandez, A.M. Alvarez, A. Sanchez, C. Roncero, J.J. Ventura, A.M. Valverde, M. Benito, *Hepatology* **2000**; *32*: 528–535.
142 I. Fabregat, A. Sanchez, A.M. Alvarez, T. Nakamura, M. Benito, *FEBS Lett* **1996**; *384*: 14–18.
143 G. Pelicci, L. Lanfrancone, F. Grignani, J. McGlade, F. Cavallo, G. Forni, I. Nicoletti, T. Pawson, P.G. Pelicci, *Cell* **1992**; *70*: 93–104.
144 E. Migliaccio, S. Mele, A.E. Salcini, G. Pelicci, K.M. Lai, G. Superti Furga, T. Pawson, P. P. di Fiore, L. Lanfrancone, P.G. Pelicci, *EMBO J* **1997**; *16*: 706–716.
145 E. Migliaccio, M. Giorgio, S. Mele, G. Pelicci, P. Reboldi, P.P. Pandolfi, L. Lanfrancone, P.G. Pelicci, *Nature* **1999**; *402*: 309–313.
146 C.P. Yang, S.B. Horwitz, *Cancer Res* **2000**; *60*: 5171–5178.
147 M. Foschi, F. Franchi, J. Han, G. La Villa, A. Sorokin, *J Biol Chem* **2001**; *276*: 26640–26647.
148 T.H. Morgan, *Roux's Arch Dev Biol* **1897**; *5*: 570–586.
149 H. Spemann H, *Embryonic Development and Induction*. New Haven, CT: Yale University Press, **1938**.
150 H. Sebag-Montefiore, *ENIGMA The Battle for the Code*, John Wiley & Sons, **2000**.
151 A.M. Turing, *Philos Trans R Soc Lond (B)* **1952**; *237*: 32–72.
152 D. Warburton, R. Seth, L. Shum, P.G. Horcher, F.L. Hall, Z. Werb, H.C. Slavkin, *Dev Biol* **1992**; *149*: 123–133.
153 L. Wolpert, *Curr Biol* **2002**; *12*: R628–R630.
154 J. Aubin, M. Lemieux, M. Tremblay, J. Berard, L. Jeannotte, *Dev Biol* **1997**; *192*: 432–445.

22
Molecular Genetics of Liver and Pancreas Development

Tomas Pieler, Fong Cheng Pan, Solomon Afelik, and Yonglong Chen

22.1
Introduction

Liver and pancreas are two major internal organs that originate from the endodermal germ layer. Similar to the lung, they are both derived from gut primordia (Figs. 22.1 and 22.2). The liver serves a broad range of metabolic functions, including processing of ingested nutrients, detoxification of a variety of different compounds and maintenance of physiological metabolite and protein concentrations in the blood. The pancreas consists of two major functional entities that exert exocrine and endocrine activities, respectively. The exocrine cells produce and secrete various digestive enzymes and they constitute more than 95 % of the total pancreatic cell mass. The four different endocrine cell types are each responsible for the production of one pancreatic hormone: glucagon is made in the α-cells, insulin in the β-cells, somatostatin in the δ-cells and pancreatic polypeptide in the PP-cells.

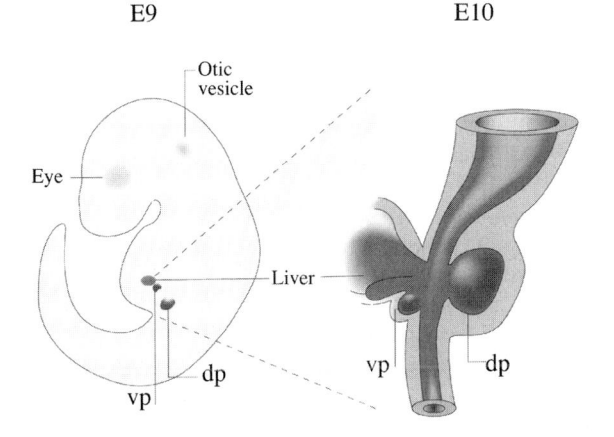

Fig. 22.1
Schematic representation of the liver as well as of the dorsal and ventral pancreatic buds evaginating from the primitive gut epithelium at mouse embryonic day E9 and E10 (after [3]). dp, dorsal pancreas; vp, ventral pancreas.

Cell Signaling and Growth Factors in Development. Edited by K. Unsicker and K. Krieglstein
Copyright © 2006 WILEY-VCH Verlag GmbH & Co. KGaA, Weinheim
ISBN 3-527-31034-7

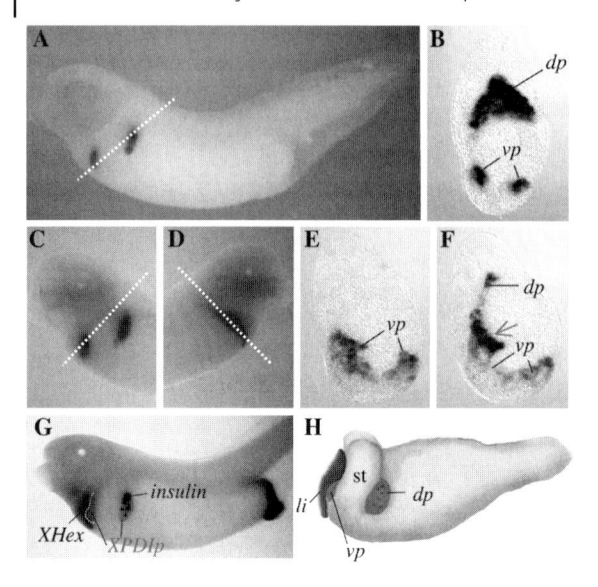

Fig. 22.2
Xenopus embryonic liver and pancreas. (A and B) Whole-mount *in situ* hybridization reveals that exocrine marker gene, XPDIp, is exclusively expressed in the dorsal and two ventral pancreatic anlagen of a stage–39 embryo. Panel B shows a transversal section across the white dashed line, as indicated in panel A. (C-F) During subsequent development (stage 40), XPDIp expression reflects the fusion process of dorsal and ventral pancreatic buds, first at the right side of the embryo (red arrow in F). Panels C and D are left and right side views of the same stage–40 embryo. Panels E and F represent transversal sections across the white dashed lines indicated in panels C and D, respectively. (G) Left side view of a stage–40 Xenopus embryo after whole-mount *in situ* hybridization analysis of liver (XHex) and pancreas (insulin and XPDIp) marker gene expression. Note that the liver (XHex expression in dark blue) is adjacent to the ventral pancreatic buds (XPDIp expression in red). Insulin expression (dark blue) is restricted to the dorsal pancreatic bud at this stage of development. The white dashed line demarcates the boundary between the liver and the ventral pancreatic bud. (H) A schematic drawing reflecting the situation depicted in A (modified from [81]). The blue dots in the dorsal pancreatic bud represent insulin expression. dp, dorsal pancreatic bud; li, liver anlage; st, stomach; vp, ventral pancreatic bud. (This figure also appears with the color plates.)

In mammals and amphibia, the pancreas develops from two different anlagen, one dorsal and two ventral pancreatic buds. Formation of the dorsal pancreatic primordium is under the control of signals from the notochord and dorsal aorta, while formation of the ventral pancreatic buds is directed by signals from the ventral vein and closely coordinated with the process of liver development. In the course of embryonic development, one of the two ventral buds regresses, while the other continues to grow. During gut rotation, the dorsal bud is translocated to the ventral side of the embryo, where the pancreatic primordia fuse [1]. Regulation of pancreas and liver development has been studied at the molecular level in a variety of vertebrate animal models, including mouse, chicken, frog and fish (reviewed in [2–6]).

22.2
From the Fertilized Egg to Primitive Endodermal Precursor Cells

Due to historical and technical reasons, early embryonic patterning events that lead to the formation of the three different germ layers have been studied in greatest detail with amphibians such as *Xenopus laevis*. The endodermal germ layer is derived from cells that localize to the vegetal hemisphere of the early Xenopus embryo. Single-cell transplantation experiments have revealed that vegetal cells are not committed to an endodermal fate prior to early gastrula stages [7, 8]. VegT, a T-box transcription factor that is encoded by a maternal mRNA localizing to the vegetal cortex of Xenopus oocytes/eggs, is essentially required for endoderm development. It promotes the expression of nodal related (TGFβ-type) signaling molecules and downstream endodermal transcription factors (such as Mixer, GATA5, Bix1, Bix4 and Sox17α) in a cell-autonomous manner at the mid-blastula transition. Subsequently, cell-cell communication, at least in part mediated by graded nodal signaling, is required to pattern endoderm-specific gene expression during late blastula and gastrula stages of development [9–14]. Dorsal and ventral pancreas, together with the liver, are derived from the dorsal endoderm of gastrula-stage embryos, which forms under the influence of the canonical Wnt-signal transduction cascade (Fig. 22.3) [15, 16].

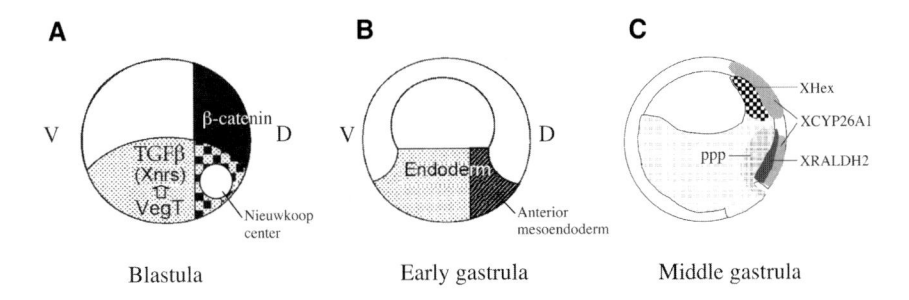

Fig. 22.3
Schematic drawings showing some of the early signaling pathways that are involved in the pre-patterning of the liver and the pancreas precursor cells in Xenopus. The VegT-activated TGFβ (Xnrs) signaling activities and the canonical Wnt pathway mediated by β-catenin are thought to coordinate the formation of an early signaling center, the so-called Nieuwkoop center, in the presumptive dorsal vegetal hemisphere of blastula stage embryos, which will give rise to the anterior mesoendoderm during gastrulation (A and B after [16]). (C) The anterior mesoendodermal cells give rise to liver, as reflected in XHex expression. The putative pancreatic precursor cells (ppp) are likely to be under the influence of the RA signaling center that is created by the RA-generating (XRALDH2) and -degrading (XCYP26A1) enzymes during gastrulation. D, dorsal; V, ventral. (This figure also appears with the color plates.)

22.3
Commitment to Pancreas and Liver Fates in Xenopus

On the basis of results obtained with vegetal explants from blastula-stage Xenopus embryos, it has been suggested that the embryonic endoderm is already pre-patterned along the dorsal-ventral body axis in these early developmental stages, as evident from the differential expression of endodermal marker genes, such as Xlhbox8 [17, 18], as well as XHex and cerberus [16], in dorsal but not ventral vegetal explants. The molecular identity of the signals involved in these endodermal pre-patterning events is not exactly defined, but TGFβ signaling has been found to exert a positive regulatory function in the vegetal explant system [16–18]. A more recent report provides strong evidence that the specification of the endoderm along the antero-posterior body axis relies on the inductive activity of adjacent mesodermal tissue in tailbud-stage embryos (from stage 25 onwards); such a signaling activity may also be produced by mesodermal cells found to be present in vegetal explants from blastula-stage embryos [19]. Nevertheless, and as already pointed out in the same study, these observations do not rule out the possibility that an early labile specification within the endoderm does exist and may confer competence to respond to later interactions with mesoderm or with other neighboring tissues.

Retinoic acid (RA) is long known to play a central role in antero-posterior patterning of the central nervous system in Xenopus embryos. More recently, we have been able to demonstrate that RA signaling also serves a regulatory function for endodermal patterning in Xenopus (Fig. 22.3). Inhibition of RA signaling at gastrula stages results in a loss of dorsal exo- and endocrine pancreas development. The ventral pancreas is only moderately affected under these conditions. Conversely, application of excessive RA during gastrulation expands the endocrine cell population at the expense of exocrine cells in the dorsal pancreas. In contrast, increased RA levels enhance exocrine marker gene expression in the ventral pancreas, while development of the liver is inhibited [20].

22.4
Liver and Pancreas Specification in Mouse, Chicken and Zebrafish

Two recent studies provide further support for the idea of a regulatory function for RA signaling in the context of early endodermal patterning. Kumar et al. [21] have used explants from chicken and quail embryos to demonstrate that lateral plate mesoderm exerts an instructive signaling activity on the adjacent endoderm in respect of the expression of various pancreas marker genes. RA signaling was found to be sufficient to induce Pdx1 expression in anterior explants that contained both endoderm as well as mesoderm and would normally not develop into pancreatic tissue; however, RA was not sufficient to induce endodermal Pdx1 expression in the absence of mesoderm. Thus, RA may work indirectly via the mesoderm rather than directly on the endoderm, or it may act directly in concert with mesodermally-derived signals.

Stafford and Prince [22] have analyzed the function of RA signaling for pancreas development in the zebrafish. While some of the effects observed in the fish equate with those obtained in the frog system, others do not. In both systems, inhibition of RA signaling during gastrulation results in a dramatic reduction of pancreas gene expression. However, application of excess RA resulted in an expansion of both exo- and endocrine cells in the fish, while it resulted in an expansion of the endocrine cell population at the expense of the exocrine cells in the dorsal pancreas of the frog. Furthermore, while inhibition of RA signaling correlated with a loss of liver in the fish, liver development in the frog was unaffected in the same situation. The pancreas develops from two separate domains in both Xenopus and zebrafish; we note, however, that their response to a modulation of RA signaling is indistinguishable in the fish, while dorsal and ventral pancreas in the frog differ significantly in several aspects of pancreas gene expression. It has been pointed out [23] that the pancreatic primordia in the zebrafish embryo are not truly equivalent to the pancreatic buds, as they develop in other vertebrates including Xenopus. Gastrulation in zebrafish results in the formation of a sheet of endodermal cells rather than a proper endodermal tube. Therefore, the discrepancies of results obtained in a comparison of both systems in respect to pancreas and liver development may, at least in part, reflect the differences in respect to gut tube formation in fish and frog embryos.

In the mouse, an endodermal pre-pattern that defines competence to respond to pancreas-inducing signals from the notochord seems to be established during gastrulation by signals derived from the overlying mesoderm and ectoderm [24]. In a slightly later phase, FGF and activin signals from the notochord are thought to induce pancreatic gene activity by suppressing sonic hedgehog expression in the dorsal pancreatic epithelium (Fig. 22.4) [25, 26]. A significantly different situation has been described in the zebrafish; sonic hedgehog was reported to be required for the establishment of the endocrine lineage in this system [27, 28].

Dorsal foregut endoderm

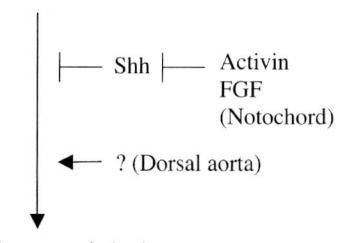

⊢ Shh ⊢ Activin
           FGF
           (Notochord)

← ? (Dorsal aorta)

Dorsal pancreatic bud

Fig. 22.4
Signals involved in dorsal pancreas specification. Studies in chicken and mouse indicate that initial signals from the notochord (likely FGF2 and activin), which inhibit Shh expression, as well as later unknown signals from the dorsal aorta, are required for dorsal pancreas specification [26, 29].

The multipotent ventral endoderm, that will give rise to both liver and pancreas, forms under the control of BMPs from the septum transversum mesenchyme. FGFs produced by the cardiac mesoderm promote liver formation and inhibit pancreas development (Fig. 22.5; reviewed in [2]). Pancreatic growth is initiated at sites in the foregut endoderm, precisely where the endodermal epithelium contacts the endothelium of major blood vessels (dorsal aorta and vitelline veins). More recently, it has been reported that the vascular as well as the vasculogenic endothelium act as sources for signals that are positively involved in both pancreas and liver differentiation [29, 30]. The molecular identity of these signal(s), however, remains to be established.

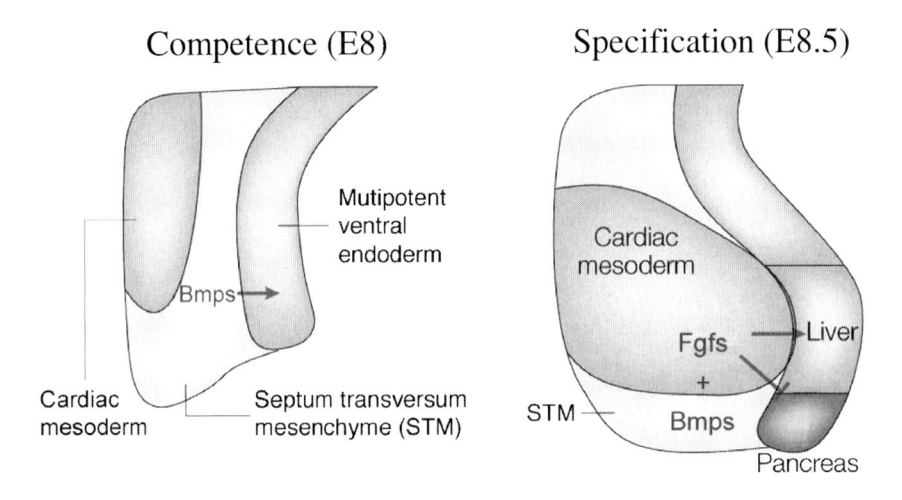

Fig. 22.5
Specification of liver and ventral pancreas in the mouse. The default state of the ventral foregut endoderm is pancreas. Signaling molecules, such as Bmps from septum transversum mesenchyme and Fgfs from cardiac mesoderm, direct subpopulations of the ventral foregut endoderm to form liver (after [2]). (This figure also appears with the color plates.)

In the zebrafish, the position and extent of the pancreatic domain were found to depend on the level of BMP signaling at or just after gastrulation [31]. Finally, in the chicken embryo, the endoderm seems to be patterned in a posterior-dominant manner with the lateral plate mesoderm acting as a source for instructive signals that induce pancreatic differentiation. BMP, activin and, as already mentioned above, RA signaling were found to mimic aspects of this inductive process in combined endoderm/mesoderm explants, even though they were not sufficient on their own in the absence of mesoderm [21].

22.5
Proliferation and Differentiation of Functionally Distinct Pancreatic and Hepatic Cell Populations

Several transcription regulators (such as Pdx1, Hlxb9, p48/Ptf1a and others) have been described, which identify pancreatic progenitor cells before expression of the pancreas differentiation markers can be detected (reviewed in [3]). A critical combination of multiples of these transcription regulators seems to be necessary to control pancreas specification, as some of these factors are also expressed outside of the pancreatic anlagen. A second, partially overlapping wave of transcription factor activity operates in the differentiation of the individual exocrine and endocrine cell types. Several of these factors (such as p48/Ptf1a, Hes1, ngn3 and NeuroD) are directly linked to the Notch signal transduction cascade.

Notch signaling is involved in both proliferation and differentiation of pancreatic precursor cells. In mice, all four vertebrate Notch genes (Notch 1 to 4), as well as the Notch ligands Serrate 1 and 2 have been found to be differentially expressed in the mesenchymal and epithelial pancreatic cells along with the Notch target gene Hes1 [32]. Blocking Notch receptor activation in early pancreatic progenitors results in high neurogenin 3 (ngn3) expression, which promotes early endocrine cell differentiation at the expense of pancreatic cell proliferation [33–35]. Conversely, increasing Notch activity prevents differentiation of pancreatic acinar cells and attenuates endocrine development [36–38]. Therefore, Notch activity, which operates via the transcription repressor Hes1, seems not only involved in the choice between endocrine and exocrine cell fates, but it may also promote cell proliferation by maintaining pancreatic cells in a progenitor state. The modulation of Notch activity could thus be crucial for the *in vitro* generation of proliferating precursor cell populations.

A large set of transcription factors has been identified that operate in the differentiation of the diverse endo- and exocrine cell types (reviewed in [3]). These include activities that promote the formation of individual endocrine cell types or acinar cells (such as Pdx1, Hlxb9, Nkx6.1, Mist1 and Hnf1a), as well as those that are crucial for multiple different endocrine cell types (such as Nkx2.2, Isl1, Pax4, Pax6, NeuroD, Foxa1, Foxa2 and Hnf6). How the activity of these different transcription regulators is interconnected and how they merge to define different programs of gene activity is not fully understood; this however, may be particularly relevant to attempts to generate specific pancreatic cell populations *in vitro* from endodermal precursor cells.

Appropriate molecular markers that allow the early events of pancreas differentiation to be monitored have been defined in the model systems employed in studying pancreas development, including Xenopus. Interestingly, early insulin expression in Xenopus remains confined to the dorsal pancreas until fusion and is first detectable at stage 35, i.e. when the formation of the dorsal bud is becoming visible [39, 40]. In contrast, PDI (protein disulfide isomerase) is an exocrine pancreatic marker gene that is expressed in both dorsal and ventral buds from stage 39 onwards (Fig. 22.2) [41]; the ventral pancreatic primordium becomes morphologically detectable at stage 37/38 [39].

The undifferentiated hepatic precursor cells, also referred to as hepatoblasts, differentiate to give rise to either hepatocytes with their characteristic epithelial morphology, or to the bile duct cells [42, 43]. Several transcription factors (including Hnf1α, Hnf1β, Hnf4 and Hnf6) have been identified to be involved in this process (reviewed in [2]).

22.6
Transdifferentiation of Pancreas and Liver

As already pointed out above, the ventral pancreas is derived from an embryonic precursor cell population that will also give rise to the liver. It has been reported that ectopic hepatocytes can occur in the pancreas after pancreatic damage [44]; such pancreatic hepatocytes are also obtained upon ectopic expression of an FGF-type signaling molecules [45]. Moreover, application of dexamethasone to either a pancreatic cell line or to pancreatic buds *in vitro* is associated with a transformation of these pancreatic cells to hepatocytes [46, 47].

The converse process has also been described; most remarkably, it was recently demonstrated that a single activated pancreatic transcription factor (Pdx1–VP16) is sufficient to transform the liver into ectopic pancreatic tissue in Xenopus [48]. Furthermore, it was reported that adenovirus-mediated Pdx–1 gene transfer into the adult liver induces both endo- and exocrine pancreatic gene expression in the mouse [49–51]. It is not entirely clear if these phenomena reflect an ability of pancreatic and/or hepatic cells to transdifferentiate, or if they are the result of differentiation from hepato-pancreatic precursor cells [52]; however, these findings certainly reflect the close anatomical relationship of pancreatic and hepatic cells, as is evident from their common origin in the ventral endoderm during early embryogenesis (summarized in Fig. 22.6).

22.7
Pancreas and Liver Regeneration

The pancreas also has a limited ability to regenerate. Upon partial pancreatectomy, new islets have been found to differentiate from ductular epithelium, recapitulating islet formation as it occurs during embryogenesis and suggesting the existence of multipotent precursor cells in the adult that can be activated during regeneration [53] (reviewed in [54, 55]). A more recent study using a genetic lineage-marking approach indicates that pre-existing β-cells (i.e. insulin gene transcribing cells), rather than multipotent stem cells, are the major source of new β-cells during adult life and after partial pancreatectomy in mice [56]. These observations do not exclude the possibility that a small portion of the insulin-positive cells serve as multipotent precursor cells with a significant proliferative capacity, rather than having the identity of fully differentiated β-cells.

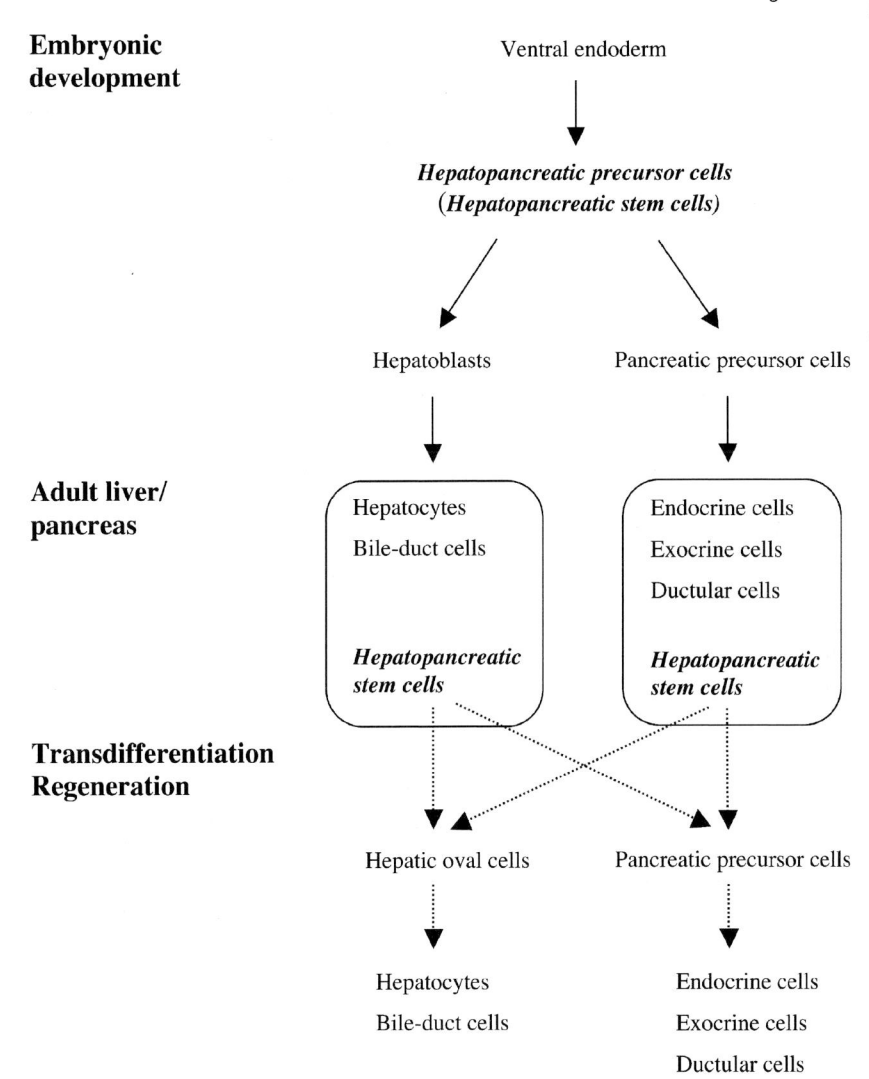

22.7 Pancreas and Liver Regeneration | 831

Embryonic development

Ventral endoderm
↓
Hepatopancreatic precursor cells
(Hepatopancreatic stem cells)
↙ ↘
Hepatoblasts   Pancreatic precursor cells
↓   ↓

Adult liver/ pancreas

Hepatocytes	Endocrine cells
Bile-duct cells	Exocrine cells
	Ductular cells
Hepatopancreatic stem cells	*Hepatopancreatic stem cells*

Transdifferentiation Regeneration

Hepatic oval cells   Pancreatic precursor cells
↓   ↓
Hepatocytes   Endocrine cells
Bile-duct cells   Exocrine cells
                                      Ductular cells

Fig. 22.6
Liver-pancreas transdifferentiation might partially recapitulate the events of early embryonic development. To explain the current liver-pancreas transdifferentiation data, we propose the model indicated. Adult liver and pancreas possess a small population of common hypothetical hepatopancreatic stem cells [52], which are equivalent to the hepatopancreatic precursor cells derived from ventral foregut endoderm during early embryogenesis. These cells can give rise to both hepatic oval cells and pancreatic precursor cells and further differentiate into liver- and pancreas-specific cell types. The same cells could account for liver to pancreas and pancreas to liver transdifferentiation under various experimental and pathological conditions. They could also account for progenitor-dependent liver and pancreas regeneration (dashed arrow lines).

Interestingly, a corresponding cell population of multipotent precursor cells has also been shown to exist in the liver; hepatic oval cells, which reside within the canal of Hering, have the potential to either differentiate into bile duct cells or, in the context of liver regeneration, into hepatocytes [57, 58] (reviewed in [59]). The protocol that allows for oval cell activation entails partial hepatectomy and simultaneous inhibition of hepatocyte proliferation. In the absence of such inhibition, liver regeneration is achieved via differentiated hepatocytes resuming mitotic activity (Fig. 22.7). In the context of this latter situation, vascular endothelium produces signals such as IL–6 and HGF, which promote proliferation and survival of adjacent hepatocytes [60, 61]. HGF is also known as a signaling molecule with a function in the embryonic development of the liver, where it exerts an anti-apoptotic effect on the proliferating hepatoblasts [62].

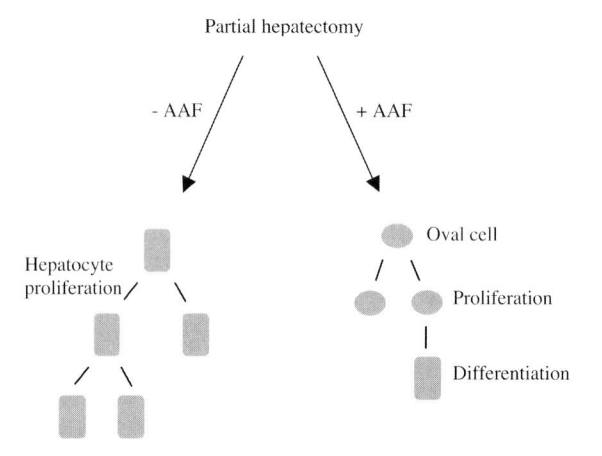

Fig. 22.7
Rat liver can regenerate via hepatocyte proliferation or oval cell proliferation and differentiation. After partial hepatectomy (PH), hepatocyte proliferation may be responsible for liver growth. In the AAF (acetylaminofluorene)/PH model, AAF treatment blocks hepatocyte proliferation after PH. The oval cells, that might be derived from the stem cell niche in the canals of Hering, are believed to first proliferate and then differentiate into hepatocytes.

22.8
Generation of Pancreatic and Hepatic Cells from Pluripotent Embryonic Precursor Cells

Due to the high prevalence of diseases correlating with pancreatic malfunction, it is of obvious medical relevance to seek to establish protocols that allow for the application of pancreatic cell-replacement therapies. Several protocols have been developed to generate insulin-producing cells from murine embryonic stem cells *in vitro* [63–68] (reviewed in [69]). Even though these pioneering studies seem highly rel-

22.8 Generation of Pancreatic and Hepatic Cells from Pluripotent Embryonic Precursor Cells

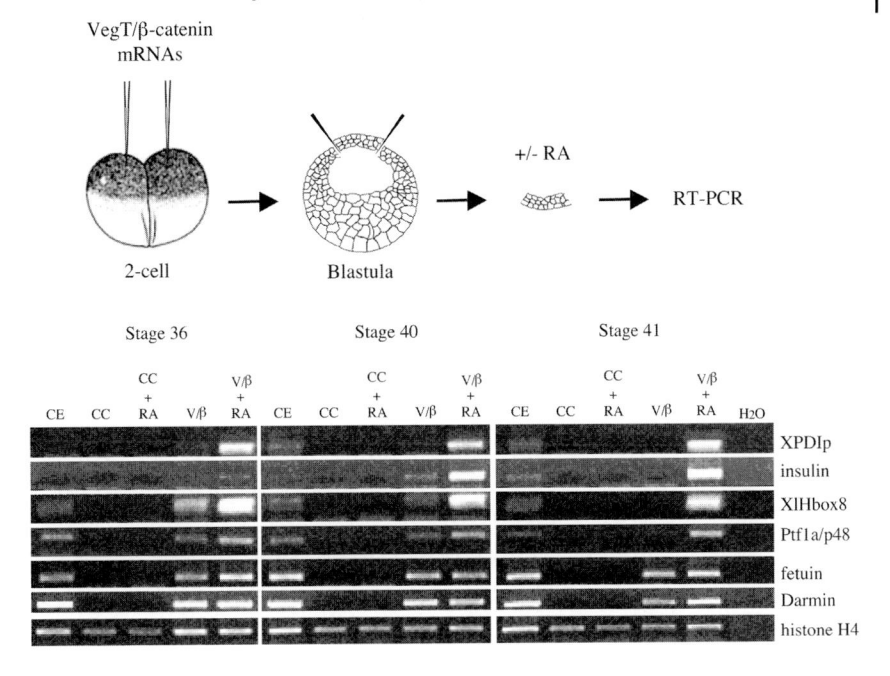

Fig. 22.8
Activation of liver and pancreas marker gene expression in pluripotent embryonic precursor cells by VegT, β-catenin and retinoic acid. Two-cell stage embryos were injected with VegT and β-catenin mRNAs in both blastomeres from the animal pole. Ectodermal explants (animal caps) were harvested at stage 8/9, treated with RA at stage 11 for 1 h and cultivated to the equivalent of embryonic stages 36, 40 and 41. Expression of the liver (fetuin), pancreas (XPDIp, insulin, XlHbox8 and Ptf1a/p48) and intestine (Darmin) marker genes was analyzed by RT-PCR. Histone H4 was used for RNA loading control. CE, control embryo; CC, control explants (caps); CC + RA, control caps treated with RA at the equivalent of stage 11; V/β, VegT/β-catenin injected caps; V/β + RA, VegT/β-catenin injected caps treated with RA at the equivalent of stage 11.

evant, the protocols employed were largely empirical, the data obtained by McKay and colleagues [63] have been questioned [70] and criteria to determine what exactly constitutes a functional pancreatic β-cell at the molecular level remain to be defined. A detailed analysis of the molecular mechanisms that allow for pancreas formation during embryonic development should help to generate the different pancreatic cell types *in vitro* by applying the appropriate combination of regulatory activities as they occur in the context of embryogenesis.

Ectodermal explants from early Xenopus embryos, which would normally develop into atypical epidermis, express a panel of endodermal marker genes if injected with a combination of VegT and β-catenin encoding mRNAs. However, in contrast to liver genes, pancreas differentiation markers were not found to be expressed [71].

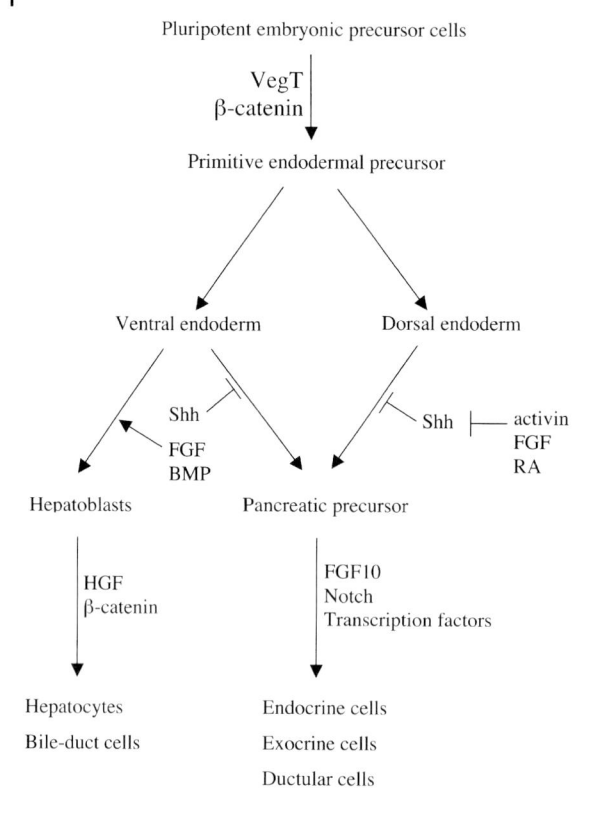

```
Pluripotent embryonic precursor cells
           │ VegT
           │ β-catenin
           ▼
    Primitive endodermal precursor
         ╱        ╲
        ╱          ╲
       ▼            ▼
Ventral endoderm   Dorsal endoderm
   ╱    ╲            ╱    ╲
  Shh    ╲          Shh    ─── activin
   FGF    ╲           FGF
   BMP     ╲           RA
    ▼       ▼           ▼
Hepatoblasts   Pancreatic precursor

    │ HGF              │ FGF10
    │ β-catenin        │ Notch
    │                  │ Transcription factors
    ▼                  ▼
Hepatocytes        Endocrine cells
Bile-duct cells    Exocrine cells
                   Ductular cells
```

Fig. 22.9.
Specific liver and pancreatic cell types could be induced from pluripotent embryonic precursor cells by sequentially modulating the activities as they govern normal embryonic development. Most of the regulators have been discussed in the text. A recent study in chicken indicates that β-catenin-mediated canonical Wnt signaling controls the size of the developing liver [82].

Asashima and coworkers have very systematically analyzed, what type of tissue develops from explanted Xenopus dorsal lips and activin-treated animal caps upon transient treatment with the signaling molecule retinoic acid (RA). It turns out, that not all, but a significant number of these RA-treated explants contain some pancreatic cells with insulin expression [72, 73]. We have been able to demonstrate that RA treatment of VegT/β-catenin injected animal caps induces the formation of clusters of cells positive for exocrine pancreas markers, which are associated with secondary clusters of cells positive for endocrine pancreas gene expression (Fig. 22.8) [20].

A number of different studies have also suggested that hematopoietic stem cells have the ability to transdifferentiate into hepatocytes [74, 75]. However, it is not entirely clear whether these phenomena reflect the ability of such adult stem cells to respond to the microenvironment of the liver by activating the hepatocyte-specific gene program, or whether they are the result of a fusion of hepatocytes with the invading hematopoietic stem cells [76, 77]. Embryonic stem cells also gain hepatocyte-specific features upon transplantation into the liver [78]. Hepatocyte-specific properties can also be induced *in vitro*, by use of adult and embryonic stem cells. HGF and FGF4 induce hepatocyte-specific characteristics in a subpopulation of hematopoietic stem cells known as MAPCs (multipotent adult progenitor cells) [79]. Similarly, use of aFGF, HGF, OSM and Dex was found to drive murine embryonic stem cells in the direction of hepatocyte differentiation [80]. Finally, the hepatocyte-specific gene program could also be activated in multipotent embryonic precursor cells from the frog by applying a combination of the transcription factor VegT and β-catenin, recapitulating regulatory events of the primitive dorsal endoderm that will give rise to the liver in this system [71] (Fig. 22.9).

References

1 Slack, J. M.W. Developmental biology of the pancreas. *Development* 1995; *121*: 1569–1580.
2 Zaret, K.S. Regulatory phases of early liver development: paradigms of organogenesis. *Nat Rev Genet* 2002; *3*: 499–512.
3 Edlund, H. Pancreatic organogenesis – developmental mechanisms and implications for therapy. *Nat Rev Genet* 2002; *3*: 524–532.
4 Kumar, M. and Melton, D. Pancreas specification: a budding question. *Curr Opin Genet Dev* 2003; *13*: 401–407.
5 Murtaugh, L. C. and Melton, D.A. Genes, signals, and lineages in pancreas development. *Annu Rev Cell Dev Biol* 2003; *19*: 71–89.
6 Jensen, J. Gene regulatory factors in pancreatic development. *Dev Dyn* 2004; *229*: 176–200.
7 Heasman, J., Wylie, C.C., Hausen, P. and Smith, J.C. Fates and states of determination of single vegetal pole blastomeres of X. laevis. *Cell* 1984; *37*: 185–194.
8 Wylie, C.C., Snape, A., Heasman, J. and Smith J.C. Vegetal pole cells and commitment to form endoderm in Xenopus laevis. *Dev Biol* 1987; *119*: 496–502.
9 Zhang, J., Houston, D.W., King, M.L., Payne, C., Wylie, C. and Heasman, J. The role of maternal VegT in establishing the primary germ layers in Xenopus embryos. *Cell* 1998; *94*: 515–524.
10 Clements, D., Friday, R.V., and Woodland, H.R. Mode of action of VegT in mesoderm and endoderm formation. *Development* 1999; *126*: 4903–4911.
11 Yasuo, H. and Lemaire, P. A two-step model for the fate determination of presumptive endodermal blastomeres in Xenopus embryos. *Curr Biol* 1999; *9*: 8609–8879.
12 Chang, C., and Hemmati-Brivanlou, A. A post-mid-blastula transition requirement for TGFbeta signalling in early endodermal specification. *Mech Dev* 2000; *90*: 227–235.

13 Engleka, M.J., Craig, E. J. and Kessler, D.S. VegT activation of Sox17 at the midblastula transition alters the response to nodal signals in the vegetal endoderm domain. *Dev Biol* **2001**; *237*: 159–172.

14 Xanthos, J.B., Kofron, M., Wylie, C. and Heasman, J. Maternal VegT is the initiator of a molecular network specifying endoderm in *Xenopus laevis. Development* **2001**; *128*: 167–180.

15 Jones, C.M., Broadbent, J., Thomas, P.Q., Smith, J.C., and Beddington, R.S. An anterior signalling centre in Xenopus revealed by the homeobox gene Xhex. *Curr Biol* **1999**; *9*: 946–954.

16 Zorn, A.M., Butler, K. and Gurdon, J.B. Anterior endomesoderm specification in Xenopus by Wnt/beta-catenin and TGF-beta signalling pathways. Dev Biol **1999**; *209*: 282–297.

17 Gamer, L.W. and Wright, C.V. Autonomous endodermal determination in Xenopus: regulation of expression of the pancreatic gene XlHbox 8. *Dev Biol* **1995**; *171*: 240–251.

18 Henry, G.L., Brivanlou, I.H., Kessler, D.S., Hemmati-Brivanlou, A. and Melton, D.A. TGF-beta signals and a pattern in *Xenopus laevis* endodermal development. *Development* **1996**; *122*: 1007–1015.

19 Horb, M.E. and Slack, J.M. Endoderm specification and differentiation in Xenopus embryos. *Dev Biol* **2001**; *236*: 330–343.

20 Chen, Y., Pan, F.C., Brandes, N., Afelik, S., Sölter, M. and Tomas Pieler, T. Retinoic acid signalling is essential for pancreas development and promotes endocrine at the expense of exocrine cell differentiation in *Xenopus Dev Biol* **2004**; *271*: 144–160.

21 Kumar, M., Jordan, N., Melton, D. and Grapin-Botton, A. Signals from lateral plate mesoderm instruct endoderm toward a pancreatic fate. *Dev Biol* **2003**; *259*: 109–122.

22 Stafford, D. and Prince, V.E. Retinoic acid signaling is required for a critical early step in zebrafish pancreatic development. *Curr Biol* **2002**; *12*: 1215–1220.

23 Biemar, F., Argenton, F., Schmidtke, R., Epperlein, S., Peers, B. and Driever, W. Pancreas development in zebrafish: early dispersed appearance of endocrine hormone expressing cells and their convergence to form the definitive islet. *Dev Biol* **2001**; *230*: 189–203.

24 Wells, J.M. and Melton, D.A. Early mouse endoderm is patterned by soluble factors from adjacent germ layers. *Development* **2000**; *127*: 1563–1572.

25 Kim, S.K., Hebrok, M., and Melton, D.A. Notochord to endoderm signalling is required for pancreas development. *Development* **1997**; *124*: 4243–4253.

26 Hebrok, M., Kim, S.K., and Melton, D.A. Notochord repression of endodermal Sonic hedgehog permits pancreas development. *Genes Dev* **1998**; *12*: 1705–1713.

27 Roy, S., Qiao, T., Wolff, C. and Ingham, P.W. Hedgehog signalling pathway is essential for pancreas specification in the zebrafish embryo. *Curr Biol* **2001**; *11*: 1358–1363.

28 diIorio, P.J., Moss, J.B., Sbrogna, J.L., Karlstrom, R.O. and Moss, L.G. Sonic hedgehog is required early in pancreatic islet development. *Dev Biol* **2002**; *244*: 75–84.

29 Lammert, E., Cleaver, O. and Melton, D. Induction of pancreatic differentiation by signals from blood vessels. *Science* **2001**; *294*: 564–567.
30 Matsumoto, K., Yoshitomi, H., Rossant, J. and Zaret, K.S. Liver organogenesis promoted by endothelial cells prior to vascular function. *Science* **2001**; *294*: 559–563.
31 Tiso, N., Filippi, A., Pauls, S., Bortolussi, M. and Argenton, F. BMP signalling regulates anteroposterior endoderm patterning in zebrafish. *Mech Dev* **2002**; *118*: 29–37.
32 Lammert, E., Brown, J. and Melton, D.A. Notch gene expression during pancreatic organogenesis. *Mech Dev* **2000**; *94*: 199–203.
33 Gu, G., Dubauskaite, J. and Melton, D.A. Direct evidence for the pancreatic lineage: NGN3+ cells are islet progenitors and are distinct from duct progenitors. *Development 129:* **2002**; 2447–2457.
34 Jensen, J., Pedersen, E.E., Galante, P., Hald, J., Heller, R.S., Ishibashi, M., Kageyama, R., Guillemot, F., Serup, P. and Madsen, O.D. Control of endodermal endocrine development by Hes–1. *Nat Genet* **2000**; *24*: 36–44.
35 Apelqvist, A., Li, H., Sommer, L., Beatus, P., Anderson, D.J., Honjo, T., Hrabe de Angelis, M., Lendahl, U. and Edlund, H. Notch signaling controls pancreatic cell differentiation. *Nature* **1999**; *400*: 877–881.
36 Hald, J., Hjorth, J.P., German, M.S., Madsen, O.D., Serup, P. and Jensen, J. Activated Notch1 prevents differentiation of pancreatic acinar cells and attenuate endocrine development. *Dev Biol* **2003**; *260*: 426–437.
37 Hart, A., Papadopoulou, S. and Edlund, H. Fgf10 maintains notch activation, stimulates proliferation, and blocks differentiation of pancreatic epithelial cells. *Dev Dyn* **2003**; *228*: 185–193.
38 Norgaard, G.A., Jensen, J. N. and Jensen, J. FGF10 signalling maintains the pancreatic progenitor cell state revealing a novel role of Notch in organ development. *Dev Biol* **2003**; *264*: 323–338.
39 Kelly, O.G. and Melton, D.A. Development of the pancreas in *Xenopus laevis*. *Dev Dyn* **2000**; *218*: 615–627.
40 Horb, M.E. and Slack, J.M. Expression of amylase and other pancreatic genes in Xenopus. *Mech Dev* **2002**; *113*: 153–157.
41 Afelik, S., Chen, Y. and Pieler, T. Pancreatic protein disulfide isomerase (XPDIp) is an early marker for the exocrine lineage of the developing pancreas in *Xenopus laevis* embryos. *Gene Expr. Patterns* **2004**; *4*: 71–76.
42 Shiojiri, N. The origin of intrahepatic bile duct cells in the mouse. *J Embryol Exp Morphol* **1984**; *79*: 25–39.
43 Germain, L., Blouin, M. J. and Marceau, N. Biliary epithelial and hepatocytic cell lineage relationships in embryonic rat liver as determined by the differential expression of cytokeratins, alpha-fetoprotein, albumin, and cell surface-exposed components. *Cancer Res* **1988**; *48*: 4909–4918.
44 Rao, M.S., Scarpelli, D.G. and Reddy, J.K. Transdifferentiated hepatocytes in rat pancreas. *Curr Top Dev Biol* **1986**; *20*: 63–78.

45 Krakowski, M.L., Kritzik, M.R., Jones, E.M., Krahl, T., Lee, J., Arnush, M., Gu, D. and Sarvetnick, N. Pancreatic expression of keratinocyte growth factor leads to differentiation of islet hepatocytes and proliferation of duct cells. *Am J Pathol* **1999**; 154: 683–691.

46 Shen, C.N., Slack, J.M. and Tosh, D. Molecular basis of trans-differentiation of pancreas to liver. *Nat Cell Biol* **2000**; *2*: 879–887.

47 Tosh, D., Shen, C.N. and Slack, J.M. Conversion of pancreatic cells to hepatocytes. *Biochem Soc Trans* **2002**; *30*: 51–55.

48 Horb, M.E., Shen, C.N., Tosh, D. and Slack, J.M. Experimental conversion of liver to pancreas. *Curr Biol* **2003**; *13*: 105–115.

49 Ferber, S., Halkin, A., Cohen, H., Ber, I., Einav, Y., Goldberg, I., Barshack, I., Seijffers, R., Kopolovic, J., Kaiser, N. and Karasik, A. Pancreatic and duodenal homeobox gene 1 induces expression of insulin genes in liver and ameliorates streptozotocin-induced hyperglycemia. *Nat Med* **2000**; *6*: 568–572.

50 Ber, I., Shternhall, K., Perl, S., Ohanuna, Z., Goldberg, I., Barshack, I., Benvenisti-Zarum, L., Meivar-Levy, I. and Ferber, S. Functional, persistent, and extended liver to pancreas trans-differentiation. *J Biol Chem* **2003**; *278*: 31950–31957.

51 Kojima, H., Fujimiya, M., Matsumura, K., Younan, P., Imaeda, H., Maeda, M. and Chan, L. NeuroD-betacellulin gene therapy induces islet neogenesis in the liver and reverses diabetes in mice. Nat Med **2003**; *9*, 596–603.

52 Grompe, M. Pancreatic-hepatic switches *in vivo*. *Mech Dev* **2003**; *120*: 99–106.

53 Bonner-Weir, S., Baxter, L.A., Schuppin, G.T. and Smith, F.E. A second pathway for regeneration of adult exocrine and endocrine pancreas. A possible recapitulation of embryonic development. *Diabetes* **1993**; *42*: 1715–1720.

54 Bonner-Weir, S. and Sharma, A. Pancreatic stem cells. *J Pathol* **2002**; *197*: 519–526.

55 Holland, A.M., Gonez, L.J., and Harrison, L.C. Progenitor cells in the adult pancreas. *Diabetes Metab Res Rev* **2004**; *20*: 13–27.

56 Dor, Y., Brown, J., Martinez, O.I., and Melton, D.A. Adult pancreatic beta-cells are formed by self-duplication rather than stem-cell differentiation. *Nature* **2004**; *429*: 41–46.

57 Fausto, N., Lemire, J.M. and Shiojiri, N. Cell lineages in hepatic development and the identification of progenitor cells in normal and injured liver. *Proc Soc Exp Biol Med* **1993**; *204*: 237–241.

58 Sirica, A.E. Ductular hepatocytes. *Histol Histopathol* **1995**; *10*: 433–456.

59 Fausto, N. and Campbell, J.S. The role of hepatocytes and oval cells in liver regeneration and repopulation. *Mech Dev* **2003**; *120*: 117–130.

60 LeCouter, J., Moritz, D.R., Li, B., Phillips, G.L., Liang, X.H., Gerber, H.P., Hillan, K. J. and Ferrara, N. Angiogenesis-independent endothelial protection of liver: role of VEGFR–1. *Science* **2003**; *299*: 890–893.

61 Davidson, A. J. and Zon, L.I. Biomedicine. Love, honor, and protect (your liver). *Science* **2003**; *299*: 835–837.

62 Schmidt, C., Bladt, F., Goedecke, S., Brinkmann, V., Zschiesche, W., Sharpe, M., Gherardi, E. and Birchmeier, C. Scatter factor/hepatocyte growth factor is essential for liver development. Nature **1995**; 373: 699–702.

63 Lumelsky, N., Blondel, O., Laeng, P., Velasco, I., Ravin, R. and McKay, R. Differentiation of embryonic stem cells to insulin-secreting structures similar to pancreatic islets. *Science* **2001**; *292*: 1389–1394.

64 Hori, Y., Rulifson, I.C., Tsai, B.C., Heit, J.J., Cahoy, J.D., and Kim, S.K. Growth inhibitors promote differentiation of insulin-producing tissue from embryonic stem cells. *Proc Natl Acad Sci USA* **2002**; *99*: 16105–16110.

65 Blyszczuk, P., Czyz, J., Kania, G., Wagner, M., Roll, U., St-Onge, L., and Wobus, A.M. Expression of Pax4 in embryonic stem cells promotes differentiation of nestin-positive progenitor and insulin-producing cells. *Proc Natl Acad Sci USA* **2003**; *100*: 998–1003.

66 Kania, G., Blyszczuk, P., Czyz, J., Navarrete-Santos, A., and Wobus, A.M. Differentiation of mouse embryonic stem cells into pancreatic and hepatic cells. *Methods Enzymol* **2003**; *365*: 287–303.

67 Kahan, B.W., Jacobson, L.M., Hullett, D.A., Ochoada, J.M., Oberley, T.D., Lang, K.M., and Odorico, J.S. Pancreatic precursors and differentiated islet cell types from murine embryonic stem cells: an *in vitro* model to study islet differentiation. *Diabetes* **2003**; *52*: 2016–2024.

68 Stoffel, M., Vallier, L., and Pedersen, R.A. Navigating the pathway from embryonic stem cells to beta cells. *Semin Cell Dev Biol* **2004**; *15*: 327–336.

69 Hussain, M.A., and Theise, N.D. Stem-cell therapy for diabetes mellitus. *Lancet* **2004**; *364*: 203–205.

70 Rajagopal, J., Anderson, W.J., Kume, S., Martinez, O.I. and Melton, D.A. Insulin staining of ES cell progeny from insulin uptake. *Science* **2003**; *299*: 363.

71 Chen, Y., Jürgens, K., Hollemann, T., Claußen, M., Ramadori, G. and Pieler, T. Cell-autonomous and signal-dependent expression of liver and intestine marker genes in endodermal precursor cells from Xenopus embryos. *Mech Dev* **2003**; *120*: 277–288.

72 Moriya, N., Komazaki, S. and Asashima, M. In vitro organogenesis of pancreas in *Xenopus laevis* dorsal lips treated with retinoic acid. *Dev Growth Differ* **2000**; *42*: 175–185.

73 Moriya, N., Komazaki, S., Takahashi, S., Yokota, C. and Asashima, M. *In vitro* pancreas formation from Xenopus ectoderm treated with activin and retinoic acid. *Dev Growth Differ* **2000**; *42*: 593–602.

74 Alison, M.R., Poulsom, R., Jeffery, R., Dhillon, A.P., Quaglia, A., Jacob, J., Novelli, M., Prentice, G., Williamson, J. and Wright, N.A. Hepatocytes from non-hepatic adult stem cells. *Nature* **2000**; *406*: 257.

75 Lagasse, E., Connors, H., Al-Dhalimy, M., Reitsma, M., Dohse, M., Osborne, L., Wang, X., Finegold, M., Weissman, I. L. and Grompe, M. Purified hematopoietic stem cells can differentiate into hepatocytes *in vivo*. *Nat Med* **2000**; *6*: 1229–1234.

76 Wang, X., Willenbring, H., Akkari, Y., Torimaru, Y., Foster, M., Al-Dhalimy, M., Lagasse, E., Finegold, M., Olson, S. and Grompe, M. Cell fusion is the principal source of bone-marrow-derived hepatocytes. *Nature* **2003**; *422*: 897–901.

77 Vassilopoulos, G., Wang, P. R. and Russell, D.W. Transplanted bone marrow regenerates liver by cell fusion. *Nature* **2003**; *422*: 901–904.

78 Theise, N.D., Badve, S., Saxena, R., Henegariu, O., Sell, S., Crawford, J. M. and Krause, D.S. Derivation of hepatocytes from bone marrow cells in mice after radiation-induced myeloablation. *Hepatology* **2000**; *31*: 235–240.

79 Schwartz, R.E., Reyes, M., Koodie, L., Jiang, Y., Blackstad, M., Lund, T., Lenvik, T., Johnson, S., Hu, W. S. and Verfaillie, C.M. Multipotent adult progenitor cells from bone marrow differentiate into functional hepatocyte-like cells. *J Clin Invest* **2002**; *109*: 1291–1302.

80 Hamazaki, T., Iiboshi, Y., Oka, M., Papst, P.J., Meacham, A.M., Zon, L. I. and Terada, N. Hepatic maturation in differentiating embryonic stem cells *in vitro*. *FEBS Lett* **2001**; *497*: 15–19.

81 Zorn, A.M. and Mason, J. Gene expression in the embryonic Xenopus liver. *Mech Dev* **2001**; *103*: 153–157.

82 Suksaweang, S., Lin, C.M., Jiang, T.X., Hughes, M.W., Widelitz, R.B. and Chuong, C.M. Morphogenesis of chicken liver: identification of localized growth zones and the role of beta-catenin/Wnt in size regulation. *Dev Biol* **2004**; *266*: 109–122.

23
Molecular Networks in Cardiac Development

Thomas Brand

23.1
Introduction

In recent years the molecular analysis of heart development has received increasing attention. One of the reasons is its medical importance. Congenital heart disease (CHD) is a severe clinical problem that has significant socio-economic implications. The incidence of moderate and severe forms of CHD is about 6/1000 live births and increases to 75/1000 live births if minor ventricular septation defects (VSD) and other small lesions are included [1]. In addition prenatal death due to CHD probably amounts to ⟩ 5 % of conceptuses that do not survive to term [2]. While surgical intervention in neonates with malformed hearts results in a high survival rate, re-surgery and cardiological interventions are common however, in adults with CHD [3–6]. For any further improvement to this situation it is therfore important to fully define cardiovascular development at the molecular and cellular level. It probably will not be possible in the near future to treat CHD causally (i.e. with the help of gene therapy), nonetheless, understanding normal cardiac development at the molecular level will be the entry point into causal understanding of CHD and will facilitate new diagnostic classifications of CHD, improved genetic counselling and prevention of CHD.

23.2
Implications of Studying Heart Development for Adult Cardiology

The study of cardiac development will also lead to a more thorough understanding of the regulatory loops that are activated in the diseased adult heart. Transcriptional programs of contractile proteins for example that are active in the fetal myocardium are silenced during postnatal maturation of the heart. However, these genes become reactivated in the course of the development of a pathological hypertrophy [7]. While it is believed that reactivation of this set of "fetal" genes might be of adaptive value, ultimately the heart is unable to prevent progression to the state of terminal failure.

Cell Signaling and Growth Factors in Development. Edited by K. Unsicker and K. Krieglstein
Copyright © 2006 WILEY-VCH Verlag GmbH & Co. KGaA, Weinheim
ISBN 3-527-31034-7

We are still far from being able to fully describe the progression in molecular terms, however, several key molecules that activate the gene expression programs in pathological hypertrophy such as NFAT, GATA4, or Mef2 have initially been defined in the context of studying cardiac development [8–10]. Likewise, coronary artery development has now entered center stage and several genes were identified that are essential for the differentiation of precursor cell populations [11, 12]. Recent analysis of the infarcted rat heart demonstrated that for example the Wilms' tumor gene, WT1, a zinc-finger transcription factor which is important for coronary artery development, becomes upregulated in endothelial and smooth muscle cells of coronary arteries in the border zone adjacent to the necrotic myocardium [13]. Thus, genes that are important for generating the coronary vasculature are re-expressed in the ischemic heart. As will become apparent in the next section, heart development is studied in different species and we might be able to learn from comparative analysis why for example a fish but not a mammalian heart is able to fully regenerate ventricular myocardium [14, 15].

23.3
Model Organisms

Cardiac development is studied in various animal models that are also commonly used in other organ systems and include *Drosophila* [16], zebrafish [17, 18], *Xenopus* [19], chick [20], and mouse [21].

23.3.1
The Mouse Embryo (*Mus musculus*)

While mice are the most preferred model organism for clinically-oriented research, the elegant genetic approach of generating null mutations through homologous recombination is hampered by the fact that during vertebrate evolution the genome has been duplicated and thus gene ablation studies are usually complicated by the fact that two or more genes with a similar function are present. Thus, at least in the case of gene families, often the functional definition of a gene requires the lengthy generation of several null mutations that need to be interbred [22, 23]. In the mouse a large spectrum of genetic modifications are possible including heart-specific ablation at different time points during heart development. For this purpose several mouse strains have been generated, that express Cre-recombinase under the control of cardiac-restricted promoter constructs such as Nkx2.5 [24], Mlc2 [25], αMHC [26], or ANP [27]. This approach has recently been further fine-tuned via hormonal control of tissue-specific ablation, thereby heart-specific ablation can be induced at any time during embryonic or adult life [28]. Cell lineage-specific expression of Cre-recombinase is also sucessfully used to fate-map cardiac cell lineages such as the cardiac neural crest [29, 30], or derivatives of the anterior heart field [31].

23.3.2
The Chick Embryo (*Gallus gallus*)

The chick is the classic organism for research on the development of the cardiovascular system and has provided insight for example, into the process of cardiac induction [32], left-right asymmetry [33], endocardium and valve formation [34], the role of the neural crest [35] and the epicardium in heart development [36]. Experiments in the chick were also fundamental in establishing the existence of an anterior heart field that generates right ventricular and outflow tract myocardium [37]. The chick lacks the option to analyze heart development by genetic analysis. Gene expression can however be manipulated by retroviral and adenoviral infection [38], electroporation [39], antisense oligonucleotides [33], morpholino [40] and RNAi [41]. Moreover, recently an EST project [42] has been started and the sequencing of the chick genome is well underway. Thereby transcriptomic analysis in combination with embryological experiments is possible and may provide further insight into heart development.

23.3.3
The Frog Embryo (*Xenopus laevis*)

Xenopus was instrumental in establishing the role of several signaling molecules for cardiac induction [43, 44]. However, amphibians differ from amniotes in several respects. For example, *Xenopus* has only a single ventricle and therefore several genes with a chamber-restricted expression pattern such as HAND1 are not present in the frog heart [45, 46], while others such as Nkx2.3 are expressed in the amphibian but not the mammalian heart [47, 48]. The most impressive use of *Xenopus* as a model is reflected in recent attempts to generate a fully functional beating heart from stem cells [49, 50].

23.3.4
The Zebrafish Embryo (*Danio rerio*)

Zebrafish, like the mouse is an organism that allows the utilization of the powerful techniques of genetics to screen for mutants with defects in cardiac development. Approximately 50 mutants have been characterized until now with specific defects in heart formation [51]. Through positional cloning this has led to the isolation of several key players in cardiac development which have not yet been identified in other systems [52–54]. Functional analysis in the zebrafish can also be approached by reverse genetics through the use of morpholino oligonucleotides that block either translation or splicing [55].

23.3.5
Lower Chordates

Recent additions to the list of model organisms for the molecular analysis of heart development are primitive chordates such as the acrania *Branchiostoma lanceolatum* or the tunicate *Ciona intestinalis* [56, 57]. These organism have heart-like organs with a very simple structural organization, nonetheless apparently the basic molecular machinery which is utilized to generate a heart is similar to the vertebrate heart [56, 57]. Embryos of *Ciona intestinalis* can be successfully treated with morpholino oligonucleotides and also transgenic overexpression seems to be possible [58, 59]. The genomes of primitive chordates are much smaller, yet these organisms are closely related to vertebrates, therefore work with these organims may define a gene's function more quickly than experimentation in higher vertebrates.

23.3.6
Drosophila melanogaster

Drosophila has contributed enormously to our understanding of cardiac development. The role of Nkx2.5 has been initially established in *Drosophila* through the identification of the *tinman* gene [60], likewise the role of Dpp as a molecular signal for heart development was first identified in this fly [61]. Early development seems to be very similar between vertebrates and flies as suggested by the similar role of GATA4 and *pannier* in heart development [62]. However, unexpectedly, more specialized functions such as ostia formation in the dorsal vessel and inflow tract development in the mouse also seem to be controlled by a similar set of transcription factors [63, 64]. Even more surprising is the dependence of the *Drosophila* outflow tract region on an extra-cardiac cell population, a situation, which is reminiscent of the function of the neural crest cell population in the vertebrate heart [65]. Recently, the initial results of a genome-wide screen of genes involved in heart development have been reported which will probably be a rich source of information for the identification of novel genes involved in heart formation [66].

23.3.7
Homo sapiens

Congenital malformations in man also provide the basis for the identification of genes involved in heart development. For example the role of T-box genes in heart development was initially identified through the analysis of the Holt-Oram and DiGeorge syndromes [67–69]. Recently, the first large-scale gene analysis of hearts with CHD resulted in the identification of several genes which may be causally linked to the development of the disease, and which will also be informative with regard to their normal function during heart development [70]. Mutations in several cardiac transcription factors and signaling molcules are now known to be responsible for the development of CHD [68, 69, 71–74].

23.4
Anatomical Description of Heart Development

The heart develops from two primordia that are derivatives of the anterior LPM [75–77]. As early as Hamburger Hamilton (HH) stage 4 in the chick and E7.25 in the mouse, the heart fields can be defined by fate mapping [78]. These heart primordia migrate anteriorly as a sheet. Two cell populations, the myocardial and endocardial cell layers are formed and separated by a thick layer of extracellular matrix, the so-called cardiac jelly [79]. In the process of head formation and descensus of the anterior intestinal portal (AIP) the heart fields meet at the ventral midline in the front of the foregut and form the tubular heart which shortly thereafter starts to beat [80].

Subsequently, the initially straight heart tube transforms into a c-shaped loop which becomes displaced towards the right of the body [81]. This is the first prominent morphological sign of the developing left-right axis. The outflow tract, and almost the entire right ventricle are derivatives of a secondary or anterior heart field [82]. The mesoderm that forms the anterior heart field is intitially present in a medial position to the primary heart field. However, after tubular heart formation its position becomes dorsoanterior to the heart tube. For an extended period of time mesodermal precursors are continuously added to the anterior end of the developing heart. The anterior heart field also contributes cells to the left ventricle and atria [83].

Transformation of the tubular heart into an organ with four chambers is mediated by further steps of positional and morphological changes, such as formation of cardiac septa and the process of growth of the cardiac chambers that has been termed ballooning [84]. A further process that is of utmost importance for the correct connection of the ventricles with the developing great arteries is the remodelling of the inner curvature of the heart loop [85]. In the atriventricular canal (AVC) and in the outflow tract, cushions are formed that contain large amounts of extracellular matrix, which subsequently become populated by endocardial cells after EMT [34]. These cushions are primitive valve-like structures that guarantee unidirectional blood flow.

Several cell populations migrate into the cardiac cushions that are remodelled to form the mature valves [86]. The outflow tract becomes invaded by an extracardiac cell population, the neural crest cells, which participates in the formation of smooth muscle cells that substitute for the myocardial wall of the outflow tract [87]. In addition the cardiac neural crest is essential for the development of the aortico-pulmonary septum, which divides the distal outflow tract into the aortic and pulmonary flow pathways.

Another cell population that participates in myocardial development but originates outside of the heart fields is the proepicardial serosa [11]. This cell population develops from the coelomic mesothelium adjacent to the sinus venosus and migrates onto the surface of the heart forming the epicardium and subepicardial connective tissue. Some cells invade the myocardial layer and differentiate into the cellular elements of the coronary vascular system, connective tissue, and even a small number of endocardial cells [88]. The epicardium and the endocardium appear to be essential signaling centers for the development of the myocardium of the ventricular wall [89, 90]. Two different populations of myocardial cells are present in the primitive ventricular chamber, i.e. (1) the compact layer, which contains immature myocytes with a high level of proliferative activity and depends on signals from the adjacent epicardium [91] and (2) the trabecular layer, which is formed by differentiated cardiac myocytes [92]. Survival of trabecular cardiac myocytes depends on signals from the adjacent endocardium [93–95].

This short and simplified description of the major steps in heart development makes it clear that cardiogenesis is a complicated process and a large fraction of the vertebrate genome is utilized in this process. In the following section I will review the molecular networks that are active during the various steps of cardiac development mentioned above.

23.5
Cardiac Induction

In the chick embryo at HH stage 3, cells within the anterior part of the primitive streak are fated to become heart muscle [96]. In addition some prospective heart precursors are present within the epiblast adjacent to the primitive streak [97]. Cardiac progenitor cells, that lie rostral will form the anterior structures of the tubular heart, while more caudally positioned cells will form posterior structures [96]. CITED2 a member of the CITED family of transcriptional co-activators is expressed at HH stage 2–4 in an anterior domain in the elongating primitive streak. This expression domain might label the cells which are fated to become heart and thus CITED2 represents one of the earliest markers for cardiac progenitor cells [98]. A gene which acting upstream of cardiac induction is *Ehox*, a paired-like homeobox-containing gene, that seems to be important for ES cell differentiation [99]. Among other cell types, antisense *Ehox* blocked, while over-expression accelerated myocardial differentiation. However, no mechanistic link between *Ehox* and myocardial gene expression has been established.

Transplantation of primitive streak cells of the chicken embryo into locations other than the LPM showed that these cells are not specified yet and have a far greater developmental potential [100]. Nonetheless, there is evidence that myocardial specification is already underway when cells are localized within the posterior epiblast [101, 102]. Explants of epiblast differentiate into cardiac mesoderm if cultured in the presence of hypoblast, which can be substituted by activin, or TGFβ [102]. High doses of activin have also been shown in Xenopus animal cap experiments to induce cardiac muscle formation [43, 50]. At present, it is difficult to distinguish between a specific role for activin-like signals for cardiac specification, or a more general role as a mesendoderm inducers, which secondarily and indirectly enhances the formation of cardiac mesoderm. In support of a role for activin-like signals as an early inducer is the finding that the activin-related factor Nodal is involved in cardiac specification in zebrafish. The zebrafish gene *one eyed pinhead* (*oep*) codes for a competence factor for Nodal signaling and *oep* mutants have a small heart [103]. Moreover, in embryos ectopically expressing low levels of lefty1, a Nodal antagonist, variable reductions of myocardium are exhibited, with a preferential loss of ventricular myocardium [104]. A more severe mutant phenotype is observed in double mutants of *oep* and the T-box transcription factor *spadetail*, which lack a heart altogether [105]. A structure with homologous function to the hypoblast in the chick is the anterior visceral endoderm in the mouse. Cardiac mesoderm explants of E7.25 embryos differentiate into cardiac muscle, if both visceral embryonic endoderm and primitive streak are present [106]. Thus, the hypoblast in chicken and the visceral endoderm in mouse seem to have a similar function in cardiac induction.

Between HH stage 3 and 4 in the chick, the precardiac mesoderm emigrates from the primitive streak in an anteriolateral direction. Subsequently two heart fields on either side of the primitive streak are formed. Migration of the precardiac mesoderm cells requires the function of the basic helix-loop-helix (bHLH) transcription factors Mesp1 and Mesp2 [107]. In ascidians, Cs-Mesp, the sole ortholog of vertebrate Mesp genes is essential for heart precursor cell specification [108]. Whether Cs-Mesp also controls cell migration is currently unexplored. Cell migration also involves fibroblast growth factor signaling as Fgfr1 mutants show a specific defect of EMT during gastrulation [109]. FGF4 appears to be controlled by Mesp1 and Mesp2 genes since it is downregulated in double mutant mice [107]. The migratory function of FGF is evolutionarily conserved since the FGF receptor *heartless* is required for migration of cardiac mesoderm in Drosophila [110, 111]. The heart fields harbors the progenitors of the atrial and ventricular cell lineage, cells that will form proximal outflow tract tissue as well as endocardial progenitor cells. All cell lineages have already separated by the time the progenitors reside within the primitive streak in the chick and at midblastula stage in zebrafish [112–115]. Cell-fate analysis in the chick suggests that atrial precursors are present in the caudal part, while ventricular progenitor cells are mainly localized to the anterior region of the heart field [75, 77, 116]. In the zebrafish blastula ventricular myocardial progenitors are positioned closer to the margin and to the dorsal midline than are atrial myocardial progenitors [104].

23.5.1
The Role of BMP2 in Heart Induction

Anterior endoderm is an important regulator of myocardial induction [117]. Ablation of endoderm results in a loss of myocardial specification in amphibian and chick embryos [118–120]. Endoderm secretes a variety of signaling molecules including various FGFs, activin, insulin-like growth factor II which all appear to promote cell survival and proliferation of cardiogenic cells and differentiating myocytes [121–125]. BMP2 is part of the heart-inducing activity that is derived from pharyngeal endoderm (Fig. 23.1) [32, 102, 126, 127]. In the chick embryo, BMP2 is expressed at HH stages 4 and 5 in anteriolateral mesendoderm and during HH stages 6–9 in pharyngeal endoderm underlying cardiogenic mesoderm [32]. Subsequently,

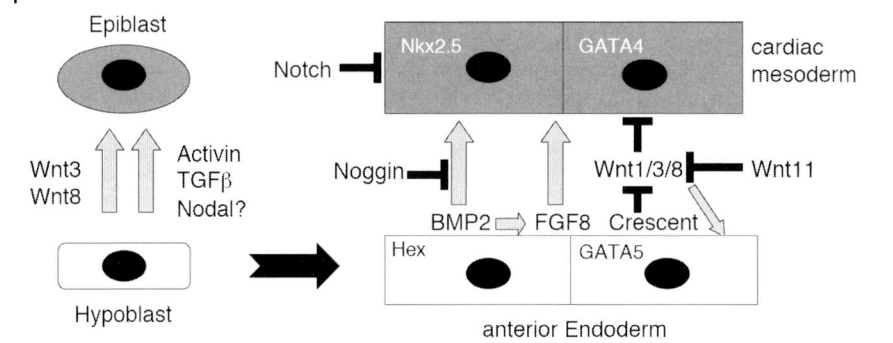

Fig. 23.1
The dual signaling model during heart induction. Data in the chick suggest that a signal from the hypoblast which can be substituted by activin or TGFβ induces heart formation in the epiblast of pre-streak embryos (left-hand side). Since activin, TGFβ and Nodal utilize the same signal transduction pathway it is also formally possible that Nodal is the signal that is active in the embryo. Data in EC P19 cells suggest that Wnt3 and Wnt8 are inducing factors during an early phase of cardiac induction. Subsequently during gastrulation (right-hand side) the anterior endoderm is a signaling center secreting BMP2, FGF8 and Crescent which all induce, directly or indirectly, cardiac transcription factors such as Nkx2.5 and GATA4. Surrounding tissues secrete or express inhibitors such as Noggin, Wnt1,– 3, –8, or Notch. Wnt11, which is expressed in the mesoderm itself acts as an inhitor of canonical Wnt signaling. Wnt1, –3, and –8 are probably required for endoderm formation in the mouse. There seems to be an anterior endoderm-specific gene expression program of which GATA5 and Hex are shown here.

during HH stages 10 and 11, BMP2 remains present in the endoderm of the sino-atrial region but is absent from the tubular heart [32]. BMP2 is able to induce the expression of myocardial lineage markers in ectopic locations [32, 102, 126, 127]. Moreover, noggin, which binds BMPs with high affinity, prevents myocardial differentiation of lateral mesendoderm cultured *in vitro* [102, 127, 128]. Multiple cardiac-restricted transcription factors are induced by BMP2 with distinct kinetics [128]. In heart field explants, noggin inhibits myocardial marker gene expression. Implantation of noggin-expressing cells into the heart field *in vivo* also interferes with myocardial determination [128]. Myocardial induction is concentration dependent and is maximal at 50 ng/ml, when tested in explant cultures [129]. Too much BMP2 is no longer cardiogenic and induces a cell type with high alkaline phosphatase activitvity. Thus, BMP signaling needs to be tightly controlled in order to ensure myocardial differentiation.

Smad6, an inhibitory Smad, which interferes with phosphorylation of receptor-regulated Smads, is expressed in the heart field and is positively activated by BMP2 signaling [130]. Thus, Smad6 is part of a negative regulatory loop, which probably limits the number of mesodermal cells that will actually enter the cardiogenic pathway. Consistent with this view is the observation that in Smad6 null mutant hearts the outflow tract becomes ossified [131].

The dependence of myocardial specification on BMP2 signaling appears to be conserved in vertebrates. In zebrafish the BMP2 mutant *swirl* lacks cardiogenic mesoderm [132]. Xenopus embryos with ectopic expression of dominant-negative BMPRII and ALK3 mutations do not form cardiac mesoderm and display an inability to maintain Nkx2.5 expression [133]. In contrast to the observations in the chick, interfering with BMP signaling does not affect initial Nkx2.5 expression, suggesting that in amphibians BMP2 is not required for cardiac induction but for maintaining cardiac differentiation [134].

Murine *BMP2* null mutants display abnormal heart development but cardiac mesoderm is formed [135]. Several members of the BMP family with overlapping functions are expressed in murine cardiogenic mesoderm and the essential function in heart formation might only become apparent in double or triple null embryos [136, 137]. It has been recently demonstrated that myocardial differentiation in murine P19 embryonic teratocarcinoma cells is dependent on a functional BMP signal transduction pathway and is mediated by Smad and a parallel signaling pathway involving TAK1 and ATF2 [138–140]. Addition of both, BMP2 and TGFβ to differentiating ES cell cultures enhanced myocardial differentiation [141]. Evidence for an important role of BMP signaling in mammalian heart induction also stems from the analysis of the Nkx2.5 promoter. There are several functionally important Smad binding sites within the enhancer that drive expression in the cardiac crescent [142, 143]. Recently, another Nkx2.5 enhancer was found that contains multimerized Smad-binding sites and adjacent GATA elements, which also seem to be active during early heart development [144]. The chick Nkx2.5 gene contains at least three cardiac enhancer regions. One of which is present in the 3′ UTR and directs robust cardiac transgene expression. This enhancer binds GATA4/5/6, YY1 and SMAD1/4 and all are necessary for BMP-mediated induction and heart-specific expression [145]. Thus, the observed function of Dpp in Drosophila, to maintain *tinman* expression seems evolutionarily conserved and also utilized in vertebrate heart formation.

Not all myocardial genes are however dependent on BMP-signaling: troponin T for example is expressed in the cardiac crescent beginning at HH stage 5 in the chick [146]. Removal of endoderm at HH stage 5 or application of noggin has no effect on cTNT mRNA levels. Thus an additional signaling pathway independent of BMP might exist that controls the expression of certain myocardial genes.

23.5.2
Canonical Wnt Signaling Interferes with Heart Formation in Vertebrates

In Drosophila, *wg* and *dpp* are both required for *tinman* expression [147]. A subtractive screen, which was designed to identify mesendodermal genes present in the anterior half of the chick embryo resulted in the isolation of the secreted wnt inhibitor Crescent [148]. Crescent binds to several wnt factors such as Wnt–1, Wnt–3a and Wnt–8c that are present in the neural ectoderm and are co-expressed with BMP2 in the LPM. Ectopic expression of Wnt–1 in the heart field blocks myocardial differentiation and induces blood formation [148]. Conversely, explants of posterior lateral mesendoderm are induced to form beating cardiac mesoderm in the presence of Crescent at the expense of globin expression. *In vivo* Crescent is unable to induce ectopic cardiogenesis in posterior mesoderm. Crescent's activity can also be mimicked by other secreted wnt inhibitors such as Dkk–1, and Frzb. Implantation of pellets of Frzb- and BMP4–expressing cells into paraxial mesoderm leads to cardiac mesoderm formation and subsequently to the migration of these cells into the heart tube [149]. These data suggest that BMP2 and inhibition of wnt signals are necessary and sufficient components to induce specification of cardiac mesoderm in explant cultures [149, 150]. The ability of Dkk1 and Crescent to promote cardiogenesis has also been demonstrated in ventral marginal zone explants from Xenopus embryos [151]. Interestingly in the mouse embryo the wnt-antagonist Sfrp5 is expressed in the visceral endoderm and may have the same biological activity as Crescent in the chick for which no homolog is present in the mouse [152].

Wnt–11, which is expressed around the posterior edge of the heart field in the chick can induce cardiogenesis in posterior non-precardiac mesoderm [153]. Similar observations were made in Xenopus suggesting an important role for Wnt–11 in heart formation [44]. Application of Wnt–11–condition medium to EC P19 cells resulted in efficient induction of cardiac myocyte differentiation, which suggests that in mammals Wnt–11 may also have cardiogenic activity [44]. Interestingly, Wnt–11 belongs to a different class of Wnt factors that differ in signal transduction and activate PKC and CamKII in a G-protein-dependent manner [154]. As in other systems, Wnt–11 might promote cardiogenesis by antagonizing Wnt–3 and Wnt–8 signal transduction [155].

Multiple hearts are formed in mouse embryos with endodermal ablation of β-catenin [156]. At first sight this suggests that the canonical wnt inhibition is also operative in mammals. However, in chimeric embryos β-catenin-deficient cells revealed an inherent inability to populate the endoderm and instead preferentially formed cardiac mesoderm. Canonical wnt signaling may therefore also be involved in the separation of endoderm and mesodermal cell lineages in addition to its inhibitory influence on cardiac mesoderm formation as found in chick and frog embryos. More confusing are recent results in EC P19 cells, which showed that early during cardiac induction Wnt–8 and Wnt–3 may have cardiac-inducing activity [157]. Thus, like BMP2, which initially during hypoblast-epiblast interaction is inhibitory for heart formation [102], canonical wnt signals may have dual activity for cardiac induction being first an inducing factor and subsequently inhibitory. This dual step model (Fig. 23.1) however, needs to be analyzed further, both *in vivo* and *in vitro*.

23.5.3
FGF Cooperates with BMP2

Additional cardiogenic signals have recently been identified including FGF8, which is mutated in the zebrafish mutant *acerebellar* (*ace*) [158]. In the zebrafish FGF8 is expressed in the cardiogenic mesoderm itself. Both Nkx2.5 and GATA4 require the presence of FGF8 while GATA6 does not. In the chick embryo FGF8 is found in the pharyngeal endoderm [120]. FGF8 is able to substitute for pharyngeal endoderm and probably cooperates with BMP2 to induce cardiac mesoderm. Interestingly, FGF8 was able to ectopically activate the expression of Nkx2.5 and Mef2C but was unable to induce GATA4, suggesting that GATA and Nkx2.5 expression are independently regulated. FGF8 deletion mutants in the mouse display aberrant left-right asymmetry and various cardiac malformations have been observed, however, early heart formation seems to be unaffected [159, 160]. Chick posterior mesoderm explants respond to the addition of a combination of BMP2 and FGF4 with cardiac muscle formation, while neither factor alone can do so [126, 129]. Thus, FGF and BMP signaling seem to synergize to drive mesodermal cells into myocardial differentiation.

23.5.4
Cripto

The founding member of the EGF-CFC family of EGF-like signaling molecules, Cripto, has an essential function in mediating Nodal signaling [161–163]. Nodal unlike activin is unable to bind to the signaling receptor complex only in the absence of Cripto. Cripto null mutants display a deficit in the conversion of the proximodistal axis into the embryonic anterio-posterior axis [164]. A related gene, Cryptic, is also involved in Nodal signal transduction and Cryptic null embryos display aberrant left-right asymmetric morphogenesis [165, 166]. Interestingly, Cripto homozygous embryonic stem cells are deficient in myocardial differentiation, while expression of cardiac-restricted transcription factors such as Nkx2.5, Mef2, or GATA4 appears to be unaltered suggesting that Cripto is involved in the transition from myocardial specification to differentiation [167, 168]. A more detailed analysis revealed that Cripto is also required during myocardial induction and that it mediates Nodal signaling via the Alk4/ActIIB/Smad2 signal transduction pathway [169]. In the mouse, Cripto is present in the nascent mesoderm, in the heart fields, the tubular heart, and between E9.5 and E10.5 in the truncus arteriosus [170]. This expression pattern is compatible with the view that Cripto is required for development of both, the primary and secondary heart fields. In the chick a related factor, chick CFC, is also expressed in the heart fields and in the tubular heart [33, 171]. While chick CFC appears to be an important factor in generating the L-R axis (see below), no functional data are currently available for a role in heart formation in the chicken embryo. Embryos lacking zygotic *oep*, the EGF-CFC factor in zebrafish, display *cardia bifida* and a reduced or absent Nkx2.5 expression [103]. The *cardia bifida* phenotype is probably related to the essential function of Nodal signaling for endoderm formation. *Oep* seems to act together with Bmp2 (*swr/bmp2b*) to control the expression of GATA5, which is essential for heart formation in the zebrafish. Another factor that genetically interacts with *oep* and is required for heart formation is the T-box factor *spadetail* [105]. Nodal signaling is known to be required for endoderm formation [172] and in zebrafish works through *bonnie and clyde* (*bon*) a Mixer type homeobox gene [173], *casanova* (*cas*) a Sox-related factor [174, 175] and *faust* (*fst*) [175]. Each of these mutants displays defects in endoderm formation. Thus, Nodal may not directly activate cardiac mesoderm but may act via endoderm formation similar to the role proposed for canonical wnt signaling. Surprisingly, overexpression of the Nodal antagonist *Cerberus* in Xenopus results in the formation of ectopic heads and often, duplicated hearts [176]. *Cerberus* is a member of the Dan family of secreted antagonists that interfere with Nodal, Wnt and BMP signaling [177]. Microsurgical ablation studies in Xenopus demonstrated that the *Cerberus* expression domain in the anterior endoderm demarcates cells that have heart-inducing activity [178]. A *Cerberus* homolog, *Cerr1*, has been isolated in the mouse but has no essential function in development [179–181]. In the chick the *Cerberus* homolog *Caronte* (*Car*) is involved in left-right axis formation acting as a BMP antagonist in the left LPM [182–184]. In contrast to *Crescent*, Car fails to promote cardiogenesis in cultured posterior LPM [150]. Thus, at present it is unclear whether Cerberus-like factors are important for heart development in higher vertebrates.

23.5.5
Shh

In the mouse, the double mutant of *Shh* and *Ihh*, or the receptor gene *smoothened* (*Smo*) lack early cardiac Nkx2.5 and Wnt–2 expression while mutant embryos lacking the inhibitory receptor *patched* (*Ptc*) have an enlarged and enhanced expression domain of Nkx2.5 [185]. Both *Shh* and *Ihh* are expressed in the pharyngeal endoderm at the time of cardiac crescent formation. The fact that heart formation, although retarded, still occurs in *Smo* mutant embryos suggests that Shh/Ihh may function as permissive rather than inductive factors in mammalian cardiogenesis.

23.5.6
Notch Signaling Interferes with Myocardial Specification

Fate map analysis in Xenopus, revealed that at neural tube stage the lateral and dorsal portions of the Nkx2.5 expression domain form the dorsal mesocardium and roof of the pericardial cavity while only the ventral Nkx2.5–expressing region develops into the heart [76]. The suppression of cardiogenesis in the lateral portion of the Nkx2.5 expression domain is mediated by a Serrate1–dependent Notch1 signaling pathway [186]. There is growing evidence for a conserved function of Notch signaling as a negative regulator of heart formation. Morpholino treatment of zebrafish embryos directed against Su(H) strongly affects myocardial development, the heart develops large edemas [187], a phenotype that is also observed in case of the RBPJ$\varkappa$ null mutant in mice [188]. Furthermore, *in vitro* differentiation of homozygous mutant RBP-J$\varkappa$ cells revealed an enhanced formation of beating heart muscle cells, suggesting that Notch signaling may limit the number of cells that are recruited into the heart cell lineage [189]. Notch signaling may also induce myocardial dedifferentiation during regeneration in the adult zebrafish heart [190]. The interpretation of the role of Notch signaling during heart development is however complicated by the fact that Notch signaling is also involved in left-right axis formation (see below).

23.6
Transcription Factor Families Involved in Early Heart Induction

23.6.1
The NK Family of Homeobox Genes

There are several cardiac transcription factors that are activated at the time of heart field formation. One of the first transcription factors that is expressed in cardiac progenitor cells within the heart fields is *Nkx2.5* [191, 192]. The *tinman* gene in Drosophila, which is related to *Nkx2.5*, is essential for heart formation [60]. In contrast in the mouse, *Nkx2.5* is dispensable for early recruitment of cells to the cardiac cell lineage and heart development is arrested only at the beginning of cardiac looping [193]. Recent detailed analysis suggests that *Nkx2.5* is specifically required for left ventricular chamber development [194]. In agreement with this genetic data, a fate map analysis in the chick suggests that at HH stage 5 *Nkx2.5* is expressed only in cells with a ventricular fate [75]. In addition to *Nkx2.5*, *Nkx2.6* is expressed in the mouse heart [195]. A possible explanation for the lack of an early mutant phenotype in mouse would therefore be the functional redundancy between *Nkx2.5* and *Nkx2.6*. However, only malformation of the pharynx and some mild effects on atrial chamber development was noted in the case of the double mutant [196]. Experiments in Xenopus using dominant-negative Nkx2.3 and Nkx2.5 point mutants or engrailed fusion proteins block heart formation [47, 197]. Initially these results were interpreted as being evidence for an essential function of NK homeobox genes during early cardiogenesis. However, since NK homeobox proteins interact with members of other transcription factor families it is likely that the dominant-negative effect was a result of titrating out other essential cardiac transcription factors. The Nkx2.5 null mutant was utilized to search for putative target genes. A number of genes are specifically downregulated in the *Nkx2.5* null mutant heart including ANP, BNP, Mlc2v, Irx4, CARP and Chisel, which all are genes that are specifically expressed during cardiac chamber formation [198–200]. An RNA helicase entitled CSM was recently identified by a subtractive hybridization approach, which is downregulated in the *Nkx2.5* null mutant [201]. The transcription factors HAND1, involved in left ventricle formation [199, 202] and the homeobox gene Hop and myocardin, both modulators of SRF activity [203–205] are also affected by the loss of *Nkx2.5*. A functionally important NKE element has been found in the chick and mouse GATA6 promoter [206, 207]. In the fly expression of D-Mef–2 in striated and visceral muscle is under the control of the *tinman* gene [208]. Maintenance of expression of Pitx2 in the lateral plate mesoderm is believed to be under the control of Nkx2.5 [209]. Thus, multiple transcription factors and differentiation genes are regulated by Nkx2.5.

23.6.2
The GATA Family of Zinc-finger Transcription Factors

The GATA family of zinc-finger transcription factors is an important gene family involved in heart formation. In Drosophila a single GATA gene, *pannier* is essential for cardiogenesis [62]. In vertebrates a total of three GATA genes, GATA4–6, are expressed in the heart [210]. Mouse embryos that lack GATA4 have bilateral heart tubes (*cardia bifida*) and a reduced number of cardiac myocytes [8, 211]. GATA5 null mutants are viable [212], and a specific function for GATA5 in endocardial differentiation has been proposed recently [213]. Essential functions of GATA5 for endoderm formation have been described in both zebrafish and Xenopus [214, 215]. Like GATA4 in the mouse, the zebrafish GATA5 mutant has a *cardia bifida* phenotype [214]. At present the role of GATA6 has not been fully defined due to the fact that GATA6$^{-/-}$ embryos die soon after implantation [216]. GATA6 has a specific function in endoderm formation since homozygous GATA6 mutants lack visceral endoderm and severely attenuate, or fail to express, genes encoding early and late endodermal marker genes [217]. Xenopus and zebrafish morphants for GATA6 display deficient myocardial differentiation, which probably relates to the inability to maintain Nkx2.3, Nkx2.5, and Nkx2.10 as well as BMP4 expression [218].

Pannier in Drosophila functions as a cardiac identity gene, since forced expression results in supernumerary cardiac cells at the expense of other derivatives of the dorsal mesoderm [62]. *Pannier* works synergistically with *tinman* and ectopic co-expression of both factors results in expanded cardiogenic gene expression domains in ventral and dorsal mesoderm [62]. Ectopic expression of GATA4 in Xenopus animal cap tissue is sufficient to induce beating heart tissue [219]. Likewise, overexpression of GATA5 in the zebrafish embryo leads to ectopic *Nkx2.5* expression and formation of ectopic beating cell clusters [214]. Thus there is ample evidence that GATA factors are able to ectopically activate the cardiogenic program at least in insects and lower vertebrates. In mammals heart formation is probably more complex, nonetheless forced expression of GATA4 in embryonic stem cells results in enhanced cardiac differentiation [220].

Promoter studies reveal a mutually reinforcing regulatory network of Nkx2.5 and GATA transcription factors during cardiogenesis. The GATA6 promoter in both mouse and chick contain functionally important Nkx2.5 binding sites [206, 207]. Similarly, the murine *Nkx2.5* promoter contains GATA sites that are involved in early heart field expression [144, 221–223]. A downstream target gene of GATA seems to be the HAND2 gene for which GATA elements have been documented in an enhancer that drives expression in the right ventricle [224]. Numerous myocardial structural genes, among these are troponin C [225] and α-myosin heavy chain [226], have essential GATA binding sites in their promoters.

23.6.3
Serum Response Factor

Serum response factor (SRF) is a member of the MADS-box family to which the MEF2 genes also belong. SRF mediates the rapid transcriptional response to various extracellular stimuli. SRF null mutants have a severe gastrulation defect and are unable to form mesoderm [227, 228]. Binding sites for SRF (CarG boxes) are present in many cardiac genes. Because SRF is not muscle specific, it has been postulated to activate muscle genes by recruiting myogenic accessory factors. A highly potent transcription factor, named myocardin, that is expressed in cardiac and smooth muscle cells, has been recently isolated [229]. Myocardin belongs to the SAP domain family of nuclear proteins and activates cardiac muscle promoters by associating with SRF. Expression of a dominant negative mutant of myocardin in Xenopus embryos interferes with myocardial cell differentiation, however, myocardial development was unaffected in the null mutant in mouse [230]. Ectopic activation of myocardin expression in various non-muscle cell types results in the induction of the transcriptional program of smooth muscle and consistent with these *in vitro* observations, the myocardin null mutant is deficient in smooth muscle development, however myocardial development is apparently normal [230–233]. There are two additional myocardin-related transcription factors (MRTF-A and -B) present in the mammalian genome [234]. Functional studies indicate that MRTF-B is an important mediator of SRF-dependent gene expression in skeletal muscle cells [235] and MRTF-A was found to associate with serum response factor to mediate serum-induced transcriptional regulation [236, 237]. It will be interesting to generate double and triple null mutations to unravel the exact role of myocardin and its related family members during heart development. SRF is negatively regulated by a small homeobox gene, *Hop* that has a divergent homeodomain and is unable to bind to DNA [203, 204]. Expression of Hop is regulated by Nkx2.5 and mice homozygous for a *Hop* null allele have two different phenotypes: some mutants develop early lethality due to a deficiency in myocyte proliferation, while the majority of null mutants develop a hyperplastic heart in the postnatal period. Therefore, Hop modulates SRF activity, which is important for balancing signaling inputs that induce growth and differentiation in the myocardium.

23.6.4
Synergistic Interaction of Cardiac Transcription Factors

By HH stage 8 many cardiac transcription factors are co-expressed in the cardiac crescent. For example *SRF*, *Nkx2.5* and *GATA4* are present in the cardiac crescent [238] and all three factors form a complex, which strongly enhances transcriptional activity of cardiac differentiation genes. Combinatorial and synergistic activation was shown for many transcription factors in the case of the ANF promoter: a combination of Nkx2.5 and Tbx5 [239], Pitx2 [240], GATA4 [241], and Mef2 [242] were shown to synergistically activate the ANF promoter. The transcription factors

GATA4 and HAND2 physically interact to synergistically activate cardiac gene expression through a p300–dependent mechanism [243]. GATA4 and SRF were found to synergistically activate expression of the muscle-specific carnitine palmitoyltransferase I beta gene in the rat heart [244]. Tbx20 was found to be able to interact with Nkx2.5 and GATA factors [245]. Although synergistic activation by cardiac transcription factors is observed in many genes that are expressed in the myocardium, it has been found that Nkx2.5 attenuates transcriptional activation of GATA4–dependent promoters, thus activation or inhibition is context dependent [246]. Taken together these results suggest the existence of cardiac enhanceosome complexes consisting of multiple transcription factors that together specify tissue-specific gene expression in the heart.

23.7
Tubular Heart Formation

After cardiac mesoderm is specified the two heart fields unite and form a single heart tube at the embryonic midline ventral to the foregut. Tubular heart formation is dependent on the descensus of the anterior intestinal portal (AIP). Preventing the descensus of the AIP, for example by incision of the AIP results in formation of two tube-like structures (*cardia bifida*) instead of a single heart tube in the midline. Both tubes acquire contractile activity and undergo some steps of cardiac morphogenesis including heart looping. Mouse mutants with a *cardia bifida* phenotype include *GATA4* [8], the *furin* proprotein convertase [247], and the helix-loop-helix transcription factor *MesP1* [248]. Homozygous mutants for *Foxp4*, a member of the Fox gene family of forkhead transcription factors also have *cardia bifida*, which is probably due to aberrant anterior foregut development [249]. In contrast to other mouse mutants with a bifid heart, the Foxp4 mutants survive until embryonic day 12.5 and the unfused heart fields display extensive cardiac morphogenesis, which appears to be independent of forming a tubular heart at the embryonic midline.

In zebrafish several mutants display *cardia bifida*, including the mutants of *gata5* [250], zebrafish *hand2* as well as several mutants that lack endodermal tissue, including *casanova, bonnie and clyde*, and *one-eyed-pinhead* (oep) [175]. Thus, the endoderm is essential for cardiac mesoderm migration towards the midline. However, tube formation also involves a cell autonomous function of cardiac mesoderm itself [248]. Analysis of the *cardia bifida* phenotype of the *miles apart* (*mil*) zebrafish mutant revealed a putative signaling function of the paraxial mesoderm to direct cardiac mesoderm to the midline [52]. Mouse embryos that lack *GATA4* have bilateral heart tubes [8, 211]. The fact that in chimeric embryos homozygous $GATA4^{-/-}$ embryonic stem cells can differentiate into contractile myocytes and the presence of GATA4 in mesoderm and endoderm suggests that the *cardia bifida* phenotype is probably related to an endoderm deficiency [251]. In the chick embryo however, evidence has been provided for a cell-autonomous defect in the GATA4–deficient mesoderm [252]. Electroporation of an RNAi construct for GATA4 into the cardiogenic mesoderm resulted in a loss of N-cadherin expression. Moreover, essential GATA elements are present in the N-cadherin promoter. Both, the mouse null mutant of N-cadherin and chick embryos treated with neutralizing N-cadherin antibodies display *cardia bifida* [253, 254]. It is therefore possible that the *cardia bifida* phenotype in the GATA4 null mutant is cell autonomous due to a loss of N-cadherin expression in the cardiac mesoderm.

23.8
Left-Right Axis Development

23.8.1
Looping Morphogenesis

Although the first morphological asymmetry in the chick embryo is present in Hensen's node by HH stage 4, the heart is the first organ that displays left-right (L-R) asymmetry [255]. Initially, at HH stage 9 a bilateral symmetric heart tube is established. Transformation of the straight heart tube into a C-shaped heart loop starts at HH stage 10 [81]. Early looping morphogenesis involves two different processes that are intertwined. The primitive ventricular region bends toward its ventral side and simultaneously rotates to the right. Thereby the original left and right sides of the straight heart tube become the ventral and dorsal sides of the post-looping ventricle. Looping morphogenesis is complex and recent simulation experiments revealed that looping involves opposite rotations of the cranial and caudal ends of the embryonic heart [256]. Thereby the heart loop acquires a helical configuration with a counter-clockwise or left-handed direction of twist. There are further steps in looping morphogenesis such as a shift of the ventricular position caudal to the atria and the positioning of the conus ventral to the right atrium. The mature myocardium displays several L-R asymmetric structures including the right and left atria, which are morphologically different [257]. The atria are connected to different vascular structures and the outflow tract shows a characteristic distribution of valves and is connected asymmetrically to the great vessels. Both atria and outflow tract are affected in asymmetry mutants in mice and man. The anatomical left and right ventricle are juxtaposed to each other only after cardiac looping is completed. The structure of the ventricular chambers is not subject to L-R asymmetry defects due to their embryological origin arising along the anterio-posterior axis [258].

23.8.1.1 Mechanisms of L-R Axis Determination
Generation of L-R asymmetry during development is an integral part of the vertebrate body plan and is especially important for the development and morphogenesis of the vertebrate heart [259–264]. Morphological asymmetries of the heart are actually at the end of a complicated process of L-R axis determination, which has its

origin much earlier than the time of looping itself. In recent years a working model has been formulated of how left-right asymmetry is established and transduced into asymmetric morphogenesis of the individual organ anlage [264]. Specification of the left-right (L-R) axis requires multiple steps: (1) generation of an initial asymmetric signal which aligns with the previously established anteroposterior and dorsoventral axes; (2) transfer of this decision to the organizer tissue; (3) transfer of asymmetric signals form the node to the LPM; (4) induction of an evolutionary conserved cascade of gene expression in the left LPM; and (5) transformation of the L-R asymmetric signals into organ-specific morphogenesis.

23.8.1.2 Generation of Initial Asymmetry

In both chick and frog embryos ionic imbalances between the right and left side are probably important early determinants [260]. Asymmetric gene expression becomes apparent in frog embryos by the two-cell stage, while in amniotes it probably develops during primitive streak formation [265]. A proposed downstream mechanism involves transport of L-R determinants through gap junction channels [266, 267]. An alternative model implicates the early activation of a Vg1 signaling pathway on the left side and an antagonistic BMP pathway on the right side in Xenopus [268]. Vg1 signaling probably occurs during early gastrulation and requires ligand presentation by ectodermal cells via syndecan–2 to the migrating mesoderm [269].

23.8.1.3 Transfer of L-R Asymmetry to the Organizer Tissue

Node transplantation experiments established that the chick organizer, Hensen's node, receives its laterality from peripheral cells. Node rotation establishes that Hensen's node is lateralized by HH stage 5 coincident with a morphological recognizable asymmetry, which is first observed by HH stage 4 [255]. Several asymmetric expression domains have been described in Hensen's node (Fig. 23.2). A right-sided asymmetric activinβB signal in the epiblast induces BMP4 on the right side of Hensen's node [270]. BMP4 antagonizes Shh expression and thereby Shh is restricted to the left [271]. BMP4 induces the expression of FGF8 and ultimately cSnR, which functions as a repressor of left identity on the right side [272]. Several additional determinants are active within Hensen's node including N-cadherin on the right side, which limits Wnt–8C-induced β-catenin signaling to the left side of Hensen's node [273, 274]. Ultimately, the TGFβ molecule Nodal is induced exclusively on the left side in the chick embryo in a domain adjacent to Hensen's node [271]. In the mouse Nodal is expressed on both sides of the node and expression is only slightly stronger on the left then on the right. Cerl–2, s Nodal antagonist plays a key role in restricting the Nodal signaling pathway toward the left side of the mouse embryo by preventing its activity in the right side [275]. Recent analysis points to an important role for the Notch pathway in determining Nodal expression adjacent to the node, which is independent of Shh [190]. Several null mutants in mouse ablating

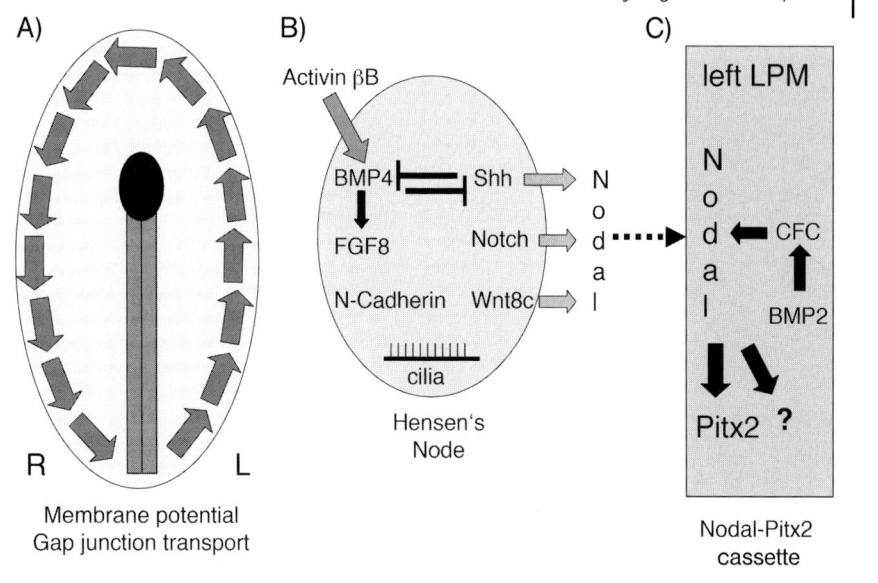

Fig. 23.2
The L-R axis is generated in several steps. (A) In the chick and Xenopus initial asymmetry is probably generated by voltage differences, which in the chick become apparent at HH stage 3 in the primitive streak. The membrane potential is probably the driving force for gap junction-mediated transport. (B) Subsequently, during gastrulation the node is lateralized from the periphery and in the chick a right-sided activin signal is believed to set in motion a cascade of signaling molecules, which ultimately generates an asymmetric Nodal domain adjacent to the organizer. In the mouse cilia are important for L-R axis formation, at present however it is unknown whether they also play a role in other species. (C) Nodal adjacent to the node induces a second domain in the lateral plate which activates Pitx2 and probably other transcription factors which mediate organ-specific L-R asymmetry. Induction of Nodal in the LPM depends on BMP2 and CFC.

genes involved in Notch signaling show abberrant heart looping and embryonic torsion. This has been reported for the Delta1 (*Dll1*) null mutant, for a double mutant of *Notch1* and *Notch2* and the effector gene *RBPjk* [190, 276, 277]. Apparently this function is conserved and also active in zebrafish and chick L-R axis determination [190].

23.8.1.4 The Nodal Flow Model

Aside from growth factor signaling, another mechanism for establishing L-R asymmetry within the organizer stems from analysis of various mouse mutants with asymmetry defects. The mouse node at E6.5 is a structure at the distal tip of the embryo consisting of a dorsal layer of ectoderm over a ventral layer of cells each having a monocilium on the apical surface [278]. These cilia rotate in a clockwise fashion generating a leftward nodal flow [279, 280]. Since many mouse mutants with defective or immotile node cilia display L-R asymmetry, it has been proposed that this nodal flow may be important for setting up L-R asymmetry. The nodal flow model was directly tested by culturing mouse embryos under the influence of an artificial rightward fluid flow that reversed the intrinsic leftward nodal flow [281]. This manipulation resulted in a reversal of the situs and altered the expression of several L-R determinants. In the *iv* mouse, a spontaneous mutation within *situs inversus*, the motor protein *left-right dynein (lrd)* is mutated resulting in immotile nodal cilia [282, 283]. Also patients with *situs inversus* harbor mutations in genes encoding dynein subunits [284]. *Lrd*-like genes are found in Hensen's node in the chick and node-like structures of Xenopus and zebrafish embryos [285]. Hensen's node in the chick lacks an equivalent of the ventral node in the mouse at its ventral surface [286], however, cilia are found in the space between the dorsal epiblast and the ventral endoderm at HH stage 4 [285]. Recent refined analysis of monocilia in the mouse node revealed that two cilia subpopulations can be distinguished, lrd localizes to a centrally located subset of node monocilia, while polycystin–2, a Ca^{2+}-channel protein [287] is found in all node monocilia [288]. Polycystin–2, which is encoded by the *pkd2* gene, is believed to act as a mechanosensor detecting the left-sided flow. Indeed, an asymmetric calcium signal appears at the left margin of the node coincident with the direction of nodal flow [288]. Another important protein involved in generating L-R asymmetry is the protein inversin (Inv), which is mutated in the *inv* mouse. The *inv* mutant is unique in that close to 100 % of the mutants displays *situ inversus totalis* [289]. Inversin homologs have been cloned from chick, Xenopus and zebrafish [290]. Expression of wild-type Inv protein in Xenopus embryos perturbed L-R determination, whereas expression of mutant Inv that lack either of the two calmodulin binding motifs did not [291]. Inv morphants in zebrafish embryos have randomized heart looping and humans with mutations in both INVS alleles have complete *situs inversus* [292]. Thus inversin has a evolutionarily conserved role in left-right axis formation. The *inv* gene encodes a protein that contains 15 tandem repeats of an ankyrin motif. Two-hybrid analysis revealed that Inv interacts with calmodulin [290, 291], the cell cycle control protein apc2 [290], catenins, N-cadherin [293] and nephrocystin. Nephrocystin couples inversin to tubulin leading to a localization of Inv to monocilia of the node [292]. At present it is unclear how Inv mechanistically interacts with lrd and/or pkd2 to determine L-R asymmetry. It is also unclear how the nodal flow and the signaling pathways which have been mentioned above interact. Analysis of the mouse mutants, *wimple (wim)* and *flexo (fxo)* revealed an unexpected link between node cilia and Shh signaling [294]. Both mutants are alleles of interflagellar transport (IFT) proteins, that in vertebrates are involved in ciliogenesis and maintenance of cilia. The *fxo* mutation is a novel hypomorphic allele of *polaris*, a protein which has been previously implicated in L-R asymmetry [295]. Interestingly analysis of the function of these IFT proteins now reveals an important function of IFT proteins for hedgehog signaling at a step downstream of the Patched1 receptor [294]. Since Shh is an important determinant of the left-sided Nodal domain adjacent to the node this finding may unify the nodal flow and the signaling pathway models.

23.8.1.5 Transfer of L-R Asymmetry to the Lateral Plate

At HH stage 8 in the chick Nodal is expressed in a large expression domain in the LPM [271]. It is thought that transfer of Nodal from the Node to the LPM involves Nodal itself and the Cerberus homolog Car [182, 183]. One model proposed that Caronte induces Nodal expression by binding and inhibiting BMP2, which is expressed bilaterally in the LPM [182, 183]. Recent data, however disproved this model and suggested that Nodal expression in the LPM actually depends on the presence of BMP2 [296–299]. BMP2 in the chick is required for the expression of the competence factor CFC in the LPM and Nodal requires the presence of CFC on the surface of responsive cells [33]. Signaling by activin, nodal, Vg1 or the related mammalian factor GDF1 is mediated through the same set of signaling receptors namely ActRIIa and ActRIIB as the constitutively active type II receptors that upon ligand binding form a complex with the type I receptor ALK 4 and 7, and activate the type I receptor through phosphorylation [300]. All of these ligands utilize the same type I and type II receptors; however, nodal and Vg1/GDF1, but not activin, require a membrane-associated EGF-CFC protein belonging to the Cripto family as a co-receptor [162, 166, 301–304]. Mutation of the Cripto family member *oep* in zebrafish leads to defective nodal/Vg1/GDF1 signaling, so that resulting embryos mimic those that lack nodal ligands [162, 304]. In the chick Nodal establishes the expression of the Nodal antagonist Lefty in the midline [33]. In the mouse Lefty1 and Lefty2 are induced by Nodal in the midline and LPM, respectively [305, 306]. A direct downstream target of Nodal signaling is the bicoid-type homeobox gene *Pitx2* [307–310].

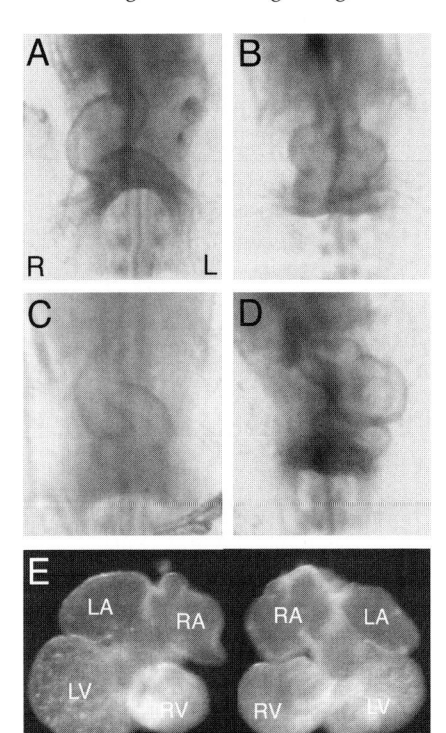

Fig. 23.3
Looping morphogenesis in the chick (A-D) and in the inv mouse (E). Ventral views of chicken embryos at HH stage 11 after treatment with CFC antisense oligonucleotides [327]. (A) The heart show normal looping, or (B-D) hearts with varying degrees of looping aberration. (E) Hearts of an E10.5 inv mutant (inv) and a wild-type embryo (WT). The outflow tract was removed to visualize the mirror image of the atrial morphology in the mutant and wild-type hearts. Also the position of the right and left ventricle are inverted, since most inv mutants have a situs inversus totalis. RA, right atrium; LA, left atrium; RV, right ventricle; LV, left ventricle.

23.8.1.6 Asymmetric Organ Morphogenesis

In contrast to our extensive insight into the signaling cascades that generate the L-R axis, we know very little about the last step, the generation of L-R-specific organ morphogenesis (Fig. 23.3). Overexpression experiments in the chick utilizing retroviruses expressing *Nodal* [311], *Pitx2* [308, 310, 312], and *Caronte* [182], as well as loss-of-function experiments utilizing antisense oligonucleotides for *SnR* [272], or *CFC* [33] affect cardiac looping morphogenesis. These data would indicate that the Nodal-Pitx2 pathway also controls cardiac looping morphogenesis. However, sidedness of *Pitx2* expression and heart looping can be experimentally and genetically uncoupled in a variety of species [313–315]. Moreover, despite severe defects in asymmetric morphogenesis of atria and outflow tract in the case of the *Pitx2* null mutant no alterations in heart looping are observed [316]. Therefore, the Nodal-Pitx2 pathway is directly affecting asymmetric morphogenesis of inflow and outflow tract, while an unknown and probably earlier acting pathway controls cardiac looping morphogenesis. Heart looping and other aspects of cardiac laterality must be coupled to allow normal heart morphogenesis to occur. Nodal is a good candidate that might regulate the different aspects of L-R asymmetry of the heart. *Nodal* can affect cardiac looping when ectopically expressed in the chick embryo and a *Nodal* hypomorph in the mouse displays aberrant looping [317]. Another signaling molecule that may play a role in heart looping is BMP4. Genetic analysis of heart formation in zebrafish reveals that mutations affecting normal cardiac looping are associated with perturbations in *BMP4* expression [318]. Loss- and gain-of-function experiments suggest that *BMP4* is a determinant of looping morphogenesis in zebrafish and Xenopus [319]. BMP signaling might also be involved in looping morphogenesis in the chick heart based on the fact that the phosphorylated form of Smad1 was found to be asymmetrically distributed in the tubular heart [320].

Several genes display asymmetric expression domains within the lateral plate mesoderm during early heart field specification. One of the earliest asymmetrically expressed genes within cardiac progenitor cells is the transcriptional co-activator *CITED2* which in the chick embryo displays transiently stronger expression in the right LPM at HH stage 5 [98]. Several extracellular matrix molecules including hLAMP and flectin on the left and JB3 on the right side are asymmetrically distributed in the precardiac mesoderm at HH stage 5 and 6 [321]. Isl–1 in the chick embryo also displays asymmetric expression in the precardiac mesoderm [322]. Popdc2 is asymmetrically expressed during early chick heart development at HH stage 7 and 8 [323]. Thus, well before *Pitx2* is present in the cardiac mesoderm, asymmetric gene expression is already found in the LPM.

In a classical fate mapping experiment Stalsberg has shown, that in the chick there is an unequal contribution of the left and right heart field along the A-P axis of the tubular heart [317]. The right heart field has a greater contribution to the rostral end and to a lesser extent to the caudal part. Recent fate mapping experiments in Xenopus also revealed regional differences in the contribution from the left and right side [324]. There also seems to be a difference in temporal progression of cardiac differentiation between the left and the right side [325]. Treatment of chick embryos with monoclonal antibodies against flectin results in randomization of heart looping suggesting that the extracellular matrix might be important for looping morphogenesis [326]. Moreover, in contrast to the sidedness of Pitx2 expression, the asymmetry of flectin strictly correlates with the direction of heart looping [327]. However, at present there is no insight into how the asymmetric expression of extracellular matrix proteins are mechanistically linked to the direction of looping. It is also unclear whether the cardiac mesoderm is affected by the underlying endoderm during lateralization. Laterality marker genes show asymmetric expression only in the LPM and Pitx2 for example is symmetrically expressed in the endoderm. Cardiac looping morphogenesis appears to be an autonomous process as the chick hearts develop a D-loop even after explantation at the tubular heart stage [328]. In the zebrafish gut the generation of L-R asymmetric morphogenesis also seems to be a cell-autonomous process of the LPM and occurs normally in the absence of gut endoderm [329]. While the bHLH genes *HAND1* and *HAND2* are symmetrically expressed in the early heart field, antisense treatment of chicken embryos results in a failure to undergo normal looping morphogenesis [330]. During looping *HAND1* expression is enhanced in the left atrial and ventricular primordia in the mouse [202] and *HAND1* null cells are underrepresented in the left caudal region of the linear heart tube [331]. Another transcription factor that has been implicated in cardiac looping is the T-box factor *Tbx20* (*hrT*), which is expressed in the early heart field [332–334]. *Tbx20* morphants in zebrafish develop normally until the looping stage at which a straight heart tube is formed [55]. In *Tbx20* morphants expression of *Tbx5* is increased while overexpression of *Tbx20* leads to a loss of *Tbx5* expression. These results indicate, that *Tbx20* negatively regulates *Tbx5* expression and that deregulated *Tbx5* expression might interfere with looping morphogenesis. Consistent with this hypothesis is the observation that *Tbx5* mutants also have unlooped hearts [335]. It has been hypothesized that cardiac looping is driven by changes of cell shape, which might involve differential contraction of the actin cytoskeleton along the heart tube [336]. Another proposed mechanism that might be involved is differential proliferative activity, however, no L-R-specific difference has been observed in the linear heart tube [337]. Recent evidence also indicates that external forces are exerted by the splanchnopleura and the omphalomesenteric veins which drive cardiac rotation [338]. At present we lack a full understanding of the pathway(s) that regulates cardiac looping and further research is required in this area.

The role of Pitx2 in generating cardiac asymmetric morphogenesis is not fully understood but currently represents the only known effector gene downstream of Nodal signaling. Pitx2 enhances cell proliferation in the outflow tract via direct activation of the cyclin D2 promoter [339]. Additional cell cycle regulators that also have functional Pitx2 binding sites are cyclin D1 and c-myc [340]. Pitx2 is a trancriptional repressor that is changed into an activator by canonical wnt signaling facilitating β-catenin interaction [341]. This interaction apparently plays an important role during outfllow tract patterning. Fate mapping studies using a Pitx2 cre recombinase knock-in allele revealed that daughters of Pitx2–expressing cells populate the right and left ventricles, atrioventricular cushions and valves and pulmonary veins [342].

23.8.2
A-P Axis Formation in the Heart

There are both morphological and molecular differences at the tubular heart stage along the anterior-posterior (A-P) axis. However, only little is known about the mechanisms that regulate the A/P axis. In Drosophila maternal genes first set up the A/P axis which subsequently through the use of different gene cascades, ultimately give an identity to the embryonic segments through the combinatorial activation of Hox genes. The Drosophila heart is composed of a simple linear heart tube, which is subdivided into an anterior positioned "aorta" and a posterior "heart", The "heart" segment has a much larger cavity than the aorta, and aortic cells do not differentiate into beating cardiac myocytes. Aorta and heart differ in the expression of homeotic genes of the *BXC* complex: the aorta largely expresses *Ubx*, while the heart expresses *abdA*. A uniform aortic configuration is present in *abdA* mutants, while overexpression of *abdA* resulted in a partial transformation of the aortic segment into a heart-like structure [343, 344]. Abd-B, which at earlier stages is expressed posteriorly to the cardiogenic mesoderm, represses cardiogenesis [345]. These data suggest that the concept of homeotic transformation by alterations of Hox gene expression can also be applied to the insect heart. Little is known about the role of Hox gene expression in vertebrate cardiogenesis, which is probably due to increased genetic complexity of vertebrates. In avian embryos several anterior Hox genes are expressed within the early heart-forming field [346]. Retinoic acid, which is known to affect the A/P axis alters Hox gene expression. Adenoviral expression of the normally posteriorly-expressed cCdxB homeobox gene leads to an activation of posterior Hox genes within the anterior heart field, however with no phenotypic consequences [347]. It may however be that the A-P axis is already laid down at the time the experiments were performed, or alternatively that Hox gene expression is not important for A/P patterning of the vertebrate heart.

23.8.3
Dorso-ventral Polarity of the Heart Tube

The dorsal side of the forming tubular heart is connected to the body wall via the dorsal mesocardium, therefore polarity along the dorsoventral (D-V) axis is morphologically apparent. Little is known about the molecular mechanisms that specify D-V polarity of the heart. The HAND1 gene is expressed exclusively along the ventral side of the tubular heart in the E8–8.5 embryo [202]. Another gene, which is expressed in a D-V-specific pattern is the ANF gene [84]. The transcriptional cofactor *Cited–2* and the muscle-specific gene *Chisel* are also selectively expressed in the ventral side of the future ventricle of the tubular heart in E8.25 embryos [198, 348].

23.9
Chamber Formation

23.9.1
Analysis of Growth Patterns in the Heart

The growth pattern of the ventricular myocardium at the tubular heart stage has been analyzed in the chick by infecting cardiac myocytes with a replication-defective retrovirus [349]. Colonies of labeled myocytes were transmural, i.e. they extended from epicardial to endocardial layers of the myocardium and generally exhibited a cone shape with the base of the cone nearest to the epicardium. This observation suggests, that mitosis in the ventricle is not random but has a defined direction [349]. This analysis has now also been performed in the mouse with the help of an inactive LacZ (nlaacZ) reporter gene under the control of the cardiac α-actin promoter [350, 351]. Upon somatic recombination, which occurs randomly at low frequency, clonal size and morphology can be analyzed retrospectively. It was found that the myocardium proliferates in two steps. The first growth phase, before E8.5, is dispersive and polarized along the axis of the primitive cardiac tube, contributing to its elongation [350]. The second growth phase at E10.5 is coherent and polarized differentially in different cardiac subregions. Right and left ventricular lineages for example, appear to differ in their growth pattern and are separated by the myocardial septum, which shows yet another growth pattern [351]. It will be interesting to apply this labeling technique to mouse models with aberrant cardiac morphognesis to analyze how gene ablation affects the clonal growth behavior.

23.9.2
Reprogramming of Gene Expression at the Onset of Chamber Development

After tubular heart formation is complete and looping has started, transformation of the expression pattern of cardiac genes is observed. Many genes that are initially expressed throughout the tubular myocardium become restricted to the presumptive atrial or ventricular primordia [352]. Several transcription factors have been identified that display chamber-specific expression patterns. Two hairy-related transcription factors, Hey1, and Hey2 are expressed in atrial and ventricular progenitor cells, respectively [353–355]. Like other hairy-related transcription factors, the Hey family is controlled by the Notch signaling pathway and these proteins act as transcriptional repressors [356]. Null mutants of *hey2* have been recently engineered and show complex cardiac phenotypes which, depending on the genetic background, either cause a massive postnatal cardiac hypertrophy [357], or display a spectrum of cardiac malformations including ventricular septal defects, tetralogy of Fallot, and tricuspid atresia [358]. The *iroquois*-related homeobox protein Irx4 labels ventricular progenitor cells in the cardiac crescent and subsequent to chamber formation Irx4 is exclusively expressed in the ventricle [359]. In the chick embryo, Irx4 seems to be involved in establishing chamber-restricted gene expression, while morphogenesis is apparently not affected by this homeobox gene [360]. Ventricle-specific expression is also observed in case of Irx4 expression in Xenopus [361]. *Irx4*–deficient mice develop normally *in utero*, however, null mutants develop a cardiomyopathy in later life [362]. Irx4 may be functionally substituted by other *Irx* homeobox genes expressed in the myocardium [362, 363]. The atrial compartment of the myocardium is labeled by GATA–4 and Tbx5. Both genes are initially expressed throughout the cardiac crescent but become localized to the posterior inflow compartment after tubulogenesis is completed. Ablation of Tbx5 in mice results in a loss of the posterior heart segments, i.e. a severe hypoplasia of the atrial and left ventricular compartments, while the right ventricle and outflow tract are not affected [239]. The orphan nuclear receptor COUP-TFII is also restricted to the inflow tract myocardium and the null mutation displays grossly retarded development of sinus venosus and atrium [64]. GATA4 and Nkx2.5 expression domains overlap during early heart formation. Later during the tubular heart stage, GATA4 is downregulated in the anterior presumptive heart field and is only maintained in the posterior presumptive atrial compartment [364]. It is believed that Nkx2.5 is involved in setting up a ventricle-specific gene expression program [359], while GATA4 is probably involved in establishing atrial identity. In retinoic acid (RA)-deficient quail embryos an oversized ventricle with deficient formation of atrium and omphalomesenteric vein is observed [365] and GATA4 expression in these animals is severely reduced [366]. Loss of GATA4 expression is accompanied by an increase in foregut apoptosis, which is probably causally related to the malformations of the inflow tract [367]. A morphogen, which has been implicated in orchestrating A/P patterning in the myocardium is retinoic acid (RA). Administration of excess RA to chick embryos truncates the anterior portion of the heart tube and enlarges its caudal portion [368, 369]. Conversely, administration of an RALDH2 antagonist diminishes expression of an atrial-specific transgene [368]. The *RALDH2* gene is specifically expressed within the posterior heart region of both, mouse and chick [77, 370]. Consistent with this role of RA as a posteriorizing reagent is the finding that in RALDH2 (*Aldh1a7*) null mutants a severe posterior truncation of the heart and a specific loss of Tbx5 expression is observed [370]. The Fwb7 F-box protein encodes an E3 ubiquitin ligase involved in ubiquitination of NICD (the notch intracellular domain) as well as cyclin E [371]. Ablation of this gene specifically affects left ventricular chamber development. Since in the mutant the concentration of NICD1 and 4 are elevated it could be that another role of the Notch pathway in the heart is the control of left ventricular chamber development.

23.9.3
The Secondary or Anterior Heart Field (AHF)

While primary cardiogenic induction is complete by HH stage 8 in the chick, classical studies suggested that at the arterial pole recruitment of splanchnic mesodermal cells to the cardiac cell lineage continues until stage 22 in the chick and E11 in the mouse [372, 373]. DiI labeling experiments also provided evidence for a progressive movement of cells from the pharyngeal mesoderm into the growing arterial pole of the mouse heart between E8.25 and E10.5 [374]. Adenoviral labeling studies demonstrate that the distal outflow tract originates anterior to the heart tube [82]. The splanchnic mesoderm adjacent to the outflow tract myocardium expresses Nkx2.5 [37], GATA4 [37], Tbx1 [375], Isl1 [83], and Nkx2.8 [376]. These cells are also labeled by a transgenic MLC2 β-galactosidase reporter gene, which has trapped an enhancer of the FGF10 gene [374] and by a LacZ knock-in mutant of the murine Nkx3.1 gene [377]. During growth of the outflow tract the arterial pole of the heart is displaced caudally and sequentially anterior cardiogenesis occurs adjacent to the arch arteries [37]. In the FGF10 enhancer trap line LacZ labeling is found in the outflow tract as well as the right ventricles, which suggests that at least in mammals most of the right ventricle might actually be derived from the anterior heart field [374]. Recent analysis of the Isl1 mutant embryo shows that the secondary heart field actually contributes not only to the outflow and right ventricle but descendants of this field also migrate into the left ventricle and atria [83]. Similar conclusions can be drawn from a Tbx1–Cre fate map experiment [375]. The distinct origin of right and left ventricle from the secondary versus primary heart field is also reflected by distinct sensitivities of the right and left ventricular chambers after ablating cardiac transcription factors. For example Tbx5 deficiency affects atrium and left ventricle while outflow tract and right ventricle are normally formed [239]. In contrast, null mutants of versican [378], HAND2 [379], and Mef2c [380] affect primarily the right ventricle. In the Isl1 mutant the outflow tract and the right ventricle are missing while left ventricle and atrium are only reduced in size [83]. Expression of Nkx2.5 in the outflow tract myocardium is controlled by an enhancer that is distinct from that which is important for expression in the primary heart field [381]. The anterior heart field enhancer of the Mef2c gene has recently been analyzed and expression depends on the binding of ISL1 and GATA4 [382] and in addition also requires binding of Foxh1 and Nkx2.5 which mediate a Smad-dependent input of a TGFβ-signal transduction pathway [383]. Signaling molecules that are present in the anterior heart field at the time of cardiogenic induction include BMP2, FGF8, FGF10, Wnt–11, Cripto, as well as Shh. BMP2 is present in the splanchnic mesoderm between HH stages 14 and 18, Shh and FGF8 are present in the pharyngeal endoderm and ectoderm, FGF10 is found within the splanchnic mesoderm, and Wnt–11 is expressed in the outflow tract mesoderm [37, 83, 374, 384]. BMP2 seems to be essential for outflow tract formation as application of noggin to splanchnic mesoderm explants impaired myocardial differentiation [37]. There is normal outflow tract and right ventricle development in the case of FGF8, FGF10, Wnt–11, or Shh null mutants in mouse, which is most likely due to the presence of additional family members expressed during anterior heart field formation. In the mouse expression of the EGF-CFC gene *Cripto* was found in the outflow tract mesoderm suggesting that Nodal class signals may either be important for outflow tract formation or are involved in mediating lateralization [170]. Indeed Nodal and Lefty2 are both present in splanchnic mesoderm of the arterial pole at E8.5 [305, 385]. Interestingly *Cripto* is under the regulation of *Mef2C* during outflow tract formation [386].

Expression of the transgenic FGF10 enhancer trap suggests that the secondary heart field is derived from LPM, which lies medial to the primary heart field [374]. In this position the cells experience a lower concentration of BMP2 while inhibitory Wnt signaling molecules and BMP antagonists are present at high concentrations and thus mesoderm in this region is probably unable to differentiate into myocardium during the time of primary heart field formation [127, 149]. Fate mapping analysis in Xenopus embryos suggests that medial Nkx2.5–expressing cells do not contribute to the heart but to the dorsal mesocardium. Dorsal mesocardial cells are prevented from differentiating into cardiac myocytes by the activation of the Notch pathway [186]. Interestingly most of the outflow tract myocardium is lost due to apoptosis during later stages of cardiogenesis and is replaced by smooth muscle cells, which originate to a large extent from the cardiac neural crest [387, 388]. It is currently not known what exactly triggers outflow tract apoptosis, however, it might be mechanistically important that the outflow tract is covered by epicardium that is not derived from the pro-epicardial serosa (PrES) but is probably derived from the anterior pericardium [36].

23.9.4
The Right Ventricle is a Derivative of the Anterior Heart Field

Right ventricular hypoplasia is apparent in the *versican* [389] and *Mef2C* mutant [380]. The search for *Mef2C*-dependent genes resulted in the identification of a heart-restricted helicase, CHAMP [386]. CHAMP probably functions as a negative regulator of cell cycle progression by upregulating the cdk inhibitor p21. In the context of *Mef2C*, CHAMP may be part of negative regulatory loop that controls right ventricular wall expansion [386, 390]. Another potential *Mef2C* target gene is the EGF-CFC factor *Cripto* that is downregulated in *Mef2C*-deficient hearts [386]. Mef2 proteins are targets of various signal transduction pathways. In particular p38a and Erk5 can phosphorylate Mef2C and Mef2A [391]. Interestingly *Mekk3* [392], *p38a* [393], and *Erk5* null mice [394] display a phenocopy of the *Mef2C* phenotype and all display deficient right ventricle formation.

The related basic helix-loop-helix transcription factor genes *HAND1* and *HAND2* are involved in ventricular chamber formation. In the mouse, both genes are co-expressed in the primary heart tube, however they differentially accumulate in the right (*HAND2*) and left (*HAND1*) ventricle [258]. Similar to Mef2C and versican, deletion of *HAND2* in mice results in hypoplasia of the right ventricular chamber [379]. Due to early lethality the function of *HAND1* can only be inferred from chimeric analysis, which suggests that *HAND1* is specifically required for the formation of the outer curvature of the left ventricular myocardium [331]. Mice homozygous for *Nkx2.5* fail to express *HAND1* in the heart, suggesting that *HAND1* acts downstream of *Nkx2.5* in controlling left ventricular chamber formation [202]. In zebrafish only one HAND gene, *dHAND*, has been identified. Mutation of *dHAND* results in loss of ventricular chamber formation [395]. Two HAND genes have also been identified in Xenopus, however only *dHAND* is utilized in the myocardium

[45]. This suggests that in lower vertebrates with a single ventricular chamber only a single HAND gene is expressed in the heart. A *hand* gene is also present in the Drosophila heart [396]. In contrast to the chamber-restricted expression pattern in higher vertebrates *hand* is expressed in all cells of the insect heart.

23.9.5
T-box Genes Pattern the Cardiac Chambers and are Involved in Septum Formation

Several genes are specifically activated in the outer curvature of the embryonic heart loop in specialized regions that demarcate the future working myocardium of atria and ventricles. Genes that label this region includes *Connexins* (*Cx40* and *Cx43*), *ANF* and *Chisel* [84]. In contrast, cells at the inner curvature of the embryonic heart loop retain the phenotype of the tubular myocardium and do not express the aforementioned genes. Cells from this region contribute to the formation of the conduction tissue and are also involved in septation. Several transcription factors have recently been implicated in the formation of the working myocardium. Tbx5 was found to cooperate with Nkx2.5 to activate expression of *ANF* and *Cx40* [239, 397]. The T-box family member *Tbx2*, encoding a transcriptional repressor is expressed in a pattern mutually exclusive to ANF. Tbx2 can form a repressive complex with Nkx2.5 on the *ANF* promoter and competes with the transactivating Tbx5–Nkx2.5 complex [239, 397]. Thus, complex formation of positively and negatively regulating factors provides a potential mechanism to generate chamber-specific gene expression. Overexpression of Tbx5 under the control of the β-MHC promoter results in the loss of anterior gene expression and retardation of ventricular chamber morphogenesis [398]. Similar experiments in the chick through viral overexpression of Tbx5 inhibited myocardial growth and formation of the trabecular layer and ventricular septum [38, 399]. In the normal heart Tbx5 is expressed in the left ventricle and Tbx20 is expressed in the right ventricle [38]. Both genes have antagonistic activities on the ANF promoter and probably also affect morphogenetic genes in the heart. Localized ectopic expression of Tbx5 in the chick right ventricle leaving only a small Tbx5–free zone, resulted in septum formation just at the border between the Tbx5–positive and negative zone [38], suggesting that Tbx5 through antagonistic interaction with Tbx20 controls muscular septum formation.

23.9.6
Cell-Cell Interaction in Chamber Formation

23.9.6.1 Formation of Compact and Trabecular Layer
Maturation of the myocardium involves the formation of two different myocardial layers within the ventricular wall, i.e. the trabecular layer and the subepicardial compact layer (Fig. 23.4). The trabecular layer is ensheathed by the endocardium, while the subepicardial compact layer is formed underneath the epicardium. Cardiac trabeculation is dependent on endocardial-myocardial interactions involving

neuregulin signaling which is mediated by a myocardial receptor complex of erbB2 and erbB4 [93–95]. The serotonin 2B (5–hydroxytryptamine, 5–HT) receptor may also be involved in trabecular layer formation since 5–HT(2B) mutant embryos exhibit a lack of trabeculae in the heart, leading to mid-gestation lethality [400]. The signaling molecule BMP10 is specifically expressed in trabeculated myocardium and is absent from the compact layer [401, 402]. The BMP10 null mutant displays ectopic and elevated expression of p57(kip2) and as a result a dramatic reduction in ventricular proliferative activity. In addition several cardiac transcription factors such as Nkx2.5 and Mef2C require the presence of BMP10 in the mid-gestational heart [403]. The trabecular and compact layer myocardium differ in several respects from each other. Initially (between E9.5 and 12.5) the trabecular myocardium contributes a larger cell number to the ventricular wall, however at later stages (E14.5–E18.5) this relationship appears to be reversed [91, 404, 405]. Moreover the transcriptional profile differs between the two layers [406]. The compact myocardium is better adapted to relaxation while the trabecular myocardium is suitable for faster contraction and has slower relaxation compared with compact myocardium. Specific loss of compact layer myocardium has an impact on the diastolic properties of the contractile cycle [407].

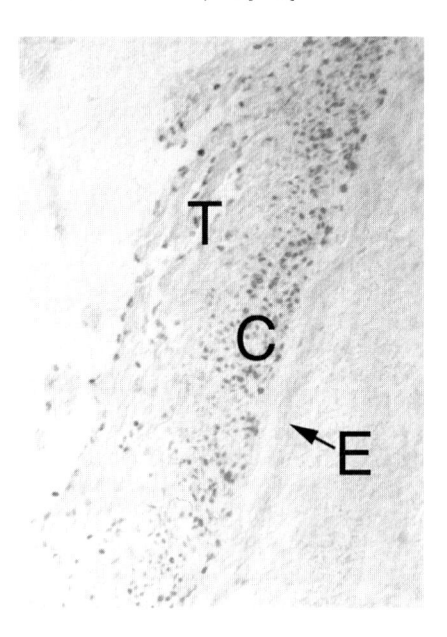

Fig. 23.4
During chamber development the ventricles generate two different compartments, i.e. the compact layer and the trabecular layer myocardium. These two compartments are shown here in a section of a 13.5 Popdc1–LacZ transgenic mouse heart with an antibody against the cell cyle inhibitor p57Kip2 [91, 404, 405] which labels the trabecular layer and by LacZ staining which visualizes the expression domain of Popdc1 in the compact layer myocardium [352]. T, trabecular layer; C, compact layer; E, epicardium. (This figure also appears with the color plates.)

23.9.6.2 The Epicardium Controls Ventricular Compact Layer Formation

The importance of the epicardium for growth control of the ventricular wall only recently became apparent. Several mouse mutants display a hypoplastic ventricular wall and die between embryonic days 13 and 16. This phenotype has been observed for mutants of the transcription factors N-myc [408], RXRα [409, 410], WT1 [411],

Tef1 [412], and FOG2 [413], the pocket protein p130 [414], the receptor gp130 [415], the receptor kinase βARK1 [416] and erythropoietin and its receptor [417]. In some cases, such as the N-myc mutant, the reduced proliferative capacity of cardiac myocytes is probably the cause of the hypoplastic phenotype. In most cases however, the gene functions in a non-cardiac cell background and the mutation is not cell autonomous. This has been shown for example for WT1 and the erythropoietin receptor. These genes are expressed in the epicardium, while the defect is apparent in the subepicardial compact layer [417, 418]. Similarly, in the RXRα mutant the mutation seems not to be cell autonomous and cannot be rescued by cardiac myocyte-specific overexpression of RXRα [419, 420]. The mutual interaction of the epicardium and the compact layer myocytes is best seen in the case of the FOG2 mutant, which has both a hypoplastic ventricular wall and a deficiency in coronary artery formation [421]. In this case both phenotypes can be rescued by myocyte-specific overexpression of FOG2. Thus, FOG2 is involved in a cell-autonomous manner in the hypoplastic growth of the compact layer, and in a non-cell autonomous manner in the differentiation of the coronary arteries. In two mutations in the mouse, VCAM and α–4 integrin, defective growth of the epicardial layer is present and in both mutations a hypoplastic ventricle has been observed [408, 422, 423]. This dependence of the ventricular compact layer on a mitogenic signal from the epicardium is also seen in the avian embryo. Ablation of the epicardium by surgical or pharmacological interventions in the chick results in a hypoplastic ventricle [424]. Interestingly, one of the retinoic acid-synthesizing enzymes, RALDH2, is expressed in the forming epicardium of mice and chicken and might be an indicator of localized RA production in the epicardium [368, 425, 426]. In the RXRα mutation a hypoplastic ventricular wall is histologically first detectable in E12 fetuses [410]. It was observed that the ventricular myocytes in the compact layer differentiate precociously [91]. While normally myocytes of the compact layer mature around day 16.5 in the absence of RXRα or in vitamin A-deficient embryos, precocious differentiation is already apparent at day 8.5 dpc. Recent evidence implicates at least two autocrine signaling loops, i.e. RA and erythropoietin that are secreted by the epicardial cells and maintain their ability to secrete another as yet unidentified mitogen that stimulates myocardial proliferation in the compact layer [89, 90].

23.9.6.3 Epicardial Cells Form the Coronary Vasculature after Epithelial Mesenchymal Transition

Some of the cells that migrate onto the heart as part of the pro-epicardium remain mesenchymal and form the subepicardial mesenchyme [427]. In addition, mesenchymal cells are formed from the epicardium after EMT [428]. The cells characteristically change from cytokeratin to vimentin expression during EMT. The genetic control of epicardial EMT has not been fully analyzed, however it seems that FGF–1, 2, 7 stimulate EMT, while TGFβ2 and TGFβ3 are inhibitory [429]. EMT transition is also controlled by α4–integrin expression since infection of epicardial cells with an adenovirus that expresses antisense chicken α4–integrin led to accelerated EMT and invasion of the ventricular wall [430]. The mesenchymal cells were however unable to seed the media of the developing coronary vasculature in the myocardial interstitium. Thus α4–integrin normally restrains EMT, invasion, and migration and is essential for correct targeting of epicardium-derived mesenchyme to the developing coronary vasculature. Another marker of epicardial development is the Bves, which is believed to be the prototype of a novel class of adhesion molecules [431]. In the epicardium Bves is localized to the plasma membrane, however, upon initiation of the EMT Bves is re-localized to a perinuclear location [432]. Another gene which is involved in this process is the zinc-finger transcription factor slug, which as in other migratory cell populations downregulates cell adhesion molecules [433]. The transcription factors Ets–1 and –2 are also essential for EMT since antisense treatment of chicken hearts block EMT [434]. Cells that migrate into the myocardium develop into coronary smooth muscle and endothelial cells. In addition perivascular and interstital fibroblast are formed by migratory epicardium-derived cells [36, 88, 435, 436]. Differentiation of epicardial cells into smooth muscle cells have been recently analyzed in a rat mesothelial cell line [437]. Induction of α-smooth muscle actin, calponin and α-smooth muscle tropomyosin were induced after cultivating the cells in serum-containing medium [438]. Differentiation into smooth muscle cells involves activation of serum response factor expression [439]. Smooth muscle cell differentiation in pro-epicardial explants is induced by serum, or by PDGF-BB, which activates rhoA-RhoK signaling which leads to re-organization of the microfilament system [440]. Similar to other cell systems differentiation of EPDC into coronary endothelial cells is probably regulated by VEGF [441].

23.10
Outflow Tract Patterning and the Role of the Neural Crest

In the chick embryo the cardiac neural crest originates at the dorsal aspect of the neural tube, between the level of the mid-otic ear placode and the third somite [87]. At HH stage 10 these cells start to migrate ventrally and laterally. By HH stage 12 cardiac neural crest cells populate the pharyngeal region and at late HH stage 13 the cells populate the third pharyngeal arch, subsequently the fourth and finally the sixth arch region. The outflow tract of the heart is reached by the fourth incubation day. The original fate map studies in the chick have been recently confirmed in the mouse through the use of the Cre/loxP system to trace the lineage of neural crest derivatives [29, 30, 442]. The cardiac neural crest is involved in the septation of the aortic sac, reorganization of the pharyngeal arteries and in conotrucal septation but is also involved in the formation of the derivatives of the three caudal pharyngeal arches, namely the great arteries of the thorax, thyroid, thymus, and parathyroids.

23.10.1
Tbx1 is Mutated in the DiGeorge Syndrome

Haploinsufficiency of chromosome 22q11 is the abnormality shared by the majority of individuals diagnosed with DiGeorge syndrome, which is therefore also known as 22q11 deletion syndrome [443]. Patients with DiGeorge syndrome often have life-threatening cardiac malformations such as persistent truncus arteriosus, Tetralogy of Fallot, and aortic arch defects [444]. Mice homozygous for mutations in the neural crest-expressed Pax3 gene (the *splotch* (Sp2h) mouse) show decreased numbers of migrating neural crest cells and a severe phenotype that includes features found in patients with DiGeorge syndrome [30, 445, 446]. The cardiovascular features in the DiGeorge syndrome are probably due to haploinsufficiency of the Tbx1 gene, which is located in the 22q11 chromosomal region [67, 68, 447]. Tbx1 is under the control of Shh and is expressed in the mesoderm and endoderm of the pharyngeal arches but not in the neural crest cells [384]. Shh acts on Tbx1 via a single cis-element upstream of Tbx1 that is recognized by winged helix/forkhead box (Fox)-containing transcription factors [448]. Shh controls the expression of Foxa2 and Foxc2 in the pharyngeal endoderm and head mesenchyme, respectively and Foxa2, Foxc1, or Foxc2 could bind and activate transcription through the critical cis-element upstream of Tbx1. Consistent with their proposed role in controlling Tbx1 expression homozygous and compound mutants of Foxc1 and Foxc2 display outflow tract abnormalities and pharyngeal patterning defects [449–451]. Tbx1 controls the expression of FGF8 in the endoderm and ectoderm of the pharyngeal arches and of FGF10 in the paraxial mesoderm of the pharyngeal arches and in the splanchnic mesoderm of the secondary heart field [452, 453]. In FGF8 hypomorphic mutants the neural crest cells migrate properly to the aortic arches and into the outflow tract region but rapidly become apoptotic. Absence of the Vegf164 isoform causes birth defects in mice, reminiscent of those found in patients with DiGeorge syndrome [454]. Vegf interacts with Tbx1, as Tbx1 expression was reduced in Vegf164–deficient embryos and knocked-down vegf levels enhanced the pharyngeal arch artery defects induced by tbx1 knockdown in zebrafish. Outflow tract malformation is also a feature of mouse embryos lacking the retinoic acid receptors RARα1 and RARβ [455]. The cardiac neural crest migrates normally in this mutant, however the specific function of these cells in forming the aorticopulmonary septum is impaired. The neural crest cells themselves do not utilize retinoid receptors and do not respond to retinoic acid and thus, an undefined tissue in the vicinity of the outflow tract responds directly to retinoic acid and permits the initiation of aorticopulmonary septation via the cardiac neural crest lineage [455]

23.10.2
The Neural Crest Cells have an Early Role in the Heart Tube

Besides its role in patterning the outflow tract, recent data indicate that neural crest cells have a role before they have even entered the outflow tract. Laser ablation of the cardiac neural crest at HH stage 9–10 affects heart development by HH stage 14, which is approximately 1 day before the neural crest cells enter the outflow tract myocardium. Ablated embryos display abnormal outflow tract looping, reduced contractility, dilatation of the ventricles and myocardial hyperplasia accompanied by myofibrillar disarray [456]. A similar range of phenotypes was found in the mouse mutant for the homeobox-containing gene Lbx1 [457]. Co-culture of pharyngeal endoderm with myocardial explants in the chick suggests a depressive influence of the pharynx on calcium mobilization and at the same time stimulation of the proliferation rate of myocardial cells [458]. FGF8 is expressed in the ectoderm and pharyngeal endoderm and FGF2 mimics the effect of pharyngeal endoderm on myocardial contraction and proliferation. The cardiac neural crest migrates between the pharyngeal endoderm and the splanchnic mesoderm separating these two cell populations and might function as a cellular barrier or alternatively may secrete an antagonist that restricts the emanation of signals from the pharyngeal endoderm. Probably the best way to explain these observations is that the cardiac neural crest might modulate the recruitment of splanchnic mesoderm to the anterior heart field. Cardiac neural crest cells are also present in zebrafish, however in contrast to amniotes in zebrafish neural crest cells participate in the development of the ventricular chamber [459, 460].

23.11
Signals Governing Valve Formation

One crucial element of cardiogenesis is the formation of valves between the developing atrial and ventricular chambers and in the outflow tract [34]. EMT is a critical event in the generation of the endocardial cushion, the primordia of the valves and septa of the adult heart. This embryonic phenomenon occurs in the outflow tract (OT) and atrioventricular (AV) canal of the embryonic heart in a spatiotemporally restricted manner.

23.11.1
The Tgfβ Superfamily and Valve Formation

AV canal endocardial cushion morphogenesis has been extensively studied in avians using an *ex vivo* assay of culturing AV canal tissue on hydrated collagen gels [461]. Use of this assay has established sequential and separate roles for TGFβ2, TGFβ3 and BMP2 and their receptors for EMT [462–466]. There are however species differences as highlighted by the fact that the TGFβ3 null mutant in mouse has no cushion defect [467] and *in vitro* experiments suggest that TGFβ2 in mouse AV canal explants is sufficient for EMT induction [468]. Several BMPs are expressed in the cushion mesenchyme or the surrounding myocardium [469], including BMP2 [470], BMP4 [471], BMP5 [472], BMP6 [136], and BMP7 [473]. While individual null mutants die too early (BMP2 and BMP4) or do not display cushion defects, the BMP6/BMP7 double mutant displays delayed OFT cushion morphogenesis with less affected AV valve formation [474]. By combining use of a BMP4 hypomorphic mutation with conditional gene inactivation in cardiac myocytes, a specific role of BMP4 during AV cushion development has been recently demonstrated [475]. BMPs are bound by the BMP type II receptor (Bmpr2). Mice carrying a hypomorphic Bmpr2 allele exhibit persistent truncus arteriosus with absent semilunar valves, however AV valve formation was normal [476]. Targeted deletion of the type I BMP receptor ALK3 in the myocardium results in hypoplastic AV cushions [477]. Overexpression of noggin in the OFT of chicken embryos induces a spectrum of of OFT defects including cushion hypoplasia, double outlet right ventricle and truncus arteriosus [478]. Thus BMPs obviously play an important role during valve formation in the heart.

23.11.2
AV Cushion Formation Requires Hyaluronic Acid

The cardiac cushions are rich in extracellular matrix components including fibronectin, fibulin1, –2, fibrillin, versican, collagens, heparan sulfate proteoglycans and glycosaminoglycans, and especially prominent is hyaluronic acid (HA). Creation of a null mutation of the gene encoding the HA-synthesizing enzyme Has2 results in embryonic lethality at E9.5 due to endocardial cushion defects in both the AV canal and the OFT [479]. Consistent with this observation in mice is the finding that in the zebrafish mutant *jekyll* no functional AV valve tissue is formed [480]. The *jekyll* gene encodes UDP-glucose dehydrogenase which is required for HA synthesis. Has2 has a specific function during cushion development mediating ErbB2 and ErbB3 signaling during valve formation [481, 482]. HB-EGF$^{-/-}$ embryos have enlarged cushions suggesting that this signaling molecule is normally a negative regulator that limits the number of endocardial cells undergoing EMT [483, 484]. HB-EGF is a membrane-bound ligand which is released from the plasma membrane by the protease TACE. Mice with a TACE null mutation exhibit hyperplastic cardiac valves which is similar to, and consistent with the phenotype of the HB-EGF null mutant [483, 485]. The matrix molecule perlecan is also required for outflow tract septation since perlecan-deficient mice have hyperplastic semilunar valves and complete transposition of the great arteries [486]. The gene Nf1 which encodes the GAP protein neurofibromin negatively regulates ras activity and is expressed in the AV cushion endocardium. Like HB-EGF, Nf1 functions as a negative regulator of EMT in the AV endocardial cushions since the NF1 null mutant has hyperplastic cushions [487].

23.11.3
NFAT2 probably Mediates VEGF Signaling during Cushion Formation

Gene targeting has implicated NFAT2 in valve formation. Mice homozygous for the NFAT2 gene display defects in cardiac valve and septum formation [488, 489]. Consistent with the observed developmental deficits, NFAT2 mRNA and protein are present in murine cardiac endothelium throughout the period of cardiac morphogenesis from a simple tube to a four-chambered heart [488, 489]. A possible local signal for NFAT activation is VEGF, which triggers nuclear import of NFAT2 in pulmonary valve endothelial cells cultured from postnatal human heart [490], just as VEGF triggers activation of NFAT1 in vascular endothelial cells [491]. VEGF and its receptor VEGF-R2/Flk–1 are present in the heart rudiment, and it is known that increased expression of VEGF at E10 in the mouse terminates EMT in the valve primordia and possibly promotes expansion and differentiation of the endothelial cell layer [492, 493]. An earlier role for NFAT signaling which defines a field of myocardial cells within the outflow tract and AV canal to initiate EMT of endocardial cells has recently been proposed [494]. In this context NFAT signaling is required to block VEGF expression within the AV canal and outflow tract myocardium and this enables the adjacent endocardial cells to respond to EMT-inducing signals.

23.11.4
Wnt and Notch/Delta Signaling Pathways and Cardiac Valve Formation

Several other signaling class molecules have also been implicated in cardiac valve formation. Loss-of-function mutations of the adenomatous polyposis coli (APC) part of the axin-containing destruction complex that phosphorylates β-catenin, tagging it for ubiquitination and degradation by the proteasome, results in excess valve formation in zebrafish [495]. In some mutant fish the entire endocardium of the ventricle has valve identity as revealed by marker gene expression such as versican or has–2 expression. There is evidence for β-catenin regulatory elements in the BMP4, versican and has2 genes, which may be downstream of β-catenin signaling during valve formation. Since *sugarless* in Drosophila is required for Wnt signaling and *jekyll* is the ortholog in zebrafish, it is possible that *jekyll* functions in the context of Wnt signaling [480]. Notch/Delta signaling is obviously also involved in controling EMT during valve formation. Both, Notch 1 and Delta4 are strongly expressed throughout the endocardium in the E8.5 to 9.5 embryonic mouse heart [496]. Null mutants of *Notch1* and of the downstream effector *RBPJx* display severe impairment of endocardial EMT. Electron microscopy and molecular marker analysis reveals that although the initiation of EMT takes place, the cells are unable to invade the cushion matrix [496]. Interestingly, there seems to be cross-talk between Notch and TGFβ signaling, since in the RBPJx mutant expression of TGFβ2, TβRII, and TβRIII are downregulated. Since TGFβ2 is an essential signal which is secreted by the AV and OFT myocardium, endocardial Notch-Delta signaling is probably required to maintain TGFβ signaling in the myocardium [496].

23.12
Epigenetic Factors

The heart is a blood-pumping organ and thus, hemodynamics could be an important morphogenetic factor which helps to sculpture the heart into its mature form of a four-chambered organ. Indeed earlier work in the chick has demonstrated that valve formation, septation and pharyngeal arch artery remodelling are all sensitive to alterations in the hemodynamic load [497]. Endothelial cells are especially sensitive to alterations in flow-induced forces, cultured endothelial cells rearrange their cytoskeletal structure and change their gene expression profiles [498, 499]. Also arterial-venous differentiation appears to be flow regulated and is mediated via modulation of ephrin2 and neuropilin 1 expression [500]. In the myocardium there is increasing evidence that the endocardium senses shear stress. In an attempt to modulate blood flow in the zebrafish heart, microbeads were implanted either in front of the sinus venosus to block blood influx into the atrium or in the back of the ventricle to block blood efflux from the heart into the aorta [501]. Interference with intracardiac hemodynamics had profound effects on cardiac valve formation, as well as chamber differentiation and organ morphology during cardiac embryogenesis. Fluid shear stress induces Vegf–2 gene expression in endothelial cells [502]. It is possible that increased flow stress interferes with valve formation via increased Vegf2 expression [492]. Flow also has some influence on the maturation of the peripheral conduction system [503]. A gadolinium-sensitive ISA channel seems to sense changes in shear stress in the endothelium. Downstream signaling might include induction of the endothelin-converting enzyme 1 (ECE1) gene [504]. Flow alterations have also been recently studied in mutants with dysfunctional atrial contraction such as the weak atrium mutation in zebrafish that harbors a mutated atrial myosin heavy chain gene [505]. Despite the fact that this gene is not expressed in the ventricle the ventricular morphology shows specific alterations: the chamber becomes more compact and the ventricular wall thickens without any increase in cell proliferation. A similar observations has been also made in the mouse lacking Mlc2a expression [506]. These mice die by E11.5 and the heart exhibits aberrant morphologies in each cardiac chamber, accompanied by overall abnormalities in looping architecture. Moreover and consistent with the experiments performed in the chick and zebrafish these mice als show defects in valve formation.

23.13
Outlook

If one views heart development as the embryonic and fetal stage of the heart life cycle it should be possible to extrapolate from heart development to heart disease. As already pointed out in the Introduction it has been demonstrated that regulatory networks utilized during heart development are also important in various forms of heart disease [507]. We now know that in pressure overload of the adult heart many of the transcription factors that are activated during different phases of cardiac development are also important regulators of pathological cardiac hypertrophy and cardiomyopathy [508–511]. A direct role for GATA4 in stretch-mediated gene regulation has been observed [512]. Both, GATA4 and Mef2 are mediators of signal transduction in cardiac hypertrophy [513, 514]. GATA4 may even act as an anti-apoptotic factor assuring survival of cardiac myocytes [515]. Another important area where cardiac development and adult cardiology have an overlapping interest is the recent finding that in adult myocardium, cardiac stem cells are obviously present [516–519]. However, the function of these cells is unclear: how are they maintained and do they become activated during cardiac repair? How similar are these cells to precardiac mesoderm cells and how are they recruited to form cardiac myocytes? While cardiac regeneration is not functional in normal mammalian myocardium, it is a normal property of the zebrafish and axolotl heart [15, 520]. The ability to regenerate after myocardial infarction may be present in a dormant state in the mammalian myocardium, since the MRL mouse strain has the surprising ability to fully regenerate necrotic myocardium [521, 522]. Recently thymosin β4a G-actin sequestering peptide was shown to be strongly expressed during cardiac chamber formation in the chick and mouse [523, 524]. Interestingly, it was found the exogenous thymosinβ4 was able to prevent cardiac tissue loss after myocardial infarction [524]. The mechanism of action is believed to involve the activation of the survival kinase Akt1, however it may also involve the mobilization of resident stem cells [525]. What are the genes and signals involved in myocardial regeneration? It would not be surprising if they were the same as those involved in cardiac development. In support of this view, the Notch pathway is apparently activated during myocardial regeneration in the zebrafish heart [14]. New technology will help us to unravel the mechanisms of cardiac induction. Interestingly, combinatorial chemistry has recently led to the identification of a novel class of small molecules that effectively stimulate the expression of cardiogenic marker genes and stimulate beating in embryonic stem cells [526]. However, classical approaches like the co-culture of anterior heart fields with mouse embryoid bodies have also revealed novel and exciting prospects which could be exploited for tissue engineering purposes [527].

It is apparent from the review in this chapter that heart development is a very complex process and that many genes such as Nkx2.5, or the Notch-Delta pathway are re-used several times during cardiac development. Thus, comparative analysis of gene function in different species as well as multiple gene targeting experiments in mice and the generation of an allelic series of hypomorphic mutations will be required to fully appreciate the different roles that a single gene may play during heart development and to appropriately position a gene in the molecular networks that are active during cardiac development.

Acknowledgments

Work in the author's laboratory is funded by the Deutsche Forschungsgemeinschaft and the German-Israeli Foundation.

References

1. Hoffman, J.I. and Kaplan, S., The incidence of congenital heart disease. *J Am Coll Cardiol* **2002**; *39*: 1890–1900.
2. Hoffman, J.I., Incidence of congenital heart disease: II. Prenatal incidence. *Pediatr Cardiol* **1995**; *16*: 155–165.
3. Alexiou, C., et al., Outcome after repair of tetralogy of Fallot in the first year of life. *Ann Thorac Surg* **2001**; *71*: 494–500.
4. Gaynor, J.W., et al., Long-term outcome of infants with single ventricle and total anomalous pulmonary venous connection. *J Thorac Cardiovasc Surg* **1999**; *117*: 506–513.
5. Schreiber, C., et al., Repair of interrupted aortic arch: results after more than 20 years. *Ann Thorac Surg* **2000**; *70*: 1896–1899.
6. Rajasinghe, H.A., et al., Long-term follow-up of truncus arteriosus repaired in infancy: a twenty-year experience. *J Thorac Cardiovasc Surg* **1997**; *113*: 869–878.
7. Feldman, A.M., et al., Selective changes in cardiac gene expression during compensated hypertrophy and the transition to cardiac decompensation in rats with chronic aortic banding. *Circ Res* **1993**; *73*: 184–192.
8. Molkentin, J.D., et al., Requirement of the transcription factor GATA4 for heart tube formation and ventral morphogenesis. *Genes Dev* **1997**; *11*: 1061–1072.
9. Molkentin, J.D., et al., A calcineurin-dependent transcriptional pathway for cardiac hypertrophy. *Cell* **1998**; *93*: 215–228.
10. Passier, R., et al., CaM kinase signaling induces cardiac hypertrophy and activates the MEF2 transcription factor *in vivo*. *J Clin Invest* **2000**; *105*: 1395–1406.
11. Männer, J., et al., The origin, formation and developmental significance of the epicardium: a review. *Cells Tissues Organs* **2001**; *169*: 89–103.
12. Reese, D.E., Mikawa, T., and Bader, D.M., Development of the coronary vessel system. *Circ Res* **2002**; *91*: 761–768.
13. Wagner, K.D., et al., The Wilms' tumor suppressor Wt1 is expressed in the coronary vasculature after myocardial infarction. *FASEB J* **2002**; *16*: 1117–1119.
14. Raya, A., et al., Activation of Notch signaling pathway precedes heart regeneration in zebrafish. *Proc Natl Acad Sci USA* **2003**; *100*(Suppl. 1): 11889–11895.
15. Poss, K.D., Wilson, L.G., and Keating, M.T., Heart regeneration in zebrafish. *Science* **2002**; *298*: 2188–2190.
16. Cripps, R.M. and Olson, E.N., Control of cardiac development by an evolutionarily conserved transcriptional network. *Dev Biol* **2002**; *246*: 14–28.

17 Glickman, N.S. and Yelon, D., Cardiac development in zebrafish: coordination of form and function. *Semin Cell Dev Biol* **2002**; *13*: 507–513.
18 North, T.E. and Zon, L.I., Modeling human hematopoietic and cardiovascular diseases in zebrafish. *Dev Dyn* **2003**; *228*: 568–583.
19 Mohun, T., Orford, R., and Shang, C., The origins of cardiac tissue in the amphibian, *Xenopus laevis*. *Trends Cardiovasc Med* **2003**; *13*: 244–248.
20 Brand, T., Heart development: molecular insights into cardiac specification and early morphogenesis. *Dev Biol* **2003**; *258*: 1–19.
21 Harvey, R.P., Patterning the vertebrate heart. *Nat Rev Genet* **2002**; *3*: 544–556.
22 Lebel, M., et al., The Iroquois homeobox gene Irx2 is not essential for normal development of the heart and midbrain-hindbrain boundary in mice. *Mol Cell Biol* **2003**; *23*: 8216–8225.
23 Bruneau, B.G., et al., A murine model of Holt-Oram syndrome defines roles of the T-box transcription factor Tbx5 in cardiogenesis and disease. *Cell* **2001**; *106*: 709–721.
24 Stanley, E.G., et al., Efficient Cre-mediated deletion in cardiac progenitor cells conferred by a 3'UTR-ires-Cre allele of the homeobox gene Nkx2–5. *Int J Dev Biol* **2002**; *46*: 431–439.
25 Chen, J., Kubalak, S.W., and Chien, K.R., Ventricular muscle-restricted targeting of the RXRalpha gene reveals a non-cell-autonomous requirement in cardiac chamber morphogenesis. *Development* **1998**; *125*: 1943–1949.
26 Agah, R., et al., Gene recombination in postmitotic cells. Targeted expression of Cre recombinase provokes cardiac-restricted, site-specific rearrangement in adult ventricular muscle i. *J Clin Invest* **1997**; *100*: 169–179.
27 de Lange, F.J., Moorman, A.F., and Christoffels, V.M., Atrial cardiomyocyte-specific expression of Cre recombinase driven by an Nppa gene fragment. *Genesis* **2003**; *37*: 1–4.
28 Minamino, T., et al., Inducible gene targeting in postnatal myocardium by cardiac-specific expression of a hormone-activated Cre fusion protein. *Circ Res* **2001**; *88*: 587–592.
29 Jiang, X., et al., Fate of the mammalian cardiac neural crest. *Development* **2000**; *127*: 1607–1616.
30 Epstein, J.A., et al., Migration of cardiac neural crest cells in Splotch embryos. *Development* **2000**; *127*: 1869–1878.
31 Brown, C.B., et al., Cre-mediated excision of Fgf8 in the Tbx1 expression domain reveals a critical role for Fgf8 in cardiovascular development in the mouse. *Dev Biol* **2004**; *267*: 190–202.
32 Andrée, B., et al., BMP-2 induces ectopic expression of cardiac lineage markers and interferes with somite formation in chicken embryos. *Mech Dev* **1998**; *70*: 119–131.
33 Schlange, T., et al., Chick CFC controls Lefty1 expression in the embryonic midline and nodal expression in the lateral plate. *Dev Biol* **2001**; *234*: 376–389.
34 Eisenberg, L.M. and Markwald, R.R., Molecular regulation of atrioventricular valvuloseptal morphogenesis. *Circ Res* **1995**; *77*: 1–6.

35 Kirby, M.L., et al., Backtransplantation of chick cardiac neural crest cells cultured in LIF rescues heart development. *Dev Dyn* **1993**; *198*: 296–311.

36 Männer, J., Does the subepicardial mesenchyme contribute myocardioblasts to the myocardium of the chick embryo heart? A quail-chick chimera study tracing the fate of the epicardial primordium. *Anat Rec* **1999**; *255*: 212–226.

37 Waldo, K.L., et al., Conotruncal myocardium arises from a secondary heart field. *Development* **2001**; *128*: 3179–3188.

38 Takeuchi, J.K., et al., Tbx5 specifies the left/right ventricles and ventricular septum position during cardiogenesis. *Development* **2003**; *130*: 5953–5964.

39 Oberg, K.C., et al., Efficient ectopic gene expression targeting chick mesoderm. *Dev Dyn* **2002**; *224*: 291–302.

40 Kos, R., et al., Methods for introducing morpholinos into the chicken embryo. *Dev Dyn* **2003**; *226*: 470–477.

41 Pekarik, V., et al., Screening for gene function in chicken embryo using RNAi and electroporation. *Nat Biotechnol* **2003**; *21*: 93–96.

42 Boardman, P.E., et al., A comprehensive collection of chicken cDNAs. *Curr Biol* **2002**; *12*: 1965–1969.

43 Logan, M. and Mohun, T., Induction of cardiac muscle differentiation in isolated animal pole explants of *Xenopus laevis*. *Development* **1993**; *118*: 865–875.

44 Pandur, P., et al., Wnt–11 activation of a non-canonical Wnt signalling pathway is required for cardiogenesis. *Nature* **2002**; *418*: 636–641.

45 Smith, S.J., et al., Xenopus hand2 expression marks anterior vascular progenitors but not the developing heart. *Dev Dyn* **2000**; *219*: 575–581.

46 Angelo, S., et al., Conservation of sequence and expression of Xenopus and zebrafish dHAND during cardiac, branchial arch and lateral mesoderm development. *Mech Dev* **2000**; *95*: 231–237.

47 Fu, Y., et al., Vertebrate tinman homologues XNkx2–3 and XNkx2–5 are required for heart formation in a functionally redundant manner. *Development* **1998**; *125*: 4439–4449.

48 Pabst, O., et al., The mouse Nkx2-3 homeodomain gene is expressed in gut mesenchyme during pre- and postnatal mouse development. *Dev Dyn* **1997**; *209*: 29–35.

49 Grunz, H., Amphibian embryos as a model system for organ engineering: *in vitro* induction and rescue of the heart anlage. *Int J Dev Biol* **1999**; *43*: 361–364.

50 Ariizumi, T., et al., Amphibian *in vitro* heart induction: a simple and reliable model for the study of vertebrate cardiac development. *Int J Dev Biol* **2003**; *47*: 405–410.

51 Yelon, D., Cardiac patterning and morphogenesis in zebrafish. *Dev Dyn* **2001**; *222*: 552–563.

52 Kupperman, E., et al., A sphingosine–1–phosphate receptor regulates cell migration during vertebrate heart development. *Nature* **2000**; *406*: 192–195.

53 Horne-Badovinac, S., et al., Positional cloning of heart and soul reveals multiple roles for PKC lambda in zebrafish organogenesis. *Curr Biol* **2001**; *11*: 1492–1502.

54 Rottbauer, W., et al., Reptin and pontin antagonistically regulate heart growth in zebrafish embryos. *Cell* **2002**; *111*: 661–672.
55 Szeto, D.P., Griffin, K.J., and Kimelman, D., HrT is required for cardiovascular development in zebrafish. *Development* **2002**; *129*: 5093–5101.
56 Dehal, P., et al., The draft genome of *Ciona intestinalis*: insights into chordate and vertebrate origins. *Science* **2002**; *298*: 2157–2167.
57 Davidson, B. and Levine, M., Evolutionary origins of the vertebrate heart: Specification of the cardiac lineage in *Ciona intestinalis*. *Proc Natl Acad Sci USA* **2003**; *100*: 11469–11473.
58 Satou, Y., Imai, K.S., and Satoh, N., Action of morpholinos in Ciona embryos. *Genesis* **2001**; *30*: 103–106.
59 Deschet, K., Nakatani, Y., and Smith, W.C., Generation of Ci-Brachyury-GFP stable transgenic lines in the ascidian *Ciona savignyi*. *Genesis* **2003**; *35*: 248–259.
60 Bodmer, R., The gene *tinman* is required for specification of the heart and visceral muscles in Drosophila. *Development* **1993**; *118*: 719–729.
61 Frasch, M., Induction of visceral and cardiac mesoderm by ectodermal Dpp in the early Drosophila embryo. *Nature* **1995**; *374*: 464–467.
62 Gajewski, K., et al., The zink finger proteins Pannier and GATA4 function as cardiogenic factors in Drosophila. *Development* **1999**; *126*: 5679–5688.
63 Lo, P.C. and Frasch, M., A role for the COUP-TF-related gene seven-up in the diversification of cardioblast identities in the dorsal vessel of Drosophila. *Mech Dev* **2001**; *104*: 49–60.
64 Pereira, F.A., et al., The orphan nuclear receptor COUP-TFII is required for angiogenesis and heart development. *Genes Dev* **1999**; *13*: 1037–1049.
65 Zikova, M., et al., Patterning of the cardiac outflow region in Drosophila. *Proc Natl Acad Sci USA* **2003**; *100*: 12189–12194.
66 Kim, Y.O., et al., A functional genomic screen for cardiogenic genes using RNA interference in developing Drosophila embryos. *Proc Natl Acad Sci USA* **2004**; *101*: 159–164.
67 Merscher, S., et al., TBX1 is responsible for cardiovascular defects in velo-cardio-facial/DiGeorge syndrome. *Cell* **2001**; *104*: 619–629.
68 Lindsay, E.A., et al., Tbx1 haploinsufficieny in the DiGeorge syndrome region causes aortic arch defects in mice. *Nature* **2001**; *410*: 97–101.
69 Basson, C.T., et al., Mutations in human TBX5 (corrected) cause limb and cardiac malformation in Holt-Oram syndrome. *Nat Genet* **1997**; *15*: 30–35.
70 Kaynak, B., et al., Genome-wide array analysis of normal and malformed human hearts. *Circulation* **2003**; *107*: 2467–2474.
71 Benson, D.W., et al., Mutations in the cardiac transcription factor NKX2.5 affect diverse cardiac developmental pathways. *J Clin Invest* **1999**; *104*: 1567–1573.
72 Bamford, R.N., et al., Loss-of-function mutations in the EGF-CFC gene CFC1 are associated with human left-right laterality defects. *Nat Genet* **2000**; *26*: 365–369.

73 Goldmuntz, E., et al., CFC1 mutations in patients with transposition of the great arteries and double-outlet right ventricle. *Am J Hum Genet* **2002**; *70*: 776–780.
74 Garg, V., et al., GATA4 mutations cause human congenital heart defects and reveal an interaction with TBX5. *Nature* **2003**; *424*: 443–447.
75 Redkar, A., Montgomery, M., and Litvin, J., Fate map of early avian cardiac progenitor cells. *Development* **2002**; *128*: 2269–2279.
76 Raffin, M., et al., Subdivision of the cardiac Nkx2.5 expression domain into myogenic and nonmyogenic compartments. *Dev Biol* **2000**; *218*: 326–340.
77 Hochgreb, T., et al., A caudorostral wave of RALDH2 conveys anteroposterior information to the cardiac field. *Development* **2003**; *130*: 5363–5374.
78 Tam, P. and Schoenwolf, G., In *Heart Development*, R. Harvey and N. Rosenthal, (Eds). Academic Press: San Diego, **1999**, p. 3–18.
79 Linask, K.K. and Lash, J.W., Early heart development: dynamics of endocardial cell sorting suggests a common origin with cardiomyocytes. *Dev Dyn* **1993**; *196*: 62–69.
80 DeRuiter, M.C., et al., The development of the myocardium and endocardium in mouse embryos. Fusion of two heart tubes? *Anat Embryol (Berl)* **1992**; *185*: 461–473.
81 Männer, J., Cardiac looping in the chick embryo: a morphological review with special reference to terminological and biomechanical aspects of the looping process. *Anat Rec* **2000**; *259*: 248–262.
82 Mjaatvedt, C.H., et al., The outflow tract of the heart is recruited from a novel heart-forming field. *Dev Biol* **2001**; *238*: 97–109.
83 Cai, C.L., et al., Isl1 identifies a cardiac progenitor population that proliferates prior to differentiation and contributes a majority of cells to the heart. *Dev Cell* **2003**; *5*: 877–889.
84 Christoffels, V.M., et al., Chamber formation and morphogenesis in the developing mammalian heart. *Dev Biol* **2000**; *223*: 266–278.
85 Harvey, R.P., Seeking a regulatory roadmap for heart morphogenesis. *Semin Cell Dev Biol* **1999**; *10*: 99–107.
86 van den Hoff, M.J., et al., Formation of myocardium after the initial development of the linear heart tube. *Dev Biol* **2001**; *240*: 61–76.
87 Kirby, M.L. and Waldo, K.L., Neural crest and cardiovascular patterning. *Circ Res* **1995**; *77*: 211–215.
88 Mikawa, T. and Gourdie, R.G., Pericardial mesoderm generates a population of coronary smooth muscle cells migrating into the heart along with ingrowth of the epicardial organ. *Dev Biol* **1996**; *174*: 221–232.
89 Stuckmann, I., Evans, S., and Lassar, A.B., Erythropoietin and retinoic acid, secreted from the epicardium, are required for cardiac myocyte proliferation. *Dev Biol* **2003**; *255*: 334–349.
90 Chen, T.H., et al., Epicardial induction of fetal cardiomyocyte proliferation via a retinoic acid-inducible trophic factor. *Dev Biol* **2002**; *250*: 198–207.

91 Kastner, P., et al., Vitamin A deficiency and mutations of RXRalpha, RXRbeta and RARalpha lead to early differentiation of embryonic ventricular cardiomyocytes. *Development* **1997**; *124*: 4749–4758.

92 Sedmera, D., et al., Developmental patterning of the myocardium. *Anat Rec* **2000**; *258*: 319–337.

93 Gassmann, M., et al., Aberrant neural and cardiac development in mice lacking the ErbB4 neuregulin receptor. *Nature* **1995**; *378*: 390–394.

94 Lee, K.-F., et al., Requirement for neuregulin receptor erbB2 in neural and cardiac development. *Nature* **1995**; *378*: 394–398.

95 Meyer, D. and Birchmeier, C., Multiple essential functions of neuregulin in development. *Nature* **1995**; *378*: 386–390.

96 Garcia-Martinez, V. and Schoenwolf, G.C., Primitive-streak origin of the cardiovascular system in avian embryos. *Dev Biol* **1993**; *159*: 706–719.

97 Lopez-Sanchez, C., Garcia-Martinez, V., and Schoenwolf, G.C., Localization of cells of the prospective neural plate, heart and somites within the primitive streak and epiblast of avian embryos at intermediate primitive-streak stages. *Cells Tissues Organs* **2001**; *169*: 334–346.

98 Schlange, T., et al., Expression analysis of the chicken homologue of CITED2 during early stages of embryonic development. *Mech Dev* **2000**; *98*: 157–160.

99 Jackson, M., et al., Cloning and characterization of Ehox, a novel homeobox gene essential for embryonic stem cell differentiation. *J Biol Chem* **2002**; *277*: 38683–38692.

100 Inagaki, T., Garcia-Martinez, V., and Schoenwolf, G.C., Regulative ability of the prospective cardiogenic and vasculogenic areas of the primitive streak during avian gastrulation. *Dev Dyn* **1993**; *197*: 57–68.

101 Yatskievych, T., Ladd, A., and Antin, P., Induction of cardiac myogenesis in avian pregastrula epiblast: the role of the hypoblast and activin. *Development* **1997**; *124*: 2561–2570.

102 Ladd, A.N., Yatskievych, T.A., and Antin, P.B., Regulation of avian cardiac myogenesis by activin/TGFbeta and bone morphogenetic proteins. *Dev Biol* **1998**; *204*: 407–419.

103 Reiter, J., Verkade, H., and Stainier, D., Bmp2b and Oep promote early myocardial differentiation through their regulation of gata5. *Dev Biol* **2001**; *234*: 330–338.

104 Keegan, B.R., Meyer, D., and Yelon, D., Organization of cardiac chamber progenitors in the zebrafish blastula. *Development* **2004**; *131*: 3081–3091.

105 Griffin, K. and Kimelman, D., One-Eyed Pinhead and Spadetail are essential for heart and somite formation. *Nat Cell Biol* **2002**; *4*: 821–825.

106 Arai, A., Yamamoto, K., and Toyama, J., Murine cardiac progenitor cells require visceral embryonic endoderm and primitive streak for terminal differentiation. *Dev Dyn* **1997**; *210*: 344–353.

107 Kitajima, S., et al., MesP1 and MesP2 are essential for the development of cardiac mesoderm. *Development* **2000**; *127*: 3215–3226.

108 Satou, Y., Imai, K.S., and Satoh, N., The ascidian Mesp gene specifies heart precursor cells. *Development* **2004**; *131*: 2533–2541.

109 Ciruna, B. and Rossant, J., FGF signaling regulates mesoderm cell fate specification and morphogenetic movement at the primitive streak. *Dev Cell* **2001**; *1*: 37–49.

110 Gisselbrecht, S., et al., Heartless encodes a fibroblast growth factor receptor (DFR1/DFGF-R2) involved in the directional migration of early mesodermal cells in the Drosophila embryo. *Genes Dev* **1996**; *10*: 3003–3017.

111 Beiman, M., Shilo, B.Z., and Volk, T., Heartless, a Drosophila FGF receptor homolog, is essential for cell migration and establishment of several mesodermal lineages. *Genes Dev* **1996**; *10*: 2993–3002.

112 Cohen-Gould, L. and Mikawa, T., The fate diversity of mesodermal cells within the heart field during chicken early embryogenesis. *Dev Biol* **1996**; *177*: 265–273.

113 Wei, Y. and Mikawa, T., Fate diversity of primitive streak cells during heart field formation *in ovo*. *Dev Dyn* **2000**; *219*: 505–513.

114 Lee, R.K., et al., Cardiovascular development in the zebrafish. II. Endocardial progenitors are sequestered within the heart field. *Development* **1994**; *120*: 3361–3366.

115 Stainier, D.Y., Lee, R.K., and Fishman, M.C., Cardiovascular development in the zebrafish. I. Myocardial fate map and heart tube formation. *Development* **1993**; *119*: 31–40.

116 Mikawa, T. and Fischman, D.A., The polyclonal origin of myocyte lineages. *Annu Rev Physiol* **1996**; *58*: 509–521.

117 Lough, J. and Sugi, Y., Endoderm and heart development. *Dev Dyn* **2000**; *217*: 327–342.

118 Jacobson, A.G. and Sater, A.K., Features of embryonic induction. *Development* **1988**; *104*: 341–359.

119 Nascone, N. and Mercola, M., An inductive role for the endoderm in Xenopus cardiogenesis. *Development* **1995**; *121*: 515–523.

120 Alsan, B.H. and Schultheiss, T.M., Regulation of avian cardiogenesis by Fgf8 signaling. *Development* **2002**; *129*: 1935–1943.

121 Kokan-Moore, N.-P., Bolender, D.L., and Lough, J., Secretion of Inhibin β_A by endoderm cultured from early embryonic chicken. *Dev Biol* **1991**; *146*: 242–245.

122 Antin, P.B., et al., Regulation of avian precardiac mesoderm development by insulin and insulin-like growth factors. *J Cell Physiol* **1996**; *168*: 42–50.

123 Parlow, M.H., et al., Localization of bFGF-like proteins as punctate inclusions in the pre-septation myocardium of the chicken embryo. *Dev Biol* **1991**; *146*: 139–147.

124 Sugi, Y. and Lough, J., Activin-A and FGF–2 mimic the inductive effects of anterior endoderm on terminal cardiac myogenesis. *Dev Biol* **1995**; *168*: 567–574.

125 Zhu, X., Sasse, J., and Lough, J., Evidence that FGF receptor signaling is necessary for endoderm-regulated development of precardiac mesoderm. *Mech Ageing Dev* **1999**; *108*: 77–85.

126 Lough, J., et al., Combined BMP–2 and FGF–4, but neither factor alone, induces cardiogenesis in non-precardiac embryonic mesoderm. *Dev Biol* **1996**; *178*: 198–202.

127 Schultheiss, T., Burch, J., and Lassar, A., A role for bone morphogenetic proteins in the induction of cardiac myogenesis. *Genes Dev* **1997**; *11*: 451–462.

128 Schlange, T., et al., BMP2 is required for early heart development during a distinct time period. *Mech Dev* **2000**; *91*: 259–270.

129 Barron, M., Gao, M., and Lough, J., Requirement for BMP and FGF signaling during cardiogenic induction in non-precardiac mesoderm is specific, transient, and cooperative. *Dev Dyn* **2000**; *218*: 383–393.

130 Yamada, M., et al., Evidence for a role of Smad6 in chick cardiac development. *Dev Biol* **1999**; *215*: 48–61.

131 Galvin, K.M., et al., A role for smad6 in development and homeostasis of the cardiovascular system. *Nat Genet* **2000**; *24*: 171–174.

132 Kishimoto, Y., et al., The molecular nature of zebrafish swirl: BMP2 function is essential during early dorsoventral patterning. *Development* **1997**; *124*: 4457–4466.

133 Shi, Y., et al., BMP signaling is required for heart formation in vertebrates. *Dev Biol* **2000**; *224*: 226–237.

134 Walters, M.J., Wayman, G.A., and Christian, J.L., Bone morphogenetic protein function is required for terminal differentiation of the heart but not for early expression of cardiac marker genes. *Mech Dev* **2001**; *100*: 263–273.

135 Zhang, H. and Bradley, A., Mice deficient for BMP2 are nonviable and have defects in amnion/chorion and cardiac development. *Development* **1996**; *122*: 2977–2986.

136 Dudley, A. and Robertson, E., Overlapping expression domains of bone morphogenetic protein family members potentially account for limited tissue defects in BMP7 deficient embryos. *Dev Dyn* **1997**; *208*: 349–362.

137 Solloway, M.J. and Robertson, E.J., Early embryonic lethality in Bmp5;Bmp7 double mutant mice suggests functional redundancy within the 60A subgroup. *Development* **1999**; *126*: 1753–1768.

138 Monzen, K., et al., Bone morphogenetic proteins induce cardiomyocyte differentiation through the mitogen-activated protein kinase kinase kinase TAK1 and cardiac transcription factors Csx/Nkx–2.5 and GATA–4. *Mol Cell Biol* **1999**; *19*: 7096–7105.

139 Monzen, K., et al., Smads, TAK1, and their common target ATF–2 play a critical role in cardiomyocyte differentiation. *J Cell Biol* **2001**; *153*: 687–698.

140 Jamali, M., et al., BMP signaling regulates Nkx2–5 activity during cardiomyogenesis. *FEBS Lett* **2001**; *509*: 126–130.

141 Behfar, A., et al., Stem cell differentiation requires a paracrine pathway in the heart. *FASEB J* **2002**; *16*: 1558–1566.

142 Liberatore, C., et al., Nkx–2.5 gene induction in mice is mediated by a Smad consensus regulatory region. *Dev Biol* **2002**; *244*: 243–256.

143 Lien, C., et al., Cardiac-specific activity of an Nkx2–5 enhancer requires an evolutionarily conserved Smad binding site. *Dev Biol* **2002**; *244*: 257–266.

144 Brown, C.O., et al., The cardiac determination factor, Nkx2–5, is activated by mutual cofactors GATA–4 and Smad1/4 via a novel upstream enhancer. *J Biol Chem* **2004**; *279*: 10659–10669.

145 Lee, K.H., et al., SMAD-mediated modulation of YY1 activity regulates the BMP response and cardiac-specific expression of a GATA4/5/6–dependent chick Nkx2.5 enhancer. *Development* **2004**; *131*: 4709–4723.

146 Antin, P., et al., Precocious expression of cardiac troponin T in early chick embryos is independent of bone morphogenetic protein signaling. *Dev Dyn* **2002**; *225*: 135–141.

147 Wu, X., Golden, K., and Bodmer, R., Heart development in Drosophila requires the segment polarity gene wingless. *Dev Biol* **1995**; *169*: 619–628.

148 Marvin, M.J., et al., Inhibition of Wnt activity induces heart formation from posterior mesoderm. *Genes Dev* **2001**; *15*: 316–327.

149 Tzahor, E. and Lassar, A.B., Wnt signals from the neural tube block ectopic cardiogenesis. *Genes Dev* **2001**; *15*: 255–260.

150 Marvin, M., et al., Inhibition of Wnt activity induces heart formation from posterior mesoderm. *Genes Dev* **2001**; *15*: 316–327.

151 Schneider, V.A. and Mercola, M., Wnt antagonism initiates cardiogenesis in *Xenopus laevis*. *Genes Dev* **2001**; *15*: 304–315.

152 Finley, K.R., Tennessen, J., and Shawlot, W., The mouse secreted frizzled-related protein 5 gene is expressed in the anterior visceral endoderm and foregut endoderm during early post-implantation development. *Gene Expr Patterns* **2003**; *3*: 681–684.

153 Eisenberg, C.A. and Eisenberg, L.M., WNT11 promotes cardiac tissue formation of early mesoderm. *Dev Dyn* **1999**; *216*: 45–58.

154 Kühl, M., et al., The Wnt/Ca^{2+} pathway: a new vertebrate Wnt signaling pathway takes shape. *Trends Genet* **2000**; *16*: 279–283.

155 Torres, M.A., et al., Activities of the Wnt–1 class of secreted signaling factors are antagonized by the Wnt–5A class and by a dominant negative cadherin in early Xenopus development. *J Cell Biol* **1996**; *133*: 1123–1137.

156 Lickert, H., et al., Formation of multiple hearts in mice following deletion of beta-catenin in the embryonic endoderm. *Dev Cell* **2002**; *3*: 171–181.

157 Nakamura, T., et al., A Wnt- and beta -catenin-dependent pathway for mammalian cardiac myogenesis. *Proc Natl Acad Sci USA* **2003**; *100*: 5834–5839.

158 Reifers, F., et al., Induction and differentiation of the zebrafish heart requires fibroblast growth factor 8 (fgf8/acerebellar). *Development* **2000**; *127*: 225–235.

159 Meyers, E.N., Lewandoski, M., and Martin, G.R., An Fgf8 mutant allelic series generated by Cre- and Flp-mediated recombination. *Nat Genet* **1998**; *18*: 136–141.

160 Meyers, E.N. and Martin, G.R., Differences in left-right axis pathways in mouse and chick: functions of FGF8 and SHH. *Science* **1999**; *285*: 403–406.

161 Salomon, D.S., Bianco, C., and De Santis, M., Cripto: a novel epidermal growth factor (EGF)-related peptide in mammary gland development and neoplasia. *Bioessays* **1999**; *21*: 61–70.

162 Gritsman, K., et al., The EGF-CFC protein one-eyed pinhead is essential for nodal signaling. *Cell* **1999**; *97*: 121–132.

163 Shen, M. and Schier, A., The EGF-CFC gene family in vertebrate development. *Trends Genet* **2000**; *16*: 303–309.

164 Ding, J., et al., Cripto is required for correct orientation of the anterior-posterior axis in the mouse embryo. *Nature* **1998**; *395*: 702–707.

165 Gaio, U., et al., A role of the cryptic gene in the correct establishment of the left-right axis. *Curr Biol* **1999**; *9*: 1339–1342.

166 Yan, Y.-T., et al., Conserved requirement for EGF-CFC genes in vertebrate left-right axis formation. *Genes Dev* **1999**; *13*: 2527–2537.

167 Xu, C., et al., Specific arrest of cardiogenesis in cultured embryonic stem cells lacking Cripto-1. *Dev Biol* **1998**; *196*: 237–247.

168 Xu, C., et al., Abrogation of the Cripto gene in mouse leads to failure of postgastrulation morphogenesis and lack of differentiation of cardiomyocytes. *Development* **1999**; *126*: 483–494.

169 Parisi, S., et al., Nodal-dependent Cripto signaling promotes cardiomyogenesis and redirects the neural fate of embryonic stem cells. *J Cell Biol* **2003**; *163*: 303–314.

170 Dono, R., et al., The murine cripto gene: expression during mesoderm induction and early heart morphogenesis. *Development* **1993**; *118*: 1157–1168.

171 Colas, J. and Schoenwolf, G.C., Subtractive hybridization identifies chick-cripto, a novel EGF-CFC ortholog expressed during gastrulation, neurulation and early cardiogenesis. *Gene* **2000**; *255*: 205–217.

172 Aoki, T., et al., Molecular integration of casanova in the Nodal signalling pathway controlling endoderm formation. *Development* **2002**; *129*: 275–286.

173 Kikuchi, Y., et al., The zebrafish bonnie and clyde gene encodes a Mix family homeodomain protein that regulates the generation of endodermal precursors. *Genes Dev* **2000**; *14*: 1279–1289.

174 Kikuchi, Y., et al., Casanova encodes a novel Sox-related protein necessary and sufficient for early endoderm formation in zebrafish. *Genes Dev* **2001**; *15*: 1493–1505.

175 Alexander, J., et al., Casanova plays an early and essential role in endoderm formation in zebrafish. *Dev Biol* **1999**; *215*: 343–357.

176 Bouwmeester, T., et al., Cerberus is a head-inducing secreted factor expressed in the anterior endoderm of Spemann's organizer. *Nature* **1996**; *382*: 595–601.

177 Glinka, A., et al., Head induction by simultaneous repression of Bmp and Wnt signalling in Xenopus. *Nature* **1997**; *389*: 517–519.

178 Schneider, V.A. and Mercola, M., Spatially distinct head and heart inducers within the Xenopus organizer region. *Curr Biol* **1999**; *9*: 800–809.

179 Belo, J.A., et al., Cerberus-like is a secreted BMP and nodal antagonist not essential for mouse development. *Genesis* **2000**; *26*: 265–270.

180 Stanley, E.G., et al., Targeted insertion of a lacZ reporter gene into the mouse Cer1 locus reveals complex and dynamic expression during embryogenesis. *Genesis* **2000**; *26*: 259–264.

181 Shawlot, W., et al., The cerberus-related gene, Cerr1, is not essential for mouse head formation. *Genesis* **2000**; *26*: 253–258.

182 Yokouchi, Y., et al., Antagonistic signaling by Caronte, a novel Cerberus-related gene, establishes left-right asymmetric gene expression. *Cell* **1999**; *98*: 573–583.

183 Rodriguez Esteban, C., et al., The novel Cer-like protein Caronte mediates the establishment of embryonic left-right asymmetry. *Nature* **1999**; *401*: 243–251.

184 Zhu, L., et al., Cerberus regulates left-right asymmetry of the embryonic head and heart. *Curr Biol* **1999**; *9*: 931–938.

185 Zhang, X.M., Ramalho-Santos, M., and McMahon, A.P., Smoothened mutants reveal redundant roles for shh and ihh signaling including regulation of l/r asymmetry by the mouse node. *Cell* **2001**; *105*: 781–792.

186 Rones, M.S., et al., Serrate and Notch specify cell fates in the heart field by suppressing cardiomyogenesis. *Development* **2000**; *127*: 3865–3876.

187 Sieger, D., Tautz, D., and Gajewski, M., The role of Suppressor of Hairless in Notch mediated signalling during zebrafish somitogenesis. *Mech Dev* **2003**; *120*: 1083–1094.

188 Oka, C., et al., Disruption of the mouse RBP-J kappa gene results in early embryonic death. *Development* **1995**; *121*: 3291–3301.

189 Schroeder, T., et al., Recombination signal sequence-binding protein Jkappa alters mesodermal cell fate decisions by suppressing cardiomyogenesis. *Proc Natl Acad Sci USA* **2003**; *100*: 4018–4023.

190 Raya, A., et al., Notch activity induces Nodal expression and mediates the establishment of left-right asymmetry in vertebrate embryos. *Genes Dev* **2003**; *17*: 1213–1218.

191 Lints, T.J., et al., Nkx–2.5: a novel murine homeobox gene expressed in early heart progenitor cells and their myogenic descendants. *Development* **1993**; *119*: 419–431.

192 Schultheiss, T.M., Xydas, S., and Lassar, A.B., Induction of avian cardiac myogenesis by anterior endoderm. *Development* **1995**; *121*: 4203–4214.

193 Lyons, I., et al., Myogenic and morphogenetic defects in the heart tubes of murine embryos lacking the homeo box gene Nkx2–5. *Genes Dev* **1995**; *9*: 1654–1666.

194 Yamagishi, H., et al., The combinatorial activities of Nkx2.5 and dHAND are essential for cardiac ventricle formation. *Dev Biol* **2002**; *239*: 190–203.

195 Tanaka, M., Yamasaki, N., and Izumo, S., Phenotypic characterization of the murine Nkx2.6 homeobox gene by gene targeting. *Mol Cell Biol* **2000**; *20*: 2874–2879.

196 Tanaka, M., et al., Nkx2.5 and Nkx2.6, homologs of Drosophila tinman, are required for development of the pharynx. *Mol Cell Biol* **2001**; *21*: 4391–4398.

197 Grow, M.W. and Krieg, P.A., Tinman function is essential for vertebrate heart development: elimination of cardiac differen-

tiation by dominant inhibitory mutants of the tinman-related genes, XNkx2–3 and XNkx2–5. *Dev Biol* **1998**; *204*: 187–196.

198 Palmer, S., et al., The small muscle-specific protein Csl modifies cell shape and promotes myocyte fusion in an insulin-like growth factor 1–dependent manner. *J Cell Biol* **2001**; *153*: 985–998.

199 Tanaka, M., et al., The cardiac homeobox gene Csx/Nkx2.5 lies genetically upstream of multiple genes essential for heart development. *Development* **1999**; *126*: 1269–1280.

200 Zou, Y., et al., CARP, a cardiac ankyrin repeat protein, is downstream in the Nkx2–5 homeobox gene pathway. *Development* **1997**; *124*: 793–804.

201 Ueyama, T., et al., Csm, a cardiac-specific isoform of the RNA helicase Mov10l1, is regulated by Nkx2.5 in embryonic heart. *J Biol Chem* **2003**; *278*: 28750–28757.

202 Biben, C. and Harvey, R.P., Homeodomain factor Nkx2–5 controls left/right asymmetric expression of bHLH gene eHand during murine heart development. *Genes Dev* **1997**; *11*: 1357–1369.

203 Chen, F., et al., Hop is an unusual homeobox gene that modulates cardiac development. *Cell* **2002**; *110*: 713–723.

204 Shin, C., et al., Modulation of cardiac growth and development by HOP, an unusual homeodomain protein. *Cell* **2002**; *110*: 725–735.

205 Ueyama, T., et al., Myocardin expression is regulated by Nkx2.5, and its function is required for cardiomyogenesis. *Mol Cell Biol* **2003**; *23*: 9222–9232.

206 Molkentin, J.D., et al., Direct activation of a GATA6 cardiac enhancer by Nkx2.5: evidence for a reinforcing regulatory network of Nkx2.5 and GATA transcription factors in the developing heart. *Dev Biol* **2000**; *217*: 301–309.

207 Davis, D.L., Wessels, A., and Burch, J.B., An Nkx-dependent enhancer regulates cGATA–6 gene expression during early stages of heart development. *Dev Biol* **2000**; *217*: 310–322.

208 Cripps, R.M., Zhao, B., and Olson, E.N., Transcription of the myogenic regulatory gene Mef2 in cardiac, somatic, and visceral muscle cell lineages is regulated by a Tinman-dependent core enhancer. *Dev Biol* **1999**; *215*: 420–430.

209 Shiratori, H., et al., Two-step regulation of left-right asymmetric expression of Pitx2: initiation by nodal signaling and maintenance by Nkx2. Mol *Cell* **2001**; *7* 137–149.

210 Molkentin, J.D., The zinc finger-containing transcription factors GATA–4, –5, and –6. Ubiquitously expressed regulators of tissue-specific gene expression. *J Biol Chem* **2000**; *275*: 38949–38952.

211 Kuo, C.T., et al., GATA4 transcription factor is required for ventral morphogenesis and heart tube formation. *Genes Dev* **1997**; *11*: 1048–1060.

212 Molkentin, J.D., et al., Abnormalities of the genitourinary tract in female mice lacking GATA5. *Mol Cell Biol* **2000**; *20*: 5256–5260.

213 Nemer, G. and Nemer, M., Cooperative interaction between GATA5 and NF-ATc regulates endothelial-endocardial differentiation of cardiogenic cells. *Development* **2002**; *129*: 4045–4055.

214 Reiter, J.F., Kikuchi, Y., and Stainier, D.Y., Multiple roles for Gata5 in zebrafish endoderm formation. *Development* **2001**; *128*: 125–135.
215 Weber, H., et al., A role for GATA5 in xenopus endoderm specification. *Development* **2000**; *127*: 4345–4360.
216 Koutsourakis, M., et al., The transcription factor GATA6 is essential for early extraembryonic development. *Development* **1999**; *126*: 723–732.
217 Morrisey, E.E., et al., GATA6 regulates HNF4 and is required for differentiation of visceral endoderm in the mouse embryo. *Genes Dev* **1998**; *12*: 3579–3590.
218 Peterkin, T., Gibson, A., and Patient, R., GATA-6 maintains BMP-4 and Nkx2 expression during cardiomyocyte precursor maturation. *EMBO J* **2003**; *22*: 4260–4273.
219 Latinkic, B.V., Kotecha, S., and Mohun, T.J., Induction of cardiomyocytes by GATA4 in Xenopus ectodermal explants. *Development* **2003**; *130*: 3865–3876.
220 Grepin, C., Nemer, G., and Nemer, M., Enhanced cardiogenesis in embryonic stem cells overexpressing the GATA-4 transcription factor. *Development* **1997**; *124*: 2387–2395.
221 Reecy, J.M., et al., Identification of upstream regulatory regions in the heart-expressed homeobox gene Nkx2–5. *Development* **1999**; *126*: 839–849.
222 Lien, C.L., et al., Control of early cardiac-specific transcription of Nkx2–5 by a GATA-dependent enhancer. *Development* **1999**; *126*: 75–84.
223 Searcy, R.D., et al., A GATA-dependent nkx–2.5 regulatory element activates early cardiac gene expression in transgenic mice. *Development* **1998**; *125*: 4461–4470.
224 McFadden, D.G., et al., A GATA-dependent right ventricular enhancer controls dHAND transcription in the developing heart. *Development* **2000**; *127*: 5331–5341.
225 Ip, H.S., et al., The GATA–4 transcription factor transactivates the cardiac muscle-specific troponin C promoter-enhancer in nonmuscle cells. *Mol Cell Biol* **1994**; *14*: 7517–7526.
226 Molkentin, J.D., Kalvakolanu, D.V., and Markham, B.E., Transcription factor GATA–4 regulates cardiac muscle-specific expression of the α-myosin heavy chain gene. *Mol Cell Biol* **1994**; *14*: 4947–4957.
227 Weinhold, B., et al., Srf(–/–) ES cells display non-cell-autonomous impairment in mesodermal differentiation. *EMBO J* **2000**; *19*: 5835–5844.
228 Arsenian, S., et al., Serum response factor is essential for mesoderm formation during mouse embryogenesis. *EMBO J* **1998**; *17*: 6289–6299.
229 Wang, D., et al., Activation of cardiac gene expression by myocardin, a transcriptional cofactor for serum response factor. *Cell* **2001**; *105*: 851–862.
230 Li, S., et al., The serum response factor coactivator myocardin is required for vascular smooth muscle development. *Proc Natl Acad Sci USA* **2003**; *100*: 9366–9370.
231 Yoshida, T., et al., Myocardin is a key regulator of CArG-dependent transcription of multiple smooth muscle marker genes. *Circ Res* **2003**; *92*: 856–864.

232 Wang, Z., et al., Myocardin is a master regulator of smooth muscle gene expression. *Proc Natl Acad Sci USA* **2003**; *100*: 7129–7134.

233 Du, K.L., et al., Myocardin is a critical serum response factor cofactor in the transcriptional program regulating smooth muscle cell differentiation. *Mol Cell Biol* **2003**; *23*: 2425–2437.

234 Wang, D.Z., et al., Potentiation of serum response factor activity by a family of myocardin-related transcription factors. *Proc Natl Acad Sci USA* **2002**; *99*: 14855–14860.

235 Selvaraj, A. and Prywes, R., Megakaryoblastic leukemia–1/2, a transcriptional co-activator of serum response factor, is required for skeletal myogenic differentiation. *J Biol Chem* **2003**; *278*: 41977–41987.

236 Cen, B., et al., Megakaryoblastic leukemia 1, a potent transcriptional coactivator for serum response factor (SRF), is required for serum induction of SRF target genes. *Mol Cell Biol* **2003**; *23*: 6597–6608.

237 Miralles, F., et al., Actin dynamics control SRF activity by regulation of its coactivator MAL. *Cell* **2003**; *113*: 329–342.

238 Sepulveda, J., et al., Combinatorial expression of GATA4, Nkx2–5, and serum response factor directs early cardiac gene activity. *J Biol Chem* **2002**; *277*: 25775–25782.

239 Bruneau, B.G., et al., A murine model of Holt-Oram syndrome defines roles of the T-box transcription factor Tbx5 in cardiogenesis and disease. *Cell* **2001**; *106*: 709–721.

240 Ganga, M., et al., PITX2 isoform-specific regulation of atrial natriuretic factor expression: synergism and repression with Nkx2.5. *J Biol Chem* **2003**; *278*; 22437–22445.

241 Lee, Y., et al., The cardiac tissue-restricted homeobox protein Csx/Nkx2.5 physically associates with the zinc finger protein GATA4 and cooperatively activates atrial natriuretic factor gene expression. *Mol Cell Biol* **1998**; *18*: 3120–3129.

242 Morin, S., et al., GATA-dependent recruitment of MEF2 proteins to target promoters. *EMBO J* **2000**; *19*: 2046–2055.

243 Dai, Y.S., et al., The transcription factors GATA4 and dHAND physically interact to synergistically activate cardiac gene expression through a p300–dependent mechanism. *J Biol Chem* **2002**; *277*: 24390–24398.

244 Moore, M.L., et al., GATA–4 and serum response factor regulate transcription of the muscle-specific carnitine palmitoyltransferase I beta in rat heart. *J Biol Chem* **2001**; *276*: 1026–1033.

245 Stennard, F.A., et al., Cardiac T-box factor Tbx20 directly interacts with Nkx2–5, GATA4, and GATA5 in regulation of gene expression in the developing heart. *Dev Biol* **2003**; *262*: 206–224.

246 Shiojima, I., et al., Context-dependent transcriptional cooperation mediated by cardiac transcription factors Csx/Nkx–2.5 and GATA–4. *J Biol Chem* **1999**; *274*: 8231–8239.

247 Roebroek, A.J., et al., Failure of ventral closure and axial rotation in embryos lacking the proprotein convertase furin. *Development* **1998**; *125*: 4863–4876.

248 Saga, Y., et al., MesP1 is expressed in the heart precursor cells and required for the formation of a single heart tube. *Development* **1999**; *126*: 3437–3447.

249 Li, S., et al., Advanced cardiac morphogenesis does not require heart tube fusion. *Science* **2004**; *305*: 1619–1622.
250 Reiter, J.F., et al., Gata5 is required for the development of the heart and endoderm in zebrafish. *Genes Dev* **1999**; *13*: 2983–2995.
251 Narita, N., Bielinska, M., and Wilson, D.B., Wild-type endoderm abrogates the ventral developmental defects associated with GATA–4 deficiency in the mouse. *Dev Biol* **1997**; *189*: 270–274.
252 Zhang, H., et al., GATA–4 regulates cardiac morphogenesis through transactivation of the N-cadherin gene. *Biochem Biophys Res Commun* **2003**; *312*: 1033–1038.
253 Nakagawa, S. and Takeichi, M., N-Cadherin is crucial for heart formation in the chicken embryo. *Dev Growth Diff* **1997**; *39*: 451–455.
254 Radice, G.L., et al., Developmental defects in mouse embryos lacking N-cadherin. *Dev Biol* **1997**; *181*: 64–78.
255 Dathe, V., et al., Morphological left-right asymmetry of Hensen's node precedes the asymmetric expression of Shh and Fgf8 in the chick embryo. *Anat Embryol* **2002**; *205*: 343–354.
256 Männer, J., On rotation, torsion, laterlization and handedness of the embryonic heart loop. New insights from a simulation model for the heart loop of chick embryos. *Anat Rec* **2004** (in press).
257 Brown, N. and Anderson, R., In *Heart Development* R. Harvey and N. Rosenthal, (Eds). Academic Press: San Diego, **1999**; p. 447–461.
258 Thomas, T., et al., The bHLH factors, dHAND and eHAND, specify pulmonary and systemic cardiac ventricles independent of left-right sidedness. *Dev Biol* **1998**; *196*: 228–236.
259 Capdevilla, J., et al., Mechanisms of left-right determination in vertebrates. *Cell* **2000**; *101*: 9–21.
260 Levin, M., Motor protein control of ion flux is an early step in embryonic left-right asymmetry. *Bioessays* **2003**; *25*: 1002–1010.
261 Mercola, M. and Levin, M., Left-right asymmetry determination in vertebrates. *Ann Rev Cell Dev Biol* **2001**; *17*: 779–805.
262 Wright, C.V., Mechanisms of left-right asymmetry: what's right and what's left? *Dev Cell* **2001**; *1*: 179–186.
263 Yost, H.J., Establishment of left-right asymmetry. *Int Rev Cytol* **2001**; *203*: 357–381.
264 Hamada, H., et al., Establishment of vertebrate left-right asymmetry. *Nat Rev Genet* **2002**; *3*: 103–113.
265 Bunney, T.D., De Boer, A.H., and Levin, M., Fusicoccin signaling reveals 14–3–3 protein function as a novel step in left-right patterning during amphibian embryogenesis. *Development* **2003**; *130*: 4847–4858.
266 Levin, M. and Mercola, M., Gap junction-mediated transfer of left-right patterning signals in the early chick blastoderm is upstream of Shh asymmetry in the node. *Development* **1999**; *126*: 4703–4714.
267 Levin, M. and Mercola, M., Gap junctions are involved in the early generation of left-right asymmetry. *Dev Biol* **1998**; *203*: 90–105.

268 Ramsdell, A.F. and Yost, H.J., Cardiac looping and the vertebrate left-right axis: antagonism of left-sided Vg1 activity by a right-sided ALK2–dependent BMP pathway. *Development* **1999**; *126*: 5195–5205.

269 Kramer, K. and Yost, H., Ectodermal syndecan–2 mediates left-right axis formation in migrating mesoderm as a cell-nonautonomous Vg1 cofactor. *Dev Cell* **2002**; *2*: 115–124.

270 Monsoro-Burq, A. and Le Douarin, N.M., BMP4 plays a key role in left-right patterning in chick embryos by maintaining Sonic Hedgehog asymmetry. *Mol Cell* **2001**; *7*: 789–799.

271 Levin, M., et al., A molecular pathway determining left-right asymmetry in chick embryogenesis. *Cell* **1995**; *82*: 803–814.

272 Isaac, A., Sargent, M.G., and Cooke, J., Control of vertebrate left-right asymmetry by a snail-related zinc finger gene. *Science* **1997**; *275*: 1301–1304.

273 Rodriguez-Esteban, C., et al., Wnt signaling and PKA control Nodal expression and left-right determination in the chick embryo. *Development* **2001**; *128*: 3189–3195.

274 Garcia-Castro, M., Vielmetter, E., and Bonner-Fraser, M., N-Cadherin, a cell adhesion molecule involved in establishment of embryonic left-right asymmetry. *Science* **2000**; *288*: 1047–1051.

275 Marques, S., et al., The activity of the Nodal antagonist Cerl–2 in the mouse node is required for correct L/R body axis. *Genes Dev* **2004**; *18*: 2342–2347.

276 Przemeck, G.K., et al., Node and midline defects are associated with left-right development in Delta1 mutant embryos. *Development* **2003**; *130*: 3–13.

277 Krebs, L.T., et al., Notch signaling regulates left-right asymmetry determination by inducing Nodal expression. *Genes Dev* **2003**; *17*: 1207–1212.

278 Sulik, K., et al., Morphogenesis of the murine node and notochordal plate. *Dev Dyn* **1994**; *201*: 260–278.

279 Nonaka, S., et al., Randomization of left-right asymmetry due to loss of nodal cilia generating leftward flow of extraembryonic fluid in mice lacking KIF3B motor protein. *Cell* **1998**; *95*: 829–837.

280 Okada, Y., et al., Abnormal nodal flow precedes situs inversus in iv and inv mice. *Mol Cell* **1999**; *4*: 459–468.

281 Nonaka, S., et al., Determination of left-right patterning of the mouse embryo by artificial nodal flow. *Nature* **2002**; *418*: 96–99.

282 Supp, D.M., et al., Mutation of an axonemal dynein affects left-right asymmetry in inversus viscerum mice. *Nature* **1997**; *389*: 963–966.

283 Supp, D.M., et al., Targeted deletion of the ATP binding domain of left-right dynein confirms its role in specifying development of left-right asymmetries. *Development* **1999**; *126*: 5495–5504.

284 Bartoloni, L., et al., Mutations in the DNAH11 (axonemal heavy chain dynein type 11) gene cause one form of situs inversus totalis and most likely primary ciliary dyskinesia. *Proc Natl Acad Sci USA* **2002**; *99*: 10282–10286.

285 Essner, J., et al., Conserved function for embryonic nodal cilia. *Nature* **2002**; *418*: 37–38.

286 Männer, J., Does an equivalent of the ventral node exist in chick embryos? A scanning electron microscopic study. *Anat Embryol (Berl)* **2001**; *203*: 481–490.

287 Pennekamp, P., et al., The ion channel polycystin–2 is required for left-right axis determination in mice. *Curr Biol* **2002**; *12*: 938–943.

288 McGrath, J., et al., Two populations of node monocilia initiate left-right asymmetry in the mouse. *Cell* **2003**; *114*: 61–73.

289 Yokoyama, T., et al., Reversal of left-right asymmetry: a situs inversus mutation. *Science* **1993**; *260*: 679–682.

290 Morgan, D., et al., Expression analyses and interaction with the anaphase promoting complex protein Apc2 suggest a role for inversin in primary cilia and involvement in the cell cycle. *Hum Mol Genet* **2002**; *11*: 3345–3350.

291 Yasuhiko, Y., et al., Calmodulin binds to inv protein: implication for the regulation of inv function. *Dev Growth Differ* **2001**; *43*: 671–681.

292 Otto, E.A., et al., Mutations in INVS encoding inversin cause nephronophthisis type 2, linking renal cystic disease to the function of primary cilia and left-right axis determination. *Nat Genet* **2003**; *34*: 413–420.

293 Nurnberger, J., Bacallao, R.L., and Phillips, C.L., Inversin forms a complex with catenins and N-cadherin in polarized epithelial cells. *Mol Biol Cell* **2002**; *13*: 3096–3106.

294 Huangfu, D., et al., Hedgehog signalling in the mouse requires intraflagellar transport proteins. *Nature* **2003**; *426*: 83–87.

295 Taulman, P.D., et al., Polaris, a protein involved in left-right axis patterning, localizes to basal bodies and cilia. *Mol Biol Cell* **2001**; *12*: 589–599.

296 Piedra, M. and Ros, M., BMP signaling positively regulates Nodal expression during left right specification in the chick embryo. *Development* **2002**; *129*: 3431–3440.

297 Schlange, T., Arnold, H.H., and Brand, T., BMP2 is a positive regulator of Nodal signaling during left-right axis formation in the chicken embryo. *Development* **2002**; *129*: 3421–3429.

298 Fischer, A., Viebahn, C., and Blum, M., FGF8 acts as a right determinant during establishment of the left-right axis in the rabbit. *Curr Biol* **2002**; *12*: 1807–1816.

299 Fujiwara, T., et al., Distinct requirements for extra-embryonic and embryonic bone morphogenetic protein 4 in the formation of the node and primitive streak and coordination of left-right asymmetry in the mouse. *Development* **2002**; *129*: 4685–4696.

300 Shi, Y. and Massague, J., Mechanisms of TGF-beta signaling from cell membrane to the nucleus. *Cell* **2003**; *113*: 685–700.

301 Yan, Y.T., et al., Dual roles of Cripto as a ligand and coreceptor in the nodal signaling pathway. *Mol Cell Biol* **2002**; *22*: 4439–4449.

302 Reissmann, E., et al., The orphan receptor ALK7 and the Activin receptor ALK4 mediate signaling by Nodal proteins during vertebrate development. *Genes Dev* **2001**; *15*: 2010–2022.

303 Bianco, C., et al., Cripto–1 activates nodal- and ALK4–dependent and -independent signaling pathways in mammary epithelial Cells. *Mol Cell Biol* **2002**; *22*: 2586–2597.

304 Cheng, S.K., et al., EGF-CFC proteins are essential coreceptors for the TGF-beta signals Vg1 and GDF1. *Genes Dev* **2003**; *17*: 31–36.
305 Meno, C., et al., Left-right asymmetric expression of the TGF beta-family member lefty in mouse embryos. *Nature* **1996**; *381*: 151–155.
306 Meno, C., et al., Lefty–1 is required for left-right determination as a regulator of lefty–2 and nodal. *Cell* **1998**; *94*: 287–297.
307 Campione, M., et al., The homeobox gene Pitx2: mediator of asymmetric left-right signaling in vertebrate heart and gut looping. *Development* **1999**; *126*: 1225–1234.
308 Logan, M., et al., The transcription factor Pitx2 mediates situs-specific morphogenesis in response to left-right asymmetric signals. *Cell* **1998**; *94*: 307–317.
309 Piedra, M.E., et al., Pitx2 participates in the late phase of the pathway controlling left- right asymmetry. *Cell* **1998**; *94*: 319–324.
310 Ryan, A.K., et al., Pitx2 determines left-right asymmetry of internal organs in vertebrates. *Nature* **1998**; *394*: 545–551.
311 Levin, M., et al., Left/right patterning signals and the independent regulation of different aspects of situs in the chick embryo. *Dev Biol* **1997**; *189*: 57–67.
312 Essner, J., et al., Mesendoderm and left-right brain, heart and gut development are differentially regulated by pitx2 isoforms. *Development* **2000**; *127*: 1081–1093.
313 Patel, K., Isaac, A., and Cooke, J., Nodal signalling and the roles of the transcription factors SnR and Pitx2 in vertebrate left-right asymmetry. *Curr Biol* **1999**; *9*: 609–612.
314 Zile, M., et al., Retinoid signaling is required to complete the vertebrate cardiac left/right asymmetry pathway. *Dev Biol* **2000**; *223*: 323–338.
315 Campione, M., et al., Pitx2 expression defines a left cardiac lineage of cells: evidence for atrial and ventricular molecular isomerism in the iv/iv mice. *Dev Biol* **2001**; *231*: 252–264.
316 Lu, M.F., et al., Function of Rieger syndrome gene in left-right asymmetry and craniofacial development. *Nature* **1999**; *401*: 276–278.
317 Stalsberg, H., The origin of heart asymmetry: right and left contributions to the early chick embryo heart. *Dev Biol* **1969**; *19*: 109–127.
318 Chen, J.N., et al., Left-right pattern of cardiac BMP4 may drive asymmetry of the heart in zebrafish. *Development* **1997**; *124*: 4373–4382.
319 Breckenridge, R.A., Mohun, T.J., and Amaya, E., A role for BMP signalling in heart looping morphogenesis in Xenopus. *Dev Biol* **2001**; *232*: 191–203.
320 Faure, S., et al., Endogenous patterns of BMP signaling during early chick development. *Dev Biol* **2002**; *244*: 44–65.
321 Smith, S.M., et al., Retinoic acid directs cardiac laterality and the expression of early markers of precardiac asymmetry. *Dev Biol* **1997**; *182*: 162–171.
322 Yuan, S. and Schoenwolf, G.C., Islet–1 marks the early heart rudiments and is asymmetrically expressed during early rota-

tion of the foregut in the chick embryo. *Anat Rec* **2000**; *260*: 204–207.
323. Breher, S.S., et al., Popeye domain containing gene 2 (Popdc2) is a myocyte-specific differentiation marker during chick heart development. *Dev Dyn* **2004**; *229*: 695–702.
324. Gormley, J.P. and Nascone-Yoder, N.M., Left and right contributions to the Xenopus heart: implications for asymmetric morphogenesis. *Dev Genes Evol* **2003**; *213*: 390–398.
325. Satin, J., Fujii, S., and DeHaan, R.L., Development of cardiac beat rate in early chick embryos is regulated by regional cues. *Dev Biol* **1988**; *129*: 103–113.
326. Linask, K.K., et al., Directionality of heart looping: effects of Pitx2c misexpression on flectin asymmetry and midline structures. *Dev Biol* **2002**; *246*: 407–417.
327. Linask, K.K., et al., Effects of antisense misexpression of CFC on downstream flectin protein expression during heart looping. *Dev Dyn* **2003**; *228*: 217–230.
328. Stalsberg, H., Development and ultrastructure of the embryonic heart. II. Mechanism of dextral looping of the embryonic heart. *Am J Cardiol* **1970**; *25*: 265–271.
329. Horne-Badovinac, S., Rebagliati, M., and Stainier, D.Y., A cellular framework for gut-looping morphogenesis in zebrafish. *Science* **2003**; *302*: 662–665.
330. Srivastava, D., Cserjesi, P., and Olson, E.N., A subclass of bHLH proteins required for cardiac morphogenesis. *Science* **1995**; *270*: 1995–1999.
331. Riley, P.R., et al., Early exclusion of Hand1–deficient cells from distinct regions of the left ventricular myocardium in chimeric mouse embryos. *Dev Biol* **2000**; *227*: 156–168.
332. Griffin, K.J., et al., A conserved role for H15–related T-box transcription factors in zebrafish and Drosophila heart formation. *Dev Biol* **2000**; *218*: 235–247.
333. Iio, A., et al., Expression pattern of novel chick T-box gene, Tbx20. Dev Genes Evol, **2001**; *211*: 559–562.
334. Kraus, F., Haenig, B., and Kispert, A., Cloning and expression analysis of the mouse T-box gene tbx20. *Mech Dev* **2001**; *100*: 87–91.
335. Garrity, D., Childs, S., and Fishman, M., The heartstrings mutation in zebrafish causes heart/fin Tbx5 deficiency syndrome. *Development* **2002**; *129*: 4635–4645.
336. Taber, L.A., Lin, I.E., and Clark, E.B., Mechanics of cardiac looping. *Dev Dyn* **1995**; *203*: 42–50.
337. Stalsberg, H., Regional mitotic activity in the precardiac mesoderm and differentiating heart tube in the chick embryo. *Dev Biol* **1969**; *20*: 18–45.
338. Voronov, D.A., et al., The role of mechanical forces in dextral rotation during cardiac looping in the chick embryo. *Dev Biol* **2004**; *272*: 339–350.
339. Kioussi, C., et al., Identification of a Wnt/Dvl/-catenin Pitx2 pathway mediating cell-type-specific proliferation during development. *Cell* **2002**; *111*: 673–685.
340. Baek, S.H., et al., Regulated subset of G1 growth-control genes in response to derepression by the Wnt pathway. *Proc Natl Acad Sci USA* **2003**; *100*: 3245–3250.

341 Hamblet, N., et al., Dishevelled 2 is essential for cardiac outflow tract development somite segmentation and neural tube closure. *Development* **2002**; *129*: 5827–5838.

342 Liu, C., et al., Regulation of left-right asymmetry by thresholds of Pitx2c activity. *Development* **2001**; *128*: 2039–2048.

343 Lovato, T., et al., The Hox gene abdominal-A specifies heart cell fate in the Drosophila dorsal vessel. *Development* **2002**; *129*: 5019–5027.

344 Ponzielli, R., et al., Heart tube patterning in Drosophila requires integration of axial and segmental information provided by the Bithorax Complex genes and hedgehog signaling. *Development* **2002**; *129*: 4509–4521.

345 Lo, P., et al., Homeotic genes autonomously specify the anteroposterior subdivision of the Drosophila dorsal vessel into aorta and heart. *Dev Biol* **2002**; *251*: 307–319.

346 Searcy, R.D. and Yutzey, K.E., Analysis of Hox gene expression during early avian heart development. *Dev Dyn* **1998**; *213*: 82–91.

347 Ehrman, L. and Yutzey, K., Anterior expression of the caudal homologue cCdx-B activates a posterior genetic program in avian embryos. *Dev Dyn* **2001**; *221*: 412–421.

348 Dunwoodie, S.L., Rodriguez, T.A., and Beddington, R.S., Msg1 and Mrg1, founding members of a gene family, show distinct patterns of gene expression during mouse embryogenesis. *Mech Dev* **1998**; *72*: 27–40.

349 Mikawa, T., et al., Clonal analysis of cardiac morphogenesis in the chicken embryo using a replication-defective retrovirus: I. Formation of the ventricular myocardium. *Dev Dyn* **1992**; *193*: 11–23.

350 Meilhac, S.M., et al., A retrospective clonal analysis of the myocardium reveals two phases of clonal growth in the developing mouse heart. *Development* **2003**; *130*: 3877–3889.

351 Meilhac, S.M., et al., Oriented clonal cell growth in the developing mouse myocardium underlies cardiac morphogenesis. *J Cell Biol* **2004**; *164*: 97–109.

352 Andrée, B., et al., Isolation and characterization of the novel popeye gene family expressed in skeletal muscle and heart. *Dev Biol* **2000**; *223*: 371–382.

353 Leimeister, C., et al., Hey genes: a novel subfamily of hairy- and Enhancer of split related genes specifically expressed during mouse embryogenesis. *Mech Dev* **1999**; *85*: 173–177.

354 Chin, M.T., et al., Cardiovascular basic helix loop helix factor 1, a novel transcriptional repressor expressed preferentially in the developing and adult cardiovascular system. *J Biol Chem* **2000**; *275*: 6381–6387.

355 Nakagawa, O., et al., HRT1, HRT2, and HRT3: a new subclass of bHLH transcription factors marking specific cardiac, somitic, and pharyngeal arch segments. *Dev Biol* **1999**; *216*: 72–84.

356 Nakagawa, O., et al., Members of the HRT family of basic helix-loop-helix proteins act as transcriptional repressors downstream of Notch signaling. *Proc Natl Acad Sci USA* **2000**; *97*: 13655–13660.

357 Gessler, M., et al., Mouse gridlock. No aorticcoarctation or deficiency, but fatal cardiac defects in Hey2 −/− mice. *Curr Biol* **2002**; *12*: 1601–1604.

358 Donovan, J., et al., Tetralogy of Fallot and other congenital heart defects in Hey2 mutant mice. *Curr Biol* **2002**; *12*: 1605–1610.

359 Bruneau, B.G., et al., Cardiac expression of the ventricle-specific homeobox gene Irx4 is modulated by Nkx2–5 and dHand. *Dev Biol* **2000**; *217*: 266–277.

360 Bao, Z.Z., et al., Regulation of chamber-specific gene expression in the developing heart by Irx4. *Science* **1999**; *283*: 1161–1164.

361 Garriock, R.J., et al., Developmental expression of the Xenopus Iroquois-family homeobox genes, Irx4 and Irx5. *Dev Genes Evol* **2001**; *211*: 257–260.

362 Bruneau, B.G., et al., Cardiomyopathy in Irx4–deficient mice is preceded by abnormal ventricular gene expression. *Mol Cell Biol* **2001**; *21*: 1730–1736.

363 Christoffels, V.M., et al., Patterning the embryonic heart: identification of five mouse Iroquois homeobox genes in the developing heart. *Dev Biol* **2000**; *224*: 263–274.

364 Jiang, Y., et al., Common role for each of the cGATA4/5/6 genes in the regulation of cardiac morphogenesis. *Dev Gen* **1998**; *22*: 263–277.

365 Dersch, H. and Zile, M., Induction of normal cardiovascular development in the vitamin A-deprived quail embryo by natural retinoids. *Dev Biol* **1993**; *160*: 424–433.

366 Kostetskii, I., et al., Retinoid signaling required for normal heart development regulates GATA–4 in a pathway distinct from cardiomyocyte differentiation. *Dev Biol* **1999**; *206*: 206–218.

367 Ghatpande, S., et al., Anterior endoderm is sufficient to rescue foregut apoptosis and heart tube morphogenesis in an embryo lacking retinoic acid. *Dev Biol* **2000**; *219*: 59–70.

368 Xavier-Neto, J., et al., A retinoic acid-inducible transgenic marker of sino-atrial development in the mouse heart. *Development* **1999**; *126*: 2677–2687.

369 Dickman, E.D. and Smith, S.M., Selective regulation of cardiomyocyte gene expression and cardiac morphogenesis by retinoic acid. *Dev Dyn* **1996**; *206*: 39–48.

370 Niederreither, K., et al., Embryonic retinoic acid synthesis is essential for heart morphogenesis in the mouse. *Development* **2001**; *128*: 1019–1031.

371 Tetzlaff, M.T., et al., Defective cardiovascular development and elevated cyclin E and Notch proteins in mice lacking the Fbw7 F-box protein. *Proc Natl Acad Sci USA* **2004**; *101*: 3338–3345.

372 Viragh, S. and Challice, C.E., Origin and differentiation of cardiac muscle cells in the mouse. *J Ultrastruct Res* **1973**; *42*: 1–24.

373 de la Cruz, M.V., et al., Experimental study of the development of the truncus and the conus in the chick embryo. *J Anat* **1977**; *123*: 661–686.

374 Kelly, R.G., Brown, N.A., and Buckingham, M.E., The arterial pole of the mouse heart forms from Fgf10–expressing cells in pharyngeal mesoderm. *Dev Cell* **2001**; *1*: 435–440.

375 Brown, C., et al., Cre-mediated excision of Fgf8 in the Tbx1 expression domain reveals a critical role for Fgf8 in cardiovascular development in the mouse. *Dev Biol* **2004** (in press).

376 Brand, T., et al., Chicken NKx2–8, a novel homeobox gene expressed during early heart and foregut development. *Mech Dev* **1997**; *64*: 53–59.

377 Schneider, A., et al., Targeted disruption of the nkx3.1 gene in mice results in morphogenetic defects of minor salivary glands: parallels to glandular duct morphogenesis in prostate. *Mech Dev* **2000**; *95*: 163–174.

378 Mjaatvedt, C.H., et al., The Cspg2 gene, disrupted in the hdf mutant, is required for right cardiac chamber and endocardial cushion formation. *Dev Biol* **1998**; *202*: 56–66.

379 Srivastava, D., et al., Regulation of cardiac mesodermal and neural crest development by the bHLH transcription factor, dHAND. *Nat Genet* **1997**; *16*: 154–160.

380 Lin, Q., et al., Control of mouse cardiac morphogenesis and myogenesis by transcription factor MEF2C. *Science* **1997**; *276*: 1404–1407.

381 Schwartz, R.J. and Olson, E.N., Building the heart piece by piece: modularity of cis-elements regulating Nkx2–5 transcription. *Development* **1999**; *126*: 4187–4192.

382 Dodou, E., et al., Mef2c is a direct transcriptional target of ISL1 and GATA factors in the anterior heart field during mouse embryonic development. *Development* **2004**; *131*: 3931–3942.

383 Von Both, I., et al., Foxh1 is essential for development of the anterior heart field. *Dev Cell* **2004**; *7*: 331–345.

384 Garg, V., et al., Tbx1, a DiGeorge syndrome candidate gene, is regulated by sonic hedgehog during pharyngeal arch development. *Dev Biol* **2001**; *235*: 62–73.

385 Collignon, J., Varlet, I., and Robertson, E.J., Relationship between asymmetric nodal expression and the direction of embryonic turning. *Nature* **1996**; *381*: 155–158.

386 Liu, Z., et al., CHAMP, a novel cardiac-specific helicase regulated by MEF2C. *Dev Biol* **2001**; *234*: 497–509.

387 Watanabe, M., Jafri, A., and Fisher, S.A., Apoptosis is required for the proper formation of the ventriculo-arterial connections. *Dev Biol* **2001**; *240*: 274–288.

388 Rothenberg, F., et al., Initiation of apoptosis in the developing avian outflow tract myocardium. *Dev Dyn* **2002**; *223*: 469–482.

389 Mjaatvedt, C.H., et al., The Cspg2 gene, disrupted in the hdf mutant, is required for right cardiac chamber and endocardial cushion formation. *Dev Biol* **1998**; *202*: 56–66.

390 Liu, Z. and Olson, E., Suppression of proliferation and cardiomyocyte hypertrophy by CHAMP, a cardiac-specific RNA helicase. *Proc Natl Acad Sci USA* **2002**; *99*: 2043–2048.

391 Zhao, M., et al., Regulation of the MEF2 family of transcription factors by p38. *Mol Cell Biol* **1999**; *19*: 21–30.

392 Yang, J., et al., Mekk3 is essential for early embryonic cardiovascular development. *Nat Genet* **2000**; *24*: 309–313.

393 Mudgett, J., et al., Essential role for p38alpha mitogen-activated protein kinase in placental angiogenesis. *Proc Natl Acad Sci USA* **2000**; *97*: 10454–10459.

394. Kegan, C., et al., Erk5 null mice display multiple extraembryonic vascular and embryonic cardiovascular defects. *Proc Natl Acad Sci USA* **2002**; *99*: 9248–9253.
395. Yelon, D., et al., The bHLH transcription factor hand2 plays parallel roles in zebrafish heart and pectoral fin development. *Development* **2000**; *127*: 2573–2582.
396. Kölsch, V. and Paululat, A., The highly conserved cardiogenic bHLH factor Hand is specifically expressed in circular visceral muscle progenitor cells and in all cell types of the dorsal vessel during Drosophila embryogenesis. *Dev Genes Evol* **2002**; *212*: 473–485.
397. Habets, P., et al., Cooperative action of Tbx2 and Nkx2.5 inhibits ANF expression in the atrioventricular canal: implications for cardiac chamber formation. *Genes Dev* **2002**; *16*: 1234–1246.
398. Liberatore, C.M., Searcy-Schrick, R.D., and Yutzey, K.E., Ventricular expression of tbx5 inhibits normal heart chamber development. *Dev Biol* **2000**; *223*: 169–180.
399. Hatcher, C.J., et al., TBX5 transcription factor regulates cell proliferation during cardiogenesis. *Dev Biol* **2001**; *230*: 177–188.
400. Nebigil, C.G., et al., Serotonin 2B receptor is required for heart development. *Proc Natl Acad Sci USA* **2000**; *97*: 9508–9513.
401. Neuhaus, H., Rosen, V., and Thies, R.S., Heart specific expression of mouse BMP–10 a novel member of the TGF-beta superfamily. *Mech Dev* **1999**; *80*: 181–184.
402. Teichmann, U. and Kessel, M., Highly restricted BMP10 expression in the trabeculating myocardium of the chick embryo. *Dev Genes Evol* **2004**; *214*: 96–98.
403. Chen, H., et al., BMP10 is essential for maintaining cardiac growth during murine cardiogenesis. *Development* **2004**; *131*: 2219–2231.
404. Toyoda, M., et al., *jumonji* downregulates cardiac cell proliferation by repressing cyclin D1 expression. *Dev Cell* **2003**; *5*: 85–97.
405. Kochilas, L.K., et al., p57Kip2 expression is enhanced during mid-cardiac murine development and is restricted to trabecular myocardium. *Pediatric Res* **1999**; *45*: 635–642.
406. Moorman, A.F. and Christoffels, V.M., Cardiac chamber formation: development genes, and evolution. *Physiol Rev* **2003**; *83*: 1223–1267.
407. Ishiwata, T., et al., Developmental changes in ventricular diastolic function correlate with changes in ventricular myoarchitecture in normal mouse embryos. *Circ Res* **2003**; *93*: 857–865.
408. Moens, C.B., et al., Defects in heart and lung development in compound heterozygotes for two different targeted mutations at the N-myc locus. *Development* **1993**; *119*: 485–499.
409. Kastner, P., et al., Genetic analysis of RXR alpha developmental function: convergence of RXR and RAR signaling pathways in heart and eye morphogenesis. *Cell* **1994**; *78*: 987–1003.
410. Sucov, H.M., et al., RXR alpha mutant mice establish a genetic basis for vitamin A signaling in heart morphogenesis. *Genes Dev* **1994**; *8*: 1007–1018.
411. Kreidberg, J.A., et al., WT–1 is required for early kidney development. *Cell* **1993**; *74*: 679–691.

412 Chen, Z., Friedrich, G.A., and Soriano, P., Transcriptional enhancer factor 1 disruption by a retroviral gene trap leads to heart defects and embryonic lethality in mice. *Genes Dev* **1994**; *8*: 2293–2301.

413 Tevosian, S.G., et al., FOG–2, a cofactor for GATA transcription factors, is essential for heart morphogenesis and development of coronary vessels from epicardium. *Cell* **2000**; *101*: 729–739.

414 LeCouter, J.E., et al., Strain-dependent embryonic lethality in mice lacking the retinoblastoma-related p130 gene. *Development* **1998**; *125*: 4669–4679.

415 Yoshida, K., et al., Targeted disruption of gp130, a common signal transducer for the interleukin 6 family of cytokines, leads to myocardial and hematological disorders. *Proc Natl Acad Sci USA* **1996**; *93*: 407–411.

416 Jaber, M., et al., Essential role of beta-adrenergic receptor kinase 1 in cardiac development and function. *Proc Natl Acad Sci USA* **1996**; *93*: 12974–12979.

417 Wu, H., et al., Inactivation of erythropoietin leads to defects in cardiac morphogenesis. *Development* **1999**; *126*: 3597–3605.

418 Moore, A.W., et al., YAC complementation shows a requirement for Wt1 in the development of epicardium, adrenal gland and throughout nephrogenesis. *Development* **1999**; *126*: 1845–1857.

419 Chen, J., Kubalak, S.W., and Chien, K.R., Ventricular muscle-restricted targeting of the RXRalpha gene reveals a non-cell-autonomous requirement in cardiac chamber morphogenesis. *Development* **1998**; *125*: 1943–1949.

420 Tran, C.M. and Sucov, H.M., The RXRalpha gene functions in a non-cell-autonomous manner during mouse cardiac morphogenesis. *Development* **1998**; *125*: 1951–1956.

421 Tevosian, S., et al., FOG–2, a cofactor for GATA transcription factors, is essential for heart morphogenesis and development of coronary vessels from epicardium. *Cell* **2000**; *101*: 729–739.

422 Kwee, L., et al., Defective development of the embryonic and extraembryonic circulatory systems in vascular cell adhesion molecule (VCAM–1) deficient mice. *Development* **1995**; *121*: 489–503.

423 Sengbusch, J.K., et al., Dual functions of [alpha]4[beta]1 integrin in epicardial development: initial migration and long-term attachment. *J Cell Biol* **2002**; *157*: 873–882.

424 Männer, J., Experimental study on the formation of the epicardium in chick embryos. *Anat Embryol* **1993**; *187*: 281–289.

425 Moss, J.B., et al., Dynamic patterns of retinoic acid synthesis and response in the developing mammalian heart. *Dev Biol* **1998**; *199*: 55–71.

426 Xavier-Neto, J., et al., Sequential programs of retinoic acid synthesis in the myocardial and epicardial layers of the developing avian heart. *Dev Biol* **2000**; *219*: 129–141.

427 Nahirney, P.C., Mikawa, T., and Fischman, D.A., Evidence for an extracellular matrix bridge guiding proepicardial cell migration to the myocardium of chick embryos. *Dev Dyn* **2003**; *227*: 511–523.

428 Munoz-Chapuli, R., et al., The epicardium as a source of mesenchyme for the developing heart. *Ital J Anat Embryol* **2001**; *106*: 187–196.

429. Morabito, C.J., et al., Positive and negative regulation of epicardial-mesenchymal transformation during avian heart development. *Dev Biol* **2001**; *234*: 204–215.
430. Dettman, R.W., et al., Inhibition of alpha4–integrin stimulates epicardial-mesenchymal transformation and alters migration and cell fate of epicardially derived mesenchyme. *Dev Biol* **2003**; *257*: 315–328.
431. Reese, D.E., et al., bves: A novel gene expressed during coronary blood vessel development. *Dev Biol* **1999**; *209*: 159–171.
432. Wada, A., Reese, D., and Bader, D., Bves: prototype of a new class of cell adhesion molecules expressed during coronary artery development. *Development* **2001**; *128*: 2085–2093.
433. Carmona, R., et al., Immunolocalization of the transcription factor Slug in the developing avian heart. *Anat Embryol (Berl)* **2000**; *201*: 103–109.
434. Lie-Venema, H., et al., Ets–1 and Ets–2 transcription factors are essential for normal coronary and myocardial development in chicken embryos. *Circ Res* **2003**; *92*: 749–756.
435. Wessels, A. and Perez-Pomares, J.M., The epicardium and epicardially derived cells (EPDCs) as cardiac stem cells. *Anat Rec* **2004**; *276A*: 43–57.
436. Dettman, R.W., et al., Common epicardial origin of coronary vascular smooth muscle, perivascular fibroblasts, and intermyocardial fibroblasts in the avian heart. *Dev Biol* **1998**; *193*: 169–181.
437. Eid, H., et al., Role of epicardial mesothelial cells in the modification of phenotype and function of adult rat ventricular myocytes in primary coculture. *Circ Res* **1992**: *71*: 40–50.
438. Wada, A.M., et al., Epicardial/Mesothelial cell line retains vasculogenic potential of embryonic epicardium. *Circ Res* **2003**; *92*: 525–531.
439. Landerholm, T.E., et al., A role for serum response factor in coronary smooth muscle differentiation from proepicardial cells. *Development* **1999**; *126*: 2053–2062.
440. Lu, J., et al., Coronary smooth muscle differentiation from proepicardial cells requires rhoA-mediated actin reorganization and p160 rho-kinase activity. *Dev Biol* **2001**; *240*: 404–418.
441. Perez-Pomares, J.M., et al., Origin of coronary endothelial cells from epicardial mesothelium in avian embryos. *Int J Dev Biol* **2002**; *46*: 1005–1013.
442. Yamauchi, Y., et al., A novel transgenic technique that allows specific marking of the neural crest cell lineage in mice. *Dev Biol* **1999**; *212*: 191–203.
443. Scambler, P.J., The 22q11 deletion syndromes. *Hum Mol Genet* **2000**; *9*: 2421–2426.
444. Yamagishi, H. and Srivastava, D., Unraveling the genetic and developmental mysteries of 22q11 deletion syndrome. *Trends Mol Med* **2003**; *9*: 383–389.
445. Franz, T., Persistent truncus arteriosus in the Splotch mutant mouse. *Anat Embryol* **1989**; *180*: 457–464.
446. Goulding, M., et al., Analysis of the Pax–3 gene in the mouse mutant splotch. *Genomics* **1993**; *17*: 355–363.
447. Jerome, L.A. and Papaioannou, V.E., DiGeorge syndrome phenotype in mice mutant for the T-box gene, Tbx1. *Nat Genet* **2001**; *27*: 286–291.

448 Yamagishi, H., et al., Tbx1 is regulated by tissue-specific forkhead proteins through a common Sonic hedgehog-responsive enhancer. *Genes Dev* **2003**; *17*: 269–281.

449 Winnier, G.E., et al., Roles for the winged helix transcription factors MF1 and MFH1 in cardiovascular development revealed by nonallelic noncomplementation of null alleles. *Dev Biol* **1999**; *213*: 418–431.

450 Kume, T., et al., The murine winged helix transcription factors, Foxc1 and Foxc2, are both required for cardiovascular development and somitogenesis. *Genes Dev* **2001**; *15*: 2470–2482.

451 Iida, K., et al., Essential roles of the winged helix transcription factor MFH–1 in aortic arch patterning and skeletogenesis. *Development* **1997**; *124*: 4627–4638.

452 Vitelli, F., et al., A genetic link between Tbx1 and fibroblast growth factor signaling. *Development* **2002**; *129*: 4605–4611.

453 Abu-Issa, R., et al., Fgf8 is required for pharyngeal arch and cardiovascular development in the mouse. *Development* **2002**; *129*: 4613–4625.

454 Stalmans, I., et al., VEGF: a modifier of the del22q11 (DiGeorge) syndrome? *Nat Med* **2003**; *9*: 173–182.

455 Jiang, X., et al., Normal fate and altered function of the cardiac neural crest cell lineage in retinoic acid receptor mutant embryos. *Mech Dev* **2002**; *117*: 115–122.

456 Waldo, K., et al., A novel role for cardiac neural crest in heart development. *J Clin Invest* **1999**; *103*: 1499–1507.

457 Schafer, K., et al., The homeobox gene Lbx1 specifies a subpopulation of cardiac neural crest necessary for normal heart development. *Circ Res* **2003**; *92*: 73–80.

458 Farrell, M.J., et al., FGF–8 in the ventral pharynx alters development of myocardial calcium transients after neural crest ablation. *J Clin Invest* **2001**; *107*: 1509–1517.

459 Li, Y.X., et al., Cardiac neural crest in zebrafish embryos contributes to myocardial cell lineage and early heart function. *Dev Dyn* **2003**; *226*: 540–550.

460 Sato, M. and Yost, H.J., Cardiac neural crest contributes to cardiomyogenesis in zebrafish. *Dev Biol* **2003**; *257*: 127–139.

461 Runyan, R.B. and Markwald, R.R., Invasion of mesenchyme into three-dimensional collagen gels: a regional and temporal analysis of interaction in embryonic heart tissue. *Dev Biol* **1983**; *95*: 108–114.

462 Potts, J.D., et al., Epithelial-mesenchymal transformation of embryonic cardiac endothelial cells is inhibited by a modified antisense oligodeoxynucleotide to transforming growth factor beta 3. *Proc Natl Acad Sci USA* **1991**; *88*: 1516–1520.

463 Boyer, A.S., et al., TGFbeta2 and TGFbeta3 have separate and sequential activities during epithelial-mesenchymal cell transformation in the embryonic heart. *Dev Biol* **1999**; *208* 530–545.

464 Brown, C.B., et al., Antibodies to the type II TGFβ receptor block cell activation and migrastion during atrioventricular cushion transformation in the heart. *Dev Biol* **1996**; *174* 248–257.

465 Brown, C.B., et al., Requirement of type III TGF-beta receptor for endocardial cell transformation in the heart. *Science* **1999**; *283*: 2080–2082.

466 Yamagishi, T., et al., Bone morphogenetic protein–2 acts synergistically with transforming growth factor-beta3 during endothelial-mesenchymal transformation in the developing chick heart. *J Cell Physiol* **1999**; *180*: 35–45.

467 Kaartinen, V., et al., Abnormal lung development and cleft palate in mice lacking TGF-beta 3 indicates defects of epithelial-mesenchymal interaction. *Nat Genet* **1995**; *11*: 415–421.

468 Camenisch, T.D., et al., Temporal and distinct TGFbeta ligand requirements during mouse and avian endocardial cushion morphogenesis. *Dev Biol* **2002**; *248*: 170–181.

469 Delot, E.C., Control of endocardial cushion and cardiac valve maturation by BMP signaling pathways. *Mol Genet Metab* **2003**; *80*: 27–35.

470 Lyons, K., Pelton, R., and Hogan, B., Organogenesis and pattern formation in the mouse: RNA distribution patterns suggest a role for bone morphogenetic protein–2A (BMP2–A). *Development* **1990**; *109*: 833–844.

471 Jones, C.M., Lyons, K.M., and Hogan, B.L., Involvement of Bone Morphogenetic Protein–4 (BMP–4) and Vgr–1 in morphogenesis and neurogenesis in the mouse. *Development* **1991**; *111*: 531–542.

472 Yamagishi, T., et al., Expression of bone morphogenetic protein–5 gene during chick heart development: possible roles in valvuloseptal endocardial cushion formation. *Anat Rec* **2001**; *264*: 313–316.

473 Lyons, K.M., Hogan, B.L., and Robertson, E.J., Colocalization of BMP 7 and BMP 2 RNAs suggests that these factors cooperatively mediate tissue interactions during murine development. *Mech Dev* **1995**; *50*: 71–83.

474 Kim, R.Y., Robertson, E.J., and Solloway, M.J., Bmp6 and Bmp7 are required for cushion formation and septation in the developing mouse heart. *Dev Biol* **2001**; *235*: 449–466.

475 Jiao, K., et al., An essential role of Bmp4 in the atrioventricular septation of the mouse heart. *Genes Dev* **2003**; *17*: 2362–2367.

476 Delot, E.C., et al., BMP signaling is required for septation of the outflow tract of the mammalian heart. *Development* **2003**; *130*: 209–220.

477 Gaussin, V., et al., Endocardial cushion and myocardial defects after cardiac myocyte-specific conditional deletion of the bone morphogenetic protein receptor ALK3. *Proc Natl Acad Sci USA* **2002**; *99*: 2878–2893.

478 Allen, S.P., et al., Misexpression of noggin leads to septal defects in the outflow tract of the chick heart. *Dev Biol* **2001**; *235*: 98–109.

479 Camenisch, T.D., et al., Disruption of hyaluronan synthase–2 abrogates normal cardiac morphogenesis and hyaluronan-mediated transformation of epithelium to mesenchyme. *J Clin Invest* **2000**; *106*: 349–360.

480 Walsh, E.C. and Stainier, D.Y., UDP-glucose dehydrogenase required for cardiac valve formation in zebrafish. *Science* **2001**; *293*: 1670–1673.

481 Erickson, S.L., et al., ErbB3 is required for normal cerebellar and cardiac development: a comparison with ErbB2–and heregulin-deficient mice. *Development* **1997**; *124*: 4999–5011.

482 Camenisch, T.D., et al., Heart-valve mesenchyme formation is dependent on hyaluronan-augmented activation of ErbB2–ErbB3 receptors. *Nat Med* **2002**; *8*: 850–855.

483 Jackson, L.F., et al., Defective valvulogenesis in HB-EGF and TACE-null mice is associated with aberrant BMP signaling. *EMBO J* **2003**; *22*: 2704–2716.

484 Iwamoto, R., et al., Heparin-binding EGF-like growth factor and ErbB signaling is essential for heart function. *Proc Natl Acad Sci USA* **2003**; *100*: 3221–3226.

485 Shi, W., et al., TACE is required for fetal murine cardiac development and modeling. *Dev Biol* **2003**; *261*: 371–380.

486 Costell, M., et al., Hyperplastic conotruncal endocardial cushions and transposition of great arteries in perlecan-null mice. *Circ Res* **2002**; *91*: 158–164.

487 Lakkis, M.M. and Epstein, J.A., Neurofibromin modulation of ras activity is required for normal endocardial-mesenchymal transformation in the developing heart. *Development* **1998**; *125*: 4359–4367.

488 Ranger, A.M., et al., The transcription factor NF-ATc is essential for cardiac valve formation. *Nature* **1998**; *392*: 186–190.

489 de la Pompa, J.L., et al., Role of the NF-ATc transcription factor in morphogenesis of cardiac valves and septum. *Nature* **1998**; *392*: 182–186.

490 Johnson, E.N., et al., NFATc1 mediates vascular endothelial growth factor-induced proliferation of human pulmonary valve endothelial cells. *J Biol Chem* **2003**; *278*: 1686–1692.

491 Armesilla, A.L., et al., Vascular endothelial growth factor activates nuclear factor of activated T cells in human endothelial cells: a role for tissue factor gene expression. *Mol Cell Biol* **1999**; *19*: 2032–2043.

492 Dor, Y., et al., A novel role for VEGF in endocardial cushion formation and its potential contribution to congenital heart defects. *Development* **2001**; *128*: 1531–1538.

493 Dor, Y., et al., VEGF modulates early heart valve formation. *Anat Rec* **2003**; *271A*: 202–208.

494 Chang, C.P., et al., A field of myocardial-endocardial NFAT signaling underlies heart valve morphogenesis. *Cell* **2004**; *118*: 649–663.

495 Hurlstone, A.F., et al., The Wnt/beta-catenin pathway regulates cardiac valve formation. *Nature* **2003**; *425*: 633–637.

496 Timmerman, L.A., et al., Notch promotes epithelial-mesenchymal transition during cardiac development and oncogenic transformation. *Genes Dev* **2004**; *18*: 99–115.

497 Hogers, B., et al., Extraembryonic venous obstructions lead to cardiovascular malformations and can be embryolethal. *Cardiovasc Res* **1999**; *41*: 87–99.

498 Davies, P.F., et al., Spatial relationships in early signaling events of flow-mediated endothelial mechanotransduction. *Annu Rev Physiol* **1997**; *59*: 527–549.

499 Topper, J.N., et al., Vascular MADs: two novel MAD-related genes selectively inducible by flow in human vascular endothelium. *Proc Natl Acad Sci USA* **1997**; *94*: 9314–9319.

500 le Noble, F., et al., Flow regulates arterial-venous differentiation in the chick embryo yolk sac. *Development* **2004**; *131*: 361–375.

501 Hove, J.R., et al., Intracardiac fluid forces are an essential epigenetic factor for embryonic cardiogenesis. *Nature* **2003**; *421*: 172–177.

502 Urbich, C., et al., Fluid shear stress-induced transcriptional activation of the vascular endothelial growth factor receptor–2 gene requires Sp1–dependent DNA binding. *FEBS Lett,* **2003**; *535*: 87–93.

503 Reckova, M., et al., Hemodynamics is a key epigenetic factor in development of the cardiac conduction system. *Circ Res* **2003**; *93*: 77–85.

504 Hall, C.E., et al., Hemodynamic-dependent patterning of endothelin converting enzyme 1 expression and differentiation of impulse-conducting Purkinje fibers in the embryonic heart. *Development* **2004**; *131*: 581–592.

505 Berdougo, E., et al., Mutation of weak atrium/atrial myosin heavy chain disrupts atrial function and influences ventricular morphogenesis in zebrafish. *Development* **2003**; *130*: 6121–6129.

506 Huang, C., et al., Embryonic atrial function is essential for mouse embryogenesis cardiac morphogenesis and angiogenesis. *Development* **2003**; *130*: 6111–6119.

507 Chien, K.R. and Olson, E.N., Converging pathways and principles in heart development and disease: CV@CSH. *Cell* **2002**; *110*: 153–162.

508 Bar, H., et al., Upregulation of embryonic transcription factors in right ventricular hypertrophy. *Basic Res Cardiol* **2003**; *98*: 285–294.

509 Thattaliyath, B.D., et al., HAND1 and HAND2 are expressed in the adult-rodent heart and are modulated during cardiac hypertrophy. *Biochem Biophys Res Commun* **2002**; *297*: 870–875.

510 Thompson, J.T., Rackley, M.S., and O'Brien, T.X., Upregulation of the cardiac homeobox gene Nkx2–5 (CSX) in feline right ventricular pressure overload. *Am J Physiol* **1998**; *274*: H1569–H1573.

511 Natarajan, A., et al., Human eHAND, but not dHAND, is down-regulated in cardiomyopathies. *J Mol Cell Cardiol* **2001**; *33*: 1607–1614.

512 Pikkarainen, S., et al., GATA–4 is a nuclear mediator of mechanical stretch-activated hypertrophic program. *J Biol Chem* **2003**; *278*: 23807–23816.

513 Liang, Q., et al., The transcription factors GATA4 and GATA6 regulate cardiomyocyte hypertrophy in vitro and in vivo. *J Biol Chem* **2001**; *276*: 30245–30253.

514 Liang, Q. and Molkentin, J.D., Divergent signaling pathways converge on GATA4 to regulate cardiac hypertrophic gene expression. *J Mol Cell Cardiol,* **2002**; *34*: 611–616.

515 Suzuki, Y.J. and Evans, T., Regulation of cardiac myocyte apoptosis by the GATA–4 transcription factor. *Life Sci* **2004**; *74*: 1829–1838.

516 Oh, H., et al., Cardiac progenitor cells from adult myocardium: homing, differentiation, and fusion after infarction. *Proc Natl Acad Sci USA* **2003**; *100*: 12313–12318.

517 Beltrami, A.P., et al., Adult cardiac stem cells are multipotent and support myocardial regeneration. *Cell* **2003**; *114*: 763–776.

518 Matsuura, K., et al., Adult cardiac Sca–1–positive cells differentiate into beating cardiomyocytes. *J Biol Chem* **2004**; *279*: 11384–11391.

519 Martin, C.M., et al., Persistent expression of the ATP-binding cassette transporter, Abcg2, identifies cardiac SP cells in the developing and adult heart. *Dev Biol* **2004**; *265*: 262–275.

520 Flink, I.L., Cell cycle reentry of ventricular and atrial cardiomyocytes and cells within the epicardium following amputation of the ventricular apex in the axolotl, Amblystoma mexicanum: confocal microscopic immunofluorescent image analysis of bromodeoxyuridine-labeled nuclei. *Anat Embryol (Berl)* **2002**; *205*: 235–244.

521 Leferovich, J.M., et al., Heart regeneration in adult MRL mice. *Proc Natl Acad Sci USA* **2001**; *98*: 9830–9835.

522 Leferovich, J.M. and Heber-Katz, E., The scarless heart. *Semin Cell Dev Biol* **2002**; *13*: 327–333.

523 Dathe, V. and Brand-Saberi, B., Expression of thymosin beta4 during chick development. *Anat Embryol (Berl)* **2004**; *208*: 27–32.

524 Bock-Marquette, I., et al., Thymosin beta4 activates integrin-linked kinase and promotes cardiac cell migration, survival and cardiac repair. *Nature* **2004**; *432*: 466–472.

525 Schneider, M.D., Regenerative medicine: Prometheus unbound. *Nature* **2004**; *432*: 451–453.

526 Wu, X., et al., Small molecules that induce cardiomyogenesis in embryonic stem cells. *J Am Chem Soc* **2004**; *126*: 1590–1591.

527 Rudy-Reil, D. and Lough, J., Avian precardiac endoderm/mesoderm induces cardiac myocyte differentiation in murine embryonic stem cells. *Circ Res* **2004**; *94*, e107–116.

24
Vasculogenesis

Georg Breier

24.1
Introduction

All blood vessels share a common function which is to deliver oxygen and nutrients and to remove metabolites, yet they display a considerable degree of morphological and functional heterogeneity. Small blood vessels only consist of endothelial cells, whereas large vessels (arteries and veins) are surrounded by one or more layers of smooth muscle cells. The endothelium of blood vessels serves as an interface between blood and tissue and regulates transport of water-soluble substances and cells from blood into tissue and vice versa. For example, blood vessels in the kidney and in the choroid plexus are highly permeable to water-soluble substances whereas capillaries in the brain are extremely impermeable and form the blood-brain barrier. Blood vessels communicate with every organ, and are involved in communication over long distances in the body.

The cardiovascular system is the first functional organ to develop in the vertebrate embryo. Genetic analyses carried out over the last 10 years have yielded important insights into the molecular and cellular mechanisms underlying the formation of such vessels. Blood vessel formation is orchestrated by a plenitude of different proteins, including cell adhesion molecules, extracellular matrix, proteases, etc. This overview focuses on the role of endothelial receptor tyrosine kinases and their ligands which serve as key signaling molecules in blood vessel formation, and on transcriptional regulators which function as upstream regulators of these endothelial signaling systems.

24.2
Modes of Blood Vessel Morphogenesis: Vasculogenesis, Angiogenesis and Arteriogenesis

In the vertebrate embryo, the vascular system develops from the lateral mesoderm that exfoliates from the primitive streak during gastrulation. The first primitive

Cell Signaling and Growth Factors in Development. Edited by K. Unsicker and K. Krieglstein
Copyright © 2006 WILEY-VCH Verlag GmbH & Co. KGaA, Weinheim
ISBN 3-527-31034-7

blood vessels form through vasculogenesis, that is, through *de novo* differentiation of endothelial progenitor cells which aggregate and form a lumen [1]. The primitive vascular plexus consists only of endothelial cells and has a regular honeycomb shape, at least in the yolk sac and in the head. It expands by the sprouting or division (intussusception) of capillaries [2, 3]. While the term angiogenesis historically referred to the sprouting of capillaries from pre-existing venules, it is now used in embryology in a more general sense to describe the complex growth and remodeling processes that transform the primary vascular plexus into the hierarchically-structured vascular tree [4]. Blood vessels undergo arteriovenous differentiation in early stage vascular development. In the course of vessel maturation, blood vessels recruit perivascular cells (pericytes and smooth muscle cells) to the endothelium, which stabilize blood vessels. Smooth muscle cells are of heterogenous origin, depending on their location [4]. They may differentiate in the mesoderm, from bone marrow, or transdifferentiate from endothelium. Large arteries form through the proliferation and enlargement of arterioles (arteriogenesis). Once the endothelial tube becomes invested with smooth muscle cells, endothelial cells stop proliferating and become resistant to apoptosis. However, the regression of certain structures of the primitive vascular system (for example the aortic arches) is also part of the developmental program, and leads to the pruning of the vascular tree.

The primitive vascular system develops before the onset of the heart beat, by mechanisms independent of perfusion [5]. After the onset of circulation, the vascular system expands rapidly and undergoes extensive remodeling. Whereas the initial pre-circulatory phase is regulated by genetic mechanisms, later on, morphogenesis of the vascular system is greatly influenced by mechanical forces and the metabolic demand of growing tissue. For example, unperfused vessels regress during embryonic development [6]. Blood flow also influences arteriovenous differentiation: although endothelial cells already express markers for arteriovenous differentiation (such as EphrinB2 or Neuropilin–1) during early development, altered blood flow in the chick embryo yolk sac can induce the re-programming of the arteriovenous blood vessel identity [7].

24.3
Endothelial Cell Differentiation and Hematopoiesis

Endothelial cells and blood cells share a common progenitor, the hemangioblast. In the yolk sac, angioblasts and hematoblasts differentiate in close proximity to each other in cell clusters called blood islands [1]. As blood islands mature, cells in the periphery flatten and differentiate into endothelial cells whereas primitive erythroblasts remain in the center. Fusion of blood islands generates the primitive vascular plexus in the yolk sac. In the embryo proper, endothelial progenitors cells (angioblasts) differentiate in the absence of hematopoiesis. Angioblasts express molecular markers characteristic of mature endothelial cells but have not formed a lumen or been integrated into a vessel [5]. They are migratory cells of mesenchymal morphology, are highly invasive, and may migrate over long distances in the body, for example to colonize the head region where they form the perineural vascular plexus. Experiments using chick-quail chimeras have shown that most developing organs contain angioblasts which may differentiate *in situ* [8], with the exception of the brain whose vasculature is completely derived from outside blood vessels and is formed in a purely angiogenic process [2].

In the embryo proper, the first hematopoietic stem cells differentiate from endothelial cells in the aorta-gonad-mesonephros (AGM) region of the dorsal aorta [9], a process that requires SCL/Tal–1 [10]. Following the first wave of hematopoiesis, the major site of hematopoiesis shifts to the fetal liver, later on to the spleen and eventually becomes localized in the bone marrow.

24.4
Adult Arteriogenesis and Vasculogenesis

Following embryonic and postnatal development, blood vessel endothelial cells cease to proliferate and may remain quiescent for several years. In the adult organism, new blood vessel formation is tightly controlled and occurs only under certain physiological and pathological conditions, such as pregnancy, wound healing, diabetic retinopathy, psoriasis, rheumatoid arthritis, or solid tumor growth. Until recently, new blood vessel formation in the adult organism was thought to occur exclusively through angiogenesis. However, the growth of collaterals in ischemic tissue (heart or extremities) does not occur by angiogenesis, but as in the embryo, by arteriogenesis [11]. During arteriogenesis, small arterioles experience increased fluid shear stress. This activates the endothelium and leads to monocyte adhesion and infiltration with the subsequent production of growth factors and proteases, needed for vessel wall enlargement.

Recent studies have suggested that in the adult organism, there are circulating endothelial progenitor cells (EPC) which are derived from bone marrow [12]. These cells are CD34, VEGFR–2/Flk–1, or AC133 antigen-positive, and may home to sites of neovascularization where they differentiate into endothelial cells *in situ*. Circulating EPC have been reported to contribute to angiogenesis in ischemic, inflamed or malignant tissue. EPC recruitment is stimulated by hypoxia, VEGF, angiopoietins, and other cytokines [13, 14]. The therapeutic potential of EPC transplantation in animal models of hind limb and myocardial ischemia has been reported, and clinical application of cell therapy has been proposed [15, 16]. However, it is not clear to what extent EPC contribute to pathological angiogenesis or regeneration. Under certain conditions, EPC do not contribute to adult neovascularization, for example during lung regeneration [17].

24.5
Endothelial Cell Growth and Differentiation Factors

Receptor tyrosine kinases (RTK) play important roles in cellular growth and differentiation processes. Several RTK are expressed in the hemangioblastic lineage and function as signaling molecules in vasculogenesis and angiogenesis (Fig. 24.1). Gene inactivation experiments in mice have revealed that virtually no functional redundancy exists between the different members of endothelial RTK families. Moreover, there is a functional hierarchy among the various endothelial RTK signaling systems. Vascular endothelial growth factor (VEGF) and the high affinity VEGF receptors (VEGFR) comprise a key signaling system that regulates vasculogenesis [18, 19]. VEGFR–2 (also known as flk–1 in the mouse and KDR in humans), is the first endothelial RTK known to be expressed in the hemangioblastic lineage [20], and its function is required for the differentiation of definitive endothelial cells in the mouse embryo [21]. VEGFR–2 induction in avian epiblast cultures, but not in mouse embryonic stem cells, requires FGF-like factors [22, 23]. Initially, VEGFR–2 is expressed broadly in mesodermal cells, but its expression becomes restricted to the hemangioblast later on [20, 24]. Once endothelial cells have formed a primary capillary plexus, the Angiopoietin/Tie system and the Ephrin/Eph receptor system come into play and cooperate with VEGF in angiogenesis, vascular remodeling, and arteriovenous differentiation [25, 26].

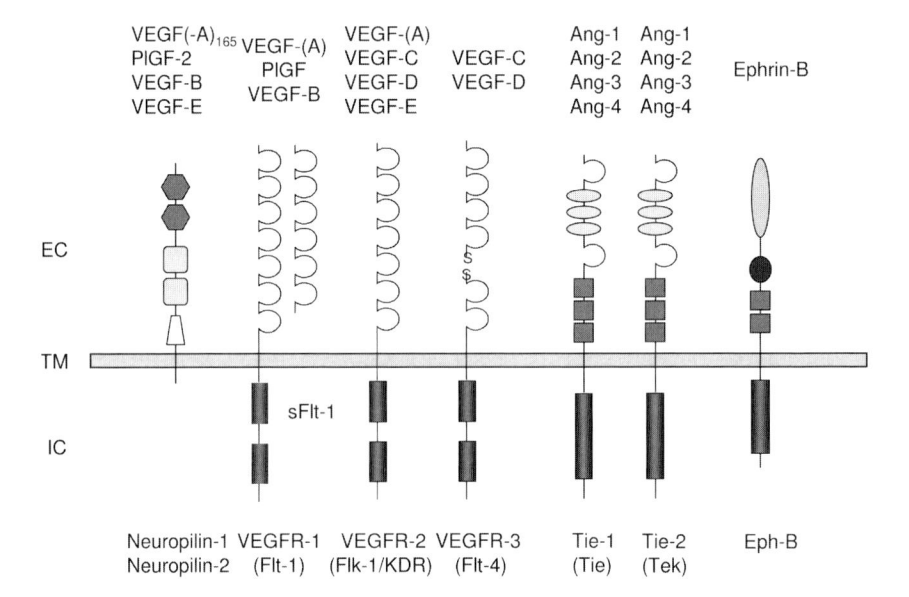

Fig. 24.1
Endothelial receptor tyrosine kinase signalling systems involved in vasculogenesis and angiogenesis. See text for a detailed explanation. EC, TM, IC refers to the extracellular, transmembrane, and intracellular domains, respectively.

24.6
VEGF and VEGF Receptors

VEGF was first identified as an endothelial cell-specific mitogen and permeability factor [19]. It is clear now that VEGF is a true multifunctional growth and differentiation factor that is involved in multiple aspects of vascular development and pathological angiogenesis. However, VEGF is also involved in hematopoiesis [19] and in certain aspects of neuronal development, including axonal growth and neural cell survival [27]. VEGF is synthesized in various splice variants that are secreted by the producer cells as dimers. Homodimeric VEGF binds to two high affinity receptor tyrosine kinases, VEGFR–1 and VEGFR–2 (Fig. 24.1). The VEGF164/165 splice variant also binds to neuropilin, a receptor for Semaphorin 3A, that has previously been implicated in axonal guidance/repulsion. The heparin-binding VEGF isoforms (primarily VEGF164/165) are required for the establishment and the maintenance of VEGF gradients in tissue [28]. In VEGFR–2 knockout mice, angioblasts differentiate but fail to assemble into vascular structures [21]. VEGF activity is balanced by the reciprocal activity of VEGFR–1 and VEGFR–2. While VEGFR–2 mediates most endothelial cell responses, such as proliferation, migration, survival, and permeability, VEGFR–1 can exert either inhibitory or stimulatory activity, depending on the context. In mouse embryos, VEGFR–1 acts as a negative regulator of vasculogenesis, which is achieved through the secretion of soluble VEGFR–1 (Fig. 24.1) by endothelial progenitor cells [29]. VEGFR–1 knockout mice show overgrowth of endothelial cells due to increased angioblast commitment, leading to abnormally enlarged blood vessels [30, 31]. The ligand binding, but not the signaling property of VEGFR–1 is required for vascular development, because the deletion of its intracellular domain does not affect embryogenesis [32]. However, VEGFR–1 signaling contributes to postnatal angiogenesis and arteriogenesis [29]. Neuropilin (Npn) is a co-receptor for VEGF in endothelial cells; its interaction with VEGFR–2 enhances VEGF signaling in endothelial cells [33]. Although Npn–1 has initially been implicated in axonal repulsion, Npn–1 knockout mice develop vascular malformation, in particular in the brain [34], consistent with its function in vascular development.

VEGF knockout mice display various defects, indicating that VEGF is involved in multiple aspects of vascular development: lumen formation, large vessel formation, yolk sac remodeling, and vascular sprouting. VEGF acts in a strictly dose-dependent manner: mice lacking a single VEGF allele die *in utero* [35, 36], and even subtle variations in VEGF levels cause abortive vascular development, and hematopoietic and endothelial differentiation [37, 38]. The remarkable ability of endothelial cells to sense VEGF dose enables them to sense growth factor gradients that guide the growth of capillary sprouts. A VEGF gradient guides the growth of new blood vessels in the developing brain [39] and in the retina [28]. In the retina, specialized tip cells are localized at the tip of vascular sprouts [40]. Tip cells express VEGFR–2 and PDGF-B and do not proliferate whereas stalk cells are PDGF-B negative and proliferate. The shape of the VEGF gradient in tissue controls the balance between tip cell migration and vessel proliferation. While the guidance of the axonal growth cone is accomplished by both attractive and repulsive signals, there is currently no firm evidence for repulsive signaling in blood vessel growth [41].

Based on the phenotype of KO embryos, it has been suggested that VEGF functions primarily during vasculogenesis. However, VEGF is also involved in the vascularization of organs. Inhibiting VEGF activity by the application of soluble VEGFR–1 impairs endochondral bone formation [42]. VEGF is also necessary for the invasion of blood vessels into the neural tube [39, 43] and for blood vessel formation in the pancreas [44]. Consistent with the expression of VEGFR–2 in the hemangioblast and in hematopoietic stem cells, VEGF is also required for primitive and definitive hematopoiesis [45].

Endothelial cell survival is important for embryonic vascular development. VEGF promotes the survival of newly formed blood vessels. This was first demonstrated in the retina [46]. VEGF survival function is dependent on the interaction between VEGFR–2, vascular endothelial (VE-) cadherin, and β-catenin, and is mediated by Akt signaling [47, 48]. VE-cadherin, a component of endothelial adherens junctions, is co-expressed with VEGFR–2 during embryonic vascular development [49], and it associates with VEGFR–2 upon stimulation of cells with VEGF. VE-Cadherin is one of the most specific markers for defining endothelial cells [50].

24.7
Other VEGF Family Members: Involvement in Angiogenesis and Lymphangiogenesis

VEGF is the prototype of a protein family with at least six members (Fig. 24.1): VEGF (also called VEGF-A), Placenta Growth Factor (PlGF), VEGF-B, VEGF-C, VEGF-D, and the Orf virus encoded VEGF-E. VEGF-A is unique with regard to its importance for early stage vascular development. PlGF, which binds to VEGFR–1 but not to VEGFR–2, is not required for embryonic vascular development but is involved in adult angiogenesis and arteriogenesis [29]. PlGF may form a heterodimer with VEGF, and potentiates its activity.

The function of VEGF-B in angiogenesis remains unclear. VEGF-C and VEGF-D are synthesized as pro-forms that activate VEGFR–3; the processed forms also bind to and activate VEGFR–2. VEGF-C and VEGF-D are involved primarily in lymphangiogenesis. The lymphatic vasculature takes up fluid from the interstitial space, thereby maintaining interstitial fluid balance and providing lymphatic clearance of macromolecules. Reduced function of lymphatic vessels and lymphedema can result from chronic inflammation, infection, or trauma [51]. There are also rare cases of primary congenital lymphedema which are associated with mutations in VEGFR–3, or the forkhead family transcription factor FOXC2 [52]. Until recently, the mechanisms of lymphangiogenesis were obscure, mainly due to the lack of suitable markers for lymphatic endothelium. Recently, several molecules that are expressed preferentially in lymphatic endothelium have been identified, including VEGFR–3 [51], the homeobox gene Prox–1 [53], the hyaluronan receptor LYVE–1 [54], and the mucin-type transmembrane glycoprotein Podoplanin [55]. Overexpression of Prox1 in blood vessel endothelial cells caused upregulation of lymphatic markers, indicating that it is a master regulator specifying lymphatic endothelial cell fate [56, 57].

Florence Sabin proposed in 1902 that the lymphatic vasculature is formed by the budding of lymph sacs at the anterior cardinal vein, and subsequent sprouting [58]. This view is supported by genetic experiments. Polarized expression of Prox1 in a subpopulation of endothelial cells in the anterior cardinal vein is necessary for budding of the lymph sac and sprouting of lymphatic vessels [53]. This process also requires the activation of VEGFR–3 by VEGF-C [59]: in vegf-c-deficient mice, endothelial cells become committed to the lymphatic lineage but do not sprout to form lymph vessels. Various other factors have been implicated in lymphatic development, including podoplanin. Podoplanin-deficient mice die at birth due to respiratory failure and have defects in lymphatic vessel pattern formation, associated with diminished lymphatic transport, congenital lymphedema and dilation of lymphatic vessels [55]. The lymphatic vasculature also serves as an important route for the dissemination of metastatic tumor cells. VEGF-C and VEGF-D are potent inducers of lymphangiogenesis in tumors and promote the metastatic spread of tumor cells via the lymphatics [60–62]. VEGFR–3 is also involved in early stage vascular development because mice lacking VEGFR–3 show defective remodeling of the primitive vascular plexus and die at mid-gestation [63]. The physiological ligand of VEGFR–3 during early stage embryonic angiogenesis is however, not known because VEGF-C knockout mice do not display a vascular phenotype.

24.7.1
Angiopoietins and Ties in Angiogenesis and Lymphangiogenesis

The angiopoietins represent another family of endothelial cell-selective angiogenic factors [26]. Ang–1 is pro-angiogenic and induces sprouting but not proliferation of endothelial cells *in vitro* [64]. Ang–1 binds to and activates the Tie2 receptor whereas Ang–2 activity is context dependent and may either stimulate or inhibit Tie–2. Ang–1 is associated with blood vessel stabilization and recruitment of perivascular cells. In contrast, Ang–2 destabilizes the vessels and this allows endothelial cells to sprout and proliferate in response to the angiogenic VEGF stimulus. Ang–1 protects the adult vasculature against plasma leakage [65]. Ang–1 KO mice develop multiple cardiovascular defects, in particular in angiogenic sprouting, vascular remodeling in the head region, and trabeculation of the heart. However, vasculogenesis is not affected by Ang–1 deficiency. In contrast, Ang–2 is not required for normal embryonic development [66], but is involved in adult angiogenesis, in particular in tumors [67, 68]. Angiopoietins also bind to the Tie–1 receptor [69]. In addition to their role in blood vessel development or growth, the angiopoietins also have a function in lymphangiogenesis. Analysis of knockout mice has shown that Ang–2 is also required for lymphatic patterning, and this role can be rescued by Ang–1 [66]. Ang–1 stimulated lymphatic endothelial cell proliferation, vessel enlargement and the generation of new sprouts [70]. Consistent with a function in lymphangiogenesis, Tie–1–deficient mice show severe edema [71].

24.7.2
Ephrins, Notch, and Arteriovenous Differentiation

Arteries and veins are morphologically and functionally different, but the mechanisms that establish this distinction are just beginning to emerge. Both genetic programs and circulatory dynamics play a major role in establishing this dichotomy [72]. It has become clear that arterial and venous endothelial cells are molecularly distinct even before the onset of the first embryonic heart beat, thus revealing the existence of genetic programs in coordinating arterial-venous differentiation. Several genes are expressed selectively in arteries or in veins, including receptor-ligand systems and transcriptional regulators. Ephrin-B2, a transmembrane ligand of the EphB RTK is specifically expressed in arteries but not veins and is essential for cardiovascular development [73, 74]. EphB4, a receptor for ephrin-B2, is exclusively expressed by vascular endothelial cells in embryos and is preferentially expressed in veins. Interaction between ephrinB2 in arteries and its EphB receptors in veins has been suggested to play a role in defining boundaries between arterial and venous domains [75]. However, recent evidence suggests that arterial-venous differentiation and patterning are controlled by hemodynamic forces, as shown by flow manipulation and *in situ* hybridization with arterial markers EphrinB2 and Npn–1 [7]. Ephrin-B2 may therefore just be a marker of arterial endothelial cells, rather than a determinant of arterial fate. Npn–1 and Npn–2 are also differentially expressed during arteriovenous differentiation. In 2–3–day-old chick embryos, Npn–1 expression is localized preferentially in arteries with an expression pattern that resembles that of ephrin-B2 whereas Npn–2 is expressed preferentially in veins [76].

Notch signaling appears to be involved in the specification of arteries and veins. By lineage tracking in zebrafish embryos it has been shown that angioblast precursors for the trunk artery and vein are spatially mixed in the lateral posterior mesoderm, however progeny of each angioblast are restricted to one of the vessels [77]. The arterial-venous decision for angioblasts is guided by gridlock (grl), an artery-restricted gene that is expressed in the lateral posterior mesoderm. Graded reduction of grl expression, by mutation or morpholino antisense progressively ablated regions of the artery, and expanded contiguous regions of the vein. This was preceded by an increase in expression of the venous marker EphB4 and reduction of expression of the arterial marker ephrin-B2. Grl is downstream of notch, and interference with notch signaling similarly reduced the artery and increased the vein regions. Thus, the notch-grl pathway seems to control assembly of the first embryonic artery [77]. The mouse homologs of gridlock, Hey1 and Hey2, also function in arterio-venous differentiation. Whereas Hey1 knockout mice show no overt vascular defects, and Hey2 knockout embryos show postnatal ventricular septum and valve defects; arterial differentiation is impaired in double knockout embryos [78, 79]. Examination of mouse Notch1 knockout embryos showed that they lack vascular Ephrin-B2, CD44 and neuropilin–1 expression, which confirms the role of notch signaling in arteriovenous differentiation [79].

24.8
Hypoxia-inducible Factors and Other Endothelial Transcriptional Regulators

Hypoxia is an important stimulus for angiogenesis in adult tissues. Hypoxia is also involved in developmental blood vessel formation, at least in the postnatal retina [46]. The cellular response to low oxygen levels is mediated by hypoxia-inducible factors (HIF), primarily by HIF–1 [80]. HIFs are heterodimeric basic helix-loop-helix transcription factors composed of an oxygen-sensitive α subunit and a constitutive β subunit (ARNT). HIF–1 and HIF–2 stimulate the expression of VEGF and of other essential angiogenesis factors such as Ang–2 following binding to short sequences (hypoxia response elements) located in promoter regions of these genes. ARNT knockout mice die of cardiovascular malformation and defective placentation [81]. HIF knockout mice also develop vascular defects, yet they die of mesenchymal cell death [82]. In contrast to HIF–1, HIF–2 is expressed preferentially in endothelial cells [83]. However, gene targeting experiments have largely failed to demonstrate a requirement for HIF–2 in vascular development, presumably because HIF–1 can compensate for the loss of HIF–2 activity. The role of endothelial HIFs was demonstrated by the endothelium-specific inhibition of HIF–1 and HIF–2. This caused impaired cardiovascular development, defective trabeculation, vascular remodeling and sprouting (Licht and Breier, unpublished data). The stability and activity of HIF-α is under the delicate control of HIF hydroxylases, a group of enzymes that require molecular oxygen, 2–oxoglutarate, and iron for their activity [80]. Hydroxylation of prolyl residues in the oxygen-dependent degradation domains of HIF-α, tags the subunit for proteasomal degradation. However, the functions of HIF hydroxylases in vascular development and angiogenesis have not as yet been elucidated.

In addition to HIFs, several other transcriptional regulators are expressed in the hemangioblastic lineage and represent candidate regulators of vascular development, for example, Ets–1 and other members of the Ets family, certain GATA-binding factors, SCL/Tal–1, HoxB5, and HIF–2 [84]. However, strictly endothelial cell-specific transcription factors have not been described. Ets–1 stimulates the expression of several endothelial RTK *in vitro* and *in vivo*, including Tie–2 and VEGFR–2 [85–87]. In contrast to targeted deficiency of endothelial RTK, inactivation of Ets–1 and of GATA binding factors affected hematopoiesis, but not vascular development. The most likely explanation for this observation is that functional redundancy exists within the transcription factor families. Yet, rescue of the embryonic lethal hematopoietic defect in SCL/Tal–1 knockout mice revealed an unsuspected role for SCL/Tal–1 in vascular development [88].

24.9
Outlook

Recent research has unraveled essential signaling systems involved in vasculogenesis, blood vessel angiogenesis, arteriovenous differentiation, and lymphatic angiogenesis. This knowledge also improved our understanding of diseases associated with aberrant angiogenesis and may open up new therapeutic avenues. However, we still know very little about the mechanisms that control organ-specific differentiation of endothelial cells, a process that is also of great relevance for the metastatic dissemination of tumor cells to specific organs. Similarly, we are just beginning to understand how the developing endothelium interacts with developing organs [89]. Recent evidence suggests that developing blood vessels are involved in complex signaling interactions needed for proper organ assembly from their constituent tissues. This is reflected for example by the recent evidence that certain signaling systems which have been implicated in the formation of the nervous system are also involved in the formation of blood vessels. Modern molecular cell biology and genetics will address these questions in the future.

References

1 Risau, W. and I. Flamme. Vasculogenesis. *Annu Rev Cell Dev Biol* **1995**; *11*: 73–91.
2 Risau, W. Mechanisms of angiogenesis. *Nature* **1997**; *386*: 671–674.
3 Burri, P.H., R. Hlushchuk, and V. Djonov. Intussusceptive angiogenesis: its emergence, its characteristics, and its significance. *Dev Dyn* **2004**; *231*: 474–488.
4 Carmeliet, P. Mechanisms of angiogenesis and arteriogenesis. *Nat Med* **2000**; *6*: 389–395.
5 Flamme, I. and G. Breier. The role of vascular endothelial growth factors and their receptors during embryonic vascular development. In *Assembly of the Vasculature and its Regulation* Tomanekk, R.J. (Ed.). Birkhäuser: Boston, **2002**.
6 Heine, U.I., A.B. Roberts, E.F. Munoz, N.S. Roche, and M.B. Sporn. Effects of retinoid deficiency on the development of the heart and vascular system of the quail embryo. *Virch Arch B Cell Pathol Incl Mol Pathol* **1985**; *50*: 135–152.
7 le Noble, F., D. Moyon, L. Pardanaud, L. Yuan, V. Djonov, R. Matthijsen, C. Breant, V. Fleury, and A. Eichmann. Flow regulates arterial-venous differentiation in the chick embryo yolk sac. *Development* **2004**; *131*: 361–375.
8 Pardanaud, L., F. Yassine, and F. Dieterlen-Lievre. Relationship between vasculogenesis, angiogenesis and haemopoiesis during avian ontogeny. *Development* **1989**; *105*: 473–485.
9 Medvinsky, A. and E. Dzierzak. Definitive hematopoiesis is autonomously initiated by the AGM region. *Cell* **1996**; *86*: 897–906.
10 Endoh, M., M. Ogawa, S. Orkin, and S. Nishikawa. SCL/tal–1–dependent process determines a competence to select the definitive hematopoietic lineage prior to endothelial differentiation. *EMBO J* **2002**; *21*: 6700–6708.
11 Schaper, W. and D. Scholz. Factors regulating arteriogenesis. *Arterioscler Thromb Vasc Biol* **2003**; *23*: 1143–1151.

12 Asahara, T., H. Masuda, T. Takahashi, C. Kalka, C. Pastore, M. Silver, M. Kearne, M. Magner, and J.M. Isner. Bone marrow origin of endothelial progenitor cells responsible for postnatal vasculogenesis in physiological and pathological neovascularization. *Circ Res* **1999**; *85*: 221–228.

13 Hattori, K., S. Dias, B. Heissig, N.R. Hackett, D. Lyden, M. Tateno, D.J. Hicklin, Z. Zhu, L. Witte, R.G. Crystal, M.A. Moore, and S. Rafii. Vascular endothelial growth factor and angiopoietin–1 stimulate postnatal hematopoiesis by recruitment of vasculogenic and hematopoietic stem cells. *J Exp Med* **2001**; *193*: 1005–1014.

14 Losordo, D.W. and S. Dimmeler. Therapeutic angiogenesis and vasculogenesis for ischemic disease. Part I: angiogenic cytokines. *Circulation* **2004**; *109*: 2487–2491.

15 Asahara, T. and A. Kawamoto. Endothelial progenitor cells for postnatal vasculogenesis. *Am J Physiol Cell Physiol* **2004**; *287*: C572–C579.

16 Losordo, D.W. and S. Dimmeler. Therapeutic angiogenesis and vasculogenesis for ischemic disease: part II: cell-based therapies. *Circulation* **2004**; *109*: 2692–2697.

17 Voswinckel, R., T. Ziegelhoeffer, M. Heil, S. Kostin, G. Breier, T. Mehling, R. Haberberger, M. Clauss, A. Gaumann, W. Schaper, and W. Seeger. Circulating vascular progenitor cells do not contribute to compensatory lung growth. *Circ Res* **2003**; *93*: 372–379.

18 Breier, G. and W. Risau. The role of vascular endothelial growth factor in blood vessel formation. *Trends Cell Biol* **1996**; *6*: 454–456.

19 Ferrara, N., H.P. Gerber, and J. LeCouter. The biology of VEGF and its receptors. *Nat Med* **2003**; *9*: 669–676.

20 Yamaguchi, T.P., D. Dumont, R.A. Conlon, M.L. Breitman, and J. Rossant. flk–1, an flt-related receptor tyrosine kinase is an early marker for endothelial cell precursors. *Development* **1993**; *118*: 489–498.

21 Shalaby, F., J. Rossant, T.P. Yamaguchi, M. Gertsenstein, X.F. Wu, M.L. Breitman, and A.C. Schuh. Failure of blood-island formation and vasculogenesis in Flk–1–deficient mice. *Nature* **1995**; *376*: 62–66.

22 Risau, W., H. Sariola, H.G. Zerwes, J. Sasse, P. Ekblom, R. Kemler, and T. Doetschman. Vasculogenesis and angiogenesis in embryonic-stem-cell-derived embryoid bodies. *Development* **1988**; *102*: 471–478.

23 Flamme, I. and W. Risau. Induction of vasculogenesis and hematopoiesis *in vitro*. *Development* **1992**; *116*: 435–439.

24 Breier, G., M. Clauss, and W. Risau. Coordinate expression of vascular endothelial growth factor receptor–1 (flt–1) and its ligand suggests a paracrine regulation of murine vascular development. *Dev Dyn* **1995**; *204*: 228–239.

25 Adams, R.H. and R. Klein. Eph receptors and ephrin ligands: essential mediators of vascular development. *Trends Cardiovasc Med* **2000**; *10*: 183–188.

26 Yancopoulos, G.D., S. Davis, N.W. Gale, J.S. Rudge, S.J. Wiegand, and J. Holash. Vascular-specific growth factors and blood vessel formation. *Nature* **2000**; *407*: 242–248.

27 Carmeliet P. Blood vessels and nerves: common signals, pathways and diseases. *Nat Rev Genet* **2003**; 710–720.

28 Ruhrberg, C., H. Gerhardt, M. Golding, R. Watson, S. Ioannidou, H. Fujisawa, C. Betsholtz, and D.T. Shima. Spatially restricted patterning cues provided by heparin-binding VEGF-A control blood vessel branching morphogenesis. *Genes Dev* **2002**; *16*: 2684–2698.

29 Carmeliet, P., L. Moons, A. Luttun, V. Vincenti, V. Compernolle, M. De Mol, Y. Wu, F. Bono, L. Devy, H. Beck, D. Scholz, T. Acker, T. DiPalma, M. Dewerchin, A. Noel, I. Stalmans, A. Barra, S. Blacher, T. Vandendriessche, A. Ponten, U. Eriksson, K.H. Plate, J.M. Foidart, W. Schaper, D.S. Charnock-Jones, D.J. Hicklin, J.M. Herbert, D. Collen, and M.G. Persico. Synergism between vascular endothelial growth factor and placental growth factor contributes to angiogenesis and plasma extravasation in pathological conditions. *Nat Med* **2001**; *7*: 575–583.

30 Fong, G.H., J. Rossant, M. Gertsenstein, and M.L. Breitman. Role of the Flt–1 receptor tyrosine kinase in regulating the assembly of vascular endothelium. *Nature* **1995**; *376*: 66–70.

31 Fong, G.H., L. Zhang, D.M. Bryce, and J. Peng. Increased hemangioblast commitment, not vascular disorganization, is the primary defect in flt–1 knock-out mice. *Development* **1999**; *126*: 3015–3025.

32 Hiratsuka, S., O. Minowa, J. Kuno, T. Noda, and M. Shibuya. Flt–1 lacking the tyrosine kinase domain is sufficient for normal development and angiogenesis in mice. *Proc Natl Acad Sci USA* **1998**; *95*: 9349–9354.

33 Neufeld, G., T. Cohen, N. Shraga, T. Lange, O. Kessler, and Y. Herzog. The neuropilins: multifunctional semaphorin and VEGF receptors that modulate axon guidance and angiogenesis. *Trends Cardiovasc Med* **2002**; *12*: 13–19.

34 Kawasaki, T., T. Kitsukawa, Y. Bekku, Y. Matsuda, M. Sanbo, T. Yagi, and H. Fujisawa. A requirement for neuropilin–1 in embryonic vessel formation. *Development* **1999**; *126*: 4895–4902.

35 Carmeliet, P., V. Ferreira, G. Breier, S. Pollefeyt, L. Kieckens, M. Gertsenstein, M. Fahrig, A. Vandenhoeck, K. Harpal, C. Eberhardt, C. Declercq, J. Pawling, L. Moons, D. Collen, W. Risau, and A. Nagy. Abnormal blood vessel development and lethality in embryos lacking a single VEGF allele. *Nature* **1996**; *380*: 435–439.

36 Ferrara, N., K. Carver-Moore, H. Chen, M. Dowd, L. Lu, K.S. O'Shea, L. Powell-Braxton, K.J. Hillan, and M.W. Moore. Heterozygous embryonic lethality induced by targeted inactivation of the VEGF gene. *Nature* **1996**; *380*: 439–442.

37 Miquerol, L., B.L. Langille, and A. Nagy. Embryonic development is disrupted by modest increases in vascular endothelial growth factor gene expression. *Development* **2000**; *127*: 3941–3946.

38 Damert, A., L. Miquerol, M. Gertsenstein, W. Risau, and A. Nagy. Insufficient VEGFA activity in yolk sac endoderm compromises haematopoietic and endothelial differentiation. *Development* **2002**; *129*: 1881–1892.

39 Raab, S., H. Beck, A. Gaumann, A. Yuce, H.P. Gerber, K. Plate, H.P. Hammes, N. Ferrara, and G. Breier. Impaired brain angiogenesis and neuronal apoptosis induced by conditional homozygous inactivation of vascular endothelial growth factor. *Thromb Haemost* **2004**; *91*: 595–605.

40 Gerhardt, H., M. Golding, M. Fruttiger, C. Ruhrberg, A. Lundkvist, A. Abramsson, M. Jeltsch, C. Mitchell, K. Alitalo, D. Shima, and C. Betsholtz. VEGF guides angiogenic sprouting utilizing endothelial tip cell filopodia. *J Cell Biol* **2003**; *161*: 1163–1177.

41 Gerhardt, H. and C. Betsholtz. How do endothelial cells orientate? *Exs* **2005**; *94*: 3–15.

42 Gerber, H.P., T.H. Vu, A.M. Ryan, J. Kowalski, Z. Werb, and N. Ferrara. VEGF couples hypertrophic cartilage remodeling, ossification and angiogenesis during endochondral bone formation. *Nat Med* **1999**; *5*: 623–628.

43 Haigh, J.J., P.I. Morelli, H. Gerhardt, K. Haigh, J. Tsien, A. Damert, L. Miquerol, U. Muhlner, R. Klein, N. Ferrara, E.F. Wagner, C. Betsholtz, and A. Nagy. Cortical and retinal defects caused by dosage-dependent reductions in VEGF-A paracrine signaling. *Dev Biol* **2003**; *262*: 225–241.

44 Lammert, E., G. Gu, M. McLaughlin, D. Brown, R. Brekken, L.C. Murtaugh, H.P. Gerber, N. Ferrara, and D.A. Melton. Role of VEGF-A in vascularization of pancreatic islets. *Curr Biol* **2003**; *13*: 1070–1074.

45 Gerber, H.P., A.K. Malik, G.P. Solar, D. Sherman, X.H. Liang, G. Meng, K. Hong, J.C. Marsters, and N. Ferrara. VEGF regulates haematopoietic stem cell survival by an internal autocrine loop mechanism. *Nature* **2002**; *417*: 954–958.

46 Alon, T., I. Hemo, A. Itin, J. Pe'er, J. Stone, and E. Keshet. Vascular endothelial growth factor acts as a survival factor for newly formed retinal vessels and has implications for retinopathy of prematurity. *Nat Med* **1995**; *1*: 1024–1028.

47 Gerber, H.P., A. McMurtrey, J. Kowalski, M. Yan, B.A. Keyt, V. Dixit, and N. Ferrara. Vascular endothelial growth factor regulates endothelial cell survival through the phosphatidylinositol 3'-kinase/Akt signal transduction pathway. Requirement for Flk-1/KDR activation. *J Biol Chem* **1998**; *273*: 30336–30343.

48 Carmeliet, P., M.G. Lampugnani, L. Moons, F. Breviario, V. Compernolle, F. Bono, G. Balconi, R. Spagnuolo, B. Oostuyse, M. Dewerchin, A. Zanetti, A. Angellilo, V. Mattot, D. Nuyens, E. Lutgens, F. Clotman, M.C. de Ruiter, A. Gittenberger-de Groot, R. Poelmann, F. Lupu, J.M. Herbert, D. Collen, and E. Dejana. Targeted deficiency or cytosolic truncation of the VE-cadherin gene in mice impairs VEGF-mediated endothelial survival and angiogenesis. *Cell* **1999**; *98*. 147–157.

49 Breier, G., F. Breviario, L. Caveda, R. Berthier, H. Schnurch, U. Gotsch, D. Vestweber, W. Risau, and E. Dejana. Molecular cloning and expression of murine vascular endothelial- cadherin in early stage development of cardiovascular system. *Blood* **1996**; *87*: 630–641.

50 Hirai, H., M. Ogawa, N. Suzuki, M. Yamamoto, G. Breier, O. Mazda, J. Imanishi, and S. Nishikawa. Hemogenic and non-

hemogenic endothelium can be distinguished by the activity of fetal liver kinase (Flk)-1 promotor/enhancer during mouse embryogenesis. *Blood* **2003**; *101*: 886–893.

51 Alitalo, K. and P. Carmeliet. Molecular mechanisms of lymphangiogenesis in health and disease. *Cancer Cell* **2002**; *1*: 219–227.

52 Saaristo, A., M.J. Karkkainen, and K. Alitalo. Insights into the molecular pathogenesis and targeted treatment of lymphedema. *Ann NY Acad Sci* **2002**; *979*: 94–110.

53 Wigle, J.T. and G. Oliver. Prox1 function is required for the development of the murine lymphatic system. *Cell* **1999**; *98*: 769–778.

54 Banerji, S., J. Ni, S.X. Wang, S. Clasper, J. Su, R. Tammi, M. Jones, and D.G. Jackson. LYVE–1, a new homologue of the CD44 glycoprotein, is a lymph-specific receptor for hyaluronan. *J Cell Biol* **1999**; *144*: 789–801.

55 Schacht, V., M.I. Ramirez, Y.K. Hong, S. Hirakawa, D. Feng, N. Harvey, M. Williams, A.M. Dvorak, H.F. Dvorak, G. Oliver, and M. Detmar. T1alpha/podoplanin deficiency disrupts normal lymphatic vasculature formation and causes lymphedema. *EMBO J* **2003**; *22*: 3546–3556.

56 Hong, Y.K., N. Harvey, Y.H. Noh, V. Schacht, S. Hirakawa, M. Detmar, and G. Oliver. Prox1 is a master control gene in the program specifying lymphatic endothelial cell fate. *Dev Dyn* **2002**; *225*: 351–357.

57 Petrova, T.V., T. Karpanen, C. Norrmen, R. Mellor, T. Tamakoshi, D. Finegold, R. Ferrell, D. Kerjaschki, P. Mortimer, S. Yla-Herttuala, N. Miura, and K. Alitalo. Defective valves and abnormal mural cell recruitment underlie lymphatic vascular failure in lymphedema distichiasis. *Nat Med* **2004**; *10*: 974–981.

58 Sabin, F.R. On the origin of the lymphatic system from the veins and the development of the lymph hearts and thoracic duct in the pig. *Am J Anat* **1902**; *1*: 367–391.

59 Karkkainen, M.J., P. Haiko, K. Sainio, J. Partanen, J. Taipale, T.V. Petrova, M. Jeltsch, D.G. Jackson, M. Talikka, H. Rauvala, C. Betsholtz, and K. Alitalo. Vascular endothelial growth factor C is required for sprouting of the first lymphatic vessels from embryonic veins. *Nat Immunol* **2004**; *5*: 74–80.

60 Karpanen, T., M. Egeblad, M.J. Karkkainen, H. Kubo, S. Yla-Herttuala, M. Jaattela, and K. Alitalo. Vascular endothelial growth factor C promotes tumor lymphangiogenesis and intralymphatic tumor growth. *Cancer Res* **2001**; *61*: 1786–1790.

61 Skobe, M., T. Hawighorst, D.G. Jackson, R. Prevo, L. Janes, P. Velasco, L. Riccardi, K. Alitalo, K. Claffey, and M. Detmar. Induction of tumor lymphangiogenesis by VEGF-C promotes breast cancer metastasis. *Nat Med* **2001**; *7*: 192–198.

62 Stacker, S.A., C. Caesar, M.E. Baldwin, G.E. Thornton, R.A. Williams, R. Prevo, D.G. Jackson, S. Nishikawa, H. Kubo, and M.G. Achen. VEGF-D promotes the metastatic spread of tumor cells via the lymphatics. *Nat Med* **2001**; *7*: 186–191.

63 Dumont, D.J., L. Jussila, J. Taipale, A. Lymboussaki, T. Mustonen, K. Pajusola, M. Breitman, and K. Alitalo. Cardiovascular failure in mouse embryos deficient in VEGF receptor–3. *Science* **1998**; *282*: 946–949.

64 Koblizek, T.I., C. Weiss, G.D. Yancopoulos, U. Deutsch, and W. Risau. Angiopoietin–1 induces sprouting angiogenesis in vitro. *Curr Biol* **1998**; *8*: 529–532.

65 Thurston, G., J.S. Rudge, E. Ioffe, H. Zhou, L. Ross, S.D. Croll, N. Glazer, J. Holash, D.M. McDonald, and G.D. Yancopoulos. Angiopoietin–1 protects the adult vasculature against plasma leakage. *Nat Med* **2000**; *6*: 460–463.

66 Gale, N.W., G. Thurston, S.F. Hackett, R. Renard, Q. Wang, J. McClain, C. Martin, C. Witte, M.H. Witte, D. Jackson, C. Suri, P.A. Campochiaro, S.J. Wiegand, and G.D. Yancopoulos. Angiopoietin–2 is required for postnatal angiogenesis and lymphatic patterning, and only the latter role is rescued by Angiopoietin–1. *Dev Cell* **2002**; *3*: 411–423.

67 Holash, J., P.C. Maisonpierre, D. Compton, P. Boland, C.R. Alexander, D. Zagzag, G.D. Yancopoulos, and S.J. Wiegand. Vessel cooption, regression, and growth in tumors mediated by angiopoietins and VEGF. *Science* **1999**; *284*: 1994–1998.

68 Vajkoczy, P., M. Farhadi, A. Gaumann, R. Heidenreich, R. Erber, A. Wunder, J.C. Tonn, M.D. Menger, and G. Breier. Microtumor growth initiates angiogenic sprouting with simultaneous expression of VEGF, VEGF receptor–2, and angiopoietin–2. *J Clin Invest* **2002**; *109*: 777–785.

69 Saharinen, P., K. Kerkela, N. Ekman, M. Marron, N. Brindle, G.M. Lee, H. Augustin, G.Y. Koh, and K. Alitalo. Multiple angiopoietin recombinant proteins activate the Tie1 receptor tyrosine kinase and promote its interaction with Tie2. *J Cell Biol* **2005**; *169*: 239–243.

70 Tammela, T., A. Saaristo, M. Lohela, T. Morisada, J. Tornberg, C. Norrmen, Y. Oike, K. Pajusola, G. Thurston, T. Suda, S. Yla-Herttuala, and K. Alitalo. Angiopoietin–1 promotes lymphatic sprouting and hyperplasia. *Blood* **2005**; *105*: 4642–4648.

71 Sato, T.N., Y. Tozawa, U. Deutsch, K. Wolburg-Buchholz, Y. Fujiwara, M. Gendron-Maguire, T. Gridley, H. Wolburg, W. Risau, and Y. Qin. Distinct roles of the receptor tyrosine kinases Tie–1 and Tie–2 in blood vessel formation. *Nature* **1995**; *376*: 70–74.

72 Torres-Vazquez, J., M. Kamei, and B.M. Weinstein. Molecular distinction between arteries and veins. *Cell Tissue Res* **2003**; *314*: 43–59.

73 Gerety, S.S., H.U. Wang, Z.F. Chen, and D.J. Anderson. Symmetrical mutant phenotypes of the receptor EphB4 and its specific transmembrane ligand ephrin-B2 in cardiovascular development. *Mol Cell* **1999**; *4*: 403–414.

74 Gerety, S.S. and D.J. Anderson. Cardiovascular ephrinB2 function is essential for embryonic angiogenesis. *Development* **2002**; *129*: 1397–1410.

75 Adams, R.H., G.A. Wilkinson, C. Weiss, F. Diella, N.W. Gale, U. Deutsch, W. Risau, and R. Klein. Roles of ephrinB ligands and EphB receptors in cardiovascular development: demarcation of arterial/venous domains, vascular morphogenesis, and sprouting angiogenesis. *Genes Dev* **1999**; *13*: 295–306.

76 Herzog, Y., C. Kalcheim, N. Kahane, R. Reshef, and G. Neufeld. Differential expression of neuropilin–1 and neuropilin–2 in arteries and veins. *Mech Dev* **2001**; *109*: 115–119.

77 Zhong, T.P., S. Childs, J.P. Leu, and M.C. Fishman. Gridlock signalling pathway fashions the first embryonic artery. *Nature* **2001**; *414*: 216–220.

78 Gessler, M., K.P. Knobeloch, A. Helisch, K. Amann, N. Schumacher, E. Rohde, A. Fischer, and C. Leimeister. Mouse gridlock: no aortic coarctation or deficiency, but fatal cardiac defects in Hey2 –/– mice. *Curr Biol* **2002**; *12*: 1601–1604.

79 Fischer, A., N. Schumacher, M. Maier, M. Sendtner, and M. Gessler. The Notch target genes Hey1 and Hey2 are required for embryonic vascular development. *Genes Dev* **2004**; *18*: 901–911.

80 Pugh, C.W. and P.J. Ratcliffe. Regulation of angiogenesis by hypoxia: role of the HIF system. *Nat Med* **2003**; *9*: 677–684.

81 Maltepe, E., J.V. Schmidt, D. Baunoch, C.A. Bradfield, and M.C. Simon. Abnormal angiogenesis and responses to glucose and oxygen deprivation in mice lacking the protein ARNT. *Nature* **1997**; *386*: 403–407.

82 Kotch, L.E., N.V. Iyer, E. Laughner, and G.L. Semenza. Defective vascularization of HIF–1alpha-null embryos is not associated with VEGF deficiency but with mesenchymal cell death. *Dev Biol* **1999**; *209*: 254–267.

83 Flamme, I., T. Frohlich, M. von Reutern, A. Kappel, A. Damert, and W. Risau. HRF, a putative basic helix-loop-helix-PAS-domain transcription factor is closely related to hypoxia-inducible factor–1 alpha and developmentally expressed in blood vessels. *Mech Dev* **1997**; *63*: 51–60.

84 Oettgen, P. Transcriptional regulation of vascular development. *Circ Res* **2001**; *89*: 380–388.

85 Kappel, A., V. Ronicke, A. Damert, I. Flamme, W. Risau, and G. Breier. Identification of vascular endothelial growth factor (VEGF) receptor–2 (Flk–1) promoter/enhancer sequences sufficient for angioblast and endothelial cell-specific transcription in transgenic mice. *Blood* **1999**; *93*: 4284–4292.

86 Kappel, A., T.M. Schlaeger, I. Flamme, S.H. Orkin, W. Risau, and G. Breier. Role of SCL/Tal–1, GATA, and ets transcription factor binding sites for the regulation of flk–1 expression during murine vascular development. *Blood* **2000**; *96*: 3078–3085.

87 Elvert, G., A. Kappel, R. Heidenreich, U. Englmeier, S. Lanz, T. Acker, M. Rauter, K. Plate, M. Sieweke, G. Breier, and I. Flamme. Cooperative interaction of hypoxia inducible factor (HIF)–2a and Ets–1 in the transcriptional activation of vascular endothelial growth factor receptor–2 (Flk–1). *J Biol Chem* **2003**; *278*: 7520–7530.

88 Visvader, J.E., Y. Fujiwara, and S.H. Orkin. Unsuspected role for the T-cell leukemia protein SCL/tal–1 in vascular development. *Genes Dev* **1998**; *12*: 473–479.

89 Lammert, E., O. Cleaver, and D. Melton. Induction of pancreatic differentiation by signals from blood vessels. *Science* **2001**; *294*: 564–567.

25
Inductive Signaling in Kidney Morphogenesis

Hannu Sariola and Kirsi Sainio

25.1
Early Differentiation of the Kidney

The permanent kidney or metanephros is derived from the intermediate mesoderm. Metanephric differentiation is regulated by reciprocal inductive signals between the ureteric bud, nephrogenic mesenchyme, and peritubular stroma. When the nephric or Wolffian duct fuses with the cloaca, a ureteric bud is formed from the duct and invades the metanephric mesenchyme (MM), a unique group of cells predetermined to form the nephron. The ureteric bud then induces metanephric mesenchymal cells to condense, undergo epithelial transformation, and subsequently segregate to the different segments of the secretory nephrons [1–3].

The induced metanephric mesenchymal cells signal reciprocally to the ureteric bud, as a result of which it undergoes dichotomous branching to establish the collecting duct network. At each tip of the branching ureteric tree further groups of mesenchymal cells are induced to become secretory nephrons. Mesenchymal cells that are not induced to epithelial differentiation become peritubular stroma. The stroma contributes to the control of ureteric bud branching by converting vitamin A to its active form, retinoic acid, which upregulates the receptors in bud cells that are critical for branching (e.g. Ret receptor tyrosine kinase). Stromal cells express transcription factor Foxd1 [4] and lack of it results in severe kidney hypoplasia and differentiation defects, which further underlines the role of stromal cells in kidney differentiation.

Induction is defined as a process, in which one embryonic tissue signals a second nearby tissue to cause the latter to differentiate in a direction it would otherwise not have taken [5]. Tissue recombination studies have shown that the nephrogenic mesenchymal cells are committed to nephrogenesis prior to their induction, and they are the only group of cells in the embryo that can be induced to kidney tubulogenesis [6].

Morphological and genetic evidence has shown that several consecutive steps constitute the differentiation of MM. First, a layer of cells in the MM surrounding the invading ureteric bud form a cap condensate and simultaneously escape the apoptotic default pathway. This early mesenchymal condensation process should not be confused with the pretubular condensation or aggregation, which follows cap condensation. Only a small fraction of cells in the cap condensate are induced to form pretubular aggregates that then epithelialize to become secretory nephrons [5]. Confocal microscopy has been used to trace the minimum cell number in a pretubular aggregate thus giving the number of cells from which a single secretory nephron originates; it is apparently far less than 50 in mouse [7]. The cells of pretubular aggregates proliferate rapidly, undergo epithelial transition, form first comma-shaped, then S-shaped bodies, and finally segregate to different segments of the secretory nephron: glomerulus, proximal and distal tubule (Fig. 25.1).

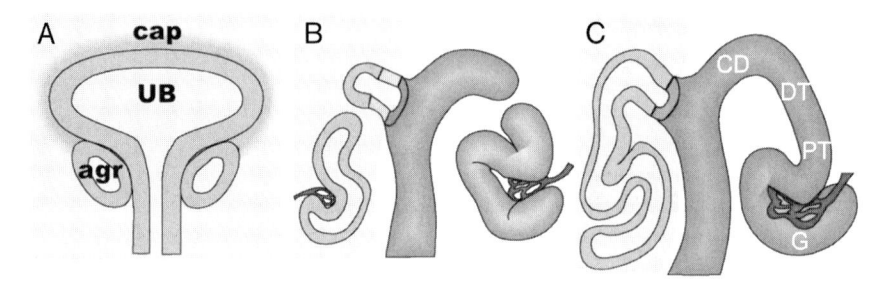

Fig. 25.1
Formation of secretory nephrons. (A) Ureteric bud (UB) induces a cap condensate (cap) from the nephrogenic mesenchyme. Subsequently, two subsets of cap cells become pretubular aggregates (agr). (B) The aggregates undergo epithelial transformation, elongate and form an S-shaped body. (C) The S-shaped body segregates into different tubule segments: glomerulus (G), proximal (PT) and distal tubule (DT). Finally, the S-shaped body fuses with the tip of a collecting duct (CD), which establishes the continuous urinary space from the glomerulus to the renal pelvis (according to [120]).

In mouse, each group of metanephric mesenchymal cells converts to nephron epithelium in 48 h. From the beginning of the organogenetic period (around embryonic day 11 (E11)), nephron induction and differentiation continues for a considerable time during embryogenesis and only ceases when the last mesenchymal cells have been induced. This occurs during late embryogenesis in human and soon after the birth in rat and mouse (Fig. 25.2). In human, kidney morphogenesis is initiated at the fifth week of embryogenesis.

Murine kidney rudiments can be cultured on porous membranes *in vitro* at a medium/gas interface (in so called Trowell-type tissue culture) and will develop numerous secretory nephrons and an extensively branched ureteric bud tree [8–10]. More than 50 secretory nephrons will form from an E11 mouse rudiment cultured for 4 days, corresponding to the same number of ureteric bud branches. In such culture, the only tissue type that does not recapitulate its normal differentiation pattern is the vasculature.

Another organotypic kidney culture is based on separation of an isolated MM and an inducer tissue by a porous filter. This so called transfilter culture model allows

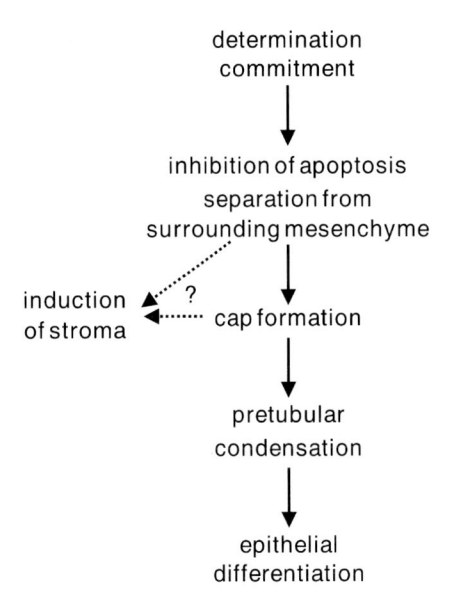

```
                determination
                commitment
                    │
                    ▼
            inhibition of apoptosis
                separation from
            surrounding mesenchyme
                    ┊
                    ▼
induction    ┄? 
of stroma  ◄┄┄┄┄ cap formation
                    │
                    ▼
                pretubular
                condensation
                    │
                    ▼
                 epithelial
               differentiation
```

Fig. 25.2
Developmental stages in nephron differentiation.

separate analyses of the responding and inducing tissues, and has been widely been used for studies on the kinetics of MM differentiation and the nature of the inducing event [11–16]. Furthermore, isolated MMs can be cultured on top of a filter for a short period of time (not more than 1 or 2 days), which has allowed the analysis of the inducing capacity of tissue extracts and purified molecules [17–19]. Culture of isolated ureteric buds has also been successful [20]. Tissue culture techniques have been helpful in identification of signals from the MM that lead to the branching of the ureteric bud, such as Glial cell line-derived neurotrophic factor (GDNF), and on the other hand, resolving signals that induce epithelial conversion of MM, such as fibroblast growth factor 2 (FGF2), leukemia inhibitory factor (LIF), and transforming growth factor β2 (TGFβ2) (reviewed in [21]).

These *in vitro* observations imply that soluble factors can play a greater role in nephron induction than had been originally thought. However, the controversy is that the roles of the tubule-inducing molecules identified until now have not been substantiated by *in vivo* gene ablation studies, or they are not expressed in the embryonic kidney at the right place and the right time. Typically, null alleles created by gene targeting have not recapitulated the expected disruption of nephrogenesis. For instance, the cytokine LIF [22] induces full epithelial differentiation of isolated rat MM growing in organ culture, but it is unlikely to play a unique role as an inducer *in vivo* as the LIF receptor-deficient mouse has functional kidneys, although they are smaller than those of wild-type mice [23]. Some of the "molecular confusion" concerning the inducing signals probably reflects redundancy and interplay of the receptors at the cell surface, but apparently also reflects the fact that the signaling network responding to induction is complex and can be activated at several points. This is best exemplified by the experiments with Wnt-proteins. Wnt4 is upregulated in the pretubular aggregates and gene ablation studies have shown its necessity for epithelial conversion of secretory nephrons [24]. The kidneys of Wnt4–deficient mice form cap condensates, but do not normally form pretubular aggregates or undergo epithelial transformation. In tissue recombination of MM with a heterologous inducer, the spinal cord, the mesenchyme undergoes epithelial differentiation (reviewed in [5]). Wnt4 together with several other Wnts are produced by the spinal cord and they induce autocrine Wnt4 production in MM. The same phenomenon is seen in recombination experiments between Wnt-expressing cells, including Wnt4, and non-induced MM indicating a common autocrine role for canonical Wnt:s [25]. In tissue recombination experiments, Wnt-proteins expressed by the spinal cord apparently mimic the action of the real, yet unidentified ureteric bud-derived inducers. Thus, Wnt4 is clearly an essential autocrine factor for epithelial conversion of MM [26].

The best assumption at the moment is that the induction of secretory nephrons *in vivo* involves several interrelated and consecutive inductions (leading to the escape from apoptosis, cap condensation around the duct tips, further condensation to form the pretubular aggregate, epithelialization of the aggregate, polarisation of epithelial cells etc.) that may involve redundancy and unexpected interplay at the receptor level.

25.2
Regulation of Ureteric Bud Branching

Collecting ducts drain urine from the secretory nephrons to renal pelvis. They form a dichotomously branched network, which is derived from the ureteric bud that evolves from the Wolffian duct. Ureteric budding and branching are controlled by the MM-derived factors, of which the role of GDNF is essential and best characterized.

Ureteric budding is negatively controlled by Bone morphogenetic protein (Bmp) signaling. Bmp4 is expressed by the peritubular mesenchyme around the Wolffian duct. Bmp4–deficient mice are early embryonic lethal. Heterozygous Bmp4–null mice display renal abnormalities with high penetrance, these include hypodysplastic kidneys, hydroureter, ectopic ureterovesical junction, and double collecting system, and Bmp4 mRNA is locally downregulated at the budding site [27]. Bmp signaling in the kidney is positively regulated by Sonic hedgehog (SHH) [28] and inhibited by several antagonists such as Noggin, Gremlin, and BAMBI. Similarly to GDNF, exogenous Noggin induces supernumerary budding from Wolffian duct in explant culture (our unpublished data). In accordance with these findings, mice deficient for gremlin develop supernumerary buds from the Wolffian duct [29]. Thus, Bmp and GDNF signaling are counteracting regulators of ureteric budding. Although it is apparent that Bmp signaling negatively regulates ureteric budding, it remains unresolved whether the primary event is downregulation of Bmp4 or upregulation of its antagonist at the budding site. Until now, only antagonists of Bmp signaling have been identified, but it is plausible to assume that such molecules are to be found in GDNF signaling as well (Fig. 25.3).

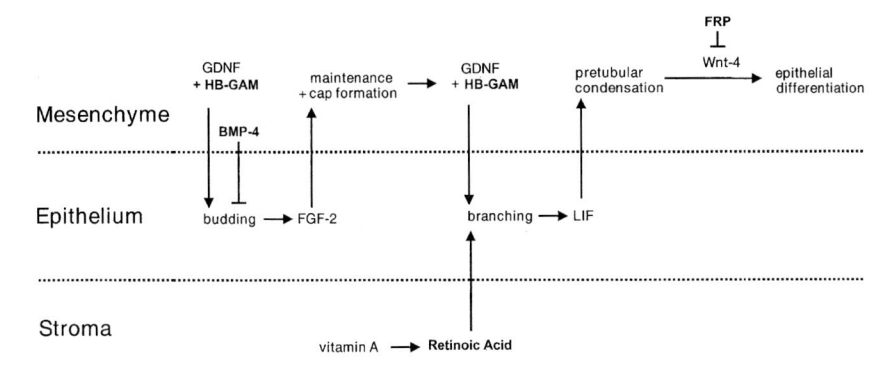

Fig. 25.3
The sequence of inductive signals during nephrogenesis: a hypothetical model. The first signal is derived from the mesenchyme, which induces budding from the ureteric epithelium which is positive for Ret receptor tyrosine kinase. Its expression is controlled by retinoic acid, which is converted from vitamin A by the stromal cells. The fourth tissue type possibly involved in the inductive signaling in nephrogenesis may be the kidney capillaries, but their role is still poorly defined. GDNF, glial cell line-derived neurotrophic factor; HB-GAM, heparin-binding growth-associated molecule; BMP–4, bone morphogenetic protein–4; FGF–2, fibroblast growth factor–2; LIF, leukemia inhibitory factor; FRP, frizzled-related protein (a Wnt antagonist).

25.3
Genes Affecting Early Nephrogenesis

Expression studies, antibody inhibition experiments, growth factor supplementation in organ culture, overexpressing and gene targeted transgenic mice have pinpointed a number of signaling molecules, their inhibitors, and transcriptional regulators that affect metanephric induction or ureteric branching morphogenesis [26, 30–35]. The complexity of the inductive networks in nephrogenesis has often made it difficult to identify precisely the target processes of the genes. It is also difficult to judge the *in vivo* significance of molecules if the tissue culture data do not gain support from *in vivo* models.

While many of the regulatory genes are highly upregulated in the induced nephrogenic mesenchyme, it is important to realize that the transition of gene expression profiles from a non-induced to induced MM is generally not black-and-white. Careful analysis of the expression patterns of many genes originally supposed to appear in MM by induction have shown that these genes are already expressed by the non-induced mesenchyme [36, 37] possibly reflecting the commitment to MM. Typically, the molecules are highly upregulated rather than appearing upon induction. Nevertheless, this raises the question of the true nature of induction. The molecules listed here are a relatively small subset of the approximately 400 genes known to be expressed in the developing kidney. The list is therefore clearly subjective and incomplete. A complete list of genes known to be expressed in the developing kidney with a short description of their proposed function has been collected on an Internet database [38].

25.4.
Signaling Molecules, their Receptors, and the Integrins

Significant progress has been made during recent years on the signaling molecules, which are expressed and function in the embryonic kidney. Tissue culture data originally suggested that induction of the kidney tubules is based on cell-to-cell contacts between MM and the ureteric bud (for review see [5]), but now it is clear that purified signaling molecules and their combination efficiently induce kidney tubule formation in MM in vitro. As mentioned, genetic studies do not support a role for these factors in normal nephrogenesis, and therefore the nature of *in vivo* inducers of tubulogenesis and their mode-of-action still remains unresolved.

25.4.1
Glial Cell Line-derived Neurotrophic Factor

GDNF [39] and the related factors, artemin (ARTN), neurturin (NRTN), and persephin (PSPN), form the GDNF family of neurotrophic factors (GFLs). GDNF maintains several neuronal sets of the nervous system, controls ureteric bud branching in kidney development (Fig. 25.4) and regulates spermatogonial stem cell fate decisions [21, 40–42].

The cellular responses to GFLs are mediated by a multi-component receptor complex consisting of Ret receptor tyrosine kinase and a glycosyl phosphatidylinositol (GPI)-linked ligand-binding subunit, known as GDNF family receptor α (GFRα) [43]. GDNF binds GFRα1, NTRN GFRα2, ARTN GFRα3, and PSPN GFRα4, but all GFLs use Ret as their signaling receptor. Mice lacking Ret, GFRα1, or GDNF show quite similar phenotypes [42]. They have renal aplasia or severe hypodysplasia, and lack enteric innervation below the stomach. GDNF-deficient mice show more severe renal malformations than Ret- or GFRα1–deficient mice [44, 45], although it is not known whether this reflects the mouse strain or biochemical differences. It is notable that GFRα:s are more broadly expressed in embryonic tissues than Ret suggesting Ret-independent roles for the α-receptors. Recent *in vitro* findings indeed demonstrate that GFLs also signal independently of Ret [46–48], and in particular through neural cell adhesion molecule (NCAM) [49]. However, the role of Ret-inde-

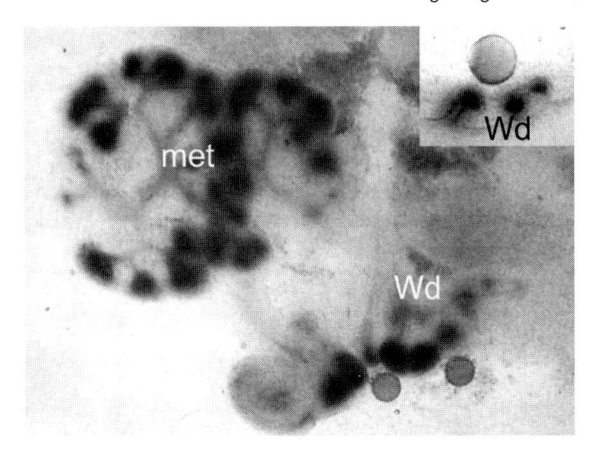

Fig. 25.4
GDNF regulates ureteric budding and branching in the metanephros (Met). Experimentally, GDNF-releasing beads promote supernumerary budding from the Wolffian duct (Wd). The branches of the ureteric buds and the ectopic GNDF-induced buds from the Wd are shown in the whole mount of *in situ* hybridization by GFRα1 mRNA. The inset shows a GDNF-releasing bead and supernumerary budding from Wd at high magnification (courtesy of Marjo Hytönen). (This figure also appears with the color plates.)

pendent signaling of GDNF may be dispensable for normal organogenesis because double transgenic mice expressing GFRα1 in cells together with Ret, but not in cells which lack Ret, show normal development of the kidney and enteric nervous plexus [50].

GDNF may act on the tip of the ureteric bud together with other mesenchyme-derived factors, such as Heparin-binding growth-associated molecule (HB-GAM) and Scatter factor/Hepatocyte growth factor (HGF) [51]. Their role in ureteric branching is based only on *in vitro* cultures and again the genetic studies have not confirmed their physiological significance. Also, Wnt11 may contribute to ureteric branching morphogenesis [52]. Wnt11 is a non-canonical Wnt and is expressed by the ureteric bud tips; its expression is reduced in the absence of GDNF signaling and Wnt11-deficient mice show renal hypoplasia [52]. Consistent with the idea that Wnt11 and GDNF signaling co-operate in the regulation the branching process, Wnt11 and Ret mutations synergistically affect ureteric branching morphogenesis and form an autoregulatory loop in the tips of the ureteric bud [52].

Heparan sulfate proteoglycans may also be important for GDNF signaling in embryonic kidneys [53–56] as the effect of depriving kidneys of heparin sulfates is very similar to that of knocking out of GDNF or Ret [45]. In accordance, heparan sulfate 2–O-sulphotransferase, an enzyme required in processing heparan sulfates, is essential for nephrogenesis and mice lacking it show frequent renal aplasia [56, 57]. Also, mice deficient for Glypican–3, a heparan sulfate core protein, exhibit cystic and dysplastic kidneys [58], but in addition to binding GDNF, glypican–3 may act through Bmp:s and Fgf:s.

Transcription factors Pax2 and Eya1 control GDNF expression in differentiating nephrogenic mesenchymal cells [59, 60], while Ret expression by ureteric bud cells is indirectly regulated by retinoic acid [61, 62]. NRTN, GFRα2, and NCAM are also expressed in the developing kidney but they have no renal phenotype when knocked-out [63–65]. Therefore, their *in vivo* role in renal differentiation remains unclear.

25.4.2
Fibroblast Growth Factors

MM is rescued from apoptosis by Fgf–2 in tissue culture [18, 19]. This may be the first step in the multi-step model of nephron differentiation. Interestingly isolated rat but not mouse MMs undergo full epithelial differentiation if Fgf–2 is combined with a conditioned medium from ureteric bud cell cultures [16] or if the Fgf–2 treatment is followed by supplementation of the medium with Lif [22]. Fgf and Bmp signaling act synergistically during nephrogenesis to maintain the competence of MM [37] and Fgf–7 modulates kidney growth [66]. The expression profiles of different Fgf:s and their receptors (FgfR:s) have been carefully mapped in the embryonic metanephric kidney [67]. Most Fgf:s (Fgf–1 through –10 except Fgf–6) and their receptors (FgfR–1 through –4) are expressed by MM, ureteric epithelium, or both. Thus, Fgf:s may act in concert and have multiple functions in nephrogenesis. Gene targeting of FgfR:s are mainly early embryonic lethal. Fgf–7–deficient mice develop hypoplastic kidneys and overexpression results in hyperplastic dilated collecting ducts [66, 68]. Fgf signaling rapidly upregulates intracellular Sprouty proteins which act as general antagonists of receptor tyrosine kinase signaling including that of Fgf and epidermal growth factor receptors [69]. The embryonic kidney expresses all four splice variants of Sprouty [70]. Targeted expression of human sprouty 2 (SPRY2) in the ureteric bud leads to postnatal death resulting from kidney failure manifested as unilateral agenesis, lobularization of the organ or reduction in organ size because of inhibition of ureteric branching [71].

25.4.3
Leukemia Inhibitory Factor

After Fgf–2 priming, microsurgically isolated rat MM:s undergo epithelial differentiation in vitro, when exposed to Lif, a member of the interleukin–6 cytokine family [22, 72]. This sequential action of inductive signals promotes nephron differentiation efficiently in rat, but not in mouse suggesting that other members of the interleukin–6 family such as oncostatin M could represent tubule inducers in mouse. The identification of Lif as a tubule inducer is rather unexpected because earlier organ culture experiments showed that Lif inhibits morphogenesis of whole kidney rudiments growing in organ culture [73], and Lif-deficient mice show only mild renal hypoplasia (see above). Thus, the *in vivo* significance of Lif or other interleukin family members in kidney morphogenesis remains to be solved.

25.4.4
Wnt Proteins

The Wnt proteins constitute a large family of extracellular signaling molecules that are found throughout the animal kingdom and are important for a wide variety of developmental and pathological processes [74]. Gene targeting of the Wnt4 gene, a murine homolog of the Drosophila Wingless, has indicated that Wnt4 acts as an autocrine epithelializing signal in the induced MM [24]. Kidney morphogenesis is stopped in Wnt4-deficient mice just before the pretubular aggregation stage, when several markers of nephron differentiation have already upregulated and the ureteric bud has undergone several branches. This mutant phenotype and the experiments described above strongly suggest that Wnt4 acts as an autocrine epithelializing signal of the induced pretubular condensates. Organ culture experiments with cell lines transfected with various members of the Wnt-family have shown that all canonical Wnt:s upregulate Wnt4 and induce differentiation of MM [25, 75] raising the possibility that an as yet unrecognized canonical Wnt may act as a ureteric bud-derived inducer of MM. The Wnt proteins expressed by the ureteric epithelial cells, Wnt7B and Wnt11, signal through the non-canonical Wnt pathway and do not induce nephrogenic mesenchyme in cell culture assays [76].

25.4.5.
Bone Morphogenetic Proteins

Bone morphogenetic proteins (Bmp:s) serve as pleiotropic signaling molecules in various developing tissues including the metanephric kidney [77]. mRNA expression studies have shown that at least Bmp2, –4, –5, and –7 are expressed either by the ureteric bud tree or MM [78]. Bmp2– and –4–deficient mice are early embryonic lethal, before initiation of the organogenetic period. Bmp7–deficient mice develop hypoplastic kidneys [79–81] and Bmp5–deficient mice possess variable renal defects including hydroureters [82]. Organ culture experiments suggested initially that Bmp7 would be an inducer of kidney tubulogenesis [83]. Later studies have shown that it maintains the nephrogenic cells by preventing apoptosis rather than inducing nephron differentiation [37]. Bmp4 triggers Wolffian duct development from the intermediate mesoderm [84]. Bmp4 also maintains isolated MM:s in culture, but their developmental fate is changed and they can no longer be induced to tubulogenesis – even after a short exposure to Bmp4 [85]. Another method to maintain isolated MMs *in vitro* is to treat them with tissue metalloproteinase inhibitor–2 (TIMP–2). It may act in a similar manner to Bmp4 because TIMP 2 treatment also inhibits tubulogenesis in cultures of whole kidneys [86]. The best characterized function of Bmp signaling in kidney morphogenesis is apparently its role in ureteric budding where it counteracts the effect of GDNF signaling (see above).

25.4.6.
Sonic Hedgehog

The signaling molecule Shh regulates several inductive events during embryogenesis [87]. Shh is expressed by the ureteric bud and its targeted disruption leads to renal aplasia in addition to severe defects in other organs including the midline defects that are often associated with kidney abnormalities [88]. The conditional loss-of-function approach has however revealed that Shh produced by the ureteric stalk epithelium is needed for the proliferation and differentiation of the smooth muscle precursors surrounding the stalk and that Shh regulates Bmp4–expression in the peritubular cells [28].

25.4.7
Formins

Formins are signaling molecules regulating limb and kidney morphogenesis and they are expressed early during nephrogenesis by MM [89]. Mice homozygous for the spontaneous *limb deformity* (*ld*) mutation lack formins and develop limb deformities and renal defects. The *ld* locus encodes a complex family of mRNA and protein isoforms of formins. Homozygous mice bearing the formin isoform IV disruption display incompletely penetrant renal agenesis but unlike the *ld* mice, no limb defects [90]. The most common renal defect in *ld* mice is renal aplasia which is thought to result from failure of ureteric bud outgrowth. Analysis of *ld* mice has suggested that formins regulate Fgf–4 and Shh expression during limb morphogenesis [90]. Although this regulatory pathway has not been analyzed in the kidney it might be speculated that in the ureteric bud formins could upregulate Shh and Fgf:s, which in turn signal back to the mesenchyme to promote its maintenance and commitment to nephrogenesis.

25.4.8
Hepatocyte Growth Factor and Met Receptor Tyrosine Kinase

Both the ureteric bud and MM express Hgf and its Met receptor. Antibody inhibition of Hgf signaling disrupts kidney morphogenesis in organ culture [91, 92]. Hgf- or Met-deficient mice do not however show defects in kidney morphogenesis [93] indicating that either they are redundant or not critical for normal nephrogenesis. While the significance of Hgf signaling in normal kidney morphogenesis does not have genetic support, it is interesting that Hgf promotes ureteric branching in recombination cultures with lung mesenchyme [94]. We have recently shown that GDNF binding to GFRα1 indirectly through Src, leads to Met phosphorylation and branching tubule formation in MDCK cells transfected with GFRα1 gene and lacking Ret [48]. The *in vivo* significance of this receptor cross-talk is still unknown.

25.4.9. Integrins

Integrin α8β1 is expressed at high levels in MM in particular by the mesenchymal cells adjacent to the ureteric bud and is critical for the regulation of kidney morphogenesis. The phenotype in integrin α8β1–deficient mice is kidney specific and these mice consistently show renal aplasia [95]. A ligand for integrin α8β1 is osteopontin, an abundant component of cartilaginous and bone matrix. It shows overlapping expression pattern with integrin α8β1, but its role in nephrogenesis is unknown [96]. Also, integrin α3β1 affects kidney morphogenesis [97] but obviously later than α8β1. Integrin α3β1–deficient kidneys display retarded branching of collecting ducts. Proximal tubules and glomeruli exhibit distinct abnormalities. The epithelial cells become microcystic and the extent of branching of glomerular capillary loops is decreased with capillary lumina being wider than normal. Consistent with this data, laminin α5, the ligand of α3β1–intergin, is expressed by the embryonic kidney and mice deficient for laminin α5 show severe defects in glomerulogenesis and occasional uni- or bilateral kidney aplasia [98].

25.4.10. Other Molecules

A number of other molecules have been implicated in morphogenesis of MM including G proteins, cadherins, extracellular matrix molecules, and various neural cell adhesion molecules [99–101]. Their putative roles and expression patterns in the embryonic kidney are listed in the Kidney Development Database [38].

25.5 Transcription Factors

It is clear that the critical signaling genes and their receptors are regulated by transcription factors but relatively little is known about their targets in embryonic kidneys. This data are however rapidly evolving.

25.5.1 Wilms' Tumor Gene 1

The product of Wilms' tumor gene–1 (WT1) is a helix-loop-helix transcription factor and an early marker of MM. It was originally identified as a gene involved in the pathogenesis of Wilms' tumour, a pediatric renal cancer, where the gene is relatively commonly inactivated [102]. The role of WT1 in the epithelial differentiation of nephrons has been verified by various experimental means: gene targeting, transfection of cell lines, and organ culture. WT1–deficient mice consistently lack kidneys, the MM undergoes apoptosis and the mice even fail to develop ureteric buds [103]. WT1 regulates many growth factors involved in kidney morphogenesis, such as Fgf–2 and interleukin II (IL–11). One of the target genes, amphiregulin, was identified by microarray screening [104]. Amphiregulin is a ligand of the epidermal growth factor receptor and it is unique in exhibiting bifunctional properties: enhancing the proliferation of some epithelial cells while inhibiting this process in many cancer cell lines [105]. However, mice deficient for amphiregulin or any of the other EGF family members have normal kidneys [106]. Thus, it is unlikely that the lack of amphiregulin could cause the severe kidney phenotype seen in WT1–deficient mice. NIH–3T3 fibroblastic cells transfected with the WT1 gene partially epithelialize and these cells show some features of renal epithelial differentiation [107]. Taken together, the current data indicate that WT1 has an early function in MM commitment and survival, and a later role in epithelial cell differentiation.

25.5.2
Pax–2

Pax–2 is a transcription regulator that is the product of a paired-box gene. Interestingly, Pax–2 characterizes many different cell types in the embryonic kidney, in both ureteric epithelium (and Wolffian ducts) and MM [108]. Pax-family members control differentiation of several organs in a complex with Eya and Six transcriptional regulators. In the developing eye for instance Pax–6 is the master regulatory gene. Likewise, Pax–2 is essential for the entire development of the urogenital system and Pax–2–deficient mice lack not only kidneys but all derivatives of the intermediate mesoderm [109]. Overexpression of Pax–2 leads to cystic growth of nephrons [110]. Although Pax–2 is strongly upregulated during kidney tubule induction, it is already expressed by the non-induced MM, possibly reflecting its commitment. The downstream targets of Pax–2 are still poorly understood but there is some indication that GDNF could be one its molecular targets [60].

25.5.3
Six-Eya Complex

Haploinsufficiency of human Eya1, a homolog of the *Drosophila melanogaster* gene *eyes absent* (eya), causes the dominantly inherited branchio-oto-renal and branchio-oto syndromes (OMIM 113650 http://www.genetests.org/query?mim=113650), which are characterized by craniofacial abnormalities and hearing loss with or without kidney defects. In mice [59], Eya1 heterozygotes show renal abnormalities and a conductive hearing loss, whereas Eya1 homozygotes lack ears and kidneys due to defective inductive tissue interactions and apoptotic regression of the organ primordia. The lack of Eya1 results in the absence of ureteric buds and a subsequent failure of metanephric induction. Eya molecules have been shown to function synergistically with other transcriptional co-activators, namely the mouse homologs of Drosophila *sine oculis* (so) six-family. Six–1 and possibly Six–2 [111] together with Eya–1 co-regulate GDNF expression in the developing metanephric mesenchyme. Six–1 mutants lack kidneys and recapitulate the Eya–1–deficient phenotype but could be downstream of Eya–1, which shows a normal expression pattern in the Six–1 null mutants [112, 113]. Both Eya–1 and Six–1 are expressed in Pax–2 null embryos, which indicates that these three signaling molecules act in the sequence Eya 〉 Six 〉 Pax. In addition to GDNF their molecular target in the MM may be the zinc-finger protein Sall–1 [113]. Accordingly, Sall–1 mutant mice show defects in nephrogenesis because of incomplete ureteric bud outgrowth [114].

25.5.4
Emx–2

Emx–2 is a murine homolog of the Drosophila *empty spiracles* homeotic gene product. Targeted disruption of Emx–2 results in a severe renal phenotype [115]. A ureteric bud evolves from the Wolffian duct but the MM rapidly undergoes apoptosis. Tissue recombination of a mutant bud with wild-type MM and vice versa demonstrated that the defect is in the ureteric bud, which fails to induce the mesenchyme. The target genes of Emx–2 in the kidney are still unknown, but these recombination experiments suggest that it regulates the molecules capable of inducing MM.

25.5.5
Lim–1

Lim-domain homeotic genes control patterning of the embryo. Lim–1 is expressed by the lateral plate and intermediate mesoderm in all species from frog to human. In mouse, the expression of Lim–1 is restricted to intermediate mesoderm and its derivatives by E10.5. After the differentiation of MM has been induced, Lim–1 is soon lost from the ureteric epithelium and is upregulated in the pretubular aggregates [116]. Gene targeted mice deficient for Lim–1 die frequently at the initiation of organogenesis [117]. However, the few survivors show consistent renal aplasia.

25.5.6
Fox Genes

The large family of Fox genes is involved in a number of morphogenetic events. Typically, they have very restricted tissue distribution and some Fox genes are expressed organ-specifically. Fox genes encode transcription factors, of which at least Foxd1 **(**previously known as BF2**)**, Foxc1, and Foxc2 are involved in kidney morphogenesis [4, 118]. They are expressed in the mesodermal mesenchyme and both spontaneous mutations (called *congenital hydrocephalus*) and targeted disruption of *Foxd1* and *Foxc1* each result in defects in kidney morphogenesis. Moreover, *Foxc1*– and *Foxc2*–combined heterozygotes develop severe defects in metanephric differentiation [118].

25.6
Future Perspectives

In spite of the rapid progress in recent years some key events in the regulation of nephrogenesis remain puzzling. In particular the identities and mode of action of the *in vivo* tubule inducers are still unknown or controversial. Further, the contribution of the renal stroma to ureteric branching and its developmental origin are poorly understood. Most confusing has been the recognition of the completely different responses of rat and mouse to the *in vitro* inducers of MM. Rat MM is readily inducible whereas in exactly similar conditions the mouse MM is not. This difference may rather reflect the differences in the explanted material than true molecular differences in the induction process of these closely related murine species. Therefore the search should be for unexpected or unorthodox explanations for this species difference. The answer may come from the developmental stage of the microdissected MM about which only little is known. However, at least one difference is clear. The rat MM already contains vascular precursors whereas mouse MM does not. The endothelial cells may not only regulate glomerular differentiation but may critically affect tubule induction [119]. The number of known inductive partners in kidney morphogenesis increased from two to three when the role of stromal cells was noticed, and may increase to four, if vessels are shown to affect epithelial differentiation. Maybe the interplay between MM and endothelial cells will be the final missing link in understanding tubule induction.

A long-standing open question concerns the number of stem cells or precursor cells in MM. If secretory nephrons, stromal cells and blood vessels are all derivatives of the same cell in MM, the metanephric mesenchymal cells could indeed be considered as stem cells. If these cell lineages all have separate origins they should rather be called precursors. At the moment very little is known about the origins of different cell types in MM.

References

1 Cho EA, Dressler GR. Formation and development of nephrons. In *The Kidney: From Normal Development to Congenital Disease*, Vize P, Woolf AS, Bard JBL (Eds). pp. 195–210. Academic Press, **2003**.

2 Sariola H, Sainio K, Bard J. Fates of the metanephric mesenchyme. In *The Kidney: From Normal Development to Congenital*

Disease, pp. 181–194. Vize P, Woolf AS, Bard JBL (Eds). Academic Press: London, **2003**.
3. Vize PD, Woolf AS, Bard JBL (Eds). *The Kidney: From Normal Development to Congenital Disease*, pp. 1–519. Academic Press: London, **2003**.
4. Hatini V, Huh SO, Herzlinger D, Soares VC, Lai E. Essential role of stromal mesenchyme in kidney morphogenesis revealed by targeted disruption of Winged Helix transcription factor BF-2. *Genes Dev* **1996**; *10*: 1467–1478.
5. Saxén L. *Organogenesis of the Kidney*. Cambridge University Press: Cambridge, **1987**.
6. Saxén L. Failure to show tubule induction in a heterologous mesenchyme. *Dev Biol* **1970**; *23*: 511–523.
7. Bard JB, Gordon A, Sharp L, Sellers WI. Early nephron formation in the developing mouse kidney. *J Anat* **2001**; *199*: 385–392.
8. Grobstein C. Inductive epithelio-mesenchymal interaction in cultured organ rudiments of the mouse. *Science* **1953**; *118*: 52–55.
9. Grobstein C. Inductive interaction in the development of the mouse metanephros. *J Exp Zool* **1955**; *130*: 319–340.
10. Grobstein C. Mechanisms of organogenetic tissue interaction. *Natl Cancer Inst Monogr* **1967**; *26*: 107–119.
11. Lombard M-N, Grobstein C. Activity in various embryonic and postembryonic sources for induction of kidney tubules. *Dev Biol* **1969**; *19*: 41–51.
12. Wartiovaara J, Lehtonen E, Nordling S Saxén L. Do membrane filters prevent cell contacts? *Nature* **1972**; *238*: 407–408.
13. Lehtonen E, Wartiovaara J, Nordling S, Saxen L. Demonstration of cytoplasmic processes in Millipore filters permitting kidney tubule induction. *J Embryol Exp Morphol* **1975**; *33*: 187–203.
14. Saxén L, Lehtonen E, Karkinen-Jaaskelainen M, Nordling S, Wartiovaara J. Are morphogenetic tissue interactions mediated by transmissible signal substances or through cell contacts? *Nature* **1976**; *259*:662–663.
15. Saxén L, Lehtonen E. Transfilter induction of kidney tubules as a function of the extent and duration of intercellular contacts. *J Embryol Exp Morphol* **1978**; *47*: 97–109.
16. Karavanova ID, Dove LF, Resau JH, Perantoni AO. Conditioned medium from a rat ureteric bud cell line in combination with bFGF induces complete differentiation of isolated metanephric mesenchyme. *Development* **1996**; *122*: 4159–4167.
17. Davies JA, Garrod DR. Induction of early stages of kidney tubule differentiation by lithium ions. *Dev Biol* **1995**; *167*: 50–60.
18. Barasch J, Qiao J, McWilliams G, Chen D, Oliver J, Hezlinger D. Ureteric bud cells secrete multiple factors incuding bFGF which rescue renal progenitors from apoptosis. *Am J Physiol* **1997**; *273*: F757–F767.
19. Perantoni AO, Dove LF, Karavanova I. Basic fibroblast growth factor can mediate the early inductive events in renal development. *Proc Natl Acad Sci USA* **1995**; *92*: 4696–4700.
20. Qiao J, Sakurai H, Nigam S.K. Branching morphogenesis independent of mesenchymal-epithelial contact in the developing kidney. *Proc Natl Acad Sci USA* **1999**; *96*: 7330–7335.

21 Saarma M, Sariola H. Other neurotrophic factors: glial cell line-derived neurotrophic factor (GDNF). *Microsc Res Tech* **1999**; *45*: 292–302.

22 Barasch J, Yang J, Ware CB, Taga T, Yoshida K, Erdjument-Bromage H, Tempst P, Parravicini E, Malach S, Aranoff T, Oliver JA. Mesenchymal to epithelial conversion in rat metanephros is induced by LIF. *Cell* **1999**; *99*: 377–386.

23 Yoshida K, Taga T, Saito M, Suematsu S, Kumanogoh A, Tanaka T, Fujiwara H, Hirata M, Yamagami T, Nakahata T.K. et al. Targeted disruption of gp130 a common signal transducer for the interleukin–6 family of cytokines leads to myocardial and hematological disorders. *Proc Natl Acad Sci USA* **1996**; *93*: 407–411.

24 Stark K, Vainio S, Vassileva G, McMahon AP. Epithelial transformation of metanephric mesenchyme in the developing kidney regulated by Wnt–4. *Nature* **1994**; *372*: 679–683.

25 Kispert A, Vainio S, McMahon AP. Wnt–4 is a mesenchymal signal for epithelial transformation of metanephric mesenchyme in the developing kidney. *Development* **1998**; *125*: 4225–4234.

26 Carroll TJ, McMahon AP. Secreted molecules in metanephric induction. *J Am Soc Nephrol* **2000**; 16: S116–S119.

27 Miyazaki Y, Oshima K, Fogo A, Hogan BL, Ichikawa I. Bone morphogenetic protein 4 regulates the budding site and elongation of the mouse ureter. *J Clin Invest* **2000**; *105*: 863–873.

28 Yu J, Carroll TJ, McMahon AP. Sonic hedgehog regulates proliferation and differentiation of mesenchymal cells in the mouse metanephric kidney. *Development* **2002**; *129*: 5301–5312.

29 Michos O, Panman L, Vintersten K, Beier K, Zeller R, Zuniga A. Gremlin-mediated BMP antagonism induces the epithelial-mesenchymal feedback signaling controlling metanephric kidney and limb organogenesis. *Development* **2004**; *131*: 3401–3410.

30 Davies JA, Bard JBL. The development of the kidney. *Curr Topics Dev Biol* **1998**; *39*: 245–301.

31 Schofield PN, Boulter CA. Growth factors and metanephrogenesis. *Exp Nephrol* **1996**; *4*: 97–104.

32 Dressler GR. Genetic control of kidney development. *Adv Nephrol Necker Hosp* **1997**; *26*: 1–17.

33 Woolf AS, Cale CM. Roles of growth factors in renal development. *Curr Opin Nephrol Hypertens* **1997**; *6*: 10–14.

34 Orellana SA, Avner ED. Cell and molecular biology of kidney development. *Semin Nephrol* **1998**; *18*: 233–243.

35 Burrow CR. Regulatory molecules in kidney development. *Pediatr Nephrol* **2000**; *14*: 240–253.

36 Donovan MJ, Natoli TA, Sainio K, Amstutz A, Jaenisch R, Sariola H, Kreidberg JA. Initial differentiation of the metanephric mesenchyme is independent of WT1 and the ureteric bud. *Dev Genet* **1999**; *24*: 252–262.

37 Dudley AT, Godin RE, Robertson EJ. Interaction between FGF and BMP signaling pathways regulates development of metanephric mesenchyme. *Genes Dev* **1999**; *13*: 1601–1613.

38 Davies JA, Brändli AW: »The kidney development database« http://mbisg2.sbc.man.ac.uk/kidbase/kidhome.html and http://www.ana.ed.ac.uk/anatomy/kidbase/kidhome.html. **1997**.

39 Lin LF, Doherty DH, Lile JD, Bektesh S, Collins F. GDNF: a glial cell line-derived neurotrophic factor for midbrain dopaminergic neurones. *Science* **1993**; *260*: 1130–1132.

40 Meng X, Lindahl M, Hyvönen ME, Parvinen M, de Rooij DG, Hess MW, Raatikainen-Ahokas A, Sainio K, Rauvala H, Lakso M, Pichel JG, Westphal H, Saarma M, Sariola H. Regulation of cell fate decision of undifferentiated spermatogonia by GDNF. *Science* **2000**; *287*: 1489–1493.

41 Airaksinen MS, Saarma M. The GDNF family: signalling biological functions and therapeutic value. *Nature Rev Neurosci* **2002**; *3*: 383–394.

42 Sariola H, Saarma M. Novel functions and signalling pathways for GDNF. *J Cell Sci* **2003**; *116*: 3855–3862.

43 Jing S, Wen D, Yu Y, Holst PL, Luo Y, Fang M, Tamir R, Antonio L, Hu Z, Cupples R, Louis JC., Hu S, Altrock BW, Fox GM.. GDNF-induced activation of the ret protein tyrosine kinase is mediated by GDNFR-a a novel receptor for GDNF. *Cell* **1996**; *85*: 1113–1124.

44 Pichel JG, Shen L, Sheng HZ, Granholm AC, Drago J, Grinberg A, Lee EJ, Huang SP, Saarma M, Hoffer BJ, Sariola H, Westphal H. Defects in enteric innervation and kidney development in mice lacking GDNF. *Nature* **1996**; *382*: 73–76.

45 Schuchardt A, D'Agati V, Larsson-Blomberg L, Costantini V, and Pachnis V. Defects in the kidney and enteric nervous system of mice lacking the tyrosine kinase receptor Ret. *Nature* **1994**; *367*: 380–383

46 Poteryaev D, Titievsky A, Sun YF, Thomas-Crusells J, Lindahl M, Billaud M, Arumäe U, Saarma M. GDNF triggers a novel ret-independent Src kinase family-coupled signaling via a GPI-linked GDNF receptor alpha1. *FEBS Lett* **1999**; *463*: 63–66.

47 Trupp M, Scott R, Whittemore SR, Ibañéz CF. RET-dependent and -independent mechanisms of glial cell line-derived neurotrophic factor signaling in neuronal cells. *J Biol Chem* **1999**; *274*: 20885–20894.

48 Popsueva A, Poteryaev D, Arighi E, Meng X, Angers-Loustau A, Kaplan D, Saarma M, Sariola H. GDNF promotes tubulogenesis of GFRalpha1–expressing MDCK cells by Src-mediated phosphorylation of Met receptor tyrosine kinase. *J Cell Biol* **2003**; *161*: 119–129.

49 Paratcha G, Ledda F, Ibañéz C.F. The neural cell adhesion molecule NCAM is an alternative signaling receptor for GDNF family ligands. *Cell* **2003**; *113*: 867–879.

50 Enomoto H, Hughes I, Golden J, Baloh RH, Yonemura S, Heuckeroth RO, Johnson EM Jr, Milbrandt J. GFRalpha1 expression in cells lacking RET is dispensable for organogenesis and nerve regeneration. *Neuron* **2004**; *44*: 623–636.

51 Sakurai H, Bush KT, Nigam SK. Identification of pleiotrophin as a mesenchymal factor involved in ureteric bud branching morphogenesis. *Development* **2001**; *128*: 3283–3293.

52. Majumdar A, Vainio S, Kispert A, McMahon J, McMahon AP. Wnt11 and Ret/Gdnf pathways cooperate in regulating ureteric branching during metanephric kidney development. *Development* **2003**; *130*: 3175–3185.

53. Barnett MW, Fisher CE, Perona-Wright G, and Davies JA. Signalling by glial cell line-derived neurotrophic factor (GDNF) requires heparan sulphate glycosaminoglycan. *J Cell Sci* **2002**; *115*: 4495–4503.

54. Tanaka M, Xiao H, Kiuchi K. Heparin facilitates glial cell line-derived neurotrophic factor signal transduction. *NeuroReport* **2002**; *13*: 1913–1916.

55. Steer DL, Shah MM, Bush KT, Stuart RO, Sampogna RV, Meyer TN, Schwesinger C, Bai X, Esko JD, Nigam SK. Regulation of ureteric bud branching morphogenesis by sulfated proteoglycans in the developing kidney. *Dev Biol* **2004**; *272*: 310–327.

56. Bullock SL, Fletcher JM, Beddington RS, and Wilson VA. Renal agenesis in mice homozygous for a gene trap mutation in the gene encoding heparan sulfate 2–sulfotransferase. *Genes Dev* **1998**; *12*: 1894–1906.

57. Wilson VA, Gallagher JT, Merry CL. Heparan sulfate 2–O-sulfotransferase (Hs2st) and mouse development. *Glycoconjug J* **2002**; *19*: 347–354.

58. Cano-Gauci DF, Song HH, Yang H, McKerlie C, Choo B, Shi W, Pullano R, Piscione TD, Grisaru S, Soon S, Sedlackova L, Tanswell AK, Mak TW, Yeger H, Lockwood GA, Rosenblum ND, Filmus J. Glypican–3–deficient mice exhibit developmental overgrowth and some of the abnormalities typical of Simpson-Golabi-Behmel syndrome. *J Cell Biol* **1999**; *146*: 255–264.

59. Xu P-X, Adams J, Peters H, Brown MC, Heaney S, Maas R. Eya1–deficient mice lack ears and kidneys and show abnormal apoptosis of organ primordia. *Nature Genet* **1999**; *23*: 113–117.

60. Brophy PD, Ostrom L, Lang KM, Dressler GR. Regulation of ureteric bud outgrowth by Pax2–dependent activation of the glial-derived neurotrophic factor gene. *Development* **2001**; *128*: 4747–4756.

61. Moreau E, Vilar J, Lelievre-Pegorier M, Merlet-Benichou C, Gilbert T. Regulation of c-ret expression by retinoic acid in rat metanephros: implication in nephron mass control. *Am J Physiol* **1998**; *275*: F938–F945.

62. Batourina E, Gim S, Bello N, Shy M, Clagett-Dame M, Srinivas S, Costantini F, and Mendelsohn C. Vitamin A controls epithelial/mesenchymal interactions through RetRET expression. *Nature Genet* **2001**; *27*: 74–78.

63. Cremer H, Lange R, Christoph A, Plomann M, Vopper G, Roes J, Brown R, Baldwin S, Kraemer P, Scheff S, Barthels D, Rajewsky K, Wille W. Inactivation of the N-CAM gene in mice results in size reduction of the olfactory bulb and deficits in spatial learning. *Nature* **1994**; *367*: 455–459.

64. Heuckeroth RO, Enomoto H, Grider JR, Golden JP, Hanke JA, Jackman A, Molliver DC, Bardgett ME, Snider WD, Johnson EM, Milbrandt J. Gene targeting reveals a critical role for neur-

turin in the development and maintenance of enteric sensory and parasympathetic neurons. *Neuron* **1999**; *22*: 253–263.

65 Rossi R, Luukko K, Poteryaev D, Laurikainen A, Sun YF, Laakso T, Eerikäinen S, Tuominen R, Lakso M, Rauvala H, Arumäe U, Pasternack M, Airaksinen MS. Retarded growth and deficits in the enteric and parasympathetic nervous system in mice lacking GFRα2 a functional neurturin receptor. *Neuron* **1999**; *22*: 243–252.

66 Qiao J, Uzzo R, Obara-Ishihara T, Degenstein L, Fuchs E, Herzlinger D. FGF–7 modulates ureteric bud growth and nephron number in the developing kidney. *Development* **1999**; *126*: 547–554.

67 Cancilla B, Ford-Perriss MD, Bertram JF. Expression and localization of fibroblast growth factors and fibroblast growth factor receptors in the developing rat kidney. *Kidney Int* **1999**; *56*: 2025–2039.

68 Nguyen HQ, Danilenko DM, Bucay N, DeRose ML, Van GY, Thomason A, Simonet WS. Expression of keratinocyte growth factor in embryonic liver of transgenic mice causes changes in epithelial growth and differentiation resulting in polycystic kidneys and other organ malformations. *Oncogene* **1996**; *10*: 2109–2119.

69 Reich A, Sapir A, Shilo B. Sprouty is a general inhibitor of receptor tyrosine kinase signaling. *Development* **1999**; *126*: 4139–4147.

70 Minowada G, Jarvis LA, Chi CL, Neubuser A, Sun X, Hacohen N, Krasnow MA,. Martin GR. Vertebrate Sprouty genes are induced by FGF signaling and can cause chondrodysplasia when overexpressed. *Development* **1999**; *126*: 4465–4475.

71 Chi L, Zhang S, Lin Y, Prunskaite-Hyyrylainen R, Vuolteenaho R, Itaranta P, Vainio S. Sprouty proteins regulate ureteric branching by coordinating reciprocal epithelial Wnt11 mesenchymal Gdnf and stromal Fgf7 signalling during kidney development. *Development* **2004**; *131*: 3345–3356.

72 Plisov SY, Yoshino K, Dove LF, Higinbotham KG, Rubin JS, Perantoni AO. TGF beta 2 LIF and FGF2 cooperate to induce nephrogenesis. *Development* **2001**; *128*: 1045–1057.

73 Bard JB, Ross AS. LIF the ES-cell inhibition factor reversibly blocks nephrogenesis in cultured mouse kidney rudiments. *Development* **1991**; *113*: 193–198.

74 Wodarz A, Nusse R. Mechanisms of Wnt signaling in development. *Annu Rev Cell Dev Biol* **1998**; *14*: 59–88.

75 Herzlinger D, Qiao J, Cohen D, Ramakrishna N, Brown AM. Induction of kidney epithelial morphogenesis by cells expressing Wnt–1. *Dev Biol* **1994**; *166*: 815–818.

76 Patapoutian A, Backus C, Kispert A, Reichardt LF. Regulation of neurotrophin–3 expression by epithelial-mesenchymal interactions: the role of Wnt factors. *Science* **1999**; *283*: 1180–1183.

77 Godin RE, Robertson EJ, Dudley AT. Role of BMP family members during kidney development. *Int J Dev Biol* **1999**; *43*: 405–411.

78 Dudley AT, Robertson EJ. Overlapping expression domains of bone morphogenetic protein family members potentially ac-

count for limited tissue defects in BMP7 deficient embryos. *Dev Dyn* **1997**; *208*: 349–362.

79. Dudley AT, Lyons KM, Robertson EJ. A requirement for bone morphogenetic protein 7 during development of the mammalian kidney and eye. *Genes Dev* **1995**; *9*: 2795–2807.

80. Jena N, Martin-Seisdedos C, McCue P, Croce CM. BMP7 null mutation in mice: developmental defects in skeleton kidney and eye. *Exp Cell Res* **1997**; *230*: 28–37.

81. Luo G, Hofmann C, Brunckens ALJ, Sohochi M, Bradley A, Karsenty G. BMP7 is an inducer of nephrogenesis and is also required for eye development and skeletal patterning. *Genes Dev* **1995**; *9*: 2808–2820.

82. King JA, Marker PC, Seung KJ, Kingsley DM. BMP5 and the molecular, skeletal, and soft-tissue alterations in short ear mice. *Devel Biol* **1994**; *166*: 112–122.

83. Vukicevic S, Kopp JB, Luyten FP, Sampath TK. Induction of nephrogenic mesenchyme by osteogenic protein 1 (BMP–7). *Proc Natl Acad Sci USA* **1996**; *93*: 9021–9026.

84. Obara-Ishihara T, Kuhlman J, Niswander L, Herzlinger D. The surface ectoderm is essential for nephric duct formation in intermediate mesoderm. *Development* **1999**; *126*: 1103–110.

85. Raatikainen-Ahokas A, Hyvönen M, Tenhunen A, Sainio K, Sariola H. BMP–4 affects the differentiation of metanephric mesenchyme and reveals an early anterior-posterior axis of the embryonic kidney. *Dev Dyn* **2000**; *217*: 146–158.

86. Barasch J, Yang J, Ware CB, Taga T, Yoshida K, Erdjument-Bromage H, Tempst P, Parravicini E, Malach S, Aranoff T, Oliver JA. Mesenchymal to epithelial conversion in rat metanephros is induced by LIF. *Cell* **1999**; *99*: 377–386.

87. Weed M, Mundlos S, Olsen BR. The role of sonic hedgehog in vertebrate development. *Matrix Biol* **1997**; *16*: 53–58

88. Chiang C, Litingtung Y, Lee E, Young KE, Corden JL, Westphal H, Beachy PA. Cyclopia and defective axial patterning in mice lacking Sonic hedgehog gene function. *Nature* **1996**; *383*: 407–413.

89. Leader B, Leder P. Formin–2 a novel formin homology protein of the cappuccino subfamily is highly expressed in the developing and adult central nervous system. *Mech Dev* **2000**; *93*: 221–231.

90. Wynshaw-Boris A, Ryan G, Deng CX, Chan DC, Jackson-Grusby L, Larson D, Dunmore JH, Leder P. The role of a single formin isoform in the limb and renal phenotypes of limb deformity. *Mol Med* **1997**; *3*:372–384.

91. Santos OFP, Barros EJG, Yang X-M, Matsumoto K, Nakamura T, Park M, Nigam SK. Involvement of hepatocyte growth factor in kidney development. *Dev Biol* **1994**; *163*: 525–529.

92. Woolf AS, Kolatsi-Jjoannou M, Harman P, Andermarcher E, Moorby C, Fine LG, Jat PS, Noble MD, Gherardi E. Roles of hepatocyte growth factor/scatter factor and the met receptor in the early development of the metanephros. *J Cell Biol* **1995**; *128*: 171–184. 113–117.

93. Birchmeier C, Gherardi E. Developmental roles of HGF/SF and its receptor, the c-Met tyrosine kinase. *Trends in Cell Biol* **1998**; *8*: 404–410.

94 Sainio K, Suvanto P, Davies J, Wartiovaara K, Saarma M, Arumäe U, Meng X, Lindahl M, Pachnis V, Sariola H. Glial-cell-line-derived neurotrophic factor is required for bud initiation from ureteric epithelium. *Development* **1997**; *124*: 4077–4087.

95 Muller U, Wang D, Denda S, Meneses JJ, Pedersen RA, Reichardt LF. Integrin alpha8beta1 is critically important for epithelial-mesenchymal interactions during kidney morphogenesis. *Cell* **1997**; *88*: 603–613.

96 Denda S, Reichardt LF, Muller U. Identification of osteopontin as a novel ligand for the integrin alpha8 beta1and potential roles for this integrin-ligand interaction in kidney morphogenesis. *Mol Biol Cell* **1998**; *9*: 1425–1435.

97 Kreidberg JA, Donovan MJ, Goldstein SL, Rennke H, Shepherd K, Jones RC, Jaenisch R. Alpha 3 beta 1 integrin has a crucial role in kidney and lung organogenesis. *Development* **1996**; *122*: 3537–3547.

98 Miner JH, Li C. Defective glomerulogenesis in the absence of laminin alpha5 demonstrates a developmental role for the kidney glomerular basement membrane. *Dev Biol* **2000**; *217*: 278–289.

99 Kanwar YS, Carone FA, Kumar A, Wada J, Ota K, Wallner EI. Role of extracellular matrix growth factors and proto-oncogenes in metanephric development. *Kidney Int* **1997**; *52*: 589–606.

100 Lechner MS, Dressler GR. The molecular basis of embryonic kidney development. *Mech Dev* **1997**; *62*: 105–120.

101 Togawa A, Miyoshi J, Ishizaki H, Tanaka M, Takakura A, Nishioka H, Yoshida H, Doi T, Mizoguchi A, Matsuura N, Niho Y, Nishimune Y, Nishikawa Si, Takai Y. Progressive impairment of kidneys and reproductive organs in mice lacking Rho GDI-alpha. *Oncogene* **1999**; *18*: 5373–5380.

102 Bard JB, Armstrong JF, Bickmore WA. WT1 a Wilms' tumour gene. *Exp Nephrol* **1993**; *1*: 218–23.

103 Kreidberg JA, Sariola H, Loring JM, Maeda M, Pelletier J, Housman D, Jaenisch R. WT–1 is required for early kidney development. *Cell* **1993**; *74*: 679–691.

104 Lee SB, Huang K, Palmer R, Truong VB, Herzlinger D, Kolquist KA, Wong J, Paulding C, Yoon SK,. Gerald W, Oliner JD, Haber DA. The Wilms tumor suppressor WT1 encodes a transcriptional activator of amphiregulin. *Cell* **1999**; *98*: 663–673.

105 Shoyab M, McDonald VL, Bradley JG, Todaro GJ. Amphiregulin: a bifunctional growth-modulating glycoprotein produced by the phorbol 12–myristate 13–acetate-treated human breast adenocarcinoma cell line MCF–7. *Proc Natl Acad Sci USA* **1998**; *85* 6528–6532.

106 Luetteke NC, Qiu TH, Fenton SE, Troyer KL, Riedel RF, Chang A, Lee DC. Targeted inactivation of the EGF and amphiregulin genes reveals distinct roles for EGF receptor ligands in mouse mammary gland development. *Development* **1999**; *126*: 2739–2750.

107 Hosono S, Luo X, Hyink DP, Schapp LM, Wilson PD, Burrow CR, Reddy JC, Atweh GF, Licht JD. WT1 expression induces features of renal epithelial differentiation in mesenchymal fibroblasts. *Oncogene* **1999**; *18*: 417–427.

108. Dressler GR, Woolf AS. Pax2 in development and renal disease. *Int J Dev Biol* **1999**; *43*: 463–468.
109. Torres M, Gomez-Pardo E, Dressler GR. Gruss P. Pax–2 controls multiple steps of urogenital development. *Development* **1995**; *121*: 4057–4065.
110. Dressler GR, Wilkinson JE, Rothenpieler UW, Patterson LT, Williams-Simons L, Westphal H. Deregulation of Pax–2 expression in transgenic mice generates severe kidney abnormalities. *Nature* **1993**; *362*: 65–67.
111. Brodbeck S, Besenbeck B, Englert C. The transcription factor Six2 activates expression of the Gdnf gene as well as its own promoter. *Mech Dev* **2004**; *121*: 1211–1222.
112. Laclef C, Souil E, Demignon J, Maire P. Thymus kidney and craniofacial abnormalities in Six 1 deficient mice. *Mech Dev* **2003**; *120*: 669–679.
113. Xu PX, Zheng W, Huang L, Maire P, Laclef C, Silvius D. Six1 is required for the early organogenesis of mammalian kidney. *Development* **2003**; *130*: 3085–3094.
114. Nishinakamura R, Matsumoto Y, Nakao K, Nakamura K, Sato A, Copeland NG, Gilbert DJ, Jenkins NA, Scully S, Lacey DL, Katsuki M, Asashima M, Yokota T. Murine homolog of SALL1 is essential for ureteric bud invasion in kidney development. *Development* **2001**; *128*: 3105–3115.
115. Miyamoto N, Yoshida M, Kuratani S, Matsuo I, Aizawa S. Defects of urogenital development in mice lacking Emx2. *Development* **1997**; *124*: 1653–1664.
116. Karavanov AA, Karavanova I, Perantoni A, Dawid IB. Expression pattern of the rat Lim–1 homeobox gene suggests a dual role during kidney development. *Int J Dev Biol* **1998**; *42*: 61–66.
117. Shawlot W, Behringer RR. Requirement for Lim1 in head-organizer function. *Nature* **1995**; *374*: 425–430.
118. Kume T, Deng K, Hogan BLM. Murine forkhead/winged helix genes Foxc1 (Mf1) and Foxc2 (Mfh1) are required for the early organogenesis of the kidney and urinary tract. *Development* **2000**; *127*: 1387–1395.
119. Oliver JA, Al-Awqati Q. An endothelial growth factor involved in rat renal development. *J Clin Invest* **1998**; *102*: 1208–1219.
120. Sariola H, Frilander M, Heino T, Jernwall J, Partanen J, Sainio K, Salminen M, Thesleff I. *Solusta yksilöksi: kehitysbiologia (From Cell to Organism: Developmental Biology)*. Duodecim: Helsinki, **2003** (in Finnish).

26
Molecular and Cellular Pathways for the Morphogenesis of Mouse Sex Organs

Humphrey Hung-Chang Yao

26.1
Introduction

How sex is determined has fascinated scientists and philosophers alike for thousands of years. The great philosopher, Aristotle, claimed that the sex of an embryo was determined by the heat of the male partner during intercourse. It was a constant battle between the heat of man's semen and the coldness of the woman's womb that decided the sex of the offspring. Indeed in many other vertebrate species such as reptiles and birds, sex determination is sensitive to environmental factors (temperature) or hormones (steroids). However, mammalian sex determination, at least in eutherian species, is controlled by genes and independent of the environment or, as Aristotle put it, "the moment of the passion". In the 1950s, Alfred Jost proposed the classic paradigm of mammalian sex determination describing this process as a sequential, three-step event. The genetic sex or chromosomal sex is first set at the time of fertilization dependent upon the chromosomal composition of the sperm (X or Y). The chromosomal sex then decides the gonadal sex (ovary for the XX or testis for the XY), which eventually sculptures the phenotypic sex. In the male or XY individual, the presence of the Y chromosome (genetic sex) determines the gonadal sex by triggering the formation of the testis (gonadal sex). Once the testis is formed, it produces androgens that induce differentiation of the internal reproductive tract, external genitalia, and male secondary sexual characteristics (phenotypic sex). In the female or XX individual, the absence of a Y chromosome leads to ovary formation and subsequent feminization. After almost 60 years of scrutiny, this classic paradigm still holds true. The goals of this chapter are to summarize the current knowledge on development of sex organs and to add molecular and cellular insights into the classic sex determination paradigm. Specific focuses will be made on the mouse model because its genetic regulation, cell fate decision, and cell-cell interaction during this dynamic event are the best characterized.

Cell Signaling and Growth Factors in Development. Edited by K. Unsicker and K. Krieglstein
Copyright © 2006 WILEY-VCH Verlag GmbH & Co. KGaA, Weinheim
ISBN 3-527-31034-7

26.2
Building the Foundation: Establishment of the Urogenital Ridge

The urogenital system originates from the intermediate mesoderm along the anterior-posterior axis of the embryo. In 9–day-old mouse embryos (E9), the intermediate mesoderm starts to differentiate into a pair of urogenital ridges, which runs longitudinally along the dorsal aspect of the coelomic cavity. Within the primitive urogenital ridges (Fig. 26.1), three kidney units arise: pronephros, mesonephros, and metanephros [1]. The pronephros, which is at the anterior end of the urogenital ridge, is rudimentary in mammals. The posteriorly located metanephros interacts with the outgrowing ureteric bud to form the definitive kidney. In the central region of the urogenital ridge lies the mesonephros, which is the only unit among the three that contributes to the reproductive system. The mesonephros harbors the Wolffian and Müllerian ducts, which are progenitors of male and female reproductive tracts, respectively (see Section 26.4). It is on the surface of the mesonephros where the bipotential gonad arises.

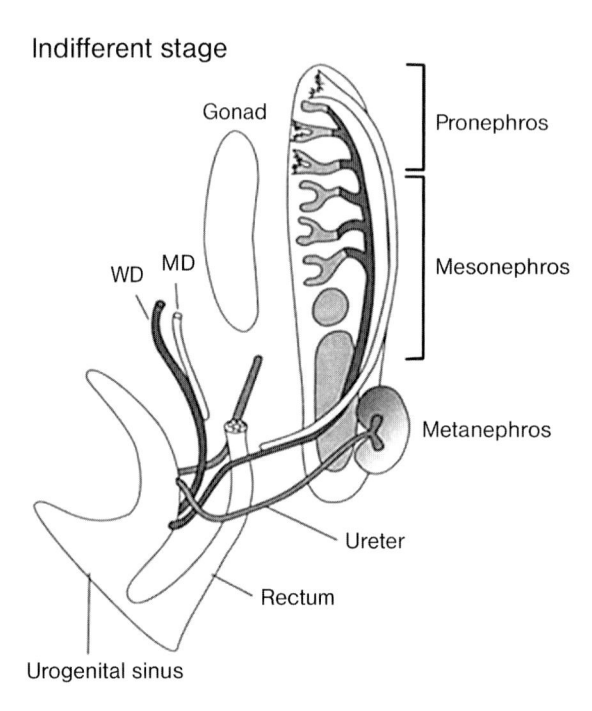

Fig. 26.1
Structure of the urogenital ridge before sex determination. MD, Müllerian duct; WD, Wolffian duct (from [140]).

26.2.1
Gonadogenesis in Mice

The mouse gonad arises on the surface of the mesonephros around midway through gestation (E9–10). The coelomic epithelium along the ventro-medial surface of the mesonephros starts to thicken and becomes the gonadal primordium or genital ridge. The gonadal primordium initially consists of somatic cell lineages derived from the coelomic epithelium and possibly the underlying mesonephros. The coelomic epithelium proliferates, delineates from the surface, and moves to the interior of the gonadal primordium. At the same time, the mesonephros-derived cells migrate into the gonadal primordium. As development progresses, primordial germ cells, which originate and migrate from the extraembryonic mesoderm at the base of the yolk sac, infiltrate the mesenchyme of the gonadal primordium and become associated with somatic cells. Initial formation of the gonad is induced by somatic cell lineages and does not require the presence of primordial germ cells. In embryos in which primordial germ cells are lost before they reach the gonads, early gonad formation occurs normally [2, 3].

26.2.2
Molecular Mechanisms of Gonadogenesis

The initial stage of gonadogenesis appears to be identical in male and female embryos. Genetic experiments have identified several genes essential for early gonad development. Null mutations of *Emx2*, *Wilms tumor 1 (Wt1)*, *Lhx9*, or *Steroidogenic factor 1 (Sf1)* have similar phenotypes with degeneration of gonads in both sexes by E13.5, suggesting that these molecules act together to form or maintain the early gonads.

The homeobox gene *Emx2* is the mouse homolog of the Drosophila head gap gene *ems*. *Emx2* null mice lack gonads and other urogenital organs in both sexes (kidney, ureters, gonads and the genital tracts except bladder). *Emx2* is expressed in the coelomic epithelium and its underlying mesenchyme in developing gonads starting at E10.5. Gonad agenesis is apparent at E11.5 with a lack of thickening of the coelomic epithelium. By E13.0, *Emx2* null gonads are completely degenerated [4].

Wilms tumor-associated gene (*Wt1*) encodes a group of proteins that result from alternate translation, RNA editing, and alternate splicing events [5]. In humans, WT1 was originally identified as a tumor-suppressor gene that is inactivated in a subset of Wilms tumors. Mutations of *WT1* are also associated with urogenital abnormality in patients with WAGR, Denys-Drash, or Frasier syndromes. The necessity for *Wt1* in gonad development has been experimentally examined in mice. Genital ridges from *Wt1* null embryos had a reduced thickening of the coelomic epithelium at E11.5, and by E14, the gonads had completely regressed [6].

Steroidogenic factor 1 (*Sf1*) or adrenal 4–binding protein (*Ad4BP*) is an orphan nuclear receptor whose ligand is currently unknown. Sf1 is expressed in the urogenital ridge of both sexes starting at E9–9.5. *Sf1* expression remains in gonads of both sexes until after E12.5, when it persists in the male but decreases in the female gonad. Targeted disruption of the *Sf1* gene causes complete loss of adrenal glands and gonads in newborn mice. In both sexes, gonads undergo degeneration by E12 with characteristics of apoptosis in both epithelium and mesenchyme of the gonadal primordium and are completely absent at E12.5 [7].

Lhx9, a member of the LIM homeobox domain gene family of transcription factors, is expressed in mouse gonadal primordium at E9.5 in both sexes. The expression seems to be restricted to the coelomic epithelium and its underlying mesenchyme. Null mutation of *Lhx9* leads to degeneration of gonads in both sexes by E13.5. At E11.5, *Lhx9* null and wild-type gonads were morphologically indistinguishable even though cell proliferation in the *Lhx9*–/– gonads is significantly attenuated. This lack of proliferation leads to gonadal degeneration by E12 and the disappearance of gonads by E13.5. Unlike the gonads of *Wt1* and *Sf1* knockout mice which exhibit histological evidence of apoptosis at E11.5–12.5, *Lhx9*–/– gonads do not show signs of apoptotic DNA fragmentation at similar stages. Instead *Lhx9* appears to be responsible for the proliferation of precursor cells which are essential for gonad formation [8].

Recently, essential relationships have emerged between a subset of these genes. Biochemical and genetic studies show that WT1 and LHX9 bind regulatory sequences within the *Sf1* promoter and activate transcription of the gene [9]. These data indicate that *Wt1* and *Lhx9* lie upstream of *Sf1* and are both required for normal expression of *Sf1* in the gonad. Whether Wt1 and Lhx9 interact directly or indirectly to transactivate the *Sf1* promoter is not known. It is clear that *Wt1* expression is not under the control of *Lhx9* because inactivation of *Lhx9* does not affect *Wt1* expression in the gonad. The similar roles of these genes in early gonadogenesis indicate that they either act synergistically or form an hierarchical pathway to initiate and/or to maintain the differentiation of somatic cell lineages in the gonadal primordium. Importantly, this process is not sex specific.

26.3
Parting the Way: Sexually Dimorphic Development of the Gonad

The drama of sexually dimorphic development begins immediately after the bipotential gonadal primordium forms. Although the gonads are indistinguishable morphologically between E10.5 and 12.0, divergent changes at the molecular level already exist. At ~ E10.5–11.0, the XY gonad starts to express *Sry* (Sex-determining region of the Y chromosome), the only gene on the Y chromosome required for initiation of testis development [10, 11]. *Sry* triggers differentiation of the Sertoli cell lineage. Sertoli cells then produce morphogenic factors to orchestrate the establishment of other somatic cell lineages (Leydig cells, peritubular myoid cells, and other interstitial cells) and subsequent testis organization. In the XX gonad where the *Sry* gene is absent, somatic cells become follicle cells that interact with oocytes to form follicles later in development.

26.3.1
Embryonic Testis

26.3.1.1 Sry: The Master Switch from the Y Chromosome

Sry is the only gene on the Y chromosome essential for mammalian testis determination. *Sry* is located on the short arm of the Y chromosome and its conserved homologs have been found on the Y chromosome of most eutherian mammals [10, 12, 13]. The correlation of Sry with testis determination has been clinically demonstrated by male to female sex reversed patients that are explained by mutations or deletion of the *SRY* gene [11, 14–17]. The functional role of Sry was first experimentally confirmed by introducing the genomic DNA containing the *Sry* locus into XX mouse embryos. The presence of the *Sry* gene resulted in testis formation in genetically female embryos [18].

In the mouse XY gonad, *Sry* is expressed in Sertoli cells and is restricted to a very narrow window of time: beginning at E10.5, reaching a peak at E11.5, and becoming undetectable at, and after E12.5 [19, 20]. This is in contrast to human, pigs, and sheep where gonadal *Sry* is expressed at most developmental stages [21–23]. The temporal expression pattern of *Sry* is also accompanied by a unique spatial arrangement. Sry mRNA expression first appears either from the center or simultaneously from the central and anterior regions of the XY gonad and then spreads to the rest of the XY gonad between E11 and 11.5. After E11.5 *Sry* expression is extinguished in an anterior to posterior wave [20]. A similar center-to-pole expression pattern is found in transgenic mice carrying an EGFP reporter gene under the control of the mouse *Sry* promoter [24]. It is not known how these temporal and spatial expression patterns of *Sry* are established in mouse gonad and therefore no conclusions can be drawn about their importance.

Although *Sry* is only produced in the XY gonad, the molecular mechanisms that control *Sry* expression appear to be present in the gonads of both sexes. The XX gonad is able to express *Sry* indicating that other genes on the Y chromosome are not required for *Sry* expression. Introduction of the *Sry* gene in a 14–kb genomic fragment into an XX embryo causes sex reversal and normal testis organogenesis [18]. This non-sex-specific transcriptional control of *Sry* gene is also supported by transgenic studies [25]. In XX embryos carrying the EGFP reporter with the *Sry* promoter, EGFP is found in the somatic cell lineage of the XX gonad. Interestingly, several transcription factors that are found in bipotential gonads of both sexes have been implicated in the regulation of Sry expression. *Wt1* and *Sf1*, which are required for early gonadogenesis, can bind and activate the human *SRY* promoter in transient transfection experiments [26–29]. Genetic analysis reveals additional components including *Fog2*, *Gata4*, *M33*, and insulin receptor family in regulation of *Sry* expression [30–32]. It remains unclear how these transcription factors and signaling pathways interact to induce *Sry* expression (Fig. 26.2).

Sry expression is also under the regulation of *Dax1* (dosage sensitive sex-reversal-adrenal hypoplasia congenital-critical region of the X chromosome gene1) [33, 34]. In humans, duplication of the portion of the X chromosome that includes *DAX1* can cause male to female sex reversal, indicating that the dosage of *DAX1* affects normal

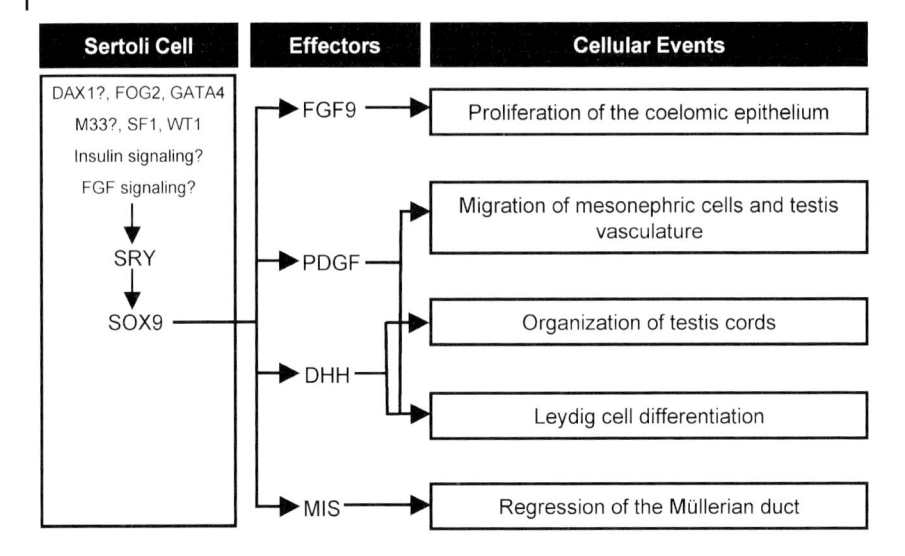

Sertoli Cell	Effectors	Cellular Events
DAX1?, FOG2, GATA4 M33?, SF1, WT1 Insulin signaling? FGF signaling? ↓ SRY ↓ SOX9	→ FGF9 →	Proliferation of the coelomic epithelium
	→ PDGF →	Migration of mesonephric cells and testis vasculature
	→ DHH →	Organization of testis cords
		Leydig cell differentiation
	→ MIS →	Regression of the Müllerian duct

Fig. 26.2
A current model showing the molecular and cellular events in the determination of the mouse testis.

Sry functions or its expression in humans. In contrast, however, over-expression of *Dax1* in mouse did not produce sex reversal phenotypes in XY embryos from the common C57BL6 strain. Sex reversal by *Dax1* over-expression only occurs in the C57BL6 strain carrying the *Mus domesticus poschiavinus* Y chromosome (Y^{pos}), which is known to carry a weaker allele of the *Sry* gene [35]. Based on these findings, Dax1 was initially labeled as an "anti-testis" gene. However, recent genetic experiments demonstrate that *Dax1* is a positive regulator of testis development. Null mutation of *Dax1* results in testicular dysgenesis with severe defects in proliferation of peritubular myoid cells, formation of testis cords, and proper Leydig cell differentiation [36, 37]. Furthermore, when the *Dax1* null allele is crossed to *Mus domesticus poschiavinus* (Y^{pos}), a complete male to female sex reversal arises [38]. Because *Dax1* is present in the gonadal primordium before sexually dimorphic development, it could either interact synergistically with other factors to regulate Sry expression or have specific functions in establishing other somatic cell lineages. Indeed, testicular phenotypes in *Dax1* null animals are improved, but not completely rescued, by the introduction of the human *DAX1* transgene under the control of either Sertoli cell- (Mis or Müllerian inhibiting substance) or Leydig cell- (Luteinizing hormone receptor) specific promoters [39, 40]. Thus, *Dax1* appears to be a pro-testis factor in the normal course of gonad development and its anti-testis characteristic is only manifest under pathological conditions.

How *Sry* triggers testis differentiation is a mystery. The conserved HMG box domains of SRY proteins across mammalian species suggests that SRY is a DNA-binding protein that either activates or suppresses expression of its targets [41]. *In vitro*, the HMG domain of SRY is shown to recognize specific DNA motifs and bend the DNA [42–44]. When the HMG domain of *Sry* is replaced by the HMG domain of *Sry*-related proteins such as *Sox9* or *Sox3*, the recombinant SRY is capable of inducing sex reversal in XX transgenic mice [45], indicating the conserved function of the HMG domain. Although the molecular targets of SRY are still under investigation, emerging genetic evidence indicates that SRY elicits its sex-determining wizardry probably through a single molecule, *Sox9* (see below).

26.3.1.2 Sox9: The Master Switch Downstream of Sry

Unlike *Sry*, whose role in testis determination is restricted only in mammals, *Sox9* (Sry-related HMG box gene 9) is a pivotal sex-determining gene that is highly conserved among vertebrates. *Sox9* is expressed at the critical period of sex determination in American alligator, domestic chicken, Leopard gecko, mouse, and human [21, 46–50]. In mouse, *Sox9* is expressed in the gonadal primordium in both sexes at E10.5. Its expression in the XY gonad is upregulated following *Sry* expression. At the same time, *Sox9* expression diminishes in the XX gonads [47, 50, 51]. *Sox9* is expressed in Sertoli cells, the same cell type that expresses *Sry*. This sex, cell type, and time specificity of *Sox9* expression suggests that it may be downstream of *Sry* in testis development. In fact, introduction of *Sox9* into an XX embryo induces testis formation [52]. Failure to downregulate *Sox9* expression in the Odsex XX gonads also leads to testis development in the absence of *Sry* [53]. Loss-of-function analysis further shows that deletion of *Sox9* in XY gonads interferes with testis development and leads to activation of the ovarian pathway [54–57]. In humans, mutations in *SOX9* cause male to female sex reversal in patients with CD/SRA1 (Campomelic Dysplasia/Autosomal Sex Reversal) [58]. These observations suggest that *Sox9* is the downstream effector of *Sry*. *Sox9* can substitute for the *Sry* in orchestrating the entire course of testis determination.

Another Sox-related gene, *Sox8*, is also involved in mammalian sex determination. *Sox8*, just like *Sox9*, encodes a transcription factor with the ability to bind conserved DNA sequences and to activate transcription *in vitro* [56]. These two genes also have similar XY-specific expression patterns in gonad development. *Sox9* appears to activate the expression of *Sox8* based on the loss of *Sox8* expression in the absence of *Sox9* [54]. However, null mutations of *Sox8* in mice have no obvious gonadal phenotypes [55].

In contrast to *Sry*, whose mode of action is unclear, SOX9 is known to directly activate the expression of *Mis* (or *Amh*, anti-Müllerian hormone) and possibly *Sf1* in Sertoli cells. *Mis* expression does not appear in the gonadal primordium until E11.5, when *Sox9* becomes male specific. The *Mis* promoter contains the conserved Sox-like binding site that allows SOX9 to bind and activate transcription [29, 59]. Mutation of this Sox-like binding site in the *Mis* promoter abolishes normal *Mis* expression in affected mouse embryos. Another potential target of SOX9 is *Sf1*, which shares a similar expression pattern with *Sox9*. The mRNA of both genes is initially present in the gonadal primordium of both sexes at E10.5. At E11.5, both *Sox9* and

Sf1 become upregulated in the XY gonads. The *Sf1* promoter region also contains a Sox-like binding site, which can be transactivated by SOX9 *in vitro* [57]. Genetic analysis is needed to confirm that Sox9 contributes to the XY-specific expression of *Sf1* after E12.5.

26.3.1.3 Specification of Sertoli Cell Lineage

Sertoli cells are now proven to be the only somatic cell lineage in the testis that expresses *Sry*. Palmer and Burgoyne performed an elegant experiment by generating mosaic embryos that contained both XX and XY cells. In the mosaic mice, most cell lineages in the testis consisted of XX and XY cells in a 50 : 50 ratio with the exception of the Sertoli cells. More than 90 % of the Sertoli cell population in the mosaic testis is XY derived, indicating a strong selection force in cells containing a Y chromosome to become Sertoli cells [60–62]. Thus, they proposed that the Sertoli cell lineage is probably the only cell type that *Sry* acts upon. But it was only recently that this hypothesis was finally tested by genetic experiments. A transgenic mouse carrying EGFP reporter driven by the Sry promoter was generated. EGFP is found only in Sertoli cells confirming that the Sertoli cell lineage is the sole cell type in the embryonic testis that expresses *Sry* [24]. The pattern of *Sry-EGFP* transgene expression is consistent with that of the *Sry* mRNA [20]. Furthermore, the *Sry-EGFP* transgene is also expressed in the granulosa cell lineage of the ovary, indicating that Sertoli cells in the testis and granulosa cells in the ovary derive from a common origin.

The origin of the Sertoli cell lineage has been the center of debate for years. Various theories have been proposed that the Sertoli cell lineage derives from the coelomic epithelium, the mesonephros, or both [63]. By labeling the coelomic epithelium with lineage tracing dye, Karl and Capel demonstrated that the coelomic epithelium moves into gonads and becomes Sertoli cells and other cell types in the gonad [64]. However, only the coelomic epithelium from gonads younger than E11.5 is able to contribute to the Sertoli cell lineage. After E11.5, the coelomic epithelium only gives rise to somatic cells located outside testis cords. Interestingly, the coelomic epithelium is also a source of granulosa cell lineage in the XX gonad. These lineage-tracing experiments do not rule out the possibility that the mesonephros also contributes to the Sertoli cell lineage [65]. There is evidence that mesonephric cells migrate into the XY gonad after E11.5. However, the migrating mesonephric cells never become Sertoli cells [66]. If the mesonephros contributes to Sertoli cell precursors, it must do so before E11.5. It is worth noting that neither the coelomic epithelium nor the mesonephros is positive for *Sry* expression. If Sertoli cell precursors derive from these two sources, they likely start to express *Sry* only after they move into the gonad.

How *Sry* triggers Sertoli cell differentiation is still not understood. As noted previously, *Sry* likely stimulates or maintains *Sox9* expression, which itself is sufficient to cause Sertoli differentiation and testis organogenesis [52]. A known target of *Sox9* includes *Mis*, a marker for differentiated Sertoli cells [29, 59, 67–70]. *Mis* is critical for the regression of the Müllerian duct in the mesonephros but is apparently not an effector downstream of Sry or Sox9 in testis organogenesis. Null mutation of *Mis* did not affect testis formation. However, while MIS is critical for the regression of the Müllerian duct, it does not appear to play a role in testis organization. Null mutation of *Mis* does not affect testis formation [71].

26.3.1.4 Cellular Events Triggered by *Sry*

Approximately 24 h after the onset of *Sry* expression (E12.0), the amorphic gonadal primordium undergoes dynamic morphological transition in the XY embryo. Sertoli cells, the only cell type that expresses *Sry*, produce morphogenic factors to coordinate the complex processes of cell fate decision, cell migration, and tissue reorganization. Different somatic cell lineages such as Leydig cells and peritubular myoid cells start to arise and structures unique to the testis began to take shape. Eventually, Sertoli cells and germ cells are sequestered inside the testis cords to create a unique environment for spermatogenesis to occur. Outside of the testis cords, Leydig cells produce androgens that are transported out of the testis via a specialized vasculature network. Although it is still a mystery how a single gene like *Sry* can change the fate of the entire organ, we have begun to understand how *Sry* functions by dissecting the cellular processes induced by *Sry*. These processes are: proliferation of the coelomic epithelium, migration of the mesonephric cells, organization of testis cords, and specification of the Leydig cell lineage (Fig. 26.2).

Proliferation of the coelomic epithelium
Proliferation of the coelomic epithelium is the earliest cellular landmark in testis organogenesis. The increase in proliferation in the coelomic epithelium begins less than 24 h after the onset of *Sry* expression and before the appearance of any other testis-specific morphological changes. This *Sry*-dependent proliferation of the coelomic epithelium occurs in two distinct stages: during the first stage (before E11.5) proliferating cells contribute to Sertoli and other somatic cell populations. During the second stage (after E11.5), proliferating cells give rise to other somatic cells in the testis except Sertoli cells [72]. Inhibition of proliferation during the first, but not the second stage, blocks testis cord formation, indicating coelomic epithelium proliferation before E11.5 is critical for the establishment of the Sertoli cell population and subsequent testis organization [73]. Interestingly, *Sry* is expressed in somatic cells inside the XY gonad but not in the coelomic epithelium. *Sry* is thought to trigger production of unknown proliferation-inducing factors that act on the coelomic epithelium. One possible candidate for this proliferation-inducing factor is *fibroblast growth factor 9* (*Fgf9*). Null mutation of *Fgf9* causes a decrease in proliferation of the coelomic epithelium in the XY gonad, leading to XY to XX sex reversal [74]. The molecular connection between *Sry* and *Fgf9* remains to be established.

Migration of mesonephric cells and testis vasculature

Only the XY gonad has the ability to induce mesonephric cell migration, and this ability is dependent upon the presence of *Sry* [66, 75]. The mechanism of mesonephric cell migration was characterized by a recombinant organ culture system with a wild-type gonad cultured on top of a mesonephros from embryos that ubiquitously express either the LacZ or EGFP transgene. Migration of the mesonephric cells (LacZ- or EGFP-positive) occurs only when the sex of the gonad is XY. The migrating mesonephric cells contribute to peritubular myoid cells and other interstitial cells but not to Sertoli cells or germ cells in the XY gonad [66, 76–79]. Although the XX gonad is not able to induce mesonephric migration, it is receptive to the mesonephric cells. Mesonephric cells can be induced to migrate into the XX gonad when it is sandwiched between the XY gonad and mesonephros [79]. The chemoattractant(s) from the XY gonad operates over a long distance (across the XX gonad) to induce mesonephric cell migration into the XX gonad. Numerous signaling pathways have been shown to be involved in mediating mesonephric cell migration. Ectopic mesonephric migration can be induced in the XX gonad by treatment with neurotropin 3, MIS, BMP2, or BMP4 [80, 81]. Among these molecules, MIS was thought to be the candidate for the endogenous chemoattractant because MIS, but not the others, is expressed specifically in the XY gonad. However, null mutation of *Mis* had no effects on mesonephric migration. It is possible that multiple and/or redundant pathways are responsible for this event. Indeed, defects in mesonephric migration are also observed in the Pdgf receptor alpha null XY gonad [82], and in XY gonads treated with inhibitors of the phosphotidylinositol 3–kinase pathway [80, 83, 84] or inhibitors of matrix metalloproteinases [85]. Because mesonephric cell migration is such a critical event for the organization of testis cords, it seems reasonable to have redundant pathways as "safety nets" to ensure its occurrence.

One major cell type that migrates from the mesonephros into the XY gonad is the endothelial cell. These migrating endothelial cells contribute to the formation of the coelomic vessel and elaboration of the arterial system in the XY gonads. This XY-specific arterial system establishes a new pattern of blood flow, which is important for transportation of hormones and possibly organization of the testis cords [86]. Recent evidence indicates that this endothelial migration is inhibited in the XX gonad via the *Wnt4* pathway [87]. In the *Wnt4* null ovary, the endothelial migration leads to ectopic formation of the coelomic vessel (also see Section 26.3.2.1).

Organization of testis cords

Formation of testis cords is one ultimate end-product of *Sry* action. Testis cords harbor Sertoli cells and male germ cells that are separated from the interstitium by the basal membrane and a single layer of squamous peritubular myoid cells, thereby creating a unique niche for male germ cell maturation. Formation of testis cords requires coordinated cellular processes including proliferation of the coelomic epithelium and mesonephric cell migration. The coelomic epithelium proliferates to expand the Sertoli cell population and at the same time, precursors of peritubular myoid cells migrate from the mesonephros. If any of these processes is interrupted, testis cords fail to form properly [73, 88].

How do these different cell types organize the testis cords? It is logical to assume that the *Sry*-expressing Sertoli cells must produce morphogenic molecule(s) to coordinate cell-cell interactions in this process. One candidate for such a morphogenic molecule is desert hedgehog (*Dhh*). *Dhh* gene encodes a secreted morphogen that is expressed exclusively in Sertoli cells [89, 90]. On the other hand, its receptor, *Patched 1* (*Ptch1*), is expressed in peritubular myoid cells and other cells in the interstitium [91]. Null mutation of *Dhh* causes malformed testis cords with incomplete deposition or absence of the basal membrane [92, 93]. In addition, a complete failure of testis cord organization occurs when signaling is inhibited by treatment of a general hedgehog inhibitor, cyclopamine, in culture [94]. *Dhh* does not participate in inducing mesonephric migration; instead, it is thought to act on the interstitial cells to trigger the differentiation of peritubular myoid cells. Sertoli cells then start to deposit the basal membrane at the interface where they are in contact with the peritubular myoid cells. Testis cord defects are also observed in *Pdgfra* and *Dax1* null embryos. These two molecules are not directly involved in the cellular organization of the testis cords. They likely regulate the proliferation and migration of the peritubular myoid cell precursors [37, 82].

Specification of the Leydig cell lineage
Fetal Leydig cells produce androgens that maintain the Wolffian duct and induce male secondary sex characteristics. This XY-specific somatic cell lineage starts to appear in the interstitium outside of testis cords at E12.5, 24 h after the peak of *Sry* expression. It is known that Leydig cell precursors and steroid-producing adrenal precursors share a common origin that expresses *Sf1* before the separation of gonadal and adrenal primordia [95–97]. However, this common origin has not as yet been identified. The mesonephros is one possible origin of the Leydig cell precursors. Migration experiments using recombinant gonad culture (see Section 26.3.1.3) reveal that a small percentage of migrating mesonephric cells differentiates into Leydig cells [77, 98]. When the E11.5 XY gonad is cultured without the mesonephros attached, some Leydig cells still appear, suggesting that if precursors of Leydig cells migrate from the mesonephros, they must do so before E11.5 [76]. Another possible source of Leydig cell precursors is the coelomic epithelium. Lineage tracing experiments using DiI labeling indicate that the coelomic epithelium contributes to interstitial cells after E11.5, but it is not clear whether these interstitial cells actually differentiate into Leydig cells [82, 99].

Although Leydig cells are specific to XY gonads, they never express the testis-determining gene *Sry*. Leydig cell differentiation appears to be indirectly regulated by *Sry*, likely through the production of morphogens by the *Sry*-expressing Sertoli cells. The signaling pathway induced by *Dhh* and its receptor *Ptch1* is required for normal Leydig cell differentiation. *Dhh* is expressed in Sertoli cells whereas its receptor *Ptch1* is expressed in interstitial cells outside of testis cords. Embryos carrying a null mutation of *Dhh* have a severely reduced Leydig cell population, leading to pseudohermaphrodite phenotypes such as undescended testes, atrophic male internal accessory organs, and female external genitalia [90, 92]. DHH is shown to act as a paracrine factor to trigger differentiation of Leydig cell precursors in the interstitium. Once Leydig cells differentiate, they no longer require the presence of the DHH signaling [91]. Decreases in Leydig cell numbers are also observed in XY embryos carrying a null mutation for Pdgf receptor alpha [82]. The PDGFα signaling is required for specification and/or expansion of the Leydig cell population. The DHH and PDGF signaling appear to operate in parallel in regulating Leydig cell differentiation based on findings that the components of PDGFα signaling remains unchanged in *Dhh* null mutants (and vice versa). Symptoms of fetal Leydig cell defects were also reported in mouse and human carrying mutations in X-linked aristaless-related homeobox gene [100]. Together, these observations suggest that the Leydig cell lineage is derived from multiple origins and its differentiation is regulated by a network of signaling pathways.

Fetal Leydig cells produce testosterone and insulin-like factor 3 (INSL3 or relaxin-like factor), which regulate differentiation of the Wolffian duct (see Section 26.4.2) and external genitalia (see Section 26.5), and testis descent (discussed here). Descent of testes into the scrotum is critical because proper spermatogenesis requires a lower temperature. Testis descent is a two-stage event including the transabdominal migration (INSL3 dependent) and inguino-scrotal descent (testosterone dependent) [101]. Before the production of INSL3 and testosterone, the bipotential gonads are loosely supported by dorsal ligaments (or cranial suspensory ligament or CSL) and ventral ligaments (later develops into gubernaculums). In the male, testosterone causes involution of the CSL whereas INSL3 induces massive growth of the gubernaculums via the LGR8 receptor during the transabdominal migration [102, 103]. The thickening gubernaculums retain the testes close to the inguinal region while the rest of the abdominal contents grow dorsally. In the second migration step, the inguinal canal and scrotum are formed under the influence of testosterone, thus facilitating the passage of the testes into the scrotum [104]. In the female embryo this type of testosterone and INSL3 are not produced, and the CSL continues to develop whereas the gubernaculums involutes, maintaining the ovary close to the kidney [105]. Inactivation of either testosterone or the INSL3/LGR8 pathway leads to cryptorchidism [106–108]. Furthermore, when the INSL3/LGR8 pathway is ectopically activated in the female embryo, trasabdominal migration of the ovary, but not the inguino-scrotal descent, occurs [105]. These observations indicate that a complete descent of testes requires both testosterone and INSL3 signaling. Deletion of the homeobox gene *Hoxa10* also causes cryptorchidism but its molecular connection with testosterone or INSL3 is not understood [109].

26.3.2
Embryonic Ovary

26.3.2.1 The Quest for the Ovary-determining Gene

Sex determination in most mammals is a male-dominant event, dictated by the presence or absence of *Sry*. SRY turns on the male pathway and transform the amorphic bipotential gonad into a testis. Without *Sry*, the bipotential gonad has amorphic status until right before birth, when formation of the ovarian follicles occurs. Scientists have been searching for the ovarian counterpart of *Sry* without any success. Genetic screening or sex reversal cases have not been fruitful in providing candidates for ovary determination. An alternative theory also proposed that *Sry* is a suppressor of the putative "Z" gene, whose function is to activate the ovarian pathway while inhibiting the "default" testis development [110]. Many genes exhibit an ovary-specific expression pattern but none of them is implicated in the initial formation of the ovary (see below).

Dax1 was initially considered as the candidate for the ovary-determining gene because of its location on the X chromosome and its ovary-specific expression [111]. However, knockout and transgenic studies in mice indicate that *Dax1* is not required for normal ovary development [112]. Ironically, it is critical for testis development (see Section 26.3.1.1). *Foxl2*, a putative forkhead transcription factor, is proposed to be an early regulator of ovary development based on its conserved ovary-specific expression pattern in vertebrates [113], its implication in premature ovarian failure in human [114–116] and intersex syndrome in the XX sex-reversed goat [117, 118]. Disappointingly, null mutation of *Foxl2* in mice does not affect ovary formation although defects in granulosa cell differentiation and ovary maintenance at a later stage have been reported [119, 120]. Other genes such as caveolin1 and *Adamt19* are also expressed in an XX-specific manner at the time of sex determination [121, 122]. Their functional roles in ovary development remain unknown.

At present, defects in early ovary development are only observed in mouse embryos carrying null mutations in *Wnt4* and *follistatin*. WNT4 and follistatin are secreted molecules expressed exclusively in the XX, but not XY gonad, at the time of sex determination [121, 123]. Inactivation of *Wnt4* or *follistatin* does not affect ovary formation but causes unique defects in the XX gonads. A large artery, which is referred to as the coelomic vessel, appears on the coelomic surface of the *Wnt4* or *follistatin* null XX gonad [87]. This coelomic vessel is a structure found only in the XY gonad and is an early morphological landmark of testis development. The coelomic vessel in *Wnt4* and *follistatin* null XX gonads is located underneath the coelomic epithelium as in normal XY gonads, except that it lacks any branching. In the normal developmental course of the XX gonad, WNT4 and FST appear to inhibit a specific aspect of testis character, the coelomic vessel. At least for *Wnt4*, it has been shown that WNT4 suppresses endothelial cell migration from the mesonephros, preventing the formation of the coelomic vessel [87]. Because mesonephric cell migration is a testis-specific event downstream of *Sry* (see Section 26.3.1.4), it would be reasonable to assume that components of the *Sry* pathway are activated in the absence of *Wnt4* or *follistatin*, and would therefore induce formation of the testis-specific vessel. However, *Sox9*, the putative downstream effector of *Sry* (see Section 26.3.1.2), is not expressed in *Wnt4* or *follistatin* null XX gonads. This observation implies that formation of the coelomic vessel does not depend upon the presence of *Sry* and activation of its downstream components. There is no doubt that further research is required to support this hypothesis.

Wnt4 was also claimed to inhibit Leydig cell differentiation in the ovary. This is based on the initial finding that androgen-producing Leydig cells appear ectopically in the *Wnt4* null XX gonad. *Wnt4* is therefore labeled as an "anti-male" factor [123].

However, further investigation indicated that the steroid-producing cells in *Wnt4* null XX gonads are not Leydig cells; instead, they express markers specific for adrenal cells. Furthermore, these ectopic adrenal cells are found not only in XX but also XY gonads in the absence of *Wnt4* [124]. Adrenal cells and Leydig cells share common precursors in the primitive adrenal-gonadal primordium. *Wnt4* is likely to be critical for the proper separation of adrenal cells from the gonadal primordium but is not a repressor of fetal Leydig cell differentiation [87, 125].

26.3.2.2 Female Germ Cells: The Key to Femaleness

The process of sex determination ensures the creation of two distinct somatic environments that support the maturation of sperm and oocytes. Unlike the testis, where morphological organization occurs immediately after the onset of sex determination, the ovarian structure does not appear until birth. The most striking difference between testis and ovary differentiation is the involvement of germ cells. Germ cells are not essential for testis determination, and testis organization progresses normally in the absence of germ cells [3]. However, when germ cells are absent in the XX gonad, ovarian follicles, the functional units in the ovary, never form [2]; or in the case when germ cells are lost after formation of follicles, follicles rapidly degenerate and somatic cells transdifferentiate into Sertoli-like cells [126–128]. Germ cells in the ovary appear to be a critical regulator in organizing and maintaining the functional structure. This distinct role of germ cells in gonad development may result from the dimorphic development pattern of the germ cells. In the XY gonad, germ cells are sequestered to the testis cords and prevented from entering the first meiosis. On the other hand, germ cells in the XX gonad enter meiosis around E13.5 and arrest in prophase I of meiosis by birth [2]. Once germ cells enter meiosis, they are destined to become oocytes. Furthermore, these meiotic germ cells ensure the progression of ovarian development by inhibiting the occurrence of the certain testis events in the XX gonad. Migration of mesonephric cells and formation of testis cords can be induced in XX gonads only before germ cells enter meiosis. But once germ cells enter meiosis, the XX gonad loses its plasticity and no longer allows these two events to occur [129, 130]. These observations suggest that meiotic germ cells hold the key to the path of ovarian development.

Understanding the mechanism of female germ cell meiosis is the first step in solving the mystery of ovary organogenesis. Female germ cells enter meiosis in a cell-autonomous manner. When germ cells develop in ectopic regions such as adrenal glands or mesonephros [131], or when they are assembled in lung aggregates in culture [132], they still enter meiosis around E13.5, the same time as they do in the ovary. This indicates that female germ cells follow an intrinsic clock to enter meiosis which is independent of the somatic environment. Once female germ cells enter the first meiosis and commit to the ovarian fate, they become oogonia and express a putative transcription *Figα* (factor in germline α). FIGα regulates the production of the zona pellucida and recruitment of granulosa cells to form primordial follicles, the basic units that nurture the oocyte [133].

There is no doubt that female germ cells control the organization of the ovary, which in fact, makes perfect sense because germ cells are the reason for the existence of the gonads. However, it is not to say that somatic cells are insignificant in this process and passively receive organizing signals from the oocyte. Finely-tuned cross-talk between oocyte and its surrounding somatic cells is required for proper formation of the ovarian follicle. Null mutations of *Wnt4* or *follistatin*, which are somatic cell-derived factors, cause loss of meiotic germ cells, and follicles never form [123, 134]. Inactivation of *Foxl2*, a somatic cell-specific transcription factor, prevents further differentiation of granulosa cells in the primordial follicle, leading to oocyte atresia (see Section 26.3.2.1). These observations raise the possibility that the long-sought ovary-determining gene may not exist. Organogenesis of the ovary is actually the product of a coordinated communication between germ cells and somatic cells.

26.4
Dimorphic Development of the Reproductive Tracts

Mesonephros, the middle nephric unit of the urogenital ridge (see Section 26.2 and Fig. 26.1), is where the primitive male (Wolffian or mesonephric duct) and female (Müllerian or paramesonephric duct) reproductive tracts arise. The Wolffian duct is originally derived from the pronephros, whose ductal derivative elongates posteriorly through the mesonephros and extends to the cloaca. The pronephros eventually degenerates but its duct derivative remains in the mesonephros and becomes the Wolffian duct between E9 and E10 (Fig. 26.3). At E11.5, the Müllerian duct starts to form by invagination of the coelomic epithelium of the anterior mesonephros, and extend posteriorly until it reaches the cloaca at E13.5 (Fig. 26.3). These two ductal systems are initially present in both male and female embryos, meaning that the embryo has to maintain the system that matches its own sex and at the same time, remove the other of the opposite sex. Alfred Jost first characterized the mechanisms for this decision-making process in the rabbit embryos [135, 136]. He found that when the male embryo was castrated, the Wolffian duct regressed whereas the Müllerian duct persisted. Removal of ovaries did not affect normal female development. Jost then transplanted testes into a female embryo and found that the female embryo underwent the same structural changes as a male embryo with regression of the Müllerian duct and maintenance of the Wolffian duct. Finally, when androgen, instead of the entire testis, was replaced in a male or female embryo without gonads, Wolffian development was supported but the Müllerian system failed to regress. Jost therefore postulated that the female or Müllerian duct is a "default" system whose development does not need the presence of the gonad. In the male, this default system is removed by a testis-derived factor called "L'hormone inhibitrice", later named MIS or AMH. On the other hand, the Wolffian duct requires the presence of testes to maintain its development. The production of MIS and testosterone in the male embryo eventually leads to Müllerian duct regression and Wolffian duct differentiation, which then becomes the epididymis, vas deferens, and seminal vesicles, whilst in the female, the absence of both testosterone and MIS results in Wolffian duct degeneration and maintenance of the Müllerian duct which eventually differentiates into the oviduct, uterus, and the upper region of the vagina.

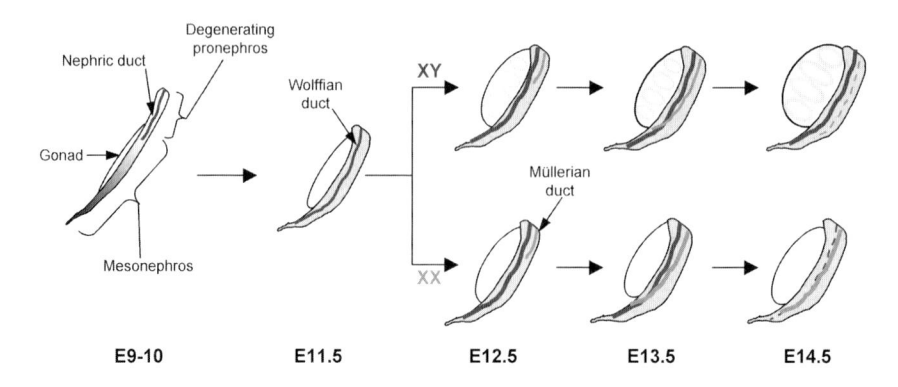

Fig. 26.3
Sexually dimorphic development of the Wolffian and Müllerian ducts in mouse between E9 and E14.5. (This figure also appears with the color plates.)

26.4.1
Initial Formation of the Wolffian and Müllerian Ducts

Initial formation of the Wolffian and Müllerian ducts is independent of the sexes of the gonad or hormones derived from the gonads. Instead, a network of transcription factors and signaling molecules control the establishment of the ducts. Both Wolffian and Müllerian ducts are believed to derive from the mesonephric mesenchyme, which undergoes transformation to differentiate into the epithelial tubes of the ducts. The mechanisms underlying this mesenchymal-epithelial transition are not well understood but current evidence suggests that this transformation is cell autonomous. Several transcription factors such as *Pax2*, *Pax8*, *Lim1*, and *Emx2* are expressed in the epithelium of the both Wolffian and Müllerian ducts. Null mutations of these genes affect different aspects of epithelial differentiation but eventually lead to a similar phenotype: degeneration of both Wolffian and Müllerian ducts by E13.5 [4, 137–139]. Despite the fact that molecular interactions among these factors are still obscure, a simplified model is proposed here based on knockout and epistatic analysis. For the Wolffian duct, both *Pax2* and *Pax8* and possibly *Lim1* are required for the epithelium transition in the pronephros at E9. As the Wolffian duct (or nephric duct at the time) elongates and extends posteriorly toward the mesonephros between E10 and E11, *Pax8* becomes indispensable and only *Pax2* is required for this elongation process. After E11.5, another transcription factor *Emx2* become essential for the maintenance of the epithelium structure. Once the Wolffian duct structure is stabilized, its further maintenance becomes hormone dependent, specifically on testosterone from the Leydig cells of the developing XY gonads [140].

Factors involved in the initial establishment of the Wolffian duct also play roles in Müllerian duct development with a few exceptions. When the Müllerian duct starts to form by invagination of the anterior coelomic epithelium into the mesonephric mesenchyme, it requires *Pax2*, *Lim1*, and *Emx2* in the epithelium to complete the transformation and for maintenance (the role of *Pax8* is not clear). Tubule formation also needs *Wnt4*, a mesenchyme-derived signaling molecule. Without *Wnt4*, the Müllerian epithelium is absent in spite of the presence of the epithelial precursors [139]. Signaling through the retinoic acid receptor family (including RARα1, RXRα1, RARβ2, and RARγ) is also essential for Müllerian development. Compound knockouts of the RAR gene family cause the complete absence of the female reproductive tract [141–143]. It is not clear how RAR signaling coordinates with other factors in this process.

26.4.2
Sexually Divergent Development of Reproductive Tracts

The sex-specific differentiation of the reproductive tract starts around E12.5 in mouse embryos. In the male embryo, the Müllerian duct, precursor of the female reproductive tract, is degraded by Sertoli cell-derived MIS [144, 145]. MIS diffuses into the neighboring mesonephros and acts on the Müllerian mesenchyme, where the receptors for MIS are located. MIS is a TGFβ family protein that requires heterodimeric interaction of its type 2 (ligand binding) and type 1 (signaling transduction) receptors. MIS activates the signaling transduction pathway mediated by MIS receptor 2 (MISR2) and ALK3 (the type 1 receptor, also know as BMPRIa) to induce apoptosis of the Müllerian epithelium and disruption of the ductal structure [146–151]. The functional links among MIS, MISR2, and ALK3 in Müllerian regression have been elegantly established by genetic analysis. Inactivation of any one of these genes leads to the identical phenotype: failure to trigger Müllerian degeneration in the affected male [71, 148, 149]. Another important player in Müllerian degeneration is *Wnt7a*. *Wnt7a* is expressed in the Müllerian epithelium and is essential for the induction of *Misr2* in the Müllerian mesenchyme [152]. Without *Wnt7a*, *Misr2* expression is absent leading to the inability of MIS to trigger Müllerian duct degeneration. Further research is still needed to understand how the activation of MIS signaling leads to the disintegration of the Müllerian duct.

While MIS triggers the Müllerian degeneration, testosterone produced by the Leydig cells keeps the Wolffian duct intact. Testosterone is a steroid hormone whose action is mediated via its receptor (androgen receptor or AR) inside the target cells. Once testosterone enters the cells and binds to the AR, the AR recognizes specific sequences in the promoter region of the target genes and activates or suppresses gene transcription. AR is expressed in both epithelium and mesenchyme of the Wolffian duct and it is assumed that testosterone acts on both compartments [153]. The importance of AR in mediating testosterone-induced signaling is best demonstrated in AR null mice [154]. The male reproductive tract and its accessory glands are completely absent in the AR null male, which is identical to the effects of castration in the original Jost experiments. It is not clear how the activation of AR by testosterone maintains the Wolffian duct. *In vitro* experiments show that epidermal growth factor (EGF) can mimic the effects of testosterone in maintaining the Wolffian duct in the female mesonephros. Treatment with anti-EGF antibody prevents the normal Wolffian duct morphogenesis in the male mesonephros, which can be reversed by EGF supplementation [155]. EGF could be a downstream target of testosterone in Wolffian duct development but genetic evidence is needed to confirm its role.

In contrast to the decisive role of the testis in reproductive tract development, the ovary is dispensable in this process as demonstrated by Jost. In the female embryo, neither MIS nor testosterone is produced and therefore the Müllerian duct is maintained while the Wolffian duct degenerates [156]. How the Müllerian duct remains intact in the female is not understood due to lack of genetic models. It is postulated that factors essential for initial formation of the Müllerian duct (such as *Pax2*, *Lim1*, and *Emx2*) are also important for the maintenance of the Müllerian duct. However, this hypothesis cannot be tested because the Müllerian duct never forms in these knockout embryos.

26.4.3
Patterning of the Reproductive Tracts

How does a single Müllerian duct differentiate into oviduct, uterus, cervix, and upper part of the vagina in an exact anterior to posterior pattern? The same question can be posed for the Wolffian duct, which is patterned to become the epididymis, vas deferens, and seminal vesicle. The molecular mechanisms for this dynamic patterning have begun to emerge just recently. Not surprisingly, this process involves a complex network of cell-autonomous differentiation, epithelium-mesenchyme interaction, and hormonal regulation. The patterning of the Müllerian and Wolffian ducts utilizes common components as well as divergent pathways. In the female, the patterning of the Müllerian duct does not occur until 5 days after birth [157, 158]. Taking advantage of this long quiescent period, Cunha pioneered a recombinant graft system to study epithelium-mesenchyme interaction in Müllerian patterning. During this period, the mesenchyme seems to provide a guiding force to direct epithelial differentiation. When the epithelium from the donor portion (i.e. presumptive vagina) is grafted to the mesenchyme from the host (i.e. presumptive uterus), the donor epithelium develops characteristics of the host (vaginal to uterine transformation in this case). The donor epithelium is only able to respond to the host mesenchyme before day 5 after birth and this receptivity is lost after that. These elegant experiments indicate that the Müllerian mesenchyme directs epithelium patterning during a brief window of time (first 5 days after birth). Once the epithelial identity has been established by the mesenchymal signal(s), the epithelial fate is set and cannot be reversed [159, 160]. Furthermore, proper mesenchyme differentiation also depends on signals from the epithelium. Without the presence of the epithelium, the mesenchyme fails to differentiate [161]. The nature of the interaction between Müllerian epithelium and mesenchyme (cell-cell contact or diffusible factors) is not understood at the molecular level.

In contrast to the late onset of Müllerian patterning, morphological transformation of the Wolffian duct occurs as early as E15 in the male. Because of the practical difficulties of grafting experiments in embryonic tissues, the interaction between Wolffian epithelium and mesenchyme remains unknown. Based on morphological analysis, the anterior or upper portion of the Wolffian duct adjacent to the testis elongates and folds into the epididymis. Meanwhile, the mesonephric tubules differentiate into efferent ducts that eventually connect the rete testis and epididymis. The middle portion of the Wolffian duct remains as a simple tube to form the vas deferens. The posterior or caudal portion of the Wolffian duct dilates, elongates cranially, and eventually form a distinct diverticulum [162].

Homeobox transcription factors (*Hox*) are appealing candidates for the establishment of reproductive tracts because of their roles in patterning of embryos and other organs. In fact, *Hoxa* and *Hoxd*, two clusters of *Hox* genes, have partially overlapping expression in the mesenchyme of the female reproductive tract (the patterns in the male tract are not known). In late gestation, *Hoxd10* and *Hoxd11* are expressed in the oviduct, *Hoxd12* is expressed in the uterus, and *Hoxd13* is expressed in the posterior uterus and in the vagina [163, 164]. After birth, *Hoxa9* is expressed in the oviduct, *Hoxa10* in the uterus, *Hoxa11* in the uterus and cervix, and *Hoxa13* in the cervix and upper vagina [165, 166]. Genetic analyses have identified functional roles for several *Hox* genes in accordance with their unique anterior to posterior expression. Null mutation of *Hoxa10* causes patterning defects in specific regions of the reproductive tract that express *Hoxa10*. In the affected female, the anterior uterus loses its identity and transform into oviduct [167]. A similar homeotic transformation is found in male embryos with the anterior vas deferens becoming the epididymis. Defects in accessory gland development also arise in *Hoxa* null embryos [168]. The importance of *Hox* genes in defining tissue boundaries in the reproductive tract is also highlighted by ectopic gain-of-function experiments. When the *Hoxa11* gene (uterus specific) is replaced by *Hoxa13* (cervix and vagina specific), the uterus is transformed into cervix and vagina [169]. Some *Hox* genes may have redundant functions because of their overlapping expression patterns. One example is the partial overlap of *Hoxd13* and *Hoxa13* expression. Null mutation of *Hoxd13* (vaginal expression) does not affect vaginal patterning. However, compound knockouts of *Hoxd* and *Hoxa* (*Hoxd13*–/– *Hoxa13*+/–) display multiple vaginal anomalies [165].

In addition to the *Hox* genes, the *Wnt* family proteins including *Wnt5a* and *Wnt7a* are also implicated in patterning of the reproductive tracts, specifically in the female. *Wnt5a* is produced in the Müllerian mesenchyme and is shown to regulate patterning of the posterior Müllerian duct (cervix and vagina) [170]. On the other hand, *Wnt7a* is derived from the Müllerian epithelium and is essential for patterning of the anterior Müllerian duct (oviduct and uterus) [171]. *Wnt5a* and *Wnt7a* together also regulate uterine glandulargenesis. There is also evidence indicating possible crosstalk between *Wnt* and *Hox* signaling. The similar vaginal phenotype in *Wnt5a* knockout and *Hoxd13;Hoxa13* compound knockouts suggests that these three factors may act in a common pathway to regulate posterior differentiation of the Müllerian duct [165, 170]. Epistatic analysis also reveals that maintenance of *Wnt7a* expression requires *Hoxa10/Hoxa11* and vice versa. *Wnt7a* and *Hoxa10/11* together direct the patterning of the uterine horns and stromal/epithelial differentiation [172, 173].

26.5
Morphogenesis of the External Genitalia

Initial formation and patterning of the external genitalia is independent of hormones and identical in male and female embryos. The outgrowth of the genital tubercle (GT) is the first morphological landmark of the external genitalia, occurring around E10.5 in mouse [174]. Several transcription factors and growth factors have been implicated in this process. *Hoxa13;Hoxd13* compound null mice fail to form a GT probably due to defects in posterior patterning [165]. Sonic hedgehog (*Shh*), a secreted morphogen, also regulates GT outgrowth and deletion of *Shh* causes penile agenesis [175, 176]. Absence of GT outgrowth is also observed in *Wnt5a* null mice, where *Wnt5a* is shown to be a proliferation-stimulating factor. Interestingly, null mutation of Noggin, a BMP antagonist, results in hypertrophy of the GT, indicating that GT development is a balanced act between positive and negative regulators [177]. A defined molecular picture of GT development is still obscure at present.

Sexually dimorphic external sex features do not become apparent until several days (E14–15) after the initial formation of the gonads. The primitive external genitalia consist of the genital tubercle (GT), the urogenital folds, and the bilateral genital swellings (Fig. 26.4). In the male under the influence of androgens, the GT elongates to form the penis, while the genital swellings migrate posteriorly and fuse in the midline to form the scrotum. The urogenital folds extend ventrally and fuse with each other, thus forming the tubular urethra. In the female where androgens are absent the GT undergoes minimal growth and becomes the clitoris. The urogenital folds and genital swellings remain separated and develop into the labia minora and majora, respectively. The urogenital folds (labia minora) outline the vestibule of the vagina where the urethral and vaginal openings arise.

Androgen receptor signaling is responsible for sculpturing the external genitalia. In the developing GT, the enzyme 5α-reductase in the mesenchyme converts testosterone into 5α-dihydrotestosterone (DHT), which is the masculinizing force in the external genitalia [178]. Androgen receptors are also present in the mesenchymal cells of the GT suggesting the autocrine/paracrine action of androgens [179]. It is worth noting that this androgen signaling system is present in the external genitalia of both sexes. However, testosterone is only produced in male but not in female embryos. Recent evidence demonstrates that coordinated cross-talk among FGF, WNT, and BMP pathways is crucial for the normal development of external genitalia [174, 177, 180]. It remains to be determined how these signaling pathways integrate with the androgen receptor pathway.

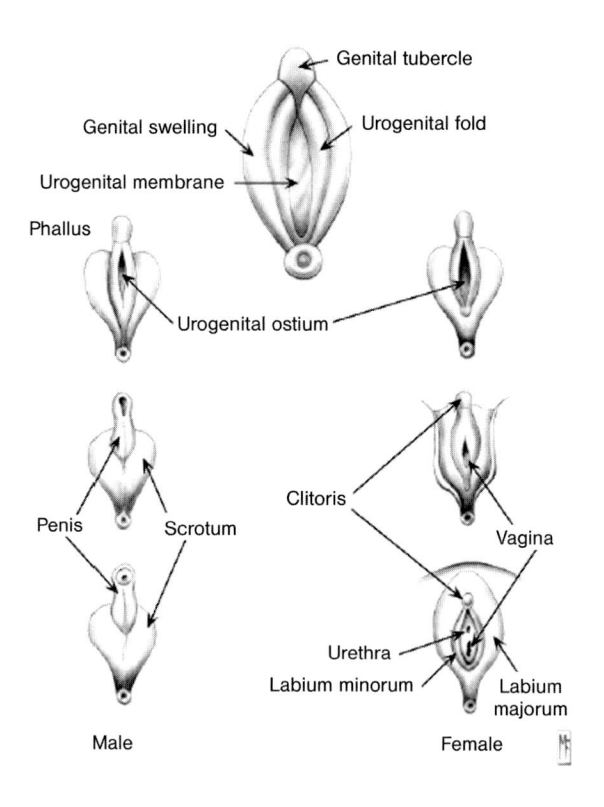

Fig. 26.4
Sex differentiation of the external genitalia (from [174]).

26.6
Summary

Understanding the development of sex organs is not just a matter of satisfying human curiosity about the origin of the sexes, but also a critical issue related to human welfare. Reproductive defects have become a rising medical problem in recent years. Infertility affects about 20 % of couples wishing to have children. Anomalies of the urogenital system such as intersexes and cryptorchidism are the third most common birth defect after circulatory and musculoskeletal defects, affecting approximately one in 1000 newborns. Although infertility and congenital disorders in urogenital systems are rarely fatal, their impact on the welfare of patients and society is enormous. The power of the mouse genetic models and the availability of genomic information have greatly advanced our knowledge regarding the molecular and cellular mechanisms of development of the sex organs. The story of sex determination and differentiation is far from complete but suffice to say that the pieces of this intriguing puzzle are gradually coming into place.

References

1 A. Staack, A.A. Donjacour, J. Brody, G.R. Cunha, P. Carroll, Mouse urogenital development: a practical approach. *Differentiation* **2003**; *71*: 402–413.

2 A. McLaren, Meiosis and differentiation of mouse germ cells. *Symp Soc Exp Biol* **1984**; *38*: 7–23.

3 H. Merchant, Rat gonadal and ovarian organogenesis with and without germ cells. An ultrastructural study. *Dev Biol* **1975**; *44*: 1–21.

4 N. Miyamoto, M. Yoshida, S. Kuratani, I. Matsuo, S. Aizawa, Defects of urogenital development in mice lacking *Emx2*. *Development* **1997**; *124*: 1653–1664.

5 V. Vidal, A. Schedl, Requirement of WT1 for gonad and adrenal development: insights from transgenic animals. *Endocr Res* **2000**; *26*: 1075–1082.

6 J.A. Kriedberg et al., Wt–1 is required for early kidney development. *Cell* **1993**; *74*: 679–691.

7 X. Luo, Y. Ikeda, K. Parker, A cell-specific nuclear receptor is essential for adrenal and gonadal development and sexual differentiation, *Cell* **1994**; *77*: 481–490.

8 O. Birk et al., The LIM homeobox gene *Lhx9* is essential for mouse gonad formation. *Nature* **2000**; *403*: 909–913.

9 D. Wilhelm, C. Englert, The Wilms tumor suppressor WT1 regulates early gonad development by activation of Sf1. *Genes Dev* **2002**; *16*: 1839–1851.

10 J. Gubbay et al., Inverted repeat structure of the *Sry* locus in mice. *Proc Natl Acad Sci. USA* **1992**; *89*: 7953–7957.

11 J.R. Hawkins et al., Mutational analysis of *SRY*: Nonsense and missense mutations in XY sex reversal. *Hum Genet* **1992**; *88*: 471–474.

12 J. Foster et al., Evolution of sex determination and the Y chromosome: SRY related sequences in marsupials. *Nature* **1992**; *359*: 531–533.

13 A.H. Sinclair et al., A gene from the human sex-determining region encodes a protein with homology to a conserved DNA-binding motif. *Nature* **1990**; *346*: 240–244.

14 P. Berta et al., Genetic evidence equating SRY and the testis-determining factor. *Nature* **1990**; *348*: 448–450.

15 J.R. Hawkins et al., Evidence for increased prevalence of *SRY* mutations in XY females with complete rather than partial gonadal dysgenesis. *Am J Hum Genet* **1992**; *51*: 979–984.

16 R.J. Jager, R.A. Pfeiffer, G. Scherer, An amino acid substitution in SRY leads to conditional XY gonadal dysgenesis. *Int Cong Hum Genet* **1991**; *8*: 12–13.

17 K. McElreavey et al., XY sex reversal associated with a deletion 5' to the SRY HMG box in the testis-determining region. *Proc Natl Acad Sci USA* **1992**; *89*: 11016–11020.

18 P. Koopman, J. Gubbay, N. Vivian, P. Goodfellow, R. Lovell-Badge, Male development of chromosomally female mice transgenic for Sry. *Nature* **1991**; *351*, 117–121.

19 A. Hacker, B. Capel, P. Goodfellow, R. Lovell-Badge, Expression of *Sry*, the mouse sex determining gene. *Development* **1995**; *121*: 1603–1614.

20 M. Bullejos, and Koopman, P., Spatially dynamic expression of Sry in mouse genital ridges. *Dev Dyn* **2001**; *221*: 201–205.
21 N.A. Hanley et al., SRY, SOX9, and DAX1 expression patterns during human sex determination and gonadal development. *Mech Dev* **2000**; *91*: 403–407.
22 P. Parma, E. Pailhoux, C. Cotinot, Reverse transcription-polymerase chain reaction analysis of genes involved in gonadal differentiation in pigs. *Biol Reprod* **1999**; *61*: 741–748.
23 E. Payen et al., Characterization of ovine SRY transcript and developmental expression of genes involved in sexual differentiation. *Int J Dev Biol* **1996**; *40*: 567–575.
24 K.H. Albrecht, E.M. Eicher, Evidence that Sry is expressed in pre-Sertoli cells and Sertoli and granulosa cells have a common precursor. *Dev Biol* **2001**; *240*: 92–107.
25 M. Ito et al., In vitro Cre/loxP system in cells from developing gonads: investigation of the Sry promoter. *Dev Growth Differ* **2002**; *44*: 549–557.
26 R. Shimamura, G.C. Fraizer, J. Trapman, C. Lau Yf, G.F. Saunders, The Wilms' tumor gene WT1 can regulate genes involved in sex determination and differentiation: SRY, Mullerian-inhibiting substance, and the androgen receptor. *Clin Cancer Res* **1997**; *3*: 2571–2580.
27 A. a. S. Hossain, G.F., The human sex-determining gene SRY is a direct target of WT1. *J Biol Chem* **2001**; *276*: 16817–16823.
28 P. de Santa Barbara et al., Steroidogenic factor–1 contributes to the cyclic-adenosine monophosphate down-regulation of human SRY gene expression. *Biol Reprod* **2001**; *64*: 775–783.
29 P. De Santa Barbara et al., Direct interaction of SRY-related protein SOX9 and steroidogenic factor 1 regulates transcription of the human anti-Mullerian hormone gene. *Mol Cell Biol* **1998**; *18*: 6653–6665.
30 S.G. Tevosian et al., Gonadal differentiation, sex determination and normal Sry expression in mice require direct interaction between transcription partners GATA4 and FOG2. *Development* **2002**; *129*: 4627–4634.
31 S. Nef et al., Testis determination requires insulin receptor family function in mice. *Nature* **2003**; *426*: 291–295.
32 Y. Katoh-Fukui et al., Male to female sex reversal in M33 mutant mice. *Nature* **1998**; *393*: 688–692.
33 E. Zanaria et al., An unusual member of the nuclear hormone receptor superfamily responsible for X-linked adrenal hypoplasia congenita. *Nature* **1994**; *372*: 635–641.
34 F. Muscatelli et al., Mutations in the DAX–1 gene give rise to both X-linked adrenal hypoplasia congenita and hypogonadotropic hypogonadism. *Nature* **1994**; *372*: 672–676.
35 A. Swain, V. Narvaez, P. Burgoyne, G. Camerino, R. Lovell-Badge, Dax1 antagonizes Sry action in mammalian sex determination. *Nature* **1998**; *391*: 761–767.
36 B. Jeffs et al., Blockage of the rete testis and efferent ductules by ectopic Sertoli and Leydig cells causes infertility in Dax1–deficient male mice. *Endocrinology* **2001**; *142*: 4486–4495.
37 J.J. Meeks et al., Dax1 regulates testis cord organization during gonadal differentiation. *Development* **2003**; *130*: 1029–1036.

38 J.J. Meeks, J. Weiss, J.L. Jameson, Dax1 is required for testis determination. *Nat Genet* **2003**; *34*: 32–33.

39 B. Jeffs et al., Sertoli cell-specific rescue of fertility, but not testicular pathology, in Dax1 (Ahch)-deficient male mice. *Endocrinology* **2001**; *142*: 2481–2488.

40 J.J. Meeks et al., Leydig cell-specific expression of DAX1 improves fertility of the Dax1–deficient mouse. *Biol Reprod* **2003**; *69*: 154–160.

41 S. Whitfield, R. Lovell-Badge, P.N. Goodfellow, Rapid sequence evolution of the sex determining gene SRY. *Nature* **1993**; *364*: 713–715.

42 K. Giese, J. Pagel, R. Grosschedl, Distinct DNA-binding properties of the high mobility group domain of murine and human SRY sex-determining factors. *Proc Natl Acad Sci USA* **1994**; *91*: 3368–3372.

43 M. Desclozeaux et al., Characterization of two Sp1 binding sites of the human sex determining SRY promoter. *Biochim Biophys Acta* **1998**; *1397*: 247–252.

44 V. Harley, R. Lovell-Badge, P. Goodfellow, Definition of a consensus DNA binding site for SRY. *Nucleic Acid Res* **1994**; *22*: 1500–1501.

45 D.E. Bergstrom, M. Young, K.H. Albrecht, E.M. Eicher, Related function of mouse SOX3, SOX9, and SRY HMG domains assayed by male sex determination. *Genesis* **2000**; *28*: 111–124.

46 P. Western, H. JL, J. Graves, A. Sinclair, Temperature-dependent sex determination: upregulation of SOX9 expression after commitment to male development. *Developmental Dynamics* **1999**; *214*: 171–177.

47 J. Kent, S.C. Wheatley, J.E. Andrews, A.H. Sinclair, P. Koopman, A male-specific role for SOX9 in vertebrate sex determination. *Development* **1996**; *122*: 2813–2822.

48 N. Moreno-Mendoza, V.R. Harley, H. Merchant-Larios, Differential expression of SOX9 in gonads of the sea turtle *Lepidochelys olivacea* at male- or female-promoting temperatures. *J Exp Zool* **1999**; *284*: 705–710.

49 C.A. Smith, M.J. Smith, A.H. Sinclair, Gene expression during gonadogenesis in the chicken embryo. *Gene* **1999**; *234*: 395–402.

50 S. M. da Silva et al., *Sox9* expression during gonadal development implies a conserved role for the gene in testis differentiation in mammals and birds. *Nat Genet* **1996**; *14*: 62–68.

51 N. Moreno-Mendoza, V. Harley, H. Merchant-Larios, Cell aggregation precedes the onset of Sox9–expressing preSertoli cells in the genital ridge of mouse. *Cytogenet Genome Res* **2003**; *101*: 219–223.

52 V.P. Vidal, M.C. Chaboissier, D. G. de Rooij, A. Schedl, Sox9 induces testis development in XX transgenic mice. *Nat Genet* **2001**; *28*: 216–217.

53 C. Bishop et al., A transgenic insertion upstream of *Sox9* is associated with dominant XX sex reversal in the mouse. *Nat Genet* **2000**; *26*: 490–494.

54 M.C. Chaboissier et al., Functional analysis of Sox8 and Sox9 during sex determination in the mouse. *Development* **2004**; *131*: 1891–1901.

55. S. Takada, P. Koopman, Origin and possible roles of the Sox8 transcription factor gene during sexual development. *Cytogenet Genome Res* **2003**; *101*: 212–218.
56. G. Schepers, M. Wilson, D. Wilhelm, P. Koopman, SOX8 is expressed during testis differentiation in mice and synergizes with SF1 to activate the Amh promoter *in vitro*. *J Biol Chem* **2003**; *278*: 28101–28108.
57. J.H. Shen, H.A. Ingraham, Regulation of the orphan nuclear receptor steroidogenic factor 1 by Sox proteins. *Mol Endocrinol* **2002**; *16*: 529–540.
58. K.C. Knower, S. Kelly, V.R. Harley, Turning on the male – SRY, SOX9 and sex determination in mammals. *Cytogenet Genome Res* **2003**; *101*: 185–198.
59. N. Arango, R. Lovell-Badge, R. Behringer, Targeted mutagenesis of the endogenous mouse Mis gene promoter: *in vivo* definition of genetic pathways of vertebrate sexual development. *Cell* **1999**; *99*: 409–419.
60. P.S. Burgoyne, M. Buehr, P. Koopman, J. Rossant, A. McLaren, Cell autonomous action of the testis-determining gene: Sertoli cells are exclusively XY in XX XY chimaeric mouse testes. *Development* **1988**; *102*: 443–450.
61. P.S. Burgoyne, M. Buehr, A. McLaren, XY follicle cells in ovaries of XX-XY female mouse chimaeras. *Development* **1988**; *104*: 683–688.
62. S. Palmer, P.S. Burgoyne, XY follicle cells in the ovaries of XO/XY and XO/XY/XYY mosaic mice. *Development* **1991**; *111*: 1017–1020.
63. A.G. Byskov, Differentiation of mammalian embryonic gonad. *Physiol Rev* **1986**; *66*: 71–117.
64. J. Karl, B. Capel, Sertoli cells of the mouse testis originate from the coelomic epithelium. *Dev Biol* **1998**; *203*: 323–333.
65. H. Wartenberg, Rodemer-Lenz, E., and C.H. Viebahn, The dual Sertoli cell system and its role in testicular development and in early germ cell differentiation (prespermatogenesis). *Reprod Biol Med* **1986**; 44–57.
66. J. Martineau, K. Nordqvist, C. Tilmann, R. Lovell-Badge, B. Capel, Male-specific cell migration into the developing gonad. *Curr Biol* **1997**; *7*: 958–968.
67. M. Nachtigal et al., Wilms'Tumor1 and Dax1 modulate the orphan nuclear receptor Sf1 in sex-specific gene expression. *Cell* **1998**; *93*: 445–454.
68. J.J. Tremblay, N.M. Robert, R.S. Viger, Modulation of endogenous GATA–4 activity reveals its dual contribution to Mullerian inhibiting substance gene transcription in Sertoli cells. *Mol Endocrinol* **2001**; *15*: 1636–1650.
69. R. Viger, C. Mertineit, J. TRasler, M. Nemer, Transcription factor GATA–4 is expressed in a sexually dimorphic pattern during mouse gonadal development and is a potent activator of the Mullerian inhibiting substance promoter. *Development* **1998**; *125*: 2665–2675.
70. K. Watanabe, T.R. Clarke, A.H. Lane, X. Wang, P.K. Donahoe, Endogenous expression of Mullerian inhibiting substance in early postnatal rat Sertoli cells requires multiple steroidogenic

factor–1 and GATA–4–binding sites. *Proc Natl Acad Sci USA* **2000**; *97*: 1624–1629.

71 R.R. Behringer, M.J. Finegold, R.L. Cate, *Müllerian* inhibiting substance function during mammalian sexual development. *Cell* **1994**; *79*: 415–425.

72 J. Schmahl, E.M. Eicher, L.L. Washburn, B. Capel, Sry induces cell proliferation in the mouse gonad. *Development* **2000**; *127*: 65–73.

73 J. Schmahl, B. Capel, Cell proliferation is necessary for the determination of male fate in the gonad. *Dev Biol* **2003**; *258*: 264–276.

74 J.S. Colvin, R.P. Green, J. Schmahl, B. Capel, D.M. Ornitz, Male-to-female sex reversal in mice lacking fibroblast growth factor 9. *Cell* **2001**; *104*: 875–889.

75 B. Capel, K.H. Albrecht, L.L. Washburn, E.M. Eicher, Migration of mesonephric cells into the mammalian gonad depends on *Sry*. *Mech Dev* **1999**; *84*: 127–131.

76 H. Merchant-Larios, N. Moreno-Mendoza, M. Buehr, The role of the mesonephros in cell differentiation and morphogenesis of the mouse fetal testis. *Int J Dev Biol* **1993**; *37*: 407–415.

77 K. Nishino, K. Yamanouchi, K. Naito, H. Tojo, Characterization of mesonephric cells that migrate into the XY gonad during testis differentiation. *Exp Cell Res* **2001**; *267*: 225–232.

78 K. Nishino et al., Establishment of fetal gonad/mesonephros coculture system using EGFP transgenic mice. *J Exp Zool* **2000**; *286*: 320–327.

79 C. Tilmann, B. Capel, Mesonephric cell migration induces testis cord formation and Sertoli cell differentiation in the mammalian gonad. *Development* **1999**; *126*: 2883–2890.

80 A.S. Cupp, L. Tessarollo, M.K. Skinner, Testis developmental phenotypes in neurotropin receptor trkA and trkC null mutations: role in formation of seminiferous cords and germ cell survival. *Biol Reprod* **2002**; *66*: 1838–1845.

81 A.J. Ross, C. Tilman, H. Yao, D. MacLaughlin, B. Capel, AMH induces mesonephric cell migration in XX gonads. *Mol Cell Endocrinol* **2003**; *211*: 1–7.

82 J. Brennan, C. Tilmann, B. Capel, Pdgfr-alpha mediates testis cord organization and fetal Leydig cell development in the XY gonad. *Genes Dev* **2003**; *17*: 800–810.

83 M. Uzumcu, K.A. Dirks, M.K. Skinner, Inhibition of platelet-derived growth factor actions in the embryonic testis influences normal cord development and morphology. *Biol Reprod* **2002**; *66*: 745–753.

84 E. Levine, A.S. Cupp, M.K. Skinner, Role of neurotropins in rat embryonic testis morphogenesis (cord formation). *Biol Reprod* **2000**; *62*: 132–142.

85 K. Nishino, K. Yamanouchi, K. Naito, H. Tojo, Matrix metalloproteinases regulate mesonephric cell migration in developing XY gonads which correlates with the inhibition of tissue inhibitor of metalloproteinase–3 by sry. *Dev Growth Differ* **2002**; *44*: 35–43.

86 J. Brennan, J. Karl, B. Capel, Divergent vascular mechanisms downstream of Sry establish the arterial system in the XY gonad, *Dev Biol* **2002**; *244*: 418–428.

87 K. Jeays-Ward et al., Endothelial and steroidogenic cell migration are regulated by WNT4 in the developing mammalian gonad. *Development* **2003**; *130*: 3663–3670.
88 M. Buehr, S. Gu, A. McLaren, Mesonephric contribution to testis differentiation in the fetal mouse. *Development* **1993**; *117*: 273–281.
89 M.J. Bitgood, A.P. McMahon, Hedgehog and Bmp genes are coexpressed at many diverse sites of cell-cell interaction in the mouse embryo. *Dev Biol* **1995**; *172*: 126–138.
90 M.J. Bitgood, L. Shen, A.P. McMahon, Sertoli cell signaling by Desert hedgehog regulates the male germline. *Curr Biol* **1996**; *6*: 298–304.
91 H.H. Yao, W. Whoriskey, B. Capel, Desert Hedgehog/Patched 1 signaling specifies fetal Leydig cell fate in testis organogenesis. *Genes Dev* **2002**; *16*: 1433–1440.
92 A.M. Clark, K.K. Garland, L.D. Russell, Desert hedgehog (Dhh) gene is required in the mouse testis for formation of adult-type Leydig cells and normal development of peritubular cells and seminiferous tubules. *Biol Reprod* **2000**; *63*: 1825–1838.
93 F. Pierucci-Alves, A.M. Clark, L.D. Russell, A developmental study of the desert hedgehog-null mouse testis. *Biol Reprod* **2001**; *65*: 1392–1402.
94 H.H. Yao, B. Capel, Disruption of testis cords by cyclopamine or forskolin reveals independent cellular pathways in testis organogenesis. *Dev Biol* **2002**; *246*: 356–365.
95 O. Hatano, A. Takakusu, M. Nomura, K. Morohashi, Identical origin of adrenal cortex and gonad revealed by expression profiles of Ad4BP/SF–1. *Genes Cells* **1996**; *1*: 663–671.
96 O. Hatano et al., Sex-dependent expression of a transcription factor, Ad4BP, regulating steroidogenic P–450 genes in the gonads during prenatal and postnatal rat development. *Development* **1994**; *120*: 2787–2797.
97 K. Morohashi, The ontogenesis of the steroidogenic tissues. *Genes Cells* **1997**; *2*: 95–106.
98 H. Merchant-Larios, N. Moreno-Mendoza, Mesonephric stromal cells differentiate into Leydig cells in the mouse fetal testis. *Exp Cell Res* **1998**; *244*: 230–238.
99 J. Karl, B. Capel, Three-dimensional structure of the developing mouse genital ridge. *Phil Tran Royal Soc Lond* **1995**; *350*: 235–242.
100 K. Kitamura et al., Mutation of ARX causes abnormal development of forebrain and testes in mice and X-linked lissencephaly with abnormal genitalia in humans. *Nat Genet* **2002**; *32*: 359–369.
101 R. Ivell, S. Hartung, The molecular basis of cryptorchidism. *Mol Hum Reprod* **2003**; *9*: 175–181.
102 Y. Kubota et al., Leydig insulin-like hormone, gubernacular development and testicular descent. *J Urol* **2001**; *165*: 1673–1675.
103 J. Kumagai et al., INSL3/Leydig insulin-like peptide activates the LGR8 receptor important in testis descent. *J Biol Chem* **2002**; *277*: 31283–31286.

104 J.M. Hutson, S. Hasthorpe, C.F. Heyns, Anatomical and functional aspects of testicular descent and cryptorchidism. *Endocr Rev* **1997**; *18*: 259–280.

105 P. Koskimies et al., Female mice carrying a ubiquitin promoter-Insl3 transgene have descended ovaries and inguinal hernias but normal fertility. *Mol Cell Endocrinol* **2003**; *206*: 159–166.

106 I.M. Adham, J.M. Emmen, W. Engel, The role of the testicular factor INSL3 in establishing the gonadal position. *Mol Cell Endocrinol* **2000**; *160*: 11–16.

107 S. Nef, L.F. Parada, Cryptorchidism in mice mutant for Insl3. *Nat Genet* **1999**; *22*: 295–299.

108 S. Zimmermann et al., Targeted disruption of the Insl3 gene causes bilateral cryptorchidism. *Mol Endocrinol* **1999**; *13*: 681–691.

109 I. Satokata, G. Benson, R. Maas, Sexually dimorphic sterility phenotypes in Hoxa10–deficient mice. *Nature* **1995**; *374*: 460–463.

110 K. McElreavey, E. Vilain, N. Abbas, I. Herskowitz, M. Fellous, A regulatory cascade hypothesis for mammalian sex determination: SRY represses a negative regulator of male development. *Proc Natl Acad Sci USA* **1993**; *90*: 3368–3372.

111 A. Swain, E. Zanaria, A. Hacker, R. Lovell-Badge, G. Camerino, Mouse *Dax1* expression is consistent with a role in sex determination as well as in adrenal and hypothalamus function. *Nat Genet* **1996**; *12*: 404–409.

112 R. Yu, T. Ito, S. Saunders, S. Camper, L. Jameson, Role of Ahch in gonadal development and gametogenesis. *Nat Genet* **1998**; *20*: 353–357.

113 K.A. Loffler, D. Zarkower, P. Koopman, Etiology of ovarian failure in blepharophimosis ptosis epicanthus inversus syndrome: FOXL2 is a conserved, early-acting gene in vertebrate ovarian development. *Endocrinology* **2003**; *144*: 3237–3243.

114 L. Crisponi et al., The putative forkhead transcription factor FOXL2 is mutated in blepharophimosis/ptosis/epicanthus inversus syndrome. *Nat Genet* **2001**; *27*: 159–166.

115 D. Fuhrer, Lessons from studies of complex genetic disorders: identification of FOXL2 – a novel transcription factor on the wing to fertility. *Eur J Endocrinol* **2002**; *146*: 15–18.

116 K. Kosaki, T. Ogata, R. Kosaki, S. Sato, N. Matsuo, A novel mutation in the FOXL2 gene in a patient with blepharophimosis syndrome: differential role of the polyalanine tract in the development of the ovary and the eyelid. *Ophthal Genet* **2002**; *23*: 43–47.

117 J. Cocquet et al., Evolution and expression of FOXL2. *J Med Genet* **2002**; *39*: 916–921.

118 E. Pailhoux et al., A 11.7–kb deletion triggers intersexuality and polledness in goats. *Nat Genet* **2001**; *29*: 453–458.

119 D. Schmidt et al., The murine winged-helix transcription factor Foxl2 is required for granulosa cell differentiation and ovary maintenance. *Development* **2004**; *131*: 933–942.

120 M. Uda et al., Foxl2 disruption causes mouse ovarian failure by pervasive blockage of follicle development. *Hum Mol Genet* **2004**; *13*: 1171–1181.

121 D. Menke, D. Page, Sexually dimorphic gene expression in the developing mouse gonad. *Gene Expr Pattern* **2002**; *2*: 359–367.
122 M. Bullejos, J. Bowles, P. Koopman, Extensive vascularization of developing mouse ovaries revealed by caveolin–1 expression. *Dev Dyn* **2002**; *225*: 95–99.
123 S. Vainio, M. Heikkila, A. Kispert, N. Chin, A. McMahon, Female development in mammals is regulated by Wnt–4 signaling. *Nature* **1999**; *397*: 405–409.
124 M. Heikkila et al., Wnt–4 deficiency alters mouse adrenal cortex function, reducing aldosterone production. *Endocrinology* **2002**; *143*: 4358–4365.
125 B.K. Jordan et al., Up-regulation of WNT–4 signaling and dosage-sensitive sex reversal in humans. *Am J Hum Genet* **2001**; *68*: 1102–1109.
126 R.R. Behringer, R.L. Cate, G.J. Froelick, R.D. Palmiter, R.L. Brinster, Abnormal sexual development in transgenic mice chronically expressing Müllerian inhibiting substance. *Nature* **1990**; *345*: 167–170.
127 J.F. Couse et al., Postnatal sex reversal of the ovaries in mice lacking estrogen receptors alpha and beta. *Science* **1999**; *286*: 2328–2331.
128 N. Hashimoto, R. Kubokawa, K. Yamazaki, M. Noguchi, Y. Kato, Germ cell deficiency causes testis cord differentiation in reconstituted mouse fetal ovaries. *J Exp Zool* **1990**; *253*: 61–70.
129 K. Tilmann, B. Capel, Mesonephric cell migration induces testis cord formation and Sertoli cell differentiation in the mammalian gonad. *Development* **1999**; *126*: 2883–2890.
130 H.H. Yao, L. DiNapoli, B. Capel, Meiotic germ cells antagonize mesonephric cell migration and testis cord formation in mouse gonads. *Development* **2003**; *130*: 5895–5902.
131 L. Zamboni, S. Upadhyay, Germ cell differentiation in mouse adrenal glands. *J Exp Zool* **1983**; *228*: 173–193.
132 A. McLaren, D. Southee, Entry of mouse embryonic germ cells into meiosis. *Dev Biol* **1997**; *187*: 107–113.
133 L. Liang, S.M. Soyal, J. Dean, FIGalpha, a germ cell specific transcription factor involved in the coordinate expression of the zona pellucida genes. *Development* **1997**; *124*: 4939–4947.
134 H.H. Yao et al., Follistatin operates downstream of Wnt4 in mammalian ovary organogenesis. *Dev Dyn* **2004**; *230*: 210–215.
135 A. Jost, Recherches sur la differenciation sexuelle de l'embryon de lapin. *Arch Anat Microsc Morph Exp* **1947**; *36*: 271–315.
136 A. Jost, Problems of fetal endocrinology: the gonadal and hypophyseal hormones. *Recent Prog Horm Res* **1953**; *8*: 379–418.
137 M. Torres, E. Gomez-Pardo, G.R. Dressler, P. Gruss, *Pax–2* controls multiple steps of urogenital development. *Development* **1995**; *121*: 4057–4065.
138 M. Bouchard, A. Souabni, M. Mandler, A. Neubuser, M. Busslinger, Nephric lineage specification by Pax2 and Pax8. *Genes Dev* **2002**; *16*: 2958–2970.
139 A. Kobayashi, W. Shawlot, A. Kania, R.R. Behringer, Requirement of Lim1 for female reproductive tract development. *Development* **2004**; *131*: 539–549.
140 U. Drews, Local mechanisms in sex specific morphogenesis. *Cytogenet Cell Genet* **2000**; *91*: 72–80.

141 P. Kastner et al., Vitamin A deficiency and mutations of RXRalpha, RXRbeta and RARalpha lead to early differentiation of embryonic ventricular cardiomyocytes. *Development* **1997**; *124*: 4749–4758.

142 C. Mendelsohn et al., Retinoic acid receptor beta 2 (RAR beta 2) null mutant mice appear normal. *Dev Biol* **1994**; *166*: 246–258.

143 C. Mendelsohn et al., Function of the retinoic acid receptors (RARs) during development (II). Multiple abnormalities at various stages of organogenesis in RAR double mutants. *Development* **1994**; *120*: 2749–2771.

144 N. Josso, Action de la testosterone sur le canal de Wolff du foetus de rat en culture organotypique. *Arch Anat Microsc* **1970**; *59*: 37–50.

145 N. Josso, J.Y. Picard, B. Vigier, Purification de l'hormone anti-Müllerian bovine a l'aide d'un anticorps monoclonal. *CRAcadSci* **1981**; *293*: 447–450.

146 T.R. Clarke et al., Mullerian inhibiting substance signaling uses a bone morphogenetic protein (BMP)-like pathway mediated by ALK2 and induces SMAD6 expression. *Mol Endocrinol* **2001**; *15*: 946–959.

147 L. Gouedard et al., Engagement of bone morphogenetic protein type IB receptor and Smad1 signaling by anti-Mullerian hormone and its type II receptor. *J Biol Chem* **2000**; *275*: 27973–27978.

148 S.P. Jamin, N.A. Arango, Y. Mishina, M.C. Hanks, R.R. Behringer, Requirement of Bmpr1a for Mullerian duct regression during male sexual development. *Nat Genet* **2002**; *32*: 408–410.

149 Y. Mishina et al., Genetic analysis of the Mullerian-inhibiting substance signal transduction pathway in mammalian sexual differentiation. *Genes Dev* **1996**; *10*: 2577–2587.

150 J.A. Visser, AMH signaling: from receptor to target gene. *Mol Cell Endocrinol* **2003**; *211*: 65–73.

151 J.A. Visser et al., The serine/threonine transmembrane receptor ALK2 mediates Mullerian inhibiting substance signaling. *Mol Endocrinol* **2001**; *15*: 936–945.

152 B.A. Parr, A.P. McMahon, Sexually dimorphic development of the mammalian reproductive tract requires Wnt-7a. *Nature* **1998**; *395*: 707–710.

153 U. Drews, O. Sulak, M. Oppitz, Immunohistochemical localisation of androgen receptor during sex-specific morphogenesis in the fetal mouse. *Histochem Cell Biol* **2001**; *116*: 427–439.

154 T. Matsumoto, K. Takeyama, T. Sato, S. Kato, Androgen receptor functions from reverse genetic models. *J Steroid Biochem Mol Biol* **2003**; *85*: 95–99.

155 C. Gupta, S. Siegel, D. Ellis, The role of EGF in testosterone-induced reproductive tract differentiation. *Dev Biol* **1991**; *146*: 106–116.

156 W.J. Dyche, A comparative study of the differentiation and involution of the Mullerian duct and Wolffian duct in the male and female fetal mouse. *J Morphol* **1979**; *162*: 175–209.

157 G.R. Cunha, Epithelial-stromal interactions in development of the urogenital tract. *Int Rev Cytol* **1976**; *47*: 137–194.

158 G.R. Cunha, Stromal induction and specification of morphogenesis and cytodifferentiation of the epithelia of the Mullerian ducts and urogenital sinus during development of the uterus and vagina in mice. *J Exp Zool* **1976**; *196*: 361–370.

159 R.M. Bigsby, P.S. Cooke, G.R. Cunha, A simple efficient method for separating murine uterine epithelial and mesenchymal cells. *Am J Physiol* **1986**; *251*: E630–E636.

160 R.M. Bigsby, G.R. Cunha, Estrogen stimulation of deoxyribonucleic acid synthesis in uterine epithelial cells which lack estrogen receptors. *Endocrinology* **1986**; *119*: 390–396.

161 C. Miller, A. Pavlova, D.A. Sassoon, Differential expression patterns of Wnt genes in the murine female reproductive tract during development and the estrous cycle. *Mech Dev* **1998**; *76*: 91–99.

162 J.H. Lipschutz et al., Clonality of urogenital organs as determined by analysis of chimeric mice. *Cells Tissues Organs* **1999**; *165*: 57–66.

163 P. Dolle, J.C. Izpisua-Belmonte, E. Boncinelli, D. Duboule, The Hox-4.8 gene is localized at the 5′ extremity of the Hox-4 complex and is expressed in the most posterior parts of the body during development. *Mech Dev* **1991**; *36*: 3–13.

164 P. Dolle, J.C. Izpisua-Belmonte, J.M. Brown, C. Tickle, D. Duboule, HOX-4 genes and the morphogenesis of mammalian genitalia. *Genes Dev* **1991**; *5*: 1767–1767.

165 X. Warot, C. Fromental-Ramain, V. Fraulob, P. Chambon, P. Dolle, Gene dosage-dependent effects of the Hoxa-13 and Hoxd-13 mutations on morphogenesis of the terminal parts of the digestive and urogenital tracts. *Development* **1997**; *124*: 4781–4791.

166 H.S. Taylor, G.B. Vanden Heuvel, P. Igarashi, A conserved Hox axis in the mouse and human female reproductive system: late establishment and persistent adult expression of the Hoxa cluster genes. *Biol Reprod* **1997**; *57*: 1338–1345.

167 G.V. Benson et al., Mechanisms of reduced fertility in Hoxa-10 mutant mice: uterine homeosis and loss of maternal Hoxa-10 expression. *Development* **1996**; *122*: 2687–2696.

168 C.A. Podlasek et al., Hoxa-10 deficient male mice exhibit abnormal development of the accessory sex organs. *Dev Dyn* **1999**; *214*: 1–12.

169 Y. Zhao, S.S. Potter, Functional specificity of the Hoxa13 homeobox. *Development* **2001**; *128*: 3197–3207.

170 M. Mericskay, J. Kitajewski, D. Sassoon, Wnt5a is required for proper epithelial-mesenchymal interactions in the uterus. *Development* **2004**; *131*: 2061–2072.

171 L. Carta, D. Sassoon, Wnt7a is a suppressor of cell death in the female reproductive tract and is required for postnatal and estrogen-mediated growth. *Biol Reprod* **2004**; *71*: 444–454.

172 W.W. Branford, G.V. Benson, L. Ma, R.L. Maas, S.S. Potter, Characterization of Hoxa-10/Hoxa-11 transheterozygotes reveals functional redundancy and regulatory interactions. *Dev Biol* **2000**; *224*: 373–387.

173 C. Miller, D.A. Sassoon, Wnt-7a maintains appropriate uterine patterning during the development of the mouse female reproductive tract. *Development* **1998**; *125*: 3201–3211.

174 G. Yamada, Y. Satoh, L.S. Baskin, G.R. Cunha, Cellular and molecular mechanisms of development of the external genitalia. *Differentiation* **2003**; *71*: 445–460.

175 C.L. Perriton, N. Powles, C. Chiang, M.K. Maconochie, M.J. Cohn, Sonic hedgehog signaling from the urethral epithelium controls external genital development. *Dev Biol* **2002**; *247*: 26–46.

176 R. Haraguchi et al., Unique functions of Sonic hedgehog signaling during external genitalia development. *Development* **2001**; *128*: 4241–4250.

177 K. Suzuki et al., Regulation of outgrowth and apoptosis for the terminal appendage: external genitalia development by concerted actions of BMP signaling (corrected). *Development* **2003**; *130*: 6209–6220.

178 D.W. Russel, J.D. Wilson, Steroid 5 alpha-reductase: two genes/two enzymes. *Annu Rev Biochem* **1994**; *63*: 25–61.

179 K.S. Kim et al., Expression of the androgen receptor and 5 alpha-reductase type 2 in the developing human fetal penis and urethra. *Cell Tissue Res* **2002**; *307*: 145–153.

180 K. Suzuki, K. Shiota, Y. Zhang, L. Lei, G. Yamada, Development of the mouse external genitalia: unique model of organogenesis. *Adv Exp Med Biol* **2004**; *545*: 159–172.

Index

a

abortion 91
ace 200, 850
acerebellar, see ace
achaete-scute complex 231
achondroplasia 628 ff
acquired immune deficiency syndrome, see AIDS
actin 119, 679, 689, 693, 695, 730, 773, 863
α-actinin 330, 333, 335
activin 18 ff., 44, 49 f., 52, 119, 123, 125, 248, 371, 416, 726, 729, 734, 738 ff., 793, 846 f., 861
activin βA 724, 726, 729
Ad4BP (adrenal 4–binding protein) 949
ADAM 84, 693
adherens junctions 86, 241, 679, 685, 689, 773, 914
adipocytes 698, 799
ADMP (anti-dorsalizing morphogenetic protein) 152, 292
adrenal 4–binding protein, see Ad4BP
adrenal primordium 957
adult stem cells 3 f., 510, 834
AE (anterior endoderm) 143, 146, 169, 847 f.
aggrecan 621
AGS (Alagille syndrome) 18
AHF (anterior heart field) 867 f., 874, 878
AIDS (acquired immune deficiency syndrome) 656
Alagille syndrome, see AGS
allantois 41, 77 ff., 88 ff.
amacrine cells 460, 464, 466 f.
AME (axial mesoderm) 143 f., 146, 149, 152, 155 f., 159 f., 164 ff., 169
ameloblastin 735, 739 f.
ameloblasts 720, 723, 725 f., 730, 733 ff., 738 f., 742 f.
Amphiregulin 7, 936
amygdala 190
anastomosis 699
angiogenesis 699, 797, 801, 806, 909 ff., 917 f.

angiopoietins 911 f., 915
anhidrotic ectodermal dysplasia 701
animal caps 116, 119, 293, 834
animal-vegetal axis 288
anoikis 778 ff.
anterioposterior 525
anterior-posterior axis 76, 11 ff., 117, 122, 196, 200 f., 211, 293, 371, 450, 507, 948
anterior endoderm, see AE
anterior heart field, see AHF
anterior neural ridge 295
anterior visceral endoderm, see AVE
anti-dorsalizing morphogenetic protein, see ADMP
aortic arch 907, 910
aortic sac 872
Apert syndrome 621
apoptosis 21, 23, 44, 46, 56 f., 75, 79 f., 85, 240, 251, 347 ff., 379, 455, 528 ff., 533 f., 539, 542 ff., 546, 563 f., 566, 568, 570, 576 ff., 623, 629, 654 f., 659 f., 662, 686, 692, 720 f., 735, 778 ff., 810, 868, 932 f., 96 f., 950, 963
arteries 845, 872, 875, 909 f., 916
arteriogenesis 663, 665, 909 ff., 913 f.
astrogliogenesis 250
asymmetric cell division 6, 231, 260 ff., 468
atonal 231, 243, 465
atonal homologs 299
autonomic neurogenesis 11, 21, 404
autonomic neuron 11, 21, 407 f., 412
autopod 524, 546 f., 568, 570, 574 f., 576 f.
AVE (anterior visceral endoderm) 76, 114, 146, 169, 451, 847
axial mesoderm, see AME
axin1 159, 294, 451

b

Barhl1 508
Barx1 724, 734
basal cell carcinomas, see BCCs
basal ganglia 190, 250, 380
blastoporal groove 109

Cell Signaling and Growth Factors in Development. Edited by K. Unsicker and K. Krieglstein
Copyright © 2006 WILEY-VCH Verlag GmbH & Co. KGaA, Weinheim
ISBN 3-527-31034-7

bazooka 261 f., 266
BCCs (basal cell carcinomas) 25, 702
Bcl–2 proteins 348
BDNF (brain-derived neurotrophic factors) 46, 252 f., 506 ff., 510, 734
B-FABP (brain fatty acid binding protein) 419, 421, 423
bile duct cells 830, 832
bipolar cells 462 ff., 467
Bix1 825
bladder 949
blastocoel cavity 74 ff.
blastocoel floor 109
blastocoel roof 109 f., 125
blastoderm 287 ff., 293, 296 f.
blastomeres 74, 123, 165
blastopore lip 109 ff., 141, 290
blastula 107 f., 110, 126, 148, 292 f., 296, 450, 825 f.
Blimp1 164
BMP (bone morphogenetic protein) pathway 5, 17 ff., 150 ff., 241, 244, 251, 365 f., 535, 542 f., 565, 858
BMP2 16, 20 ff., 41, 83, 88, 250 f., 351, 398 f., 407, 544, 570, 627, 648, 702, 719, 738, 847 ff., 956
BMP2b 292, 304, 365, 851
BMP3 52, 152, 726, 795
BMP4 20 f., 41, 52 f., 55, 88, 241, 250 f., 379, 402 f., 493, 507, 544 f., 560, 577, 642, 647 ff., 701 f., 726 ff., 733 ff., 739 f., 793 ff., 808, 850, 858, 862, 874 ff., 928
BMP5 88, 312, 545, 577, 726, 795, 874
BMP6 47, 51 f., 545, 627, 726, 874
BMP7 20, 52, 88, 365, 410, 457, 545, 627, 701, 720, 726 f., 736, 795, 874 f.
BMP8 41, 55
BMP15 47, 51 f.
BNP (brain natriuretic peptide) 631, 853
Bnl (branchless) 799 f.
body axis 110, 141, 189, 525, 826
bon 851, 856
bone marrow 3 f., 14 f., 20, 623, 664 f., 911
bone morphogenetic proteins, see BMPs
bonnie and clyde, see bon
bozozok 146, 169, 291
BPD (Bronchopulmonary Dysplasia) 795, 803
brachydactyly 537, 545
brachydactyly sydromes 578
Brachyury 88, 117, 121 f., 559
brain fatty acid binding protein, see B-FABP
brain natriuretic peptide, see BNP

brain vesicles 189
brain-derived neurotrophic factors, see BDNF
branchial arches 190 ff., 211, 313, 652
branching morphogenesis 78, 89 f., 496, 692, 791 ff., 796 ff., 815, 929, 931
branchless, see bnl
breathless, see btl
Brn3a 412, 414, 466, 505
Bronchopulmonary Dysplasia, see BPD
brush border 771, 773 f., 776
btl (breathless) 799 f.

c

cachexia 659 f.
CADASIL (Cerebral Autosomal Arteriopathy with Subcortical Infarcts and Leucoencephalopathy) 18
cadherins 8, 123, 240, 313, 509, 538, 557, 574, 689, 720, 935
calponin 872
cAMP responsive element binding protein, see CREB
campomelic dysplasia 621, 953
cardia bifida 851, 854, 856 f.
cardiac cushions 845, 875
cardiogenesis 648, 846, 850, 852 f., 854, 867 f., 874
Caronte 852, 861 f.
cartilage oligomeric matrix protein, see COMP
cas (casanova) 852, 856
casanova, see cas
Castor 814
catecholestrogen 82
β-catenin 7 ff., 15, 115 ff., 119 f., 124, 158 ff., 165, 167 ff., 238, 240 f., 291, 294 f., 365, 397 f., 408, 538 ff., 552, 557 ff., 643 f., 647, 649, 686 ff., 700 ff., 730, 736, 770, 772, 774 f., 807 f., 833 ff., 850, 858, 863, 879, 914
caudal diencephalon 190, 239
caudal homolog 3 153
cavitation 74, 81, 350,
CCAAT-displacement protein, see CDP
CCR7 685
CD34 661, 911
CD71 703
cdc42 115 f., 119, 339 ff., 343
CDK (cyclin dependent kinase) 258, 653 f.
CDK inhibitors, see CKI 23, 258, 654
CDP (CCAAT-displacement protein) 778
α-cell 823
β-cells 823
δ-cells 823

cell adhesion molecules 114, 123, 207f., 397, 551f., 872, 935
cell cycle 25, 57ff., 126, 231f., 236f., 240, 242, 254ff., 262f., 300, 404f., 460ff., 623, 653ff., 770f., 778, 803
cementoblast 721
Cer (Cerberus) 147f., 150, 156f., 159, 162, 165, 169
Cerberus, see Cer
cerebellar primordium 16
cerebellum 25, 190, 200, 247, 249, 256
cerebral cortex 190, 247, 314, 380,
chemoattractant 44, 652, 793, 956
chemokine receptor, see CCR7 699
chief cells 756, 761f., 764
Chisel 853, 864, 869
chondrodysplasia 624, 626
chondrodysplastic syndrome 579
CHOP 771
Chordin 150, 152, 156, 162, 164f., 169, 251, 290, 293, 446, 543
chorio-allantoic 76ff., 85, 88ff.
chorion 77f., 80, 88f., 91
Chx10 373, 378, 462f., 467
chylomicron 763
ciliary epithelium 469
ciliary ganglion 410, 413, 416
ciliary neurotrofic factor, see CNTF
cingulin 81
CITED2 846, 862
CKI 258, 654
c-kit 42, 53ff., 58, 397, 407
Clara cell protein 10 802
cleidocranial dysplasia 620
cleidocranial dysplasia syndrome 741
cloaca 925, 961
c-Met 90f., 651f., 662
c-myc 8, 10, 23, 400, 704, 778, 863
CNP (C-type natiuretic peptide) 628, 631
CNTF (ciliary neurotrofic factor) 43, 253, 417
cochlea 352, 487f., 492f., 496, 498, 500, 503f., 506ff., 510
Coco 156, 290
coelomic cavity 760, 948
coelomic epithelium 949f., 954ff., 959, 961, 963
cofilin 330ff., 340
colgate 160
collagens 86, 125, 698f., 769, 875
coloboma 377, 379, 458f., 460,
colon carcinoma 13
colony stimulating factor–1, see CSF–1

commissural axons 371
COMP (cartilage oligomeric matrix protein) 621, 624
condensation 59f., 348, 540f., 544, 551f., 570, 574f., 620, 686ff., 925f., 928
congenital lymphedema 914f.
connexins 81, 869
Contactin 301
cortactin 332
cortical plate 239
cyclooxygenase–2 (COX–2) 52, 83
cranial nerves 190f., 201, 296
Craniosynostosis syndrome 621
CREB (cAMP responsive element binding protein) 242, 410, 539
Crescent 160, 849f.
crinkled 732
Cripto 851, 861, 867f.
crypt 5, 758f., 768ff., 775f., 779ff.
crypt cells 769, 771, 758, 775, 779f.
crypt villus axis 756, 758ff., 767f., 779
Cryptic 851
cryptorchidism 958, 967
CSF–1 (colony stimulating factor–1) 80f.
c-Ski 19
C-type natiuretic peptide, see CNP
Cubitus Interruptus 549
Cux 559
cyclins 258
cyclin A 778, 804
cyclin D1 240, 460, 580, 626, 772, 862
cyclin D2 240, 863
cyclin D3 56, 85
cyclin E 56
cyclin dependent kinase, see CDK
cyclin-dependent kinase inhibitor p21 15, 85, 251, 257f., 533, 628, 654f., 778, 865
cyclooxygenase–2, see COX–2
cyclopia 146, 157, 159f., 456ff.
cyclops 155, 288, 456
cytokinesis 122

d

Daam1 538
dark skin, see Dsk
Darmin 833
Dax1 951f., 957, 959
DBH (dopamine-β-hydroxylase) 408, 413
DBL (dorsal blastopore lip) 141, 143f., 146, 151, 166
deciduum 79, 85f.
decorin 699, 803f.
delamination 32, 261f., 352, 366f., 401ff., 410, 651f.

Delta 13, 15f., 233ff., 246, 25f., 300, 302f., 701, 704, 733, 876
dendritic cells 683, 685
dental epithelium 719ff., 726, 728ff., 739, 742
dental follicle 719, 721, 726, 733, 739
dental placodes 719, 734, 736ff., 742
dentate gyrus 24f., 247ff.
Denys-Drash 949
dermis 641, 679, 681ff., 686, 695ff., 700, 702, 707
dermomyotome 641ff., 648, 652
desert hedgehog, see Dhh
desmoglein 81, 682, 689
desmoplakin 81, 689
desmosomes 81, 322, 689
dHand 413f., 555, 561f., 571, 868
dharma, see bozozok
Dhh (desert hedgehog) 419, 423, 547, 957f.
Dickkopf1, see Dkk1
DiGeorge syndrome 873
digit formation 352, 549
digits 352, 524ff., 537, 542, 546, 548f., 567f., 570f., 574, 576
discs large 263
Dishevelled, see Dsh
Disp (dispatched) 548
dispatched, see Disp
Dkk1 (Dickkopf1) 159, 162, 290, 295, 304, 539, 544, 559f., 567, 576ff., 802, 850
dopamine-β-hydroxylase, see DBH
dorsal aorta 21, 408, 410, 824, 828, 911
dorsal blastopore lip, see DBL
dorsal neural tube 11, 306, 362ff., 369f., 378, 402f., 405, 642ff.
dorsal pallium 381, 383
dorsal-ventral axis 290, 298
dorsal-ventral patterning 304, 310
downless 701, 732
DRG11 412
Dsh (Dishevelled) 7, 115f., 158, 160, 238, 537f.
Dsk (dark skin) 683
dystroglycan 86, 693, 695
Dystrophic Epidermolysis Bullosa 698

e

E2A 14, 231
E3 ubiquitin ligase 19, 303, 866
E-cadherin 74, 80, 120, 122, 685, 689f., 773
ECM (extracellular matrix) 86, 124f., 322, 334, 344, 763, 766
ectoderm 76f., 87, 107, 109f., 118ff., 123, 151, 169ff., 250, 287ff., 292f., 305, 363, 365ff., 378, 399, 449, 454ff. 492ff., 497, 526f., 530f., 533ff., 547, 551ff., 566, 642f., 687f., 732f., 736, 757, 799f.
Ectodermal Dysplasia Syndrome 732, 736
ectodin 726, 734, 738
ectodysplasin 701, 731f., 738
ectopic pregnancies 91
EEC syndrome 682, 705
EGF (epidermal growth factor) 80, 82ff., 91, 244ff., 339, 692, 733, 761ff.,770ff., 803, 809f., 875, 936, 964
egg cylinder 114, 147
EGL 25, 348
Ehox 846
EMT, see epithelial to mesenchymal transition
Emx2 23, 380f., 949, 962ff.
En1 (Engrailed–1) 556
enamel organ 719f., 723, 730
endoderm 75ff., 81f., 86f., 107, 109f., 113f., 155, 757ff., 777, 782, 794ff., 802, 825ff., 847ff., 854ff., 860, 863, 867, 873f.
endoglin 797, 804
endothelin 397, 411, 423, 425
endothelin-converting enzyme 877
Engrailed–1, see En1
enterocolitis 784
Eph receptors 207f., 210, 552, 570, 574
epiblast 41, 75f., 111ff., 117, 147, 155f., 371, 846, 848, 850
epibranchial placode 411f.
epicardium 843, 845f., 869ff.
epidermal fragility syndrome 705
epidermal growth factor, see EGF
epidermis 10, 15, 107, 151, 288, 306, 364, 399, 679ff., 692, 695, 699ff., 833
epididymis 961, 964f.
epimorphin 757, 760
epithalamus 190
epithelial to mesenchymal transition (EMT) 23, 76, 113, 120, 648, 653
Ephrin 207, 380, 552, 574f., 912, 916
ErbB1 82
esophagus 411, 756
estrogen 81f., 85
external germinal layer 25, 242, 250
extracellular matrix, see ECM
Extradenticle/Homothorax-like homeobox gene 204
eye development 375, 377, 449ff., 460ff.

f

FADD (Fas-associated death domain) 349

FAK (focal adhesion kinases) 125, 336
Fas-associated death domain, see FADD
fascin 333
fast muscle fibers 656, 658
faust, see fst
feather 681, 701, 719, 736
femur 524 f., 630
fertility 39, 51
fetal liver kinase–1, see FLK–1
fetal liver tyrosinase–1, see FLT–1
fetal ovary 40, 45
fetuin 833
FGF (fibroblast growth factor) 80 f., 87, 90, 114, 119 f., 153 f., 161, 199 ff., 203 f., 244 ff., 251, 291, 311 ff., 377 f., 495 f., 627 f., 631, 643, 652, 662, 682, 687, 701, 723 ff., 728, 734, 793 f., 798 ff., 812 ff., 850 f.
FGF-gradient 296 f., 798
FGFR1 (fibroblast growth factor receptor 1) 120 ff., 201, 496 f., 531, 533, 621, 650 f., 662 ff., 723, 798 f., 847
fibroblast growth factor, see FGF
fibroblast growth factor receptor 1, see FGFR1
fibronectin 86, 91, 110 f., 125, 334, 769, 795, 808, 811
fibrosis 795, 797 f., 804 f.
fibula 524, 544
filaggrin 683
filamin 333, 335
filipodia 333
fimbrin 333 f.
flamingo 117 f.
flexo 551, 860
FLK–1 (fetal liver kinase–1) 805, 876, 911 f.
floating head homeobox gene 124
FLT–1 (fetal liver tyrosinase–1) 805 f.
focal adhesion kinases, see FAK
follicular cells 50
folliculogenesis 46, 50 ff.
follistatin 150, 251, 290, 449, 701, 726 f., 738 ff., 959, 961
forebrain development 152
foregut 147, 757, 805, 828, 831, 845, 856, 866
formins 333 f., 340, 934
Fox 366 f., 400, 499, 856, 873, 937
FoxA2 164 ff., 371 f., 873
Foxi1 494
Foxl1 761
FRAP 43
Frasier syndrome 949
Fringe 14, 302, 552
frizzled 115 f., 119, 158, 238 f., 537, 539, 643, 645, 730,f., 772, 806 f.

frizzled receptors 7, 9, 115, 239, 731, 806
frizzled-related proteins 539, 645
FSH 46 f., 49, 52 f., 55, 57, 59
Fst (faust) 852
furin 234, 301
furin proprotein convertase 856

g

GABAergic neurons 244
gametogenesis 39 f.
gap junctions 22, 81
Gas1 (growth arrest specific gene 1) 559, 645
gastrointestinal tract 568, 755 ff., 760 f.
gastrula organizer 122, 143 ff., 149, 160 ff., 168
gastrulation 76, 87, 89, 107 ff., 117 ff., 143 f., 146 ff., 153, 156 f., 160 f., 165 f., 168 f., 189, 193, 196 f., 240, 287, 292 ff., 301, 401
Gata2/3 409, 413
GBX 295
GBX2 166 f., 293, 295, 456
GCNF (germ cell nuclear factor) 51
GDNF 53 ff., 312, 403, 411, 415 ff., 734, 927 ff., 936 f.
Geminin 461 ff.
geniculate 397
genitalia 947, 957, 966 f.
germ cell nuclear factor, see GCNF
GFAP 419, 422
GFRαs 411, 415 ff., 930 f., 934
GGF (glial growth factor) 227
GI cancer 579, 784
Gli 23 ff., 36 f., 242, 244, 381, 549 ff., 626, 646 f., 737, 729
glial growth factor, see GGF
gliomas 25, 579
glucagon 44, 771, 823
glucocorticoids 765
glycogen synthase kinase 3β, see GSK3β
glycosaminoglycans 875
Glypican–3 931
GM-CSF (granulocyte-marcophage colony stimulating factor) 80
gonadal primordium 949 ff., 960
gonadogenesis 949 ff.,
gonadotropin 46, 52, 82
gonocytes 39
goosecoid, see gsc
Gorlin's syndrome 25, 759
gp130 43 f., 82, 871
granule neurons 25
granulocyte-marcophage colony stimulating factor, see GM-CSF

granulosa cells 40, 47 ff., 954, 960 f.
Groucho 7, 167, 234 f., 291, 310, 374, 461, 469
growth arrest specific gene 1, see GAS1
growth cone 312, 321, 339, 341 ff., 414 ff., 913
gsc (goosecoid) 124, 165
GSK3β (glycogen synthase kinase 3β) 7 f., 116, 158 ff., 238, 300, 550
GTPase 116, 118 f., 336, 399 ff., 343, 538
gut 288, 410 f., 414, 417, 757 ff., 766, 769, 775 f., 780 f., 823 f., 827

h

hair cells 118, 352, 492 f., 496 f., 500 ff.
hair follicle 10, 15, 681, 686 ff., 693, 697, 701 ff.
Hairy/Enhancer of Split, see HES
Hand1 87 f., 91, 843, 853, 863, 868
HDAC (histone deacetylase) 778
HDL (high density lipoprotein) 766
head formation 114, 126, 141, 146 f., 152, 158 f., 161, 165 f., 168, 451 f., 845
headless 152 f., 159, 164 f., 167, 451
heart disease 841, 877
heartless 847
hematopoiesis 14, 24, 77, 910 f., 913 f., 917
hematopoietic stem cell 8
hemangioblastic lineage 912, 917
hemidesmosomes 679, 684, 690, 692 f.
Hensen's node 108, 113, 857 f., 860
heparan sulfate proteoglycans, see also HSPG 244, 530, 695 f., 723, 875, 931
heparin-binding (HB)-EGF 82
hepatocyte growth factor, see HGF
hepatocyte nuclear factor 1, see HNF1
hepatocyte nuclear factor 3β, see HNF3β
hepatocytes 4, 651, 809, 830 ff.
hereditary hemorrhagic telangiectasia type 1, see HHT1
HES (Hairy/Enhancer of Split) 13, 16, 234 f., 300, 509, 733, 775
Hex 147 ff., 165
HGF (hepatocyte growth factor) 90 f., 651 ff., 662, 759, 761, 765, 792, 815, 832, 835, 931
HHT1 (hereditary hemorrhagic telangiectasia type 1) 797
HIF–1 (hypoxia-inducible factor–1) 88 f., 917
high density lipoprotein, see HDL
hindbrain development 191, 193 f., 202, 205 f. 209, 374
hindgut 41, 43, 757, 759
hippocampus 24 f., 190, 239, 381
His 395
histone deacetylase, see HDAC
HNF1 (hepatocyte nuclear factor 1) 777, 779
HNF3β (hepatocyte nuclear factor 3β) 164
HNK–1 397
Hop 853, 855
horizontal cells 464, 467 f.
Hox gene 193 ff., 200 f., 203 ff., 210 f., 213, 296 ff., 314, 405, 560, 575, 814, 864
Hox transcription factor 193
humerus 524 f., 568
Hunchback 314
HSPG 244, 530, 695 f., 723, 875, 931
hyaluronic acid 52, 875
hydroureter 928
hypoblast 114, 142, 147 ff., 157, 168, 846 ff., 850
hypodysplastic kidney 928
hypohidrotic ectodermal dysplasia 701, 732, 739
hypothalamus 190, 374, 457
hypoxia-inducible factor–1, see HIF–1

i

ICAT (inhibitor of β-catenin and TCF) 159
Id 20, 236
IGF–1 (insulin growth factor–1) 45 ff., 51, 654 f., 734
Iguana 550
IL, see interleukins
implantation 73 ff., 79 ff., 89 ff., 90 f., 350, 405
infertility 967
inflammatory bowel disease 780
inhibitor of β-catenin and TCF, see ICAT
inner ear 352, 487, 489, 499 f., 504, 510
Inscuteable 262
insulin 762, 765, 781 f., 823 f., 829 f., 832 ff.
insulin growth factor–1, see IGF–1
integrins 7, 84, 110, 124, 334 ff., 338 f., 341, 693, 755, 763, 766 ff., 782 f., 930, 935
interleukin–1 46, 85, 660, 685, 771
interleukin–3 9
interleukin–6 43 f., 85, 929
interleukin–11 936
interleukin receptor 46, 48
intermediate filaments 324, 327, 338
intersexes 967
intestinal epithelium 13, 766 f., 769 ff., 775 ff., 779, 782
intestinal villus 5
intestine 755 ff., 760, 765 ff., 771 f., 775, 777 ff.

inversin 860
Irx3 167, 295, 308 ff., 373, 459
Irx4 853, 865 f.
islet1 312, 373
isthmus 200, 761

j

jagged 13, 18, 234, 302, 552, 733, 736
Janus kinase (JAK) 8, 42, 533
jekyll 875 f.
JNK pathway 115 f.
junctional epidermolysis bullose 691, 696
juvenile polyposis syndrome 756

k

keratinization 682 f., 688
keratinocyte 10, 15, 679 ff., 690 ff., 695 ff., 704
keratinocyte growth factor (KGF) 685, 766
KGF, see keratinocyte growth factor
kidney 351, 777, 925 ff., 948 f.
kit ligand (Kl), see also stem cell factor (SCF) 9, 42, 53
Knypek 117, 537
Kreisler 200 ff., 211
Kremen 159, 536
Krox20 200 ff., 206 f., 211
Kruppel 314
Kruppel-like transcription factors 776

l

labyrinth 78 ff., 90 f., 487, 492, 503
lamellipodia 119, 322 ff., 328 ff., 336 f., 339 f.
lamellipodial membrane protrusion 322
laminins 7, 84, 26 f., 89 f., 242, 334, 690, 692 f., 695 ff., 762 f., 676 f., 935
Langerhans cells 683, 685, 690
laryngo-tracheal 791
lateral ganglionic eminence 381, 383
lateral inhibition 233, 235, 300, 302, 304, 310
lateral plate mesoderm 151, 524 ff., 530 f., 553 ff., 557, 563, 620, 642, 648, 826, 828, 862
lazarus, see lrz
LEF (lymphoid enhancer factor) 7, 10, 158 f., 238 ff., 291, 365, 538 ff., 553 ff., 567 f., 575, 643 f., 649, 683, 687, 700 ff., 704, 730 f., 734, 807
left-right axis 845, 852 f., 860
left-right dynein, see lrd
LEKTI 683
lens 449, 454, 458 f., 464
lens placode 376 ff., 459
lethal giant larvae 263

leukemia inhibitory factor, see LIF
leukocytes 321, 328
Leydig cells 53, 58, 950, 957 ff., 962 f.
LH 46 f., 49, 52 f.
LIF (leukemia inhibitory factor) 8, 20, 43 ff., 49, 82 f., 86, 90 f., 253, 927, 932
LIFR (leukemia inhibitory factor receptor) 43, 253
LIM kinase 86
Lim1 124, 165 f., 308, 312, 414, 962 ff.
limb development 520, 528 f., 534 f., 537 ff., 541, 549, 551 f., 570, 578 ff.
limb deformity 569, 934
LIV1 122
liver 77, 499, 757, 777, 823 ff., 830 ff., 911
LMX1B 368, 556, 566 f., 688
loop-tail mice 118
loricrin 683, 705
lrd (left-right dynein) 860
LRP 238, 539, 730, 806
Lrz (lazarus) 205
lunatic fringe 507, 773
lung epithelium 797 ff., 803, 807 f.
lymphangiogenesis 914 f.
lymphoid enhancer factor, see LEF
LYNX 303
LYVE–1 700, 914

m

macrophages 43, 85, 324, 342, 344
mafB 200, 202 ff.
mammalian neurogenesis 15, 259
MAPK (mitogen-activated protein kinase) 20, 23, 42 f., 59, 153 ff., 241, 533 f., 577, 690, 763 f., 771, 773 f., 800
MARCKS (myristoylated alanine-rich C kinase substrate) 119
marcophage stimulating protein, see MSP
masculinizing force 966
Mash1 14, 16, 20 f., 244, 256, 311, 314, 370, 467 ff.
masterblind, see mbl
Math3 236, 467 f.
matrilins 621
matrix metalloproteinases, see MMPs
MBT (midblastula transition) 158
Mbl (masterblind) 294, 451 f.
MCP 685
mechanosensitive sensory neuron 408
medial ganglionic eminence 242, 381
medulla oblongata 190
medulloblastomas 579
meiosis 40 f., 45 f., 51 ff., 57 f., 74, 960

Meis 204f., 565f., 572
MEK (mitogen-activated proteinkinase kinase) 42f., 57, 59, 245, 655, 770f., 773f., 809
melanocyte 407f., 421, 683
Merkel cells 683, 685
mesencephalon, see also midbrain 189
mesenchyme 681f., 684, 688, 695
mesendodermal cells 122, 126
mesodermal cells 76, 113, 122f., 290, 641, 800, 826, 848, 912
mesodermal tissue 110, 826
metanephric mesenchyme 925, 937
metencephalon 190
mid-blastula transition 825
MHB (midbrain-hindbrain boundary) 154, 166, 239, 245, 294f., 305
microcephaly 159
microfilaments 322, 332
midblastula transition, see MBT
midbrain 143f., 159f., 165ff., 189ff., 200, 229, 234, 239, 245, 247, 287, 294f., 371, 455
midbrain-hindbrain boundary, see MHB
midgut 757, 802
migratory pathways 395f.
mil (miles apart) 856
miles apart, see mil
mind bomb 301
microtubules 111, 262, 323ff., 327, 338, 341, 343, 550, 801
Miranda 262
MIS (Müllerian inhibitory substance) 49f., 52, 796, 952ff., 961ff.
Mitf 377f., 683
mitogen-activated proteinkinase kinase, see MEK
mitogen-activated protein kinase, see MAPK
Mix.1 124
Mixer 825, 851
MJR 88f.
MKP3 120, 315, 533ff., 565
MMPs (matrix metalloproteinases) 84, 322, 629, 696, 956
MNs 307
morphogens 23, 144, 238, 288, 313f., 758f., 761, 811, 957
morpholino 117f., 121, 195, 200, 399ff., 458, 843f., 852, 916
morula 74f., 81
motor neuron 194, 206, 310ff., 361, 373, 424
motor neuron progenitor cells 301, 310
MSP (marcophage stimulating protein) 416
mTOR 42f., 86
Mucin1 83

Muller glia 460
Müllerian duct 948, 954f., 961ff.
Müllerian inhibitory substance, see MIS
muscle fiber hypertrophy 653, 655
muscle hypoplasia 655
myelencephalon 190
myoblasts 328, 645, 648ff., 653ff., 659ff.
myocardin 853, 855
myocardium 841ff., 846f., 855ff., 865ff., 872, 874f., 876ff.
MyoD 14, 20, 491, 537, 641, 643ff., 648f., 651, 654f., 659f.
myofibers 655ff., 660, 664
myogenesis 539, 641ff., 648ff., 653f., 657, 660, 663ff.
Myogenin 641, 654f., 659, 663
myosins 327, 333, 343
Myostatin 653
myristoylated alanine-rich C kinase substrate, see MARCKS

n

Na^+/K^+-ATPase 81
nail-patella syndrome 567, 578, 688
nails 688
naked cuticle, see Nkd
Nanog 8, 41, 81
NDF (neu differentiation factor) 247
necrosis 347f.
Nek2 59
nemo-like kinase, see Nlk
neocortex 24, 243ff., 381
nephrogenesis 925, 927ff., 934f., 937f.
nephron 925ff., 932f., 935f., 938
Netherton syndrome 702f.,
Netrin–1 371
netrins 343
neu differentiation factor, see NDF
neural crest formation 365, 367
neural induction 141, 146, 149, 151, 154, 156, 160f., 169, 293, 449
neural plate 107, 109f., 144f., 147, 149, 151ff., 159f., 166, 189, 191, 264, 287f., 290, 301ff., 308f., 311, 361, 363ff.
neural retina, see NR
neural tube 11, 21, 23, 107, 118, 189, 191, 229, 232, 236, 239f., 242, 250, 257, 263f.
neuregulins 11, 246ff., 421f., 424, 759, 870
NeuroD 236, 300, 373, 409, 412, 467ff., 504, 829
neuroectoderm 117, 119, 145, 151ff., 15, 159f., 166f., 191, 233, 246, 262, 293ff., 304f.
neurofilament 327, 410

neurogenesis 21 f., 24, 208, 229 ff., 242 ff., 287, 296, 298 ff., 303, 305, 310 f., 407 f., 422, 465 f.
Neurogenins 14, 231, 397, 409, 412
neurulation 107 ff., 113, 115, 118, 123 f., 189, 287, 305, 399
NeuroM 236, 299, 310 f., 373, 412
neuro-muscular junction 418
neuronal circuits 370
neuropilin–2 397
neurosphere 11, 16 f., 24 f.
neurotrophic theory 349, 417
neurotrophins 46, 252 f., 343, 397, 409, 506 f.
neutrophils 321, 338 f.
Newth 395
NEXT (Notch extracellular truncation) 302
NF-protocadherin 123
niche 5 f., 10, 18, 153, 469, 723, 733, 741 f., 956
nidogens 696
nieuwkoid, see bozozok
NIMA 59
nitric oxide 85, 662
Nkd (naked cuticle) 115
Nkx 499 f., 761
Nlk (nemo-like kinase) 117
Nodal-related 3 150
nodose ganglion 397
Noelin 397, 401, 404
Noggin 21, 149 f., 152, 156, 162, 169, 244, 250 f., 290, 293, 379, 402 ff., 410, 449, 543 ff., 556 f., 560, 575 f., 620, 642, 644 f., 648 f., 701, 739 f., 794, 848 f., 867, 928, 966
noradrenergic neuron 220, 405
Not 124
Notch signaling 13 ff., 208 f., 233 ff., 256, 297 f., 300 ff., 310, 364, 422 f., 469, 701, 733 f., 775, 829, 852 f., 859, 865, 916
Notch extracellular truncation, see NEXT
notochord 23, 111, 113 f., 124, 242, 288, 290, 308 f., 311, 358, 371 ff., 408, 410, 411, 488, 641 f., 644, 646, 824, 827
NR (neural retina) 375 ff., 455, 466
NRH1 117 f.
NSCL–1 236, 409, 412
Numb 14, 16, 229, 262, 266, 303 f.

o
occludin 81
Oct4 8, 41, 81
odontoblasts 720, 723, 725 f., 730, 733, 735, 738 ff., 742
oep (one-eyed pinhead) 155, 157, 290, 451, 456, 465, 847, 851, 856, 861

olfactory bulb 248
olfactory placodes 381
Olig2 244, 301, 309 ff., 373
oligodendrocyte 16, 229, 242 ff., 247, 250 f., 253, 258, 301, 304
oligozeugodactyly 527
omphalomesenteric veins 863
Oncostatin M, see OSM
one-eyed pinhead, see oep
oocyte 40 f., 45 ff., 290, 825, 950, 960 f.
oogonia 39 f., 46, 58, 960
optic disc 379 f.
optic stalk 375 ff., 449 f., 456 ff., 465
optix 455
organ of Corti 352, 498, 508, 510
OSM (Oncostatin M) 43, 835, 932
ossification 498, 502, 524, 574, 576, 620 ff., 625 ff.
osteogenesis imperfecta 624
osteonectin 769
osteopontin 84, 502, 768
osteoporosis 578
otic placode 352, 491, 493 ff., 498 f.
otic vesicle 143, 194, 352, 497 f.
otoconia 487, 500 ff.
otocyst 491 ff., 499, 504, 506 f.
otogelin 502
otopetrin 502 f.
OTX 166 f., 293 f., 375, 377, 379, 455 f., 468, 496
Otx2 166 f., 297, 375, 379, 455 f., 468
ovarian follicles 958, 960
ovary 46, 48, 50, 52, 58, 81, 947, 956, 958 ff., 964
oviduct 74, 112, 962, 964 f.

p
P0 407
p15 23
p63 535, 542, 557 ff., 682, 688, 704 ff., 736
p73 258
P75 117, 252, 349 f., 397, 407 f., 506, 702
PACAP 44
palladin 86
pancreas 489, 775, 777, 823 ff., 914
pancreatic insufficiency 765, 784
pancreatic polypeptide 823
pancreatic transcription factor, see Pdx1
pannier 499, 844, 854
PAPC 118, 124
parachute 124
parasympathetic 408 ff., 413, 415 ff.
parathyroid hormone-related protein, see PTHrP

parenchymal cells 321
Parkinson's disease 3
Pas 88
patched 23, 25. 242f., 469, 548, 626, 646, 702, 705, 729, 793, 802, 852, 860, 957
paternally expressed 3, see Peg3 or PW1
pbl 122
Pbx 204f., 622
PC12 cells 253, 258
P-cadherin 86
PCP (planar cell polarity pathway) 115ff.,124, 538, 807
PDGF (platelet-derived growth factor) 125f., 229, 249, 253f., 339, 655, 700ff., 792, 795, 803, 808, 814f., 913, 957f.
Pdx1 826, 829f.
pebble, see pbl
Peg3 (paternally expressed 3) 660
pentraxin3 52
periderm 682, 697
peripheral nerves 419, 421, 425
perlecan 84ff., 695, 767, 875
petrosal 397
Pfeiffer syndrome 621
phalanges 524, 545f., 574f.
pharyngeal arches 401, 491, 873
phospholipase-a 500
photoreceptor 248, 464, 466ff.
Phox2 257, 408f., 411ff.
PI3–K 42f., 46, 52, 56f., 781f.
pipetail 117
PKC-alpha 119
placenta 73, 76ff., 82, 84f., 88f., 91f., 914
placenta growth factor, see PlGF
plakoglobin 81
plakophilin 70f.
planar cell polarity pathway, see PCP
plasminogen activators, see Pas
platelet-derived growth factor, see PDGF
pleura 803, 810
PlGF (placenta growth factor) 914
pluripotent stem cell 4, 44
PME 143, 145ff., 151
Podoplanin 914f.
polaris 551, 560
polycystin–2 860
polydactyly 544ff., 548ff., 568, 574, 578
pons 190
post-mitotic state 230, 254
Pou domain factor, see Pou4f and also Brn3a 505f., 508
PP-cells 823
prechordal mesoderm 641

premature neurogenesis 15, 229
presenilin 234
prespermatogonia 39ff., 52
prickle 117
primitive streak 76, 209, 122ff., 120f., 146, 155, 157, 189, 193, 296, 846f., 858f., 909
profiling 326, 329f., 333
progesterone 52, 74, 81ff.
proprioceptive 306, 308, 408ff.
prosencephalon, see also forebrain development 189, 375
Prospero 262, 468
prostaglandins 83, 770
protein tyrosine kinase 7, see PTK7
Prox1 468, 915
proximal long bone 524
pseudopods 328
pseudostratified epithelial structure 395
PTHrP (parathyroid hormone-related protein) 626ff., 631, 734
PTK7 (protein tyrosine kinase 7) 118
Purkinje neurons 25, 242
PW1 66

r

Rab23 550f.
rac 86, 115f., 119, 124, 339ff., 343
radial glial cells 251, 256, 263
radical fringe 208, 552, 559
radius 293, 524f., 568
RALDH2 157, 196, 198, 201, 203, 311f., 379, 529f., 535, 554f., 558, 565f., 574, 866, 871
RAR (retinoic acid receptor) 157, 167f., 197f., 201, 203, 529, 963
Ras 23, 42, 243, 339, 533, 690, 692, 770ff., 800f., 809f., 875
reproductive tract 80, 83, 947, 963ff.
Ret 54, 411, 414ff., 925, 930ff., 934
retina 375ff., 449f., 454, 458ff., 462ff., 913f., 917
retinoic acid receptor, see RAR
retinoids 197ff., 204f., 374, 500, 528f.
retino-tectal projections 380
rfng 208f.
rho 115f., 118f., 122ff., 338ff., 343, 538
rhombencephalon, see also hindbrain development 189, 296
rhombomere 4 190, 299
ribbon synapse 507
RNA interference (RNAi) 398, 843, 856
Robos 397
ROCK 338, 340
roof plate 283, 308, 363, 367ff., 377f.

Ror2 537, 541
Runt/Runx3 412, 620 ff., 724 f., 741
Ryk 537

S

S100 419, 423
SCAR 332
SCF (stem cell factor) 9, 42, 53
SCG10 410
sclerosteosis 578
sclerotome 641 ff., 645 f.
scribble 118, 263
sebaceous glands 687, 703
secreted frizzled-related proteins, see Sfrps
Sef 120, 297, 534
segmentation 189 ff., 195 f., 200 f., 203, 206, 208 ff., 535, 540, 546, 565, 573 ff., 620, 643
seminal vesicles 961
seminiferous epithelium 53, 55, 57 f.
septation 802, 841, 869, 872 f., 875, 877
serotonin 2B receptor 870
Serrate 13, 302, 552, 733, 829
Sertoli cells 40, 52 ff., 57 ff., 950 f., 953 ff.
serum response factor, see SRF
sex determination 947 f., 953, 958 ff., 967
sex-determining region of the Y chromosome, see Sry
Sfrps (secreted frizzled-related proteins) 645
Shh 7, 23 ff., 118, 241 ff., 251, 256, 371 ff., 379 ff., 456 ff., 465, 469, 491, 535, 542 ff., 560 ff., 567 ff., 575, 642 ff., 652, 656 f., 702, 726, 728 ff., 734, 737 f., 758 f., 793 f., 802, 813, 852, 860, 873, 934, 966
siamois 115, 146, 291
signal transducer and activator of transcription-3, see Stat-3
silberblick 116 f., 160
Sim1 373, 648
sine oculis 455, 497, 937
situs inversus 860 f.
Six3 167, 295, 375, 453 ff., 458 f., 461 ff., 467, 469
skeleton 578, 619 f., 621, 625, 627 f., 641
slow muscle fibers 656 ff.
Smad 19 f., 23, 55, 244, 248, 251, 542 f., 573, 577, 725, 729, 803 ff., 848 f., 867
Smad ubiquitination-related factor 1, see Smurf1
smad5 19, 41, 44, 55, 365, 557, 810
Smoothened 23 ff., 549 f., 573, 646 f., 729, 793, 852
SMP 423
Smurf (Smad ubiquitination-related factor 1) 19, 801

snail 86, 120, 122, 126, 366F., 397, 399 ff.
snailhouse 365
SnoN 19
SOCS2 254
somatostatin 823
somitabun 292, 365
somites 108, 151, 196, 206, 246, 298, 311, 402 ff., 488, 524, 539, 565, 620, 641, 643 ff., 651 f., 656
somitogenesis 246, 293, 296 ff., 311
Sox2 8, 81, 293
spadetail 122, 124, 851
Spalt 159
SP-C (surfactant protein C) 795 f., 798, 800, 802 f., 805, 808, 811
Spemann's organizer 108, 144 f.
sperm 109, 322, 947, 960
spermatocytes 40, 53, 57 ff.
spermatogenesis 40, 52 ff., 955, 958
spermatogonia 40, 52 ff.
spermatogonial stem cells 40, 53, 55 f.
spermatozoa 40, 53, 59
spinal cord injury 3
SPINK-5 705
spiral limbus 498
Splotch 403, 422, 873
spred 120
sprouty 120, 534, 72, 800 f., 932
Spry 800 ff., 812
squint 155, 288, 304, 456
Src 42 f., 336, 934
SRF 853, 855 f.
Sry 950 ff.
Stat-3 (signal transducer and activator of transcription-3) 8
Staufen 262
stem cells 3 ff., 10 ff., 15, 20, 24 f. 40
stereocilia 118, 328, 500, 503, 507 ff.
steroidogenesis 52
steroidogenic factor1 949
stomach 411, 755 ff., 759 f., 930
strabismus 117 f.
striatum 247, 252
stroke 3
subthalamus 190
sugarless 876
surfactant protein C, see SP-C
swirl 292, 365, 849
sympathetic ganglia 311, 408, 413, 415, 423 f.
synaptic plasticity 244, 417
syncytiotrophoblast 79, 89
syndecan 693, 723, 858
syntaxin 760

t

Tabby 701, 731f., 738f.
TACE (TNFα converting enzyme) 302, 693, 875
TAF1 123
Tak1 (TGFβ activated kinase 1) 533f., 542f., 573, 577f., 849
Talin 335, 337
Tbx6 121
TCF (ternary complex factor) 7, 10, 13, 117, 121, 158f., 163, 167, 238, 240f., 291, 294f., 365, 381, 451, 455, 538ff., 553ff., 567f., 575, 643f., 649, 683, 687, 700, 704, 730f., 772, 775, 807
TDGF1 157
tectorial membrane 500, 502f.
α-tectorin 501ff.
β-tectorin 501ff.
telencephalon 17, 152, 189f., 242f., 253, 294f., 350, 363, 75, 380f., 383, 451f.
Tenascin-C 399, 768f.
ternary complex factor, see TCF
testis 40, 43, 52f., 55ff., 947, 950ff., 964f.
testosterone 958, 961ff., 966
tetralogy of Fallot 865
TGFα (transforming growth factor α) 50, 80, 762f., 765, 772
TGFβ (transforming growth factor β) 5, 7, 11, 18f., 44, 50f., 80, 83, 150, 155, 157, 249, 288, 290, 308, 364, 368f., 377, 417, 449, 541f., 642, 646, 648f., 653, 724ff., 738, 765, 793, 795ff., 802ff., 810, 825f., 846, 848f., 858, 867, 874, 876, 963
TGFβ activated kinase 1, see Tak1
TGIF 159, 542, 810
TH (tyrosine hydroxylase) 408ff., 413
thalamus 190, 295
thymocytes 24, 685
tibia 524, 630
tight junctions 81, 86, 264
TIMPs 86
tinman gene 844, 853
TIS21 259, 264f.
TNF (tumour necrosis factor) 80, 85, 349, 353, 655, 659ff., 685, 693, 701, 719, 729, 731ff., 742, 771
TNFRSF19 732
TNFα converting enzyme, see TACE
tooth 719ff.
Torso 800
TRAF 349
transferrin receptor, see CD71
transforming growth factor β, see TGFβ
tricuspid atresia 865
trigeminal 190, 194f., 198, 201, 205f., 397
Trk 46ff., 252, 397, 407ff., 505f., 702
trophectoderm 74ff., 79, 81f., 84ff., 114
trophoblast 74ff., 78f., 81f., 84ff.
Tsukushi 150
TUJ1 419
tumour necrosis factor, see TNF
Twist 399f., 857
tyrosine hydroxylase, see TH

u

U-boot 656
ulna 524f., 544
ureter 949
ureteric bud 925ff., 930ff., 937, 948
urogenital ridge 948f., 961
uterus 73f., 79, 81f., 85, 91f., 962, 964f.
utrophin 695

v

vagina 962, 964ff.
Valentino 202f.
van gogh 117f.
Vangl2 118
vascular cell-adhesion molecule–1, see Vcam1
vas deferens 961, 965
Vax 377, 379f., 459
Vcam1 (vascular cell-adhesion molecule–1) 88ff., 871
VEGF 623f., 627ff., 699f., 793, 801, 805f., 814f., 872f., 876f., 911ff., 197
VegT 290, 825, 833ff.
ventral neural tube 371f., 410, 457, 642
ventricular septal defects 865
versican 867f., 875f.
very low density lipoprotein, see VLDL
vestibulo-cochlear 397
vestigial tail 159
vibrissae 701
villus 5, 757ff., 768f., 776, 779ff.
vimentin 327, 338, 871
vinculin 335
vitelline veins 828
VLDL (very low density lipoprotein) 766
VP16, see Pdx–1

w

WAGR 949
WAVE 332
Wif 238, 539
Wilm's tumor gene, see WT1
wimple 860

winged helix/forkhead box, *see also* Fox 164, 371, 400, 761, 873
Wiskott-Aldrich syndrome 332
WISE (Wnt modulator in surface ectoderm) 159, 170, 399, 539
Wnt modulator in surface ectoderm, *see* WISE
Wnt proteins 238, 364, 368, 451, 536, 647, 658, 806, 928, 936
Wolffian duct 925, 928, 933, 937, 957f., 961ff.
WT1 (Wilm's tumor gene) 842, 870f., 935f., 949ff.

X
X-linked aristaless-related homoebox gene 958

Z
zath1 299
ZO–1 81
zona pellucida 48, 74, 81, 960